The Avocado

Botany, Production and Uses
3rd Edition

This book is dedicated to the memory of Medora Krome, Andre A. Ernst, Reuben Hofshi, and Samuel Salazar-Garcia in recognition of their lifetime contributions to advancing the avocado industry.

The Avocado

Botany, Production and Uses
3rd Edition

Edited by

Daniel Carrillo

Tropical Research and Education Center, University of Florida, USA

Bruce Schaffer

Tropical Research and Education Center, University of Florida, USA

Anthony W. Whiley

Sunshine Horticultural Services, Queensland, Australia

B. Nigel Wolstenholme

University of KwaZulu-Natal, South Africa

CABI

CABI is a trading name of CAB International

CABI
Nosworthy Way
Wallingford
Oxfordshire OX10 8DE
UK

Tel: +44 (0)1491 832111
E-mail: info@cabi.org
Website: www.cabi.org

CABI
200 Portland Street
Boston
MA 02114
USA

Tel: +1 (617)682-9015
E-mail: cabi-nao@cabi.org

A catalogue record for this book is available from the British Library, London, UK.

ISBN-13: 9781800621800 (hardback)
9781800621817 (ePDF)
9781800621824 (ePub)

DOI: 10.1079/9781800621824.0000

Commissioning Editor: Rebecca Stubbs
Editorial Assistant: Emma McCann
Production Editor: James Bishop

Typeset by Straive, Pondicherry, India
Printed and bound in the UK by CPI Group (UK) Ltd, Croydon CR4 0YY

Contents

List of Contributors

M. Aluja, Instituto de Ecologia, Km. 2.5 Antigua Carretera a Coatepec No. 357, Congregación El Haya CP 91070, Xalapa, México.

M.L. Arpaia, Department of Botany and Plant Sciences, University of California, Riverside, CA 92521, USA.

C.F. Balerdi, University of Florida/Miami-Dade County Cooperative Extension Service, 18710 SW 288 Street, Homestead, FL 33030, USA.

A. Barceló-Muñoz, IFAPA, Centro de Churriana, Cortijo de la Cruz s/n. 29140 Churriana, Málaga, Spain.

A.F. Barrientos-Priego, Departamento de Fitotecnia, Universidad Autónoma Chapingo, Km. 38.5 carr. México-Texcoco, Chapingo, Estado de México, CP 56230, México.

G.S. Bender, University of California Cooperative Extension Service, San Diego County, 555 Overland Avenue, San Marcos, CA 92123, USA.

J.B. Bost, Sustainable Development Department, Appalachian State University, North Carolina, USA.

J. Bower, 33361 Seventh Avenue, Mission, British Columbia V2V 2E4, Canada.

T. Cantuarias-Avilés, Luiz de Queiroz College of Agriculture, University of São Paulo, Brazil.

D. Carrillo, Tropical Research and Education Center, University of Florida, 18905 SW 280 Street, Homestead, FL 33031, USA

R. Castillo-Argaez, Department of Agronomy, University of Florida, Gainesville, FL 32611, USA.

A.S. Chanderbali, Florida Museum of Natural History, University of Florida, Gainesville, FL 32611, USA

J.H. Crane, Tropical Research and Education Center, University of Florida, 18905 SW 280 Street, Homestead, FL 33031, USA

E.K. Dann, Queensland Alliance for Agriculture and Food Innovation, University of Queensland, Ecosciences Precinct, Dutton Park, Queensland 4102, Australia.

B.G. Defilippi, Instituto de Investigaciones Agropecuarias (INIA), Santiago, Chile.

B.A. Faber, University of California Cooperative Extension, Ventura County, 669 County Square Drive, Ventura, CA 93003, USA.

E.D. Focht, Department of Botany and Plant Sciences, University of California, Riverside, CA 92521, USA.

R. Folgado, The Huntington Library, Art Museum and Botanical Gardens, San Marino, CA 91108, USA.

C. Fuentealba, Escuela de Agronomía, Facultad de Ciencias Agronómicas y de los Alimentos, Pontificia Universidad Católica de Valparaíso, Valparaíso, Chile.

M.E. Galindo-Tovar, Facultad de Ciencias Biológicas y Agropecuarias, Universidad Veracruzana, Veracruz, México

M. García-Rojas, Instituto de Investigaciones Agropecuarias (INIA), Santiago, Chile.

F.J. Gardiazabal, GAMA, Sociedad Gardiazabal y Mena Ltda. Quillota, Chile.

L.C. Garner, Plant Sciences Department, California Polytechnic State University, San Luis Obispo, CA 93407, USA.

R. Gazis, Tropical Research and Education Center, University of Florida, 18905 SW 280 Street, Homestead, FL 33031, USA.

P.M. Gil, Departamento de fruticultura y Enología, Facultad de Agronomía e Ingeniería Forestal, Pontificia Universidad Católica de Chile, Casilla 306–22, Santiago, Chile.

I. Hernández, Escuela de Agronomía, Facultad de Ciencias Agronómicas y de los Alimentos, Pontificia Universidad Católica de Valparaíso, Valparaíso, Chile.

M.S. Hoddle, Department of Entomology, University of California, Riverside, CA 92521, USA.

P.J. Hofman, Maroochy Research Station, Department of Agriculture, Fisheries and Forestry, PO Box 5083 SCMC, Nambour, Queensland 4560, Australia.

J.I. Hormaza, Instituto de Hortofruticultura Subtropical y Mediterránea "La Mayora", Consejo Superior de Investigaciones Científicas – Universidad de Málaga (IHSM la Mayora — CSIC-UMA), Universidad de Málaga,-29071 Málaga, Spain

E. Lahav, Institute of Horticulture, ARO, The Volcani Center, PO Box 6, Bet Dagan 50250, Israel.

U. Lavi, Institute of Horticulture, ARO, The Volcani Center, PO Box 6, Bet Dagan 50250, Israel.

R.E. Litz, Tropical Research and Education Center, University of Florida, 18905 SW 280 Street, Homestead, FL 33031, USA.

R. López-Gómez, Laboratorio de Fisiología Molecular de Plantas, Universidad Michoacana de San Nicolás de Hidalgo, Edificio A1' Ciudad Universitaria, 58060 Morelia, Michoacán, México.

C.J. Lovatt, Department of Botany and Plant Sciences, University of California, Riverside, CA 92521, USA.

P. Manosalva, Department of Microbiology and Plant Pathology, University of California, Riverside, CA 92521, USA.

F.J. Mena, GAMA, Sociedad Gardiazabal y Mena Ltda. Quillota, Chile.

J.A. Mercado, Instituto de Hortofruticultura Subtropical y Mediterránea "La Mayora," Universidad de Málaga-Consejo Superior de Investigaciones Científicas (IHSM-UMA-CSIC), Departamento de Biología Vegetal, Universidad de Málaga, 29071 Málaga, Spain.

F. Meza, Departamento de fruticultura y Enología, Facultad de Agronomía e Ingeniería Forestal, Pontificia Universidad Católica de Chile, Casilla 306–22, Santiago, Chile.

M.V. Mickelbart, Department of Botany & Plant Pathology and Department of Horticulture & Landscape Architecture, Purdue University, 625 Agriculture Mall Drive, West Lafayette, IN 47907, USA.

I. Narváez, Instituto de Hortofruticultura Subtropical y Mediterránea "La Mayora," Universidad de Málaga-Consejo Superior de Investigaciones Científicas (IHSM-UMA-CSIC), Departamento de Biología Vegetal, Universidad de Málaga, 29071 Málaga, Spain.

D. Olivares, Instituto de Investigaciones Agropecuarias (INIA), Santiago, Chile.

E. Palomo-Ríos, Instituto de Hortofruticultura Subtropical y Mediterránea "La Mayora," Universidad de Málaga-Consejo Superior de Investigaciones Científicas (IHSM-UMA-CSIC), Departamento de Biología Vegetal, Universidad de Málaga, 29071 Málaga, Spain.

J.E. Peña, Tropical Research and Education Center, University of Florida, 18905 SW 280 Street, Homestead, FL 33031, USA.

C. Pliego, IFAPA, Centro de Churriana, Cortijo de la Cruz s/n. 29140 Churriana, Málaga, Spain.

F. Pliego-Alfaro, Instituto de Hortofruticultura Subtropical y Mediterránea "La Mayora," Universidad de Málaga-Consejo Superior de Investigaciones Científicas (IHSM-UMA-CSIC), Departamento de Biología Vegetal, Universidad de Málaga, 29071 Málaga, Spain.

J.G. Ramirez-Gil, Universidad Nacional de Colombia sede Bogotá-Facultad de Ciencias Agrarias-Departamento de Agronomía, Bogotá, Colombia

B. Schaffer, Tropical Research and Education Center, University of Florida, 18905 SW 280 Street, Homestead, FL 33031, USA

D.E. Soltis, Department of Biology, University of Florida, Gainesville, FL 32611, USA.

P.S. Soltis, Florida Museum of Natural History, University of Florida, Gainesville, FL 32611, USA.

L.A. Valbuena-Gaona, Universidad Nacional de Colombia sede Bogotá-Facultad de Ciencias Agrarias-Departamento de Agronomía, Bogotá, Colombia.

A.W. Whiley, Sunshine Horticultural Services, Pty Ltd, 287 Dulong Road, Nambour 4560, Queensland, Australia

B.N. Wolstenholme, University of KwaZulu-Natal, 30 Wavell Drive, Wembley, Pietermaritzburg 3201, South Africa.

A. Woolf, New Zealand Institute for Plant and Food Research Limited, Mt Albert Private Bag 92169, Auckland Mail Centre 1142, Auckland, New Zealand.

Preface

It has been 20 years since publication of the first edition of this textbook. In view of the increased interest in avocado cultivation worldwide, and the success of the previous editions, in 2022, the publishers contacted Dr Bruce Schaffer, the lead editor of the second edition, to discuss the possibility of a third edition. Dr Schaffer saw the need for a generational change and invited Dr Daniel Carrillo, an entomologist with active avocado research, to lead the new edition. Dr Anthony Whiley and Dr Nigel Wolstenholme were also invited to join the editorial team in recognition of their valuable contribution to the previous editions.

Considering the volume of new literature since 2013, more authors were invited to contribute compared with the previous editions, and most of the new authors are from countries with growing avocado industries. Authors were chosen based on their current avocado research participation, scientific expertise, and familiarity with the chapter topics. Also, because of the significant expansion in avocado consumption in the last decade, the third edition includes a new chapter focused on avocado uses and human nutrition.

This book presents state-of-the-art information to avocado researchers, students, teachers, and growers. The editors helped compile and share this information to promote improved production systems and the continued growth of the avocado industry worldwide. The editors thank the authors who have given their time, without any financial reward, to contribute to this edition, as well as authors of the first and second editions, who laid the basic foundation for this important book. We also acknowledge the assistance of several scientists who reviewed the chapters and made valuable contributions to improve the end product.

Daniel Carrillo, Bruce Schaffer, Anthony Whiley, and Nigel Wolstenholme

1 Introduction

D. Carrillo[1] and B. Schaffer[1]
[1]*Tropical Research and Education Center, University of Florida, Homestead, Florida*

The third edition of 'The Avocado: Botany, Production and Uses' was written with specific goals in mind. The first was to document the remarkable expansion of avocado cultivation during the last decade. Avocado production nowadays supports a multibillion-dollar industry that has grown exponentially, reaching all continents except Antarctica and with prospects of even faster growth in the years to come. Global avocado production has almost doubled recently, with a significant expansion in areas planted, especially in Latin America, but also in Africa and Asia. Mexico continues to be the largest avocado producer, with approximately a quarter of the global production, estimated at 9 million metric tonnes in 2022 (FAOSTAT, 2024). Mexico produces nearly 2.5 times more avocados than Colombia, the country with the second-largest production. Peru, the Dominican Republic, Chile, Brazil and Ecuador are other important avocado-producing countries in Latin America. Commercial avocado production has also greatly expanded outside the Americas, especially in Indonesia and Kenya, with nearly 8% and 5% of the global output in 2022 from these two countries, respectively. In other traditional avocado-producing countries, such as Mexico, the USA, Israel, Chile, South Africa, New Zealand, Australia and Spain, production has not increased as much in recent years, but these countries sustain vibrant industries and produce a good part of the research and technology used in avocado production systems worldwide.

This book compiles and updates the technological advances and knowledge that sustain current avocado production. Chapter 2 encompasses the history of avocado, from its origins traced back to the Pangea era, through its domestication by Indigenous people in different areas of Mesoamerica, to its recent dispersion as a modern crop with established productive systems worldwide. Chapter 3 discusses the placement of avocado within the large and diverse Lauraceae plant family, which was prominent in tropical forests, where avocados were domesticated. Separate domestication events in different geographic areas are believed to have produced the recognized horticultural races of avocado: Mexican, Guatemalan and West Indian, ultimately resulting in the development of the most commercially important commodity within the Lauraceae family. Chapter 4 offers a comprehensive overview of recent advancements in avocado genetics and breeding, emphasizing its promising future shaped by evolving technologies and available genetic diversity, particularly in the centre of origin of the species. Those advances have relied

Corresponding author: dancar@ufl.edu

 The Avocado: Botany, Production and Uses (eds D. Carrillo, B. Schaffer, A.W. Whiley and B.N. Wolstenholme)
DOI: 10.1079/9781800621824.0001

on traditional breeding methods, but a transition towards molecular breeding is already taking place and will likely accelerate in the coming years to produce avocado varieties and rootstocks with improved agronomic traits.

Much of the avocado production occurs in the tropics and subtropics, as well as in regions characterized by a Mediterranean climate. Chapter 5 analyses the biomes supporting avocado cultivation and addresses the ecological, edaphic and climatic requirements for optimal production. Also discussed are predicted changes in the areas suitable for avocado production because of evolving environmental changes and climate variability. Increased temperatures are expected to result in the establishment of avocado orchards at higher altitudes and latitudes, changing the areas considered suitable for avocado production. Chapter 6 describes the most critical periods in the avocado production cycle, the flowering and fruit development periods. All endogenous and exogenous factors affecting these phenological stages and necessary for achieving optimal yields are discussed. Chapter 7 addresses the whole-plant physiology of avocado trees and how they respond to environmental factors, including light, temperature, humidity, drought, flooding, wind, salinity, elevated atmospheric CO_2, air pollution and the physiological effects of important plant diseases. Some of the available methods to reduce or alleviate plant stress due to these factors are also discussed.

Currently, more than 90% of the global avocado trade, and the vast majority of the recent growth in cultivated areas, is based on the exceptional 'Hass' scion cultivar. The overreliance on a single cultivar is concerning; therefore, efforts have increased to select scion and rootstock cultivars better adapted to different agroecological conditions. Chapter 8 contains updated information on scions and rootstocks available worldwide, which can be critical for mitigating the adverse effects of environmental stressors and avocado diseases, and for the better use of natural resources. Chapter 9 describes the propagation techniques used to obtain superior rootstock and scion cultivars, from when trees were planted as self-rooted seedlings to the most recent development of *in vitro* rootstock propagation. This chapter also reviews nursery production protocols and includes grafting techniques in the nursery and in the field when top-working older orchards. Biotechnology offers a promising alternative to conventional propagation. Chapter 10 describes new genomic tools that provide a more efficient framework for avocado selection and breeding, and a better outlook for developing new avocado cultivars and rootstocks tailored to the conditions and needs of the different production areas. The recent annotation of avocado genomes, together with the development of genetic markers, will allow assessing the genetic diversity of avocado germplasm. These new tools and significant advances in biotechnology allow faster and more efficient selection of superior cultivars than the current traditional breeding programmes.

This book compiles state-of-the-art information that could be used to promote more sustainable production systems with reduced environmental impacts and better preparedness to face the significant challenges that will likely shape the future of avocado production. Erratic weather patterns, due to climate change, are already affecting avocado-producing regions and are likely to affect multiple aspects of avocado production, which are discussed in various chapters of this book. Chapter 11 discusses water use in avocado production, which is a highly contentious issue that will probably shape the future of the industry. There is an increased public misperception that avocado has extremely high water demand in fact, the water requirement is average when compared with other tree crops. Water conflicts involving avocados are not caused by high water demand but rather because avocado orchards are planted in areas with water deficits, where agriculture competes with natural ecosystems and humans for this precious resource. This is the case in many areas with a Mediterranean climate, where not only conditions for avocado production are generally favourable but also where water is often a limited natural resource. More careful planning is needed to establish avocado orchards in areas without water conflicts, and cultivars less demanding in terms of water consumption and more tolerant to drought need to be used. The escalating adoption of biofertilizers, organic fertilizers and tree row mulching in avocado orchards, together with significant transformations in avocado nutrient management, is also discussed in Chapter 11.

Extreme weather events (i.e. freezes, low or high temperatures) are expected to increase in frequency and intensity and are likely to aggravate problems associated with alternate bearing, which will require implementing specific crop management tactics, discussed in Chapter 12, to avoid irregular cropping cycles and producing fruit not up to quality standards. In many countries, climate change is reducing the availability of arable land and increasing the cost of high-quality irrigation water; therefore, high-density plantings pioneered in Chile will likely be adopted in other countries.

Pests and diseases are another factor that will continue shaping the future of avocado production. Chapter 13 addresses avocado diseases, from which Phytophthora root rot continues to be the most limiting disease worldwide. Years of research and technological innovation have led to the development of effective treatments and tolerant rootstocks to manage this disease. However, with the continued expansion of avocado production to new geographical areas and the increased movement of goods between countries in a globalized world, new diseases are likely to emerge. New pathogen associations can be devastating because of the lack of evolved defences against them. This is the case of laurel wilt, a lethal vascular disease so far only known to be present in Florida and Myanmar but with the potential to spread and cause significant damage in other production areas. Chapter 14 describes the arthropod pest complex affecting avocados in different regions. The majority of specialized pests of avocado are still restricted to the native range of the species in Mexico and Central and South America. Strict regulatory actions are in place to prevent their spread to the many new areas, where avocado is now cultivated. Despite these restrictions, new pests are emerging in new cultivation areas, which require additional quarantine actions to prevent their spread to other areas.

Avocado production represents a vital economic activity for many developing countries. Production in these countries is usually export-oriented, targeting consumers in developed economies, where demand for avocados continues to grow, and production still cannot keep up with demand. Chapter 15 reviews the available information on postharvest technology and practices (including processing) for avocado fruit. This chapter provides updated information on quality parameters, sensory perception and maturity, production and postharvest factors affecting quality, disinfestation and storage (including preconditioning), prediction of out-turn quality and processing. Avocado is marketed as a 'superfood' and is considered a source of desirable nutrients linked to the prevention of diseases. It is also believed to offer several health benefits beyond its nutritional value, as discussed in Chapter 16. This chapter describes the traditional uses of avocado and the increasing processing into products such as dehydrated pulp, frozen pulp and guacamole, driven by the need to utilize surplus and damaged fruit effectively.

The apparently insatiable demand for avocados will likely drive the continued expansion of the worldwide avocado industry. Avocados will likely be embedded in the socioeconomic development of several countries, representing an important source of income and an opportunity for rural development. The biggest challenge is to achieve better coordination and planning for this expansion, taking advantage of the available genetic resources, crop management advances, knowledge of the suitability of environments for avocado production and increased preparedness to combat existing pests and diseases and mitigate new phytosanitary problems. International organizations and institutions could play a major role in technology transfer and coordination efforts to structure the international avocado trade. New growing areas need to develop research capabilities and establish breeding programmes for the development of superior rootstocks and scion cultivars tailored to their agroecological conditions. Local governments need to regulate land use to prevent further expansion of avocado cultivation in areas with water deficits and environmental conflicts, where natural habitats should be conserved instead. Cooperation and collaboration at different levels will be required to improve the sustainability of the worldwide avocado industry. The third edition of 'The Avocado: Botany, Production and Uses' is a step in that direction. Scientists from 11 countries with prominent avocado industries contributed to this book by sharing their experience studying multiple aspects of this valued plant, including more contributing authors, since the previous editions, from countries where avocado cultivation has greatly expanded in the last decade.

Reference

FAOSTAT (Food and Agriculture Organization of the United Nations) (2024) Avocado: post-harvest operations. Available at: https://www.fao.org/fileadmin/user_upload/inpho/docs/Post_Harvest_Compendium_-_Avocado.pdf (accessed 5 June 2024).

2 History, Distribution and Uses

M.E. Galindo-Tovar[1,]* and J.B. Bost[2]
[1]*Facultad de Ciencias Biológicas y Agropecuarias, Universidad Veracruzana, Xalapa, México;* [2]*Sustainable Development Department, Appalachian State University, Boone, North Carolina*

Abstract
This chapter discusses the history of the avocado from its origin, which is traced back to the Pangea times, to its current worldwide distribution. Also discussed are avocado archaeological records, evidence of its domestication by Indigenous people in Mesoamerica, its early dispersal in the Americas, its further dispersal and introduction to Europe during the colonial period, its cultural importance, modern history and current world production.

Introduction

Humans have consumed avocado fruit since ancient times. Nowadays, it is one of the most consumed fruits and the world's second tropical fruit in terms of tonnes of fruit produced. The origin, evolution, dispersion and domestication of avocado have been documented by integrating archaeological, historical, ethnobotanical and molecular data. Avocado ancestors have been placed in the Sierra Nevada (California) about 5 to 8 million years ago. Avocado trees dispersed north to south, establishing in different climates and adapting to different ecological conditions. The contact with Mesoamerican cultures resulted in different domestications that originated the three recognized horticultural varieties that have been classified as Mexican avocado, *Persea americana* var. drymifolia (Schlecht *et al.* Cham. Blake); Guatemalan avocado, *P. americana* var. guatemalensis L. Wms.; and West Indian avocado, *P. americana* Mill. var. *americana* (Scora and Bergh, 1992). In addition, diverse avocado groups, derived from genetic differentiation of the horticultural varieties, have developed in other places of the Americas, resulting in almost 1000 varieties currently registered for this species. Avocado is a tree with tradition and history. Its relationship with different cultures and its dispersion during colonial and modern times have made it one of the most important foods the Americas have given to the world. Currently, 64 countries produce avocado worldwide, with México as the top producer, followed by Colombia, Dominican Republic, Perú and Indonesia as the five major producers.

Avocado Origin

The origin of avocado has been traced back to Pangea times, when ancient plants originated in Gondwana and migrated to North America via Laurasia (Scora and Bergh, 1992). Avocado belongs

*Corresponding author: megalindo@uv.mx

 The Avocado: Botany, Production and Uses (eds D. Carrillo, B. Schaffer, A.W. Whiley and B.N. Wolstenholme)
DOI: 10.1079/9781800621824.0002

to the Lauraceae, one of the most ancient and largest plant families in the Magnoliales order. In this regard, archaeological data indicate that the Lauraceae family, as all angiosperms, were part of Gondwana vegetation during the Paleocene and in the Tertiary period migrated towards Laurasia via Tethys, when Africa and Laurasia were joined (Raven and Axelrod, 1974). According to Renner (2004), during this migration the first radiation occurred, deriving in the basal lineages of the Lauraceae family. These data agree with those reported by Chanderbali *et al.* (2001), who, using chloroplast and nuclear DNA sequences, estimated that this radiation occurred 174 + 32 million years ago, in the middle of the Mesozoic era. Renner (2004) estimated that the Lauraceae family was diverse and widely dispersed in the early Cretaceous. During the Eocene-Oligocene (55.8–23 million years ago), when the climate was tropical in the North Atlantic Ocean and Laurasia (part of the Pangea supercontinent), another radiation occurred in North America and Asia, probably caused by weather changes in the Boreotropical area during this period. Chanderbali *et al.* (2001) situated this radiation 44 + 7 million years ago.

How avocado ancestors migrated from Gondwana to North America has not been clearly understood. However, it is presumed that the initial dispersal of the Lauraceae family occurred accidentally through the transport of seeds or propagules with soil or water, in the same way that has been proposed for other plants (MacDougall, 2003; Renner, 2004). Later, animals seem to have been important avocado seed dispersers, and megafauna extinctions explain population fragmentation and genetic heterogeneity. Afterwards, humans became important seed dispersers and domesticators of these fragmented and isolated populations (Spengler *et al.*, 2021).

It has been challenging to assign fossils to the Lauraceae family; however, well-preserved flowers in Eocene deposits indicate the presence of the *Persea* genus in North America (Taylor, 1988). In addition, six fossils assigned to the *Persea* genus, virtually identical to the extant species, were found in the eastern USA and have been chronologically placed in the late Cretaceous (93.5–65 million years ago) (Tupper, 1927; Eklund, 2000). Other leaf fossils found in Nevada and California suggest the presence of modern avocado progenitors in these places since the Tertiary (60–10 million years ago), when the tropical climate prevailed in this region. It has been suggested that 68% of the preserved specimens in the Palmdale, California, area were avocados, indicating the presence of prehistoric avocados in California around 50 million years ago (Schröeder, 1968; Millar, 1996). Archaeological evidence documented by Schröeder (1968) indicates that avocados were a common plant in this area. In light of these data, Galindo and Arzate (2010a) proposed that the ancient avocado origin was in the area currently occupied by the Sierra Nevada in California, which during the early Tertiary was a plain, with hot, humid weather and high plant diversity. When the Sierra Nevada emerged during the Cenozoic era, about five to eight million years ago, the climate changed to dry and cold, and ancient avocados migrated towards the south and went extinct in the area, leaving plant fossil evidence. Even though these plants did not survive when the climate became semiarid, *Persea* fossils found in the semiarid areas of northern México (Sierra Madre Occidental, Durango and Sonora) indicate the relationship of these specimens with those of California and their migration to the south (Schröeder, 1968). In this migration, large animals carried the fruit to other places, playing an important role in the dispersal of the avocado (Spengler *et al.*, 2021). It is possible that large mammals, such as ground sloths, were avocado dispersers, especially during the ice age (Barlow, 2002). Although the edible portion of the fruit was very thin, its oil content and flavour likely attracted these mammals, which had relatively small and blunt teeth and could swallow the fruit and excrete ready-to-germinate seeds (Barlow, 2002; Diamond, 2017). Early avocado dispersal has been documented by Bost *et al.* (2013). Later, 17,000 to 12,000 years ago, when these mammals became extinct, humans became the main avocado dispersers (Spengler *et al.*, 2021).

During migration, avocado trees had to adapt to different ecological conditions such as daily light and temperature variations, and seasons. New avocado populations developed where avocados found appropriate habitats and crossing opportunities. However, the populations that colonized new habitats were small, resulting in bottlenecks that contributed to the species diversification. Over time and adaptation to different environments, avocados developed characteristics that originated diverse types. North-to-south avocado migration has been documented with archaeological and ecological data. The presence of wild avocados in the Nuevo León Sierra in México, according to Gutierrez *et al.* (2009), is indicative of a migration route to south-east.

Based on the diversity and presence of some primitive avocado types, it has been proposed that the primitive avocado gave rise to divergent groups that gradually formed the three horticultural modern varieties, and other types that can be identified based on morphological and molecular characters, and the presence of essential oils. Scora and Bergh (1992) identified the Mexican race by the presence of an estragole monoterpene, while the sesquiterpene caryophyllene identified varieties of the West Indian and Guatemalan races. Furthermore, the botanical races have been studied in the last three decades using molecular techniques to distinguish them (Furnier *et al.*,1990; Bufler and Ya'acov, 1992; Davis *et al.*, 1998; Fielder *et al.*, 1998; Ashworth and Clegg, 2003; Lavi *et al.*, 2003; Schnell, 2003; Ge *et al.*, 2019). In light of new evidence integrating palaeoecological, anthropological and molecular data, it is important to reconsider the origin centres of the three traditionally recognized botanical races and a wide range of physiological varieties that undoubtedly originated by adaptation and domestication not only in Mesoamerica but also in Central and South America, as reported by Berdugo-Cely *et al.* (2023).

Early Dispersion

Archaeological and molecular data indicate that an early radiation could have occurred during the avocado dispersion from the Sierra Nevada to the south. This first split was dated about 1.3 million years ago by Solares *et al.* (2023). One migration took place towards Sierra Madre Occidental, Durango and Sonora, as indicated by archaeological data (Schröeder, 1968), and the other to the south-east, as indicated by the presence of wild avocados in the Nuevo León Sierra in México (Gutierrez *et al.*, 2009). This geographical separation gave rise to two sister clades, one for the Mexican group and the other for the Guatemalan and Lowland lineages, which separated about 440,000 years ago. Later, different domestications derived in the three horticultural races (Solares *et al.*, 2023).

According to Buckler *et al.* (1998), when avocados dispersed from Sierra Nevada to the south, they arrived in the central part of México. During the glaciation until 10,000 BC (before Christ), the climate was more humid and colder than it is today, and the Tehuacán Valley was hospitable to avocados. The tree adapted to this climate and proliferated even before humans arrived in Mesoamerica (about 15,000 years ago). Avocado fruit has been consumed since the first seminomadic settlers arrived at the Tehuacán Valley (Goebel *et al.*, 2008). This suggests that the first human groups that settled near the forests and jungles collected avocado, probably systematically and selectively, rotating the collection areas, as has been documented by Mac Neish (1964). From this point of view, probably the first human groups who arrived at Mesoamerica began gathering avocado and other abundant plants that required less processing to be consumed (Betz, 1999). Thus, avocado domestication began unintentionally and even before that of other short-cycle plants, when humans foraged plants and selecting the best ones. This initial domestication was initiated in the Coaxcatlan Valley between 500 BC and 1500 AD, where avocado was selected for fruit size, resulting in larger fruit (Smith, 1966; Fuller, 2018). In this context, domestication initiated when humans started to care for and protect useful plants and began the *in situ* management of different trees in forests. According to Casas *et al.* (2007), agroforest systems became common in Mesoamerica, one of the main domestication centres in the world. Once humans began to use the avocado fruit, its widespread dissemination and domestication began (Chen *et al.*, 2009). This process began when the first hunter-gatherer groups arrived in Mesoamerica (18,000–10,000 BC) and found the avocado in central México.

According to Smith (1966), the avocados found in Coaxcatlan probably came from the Sierra de Zongolica, an area in which the avocado with an anis-like leaf odour is distributed naturally and traditionally cultivated in backyards and used as live fences or for shade. Since ancient times, the fruit has been consumed locally in the municipalities of Zongolica, Aquila and

Maltrata in Veracruz. The leaves are used as a condiment and, together with the fruit, are consumed and sold locally in the tianguis (open-air market). In the Sierra de Zongolica, Veracruz, avocado leaves are used together with the tortilla, to wrap and consume edible insects (Landero *et al.*, 2012).

Later, a large part of the cultures that inhabited Mesoamerica since pre-Columbian times continued to consume avocado. Among these cultures, we can include the Mokaya, the Olmec, the Aztec and the Maya. In this respect, avocado domestication has been placed in Mesoamerica (Smith, 1966, 1969; Knight, 2002), an area with great biological and cultural diversity (Toledo *et al.*, 2001).

When the climate became dry and variable (8000–2000 BC) in the central area of México, it was necessary to cultivate avocado trees to preserve a fruit that was surely already valued by the inhabitants of the area (Buckler *et al.*, 1998). Human groups learned that a new plant would emerge by dropping a seed on the ground (Mac Neish, 1964). Therefore, starting avocado cultivation should not have been a difficult task. Changing avocado's natural growth and reproduction to cultivation conditions meant continuing the selection of trees with useful characteristics, which unconsciously created a different environment in which the trees changed morphologically and genetically to adapt to this new environment. According to Gepts (2005), the process of selecting the best trees and their adaptation to cultivation gave rise to the first domesticated trees. In the case of the avocado, its domestication likely began between 8000 and 2000 BC, mainly as a form of conservation (Ellstrand and Marshall, 1985). However, the low selection pressure, as well as the long life cycle of the avocado, must have been an important factor in its long and complex process of domestication, which, according to Gómez-Pompa *et al.* (1990), Gama and Gómez (1992), and Clegg *et al.* (1993), has occurred several times, involving different cultures that have inhabited Mesoamerica.

The oldest culture considered in Mesoamerica is the Mokaya. This culture inhabited the Soconusco area in 1800 BC, as reported by Taube (2004). Even though its name means maize, in the beginning, the Mokaya did not consume maize. Evidence of this is the small and unproductive maize ears recovered in the area. Furthermore, chemical analyses performed on human bone collagen indicate that plants with C4 photosynthesis (such as maize) were not part of the local diet. By contrast, avocado was a common tree in the area inhabited by this culture and was likely a part of their diet, representing a source of energy and nutrition. Through caring for, propagating and selecting avocado plants, they continued the domestication process started by the first settlers, who later passed their knowledge to other cultures that inhabited this area. One of these cultures was the Olmec. The oldest data on this culture date from 1600 to 1500 BC (Ortiz and Rodríguez, 2000). According to Taube (2004), they probably inherited the knowledge of avocado cultivation from the Mokaya culture. The Olmec culture must also be among the first domesticators of the avocado tree. In addition to having developed in an area where avocados grew naturally, their daily activities, such as farming, constructing, playing ball games, weaving and making pottery, required energetic and nutritious foods such as the avocado fruit. The domestication later gave origin to the Mexican avocado race, an avocado with a high oil content, a thin skin and leaves that smell like anise. According to Williams (1977), among the three avocado races, the oldest seems to be the Mexican, since it has the oldest archaeological references. For this race, Storey *et al.* (1986) proposed a triangular-shaped centre of origin, whose vertices are the southern parts of the state of Nuevo León, Pico de Orizaba (Veracruz) and the upper parts of the state of Michoacán. In addition, the Mexican race is the one that has been distributed furthest to the north and, therefore, the closest to the area of natural dispersion of the species. Furthermore, according to Mhameed *et al.* (1997) and Schnell *et al.* (2003), the Mexican race has the greatest genetic variability, indicating proximity to the putative refuge. Along with these studies, Corona *et al.* (2016) compared genetic diversity among avocados from Zaragoza in Nuevo León, Aquila in Veracruz and Chicola in Yucatán state, México, and La Antigua in Guatemala. The avocados located in the northernmost regions (Zaragoza and Aquila), in the area proposed to be the centre of origin for the Mexican race, exhibited the highest molecular diversity. These results further support the conclusion that the Mexican race is the oldest avocado race and strengthens the proposal that the centre of origin of this race is in central México.

Other Dispersions

Avocados from central México were taken to different places, where other domestications occurred. Smith (1966) proposed that avocados already selected in the Tehuacán Valley were taken to the central valleys of Oaxaca, located between the Mixteco Nudo, the Sierra de Juarez and the Sierra Madre Sur. Archaeological evidence shows that improved fruit found in the Oaxaca valleys (1200 BC) have similar characteristics to those found in the Tehuacán Valley (Smith, 1966). Furthermore, ethnobotanical data indicate that anise-scented avocado leaves have been used since ancient times by cultures such as the Zapotecs, who used them as a condiment for broths, tamales and barbecue (Messer, 2003).

The domestication that originated the Guatemalan race has been placed in the Chiapas highlands. However, how avocados arrived in this area is not clear. One hypothesis is the dispersion route from Sierra de Zongolica, Tehuacán, Oaxaca and the highlands of Chiapas, which is supported by Mhameed *et al.* (1997). They described a close relationship between the Mexican and Guatemalan avocado races, with phenotypic similarities, such as their cold tolerance and oil content. Furthermore, molecular data (Corona *et al.*, 2016) showed a close relationship among avocados from Veracruz (Sierra de Zongolica) and Guatemala (La Antigua). An alternative hypothesis suggests that the Mayans domesticated avocados that grew naturally in the area, since avocados grew naturally in the highlands of Chiapas and Guatemala since ancient times (Fedick, 1995). This is another area with great diversity and quantity of avocados, suggesting their ancient presence. Rosenswig *et al.* (2015) also reported avocado pit fragments during the Early Formative Period (2000 to 1000 BC) at Cuauhtemoc, Chiapas. However, low frequencies of the genotypes that distinguished the localities in México and Guatemala do not allow these localities to be clearly distinguished, suggesting a reticulated distribution. Nevertheless, the presence of different loci in the two regions and the different frequencies of the shared loci suggest different domestications (Corona *et al.*, 2016). Solares *et al.* (2023) proposed that Mexican and Guatemalan groups are sister clades; therefore, the more likely hypothesis is that avocados migrated from Sierra Nevada to the south, via Sierra Madre Occidental, before humans arrived in the Chiapas highlands. When humans arrived in this area, with an inherited knowledge from other cultures, they domesticated another type (race) of avocado: the Guatemalan. Later, avocados from Oaxaca were taken to Chiapas and Guatemala, as a part of the trading network of that time, giving rise to the reticulated characteristics of avocados.

The next question is as follows: How did avocados arrive in Yucatán to be domesticated as the West Indian race? Colunga and Zizumbo (2004) proposed that avocados domesticated in central México were brought to the Mayan lowlands (Yucatán and Belize) around 3400 BC, when the first human groups arrived, and they continued the domestication process, mainly in Mayan backyard gardens. However, Solares *et al.* (2023) propose that Guatemalan and West Indian avocados are sister clades, suggesting that avocados from Mayan highlands in Chiapas, Guatemala, El Salvador, Honduras and Nicaragua were brought to Yucatán via Central America, as Kennett *et al.* (2022) have proposed.

It is assumed that the avocado domesticated in Yucatán and Belize acquired its characteristics by adapting to the Yucatán Peninsula's climate and calcareous soils. Consequently, the West Indian avocado is characterized by its adaptation to tropical conditions and its resistance to saline soils and soils with low-iron content (Galindo and Arzate, 2010b). In addition, Dine *et al.* (2019) reported avocado among the most frequent species found in Mayan lowlands called *rejoyadas* (shallow sinkholes). The Mayans have been recognized as one of the main avocado domesticators (Gama and Gómez, 1992). They inherited their knowledge from the first settlers in the area and the Olmecs.

It is also interesting that New Spain chroniclers such as Landa ([1590] 2014), in his book 'Relationship of Yucatán Things', described an avocado with characteristics of the West Indian race in Yucatán as follows:

> *. . . a very large and cool tree which the Indians call on; it bears a fruit like large courgettes of great softness that seems to have a buttery flavor and is buttery, and it is of very high maintenance and substance. It has a large stone and a delicate skin, and it is eaten cut into slices like melon and with salt.*

Cobo ([1653] 2008) in his 'History of the New World' also described the avocado of Yucatán:

> *The varieties from the Yucatán province are as big as a small pumpkin or a large grapefruit. The avocado has a thin skin, softer and more flexible than the Ceuta lemon, green on the outside, and when the fruit is very ripe, it is easily peeled. It has the largest seed I have ever seen in any fruit, even in the Indies of Europe . . . Between the seed and the shell is the flesh, a little thicker than a finger except at the neck where it is thickest. It is a whitish-green color, delicate, buttery, and very smooth. Some people eat it with sugar and salt, others directly from the tree, it tastes so good that it does not require seasoning.*

Other evidence that the West Indian avocados were domesticated in the lowlands of Yucatán relies on linguistic data. Given the use of the plant and its cultural importance since ancient times, the Mayans named the avocado with the word *on* (Gama and Gómez, 1992). Undoubtedly, the first groups of settlers in Mesoamerica began the process of avocado domestication, first selecting trees in the forests and later dispersing their improved fruit and passing their knowledge on to the first cultures that developed in this area, thereby originating the first avocado cultivars. At the same time that avocado domestication was initiated in Mesoamerica, it was dispersed to Central and South America in different ways. There is archaeological evidence of the presence of avocado in Honduras. Kennett *et al.* (2017) reported that avocado, among other plants, was utilized by occupants of the 'El Gigante' Rockshelter, about 11,000 and 10,000 years ago. Also, the Olmecs traded avocado with the Papayecas around 1200 BC, which coincides with seed fossils found in Río Claro, Honduras (the area inhabited by the Papayecas) (Healy, 1978). This suggests that the Olmecs commenced transporting avocado by sea or land.

Pope *et al.* (2001) reported the presence of *Manihot* sp. in Tabasco from 4600 BC, indicating the indirect contact of inhabitants of this area with those of the Amazon basin. From this point of view, the presence of the avocado in the northern part of South America can be explained through the exchange between these groups. Most likely, avocado trees from Yucatán and Belize were taken by sea to Colombia, where the avocados were described first when Fernández de Enciso arrived in Yaharo, Colombia, in 1519 (Fernández de Enciso, [1519] 2014). In this regard, according to Berdugo-Cely *et al.* (2023), the different groups of Colombian avocados possibly originated from the West Indian race, even when a less parsimonious alternative could be a radiation of the West Indian race from the Colombian group. Later, these avocados were taken to Ecuador, where West Indian-type avocados were described by Juan and Ulloa ([1748] 1970) around Quito:

> *This figure resembles that of the gourds that usually serve as snuffboxes: that is, in the circular or round lower part; and then it extends through the upper part forming a neck, which will end at the nipple area; from which to its base they are regularly 3 to 5 inches long; a very thin shell covers it, which when the fruit is ripe, easily detaches from the flesh, and on the outside it is shiny, as before seasoning, it is always green, although clearer at the end, than when it has not been perfected: the meat is of some consistency, but not so much, that it stops falling apart when pressed with the fingers; white, turning to greenish; and not at all sweet in taste; and it is necessary to put some salt on it, to make it stand out in it; it is somewhat fibrous, which is less noticeable in those of good quality, than in those that are not so outstanding. The tree is very tall; copious; and very leafy, its leaf somewhat larger than that of the cherimoya, and with some difference in the figure.*

Palaeographic data indicate that avocado was distributed from Ecuador to other locations. One of these locations was undoubtedly Perú. According to Wolters (1999), the Valdivia culture that inhabited Ecuador travelled by boat from Ecuador to Perú and Central America around 2200 BC, transporting avocado and other plants between Central America and northern South America. This explains the presence of avocados on the Pacific coasts, and according to Berdugo-Cely *et al.* (2023), the presence of avocado in the Andes may reflect the different local adaptations after the initial colonization of Mesoamerica. Slotten and Lentz (2021) identified avocado fragments during the Late Classic Period in Joya de Ceren, El Salvador. Avocado seed fossils dated from 2500 to 1800 BC were also found in the Moche Valley, Perú (Pozorski, 1979); on the Peruvian Pacific coast around

1500 BC (Heiser, 1979); in the Supé Valley, Perú, in 1200 BC (Solis, 2001; Skidmore, 2005); and in Huaca Gallinazo and Huaca Santa Clara at Virú Velley, Perú, in 200 BC (Mansur *et al.*, 2018). De la Vega provides another interesting piece of avocado dispersion history in his book 'Real Comments of the Incas'. He recounts how the avocado was brought to the Inca valleys between 1450 and 1475 AD, when King Yupac Inca Yupanqui conquered the province of Palta in Ecuador (Vega, [1604] 2017).

Colonial History

During colonial times, the flora and fauna of the New World attracted the chroniclers' attention and curiosity. When the Spaniards arrived in America, the avocado had already been domesticated and introduced in northern South America, and diverse types that received names in different languages were recognized. The Spanish chroniclers who arrived in New Spain made the first written descriptions of the avocado. Sahagún ([1590] 2019), in his book 'Historia General de las cosas de la Nueva España' [*General History of the Things of New Spain*], relates that this species:

> *... was appreciated and used by the different cultures that populated the New Spain. The ahuacatl or ahuacacahuitl, was a tree with dark green leaves, with a fruit, black in the outside and white and green in the inside, with a heart shaped seed; other avocado was the tlacazolahuacatl, a big tree, as the former; and the quilahuacatl, with a fruit green in the outside, and very good to eat.*

The oldest known description of the avocado is found in the first document written in America: 'Suma de Geografía' [*The Sum of Geography*]. This document is attributed to Martín Fernández de Enciso and was published in 1519 in Seville, Spain (Fernández de Enciso, [1519] 2014). He described the avocado as follows:

> *Before reaching Santa Marta is Yaharo, which is in the snowy mountains. Yaharo is a good port and good land and here there are trees with many fruits to eat and among others there is one that looks like orange, and when it is seasoned to eat it turns yellow; what is inside is like butter and has a wonderful flavor and leaves a taste so good and so soft that it is a wonderful thing.*

The avocado was also described by Gónzalo Fernández de Oviedo ([1526] 1999) in the 'Sumario de la Natural Historia de las Indias' [*Summary of the Natural History of the Indies*]:

> *On the mainland there are some trees called pear trees, but they are not like those of Spain, but there are others of no less esteem; they are of such a fruit, that they give a lot of advantage to the pears here. These are large trees, and the leaves are broad and somewhat similar to that of the laurel, but it is larger and greener. This tree gives some pears weighing a pound and much larger, and some less; but commonly they are a pound, a little more or less, and the color and size are true pears, and the rind is somewhat thicker, but softer, and in the middle it has a seed like a grafted chestnut, peeled; but it is very bitter, as was said earlier about the mamey, except that this is in one piece, and that of the mamey is three, but in itself bitter and in the same way, and on top of this nugget there is a very thin telica, and between it and the peel is what is to eat, which is of a liquor or paste very similar to butter and very good delicacy, and very good flavor, and such that those who may have it keep and appreciate it; and they are wild trees, this one as well as all those that are mentioned, because the main gardener is God, and the Indians do not put any work on these trees. These pears taste very good with cheese, and pick them early, before they ripen, and store them, and after they are ready, which ones are best to eat, they lose if you delay and let them overripen when they are good to eat.*

Cervantes de Salazar ([1594] 1985), in his 'Crónica de la Nueva España' [*Chronicle of New Spain*], described the avocado as a fig, but larger. Fray Diego de Landa ([1590] 2014), in the 'Relación de las cosas de Yucatán' [*Relation of the Things of Yucatan*], mentioned the avocado as a large and fresh tree with fruit resembling a zucchini. Garcilazo de la Vega, ([1604] 2017), in 'Los Comentarios Reales de los Incas' [*The Royal Commentaries of the Incas*], referred to the avocado as a tasty and healthy fruit for sick people, and with sugar it is like a preserve. Fray Francisco Ximénez ([1615] 2001), in his four books 'Naturaleza y Virtudes de las Plantas y Animales de uso Medicinal en la Nueva España' [*Nature and Virtues of Plants and Animals for Medicinal Use in New Spain*], described the *ahuacaquahuitl* tree as an oak with orange leaves and fruits such as figs. Later, Juan and Ulloa ([1748] 1970), in 'Relación Histórica del Viaje a la América Meridional' [*The Historical Relationship of the Journey to*

South America], described the avocado as a tall, bushy and very leafy tree with fruit like gourds. The avocado tree and its fruit were also described by Carvajal ([1542] 1999) in 'Relación del Nuevo Descubrimiento del Rio Grande de las Amazonas' [*Account of the New Discovery of the Rio Grande of the Amazonas*]; Fray Bernardino de Sahagún ([1590] 2019) in the 'Historia General de las cosas de la Nueva España' [*General History of the Things of New Spain*]; Father José Acosta ([1590] 2002) in the 'Historia Natural y Moral de las Indias' [*Natural and Moral History of the Indies*]; Antonio de Ciudad Real ([1590] 1993), in his 'Tratado Curioso y Docto de las Grandezas de la Nueva España' [*Curious and Learned Treatise on the Greatness of New Spain*]; and Fray Bernabé de Cobo ([1653] 2008) in 'Historia del Nuevo Mundo' [*History of the New World*].

In 1657, 2 years after the British took Jamaica from the Spanish, 'A Book of the Continuation of Foreign Passages' mentions avocado as 'avocatas'. Subsequent reports from Jamaica refer to avocado as albecato, Spanish pear, abacado, avocado and alligator pear (Popenoe and Zentmyer, 1997). This report is the first mention of avocado in the West Indies, although it is assumed to have been brought to the West Indies earlier, probably first to Hispañola (Dominican Republic and Haiti) and/or Cuba, by the Spanish (Popenoe, 1939).

At the end of the 16th century, Philip II, King of Spain and the Indies, asked Francisco Hernández for a description of the plants, animals and minerals of New Spain, paying particular attention to their uses and virtues. Hernández wrote 'Historia de las Plantas de la Nueva España' [*History of the Plant of New Spain*], in which he described 'the ahoacaquahuit tree as an oak with leaves that smell of anise and hanging black fruits like figs' (Hernández, [1605] 2007). At that time, Samuel Champlain also described the avocado in his manuscript 'Narrative of a Voyage to West Indies and México in the Year 1599–1602, with Maps and Illustrations' (Webber, 1936). The Spaniards soon appreciated the avocado fruit, and the Guatemalan and West Indian varieties were considerably improved and subsequently dispersed (Knight *et al.*, 1999a, b). They took it to Spain, other colonies in America and even other places such as Indonesia in 1750 (Knight, 2002).

Historically, systematic botanists and ethnobotanists agree that the avocado was unknown in the European continent before pre-Columbian times (Scora and Bergh, 1992). In Europe, the first record of the avocado is found in 'Rariorum Plantarum Historia', where Clusius in 1601 described them as trees of Mexican origin and type developed in Valencia, Spain (Popenoe, 1963). Later, in 1753, Carl Linnaeus described the avocado in 'Species Plantarum' as a plant with pear-shaped fruit and a habitat in warm America and gave it the name *Laurus persea*. Later, in 1754, Miller published a description of the avocado under the genus and species name *Persea americana*, which has been preserved to date (Kopp, 1966).

As the colonial era progressed, avocado found its way around the world. Some of the earliest long-distance movements were to the Philippines by the Spanish, perhaps as early as the end of the 16th century, on the trading galleons sailing from Acapulco to Chile by Spanish clergy during the early colonial times. Avocado dispersion by the Spanish has been well documented by Morton (1987), Smith *et al.* (1992) and Knight (2002).

Cultural Importance

This journey through avocado history allows us to perceive its cultural importance. Before the arrival of the Spanish, iconographic data provide a symbolic representation of the relationship of the avocado with different cultures. In addition, linguistic evidence indicates that the avocado was valued by different cultures that inhabited Mesoamerica. For example, avocado was called *Oj, oh, um, un* and *on* by Mayans in the highlands; and *on, oon* and *matuuloj* in -lowlands; cupanda by Tarascans (Mich.); *ahuacatl* and *pahua* by Aztecs; *yaus, yashu, ishu, isu* by Zapotecs (Oax.); and *nttzani* by Otomies (SLP).

The Mayans in the Haab calendar, developed between 300 and 800 BC, named the months based on the seasons and agricultural events; for example, the 14th month was represented by the avocado glyph that the Mayans called *Uniw, Uniiw* which corresponds to *K'ank'in* from the Yucatecan language of the 16th century. This word comprises two morphemes: *Kan*, which means yellow or ripe, and *kin*, which means day, sun, and time (Kettunen y Helmke,

2020). Another important reference to the interaction of the avocado tree with the Mayan culture is found in the inscriptions of the tomb of King Pakal in Palenque, Chiapas, México. On the sides of the sarcophagus, dated 650 BC, there are ten figures representing King Pakal's ancestors. Nine are believed to be male and one female. Each figure emerges from the Earth, and behind each one there is a tree representing a fruit, among which are cocoa, soursop, sapodilla and avocado (Schele, 1974). The pattern of the leaves seems to be associated with figures of gods. The fruit trees represent those planted by the Mayans and used as a food source, which they cared for around their houses, because they believed that their ancestors were reincarnated in them (Schele and Mathews, 1998). The female figure, Lady Onal, emerges with the avocado tree.

Avocado also had cultural importance for the Aztecs. In the Mendoza codex, a manuscript painted by a tlacuilo at the time of Viceroy Don Antonio de Mendoza ([1535–1550] 1989), the life and customs of the peoples who inhabited the area of influence of the Aztecs are described. The town of Ahuacatlán is mentioned as the place where avocado abounded (Teliz, 2000). Avocado also has been described among the fruits Aztecs exchanged in the Tlatelolco market (Cervantes de Salazar, [1594] 1985), and it is known that it was consumed by the 'Señores Aztecas' (Sahagún, [1590] 2019). Interestingly, the Nahua legend describes how Yaotl was turned into an avocado grasshopper (ahuacachapulin) as a deserved punishment for coursing the goddess Tehuehuetl (Robelo, 1951). In addition, avocado was valued as tribute. Fray Diego de Durán ([1867–1880] 1967), in his book 'Historia de las Indias de Nueva España' [*History of the Indies of the New Spain*], mentioned the avocado among the tributes and riches that entered Tenochtitlan from the provinces and cities that the Aztecs had held due to the war.

Avocado history is also related to how it was used by the different cultures that inhabited Mesoamerica in pre-Hispanic times. Cervantes de Salazar ([1594] 1985) documented in his 'Crónica de la Nueva España' [*Chronicle of New Spain*] that the avocado was a fruit that helped digestion (warm fruit) and that an oil or butter was made from the seed and bleach from the flower, which barbers used because it was fragrant. Garcilazo de la Vega ([1604] 2017) mentioned that the fruit was healthy for sick people. Fray Francisco Ximénez ([1615] 2001) mentioned that the fruit was used in lavatories, to relieve instep pain, cure blood chambers due to their astringency and prevent damage to the hair ends. The seed was used to mark cloth, and the mark was not removed. Labat (1742), in his book 'Noveau Voyage Aux Isles de L'Amerique', mentioned the avocado as warm and nutritious, and good for the stomach. Buds and branches in herbal tea were used to relieve the ailments of smallpox. The extract, when taken on an empty stomach, induces vomiting and expulsion of blood clots. The use of this fruit stops diarrhoea and dysentery and, being a warm fruit, stimulates sexual appetite. In the island of Santo Domingo, Labat (1742) mentioned that sangliers (a type of pig) eat the ripe fruit that is on the ground. These animals grow fat, and their meat acquires an excellent flavour.

Modern History

Currently, avocado is cultivated in most of the world where climatic conditions are favourable. When the Spanish arrived in the Americas, avocados were taken to Europe and Asia. Dates of introduction to specific locales are unclear due to the paucity of written documentation. A few early introductions amounted to much, mainly serving as curiosities in botanical gardens and not being widely adopted into local diets. An exception to this is in the Caribbean, where avocado quickly became adopted into the diet and spread widely and abundantly, so much so that early botanists believed it to be native to the islands. In general, these early dispersals of avocado were haphazard, and in many cases, the races distributed were not appropriate to the climate of their new habitats.

Avocado selections have been introduced into Central America, the Caribbean, South America, Africa, Oceania, Europe and other areas of Asia (Knight, 2002). By the 19th century, avocados were taken to the Canary Islands and Hawaii in 1855. In Brazil, West Indian avocados were introduced in 1809 and Guatemalan avocado later. This fruit was taken to the Philippines in 1890 and later to the Middle East (Israel). In 1908, French monks in the Laturn Monastery planted the first avocado trees in former Palestine. In 1924, grafted avocados from California were planted at Mique Israel Agricultural School. Nowadays, Israel produces avocados year-round, cultivating varieties with different ripening seasons. Table 2.1 shows a historical review of avocado introduction all over the world.

In North America, avocado was introduced into Florida and later into California. There are reports that Henry Perrine introduced the avocado (alligator pears) in Florida in 1833. It is unknown whether avocados were ever cultivated sporadically by Spanish colonialists in St. Augustine, Florida, or by Indigenous or Seminole residents of southern Florida. The literature does not address this topic, though it is considered likely (Hodgson, 1926; Popenoe, 1936).

In southern Florida, seedling trees from Cuban avocados (recognized as the best in the West Indies) were planted and succeeded to varying degrees in the early 1900s as American settlers arrived with the railroad's completion. One of the early pioneers in southern Florida was George B. Cellon, who began to bud avocados in 1901 (Popenoe, 1936). The practice of budding developed by Cellon allowed, for the first time, the vegetative propagation of *Persea*, something entirely new and revolutionary in its evolutionary history. In contrast to the age-old practice of seed propagation, humans could quickly and reliably perpetuate high-quality genotypes true-to-type.

It is through this same practice that almost all of the avocado cultivars currently in production have come into cultivation. Two West Indian race cultivars, originally propagated by Cellon and named 'Trapp' and 'Pollock', became important cultivars in tropical countries (Popenoe, 1936). The work of early pioneers in Florida was continued by David Fairchild, Edward Simmonds of the USDA, Will Booth and William J. Krome, who was an early commercial grower in Homestead, among others (Popenoe, 1936; Knight and Campbell, 1999b). Most of the Florida crop was then, as now, bound for the markets of the northern cities, principally New York. As time progressed, it was recognized that adding Guatemalan, Guatemalan–West Indian hybrids and Guatemalan–Mexican hybrids to the repertoire of cultivars could extend the season in Florida. Many popular varieties still in use today were selected, such as 'Lula', a West Indian × Guatemalan hybrid (Hodgson, 1926). Due to the overdevelopment of southern Florida, resulting in high land prices, competition from Cuba (until the embargo) and later the Dominican Republic, hurricanes and less-than-ideal edaphic conditions, the avocado industry never grew to the extent of the California industry. That said, it is recognized that the earlier beginning of the Florida industry inspired the southern California growers. The cultivars and techniques developed in southern Florida have greatly influenced and continue to influence the avocado industry throughout the Caribbean basin and the tropical world.

Strangely, evidence of avocado presence in California during the Spanish colonial missions is lacking. Some have suggested, especially in American Hispanic folklore, that because avocados are aphrodisiacs, the Franciscans rejected the fruit. The earliest verified introduction of avocado to California was in 1856 on the property of Dr Thomas J. White near San Gabriel, who had obtained various tropical fruit species from Nicaragua (Condit, 1916), probably taken to California by a Nicaraguan gold prospector. Most likely, this introduction was unsuccessful, as the tree(s) were probably of West Indian descent. The first successful introduction of cold-hardy avocados from México was that of Judge R. B. Ord of Santa Barbara in 1871; two of these trees fruited and encouraged interest in avocado culture in southern California (Popenoe, 1936).

As the 19th century came to a close, avocados of not particularly high quality existed in California. It was in the early 20th century that the California industry (and by extension and influence most of the commercial world industry) was born. As more and more seedling trees came into bearing in southern California, the variability of quality was noted, and the hunt for superior cultivars commenced. Taking the cue from Florida, growers began the budding of superior chance seedlings in California around 1905 (Popenoe, 1936). Some of the seedlings were the progeny of local trees, while others contained genes from the seedlings grown from the avocados and seeds constantly and increasingly being imported from México. A small nursery industry began to grow in southern California,

Table 2.1. Introduction of avocado after the late 16th century by region, country and date.

Country	Date of introduction	Citation
North America[a]		
USA (mainland)	1825	Condit, 1916; Fairchild, 1945; Purseglove, 1968; Popenoe, 1974
Hawaii	1833 (1835)	Gustafson, 1976; Morton, 1987
Florida	1833 (1835)	Wolfe *et al.*, 1942; Knight, 2010
California	1856 (1871)	Condit, 1916; Popenoe, 1936
South America[b]		
Brazil	1809	Simão, 1971 (Wiltbank, 1977)
Chile	17th century	Schmidt, 1965
Oceania		
Australia	Mid- to late 19th century	Alexander, 1978; Whiley, 1982, 1983; Moore, 2001
New Zealand	1920	White, 2001
Asia		
China	1924	Liu and Zhou, 2000
India (Bangalore)	1892; <1906	Raman and Balaram, 1967; Morton, 1987
Indonesia	18th century (~1750)	Morton, 1987; Harjadi, 2000
Malaysia	1830–1840	Morton, 1987; Gaillard and Godefroy, 1995
Myanmar (Burma)	<1940	Soe, 2000
Philippines	Late 16th century? (1890)	Sotto, 2000; Knight, 2002
Sri Lanka	1883; 1927	Schröeder and Schröeder, 1982; Dionysius, 2000
Taiwan	1918	Liu and Zhou, 2000
Thailand	*c.*1910	Babpraserth and Subhadrabandhu, 2000
Vietnam	1940	Chau and Truyen, 2000
Caribbean		
Dutch West Indies (part of Lesser Antilles)	<1750	Gaillard and Godefroy, 1995
Jamaica	~1650 (1696)	Popenoe, 1974; Morton, 1987; Gaillard and Godefroy, 1995
Martinique	18th century	Gaillard and Godefroy, 1995
Africa		
Algeria	18th century	Gaillard and Godefroy, 1995
La Réunion	18th century	Gaillard and Godefroy, 1995
Madagascar	18th century	Gaillard and Godefroy, 1995
Mauritius	1780	Morton, 1987
Morocco	1931	Gaillard and Godefroy, 1995
South Africa	late 19th century (West Indian Seedlings); >1920	Le Roux, 1940; van Zyl and Ferreira, 1995
Zanzibar (Tanzania)	1892	Ghosh, 2000
Western Europe		
Canary Islands	19th century	Rodriguez-Garcia, 1979; Galán-Saúco, 1990
Corsica	1958	Vullin, 1982
Italy	1909	Calabrese *et al.*, 1995
Spain	1601	Popenoe, 1963; Purseglove, 1968
Middle East		
Egypt	1870–1880	Milad, 1936
Israel	1908	Shachar, 1982; Homsky, 1995
Palestine	1924	Ryerson, 1929

where superior avocados were propagated by budding. One of these nurseries was West India Gardens founded by F.O. Popenoe in Altadena, California. His son, Wilson Popenoe, would become one of the most influential tropical fruit horticulturists of his time. Wilson Popenoe (1936) wrote of this time in California:

> *Notwithstanding the considerable number of local seedlings which had come upon the scene, vigorously recommended by their backers, it was still evident that the ideal commercial avocado for California conditions had not put in its appearance. You will say, of course, that we should have recognized, in the first place, that no one variety would be found suitable to the diverse conditions of climate and soil which are represented in California. But we did not realize any such thing. We thought of the Washington Navel orange, and more particularly of the Trapp avocado of Florida. We wanted a California Trapp; an -avocado of good commercial characteristics which could be grown from San Diego to Santa Barbara.*

In an effort to find superior material to grow in California, West India Gardens commissioned Carl B. Schmidt to travel and collect avocado budwood in México. He sent back 41 selections, one of which was from Atlixco, Puebla. The first 'commercial' avocado orchard of budded trees was planted in 1908 by William Hertrich in San Marino, Los Angeles County, but the orchard was heavily damaged by the catastrophic '1913 Freeze'. The trees that survived the freeze were named 'Fuerte', after the Spanish word for strong, and were planted in Orange County in 1914, and later introduced into many other countries. But this crop had short harvest seasons and erratic yields. At that time, the California Avocado Association was formed (1915), and the common name was changed from the common Mexican name 'ahuacate' used in California and 'alligator pear' in Florida to 'avocado' by the association (Shepherd and Bender, 2001).

This cold-hardy, superb-tasting cultivar was released by West India Gardens in 1911, and by 1940 it occupied nearly 75% of the orchard area in California, while it was spread around the globe, birthing new avocado industries in other subtropical Mediterranean and summer-rainfall climates, including Israel, South Africa, Australia and Chile (Knight, 2002).

While 'Fuerte' continues to be an important variety around the world, it has been eclipsed by another chance seedling that arose in southern California, the 'Hass' avocado, by far the most widely grown cultivar of *P. americana* in the world today. 'Hass', originated around 1926 in La Habra Heights, California, when a farmer from Wisconsin, named Rudolph Hass, purchased some Guatemalan avocado seedlings and grafted them to the 'Fuerte' variety. After several attempts, he obtained a fruit with creamy flesh, nutty taste, lack of fibre, 18% oil content and a long harvest season. This selection was named after Rudolph Hass, and a patent was obtained in 1935 (Smith *et al.*, 1992; Shepherd and Bender, 2001; Chen *et al.*, 2009). The genetic provenance of 'Hass' is not fully resolved; however, this variety falls predominantly into the Guatemalan group (Solares *et al.*, 2023). By 1990, 'Hass', though slowly accepted by consumers, had become the industry norm, accounting for 83% of California's avocado production, while 'Fuerte' had faded to only 2% (Shepherd and Bender, 2001). Though no doubt an economic (and gustatory) success, some are concerned that the overreliance on 'Hass' makes the industry vulnerable and is leading to a drastic genetic narrowing of avocado production worldwide (Smith *et al.*, 1992).

In addition to this cultivar, other ones have been developed. 'Gwen', as an example, is a descendant of 'Hass'. In 1956, 'Gwen' emerged at the Citrus Experiment Station, in Riverside, California, and in the 1980s was introduced into the market. At that time, avocado improvement via interracial hybridization between Guatemalan and Mexican germplasm in California and Guatemalan and West Indian in Florida resulted in complex and inaccurately characterized modern cultivars (Chen et *al.*, 2009). Avocado-producing countries around the world have embraced cultivars and horticultural techniques that emerged from these pioneering avocado-growing areas.

As the selected cultivars from California and Florida spread around the globe during the middle of the 20th century, fledgling industries were boosted and new industries were founded (Smith *et al.*, 1992). During this time, commercial orchards based on imported cultivars, and later on some local selections, were planted in Israel (Homsky, 1995), Chile (Hofshi, 2002), Brazil, South Africa (Witney, 2002), Australia (Whiley, 2000), New Zealand (White, 2001), the Dominican Republic and in the highlands of Kenya (Sauer, 1993).

Newer production areas with rapidly growing harvests include China, Cameroon, the Democratic Republic of Congo, Ethiopia (Statista, 2004), the Dominican Republic and Perú. In these newer areas and others, avocado has become an important part of local agriculture, even though this importance is not necessarily reflected in the literature. In some countries, industries are nascent, as in much of southeast Asia. In many other areas, organized industries do not exist, yet avocado is an important part of the local diets and markets. For example, it is noted that in Cameroon, avocado is perhaps the most abundant fruit tree in home gardens and agroforests (G. Coppens, Montpellier, France, 2011, personal communication), while in the island of Mare in the Loyalty Islands, New Caledonia, an annual avocado festival is held each May to celebrate the avocado (Logan and Cole, 2001). Thus, as avocado has spread, its importance in local agriculture systems and food security is not always reflected in statistics focused on commercial-scale production, mainly in subtropical cultivars as opposed to tropical cultivars. In the USA, avocado became a commercial crop in the 1950s. Recently, the avocado has become the second most important tropical fruit produced in the world (FAO, 2023).

Since the middle of the 20th century, México has been the leading producer in the world. In the 1960s, the production was around 108,000 tonnes, which has since progressively increased reaching 2,699,738.52 tonnes by 2020 (Statista, 2024). According to SIAP (2021), México is the top worldwide producer with 2.4 million tonnes and the main exporter, with a production that increased by 13.9% between 2019 and 2020 (FAOSTAT, 2022). The Dominican Republic was the second producer with 372,117 tonnes, Colombia the third with 335,977 tonnes, and Perú the fourth with 308,593 tonnes from 2000 to 2022. However, ample global demand and high export prices continue to be critical growth drivers in this dynamic sector, stimulating substantial investments in area expansion in emerging producing countries (FAOSTAT, 2024).

In regard to exports, ample global demand and high export prices propitiate avocado production expansion in emerging producing countries. Even when México produces avocado in all seasons, data from the last years indicate a fall in shipments. On the other hand, the harvested area in Perú has expanded from 2017 to 2021, consolidating this country as a major global exporter. Kenya is also an emerging supplier to European Union markets. Chile and South Africa remain significant exporters, which continue expanding this industry.

Regarding imports, the two major import markets are the USA and the European Union, accounting for 45% and 25% of the international avocado trade, respectively.

World Production

The global avocado market size has been estimated at USD 14.85 billion in 2022 and was expected to reach USD 15.83 billion in 2023. The global avocado market could reach USD 26.04 billion by 2030, assuming an annual growth rate of 7.3%. The demand for nutritious food is a factor driving the consumption of fresh avocados. The improvement in living standards, increases in health expenditures and disposable income and the large-scale promotion of processed avocado are factors driving the avocado market (GBR, 2024).

As of 2022, 64 countries produced 898 million metric tonnes of avocado. Notably, the global production of avocado has increased in recent years from 2,699,738.52 tonnes in 2000 to 8,978,275.2 tonnes in 2022, with the Americas accounting for 71.2% of global production, Africa for 13%, Asia for 12.3%, Europe for 1.9% and Oceania for 1.6%. México, Dominican Republic, Colombia, Perú, Indonesia, the USA, Brazil, Kenya, Chile and Vietnam were the top ten producers on average from 2000 to 2022 (FAO, 2024).

As more countries produce avocado for export or become familiar with the fruit, the increase in production area, production and consumption is likely to continue. Although avocado is primarily consumed as fresh fruit, substantial gains in the use of 'value-added' avocado products (e.g. guacamole) and oil processing for cosmetics and culinary use also suggest further market growth.

References

Acosta, J. [1590] (2002) *Natural and moral history of the indies*. Duke University Press. Durham, North Carolina.

Alexander, D. McE. (1978) *Some Avocado Varieties for Australia*. CSIRO, Australia.

Ashworth, V.E.T.M. and Clegg, M.T. (2003) Microsatellite markers in Avocado (*Persea americana* Mill.): genealogical relationships among cultivated Avocado genotypes. *Journal of Heredity* 94, 407–415.

Babpraserth, C. and Subhadrabandhu, S. (2000) Avocado production in Thailand. In: Papademetriou, M.K. (ed.) *Avocado Production in Asia and the Pacific*. FAO/RAP Publication 2000/09, FAO Publisher, Bangkok, pp. 57–64.

Barlow, C. (2002) *The Ghosts of Evolution*. Perseus Books Group. USA.

Berdugo-Cely, J.A., Cortés, A.J., López-Hernández, F., Delgadillo-Durán, P., Cerón-Souza, I. *et al.* (2023) Pleistocene-dated genomic divergence of avocado trees supports cryptic diversity in the Colombian germplasm. *Tree Genetics & Genomes* 19:42.

Betz, V.M. (1999) Cotton, maize and chocolate: Plant cultivation in prehistoric Mesoamerica. *Athena Review* 2, 24–31.

Bost, J.B, Smith, N.J.H. and Crane J.H. (2013) History distribution and uses. In: Bruce Schaffer, B. Nigel Wolstenholme and Anthony William Whiley (eds) *The Avocado: Botany, Production and Uses,* 2nd edn. CAB International, Wallingford, UK, pp. 10–33.

Buckler IV, E.S., Pearsall, D.M. and Holtsford, T.P. (1998) Climate, plant ecology, and Central Mexican archaic subsistence. *Current Anthropology* 39, 152–164.

Bufler, G. and Ya'acov, A.B. (1992) *A study of avocado germplasm resources, 1988–1990. III. Ribosomal DNA repeat unit polymorphism in avocado*. Second World Avocado Congress. California University, Orange, California, pp. 545–550.

Calabrese, F., Azzena, M., Barone, F., Continella, G., D'Ascanio, R. *et al.* (1995) Research on avocado in Italy. *Proceedings of the Third World Avocado Congress.* Tel Aviv, Israel, pp. 457–460.

Carvajal, G. de. [1542] (1999) *Relación del nuevo descubrimiento del famoso Río Grande de las Amazonas.* Fondo de Cultura Económica. México.

Casas, A., Otero, A., Pérez, E. and Valiente, A. (2007) In situ management and domestication of plants in Mesoamerica. *Annals of Botany* 100, 1101–1115.

Cervantes de Salazar, F. [1594] (1985) *Crónica de la Nueva España.* Editorial Porrúa, S.A. México.

Chanderbali, A., van der Werff, and Renner, S. (2001) Phylogeny and historical biogeography of Lauraceae: evidence from chloroplast and nuclear genomes. *Annals of the Missouri Botanical Garden* 88, 104–134.

Chau, N.M. and Truyen, V.T. (2000) Avocado production in Vietnam. In: Papademetriou, M.K. (ed.) *Avocado Production in Asia and the Pacific*. FAO Publisher, Bangkok, pp. 65–72.

Chen, H., Morell, P.L., Ashworth, V.E.T.M., De la Cruz, M. and Clegg, M.T. (2009) Tracing the geographic origins of mayor avocado cultivars. *Journal of Heredity* 100, 56–65.

Ciudad Real, A. [1590] (1993) *Tratado curioso y docto de las grandezas de la Nueva España*. Universidad Nacional Autónoma de México. México.

Clegg, M., Brandon, G., Duvall, M. and Davis, J. (1993) Inferring plant evolutionary history from molecular data. *New Zealand Journal of Botany* 3,307–316.

Cobo, B. [1653] (2008) *Historia del Nuevo Mundo*. Biblioteca de Autores Españoles. Sevilla, Spain.

Colunga, P. and Zizumbo, D. (2004) Domestication of plants in Maya lowlands. *Economic Botany* 58, S101–S110.

Condit, I.J. (1916) History of the avocado and its varieties in California with a checklist of all named varieties. *California Avocado Association 1916 Annual Report* 2, 105–144.

Corona, E., Galindo, M.E., Lee, H. and Landero, I. (2016) Diversidad genenética del aguacate (*Persea americana* Mill.) en cuatro zonas de su área de dispersión natural. *Agroproductividad* 9, 80–85.

Davis, J., Kobayashi, H. and Clegg, M. (1998) Genealogical relationships among cultivated avocado as revealed through RFLP analyses. *The American Genetic Association* 89, 319–323.

Diamond, J. (2017) *Guns, germs, and steel. The fates of human societies*. W. W. Norton and Company, Inc, USA.

Dine, H., Arden, T., Bascopé, G. and Báez, C.G. (2019) Famine Foods and Food Security in the Northern Maya Lowlands: modern Lessons for Ancient Reconstructions. *Ancient Mesoamerica* 30, 517–534.

Dionysius, M. (2000) Avocado production in Sri Lanka. In: Papademetriou, M.K. (ed.) *Avocado Production in Asia and the Pacific*. FAO Publisher, Bangkok, pp. 49–56.

de Durán, F.D. [1867–1880] (1967) *Historia de las Indias de Nueva España e Islas de Tierra firme*. Editorial Porrúa, México.

Eklund, H. (2000) Lauraceous flowers from the Late Cretaceous of North Carolina, U.S.A. *Botanical Journal of the Linnean Society* 132, 397–428.

Ellstrand, N. and Marshall, D. (1985) The impact of domestication on distribution of allozyme variation within and among cultivars of radish, *Raphanus sativus* L. *Theoretical Applied Genetics* 69, 393–398.

Fairchild, D. (1945) Personal recollections of George B. Cellon, horticultural pioneer of South Florida. *Proceedings of the Florida State Horticultural Society* 58, 205–209.

FAO (2023) Major tropical fruits market review – preliminary results 2022. Rome. https://openknowledge.fao.org/server/api/core/bitstreams/852265a4-9006-4d54-a792-51f1d9c44673/content (accessed 22 September 2024).

FAOSTAT (2022) Selected indicators. Available at: https://www.fao.org/faostat/en/#country (accessed 12 August 2023).

FAOSTAT (2024) Crops and livestock products. Available at: https://www.fao.org/faostat/en/#data/QCL/visualize (accessed 22 September 2024).

Fedick, S. (1995) Indigenous agriculture in the Americas. *Journal of Archaeological Research* 3, 257–303.

Fernández de Enciso, M. [1519] (2014) *Suma de Geografía que trata de todas las partidas y provincias del mundo, en especial de las indias y trata largam ete del arte del marear ju tamente*. Nuevo Mundo Paraiso, México.

Fernández de Oviedo, G. [1526] (1999) *Sumario de la Natural Historia de las Indias*. Fondo de Cultura Económica, México.

Fielder, J., Buffer, G. and Bangerth, F. (1998) Genetic relations of avocado (*Persea americana* Mill.) using RAPD markers. *Euphytica* 101, 249–255.

Fuller, D.Q. (2018) Long and Attenuated: comparative Trends in the Domestication of Tree Fruits. *Vegetation History and Archaeobotany* 27,165–176.

Furnier, G., Cummings, M. and Clegg, M. (1990) Evolution of the avocados as revealed by DNA restriction fragment variation. *Journal of Heredity* 81, 183–188.

Gaillard, J.P. and Godefroy, J. (1995) *Avocado*. Technical Centre for Agriculture and Rural Cooperation, Wageningen, The Netherlands, pp. 1–3.

Galán-Saúco, V. (1990) Aguacate (*Persea americana* Mill.) In: Galán-Saúco, V. (ed.) *Los Frutales Tropicales en Los Subtropicos, I. Aguacate-Mango-Litchi y Longan*. Ediciones Mundi-Prensa, Madrid, pp. 25–58.

Galindo, M. and Arzate A. (2010a) Consideraciones sobre el origen y primera dispersión del aguacate (*Persea americana* Lauraceae). *Cuadernos de Biodiversidad* 33, 11–15.

Galindo, M. and Arzate A. (2010b) West Indian Avocado: where did it originate? Phyton. *International Journal of Experimental Botany* 79, 203–207.

Gama, L. and Gómez. A. (1992) *An ethnoecological approach for the study of Persea: a case study in the Maya area*. Second World Avocado Congress, University of California, Orange, California, pp. 11–17.

GBR (2024) Avocado Market Size, Share & Trends Analysis Report. In: *Avocado Market Report*. Globalnewswire. Available at: https://www.grandviewresearch.com/industry-analysis/fresh-avocado-market-report/to (accessed 22 September 2024).

Ge, Y., Dong, X., Wu, B., Wang, N., Chen, D. *et al.* (2019) Evolutionary Analysis of Six Chloroplast Genomes from Three *Persea americana* Ecological Races: into Sequence Divergences and Phylogenetic Relationships. *PLOS ONE* 14(9): e0221827.

Gepts, P. (2005) Plant and animal domestication as human-made evolution. In: Cracaft, J. and Bybee, R. (eds) *Evolutionary Science and Society: Educating a new generation*. BSCS, AIBS symposium. NABT Convention Chicago, IL. BSCS, Colorado Springs/USA.

Ghosh, S.P. (2000) Avocado production in India. In: Papademetriou, M.K. (ed.) *Avocado Production in Asia and the Pacific*. FAO Publisher, Bangkok, pp. 24–30.

Goebel, T., Waters M. and O'Rourke, D. (2008) The late Pleistocene dispersal of modern humans in the Americas. *Science* 319, 1497–1502.

Gómez-Pompa, A., Flores, J. and Fernández, M. (1990) The sacred cacao groves of the Maya. *Latin American Antiquity* 1, 247–257.

Gustafson, D. (1976) World avocado production. *Proceedings of the First International Tropical Fruit Short Course, The Avocado*. Fruit Crops Department, Florida Cooperative Extension Service, University of Florida, IFAS, Gainesville, Florida, pp. 1–15.

Gutierrez, A., Martínez, J., García, E., Iracheta, L., Ocampo, J. and Cerda, I. (2009) Estudio de la diversidad genética del aguacate nativo en Nuevo León, México. *Revista Fitotecnia Mexicana* 32, 9–18.

Harjadi, S.S. (2000) Avocado production in Indonesia. In: Papademetriou, M.K. (ed.) *Avocado Production in Asia and the Pacific*. FAO Publisher, Bangkok, pp. 31–35.
Healy, P. (1978) Excavations at Rio Claro, Northeast Honduras: preliminary report. *Journal of Field Archaeology* 5, 16–28.
Heiser, C. (1979) Origins of some cultivated new world plants. *Annual Review of Ecological Systems* 10, 309–326.
Hernández, F. [1605] (2007) *Historia natural de la Nueva España*. Imprenta Universitaria. UNAM. México.
Hodgson, R.W. (1926) The Florida Avocado industry – a survey. *California Avocado Association Annual Report 1925–26 Yearbook* 10, 60–71.
Hofshi, R. (2002) The Chilean Avocado industry: an overview. *AvoResearch* 2, 1–12.
Homsky, S. (1995) The avocado industry in Israel – an overview. *Alon Hanotea* 49, 479–488.
Juan, J. and de Ulloa A. [1748] (1970) Relación histórica del viaje a la América Meridional, hecho de orden de S. Mag. para medir algunos grados de meridiano terrestre, y venir por ellos en conocimiento de la verdadera figura, y magnitud de la Tierra, con otras observaciones astronómicas, y físicas. Biblioteca Digital AECID. Available at: https://bibliotecadigital.aecid.es/bibliodig/es/consulta/registro.cmd?id=1084 (accessed 22 September 2024).
Kennett, D.J., Lipson, M., Prufer, K.M. Mora-Marín, D., George, R.J. *et al.* (2022) South-to-north migration preceded the advent of intensive farming in the Maya region. *Nature Communications* 13, 1530. doi:10.1038/s41467-022-29158-y.
Kennett, D.J., Thakar, H.B., VanDerwarker, A.M., Webster, D.L., Culleton, B.J. *et al.* (2017) High-Precision Chronology for Central American Maize Diversification from El Gigante Rockshelter, Honduras. *Proceedings of the National Academy of Sciences* 114(34), 9026–9031.
Kettunen, H. and Helmke, C. (2020) Introduction to Maya hieroglyphs. Available at: https://www.mesoweb.com/resources/handbook/IMH2020.pdf (accessed 22 September 2024).
Knight, R. (2002) History, Distribution and Uses. In: Whiley, A. W., Schaffer, B. and Wolstenholme, B.N. (eds) *The Avocado: Botany, Production and Uses*. CAB International, Wallingford, UK.
Knight, R.J. (2010) The Florida experience: importation, breeding, and exportation. Available at: http://www.quisqualis.com/avoflknight.html (accessed 14 February 2024).
Knight, R.J. and Campbell C. (1999a) Ecological adaptation and the evolution of modern avocado cultivars. *Revista Chapingo. Serie Horticultura* 5, 49–54.
Knight, R.J. and Campbell, C.W. (1999b) Florida's contribution to the world avocado industry. *Proceedings of the Florida State Horticultural Society* 112, 233–236.
Kopp, L.E. (1966) A taxonomic revision of the genus Persea in the Western hemisphere (Persea-Lauraceae). *Memoirs of the New York Botanical Garden* 14, 1–120.
Labat, R.P. (1742) *Noveau Voyage Aux Isles de L'Amerique*. Available at: https://ia804501.us.archive.org/31/items/nouveauvoyagea04laba/nouveauvoyagea04laba_bw.pdf (accessed 22 September 2024).
Landa, F. [1590] (2014) *Relación de las Cosas de Yucatán*. Editorial Linkgua. España.
Landero I., Oliva, H., Galindo M., Balcazar M., Murguía J. and Ramos J. (2012) Uso de la larva Arsenura Armida Armida (Cramer, 1779) (Lepidoptera:Saturniidae), 'cuecla' en Ixtahuapa, Veracruz, México. Cuadernos de Biodiversidad 38, 4–8.
Lavi, U., Sa'ada, D., Ragev I. and Lahav, E. (2003) Avocado genetics and breeding-present and future. *World Avocado Congress V* 42, 134–135.
Le Roux, J.C. (1940) *The Avocado in South Africa*. Department of Agriculture and Forestry Pamphlet no. 217. Pretoria, South Africa.
Liu, K. and Zhou, J. (2000) Avocado production in China. In: Papademetriou, M.K. (ed.) *Avocado Production in Asia and the Pacific*. FAO Publisher, Bangkok, pp. 15–23.
Logan, L. and Cole, G. (2001) *Lonely Planet New Caledonia.* Lonely Planet Publications, London.
Mac Neish, R.S. (1964) Ancient mesoamerican civilization. *Science* 143, 531–537.
MacDougall, A. (2003) Did native Americans influence the northward migration of plants during Holocene? *Journal of Biogeography* 30, 633–647.
Mansur, L.J., Millaire, J. and Blake, M. (2018) Peanuts and Power in the Andes: the Social Archaeology of Plant Remains from the Virú Valley, Perú. *Journal of Ethnobiology* 38(4), 589–609.
Mendoza, E.C. (1989) Colección de Mendoza o Códice Mendocino: documento mexicano del siglo XVI que se conserva en la biblioteca de Bodleiana de Oxford, Inglaterra. /facsímile fototípico dispuesto por Francisco del Paso y Troncoso; edición acom- pañada de una introducción, anotaciones y comen- tarios por J. Galindo y Villa. Editorial Innovación, México.

Messer, E. (2003) Plantas alimenticias Zapotecas: transformación de dos culturas. In: Long, J. (ed.) *Conquista y comida. Consecuencias del encuentro de dos mundos*. Instituto de Investigaciones Históricas UNAM, México, pp. 312–337.

Mhameed, S., Sharon, D., Kaufman, D., Lahav, E., Hillel, J., Degani, C. and Lavi, U. (1997) Genetic relationships within avocado (*Persea americana* Mill.) cultivars and between Persea species. *Theoretical Applied Genetics* 94, 279–286.

Milad, Y. (1936) Avocado culture in Egypt. *California Avocado Society Yearbook* 21, 76–78.

Millar, C. (1996) Tertiary vegetation history. In: *Sierra Nevada Ecosystem Proyect: Final report*. University of California, Davis, California, pp. 71–122.

Moore, F. (2001) Australian avocado industry update for 2001: the history and development of the Australian avocado industry. *California Avocado Society Yearbook* 85, 65–75.

Morton, J.F. (1987) *Fruits of Warm Climates*. Wipf and Stock Publishers, Eugene, Oregon, pp. 91–102.

Ortiz, P. and Rodríguez, M.C. (2000) The sacred hill of El Manatí: a preliminary discussion on the site's ritual paraphernalia. In: Clark, J.E and Pye M.E. (eds) *Olmec Art and Archeology in Mesoamerica*. National Gallery of Art. USA.

Pope, K., Pohl, M., Jones, J., Lentz, D., von Nagy, C., Vega, F. and Quitmyer, I. (2001) Ancient Agriculture in the Lowlands of Mesoamerica. *Science* 292, 1370–1373.

Popenoe, W. (1936) Looking back. *California Avocado Society Yearbook* 21, 51–59.

Popenoe, W. (1939) *Manual of Tropical and Subtropical Fruits*. MacMillan, New York.

Popenoe, W. (1963) Early history of avocado. *California Avocado Society 1964 Yearbook* 47, 19–24.

Popenoe, W. (1974) *Manual of Tropical and Subtropical Fruits*, 2nd edn. Hafner Press, New York, USA.

Popenoe, W. and Zentmeyer, G.A. (1997) Early story of avocado. *California Society 1997 Yearbook* 81, 163–171.

Pozorski, S. (1979) Prehistoric diet and subsistence of the Moche Valley, Perú. *World Archaeology* 2, 163–185.

Purseglove, J.W. (1968) *Tropical Crops, Dicotyledons*. Wiley, New York, pp. 192–198.

Raman, T. and Balaram, P. (1967) Avocado culture in India. *California Avocado Society Yearbook* 51, 97–106.

Raven, P. and Axelord. D. (1974) Angiosperm biogeography and past continental movements. *Annals of the Missouri Botanical Garden* 61, 539–673.

Renner, S. (2004) Variation in diversity among Laurales: Early Cretaceous to present. *Biologiske Skrifter* 55, 441–458.

Robelo, C.A. (1951) *Diccionario de Mitología Nahuatl*. Ediciones Fuente Cultural. México.

Rodriguez-Garcia, V. (1979) La historia del Jardin Botánico de Tenerife en el siglo XVIII. Las fuentes documentales del Archivo General de Indias en Semilla. EXMO. Cabildo Insular de Gran Canaria. Separata aparte del Libro II Coloquio de Historia Canaria Americana, 1977, Tomo II.

Rosenswig, R.M., VanDerwarke, A.M., Culleton, B.J. and Kennett, D.J. (2015) Is it Agriculture yet? Intensified Maize-use at 1000 cal BC in the Soconusco and Mesoamerica. *Journal of Anthropological Archaeology* 40, 89–108.

Ryerson, K. (1929) Avocado notes from foreign countries Haiti and Palestine. *California Avocado Society Yearbook* 14, 96–99.

Sahagún, F. [1590] (2019) *Historia general de las cosas de la Nueva España*. Cien de México. México.

Sauer, J.D. (1993) *Historical Geography of Crop Plants: A Select Roster*. CRC Press, Boca Raton, Florida.

Schröeder, C.A. (1968) Prehistoric avocados in California. *California Avocado Society Yearbook* 52, 29–34.

Schröeder, C.A. and Schröeder, M.R. (1982) The Avocado in Sri Lanka. *California Avocado Society Yearbook* 66, 109–144.

Schele, L. and Mathews, P. (1998) *The code of the kings: The language of seven sacred Maya temples and tombs*. Scribner. New York. USA. p. 432.

Schele, L. (1974) *Observations on the cross motif at Palenke*. Mesa Redonda de Palenke. Pebble Beach, California.

Schmidt, M. (1965) Avocado growing in Chile. *California Avocado Society Yearbook* 49, 45–46.

Schnell, R., Brown, J., Olano, C., Power, E, and Krol. C. (2003) Evaluation of avocado germplasm using microsatellites markers. *Journal of the American Society of Horticultural Science* 128, 881–889.

Scora, R. and Bergh. B. (1992) *Origin of and taxonomic relationships within the genus Persea*. Second World Avocado Congress, California University, Orange, California, pp. 505–514.

Shachar, Z. (1982) The avocado in Israel. *California Avocado Society Yearbook* 66, 103–108.

Shepherd, J. and Bender, G. (2001) A history of the avocado industry in California. *California Avocado Society Yearbook* 85, 29–50.

SIAP (2021) Aguacate mexicano. Otro protagonista del Superbowl. SIAP. México. Available at: https://www.gob.mx/cms/uploads/attachment/file/614377/aguacate_brochure.pdf (accessed 22 September 2024).
Simão, S. (1971) *Manual de Fruticultura*. Editora Agronómica Ceres, Sao Paulo, Brazil.
Skidmore, J. (2005) Earlist Complex culture in the Americas. Mesoweb. Reports and News. Available at: https://www.mesoweb.com/reports/caral2.html (accessed 22 September 2024).
Slotten, V. and Lentz, D.L. (2021) Trees, Shrubs, and Forests at Joya de Cerén, a Late Classic Mesoamerican Village. *Quaternary International* 593–594, 270–283.
Smith, C. (1966) Archeological evidence for selection in avocado. *Economic Botany* 20, 169–175.
Smith, C. (1969) Additional notes on pre-conquest avocados in México. *Economic Botany* 23,135–140.
Smith, N.J.H., Williams, J.T., Plucknett, D.L. and Talbot, J.P. (1992) *Tropical Forests and Their Crops*. Cornell University Press, Ithaca, New York.
Soe, U. (2000) Avocado production in Myanmar. In: Papademetriou, M.K. (ed.) *Avocado Production in Asia and the Pacific*. FAO Publisher, Bangkok, pp. 36–38.
Solares, E., Morales-Cruz, A., Figueroa Balderas, R., Focht, E., Ashworth, V.E.T.M., *et al.* (2023) Insights into the domestication of avocado and potential genetic contributors to heterodichogamy. *Plant Genetics and Genomics* 13(2). DOI: 10.1093/g3journal/jkac323.
Solis, R., Hass, J. and Creamer, W. (2001) Dating Caral, a preceramic site in the Supé Valley on the Central Coast of Perú. *Science* 292, 723–726.
Sotto, R.C. (2000) Avocado production in the Philippines. In: Papademetriou, M.K. (ed.) *Avocado Production in Asia and the Pasific*. FAO Publisher, Bangkok, Thailand, pp. 39–48.
Spengler, R.N., Petraglia, M., Roberts, P., Ashastina, K., Kistler, L., Mueller, N.G. and Boivin, N. (2021) Exaptation Traits for Megafaunal Mutualisms as a Factor in Plant Domestication. *Frontiers in Plant Science* 12, 649394.
Statista (2024) Global avocado production by country. Available at: https://www.statista.com/statistics/593211/global-avocado-production-by-country/ (accessed 22 September 2024).
Storey, W., Bergh. B. and Zentmyer, G. (1986) The origin, Indigenous range, and dissemination of the avocado. *California Avocado Society Yearbook* 70, 127–133.
Taube, K. (2004) *Olmec Art*. Dumbarton Oaks, USA.
Taylor, D. (1988) Eocene floral evidence of Lauraceae: corroboration of the North American megafossil record. *American Journal Botany* 75, 948–957.
Telíz, D. (2000) *El aguacate y su manejo integrado*. Mundi Prensa. México.
Toledo, M., Alarcón, P., Moguer, P., Olivo, M., Leyquien, E., Cabrera, A. and Rodríguez. A. (2001) Atlas etnoecológico de México y Centroamérica: fundamentos, métodos y resultados. *Etnoecológica* 6, 1–41.
Tupper, W. (1927) A comparative study of Lauraceous woods. *American Journal of Botany* 14, 520–527.
Van Zyl, J.L. and Ferreira, S.G. (1995) An overview of the avocado industry in South Africa. *South African Avocado Growers' Association Yearbook* 18, 23–30.
Vega, I.G. de la [1604] (2017) *Cometarios Reales de los Incas*. Editorial Gradiente. Salamanca, Spain.
Vullin, G. (1982) Le greffage de l'avocatier en Corse, Essai d'une nouvelle technique. *Fruits* 37, 295–200.
Webber, H.J. (1936) An early illustration of the avocado. *California Avocado Association 1936 Yearbook* 21, 39–41.
Whiley, A.W. (1982) The Avocado – *Persea americana* Miller. *Australian Horticulture* 80(3), 80–89.
Whiley, A.W. (1983) Lauraceae. In: Page, P.E. (ed.) *Tropical Fruit Tree Fruits for Australia*. Queensland Department of Primary Industries, Brisbane, Australia, pp. 70–77.
Whiley, A.W. (2000) Avocado production in Australia. In: Papademetriou, M.K. (ed.) *Avocado Production in Asia and the Pacific*. FAO Publisher, Bangkok, pp. 5–14.
White, J. (2001) The avocado industry in New Zealand. *California Avocado Society Yearbook* 85, 77–91.
Williams, L. (1977) The avocados, a synopsis of the genus *Persea*, sub. *Persea*. *Economic Botany* 31, 315–320.
Wiltbank, W.J. (1977) Mango and avocado cultivars in Brazil. *Proceedings of the Florida State Horticultural Society* 90, 243–244.
Witney, G.W. (2002) An overview of the South African avocado industry. *AvoResearch* 2, 5–9.
Wolfe, H.S., Toy, L.R. and Stahl, A. (revised by Wolfe, H.S.) (1942) *Avocado production in Florida*. Bulletin 112. Agricultural Extension Service, University of Florida and USDA, Gainesville, Florida, pp. 6–9.
Wolters, B. (1999) Dispersion and ethnobotany of the cacao tree and other Amerindian crop plants. *Angewandte Botanik* 73, 128–137.
Ximénez, F.F. [1615] (2001) *Cuatro Libros de la Naturaleza y Virtudes de las Plantas y Animales, de Uso Medicinal en la Nueva España*. Imprenta de Juan Pablos, S.A., México.

3 Taxonomy and Botany

A.S. Chanderbali[1], D.E. Soltis[2], P.S. Soltis[1] and B.N. Wolstenholme[3]
[1]Florida Museum of Natural History, University of Florida, USA; [2]Department of Biology, University of Florida, USA; [3]University of KwaZulu-Natal, South Africa

Abstract

This chapter discusses the taxonomic placement of avocado within the *Persea* group belonging to the large and diverse plant family Lauraceae. Also discussed are the morphology, architecture and growth habits of the cultivated avocado tree, emphasizing differences among the three recognized horticultural races of avocado: the Mexican, Guatemalan and West Indian.

Introduction

The avocado, *Persea americana* Mill., is a member of the Lauraceae, a pantropical family of 57 genera and 3326 species (WFO, 2022). Lauraceae are among the most prominent components of tropical forests in the Neotropics. They replace Fabaceae (legumes) as the dominant plant family on the lower Andean slopes (Gentry, 1988; Tagawa, 1995; Song and Da, 2016). They also have a remarkably rich fossil record, consisting of flowers, fruits, leaves and wood, distributed across North America, Europe and Asia from the Mid-Cretaceous onwards (Drinnan *et al.*, 1994; Herendeen *et al.*, 1994; Eklund and Kvaček, 1998; Eklund, 2000; Chambers *et al.*, 2012; Takahashi *et al.*, 2014, 2001; Atkinson *et al.*, 2015; Poinar, 2017). This paleobotanical evidence indicates that Lauraceae were part of the major early diversification of angiosperms that occurred at roughly the same time as the initial diversification of eudicots and monocots and were likely among the most widespread and ecologically significant of the Cretaceous angiosperms (Takahashi *et al.*, 2014). In more modern times, the Greeks offered leaves of *Laurus nobilis* L., the commercial bay laurel, to reward accomplishment, a practice echoed by the term 'laureate' in contemporary accolades (e.g. Nobel Laureate; baccalaureate). Other uses include cinnamon and camphor from *Cinnamomum* Schaeff., perfume oils from *Aniba* Aubl. and durable timber from *Chlorocardium* Rohwer, H.G. Richt. & van der Werff, *Eusideroxylon* Teysmann & Binn. and *Mezilaurus* Kuntze ex Taub (Chanderbali *et al.*, 2008). Avocado, however, is undoubtedly the most important commodity from the family.

The Lauraceae are largely characterized by trimerous flowers, bi- or quadrilocular anthers with apical valvate dehiscence, and an unicarpellate gynoecium containing a single anatropous, apically attached ovule. Fruit are fleshy drupes, typically black at maturity and usually

achander@ufl.edu

(eds D. Carrillo, B. Schaffer, A.W. Whiley and B.N. Wolstenholme)
DOI: 10.1079/9781800621824.0003

borne on enlarged receptacles that can form cupulate (hemispherical) to fully enclosing structures surrounding the drupe (Rohwer, 1993). The monophyly of the family is well established (Renner, 1999), and intergeneric relationships are becoming clearer (Rohwer, 2000; Chanderbali *et al.*, 2001; Li *et al.*, 2020; Penagos Zuluaga *et al.*, 2021; Rohwer *et al.*, 2014), although current generic circumscriptions are often artificial (Chanderbali *et al.*, 2001). *Persea* belongs to a phylogenetically unresolved complex of 400 to 450 species in eight currently recognized genera, the *Persea* group, that originated and radiated in the early Eocene on the ancient northern supercontinent of Laurasia (Chanderbali *et al.*, 2001; Li *et al.*, 2011). The *Persea* group is now primarily distributed in tropical to subtropical Asia, about 80% of the species, whereas ~20% (including avocado) are native to warm temperate to tropical regions of the New World. Two species, *Apollonias barbujana* and *Persea indica*, are distributed in the Macaronesian Islands (Canary Islands and Madeira, with *P. indica* also on the Azores). This tropical amphi-Pacific disjunction in the *Persea* group has been credited to the disruption of the ancestral Boreotropical ranges by Late Eocene climatic cooling (Chanderbali *et al.*, 2001; Li *et al.*, 2011), with the Macaronesian species representing relicts of the European-Mediterranean Tertiary laurel forests (Rohwer *et al.*, 2009).

Generic delimitation within the *Persea* group has always been controversial and somewhat inconsistent. All 80–90 New World species (plus *P. indica*) have been placed in a broadly defined genus *Persea*, regardless of variation in characters such as the number of pollen sacs (locules) per anther, uniformity or dimorphism in tepal morphology and the relative persistence of tepals in fruit. In contrast, these characters have been used to delimit seven genera in the Old World (*Alseodaphne* [~50 spp.], *Alseodaphnopsis* [nine spp.], *Apollonias* [two spp.], *Dehaasia* [~35 spp.], *Machilus* [~100 spp.], *Nothaphoebe* [~40 spp.] and *Phoebe* [~100 spp.]). The American *Persea* was last revised taxonomically by Kopp (1966), who recognized two subgenera. The subgenus *Persea*, which includes avocado, is characterized by having equal or subequal tepals, these reflexed and usually deciduous in fruit, quadrilocular anthers, stipitate glands at the base of the stamens of the third whorl and a pubescent gynoecium. Subgenus *Eriodaphne* Nees is recognized by its unequal tepals (the inner three clearly longer than the outer three) that are persistent and patent in fruit, anthers with a variable number of locules (zero, two or four), sessile glands at the base of the filaments of the inner three stamens and a glabrous or pubescent gynoecium. The main differences between the two subgenera (equal tepals, which are either reflexed or usually deciduous in fruit vs unequal tepals that are patent and persistent in fruit) are easy to observe, and most species can be easily placed in a subgenus, with a few exceptions (van der Werff, 2002). For example, tepals are equal and persistent in several species, including the sole Old World species of *Persea*, *P. indica* from the Canary Islands. Subgenus *Persea*, in which avocado is the nomenclatural type, is a primarily Mesoamerican entity separate from the primarily South American subgenus *Eriodaphne*.

Several recent molecular phylogenetic studies have provided insights into evolutionary relationships in the *Persea* group (Rohwer *et al.*, 2009; Li *et al.*, 2011; Mo *et al.*, 2017; Xiao *et al.*, 2022). The two subgenera of *Persea* are well supported as monophyletic groups but are not always sister groups to each other. The Macaronesian *A. barbujana* has been found to be sister to either subgenus *Persea* or *Eriodaphne*, prompting the new combination *Persea barbujana* (Cav.) Mabb. & Nieto Fel. made in Mabberley's Plant-Book, ed. 4: 1102 (2017). The second Macaronesian species of *Persea*, *P. indica*, can be included in the subgenus *Eriodaphne*. Remarkably, subgenus *Persea*, along with *A. barbujana*, is sometimes nested among Asian genera of the *Persea* group in some molecular phylogenies, albeit with poor statistical support (Rohwer *et al.*, 2009; Li *et al.*, 2011). Several neotropical species currently placed in *Persea*, but which do not meet the morphological characters of any of the two subgenera (e.g. *P. sphaerocarpa*), appear nested among species currently placed in *Alseodaphne* (Asian). Along with these American *Persea* species, *Dehaasia* and *Nothaphoebe* (both Asian) are also nested among species of *Alseodaphne*, in a clade that includes the type of *Alseodaphne* (Li *et al.*, 2011). However, because of weak statistical support for these relationships, any nomenclatural changes should await the results of investigations with a much denser sampling and new molecular markers with higher resolution (Li *et al.*, 2011). A separate *Alseodaphne* clade has been recently recognized as a new genus, *Alseodaphnopsis* (Mo *et al.*, 2017), which differs from all other *Persea* group genera in having unisexual flowers (van der Werff, 2019). The Asian genus *Machilus* cannot be distinguished from American *Persea* on the basis of floral morphology but appears to be monophyletic and more closely related to other Asian genera of the *Persea* group. *Phoebe* may also be monophyletic, although it is poorly supported in most data sets so far. Notably, these molecular phylogenies are primarily based on just two markers, the internal transcribed spacer (ITS) of nuclear ribosomal DNA and the second intron of the single-copy nuclear gene, *LEAFY*. Furthermore, basic taxonomic knowledge of *Persea* has not advanced appreciably since Kopp's (1966) treatment, while monographs are lacking for all Old World genera (Rohwer *et al.*, 2009). Classical morphological work and genome-scale molecular phylogenetics with species-level taxonomic sampling are both needed to improve our understanding of the *Persea* group.

Persea Subgenus *Persea*: Species and Infraspecific Concepts

Taxonomic treatments of subgenus *Persea* differ dramatically in species and infraspecific concepts, especially regarding the circumscription and infraspecific entities of *P. americana*. Kopp (1966) recognized four species in subgenus *Persea*: *P. floccosa* Mez, *P. schiedeana* Nees, *P. steyermarkiana* C.K. Allen and *P. americana* with three botanical varieties: var. *americana*, var. *drymifolia* (Schlechtendal & Chamisso) Blake and var. *nubigena* (Williams) Kopp. Williams' (1977) concept of the subgenus included *P. floccosa*, *P. schiedeana*, *P. steyermarkiana* and *P. americana* with two varieties (var. *americana* and var. *drymifolia*), *Persea nubigena* with two botanical varieties (var. *nubigena* Williams and var. *guatemalensis* Williams) and two new species: *P. parviflora* Williams and *P. primatogena* Williams and Molina. The latter name is a synonym for *Beilschmiedia riparia* Miranda (Nishida, 1999), and *P. parviflora* has recently been synonymized under *P. pallescens* (Mez) Lorea-Hernández, a new combination in the subgenus *Persea* (Lorea-Hernández, 2009). Van der Werff (2002) published the broadest concept of *P. americana* to date, including *P. drymifolia*, *P. floccosa*, *P. steyermarkiana* and *P. nubigena* as synonyms without formal infraspecific varieties. Scora *et al.* (2002) proposed a similarly broad concept of *P. americana* but kept the three botanical varieties recognized by Kopp (1966) and proposed five new varieties based on *P. floccosa*, *P. nubigena* var. *guatemalensis*, *P. steyermarkiana*, *P. tolimanensis* and *P. zentmyerii*. The original publication of *P. tolimanensis* and *P. zentmyerii* lacked Latin diagnoses and designated type specimens (see Zentmyer and Schieber, 1990; Schieber and Bergh, 1987, respectively), and Scora's varieties lacked complete basionym citation (Scora *et al.*, 2002). Hence, all five of Scora's new varietal names were not published in accordance with current nomenclatural rules (McNeill *et al.*, 2006) and are listed currently as 'nom. inval.' in the International Plant Names Index (2011).

Beyond the field of classical taxonomy, there has been great interest in the infraspecific entities of *P. americana*, conceivably stemming from the long-established view that geographically separated domestication events produced the three horticultural races of avocado: the Mexican, the Guatemalan and West Indian (or Lowland, sometimes referred to as Antillean). Racial differences relate primarily to ecological preferences and fruit characteristics (Bergh and Ellstrand, 1986). Thus, Mexican avocados are characterized by adaptation to a Mediterranean climate, relatively good cold tolerance and early maturing fruit with a thin, usually purplish-black skin. The Guatemalan avocados are somewhat cold-tolerant due to their origin in a tropical highland climate but conspicuous primarily because of their thick, tough skin, which remains green until maturity. West Indian avocados are adapted to humid tropical lowland conditions and are hence very cold-sensitive but show tolerance to soil salinity and other adverse edaphic conditions. Their fruit have a thin yellowish-green skin and lower oil and higher sugar contents than the other two races. Bergh *et al.* (1973) recognized a need to integrate these racial concepts with formal taxonomy:

> The question arises as to how the taxonomy of the three more or less equivalent horticultural races can best be delineated ... One possibility is to treat the horticultural races simply as one

polytypic species, *P. americana*, or as a variety of that species in contrast to other possible varieties (see below). At the opposite pole from such lumping would be a splitting into three separate species. We have concluded that neither extreme is justified by the available data and that the taxonomic situation is best reflected by a recognition of three botanical varieties. Thereby, the Mexican race remains *P. americana* var. *drymifolia*. The Guatemalan and West Indian races are then separated. The West Indian race [actually now known to be indigenous to central and northern South America and to have been introduced into the West Indies only in post-Columbian times (Bergh, 1969)] probably has priority on the designation '*americana*', since it is the form prevalent along the seacoast and so was the first form described botanically; it retains the designation *P. americana* var. *americana*. The Guatemalan race therefore requires a new varietal name. Considering its common racial name, which accurately reflects its centre of indigenousness, it logically becomes *P. americana* var. *guatemalensis*.

(Bergh *et al.*, 1973)

Nomenclatural rules do not govern the association of published scientific names with organisms or populations thereof, as this is the practice of identification rather than taxonomy. Bergh *et al.*'s (1973) identification of the Mexican and West Indian races as representatives of *P. americana* var. *drymifolia* and *P. americana* var. *americana*, respectively, seems entirely reasonable. Unfortunately, the name *P. americana* var. *guatemalensis* for the Guatemalan race was given without a Latin diagnosis or a type specimen, both required for valid publication after 1958 (McNeill *et al.*, 2006). Yet, Bergh *et al.*'s (1973) names have been widely used to refer to the three races in the avocado research and horticultural literature. Williams (1977) subsequently described the Guatemalan race in a species separate from *P. americana*, as *P. nubigena* var. *guatemalensis* and explicitly linked his concept of this taxon to the Guatemalan race: 'This is the avocado which is known to horticulturists as the Guatemalan type, Guatemalan race, or simply as Guatemalans' (Williams, 1977). The varietal names associated with the three avocado races are therefore placed in different species, and the formal transfer of var. *guatemalensis* from *P. nubigena* to *P. americana* is yet to be made [*P. americana* var. *guatemalensis* (Williams) Scora is nom. inval.]. An additional complicating factor, as noted by Bergh *et al.* (1973), Williams (1977) and Furnier *et al.* (1990), is that additional botanical varieties would be needed to fully circumscribe *P. americana*:

Persea nubigena should probably be relegated to varietal status: *P. americana* var. *nubigena*. This seems reasonable also for *P. floccosa*, which would then be *P. americana* var. *floccosa*.

(Bergh *et al.* 1973)

P. nubigena was reduced by Kopp (1966) to varietal status: *P. americana* var. *nubigena*. Our leaf-oil results support this classification and suggest an affinity to *P. americana* var. *guatemalensis*. In fact, morphologically, its thick fruit skin especially is like the Guatemalan race to a degree unique among all of the 80 or so known *Persea* species, except for *P. schiedeana* and, to a lesser extent, *P. floccosa*. *Persea gigantea* (Williams, 1953) may be a distinct botanical variety, but Kopp's (1966) relegation of it to a mere variant form of *P. americana* var. *nubigena* seems reasonable. Kopp (1966) retained *P. floccosa* as a valid species, but she acknowledged that *P. americana var. nubigena* 'links *P. americana* var. *americana* with *P. floccosa*'. She noted further that '*P. floccosa* exhibits the typical floral and inflorescence structure of *P. americana* ...' Our unpublished results repeatedly indicated that there are no sterility barriers between *P. floccosa* and *P. americana* varieties: F2 and backcross segregations have been typical of quantitative inheritance, involving complete fertility. The morphological differences are somewhat greater than those between most *P. americana* botanical varieties but hardly significantly so. Varietal status, *P. americana* var. *floccosa*, would seem to reflect best the biological data. The only remaining taxon in *Persea* subg. *Persea*, the *steyermarkii* form, could conceivably represent an early step in the evolution of the *americana* variety; more likely it is a slight divergence from a common ancestor of the two forms.

(Williams, 1977)

If *P. americana* is to be maintained as a species containing var. *americana*, var. *drymifolia* and var. *guatemalensis*, then our data suggest that it should also contain varieties corresponding to *P. floccosa*, *P. nubigena* and *P. steyermarkii*.

(Furnier *et al.*, 1990)

Recent whole-genome sequences for individuals representing *P. americana* var. *drymifolia* (Mexican race) and the popular commercial cv. 'Hass', as well as genome resequencing data for multiple Guatemalan, Mexican and West Indian varieties and the closely related *P. schiedeana* (Rendón-Anaya *et al.*, 2019), also support a broad species concept for *P. americana*. The genome-scale data sets suggest poor population structuring within avocado as a whole, with a closer relationship between West Indian and Guatemalan races than with the Mexican race. They are also beginning to reveal cryptic variation among avocados and potentially support a fourth race among Colombian avocados related to the West Indian race (Berdugo-Cely *et al.*, 2023). Further, the 'Hass' genome is intermediate between the Guatemalan and Mexican subpopulations, in agreement with the previously noted hybrid nature of this variety (e.g. Chen *et al.*, 2009) and earlier estimates of population structure in avocado (e.g. Furnier *et al.*, 1990; Bufler and Ben-Ya'acov, 1992; Mhameed *et al.*, 1997; Davis *et al.*, 1998). Additionally, the 'Hass' genome seems to be derived from Guatemalan source regions introgressed into a Mexican race background, with introgressed genomic blocks of very large sizes (e.g. whole-chromosome arms) supporting a very recent (20th century) origin of this cultivar. A second genome assembly of 'Hass' with significantly improved contiguity (N50 = 26 Mb vs 296 kb) and completeness (96.6% vs 85% of 1440 conserved embryophyte single-copy genes) relative to the assembly of Rendón-Anaya *et al.* (2019) has been recently produced (Nath *et al.*, 2022) and has been analysed in the context of avocado relationships at broader (angiosperm-wide) phylogenetic scales. In this regard, it has become clear that the avocado genome carries evidence of two ancient whole-genome duplication (WGD) events in its deep evolutionary history, both of which are independent of those that are associated with the large eudicot and monocot clades of flowering plants (Rendón-Anaya *et al.*, 2019; Nath *et al.*, 2022). The earlier avocado WGD appears to be shared with members of the Magnoliales/Laurales clade of magnoliid angiosperms, while the more recent WGD may be specific to the derived Perseeae-Laureae clade of the Lauraceae (Chaw *et al.*, 2019; Rendón-Anaya *et al.*, 2019; Chen *et al.*, 2020; Qin *et al.*, 2021; Shen *et al.*, 2021; Nath *et al.*, 2022).

P. americana is clearly a highly polymorphic species, and the sentiment that 'The whole *P. americana* complex needs much genetic and field study before many of the problems can be solved' (Kopp, 1966) seems as relevant today as when stated more than half a century ago. Until new data emerge to the contrary, it seems best to refrain from an infraspecific classification that arbitrarily elevates all of avocado's variations to formally named botanical varieties. This position may have been embraced by the late pioneer in avocado botany, Wilson Popenoe,

> So far as my own conception is concerned, I have reached the opinion that all cultivated avocados are horticultural forms of *P. americana*; and their wild prototypes (of which more later) are geographical forms of the same species. (Popenoe, 1935)

Taxonomic Treatment

Persea Mill., Gard. Dict. ed. 4, 1754. TYPE: *Persea americana* Mill.

Tamala Rafinesque, Sylva Telluriana 136. 1838. TYPE: not designated.
Mutisiopersea Kostermans, Rheedea 3: 133. 1993. TYPE: *Mutisiopersea mutisii* (HBK) Kostermans.

Trees and shrubs. Leaves alternate, evenly distributed along the twigs or clustered, pinnately veined, axillary tufts of hairs or domatia lacking. Inflorescences in the axils of leaves or cataphylls, paniculate-cymose, lateral flowers of a cyme always strictly opposite. Flowers bisexual; tepals six, equal or unequal; if unequal, outer three tepals smaller than the inner three; tepals deciduous in fruit or persistent, if persistent, usually remaining entire, rarely with the tips of the inner tepals breaking off; stamens nine or six; if six, stamens of whorl III staminodial, anthers usually four celled, rarely two celled; inner three stamens with two glands at the base; staminodia of whorl IV present, each with a glandular apex. Fruit with or without persistent tepals at the base but never with a cupule.

Persea *Mill. subgenus* Persea

Inflorescences subterminal; flowers with outer perianth segments on both surfaces pubescent, lanceolate to elliptic, subequal to the inner segments; stamens of series I and II with quadrilocular anthers, those of series III having quadrilocular anthers and each filament with two stipitate glands attached to the base. Gynoecium pubescent. Perianth segments subtending the fruit and reflexed, mostly deciduous.

Key to species of *Persea* subgenus *Persea*

1a. Pubescence of abaxial leaf surface made of minute, straight, appressed hairs and scattered (better seen on young leaves) .... ***P. pallescens***
1b. Pubescence of abaxial leaf surface made of rather erect, sinuous hairs, mostly noticeable by the naked eye (particularly on young leaves), scattered to dense **2**
2a. Bracts at the base of the inflorescences widely ovate to depressed ovate; staminode apex elliptic to lanceolate in outline .... ***P. schiedeana***
2b. Bracts at the base of the inflorescences lanceolate to ovate; staminode apex trullate to ovate in outline .... ***P. americana***

Persea pallescens (Mez) Lorea-Hernández, Novon 19: 201. 2009. Basionym: *Phoebe pallescens* Mez, Jahrb. Königl. Bot. Gart. Berlin 5: 218. 1889. *Cinnamomum pallescens* (Mez) Kostermans, Reinwardtia 6: 22. 1961. TYPE: Mexico. Veracruz: near Orizaba, 1853, *Müller 90.*

Persea perglauca Lundell, Wrightia 5: 147. 1975. TYPE: Guatemala. Baja Verapaz: Unión Barrios, 16 Apr. 1975, *Lundell & Contreras 19217.*

Persea parvifolia Williams, Econ. Bot. 31: 320. 1977. TYPE: Mexico. Veracruz: above Tetla, Mun. Chocamán, 17 Oct. 1947, *Williams & Popenoe 13511.*

Trees to 7 m tall; branchlets glabrous. Leaves coriaceous, elliptic to elliptic-lanceolate, leaf blades 4–11 cm long, 1–3 cm wide, apices acute or short acuminate, bases acute, 6–8 pairs of secondary veins divergent at 45° inconspicuously impressed; petioles 0.4–2 cm long. Inflorescences lateral, thyrses to about 10 cm long. Flowers small, about 2 mm, greenish, obscurely puberulent; ovary glabrous, style 2 mm long. Fruit obovoid-globose, broadest above middle, about 3.5 cm long and 2.5 cm in diameter, rind very thin, bright green, lustrous, flesh about 4 mm thick and greenish; seed about 1.7 cm in diameter.

Typical features include whitish twigs contrasting with the dark petioles, clustered leaves, which are glaucous on the lower surface, the impressed midrib and the small flowers with tepals 2 mm long. This taxon is the smallest leaved of subgenus *Persea*. From the cloud forest at an altitude of 2300–2600 m in Veracruz, Mexico, on high-fertility soils in the warm, humid climate of the Neovolcanic Mountain Chain. Vernacular name: 'aguacate cimarrón' and 'aquacatillo'.

Persea schiedeana Nees, Syst. Laurin. 130. 1836. *Persea gratissima* var. *schiedeana* (Nees) Meissner, DC. Prodr. 15(1): 53. 1864. TYPE: Mexico. Misantla, *Schiede 1141.*

Persea pittier Mez, Bot. Jahrb. Syst. 30, Beibl. 67: 15. 1901. TYPE: Costa Rica. *Pittier 1156.*

Trees can grow up to 30 m tall. The crown is broad, and the branchlets are thick, roughened by leaf and bud scale scars and densely ferruginous villous. The leaves are chartaceous to subcoriaceous; leaf blades are 8–35 cm long and 4–25 cm wide. The leaves are broadly elliptic to obovate, with rounded to obtuse apices and obtuse to rounded bases. The adaxial surfaces are glabrescent, and the abaxial surfaces are moderately to densely ferruginous villous, especially when young. The leaves have 6–15 pairs of secondary veins that are divergent at 40–70° and are prominently impressed at the abaxial surface. The petioles are 2–5 cm long, slender and densely ferruginous villous. Inflorescences are subterminal thyrsoid, multiple, subtended by scarious margined, broad, round-tipped bud scales, which are up to 2 cm long with the abaxial surface being ferruginous villous. Flowers are 6–10 mm long and tomentose. The filaments of series I and II are tawny tomentose, while those of series III are densely so. The anthers of series I, II and III are oblong, quadrilocular, while the staminodia of series IV are linear. The gynoecium is densely pubescent with short, straight hairs. There are one to two fruits per branch terminal, ranging from long necked to small pyriform to almost round in shape, reaching 5–10 cm at maturity, with the pubescence persisting at the fruiting stage. The thickness of the exocarp is less than 3 mm. The colour ranges from light or dark green, yellowish and brown to almost black purple. The seed is variable, ranging from spheroid to ovate in shape, and is often large.

P. schiedeana is readily identifiable by its large flowers with long pedicels and its large somewhat clustered leaves. Other distinguishing characters are the broad scarious-margined bud scales and the short, straight, stiff and persistent pubescence of the gynoecium. There exists a great deal of morphological variability within this species, as evidenced by the obovoid fruit and broadly elliptic leaves in the montane regions and elongated fruit and leaves in the lowlands. The trees are tall and long-lived and cultivated on a limited scale. This taxon is sometimes used as shade trees in coffee plantations. It is found in cloud and coniferous forests in warmer and more humid locations compared to other wild avocados, at altitudes ranging from 250 to 2500 m. It is concentrated in the eastern and southern parts of Mexico and Panama and grows on a variety of soil types, including those with poor drainage.

Distribution: It is found in forests and meadows at 90–2000 m altitude in southern Mexico and Central America; in the Mexican states of Chiapas, Michoacán, Oaxaca, Quinta Roo, Tabasco and Veracruz; in Guatemala at Alta Verapaz, Huehuetenango and Izabal; in Honduras at Chicinul, Morazan, Tegucigalpa and Zacapa; in Costa Rica at Alajuela, Cartago, Heredia, Puntarenas, San Jose and Yoro; in Panama at Alajuela, Cartago, Chiriqui, Guanacaste and Heredia; in Honduras at Atlantida, Olando and Yoro; and in El Salvador and Colombia. Vernacular names: 'Aguacate' in Belize; 'Yas' in Costa Rica; 'Coyo', 'Coyou', 'Chalte' and 'Chucte' in Guatemala; 'Guaco' in Honduras; 'Chininni', 'Hip' and 'Zucte' in Mexico; and 'Aguacaton' in Panama.

Persea americana Mill., Gard. Dict. ed. 8. 1768. *Laurus persea* L. TYPE: Plate 222, Fig. 2 in Sloane (1725; lectotype).

Persea americana var. *angustifolia* Miranda Anal. Inst. Biol. Mexico. 17: 129. 1946. TYPE: Mexico. Puebla, *Miranda 3482*.

Persea drymifolia Schlechtendal & Chamisso, Linnaea 6: 365. 1831. *Persea americana* var. *drymifolia* (Schlechtendal & Chamisso) Blake, J. Wash. Acad. Sci. 10: 15. 1920. *Persea gratissima* var. *drymifolia* (Schlechtendal & Chamisso) Mez, Jahrb. Königl. Bot. Gart. Berlin 5: 147. 1889. TYPE: Mexico, *Schiede & Deppe 1140*.

Persea edulis Rafinesque, Sylva Tell. 134. 1838. Nom. superfl. for *Persea americana* Mill.

Persea floccosa Mez, Jahrb. Königl. Bot. Gart. Berlin 5: 148. 1889. TYPE: Mexico. *Liebmann 85*.

Persea gigantea L.O. Williams, Ceiba 4: 39. 1953. TYPE: Honduras. *Williams & Molina 12629*.

Persea gratissima Gaertner, Fruct. Sem. Pl. 3: 222. 1807. Nom. superfl. for *Persea americana* Mill.

Persea gratissima var. *oblonga* Meissner DC Prodr 15(1): 53. 1864. SYNTYPES: Mexico. *Aschenborn 110, Galeotti 7010*; Peru. *Ruiz & Pavon s.n.*; Brazil. *Riedel 106, Martius s.n.*; Java, *Zollinger 457 & 3101 ex parte*.

Persea gratissina var. *praecox* Nees, Syst. Laurin. 129. 1836. *Persea gratissima* var. *macrophylla* Meissner, DC. Prodr. 15(1): 53. 1864. TYPE: Peru. *Poeppig 2446*.

Persea leiogyne Blake, J. Wash. Acad. Sci. 10: 19. 1920. TYPE: U.S.A. Florida, *Popenoe 219*.

Persea nubigena Williams, Ceiba 1: 55. 1950. *Persea americana* var. *nubigena* (Williams) Kopp, Mem. New York Bot. Gard. 14: 19. 1966. TYPE: Guatemala. *Williams & Molina 16833*.

Persea nubigena var. *guatemalensis* Williams, Econ. Bot. 31: 318. 1977. TYPE: Guatemala. *Molina & Williams s.n.*

Persea paucitriplinervia Lundell, Wrightia 5: 146. 1975. TYPE: Guatemala. *Lundell & Contreras 19177*.

Persea steyermarkiana Allen, J. Arnold Arbor. 2. 6: 286. 1945. TYPE: Guatemala. *Steyermark 37061*.

Trees can grow up to 30 m tall. The branchlets are scarcely to moderately pubescent with tawny to subferruginous hairs. Leaves are chartaceous, narrowly to broadly elliptic to subobovate in shape. The leaf blades are 10–30 cm long, 3–19 cm wide. The apices are acute to acuminate, with acute to obtuse to rounded bases. The adaxial surfaces are glabrescent, while the abaxial surfaces are occasionally pruinose and sparsely to moderate tawny-pubescent. The leaves have six to nine pairs of secondary veins, diverging at 40–60°, which are slightly to moderately impressed on the abaxial surface. The petioles are 1–6 cm long, slender and canaliculate. The inflorescences are axillary, subterminal, multiple, thyrsoid and densely pubescent. Flowers are 3–6 mm long. The perianth lobes are 3–6 mm long and 2–3 mm wide, with a yellowish-green colour, and are densely silky tomentose on both surfaces. Tips are acute, stamens are 2–3 mm long and filaments of series

I and II are pubescent. The anthers are quadrilocular, and the staminodia are sagittate. The gynoecium is moderately to densely pubescent. There are one to three fruits per branch terminal. The fruits are 10–25 cm long, broadly or narrowly pyriform in shape. The exocarp is light or dark green and brownish to purple in colour; the flesh of the fruit is ivory coloured. Seeds are ovate and variable in size with a rough surface.

Distribution: It is indigenous to Mexico and Mesoamerica and is now widely cultivated below 1200 m in the tropics and subtropics of the Old and New Worlds.

Vernacular names: 'aguacate' as general term in Mesoamerica and 'Aguacate oloroso'.

As currently circumscribed, *P. americana* includes several species of earlier classifications. The type of *P. leiogyne* is from a cultivated tree known as the 'Trapp avocado' (Kopp, 1966), and it can be assumed that the types of *P. americana* and *P. gratissima* also come from cultivated trees, based on their origin outside of the Central American highlands. It is unclear whether the type of *P. drymifolia* was cultivated or collected from the wild. In addition to the horticultural characters noted for the Mexican race, it is recognizable by the arachnoid appearance of the longer appressed pubescence of flowers and branchlet tips, granular waxy deposits on various specimens, and the edible fruit with reflexed, more persistent perianth segments. Names applied to clearly noncultivated forms, and plants therefore considered 'wild' avocados, are *P. floccosa*, *P. nubigena* and *P. steyermarkii*, all from the Central American highlands. *P. floccosa* has been considered distinct from *P. americana* on account of its relatively small flowers, more globose fruit and the floccose pubescence to which its specific epithet refers. The features used to distinguish *P. steyermarkii* were the smaller glabrous leaves with widely divergent secondary veins and the development of cork on two-year-old branchlets. Williams (1976, 1977) regarded *P. nubigena* as the wild progenitor of the Guatemalan race, distinguished by its smaller fruit with thin flesh, while Kopp (1966) noted the coarser texture of its leaves due to more prominent reticulation and impressed upper reticulation relative to *P. americana* var. *americana*. *P. americana* var. *angustifolia* was always recognized as a narrow leaf variant of *P. americana*. The taxonomic significance of these morphological characters remains to be determined in future efforts to resolve the infraspecific classification of *P. americana*.

Botanical Overview of the Avocado Tree

Apart from the taxonomic overview of *Persea* spp. above, various aspects of the botany of cultivated avocado trees are dealt with elsewhere (see Hormaza *et al.*, Chapter 4; Hormaza *et al.*, Chapter 6 and Crane *et al.*, Chapter 8, this volume). To reduce duplication, this overview is a broad summary of the gross morphology, architecture and growth habit of the cultivated avocado tree, as a basis for a better understanding of tree management and manipulation.

Horticultural races: horticultural aspects

The postulated botanical varieties or horticultural races of *P. americana* Mill. have been previously described (see Galindo-Tovar *et al.*, Chapter 2, this volume and the 'Taxonomy' section of this chapter). The horticultural features of the three botanical varieties or races of cultivated avocados were summarized by Bergh (1975, 1992) and Bergh and Lahav (1996). Bergh recognized distinct differences in climatic tolerance, tree form and leaf, flower and fruit characteristics. In general, Guatemalan cultivars have the most useful horticultural genes. They dominate the germplasm of the world's 'subtropical' avocado cultivars, which dominate world trade, and better types are recognized for desirable fruit quality, small seeds and late fruit maturity. Good-quality pure Mexican cultivars are rare, but they have contributed genes for early maturity and cold tolerance. West Indian types, or their hybrids with Guatemalan types, dominate the world's tropical and semi-tropical industries, still largely based on seedling trees in developing countries (excluding Florida, USA). West Indian avocado fruit are renowned for relatively low oil but relatively high sugar content, and a distinctly less 'nutty' flavour than that found in the subtropical cultivars. In spite of their great seedling variability, they play an important role in upgrading the diets of mostly poor people in less developed tropical lowland countries.

The avocado tree

The seedling trees of Guatemalan and West Indian race ecotypes, especially in their native rainforest environments, can reach heights exceeding 30 m, while Mexican seedling trees are shorter, reaching a height of about 15 m. Grafted trees, however, are dwarfed to a varying extent, depending on rootstock vigour, graft compatibility and growing conditions. Orchard trees in the subtropics can reach more than 10–15 m in height (Fig. 3.1) but are usually kept to no more than 4–5 m due to difficulty of pest management and fruit harvest. In the non-invigorating Chilean environments, high-density plantings are found, with tree height limited to 2–3 m. Cultivated avocado trees are mostly evergreen, in spite of the surprisingly short leaf longevity of 10–12 months (Whiley and Schaffer, 1994). Some cultivars are more prone to defoliation just before flowering, especially the previous spring flush leaves in environments that impose stress, for example, winter cold and drought, soil salinity and root infection by *Phytophthora cinnamomi*. Severe photoinhibition of leaves in the dry winter of summer rainfall areas may also hasten leaf senescence, aggravated by the loss of feeder roots accompanying heavy flowering and cropping (Whiley, 1994).

Fig. 3.1. A large, grafted 'Bacon' tree growing at Lindcove Field Station, Tulare County, California.

Avocado trees are capable of rapid growth in height and spread, with 1 m per year not unusual in young trees in the moist subtropics. The wood is rather spongy due to the relatively thin-walled fibres caused by a rapid increase in branch thickness. Branches bend easily under the weight of fruit, and for a given thickness, are not nearly as strong as those of citrus (Chandler, 1958).

Chandler (1958) noted that avocado trees in California store considerable amounts of starch in wood and roots, more than citrus trees in similar environments. Whiley (1994) and Whiley *et al.* (1996a,b) also reported on starch storage in trunk and branch tissues as well as leaves in subtropical Queensland but noted that the seasonal flux was not as great as in the more stressful environment of temperate, semi-arid southern Australia (Scholefield *et al.*, 1985). They suggested that in the latter type of environment, avocado trees are more likely to be semi-deciduous, store more carbohydrates in autumn and winter but use more in spring, and be more reliant on stored carbohydrate reserves in relation to current photosynthate than in the warm, moist subtropics. Goldschmidt and Koch (1996), when reviewing evidence for 'source-limitation' of yield in citrus, noted that high starch levels in citrus organs do not imply a carbohydrate surplus. The accumulation of reserve carbohydrates seems to have high priority in citrus as part of its survival strategy. Higher carbohydrate reserves in avocado trees in stressful environments suggest a similar need for a greater storage buffer in such circumstances.

Vegetative growth flushes

Free-standing avocado trees typically have a rounded canopy with dense foliage. The growth form varies, however, from upright (i.e. 'Bacon', 'Edranol', 'Reed' and most seedlings), to rounded (i.e. 'Hass') to spreading (i.e. 'Fuerte'). As for most polyaxial evergreens, shoot growth in mature trees is synchronized into flushes of varying vigour, duration and extent (Verheij, 1986). In the moist, summer-rainfall subtropics, bearing trees are characterized by a spring growth flush that starts during flowering and peaks in early summer. The proportion of terminal and subterminal shoots in active growth then reduces to a low level, to be followed by a second (summer and early autumn) growth flush (Whiley *et al.*,

1988b). In climates such as those of southern California (and coastal Israel and Chile), the spring flush may end earlier, allowing an early summer flush and an autumn flush (Chandler, 1958). In the subhumid, semi-warm subtropical climate of Nayarit state, Mexico, Salazar-García *et al.* (2006) reported one winter and three summer flushes, all contributing to varying degrees to the return winter bloom. In contrast, from a 3-year study in a lowland and semi-tropical part of Florida, Davenport (1982) noted that trees were almost constantly growing throughout summer although not all branches were growing at the same time. Each branch flushes two to three times a year, except for the shaded or bearing shoots, which usually only flush in spring. Differences in flushing synchrony among cultivars were also found. In general, new flushes tend to be from the tree periphery, from non-fruit-bearing shoots, continually adding to tree size.

Different flushes on a particular branch are easily recognizable, as the approach of a period of quiescence is marked by shorter internodes and then a ring of closely spaced buds (the intercalation) immediately distal to the resting terminal bud (Fig. 3.2) (Davenport, 1990). The avocado tree abscises, perhaps through activity of the periderm, most of its lateral buds at about 1 year of age or sooner (Chandler, 1958), except at the intercalation. A pruning cutback made to the previous intercalation in autumn will release growth of the ring of buds and increase shoot complexity (Cutting *et al.*, 1994). The loss of axillary buds affects graft wood selection and is a reason for the widespread use of grafting rather than budding for avocado propagation. The persisting intercalary buds at the 'bud ring' between shoot flushes grow sufficiently each year to keep their meristems at the bark surface. They may subsequently sprout after severe pruning or branch bending (Chandler, 1958).

Fig. 3.2. The intercalation zone on an avocado shoot showing the 'ring of buds'.

Avocado leaves expand to full size in about 30 days. A net carbon loss was found in Florida in 'Booth 8' leaves until they reached about 72% of full size (Schaffer *et al.*, 1991) while for 'Hass' the sink-to-source transition was at 80% of leaf expansion (Whiley, 1990). There are 350–510 stomates mm^{-2} on the abaxial surface, while stomatal efficiency in older leaves appears to be reduced by dense epicuticular wax, which also covers the stomates (Blanke and Lovatt, 1993).

Tree architecture and growth modules

The avocado tree conforms to Rauh's architectural model (Hallé *et al.*, 1978). The monopodial trunk grows rhythmically and develops tiers of branches, which are morphogenetically identical to the trunk. Thus, a tree can be considered as comprising three to five 'mini-tree' modules, which can be rotated and renewed over a period of years by selectively cutting back one major branch per year, thus also limiting tree size and improving light relations. The remarkable tolerance of the avocado tree to severe pruning or even staghorning can also be explained by the Rauh model. Hallé *et al.* (1978) call this development of shoots outside the normal expression of the architectural model 'reiteration', which is strongly developed in the avocado tree. The model is characterized by pseudo-terminal inflorescences developing from lateral buds close to the terminal vegetative bud. Inflorescences, although commonly called panicles, are determinate thyrses (with a terminal vegetative bud) with branches terminating in flowers. Early and vigorous shoot elongation from the subtending terminal vegetative bud of the inflorescence can have profound adverse implications for fruit set due to unwanted vegetative–reproductive competition while leaves are still physiological sinks. In 'Fuerte', from 5% to 20% of the terminal buds of the inflorescence can abort, and this proportion can increase with heavier flowering (Schroeder, 1944). In such cases, there is no competition with adjacent vegetative growth, leading to better fruit set and increased fruit size (Whiley, 1994), although such fruit are more subject to sunburn due to reduced protective leaf cover.

The avocado literature confuses the concepts of determinate and indeterminate growth in shoots and inflorescences. In avocados, all branches are monopodial and, therefore, indeterminate and ending in a vegetative bud (Fig. 3.3d). All inflorescences, at least initially, are also subtended by a terminal vegetative bud, which may abort or grow out to compete with setting fruitlets (Fig. 3.3e). Schroeder (1944) thus referred to 'determinate flowering shoots' (where the terminal vegetative bud aborts or does not grow out) and 'indeterminate flowering shoots', where it does grow. This reference to 'determinate' and 'indeterminate' inflorescences is inaccurate and confusing. In the interpretation of the inflorescence accepted below, the terminal vegetative bud is not part of the inflorescence, and all avocado inflorescences (thyrses) are determinate and subterminal. Horticultural usage of 'functionally determinate' and 'functionally indeterminate' flowering shoots would be acceptable.

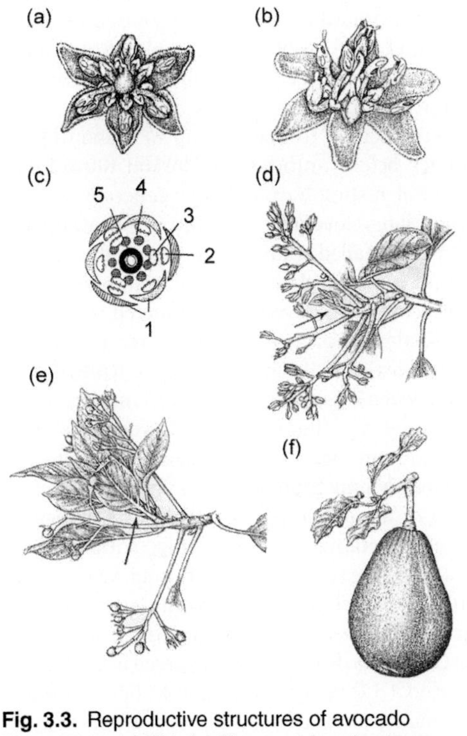

Fig. 3.3. Reproductive structures of avocado (*P. americana* Mill.). (a) Flower at female stage; (b) flower at male stage (pollen releasing); (c) floral diagram where 1 = sepal (outer) and petals (inner) or more correctly tepals; 2 = stamen of mid-whorl; 3 = stamen of inner whorl; 4 = glandular staminode; 5 = pistil. See Salazar-García *et al.*, Chapter 6, this volume for further information. (Drawings by P. Fawcett in Tomlinson (1980), with permission.)

Thorp and Sedgley (1993b, 1994) proposed a terminology for avocado shoot growth modules. They distinguished between proleptic shoot modules (axillary shoots grow out from dormant buds), which can be vegetative or reproductive and in which nodes were preformed, and sylleptic shoots. In the latter, there is contemporaneous development of a lateral shoot with its parent axis, without a period of dormancy, and shoots are invariably vegetative and not preformed. They showed that the indeterminate growth of Rauh's model is predominant in avocado. In another study, Thorp and Sedgley (1993a) found that tree form could be related to the relative dominance of proleptic and sylleptic axillary shoots, in turn controlled by the interaction between apical dominance and apical control (acrotony). 'Fuerte' had weak apical dominance and produced many sylleptic shoots and displayed weak acrotony. 'Sharwil' showed strong apical dominance with few sylleptic shoots but strong acrotony (numerous proleptic shoots). 'Hass' was intermediate. They suggested that yield efficiency could be improved by manipulating tree structure through increasing sylleptic shoot growth. Further studies are needed to relate such findings to productivity and fruit quality. An example of tree manipulation in Chilean high-density plantings is the selection of sylleptic shoots, and removal of proleptic shoots, in training young 'Hass' trees in high-density plantings, until trees reach 2 m in height (Gardiazabal *et al.*, 2011).

An interesting practical problem related to tree architecture is the extent of translocation and redistribution of chemicals injected into the trunk and major branches. Guest *et al.* (1995) noted that there is little lateral redistribution of injected phosphonate across avocado trunk tissues, necessitating multiple injections around the trunk. This is in marked contrast to cocoa stems and coconut, where a single injection site suffices.

The implications for fruit tree management of tree architecture were reviewed by Costes *et al.* (2006), with emphasis on tree training concepts. Thorp and Stowell (2001) discussed pruning height and limb removal in avocado to partition more biomass into fruiting shoots.

The root system

The avocado root system is invariably described as relatively shallow and not spreading much beyond the drip line of the tree canopy. Bergh (1992) believed that three aspects of evolution have shaped avocado roots: first, the frequent heavy rains, as found in the indigenous rainforest habitat; second, rapidly draining soils, as exemplified by the high oxygen requirement of roots and their sensitivity to poor drainage; and finally, the presence of a rich surface organic mulch, as shown by the tendency of healthy feeder roots to grow into any decomposing litter layer. Broadbent and Baker (1974), in subtropical Australia, first championed the benefits of organic mulches in helping create more 'suppressive' soils for Phytophthora root rot control. Other beneficial effects have been found in alleviating physiological stress, increasing 'Hass' fruit size and yield, and reducing the incidence of pedicel 'ring-neck' in cool subtropical South Africa, apparently through promoting root growth and health in the 'litter zone' (Moore-Gordon and Wolstenholme, 1996; Wolstenholme *et al.*, 1996). However, reinforcement of the natural leaf mulch may be counterproductive in very wet soils, on very heavy clays, and where soil salinity is excessive. The greatest benefits of mulching will be found in stressful soil and environmental conditions, for example, sandy soils low in organic matter, in hot, semi-arid areas (see Mena *et al.*, Chapter 12, of this volume).

Comprehensive surface feeder root (Whiley *et al.*, 1988b) and rhizotron studies (Whiley, 1994) in deep-red, high-organic-content krasnozem soils in cool subtropical Queensland, Australia, have shown considerable feeder root growth as deep as 1 m, and extension of the second (autumn) root growth flush well into winter. Nevertheless, the majority of white, unsuberized feeder roots are found in the top 600 mm of soil and at even shallower depths in heavy clay soils. Avocado roots have few or no root hairs. Vesicular-arbuscular mycorrhizae are commonly associated with roots in Israeli orchards (Ginsburg and Avizohar-Hershenson, 1980), and isolates of *Glomus fasciculatus* applied to sterilized growing media improved the growth and nutrition of avocado seedlings (Menge *et al.*, 1980). Whiley (1994) found pronounced attrition of feeder roots in south-eastern Queensland coincident with flowering in spring, which, together with the loss of photosynthetic capacity due to winter photoinhibition, reduced the capacity to supply water, minerals and metabolites to setting fruit. Heavy cropping, as in the 'on' season of the alternate-bearing cycle, will greatly reduce root growth.

Flowers and fruit

Flower primordia can usually be found in shoot terminal and subterminal buds, from both spring- and summer-grown shoots, in the humid subtropics, plus autumn and other flushes elsewhere, from 4 to 5 weeks up to 2 months or longer before anthesis. In Florida, floral initiation of resting terminal buds generally occurs from late November to January (Davenport, 1982). Detailed studies of the time when shoot apical buds reach irreversible commitment to flowering (ICF) have been conducted in California (Salazar-García *et al.*, 1999) and in Nayarit state, Mexico (Salazar-García *et al.*, 2006). Such knowledge is essential for orchard management strategies to either increase flowering (at the start of an 'off' bearing season) or decrease flowering and promote shoot flushing (at the start of an 'on' cropping season). In California, three vegetative flushes (spring, summer and autumn) are typical. The subhumid, semi-warm subtropical climate of Nayarit leads to four growth flushes (one winter and three summer). In California, ICF in 'Hass' occurred from 30 August to 15 October (equivalent to late February to mid-April in the southern hemisphere). In Nayarit, irrespective of shoot age, ICF occurred by 15 October (mid-April in the southern hemisphere) after an average of 27.5 chilling days (temperature ≤19°C). In humid subtropical climates in the southern hemisphere, anatomical evidence of floral 'initiation' is typically evident during March, and many buds in winter will be swollen due to the growth of flower initials prior to anthesis in spring. Peak anthesis varies seasonally by up to 4 weeks in a particular locality, timing being a function of complex interacting factors. Furthermore, flowering is less well-synchronized than, for example, in citrus trees. Flowering in a particular cultivar may extend over a period of 6 weeks or more, especially in warm climates, with a new cohort of flowers opening for the first time each day. To generalize, Mexican cultivars flower earliest, West Indian cultivars next and pure Guatemalan cultivars last.

Avocado flowering in relation to the unique 'protogynous dichogamy' opening behaviour, and fruit set, are described by Hormaza *et al.* (see Chapter 6, this volume). Other reviews include those of Bergh (1986) and Davenport (1986). Basically, under favourable conditions, each flower opens twice over a 24-h period, first as functionally female (Fig. 3.3a) and second as functionally male (Fig. 3.3b). This favours cross-pollination in a wild, seedling population but with a fail-safe mechanism, ensuring some self-pollination in most situations.

The inflorescence is most often simply called a much-branched panicle (Fig. 3.3d). A single tree may have hundreds of panicles (more correctly thyrses), each potentially with hundreds or thousands of flowers. Thus, a mature tree may have a million flowers (Bergh, 1986), a reflection of a primitive feature associated with an ancient genus. Inflorescences arise most often from shoot terminal buds but also from some subterminal buds in vigorous shoots. Confusion as to the ultimate structure of the panicle subunits arises partly from the presence or absence (by abortion) of the terminal vegetative bud on inflorescences. As noted earlier, we regard all inflorescences as determinate, with the vegetative bud not part of the inflorescence. One interpretation is that the inflorescence is a panicle of cymes (Reece, 1942; Bergh, 1986; van der Werff and Richter, 1996 for *Persea*). Osche *et al.*'s (1961) view on the panicle of racemes is incorrect. Alternatively, Kopp (1966) called the inflorescence a panicle in which the ultimate units were bidichasia, that is, the basic structure is a dichasium. In reviewing this chapter in the first edition, P.J. Robbertse (Pretoria, South Africa, 1999, personal communication) suggested that the inflorescence is a thyrse, that is, thyrsoid, with the primary axis paniculate (and ending in a vegetative bud) and the secondary axes dichasial. It is clear that more research is needed.

The basic flower structure has been summarized earlier in this chapter (see '*Persea* subgenus *Persea*' section) and is illustrated in Fig. 3.3. The avocado flower is hypogynous, regular, complete and trimerous, 3–7 mm in length. There are six greenish-yellow perianth segments, in two alternating whorls of three, described variously as either tepals (the correct interpretation – see the Taxonomy section of this chapter), three outer sepals and three inner petals (Fig. 3.3c), the latter slightly longer and 5–6 mm in length, the former slightly more reflexed, both whorls pubescent. There are typically four androecial whorls, the outer two whorls consist of (series I and II) stamens 2–3 mm long, with quadrilocular anthers and valves hinged at the top, which are introrse; the innermost (series III) whorl consists of three longer stamens, which are extrorse, each with a pair of orange nectaries or stipitate glands attached near the base of the pubescent filament, secreting nectar at the second (staminate) flower opening; the innermost androecial (series IV) whorl comprises three short, sagittate staminodes alternating with the series III stamens, secreting nectar at the first (pistillate) flower opening. The gynoecium is unicarpellate, pubescent; it consists of a simple pistil with a superior ovary that is unilocular and contains a single pendulous, anatropous, bitegmic, and crassinucellate ovule. The style is slender and hirsute, leading to a simple, slightly lobed papillate stigma; the pedicel is yellowish-green, pubescent and attached to the peduncle (slightly modified from Purseglove, 1968; Bergh, 1986).

Again, there is controversy as to some aspects of flower morphology, particularly for perianth segments. These have been incorrectly interpreted as two calyx whorls with no corolla (Calvino, 1939; Osche *et al.*, 1961; Purseglove, 1968) or the reverse (Malan and van der Meulen, 1954). Reece (1939) found that the perianth traces rise at different levels and concluded that there were three outer sepals and three inner petals, an interpretation that was accepted by Schroeder (1952) and Bergh (1975, 1986) (Fig. 3.3c). However, more recent studies concluded that the two perianth whorls could not be separated at the differentiation stage, or by position in the flower, visual gross morphology or microscopic surface structure, and proposed using the term 'tepals' (Inoue and Takahashi, 1989; Blanke and Lovatt, 1993). Estimates of the number of stamens and staminodes also varied in different studies. There is much evidence for variability in flower structure and a high percentage of abnormal and even sterile flowers (Tomer and Gottreich, 1978; Bergh, 1986).

Recent evidence from the relatively young field of evolution and development (often referred to as the evolution of development, or 'evo-devo') indicates that the tepals of *Persea* are staminodial in origin (Chanderbali *et al.*, 2006). The genetic regulation of floral development has underlying framework that is widely known as 'the ABCE model' of floral organ identity. In elegant simplicity, the ABCE model posits that four functions (A, B, C and E) act in a combinatorial manner to regulate the development of the four primary floral organs: A + E = sepals, A + B + E = petals, B + C + E = stamens and C + E = carpels (Soltis *et al.*, 2006). Remarkably, B, C and E genes are all expressed in avocado tepals, suggesting that the B + C + E equation for stamens also exists in these organs. The presence of B + C + E expression in avocado tepals might represent the genetic footprints of stamen ancestors, a vestige of a staminal past. In support of a 'staminodial' scenario for the evolution of avocado perianth, tepaloid staminodes have replaced outer stamen whorls in other Lauraceae, including *Dicypellium*, *Phyllostemonodaphne*, *Eusideroxylon* and *Endlicheria*, perhaps indicating a genetic predisposition that underlies the evolutionary origin of Lauraceae tepals in general (Chanderbali *et al.*, 2006).

Although a large, heavily flowering tree may have over one million flowers (Lahav and Zamet, 1999), a high percentage of flowers are likely to be abnormal or sterile (Tomer and Gottreich, 1978). Although under favourable conditions a high number of fruitlets may set (Fig. 3.4), a subsequent heavy fruit drop is a normal occurrence. A large crop can be obtained from as few as 200 (small trees at high density) and up to 500 or more mature fruit per tree. In subtropical climates, there are two periods of intense abscission, in late spring/early summer and in mid-summer (Whiley *et al.*, 1988b). Although flowers (2.8–3.4 mm^{-2}) and young fruit (50–75 mm^{-1}) have few stomates relative to leaves (Blanke and Lovatt, 1993), it has been shown that heavy flowering substantially increases total transpiration (Whiley *et al.*, 1988a).

Fig. 3.4. A heavy fruit set on a 'functionally determinate' flowering shoot where the terminal vegetative bud has aborted. Initial fruit retention can be high in the absence of competition from shoot growth in the immediate proximity.

The avocado fruit is botanically a one-seeded berry (Fig. 3.3f) and is highly variable in size (50 g to nearly 2 kg), shape (round, oval or pyriform), rind characteristics (thickness, surface features and colour), flesh and seed characteristics (i.e. size and tightness in cavity). Cummings and Schroeder (1942) described the basic fruit anatomy. The exocarp (rind or skin) has a thin waxy cuticle, a single-layered epidermis, a hypodermis of one to three cell layers, several layers of parenchyma cells and an interrupted layer of sclerenchyma tissue. The latter is more evident in thick-rinded cultivars, where lignified stone cells occur. Stomata are prominent in young fruit. In older fruit, the stomata may degenerate due to lenticel formation, producing white or grey specks on the rind surface (e.g. 'Fuerte'), especially in semi-arid climates. Lenticels or rind nodules can become corky and rough in some cultivars and environments, where the epidermis can be ruptured. This corkiness may blacken and become unsightly in moist climates, resulting in downgrading of fruit ('lenticel damage' – see Hofman *et al.*, Chapter 15, this volume).

The edible flesh of the avocado fruit, as usually interpreted, is mostly parenchymatous mesocarp tissue. The vacuoles of these cells contain many small oil droplets. Scattered among these cells are larger, specialized oil-containing cells called ideoblasts, which comprise about 2% of the flesh volume. Ideoblasts have thicker, more complex cell walls and a single, large oil-filled vacuole (Cummings and Schroeder, 1942; Scott *et al.*, 1963). Vascular bundles from the pedicel enter the fruit as a solid cylinder, and break up into six major strands that divide further to permeate the mesocarp and endocarp. Most vascular bundles then coalesce near the fruit apex and enter the seedcoat, where they again anastomose in the pachychalaza (Steyn *et al.*, 1993). The innermost flesh is a rather indistinct endocarp, made up of a few rows of smaller, more flattened parenchyma cells. It is botanically incorrect to refer to the flesh as 'mesocarp' as, by definition, the flesh of a berry fruit comprises mesocarp plus endocarp. There have been several ultrastructural studies of oil cells and mesocarp (Platt-Aloia and Thomson, 1981, 1992; Platt-Aloia *et al.*, 1983). The flesh recovery of superior cultivars is greater than 70% of the total fruit weight. The high oil content of the fruit flesh, together with the large, nutrient-dense seed, partly explains the relatively low average yields of contemporary avocado orchards, where good growers achieve 20–30 t ha^{-1} under favourable conditions (Wolstenholme and Whiley, 1999).

Fruit development

Several studies have been conducted on the morphology and anatomy of avocado fruit development, which may last from 6 to more than 12 months, depending on the cultivar and growing conditions. The review by Bower and Cutting (1988) on avocado fruit development and ripening focuses on ripening physiology, and that by Cowan *et al.* (2001) mainly deals with the biochemical aspects of fruit growth. The fruit growth curve is sigmoidal when measured by increase in mass or volume, although growth may slow down in winter and resume in spring in cool climates. As for fleshy fruit, most cell division occurs early in the life of the fruit, and cell enlargement follows. The avocado fruit is unusual, however, in that cell division continues, albeit at a slower rate, as long as the fruit is held on the tree (Schroeder, 1953; Valmayor, 1967; Cowan *et al.*, 1997; Moore-Gordon, 1997). Differences in fruit size appear to result mainly from cell number rather than cell size. The fruit is dependent on its seed for longer than most fleshy fruits, and only with the degeneration and browning of the seedcoat, usually close to horticultural maturity, is the seed mature and anatomically isolated from the flesh. Small, seedless cucumber-shaped fruit ('cukes') may form when seeds abort very soon after ovule fertilization (stenospermocarpy) in some cultivars (Blumenfeld and Gazit, 1974; Tomer *et al.*, 1980). The 'Hass' small fruit phenotype results from premature seedcoat abortion much later in fruit development and is associated with tree stress in hot environments (Cowan *et al.*, 1997; Moore-Gordon, 1997).

Sound avocado fruit will not ripen while held on the tree, a still unexplained manifestation of 'physiological juvenility'. On-tree fruit storage after attainment of legal maturity is, therefore, possible, especially in cool, low-stress environments. Delayed harvest from 3 (West Indian cultivars) to 6 months (Guatemalan race cultivars) is possible, particularly if the latter are grown in the cool, mesic subtropics (Kaiser and Wolstenholme, 1994; Whiley *et al.*, 1996a,b). Whiley *et al.* (1996a,b) demonstrated the accompanying trade-offs, however, of reduced subsequent yield and accentuation of alternate bearing (see Mena *et al.*, Chapter 12, this volume), while shelf life of late-harvested fruit is also substantially reduced.

There have been a number of informative studies comparing the growth of normal and seedless (cukes) fruit (Blumenfeld and Gazit, 1974; Tomer *et al.*, 1980; Steyn *et al.*, 1993), as well as studies on pollination and fruit set problems (Tomer and Gazit, 1979; Sedgley, 1987), which help explain the requirements for normal fruit growth. The importance of normal seed growth is underlined by the fact that seedless fruit (the result of stenospermocarpy) are ten times smaller than normal fruit and also exhibit a different shape (Fig. 3.5). Blumenfeld and Gazit (1974) highlighted the role of the seedcoat (not a true testa) in fruit development. In young fruit, the seedcoat is thick, white, fleshy and a rich source of promotive growth substances. With browning and degeneration of the seedcoat, vascular connections are lost, and seed growth stops. Premature seedcoat death is invariably

Fig. 3.5. Seedless avocado fruit (the result of stenospermocarpy) often set but do not reach the normal size of fruit with seeds. The seedless fruit are usually curved and elongated and are commonly known as 'cukes' or 'cocktail avocados'.

associated with the second fruit drop accompanying the summer growth flush, as well as with small fruit (and small seeds) if fruit are retained. Recent physiological studies suggest that this is stress-related and, along with pedicel 'ring-neck', can be partially alleviated by mulching and improved root health (Cowan *et al.*, 1997). The review by Cowan *et al.* (2001) summarizes recent physiological work on small as opposed to normal-sized 'Hass' fruits.

The role of previously unreported tissues in avocado fruit development has been outlined by Steyn *et al.* (1993). They note that the vascularized part of the seedcoat is actually a pachychalaza, with the testa (the contribution of the integuments) representing only a very small non-vascularized portion. It is the former which supplies photosynthate, minerals and water to the enlarging seed during its long development period. They found that cukes arose from female sterile (cryptically male) flowers in consistently low-yielding 'Fuerte' trees, in which endosperm development ceased early, and the embryo stopped growing.

The relatively large avocado seed, well endowed with energy and minerals, has undoubtedly been selected against over centuries by people seeking a higher proportion of edible flesh. The ecological importance of large seeds is probably related to germination and survival in the native rainforest, typically under dense shade awaiting a 'canopy gap' and being structured for dispersal from the parent tree. The latter was thoroughly explored in Barlow's (2000) book, *The Ghosts of Evolution*, in which wild avocado fruit feature prominently. The central argument is that many fruits coevolved with large dispersal agents, which, in most parts of the world (excluding Africa), have become extinct. In the neotropical forests of Central America, mastodons (forest elephants), giant ground sloths, glyptodonts, native horses, and other large mammals are believed to have been the 'megafaunal dispersal syndrome' for wild avocado but became extinct *c.*13,000 years ago, following climate change and the arrival of humans (see Galindo-Tovar and Bost, Chapter 2, this volume). The result in Central and South America, as well as in other continents, is the presence of many fruits considered 'anachronistic' in today's world (Janzen and Martin, 1982).

The ripe avocado, according to Barlow (2000), is an 'overbuilt' and extravagant fruit in today's world. It is large, with an oily, energy-rich and highly nutritious pulp, enveloping a seemingly unnecessarily large, inedible, bitter-tasting and mildly poisonous seed, packed with storage reserves and ready to germinate. Wolstenholme and Whiley (1999) also noted that the seed is physiologically 'recalcitrant', that is, sensitive to drying out and suited to competing in a low-light forest floor environment. In summary, the fruit has an attractive pulp with a chemically protected seed, suited to a world that no longer exists, wasteful of energy and seriously hindering seed dispersal now. Today's dispersal agents for wild avocados, including tapirs, agoutis, monkeys, squirrels and large birds, are 'pulp thieves' and/or seed predators, 'under-abundant, ill-fitted or fickle' (Barlow, 2000). Only an occasional jaguar actually swallows the seed (Cook, 1982). Wild avocados in neotropical forests are, accordingly, far less common today. It is mainly humans who have dispersed avocados to villages and orchards since the megafaunal extinction (see Galindo-Tovar *et al.*, Chapter 2, this volume).

The avocado fruit, therefore, evolved with a now-extinct megafauna, to be swallowed whole. From a functional and evolutionary perspective, the reward provided by the pulp aided the fast passage through the gut of a large animal of the large, dense, slippery seed, to be deposited in a fertile pile of dung far from the parent tree (Barlow, 2000). Wolstenholme and Whiley (1999) discussed these and other traits of the avocado fruit and tree, which help to explain the low-yield problem and the exceptional fruit nutrient density, making it arguably the world's most nutritious fruit. To these can be added the seed germination strategy. Awareness of evolutionary ecology in a relatively poorly domesticated fruit, in other words, helps to guide orchard management and breeding strategies.

Conclusions

Subgenus *Persea* consists of *P. schiedeana*, *P. pallescens* and *P. americana*. This latter species includes the three horticultural avocado races and several related entities that have been treated as separate species, botanical varieties or both over time. The contribution of molecular phylogenetics to avocado taxonomy is increasing and will complement the morphological techniques in resolving these taxonomic problems in the future. An overview of avocado taxonomy and morphology has brought to light certain unresolved aspects of inflorescence, flower and fruit structure, indicating the necessity for further research.

An understanding of avocado botany is essential for guiding tree and orchard manipulation for improved crop yield and fruit quality. Aspects of tree architecture explain the remarkable tolerance to severe pruning and other manipulatory techniques. Similarly, many aspects of flowering, fruit set and development are unique in cultivated fruit tree orchard crops.

References

Allen, C.K. (1945) Studies in the Lauraceae. VI. Preliminary survey of the Mexican and Central American species. *Journal of the Arnold Arboretum* 26, 282–434.

Atkinson, B.A., Stockey, R.A., Rothwell, G.W., Mindell, R.A. and Bolton, M.J. (2015) Lauraceous flowers from the Eocene of Vancouver Island: *Tinaflora beardiae* gen. et sp. nov. (Lauraceae). *International Journal of Plant Sciences* 176, 567–585.

Barlow, C. (2000) *The Ghosts of Evolution*. Basic Books, New York.

Berdugo-Cely, J.A., Cortés, A.J., López-Hernández, F., Delgadillo-Durán, P., Cerón-Souza I. *et al.* (2023) Pleistocene-dated genomic divergence of avocado trees supports cryptic diversity in the Colombian germplasm. *Tree Genetics & Genomes* 19:42.

Bergh, B.O. (1969) Avocado (*Persea americana* Miller). In: Ferwerda, F.P. and Wit, F. (eds) *Outlines of Perennial Crop Breeding in the Tropics*. Landbouwhogeschool, Wageningen, The Netherlands, Misc. Paper 4, pp. 23–51.

Bergh, B.O. (1975) Avocados. In: Janick, J. and Moore, J.N. (eds) *Advances in Fruit Breeding*. Purdue University Press, West Lafayette, Indiana, pp. 541–567.

Bergh, B.O. (1986) *Persea americana*. In: Halevy, A.B. (ed.) *CRC Handbook of Flowering*. Vol.5. CRC Press, Boca Raton, Florida, pp. 253–268.

Bergh, B.O. (1992) The origin, nature, and genetic improvement of the avocado. *California Avocado Society Yearbook* 76, 61–75.

Bergh, B.O. and Ellstrand, N. (1986) Taxonomy of the avocado. *California Avocado Society Yearbook* 70, 135–145.

Bergh, B.O. and Lahav, E. (1996) Avocados. In: Janick, J. and Moore, J.N. (eds) *Fruit Breeding,* Volume I*: Tree and Tropical Fruits*. Wiley, West Lafayette, Indiana, pp. 113–166.

Bergh, B.O., Scora, R.W. and Storey, W.B. (1973) A comparison of leaf terpenes in *Persea* subgenus *Persea*. *Botanical Gazette* 134, 130–134.

Blake, S.F. (1920) A preliminary revision of the North American and West Indian avocados (*Persea* spp.). *Journal Washington Academy of Sciences* 10, 9–21.

Blanke, M.M. and Lovatt, C.J. (1993) Anatomy and transpiration of the avocado inflorescence. *Annals of Botany* 71, 543–547.

Blumenfeld, A. and Gazit, S. (1974) Development of seeded and seedless avocado fruits. *Journal of the American Society for Horticultural Science* 99, 442–448.

Bower, J.P. and Cutting, J.G.M. (1988) Avocado fruit development and ripening physiology. *Horticultural Reviews* 10, 229–271.

Broadbent, P. and Baker, K.F. (1974) Behaviour of *Phytophthora cinnamomi* in soils suppressive and conducive to root rot. *Australian Journal of Agricultural Research* 25, 121–137.

Bufler, G. and Ben-Ya'acov, A. (1992) A study of the avocado germplasm resources, 1988–1990. 3. Ribosomal DNA repeat unit polymorphism in avocado. In: Lovatt, C., Holthe, P.A. and Arpaia, M.L. (eds) *Proceedings of the Second World Avocado Congress*. University of California, Riverside, California, 2, pp. 545–550.

Calvino, E.M. (1939) Biologia florale della *Persea drymifolia* (aguacate) in Sanremo. *Stazione Sperimentale di Floricoltura* 'O Raimondo' N.31. (Transl.) *California Avocado Society Yearbook* 24, 79–86 (cited by Bergh, 1986).

Chambers, K.L., Poinar, G.O. and Chanderbali, A.S. (2012) *Treptostemon* (Lauraceae), a new genus of fossil flower from mid-tertiary Dominican amber. *Journal of the Botanical Research Institute of Texas* 6, 551–556.

Chanderbali, A.S., van der Werff, H. and Renner, S.S. (2001) Phylogeny and historical biogeography of Lauraceae: evidence from the chloroplast and nuclear genomes. *Annals of the Missouri Botanical Garden* 88, 104–134.

Chanderbali, A.S, Kim, S., Buzgo, M., Zheng, P., Oppenheimer, D.G., Soltis, D.E. and Soltis, P.S. (2006) Genetic footprints of stamen ancestors guide perianth evolution in *Persea* (Lauraceae). *International Journal of Plant Sciences* 167, 1075–1089.

Chanderbali, A.S., Albert, V.A., Ashworth, V.E., Clegg, M.T., Litz, R.E., Soltis, D.E. and Soltis, P.S. (2008) *Persea americana* (avocado): bringing ancient flowers to fruit in the genomics era. *BioEssays* 30, 386–396.

Chandler, W.H. (1958) *Evergreen Orchards*, 2nd edn. Lea and Febiger, Philadelphia, Pennsylvania.

Chaw, S.-M., Liu, Y.-C., Wu, Y.-W., Wang, H.-Y., Lin, C.-Y.I. *et al.* (2019) Stout camphor tree genome fills gaps in understanding of flowering plant genome evolution. *Nature Plants* 5, 63–73.

Chen, H., Morrell, P.L., Ashworth, V.E.T.M., de la Cruz, M. and Clegg, M.T. (2009) Tracing the geographic origins of major avocado cultivars. *Journal of Heredity* 100, 56–65.

Chen, Y.-C., Li, Z., Zhao, Y.-X., Gao, M., Wang, J.-Y. *et al.* (2020) The *Litsea* genome and the evolution of the laurel family. *Nature Communications* 11:1675.

Cook, R.E. (1982) Attractions of the flesh. *Natural History* 91, 20–24.

Costes, E., Lauri, P.É. and Regnard, J.L. (2006) Analyzing fruit tree architecture: implications for tree management and fruit production. *Horticultural Reviews* 32, 1–61.

Cowan, A.K., Moore-Gordon, C.S., Bertling, I. and Wolstenholme, B.N. (1997) Metabolic control of avocado fruit growth: isoprenoid growth regulators and the reaction catalysed by 3-hydroxyl-3-methylglutaryl coenzyme A reductase. *Plant Physiology* 114, 511–518.

Cowan, A.K., Cripps, R.F., Richings, E.W. and Taylor, N.J. (2001) Fruit size: towards an understanding of the metabolic control of fruit growth using avocado as a model system. *Physiologia Plantarum* 111, 127–136.

Cummings, K. and Schroeder, C.A. (1942) Anatomy of the avocado fruit. *California Avocado Society Yearbook* 27, 56–64.

Cutting, J.G.M., Cocker, B. and Wolstenholme, B.N. (1994) Time and type of pruning cut affects shoot growth in avocado *Persea americana* (Mill.). *Journal of Horticultural Science* 69, 75–80.

Davenport, T.L. (1982) Avocado growth and development. *Proceedings of the Florida State Horticultural Society* 95, 92–96.

Davenport, T.L. (1986) Avocado flowering. *Horticultural Reviews* 8, 257–289.

Davenport, T.L. (1990) Citrus flowering. *Horticultural Reviews* 12, 349–408.

Davis, J., Henderson, D., Kobayashi, M. and Clegg, M.T. (1998) Genealogical relationships among cultivated avocado as revealed through RFLP analysis. *Journal of Heredity* 89, 319–323.

Drinnan, A.N., Crane, P.R. and Hoot, S.B. (1994) Patterns of floral evolution in the early diversification of non-magnoliid dicotyledons (eudicots). In: Endress, P.D.P.K. and Friis, P.D.E.M. (eds) *Plant Systematics and Evolution Supplement 8: Early Evolution of Flowers*. Springer, Vienna, pp. 93–122.

Eklund, H. (2000) Lauraceous flowers from the Late Cretaceous of North Carolina, U.S.A. *Botanical Journal of the Linnean Society* 132, 397–428.

Eklund, H. and Kvaček, J. (1998) Lauraceous inflorescences and flowers from the cenomanian of bohemia (Czech Republic, Central Europe). *International Journal of Plant Sciences* 159, 668–686.

Furnier, G.R., Cummings, M.P. and Clegg, M.T. (1990) Evolution of the avocados as revealed by DNA restriction site variation. *Journal of Heredity* 81, 183–188.

Gaertner, J. (1807) *De Fructibus et Seminibus Plantarum*, Vol. 3. Richter, Leipzig, part 22, plate 221.

Gardiazabal, F., Magdahl, C. and Mena, F. (2011) *New avocado planting and management systems in Chile*. Internal document, Sociedad Gardiazabal Mena Ltd, Chile.

Gentry, A.H. (1988) Changes in plant community diversity and floristic composition on environmental and geographical gradients. *Annals of the Missouri Botanical Garden* 75, 1–34.

Ginsburg, O. and Avizohar-Hershenson, Z. (1980) Observations on vesicular-arbuscular mycorrhiza associated with avocado roots in Israel. *Transactions of the British Mycological Society* 48, 101–104.

Goldschmidt, E.E. and Koch, K.E. (1996) Citrus. In: Zamski, E. and Schaffer, A.A. (eds) *Photoassimilate Distribution in Plants and Crops: Source-Sink Relationships.* Marcel Dekker Inc., New York, pp. 797–823.

Guest, D.I., Pegg, K.G. and Whiley, A.W. (1995) Control of *Phytophthora* diseases of tree crops using trunk-injected phosphonates. *Horticultural Reviews* 17, 299–330.

Hallé, F., Oldeman, R.A.A. and Tomlinson, P.B. (1978) *Tropical Trees and Forests: An Architectural Analysis*. Springer, Berlin, Germany.

Herendeen, P.S., Crepet, W.L. and Nixon, K.C. (1994) Fossil flowers and pollen of Lauraceae from the Upper Cretaceous of New Jersey. *Plant Systematics and Evolution* 189, 29–40.

Inoue, H. and Takahashi, B. (1989) Differentiation and development of avocado flower buds in Japan. *Journal of the Japanese Society for Horticultural Science* 58, 105–111.

International Plant Names Index (2011) Available at: http://www.ipni.org (accessed 29 August 2011).

Janzen, D.H. and Martin, P.S. (1982) Neotropical anachronisms: the fruits the gomphotheres ate. *Science* 215, 19–27.

Kaiser, C. and Wolstenholme, B.N. (1994) Aspects of delayed harvest of 'Hass' avocado (*Persea americana* Mill.) fruit in a cool subtropical climate. II. Fruit size, yield, phenology and whole-tree starch cycling. *Journal of Horticultural Science* 69, 447–457.

Kopp, L.E. (1966) A taxonomic revision of the genus *Persea* in the western hemisphere (Persea-Lauraceae). *Memoirs of the New York Botanical Garden* 14, 1–120.

Kostermans, A.J.G.H. (1961) The New World species of *Cinnamomum* Trew (Lauraceae). *Reinwardtia* 6, 17–24.

Kostermans, A.J.G.H. (1993) *Mutisiopersea* Kostermans, a new genus in Lauraceae. *Rheedia* 3, 132–135.

Lahav, E. and Zamet, D. (1999) Flowers, fruitlets and fruit drop in avocado trees. *Revista Chapingo Serie Horticultura Num. Especial* V, 95–100.

Li, L., Li, J., Rohwer, J.G., van der Werff, H., Wang, Z.-H. and Li, H.-W. (2011) Molecular phylogenetic analysis of the *Persea* group (Lauraceae) and its biogeographic implications on the evolution of tropical and subtropical Amphi-Pacific disjunctions. *American Journal of Botany* 98, 1520–1536.

Li, H., Liu, B., Davis, C.C., and Yang, Y. (2020) Plastome phylogenomics, systematics, and divergence time estimation of the *Beilschmiedia* group (Lauraceae). *Molecular Phylogenetics and Evolution* 151, 106901.

Lorea-Hernández, F.G. (2009) *Persea pallescens*, a new combination for *Phoebe pallescens* (Lauraceae, Perseeae), a mistaken taxon of Mexico and Guatemala. *Novon* 19, 201–203.

Lundell, C.L. (1975) Studies of American plants, IX. *Wrightia* 5, 146–148.

Mabberley, D.J. (2017) *Mabberley's Plant-book: A Portable Dictionary of Plants, Their Classification and Uses*, 4th edn. Cambridge University Press, Cambridge, UK.

Malan, E.F. and van der Meulen, A. (1954) Avocado culture in South Africa. *Farming in South Africa* 29, 309–315.

McNeill, J., Barrie, F.R., Burdet, H.M., Demoulin, V., Hawksworth, D.L. *et al.* (eds) (2006) International Code of Botanical Nomenclature (Vienna Code). *Regnum Vegetabile* 146, 1–568.

Meissner, C.F. (1864) Lauraceae. In: De Candolle, A.P. (ed.) *Prodromus*. Sumptibus Sociorum Treuttel et Würtz, Parisii. Vol. 15, pp. 1–260.

Menge, J.A., La Rue, J., Labanauskas, C.K. and Johnson, E.L.V. (1980) The effect of two mycorrhizal fungi upon growth and nutrition of avocado seedlings grown with six fertilizer treatments. *Journal of the American Society for Horticultural Science* 105, 400–404.

Mez, C. (1889) Lauraceae Americanae: monographice Descripsit. *Jahrbuch Koeniglicher Botanischer Garten Berlin* 5, 1–566.

Mez, C. (1901) Bromeliaceae et Lauraceae novae vel adhuc nonsatis cognitae. *Engler Botanische Jahrbuecher Fuer Systematic, Pflanzengeschichte and Pflanzengeographie* 30(Suppl. 67), 15.

Mhameed, S., Sharon, D., Kaufman, D., Lahav, E., Hillel, J., Degani, C. and Lavi, U. (1997) Genetic relationships within avocado (*Persea americana* Mill.) cultivars and between *Persea* species. *Theoretical and Applied Genetics* 94, 279–286.

Miller, P. (1754) *The Gardener's Dictionary*, 4th edn. London.

Miller, P. (1768) *The Gardener's Dictionary*, 8th edn. London.

Miranda, F. (1946) Algunas Plantas Notables Del Declive Oriental de La Mesa de Anahuac. *Anales del Instituto de Biologia (Mexico)* 17, 129.

Mo, Y., Li, L., Li, J., Rohwer, J.G., Li, H. and Li, J. (2017) *Alseodaphnopsis*: a new genus of Lauraceae based on molecular and morphological evidence. *PLOS ONE* 12:e0186545.

Moore-Gordon, C. and Wolstenholme, B.N. (1996) The Hass small fruit problem: role of physiological stress and its amelioration by mulching. *South African Avocado Growers' Association Yearbook* 19, 82–86.

Moore-Gordon, C.S. (1997) Aspects of avocado fruit growth and development: towards understanding the 'Hass' small fruit syndrome. PhD thesis, Department of Horticultural Science, University of Natal, Pietermaritzburg, South Africa.

Nath, O., Fletcher, S.J., Hayward, A., Shaw, L.M., Masouleh, A.K. *et al.* (2022) A haplotype resolved chromosomal level avocado genome allows analysis of novel avocado genes. *Horticulture Research* 9:uhac157.

Nees, C.G.D. (1836) *Systema Laurinarum*. Veit, Berolini, Germany.

Nishida, S. (1999) Revision of *Beilschmiedia* (Lauraceae) in the Neotropics. *Annals of the Missouri Botanical Garden* 86, 657–701.

Osche, J.J., Soule, M.J. Jr., Dijkman, M.J. and Wehlburg, C. (1961) *Tropical and Subtropical Agriculture*. Macmillan, New York, pp. 617–642.

Penagos Zuluaga, J.C., van der Werff, H., Park, B., Eaton, D.A.R., Comita, L.S. *et al.* (2021) Resolved phylogenetic relationships in the *Ocotea* complex (Supraocotea) facilitate phylogenetic classification and studies of character evolution. *American Journal of Botany* 108, 664–679.

Platt-Aloia, K.A. and Thomson, W.W. (1981) Ultrastructure of the mesocarp of mature avocado fruit and changes associated with ripening. *Annals of Botany* 48, 451–465.

Platt-Aloia, K.A. and Thomson, W.W. (1992) Ultrastructure of avocados: ripening, chilling injury and isolation of ideoblast oil cells. In: Lovatt, C., Holthe, P.A. and Arpaia, M.L. (eds) *Proceedings of the Second World Avocado Congress*. University of California, Riverside, California, pp. 417–425.

Platt-Aloia, K.A., Oross, I.W. and Thomson, W.W. (1983) Ultrastructure and development of oil cells in the mesocarp of avocado fruit. *Botanical Gazette* 144, 49–465.

Poinar, G. (2017) A mid-Cretaceous Lauraceae flower, *Cascolaurus burmitis* gen. et sp. nov., in Myanmar amber. *Cretaceous Research* 71, 96–101.

Popenoe, W. (1935) Origin of the cultivated races of avocado. *California Avocado Society Yearbook* 20, 184–194.

Purseglove, J.W. (1968) *Tropical Crops: Dicotyledons I*. Longman, London, pp. 192–198.

Qin, L., Hu, Y., Wang, J., Wang, X., Zhao, R. *et al.* (2021) Insights into angiosperm evolution, floral development and chemical biosynthesis from the *Aristolochia fimbriata* genome. *Nature Plants* 7, 1239–1253.

Rafinesque, C.S. (1838) *Sylva Telluriana* 136. Rafinesque, Philadelphia, Pennsylvania.

Reece, P.C. (1939) The floral anatomy of the avocado. *American Journal of Botany* 26, 429–433.

Reece, P.C. (1942) Differentiation of avocado blossom buds in Florida. *Botanical Gazette* 104, 323–328.

Rendón-Anaya, M., Ibarra-Laclette, E., Méndez-Bravo, A., Lan, T., Zheng, C. *et al.* (2019) The avocado genome informs deep angiosperm phylogeny, highlights introgressive hybridization, and reveals pathogen-influenced gene space adaptation. *Proceedings of the National Academy of Sciences of the United States of America* 116, 17081–17089.

Renner, S.S. (1999) Circumscription and phylogeny of Laurales: evidence from molecular and morphological data. *American Journal of Botany* 86, 1301–1315.

Rohwer, J.G. (1993) Lauraceae. In: Kubitzki, K., Rohwer, J. and Bittrich, V. (eds) *The Families and Genera of Flowering Plants.* Springer, Berlin, Germany, pp. 426–437.

Rohwer, J.G. (2000) Toward a phylogenetic classification of the Lauraceae: evidence from matK sequences. *Systematic Botany* 25, 60–71.

Rohwer, J.G., Li, J., Rudolph, B., Schmidt, S.A., van der Werff, H. and Li, H. (2009) Is *Persea* (Lauraceae) monophyletic? Evidence from nuclear ribosomal ITS sequences. *TAXON* 58, 1153–1167.

Rohwer, J.G., Moraes, P.L.R.D., Rudolph, B. and van der Werff, H. (2014) A phylogenetic analysis of the *Cryptocarya* group (Lauraceae), and relationships of *Dahlgrenodendron*, *Sinopora*, *Triadodaphne*, and *Yasunia*. *Phytotaxa* 158, 111–132.

Salazar-Garcia, S., Lord, E.M. and Lovatt, C.J. (1999) Inflorescence development of the 'Hass' avocado: commitment to flowering. *Journal of the American Society for Horticultural Science* 124, 478–482.

Salazar-Garcia, S., Cossio-Vargas, L.E., Lovatt, C.J., Gonzales-Duran, I.J.L. and Perez-Barraza, M.H. (2006) Crop load affects vegetative growth flushes and shoot age influences irreversible commitment to flowering of 'Hass' avocado. *Journal of the American Society for Horticultural Science* 41, 1541–1546.

Schaffer, B., Whiley, A.W. and Kholi, R.R. (1991) Effects of leaf age on gas exchange characteristics of avocado (*Persea americana* Mill.). *Scientia Horticulturae* 48, 21–28.

Schieber, E. and Bergh, B. (1987) *Persea zentmyerii*: a new species from Guatemala. *California Avocado Society Yearbook* 71, 199–203.
Schlechtendal, D.F.L., von and Chamisso, A. (1831) Plantarum Mexicanarum. *Linnaea* 6:365.
Scholefield, P.B., Sedgley, M. and Alexander, D. (1985) Carbohydrate cycling in relation to shoot growth, floral initiation and development and yield in the avocado. *Scientia Horticulturae* 25, 99–110.
Schroeder, C.A. (1944) The avocado inflorescence. *California Avocado Society Yearbook* 29, 39–40.
Schroeder, C.A. (1952) Floral development, sporogenesis, and embryology in the avocado *Persea americana*. *Botanical Gazette* 114, 270–278.
Schroeder, C.A. (1953) Growth and development of the Fuerte avocado fruit. *Proceedings of the American Society for Horticultural Science* 61, 103–109.
Scora, R.W., Wolstenholme, B.N. and Lavi, U. (2002) Taxonomy and botany. In: Whiley, A.W., Schaffer, B. and Wolstenholme, B.N. (eds) *The Avocado: Botany, Production and Uses*. CAB International, Wallingford, UK, pp. 15–37.
Scott, F.M., Bystrom B.G. and Bowler, E. (1963) *Persea americana*, mesocarp cell structure, light and electron microscope study. *Botanical Gazette* 124, 423–428.
Sedgley, M. (1987) Flowering, pollination and fruit-set of avocado. *South African Avocado Growers' Association Yearbook* 10, 42–43.
Shen, Z., Li, W., Li, Y., Liu, M., Cao, H. *et al.* (2021) The red flower wintersweet genome provides insights into the evolution of magnoliids and the molecular mechanism for tepal color development. *The Plant Journal* 108, 1662–1678.
Sloane, H. (1725) *A Voyage to the Islands Madera, Barbados, Nieves, S. Christophers and Jamaica*. Vol. 2. Privately printed, London.
Soltis, P.S., Soltis, D.E., Kim, S., Chanderbali, A.S. and Buzgo, M. (2006) Expression of floral regulators in basal angiosperms and the origin and evolution of ABC-function. *Advances in Botanical Research* 44, 483–506.
Song, Y.C., and Da, L.J. (2016) *Vegetation structure and function at multiple spatial, temporal and conceptual scales. Part 2: evergreen broad-leaved forest of East Asia*. Springer International Publishing, Cham, Switzerland.
Steyn, E.M.A., Robbertse, P.J. and Smith, D. (1993) An anatomical study of ovary-to-cuke development in consistently low-producing trees of the 'Fuerte' avocado (*Persea americana* Mill.) with special reference to seed abortion. *Sexual Plant Reproduction* 6, 87–97.
Tagawa, H. (1995) Distribution of Lucidophyll Oak-Laurel forest formation in Asia and other areas. *Tropics* 5, 1–40.
Takahashi, M., Herendeen, P.S. and Crane, P.R. (2001) Lauraceous fossil flower from the Kamikitaba locality (Lower Coniacian; Upper Cretaceous) in northeastern Japan. *Journal of Plant Research* 114, 429–434.
Takahashi, M., Herendeen, P.S., Xiao, X. and Crane, P.R. (2014) Lauraceous fossil flowers from the Kamikitaba assemblage (Coniacian, Late Cretaceous) of northeastern Japan (Lauraceae). *Systematic Botany* 39, 715–724.
Thorp, T.G. and Sedgley, M. (1993a) Architectural analysis of tree form in a range of avocado cultivars. *Scientia Horticulturae* 53, 85–98.
Thorp, T.G. and Sedgley, M. (1993b) Manipulation of shoot growth patterns in relation to early fruit set in 'Hass' avocado (*Persea americana* Mill). *Scientia Horticulturae* 56, 147–156.
Thorp, T.G. and Sedgley, M. (1994) Preformation of node number in vegetative and reproductive proleptic shoot modules of *Persea* (Lauraceae). *Annals of Botany* 73, 13–22.
Thorp, T.G. and Stowell, B. (2001) Pruning height and selective limb removal affect yield of large 'Hass' avocado trees. *HortScience* 36, 699–702.
Tomer, E. and Gazit, S. (1979) Early stages in avocado *Persea americana* (Mill.) fruit development: anatomical aspects. *Botanical Gazette* 140, 304–309.
Tomer, E. and Gottreich, M. (1978) Abnormalities in avocado (*Persea americana* Mill.) ovule development. *Botanical Gazette* 139, 81–86.
Tomer, E., Gazit, S. and Eisenstein, D. (1980) Seedless fruit in 'Fuerte' and 'Ettinger' avocado. *Journal of the American Society for Horticultural Science* 105, 341–346.
Tomlinson, P.B. (with illustrations by P. Fawcett) (1980) *The Biology of Trees Native to Tropical Florida*. Harvard University Printing Office, Allston, Massachusetts, pp. 188–189.
Valmayor, R.V. (1967) Cellular development of the avocado from blossom to maturity. *Philippine Agriculturist* 50, 907–976.
van der Werff, H. (2002) A synopsis of *Persea* (Lauraceae) in Central America. *Novon* 12, 575–586.

van der Werff, H. (2019) *Alseodaphnopsis* (Lauraceae) revisited. *Blumea–Biodiversity, Evolution and Biogeography of Plants* 64, 186–189.
van der Werff, H. and Richter, H.G. (1996) Towards an improved classification of Lauraceae. *Annals of the Missouri Botanical Garden* 83, 409–418.
Verheij, E.W. (1986) Towards a classification of tropical tree fruits. *Acta Horticulturae* 175, 137–150.
WFO (2022) World flora online. Available at: http://www.worldfloraonline.org/ (accessed 25 July 2022).
Whiley, A.W. (1990) CO_2 assimilation of developing fruiting shoots of cv. Hass avocado (*Persea americana* Mill.). *South African Avocado Growers' Association Yearbook* 13, 28–30.
Whiley, A.W. (1994) Ecophysiological studies and tree manipulation for maximisation of yield potential in avocado (*Persea americana* Mill.). PhD thesis, Department of Horticultural Science, University of Natal, Pietermaritzburg, South Africa.
Whiley, A.W. and Schaffer, B. (1994) Avocado. In: Schaffer, B. and Andersen, P.C. (eds) *Handbook of Environmental Physiology of Fruit Crops. Volume* II: *Sub-Tropical and Tropical Crops.* CRC Press, Boca Raton, Florida, pp. 3–35.
Whiley, A.W., Chapman, K.R. and Saranah, J.B. (1988a) Water loss by floral structures of avocado (*Persea americana* cv. Fuerte) during flowering. *Australian Journal of Agricultural Research* 39, 457–467.
Whiley, A.W., Saranah, J.B., Cull, B.W. and Pegg, K.G. (1988b) Manage avocado tree growth cycles of productivity gains. *Queensland Agricultural Journal* 4, 29–36.
Whiley, A.W., Rasmussen, T. S., Saranah J.B. and Wolstenholme, B.N. (1996a) Delayed harvest effects on yield, fruit size and starch cycling in avocado in subtropical environments. I. The early maturing cv. Fuerte. *Scientia Horticulturae* 66, 23–34.
Whiley, A.W., Rasmussen, T.S., Saranah, J.B. and Wolstenholme, B.N. (1996b) Delayed harvest effects on yield, fruit size and starch cycling in avocado (*Persea americana* Mill.) in subtropical environments. II. The late-maturing cv. Hass. *Scientia Horticulturae* 66, 35–49.
Williams, L.O. (1950) Two new *Perseas* from Central America. *Ceiba* 1: 55.
Williams, L.O. (1953) New Central American plants. *Ceiba* 4: 39.
Williams, L.O. (1976) The botany of the avocado and its relatives. In: Sauls, J.W. (ed.) *Proceedings of the First International Tropical Fruit Short Course. The Avocado.* Miami Beach, Florida, pp. 9–15.
Williams, L.O. (1977) The avocado, a synopsis of the genus *Persea*, subg. *Persea. Economic Botany* 31, 315–320.
Wolstenholme, B.N. and Whiley, A.W. (1999) Ecophysiology of the avocado (*Persea americana* Mill.) tree as a basis for pre-harvest management. *Revista Chapingo Serie Horticultura Núm. Especial* V, 77–88.
Wolstenholme, B.N., Moore-Gordon, C. and Ansermino, S.D. (1996) Some pros and cons of mulching avocado orchards. *South African Avocado Growers' Association Yearbook* 19, 87–91.
Xiao, T.-W., Yan, H.-F., Ge, and X.-J. (2022) Plastid phylogenomics of tribe *Perseeae* (Lauraceae) yields insights into the evolution of East Asian subtropical evergreen broad-leaved forests. *BMC Plant Biology* 22: 32.
Zentmyer, G.A. and Schieber, E. (1990) *Persea tolimanensis*: a new species for Central America. *Acta Horticulturae* 275, 383–386.

4 Genetics and Breeding

E. Lahav[1], U. Lavi[1] and J.I. Hormaza[2]
[1]*Institute of Horticulture, ARO, The Volcani Center, Rishon LeTsiyon, Israel;*
[2]*Instituto de Hortofruticultura Subtropical y Mediterránea La Mayora, Consejo Superior de Investigaciones Científicas – Universidad de Málaga (IHSM la Mayora – CSIC-UMA), Málaga, Spain*

Abstract

This chapter offers a comprehensive overview of recent advancements in avocado genetics and breeding, emphasizing its promising future shaped by evolving technologies and the yet-to-be fully explored genetic diversity, particularly in the centre of origin of the species. Those advances have so far relied on traditional breeding methods, beginning with the evaluation of seedling progeny based on fruit traits, followed by subsequent selection of promising candidates for further testing. Nevertheless, as in other crops, a transition towards molecular breeding, based on advancements in sequencing, genomics and other omics approaches, is taking place in avocado, and it will accelerate in the coming years. The integration of molecular breeding with traditional approaches is envisioned to expedite breeding programmes and produce avocado varieties and rootstocks with improved agronomic traits as well as to provide appropriate material for the generation of DNA-edited plants with desired traits.

Introduction

The avocado (*Persea americana* Mill.) is a subtropical evergreen woody perennial tree crop within the genus *Persea*, a member of the family Lauraceae in the order Laurales, which, together with the Canellales, Magnoliales and Piperales, forms the Magnoliid clade, considered as a sister clade to the eudicot and monocot angiosperms (Angiosperm Phylogeny Group, 2016). The genus *Persea* encompasses approximately 150 recognized species, with 70 distributed in America, 80 in Southeastern Asia and a single species, *P. indica*, endemic to the Macaronesian islands in the Atlantic Ocean. Three subgenera are usually considered in the genus *Persea*: *Machilus*, *Persea* and *Eriodaphne*, although *Machilus*, limited to Asia, is often treated as an independent genus by many authors. The 70 American species of *Persea* are divided into two subgenera (Kopp, 1966): *Eriodaphne* (primarily with a South American origin) and *Persea* (primarily with a Central American origin). Subgenus *Persea* includes at least three species (*Persea pallescens*, *Persea schiedeana* and *P. americana* – the avocado), although additional species are added by some authors (Chanderbali *et al.*, 2013). *P. americana* is a polymorphic species with at least eight botanical varieties or subspecies that have evolved under different edaphoclimatic conditions and geographical isolation. Three of them, also referred to as horticultural races, have agronomic importance (Bergh and Ellstrand, 1986). These include *P. americana* var. *drymifolia*, *P. americana*

Corresponding author: ihormaza@eelm.csic.es

 The Avocado: Botany, Production and Uses
(eds D. Carrillo, B. Schaffer, A.W. Whiley and B.N. Wolstenholme)
DOI: 10.1079/9781800621824.0004

var. *guatemalensis* and *P. americana* var. *americana* recognized as the Mexican, Guatemalan and West Indian (Lowland or Antillean) horticultural races, respectively, based on their presumed centres of origin (Popenoe, 1941). The West Indian race originated in the lowland tropics, the Guatemalan race in the valleys of the Central American mountain ranges and the Mexican race in the tropical highlands of Southern Mexico. Consequently, these horticultural races primarily differ in botanical characteristics and edaphoclimatic preferences. The Guatemalan and Mexican subspecies exhibit better adaptation to cooler subtropical climates, with the former being more susceptible to low temperatures than the latter. Conversely, the West Indian subspecies is more adapted to tropical climates. The number of chromosomes in avocado is $2n = 24$ (Garcia, 1975).

No sterility barriers have been reported among the three races or any taxa classified under *P. americana*. Consequently, hybridization may occur wherever trees of different races grow in proximity, whether in their native habitat (Popenoe and Williams, 1947) or under cultivation (Bergh, 1969). Most of the currently cultivated avocado commercial varieties in subtropical and Mediterranean climates are interracial Mexican × Guatemalan hybrids. Examples include 'Fuerte', a cultivar that historically dominated production in California and most other Mediterranean/subtropical growing regions, and 'Hass', the current predominant cultivar in most of these regions. Recent genome sequencing data reveal that approximately 40% of the 'Hass' genome is from the Guatemalan race introgressed into a Mexican-race background (Rendón-Anaya *et al.*, 2019). Guatemalan × West Indian hybrids are currently the leading cultivars in semitropical Florida and look promising for future production in more tropical areas.

Avocado Domestication and Selection

Most fruit crops are propagated clonally by grafting. Consequently, fruit tree improvement typically encompasses two main steps: first, the identification and selection of superior genotypes, followed by their vegetative propagation. Avocado is characterized by a unique flowering mechanism known as synchronous protogynous heterodichogamy (Stout, 1923). This mechanism promotes cross-pollination, resulting in a high heterozygosity with a highly variable progeny. The earliest known instance of avocado grafting dates back to Florida before 1900 (Ruehle, 1963).

The centre of origin of avocado can be located in Central America, spanning a broad geographical region that extends from the highlands of Southern Mexico, Guatemala and Honduras to the Pacific coast (Kopp, 1966). Archaeological evidence indicates that native cultures in Mexico were already consuming avocados over 10,000 years ago, as remnants dating back to that period have been found in the Coxcatlan Cave, in Tehuacan, located in the state of Puebla (Smith, 1966; see Chapter 2, this volume, for a detailed discussion of the history of avocado). From its centre of origin in Central America, avocados likely spread to South America during pre-Columbian times. In fact, the first recorded description of avocados after the European arrival to America was made in Colombia in 1519 (Fernandez de Enciso, 1519). Selection of horticulturally improved avocados occurred long before they were asexually propagated. Extensive pre-Columbian avocado selection is evident from the high horticultural quality already present in avocado in pre-Columbian times. This likely involved the laborious process of selecting and propagating superior cultivars from small-fruited wild forms found in the forests of Mexico and Central America. Popenoe (1919) suggested selection methods such as eliminating inferior seedlings, planting seeds from superior seedlings and marketing selected fruit, thereby disseminating the better genotypes. Further selection occurred during the last century, with superior cultivars being preserved through vegetative propagation (Popenoe, 1952). The initial stages of modern avocado breeding took place in California following the introduction of material (seeds and grafting wood) from Mexico. An illustrative example is the variety 'Fuerte', which, after its introduction as grafting wood from Atlixco, Mexico, in 1911, quickly became the dominant variety, first in California and later in other regions. Additional introductions, mainly of seeds, were made in previous and following years, increasing germplasm diversity. Consequently, most of the later cultivars in California (with the exception of 'Fuerte') are local random selections derived from open-pollinated seedlings.

Avocado breeding and selection also took place in Florida, particularly focusing on the West Indian germplasm and playing a crucial role in shaping the industry. All current important Florida cultivars were selected from locally grown seedlings derived from open pollination. These seedlings predominantly originated from West Indian avocados and, more recently, from West Indian × Guatemalan hybrids. The advancement of avocado cultivation in other tropical regions has also been achieved by selecting and vegetatively propagating superior local seedlings, underscoring the importance of adapting avocado production to regional conditions.

A high level of still-unexplored germplasm is available in avocado (Fig. 4.1), and successful avocado breeding will be closely related to the conservation and phenotyping of extant avocado germplasm.

Breeding Objectives

Avocado breeding combines a range of general and specific objectives. General breeding goals include achieving high and consistent yields, minimizing alternate bearing, improving fruit quality, extending the harvest season and prolonging shelf life. In most breeding programmes, since 'Hass' is currently the dominant variety in the global avocado market, there is a preference for fruit that turn black when ripened. In addition, some more specific objectives are usually found in avocado rootstock breeding, mainly related to enhanced tolerance to soil fungi, salinity or drought conditions. The following are the main objectives of avocado breeding.

Fig. 4.1. Example of the diversity of avocados found in germplasm collections. In this case, part of the avocado germplasm collection at the IHSM La Mayora in Málaga, Spain.

Rootstocks

Phytophthora root rot

Phytophthora root rot (PRR), caused by the soil-borne oomycete pathogen *Phytophthora cinnamomi* Rands., is one of the most important limiting factors for avocado production worldwide (Coffey, 1987; Ben-Ya'acov and Michelson, 1995; see Chapter 13, this volume, for a detailed discussion of avocado diseases). No resistance to this pathogen has been found in *P. americana* or any other species of the subgenus *Persea*. While resistance has been found in species belonging to the subgenus *Eriodaphne*, they are sexually and graft-incompatible with species in the subgenus *Persea* (Zentmyer, 1980; Bergh and Ellstrand, 1986). Consequently, attempts to hybridize avocado with *Persea* species of the subgenus *Eriodaphne* that are resistant to the disease have failed. An alternative approach suggested by Witjaksono and Litz (1998) involves protoplast fusion and somatic hybridization, but further data on the success of this approach are limited (Pliego-Alfaro *et al.*, 2020). Through this technology, sexually incompatible and graft-incompatible species such as *Persea borbonia*, *P. caerulea* and *Machilus* spp. may be hybridized with *P. americana*. The assumption underlying this method is that cytoplasmic genes control the resistance. Regeneration of somatic hybrids remains a limiting factor in this approach. Limited resistance to PRR is known in certain lines of avocado and the closely related *P. schiedeana*. In California, several rootstocks have been selected for PRR tolerance, and they have been clonally propagated to maintain their resistance levels. The most relevant include 'Duke' and its progeny ('Duke 7', 'Barr Duke' and 'D9'), 'Thomas' and 'Toro Canyon'. Over the years, Zentmyer (1972) and his successors have explored Mexico and Central America for wild sources of root rot resistance. The Mexican-race 'G6' accession achieved some commercial use. Nevertheless, some of the tolerant selections have serious limitations. An example is 'Martin Grande' (G755), a cross involving *P. schiedeana* and *P. americana* (Guatemalan race) (Ellstrand *et al.*, 1986), that produces trees with a low yield due to excessive vigour. Recent releases from this breeding programme include 'Steddom', 'Uzi' and 'Zentmyer' (Menge *et al.*, 2012). They are relatively tolerant to root rot caused by *P. cinnamomi* and are claimed to produce twice as much fruit as 'Thomas' (Douhan, 2009).

In South Africa, 'Dusa', of Guatemalan × Mexican origin, selected from a survivor tree, has shown excellent performance in soils infested with *P. cinnamomi* (Kremer-Köhne and Köhne, 2007). 'Dusa' also shows a good level of tolerance to cold winters and to high-salinity conditions (Kremer-Köhne and Köhne, 2007). Another rootstock, 'Latas', has also shown good tolerance to root rot and salinity conditions (Douhan, 2009). Additionally, 'Bounty', selected from another survivor tree, is also used in South Africa.

Notably, Tsao *et al.* (1992) found that some rootstocks tolerant to *P. cinnamomi* are sensitive to *Pytophthora citricola*, a serious pathogen of avocado trees.

White root rot

Another significant soil-borne pathogen affecting avocados is the ascomycete *Rosellinia necatrix* Prill, the causal agent of white root rot disease. Whereas *P. cinnamomi* affects avocado orchards worldwide, *R. necatrix* is mostly present in temperate regions. It is a serious problem in Spain, where it has been extensively studied (López-Herrera *et al.* 1998; López-Herrera and Zea-Bonilla, 2007; Moreno-Perez *et al.*, 2023). More recently, *R. necatrix* has also been detected in South Africa (Van den Bergh *et al.*, 2018) and Italy (Fiorenza *et al.*, 2021), highlighting its potential impact in regions beyond its traditional range. A breeding programme at IFAPA-Centro de Málaga, Spain, has been actively selecting avocado material with tolerance to *R. necatrix* (Barceló-Muñoz *et al.*, 2007). Several promising selections from this breeding programme are currently being tested in plots infested with *R. necatrix* at IHSM La Mayora Experimental Station in Algarrobo-Costa, Málaga. The expectation is that these selections will be released in the coming years, offering avocado growers in affected regions a valuable tool for managing and mitigating the impact of white root rot disease.

Other soil fungi

Among other fungal soil-borne pathogens affecting avocados, *Verticillium dahliae*, causing *Verticillium* wilt, is a significant concern, especially in soils where previous crops were grown (Haberman *et al.*, 2020). Guatemalan rootstocks appear to be more sensitive to this disease than Mexican rootstocks (Zentmyer *et al.*, 1965). Ben-Ya'acov and Frenkel (1974) found significant differences in sensitivity to *Verticillium* wilt among different West Indian rootstocks. Recent work has also shown differences among rootstocks growing in *V. dahliae*-infested orchards (Haberman *et al.*, 2020). This emphasizes the importance of selecting rootstocks with resistance or tolerance to specific soil-borne pathogens, such as *V. dahliae*, to effectively manage diseases and ensure the health of avocado orchards.

Soil quality and climatic tolerance

Soil quality and climatic tolerance are critical factors influencing the successful cultivation of avocados. Examples include regions characterized by high salinity or limited water supply, such as California (USA), Spain, Israel or Australia. In such environments, developing clonal rootstocks capable of using brackish water for irrigation becomes a highly desirable objective. Resistance to salinity is highest in West Indian-race cultivars and lowest in Mexican-race cultivars. Nonetheless, considerable variation is present within each of the three races (Kadman and Ben-Ya'acov, 1976) and even among seedlings from the same tree (Kadman, 1968). Resistance to Ca^{2+}-induced chlorosis is highest in West Indian cultivars, although there is also considerable variation in this trait within seedling populations (Ben-Ya'acov, 1972). Although usually West Indian seedling rootstocks perform poorly in heavy soils and under waterlogged conditions (Ben-Ya'acov *et al.*, 1974), observations with 'Simmonds' grafted onto either West Indian, Guatemalan or Mexican clonal rootstocks showed that those grafted on the West Indian rootstocks were more tolerant to flooding in terms of physiological variables and survival (Yin and Schaffer, unpublished data). Because of these challenges, efforts have been directed towards breeding rootstocks with enhanced tolerance to both salinity and chlorosis, as reported by Sánchez-Colin and Barrientos-Priego (1987). The South African selection 'Dusa' has shown a good level of tolerance to saline irrigation water in California (Crowley and Arpaia, 2002; Celis *et al.*, 2018).

Pure West Indian rootstocks perform poorly in winter-cold California soils, but hybrids, especially with Mexican types, seem to be well adapted. While the development of a rootstock that improves the cold hardiness of the scion is highly desirable in frost-prone areas, studies on chill-tolerant Mexican rootstocks have not shown a successful transmission of their hardiness to the scion (Halma and Smoyer, 1951; Ben-Ya'acov, 1987, 1998). Addressing this limitation remains an area of interest and potential improvement in avocado cultivation, especially in regions prone to winter frosts.

Dwarfing

Sánchez-Colin and Barrientos-Priego (1987) reported a significant dwarfing effect by their 'Colin V-33' selection, whether used as an inter-stock or a conventional rootstock. However, it is noteworthy that this selection tested positive for the avocado sunblotch viroid (ASBVd). Consequently, its use for dwarfing purposes remains unclear, as the presence of the viroid may introduce additional considerations. The prospect of a dwarfing rootstocks holds potential advantages for most avocado producers. In Israel, the West Indian rootstocks 'Nahlat' (Ben-Ya'acov *et al.*, 1979) and 'Maoz' (Kadman and Ben-Ya'acov, 1980) were found to have some dwarfing effects. In other studies, trees grafted onto Mexican rootstocks were smaller than comparable trees grafted onto West Indian types (Ben-Ya'acov, 1976).

Cultivars

Fruit traits

Skin colour

Since the 1970s, the global market standard for avocados has been 'Hass', which produces fruit with a black and rough epicarp, followed, to a much lesser extent, by varieties such as 'Fuerte', which produces fruit with a green and smooth epicarp. As 'Hass' has become dominant, cultivars with green skins usually sell for less. Thus, in most breeding programmes, one of the goals is to develop new cultivars that resemble 'Hass'. However, this has not always been the case, and historical market dynamics tell a different story. 'Fuerte' was the leading avocado variety in the international markets until the 1970s. Given the historical variability in preferred skin colour and the evolving avocado market, breeding programmes are increasingly focusing on overall fruit-quality traits. Fruit with mixed green and purple skin (seen in a considerable proportion of selfed 'Nabal' progeny) are usually less attractive. Glossy skin surfaces, as those present in 'Ettinger', are commonly considered attractive. In California, the severe russeting of 'Regina' and the end-spotting of 'Zutano' were found to be present significantly in their respective progeny. This highlights the importance of addressing not only colour but also texture and appearance traits to meet diverse market preferences.

Size

The current optimal fruit size for most markets falls within the range of 250–350 g. For European and US markets; fruit outside the 150–400 g range are generally unacceptable. Various factors influence the size of avocados, including the variety, crop load, proximity to other fruit on the tree, stage of maturity, cultural practices, climatic conditions and other unknown factors (Lahav and Kalmar, 1977; Whiley and Schaffer, 1994). Usually, avocados are sized into categories in the packing house based on the weight of individual fruit relative to the box in which they are packed. For the US market, sizes range from 32 (32 units per box, each weighing more than 330 g) to 84 (84 units per box, each higher than 120–150 g).

In the EU, sizes range from 12 (12 units, each higher than 330 g) to 26 (26 units, each higher than 120–150 g).

Shape

Fruit shape segregates extensively among most self-progenies. Desirable shapes include the squat-pyriform shape of 'Hass', the ovate form of 'Bacon' and the thick-ovate form of 'Gwen'. Excessively elongated fruits are quite common among progeny of 'Hass' (Bergh and Whitsell, 1974). The excessive elongated shape of 'Galil' (selected by the Israeli breeding project) or 'Pinkerton' can be shortened by treatments with growth retardants (Lahav *et al.*, 1998) such as uniconazole (Adato, 1990).

Skin thickness

The leathery easy-peeling type of 'Fuerte' or the thicker skin of 'Hass' (ideal for spoon eating) are usually preferred. While thin-skinned fruit is prone to damage, excessively thick skin hinders the determination of ripening time. Striking a balance in the level of skin thickness is essential to ensure both durability during handling and the ability to assess ripeness accurately.

Seed size

Variability in seed size is very common within the same progeny. A small seed, relative to fruit size, which is tight in the pulp cavity, is a superior attribute of many Guatemalan lines. 'Irving', with its later-maturing fruit and slightly thicker skin indicating more Guatemalan alleles, has an exceptionally low seed/flesh ratio. Among the West Indian varieties, 'Ruehle' stands out for its small seed size. Elongated fruits, such as the relatively long-necked 'Pinkerton', generally have a smaller seed. It is highly desirable that both seedcoats remain attached to the embryo instead of the flesh (loose seeds are frequently found in Mexican types).

Ripening

In most cultivars, the fibrovascular bundles tend to become more prominent with increasing maturity, although this undesirable trait is less pronounced in most Guatemalan lines. Uniform and adequate softening of fruit as it ripens is independent of race. For instance, 'Jalna' (Mexican) and 'Pinkerton' (Guatemalan hybrid) are sometimes characterized by uneven ripening (Piccone and Whiley, 1986). A long period from harvest to softening is advantageous, especially when fruit is transported over long distances. Also highly desirable is a longer time from softening to flesh deterioration. Avocado has a relatively short shelf life compared with other fruits, although it varies between very short in 'Anaheim' to longer in 'Hass', 'Fuerte' and others. West Indian lines have much lower oil content than those of the other two races, but there is considerable intra-race variability. Guatemalan and especially Mexican cultivars can reach oil content well over 20%, while 'Anaheim' and 'Mayo' cultivars of these two respective races may have an oil content below 8% when they begin dropping or deteriorating.

Flavour

The rich, slightly nutty taste of 'Hass', 'Fuerte' or 'Benik' is generally preferred over bland flavours. However, population preferences do exist, and the spicy or anise-like flavour of Mexican types such as 'Mexicola' or 'Duke' is more desirable to some consumers. The mild, sweeter taste typical of West Indian cultivars is preferred by other populations, such as those in Central America and the Caribbean. These diverse flavour profiles cater to varied consumer preferences and highlight the importance of considering regional taste preferences in avocado-breeding programmes.

Shelf life

Mexican-type avocado fruit and hybrids are strongly climacteric. Following its initiation, ripening cannot be arrested. Fruit of Guatemalan and Mexican cultivars can be stored on the tree during the lag phase for 2–4 months after reaching maturity, and they accumulate oil during this period (Whiley and Schaffer, 1994). Ripening begins only after the fruit is picked. In contrast, West Indian and West Indian × Guatemalan cultivars cannot be stored on the tree in the same manner and must be picked at maturity. On-tree storage, combined with different climatic conditions, has enabled producers in many areas to concentrate production on a single cultivar, 'Hass'. To ensure year-round production of West Indian and West Indian × Guatemalan fruit in the tropics, several cultivars with different maturity dates must be grown (Crane *et al.*, 1996). Consequently, there is no single market standard for tropical avocados, emphasizing the need for diversified cultivation strategies.

Fruit diseases and insect pests

In the humid tropics and subtropics, fruit diseases are a serious problem. Genetic resistance is desirable, as fungicide treatments are expensive, not fully effective and can be detrimental to the environment. Ruehle (1963) lists the following relative cultivar susceptibilities to various diseases: *Cercospora* spot or blotch (caused by *Pseudocercospora purpurea* Cooke) is much less severe on 'Collinson', 'Fuchsia' and 'Pollock'; 'Fuerte' and other Mexican types are highly susceptible to anthracnose or black spot (caused by *Colletotrichum gloeosporioides* Penz.). 'Fuchsia', 'Pollock', 'Booth 1' and 'Waldin' are moderately resistant, while 'Lula' is highly susceptible to avocado scab (caused by *Sphaceloma perseae* Jenkins). 'Ardith', selected in the California breeding programme, but only commercialized in Israel, is highly susceptible to orchid thrips (*Chaetanophothrips orchidii*). Rootstocks seem to clearly affect resistance to anthracnose, probably because fruit calcium content plays a key role in tolerance to this disease (Coates *et al.*, 2011).

Tree characteristics

Yield

The most important tree characteristics for avocado cultivation include precocity with consistent and high yields. The consistency of production from year to year may be as important as achieving overall heavy production (Bergh, 1961). Given the considerable differences in tree size, yield is more accurately assessed as tonnes per hectare at the respective tree spacing. A more subjective evaluation involves considering the fruit number per tree volume or the 'covering area' of the tree canopy (Kadman *et al.*, 1976). It is desirable that the majority of fruit reach commercial maturity around the same time, particularly for earlier-season cultivars that have a short life on the tree. The heritability of this trait, which is typically assessed in the second selection stage, is currently unknown. Environmental conditions have a strong influence on this matter, especially in tropical regions, where several flowering flushes can occur, resulting in fruit from different flowering periods coexisting on the same tree.

Architecture

The architecture of avocado trees varies widely, ranging from erect forms exemplified by 'Reed', to spreading types like 'Fuerte' and even weeping such as 'Wurtz'. A spread approximately equal to tree height is considered desirable, and dwarf or semidwarf erect trees are considered ideal. Very tall trees such as 'Bacon' or 'Ettinger' are difficult to manage and make fruit harvest expensive. Excessive vigour and fruitfulness are not entirely compatible with exceptionally robust seedlings, often bearing little or no fruit. According to Bergh and Lahav (1996), tree vigour declines with inbreeding and is restored when inbred lines are crossed. Maintaining sufficient vigour is crucial for ensuring good tree health and high productivity.

Cold tolerance

Most of the major world avocado regions are subject to occasional frost damage, and, consequently, cold tolerance becomes a highly advantageous trait for both the fruit and the tree. Cold-hardiness is primarily associated with the Mexican race, while West Indian-race cultivars may show damage even in above-freezing temperatures. Variability in cold tolerance exists within as well as between races. For example, among cultivars believed to be pure Guatemalan, 'Nabal' and 'Reed' are cold-hardy whereas 'Anaheim' is sensitive to low temperatures. 'Yama' is considered one of the hardiest Mexican cultivars, withstanding temperatures as low as −8°C without severe injury, making it a favourable candidate for breeding programmes focused on enhancing cold-hardiness. Promising results with Mexican seedlings in Florida indicate potential for generating commercially viable selections with improved cold-hardiness to withstand periodic freeze conditions (Knight, 1971).

Heat tolerance

Considerable variability in heat tolerance is observed among avocado cultivars. Generally, the

Mexican race shows greater average heat tolerance than other races, as exemplified by 'Mexicola', 'Mayo' or 'Indio'. Moreover, because Mexican-race cultivars bloom earlier, their fruit is usually more advanced in development and, thus, less vulnerable to potential damage from spring heatwaves.

Disease tolerance

LAUREL WILT DISEASE. Laurel wilt is a vascular disease caused by the pathogenic fungus *Harringtonia lauricola* and transmitted by certain species of ambrosia beetles. In the south-eastern USA, the invasive redbay ambrosia beetle, *Xyleborus glabratus*, is the primary vector of this pathogen in forests with wild *Persea* species but is rarely found in avocados (Carrillo *et al.*, 2012). However, several ambrosia beetles that inhabit avocado orchards have acquired the pathogen and act as laurel wilt vectors (Carrillo *et al.*, 2014). The disease was first reported in the state of Georgia in the USA in 2002, affecting *P. borbonia*, the redbay (Harrington *et al.*, 2008), and has spread to the avocado-producing areas of south Florida, leading to considerable damage. It has expanded as far west as eastern Texas and as far north as Tennessee (Castillo-Argaez *et al.*, 2021a). There is currently no known resistance to laurel wilt within *P. americana*. Nevertheless, studies involving *P. borbonia*, *Persea palustris* and *P. americana* (Castillo-Aragez *et al.*, 2021b) show that although all three species are susceptible to laurel wilt, differences among species in the extent of symptom development are observed. Interestingly, the West Indian avocado 'Waldin' exhibited less symptom development than genotypes of the other two species, possibly due to a greater density of wider vessels, opening the way for selecting laurel wilt-tolerant avocado plants. Indeed, West Indian cultivars appear to be generally more susceptible to laurel wilt than Mexican or Guatemalan, but differences in susceptibility among cultivars within each race have been reported (Ploetz *et al.*, 2012). In addition, the scion/rootstock combination affects susceptibility to laurel wilt (Castillo-Argaez *et al.* 2021a).

SUNBLOTCH DISEASE. Avocado sunblotch disease is caused by the ASBVd, a member of the Avsunviroidae family of viroids. ASBVd causes symptoms on avocado fruit and leaves, and the extent of infection within the plant can vary depending on the tissue origin and environmental conditions. Some trees can be asymptomatic (Semancik and Szychowsky, 1994), and seemingly healthy trees can carry the viroid, indicating unequal distribution of the viroid, low efficiency of the indexing technique or host silencing mechanisms (Wallace, 1967). The most effective control measure is the eradication of affected plants. This underscores the importance of implementing measures to prevent the spread of the viroid through various vectors and maintaining a rigorous approach to plant health management. Although no clear differences are observed among avocado varieties in their susceptibility to sunblotch, early work showed that 'Zutano' apparently showed a slower development of symptoms. In fact, varieties of the Mexican race seem to manifest symptoms somewhat earlier after experimental inoculation than varieties of the Guatemalan race (Wallace, 1958).

BRANCH CANKERS AND LIMB DIEBACK. Branch cankers and limb dieback of avocado are attributed to a complex of fungal species within the family Botryosphaeriaceae. The prevalence of Botryosphaeria branch canker and dieback in avocado orchards has increased in recent years in most avocado-growing areas. Interestingly, avocados grafted onto Mexican rootstocks appear to exhibit greater tolerance to this disease compared to those grafted onto Guatemalan rootstocks (Halma and Zentmyer, 1953).

Floral Biology

Avocado flowers are hermaphroditic with a single pistil and nine stamens in two whorls (Schroeder, 1944). The avocado is self-compatible (Sedgley, 1979). Avocado flowers are arranged in terminal inflorescences, forming panicles of cymes with multiple secondary and tertiary axes, highly compound cymes or thyrses containing dozens to hundreds of flowers each (see Chapter 6, this volume). Consequently, a single tree may produce a million flowers during one blooming period, although only 100 to 200 develop into mature fruit (Lahav and Zamet, 1999). This low rate of fruit set makes hand pollination impractical for commercial purposes, although it remains feasible for research purposes (Alcaraz and Hormaza, 2014).

Understanding avocado floral biology is crucial for the generation of both selfed and crossed progeny. Avocado flowering has been thoroughly reviewed by Bergh (1986), Davenport (1986) and Hormaza *et al.* (see Chapter 6, this volume). Avocado exhibits a synchronous protogynous dichogamy system, where each hermaphrodite avocado flower opens twice during two consecutive days. On the first day, it functions as a female flower and, during the second opening, as a male flower. These two openings are separated by at least one overnight period. Avocado genotypes have been traditionally classified into two different types (A or B) based on their flowering behaviour. Type A cultivars have flowers that open in the female stage in the morning on the first day of the flowering cycle, while type B cultivars have flowers that open in the afternoon (Lahav and Gazit, 1994):

1. Group A: The first (female) opening starts in the morning and ends before noon. The second (male) opening occurs in the afternoon of the next day.
2. Group B: The reverse pattern: the female opening occurs in the afternoon and the male opening happens the next morning or two days after.

The two groups are reciprocal, favouring cross-pollination, since self-pollination (within the flower) or close pollination (between neighbouring flowers) is limited to the specific hours when closing female flowers on the first day of the flowering cycle overlap with opening male flowers on the second day of the flowering cycle. This overlap can occur within the same inflorescence, tree or cultivar. Temperature and the specific avocado genotype significantly influence the timing of flower opening and closing. Under cooler temperatures, the process is delayed, and more than one night may elapse between the first and second openings, affecting the duration of the overlap (see Chapter 6, this volume). While the genetic basis of heterodichogamy remains unknown in angiosperms, recent molecular and genomic studies have shed light on this aspect in avocados. Ashworth *et al.* (2019) identified markers associated with a quantitative trait loci (QTLs) region for flowering type, and Solares *et al.* (2023) identified putative candidate genes that could be involved in avocado type A and B flower development.

The length of avocado flowering varies with cultivar and climate, with cooler temperatures generally leading to a longer flowering period. In most avocado-producing regions, pure Mexican cultivars and Mexican × Guatemalan hybrids have been reported to bloom early (Schroeder, 1951). In California, generally, the Mexican and Mexican × Guatemalan hybrids are the earliest to begin bloom (January). West Indian cultivars typically begin flowering in mid-January to February. Bloom length is cultivar- and temperature-dependent; consequently, bloom length can vary significantly from year to year for a given cultivar (Levin, 1981; Garner and Lovatt, 2008). Usually, each avocado tree flowers continuously for about 1–3 months. In some climates, such as in Michoacan (Mexico), up to four bloom flushes from July to April can be produced; they are named, respectively, 'loca' (crazy), advanced, normal and marceña (from March) (Salazar-García *et al.*, 2005). Flowering time can be manipulated to achieve overlap of desired parents. For example, flowering can be advanced as much as 2–3 months by girdling (Lahav *et al.*, 1972) or by placing container-grown trees in a greenhouse to promote early flowering during winter. Conversely, flowering can be delayed by holding container-grown trees in a cool growth chamber or by storing scions with differentiated flower buds at around 4°C until grafting time is convenient for the breeder. This approach allows for an extended pollination period of up to 6 months (Sedgley and Alexander, 1983). Growth retardants have had limited success in controlling the flowering period (Levin, 1981).

Genetics of the Juvenile Phase

Avocado, like all vascular plants, undergoes two distinct developmental phases, juvenile and adult. During the juvenile phase, plants are unable to respond to floral inductive signals that initiate floral development. The length of the juvenile phase is a major factor affecting the efficiency of fruit tree breeding programmes in general and in avocado in particular. Compared to other fruit trees, the juvenile phase in avocado is quite long and can extend beyond 10 years depending on the genotype (Bergh and Lahav, 1996). The duration of the juvenile phase is inversely related to the efficiency of genetic improvement in breeding programmes. Thus, shortening the juvenile phase is essential to enhance the effectiveness of breeding efforts. In breeding programmes based on controlled crosses, it may take about 18 years before a selection is released for small-scale commercial evaluation (Table 4.1). However, in breeding programmes based on open pollination with a fast release of selected progeny, the cycle can be reduced to around 10 years.

Similar to other fruit crops, the length of the juvenile period in avocado is under genetic control (Johnson, 1940; Visser *et al.*, 1976). The juvenile period can be significantly shortened by selecting appropriate parents (Lavi *et al.*, 1992). Cultivars such as 'Pinkerton' and 'Gwen' (Bergh and Lahav, 1996) and 'Arad' (Lahav *et al.*, 2005) have been found to produce very precocious offspring, sometimes bearing fruit just 2 years after grafting. In contrast, Lavi *et al.* (1992) reported that the mean flowering age ranged from 4.3 years for the progeny of 'Rosh-Hanikra II' × 'Ettinger' to 8.6 years for the self-pollinating progeny of 'Nabal'. Among the 11 progeny populations investigated, flowering age ranged from 3 to 11 years (with some seedlings taking up to 14 years). Fruiting age also varied from 3 to 11 years (up to 14 years in one case). No seedling flowered or fruited earlier than 3 years after planting, and only 4.2% did so at the age of 3 years. A substantial portion of each progeny population did not flower or fruit during the 10 years of the experiment. Although some variations were observed between different years, statistical analysis revealed significant differences in flowering and fruiting ages among various progeny populations, indicating quantitative genetic control. A difference by a factor of two was observed between the earliest and the latest mean flowering ages, likely depending on genetic factors that are quantitatively controlled. No differences were detected between self- and cross-pollinations, suggesting that heterozygosity levels do not significantly affect the juvenile period (Lavi *et al.* 1992). It is noteworthy that the majority of seedlings that flowered in a certain year also fruited the same year or the following year, suggesting that the main constraint in an avocado-breeding programme is the time until the first flowering. The conclusion drawn was that the choice of parents might influence the length of the juvenile period and, consequently, the efficiency of the breeding project. However, the use of this tool is limited since the choice of parents in breeding programmes is influenced by other important factors.

Recent advances in understanding the regulation of phase transition in the model plant *Arabidopsis thaliana* have provided insights that could be applied to avocados. Genes differentially expressed in the juvenile and adult phases of avocados have been identified, and they could serve as markers to screen for factors regulating

Table 4.1. Tentative timetable for avocado breeding.

Year	Stage	Controlled crosses	Year	Open pollination
1	I	Preparation of parents[a]	1	Pollination
2		Continued preparation of parents	2	Production of plantlets
3		Pollination (in cages or manual)	3	Planting in field
4		Production of plantlets	4	Continued planting in field
5		Planting in field	5	Girdling, if practised
6		Continued planting in field	6	Evaluation and selection
7		Girdling, if practised		
8		Evaluation and selections		
9	II	Top-working	7	Top-working
10		Intense evaluation	8	Intense evaluation
11		"	9	"
12		"	10	"

[a]Comments: (i) Parents are prepared to be close to each other, either by planting them side by side or by grafting one parent on top of the other. (ii) The actual timetable depends on the local conditions. (iii) Following stage II, selected types are grafted and evaluated on a semi-commercial scale.

phase transition (Ahsan *et al.* 2019). New molecular approaches could allow the development of transgenic avocado lines with a reduced length of the juvenile phase. These advancements could enable the implementation of techniques such as the 'FasTrack' breeding system developed for other fruit crops such as plum (*Prunus domestica*). The 'FasTrack' system involves overexpression of the *FLOWERING LOCUS T* (*FT*) gene, allowing for a more rapid evaluation of phenotypic traits of interest (Petri *et al.* 2018). This approach holds promise for accelerating avocado breeding programmes and enhancing the efficiency of trait selection.

Breeding Techniques

Shortening the juvenile period

Several horticultural practices have been applied to shorten the juvenile period in avocado. One such method is top-working, which has been used on a limited scale in Australia (Sedgley and Alexander, 1983), Mexico (Sánchez-Colin and Barrientos-Priego, 1987), California and Israel. Top-working involves grafting or budding desired scions onto established rootstocks or trees. This technique allows for the rapid introduction of new cultivars or selections, potentially reducing the time it takes for a tree to start producing fruit.

Attempts have been made in California and Israel to induce earlier fruiting by bending seedling branches into a horizontal position, but these attempts did not result in significant improvements. In Mexico, a shorter juvenile period has been reported by training seedlings as a single stem and allowing them to branch when they reach a height of 2 m (Barrientos-Priego *et al.*, 1991). Additionally, breeders in Israel and Mexico have observed that larger planting distances appear to shorten the juvenile period, probably due to better illumination.

The use of growth retardants has been considered as a potential method for promoting fruitfulness by altering the balance between vegetative and reproductive growth in avocado trees. However, in Israel, no significant effect on the juvenile period was observed after application of various growth retardants, suggesting that the impact of growth retardants on avocado trees may vary, and their effectiveness in shortening the juvenile period might be limited.

Girdling, a horticultural practice involving the removal of a strip of bark around the trunk of the tree, was suggested as a method to potentially shorten the juvenile period in avocado (Lahav *et al.*, 1986). The earliest autumn girdling was found to increase the proportion of seedlings that flowered from 47% to 100%, significantly intensifying flowering and increasing the proportion of seedlings setting fruit. Most importantly, this resulted in a sevenfold increase in the number of fruit per tree (1.3 vs 9.3), allowing a reliable seedling evaluation. However, the success of girdling varied, being effective in some cases but not in others.

Pollination

Avocado pollen is sticky (Stout, 1933), and successful pollination requires the presence of large flying insects. In the avocado region of origin in Central America, various native pollinators have been identified, including stingless bee species (Apidae, Meliponinae) (Ish-Am *et al.*, 1999; Can-Alonzo *et al.*, 2005), the Mexican honey wasp (*Brachygastra mellifica* Say) and flies (*Chrysomya megacephala*) (Ish-Am *et al.*, 1999; Pérez-Balam *et al.*, 2012). However, as in most crops, commercial avocado production worldwide relies mainly on European honeybees (*Apis mellifera* L.), which were introduced to the Americas by the Europeans (Dymond *et al.*, 2021). While honeybees are commonly used for avocado pollination, they may not be the most effective avocado pollinators because they are more attracted to flowers with strong fragrances (Ish-Am and Eisikowitch, 1993, 1998). In addition to honeybees, many different pollinating insects are frequently observed visiting avocado flowers worldwide (Pattemore *et al.*, 2018; Dymond *et al.*, 2021; Okello *et al.*, 2021; Muñoz *et al.*, 2021; Sagwe *et al.*, 2022). Recent work in Israel (Stern *et al.*, 2021) has shown that pollination by bumblebees (*Bombus terrestris* L.) can significantly increase the yield of 'Hass' avocados. Bumblebees show several advantages over honeybees, including their ability to pollinate at temperatures below 14°C (Stern *et al.*,

2021). This makes them potentially more effective in environments where temperature conditions may not be optimal for honeybee pollination.

Pollen grain germination and pollen tube growth are sensitive to temperature. The optimal temperature range for *in vitro* avocado pollen germination is 20–30°C (Loupassaki *et al.*, 1997; Alcaraz *et al.*, 2011), although the optimal temperatures can vary among cultivars (Alcaraz *et al.*, 2011). Avocado pollen requires approximately 2 h to grow from the stigma to the ovary, with the actual time depending on the prevailing temperature (Schroeder, 1942; Papademetriou, 1974; Sedgley, 1977). Ovule penetration typically occurs 18–24 h after pollination, and fertilization of the polar nuclei and the egg cell takes place within 48 h of pollination (Sedgley, 1979). After fixation of the flowers, pollen tube progress in the style can be followed by fluorescence microscopy after staining with aniline blue (Tomer and Gottreich, 1975; Alcaraz and Hormaza, 2014).

Hand pollination in avocados is a feasible method for producing a few fruit, but it becomes challenging when large amounts of seedlings are needed in breeding programmes. The process involves collecting male-stage flowers with dehisced anthers and applying the pollen to the stigmas of female-stage flowers. For controlled self-pollination, male flowers should be collected just before anther dehiscence and could be stored for several days at 4°C, until female-stage flowers are available (Alcaraz and Hormaza, 2014). Avocado pollen tends to stick in clumps to the opened valve of the anther until it is removed by insects or drops with the flower. Methods of pollen collection, such as vacuum devices, have not been successful for pollen collection in avocado. Storage viability can be extended by reducing both temperature and humidity. Sedgley (1981) successfully stored pollen for 1 month at 4°C and 1–23% relative humidity, and for 1 year at −196°C and 0% relative humidity, in liquid nitrogen. This procedure allows crosses between a wide range of avocado cultivars with various flowering times. However, due to the gradual loss of pollen viability over time, it is advisable to use fresh pollen whenever possible (Alcaraz and Hormaza, 2014). Fruit set increases when pollen application is performed on flowers within determinate inflorescences or within indeterminate inflorescences where vegetative growth has been removed, enhancing the chances of successful fertilization and fruit development (Alcaraz and Hormaza, 2014).

Controlled pollination

Controlled pollination in avocado offers the advantage of knowing the identity of the parents, although the cost per seedling is much higher. Self-pollinated or cross-pollinated seedlings can be obtained through three basic procedures:

1. Natural pollination. For self-pollinations, fruit can be harvested from a tree at a sufficient distance from avocado trees of other genotypes. However, cross-pollination by bees or other insects can occur over long distances (Torres and Bergh, 1978; Vrecenar-Gadus and Ellstrand, 1985; Degani *et al.*, 1989). The further away from potentially contaminating pollen, the greater the likelihood of selfing. For cross-pollinations, two different cultivars can be planted in close proximity but at a distance from other avocado trees to allow them to cross-pollinate. The likelihood of successful crossing increases with the proximity of the cultivars. However, the effectiveness of cross-pollination can vary between different cultivars (Degani *et al.*, 1990), and progeny have to be identified by molecular markers to ensure that they derive from the desired cross.

2. Caged pollination. The tree can be enclosed within an insect-proof net, and beehives, bumblebees or flies can be introduced to facilitate pollination (Degani *et al.*, 2003). For cross-pollinations, flowering branches or trees of the pollenizer can be enclosed in a cage with bees or other pollinating insects. This method will produce a variable mixture of crosses and self-pollinations (Degani and Gazit, 1984), and molecular markers can be used to distinguish among them.

3. Hand pollination. Hand pollination is challenging when a high number of progeny is needed for breeding purposes but can be used to generate a relatively small number of fruit. For successful fruit set in crosses between different floral groups, the male flowers should be collected at anther dehiscence, and the pollen should be transferred by direct contact of anthers with the stigma of functionally female flowers. During the process, the flowers could be stored in a box with high relative humidity in the orchard. When the crosses involve cultivars of the same floral group, the male flowers should be collected just before anther dehiscence and could be stored for several days at 4°C. In both cases, the pollen application can be done during the entire female stage, preferably using determinate inflorescences. Pollinating about ten flowers per inflorescence and removing the rest of the flowers or undesired fruitlets is recommended for optimal results (Alcaraz and Hormaza, 2014).

None of the above methods can ensure the identity of the progeny, which would mostly be mixtures of selfing and crossing. In scenarios where accurate knowledge of seedling identity is crucial, especially in breeding programmes aiming for specific traits, molecular markers become essential. These markers help identify and verify the genetic makeup of the progeny. The only exception is truly isolated trees (or large blocks of a single cultivar), in which the resulting seedlings are primarily the result of self-pollination, simplifying the genetic composition without the need for molecular markers in such cases.

Open pollination

In scenarios where the breeding objectives for avocados are considered 'general', controlled pollinations or identification of progeny by molecular markers might not be necessary. In these cases, the primary focus is ensuring the maximum genetic variation within the seedling population, and this is best achieved by open pollination. However, progeny identification is required for particular genetic studies that will eventually be important to avocado breeding.

For open pollination, seeds are collected from the selected tree(s), and depending on the degree of isolation and the cultivar, various proportions of out-crossed progeny can be produced. The advantage of this approach is its low cost, allowing the rapid evaluation of many seedlings. However, the disadvantage is that the inheritance of traits is limited, and if specific crosses are desired, molecular markers have to be used to check for paternity.

Increasing seed set

Hand pollination results in a low fruit set and, consequently, can be impractical when a high number of seedlings are required. Different rates of success have been reported after hand pollination. For instance, Schroeder (1958) cross-pollinated over 10,000 'Fuerte' flowers, and only four mature fruits were obtained. Eisenstein and Gazit (1989) reported a yield of 1.3%–27.2% fruitlets with hand pollination of various crosses, while Alcaraz and Hormaza (2014) reported 2.43% of fruit set in 'Hass' flowers pollinated with pollen from 'Fuerte', compared to 0.15 fruit set with open pollination.

A number of techniques can increase the fruit set in hand pollination:

I. Optimal tree location and management. Place trees in an optimal location, and provide appropriate management practices, including irrigation, nutrition, disease and insect control, and wind protection (Bergh, 1967).

II. Shading control. Avoid excessive shading in the cages by using screen material that transmits as much light as possible, and remove shading limbs to improve light exposure (Lahav, 1970).

III. Use open anthers. Pollinate the stigmas using open anthers of the male parent (Alcaraz and Hormaza, 2014).

IV. Timing of pollination. Hybridize or self-pollinate in the productive ('on') year for that particular tree, considering that avocado trees exhibit alternate bearing to some degree.

V. Girdling after fruit set. Girdling after fruit set may increase the chances of fruitlet survival (Lahav *et al.*, 1971, 1972).

VI. Select heavy-setting cultivars. If possible, try to choose heavy-setting cultivars as the mother parent to increase the likelihood of successful fruit set.

VII. Suitable pollen parent. If possible, select a suitable pollen parent based on pollen performance (Lahav and Gazit, 1994).

VIII. Reduce vegetative competition. Reduce the competition between vegetative growth and fruit development by employing techniques such as young shoot tipping (Biran, 1979) or growth retardants (Adato, 1990; Wolstenholme *et al.*, 1990).

IX. Inflorescence management. Use determinate inflorescences or remove the vegetative growth of indeterminate inflorescences to focus energy on fruit set and development (Alcaraz and Hormaza, 2014).

X. Selective pollination. Pollinate about ten flowers per inflorescence, and remove undeveloped flowers and fruitlets to concentrate resources on the desired fruit (Alcaraz and Hormaza, 2014).

Seed Germination and Growth of the Seedlings

Fruit should be harvested when they reach physiological maturity (having the ability to ripen after harvest), ensuring the seeds are fully mature for optimal germination. Harvesting undeveloped fruit can result in poor germination. Avocado seeds have physiologically recalcitrant germination. Avocado seeds left at room temperature and humidity remain viable for only a few days after their removal from the fruit, as the embryo has no protection against desiccation (Storey *et al.*, 1986). However, avocado seeds can remain viable for up to 15 months when stored at 5°C with high humidity (Halma and Frolich, 1949). Humidity can be easily maintained during storage by placing the seeds in sealed polyethylene bags. Under optimal growth conditions (night–day temperature range of about 23°C–25°C), most viable seeds will germinate within a month (Alexander, 1977). Vigorous seedlings will attain a height of about 1 m in 3 months, while less vigorous seedlings may require 6 months to reach a suitable size for field planting. Cooler temperatures significantly retard seedling growth.

Growing seedling progeny on their own roots is cost-effective, but in some conditions (such as irrigation with saline water, calcareous soil or root rot), grafting onto resistant or tolerant rootstocks may be necessary. Although in most breeding programmes in the first stage avocado seedlings are grown on their own roots, most of the progeny of the Israeli breeding programme are grafted due to calcareous soil and saline water. Avocado seedlings and their grafted duplicates showed no significant differences in their performance, productivity or the length of their juvenile period (Lahav *et al.*, 1995).

The most efficient seedling spacing is the lowest that permits adequate fruitfulness. This will vary with the average vigour of each seedling group. Suitable planting distances may be 1–2 m in the row and 4–6 m between rows. Generally, closer planting distances may increase the juvenile period and should therefore be avoided. At closer spacing, some of the smaller trees may have their fruiting delayed by shading, but this can be addressed by gradually reducing tree density by removing inferior seedlings. Typically, breeding blocks can be replaced with new seedlings after 6–8 years, providing ample time for individual seedlings to exhibit precocity while allowing a safety margin to cover delayed fruiting due to genetic or environmental factors. Recommended management techniques for nursery and field-planted seedlings were summarized by Bergh and Lahav (1996).

Assessment of the Seedlings

A detailed list of avocado descriptors was prepared by the International Plant Genetic Resources Institute (1995). A developmental scale was established according to the Biologische Bundesanstalt, Bundessortenamt und Chemische Industrie (BBCH) code by Alcaraz *et al.* (2013), facilitating exchange of scientific information among avocado experiments in different environmental conditions. In the California breeding programme, evaluations typically focus on the seedlings selected for further testing, emphasizing crucial commercial traits. Tree assessment considers size, shape, productivity and flowering characteristics. Fruit evaluations include size, shape, colour, lack of damage, attractiveness, time of maturity, skin thickness and roughness, seed size and tightness in the cavity, flesh attractiveness and presence of fibre in the flesh. Ripened fruit undergo additional laboratory evaluations, emphasizing flavour (nuttiness, sweetness, bitterness or defects), peeling ability and the time from harvest to softening (eating-ripe).

The Israeli breeding programme evaluates every seedling, recording numerous parameters for genetic studies, including all of those assessed in California and adding the following:

1. *Tree:* The distance between buds, leaf and flush colour, flush lenticels, as well as leaf size, shape, habit, margin waviness and anise scent.

2. *Flowering:* Time of year, intensity and flowering group.
3. *Fruit:* Length of inflorescence (peduncle) and fruit stalks (pedicel), thickness of fruit stalk, its attachment position, suitability for snap picking (instead of stalk cutting), skin gloss, seed surface, flesh texture, oxidation of cut fruit and shelf life.

While some traits are quantitatively measured, most are estimated visually. Evaluation of economic traits is summarized in Table 2 (Lahav *et al.*, 1995). For genetic studies, many traits can be assessed, but for selecting new cultivars, the evaluation is usually limited to a few economically significant traits.

The recommended approach involves two selection stages. The first stage focuses on fruit assessment and is conducted in the first year of cropping. In the second stage, selected seedlings are grafted onto two to four mature trees and subsequently assessed for all fruit and tree characteristics under commercial conditions. The selected seedlings are then re-evaluated in a semi-commercial orchard (about 1/10 ha). This two-stage selection process ensures faster and more efficient breeding outcomes and helps minimize 'type 2 errors' (the incorrect commercial release of a seedling).

The genetic basis of fruit abscission

Fruit abscission, which usually takes place in the weeks following anthesis, is a major issue in avocado and can be attributed to several factors. One major cause is the inadequate embryo development, often associated with unsuccessful fertilization (Garner and Lovatt, 2016). Although pollination without fertilization can briefly stimulate ovary enlargement, the resulting parthenocarpic fruit enlarges to approximately 2–3 mm or less in diameter (Sedgley, 1980) and approximately 4 mm or less in length (Tomer and Gazit, 1979) before abscising 2–4 weeks after flowering. In some cases, fertilization takes place, but due to stenospermocarpy, embryo abortion occurs, resulting in seedless fruits, known as 'cukes' (Tomer *et al.*, 1980; Steyn *et al.*, 1993). Various internal and external factors contribute to fruit abscission in fertilized fruit in the weeks following anthesis. Internal factors include limited nutrient and hormone

Table 4.2. Evaluation of economic avocado traits. Adapted from Lahav *et al.* (1995).

Traits	Evaluation criteria
Tree size	Very big (1); Big (2); Medium (3); Small (4); Dwarf (5)
Flowering intensity	Profuse (1); High (2); Medium (3); Light (4); Very light (5); None (6)
Flowering time	Precocious (1); Early (2); Early-midseason (3); Late-midseason (4); Late (5); Very late (6)
Fruit weight	In grams
Fruit size uniformity	High (1); Medium (2); None (3)
Fruit shape uniformity	High (1); Medium (2); Slight (3)
Fruit density on the tree	Dense (1); Medium (2); Light (3); Very light (4)
Damage by snap picking	Minimal (1); Slight (2); Medium-severe (3)
Skin thickness	Mexican type (1); Like 'Fuerte' (2); Like 'Tova' (3); Like 'Hass' (4); Like 'Nabal' (5); West Indian type (6)
Ease of peeling	Excellent (1); Good (2); Medium (3); Difficult (4); Impossible (5)
Separation of seed from flesh	Easily (1); With some difficulty (2); Impossible (3)
Seed weight (% of fruit weight)	<6% (1); 6%–10% (2); 11%–15% (3); 16%–20% (4); 21%–25% (5); 26%–30% (6); >30% (7)
Taste evaluation	Excellent (1); Very good (2); Good (3); Poor (4); Bad (5)
Darkening of cut surface (6 h after cutting)	None (1); Slight (2); Severe (3)
Harvest to softening time (room temperature at about 20°C)	<6 days (1); 6–10 days (2); 11–15 days (3); 16–20 days (4); 21–25 days (5); 26–30 days (6); >30 days (7)
Shelf life	In days

levels, while external factors encompass wind, low humidity and low or high temperatures (Garner and Lovatt, 2016). Fruit abscission is a major problem significantly affecting yield and is a major factor contributing to the low success rate of hand pollination. Despite efforts, most horticultural practices have not been highly successful in significantly reducing fruit drop. Some approaches, such as shoot tipping in spring (Biran, 1979) and mid-anthesis foliar application of growth retardants (Wolstenholme *et al.*, 1990), have shown partial success in reducing abscission. Subsequent research in Israel (Adato, 1990) confirmed that growth retardants significantly reduce fruit drop. Genetic selection also plays a crucial role in avocado fruitlet abscission. It was shown that abscised fruitlets exhibit different genotypes compared with those remaining on the tree, indicating that abscission is not a random occurrence but involves selection based on seed genotype (Degani *et al.*, 1986).

Classical Breeding

Selection of naturally occurring superior variants

Many commercially successful avocado cultivars, such as 'Fuerte' and 'Hass', originated as random seedlings. 'Fuerte' was, for decades, the leading cultivar in Mediterranean and subtropical regions worldwide. It was introduced into California in 1911 as budwood from a seedling tree in Atlixco (State of Puebla), Mexico. Similarly, 'Hass', currently the most cultivated avocado worldwide, derives from a single seedling grown from a rootstock of unknown origin planted by Rudolph Hass in 1926 in La Habra Heights, California. The variety was patented in 1935 and became the leading variety in California in the 1970s.

Most of the early Californian cultivars originated in Mexico or Central America, either through asexual propagation of selections or from imported seeds (Bergh, 1957). Guatemalan cultivars such as 'Benik', 'Itzamna' and 'Nabal' were introduced as budwood, while 'Dickinson' originated from imported seed. Later, introductions to California primarily focused on searching for resistance to PRR (Coffey, 1987), mainly from Central America and neighbouring regions where *Persea* species abound.

Seeds imported from California, Mexico and Central America formed the basis for the Florida avocado industry (Ruehle, 1963), later serving as a germplasm source for tropical areas, mirroring California's role for subtropical regions. Florida cultivars, including some pure West Indian race and Guatemalan × West Indian hybrids, have been successful in various countries of Central America, South America, the West Indies, coastal Mexico and tropical regions worldwide.

Several hundred random seedlings selected in California have been named, and some of them ('Bacon', 'Edranol', 'Fuerte', 'Hass', 'Pinkerton', 'Reed', 'Shepard' and 'Zutano') are currently grown on a commercial scale either as main producing varieties or as pollinizers for 'Hass' in different parts of the world. Similar local, randomly selected seedlings have achieved some commercial success in other countries, notably 'Ettinger' and, to a lesser degree, 'Horshim' in Israel and 'Sharwil' in Australia and Hawaii.

Interspecific crosses

Efforts have been made to develop PRR-resistant rootstocks through interspecific hybrids of avocado with *Persea* species known to be resistant to the disease. However, all such attempts have proven unsuccessful. Both graft and cross incompatibilities appear to be complete between the subgenus *Persea*, which includes the avocado, and subgenus *Eriodaphne*, which includes all the known tolerant/resistant species.

In the subgenus *Persea*, 'Martin Grande' (G755) appears to be a natural hybrid between *P. americana* var. *guatemalensis* and *P. schiedeana* (Ellstrand *et al.*, 1986; Furnier *et al.*, 1990). While 'Martin Grande' exhibits considerable tolerance to *P. cinnamomi*, the production from trees grafted onto this invigorating rootstock has been disappointing in several countries where it has been tested (Whiley *et al.*, 1990).

Persea floccosa Mez. has the valuable trait of setting significantly larger numbers of fruit compared to other taxa in the *Persea* subgenus.

However, its fruit size is very small, and the seeds are relatively large. Despite hybridization attempts with several large-fruited, small-seeded cultivars, the resulting F1s, when selfed or 'backcrossed' to commercial cultivars, have shown a rapid loss of the heavy-setting ability. Additionally, the flavour has been mediocre at best, even as commercial quality is approached (Bergh and Lahav, 1996).

Mutations and polyploidy

Occasional spontaneous somatic mutations have been observed in avocado, leading to variations in tree shape, leaf size, shape and colour, fruit size and shape, and characteristics such as skin surface and thickness on certain 'sported' limbs of some cultivars. Several such mutations, such as 'Weisel', 'Newman' and 'de Bard', have been identified in 'Fuerte' (Hodgson, 1945). Several mutants have also been found in 'Hass'; examples include 'Carmen Hass'/'Mendez No 1', 'Jimenez-1', 'Jimenez-2' and 'Flor de Maria' in Mexico, 'Andes 3' and 'Andes 4' in Chile and 'Turner Hass' in Australia.

The 'Duke' variety (Mexican race), initially selected for its early green, good-sized fruit with cold-hardiness and wind resistance, was later found to possess tolerance to PRR (Zentmyer *et al.*, 1963). In addition to seedlings rootstocks derived from 'Duke' (such as 'Duke 6' or 'Duke 7'), in the hope of enhancing its moderate root rot tolerance, 'Duke' scions were also irradiated with fast neutrons in California. One resulting selection, 'D9', was tested as a commercial rootstock because of its considerable root rot tolerance and somewhat dwarfing impact. 'D9' was more productive than 'Martin Grande' but less than 'Borchard' or 'Duke 7' (Arpaia *et al.*, 1992). A radiation-induced mutation-breeding programme with ^{60}Co gamma rays was conducted in Mexico by De la Cruz *et al.* (1999). Some of the irradiated trees exhibited a higher number of fruit per tree. Tetraploid genotypes were generated by Bringhurst (1956) through colchicine treatments in 'Fuerte' clones and 'Mexicola' seedlings. Although a significant increase in the vegetative organs was observed, fruit set was reduced to almost nil. Recently, another tetraploid rootstock ('Canacado'), developed using colchicine, has been reported in Israel, showing higher yields and lower susceptibility to branch dieback associated with infections from Botryosphaeriaceae pathogens (Winer, 2022).

Genetic Analysis

Qualitative traits

The genetics of fruit skin colour, flowering group and anise scent were studied in breeding populations comprising 1699 seedlings by Lavi *et al.* (1993a). The study spanned over 2 years, and the three traits were systematically recorded. The parent cultivars used in this study included 'Anaheim', 'Ettinger', 'Fuerte', 'Hass', 'Horshim', 'Irving', 'Nabal', 'Pinkerton', 'Reed', 'Regina', 'Rincon', 'Rosh-Hanikra II', 'Tova' and 'Wurtz'. Isozyme analysis was used to distinguish between hybrids and self-fertilized seedlings (Degani and Gazit, 1984). However, the possibility that some seedlings were wrongly classified cannot be ruled out. In all types of crosses, the average X/Y ratio (X and Y being green or purple skin colour, A or B flowering group, and presence or absence of anise) among the progeny was one or higher, with a wide variation in the ratio between crosses. It is interesting to note that selfing in each trait (fruit skin colour: green × green or purple × purple; flowering group: B × B or A × A; and anise scent: (–) × (–) or (+) × (+)) resulted in more progeny of the first phenotype (green skin, flowering group B and no anise scent) (Lavi *et al.*, 1993a). These results rule out the possibility that a single gene controls these traits. The results could be explained by the assumption that the traits are coded by several loci with several alleles in each, while the various phenotypes may result from various heterozygous combinations. Furthermore, it was suggested that the inheritance of these traits is based on a threshold value beyond which the phenotype shifts from one phase to another. However, in the case of flowering type, more recent research seems to indicate that it can be controlled by a one-gene Mendelian model (Ashworth *et al.*, 2019).

Quantitative traits

Most traits of interest in avocados are inherited quantitatively, as they are controlled by many genes, each with a small effect and usually under strong environmental influence. Hybridization is considered the most efficient way to combine complementary desirable features of different cultivars and obtain a desirable intermediate trait when the available breeding materials have extreme phenotypes. For example, the commercially important 'Booth' selections in Florida, which are natural hybrids of the Guatemalan and West Indian races, are intermediate in harvesting season and have other useful traits. Interracial hybridizations between Guatemalan and West Indian types have also led to significant intermediate cultivars in Hawaii. The underlying assumption in hybridization is that most of the genetic variance is additive; therefore, combining alleles from two parents will result in offspring having the desired performance (Hansche, 1983). An analysis of several quantitative traits in avocado was conducted to estimate the 'variance components' and their heritability. The study demonstrated (Lavi *et al.*, 1993b) that genetic variance (both additive and nonadditive) is substantial for most avocado traits. In four of the nine traits studied (tree size, flowering intensity, fruit density and inflorescence length), relatively large nonadditive variances were detected. In a more detailed stage of the study, 14 cultivars were used to perform 12 crosses and 12 selfings. The number of progeny varied from 2 to 431 seedlings per each 'selfing' or cross, and the total population consisted of 1938 seedlings. The parent cultivars represented a broad spectrum of the variation that exists within avocado. Parentage verification was based on isozyme analyses. Only seedlings of known parentage were used in that study, and the traits were assessed by measurements or by visual scoring with results averaged over 2–5 years. The analysis of avocado traits revealed that the nonadditive genetic variance significantly exceeded zero for all traits studied (anise scent, fruit density, flowering intensity, fruit weight, harvest duration, inflorescence length, seed size, fruit softening time and tree size). These values ranged from 36.3% in time to softening to 49.3% in harvest duration. In contrast, the additive genetic variance was found to be non-significant for all traits. However, significant environmental variance was also present, and except for flowering intensity, its estimate contributed to 33%–35% of the total phenotypic variance. The values of the narrow sense heritability varied among traits, ranging from zero for fruit density and flowering intensity to 0.5 for seed size and 0.48 for softening time (Lavi *et al.*, 1993b).

These estimates indicate that nonadditive genetic variance is a major component of the total genetic variance in avocado traits and is significantly greater than zero for all studied traits. The presence of a high level of heterozygosity in avocado (Lavi *et al.*, 1991b) may explain the prevalence of a large nonadditive (dominant) genetic variance (Fisher, 1930). Thus, these results explain the common frustration after crossing two cultivars and obtaining a wide range of phenotypes among the progeny (rather than the naive expectation of combining traits from both parents). It must be emphasized that these conclusions are limited to the set of cultivars used in that study.

The importance of the broad sense heritability (h^2b) originates from the fact that the best-performing seedlings in the selection plots could result from either environmental or genetic factors. The higher the value of the h^2b, the greater the confidence that genetic factors control the tested traits. In such cases, generating a few grafted trees from each selected seedling becomes necessary for the next evaluation phase. The choice of parents in avocado breeding should align first of all with the breeding objectives, aiming to obtain new cultivars that are better than those currently available. Since nonadditive genetic variance plays a major role in the traits essential to breeders, selecting parents to maximize the genetic variance in the progeny becomes crucial. Thus, the chance of obtaining the desired combination of genes and alleles increases. This should be achieved by choosing a broad spectrum of parent cultivars and may even include some cultivars with potentially inferior performance.

Other studies have used quantitative genetics to assess the phenotypic value of different quantitative traits in avocado. For instance, Chen *et al.* (2007) showed in a progeny of 'Gwen' that traits related to growth, such as plant height, trunk diameter and canopy size, are under genetic control, showing sufficient heritability to respond to selection. Calderón-Vázquez *et al.* (2013) used a segregating population of the cross 'Gwen' × 'Fuerte' to show that some compounds in the fruit (such as β-sitosterol, carotenoids and vitamin E) can also respond to selection.

Genetic Markers

The use of molecular markers provides a more precise and efficient way to analyse the genetic makeup of avocado varieties. Molecular tools have been used in avocado for genotype fingerprinting, differentiation among the main races, diversity and phylogenetic studies, linkage map construction and screening for traits of interest. As in other crops, isozymes were the first genetic markers used in avocado and used in the 1980s and 1990s mainly for fingerprinting and to assess cross-pollination levels by different authors (Degani *et al.*, 1997, Vrecenar-Gadus and Ellstrand, 1985; Goldring *et al.*, 1987, Degani and Gazit, 1984, Torres and Bergh, 1978; Torres *et al.*, 1978).

The development of DNA markers (Botstein *et al.*, 1980) revolutionized genetic research and opened up various applications in plant genetics. Several classes of DNA markers have been applied in avocados to assess polymorphism among individuals of the same species as well as between species. In the 1990s, DNA markers used in avocado research were primarily restricted to restriction fragment length polymorphisms (RFLPs), Random Amplified Polymorphic DNA (RAPD) and minisatellites. A limited number of studies (Furnier *et al.*, 1990; Davis *et al.*, 1998) have used RFLPs in avocados mainly to classify the avocado botanical races. Thus, Furnier *et al.* (1990) could distinguish the three botanical races and proposed that the Guatemalan race originated as a hybrid between *P. steyermarkii* and *P. nubigena*. Davis *et al.* (1998) analysed 36 genotypes using RFLPs and successfully separated them into the three major botanical races, along with identifying smaller clusters representing interracial hybrids. Twenty-two RAPD markers were used by Fiedler *et al.* (1998) to study the genetic differences among 16 avocado genotypes representing the three races, and they were able to separate the races by identifying some putative race-specific markers. Lavi *et al.* (1991a) used three minisatellites to identify 14 avocado cultivars and assigned them to their respective races, whereas Mhameed *et al.* (1997) used minisatellite markers to analyse 24 genotypes and were able to confirm racial assignment for 19 of them. These studies demonstrate the versatility of DNA markers in avocado research, allowing researchers to explore the genetic diversity, relationships among different varieties and the classification of botanical races. As technology advanced, additional DNA markers became more widely utilized for their higher resolution and efficiency in genetic analyses.

More recently, the development of microsatellite or simple sequence repeat (SSR) markers has advanced in avocado research, providing valuable tools for various applications (Sharon *et al.*, 1997; Ashworth *et al.*, 2004; Borrone *et al.*, 2007; Gross-German and Viruel, 2013). They have been used for linkage map construction (Sharon *et al.* 1997; Borrone *et al.* 2009), paternity analyses (Borrone *et al.*, 2008; Ying *et al.*, 2009) and molecular characterization of germplasm and studies of diversity in different countries (Ashworth and Clegg, 2003; Schnell *et al.*, 2003; Alcaraz and Hormaza, 2007; Acheampong *et al.*, 2008; Gross-German and Viruel, 2013; Ferrer Pereira *et al.*, 2017; Guzmán *et al.*, 2017; Boza *et al.* 2018). Ashworth and Clegg (2003) analysed 35 avocado cultivars and two wild relatives using 25 microsatellite loci, suggesting the presence of ancient hybridization or a recent origin of the avocado races. Schnell *et al.* (2003) studied the genetic diversity among 254 accessions with 14 SSR loci, clustering the genotypes according to the three races. Alcaraz and Hormaza (2007) analysed 75 avocado genotypes from different origins with 16 microsatellite loci, clustering most of them into the three races, although bootstrap support was low, probably due to the presence of several interracial hybrids. Acheampong *et al.* (2008) genotyped 172 Ghanaian accessions with 12 SSR loci, showing that most of the Ghanaian accessions studied were probably West Indian. Ying *et al.* (2009) used 17 SSR loci for paternity analyses. Gross-German and Viruel (2013) developed 47 SSR and expressed sequence Tags-SSR (EST-SSR) markers and used them to analyse 42 accessions that clustered into the three races, finding markers that could be used to distinguish the races. Ferrer Pereira *et al.* (2017) analysed 49 genotypes from the Venezuelan germplasm collection with 13 SSR loci, grouping the genotypes according to their racial composition. Guzmán *et al.* (2017) analysed 318 accessions from the National Avocado Genebank in Mexico with 28 SSR loci. Their results did not allow a clear differentiation among the three races, but two different groups, corresponding to two different ecological regions, were found. Boza *et al.* (2018) used 55 SSR loci to analyse 354 accessions of different *Persea* species (327 belonging to *P. americana*) from different origins, grouping the *P. americana* accessions according to the three races. These studies demonstrate the versatility of SSR markers in avocado research, enabling researchers to study genetic relationships, diversity and population structure among avocado cultivars and accessions.

The development of high-throughput new-generation sequencing technologies has enabled the use of single-nucleotide polymorphism (SNP) markers in genetic studies across various crops. SNPs are increasingly being considered as the preferable type of marker for genetic studies for diverse objectives such as the construction of genetic linkage maps, the characterization of QTLs, association studies with traits of agronomic interest, marker-assisted selection (MAS) or genomic selection (GS). SNPs show several advantages, particularly relevant in woody perennial crops, such as avocado, in which breeding programmes are characterized by long generation times: the possibility of developing a large number of markers at increasingly reduced costs, their bi-allelism that results in high accuracy in the analyses, or their high replicability that ensures consistency of the results. SNPs arise from point mutations, involving a single nucleotide change at any place in the genome for which a short flanking sequence is known. Changes can be either the substitution of one nucleotide with another or the insertion/deletion (indels) of a single nucleotide. SNPs are codominant single-locus markers with mostly two alleles per locus, resulting in a relatively low level of polymorphism. The first studies with SNP markers in avocado were performed by Chen *et al.* (2008, 2009). Chen *et al.* (2008) analysed 21 wild avocado accessions from Mexico, Costa Rica, Ecuador and the Dominican Republic, showing a relatively low genetic variation in the material studied. Further work with the same wild genotypes and markers plus 33 cultivars (Chen *et al.*, 2009) allowed performing population structure analyses and showed substantial genetic differentiation among three geographic groups of the wild germplasm that corresponded to the three avocado races. More recently, Khun *et al.* (2019a, b) described a set of SNPs derived from leaf- and flower-expressed genes using RNA sequences and selected 6000 SNPs to design an SNP genotyping array. Talavera *et al.* (2019) developed 7108 SNPs and analysed genetic diversity and population structure in 71 avocado accessions, classifying them according to their botanical race into four main groups: Mexican, Guatemalan, West Indian and an additional group of Guatemalan × Mexican hybrids. Similar results were reported by Ge *et al.* (2019a) and Rubinstein *et al.* (2019) with different sets of avocado accessions.

Genetic Linkage Maps

The construction of genetic linkage maps is crucial for advancing breeding programmes in various crops, enabling the identification of important traits, MAS and understanding the genetic architecture of different varieties. In the case of avocados, the development of linkage maps has been relatively limited. The first avocado genetic linkage map was developed by analysing 50 offspring of the cross 'Ettinger' × 'Pinkerton' with 93 SSR markers (of which 51 were found to be polymorphic and reliable), 23 minisatellite DNA fingerprints (DFPs) and 17 RAPD markers (Sharon *et al.*, 1997). Two-point analysis resulted in 29 linked marker loci (logarithm of the odds (LOD) score ≥ 3). The map represented the 12 avocado linkage groups having 2–5 markers per group (a total of 35 markers), covering 357.2 cM. Comparison of the parental with the maternal map showed that the crossing-over frequency was on average 21% higher in the maternal parent ('Pinkerton'). No linkage was detected between SSR and DFP markers and within RAPD markers. Based on the Poisson distribution, the markers were found to be randomly allocated to the linkage groups.

Later, in Florida, a new linkage map was developed by Borrone *et al.* (2009) with 715 F1 individuals (456 from the cross 'Tonnage' × 'Simmonds' and 259 from the reciprocal cross

'Simmonds' × 'Tonnage') using 163 loci (112 EST-SSRs and 23 SSRs). The 12 avocado linkage groups were also detected, and the total map length was longer (1087.4 cM). Linkage groups ranged from 157.3 (LG2) to 2.4 cM (LG12), with 2 (LG12) to 29 (LG1) markers per linkage group.

More recently, a linkage map was constructed using SSRs and SNPs from several sources using 50 individuals of a full-sib progeny of a 'Gwen' × 'Fuerte' cross to develop a dense genetic linkage map and to analyse QTL related to seven phenotypic traits (Ashworth *et al.*, 2019). The materials were grafted on Duke 7 rootstocks and studied in two locations in two successive years. The resulting linkage map comprised 1346 markers distributed along the 12 avocado linkage groups covering 1044.7 cM. Numerous molecular markers located in linkage group 10 were associated with a QTL for flowering type, one marker on linkage group 1 was associated with a QTL for β-sitosterol content in fruit, and a region on linkage group 3 was associated with vitamin E (α-tocopherol) fruit levels.

Although those genetic maps are useful resources for avocado breeding, new efforts are needed to develop more saturated genetic maps in this crop. Challenges such as the long juvenile period, low yields obtained after natural or hand pollination, and the dichogamy system have limited the scale and efficiency of genetic mapping studies in avocados. Larger populations, increased molecular resources and extensive phenotypic data will be required in order to detect QTLs and develop marker-assisted breeding programmes in avocado.

Non-classical Breeding Methodologies

Similar to other crops, various biotechnological approaches could play a pivotal role in future avocado breeding programmes. These include *in vitro* techniques such as shoot, embryo and calli culture, somatic embryogenesis, protoplast isolation and plant regeneration, genetic transformation, *in vitro* conservation, and cryopreservation, which are covered in detail in Chapter 10 of this volume. Regarding genetic transformation, success has been accomplished in a few cases to study some genes influencing different horticultural traits using *Agrobacterium rhizogenes* and *Agrobacterium tumefaciens* (Pliego-Alfaro *et al.*, 2020). Transient gene expression has also been accomplished using biolistics in avocado embryogenic cultures (Chaparro-Pulido *et al.*, 2014). Furthermore, as in other crops, the development of clustered regularly interspaced palindromic repeats/Cas9 CRISPR/Cas methodologies can potentially revolutionize gene editing in avocado. However, the current limitation mainly lies in the absence of efficient regeneration protocols from avocado protoplasts, a prerequisite for the success of this approach.

Linkage Between DNA Markers and Loci Controlling Important Traits

Establishing genetic linkage between DNA markers and genes controlling important traits can significantly improve avocado breeding programmes. This is accomplished through MAS, where emphasis is placed on selecting for the marker rather than the trait itself and, thus, making breeding more efficient by saving valuable time and space. In the broader perspective, the identified linkage can serve as the foundation for isolating and cloning the genes responsible for these traits (Tanksley *et al.*, 1995).

Only a few reports of linkage between DNA markers and loci controlling important traits are available for avocado, which is lagging behind other crops. The association between DNA fingerprint (DFP) fragments and 16 avocado trait loci was tested in two families by one-way analysis of variance and multiple regression (Sharon *et al.*, 1998). The DFP fragments P4, P8, E2 and E5 in a 'Pinkerton' × 'Ettinger' progeny were found to be associated with harvest duration, skin colour, skin thickness and skin surface, respectively. The fragments P1, P8, B1 and B4 in a 'Pinkerton' × 'Bacon' progeny (half-sibs of the first population) were found to be associated with fruit weight, skin colour, seed size and peeling, respectively. Based on the two populations, fragment P8 was found to be associated with the black-purple fruit skin colour. The higher intensity of this fragment in the DNA pools of progeny with green skin colour compared to those with black-purple skin colour supported this association. These results are interpreted as an association and maybe genetic linkage between the DNA fingerprint fragment P8 and the locus regulating avocado fruit skin colour.

A study involving 60 progeny from a 'Pinkerton' × 'Ettinger' cross aimed to identify linkages with loci coding for agriculturally important traits (Sharon *et al.*, 1998). One-way analysis of variance revealed significant associations ($P \leq 0.01$) with seven out of the nine analysed traits. Skin gloss was linked to two SSR markers ($P = 0.0014$); seed size was linked to one SSR marker ($P = 0.0006$); and the amount of fibre in the flesh was linked to SSR markers on linkage group 3 (especially with the marker AVAO4) ($P = 0.00001$). The application of interval mapping to allocate loci coding for these traits resulted in three cases where the LOD score was ≥2 (skin gloss in linkage group 6; skin surface in linkage group 9; and fibres in linkage group 3). Further analysis revealed allelic interaction in the locus controlling fibres in the flesh, with a high level of significance observed in the marker locus AVAO4 in specific genotypes (Sharon *et al.*, 1998). These findings underscore the utility of SSR markers in genetic studies and highlight the advantages of employing this technology in breeding fruit tree crops, such as avocado.

As described previously, using 50 individuals of a full-sib progeny of a 'Gwen' × 'Fuerte' cross, Ashworth *et al.* (2019) found numerous molecular markers in linkage group 10 associated with a QTL for flowering type, one marker on linkage group 1 associated with a QTL for β-sitosterol content in fruit and a region on linkage group 3 associated with vitamin E (α-tocopherol) fruit levels.

It is noteworthy that despite the high expectations for MAS, only a few examples of this approach are applied to fruit breeding, mainly for resistance to various diseases. The primary impediment lies in the lack of readily available markers closely linked to essential traits. It is anticipated that advancements in new genomic technologies (see below) will eventually find application in avocado breeding, thereby enhancing efficiency in the breeding process.

Avocado Genomics

New-generation sequencing approaches in avocado have primarily been focused on transcriptome analyses and SNP development. These applications have been predominantly employed in studying disease resistance, plant development, stress tolerance and genetic diversity (Ibarra-Laclette *et al.*, 2015; Kilaru *et al.*, 2015; Vergara-Pulgar *et al.*, 2019; Ge *et al.*, 2019a, 2019b, 2019c; Kuhn *et al.*, 2019a, 2019b; Rubinstein *et al.*, 2019; Talavera *et al.*, 2019; Chabikwa *et al.*, 2020; Kämper *et al.*, 2021; Pérez-Torres *et al.*, 2021; Fick *et al.*, 2022; Hernández *et al.*, 2022).

In recent years, several nuclear genome sequences of avocado have been published. The first avocado genome was reported by Rendón-Anaya *et al.* (2019) in drafts of 'Hass' and a wild Mexican-race accession. The 'Hass' assembly included 8135 contigs, with 915 scaffolds anchored to a genetic map covering approximately 46.2% of the estimated avocado genome. Furthermore, ten genomes representing the three avocado subspecies or races, as well as 'Hass'-related cultivars and *P. schiedeana*, were resequenced to help better understand the genetic diversity and the evolutionary history of avocado. Subsequently, Sharma *et al.* (2021) developed another 'Hass' genome reference, resulting in an assembly of 788 Mb representing ca. 88% of the avocado genome size although not anchored into scaffolds. Nath *et al.* (2022) published a chromosome-level genome assembly for 'Hass' with 913 Mb, with an N50 contig of 84 Mb. Contigs assigned to the 12 chromosomes represented 874 Mb with 62.6% repeat elements. Annotation of protein-coding sequences identified 48,915 avocado genes, of which 39,207 could correspond to ascribed functions. Solares *et al.* (2023) sequenced the 'Gwen' genome, resulting in an N50 of 3.37 Mb organized in 12 pseudo-chromosomes with a total of 49,450 genes. Berdugo-Cely *et al.* (2023) sequenced 144 accessions from the Colombian National genebank and 240 accessions from local commercial orchards in the Colombian north-western Andean region. Merging the resulting reads with sequences of reference genotypes led to the discovery of 4931 SNPs. The analysis of population structure and phylogenetic diversity revealed the existence of a new Colombian group closely related to the West Indian race, in addition to the three well-recognized races (Mexican, Guatemalan and West Indian).

Avocado genomic advances need adequate bioinformatic platforms to host avocado omics data. A recent example is Avobase (https://www.avocado.uma.es), a web portal dedicated to avocado genomics, providing multiple interactive bioinformatics tools, sequences and annotations for analysing, visualizing and downloading avocado omics data.

Conclusions

This chapter summarized relevant information on avocado genetics and breeding, highlighting that the future of avocado breeding appears promising, as new technologies are being developed and since current breeding programmes have only scratched the surface of the available genetic variation in this crop, particularly in most tropical regions where a high unexplored genetic diversity is still available.

Effective breeding starts with ensuring ample genetic diversity in the parent plants. Due to the high level of heterozygosity, and since a large number of seeds is required, open pollination is often favoured over controlled pollination due to its cost-effectiveness. However, when possible, and when a specific breeding target is the goal, and especially when the trait is controlled by single gene, controlled pollination has an advantage.

The breeding process is composed of two selection stages: The first one, ideally initiated at the first year of fruiting, focuses on evaluating the seedling progeny population (derived from either controlled crosses or open pollination), based on fruit traits only. At this stage, seedlings are planted at a condensed spacing, allowing a large number of seedlings per area (it is noteworthy that, in our experience, this did not affect the length of the juvenile period). At the second stage, potential selected seedlings (about 1%) are grafted on several mature trees, in several locations, and grown under commercial management protocols. Selection at this stage focuses on yield, shelf life, disease and pest sensitivity, and adaptation to the environment. This structure of the breeding project results in significant cutting of the cost and shortening of the project time.

While classical breeding approaches have dominated avocado breeding, we believe that molecular breeding (already available in various crops) will be applied to more fruit trees, including avocado. The production of transgenic plants requires the identification of DNA sequences controlling the trait of interest and the technology to introduce them to the target tree. The technology of transformation is being developed for avocado, and the subject is discussed in Chapter 10 of this book. Identification of the important DNA sequences is based on the recent advancements in sequencing, transcriptomics, linkage analysis and genome-wide association. Discoveries of additional genes governing agronomically significant traits are expected in the coming years. This knowledge is crucial to use molecular breeding to complement classical breeding approaches as well as to provide appropriate material for the generation of DNA-edited plants with desired traits.

Acknowledgement

This chapter is a revised and updated version of the chapter 'Genetics and Breeding' published in the previous edition of this book.

References

Acheampong, A.K., Akromah, R., Ofori, F.A., Takrama, J.F., Saada, D., Bitton, I. and Lavi, U. (2008) Genetic characterization of Ghanaian avocados using microsatellite markers. *Journal of the American Society for Horticultural Science* 133, 801–809.

Adato, I. (1990) Effects of paclobutrazol on avocado (*Persea americana* Mill.) cv. Fuerte. *Scientia Horticulturae* 45, 105–115.

Ahsan, M.U., Hayward, A., Alam, M., Bandaralage, J.H., Topp, B., Beveridge, C.A. and Mitter, N. (2019) Scion control of miRNA abundance and tree maturity in grafted avocado. *BMC Plant Biology* 19: 382.

Alcaraz, M.L. and Hormaza, J.I. (2007) Molecular characterization and genetic diversity in an avocado collection of cultivars and local Spanish genotypes using SSRs. *Hereditas* 144, 244–253.

Alcaraz, M.L. and Hormaza, J.I. (2014) Optimization of controlled pollinations in avocado (*Persea americana*, Lauraceae). *Scientia Horticulturae* 180, 79–85.

Alcaraz, M.L., Montserrat, M. and Hormaza, J.I. (2011) In vitro pollen germination in avocado (*Persea americana* Mill.): optimization of the method and effect of temperature. *Scientia Horticulturae* 130, 152–156.

Alcaraz, M.L., Thorp, T.G. and Hormaza, J.I. (2013) Phenological growth stages of avocado (*Persea americana*) according to the BBCH scale. *Scientia Horticulturae* 164, 434–439.

Alexander, D.McE. (1977) Propagation. In: Allen, R.N. (ed.) *Proceedings of the Australian Avocado Research Workshop*. New South Wales Department of Agriculture, Wollongbar, pp. 21–25.

Angiosperm Phylogeny Group, Chase, M.W., Christenhusz, M.J.M., Fay, M.F., Byng, J.W., Judd, W.S. *et al.* (2016) An update of the Angiosperm Phylogeny Group classification for the orders and families of flowering plants: APG IV. *Botanical Journal of the Linnean Society* 181(1), 1–20.

Arpaia, M.L., Bender, G.S. and Witney, G.W. (1992) Avocado clonal rootstock production trial. In: Lovatt, C., Holthe, P.A. and Arpaia, M.L. (eds) *Proceedings of the Second World Avocado Congress*. University of California, Riverside, California 1, 305–310.

Ashworth, V.E.T.M. and Clegg, M.T. (2003) Microsatellite markers in avocado (*Persea americana* Mill.): genealogical relationships among cultivated avocado genotypes. *Journal of Heredity* 94(5), 407–415.

Ashworth, V.E.T.M., Kobayashi, M.C., De la Cruz, M. and Clegg, M.T. (2004) Microsatellite markers in avocado (*Persea americana* Mill.): development of dinucleotide and trinucleotide markers. *Scientia Horticulturae* 101, 255–267

Ashworth, V.E.T.M., Chen, H., Calderón-Vázquez, C.L., Arpaia, M.L., Kuhn, D.N. *et al.* (2019) Quantitative Trait Locus analysis in avocado: The challenge of a slow-maturing horticultural tree crop. *Journal of the American Society for Horticultural Science* 144, 352–362.

Barceló-Muñoz, A., Zea-Bonilla, T., Jurado-Valle, I., Imbroda-Solano, I. Vidoy-Mercado, I., Pliego-Alfaro, F., and Lopez-Herrera, C.J. (2007) Selection program for avocado rootstocks tolerant to white rot caused by *Rosellinia necatrix* in Southern Spain (1995–2007). In: *Proceedings of the Sixth World Avocado Congress,* Viña del Mar, Chile. Available at: https://www.avocadosource.com/wac6/en/extenso/1a-9.pdf (accessed 23 September 2024).

Barrientos-Priego, A.F., Barrientos-Perez, F., Sánchez-Colin, S. and Aguilar-Melchor, J.J. (1991) Utilizacion del huerto-vivero para reducir la fase juvenil de plantulas de aguacate (*Persea americana* Mill.). *Proceedings of the Interamerican Society of Tropical Horticulture* 35, 18–22. (in Spanish.)

Ben-Ya'acov, A. (1972) Avocado rootstock-scion relationship: a long-term, large-scale field research project. *California Avocado Society Yearbook* 55, 158–161.

Ben-Ya'acov, A. (1976) Avocado rootstock-scion relationships: a long-term, large-scale field research project. V. Final report on some orchards planted during the years 1960–1964. *California Avocado Society Yearbook* 59, 122–130.

Ben-Ya'acov, A. (1987) Avocado rootstock-scion relationships. *South African Avocado Growers' Association Yearbook* 10, 30–32.

Ben-Ya'acov, A. (1998) The taxonomy of the avocado: a proposed new classification of the *Persea*, subgenus *Persea*. In: *Proceedings of the Third World Avocado Congress*. Tel Aviv, Israel (Abstract), p. 75.

Ben-Ya'acov, A. and Frenkel, M. (1974) Sensitivity of avocado rootstocks to *Verticillium* wilt. *California Avocado Society Yearbook* 57, 111–113.

Ben-Ya'acov, A. and Michelson, E. (1995) Avocado rootstocks. *Horticultural Reviews* 17, 381–429.

Ben-Ya'acov, A., Michelson, E. and Sela, I. (1974) Avocado rootstock-scion relationships: a long-term, large-scale field research project. Sensitivity of avocado rootstocks to inadequate soil aeration. *California Avocado Society Yearbook* 57, 108–113.

Ben-Ya'acov, A., Michelson, E. and Sela, I. (1979) Field research for selection of avocado rootstocks and scion in commercial plantations: report of the first decade of research, 1968–1978. The Division of Subtropical Horticulture, Agricultural Research Organization, Israel. (in Hebrew.)

Berdugo-Cely, J.A., Cortés, A.J., López-Hernández, F., Delgadillo-Durán, P., Cerón-Souza, I. *et al.* (2023) Pleistocene-dated genomic divergence of avocado trees supports cryptic diversity in the Colombian germplasm. *Tree Genetics & Genomes* 19: 42.

Bergh, B.O. (1957) Avocado breeding in California. *Proceedings of the Florida State Horticultural Society* 70, 284–290.

Bergh, B.O. (1961) Breeding avocados at C.R.C. *California Avocado Society Yearbook* 45, 67–74.

Bergh, B.O. (1967) Reasons for low yields of avocados. *California Avocado Society Yearbook* 51, 161–172.

Bergh, B.O. (1969) Avocado (*Persea americana Miller*). In: Ferwerda, F.P. and Wit, F. (eds) Outlines of Perennial Crop Breeding in the Tropics, Landbouwhogeschool, Wageningen, The Netherlands, pp. 23–51.

Bergh, B.O. (1986) *Persea americana*. In: Halevy, A.H. (ed.) *Handbook of Flowering*, Volume 5. CRC Press, Boca Raton, Florida, pp. 253–268.

Bergh, B.O. and Ellstrand, N.C. (1986) Taxonomy of the avocado. *California Avocado Society Yearbook* 70, 135–145.

Bergh, B.O. and Lahav, E. (1996) Avocados. In: Janick, J. and Moore, J.N. (eds) *Fruit Breeding, Vol I: Tree and Tropical Fruits.* Wiley, West Lafayette, Indiana, pp. 113–166.

Bergh, B.O. and Whitsell, R.H. (1974) Self-pollinated Hass seedlings. *California Avocado Society Yearbook* 57, 118–126.

Biran, D. (1979) Fruitlet abscission and spring growth retardation – their influence on avocado productivity. MSc thesis, Hebrew University of Jerusalem, Rehovot, Israel. (in Hebrew with English summary.)

Borrone, J.W., Schnell, R.J., Violi, H.A. and Ploetz, R.C. (2007) Seventy microsatellite markers from *Persea americana* Miller (avocado) expressed sequence tags. *Molecular Ecology Notes* 7, 439–444.

Borrone, J.W., Olano, C.T., Kuhn, D.N., Brown, J.S., Schnell, R.J. and Violi, H.A. (2008) Outcrossing in Florida avocados as measured using microsatellite markers. *Journal of the American Society for Horticultural Science* 133, 255–261.

Borrone, J., Brown, J.S., Tondo, C.T., Mauro-Herrera, M., Kuhn, D.N. *et al*. (2009) An EST-SSR-based linkage map for *Persea americana* Mill (avocado). *Tree Genetics & Genomes* 5, 553–560.

Botstein, D., White, R.L., Skolnick, M. and Davis, R.W. (1980) Construction of a genetic linkage map in man using restriction fragment length polymorphisms. *American Journal of Human Genetics* 32, 314–331.

Boza, E.J., Tondo, C.L., Ledesma, N., Campbell, R.J., Bost, J., Schnell, R.J. and Gutiérrez, O.A. (2018) Genetic differentiation, races and interracial admixture in avocado (*Persea americana* Mill.), and *Persea* spp. evaluated using SSR markers. *Genetic Resources and Crop Evolution* 65, 1195–1215.

Bringhurst, R.S. (1956) Breeding tetraploid avocados. *Proceedings of the American Society for Horticultural Science* 67, 251–257.

Calderón-Vázquez, C., Durbin, M.L., Ashworth, V.E.T.M., Tommasini, L., Meyer, K.K.T. and Clegg, M.T. (2013) Quantitative genetic analysis of three important nutritive traits in the fruit of avocado. *Journal of the American Society for Horticultural Science* 138, 283–289.

Can-Alonzo, C., Quezada-Euan, J.J., Xiu-Ancona, P., Moo-Valle, H., Valdovinos-Nunez, G. and Medina-Peralta, S. (2005) Pollination of 'criollo' avocados (*Persea americana*) and the behaviour of associated bees in subtropical Mexico. *Journal of Apicultural Research* 44, 3–8.

Carrillo, D., Duncan, R.E. and Peña, J.E. (2012) Ambrosia beetles (Coleoptera: Curculionidae: Scolytinae) that breed in avocado wood in Florida. *Florida Entomologist* 95(3), 573–579.

Carrillo, D., Duncan, R.E., Ploetz, J.N., Campbell, A.F., Ploetz, R.C. and Peña, J.E. (2014) Lateral transfer of a phytopathogenic symbiont among native and exotic ambrosia beetles. *Plant Pathology* 63, 54–62.

Castillo-Argaez, R., Konkol, J.L., Vargas, A.I., Ploetz, R.C. and Schaffer, B. (2021a) Disease severity and ecophysiology of rootstock/scion combinations of different avocado (*Persea americana* Mill.) genotypes in response to laurel wilt. *Scientia Horticulturae* 287, 110250.

Castillo-Argaez, R., Vazquez, A., Konkol, J.L., Vargas, A.I., Ploetz, R.C., Etxeberria, E. and Schaffer, B. (2021b) Sap flow, xylem anatomy and photosynthetic variables of three *Persea* species in response to laurel wilt, *Tree Physiology* 41, 1004–1018.

Celis, N., Suarez, D.L., Wu, L., Li, R., Arpaia, M.L. and Mauk, P. (2018) Salt tolerance and growth of 13 avocado rootstocks related best to chloride uptake. *HortScience* 53(2), 1737–1745.

Chabikwa, T.G., Barbier, F.F., Tanurdzic, M. and Beveridge, C.A. (2020) *De novo* transcriptome assembly and annotation for gene discovery in avocado, macadamia and mango. *Scientific Data* 7: 9.

Chanderbali, A.S., Soltis, D.E., Soltis, P.S. and Wolstenholme, B.N. (2013) Taxonomy and Botany. In: *The Avocado: Botany, Production, and Uses* (eds. Schaffer B., Wolstenholme, B.N & Whiley A.W.). CAB International, Wallingford, UK.

Chaparro-Pulido, C.A., Montiel, M.M., Palomo-Ríos, E., Mercado, J.A. and Pliego-Alfaro, F. (2014) Development of an efficient transient transformation protocol for avocado (*Persea americana* Mill.) embryogenic callus. *In Vitro Cellular and Developmental Biology – Plant* 50, 292–298.

Chen, H., Ashworth, V.E.T.M., Xu, S. and Clegg, M.T. (2007) Quantitative genetic analysis of growth rate in avocado. *Journal of the American Society for Horticultural Science* 132, 691–696.

Chen, H.F., Morrell, P.L., de la Cruz M. and Clegg, M.T. (2008) Nucleotide diversity and linkage disequilibrium in wild avocado (*Persea americana* Mill.). *Journal of Heredity* 99, 382–389.
Chen, H., Morrell, P.L., Ashworth, V.E.T.M., De la Cruz, M. and Clegg, M. (2009) Tracing the geographical origins of major avocado cultivars. *Journal of Heredity* 100, 56–65.
Coates, L.M., Dann, E.K., Shuey, L.S., Smith, L.A., Dean, J.R. *et al.*(2011) Effects of rootstock on avocado fruit quality – assessment of postharvest disease, major cations and biochemical traits. *Proceedings VII World Avocado Congress* 2011, 206–214.
Coffey, M.D. (1987) *Phytophthora* rootrot of avocado: an integrated approach to control in California. *Plant Disease* 71, 1046–1052.
Crane, J.H., Balerdi, C.F. and Campbell, C.W. (1996) *The Avocado Circular 1034*. Florida Cooperative Extension Service, Institute of Food and Agricultural Sciences, University of Florida, Gainesville.
Crowley, D. and Arpaia, M.L. (2002) Rootstock selections for improved salinity tolerance of avocado. In: *Proceedings of the California Avocado Research Symposium.* University of California, Riverside, pp. 83–86.
Davenport, T.L. (1986) Avocado flowering. *Horticultural Reviews* 8, 257–289.
Davis, J., Henderson, D., Kobayashi, M., Clegg, M.T. and Clegg, M.T. (1998) Genealogical relationships among cultivated avocados as revealed through RFLP analyses. *Journal of Heredity* 89, 319–323.
De la Cruz, E.T., Ibanez, P.J., Mijares-Oviedo, P. and Garcia, J.M. (1999) Evaluacion de productividad y calidad de fruta en aguacate 'Hass' de arboles derivados de exposicion a radiacion gamma. *Revista Chapingo Serie Horticultura* 5, 373–374.
Degani, C. and Gazit, S. (1984) Selfed and crossed proportions of avocado progenies produced by caged pairs of complementary cultivars. *HortScience* 19, 258–260.
Degani, C., Goldring, A., Gazit, S. and Lavi, U. (1986) Genetic selection during the abscission of avocado fruitlets. *HortScience* 21, 1187–1188.
Degani, C., Goldring, A. and Gazit, S. (1989) Pollen parent effect on outcrossing rate in 'Hass' and 'Fuerte' avocado plots during fruit development. *Journal of the American Society for Horticultural Science* 114, 106–111.
Degani, C., Goldring, A., Adato, I., El Batsri, R. and Gazit, S. (1990) Pollen parent effect on outcrossing rate, yield, and fruit characteristics of 'Fuerte' avocado. *HortScience* 25, 471–473.
Degani, C., El-Batsri, R., Gazit, S. (1997) Outcrossing rate, yield and selective fruit abscission in 'Ettinger' and 'Ardith' avocado plots. *Journal of the American Society for Horticultural Science* 122, 813–817.
Degani, C., Lahav, E., El-Batsri, R. and Gazit, S. (2003) Caging single avocado trees with a beehive does not guarantee exclusive formation of selfed progeny. *HortScience* 38, 1433–1434.
Douhan, G.W. (2009) Current and future perspectives regarding avocado rootstock breeding at UCR. *California Avocado Society Yearbook* 92, 87–111.
Dymond, K., Celis-Diez, J., Potts, S., Howlett, B., Willcox, B. and Garratt, M. (2021) The role of insect pollinators in avocado production: a global review. *Journal of Applied Entomology* 145, 369–383.
Eisenstein, D. and Gazit, S. (1989) Initial set and fruitlet development in avocado resulting from hand pollination with self and outcross pollen. *Alon Hanotea* 43, 339–348. (in Hebrew.)
Ellstrand, N.C., Lee, J.M., Bergh, B.O., Coffey, M.D. and Zentmyer, G.A. (1986) Isozymes confirm hybrid parentage for 'G755' selections. *California Avocado Society Yearbook* 70, 199–203.
Fernandez de Enciso, M. (1519) *Suma de Geografia*. Sevilla, Spain.
Ferrer Pereira, H., Pérez Almeida, I., Raymúndez Urrutia, M. and Suárez L. (2017) Genetic relationship analysis for avocado cultivars from Venezuelan germplasm bank (INIA-CENIAP) using molecular markers. *Tree Genetics & Genomes* 13, 1–12.
Fick, A., Swar, V., Backer, R., Bombarely, A., Engelbrecht, J. and van den Berg, N. (2022) Partially resistant avocado rootstock Dusa® shows prolonged upregulation of nucleotide binding-Leucine rich repeat genes in response to *Phytophthora cinnamomi* infection. *Frontiers in Plant Science* 13, 793644.
Fiedler, J., Bufler, G. and Bangerth, F. (1998). Genetic relationships of avocado (*Persea americana* Mill.) using RAPD markers. *Euphytica* 101, 249–255.
Fiorenza, A., Aiello, D., Leonardi, G. R., Continella, A. and Polizzi G. (2021) First report of *Rosellinia necatrix* causing white root rot on avocado in Italy. *Plant Disease* 105(10).
Fisher, R.A. (1930) The evolution of dominance in certain polymorphic species. *American Nature* 64, 385–400.
Furnier, G.R., Cummings, M.P. and Clegg, M.T. (1990) Evolution of the avocados as revealed by DNA restriction site variation. *Journal of Heredity* 81, 183–188.
Garcia, A.V. (1975) Cytogenetic studies in the genus *Persea* (Lauraceae). I. Karyology of seven species. *Canadian Journal of Genetic Cytology* 17, 173–180.

Garner, L.C. and Lovatt, C.J. (2008) The relationship between flower and fruit abscission and alternate bearing of 'Hass' avocado. *Journal of the American Society for Horticultural Science* 133(1), 3–10.

Garner, L.C. and Lovatt, C.J. (2016) Physiological factors affecting flower and fruit abscission of 'Hass' avocado. *Scientia Horticulturae* 199, 32–40.

Ge, Y., Cheng, Z., Si, X., Ma, W., Tan, L. *et al.* (2019a) Transcriptome profiling provides insight into the genes in carotenoid biosynthesis during the mesocarp and seed developmental stages of avocado (*Persea americana*). *International Journal of Molecular Sciences* 20(17), 4117.

Ge, Y., Tan, L., Wu, B., Wang, T., Zhang, T. *et al.* (2019b) Transcriptome sequencing of different avocado ecotypes: de novo transcriptome assembly, annotation, identification and validation of EST-SSR markers. *Forests* 10(5), 411.

Ge, Y., Zhang, T., Wu, B., Tan, B.L., Ma, F. *et al.* (2019c) Genome-wide assessment of avocado germplasm determined from specific length amplified fragment sequencing and transcriptomes: population structure, genetic diversity, identification, and application of race-specific markers. *Genes* 10(3), 215.

Goldring, A., Gazit, S. and Degani, C. (1987) Isozyme analysis of mature avocado embryos to determine outcrossing rate in a 'Hass' plot. *Journal of the American Society for Horticultural Science* 112, 380–392.

Gross-German, E. and Viruel, M.A. (2013) Molecular characterization of avocado germplasm with a new set of SSR and EST-SSR markers: genetic diversity, population structure and identification of race-specific markers in a group of cultivated genotypes. *Tree Genetics & Genomes* 9, 539–555.

Guzmán, L.F., Machida-Hirano, R., Borrayo, E., Cortés-Cruz, M., Espíndola-Barquera, M.D. and Heredia-García, E. (2017) Genetic structure and selection of a core collection for long term conservation of avocado in Mexico. *Frontiers in Plant Science* 8: 243.

Haberman, A., Tsror (Lahkim), L., Lazare, S., Hazanovsky, M., Lebiush, S., Zipori, I., Busatn, A., Simenski, E. and Dag, A. (2020) Management of Verticillium Wilt of avocado using tolerant rootstocks. *Plants* 9(4), 531.

Halma, F.F. and Frolich, E. (1949) Storing avocado seeds and hastening germination. *California Avocado Society Yearbook* 34, 136–138.

Halma, F.F. and Smoyer, K.M. (1951) Cold-hardiness of young avocado trees on Mexican and Guatemalan rootstocks. *California Avocado Society Yearbook* 36, 75–78.

Halma, F.F. and Zentmyer, G.A. (1953) Relative susceptibility of Guatemalan and Mexican avocado varieties to Dothiorella canker. *California Avocado Society Yearbook* 38, 156–158.

Hansche, P.E. (1983) Response to selection. In: Janick, J. and Moore, J.N. (eds) *Methods in Fruit Breeding.* Purdue University Press, West Lafayette, Indiana, pp. 154–171.

Harrington, T.C., Fraedrich, S.W. and Aghayeva, D.N. (2008) *Raffaelea lauricola*, a new ambrosia beetle symbiont and pathogen on the Lauraceae. *Mycotaxon* 104, 399–404.

Hernández, I., Uarrota, V., Fuentealba, C., Paredes, D., Defilippi, B.G. *et al.* (2022) Transcriptome and hormone analyses reveals differences in physiological age of 'Hass' avocado fruit. *Postharvest Biology and Technology* 185, 111806.

Hodgson, R.W. (1945) Suggestive evidence of the existence of strains of 'Fuerte' avocado variety. *California Avocado Society Yearbook* 30, 24–26.

Ibarra-Laclette, E., Méndez-Bravo, A., Pérez-Torres, C.A., Albert, V.A., Mockaitis, K. *et al.* (2015) Deep sequencing of the Mexican avocado transcriptome, an ancient angiosperm with a high content of fatty acids. *BMC Genomics* 16(1), 1–18.

International Plant Genetic Resources Institute (1995) *Descriptors for Avocado* (*Persea* spp.). International Plant Genetic Resources Institute, Rome.

Ish-Am, G. and Eisikowitch, D. (1993) The behaviour of honey bees (*Apis mellifera*) visiting avocado (*Persea americana*) flowers and their contribution to its pollination. *Journal of Apicultural Research* 32, 175–186.

Ish-Am, G. and Eisikowitch, D. (1998) Low attractiveness of avocado (*Persea americana* Mill.) flowers to honeybees (*Apis mellifera* L.) limits fruit set in Israel. *Journal of Horticultural Science and Biotechnology* 73(2), 195–204.

Ish-Am, G., Barrientos-Priego, A.F., Castañeda-Vildózola, A. and Gazit, S. (1999) Avocado (*Persea americana* Mill.) pollinators in its region of origin. *Revista Chapingo Serie Horticultura* 5, 137–143.

Johnson, H. (1940) Heredity of precocious flowering in *Betula verrucosa* and *B. pubescens. Hereditas* 35, 112–114.

Kadman, A. (1968) Selection of avocado rootstocks suitable for use with saline irrigation water. *California Avocado Society Yearbook* 52, 145–147.

Kadman, A. and Ben-Ya'acov, A. (1976) Selection of avocado rootstocks for saline conditions. *Acta Horticulturae* 57, 189–197.

Kadman, A. and Ben-Ya'acov, A. (1980) 'Maoz' avocado rootstock selection. *HortScience* 15: 207.

Kadman, A., Ben-Ya'acov, A., Raz, Z. and Peleg, J. (1976) A method for measuring the 'covering area' of avocado trees. *California Avocado Society Yearbook* 59, 105–108.

Kämper, W., Trueman, S.J., Cooke, J., Kasinadhuni, N., Bruton, A.J. and Ogbourne, S.M. (2021). Single-nucleotide polymorphisms that uniquely identify cultivars of avocado (*Persea americana*). *Applications in Plant Sciences* 9(6): e11440.

Kilaru, A., Cao, X., Dabbs, P.B., Sung, H.J., Rahman, M.M. *et al.* (2015) Oil biosynthesis in a basal angiosperm: Transcriptome analysis of *Persea americana* mesocarp. *BMC Plant Biology* 15: 203.

Knight, R.J. (1971) Breeding for cold hardiness in subtropical fruits. *HortScience* 6, 157–160.

Kopp, L.E. (1966) A taxonomic revision of the genus *Persea* in the Western hemisphere (Persea-Lauraceae). *Memoirs of the New York Botanical Garden* 14, 1–120.

Kremer-Köhne, S. and Köhne, J.S. (2007) 25 years of avocado rootstock development in South Africa. In: *Proceedings of the Sixth World Avocado Congress.* Viña del Mar, Chile, p. 39. Available at: https://www.avocadosource.com/wac6/en/extenso/1a-8.pdf (accessed 23 September 2024).

Kuhn, D.N., Groh, A., Rahaman, J., Freeman, B., Arpaia, M.L. *et al.* (2019a) Creation of an avocado unambiguous genotype SNP database for germplasm curation and as an aid to breeders. *Tree Genetics & Genomes* 15: 71.

Kuhn, D.N., Livingstone III, D.S., Richards, J.H., Manosalva, P., Van den Berg, N. and Chambers, A.H. (2019b) Application of genomic tools to avocado (*Persea americana*) breeding: SNP discovery for genotyping and germplasm characterization. *Scientia Horticulturae* 246, 1–11.

Lahav, E. (1970) *Localization of fruit on the tree, branch girdling and fruit thinning*. In: The Volcani Institute, Agricultural Research Division Subtropical Horticulture Report 1960–1969, Israel, pp. 60–61.

Lahav, E. and Gazit, S. (1994) World listing of avocado cultivars according to flowering type. *Fruits* 49, 299–313.

Lahav, E. and Kalmar, D. (1977) Water requirement of avocado in Israel. 2. Influence on yield, fruit growth and oil content. *Australian Journal of Agricultural Research* 28, 869–877.

Lahav, E. and Zamet, D. (1999) Flowers, fruitlets and fruit drop in avocado trees. *Revista Chapingo Serie Horticultura Número Especial* V, 95–100.

Lahav, E., Gefen, B. and Zamet, D. (1971) The effect of girdling on the productivity of the avocado. *Journal of the American Society for Horticultural Science* 96, 396–398.

Lahav, E., Gefen, B. and Zamet, D. (1972) The effect of girdling on fruit quality, phenology, and mineral analysis of the avocado tree. *California Avocado Society Yearbook* 55, 162–168.

Lahav, E., Zamet, D., Gazit, S. and Lavi, U. (1986) Girdling as a means of shortening the juvenile period of avocado seedlings. *HortScience* 21, 1038–1039.

Lahav, E., Tomer, E., Gazit, S. and Lavi, U. (1995) Performance of avocado (*Persea americana* Mill.) and mango (*Mangifera indica* L.) seedlings compared with their grafted trees. *Journal of the American Society for Horticultural Science* 120, 265–269.

Lahav, E., Lavi, U., Degani, C., Sa'ada, D., Chapnik, A. *et al.* (1998) 'Galil' – a new avocado cultivar. *Alon Hanotea* 52: 287. (in Hebrew.)

Lahav, E., Israeli, T., Regev, I., Ardity, H., Lavi, U. *et al.* (2005) 'Arad' – a new avocado cultivar. *HortScience* 40: 488.

Lavi, U., Hillel, J., Vainstein, A., Lahav, E. and Sharon, D. (1991a) Application of DNA fingerprints for identification and genetic analysis of avocado. *Journal of the American Society for Horticultural Science* 116, 1078–1081.

Lavi, U., Lahav, E., Genizi, A., Gazit, S. and Hillel, J. (1991b) Quantitative genetic analysis of traits in avocado cultivars. *Plant Breeding* 106, 149–160.

Lavi, U., Lahav, E., Degani, C. and Gazit, S. (1992) The genetics of the juvenile phase in avocado and its application for breeding. *Journal of the American Society for Horticultural Science* 117, 981–984.

Lavi, U., Lahav, E., Degani, C. and Gazit, S. (1993a) Genetics of skin colour, flowering group, and anise scent in avocado. *Journal of Heredity* 84, 82–84.

Lavi, U., Lahav, E., Degani, C., Gazit, S. and Hillel, J. (1993b) Genetic variance components and heritabilities of several avocado traits. *Journal of the American Society for Horticultural Science* 118, 400–404.

Levin, A. (1981) Factors affecting inflorescence and vegetative development regulation in avocado (*Persea americana* Mill.). MSc thesis, The Hebrew University of Jerusalem, Rehovot, Israel. (in Hebrew.)

López-Herrera, C. J. and Zea-Bonilla, T. (2007) Effects of benomyl, carbendazim, fluazinam and thiophanate methyl on white root rot of avocado. *Crop Protection* 26, 1186–1192.

López-Herrera, C.J., Pérez-Jiménez, R.M., Zea-Bonilla, T., Basallote-Ureba, M.J. and Melero-Vara, J.M. (1998) Soil solarization in established avocado trees for control of *Dematophora necatrix*. *Plant Disease* 82, 1088–1092.

Loupassaki, M.H., Vasilakakis, M. and Androulakis, I.I. (1997) Effect of pre-incubation humidity and temperature treatment on the in vitro germination of avocado pollen grains. *Euphytica* 94, 247–251.

Menge, J.A., Douhan, G.W., McKee, B., Pond, E., Bender, G.S. and Faber, B. (2012) Three new avocado rootstock cultivars tolerant to Phytophthora Root Rot: 'Zentmyer', 'Uzi', and 'Steddom'. *Journal of the American Society for Horticultural Science* 47(8), 1191–1194.

Mhameed, S. Sharon, D., Kaufman, D., Lahav, E., Hillel, J., Degani, C. and Lavi, U. (1997) Genetic relationships within avocado (*Persea americana* Mill.) cultivars and between *Persea* species. *Theoretical Applied Genetics* 94, 279–286.

Moreno-Pérez, A., Zumaquero, A., Martínez-Ferri, E., López-Herrera, C., Pliego-Alfaro, F., Palomo-Ríos, E. and Pliego, C. (2023) A comparative transcriptome analysis of avocado embryogenic lines susceptible or resistant to *Rosellinia necatrix* exudate. *Agronomy* 13: 1354.

Muñoz, A.E., Plantegenest, M., Amouroux, P. and Zaviezo, T. (2021) Native flower strips increase visitation by non-bee insects to avocado flowers and promote yield. *Basic and Applied Ecology* 56, 369–378.

Nath, O., Fletcher, S.J., Hayward, A., Shaw, L.M., Masouleh, A.K. *et al.* (2022) A haplotype resolved chromosomal level avocado genome allows analysis of novel avocado genes. *Horticultural Research* 30;9:uhac157.

Okello, E.N., Amugune, N.O., Mukiama, T.K. and Lattorff, H.M.G. (2021) Abundance and community composition of flower visiting insects of avocado (*Persea americana* Mill) in the East African region. *International Journal of Tropical Insect Science* 41, 2821–2827.

Papademetriou, M.K. (1974) A study of the variability of avocado pollen under natural conditions. *California Avocado Society Yearbook* 58, 74–76.

Pattemore, D.E., Buxton, M.N., Cutting, B.T., McBrydie, H.M., Goodwin, R.M., Dag A. (2018) Low overnight temperatures delay 'Hass' avocado (*Persea americana*) female flower opening, leading to nocturnal flowering. *Journal of Pollination Ecology* 23, 127–135.

Pérez-Balam, J., Quezada-Euán, J.J.G., Alfaro-Bates, R., Medina, S., McKendrick, L., Soro, A. and Paxton, R.J. (2012) The contribution of honeybees, flies and wasps to avocado (*Persea americana*) pollination in southern México. *Journal of Pollination Ecology* 8, 42–47.

Pérez-Torres, C.A., Ibarra-Laclette, E., Hernández-Domínguez, E.E., Rodríguez-Haas, B., Pérez-Lira, A.J. *et al.* (2021) Molecular evidence of the avocado defense response to *Fusarium kuroshium* infection: a deep transcriptome analysis using RNA-Seq. *PeerJournal* 9: e11215.

Petri, C., Alburquerque, N., Faize, M., Scorza, R. and Dardick, C. (2018) Current achievements and future directions in genetic engineering of European plum (*Prunus domestica* L.). *Transgenic Research* 27(3), 225–240.

Piccone, M.F. and Whiley, A.W. (1986) Don't rush into planting Pinkerton. *Queensland Fruit and Vegetable News* 57, 26–31.

Pliego-Alfaro, F., Palomo-Ríos, E., Mercado, J.A., Pliego, C., Barceló-Muñoz, A. *et al.* (2020). Chapter 12.1. *Persea americana* Avocado. In: R.E. Litz, F. Pliego, J.I. Hormaza (eds) *Biotechnology of Fruit and Nut Crops,* 2nd Edition. CAB International, Wallingford, UK, pp. 258–281.

Ploetz, R.C., Pérez-Martínez, J.M., Smith, J.A., Hughes, M., Dreaden, T.J., Inch, S.A. and Fu, Y. (2012) Responses of avocado to laurel wilt, caused by *Raffaelea lauricola*. *Plant Pathology* 61, 801–808.

Popenoe, W. (1919) The avocado in Guatemala. *USDA Bulletin* 743.

Popenoe, W. (1941) The avocado: a horticultural problem. *Tropical Agriculture* 18, 3–7.

Popenoe, W. (1952) The avocado. *Ceiba* 1, 269–367.

Popenoe, W. and Williams, L.O. (1947) The expedition to Mexico of October 1947. *California Avocado Society Yearbook* 32, 22–28.

Rendón-Anaya, M., Ibarra-Laclette, E., Méndez-Bravo, A., Lan, T., Zheng, C. *et al.* (2019) The avocado genome informs deep angiosperm phylogeny, highlights introgressive hybridization, and reveals pathogen-influenced gene space adaptation. *Proceedings of the National Academy of Sciences* 116(34), 17081–17089.

Rubinstein, M., Eshed, R., Rozen, A., Zviran, T., Kuhn, D.N. *et al.* (2019) Genetic diversity of avocado (*Persea americana* Mill.) germplasm using pooled sequencing. *BMC Genomics* 20(1), 1–10.

Ruehle, G.D. (1963) *The Florida Avocado Industry*. University of Florida Agricultural Experiment Station Bulletin 602.

Sagwe, R.N., Peters, M.K., Dubois, T., Steffan-Dewenter, I. and Lattorff, H.M.G. (2022) Pollinator efficiency of avocado (*Persea americana*) flower insect visitors. *Ecological Solutions and Evidence* 3(4), e12178.

Salazar-García, S., Zamora-Cuevas, L. and Vega-López, R.J. (2005) Update on the avocado industry of Michoacán, México. *California Avocado Society Yearbook* 87, 31–44.

Sánchez-Colin, S. and Barrientos-Priego, A. (1987) Avocado production and breeding in Mexico. *South African Avocado Growers' Association Yearbook* 10, 24–26.

Schnell, R.J., Brown, J.S., Olano, C.T., Power, E.J., Krol, C.A., Kuhn, D.N. and Motamayor, J.C. (2003) Evaluation of avocado germplasm using microsatellite markers. *Journal of American Society for Horticultural Science* 128, 881–889.

Schroeder, C.A. (1942) Pollen germination in the avocado. *Proceedings of the American Society for Horticultural Science* 41, 181–182.

Schroeder, C.A. (1944) The avocado inflorescence. *California Avocado Society Yearbook* 28, 39–40.

Schroeder, C.A. (1951) Flower bud development in the avocado. *California Avocado Society Yearbook* 35, 159–163.

Schroeder, C.A. (1958) The origin, spread, and improvement of the avocado, sapodilla and papaya. *Indian Journal of Horticulture* 15, 116–128.

Sedgley, M. (1977) The effect of temperature on floral behaviour, pollen tube growth and fruit set in the avocado. *Journal of Horticultural Science* 52, 135–141.

Sedgley, M. (1979) Light microscope study of pollen tube growth, fertilization, and early embryo and endosperm development in the avocado varieties Fuerte and Hass. *Annals of Botany* 44, 353–359.

Sedgley, M. (1980) Anatomical investigation of abscised avocado flowers and fruitlets. *Annals of Botany* 46, 771–777.

Sedgley, M. (1981) Storage of avocado pollen. *Euphytica* 30, 595–599.

Sedgley, M. and Alexander, D.McE. (1983) Avocado breeding research in Australia. *California Avocado Society Yearbook* 67, 129–140.

Semancik J.S. and Szychowski, J.A. (1994). Avocado sunblotch viroid and persistent infection in which variants area associated with differential symptoms. *Journal of Genetic Virology* 75, 1543–1549.

Sharma, P., Al-Dossary, O., Alsubaie, B., Al-Mssallem, I., Nath, O. *et al.* (2021) Improvements in the sequencing and assembly of plant genomes. *Gigabyte* 10:2021:gigabyte24.

Sharon, D., Cregan, P.B., Mhameed, S., Hillel, J., Lahav, E. and Lavi, U. (1997) An integrated genetic linkage map of avocado. *Theoretical Applied Genetics* 95, 911–921.

Sharon, D., Hillel, J., Mhameed, S., Cregan, T.B., Lahav, E. and Lavi, U. (1998) Association between DNA markers and loci controlling avocado traits. *Journal of the American Society for Horticultural Science* 123, 1016–1022.

Smith, C.E. Jr (1966) Archaeological evidence for selection in avocado. *Economic Botany* 20, 169–175.

Solares, E., Morales-Cruz, A., Figueroa-Balderas, R., Focht, E., Ashworth, V.E.T.M. *et al.* (2023) Insights into the domestication of avocado and potential genetic contributors to heterodichogamy. *G3 Genes|Genomes|Genetics* 13(2), jkac323.

Stern, R.A., Rozen, A., Eshed, R., Zviran, T., Sisai, I. *et al.* (2021) Bumblebees (*Bombus terrestris*) improve 'Hass' avocado (*Persea americana*) pollination. *Plants* 10(7), 1372.

Steyn, E.M.A., Robbertse, P.J. and Smith, D. (1993) An anatomical study of ovary-to-cuke development in consistently low-producing trees of the 'Fuerte' avocado (*Persea americana* Mill.) with special reference to seed abortion. *Sexual Plant Reproduction* 6, 87–97.

Storey, W.B., Bergh, B.O. and Zentmyer, G.A. (1986) The origin, indigenous range and dissemination of the avocado. *California Avocado Society Yearbook* 70, 127–133.

Stout, A.B. (1923) A study in cross-pollination of avocados in Southern California. *California Avocado Association Annual Report* 7, 29–45.

Stout, A.B. (1933) The pollination of avocados. *University of Florida Agricultural Experiment Station Bulletin* 257, 4–44.

Talavera, A., Soorni, A., Bombarely, A., Matas, A.J. and Hormaza, J.I. (2019) Genome-Wide SNP discovery and genomic characterization in avocado (*Persea americana* Mill.). *Scientific Reports* 9(1), 1–13.

Tanksley, S.D., Ganal, M.W. and Martin, G.B. (1995) Chromosome landing: a paradigm for map-based gene cloning in plants with large genomes. *Trends in Genetics* 11, 63–68.

Tomer, E. and Gazit, S. (1979) Early stages in avocado (*Persea americana* Mill.) fruit development: anatomical aspects. *Botanical Gazette* 140, 304–309.

Tomer, E. and Gottreich, M. (1975) Observation on the fertilization process in avocado with fluorescent light. *Euphytica* 24, 531–535.

Tomer, E., Gazit, S. and Eisenstein D. (1980) Seedless fruit in 'Fuerte' and 'Ettinger' avocado. *Journal of the American Society for Horticultural Science* 105, 341–346.

Torres, A.M. and Bergh, B.O. (1978) Isozymes as indicators of outcrossing among 'Pinkerton' seedlings. *California Avocado Society Yearbook* 62, 103–110.

Torres, A.M., Diedenhofen, U., Bergh, B.O. and Knight, R.J. (1978) Enzyme polymorphisms as genetic markers in the avocado. *American Journal of Botany* 65, 134–139.

Tsao, P.H., Alizadeh, A. and Krausman, E. (1992) Susceptibility to *Phytophthora citricola* of certain avocado rootstock cultivars known to be tolerant to *Phytophthora cinnamomi*. In: Lovatt, C., Holthe, P.A. and Arpaia, M.L. (eds) *Proceedings of the Second World Avocado Congress*. University of California, Riverside, California, pp. 89–94.

Van den Berg, N., Hartley, J., Engelbrecht, J., Mufamadi, Z., van Rooyen, Z. and Mavuso, Z. (2018) First report of white root rot caused by *Rosellinia necatrix* on *Persea americana* in South Africa. *Plant Disease* 102: 1850.

Vergara-Pulgar, C., Rothkegel, K., González-Agüero, M., Pedreschi, R., Campos-Vargas, R., Defilippi, B.G. and Meneses, C. (2019) De novo assembly of *Persea americana* cv.'Hass' transcriptome during fruit development. *BMC Genomics* 20(1), 1–14.

Visser, T., Verhaegh, J.J. and de Vries, D.P. (1976) A comparison of apple and pear seedlings with reference to the juvenile period. I. Seedling growth and yield. *Acta Horticulturae* 56, 205–214.

Vrecenar-Gadus, M. and Ellstrand, N.C. (1985) The effect of planting design on out-crossing rate and yield in the 'Hass' avocado. *Scientia Horticulturae* 27, 215–221.

Wallace, J.M. (1958) The sunblotch disease of avocados. *California Avocado Society Yearbook* 42, 86–89.

Wallace, J.M. (1967) Infected symptomless avocado trees and their possible use in avoiding sunblotch disease. *California Avocado Society Yearbook* 51, 187–189.

Whiley, A.W. and Schaffer, B. (1994) Avocado. In: Schaffer, B. and Andersen, P.C. (eds) *Handbook of Environmental Physiology of Fruit Crops. Volume 2: Subtropical and Tropical Crops*. CRC Press, Boca Raton, Florida, pp. 165–197.

Whiley, A.W., Köhne, J.S., Arpaia, M.L. and Bender, G.S. (1990) Future prospects with new avocado cultivars and elite rootstocks. *South African Avocado Growers' Association Yearbook* 13, 16–20.

Winer, L. (2022) Performance of a tetraploid avocado rootstock 'Canacado VC-320'. *Acta Horticulturae* 1346, 627–632.

Witjaksono and Litz, R.E. (1998) Biotechnology strategies for improving avocado. *California Avocado Society Yearbook* 82, 101–118.

Wolstenholme, B.N., Whiley, A.W. and Saranah, J.B. (1990) Manipulating vegetative: reproductive growth in avocado (*Persea americana* Mill.) with paclobutrazol foliar sprays. *Scientia Horticulturae* 41, 315–327.

Ying, Z., Davenport, T.L., Zhang, T., Schnell, R.J. and Tondo, C.L. (2009) Selection of highly informative microsatellite markers to Identify pollen donors in 'Hass' avocado orchards. *Plant Molecular Biology Reporter* 27, 374–380.

Zentmyer, G.A. (1972) Expansion of avocado root rot resistance program. *California Avocado Society Yearbook* 55, 87–89.

Zentmyer, G.A. (1980) Phytophthora cinnamomic and the diseases it causes. Phytopathological Monograph 10. American Phytopathological Society. *Soil Biology and Biochemistry* 13(6), 559.

Zentmyer, G.A., Thorn, W.A. and Burns, R.M. (1963) The Duke avocado. *California Avocado Society Yearbook* 47, 28–36

Zentmyer, G.A., Paulus, A.O., Gustafson, C.D., Wallace, J.M. and Burns, R.M. (1965) Avocado diseases. *California Agriculture Experiment Station Extension Service Circular* 534.

5 Ecology: Climate, Soils, Biogeography, Climatic Variability and Climate Change

J.G. Ramírez-Gil[1]*, L.A. Valbuena-Gaona[1] and B.N. Wolstenholme[2]
[1]*Laboratorio de Agrocomputación y Análisis Epidemiológico, Departamento de Agronomía, Facultad de Ciencias Agrarias, Universidad Nacional de Colombia sede Bogotá, Bogotá, Colombia;* [2]*University of KwaZulu-Natal, Durban, South Africa*

Abstract

This chapter addresses the ecological, edaphic and climatic requirements for optimal avocado production, particularly focused on cv. 'Hass'. Advanced data analytics tools explore insights into the biomes supporting avocado cultivation in traditional production regions in Mexico, Israel, Chile, USA, Spain, South Africa, Australia and New Zealand, along with detailed edaphoclimatic profiles. Attention is also drawn to emerging avocado-growing countries such as Peru, Colombia, Ethiopia and Kenya. Climate variability and the potential impacts of climate change are discussed as critical factors influencing avocado production systems worldwide. The contentious issue of water usage in avocado cultivation is also addressed, including conflicts over this vital resource in certain production areas. Pressing challenges facing avocado cultivation are discussed, particularly sustainability and adaptation strategies required to face climate-variability and -change scenarios. This chapter synthesizes knowledge of avocado ecology and edaphoclimatic conditions and records efforts towards sustainable avocado cultivation under evolving environmental and commercial dynamics.

Introduction

Avocado (*Persea americana* Mill.) is native to America, specifically Mexico, Central America and northern parts of South America. It has been associated with three subspecies, also known as geographic races, ecotypes or systematic botanists, botanical varieties. These are Mexican (*P. americana* var. *drymifolia* (Schlect and Cham) Blake), Guatemalan (*P. americana* var. *guatemalensis* Williams) and West Indian (*P. americana* var. *americana* Mill.) (Storey *et al.*, 1986; Galindo-Tovar *et al.*, 2008; Bost *et al.*, 2013). Because of the existence of these three races and due to natural and artificial hybridization processes, avocados have been dispersed across a wide variety of climates, soils and ecosystems and can be grown commercially in different tropical and subtropical regions of the world (Storey *et al.*, 1986; Galindo-Tovar *et al.*, 2007; see Galindo and Bost, Chapter 1 of this volume). Moreover, recent studies have proposed the potential existence of a fourth race referred to as the 'Colombian genetic group', which may exhibit genetic substructures correlated with geographical origins, particularly the Andean and Caribbean regions. These findings suggest that Colombia could represent a new centre for *P. americana* diversity (Berdugo-Cely *et al.*, 2023).

In general, the Mexican, Guatemalan and West Indian races are associated with subtropical, semitropical and tropical climates. The Mexican

*Corresponding author: jgramireg@unal.edu.co

 The Avocado: Botany, Production and Uses (eds D. Carrillo, B. Schaffer, A.W. Whiley and B.N. Wolstenholme)
DOI: 10.1079/9781800621824.0005

and Guatemalan ecotypes are indigenous to elevated montane forests or 'tropical highlands'. Both rootstock and scion cultivars of these two ecotypes, especially those dominated by Guatemalan genes with at least some Mexican genes, form the basis for today's technologically advanced 'subtropical' avocado industries (e.g. Australia), as well as those of semitropical and tropical highlands in countries such as Mexico, Colombia, Peru, Dominican Republic, Indonesia, China, Kenya, Ethiopia and Guatemala (Fig. 5.1). Meanwhile, lowland tropical avocados of a predominantly West Indian ecotype are extensively grown, mainly as seedlings, in many tropical and subtropical countries, but typically with a lower technology base and by smallholder growers in dooryard settings or the wild.

Based on the assumption that the components of avocado yield, quality and nutritional composition are closely related to soil and climatic factors, this chapter gives an overview of the climatic variables that determine where avocados can be grown successfully, emphasizing cv. 'Hass', the variety that dominates avocado trade worldwide. The present work has three thematic sections: (i) edaphoclimatic characterization of areas planted with avocado crops worldwide; (ii) edaphoclimatic variables from major areas worldwide planted with cv. 'Hass'; and (iii) avocados under climate-variability, climate-change and water-conflict scenarios.

Edaphoclimatic Characterization of Areas Planted with Avocado Crops Worldwide

Geographical distribution and general description of avocado edaphoclimatic variables in areas of native avocado races and cv. 'Hass'

Mexican ecotypes are reported to be indigenous to highland areas between 19°N and 24°N, specifically at the border of subtropical to semitropical highlands, and are distributed in the elevation range of 1500–3000 m (Storey *et al.*, 1986; Bergh, 1992). Wild Guatemalan avocados range from 14°N to 16°N in the mountainous terrain corresponding to tropical highlands with an elevation range between 1000 and 2000 m (Storey *et al.*, 1986; Bergh, 1992). The most tropical ecotype, the West Indian or Lowland (Antillean) (Me and Arzate-Fernández, 2010), extends from 82°W to 92°W in lowland areas on the Pacific coast, in the hot and humid lowlands of Central or northern South America with an elevation range of 0–1000 m (Storey *et al.*, 1986; Bergh, 1992). In general, the West Indian cultivars prefer higher temperatures and humidity and do not tolerate frost, whereas Mexican races are more frost-tolerant and grow at lower temperatures and humidity. As intermediate genotypes associated with Guatemalan cultivars, they are sensitive to high temperatures (>35°C) and low humidity (<65%) and yield well under cool conditions (Knight and Campbell, 1999).

The cv. 'Hass' is commercially grown from about 40°N to 39°S in diverse and contrasting climates (Figs. 5.1, 5.2, and 5.3; Table 5.1). For example, the accumulated precipitation in its cultivation areas varies from 200 mm year^{-1} in semidesert (irrigated) areas (Peru, Morocco and Israel) to 3000 mm in tropical rainforests (some highland areas of Colombia, Central America and Brazil) (Fig 5.3; Table 5.1). The cv. 'Hass' can adapt to a wide variety of temperature conditions (Table 5.1) (Lahav and Trochoulias, 1982; Whiley and Winston, 1987), but the optimal is a range of diurnal and nocturnal temperatures of 18°C–25°C and 14°C–18°C, respectively (Lahav and Trochoulias, 1982; Whiley and Winston, 1987). Avocado generally prefers loose, deep soils with good aeration, low levels of clay and moisture, good drainage, high levels of organic matter and/or litter, without impermeable layers (Fig. 5.2C,D). The wide distribution of the cv. 'Hass' demonstrates its high plasticity, possibly due to its Guatemalan (58%) and Mexican (42%) (Chen *et al.*, 2009) genetic origin and its remarkable adaptation to subtropical climates (Fig. 5.1A) (Knight and Campbell, 1999).

Not all production areas are suitable for growing avocado crops (Figs. 5.1 and 5.2). The avocado tree is highly sensitive to both climatic and edaphic factors, which is considered one of the reasons why the productive potentials of avocado, estimated at 30 tonnes ha^{-1}, are generally not achieved (Fig. 5.1A; Table 5.2) (Wolstenholme and Whiley, 1999; Wolstenholme, 2013; Ramírez-Gil *et al.*, 2018). The main assumption to identify the best edaphoclimatic conditions is that, except for specific variations

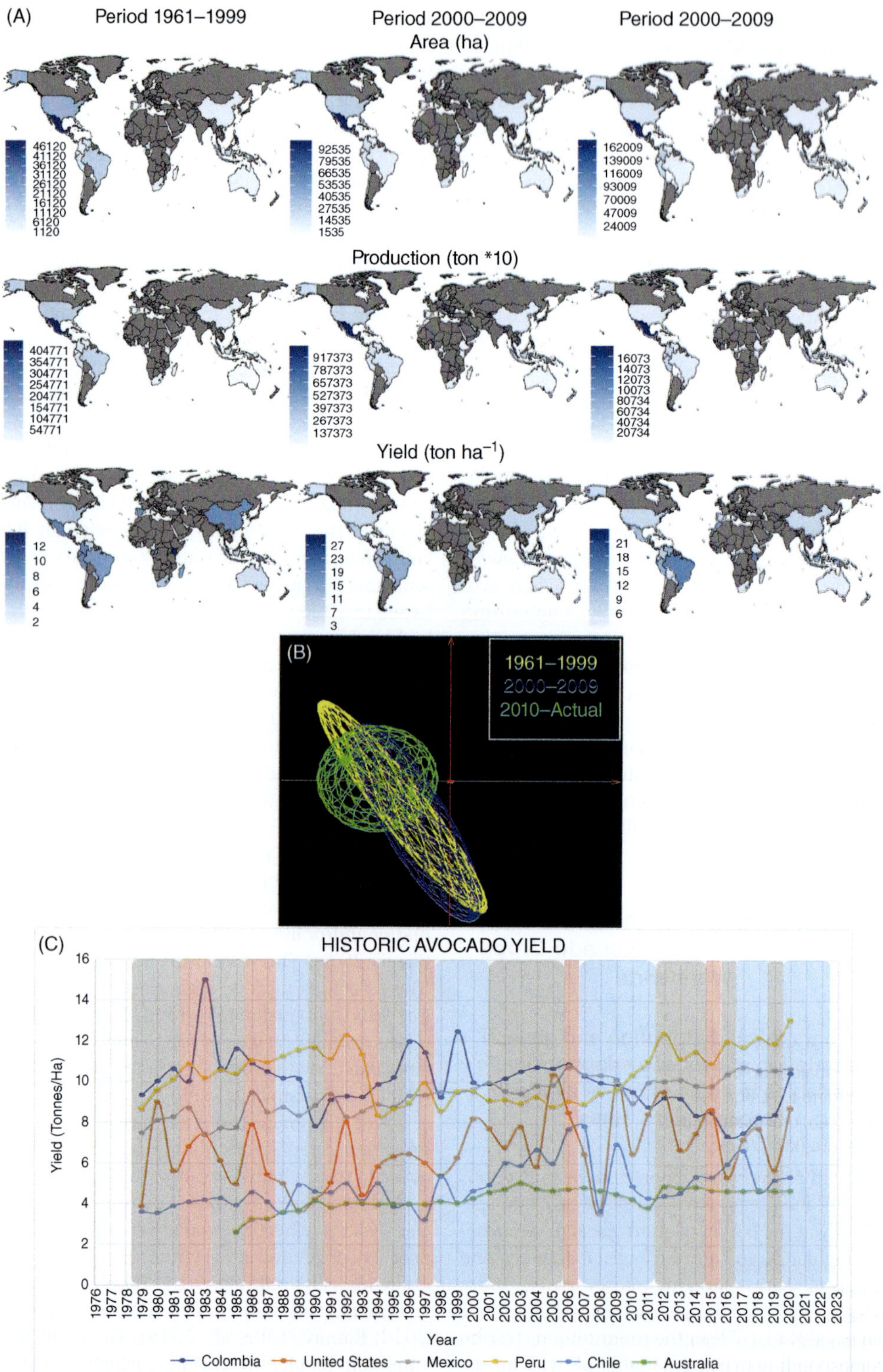
(A)
Period 1961–1999
Period 2000–2009
Period 2000–2009
Area (ha)
46120 41120 36120 31120 26120 21120 16120 11120 6120 1120
92535 79535 66535 53535 40535 27535 14535 1535
162009 139009 116009 93009 70009 47009 24009
Production (ton *10)
404771 354771 304771 254771 204771 154771 104771 54771
917373 787373 657373 527373 397373 267373 137373
16073 14073 12073 10073 80734 60734 40734 20734
Yield (ton ha^{-1})
12 10 8 6 4 2
27 23 19 15 11 7 3
21 18 15 12 9 6
(B)
1961–1999
2000–2009
2010–Actual
(C)
HISTORIC AVOCADO YIELD
Yield (Tonnes/Ha)
0 2 4 6 8 10 12 14 16
Year
1976 1977 1978 1979 1980 1981 1982 1983 1984 1985 1986 1987 1988 1989 1990 1991 1992 1993 1994 1995 1996 1997 1998 1999 2000 2001 2002 2003 2004 2005 2006 2007 2008 2009 2010 2011 2012 2013 2014 2015 2016 2017 2018 2019 2020 2021 2022 2023
Colombia
United States
Mexico
Peru
Chile
Australia

of rootstock genotypes, avocados are very susceptible to waterlogging or flooding stress (Fig. 5.1A) (Schaffer, 1998; Sanclemente *et al.*, 2014; Ramírez-Gil and Morales-Osorio, 2018; Ramírez-Gil *et al.*, 2020a; Habibi *et al.*, 2023; Yin *et al.*, 2023) (see Schaffer *et al.*, Chapter 7 in this volume for greater depth on the topic associated with flooding), and to high concentrations of salinity ions like sodium and chloride in soil (Bergh, 1992; Kourgialas and Dokou, 2021). Additionally, avocados can be susceptible to dry, windy and frosty weather, which can cause damage to fruits, branches and stems (Arpia, 1994, 2005). Furthermore, avocado trees need a healthy root system with predominant secondary roots (Fig. 5.2D,E,F), which are highly sensitive to soils with hard layers (compacted), low oxygen levels or elements that obstruct free growth (Fig. 5.2A,B) (Ramírez-Gil and Osorio, 2021).

The expansion of avocado crops over time has led to the colonization of new zones away from the centre of origin, where environmental conditions are less 'ideal' (Fig. 5.1B) (see details in the following sections). This situation requires improving environmental conditions through agronomic practices (Fig. 5.2G,H,I,J,K) or is reflected in low growth and yields (Figs. 5.1A and 5.2A,B; Table 5.2). Similarly, climate variability phenomena such as the ENSO have had a significant impact on avocado productivity, especially in countries where it has a strong influence, such as Colombia, Peru, Australia and Mexico (Fig. 5.1C) (discussed in detail below).

Macroclimatic biomes associated with the commercial production of cv. 'Hass' avocados

A biome is defined as a biogeographical unit with an ecological community characterized by dominant vegetation adapted to a physical environment as a consequence of the soil type and climatic conditions (Pennington *et al.*, 2004). Additionally, Holdridge proposed that the definition of biomes is deeply embedded in climate (Holdridge, 1967). The current cv. 'Hass' avocado-producing areas worldwide can be grouped into six biomes (Fig. 5.3; Table 5.1).

Tropical and subtropical moist broadleaf forests. This biome represents nascent avocado industries in countries such as Colombia, the Peruvian Andean mountains, the Central American region, the Dominican Republic, Brazil, China and Africa (Kenya, Ethiopia and South Africa (the highest producing areas)) (Fig. 5.3B). The climatic conditions in these regions are characterized by high levels of annual precipitation (>1200 and <3000 mm year^{-1}), wide average maximum and minimum temperature ranges of 16°C–22°C and 8°C–35°C, respectively, associated with elevations ranging from 1000 to 3000 m above sea level, and positive water balance (Figs. 5.3B and 5.4A; Table 5.1).

Tropical and subtropical coniferous forests. This biome represents the most important area planted with cv. 'Hass' worldwide, located in Mexico, mainly in the states of Michoacán, Mexico and Jalisco (Cruz-López *et al.*, 2022;

Fig. 5.1. The expansion of avocado crops from 1961 to 2021 based on planted area, production, yield and relationship with climatic suitability and the El Niño-Southern Oscillation (ENSO) phenomenon. (A): The maps were made using information from agriculture data: Crops and livestock products (FAO, 2023). (B): Visualization of the niche in the environmental space of avocado cv. 'Hass' for each of the crop periods according to A. For the niche, the environmental variables (minimum, maximum and average temperatures (°C); precipitation (mm year^{-1}); solar radiation (kJ m^{-2} day^{-1}); wind speed (m s^{-1}); water vapour pressure (kPa) (Fick and Hijmans, 2017); annual potential evapotranspiration (PET) (mm year^{-1}); and Thornthwaite aridity index, a metric of relative wetness and aridity (Title and Bemmels, 2018)) were used at a spatial resolution of ~5 km. These were subjected to dimensionality reduction and analysis using minimum-volume ellipsoids implemented with the Niche A software (Qiao *et al.*, 2016). (C): Historic avocado yield dynamics (FAO, 2023) in six important production countries influenced by the ENSO phenomenon. Red: ENSO-El Niño; blue: ENSO-La Niña; black: normal conditions. The ENSO classification was based on MEIv2: multivariate index of the ENSO cycle: (i) sea-level pressure; (ii) sea-surface temperature; (iii) zonal wind component (east–west); (iv) meridional wind component (north–south); and (v) longwave radiation (Zhang *et al.*, 2019). The MEIv2 data were obtained from the National Oceanic and Atmospheric Administration (NOAA) Application Programming Interface (API) (https://psl.noaa.gov/enso/mei/). The analyses and figures presented in this figure were created by the authors.

Fig. 5.2. General aspects associated with edaphic parameters, root architecture, and anthropic adaptation strategies associated with avocado crops. (A), (B), and (C): restrictive, acceptable, and suitable soil profiles, respectively, for avocado cultivation in the highlands of Colombia. (A): Andosol with 40% clay, 7%

Ramírez-Mejía *et al.*, 2022). The climate is cool, subhumid, with a mean temperature ranging from 17°C to 21°C, and low to medium precipitation (800–1600 mm year^{-1}) (Figs. 5.3C and 5.4B; Table 5.1). Avocados are grown at an elevation range of 1200 to 2400 m and with a water deficit in some months of the year (Ramírez-Mejía *et al.*, 2022).

Mediterranean forests, woodlands and scrub or sclerophyll forests. The avocado areas planted in this biome are located in Chile, USA (California), Spain, Morocco, Turkey, Israel, Italy and Greece, as well as small zones in Australia and South Africa (Fig. 5.3C,D,E,F). The climate is generally characterized by mild, wet winters and warm to hot and dry summers (Lionello *et al.*, 2006). It experiences a large variation in the level of precipitation (Table 5.1; Fig. 5.4C), classified as mostly dry (<700 mm year^{-1}) and wet (>1300 mm year^{-1}), depending specifically on the microclimatic conditions (high-moisture gradients at small scales) and spatiotemporal variation (specific locations and seasons of the year) of some regions, and many areas are weakly seasonal (>12% of annual precipitation during summer) (Deitch *et al.*, 2017). The temperature varies according to the geographical area and season, presenting a negative water deficit (Table 5.1; Fig. 5.4C) (Lionello *et al.*, 2006; Deitch *et al.*, 2017).

Deserts and xeric shrublands. This biome is associated with new important producer countries such as Peru (desert zones between valleys) and small parts of Israel and Morocco (Fig. 5.3B,F). The zones of this biome are characterized by a hot and dry climate with very little precipitation (<250 mm year^{-1}) (Table 5.1; Fig. 5.4D). The temperature can vary from day to night, depending on the season. Daytime temperatures can exceed 35°C in the summer, while nighttime temperatures can drop below freezing in the winter. Furthermore, this biome has high evaporation rates, solar radiation, winds and the highest water deficit (Table 5.1; Fig. 5.4D).

Temperate broadleaf and mixed forests. The main producing area of New Zealand is found in this biome (Fig. 5.3D,F). Climatic conditions are characterized by a temperate and humid climate, with mild to cool temperatures and relatively high precipitation throughout the year (Table 5.1; Fig. 5.4). The mean temperature ranges from 10°C to 20°C, with temperatures dropping to below freezing in the winter months. The precipitation ranges from 750 to 1500 mm year^{-1} (Table 5.1; Fig. 5.4).

Tropical and subtropical dry broadleaf forests. This biome is associated with avocados in borderline suitable areas for cultivation in countries such as Colombia, Ecuador, Mexico, the Central American region and Argentina, among others (Fig. 5.3). In general, these regions have different dry seasons followed by wet seasons, with annual precipitation of 500–2000 mm and temperatures of 20°C–30°C. During the dry season, these biomes experience hot and dry conditions (25°C–35°C), while during the wet season, the conditions are characterized by heavy rainfall and relatively cooler temperatures (15°C–25°C) (Table 5.1; Fig. 5.4E).

Fig. 5.2. Continued.

organic matter, 220% moisture retention, hydromorphic or redoximorphic characteristics with the presence of iron and manganese nodules and/or concretions associated with poor drainage, high water accumulation, and low oxygen levels. (B): Andosol, 20% clay, 12% organic matter, 200% moisture retention, with compacted layers associated with the use of disc ploughs in the intensive production of vegetables and livestock trampling. (C): Andosol, 18% clay, 15% organic matter, 180% moisture retention, with undisturbed structure and excellent drainage capacity. (D): Development of healthy roots in abundant vegetation cover. (E): Avocado pivoting root system without disturbance. (F): Healthy roots are affected by bifurcation in nurseries and planting in compacted soils, respectively (Fig. 5.2B). (G): The use of mulch for root induction and adaptation to climatic variability conditions, especially in drought conditions (ENSO-El Niño) (Ramírez-Gil *et al.*, 2020a). (H) and (I): Intensive drainage networks and mound planting to remove moisture from the soil profile. (J): Drip irrigation for adaptation to dry periods or planting in areas with hydric deficit. (K): Construction of artificial soils in deserts to increase organic matter and ion retention and improve structure by planting and incorporating biomass. PR: values of penetration resistance of the different types of soils in avocado areas, determined using a manual digital cone penetrometer. The analyses and figures presented in this figure were created by the authors.

Table 5.1. Average climatic variables in avocado cropping areas worldwide under current climatic conditions and associated with climate-change scenarios.

Biome	Annual Mean Temperature			Maximum Temperature			Minimum Temperature			Annual Precipitation			Annual PET	Water State
TSMBFs	18.3 ± 1.9	19.3 ± 1.9	19.5 ± 1.9	24.1 ± 2.5	25.3 ± 2.5	25.5 ± 2.5	12.5 ± 2.3	13.5 ± 2.3	13.7 ± 2.3	2075.8 ± 509.4	2149.4 ± 533.4	2108.5 ± 519.5	1394.5 ± 138.6	681.3 ± 590.0
TSCFs	17.8 ± 2.2	19.0 ± 2.2	19.2 ± 2.2	28.7 ± 2.6	30.0 ± 2.6	30.2 ± 2.6	6.1 ± 2.7	7.3 ± 2.7	7.4 ± 2.7	1037.4 ± 193.1	1048 ± 190.9	1033.5 ± 186.1	1545.8 ± 114.9	508.3 ± 250.5
MFWS	16.1 ± 1.2	17.2 ± 1.3	17.4 ± 1.3	28.0 ± 2.7	29.3 ± 2.9	29.4 ± 2.9	6.1 ± 1.1	7.0 ± 1.1	7.1 ± 1.1	359.8 ± 181.8	342.6 ± 176.5	343.0 ± 175.2	1280.2 ± 118.4	–920 ± 195.3
DXSs	19.3 ± 1.5	20.4 ± 1.5	20.6 ± 1.5	27.7 ± 2.2	28.9 ± 2.3	29.0 ± 2.3	11.7 ± 2.5	12.9 ± 2.5	13.0 ± 2.5	77.4 ± 150.9	78.1 ± 151.2	77.7 ± 50.3	1411.4 ± 125.5	–1334 ± 170.5
TBMFs	16.3 ± 2.2	17.3 ± 2.2	17.5 ± 2.2	26.5 ± 2.1	27.5 ± 2.1	27.7 ± 2.1	6.4 ± 2.3	7.2 ± 2.3	7.3 ± 2.3	1009.5 ± 531.0	1000.3 ± 542.7	989.8 ± 533.3	1236.28 ± 146.4	–226 ± 611.2
TSDBFs	20.8 ± 1.7	22.0 ± 1.7	22.2 ± 1.7	32.5 ± 1.9	33.8 ± 1.9	34.0 ± 1.9	8.5 ± 3.1	9.6 ± 3.1	9.7 ± 3.1	986.3 ± 299.6	1000.4 ± 309.7	985.9 ± 302.2	1694.8 ± 104.3	–708.5 ± 332

TSMBF: tropical and subtropical moist broadleaf forests. TSCFs: tropical and subtropical coniferous forests. MFWS: Mediterranean forests, woodlands and scrub. DXSs: deserts and xeric shrublands. TBMFs: temperate broadleaf and mixed forests. TSDBFs: tropical and subtropical dry broadleaf forests. SSP126: shared socio-economic pathways under sustainability or low-emission scenarios. SSP585: shared socio-economic pathways under fossil-fuelled development scenario model. Data for current conditions were obtained from climatological normal variables at ~1-km spatial resolution <1 km available in the WorlClim 2 free database (https://www.worldclim.org/data/monthlywth.html) (Fick and Hijmans, 2017), extracting the values associated with the coordinates of commercial lots around the world. The future climate was obtained using an ensemble model (weighted average as a function of the variance in terms of the climate condition per biome) using 20 general circulation models (GCMs) and the two SSPs (SSP126 and SSP12) at ~1-km spatial resolution and projected to 2040 using free data hosted in WorlClim 2 API (https://www.worldclim.org/data/cmip6/cmip6climate.html) (Fick and Hijmans, 2017). Annual PET: annual potential evapotranspiration. Water state: precipitation – annual PET. The date of the annual PET was obtained from ENVIREM API at ~1-km spatial resolution (https://envirem.github.io/) (Title and Bemmels, 2018). The analyses and data presented in this table were created by the authors.

Optimum climatic variables for avocado commercial crops

Based on the climates of their native habitats, it is likely that avocado trees, especially the Mexican and Guatemalan races originating in cool to warm highland environments, can be intolerant of extreme heat. In the typical warm and humid tropical and subtropical regions, where the most important avocado industry is located, climates have mean annual temperatures between 16.5°C and 22°C (Fig. 5.4A,B), the latter representing the upper limit for good cv. 'Hass' performance, including reasonable fruit size and physicochemical parameters (Lahav and Trochoulias, 1982; Carvalho *et al.*, 2015; Ramírez-Gil *et al.*, 2019b). Extreme heat could be expected to be more detrimental during critical periods, such as pollination and fruit set. Conversely, low minimum temperatures (<10°C) during flowering can lead to excessive embryo abortion or the retention of stenospermocarpic fruit in cv. 'Hass' (Sedgley and Grant, 1983), more commonly known as 'cukes', which have no commercial value. When assessing the effects of heat stress during these phenological phases, a daily maximum of 33°C can be considered the threshold value (Lomas, 1988, 1992). In contrast, temperatures <−2°C still occasionally severely damage fruit and trees (Arpia, 1994, 2005). In areas such as Mexico and Colombia, mean annual temperatures of approximately 17.8°C ± 2.2°C and 18.30°C ± 1.9°C represent the optimum for high yield and quality, respectively (Tables 5.1 and 5.2; Fig. 5.4A,B).

Avocado trees generally require 1200 to 1800 mm of water year^{-1} from well-distributed precipitation (Nakasone and Paul, 1998). In typical subtropical growing areas with summer-dominant rainfall, a total of 1000 mm water year^{-1} is considered the minimum amount desirable (Fig. 5.4F; Table 5.1). On the contrary, where annual rainfall is very high (>1800 mm year^{-1}) and there are several very wet months (>300 mm month^{-1}), the risk of waterlogging is severe, except in well-drained soils (Fig. 5.2C). In all tropical and some subtropical areas, water needs are naturally supplied, and there are excesses (Figs. 5.4A,B and 5.5B; Table 5.1); meanwhile, in dry zones, water deficit is evident, resulting in the need for irrigation (Figs. 5.4 and 5.5C,D,E,F; Table 5.1).

Irradiance (sunlight) quickly becomes a limiting factor and a key factor in achieving an acceptable avocado yield in commercial situations. There are a few guidelines with regard to the amount of sunlight or light required for avocado orchards. In general, avocado trees are reported to thrive in areas with >2000 hours of sunlight year^{-1} (Gaillard and Godefroy, 1995), but in zones with high radiation, such as tropical regions (e.g. Colombia) and low levels of leaf area, many fruits can be damaged by the sun (Ramírez-Gil *et al.*, 2020b). Another factor that can have an impact on avocado productivity is wind, especially in areas influenced by hurricanes/cyclones (e.g. subtropical coastal regions such as Florida, the USA, northern and north-eastern Australia and the east coast of Africa, as well as tropical coastal areas or islands). Wind could cause the several forms of mechanical damage that can affect quality and yield, such as branch and stem break; rubbing and chaffing of fruit, vegetative buds and flowers; wind burn; and massive fruit drop. Additionally, hail can be catastrophic, completely destroying the commercial value of the current crop (A.W. Whiley, SHS, QLD Australia, 2024, personal communication). While hail is not common, it can have a devasting impact on trees, causing significant damage to leaves, vegetative shoots, flowers and fruit under tropical conditions, for example, in Colombia (Ramírez-Gil *et al.*, 2020b). In severe cases, significant damage to bark can occur, resulting in limb-dieback or the death of young trees (<2 years old) (A.W. Whiley, SHS, QLD Australia, 2024, personal communication). Furthermore, relative humidity (RH) is a key factor in avocado growth through its role in aggravating (low RH) or alleviating (high RH) physiological stress and also influencing the incidence of pests.

Soil environmental conditions of avocado production areas

In recent years, there has been a remarkable expansion in the range of soil types that support successful orchards, often requiring increased management inputs and expenses (Fig. 5.2H,I,J,K). Very good internal drainage, aeration, light texture, sandy and an intact structure are the most important criteria in soils for good avocado

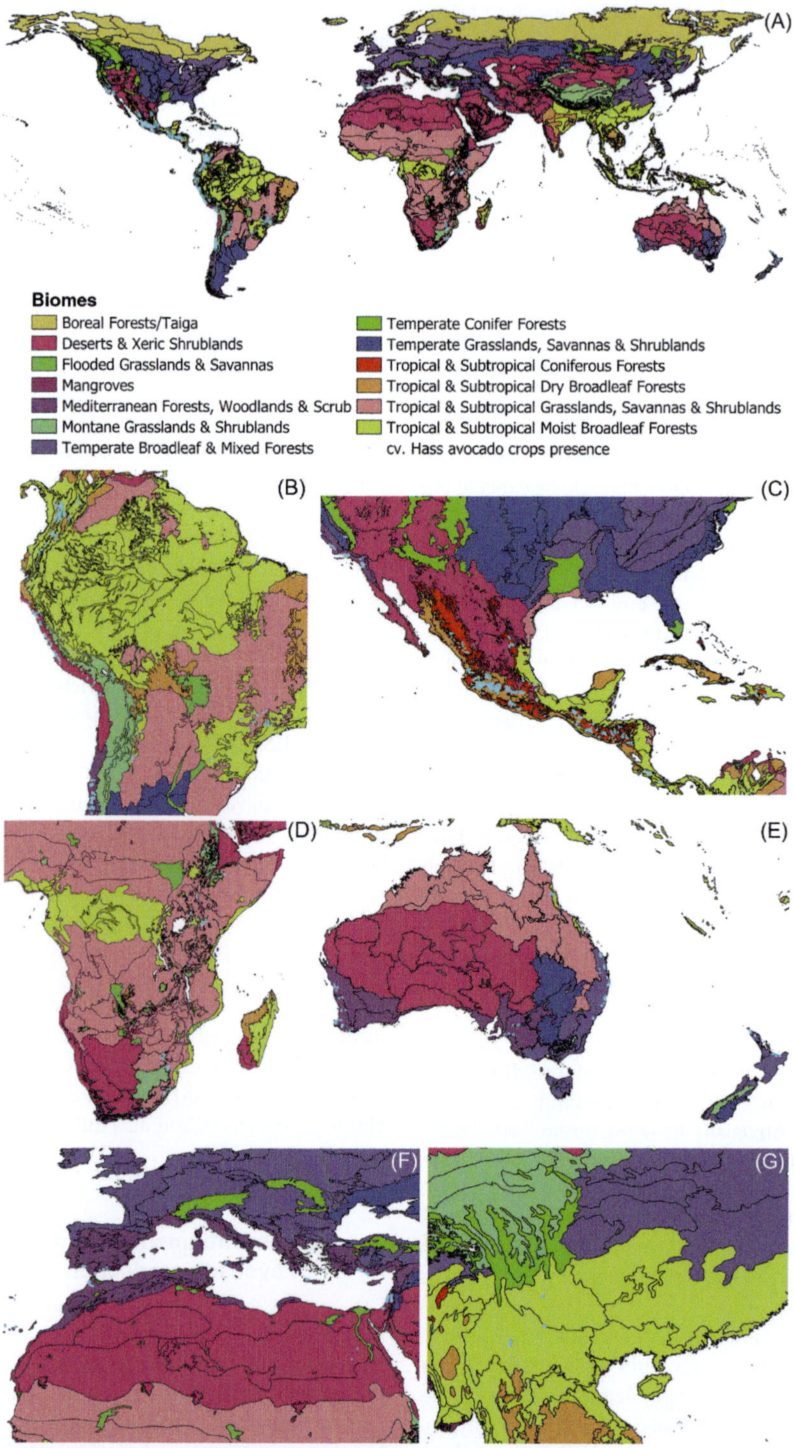

Fig. 5.3. The geographical location of the areas with planted commercial 'Hass' avocados and their relationship with different biomes. (A): worldwide biomes and commercial avocado orchards. (B), (C), (D),

growth (Fig. 5.2C). An impermeable clay subsoil, hardpan, or compacted layer, and high levels of clay near the surface can cause temporary saturation or rising, or perched water tables during periods of excess precipitation (Fig. 5.2A,B). In zones with heavy soil, inadequate drainage and/or shallow water tables, modifications must be included by ridging and mounding, with limitations in design, implementation, environmental impact and infrastructure costs (Fig. 5.2H,I,J,K). On the other hand, dry, excessively drained soils with low water retention capacity, poor organic matter and low bulk density increase dependence on adequate rainfall and/or irrigation and can require the construction of artificial soils (Fig. 5.2K).

Avocado roots have a high oxygen requirement with diffusion rates exceeding $0.20\ \mu\ cm^{-3}\ min^{-1}$ for growth (Valoras *et al.*, 1964), equivalent to >17% of air in the soil. A useful indirect measure of soil aeration is porosity. The total gas content can be determined using a mathematical ratio between porosity and soil volumetric moisture (Ramírez-Gil and Osorio, 2021). Low soil oxygen concentration, caused by waterlogging associated with soil physical properties, and/or high precipitation, induces the abiotic physiological disorder known as hypoxia-anoxia (Fig. 5.2A,B,C) (Ramírez-Gil and Morales-Osorio, 2018; Ramírez-Gil and Morales, 2019). Avocado trees are considered intolerant of flooding (Habibi *et al.*, 2023). Furthermore, a soil bulk density <0.7 is required for good penetration of avocado roots (Fig. 5.2A,B,C) (Durand and Claassens, 1987). However, root growth is absent or is greatly restricted in compacted soils with bulk densities >1.2 and/or penetration resistance values of >3 MPa (Fig. 5.2A,B,F).

The origin of avocado from highland to lowland tropical rainforest implies adaptation to moderate to highly weathered, leached, acid soils and partly with 'low fertility'. As an adaptation to these conditions, avocados have developed dense feeder roots in the top of soil or mulch cover (first 20–30 cm), corresponding to the nutrient-rich litter zone of a forest or agroecosystems. This suggests that avocados are adapted to thrive in nutrient-rich forest soils and not in nutrient-poor conditions. Avocado nutrient requirements can vary depending on their seasonal growth pattern, soil conditions, cultivar type and yield potential, among other factors (Silber *et al.*, 2018; Selladurai and Awachare, 2020). Integrated fertilizer management can be formulated based on periodic soil analysis, monitoring leaf nutrient concentration at different growth stages, total nutrient removal by harvested fruit and yield potential (Silber *et al.*, 2018; Selladurai and Awachare, 2020) (see Gil *et al.*, Chapter 11, this volume, for additional information on nutrient management).

As a general principle, volcanic ash soils, known as andosols, are regarded as optimal for avocados. This type of soil is present in the most important zone for avocado production worldwide, namely Michoacán state, Mexico (Aguilera-Montanez and Salazar-Garcia, 1991; Salazar-Garcia *et al.*, 2005), as well as in parts of the Andean mountains in Colombia (Fig. 5.2C) (Tamayo-Velez and Osorio-Vega, 2022). Most andosols have excellent physical properties, including fast drainage, good aeration, high porosity, good water retention capacity, high organic matter content and a good range of particle size distribution (structure) (Fig. 5.2C).

Edaphoclimatic variables of major areas worldwide planted with cv. 'Hass' avocados

In Mexico, the altitude across the avocado belt ranges from 1300 to 2400 m. The climate is warm, subhumid, with summer rainfall and annual mean temperatures of 10°C–24°C, annual

Fig. 5.3. Continued.

(E) and (F): biomes and 'Hass' avocado crops in South America, Central America, the USA (California), Africa, Oceania, the Mediterranean region and Asia (China), respectively. Data from the spatial worldwide distribution of biomes were obtained from a free data set from the European Environment Agency (https://www.eea.europa.eu/data-and-maps/data/biogeographical-regions-europe-3). Data on the presence of commercial lots are part of the historical database compiled by the Agrocomputing and Epidemiological Analysis Laboratory based on free data from different sources using a detailed verification and validation process. The analyses and figures presented in this figure were created by the authors.

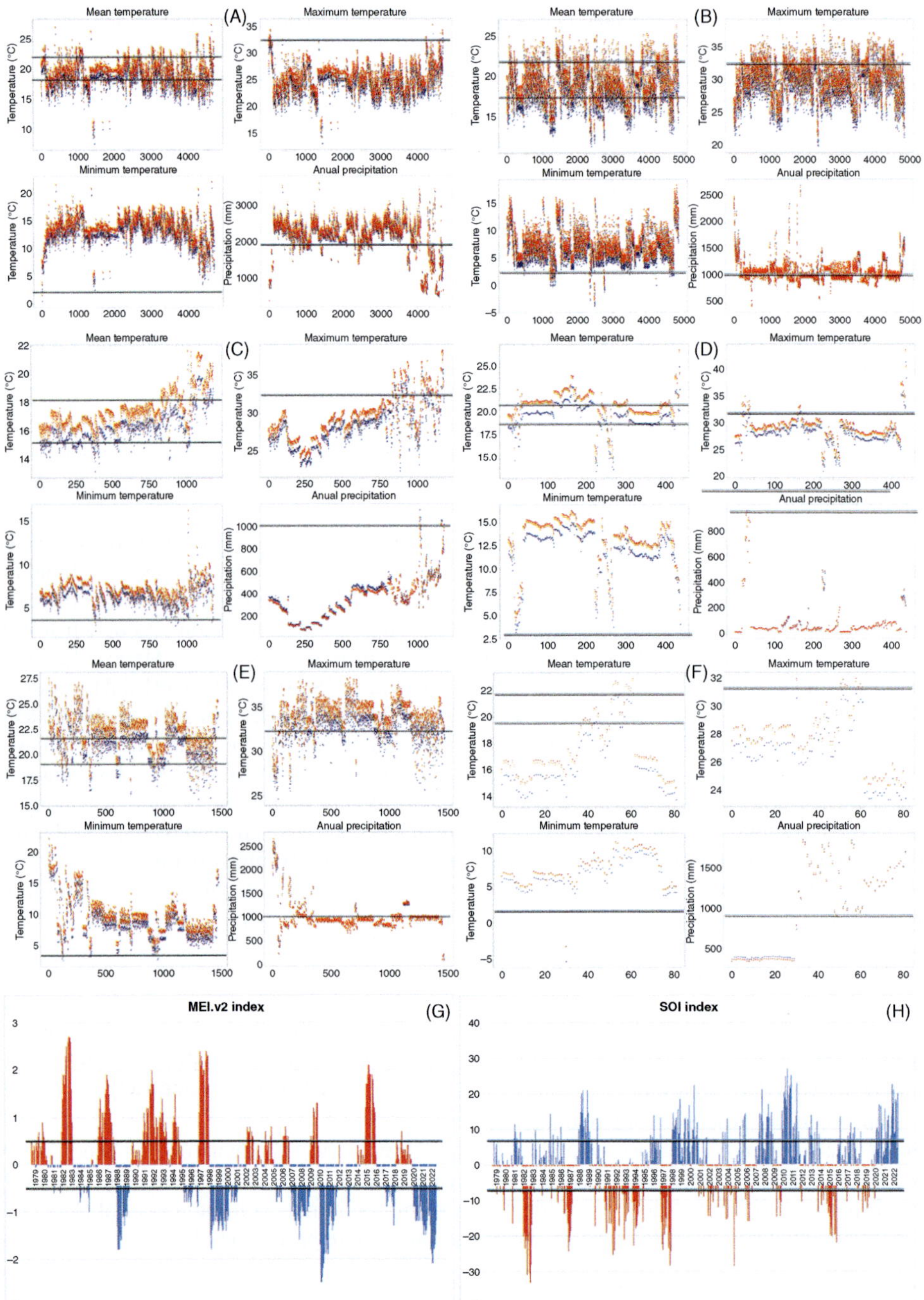

Fig. 5.4. Representation of the climatic variation in 'Hass' avocado production worldwide under scenarios of climatic variability and climate change. (A): Tropical and subtropical moist broadleaf forests. (B): Tropical and subtropical coniferous forests. (C): Mediterranean forests, woodlands and scrub. (D): Deserts and xeric shrublands. (E): Tropical and subtropical dry broadleaf forests. (F): Temperate broadleaf and mixed forests. SSP126: shared socio-economic pathways under sustainability or low emission scenarios. SSP585: shared socio-economic pathways under a fossil-fueled scenario model. Blue dots: current

precipitation of 800–1600 mm and water deficit in dry periods (Fig. 5.4B; Table 5.1). The predominant soils are andosols with high levels of organic matter (4–10%), pH of 5.5–6.9, phosphorous (P) concentration of 0.5–6 mg kg^{-1}, potassium (K) concentration of 0.5–1.5 meq 100 g^{-1}, calcium (Ca) concentration of 3–7 meq 100 g^{-1}, magnesium (Mg) concentration of 1–4 meq 100 g^{-1} and low levels of zinc and boron (Aguilera-Montanez and Salazar-Garcia, 1991; Salazar-Garcia *et al.*, 2005; Ramírez-Mejía *et al.*, 2022). The RH is between 75% and 80%, with occasional values of <60%, and >1000 sunlight hours $year^{-1}$ (Ramírez-Mejía *et al.*, 2022). Andosols have adequate physical properties for root system development and water management, as previously indicated.

In Colombia, avocado-sown areas generally have high edaphoclimatic variability, directly impacting productivity (Table 5.2) (Ramírez-Gil *et al.*, 2018). The optimal conditions that maximize yields are found at an elevation range of 1900–2400 m, with mean, minimum and maximum temperatures of 17.5°C, 6.7°C and 27°C, respectively; rainfall of 1200–2000 mm $year^{-1}$ and average radiation of 430–450 Watt m^{-2} (Table 5.2). The soils in these areas have low nutrient availability, low pH (<6.0), high levels of organic matter (>8% and <20%), high anion exchange capacity, fixation of P, low-to-high levels of Ca–Mg, medium to high levels of K, medium to high levels of Fe, Mn, Cu and Zn, and low B and Mo availability (Tamayo-Velez and Osorio-Vega, 2022). The soils are typically andosols with good performance, but this can be affected by damage to the structure, high levels of clay and location in areas with high rainfall, making them unsuitable for growing avocados (Fig. 5.2A,B; Table 5.2).

In Peru, avocados are grown in two zones, a major zone in the desert and a minor zone in the Andean highland 'Sierra Peruana'. Production in the desert uses high technology, specifically associated with drip irrigation using water from the Andean mountains (Vos and Marshall, 2017; Apaza *et al.*, 2019). The climate is dry, sunny, arid to semiarid, with mean precipitation varying from 17 to 473 mm $year^{-1}$. The temperature is stable with a range of 9.8°C to 33.3°C and a mean of 18.1°C ± 2°C; RH of 77% ± 20%; and high evapotranspiration and water deficit (Table 5.1; Fig. 5.4D) (Wei *et al.*, 2021; Felles-Leandro and García-Bendezú, 2022). In general, the physicochemical characteristics of the soil are clay loam texture, pH: 7.0–8.3, electric conductivity: 0.3–1.4 dS m^{-1}, organic matter: 0.3–1.0%, $CaCO_3$ (%): 0.5–4.0, P-Olsen (mg kg^{-1}): 7.4–16.3, K-ammonium acetate (mg kg^{-1}): 159.0–302.0 and CIC (meq 100 g^{-1}): 5.3–13.1 (Felles-Leandro and García-Bendezú, 2022). The success of avocado cultivation in desert areas in Peru is associated with low seasonal weather variability, reducing the impact of extreme conditions in summer or winter (Fig. 5.4; Table 5.1).

In Mediterranean-type zones such as the USA (California), Chile, Israel, Spain, Morocco and Turkey (among others), the climate varies according to the topographic conditions and factors at local levels. A semiarid climate with controlled summer irrigation allows the establishment of avocado orchards. In general, annual temperatures are in the range of 15°C–21°C with a mean of 16.18°C ± 1.2°C, annual precipitation is 200–700 mm; there is a high water deficit, and 3000–3500 hours of sunlight $year^{-1}$ (Fig. 5.4C; Table 5.1). In this zone, careful site selection can reduce the risk of frost damage in areas where temperatures <−2°C are likely to occur. In contrast, other climate-limiting factors occur when the mean RH is <20% in the hills and <30% in

Fig. 5.4. Continued.

climate. Orange dots: projected climate variables under the climate-change pessimistic scenario model SSP 126. Red dots: projected climate variables under the optimistic scenario model for climate change SSP585. (G): Dynamics of ENSO phenomena episodes based on the multivariate MEIv2 index used as a parameter in the zones of influence of the American continent. (G): Dynamics of ENSO phenomena episodes based on the multivariate Southern Oscillation Index (SOI) used as a parameter in the zones of influence of the Oceania continent. Blue bars: ENSO-La Niña. Red bars ENSO-El Niño. See the origin in Fig. 5.1 and Table 5.2 for details and data. For climate variables, precipitation, and temperature (mean, maximum and minimum), the upper, lower, and mean limits (black lines) were used, according to general reports of the critical ranges for avocado (Lahav and Trochoulias, 1982; Nakasone and Paul, 1998). The analyses and figures presented in this figure were created by the authors.

Table 5.2. Production and percentage of areas under cv. 'Hass' avocado in tropical conditions (Colombia) as a function of the area planted with the crop and basic characterization associated with the environmental conditions and the suitability of the agroecological zone.

Period	Zone area (%)[1]		Elevation range[2]	Temperature (°C)[2]		Precipitation (mm)[2]		ADGSR (Watt m^{-2})[2,3]		Yield (ton ha^{-1}) [4]
				A	B	A	B	A	B	
Before 2000	Adequate	90	1900–2400	17.5	6.7–27.0	1800	1200–2000	440.8	380.1–480.9	12.5
	Acceptable	7	1800–2500	17. 1	3.5–28.0	1900	1200–2300	416.5	360.4–500	10.8
	Restrictive	3	1700–2500	18.5	3.5–29.1	2000	1000–2500	400.4	330.1–550.5	6.5
2000–2010	Adequate	70	1900–2400	17.5	5.5–27.4	1750	1100–2100	430.2	370.4–475.2	14.4
	Acceptable	20	1800–2500	17. 5	3–29.0	1950	1000–2400	420,4	350.9–516.7	12.3
	Restrictive	10	1600–2550	18.5	2.7–30.0	2050	900–2550	400.4	330.1–550.5	5.5
2010 to Actual	Adequate	60	1900–2400	17.5	5–28.0	1735	1100–2100	450.5	365.3–485.6	16.7
	Acceptable	25	1800–2500	18.4	2.5–29.9	2000	1000–2450	410.4	345.4–500.1	13.2
	Restrictive	15	1400–2600	21.5	2.5–32.7	2100	900–2700	400.4	310.3–570.8	4.5

[1]Agroecological classification based on optimal edaphoclimatic conditions for the development of avocado cultivation under tropical conditions in Colombia. Adequate, acceptable and restrictive are classes based on the physical, chemical and microbiological properties of the soil (Ramírez-Gil *et al.*, 2021). [2]Climatic data obtained from the avocado-producing areas in Colombia according to the period analysed using the IDEAM (Institute of Hydrology, Meteorology and Environmental Studies, by its acronym in Spanish) hydro-climatological station data (obtained at: http://dhime.ideam.gov.co/atencionciudadano/). [3]DGSR: average daily global solar radiation. The timing of these periods was established by examining the recent history of the 'Hass' avocado in Colombia. It begins with an initial phase dedicated to establishing the crop, followed by a period of initial expansion constrained by limitations in commercialization and focused on local market objectives. The final stage is characterized by extensive expansion into regions with moderate cold climates, driven by global commercial dynamics. (A) Mean, (B) Range. [4]Yield per year (main and secondary harvest) for trees of productive age (>5 y) and assuming an average planting density of 250 plants ha^{-1}. The analyses and data presented in this table were created by the authors.

the plains, and the mean daily temperature exceeds 30°C (Lomas and Mandel, 1973). The soil in these regions generally is highly fertile, heavy-textured, calcareous (8%–40% free lime ($CaCO_3$)), with a high clay content (40%–60%), low organic matter (<3%) and a pH of 7.0–8.2, where adapted West Indian rootstocks are used (Ben-Ya'acov and Michelson, 1995).

In Australia (specifically in Queensland with an estimated 65% of the nation's production), the climate is classified as subtropical and tropical, with high rainfall (900–2000 mm year^{-1}), temperature and humidity peaks in summer, dry winter and spring, and large range in temperature (10°C–35°C), associated with intra-annual seasonality and spatial variation. Growers are advised to maintain soil pH between 5.0 and 5.5, and concentrations of $P > 80$ mg kg^{-1}, K > 0.75 meq 100 g^{-1}, Ca > 5.0 meq 100 g^{-1} and Mg > 1.6 meq 100 g^{-1} (A.W. Whiley, Nambour, Australia, 1995, personal communication).

South Africa's coastal regions suitable for avocado production have warm and humid climates with regular rainfall throughout the year. The temperature ranges from 15°C to 35°C, and the average annual rainfall is 700–1500 mm. The recommended pH is 5.5–6.5, and the recommended soil nutrient concentrations are as follows: P (Bray 1): 28–60 mg kg^{-1}, K: 150–250 mg kg^{-1}, Ca: 750–1000 mg kg^{-1}, Mg: 100–200 mg kg^{-1}, Ca/Mg ratio: 2.5–5 and Ca + Mg/K ratio: 4–5 (The ARC-ITSC (Agricultural Research Council-Institute for Tropical and Subtropical Crops), Nelspruit). The dominance of granite generally leads to less clayey soils (typically 30%–40% clay on the B horizon) and a lower organic matter content, typically 1%–2%, in the top soils. Such soils are also less fertile and require substantial liming.

Important areas with cooler maritime climates, such as New Zealand, have high temperature variations, with a mean temperature of 16.39°C ± 2.2°C (Fig. 5.4F; Table 5.1), which is the coolest in the world for 'Hass' avocados. A major risk factor in this zone is associated with night temperatures of 10°C or lower for up to five or six consecutive nights, coinciding with pollination and fruit set. In the Bay of Plenty, New Zealand, the main soil type is andosol, which has good physical and chemical properties, while in the far north, where planting has significantly increased in recent years, soils around Whangarei are red/brown, heavy clay loams of volcanic origin, and near Kaitaia, ancient marine sands overlaying peat beds are formed from prehistoric Kauri forests (A.W. Whiley, SHS, Qld, Australia, 2024, personal communication).

Avocados under Climate Variability, Climate-change, and Water-conflict Scenarios

Avocado crops are very vulnerable, with a high potential and risk (positive and negative), to climate variability and climate-change scenarios (Ramírez-Gil and Morales-Osorio, 2018; Howden and Deuter, 2005; Ramírez-Gil *et al.*, 2019a; Charre-Medellín *et al.*, 2021; Grüter *et al.*, 2022). Due to climate uncertainty, phenomena such as droughts, flooding, hurricanes, hail and extreme temperatures (low and high) could increase in frequency and intensity, potentially affecting yield, quality, production costs and reduced suitability of avocado cultivation at locations where the crop is currently grown.

The impacts of climate variability and climate change on avocados could be specific or general, depending on the phenomenon and the spatial scale on which it occurs (orchard, farm, agroecological zones, region, country, continent and global). For example, in the subtropical zone of Limpopo, South Africa, data show that an increase in temperature over time has reduced avocado yield, but no correlation was found with changes in rainfall (Randela, 2018). Similar results are reported in a historical analysis in California, where avocado was one of the most affected crops by climate trends (colder winters, hotter summers, drier periods (less available water), stronger winds, among others) (Lobell *et al.*, 2007). Meanwhile, climate variability in the Andalusia region of Spain induces water shortages and extreme heat, with a high negative impact on avocado production (Moreno-Ortega *et al.*, 2019). Another analysis reported that avocado production in New Zealand may not be as vulnerable to climate change as in Australia, as New Zealand has colder areas and microclimatic conditions in the mountains, where disturbances of climatic variables could

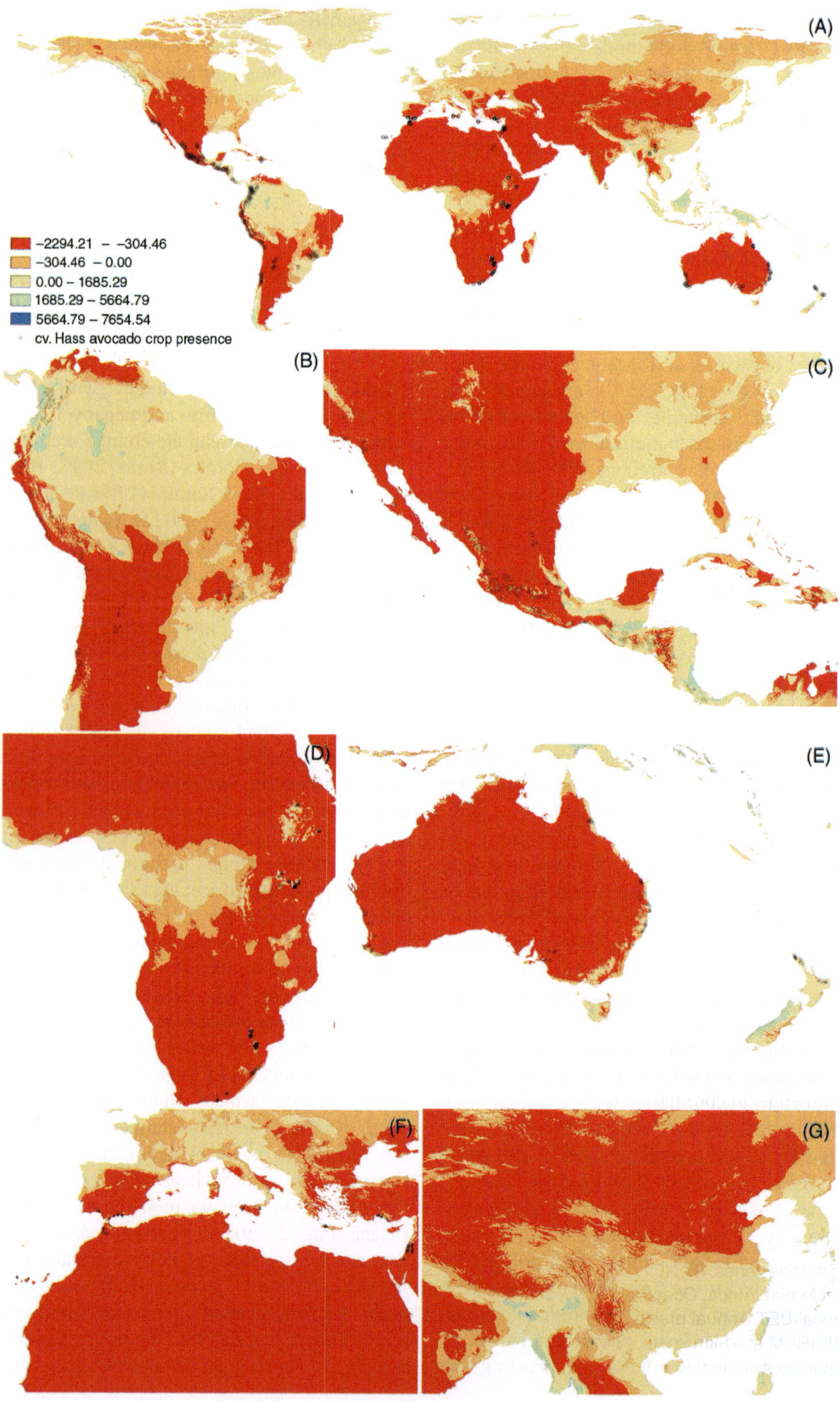
(A)
–2294.21 – –304.46
–304.46 – 0.00
0.00 – 1685.29
1685.29 – 5664.79
5664.79 – 7654.54
cv. Hass avocado crop presence
(B)
(C)
(D)
(E)
(F)
(G)

have negative impacts on growth, disease risk, fruit quality and industry location, as drivers of change in productive areas (Howden and Deuter, 2005). In contrast, in tropical zones (Colombia), climate variability has impacted the industry, causing flooding under the ENSO-La Niña, with serious consequences in terms of tree mortality and increased incidence of root rot disease (*Phytophthora cinnamomi* Rands.) (Ramírez-Gil and Morales-Osorio, 2018; Ramírez-Gil *et al.*, 2020a). To a lesser extent, increases in temperature and reductions in precipitation under ENSO-El Niño have affected production by reducing fruit size, low accumulation of unsaturated fatty acids, and inducing physiopathology of unknown origin (Carvalho *et al.*, 2015; Ramírez-Gil *et al.*, 2019b, 2020b).

Based on our macroanalysis, under climate-change scenarios, different impacts could be expected in the different avocado production areas (Fig. 5.4; Table 5.1). By 2040, an increase in temperature of between 1°C and 1.5°C is projected, which will vary depending on the scenario (optimistic or pessimistic) and the specific conditions of the biomes (Figs. 5.3 and 5.4; Table 5.1). Areas in major producing countries such as Mexico, Chile, Spain, Israel, the USA, Australia and South Africa could have limitations associated with the upper-temperature threshold (>35°C) (Fig. 5.4). The projections of precipitation in the future present more uncertainty, with current drier areas having a tendency for less rainfall (Fig. 5.4; Table 5.1). At the macro level (Fig. 5.1C), a causal relationship was found between avocado yield and ENSO-La Niña and El Niño phases, depending on the country and its predominant climatic condition. For example, in countries with excessive precipitation, such as Colombia under the La Niña, a negative impact on yield was observed, whereas an increase occurred under the El Niño pattern. Meanwhile, in countries with dominant dry climates, such as Peru, Chile and the USA (California), yield was favoured by La Niña. In countries such as Mexico, Colombia, Australia and Peru, variations in avocado production occur during both El Niño and La Niña events. These fluctuations may be influenced by local factors and the impacts associated with these weather phenomena.

It is urgent to design and apply different strategies to mitigate and adapt avocado crops to phenomena such as the ENSO-La Niña and El Niño (Ramírez-Gil *et al.*, 2020a). Zoning and determining areas of climatic risk is an alternative at the macro level (Howden and Deuter, 2005; Ramírez-Gil *et al.*, 2019a). Meanwhile, at the orchard, farm and agroecological levels, strategies that can be implemented include adapted genotypes, plant growth-promoting microorganisms, water harvesting and storage structures, increased irrigation area, the use of mulch, growth regulators and windbreaks, canopy management, earlier harvest, effective soil moisture monitoring, better drainage (e.g. higher row mounding), stricter pest monitoring and increasing soil organic matter levels (Howden and Deuter, 2005; Randela, 2018; Ramírez-Gil *et al.*, 2020a).

In recent years, controversy has arisen about the water consumption of avocado trees. The 'high' water footprint of avocado has generated water conflicts and other negative environmental and socio-economic impacts at the local level in the main production zones (Sommaruga and Eldridge, 2021). In this regard, it should be clarified that avocado is not a highly water-consuming species compared to others that can be planted under the same conditions (Apaza *et al.*, 2019). The expansion of cultivated areas in extreme climates (Fig. 5.1B), in biomes such as tropical and subtropical coniferous forests, Mediterranean, deserts, and tropical and subtropical dry broadleaf forests, in countries such as Chile, Peru, the USA, Spain and Israel, where the water balance is negative, and the minimum

Fig. 5.5. Water status of 'Hass' avocado production areas worldwide. (A): water status in cv. 'Hass' production zones. (B), (C), (D), (E), (F) and (G): 'Hass' crops in South America, Central America, USA (California), Africa, Oceania, Mediterranean and Asia (China), respectively. Water state: precipitation – annual PET. Annual precipitation and PET data were obtained from the API of WordClim 2 and ENVIREM at ~1-km spatial resolution (Fick and Hijmans, 2017; Title and Bemmels, 2018). The analyses and figures presented in this figure were created by the authors.

water needs are not supplied by precipitation, has resulted in the need for irrigation (Figs. 5.4 and 5.5; Table 5.1). According to future water projections, these regions have low water reserves or source areas to sustain avocado production (Sommaruga and Eldridge, 2021). Moreover, climate-change scenarios project precipitation reductions and temperature increases (Fig. 5.4; Table 5.1).

An example of this situation is in the Mediterranean region, where decreasing rainfall requires irrigation. In Mexico, 80% of all avocado orchards have supplemental irrigation during the dry months, and in some cases, no water is available for irrigation (Moreno-Ortega *et al.*, 2019; Sommaruga and Eldridge, 2021). Similarly, in Australia, most avocado orchards have access to irrigation water either from on-farm storage or reticulated water schemes from government-funded dams (A.W. Whiley, SHS, Qld, Australia, 2024, personal communication). In Chao, Virú, Moche and la Libertad, provinces in Peru, the avocado crop uses a mean of 16,941.2 m^3 water $hectare^{-1}$ $year^{-1}$. However, non-significant yield increases occurred when water was increased (Apaza *et al.*, 2019). Meanwhile, in the tropical areas of Colombia, the expansion of the crop to extreme climatic zones (Table 5.2) and the intra-annual and inter-annual climate variability suggests the need for irrigation in some areas. However, the water balance is positive in most areas classified as suitable for avocado production in Colombia (Erazo-Mesa *et al.*, 2021). The above indicates the need to investigate each zone, considering the water needed for satisfactory growth, development, yield and quality of avocado crops to improve and optimize the use of water resources (Erazo-Mesa *et al.*, 2021, 2022).

Conclusions

Avocados are grown worldwide in four major climatic or biome areas. The best yield potential is found in cool tropical to borderline subtropical highlands of Central America, Mexico, Colombia, Kenya, Ethiopia and others. Meanwhile, the industry is gaining importance in semiarid, cool, winter-rainfall areas such as Peru, the USA (California), Spain and Israel, where technology is used to counterbalance higher stress levels and lower yield. The large genetic variability within avocado results in broad environmental adaptation around the world. In the three (or maybe four) primary ecotypes (Mexican, Guatemalan, West Indian and perhaps a Colombian genetic group), climatic adaptability is constrained by their specific edaphoclimatic origin characteristics. However, at the cultivar level, such as with 'Hass', adaptability tends to be broader, ensuring optimal tree performance across a greater range of conditions. This is because cv. 'Hass' is a Guatemalan × Mexican hybrid, providing it with greater environmental plasticity. The choice of the avocado soil type has been dominated by the risk of waterlogging and pest incidence (i.e. *P. cinnamomi*), especially in the higher-risk humid tropics and subtropics. A mean temperature increase of 1°C–1.5°C in tropical and subtropical regions is likely in the next few decades, with a reduction in precipitation in important avocado-growing areas such as Mexico, Chile, Peru, Spain, the USA and Israel. With increasingly scarce water supplies and greater weather extremes, more stressful atmospheric and soil environments could be expected for avocado production. There will be increasing pressure to farm more sustainably, concentrate production in low-risk areas with good water balance, reduce carbon footprints and preserve soil organic matter as a major carbon sink.

References

Aguilera-Montanez, J. and Salazar-Garcia, S. (1991) The avocado industry in Michoacán, Mexico. *South African Avocado Growers' Association Yearbook* 14, 94–97.

Apaza, W., Quiroz, P. and Julca-Otiniano, A. (2019) Characterisation of avocado and asparagus farms in the Chavimochic irrigation project in La Libertad, Peru. *Peruvian Journal of Agronomy* 3(3), 91–103.

Arpia, M. (1994) Preharvest factors influencing postharvest quality of tropical and subtropical fruit. *HortScience* 29(9), 982–985.

Arpia, M. (2005) Avocado postharvest quality: an overview. In: *Session 6. Postharvest quality, outturn. New Zealand and Australia Avocado Grower's Conference '05.* Tauranga, New Zealand, p. 8.

Ben-Ya'acov, A. and Michelson, E. (1995) Avocado Rootstocks. *Horticultural Reviews* 17, 381–429.

Berdugo-Cely, J.A., Cortés, A.J., López-Hernández, F., Delgadillo-Duran, P., Ceron-Souza, I. *et al.* (2023) Pleistocene-dated genomic divergence of avocado trees supports cryptic diversity in the Colombian germplasm. *Tree Genetics & Genomes* 19:42.

Bergh, B. (1992) The origin, nature, and genetic improvement of the avocado. *California Avocado Society 1992 Yearbook* 76, 61–75.

Bost, J.B., Smith, N.J. and Crane, J.H. (2013) History, distribution and uses. In: Schaffer, B.A., Wolstenholme, B.N. and Whiley, A.W. (eds) *The Avocado: Botany, Production and Uses*. CABI International, Wallingford, UK, pp. 10–30.

Carvalho, C.P., Bernal-Estrada, J., Velásquez, M.A. and Cartagena-Valenzuela, J.R. (2015) Fatty acid content of avocados (*Persea americana* Mill. cv. Hass) in relation to orchard altitude and fruit maturity stage. *Agronomía Colombiana* 33(2), 220–227.

Charre-Medellín, J.F., Mas, J.F and Chang-Martínez, L.A. (2021) Potential expansion of Hass avocado cultivation under climate change scenarios threatens Mexican mountain ecosystems. *Crop and Pasture Science* 72(4), 291–301.

Chen, H., Morrell, P., Ashworth, V.E.T.M., de la Cruz, M. and Clegg, M. (2009) Tracing the geographic origins of major avocado cultivars. *Journal of Heredity* 100(1), 56–65. doi:10.1093/jhered/esn068.

Cruz-López, D.F., Caamal-Cauich, I., Pat-Fernández, V.G. and Reza Salgado, J. (2022) Competitiveness of Mexico's Hass avocado exports in the world market. *Revista Mexicana de Ciencias Agrícolas* 13(2), 355–362.

Deitch, M.J., Sapundjieff, M.J. and Feirer, S.T. (2017) Characterizing precipitation variability and trends in the world's Mediterranean-climate areas. *Water* 9(4): 259.

Durand, B.J. and Claassens, N.J.F. (1987) Root distribution of avocado trees in different soil types. *South African Avocado Growers' Association Yearbook* 10, 15–19.

Erazo-Mesa, E., Ramírez-Gil, J.G. and Sánchez, A.E. (2021) Avocado cv. Hass Needs Water Irrigation in Tropical Precipitation Regime: evidence from Colombia. *Water* 13(14): 1942.

Erazo-Mesa, E., Echeverri-Sánchez, A. and Ramírez-Gil, J.G. (2022) Advances in Hass avocado irrigation scheduling under digital agriculture approach. *Revista Colombiana de Ciencias Hortícolas* 16(1): e13456.

FAO (2023) FAOSTAT: food and agriculture data. [Crops and livestock products] Available at: http://www.fao.org/faostat/es/#home (accessed 5 January 2019).

Felles-Leandro, D. and García-Bendezú, S. (2022) Factores físicos y técnicos que influyen en el rendimiento del palto orgánico en Perú. *Chilean Journal of Agricultural & Animal Sciences* 38(3), 243–258.

Fick, S.E. and Hijmans, R.J. (2017) WorldClim 2: new 1-km spatial resolution climate surfaces for global land areas. *International Journal of Climatology* 37(12), 4302–4315.

Gaillard, J. and Godefroy, J. (1995) The avocado. In: Coste, R. (ed.) *The Tropical Agriculturist Series*. Education Ltd., London, p. 128.

Galindo-Tovar, M.E., Arzate-Fernández, A.M., Ogata-Aguilar, N. and Landero-Torres, I. (2007) The avocado (*Persea americana*, Lauraceae) crop in Mesoamerica: 10,000 years of history. *Harvard Papers in Botany* 12(2), 325–334.

Galindo-Tovar, M.E., Ogata-Aguilar, N. and Alzate-Fernández, A.M. (2008) Some aspects of avocado (*Persea americana* Mill.) diversity and domestication in Mesoamerica. *Genetic Resources and Crop Evolution* 55, 441–450.

Grüter, R., Trachsel, T., Laube, P. and Jaisli, I. (2022) Expected global suitability of coffee, cashew and avocado due to climate change. *PLS ONE* 17(1): e0261976.

Habibi, F., Liu, T., Shahid, M.A., Schaffer, B. and Sarkhosh, A. (2023) Physiological, biochemical, and molecular responses of fruit trees to root zone hypoxia. *Environmental and Experimental Botany* 206, 105179.

Holdridge, L. (1967) *Life Zone Ecology*. Tropical Science Center, San Jose, Costa Rica.

Howden, S.M. and Deuter, P. (2005) Climate change–risks and opportunities for the avocado industry. In: *New Zealand and Australia Avocado Grower's Conference '05.* Tauranga, New Zealand, pp. 1–19.

Knight, R.J. and Campbell, C.W. (1999) Ecological adaptation and the evolution of modern avocado cultivars. *Revista Chapingo Serie Horticultura* 5, 49–54.

Kourgialas, N.N. and Dokou, Z. (2021) Water management and salinity adaptation approaches of avocado trees: a review for hot-summer Mediterranean climate. *Agricultural Water Management* 252, 106923.

Lahav, E. and Trochoulias, T. (1982) The effect of temperature on growth and dry matter production of avocado plants. *Australian Journal of Agricultural Research* 33, 549–558.

Lionello, P., Malanotte-Rizzoli, P., Boscolo, R. and Xoplaki, E. (2006) The Mediterranean climate: an overview of the main characteristics and issues. In: Lionello, P., Malanotte-Rizzoli, P. and Boscolo, R. (eds) *Developments in Earth and Environmental Sciences*. Elsevier, Amsterdam, The Netherlands, pp. 1–26.

Lobell, D.B., Cahill, K.N. and Field, C.B. (2007) Historical effects of temperature and precipitation on California crop yields. *Climatic Change* 81(2), 187–203.

Lomas, J. (1988) An agrometeorological model for assessing the effect of heat stress during the flowering and early fruit set on avocado yields. *Journal of the American Society for Horticultural Science* 113, 172–176.

Lomas, J. (1992) Analysis of the effect of heat stress during flowering on the yield of avocado under Mediterranean climatic conditions. *Agricultural and Forest Meteorology* 59(3), 207–216.

Lomas, J. and Mandel, M. (1973) The quantitative effects of two methods of sprinkler irrigation on the microclimate of a mature avocado plantation. *Agricultural Meteorology* 12, 35–48.

Me, G.-T. and Arzate-Fernández, A.M. (2010) West Indian avocado: where did it originate? *Phyton (B Aires)* 79(1), 203–207. doi:10.32604/phyton.2010.79.203.

Moreno-Ortega, G., Pliego, C., Sarmiento, D., Barceló, A. and Martínez-Ferri, E. (2019) Yield and fruit quality of avocado trees under different regimes of water supply in the subtropical coast of Spain. *Agricultural Water Management* 221, 192–201.

Nakasone, H. and Paul, R. (1998) *Tropical Fruits*. CAB International, Wallingford, UK.

Pennington, P.T., Cronk, Q.C.B., Richardson, J.A., Woodward, F.I., Lomas, M.R. and Kelly, C.K. (2004) Global climate and the distribution of plant biomes. *Philosophical Transactions of the Royal Society of London. Series B: Biological Sciences* 359(1450), 1465–1476.

Qiao, H., Peterson, A.T., Campbell, L.P., Soberón, J., Ji, L. and Escobar, L.E. (2016) NicheA: creating virtual species and ecological niches in multivariate environmental scenarios. *Ecography* 39(8), 805–813.

Ramírez-Gil, J., Morales, J. and Peterson, A. (2018) Potential geography and productivity of 'Hass' avocado crops in Colombia estimated by ecological niche modeling. *Scientia Horticulturae* 237, 287–295.

Ramírez-Gil, J.G. and Morales, J.G. (2019) Polyphasic identification of preharvest pathologies and disorders in avocado cv. Hass. *Agronomía Colombiana* 37(3), 213–227.

Ramírez-Gil, J.G. and Morales-Osorio, J.G. (2018) Microbial dynamics in the soil and presence of the avocado wilt complex in plots cultivated with avocado cv. Hass under ENSO phenomena (El Niño—La Niña). *Scientia Horticulturae* 240, 273–280.

Ramírez-Gil, J.G. and Osorio, J.G.M. (2021) Source of inoculum of pathogens, the origin of disorders and diseases management in avocado nurseries. *Australasian Plant Pathology* 50(4), 457–468.

Ramírez-Gil, J.G., Cobos, M.E., Jiménez-García, D. and Peterson, A.T. (2019a) Current and potential future distributions of Hass avocados in the face of climate change across the Americas. *Crop and Pasture Science* 70(8), 694–708.

Ramírez-Gil, J.G., Franco, G. and Henao-Rojas, J.C. (2019b) Review of the concept of quality in Hass avocado and the pre-harvest and harvest factors that determine it under tropical conditions. *Revista Colombiana de Ciencias Hortícolas* 13(3), 359–370.

Ramírez-Gil, J.G., Henao-Rojas, J.C. and Morales-Osorio, J.G. (2020a) Mitigation of the Adverse Effects of the El Niño (El Niño, La Niña) Southern Oscillation (ENSO) Phenomenon and the Most Important Diseases in Avocado cv. Hass Crops. *Plants* 9(6): 790.

Ramírez-Gil, J.G., López, J.H. and Henao-Rojas, J.C. (2020b) Causes of Hass avocado fruit rejection in preharvest, harvest, and packinghouse: economic losses and associated variables. *Agronomy* 10(1): 8.

Ramírez-Gil, J.G., Castañeda-Sánchez, D. and Morales-Osorio, J.G. (2021) Edaphic factors associated with the development of avocado wilt complex and implementation of a GIS tool for risk visualization. *Scientia Horticulturae* 288, 110316.

Ramírez-Mejía, D., Levers, C. and Mas, J.F. (2022) Spatial patterns and determinants of avocado frontier dynamics in Mexico. *Regional Environmental Change* 22(1): 28.

Randela, M.Q. (2018) *Climate Change and Avocado Production: A Case Study of the Limpopo Province of South Africa*. University of Pretoria, Pretoria, South Africa.

Salazar-Garcia, S., Zamora-Cuevas, L. and Vega-Lopez, R. (2005) Update on the Avocado Industry of Michoacán, México. *California Avocado Society Yearbook* 87, 31–44.

Sanclemente, M., Schaffer, B., Gil, P., Vargas, A. and Davies, S.F. (2014) Pruning after flooding hastens recovery of flood-stressed avocado (*Persea americana* Mill.) trees. *Scientia Horticulturae* 169, 27–35.

Schaffer, B. (1998) Flooding responses and water-use efficiency of subtropical and tropical fruit trees in an environmentally-sensitive wetland. *Annals of Botany* 81(4), 475–481.

Sedgley, M. and Grant, J.R. (1983) Effect of low temperatures during flowering on floral cycle and pollen tube growth in nine avocado cultivars. *Scientia Horticulturae* 18, 207–213.

Selladurai, R. and Awachare, C.M. (2020) Nutrient management for avocado (*Persea americana* miller). *Journal of Plant Nutrition* 43(1), 138–147.

Silber, A., Naor, A., Cohen, H. and Assouline, S. (2018) Avocado fertilization: matching the periodic demand for nutrients. *Scientia Horticulturae* 241, 231–240.

Sommaruga, R. and Eldridge, H.M. (2021) Avocado production: water footprint and socio-economic implications. *EuroChoices* 20(2), 48–53.

Storey, W., Bergh, B. and Zentmyer, G.A. (1986) The origin, Indigenous range, and dissemination of the avocado. *California Avocado Society Yearbook* 70, 127–133.

Tamayo-Velez, A. and Osorio-Vega, N. (2022) Nutrición y fertilización. In: Jorge Alonso Bernal Estrada and Cipriano Arturo Díaz Díez AGROSAVIA (eds) *Actualización tecnológica y buenas prácticas agrícolas (BPA) en el cultivo de aguacate*. Catálogo Editorial Agrosavia (2020–2023). Bogota, Colombia, pp. 411–465.

Title, P.O. and Bemmels, J.B. (2018) ENVIREM: an expanded set of bioclimatic and topographic variables increases flexibility and improves performance of ecological niche modeling. *Ecography* 41(2), 291–307.

Valoras, N., Letey, J., Stolzy, L. and Frolich, E. (1964) The oxygen requirements for root growth of three avocado varieties. *Proceedings of the American Society for Horticultural Science* 85, 172–178.

Vos, J. and Marshall, A. (2017) Conquering the desert: drip irrigation in the Chavimochic system in Peru. In: Venot, J.-P., Kuper, M., and Zwarteveen, M. (eds) *Drip Irrigation for Agriculture. Corporate.* Taylor & Francis Group. Routledge, London, pp. 134–152.

Wei, X., Garcia-Chevesich, P., Alejo, F. and McCray, J.E. (2021) Hydrologic analysis of an intensively irrigated area in Southern Peru using a crop-field scale framework. *Water* 13(3): 318.

Whiley, A.W. and Winston, E.C. (1987) Effect of temperature on varietal productivity in some avocado-growing areas of Australia. *South African Avocado Growers' Association Yearbook* 10, 45–47.

Wolstenholme, B.N. (2013) Ecology: climate and soils. In: Schaffer, B., Wolstenholme, B.N. and Whiley, A.W. (eds) *The Avocado: Botany, Production and Uses*. CAB International, Wallingford, UK, pp. 86–117.

Wolstenholme, B.N. and Whiley, A. (1999) Ecophysiology of the avocado (*Persea americana* Mill.) tree as a basis for pre-harvest management. *Revista Chapingo Serie Horticultura* 5, 77–88.

Yin, M., Vargas, A., Fuentealba, C., Shahid, M., Bassil, E. and Schaffer, B. (2023) Differences in physiological and biochemical responses to short-term flooding among the three avocado (*Persea americana* Mill.) races. *Plant Physiology and Biochemistry* 196, 925–939.

Zhang, T., Hoell, A., Perlwitz, J. and Hamill, T.M. (2019) Towards probabilistic multivariate ENSO monitoring. *Geophysical Research Letters* 46, 10532–10540.

6 Reproductive Biology

J.I. Hormaza[1]*, L.C. Garner[2] and C.J. Lovatt[3]
[1]*Instituto de Hortofruticultura Subtropical y Mediterránea La Mayora, Consejo Superior de Investigaciones Científicas – Universidad de Málaga (IHSM la Mayora – CSIC-UMA), Málaga, SPAIN;* [2]*Plant Sciences Department, California Polytechnic State University, San Luis Obispo, California;* [3]*Department of Botany and Plant Sciences-072, University of California, Riverside, California*

Abstract

This chapter offers a comprehensive overview of the reproductive biology of avocado trees. The avocado flowering and fruit development processes are discussed in detail, including all endogenous and exogenous factors that affect them.

Introduction

Reproductive plasticity enables sessile plants to adapt to rapid changes in their environment. For *Persea americana* (Mill.), reproductive plasticity is a critical factor in the survival of the species in nature as well as the success of commercial avocado production under changing environmental conditions. In-depth knowledge of the effects of temperature and other environmental factors on key stages of avocado reproductive biology is essential for increasing production in existing orchards and for selecting appropriate sites for new orchards. This will be particularly important as part of the global expansion of avocado production required to successfully meet current and projected increases in consumer demand for avocados as both fresh fruit and value-added products.

With global changes in climate, avocado production areas worldwide have been negatively impacted by droughts of greater duration, record flooding, increased frequency of winter prebloom freezes and excessively high or low temperatures during fruit set. In many instances, the loss of crop in response to an adverse climate event initiates alternate bearing, the production of a low 'off'-crop yield followed by a high 'on'-crop yield 1–2 years later, depending upon how quickly the trees recover from the initial stress. Once initiated, alternating 'off'-/'on'-yield cycles become entrained by endogenous tree factors that ultimately impact floral intensity and, hence, yield. Alternate bearing has become a problem of significant negative economic consequence. Research on alternate bearing, under way in several avocado-producing countries, is providing fundamental information about the internal and external signals regulating floral development in avocado in relation to tree phenology, fruit development and crop load (number of fruit per tree) across the varied climatic conditions under which avocado fruit are commercially produced. It is anticipated that the results of this research will lead to practical solutions to the problem of alternate bearing and new strategies

*Corresponding author: ihormaza@eelm.csic.es

 The Avocado: Botany, Production and Uses (eds D. Carrillo, B. Schaffer, A.W. Whiley and B.N. Wolstenholme)

DOI: 10.1079/9781800621824.0006

for mitigating the negative effects of adverse climate events on reproductive development, growth and productivity.

In many countries, climate change and population growth increasingly jeopardize the availability and cost of arable land and high-quality irrigation water. Fertilizer, pesticide and labour prices are steadily increasing worldwide. Ever-increasing taxes and indirect expenses associated with new regulations further raise the cost of producing avocados. To increase net profit, it is imperative that growers reduce the cost of inputs and/or increase yield per hectare of fruit of commercially valuable sizes. Knowledge of avocado reproductive biology is fundamental to increasing flowering, fruit set, fruit size and yield and for mitigating alternate bearing to increase grower income and the sustainability of avocado commodity-based industries worldwide.

Flowering

Avocado trees grown from seeds are the product of sexual reproduction and, thus, are not genetically identical to either parent tree (i.e. they are not clones). They are instead an unknown mixture of traits from each parent. Moreover, avocado trees grown from seeds have a juvenile phase that can exceed 10 years before the seedling trees flower. *P. americana*, like all vascular plants, has two distinct developmental phases – juvenile and adult. During the juvenile phase, plants are restricted to vegetative growth. Juvenile plants are incompetent with regard to sexual reproduction and unable to respond to floral inductive signals that initiate floral development. Whereas breeding programs have produced unique progeny from seeds that flower and set fruit in dramatically shorter periods of time, for example, progeny of 'Pinkerton' and 'Gwen' flower and set fruit within 2 years of seed planting (Bergh and Lahav, 1996), identification of the cues that trigger phase transition from juvenile to adult in avocado remains unknown. However, based on the advances made in understanding the regulation of phase transition in the model plant *Arabidopsis thaliana*, Ahsan *et al.* (2019) have identified genes that are uniquely expressed in the juvenile and adult phases of *P. americana* that could potentially serve as markers to screen for factors regulating phase transition. In *P. americana*, as previously demonstrated in *A. thaliana* and other plant species, two microRNAs, miR156 and miR172, were associated with the regulation of phase transition (Ahsan *et al.*, 2019). The juvenile phase was marked by high levels of miR156, which decreased as the plant aged, resulting in the upregulation of miR172 transcription, which steadily increased in the adult phase. Expression of miR172 was associated with the activation of *APETALA1* (*AP1*) (Ahsan *et al.*, 2019), a floral meristem identity gene regulating bud determinacy. *AP1* is also a floral organ identity gene expressed in the two whorls of tepals and the stamens of the avocado flower (Chanderbali *et al.*, 2006). Ahsan *et al.* (2019) proposed that miR156 and *AP1* could serve as markers for distinguishing the juvenile and adult phases, respectively.

As in most fruit tree crops, in the establishment of commercial orchards, the problem of a prolonged juvenile period is circumvented by planting trees propagated by grafting a scion bud from an adult tree onto a seedling or clonal rootstock. Such trees are fully responsive to internal and external signals that initiate floral development and are capable of flowering and producing fruit after a short vegetative adult phase of 2–5 years, depending upon the cultivar. In the adult avocado, flowering requires multiple steps. The critical first step is floral induction (floral initiation) and transition of the vegetative shoot apical meristem (SAM) to an inflorescence meristem, which produces new lateral primordia that must acquire the identity of floral (flower) meristems (Benlloch *et al.*, 2007). Genetic control of the induction process is complex, resulting in an inflorescence meristem, formation of flower meristems within the inflorescence, initiation of floral organ primordia, floral organ specification and development of individual flowers (Ma, 1994). For avocado, our understanding is evolving regarding the regulation of the annual transition of the SAM from vegetative to floral development, resulting in inflorescence and flower formation in apical and axillary (lateral) buds along a vegetative shoot. Based on advances in the identification of the genes that regulate floral development across many plant species (Benlloch *et al.*, 2007), several key genes regulating floral development in *P. americana* have been identified (Chanderbali *et al.*, 2006,

2009; Ziv *et al.*, 2014; Acosta-Rangel *et al.*, 2021). However, information on the environmental, physiological and genetic regulation of floral development in avocado is lagging relative to that of herbaceous annuals and other woody perennials, such as apple (*Malus domestica*), sweet orange (*Citrus sinensis*) and common mandarin (*C. reticulata*).

Environmental factors that promote flowering

Floral induction resulting in the initiation of floral development is either autonomous (in response to endogenous developmental events) or regulated by environmental cues (Takeno, 2016). Environmental factors documented to induce flowering across plant species include, but are not limited to, long- and short-day length (photoperiod), high or low light intensity (LLI), shading, UV light, high and low soil moisture (LSM) (i.e. hypoxia and water deficit, respectively), or high temperature and low temperature (LT), and nutrition (Takeno, 2016). LT is used to describe both chilling (>7°C but <20°C) and freezing (<0°C) temperatures (Ritonga and Chen, 2020). In this chapter, LT refers exclusively to temperatures within the range noted for chilling. In some cases, interactions between environmental cues and endogenous developmental processes are necessary to initiate flowering, or at a minimum, the interaction influences the timing and intensity of flowering. Different environmental factors are perceived by different parts of the plant. Photoperiod, irradiance, UV light and shading are perceived mainly by mature leaves. Water availability is thought to be perceived by the root system (Davies and Zhang, 1991); however, a leaf-derived signal cannot be ruled out in the case of water-deficit stress (Chica and Albrigo, 2013b). Temperature is perceived by all plant parts, although LT is mainly perceived by the SAM (Bernier *et al.*, 1993). Since the different factors that promote flowering are perceived by different parts of the plant, it is implied that the fate of the SAM is controlled by an array of long-distance signals that are integrated within the plant to initiate a common sequence of events leading to floral determinacy, the irreversible commitment of the SAM to floral development (irreversible commitment to flowering, ICF) (Bernier *et al.*, 1993). Thus, whether a developmental event and/or an environmental cue is perceived by the leaf, the root or the bud, it must ultimately initiate the sequence of genetic events that take place within the bud to elicit determinacy.

Among tropical and subtropical tree crops, LT (>7°C but <20°C) and water deficit are both known to induce flowering in *Citrus* spp. (Southwick and Davenport, 1986; Lovatt *et al.*, 1988; Nakajima *et al.*, 1992; Chica and Albrigo, 2013a,b; Tang and Lovatt, 2022), mango (*Mangifera indica*) (Núñez-Elisea and Davenport, 1994) and litchi (*Litchi chinensis*) (Stern *et al.*, 1993; Shen *et al.*, 2016). In contrast, LT, but not LSM, results in flowering in longan (*Dimocarpus longan*) (Chen *et al.*, 2010), whereas water deficit promotes significant flowering in star fruit (*Averrhoa carambola*) (Salakpetch *et al.*, 1990). Similarly, LTs within the chilling range have been documented to induce flowering in *P. americana*, cv. 'Hass' and 'Fuerte' (Mexican × Guatemalan hybrids) (Buttrose and Alexander, 1978; Nevin and Lovatt, 1989; Chaikiattiyos *et al.*, 1994; Salazar-García *et al.*, 1999; Acosta-Rangel *et al.*, 2021). In contrast, research into the capacity of other environmental factors to induce flowering in avocado cultivars, including 'Hass', the cultivar that dominates the global avocado industry, is limited, and the results are frequently equivocal.

However, in the case of using water-deficit stress to promote avocado flowering, the results, although limited, are consistent. Neither cyclic nor even constant water-deficit stress (predawn Ψ_{leaf} decreased to −2.1 MPa in the constant-stress treatment) for 12 weeks was effective in stimulating flowering in 'Hass' avocado scions on an unknown rootstock (Chaikiattiyos *et al.*, 1994). Water deficit halted the vegetative shoot extension growth, a putative prerequisite for the transition of the SAM from vegetative to reproductive development, but the trees failed to flower subsequently. Moreover, the study documented the sensitivity of avocado to water-deficit stress. Trees subjected to water-deficit stress exhibited severe leaf drop, browning of shoot tips and necrosis of the leaf tips and leaf margins (Chaikiattiyos *et al.*, 1994). These results are consistent with more recent research. 'Hass' avocado trees grafted on clonal 'Duke 7' rootstock were maintained for 5 months under optimal growth conditions (OGCs) (14-h day, using supplemental lighting, with photosynthetic active radiation [PAR] > 900 μmol m^{-2} s^{-1} at 30°C and a 10-h night at 20°C) at a soil volumetric water content (VWC) of 20% to 25%. The trees were subsequently subjected to reduced soil VWC of 8% to 12% for 6 weeks, which significantly reduced leaf mid-day water potentials from >–1.0 MPa at the start of the experiment to ≤–2.0 MPa during the 6-week period. After 4 weeks of treatment, symptoms of water-deficit stress appeared on the developing shoots and young leaves, including shoot-tip browning and leaf necrosis. Mature shoots were not damaged but browning and abscission of older leaves occurred. The trees did not flower (Acosta-Rangel *et al.*, 2021). Although not exhaustive, the results of these two studies taken together are consistent in documenting the sensitivity of avocado to moderate water-deficit stress and in suggesting that, unlike *Citrus* spp., water-deficit stress cannot be used to replace LT to induce flowering in *P. americana*.

Davenport (1986) proposed that for avocado cultivars better adapted to warmer climates (West Indian–Guatemalan race hybrids), day length might play a more important role in inducing flowering than LT. Later studies on 'Choquette' and 'Booth-8', both West Indian–Guatemalan hybrids, demonstrated that LTs in the chilling range (≤20°C) induced flowering in the semi-warm climate of Nayarit, Mexico (21°75′N, 104°84′ W) (Cossio-Vargas *et al.*, 2007; Salazar-García *et al.*, 2008). Under controlled environmental conditions in a growth chamber, short day length combined with LT stimulated flowering in 'Fuerte' avocado (Buttrose and Alexander, 1978). The study confirmed that the absence of high temperatures (e.g. >20°C) was responsible for inflorescence initiation under a short photoperiod. In the study, the effect of short day length was not qualitatively associated with floral initiation, but it decreased the time to flowering. Under controlled environmental conditions, reducing light intensity by 50% in combination with LT did not significantly affect flower numbers compared to LT alone (Buttrose and Alexander, 1978). Under conditions characterized by long days and warm temperatures (WT) (14-h day at 30°C/10-h night at 20°C), reducing PAR from 900 μmol m^{-2} s^{-1} by 85% to <130 μmol m^{-2} s^{-1} using black net shade cloth did not induce flowering in 'Hass' avocado trees (Acosta-Rangel *et al.*, 2021).

Responses of 'Hass' avocado trees to day length or light intensity imposed under controlled environmental conditions in a growth chamber, as described above, appear to be different from the responses attained in nature. For example, in Gainesville, Florida, USA (29°39′N, 82°32′ W), LT with LLI was observed to delay and concentrate 'Hass' flowering on warm sunny days (Buzgo *et al.*, 2007). In San Vicente de Cañete, Lima, Peru (13°05′ S, 76°23′ W), coastal fog during winter-spring reduces the period of full sun to less than 3 h per day, resulting in an intense and extended blooming period, composed mostly of determinate axillary inflorescences. In contrast, in many avocado-producing regions of Mexico, such as Uruapan in Michoacán (19°25′N, 102°30′W) and in California, USA, for example, Corona (32°52′ N, 117°33′ W) and Cambria (35°37′ N, 121°40′ W), the common type of inflorescence is indeterminate, possibly the result of longer day lengths and/or greater light intensity.

The cv. 'Hass' dominates avocado production worldwide, and new orchards are being established in areas with no prior history of *P. americana* as a native species or cultivated crop. Environmental conditions that ensure flowering might prove challenging in these new areas. Data from existing avocado-producing areas indicate that a decrease in temperature is associated with the successful annual transition from vegetative growth to flowering of 'Hass' at different latitudes. Differences in annual temperature patterns between the northern and southern hemispheres are evident (Fig. 6.1). Monthly maximum temperatures in the northern hemisphere range from 17.3°C (San Luis Obispo, California; 35°16′ N, 120°39′ W) to 32.1°C (Peribán, Michoacán; 19°52′ N, 102°41′ W). Average minimum temperatures range from 5°C to less than 15°C, except for Xalisco, Nayarit (21°45′N, 104°89′ W), which has higher minimum temperatures in the summer (July–September) (Fig. 6.1a). For the southern hemisphere, the average maximum temperature ranges from 15.2°C (Pemberton, Western Australia; 34°44′ S, 116°03′ E) to 33.8°C (Renmark, South Australia; 34°17′ S, 140°74′E) (Fig. 6.1b), a range that is very similar to that of the northern hemisphere. Average minimum temperatures in the southern hemisphere range from 3.8°C (Renmark, South Australia) to 21°C (Tucumán, Argentina; 26°56′ S,

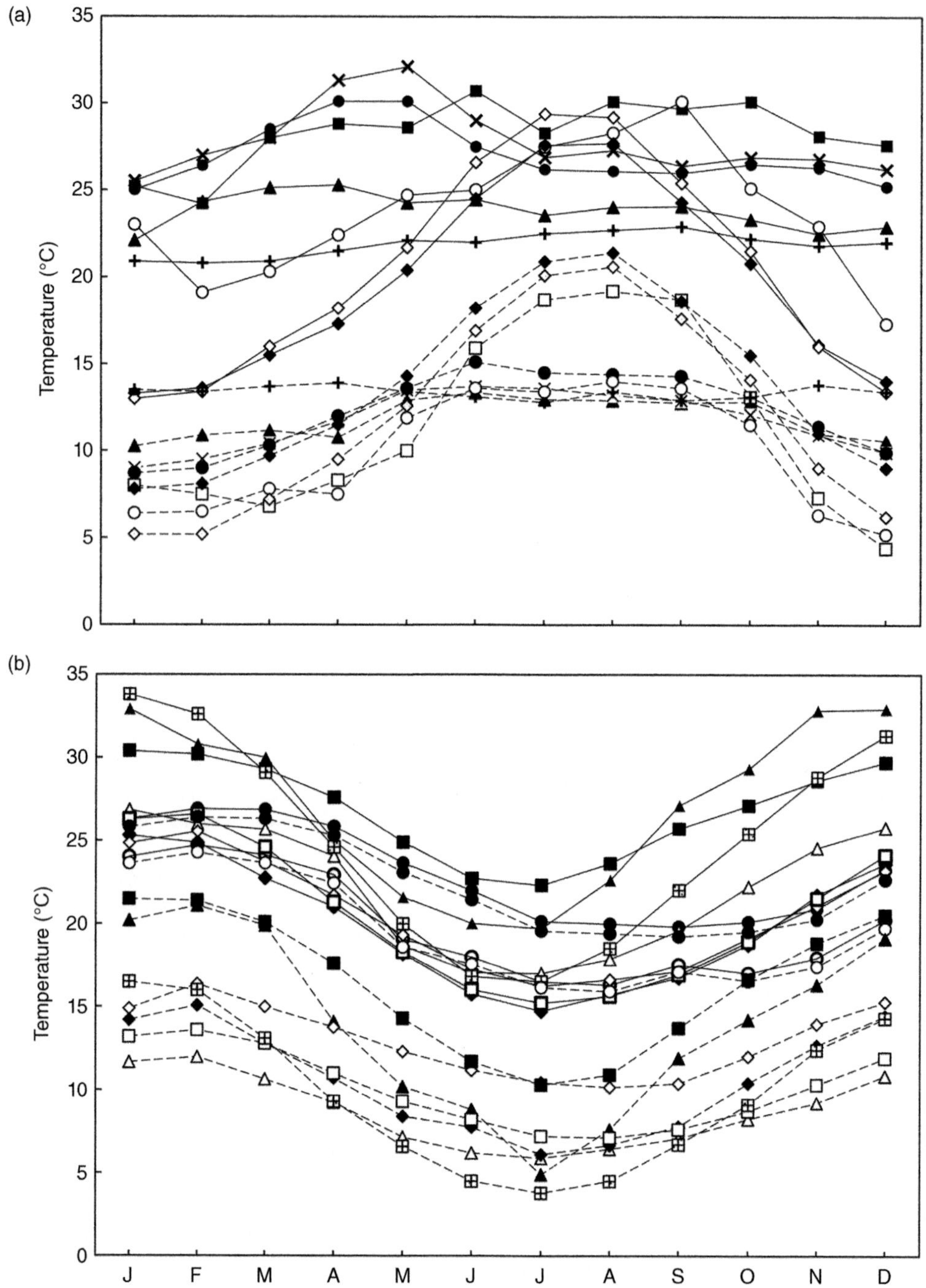

Fig. 6.1. Average maximum and minimum monthly temperatures in several 'Hass' avocado-producing regions in the northern (**a**) and southern (**b**) hemispheres. Panel (**a**): ▲ Zona de los Santos, CR; ✕ Peribán, Michoacán, MEX; ● Uruapan, Michoacán, MEX; ■ Xalisco, Nayarit, MEX; + Rionegro, Antioqía, COL; ○ San Luis Obispo, California, USA; ◆ Algarrobo, Málaga, SP; ◇ Callosa d'en Sarrià, Alicante, SP. Panel (**b**): ● Piura, Piura, PE; ○ Virú, La Libertad, PE; ▲ Tucumán, ARG; △ Quilota, CL; ■ Bunderberg, Queensland, AU; □ Pemberton, Western AU; ⊞ Renmark, South AU; ◆ Te Puke, Bay of Plentry, NZ; ◇ Kaitaia, Far North, NZ. (Data contributed by S. Salazar-García, Mexico; I. Hormaza, Spain; J. Dixon, New Zealand; Z. Jalata, Ethiopia; S. Newett, Australia; and A. Rebolledo-Roa, Colombia.)

65°17′ W). There are two avocado-producing regions in north-west Peru, Piura (5°12′ S, 80°37′W) and Virú (8°28′S, 78°45′W), which have 'unusually' WTs for 'Hass' trees. In these regions, monthly averages for maximum and minimum temperatures can vary less than 1°C. The monthly averages for minimum and maximum temperatures throughout the year fluctuate from 19.2°C to 26.9°C in Piura and 15.9°C to 24.7°C in Virú (Fig. 6.1b). Avocado, predominantly the 'Hass' cultivar, is commercially produced in two major areas of Ethiopia (8° N, 38° E). The larger avocado-producing area is the *Woyna Dega*, a cool subhumid mid-highland zone (1200–2300 meters above sea level [m.a.s.l]), having average annual temperatures ranging from 16.0°C to 20°C and moderate-to-high rainfall of 800 to 1200 mm. In sharp contrast, the second-largest avocado-growing area is the hot semi-arid lowland *Kolla* zone (500–1500 m.a.s.l.), located in the geographic peripheries of the south, south-east, west and north-eastern parts of Ethiopia. This zone has average annual temperatures between 20°C and 30°C and only 200 to 800 mm of annual rainfall (Z. Jalata, Ethiopia, 2023, personal communication). Research on the regulation of floral development of 'Hass' under the unique environmental conditions reported herein for parts of Peru and Ethiopia could provide valuable information to expand our current understanding of the plasticity of avocado reproductive biology.

Nevin and Lovatt (1989) used LT stress to promote flowering in 'Hass' avocado. They subjected trees to an LT treatment of 8-h days (500 μmol m^{-2} s^{-1}) at 15°C to 18°C/16-h nights at 10°C to 13°C. Flowering intensity was the same for 'Hass' trees receiving 4 or 8 weeks of LT treatment, and inflorescence development followed the same timetable with the exception of flower opening, which was 1 week later in trees receiving 8 weeks of LT treatment. This study and a later one carried out by Salazar-García *et al.* (1999) under controlled conditions confirmed that 'Hass' requires no more than 4 weeks of LT to induce flowering. Similarly, a 2-year study done in commercial 'Hass' orchards in the semi-warm climate of Nayarit, Mexico, confirmed that accumulation of 27.5 days at temperatures ≤19°C resulted in flowering of apical buds on winter- and summer-flush shoots (Salazar-García *et al.*, 2006). This amount of 'chilling' was equivalent to the 4-week LT treatment documented to induce flowering of 'Hass' under controlled conditions in growth chambers (Nevin and Lovatt, 1989; Salazar-García *et al.*, 1999). The basic information provided above is useful in selecting sites for new orchards but leaves open many unanswered questions related to the accumulation of 'chilling days' in an era of climate change. For example, does the interruption of the LT period with high temperatures (>20°C) before buds are determined (ICF) negate some of the 'chilling days' already accumulated or merely stop the process until LTs resume? Note that the results of Nevin and Lovatt (1989) and Salazar-García *et al.* (1999) clearly demonstrated that once buds reached ICF in response to 4 weeks of LT, transferring the trees to warm day/night temperatures of 24°C/19°C or 25°C/20°C had no negative effect on inflorescence or flower development or inflorescence number at bloom.

Transition from vegetative to reproductive growth

Annual phase transition from vegetative shoot growth to reproductive (inflorescence and flower) development is a critical first step in avocado fruit production. Salazar-García *et al.* (1998) were first to document this transition period in the primary axis meristem (PAM) of avocado buds. The research was conducted using apical buds on summer-flush shoots of 10-year-old 'Hass' avocado trees in a commercial orchard in southern California. Close to the end of the extension of the summer flush of vegetative shoots (23 July), apical buds had an external (macroscopic) pointed shape and an internal (microscopic, anatomical) convex PAM and one or two young secondary-axis inflorescence meristems in the axils of bracts. By the end of July, the buds were pointed and closed. At this stage, a low-convex PAM with separated bracts was evident in addition to one to three secondary-axis inflorescence meristems. By 30 August, bud scales started to abscise. The PAM was convex again, and four secondary-axis inflorescence meristems were present in the axils of the inflorescence bracts. A decreased rate of secondary-axis inflorescence meristem production was observed at this stage. Salazar-García *et al.* (1998) noted that the presence of a convex PAM indicated an active apex, producing either inflorescence bracts or leaf primordia. Secondary-axis inflorescence meristems form in the axils of bracts, just below the PAM. These axillary meristems typically form inflorescences. But, as reported for other species (Bernier *et al.*, 1981a), flowering in avocado is not obligatory at this point but depends on environmental conditions. After initiation of these secondary-axis meristems, the PAM becomes flattened and its activity decreases, but at no time is dormancy detected.

Flattening of the PAM during this period is associated with the potential for a full transition from vegetative to reproductive development. A flattened meristem is associated with the transition to the reproductive condition in many species (Bernier *et al.*, 1981a). If environmental conditions promoting vegetative growth occur, the growth of the two apical secondary-axis inflorescence meristems formed prior to the transition phase is suppressed by the production of new leaf primordia on the PAM. However, under optimal conditions for flowering, these meristems develop into the basal lateral cymes of the inflorescence, with additional cymes produced by new activity of the PAM (Salazar-García *et al.*, 1998). The period of transition from the vegetative to the reproductive stage was identified as S-1 (convex PAM), S-2 (low-convex PAM) and S-3 (convex PAM) of the macro- and microscopic scales designed to assess floral development in 'Hass' (Fig. 6.2) (Salazar-García *et al.*, 1998). The subsequent formation of additional secondary-axis inflorescence meristems is consistent with irreversible commitment of the PAM of the bud to floral development.

In 'Hass' avocado, the transition phase occurs early in the process of shoot development. It may take from 2 weeks to more than 4 months (Salazar-García *et al.*, 1998; Salazar-García *et al.*, 2006; Rocha-Arroyo *et al.*, 2010). In addition to orchard management practices, geographical factors such as latitude and climatic conditions affect the time of emergence of vegetative flushes, which in turn affects when the transition phase occurs. Table 6.1 illustrates the periods of occurrence of vegetative flushes in several avocado-producing regions.

Cells with vertical lines indicate the period of major vegetative shoot growth in winter or spring (usually during or after bloom) or in summer. Cells with horizontal lines or dots indicate the period of secondary and tertiary vegetative shoot flushes, respectively. Data contributed by S. Salazar-García, Mexico; I. Hormaza, Spain; J. Dixon, New Zealand; Z. Jalata, Ethiopia; S. Newett, Australia; and A. Rebolledo-Roa, Colombia.

In addition, the genetic characteristics of different avocado cultivars influence when shoots undergo phase transition. There is evidence that the transition phase might take longer for cultivars of West Indian–Guatemalan-race hybrids, which are more suited to warm climates, than for 'Hass'. For example, under the semi-warm climate of Nayarit, Mexico (21°28′ N, 104°84′ W; 440 m.a.s.l.), spring shoots of 'Choquette' required 6 months (April–October) and summer shoots 3 months (July–October) to reach the transition phase. Similarly, summer shoots of 'Booth-8' needed 4 months (July–November) to initiate annual phase transition (Cossio-Vargas, *et al.*, 2008a; Salazar-García *et al.*, 2008).

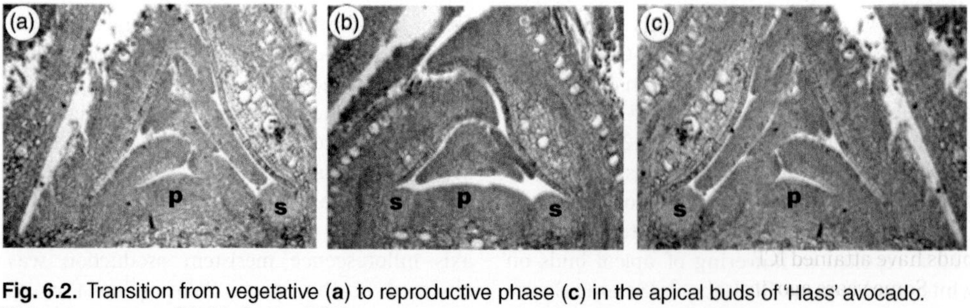

Fig. 6.2. Transition from vegetative (**a**) to reproductive phase (**c**) in the apical buds of 'Hass' avocado. Note a low-convex PAM in **b**. *P* = PAM; s = secondary-axis inflorescence meristem.

Table 6.1. Time of emergence of vegetative flushes in several 'Hass' avocado-producing regions located at different latitudes.

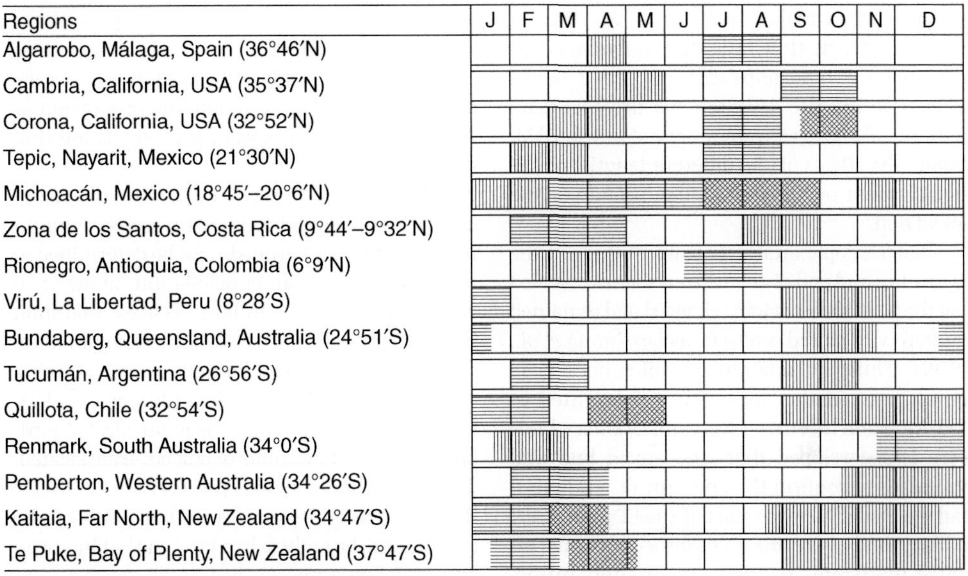

Commitment to flowering

Determinacy, ICF, is the point after the annual transition of the bud from vegetative to reproductive development, when the process is no longer reversible, and the PAM is committed to following the floral development pathway and flowering (McDaniel, 1994). In the avocado, only some apices flower; others continue the vegetative growth of the shoots; thus, the PAM plays two roles, one is to produce inflorescence bracts and the other is to produce leaf primordia.

The study of early stages of avocado flowering is critical to the development of orchard management strategies to increase floral intensity or, alternatively, to promote vegetative shoot growth and decrease flowering. For such strategies to be successful, it is essential to know the time when shoot apical buds reach ICF. To prevent an inadvertent increase in vegetative shoot growth at the expense of flowering, cultural practices that can prevent flowering, such as nitrogen fertilization, canopy pruning (hedging, topping) or application of some plant growth regulators (PGRs), for example, gibberellic acid (GA_3), must be carried out after the majority of buds have attained ICF.

Some years ago, it was assumed that a PAM had reached commitment to flowering when the two apical bracts (microscopic level) were extended over the secondary-axis inflorescence meristem, which is indicative of the initiation of the terminal flower in the triplet (Thorp *et al.*, 1994). However, the results of Salazar-García *et al.* (1998) showed that this stage occurs very late in the development of the avocado inflorescence, whereas commitment to flowering is an early event.

Temperature is a factor known to inhibit (WT) or promote (LT) floral initiation in avocado, which can be used to identify an anatomical change associated with the commitment to flowering. Container-grown 'Hass' trees were exposed to LT treatments of 10°C/7°C (day/night) to stimulate flowering and WT conditions of 25°C/20°C (day/night) to inhibit the process (Salazar-García *et al.*, 1999). This controlled-environment study showed that control trees maintained under WT conditions and trees receiving 1 or 2 weeks of LT treatment did not flower; all growth was vegetative. Four weeks at 10°C/7°C (10-h day/14-h night) were sufficient for apical buds of 'Hass' to become committed to flowering; 83% of the buds produced inflorescences and no vegetative shoot growth, with the remaining buds remaining quiescent (inactive, producing neither vegetative nor floral shoots). Anatomical sections of apical buds at the start of the experiment revealed a convex PAM that had produced a secondary-axis inflorescence meristem in the axil of each of the two inflorescence bracts. At that time, the secondary-axis inflorescence meristems did not have apical bracts. During the 4-week exposure to LT, no anatomical changes were observed in the PAM, despite the effects of the different lengths of the LT treatment on the developmental fate of the apical bud.

Macroscopic analysis of apical buds at the beginning and end of the LT treatments revealed that the buds had not visibly changed and remained pointed with closed scales (Salazar-García *et al.*, 1999). Thus, no anatomical differences were found between the apical buds committed to flowering and those that were not. Axillary buds were less developed than the apical bud and were used to confirm the anatomy of a bud that had reached ICF. At the start of the LT treatment, axillary buds bore no secondary-axis inflorescence meristems in the axils of their inflorescence bracts, whereas apical buds had two secondary-axis inflorescence meristems (apical bracts were absent). After 4 weeks of LT treatment, the axillary buds had produced one pair of secondary-axis inflorescence meristems (with no apical bracts). Thus, after 4 weeks of LT treatment, the axillary buds had caught up with the apical buds anatomically and developmentally, and they were now also committed to flowering. Hence, based on their anatomy, it is possible to distinguish axillary buds fully committed to flowering from those that are not by the presence or absence of a single pair of secondary-axis inflorescence meristems (no apical bracts). Henceforth, the criteria used to facilitate the identification of the apical buds that had attained ICF were the presence of a third, or fourth, secondary-axis inflorescence meristem (without apical bracts).

Differences in avocado cultivar, latitude, climate and shoot age are factors that influence tree phenology and affect the time and stage of development when apical buds become committed to flowering. In a 2-year study in Corona, California, apical buds of summer shoots of 'Hass' became committed to flowering by 15 October (Salazar-García *et al.*, 1998). Partial senescence, abscission and separation of the outermost bud scales of apical and axillary buds were the visual cues associated with the stage of inflorescence development when three or more secondary-axis inflorescence meristems were formed (Fig. 6.3). These visual cues correspond to Stage 3 of the visual scale developed by Salazar-García *et al.* (1998) (Fig. 6.4) and can be used in commercial avocado orchards to estimate the proportion of buds committed to flowering versus those that are not.

The stage of tree phenology and the calendar date when buds become irreversibly committed to flowering in avocado can be determined in an orchard using shoot defoliation, alone or in combination with girdling. Leaves maintain buds in a paradormant condition. Defoliation removes this growth-inhibiting factor, forcing rapid bud growth, hence, its utility as a tool for imposing timed responses. The main effect of girdling is to prevent the influx of phloem-translocated, floral-stimulating signals (Bernier *et al.*, 1981b; Bernier, 1988) produced in other parts of the tree into the branches. In a branch that has been both defoliated and girdled, buds are forced to grow and express their developmental state without the influence of incoming developmental signals. Thus, defoliation and girdling can serve as convenient tools to determine the time of ICF in avocado buds under field conditions. If flowering occurs on defoliated and girdled shoots, buds would have been irreversibly committed to flowering at the time of treatment. In that case, buds were capable of initiating floral morphogenesis using their own 'developmental program', which was upregulated prior to treatment. Factors that influence when buds become

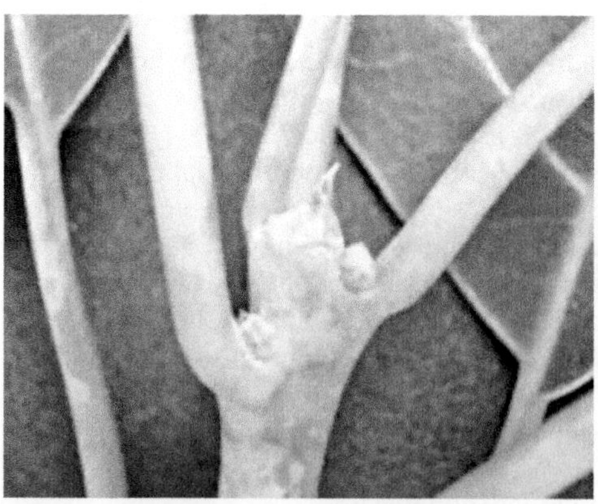

Fig. 6.3. External characteristics of 'Hass' avocado buds irreversibly committed to flowering.

Fig. 6.4. Visual scale to evaluate avocado floral development.

irreversibly committed to flowering have been identified in field studies employing shoot defoliation combined with girdling and are presented in the series of paragraphs that follow.

A shoot defoliation study conducted at two locations in California (Irvine and Somis) showed differences in the date that fall-flush shoots emerged and reached ICF. In the Irvine orchard (33°41′ N, 117°49′ W) (south of Los Angeles) fall shoots emerged 1 month earlier than in Somis (34°15′ N, 118°59′ N) (north of Los Angeles). However, in Irvine, apical buds needed 45 days more to have 30% of the buds committed to flowering, compared to Somis, which required only 15 days (Table 6.2). The lower night temperatures in Somis (annual average maximum/minimum temperature = 24.2°C/7.3°C) compared to those in Irvine (annual average maximum/minimum temperature = 24.2°C/10.3°C) could logically explain these differences (Rocha-Arroyo *et al.*, 2010). Buds that had reached ICF in both locations of California were at Stage 3 of the visual scale of Salazar-García *et al.* (1998).

Shoot defoliation and girdling were used to determine the time when apical buds, borne on the winter and summer shoots of 'Hass' avocado trees in nonirrigated orchards under the subhumid semi-warm climate of Nayarit, Mexico (annual average maximum/minimum temperature = 28.7°C/14.3°C) (21°28′ N, 104°84′ W; 440 m.a.s.l.), reached ICF (Salazar-García *et al.*, 2006). Both types of shoots were defoliated (Year 1) or defoliated and girdled (Year 2) at different stages of bud development from September to January in each case. Irrespective of the time of treatment or shoot age, ICF of apical buds occurred by 15 October, and in both years, this stage was associated with an average of 27.5 days at temperatures ≤19°C. According to the visual scale designed by Salazar-García *et al.* (1998), buds irreversibly committed to flowering were at Stage 3, which corresponds to closed and pointed buds with partial senescence of bud scales.

Shoot defoliation and girdling were used to determine the influence of climate, irrigation and shoot age (winter, spring and summer flushes) on the date apical buds of 'Hass' avocado trees reached ICF in Michoacán, Mexico (Rocha-Arroyo *et al.*, 2010). Climate influenced the date of ICF on shoots from the three flushes studied, occurring earlier in cooler climates (Table 6.3). Winter shoots (oldest) required more time after emerging to achieve ICF, compared to spring and summer shoots (youngest). Shoots from the three vegetative flushes attained ICF 29 to 42 days earlier in nonirrigated (rainfed)

Table 6.2. Type of growth produced by fall shoots of 'Hass' avocado trees in response to shoot defoliation in two orchards in California. Evaluations were done at full bloom: Irvine (17 April) and Somis (2 April). (S. Salazar-García and C.J. Lovatt, 2010, unpublished results.)

Location	Defoliation date (2003)	Bud development[y]	Type of growth (% of total shoots)[z]		
			Floral[x]	Vegetative	Inactive
Somis, Ventura County	22 October	1.0	0 c[w]	100 a	0 b
(34°16′55.78″ N,	6 November	2.8	40 b	60 b	0 b
119° 2′44.52″ W)	21 November	3.8	80 a	0 d	20 a
	Control		70 ab	30 c	0 b
Irvine, Orange County	19 September	1.0	0 b	100 a	0 a
(33°43′10.84″ N,	4 October	2.0	0 b	90 a	10 a
117°44′26.18″ W)	19 October	2.5	10 b	90 a	0 a
	3 November	2.7	10 b	90 a	0 a
	18 November	2.9	30 ab	70 ab	0 a
	3 December	4.0	50 a	50 b	0 a
	Control		50 a	50 b	0 a

[z]Average of ten tree replications for each defoliation date and 20 tree replications for the control.
[y]According to the scale of Salazar-García *et al.* (1998).
[x]All floral shoots were indeterminate.
[w]Means in vertical columns followed by different letters are significantly different at $P = 0.05$ by Duncan's multiple range test.

orchards than in irrigated orchards. Although affected by shoot age, the predominant stage of bud development at the time of ICF was between Stages 3 and 4 on the visual scale of Salazar-García *et al.* (1998). However, for summer-flush shoots, it was evident that some buds reached ICF at Stages 1 or 2 (Table 6.3). Anatomical sections of these buds showed the same developmental stages characteristic of the transition from vegetative to reproductive development, illustrated in Fig. 6.2, but this transition occurred much faster, and buds looked visually smaller, resulting in a lower number on the visual scale.

Shoot defoliation and girdling were used to establish the date when ICF occurred for apical buds of spring and summer shoots of the West Indian × Guatemalan race hybrids 'Choquette' and 'Booth-8' cultivated under rainfed conditions (annual rain 1237 mm) (Salazar-García *et al.*, 2008). The study was done in the subhumid warm climate (average temperature maximum/minimum 30.1°C/15.4°C; 442 m) of Jalcocotán, Nayarit, Mexico (21°28′N; 105°5′W). The date of ICF for 'Choquette' avocado was 15 November, and there were no differences due to shoot age (spring or summer shoots). For 'Booth-8', ICF of the summer shoots occurred by 30 November. Macroscopic characteristics of apical buds irreversibly committed to flowering were closed and pointed with partial senescence of bud scales with averages of 3.0 to 3.4 on the visual scale designed by Salazar-García *et al.* (1998), consistent with previous results with 'Hass'.

In new avocado-growing areas with no prior history of avocado production, defoliation and girdling can serve as convenient tools to determine the time of ICF in avocado buds under field conditions. This information is useful for managing the balance between vegetative and floral shoots and for avoiding the use of cultural practices at the wrong time, thereby preventing the disruption of this balance.

Floral development

The scale developed by Salazar-García *et al.* (1998) depicting both macroscopic (external) and microscopic (anatomical) features of both bud and floral shoot development is a fast and reliable way to evaluate the progress of floral development from the vegetative stage (Stage 1) to anthesis (Stage 11). This information was integrated for use in the field by avocado researchers and orchard managers (see Fig. 6.4). Stages 1 to 3 correspond to the annual transition from the vegetative to the reproductive phase. Floral initiation starts at S-4, and buds are characterized by separation of bud scales, revealing the expanding inflorescence bracts; almost all secondary-axis inflorescence meristems can be present, with the basal secondary axes showing bracts. Initiation of the flower perianth occurs at S-5, first in the terminal flowers on secondary and tertiary axes. Buds at S-6 have a round shape, and expansion of inflorescence bracts is evident. At this stage, the secondary axes of the inflorescence are fully formed, and each one bears several cymes of flowers (tertiary axis of the inflorescence). Additional features of this stage are that terminal flowers on both secondary and tertiary axes have a complete perianth, with anthers and gynoecium at early formation. At S-7, bud break occurs; the flowers are more developed, although ovule and pollen grains are not fully formed yet. In S-8 ('cauliflower stage'; Lovatt, 1994), the elongation of the secondary axes of the inflorescence occurs. Pollen and ovule formation is still ongoing. At S-9, the elongation of the tertiary axes (cymes) of the inflorescence is observed. Pollen and ovule development is complete. In the indeterminate floral shoots, the vegetative bud emerging at the tip of the inflorescence can be visible. At S-10, flowers are fully differentiated but unopened. Sexual organs are mature and ready for anthesis. S-11 corresponds to the flower at anthesis where the stigma is receptive, and pollen is ready to be shed. The terminal bud at the apex of the indeterminate floral shoots initiates vegetative shoot growth between Stages 9 and 11, depending upon the tree and environmental conditions.

In addition to the visual and anatomical scale of avocado floral development established by Salazar-García *et al.* (1998), a scale for avocado reproductive development was established according to the Biologische Bundesanstalt, Bundessortenamt und Chemische Industrie (BBCH) code by Alcaraz *et al.* (2013a). The BBCH scale for avocado is more complex and more extensive than the scale of Salazar-García *et al.* (1998), providing additional stages related to floral development from bud break to anthesis and including stages of fruit growth and vegeta-

Table 6.3. ICF in 'Hass' avocado as affected by climate and shoot age in Michoacán, Mexico. Adapted from Rocha-Arroyo *et al.* (2010).

	Type of shoot								
	Winter			Spring			Summer		
Climate/avg. temp.	DOE[z]	ICF[y]	BD[x]	DOE	ICF	BD	DOE	ICF	BD
Subhumid warm (21.4°C)	9 January	23 July (195)	3.6	14 April	26 September (165)	5.0	24 June	7 September (75)	4.1
Subhumid semi-warm (19.8°C)	15 January	29 June (165)	3.6	13 April	28 May (45)	2.9	26 July	9 September (45)	3.0
Humid semi-warm (16.6°C)	9 January	23 June (165)	3.5	24 April	7 August (105)	4.6	23 July	6 September (45)	2.3
Subhumid temperate (17.1°C)	15 January	30 May (135)	3.6	12 April	26 June (75)	3.4	13 July	28 July (15)	1.4

[z]Date of emergence.
[y]Days to ICF. Numbers in brackets are average shoot age (days) to beginning of ICF.
[x]Stage of bud development according to the visual scale of Salazar-García *et al.* (1998).

tive shoot development. Both were designed to facilitate the proper timing of the orchard management practices. Notably, the BBCH scale does not include the cauliflower stage of inflorescence development, which has proven very responsive to canopy applications of fertilizers (boron, urea and potassium phosphite, among others) and PGRs (GA_3) for increasing fruit set and total yield and fruit size (Jaganath and Lovatt, 1998; Lovatt, 2022). Moreover, the cauliflower stage of inflorescence development is used and understood by growers and referred to in PGR labels and extension publications.

Avocado produces two types of floral shoots: determinate, in which the primary axis develops into a terminal flower (Schroeder, 1944), and indeterminate, in which the primary axis continues the vegetative growth of the shoot (Fig. 6.5) (Reece, 1942). The indeterminate type of inflorescence is more abundantly produced (Schroeder, 1944; Salazar-García *et al.*, 1998); however, the determinate type is common on stressed trees and 'on'-crop trees in an alternate-bearing orchard. Both types of floral shoots consist of secondary axes (lateral panicles), producing tertiary axes (cymes), which bear a terminal flower and two lateral flowers (Reece, 1942; Salazar-García *et al.*, 1998). Development of the secondary axes within the bud proceeds in an acropetal fashion, so a developmental range occurs in a single floral shoot, with the most advanced secondary axes (i.e. at perianth differentiation) at the base

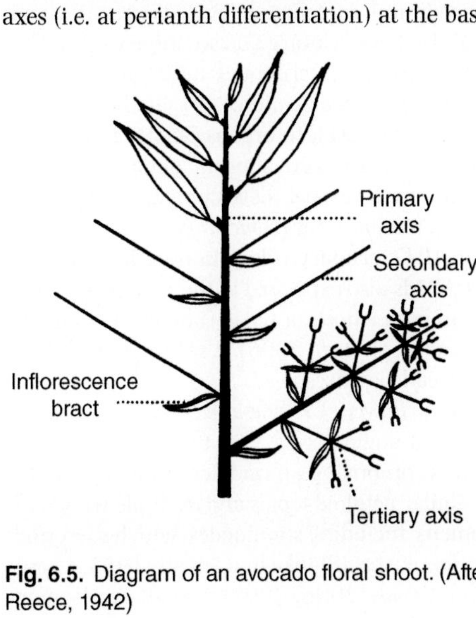

Fig. 6.5. Diagram of an avocado floral shoot. (After Reece, 1942)

and the youngest just below the PAM (Salazar-García *et al.*, 1998). However, on an individual secondary axis, development is basipetal, with the terminal flowers (lacking subtending bracts) differentiating and maturing first, as is typical in a cyme (Salazar-García *et al.*, 1998). The fate of the apex of the PAM of the floral shoot becomes determined after the secondary and tertiary axes have reached ICF (Salazar-García and Lovatt, 1999).

Fully open flowers are ~1.0 cm in width and 6–7 mm in length (Davenport, 1986). They are perfect, hypogynous, regular and trimerous (Schroeder, 1952). The perianth is formed by six tepals (Blanke and Lovatt, 1993). Inside the perianth, there are three whorls of stamens and one of staminodes. Aligned with each tepal is one stamen and one nectar-secreting, yellow staminode. Similarly, two stamens are aligned with each tepal, with the interior one having a pair of nectaries at its base. Therefore, each flower normally has a total of nine stamens (Bergh, 1986; Blanke and Lovatt, 1993). The single pistil is located in the centre. It has a superior ovary with one anatropous ovule (Schroeder, 1952; Bergh, 1986; Blanke and Lovatt, 1993).

Genetic regulation of avocado floral development

The genetic network regulating floral development is highly conserved among plant species (Benlloch *et al.*, 2007). Based on this, Acosta-Rangel *et al.* (2021) conducted a comparative analysis of the temporal patterns of expression of several core genes that might logically regulate floral timing (induction) (*FLOWERING LOCUS T* [*FT*] and *LEAFY* [*LFY*]), floral meristem identity (determinacy, ICF) (*LFY* and *AP1*) and floral organ identity (flower formation) (*APETALA2* [*AP2*]), *APETALA3* [*AP3*], *PISTILLATA.1* [*PI.1*], *AGAMOUS.1* [*AG.1*] and *AGAMOUS.3* [*AG.3*]) in the buds of 'Hass' avocado trees grown under different environmental conditions. Separate sets of trees were subjected to LT conditions known to induce flowering, OGCs known to promote vegetative shoot growth at the expense of flowering, and LSM and LLI of unknown efficacy to promote flowering. Only buds of LT-treated trees flowered (week 14). At the start of the experiment (week 0,

15 July), buds of all trees expressed *LFY*, *AP1* and *AP2*. After 4 weeks of treatment, bud *AP1* expression was greater than in week 0 and continued to increase through week 8 only in LT-treated trees, whereas by week 8, *LFY* expression increased in LT- and LSM-treated trees. At the end of 8 weeks of treatment, all trees were transferred to OGC. Two weeks after transfer (week 10), bud *FT* expression increased only in LT-treated trees, followed by activation of downstream floral organ identity genes *AP3*, *PI.1*, *AG.1* and *AG.3* by week 12 and flowering in week 14. For OGC-, LSM- and LLI-treated trees, bud *AP1*, *FT* and floral organ identity gene expression remained low or undetected at week 12. These trees did not flower. The results of Acosta-Rangel *et al.* (2021) are compelling evidence that the accumulation of both *LFY* and *AP1* transcripts to the maximum levels, which occurred in buds only after 8 weeks of LT treatment, was required and sufficient for irreversible commitment of the SAM to flowering in 'Hass' avocado. This interpretation is supported by the fact that transfer of LT-treated trees to the WTs of the OGC did not prevent flowering but instead resulted in the activation of *FT* and, subsequently, the downstream floral organ identity genes with concomitant flowering. In contrast, buds of LLI- and OGC-treated trees failed to accumulate transcripts of either *LFY* or *AP1* to the level of LT-treated trees by the end of the 8-week treatment and did not flower. Even buds of LSM-treated trees, which accumulated transcript levels of *LFY* equal to those of LT-treated trees by week 8 but expressed lower levels of *AP1* than the buds of trees in all other treatments, did not flower.

In *A. thaliana* and other plant species, *LFY* and *AP1* are expressed in early floral primordia, consistent with their role as key regulators of floral meristem identity (Blazquez *et al.*, 2006; Benlloch *et al.*, 2007; Sablowski, 2007; Hong and Jackson, 2015; Ma, 1994). Ziv *et al.* (2014) reported the expression of *LFY* and *AP1* at the start of the experiment, suggesting the induction process might have been initiated, possibly by a signal (*FT*) from the leaves or an autonomous (endogenous developmentally regulated) floral development pathway that triggered the transition from vegetative to reproductive SAM, which occurs in summer in California (Salazar-García *et al.*, 1998). Thus, Ziv *et al.* (2014) reported the first evidence to support the role of leaf-expressed *FT* as a signal in the initiation of the floral induction process in 'Hass' avocado under LT conditions, resulting in the upregulation of *LFY* and *AP1* expression in buds, analogous to the initiation of floral development in *A. thaliana*. Consistent with this possibility, the terminal buds of LSM-, LLI- and OGC-treated trees ceased vegetative shoot growth between weeks 4 and 8, and reached Stage 1 of inflorescence development according to the visual scale of Salazar-García *et al.* (1998) by week 8, but failed to develop further due to insufficient bud *LFY* and *AP1* expression prior to transfer of the trees to OGC (Acosta-Rangel *et al.*, 2021). In contrast, *LFY* and *AP1* expression continued to increase to maximum levels in response to 8 weeks of LT treatment, which was sufficient to upregulate *FT*, *AP3*, *PI.1*, *AG.1* and *AG.3* after transfer to OGC. At this time, the buds were at Stage 8 (cauliflower stage) of inflorescence development, in which all floral organs are present and during which the final stages of pollen and ovule development occur (Salazar-García *et al.*, 1998). In *A. thaliana*, *FT* has a second role, independent of its primary function in floral initiation, to maintain floral meristem determinacy and prevent floral reversion, which is essential for the full development of the inflorescence, especially one that develops acropetally, like *P. americana* (Parcy *et al.*, 2002; Müller-Xing *et al.*, 2014). Acosta-Rangel *et al.* (2021) proposed this role for *FT* in the development of the avocado inflorescence. In *A. thaliana*, *AP1* and *LFY* also have primary and secondary roles. The first is in bud determinacy and the second in floral organ specification: *AP1* in the development of the perianth organs (sepals and petals) and *LFY* in the activation of genes specifying stamen and carpel development (Sablowski, 2007). Similarly, for *P. americana*, results suggest that *AP1*'s primary role is in bud determinacy, but *AP1* is also expressed in the perianth organs (i.e. in both whorls of tepals) and in the stamens (Chanderbali *et al.*, 2006). A second role for *LFY* in avocado is unresolved.

The flower of *P. americana* is typical of early-divergent angiosperms, with the first and second whorls comprising an undifferentiated perianth of similar petaloid tepals and multiple whorls of stamens, including staminodes, which surround a single carpel (Blanke and Lovatt, 1993; Chanderbali *et al.*, 2006, 2009). Broad, overlapping patterns of gene expression specify floral organ development in *P. americana* rather than the precise floral organ specification described by the ABC model based on *A. thaliana* (Chanderbali *et al.*, 2006, 2009, 2010, 2016). In the ABC model, A-function genes *AP1* and *AP2* specify sepals, A-function genes plus the B-function genes *AP3* and *PI* specify petals, B-function genes plus the C-function gene *AG* (antagonistic to the A-function genes) specify stamens, and the C-function gene alone specifies the carpel (Bowman *et al.*, 1991; Coen and Meyerowitz, 1991; Krizek and Fletcher, 2005). In the *P. americana* flower, *AP1* is expressed in the whorls of tepals and stamens, with homologues of *AP3* and *PI.1* (putative B-function genes) also expressed in the whorls of tepals and the stamens (Chanderbali *et al.*, 2006, 2009; Soltis *et al.*, 2007a, b). Chanderbali *et al.* (2006, 2009) identified three *AG* homologues (putative C-function genes) in *P. americana*, with *AG.1* and *AG.2* expressed in outer and inner tepals, stamens and carpel, and *AG.3* expression restricted to the stamens and carpel.

Acosta-Rangel *et al.* (2021) proposed the operation of a failsafe mechanism in avocado that synchronizes flowering induced by lower fall and winter temperatures with the warmer spring temperatures. The proposal is based on their results documenting that *LFY* and *AP1* transcripts accumulated to maximum levels in the buds at the end of 8 weeks of LT treatment, whereas bud expression of *FT* and the downstream floral organ identity genes did not occur until after the transfer of LT-treated trees to OGC. Similarly, transcripts of *LFY* and *AP1* accumulated in buds under LTs during fall and winter to a level sufficient to confer floral meristem identity. However, bud *FT*, which likely plays a role in maintaining a commitment to flowering, and the downstream floral organ identity genes, *AP3*, *PI.1*, *AG.1* and *AG.3*, would only be fully expressed in 'Hass' avocado buds when spring temperatures were sufficiently warm, thereby preventing flower formation under adverse temperature conditions with potential impacts on pollination, fertilization and fruit set.

Rootstock effects on floral development

Genetic characteristics of avocado scions and rootstocks, either individually or through their combined interactions, influence floral development as well as bloom time and fruit maturation. However, expression of this effect can be hidden or modified by orchard management and environmental conditions. Generally, it is accepted that some genotypes of Mexican, Guatemalan and West Indian rootstocks might cause early, intermediate and late bloom, respectively. Results from a rootstock trial of 'Hass' on several rootstocks in the subhumid semi-warm climate of Tepic, Nayarit, Mexico, showed that the Mexican race rootstocks caused bloom to occur 4 and 8 weeks before clonal 'Hass' (Guatemalan × Mexican), grown on its own roots and on West Indian seedlings, respectively. However, the 8-week difference in time of bloom advanced fruit maturity (21.5% mesocarp, edible portion, dry matter) by only 3 weeks (S. Salazar-García, personal observation). It was hypothesized that this was because fruit that set during the late flowering period experienced warmer temperatures after ovule fertilization compared to fruit that set earlier.

Results of field trials in Queensland, Australia, provide evidence that cultivars grafted onto Mexican race rootstocks flower earlier than when grafted onto Guatemalan race rootstocks (Young, 1992). This early flowering resulted in earlier fruit maturity, between 3 to 5 weeks and 8 to 12 weeks, respectively, for 'Fuerte' and 'Hass' grafted onto Mexican race rootstocks ('Mexicola' seedlings and 'Duke 7' clonal) in comparison to the local Guatemalan × Mexican race hybrid rootstocks ('Plowman' and 'Velvick' seedlings). Flowering of 'Fuerte' and 'Hass' scions grafted onto Mexican race rootstocks was consistently earlier in a subtropical climate, intermediate in a cool temperate climate and not different from Guatemalan rootstocks in hot-humid tropical areas in northern Queensland (Young, 1992). This was probably because, after fruit set, high temperatures during the spring and summer speed up fruit development, nullifying the influence that rootstocks have on bloom time.

Root growth is affected by orchard management practices; however, the genetic background of the rootstock is important. In warm environments (such as on the Peruvian coast), root growth of Mexican race rootstocks was almost nil during the summer due to constant high soil temperatures of 25°C to >34°C (night/day) (S. Salazar-García, personal observation).

In comparison, Guatemalan and West Indian rootstocks maintained full root activity. Decreased root activity in the Mexican race rootstocks causes cessation of shoot growth conducive to earlier transition to floral development, resulting in a very intense and profuse flowering, composed predominantly of determinate inflorescences.

Interaction between the scion and the rootstock is important when managing pollinizers to ensure bloom overlap. In the warm climate of Virú, Peru (8°32′ S; 78°40′ W), there was almost perfect synchrony between 'Hass' bloom and that of pollinizers 'Ettinger', 'Zutano' and 'Fuerte' when grafted on to Mexican race 'Topa Topa' seedling rootstocks (Fig. 6.6). However, the lack of full overlap might occur if scions of either 'Hass' or the pollinizers are grafted onto rootstocks of different races. As stated above, under these conditions, bloom usually starts first on the scions grafted onto Mexican, followed by Guatemalan and West Indian rootstocks, although the last two likely show no practical differences.

Mickelbart *et al.* (2012) provided evidence that rootstock genotype had no significant effect on the flowering time or the duration of the flowering period of alternate-bearing trees. In an alternate-bearing orchard, the bloom that sets the 'on' crop is earlier, consistent with bud break being less inhibited in the presence of the 'off'-crop set the previous year and occurs over a longer period than the 'off'-crop bloom (Garner and Lovatt, 2008). 'Hass' scions grown on four clonal Mexican race rootstocks ('Thomas', 'Topa Topa', 'Duke 7', or 'D9'), in Irvine, California (33°41′ N, 117°49′ W), exhibited no difference with regard to time of flowering attributable to rootstock (Mickelbart *et al.*, 2012). Flowering was initiated with the spring vegetative shoot flush from early February through mid-April and finished in early to mid-May. Time of flowering in a given year was related to alternate bearing and not influenced by any rootstock. Over the 4-year period, flowering of 'on'-crop trees started earlier and was protracted. The mean start date for flowering in the heavy and light crop years was 28 February and 27 March, respectively, with the mean duration for the 'on'- and 'off'-bloom periods 77 and 49 days, respectively ($P < 0.001$).

Effects of the scion on floral development

Time of bloom is an important factor in commercial avocado production. In avocado, bloom occurs at the end of a protracted floral development process. Successful transition from a vegetative to reproductive SAM occurs many months prior to floral organogenesis and spring

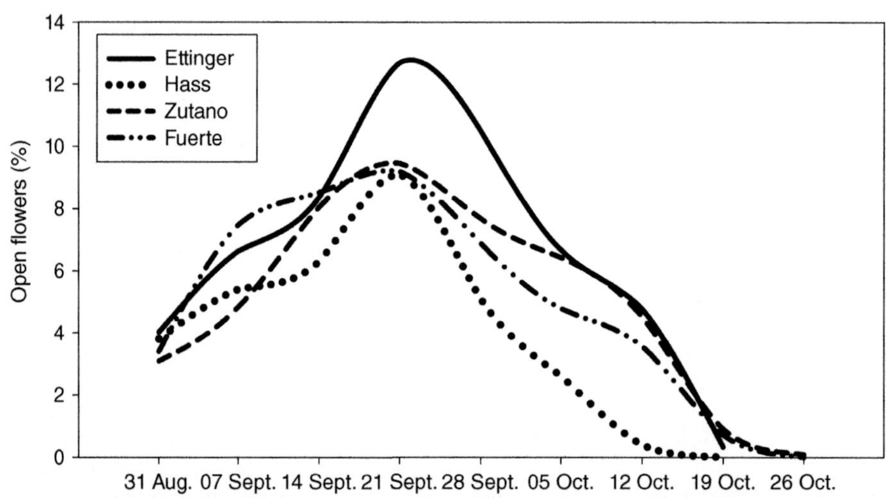

Fig. 6.6. Progression of flower opening of several avocado cultivars on 'Topa Topa' seedling rootstocks in Chao, Virú, Peru (southern hemisphere). (C. Calderón and J. Huanca, n.d., unpublished results)

bloom. Interactions between scion cultivars with different genetic backgrounds and environmental factors, rootstock and cultural management practices that occur during this period affect the time of bloom. For a seed-producing crop like avocado, flowering must be synchronized with that of the pollinizers and occur under environmental conditions that favour pollinators, if required, and support successful syngamy. Thus, selecting a scion cultivar optimal for a specific avocado-producing area is of major importance. If bloom occurs early in the winter or spring, LTs may negatively affect pollen development, pollinator activity and fruit set. Conversely, with flowering late in the spring or early in the summer, reduced or no fruit set may result due to excessively high temperatures. Depending upon the latitude, rains could occur in the winter, summer or year-round, affecting both flowering phenology and insect activity, including insects required for pollination and fruit set; the occurrence of diseases that attack flowers and young fruit is an additional issue influenced by rain.

In most avocado-producing regions, scion cultivars of Mexican race and Mexican × Guatemalan hybrids have been reported to bloom early (Schroeder, 1951). In California, generally, the Mexican and Mexican × Guatemalan hybrids are the earliest to begin bloom (January). West Indian cultivars start to flower in mid-January to February. The pure Guatemalan race cultivars start blooming from mid-February to mid-March (Bergh, 1986). In Florida, Reece (1942) and Davenport (1982) reported that the earliest evidence of floral bud development occurred in the West Indian cultivars Pollock and Simmonds and their hybrid 'Lula' from late November to early January. In the pure Guatemalan race cultivars of Nabal and Itzamna, flower development only became evident in January.

For many years, researchers and growers have been searching for cultivars that might be more productive than 'Hass' or extend the harvest season. Some cultivars have been tested, such as Pinkerton, Shepard, Reed, Gwen, Whitsell and Esther, mainly in the USA, Australia and South Africa (Whiley *et al.*, 1990). However, none fully passed all the criteria imposed by growers, packers, shippers, marketers, value-added product industries and consumers. In countries growing Mexican and Guatemalan race avocados, two cultivars are of major economic importance, Fuerte and Hass. These two have become the industry standards for avocados and comprise the majority of the international avocado trade.

As consumer market demands for 'Hass' avocado increase, Hass-like varieties are a must. 'Lamb Hass' is a promising cultivar for some countries where it has been evaluated (California, Australia, New Zealand, South Africa and Spain, among others). In some countries, 'Lamb Hass' is of great interest for extending the harvest season, since it is harvested later than 'Hass'.

An extended flowering season has been documented for 'Hass' in the state of Michoacán, Mexico. This is due to variations in climatic conditions across different altitude zones in the biggest avocado-producing region in the world. In some climates, a single tree may produce up to four bloom flushes from July to April the next year, named, respectively, Loca (Crazy), Advanced, Normal and Marceña (from March). The different bloom flushes overlap across the different altitude zones, making it possible to harvest 'Hass' throughout the year. A regional crop is composed of fruit set from the different blooms in the following proportions: Crazy = 5%–20%, Advanced = 10%–25%, Normal = 60%–80% and Marceña = 15%–30% (Salazar-García *et al.*, 2005). The origin of the name 'Crazy bloom' refers to the fact that this bloom occurs only in some climates, in some years, on some trees and on some vegetative flushes (Rocha-Arroyo *et al.*, 2011a). Fruit set from this bloom is harvested during a period of high demand (June to early September) and reduced domestic supply, selling at an extremely high price. Fruit set by the Crazy bloom is reduced due to summer rains. Also, if the Crazy bloom sets fruit, it is common for the more developed fruit from earlier blooms (i.e. Normal or Marceña) to abscise.

In the late 1980s, Mr Carlos Méndez detected a distinct avocado tree in Uruapan, Michoacán, Mexico. This tree consistently exhibited the Crazy bloom (July–September), compared with surrounding 'Hass' trees. Harvest of the fruit from this bloom coincided with the Crazy bloom crop of 'Hass'. This Hass-like avocado was locally named 'Hass-Méndez' and is now trademarked as 'Hass Carmen®'

(Illsley-Granich *et al.*, 2011). The fruit of 'Hass Carmen' is not morphologically different from the range of fruit morphologies found for the conventional 'Hass' in Mexico. This new avocado cultivar is widely grown in Mexico and is in early commercial development or commercially available in the major avocado-producing and exporting countries (Illsley-Granich *et al.*, 2011). These authors speculate that the 'off'-season-producing habit of 'Hass Carmen' will be more pronounced in semitropical climates that experience wet summers and dry, mild winters; while in more extreme Mediterranean climates with cold winters, the bloom will be more concentrated and occur during the traditional bloom period. In warm climates, the erratic occurrence of cool night temperatures in the early part of the summer in the northern hemisphere stimulates vegetative shoot growth at the expense of the Crazy bloom (S. Salazar-García, personal observation).

Additional avocado scion cultivars of interest are Gem and Luna UCR from the University of California, Riverside, breeding program and Maluma, from Allesbeste Nursery, in Tzaneen, South Africa. These 'Hass' relatives produce fruit of 'Hass'-like quality and produce smaller trees more suitable for high-density planting, with 'Maluma' also suitable for trellising. They appear to be less prone to alternate bearing than 'Hass', with 'Maluma' maturing approximately 1 month ahead of 'Hass' in South Africa and 'Gem' and 'Luna UCR' maturing slightly later than 'Hass' (see Chapter 8, this volume). Whereas 'Maluma' and 'Gem' are A-type flowering varieties, 'Luna UCR' is a B-type variety and, consequently, could be used as a pollinizer for 'Hass' and other A-type flowering varieties.

Phenology and modelling of floral development

PHENOLOGY OF FLORAL DEVELOPMENT. At the end of vegetative shoot elongation, if environmental and shoot developmental conditions are conducive to floral development, terminal and axillary buds transition to floral development. If conditions do not support floral development, the buds continue vegetative shoot growth or remain inactive. Once the buds are committed to floral development, the length of time to reach anthesis is predominantly driven by temperature (Salazar-García *et al.*, 1998; Salazar-García *et al.*, 2007a, b). Results from controlled environment and field research confirm that irrespective of the cultivar, a decrease in temperature is necessary for buds to become committed to reproductive development (Buttrose and Alexander, 1978; Nevin and Lovatt, 1989; Salazar-García *et al.*, 1999; Cossio-Vargas *et al.*, 2007; Rocha-Arroyo *et al.*, 2010; Acosta-Rangel *et al.*, 2021). This means that buds on later-developing vegetative shoots (summer and fall) will become committed to flowering faster, as they will reach the LT conditions that induce flowering in a shorter time than buds on vegetative shoots that developed earlier in winter or spring, and remain vegetative during the upcoming period of WTs. Thus, elapsed time from the end-of-shoot elongation to anthesis (S-1 to S-11) on the floral development scale of Salazar-García *et al.* (1998) can vary from 4 to more than 12 months. Winter shoots require more time to complete the process of floral development than summer and fall shoots in the subhumid semi-warm climate of Tepic, Nayarit, Mexico (Cossio-Vargas *et al.*, 2008b; Rocha-Arroyo *et al.*, 2011a).

When determining differences in avocado floral development due to natural or induced conditions, a good strategy is to use a 'biofix', a biological indicator of a developmental event. The end-of-shoot elongation has been useful for this purpose. Comparison of the progress of floral development on winter shoots in several climates of Michoacán, Mexico, illustrates the differences in the rates of development that disappear at the time of anthesis (S-11) (Fig. 6.7a). Similar results were obtained for apical buds on summer shoots in five locations of California from Cambria (35°33′ N, 121°40′ W) to Valley Center (33°14′ N, 117′00′ W); however, Cambria (coldest location) reached anthesis (S-11) 2 weeks ahead of those in the warmer locations (Fig. 6.7b).

Under normal conditions, the fate of the buds produced by the different vegetative flushes during the year is influenced by tree crop load and environmental conditions present during the first weeks of shoot growth. Buds on vegetative shoots could produce vegetative or floral shoots or remain inactive. Irrespective of the time of vegetative flushing, flowering shoots

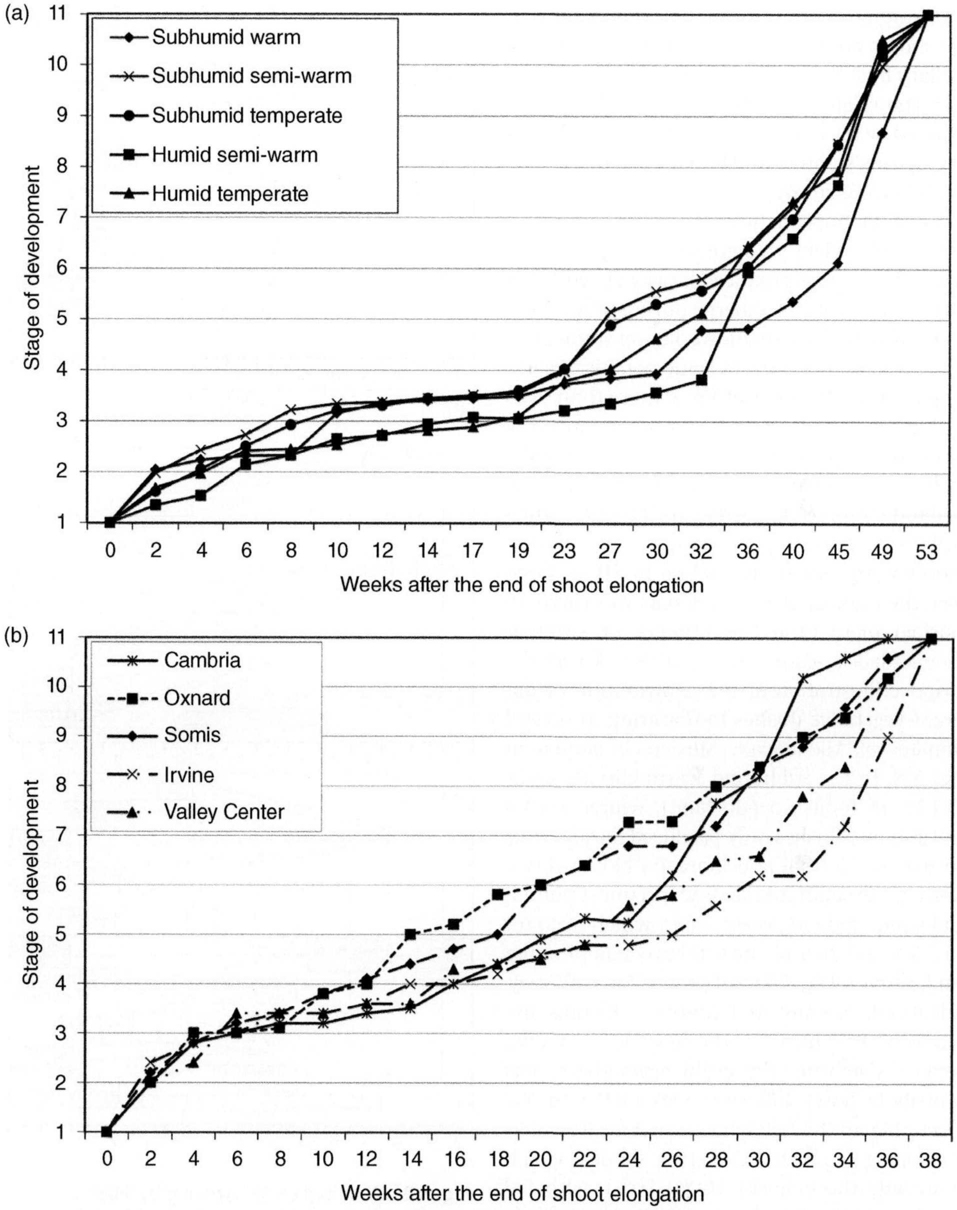

Fig. 6.7. Floral development of 'Hass' avocado: (**a**) in several climates of Michoacán, Mexico (winter shoots were born in December 2005–January 2006); (**b**) at six locations from south (Valley Center) to north (Cambria) California (summer shoots were born in July–August 2005). (S. Salazar-García, 2007, unpublished results.)

look very similar at bloom. An experienced eye might easily recognize shoot age; oldest shoots have a wider diameter and harder leaves and a higher production of axillary floral shoots, which, in prolonged winters, become the determinate type. At bloom, it is common for winter, spring and summer shoots to exhibit similar vigour. A key feature to distinguish winter, spring and summer shoots from fall shoots is that fall shoots produce only terminal floral shoots; axillary buds do not undergo bud break.

The relative number of floral shoots contributed by the different vegetative shoot flushes (e.g. spring, summer and fall) to return bloom in the following spring is valuable information to successfully implement management strategies to maximize yield. In some avocado-producing areas, flowering is produced largely by winter or spring shoots, which emerge during or after the major bloom. However, flowering on summer or fall shoots is important in some regions. For example, in coastal areas of Peru, more than 90% of the flowering shoots of 'Hass' are produced by spring-flush shoots that developed 10–12 months earlier. A similar situation occurs in New Zealand (Dixon *et al.*, 2008). In Nayarit, contributions to flowering of winter and summer shoots were very similar (close to 50%); however, the number of winter shoots was three- to fourfold more than the number of summer shoots (Salazar-García *et al.*, 2006). A very descriptive illustration of the contribution of different vegetative flushes to flowering in several climates of Michoacán, Mexico, is shown in Fig. 6.8. In the subhumid warm climate areas (0.12% of total avocado land), winter, spring and summer flushes only produced floral shoots during the Normal and Marceña bloom times. However, in subhumid semi-warm (most humid) and subhumid temperate areas, which account for 57% and 26% of the total avocado-growing land, respectively, different proportions of Crazy, Advanced, Normal and Marceña blooms are produced. In contrast, in the avocado-producing areas of California, the spring vegetative shoots contribute fewer inflorescences (30%) to the spring bloom the following year than the later-developing summer vegetative shoots, which contribute the majority (60%–70%), with fall shoots contributing the least (0%–10%), likely because the majority of the shoots develop too late in the season to undergo phase transition (Lovatt, 2011a). In the two major avocado-growing areas of Spain, Algarrobo, Málaga (36°46′ N), and Callosa d'en Sarrià, Alicante (38°39′ N), which are at the northernmost latitudes of avocado commercial production, vegetative shoots of the summer flush contribute the majority of floral shoots to the return bloom the following spring (J.I. Hormaza, Spain, 2023, personal observation). Across the avocado-producing areas of Australia, the summer flush of vegetative shoots is the major contributor of floral shoots at bloom the following spring (S. Newett,

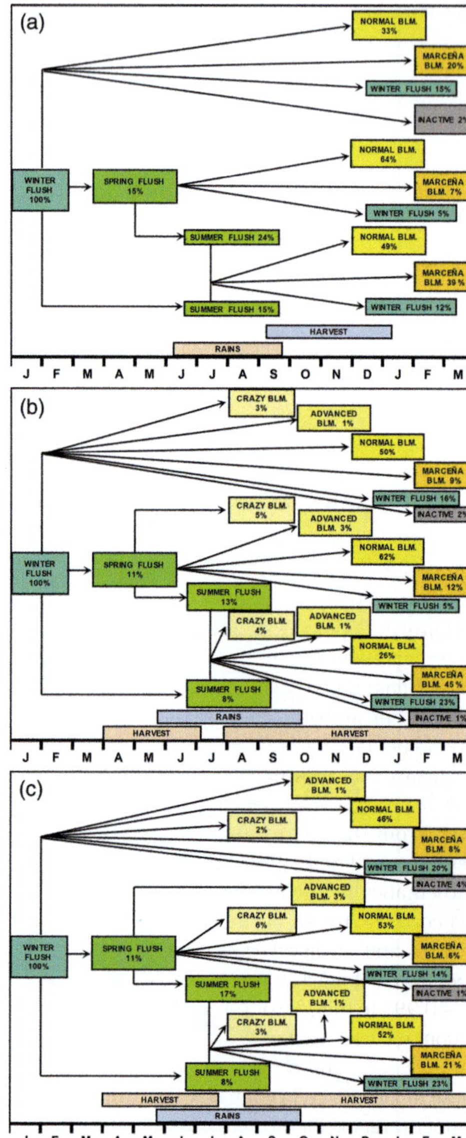

Fig. 6.8. Contribution to flowering by 'Hass' avocado shoots produced in the winter, spring and fall flushes in three contrasting climates of Michoacán, Mexico (northern hemisphere): (**a**) subhumid warm, (**b**) subhumid semi-warm (most humid), and (**c**) subhumid temperate. (Adapted from Rocha-Arroyo *et al.*, 2011b.)

Australia, 2023, personal communication). In the two major avocado-growing areas of New Zealand, Kataia, Far North (34°47′ S) and Te Puke, Bay of Plenty (37°47′ S), the summer flush and fall flush if the summer flush keeps growing, are the major source of inflorescences at return bloom (J. Dixon, New Zealand, 2023, personal communication). This information makes it possible to test treatments to alter the time of vegetative shoot flushes, to increase the number of shoots in a particular flush to have more floral shoots and greater yield the following year, or to promote the uniform intensity of the vegetative shoot flush, producing the greatest return bloom on an annual basis and even out alternate bearing.

MODELLING FLORAL DEVELOPMENT. The preceding sections of this chapter described the role of LT as a promoter of avocado floral development. This allowed developing models to predict when key floral developmental events would occur. Using ambient temperature records and field floral development data obtained from 1998 to 2006, the effect of ambient temperature on the floral development of 'Hass' in the semi-warm climate of Nayarit was quantified (Salazar-García *et al.*, 2007b). The complete process of floral development (from the end-of-shoot elongation to anthesis) was related to ambient temperature and mathematically modelled. Floral development was correlated with accumulated chilling days (ACDs), and a floral development prediction model was obtained for winter shoots, $\text{winter}_{\text{ACD} \leq 21}$ ($R^2 = 0.99$). In the case of summer shoots, two prediction models were developed, $\text{summer}_{\text{ACD} \leq 19}$, $\text{summer}_{\text{ACD} \leq 20}$, both of them with $R^2 = 0.99$. These models were validated in the same 'Hass'-producing region for their capacity to predict when buds or floral shoots reached S-8 (cauliflower stage) or S-11 (anthesis), resulting in $R^2 = 0.99$.

In another study, the prediction models generated by Salazar-García *et al.* (2007b) for 'Hass' in Nayarit were evaluated for their capacity to predict the floral bud development on winter, spring and summer shoots of trees growing in several climates of Michoacán, Mexico. The effectiveness of the Nayarit prediction models depended on the climate and shoot age (Salazar-García *et al.*, 2009). Only the model $\text{summer}_{\text{DFA} \leq 19}$ showed a strong capacity to predict floral development of summer shoots ($R^2 = 0.94$) using a data set from four climates of Michoacán, one subhumid warm, two subhumid semi-warm and one subhumid temperate, which together account for 85% of the 'Hass' avocado-producing area of the state. These results justified the development of floral development prediction models for Michoacán.

The floral development prediction models described for 'Hass' have been used to construct simulation programs available on the internet. These applications predict the date of occurrence of key floral developmental events based on daily temperature information provided by automatic weather stations situated in the 'Hass'-producing regions of Nayarit and Michoacán. These models also work in California 'Hass' orchards using temperature data from nearby California Irrigation Management Information System (CIMIS) sites. Floral development simulators are helpful for proper and effective timing of foliar sprays to control pests and diseases that coincide with anthesis and early vegetative shoot growth, increasing their benefit-to-cost ratio. More examples of the use of the prediction models are related to properly timing: (a) pruning; (b) fertilization (especially soil-applied nitrogen); (c) foliar applications of boron, urea or GA_3 at the cauliflower stage of inflorescence development; and (d) placement of beehives in the orchard at the optimal time for pollination, and other cultural practices.

Manipulating flowering

GROWTH RETARDANTS. Physiology of avocado flowering is often thought to be similar to other woody evergreen fruit trees, such as citrus and mango (Bower *et al.*, 1990). The most common factor among these tree crops is that a cessation of vegetative growth must occur in order to start the floral development process. Since in avocado absence of vegetative growth is conducive to flowering, it was assumed that the use of the growth retardant paclobutrazol {(*R**,*R**)-ß-[(4-chlorophenyl)methyl]-α-(1,1-dimethylethyl)-1*H*-1,2,4-triazole-1-ethanol}, (Cultar®; Syngenta Crop Protection, UK Limited), a triazole PGR that inhibits gibberellin (GA) biosynthesis (Davis *et al.*, 1988) and stops vegetative shoot growth, would increase flowering in avocado as it does in citrus (Muñoz-Fambuena *et al.*, 2012) and mango (Yeshitela *et al.*, 2004). Early studies on the use of paclobutrazol with avocado resulted in profuse flowering. This was the case for potted (Symons and Wolstenholme, 1989) or field-grown 'Hass' (Köhne and Kremer-Köhne, 1990) trees. Soil drenching, foliar sprays or trunk injections similarly increased flowering, but floral intensity was dose-dependent (Köhne and Kremer-Köhne, 1990). For 'Hass', the duration of the effect of paclobutrazol on flowering was 4 to 6 weeks after application as a foliar spray, whereas the suppression of shoot growth lasted 6 months when applied as a soil drench (Köhne and Kremer-Köhne, 1990). Adato (1990) reported no residual or cumulative effect of 4% Cultar® (paclobutrazol, Syngenta Crop Protection, UK Limited) sprays in 2 consecutive years of treatment for tree vigour, general appearance and yield.

The initial purpose of using paclobutrazol, uniconazole (or uniconazole-P) or other triazole PGRs that inhibit GA biosynthesis was to control avocado tree vigour and tree size. The greatest current use in regions in which these chemicals are authorized for avocado is for delaying and reducing shoot growth during the spring bloom. This is achieved by foliar sprays before or during anthesis with the goal of limiting the competition for tree resources in these critical stages of phenology by reducing vegetative shoot growth in favour of fruit set and initial fruit growth (Köhne and Kremer-Köhne, 1987; Adato, 1990; Wolstenholme *et al.*, 1990). However, careful attention must be paid to the growing conditions, current alternate-bearing cycle, doses, reapplications and stage of tree phenology. In addition, orchard management practices, such as nitrogen fertilization or pruning, which stimulate excessive vegetative growth during the spring flush, should be avoided.

GIBBERELLIC ACID (GA_3). Opposite to growth retardants that inhibit GA biosynthesis and reduce vegetative shoot growth, GAs stimulate growth (e.g. bud break, shoot elongation and fruit growth). Some research has been done to develop strategies using foliar-applied GA_3 to regulate inflorescence phenology and flowering intensity to increase yield and even out 'on'/'off' yields in alternate bearing of 'Hass' orchards.

GA_3 foliar sprays affected the type of growth produced by the apical buds. A study carried out in an alternate-bearing orchard in California reported higher production of vegetative shoots at the expense of inflorescence production for control trees carrying an 'on' crop. The opposite occurred for the 'off'-crop year, resulting in increased inflorescence production (Salazar-García and Lovatt, 2000). This research was the first to use GA_3 to alter this relationship. GA_3 (100 mg l^{-1}) applied in September, before ICF (buds at Stage 3; Fig. 6.4) in either cropping year, effectively reduced inflorescence production and increased vegetative shoot production. GA_3 (25 or 100 mg l^{-1}) applied in November (buds at Stage 4), prior to the 'on'-bloom year, reduced inflorescence production and increased the production of vegetative shoots compared to the control. GA_3 sprays in November, when buds had only one to three secondary-axis inflorescence meristems without apical bracts during the year of this study, resulted in the production of partially formed inflorescences (with less than ten secondary axes) (Salazar-García and Lovatt, 1998), which reduced flowering intensity. Stimulation of vegetative PAM growth by GA_3 arrested the growth of secondary-axis inflorescence meristems adjacent to it; the response was stronger at higher GA_3 concentrations. However, flowering was not fully inhibited, demonstrating that GA_3 did not revert secondary-axis inflorescence meristems to vegetative ones. Aborted meristems could be observed after inflorescence elongation as scars with small inflorescence bracts (Salazar-García and Lovatt, 1998). GA_3 (25 mg l^{-1}) applied in November decreased yield (50%) in the 'on'-bloom year. GA_3 treatment in January (buds at Stage 5) or in March (buds at the cauliflower stage; Stage 8) had no effect on the number of inflorescences produced in either the 'on'- or 'off'-bloom year due to the high proportion of buds at an advanced stage of development at the time of treatment. Monthly applications of GA_3 (25 mg l^{-1}) during inflorescence bud development (September to January) reduced the production of inflorescences by more than 65% in both 'off'- and 'on'-bloom years; however, there was no related effect on yield. Similar inhibitory effects of GA_3 (250 mg l^{-1}) were obtained with repeated sprays (February + March + April + two sprays in May; southern hemisphere) to young potted 'Hass', 'Fuerte' and

'Ryan' trees (Rossouw *et al.*, 2000). Untreated control trees produced more floral than vegetative shoots at spring bloom.

In addition, GA_3 treatments alter floral shoot phenology in avocado. Different from what has been reported for other fruit tree species, GA_3 sprays never delayed flowering, but hastened floral shoot development of the 'Hass' avocado. Even at a low concentration (50 mg l^{-1}), when GA_3 was applied in December or January in California 50% of the inflorescences reached the cauliflower stage 23 days earlier than the untreated control. Higher GA_3 concentrations applied to the foliage (100 or 1000 mg l^{-1}) or later application dates further accelerated the rate at which buds reached the cauliflower stage; however, the date of anthesis was not significantly affected (Salazar-García and Lovatt, 1998). This behaviour seems to be a natural mechanism in avocado to avoid anthesis when LTs are prevailing.

Application of GA_3 also alters floral shoot morphology. Determinate inflorescences were the most abundant type produced by young 'Hass' trees growing in a glasshouse. Trunk injections of GA_3 (25 or 50 mg per tree) at Stage 5 of inflorescence development significantly increased the production of indeterminate inflorescences at the expense of determinate inflorescences (Salazar-García and Lovatt, 1999). Inflorescences from both control and GA_3-treated trees had ten secondary axes, the normal number for a fully formed inflorescence (Salazar-García *et al.*, 1998). However, GA_3-treated trees produced inflorescences with longer secondary axes than control trees. Secondary axes of control inflorescences ranged from 5 to 20 cm in length, whereas those of GA_3-treated trees were 30 to 50 cm long. In that study, GA_3 injected at Stage 5 of inflorescence development caused precocious development of the vegetative shoot apex of indeterminate floral shoots. Stimulating vegetative growth at the inflorescence apex did not inhibit inflorescence development. Whereas the full number of secondary-axis meristems of the inflorescence is already formed at Stage 5, tertiary axes are still being initiated (Salazar-García *et al.*, 1998). The results provide additional evidence that secondary and tertiary axes are committed to flowering at this stage, consistent with the presence of the perianth on these axes, and that the final fate of the floral shoot apex is determined much later. Thus, potential determinate inflorescences were redirected to indeterminate floral shoots by the GA_3 treatment.

Competition between the terminal vegetative shoot and the flowers of indeterminate floral shoots at the time of fruit set has been hypothesized as one of the reasons to explain low fruit set in avocado (Zilkah *et al.*, 1987; Cutting and Bower, 1990; Whiley, 1990; Bower and Cutting, 1992). Stimulating growth of the vegetative shoot at the apex of indeterminate floral shoots was predicted to reduce fruit set and yield as a result of competition for resources between the growing shoots and developing fruit (Kalmer and Lahav, 1976). In California, vegetative growth and leaf expansion of indeterminate floral shoots of 'Hass' are usually delayed relative to the elongation of the secondary axes of the inflorescence. The rate of development of the vegetative shoot at the apex of indeterminate floral shoots was dramatically increased with GA_3 sprays (Salazar-García and Lovatt, 1998, 2000). It was hypothesized that precocious vegetative shoot development of indeterminate floral shoots caused by GA_3 treatments might result in more successful inflorescences – in terms of fruit set and yield – by eliminating the competition between developing fruit and vegetative shoot growth. Leaves of the precocious vegetative shoot of indeterminate inflorescences treated with GA_3 were sufficiently mature to be sources (based on net CO_2 assimilation rates), capable of exporting carbohydrates, amino acids and other metabolites at the time of fruit set rather than being a competing sink (Blanke and Lovatt, 1998). A 2-year field experiment testing the efficacy of foliar-applied GA_3 (0, 25 and 100 mg l^{-1}) at different stages of inflorescence development on fruit yield demonstrated that GA_3 (25 mg l^{-1}) applied in March of an 'off'-bloom year resulted in a twofold increase in the production of commercially valuable size (CVS) fruit (213–269 g per fruit) and delayed exocarp (peel) blackening of mature fruit still on the tree at the time GA_3 was applied (Salazar-García and Lovatt, 2000).

In field experiments in nine commercial 'Hass' orchards in California (Lovatt, 2022), the optimal time to apply GA_3 to significantly increase the yield of CVS fruit and total yield was between the cauliflower stage of inflorescence development and 10% inflorescence expansion

(Stages 8 and 9 on the visual scale of floral development developed by Salazar-García *et al.*, 1998) (Fig. 6.4); these stages of inflorescence development typically occur in March in California. GA_3 (61.8 g ai/ha, in sufficient volume for good canopy coverage, using an organosilicone surfactant and a final pH of 5.5–6.0) is applied as a ground or aerial (helicopter) spray when 50% of the trees in the block have 50% of their bloom between the cauliflower stage and 10% inflorescence expansion. This means that 25% of the bloom will be at an earlier stage of inflorescence development, and 25% will be approaching bloom (open flowers). If the application cannot be made at this time, being slightly late (i.e. approaching open flowers) in applying the treatment provides better efficacy than being too early (i.e. closer to bud break), but applications made at full bloom are typically ineffective. Properly applied, GA_3 resulted in an average net yield increase of 51% kg per tree, a 64% net increase in CVS fruit (178–325 g/fruit) per tree, a 68% net increase in the total number of fruit per tree and a 61% net increase in the number of CVS fruit(178–325 g/fruit) per tree, indicating that the GA_3 treatment increased both fruit set and fruit size (Lovatt, 2022). Application of GA_3 at the cauliflower stage of inflorescence development had no negative effects on fruit quality, including seed size, seed germination, vascularization of the mesocarp or days to ripen. Thus, the treatment increased fruit size (g/fruit) by increasing the mesocarp (edible portion of the fruit), not by increasing seed size. The treatment also stimulated the growth of the vegetative shoot at the apex of indeterminate inflorescences, providing expanded leaves that protected the young developing fruit from sunburn compared to indeterminate inflorescences of untreated control trees. For mature fruit on the tree at the time of treatment, GA_3 application delayed exocarp blackening for some years. No visual signs of toxicity have been observed for GA_3 up to 100 mg l^{-1} on avocado (Salazar-García *et al.*, 1998). However, concentrations of GA_3 $\geq$ 155 g a.i./ha reduced yield (C.J. Lovatt, personal observation).

The cauliflower stage of the avocado inflorescence represents an important stage in fruit production. The final stages of pollen development and ovule formation are occurring, and integuments, tissues that will later develop into the seedcoat, are forming around the ovule, which, after pollination and fertilization, will develop into the seed (Salazar-García *et al.*, 1998). Research results with other plant species document that GA_3 is necessary for the development of the stamen, the male reproductive structure of the flower, pollen formation and pollen viability (Chhun *et al.*, 2007). In addition, observations in different plant species show that GA_3 improves pollen tolerance to cold stress (Sharma and Nayyar, 2016), enhances pollen tube growth after pollination, and is important for subsequent initiation of seed development (Singh *et al.*, 2002), and plays a role in petal formation during flower development (Hu *et al.*, 2008). Whether GA_3 has a similar positive effect on these developmental processes in avocado remains to be determined.

The Mature Flower

Flower development

Avocado flowers are produced in inflorescences, which are panicles of cymes with multiple secondary and tertiary axes (Fig. 6.5) (Reece, 1942; Buzgo *et al.*, 2007). These inflorescences can be either determinate or indeterminate. Determinate inflorescences have a terminal flower, whereas the apical meristem of indeterminate floral shoots continues to grow vegetatively (Schroeder, 1944). The ratio of indeterminate to determinate inflorescences is highly variable, and it appears to be dependent on cultivar, growing region, floral shoot age and the 'on' versus 'off' bloom during alternate bearing (Schroeder, 1944; Thorp *et al.*, 1994; Salazar-García and Lovatt, 1998; Lovatt 2011a). Determinate flowering shoots appear to be more productive than indeterminate ones (Salazar-García and Lovatt, 1998; Alcaraz and Hormaza, 2014; D'Asaro *et al.*, 2021). Removal of the vegetative terminal shoot increases initial fruit set in indeterminate inflorescences (Cutting and Bower, 1990; Alcaraz and Hormaza, 2014), consistent with competition for nutrients and carbohydrates between flowers and shoots as a reason for lower fruit set by indeterminate floral shoots (Whiley, 1990). In contrast, recent research by D'Asaro *et al.* (2021) provided evidence that lower-per cent fruit set by indeterminate floral shoots of avocado was hormonally regulated and not due to competition for photosynthates caused by apical vegetative shoot growth. They demonstrated that C6 and C7 carbohydrate availability was sufficient for fruit set, but lower GA_1 and *trans*-Zeatin concentrations limited fruit set in indeterminate inflorescences. Consistent with these data, foliar-applied GA_3 increased fruit set of indeterminate floral shoots despite stimulating the growth of the vegetative shoot apex (Köhne, 1989). It is clear that further research is required to solve the problem of low fruit set by indeterminate floral shoots, and the low fruit set problem of avocado trees in general. A mature avocado tree can produce thousands of inflorescences, each containing between 100 and 500 individual flowers (Salazar-García and Lovatt, 1998; Garner and Lovatt, 2008), and, as a consequence, a mature avocado tree could produce more than a million flowers. Less than 1% of set fruit remains on the tree until harvest time. In California, on average, ≥70% of the abscising 'Hass' flowers were pollinated, but due to the failure of the pollen grains to germinate and produce pollen tubes, fertilization never occurred and the ovule deteriorated (Garner and Lovatt, 2016). Only 25% of the total number of abscised fruit per crop year was fertilized. Globally, the very low ratio of fruit harvested to flowers produced in avocado is mainly due to the lack of pollination and/or fertilization (Garner and Lovatt, 2016; Alcaraz and Hormaza, 2021).

Flowering phenology

In subtropical and Mediterranean climates, buds transition from vegetative to reproductive development in the late summer to early fall (Salazar-García *et al.*, 1998; Acosta-Rangel *et al.*, 2021), with the irreversible commitment of the buds to floral development documented to occur from the end of October through November in California and Mexico (Salazar-García *et al.*, 1998, 2006), and slightly later in Florida (Buzgo *et al.*, 2007) and Israel (Ziv *et al.*, 2014), suggesting that floral induction could be regulated by sustained LTs in autumn into winter. Bloom occurs during the following spring after WT-dependent floral organogenesis (Acosta-Rangel *et al.*, 2021). Avocado bloom length studies have been conducted in numerous growing regions, including California (Winslow and Enderud, 1955; Robinson *et al.*, 2002; Garner and Lovatt, 2008), Brazil (Falcão *et al.*, 2001), Mexico (Cossio-Vargas *et al.*, 2008b; Rocha-Arroyo *et al.*, 2011a), Israel (Levin, 1981), China (Lu *et al.*, 2004), South Africa (Kaiser and Wolstenholme, 1993) and the Mediterranean region (Papademetriou, 1976; Loupassaki *et al.*, 1995; Demirkol, 2002; Alcaraz and Hormaza, 2009). It should be noted that the criteria used to determine the beginning and end of bloom within these studies, when provided, varied in inclusiveness, from those that excluded periods when few flowers were at anthesis to those that included the period from the first flowers at anthesis until no unopened flowers were observed. Generalizations regarding the length of bloom are possible despite variations in the timing and intensity of reproductive development associated with cultivar, location and seasonal weather conditions. A precise description of the different avocado growth stages from vegetative bud dormancy to fruit harvest using the BBCH code was developed by Alcaraz *et al.* (2013a), thus allowing the standardization of observations made in different environments. Bloom is typically protracted, and, for most cultivars and growing regions, the bulk of bloom occurs over a period of 1–3 months. Bloom length is cultivar and temperature dependent; consequently, for a given cultivar, bloom length is highly variable from year to year (Winslow and Enderud, 1955; Levin, 1981; Garner and Lovatt, 2008). Though a relationship between crop load and the duration and timing of bloom has been suggested (Robinson *et al.*, 2002), studies including multiple cultivars do not demonstrate consistent relationships between crop production and bloom length (Winslow and Enderud, 1955; Garner and Lovatt, 2008). However, in circumstances in which bloom length is relatively short and often occurs in early spring, when temperatures are still low, such as in southern Spain, where the bloom season for 'Hass' lasts about 1 month, most of the crop develops on late-opening flowers when temperatures are usually higher and, consequently, more conducive to pollen germination, pollen tube growth and honeybee activity (Alcaraz and Hormaza, 2021). Thus, under those conditions, an extended and delayed bloom could increase fruit set.

Anthesis

Avocado flowers are hermaphroditic with a single pistil and nine stamens in two whorls (Schroeder, 1944). The inner whorl consists of three stamens with staminodes alternating between them. The outer whorl consists of six stamens. Each of the inner stamens has two small nectaries, one on each side of the filament at the base. However, abnormal flowers that deviate from the normal numbers of stamens, pistils and perianth parts are sometimes found (Schroeder, 1944). Avocado is characterized by a rare flowering mechanism called synchronous protogynous heterodichogamy (Stout, 1923). The functionality of the male and female organs is typically temporally separated. Each flower opens twice. During the first opening, the flowers are functionally female with a receptive stigma and undehisced anthers. At the second opening, the flowers are functionally male with a typically desiccated and unreceptive stigma and anthers that dehisce 1–2 h after flower opening. Avocado genotypes can be divided into two types based on the timing of flower openings (Stout, 1923). Type A cultivars (such as Hass) open as functionally female in the morning of the first day of anthesis and as functionally male in the afternoon of the second day of anthesis. Type B cultivars (such as Bacon, Fuerte, Zutano and Luna UCR) open as functionally female in the afternoon of the first day of anthesis and then as functionally male in the morning of the second day of anthesis or, under some conditions, one day later. Under optimal environmental conditions, this provides an elegant system for cross-pollination, because when anther dehiscence occurs in one flowering type, only the other flowering type is receptive. The entire process of anthesis is highly dependent on environmental conditions, mainly temperature, humidity and light, with cold, cloudy days tending to delay anthesis. The genetic basis of heterodichogamy is still unknown in angiosperms, but recent molecular and genomic work is giving insights into this topic in avocado. Specifically, Ashworth *et al.* (2019) identified markers associated with a quantitative trait locus (QTL) region for flowering type, whereas Solares *et al.* (2023) identified putative candidate genes that could be involved in avocado type A and type B flower development.

During the first opening in the female stage, pollen from male stage flowers of both the same cultivar and complementary flower type cultivars can adhere, hydrate and germinate successfully on avocado stigmas (Papademetriou, 1974; Sedgley, 1979a; Alcaraz and Hormaza, 2014), though typically no more than one pollen tube reaches the ovary (Sedgley, 1976). Studies in Florida suggest that, under particular conditions, the stigma could continue to be receptive during the second opening, possibly due to high humidity (Davenport *et al.*, 1994). Several studies have demonstrated that despite the fact that pollen grain germination occurs in hand-pollinated avocado flowers during the functionally male stage, pollen tube growth is limited and few pollen tubes reach the ovary within 24 h (Sedgley, 1977b; Shoval, 1987), possibly due to callose deposition within the ageing style (Sedgley, 1977b, 1979c). Results in Spain show that no embryo or endosperm development takes place when pollination occurs during the male flowering phase (Alcaraz and Hormaza, 2021).

Pollination

Since in most cases self-pollination in the male phase does not result in fruit production, for effective fertilization to take place there must be a temporal overlap in the opening of functionally male and female flowers. In the 1920s, growers were advised to interplant complementary (type A and type B) cultivars within their orchards to increase pollination and fruit set because opportunities for self-pollination (pollen movement from anther to stigma within a flower, or between flowers of the same cultivar ['close pollination']) were thought to be almost zero due to the temporal separation of functionally male and female flowers (Stout, 1923). Later work revealed that environmental conditions have a substantial effect on the overlap of functionally male and female flowers. At LTs of 17°C/12°C (day/night), flower openings in the female stage of type A trees are delayed (Sedgley and Annells, 1981). An extreme example has been reported in New Zealand, where decreasing minimum overnight temperatures were associated with a delay in the female flower opening in 'Hass', resulting in nocturnal opening of both functionally male and female flowers (Pattemore *et al.*, 2018). A significant negative linear relationship exists between the mean daily temperature and the hour of the day of flower openings (Ish-Am and Eisikowitch, 1992). Therefore, frequent or erratic changes in temperature can result in 'irregular' blooming and overlap of female and male stages within the same flowering type. These effects should allow pollen transfer within inflorescences, among inflorescences of the same tree or among trees of the same cultivar (Lesley and Bringhurst, 1951). This overlap of functionally male and female flower openings has been demonstrated not only by observational studies of the flowers (Continella *et al.*, 1992; Ish-Am and Eisikowitch, 1992; Loupassaki *et al.*, 1995; Alcaraz and Hormaza, 2009) but also by the successful fruit set in large-scale single-cultivar orchards (Hodgson, 1947). Furthermore, high rates of out-crossing between cultivars of the same flowering type have been reported (Violi *et al.*, 2009). In the case of type B avocado cultivars, LTs of 17°C/12°C (day/night) can cause the female stage of flowering to be omitted altogether, thus reducing pollination rates (Sedgley, 1977a; Sedgley and Grant, 1983).

Davenport *et al.* (2007) hypothesized that fruit that result from self-pollination must be the result of male-stage self-pollination, not close pollination, because they rarely observed extended periods of overlap of female and male stages within a cultivar. However, the length of the overlap between the female and male stages can be more than ample for pollination to occur. For example, Alcaraz and Hormaza (2009) found that in Spain, the functionally female and male stages of floral development overlapped for at least 2 h per day for seven of the ten cultivars studied. Furthermore, prolonged overlap may not be required for pollen transfer, as bees or other insects could carry pollen for an extended period.

The rates of pollination are affected by the movement of pollen through the orchard. Avocado pollen has been described as sticky (Stout, 1933), requiring a large, flying pollinator. In its region of origin in Central America, several efficient, putative native pollinators of avocado have been identified in Mexico, including stingless bee species (Apidae, Meliponinae) (Ish-Am *et al.*, 1999; Can-Alonzo *et al.*, 2005), the Mexican honey wasp (*Brachygastra mellifica* Say) and flies (*Chrysomya megacephala*) (Ish-Am *et al.*, 1999; Pérez-Balam *et al.*, 2012). However, as in most crops, avocado commercial production worldwide relies on the availability of the European honeybees (*Apis mellifera* L.), which were not present in the Americas before the arrival of Europeans in 1492 (Dymond *et al.*, 2021). However, honeybees are not very effective pollinators for avocado. They are easily attracted to more fragrant flowers, such as citrus (*Citrus* spp.) and wildflowers (Ish-Am and Eisikowitch, 1993, 1998). Avocado nectar is unattractive to honeybees (Afik *et al.*, 2006b), due potentially to its high potassium and phosphorous content (Afik *et al.*, 2006a). Furthermore, the activity of European honeybees is severely limited in cool, overcast or rainy weather (reviewed in Free, 1970). In addition to honeybees, many different pollinating insects are frequently observed visiting avocado flowers worldwide, and they could perform a significant role in avocado pollination (Pattemore *et al.*, 2018; Dymond *et al.*, 2021; Okello *et al.*, 2021; Muñoz *et al.*, 2021; Sagwe *et al.*, 2022). Recent work in Israel (Stern *et al.*, 2021) demonstrated that pollination by bumblebees (*Bombus terrestris* L.) can significantly increase 'Hass' avocado yield. Bumblebees show several advantages over honeybees, including their ability to pollinate at temperatures below 14°C (Stern *et al.*, 2021).

Although most studies show that insects are required for effective avocado pollination, fruit set and yield (Dymond *et al.*, 2021), some studies suggest that wind pollination could be relevant under some circumstances (Davenport *et al.*, 1994; Davenport, 2019; Ying *et al.*, 2009a, 2009b). However, evidence of how widespread wind pollination of avocado is and its impact on pollen viability, pollination and fertilization rates and fruit set was not demonstrated definitively in those studies. In a 7-year, five-cultivar field study in Israel, Ish-am and Lahav (2011) found that the rate of bee visitations to avocado trees was positively correlated with pollination rates, whereas wind velocity was not significantly related to pollination rates. Other researchers have found that avocado pollination and/or fruit set are typically greatly reduced when pollinators are prevented from accessing avocado inflorescences (Peterson, 1955; Papademetriou, 1976; Gazit, 1977; Vithanage, 1990; Johannsmeier and Morudu, 1999; Malerbo-Souza *et al.*, 2000; Can-Alonzo *et al.*, 2005), suggesting that honeybees and other pollinators contribute significantly to avocado pollination, and ultimately to yield. The results of these studies are consistent with Scholefield's (1982) conclusion that the nectaries and small stigmatic surface of an avocado flower are inconsistent with the typical morphology of wind-pollinated flowers. A study of 'Hass' trees in California by Hoddle *et al.* (2010) showed that open-pollinated trees, open-pollinated trees sprayed with water (spray control) and caged trees accessible to honeybees had equally high levels of fruit set. Trees caged to exclude large, flying insects, including honeybees, but accessible to small insects, such as flower thrips (*Frankliniella* spp.), and windborne pollen, had very low fruit set. Caged trees sprayed repeatedly with insecticides to eliminate even small insects so that fruit set could result only from wind pollination had no fruit. These results substantiate the very limited contribution that self-pollination in the absence of pollinators, small insects and wind make to the pollination and yield of avocado.

Field studies in Spain showed that most 'Hass' flowers did not receive pollen during the functionally female stage, and, of those that did, most received just one to five pollen grains; as a result, only approximately 8% of the flowers received more than five pollen grains on their stigmas (Alcaraz and Hormaza, 2021). The same study found a significant correlation between the number of pollen grains on the stigma and the likelihood of the flower producing a fruit.

Fertilization

The avocado flower is self-compatible (Continella *et al.*, 1992; Sedgley, 1979a). Pollen grain germination and pollen tube growth are sensitive to temperature. Temperatures within the range of 20–30°C are optimal for avocado *in vitro* pollen germination (Loupassaki *et al.*, 1997; Alcaraz *et al.*, 2011), although optimal temperatures for *in vitro* pollen germination and pollen tube growth appear to be cultivar-specific (Alcaraz *et al.*, 2011). Pollen tubes grow abnormally or fail to reach the ovule when exposed to temperature regimes of 33°C/28°C (day/night) or 17°C/12°C (day/night), respectively (Sedgley, 1977a; Sedgley and Annells, 1981). Ovule growth and viability are also inhibited at these two temperature regimes (Sedgley and Annells, 1981). Analysis of 'Hass' flowers revealed that the growth of many pollen tubes stopped mid-style, with only one pollen tube continuing to grow to the ovule (Jaganath, 1993). Based on these data, Lovatt (1997) hypothesized that 'Hass' pollen tube growth might be stopped by a mid-style barrier that breaks down in the absence of outcrossing to enable self-pollination and the success of the species. Although pollen tubes are capable of travelling the length of the stigma and style within 1–3 h (Schroeder, 1942; Papademetriou, 1974; Sedgley, 1977a), Sedgley (1979b) noted that ovule penetration does not occur until 18–24 h after pollination, with fertilization of the polar nuclei and egg cells taking place within 48 h of pollination (Sedgley, 1979b). Even when a pollen tube penetrates the ovule, fertilization might still fail to occur, a phenomenon observed more frequently in 'Fuerte' than in 'Hass' ovules (Sedgley, 1979b).

In addition to temperature, several other factors affect avocado flower abscission, including, but not limited to, the growth of fungi on pistils (Thomas *et al.*, 1994) and nutrient availability. Even when trees were supplied with nutrients in amounts considered adequate for production and had leaf nutrient concentrations in the optimal range, foliar applications of boron (B) or nitrogen (N as urea) at the cauliflower stage of inflorescence development increased the yield of 'Hass' (Lovatt, 1994). These results were consistent with the demonstrated effects of these nutrients on pollen tube growth and ovule longevity, respectively (Jaganath, 1993). In Australia, prebloom soil applications of increasing boron concentrations had no effect on the total number of fruit set until 10 weeks after full bloom. At that time, excess boron reduced fruit set by 56%, resulting in a yield reduction of 25% (Hapuarachchi *et al.*, 2022). Boron concentration had no effect on the number of fruit set from outcrossing or self-pollination. Approximately 88–92% of the mature fruit harvested were self-pollinated. Although one Australian study indicated that N availability was not limited during flower development on vigorous shoots (Thorp *et al.*, 1993), carefully timed soil applications of N in California during bloom and early fruit set increased the 4-year cumulative yield of 'Hass' (Lovatt, 2001). These results indicate that the tree utilized increased B and N during flower and early fruit set. In a 3-year study comparing foliar KH_2PO_3 and KH_2PO_4 applied at the cauliflower stage of inflorescence development with a control receiving soil-applied KH_2PO_4, foliar-applied KH_2PO_3 significantly increased the 3-year cumulative yield of 'Hass' (Gonzalez *et al.*, 2010). Studies in Spain (Alcaraz *et al.*, 2013b) and New Zealand (Boldingh *et al.*, 2016) have shown that the sink strength of avocado flowers is predetermined at anthesis and that this flower quality influences the probability of fruit set; thus, flowers with higher carbohydrate and boron content at anthesis are more likely to produce fruit after fertilization has taken place. Endogenous hormone concentrations play a critical role in establishing and maintaining the sink strength of the flowers and in the developing fruit (Cowan, 2004; Cowan *et al.*, 2005; D'Asaro *et al.*, 2021). The role of endogenous hormones in avocado flowering and fruiting warrants further investigation.

Outcrossing

Commercial scions of avocado are clonally propagated, and although single-cultivar avocado orchards can be productive (Hodgson, 1947), a study in 1966 concluded that close planting of different avocado cultivars increased yields (Bergh *et al.*, 1966). In 1978, isozyme analysis was first used to demonstrate that outcrossing occurs between neighbouring avocado cultivars in California (Torres and Bergh, 1978).

In a series of studies conducted in Israel, isozyme analysis of avocado embryos demonstrated that fruit that are the product of self-fertilization (as a result of self- or close pollination) abscised at a much higher rate than fruit that were the product of outcrossing (Degani *et al.*, 1986, 1989, 1997). High outcrossing rates have also been reported among progeny of avocado trees interplanted with or near pollinizer cultivars in Israel (Degani *et al.*, 1986, 1989, 1997; Stahl *et al.*, 2019; Stern *et al.*, 2021), Spain (Alcaraz and Hormaza, 2011), Australia (Sulaiman *et al.*, 2004), California (Vrecenar-Gadus and Ellstrand, 1985; Chen *et al.*, 2007; Garner *et al.*, 2008; Schnell *et al.*, 2009; Ying *et al.*, 2009a, 2009b) and Florida (Davenport *et al.*, 1994; Borrone *et al.*, 2008; Violi, *et al.*, 2009). The outcrossing rate is highly variable depending upon the location, tree spacing and distribution, and pollinizer cultivar (Sánchez-González *et al.*, 2019). The success of the pollen parent in promoting fruit retention is cultivar-dependent (Gazit and Gafni, 1986; Goldring *et al.*, 1987; Degani *et al.*, 1989, 1990, 1997; Sulaiman *et al.*, 2004). The contribution of outcrossing to fruit set must be due to selective abscission of the self-pollinated fruit, superior resource allocation and/or hormone content of the outcrossed fruit, or a combination of these factors. In fact, different studies have shown differences in seed and fruit characteristics between fruit derived from selfing and outcrossing (Degani *et al.*, 1990; Alcaraz and Hormaza, 2021), although this is not always the case (Kämper *et al.*, 2021). It has been hypothesized that offspring fitness of many angiosperms is increased by selective abortion of certain developing fruit (reviewed in Lee, 1984; Stephenson and Winsor, 1986). It has been suggested that inbreeding depression results in decreased embryo vigour, thus promoting the abscission of avocado fruit with embryos that are the product of self-pollination (Degani *et al.*, 1997). Inbreeding depression has also been demonstrated in studies on avocado seedlings, in which progeny that are the result of self-fertilization were found to be more susceptible to *Phytophthora cinnamomi* infection (Violi *et al.*, 2009) or more likely to have lethal mutations (Schnell *et al.*, 2009) compared to those that were the result of outcrossing. Though Alcaraz and Hormaza (2011) found that abscised fruit were typically the result of self-fertilization, they suggested that the abscised fruit had likely been produced late in bloom, when bloom overlap within the orchard was limited. These late-season fruit were unable to compete with larger, faster-growing fruit that set earlier in the season.

Despite the fact that the selective retention of outcrossed fruit is typically highly significant, the relationship between outcrossing rate and yield has only been established in a few Israeli studies, and these were often weak or cultivar dependent (Degani *et al.*, 1989, 1990, 1997). Researchers have speculated that the initial rate of outcrossing and its effect on fruit set and yield might be dependent, in part, on climate (Goldring *et al.*, 1987). Under temperate growing conditions, only weak correlations between outcrossing rates and yields have been obtained. In an isozyme study, a significant relationship between outcrossing and yield was detectable only when data from all replications and orchards were pooled, and even this relationship was extremely weak ($r^2 = 0.11$) (Vrecenar-Gadus and Ellstrand, 1985). In another study of 'Hass' orchards at three different California locations, outcrossing rates, determined by random amplified polymorphic DNA (RAPD), were only significantly correlated with yield at one location ($R^2 = 0.50$) (Kobayashi *et al.*, 2000). In an alternate-bearing orchard in California, outcrossing rates determined by microsatellites were not related to the number or kilograms of fruit harvested for either crop year or for both crop years combined ($P > 0.05$ and $R^2 < 0.01$ for number and kilograms harvested), and the degree of alternate bearing was not significantly correlated with outcrossing rate (Garner *et al.*, 2008). In Spain, Alcaraz and Hormaza (2011) found that the relationship between 'Hass' yields and outcrossing rates, as determined by microsatellites, was not significant ($P = 0.864$ and $P = 0.702$ for the 2 years of the study).

Even if interplanting of pollen donor trees contributes to increased fruit set and yield, the practice may not be horticulturally and economically practical in all growing regions. In areas where cool air temperatures often delay flower openings, substantial yields are attainable in orchards with high rates of close pollination (Alcaraz and Hormaza, 2021; Hapuarachchi *et al.*, 2022). Outcrossing rates, and the potential contribution of outcrossing to fruit persistence and yield, could also be limited by other factors, including bloom overlap with an effective pollinizer and the distance to the nearest pollinizer. In the few studies on avocado outcrossing rates in which the timing of bloom was reported, overlap between cultivars was typically extensive but incomplete (Degani *et al.*, 1997; Garner *et al.*, 2008; Alcaraz and Hormaza, 2011), suggesting that high outcrossing rates could require more than one pollinizer cultivar to guarantee consistent bloom overlap.

Though outcrossing has been detected among 'Hass' up to 80 m from the nearest pollinizer tree (Vrecenar-Gadus and Ellstrand, 1985), there is typically a negative correlation between outcrossing rate and distance to a pollinizer cultivar (Vrecenar-Gadus and Ellstrand, 1985; Degani *et al.*, 1989, 1990, 1997; Kobayashi *et al.*, 2000; Ying *et al.*, 2009a; Alcaraz and Hormaza, 2011; Kämper *et al.*, 2021). Higher outcrossing rates require close proximity to pollinizer trees, likely due to the foraging habits of honeybees (*Apis mellifera* L.) within avocado orchards. The movement of honeybees to adjacent rows decreases hyperbolically (Ish-Am and Eisikowitch, 1998; Peña and Carabalí, 2018). Ying *et al.* (2009a) have suggested that the relationship between outcrossing rates and distance from a pollinizer is also consistent with wind pollination. Garner *et al.* (2008) found that high outcrossing rates in 'Hass' trees in California were only possible when each 'Hass' tree was adjacent to a pollinizer tree and hypothesized that high densities of pollinizer trees (a minimum of 16% of the trees in an orchard planted on the square) would be required for high rates of outcrossing throughout an orchard. They suggested that because the financial return of pollinizer cultivars in California is extremely low in comparison with 'Hass', the cost of interplanting pollinizer cultivars could far exceed the monetary benefits of any potential increases in 'Hass' yields that outcrossing might provide, especially during the years when environmental conditions during bloom decrease the likelihood of outcrossing (Garner *et al.*, 2008).

Fruit Development

Often, a high number of fruit initially seem to be set under favourable conditions. However, heavy fruit abscission can occur in the weeks following anthesis due to several factors. On the one hand, developing fruit often abscise due to lack of proper embryo development (Garner and Lovatt, 2016). In some cases, those fruit have not been fertilized. Although pollination without fertilization can briefly stimulate ovary enlargement, the resulting parthenocarpic fruit enlarges to approximately 2–3 mm or less in diameter (Sedgley, 1980) and approximately 4 mm or less in length (Tomer and Gazit, 1979) before abscising 2–4 weeks after flowering. In other cases, fertilization occurs but seedless fruit, typically referred to as 'cukes', are produced. These fruit are not the result of parthenocarpy but of stenospermocarpy, in which fertilization takes place (Tomer *et al.*, 1980), but the embryo and endosperm of the fruit fail to develop properly, resulting in embryo abortion (Tomer *et al.*, 1980; Steyn *et al.*, 1993). The persistence of these seedless fruit is rare in most avocado cultivars, though it is relatively common for the Fuerte cultivar (Blumenfeld and Gazit, 1974) and occurs on girdled tree branches of 'Hass' (Garner, 2004). Seedless fruit abscission is thought to be due to the limited sink strength and availability of growth-promoting hormones typically provided by the large seed of the avocado (Blumenfeld and Gazit, 1974). On the other hand, both internal factors, such as limited nutrient and hormone levels, and external factors, such as wind, low humidity and low or high temperatures, can affect abscission rates of fertilized fruit in the weeks following anthesis (Garner and Lovatt, 2016).

The avocado fruit is botanically a pyriform or globose fleshy drupe containing one seed. The exocarp (peel), of variable thickness and colour depending upon the race and cultivar, surrounds the mesocarp, which constitutes the edible part of the fruit, with a thin endocarp next to two papery seedcoats. Exocarp colour and texture are highly variable, and the colour at ripening can be black, brown, green or purple, with intermediate intensities of each one depending upon the genotype. The fruit shape and size are also variable and can range from spherical to pear or egg-shaped.

Avocado fruit maturation can be divided into two main phases, fruit maturation on the tree and postharvest ripening. Fruit maturation that takes place on the tree can last from 6 to 18 months depending upon both the environmental conditions and the genotype. Postharvest ripening takes place after the detachment of the fruit from the tree, while the mesocarp softens and other organoleptic characteristics of the fruit change. Avocado fruit is harvested when physiological maturity (the ability to ripen after harvest) is reached. Edible ripeness takes place several days after harvesting, and the period between physiological and edible maturity depends on the variety, the time of harvesting and postharvest storage temperatures.

Basic anatomical studies on avocado embryo and endosperm development have been conducted (Cummings and Schroeder, 1942; Schroeder, 1952; Sedgley, 1979b; Tomer and Gazit, 1979). With successful fertilization, fruit enlargement occurs rapidly as a result of high rates of cell division and expansion (Stage I), but unlike other fruit, cell division continues throughout fruit development, even in mature fruit attached to the tree (Schroeder, 1953; Bower and Cutting, 1988). This increase in fruit size has generally been described as a single sigmoid curve (Bower and Cutting, 1988), but the accuracy of such a description is dependent on both cultivar and method of measurement (Piper and Gardner, 1943; Valmayor, 1967; Blumenfeld and Gazit, 1974; Mougheith *et al.*, 1978). Avocado is a climacteric fruit; thus, an increase in the respiration rate and ethylene production takes place at the onset of fruit ripening (Bower and Cutting, 1988).

It has been suggested that early increases in fruit size can be attributed to seed growth (Mougheith *et al.*, 1978). Basic seed development was described by Piper and Gardner (1943). The embryo and endosperm are surrounded by a thick integument, and the embryo grows within the gelatinous nucellus or endosperm, which is thought to nourish the developing embryo. As the seed matures, the embryo is surrounded by large cotyledons sheathed by a continuously thinning integument. At maturity, the integument, now typically referred to as the seedcoat, is brown and papery. The seedcoat provides the only vascular connection between the seed and mesocarp (Blumenfeld and Gazit, 1970b). Anatomical observations of abscising avocado fruit are often characterized by premature deterioration of the nucellus (Davenport and Manners, 1982; Garner and Lovatt, 2016) and the seedcoat (Davenport and Manners, 1982; Cowan *et al.*, 1997; Garner and Lovatt, 2016). Whether deterioration of these tissues is the cause of fruit abscission or a symptom thereof remains to be determined.

The second stage of avocado fruit growth (Stage II) is dominated by mesocarp growth (Mougheith *et al.*, 1978). The seed is surrounded by an actively dividing mesocarp that accumulates carbohydrates and proteins (Davenport and Ellis, 1959; Mougheith and Abdel-Hamid, 1978) and fatty acids, especially oleic acid (Mougheith and Abdel-Hamid, 1978; Eaks, 1990). Unlike other fruits, avocado accumulates oil instead of sugars and, during fruit growth, total mesocarp water content decreases as oil content increases, and, thus, oil content in the fruit can be used to establish the different fruit developmental stages in order to decide when physiological maturity has been reached (Lee, 1981). There is a positive correlation between oil content and dry matter (defined as the solid content of the fruit minus its water content) and, consequently, dry matter is an indirect indicator of oil content and, thus, of fruit maturity (Church and Chace, 1922; Davenport and Ellis, 1959; Eaks, 1990). Avocado fruit is unique relative to other fruit tree crops in that it accumulates high concentrations of protein (Church and Chace, 1922; Davenport and Ellis, 1959) and seven carbon (C7) sugars (e.g. mannoheptulose and perseitol) instead of C6 sugars (Davenport and Ellis, 1959; Liu *et al.*, 1999b, 2002; Cowan, 2004). The C7 sugar, D-*manno*-heptulose, might be responsible for continued sugar uptake and growth of avocado fruit by inhibiting respiration, hexokinase activity and/or damage by reactive oxygen species (Cowan, 2004; Bertling and Bower, 2005, 2006; Bertling *et al.*, 2007). Avocado fruit development is slower than that of most other fruit tree crops. For example, developing 'Hass' fruit reach maturity 14–18 months after the start of bloom in some avocado-growing regions (Bergh, 1975). Therefore, mature fruit could be present on the tree during the spring bloom and early fruit set of the next crop year. However, the fruit of many avocado cultivars, including Hass, do not ripen until harvested or until their pedicels are girdled (reviewed in Tingwa and Young, 1975) and, therefore, can be left to 'hang' on the tree for 18–19 months from the initial bloom date. The C7 sugars might play a role in the unusual ripening habit of avocado fruit (Robinson *et al.*, 2002; Cowan, 2004). Since the phloem-transported C7 sugar content of avocado fruit decreases after both harvesting and pedicel girdling, it has been suggested that C7 sugars might prevent ripening when the fruit are still attached to the tree (Liu *et al.*, 1999b, 2002). However, other biochemical changes, including changes in fruit polyamine concentrations (Winer and Apelbaum, 1986) and hormone concentrations of fruit or stems, have also been suggested as possible ripening signals (Tingwa and Young, 1975). The putative role of C7 sugars in ripening and their known antioxidant activity have led to studies on the role of C7 sugars in postharvest fruit quality (Bertling and Bower, 2005; Bertling *et al.*, 2007).

Assimilate partitioning and reproductive development

Inter- and intra-seasonal competition for limited carbohydrate resources has often been suggested as a major factor that restricts fruit production in avocado. Several studies on alternate-bearing trees demonstrated correlations between starch reserves in branches and yields the following year (Scholefield *et al.*, 1985; Van der Walt *et al.*, 1993; Whiley *et al.*, 1996a, 1996b). At the end of an 'off' year, starch reserves were high and at the end of an 'on' year, starch reserves were low, with the following years' yields positively correlated to the size of these reserves. Although the correlations were highly significant in all of these studies, there was no evidence of causation in that there was no indication that carbohydrate availability limited yields in 'off' years.

Competition for assimilates also occurs every spring and summer during reproductive and vegetative development. This competition is thought to be especially acute within indeterminate floral shoots. The leaves of indeterminate floral shoots are sinks during at least some period of early fruit set (Whiley, 1990; Finazzo *et al.*, 1994). Thus, vegetative and reproductive sinks are developing contemporaneously and in close spatial proximity. Indeterminate floral shoots have lower fruit set than determinate floral shoots (Buchholz, 1986; Lovatt *et al.*, 1994; Thorp *et al.*, 1994; Salazar-García and Lovatt, 1998). The removal of the vegetative portion of indeterminate floral shoots (Biran, 1979; Zilkah *et al.*, 1987; Bower and Cutting, 1992) and the inhibition of vegetative growth with foliar applications of paclobutrazol (Cultar®; Syngenta Crop Protection, UK Limited), a GA biosynthesis inhibitor (Wolstenholme *et al.*, 1990; Zilkah and David, 1993), have been shown to increase fruit set. It has, therefore, been suggested that the lack of competition within determinate floral shoots, which do not have a vegetative component, is responsible for their higher fruit set. However, other researchers argue that carbohydrate export from developing indeterminate floral shoots to other tree structures provides evidence that carbohydrate production does not limit fruit set on those shoots (Thorp *et al.*, 1993; Finazzo *et al.*, 1994). Furthermore, Finazzo *et al.* (1994) noted that fruit abscission continues after this competition period, suggesting that fruit abscission from indeterminate floral shoots cannot be due solely to competition between the vegetative and reproductive sinks.

Numerous studies have examined the cycling of carbohydrate reserves in avocado trees throughout their phenology in attempts to determine if and how carbohydrates limit avocado fruit set. Stem reserves are thought to be the only carbohydrate reserves utilized during spring reproductive and vegetative growth, as trunk and root supplies are not depleted at this time (Liu *et al.*, 1999a). In the studies on 'Hass' and 'Fuerte' trees in Australia and 'Hass' in California, branch starch concentrations were maximum just prior to flowering and then decreased during flowering, summer vegetative shoot growth and fruit development, with starch concentrations then increasing gradually during the winter until the next flowering period (Scholefield *et al.*, 1985; Whiley *et al.*, 1996a, 1996b; Liu *et al.*, 1999a). Total soluble solid (TSS) cycling was similar to that of starch, except that TSS concentrations peaked 2 months prior to flowering and again in the summer as flowering was ending and early fruit set was beginning (Liu *et al.*, 1999a). The authors noted that photoassimilates produced during the summer appeared to be utilized during summer growth, since storage reserves did not increase at this time. They hypothesized that the photosynthate produced during the vegetative and reproductive growth in summer might be adequate to sustain such growth (Liu *et al.*, 1999a). This suggests that limitations in fruit production are due to resource allocation, not resource availability. The idea that carbohydrates might not be limiting to yields has been hypothesized for other fruit tree crops (Bangerth, 2000). This has led many researchers to investigate the roles of endogenous hormones in the allocation of reserves to fruit and competing sinks, and the use of exogenous PGRs to increase fruit growth by manipulating sink strength.

Endogenous hormones and fruit development

The seed of developing avocado fruit contains high hormone concentrations, but as the fruit nears maturity and the seedcoat becomes desiccated, concentrations of hormones decrease dramatically. Researchers in Israel found that numerous cultivars had a high concentration of auxin in the seed, with higher concentrations found in fruit with faster mesocarp growth, suggesting that auxins influence sink strength of growing fruit (Gazit and Blumenfeld, 1972). GA synthesis might also influence the sink strength of fruit, as GA-like activity in the seedcoat decreases over time as tissue growth rates decrease (Blumenfeld and Gazit, 1972). Cytokinin (CK) concentrations are also high in avocado seeds and seedcoats, and decrease as fruit mature (Blumenfeld and Gazit, 1970a). Other researchers have found that concentrations of isopentenyladenine (IPA), a CK, are high in the developing seedcoat and decrease with fruit maturity, becoming undetectable once the seedcoat is dry (Cutting *et al.*, 1986). The concentrations of the polyamines putrescine, spermidine and spermine are high during early fruit development, suggesting that polyamines might play a vital role in early fruit growth, possibly by promoting cell division (Winer and Apelbaum, 1986; Kushad *et al.*, 1988). Polyamine concentrations are highest in the seedcoat and decrease with maturity (Winer and Apelbaum, 1986). Thus, the large seed, or more specifically its seedcoat, is thought to provide the sink strength for avocado fruit. This might help explain the low rate of persistence of seedless avocado fruit.

Phenotypically small 'Hass' fruit (<200 g) have been used as a model system to investigate the synergistic and antagonistic nature of hormones in avocado fruit growth. Cutting (1993) referred to the production of this small fruit variant in the warmest avocado-growing regions of South Africa and 'rescuing' the fruit with CK applications. The small fruit variant has since been used to model the role(s) of hormones in avocado fruit development and the interrelationship of the hormones produced by the isoprenoid pathway, which generates, among other bioactive compounds, CKs, abscisic acid (ABA) and GAs. Compared to normal fruit, the small fruit are characterized by early degeneration of the seedcoat, decreased cell numbers, high mesocarp ABA concentrations, low CK concentrations and a 70% decrease in the activity of the rate-limiting enzyme, 3-hydroxy-3-methylglutaryl-CoA reductase

(HMGR), early in the isoprenoid pathway (Cowan *et al.*, 1997). HMGR activity and/or its expression in the developing fruit are hypothesized to be regulated by sugar and ABA content (Richings *et al.*, 2000). Initially, the ratio of CKs to ABA was thought to be critical for proper fruit development (Cowan *et al.*, 1997). Moore-Gordon *et al.* (1998) confirmed that as IPA concentrations decreased, ABA concentrations increased and a significant positive linear relationship exists between the IPA to ABA ratio and fruit size. Further work by the Cowan laboratory using the small fruit variant model demonstrated that though seedcoat ABA content increases transiently during the onset of the small fruit phenotype, CK-like activity is not different between the normal and small fruit types during fruit development (Taylor and Cowan, 2001; Cowan *et al.*, 2005). However, indole-3-acetic acid (IAA) content within the developing seed was significantly higher in normal fruit compared to small fruit (Cowan *et al.*, 2005). Decreased ABA biosynthesis and increased IAA biosynthesis are correlated with decreased xanthine dehydrogenase (XDH) activity (Taylor and Cowan, 2004). Cowan *et al.* (2005) hypothesized that stress might decrease the activity of XDH, increasing the seed ABA to IAA ratio, ultimately resulting in reduced fruit growth. In addition to being a model for the role of hormones in avocado fruit development, differences in the expression of genes thought to be responsible for the regulation of cell division have also been demonstrated between the normal and small fruit phenotypes (Dahan *et al.*, 2010).

Ethylene is produced by developing avocado fruit, with the majority of it being produced in the seedcoat (Sitrit *et al.*, 1987). An increase in ethylene production by developing avocado fruit precedes their abscission and is correlated positively with fruit abscission rates and inversely with fruit size (Adato and Gazit, 1977a). However, ethylene evolution is thought to be one of the last steps in the changes preceding abscission and not the underlying cause of fruit abscission (reviewed in Bangerth, 2000). The seedcoat is also the site of high levels of pre-climacteric ethylene synthesis in harvested fruit, though this ethylene production only results in the ripening of the mature fruit (Adato and Gazit, 1977b).

Flower and Fruit Abscission

In contrast to other fruit tree crops such as apple (*Malus domestica* Borkh.) and pear (*Pyrus communis* L.), which often require thinning to reduce fruit set and increase the production of fruit of marketable size and quality (reviewed in Williams and Edgerton, 1981; Looney, 1998; Bangerth, 2000), the avocado has extremely low fruit set (<0.1%) even in healthy, well-managed orchards (Whiley and Schaffer, 1994; Garner and Lovatt, 2008; Alcaraz and Hormaza, 2021). The avocado and many other subtropical fruit tree crops, including mango (*Mangifera indica* L.) (Singh, 1960), macadamia (*Macadamia* sp.) (Sakai and Nagao, 1985) and litchi (*Litchi chinensis* Sonn.) (Stern *et al.*, 1995), are characterized by the abscission of thousands of flowers and immature fruit, with only a small percentage of these fruit persisting to harvest.

One of the first attempts to characterize flower and fruit abscission of the avocado was a branch study conducted in Trinidad utilizing numerous cultivars (Papademetriou, 1975). More than 95% of the observed flowers abscised within 1 week of the end of bloom, and fruit abscission was completed 4–12 weeks after the end of bloom, depending upon the cultivar. Final fruit set estimates ranged from 0.00% to 0.66% per branch. In studies of fruit subsamples monitored on 'Fuerte' trees in Mexico (Muñoz Pérez *et al.*, 1988) and 'Hass' trees in Israel (Adato and Gazit, 1977b), it was found that most fruit abscised within a few months of the end of the bloom, followed by a gradually declining rate of fruit abscission, with fruit abscission ceasing in October. During bloom observations in Florida, Davenport (1982) found that many commercially important cultivars, including Hass and Fuerte, were characterized by high rates of fruit abscission, whereas others had low initial fruit set rates and limited abscission. In one of the few studies designed to characterize and quantify whole-tree flower and fruit abscission of the avocado, researchers in Israel found that individual 'Fuerte' trees can lose hundreds of thousands of flowers, and thousands of developing fruit in a single season with a final fruit set of <0.02% (Lahav and Zamet, 1975). Fruit abscission was also monitored on 'Hass' and 'Ettinger' trees during this study, and for all three cultivars, fruit loss was intense during May and June but was negligible by September (northern hemisphere). In a study of two 'Fuerte' trees in South Africa, more than 400,000 flowers and 10,000 developing fruit abscised per tree within a 10-week period (Slabbert, 1981). In Japan, Fuerte, Bacon and Zutano cultivars abscised tens of thousands of flowers and hundreds of fruit per tree (Inoue and Takahashi, 1990). Cultivar differences and year effects were pronounced. The highest fruit set value in this study was <0.04%. In a small study (two trees observed for 2 years), 'Fuerte' trees in California lost hundreds of thousands of flowers per season, with a mean final fruit set of 0.07% (Cameron *et al.*, 1952). In a study of ten alternate-bearing 'Hass' trees in California, an average of 75,000 and 339,000 flowers per tree abscised in 'off'- and 'on'-crop years, respectively (Garner and Lovatt, 2008), and the majority of these flowers were unpollinated (Garner and Lovatt, 2016). Abscission of immature fruit peaked approximately 1 month after peak flower abscission, with an average of 50 and 280 fruit per tree per day in 'off'- and 'on'-crop years, respectively (Garner and Lovatt, 2008). Researchers using both open and hand pollination have noted that many fruit abscising during this time period, typically referred to as June drop, are unfertilized enlarged ovaries (Tomer and Gazit, 1979; Sedgley, 1980). Similar observations have been made in Spain (Alcaraz and Hormaza, 2011, 2021). Taken together, these studies on avocado flower and fruit abscission indicate that excessive abscission occurs in the spring and summer, the early and June fruit drop periods, respectively, after which the fruit abscission rate declines, typically reaching zero fruit per week in autumn, though the intensity, timing and longevity of the abscission periods vary with both cultivar and location. When abscission of immature fruit decreases, the abscission of mature fruit from the preceding year's crop begins, suggesting that the two abscission processes are regulated independently (Garner and Lovatt, 2008). The 'off'- or 'on'-crop status of the tree does not affect the timing of flower and fruit abscission, percentage fruit set, or the average diameter or biomass of individual fruit abscising at similar phenological stages, despite differences in the magnitude of the number of reproductive structures abscising (Garner and Lovatt, 2008). This suggests that yield is not solely resource limited and that alternate bearing is neither the cause nor the result of flower or fruit abscission.

Although the loss of large numbers of fruit is a natural thinning process for most forest species and fruit tree crops, '... from a horticultural point of view, this self-regulatory mechanism may be too strong for fruit species, such as... avocado...' (Bangerth, 2000). Wolstenholme (1990) listed the avocado among subtropical evergreen crops domesticated relatively recently and described the need for new cultural practices to maximize their productivity. Avocado trees are physiologically capable of bearing much larger crops. By determining the energy required to produce 1000 kg of apple, 'Valencia' orange (*Citrus sinensis* L.), or avocado fruit, and then comparing yields of high-intensity apple, orange and avocado production, it was estimated that a well-managed avocado orchard is theoretically capable of producing 32.5 t ha^{-1} of fruit with 17% oil content (Wolstenholme, 1986). Worldwide, avocado yields have averaged significantly less than this due to several factors including alternate bearing, though 'on' crops meet this high yield estimate in many avocado-producing countries.

Alternate Bearing

In avocado reproductive biology, alternate bearing is still an enigma. Avocado trees, especially the Hass cultivar, are prone to alternate bearing (also called biennial, irregular or uneven bearing). As described in the introduction, alternate bearing is repeating cycles of a light, low-yield 'off' crop ('off' tree, 'off' year) followed by a heavy, high yield 'on' crop ('on' tree, 'on' year). Although, in general, the 'off'/'on' cycles are biennial, in some cases, an 'on' year can be followed by two or more consecutive 'off' years, and *vice versa*. Yield reductions from the 'on-' to 'off' year can be 100%, with no fruit in the year following the 'on' crop (Whiley *et al.*, 1992). The severity of alternate bearing is evaluated by calculating the alternate bearing index (ABI) for each data tree in an orchard (or an entire orchard) for each pair of consecutive harvests using the following equation: ABI = (year 1 yield − year 2 yield [expressed as an absolute number])/(year 1 yield + year 2 yield); yield is defined as total weight (kg) of fruit at harvest per tree or orchard (Pearce and Dobersek-Urbanc, 1967).

The range for ABI is from 0 (no alternate bearing) to 1 (complete alternate bearing; crop 1 year, no crop the other year). Alternate bearing is a widespread phenomenon and appears to be inherent in polycarpic woody perennials, occurring in both deciduous and evergreen trees under commercial production and in nature, where it is called 'masting' (Monselise and Goldschmidt, 1982). It has been observed to occur on a single branch, part of a tree and over the entire tree and to be synchronized across an entire orchard and even a large geographical area (Monselise and Goldschmidt, 1982).

Alternate bearing causes significant negative economic consequences to commercial avocado production on a local scale as well as globally. The 'on' crop is characterized by a large number of small-sized fruit of reduced commercial value at harvest (Lovatt, 2011a). For the 'off' crop, fruit size is large but there are too few fruit to provide growers with a good net income. In addition, 'off'-/'on'-crop loads, especially within a single orchard, complicate orchard management, including fertilization and use of PGRs to increase yield and fruit size (Garner *et al.*, 2011). Moreover, the reduced amount of fruit in the 'off' year has a negative impact on every step in the production chain from farm to consumer, including harvesting, packinghouse operation, distribution, marketing and fruit pricing, which jeopardizes the stability and sustainability of avocado commodity-based industries. Years with low fruit volume can also compromise the development and success of value-added product industries.

Initiation of alternate bearing

Alternate bearing is typically initiated by climatic events, for example, freeze, low or high temperatures, lack of rain or hypoxia due to excessive rain or over-irrigation, which occur during the stages of floral development, anthesis or fruit set and cause floral bud abortion, reduced pollination and syngamy in seeded fruit crops, or excessive abscission of flowers or fruit (Jonkers, 1979). This results in an 'off' crop, typically followed by an 'on' crop in 1 to 2 years, depending upon how long it takes the trees to recover from the environmental stress. Conversely, optimal conditions during bloom and fruit set that preclude natural flower and fruit thinning result in an 'on' crop followed by an 'off' crop. Cultural management practices can initiate alternate bearing. Over-irrigation or fertilization, which result in excessive flower and fruit abscission, can create an 'off' crop that, with timely correction of these cultural conditions, can be followed by an 'on' crop. Excessive pruning, resulting in low flower or fruit number and vigorous vegetative shoot growth, can initiate an 'off' crop followed by heavy flowering on the new vegetative shoots and an 'on' crop. The practice of removing fruit from newly planted trees to promote vegetative shoot growth so that young trees grow into their allotted space faster during the establishment of a new orchard disrupts the natural balance between fruiting and non-fruiting shoots and increases the potential for alternate bearing. Delayed harvest has been proposed as a factor that exacerbates alternate bearing in avocado (Paz-Vega, 1997). In California, 'Hass' is harvested 11 to 16 months after bloom. Once initiated, alternating 'on'-/'off'-yield cycles are perpetuated by the effect of fruit number (crop load) on factors within the tree that ultimately impact floral intensity the following year, such that the 'on' crop reduces return bloom the following spring, whereas the 'off' crop results in an intense return bloom (Van der Walt *et al.*, 1993). Thus, once initiated, alternate bearing becomes entrained until reset by a subsequent climate event or implementation of a cultural management practice designed to reduce crop load (number of fruit per tree) in the 'on' year.

Perpetuation of 'on'-/'off'-yield cycles

In avocado, the perpetuation of alternating 'on'/'off' crops begins shortly after fruit set with the young developing fruit. The number of avocado fruit developing per tree (crop load) controls the return bloom by three basic mechanisms that have a cumulative negative effect as fruit continue to develop and mature. The mechanisms are typically discussed in terms of the negative effects of the 'on' crop. The young developing fruit of the 'off' crop have minimal effect, thus appearing to have the opposite effect of the 'on' crop. These physiological mechanisms entrain the 'on'-/'off'-yield cycles and are key determinants of avocado tree productivity.

Note that the underlying physiology of each mechanism is not yet fully known. First, the 'on' crop of young developing fruit reduces summer vegetative shoot growth, limiting both the number and length of new shoots that develop, which reduces the number of nodes that can bear buds with the potential to develop into inflorescences at return bloom. Second, the developing fruit of the 'on' crop inhibit the expression of key genes necessary for the initiation of the flowering process, inflorescence development and flower formation. Third, the 'on' crop of the maturing fruit inhibits spring bud break, further reducing the number of inflorescences at spring bloom following the 'on'-crop year. If the mature 'on' crop of fruit continues to be held on the tree through summer or fall in the year following the year the crop was set (Year 1), the 'on' crop of mature fruit will inhibit summer and fall vegetative shoot growth in Year 2, as they did as young developing fruit in Year 1, which reduces the number of floral buds at return bloom in Year 3, thus, causing back-to-back 'off' crops (Lovatt, 2011a). The negative effects of the 'on' crop in reducing the return bloom accumulate over the course of fruit development.

It is of value to note that several effects of the 'on' crop that contribute to the entrainment of the 'on'-/'off'-crop cycles in other perennial fruit and nut crops do not occur in avocado. For example, young developing avocado fruit of the 'on' crop do not cause abscission of floral buds for next year's crop during the period of exponential fruit growth. This mechanism of entraining 'on'-/'off'-crop cycles was first discovered in pistachio (*Pistacia vera*) (Crane and Nelson, 1971; Lovatt *et al.*, 2006) and was subsequently demonstrated to occur in 'Manzanillo' olive (*Olea europaea*) (Fichtner *et al.*, 2021). Results of the research designed to determine whether the yield in 1 year influenced the number of reproductive structures that abscised from 'Hass' trees the following year demonstrated that the 'off'- or 'on'-crop status of the tree did not influence flower or fruit abscission the next year. Furthermore, the number of mature fruit in the previous year's crop did not influence the per cent fruit set, average fruit diameter or biomass of individual fruit that abscised at similar developmental stages. In both years of the research, as abscission of the immature fruit decreased, abscission of the preceding year's crop (mature fruit) increased, indicating the processes were controlled independently and not driven by crop load. During the study, neither weather conditions nor tree nutrient status was associated with key abscission events. Taken together, the results of this research provided strong evidence that the previous year's yield did not influence current year flower or fruit abscission. Thus, the normal seasonal abscission of reproductive structures, in the absence of a triggering climatic or cultural event, appears to be an independent process that does not contribute to the entrainment of 'on'-/'off'-yield cycles of the 'Hass' during alternate bearing. The research provided additional evidence that the final yield of the 'Hass' avocado is strongly related to the number of floral shoots (inflorescences) at bloom (Garner and Lovatt, 2008).

Effect of crop load on vegetative shoot growth and return bloom

Typically, avocado trees produce two major vegetative shoot flushes in the spring and summer of each year, respectively (Scholefield *et al.*, 1985; Whiley *et al.*, 1988; Whiley and Schaffer, 1994), with avocado-growing areas such as California having a third, fall vegetative shoot flush (Salazar-García *et al.*, 1998) and some avocado-growing areas of Mexico having an additional (fourth) winter vegetative shoot flush (Salazar-García *et al.*, 2006) (Table 6.1). Shoot growth is generally monopodial, and axillary shoots can be proleptic (formed after a rest period of the apical meristem, typically in winter or spring) or sylleptic (formed without a rest period, typically in summer and fall). The vegetative shoots of the winter, spring, summer and fall flushes have the potential to produce floral shoots the following spring. The heavy 'on' crop suppresses both the number and length of vegetative shoots that develop (Lahav and Kalmer, 1977; Wolstenholme *et al.*, 1990; Lovatt, 2011a; Ziv *et al.*, 2014). Schaffer *et al.* (1991) reported that individual fruit influenced shoot growth. Non-fruiting shoots produced multiple vegetative shoot flushes, whereas fruiting shoots flushed only once in spring. In addition, the vegetative shoot at the apex of indeterminate floral shoots that set fruit does not produce summer or fall vegetative shoots (Davenport, 1982; Salazar-García *et al.*, 1998; Lovatt, 2011a). In contrast, avocado trees that set an 'off' crop in spring produce a significant number of summer and fall vegetative shoots that contribute floral shoots to bloom the next spring (Year 2), resulting in an 'on' bloom.

In California, for 'off'-crop 'Hass' trees (Year 1), 60% to 70% of the floral shoots at bloom (Year 2) were produced by the Year 1 summer vegetative shoots, with spring vegetative shoots contributing ~30% and fall shoots 0% to 10% of the Year 2 floral shoots (Tables 6.4 and 6.5) (Lovatt, 2011a). Summer vegetative shoots produce a greater number of floral shoots than spring and fall vegetative shoots, because floral shoots are borne in the axils of the leaves along the length of a summer vegetative shoot (Thorp *et al.*, 1993; Salazar-García and Lovatt, 1998; Lovatt, 2011a). In California, Year 1 summer shoots that develop during the 'off'-crop year produce 80% of the Year 2 determinate floral shoots, with the majority borne in axillary positions (Lovatt, 2011a). Determinate floral shoots, which are more productive than indeterminate floral shoots (Buchholz, 1986; Lovatt *et al.*, 1994; Thorp *et al.*, 1994; Salazar-García and Lovatt, 1998), are produced by both apical and axillary buds of summer-flush vegetative shoots (Salazar-García and Lovatt, 1998). Greater numbers of determinate inflorescences were reported to occur during heavy 'on' blooms following an 'off'-crop year in California (Lovatt, 2011a), New Zealand (J. Dixon, New Zealand, 2023, personal communication) and Spain (J.I. Hormaza, Spain, 2023, personal observation), typically along the axis of sylleptic shoots that developed during the summer of the previous 'off'-crop year. Fall shoots that flower produce only a single terminal indeterminate floral shoot (Thorp *et al.*, 1993; Salazar-García and Lovatt, 1998). Late-developing fall shoots remain vegetative. The 'on' crop of young developing fruit inhibits summer and fall vegetative shoot growth, thereby reducing the contribution made by these shoots to bloom the following spring, resulting in an 'off' bloom produced predominantly by the Year 1 spring flush of vegetative shoots and characterized by a significantly greater proportion of indeterminate to determinate floral shoots (Tables 6.4 and 6.5).

The importance of the summer and fall shoots to the intensity of the return bloom was confirmed by removing all summer and fall vegetative shoots or just the fall vegetative shoots from 'off'-crop 'Hass' avocado trees. The floral intensity at return bloom and the return yield of these trees were compared to untreated 'off'- and 'on'-crop control trees. At return bloom the next year, the untreated 'off'-crop control trees and 'off'-crop trees with only the fall-flush vegetative shoots removed flowered profusely, whereas the 'off'-crop trees with the summer plus fall shoots removed had a weak bloom and low return yield (2 kg per tree); 'on'-crop control trees yielded 10 kg per tree (A. Lopez-Jimenez and C.J. Lovatt, 2005, unpublished results; summarized in Lovatt, 2011a). Knowing which flush of vegetative shoots contributes the majority of inflorescences to the return bloom is critical in applying management strategies to mitigate alternate bearing and improve annual yields.

Table 6.4. Contribution of Year 1 spring and summer plus fall shoots with and without fruit on 'on'- and 'off'-crop 'Hass' avocado trees to return bloom in Year 2. Reproduced from Lovatt (2011a).

	Total	Spring	Summer + Fall
	---- no. floral shoots/4 shoots ----		
Non-fruiting shoots			
'Off'-crop tree	10.2 a[z]	1.9 ab	8.3 a
'On'-crop tree	4.2 b	2.3 a	1.9 b
Fruiting shoots			
'Off'-crop tree	1.7 bc	0.7 bc	1.0 b
'On'-crop tree	0.3 c	0.2 c	0.1 b
P-value	<0.0001	0.0076	<0.0001

[z]Values in a vertical column followed by different letters are significantly different at the *P*-value specified by Fisher's protected Least Significant Difference (LSD) test.

Whole-tree and localized effects of fruit

The effect of the 'on' fruit crop on the return bloom is both localized to shoots that set fruit (fruiting shoots) and a function of the total number of fruit per tree (whole-tree crop load

Table 6.5. Effect of the 'off' and 'on' crop and fruit removal from 'on'-crop 'Hass' avocado trees on the contribution of Year 1 spring, summer and fall vegetative shoots to the return bloom (Year 2). (C.J. Lovatt, 2009, unpublished results.)

	Spring shoots			Summer (S) and fall (F) shoots			
	Floral shoots	Vegetative shoots	Inactive buds	Total S + F shoots	Floral shoots	Vegetative shoots	Inactive buds
	No. of shoots/branch						
'Off'-crop trees	10.6 cde[z]	0.7 c	0.8 a	30.7 c	66.8 bc	4.3 bcd	0.5 a
'On'-crop trees with fruit removed							
June	14.4 abc	2.9 bc	2.3 a	42.8 ab	80.5 ab	3.3 bcd	1.1 a
July	21.8 ab	2.0 bc	1.6 a	50.1 a	119.0 a	2.4 bcd	1.3 a
August	21.8 ab	2.1 bc	0.8 a	35.6 bc	92.5 ab	2.1 bcd	0.4 a
September	23.8 a	1.4 c	0.6 a	24.6 cde	62.4 bc	1.4 cd	0.1 a
October	13.9 bcd	0.8 c	1.9 a	18.0 e	33.0 cd	0.8 d	0.4 a
November	13.8 bcd	2.1 bc	1.5 a	21.1 de	28.6 cd	9.5 abcd	1.9 a
December	5.0 cde	5.9 a	2.4 a	17.5 e	7.6 d	12.6 abc	2.6 a
January	2.9 e	2.6 bc	2.8 a	27.4 cde	19.4 d	17.5 a	2.1 a
'On'-crop trees	4.5 de	4.1 ab	3.1 a	22.6 de	19.1 d	13.3 ab	1.7 a
P-value	0.0010	0.0036	0.1388	0.0003	0.0002	0.0533	0.4319

[z]Values in a vertical column followed by different letters are significantly different at specified *P*-values by Fisher's protected LSD test.

effect). The impact of the localized versus whole-tree effects of fruit on the return bloom was determined by comparing the number of floral shoots produced by Year 1 spring shoots that set fruit (fruiting shoots) and shoots that did not set fruit (non-fruiting shoots) at the return bloom of 'on'- and 'off'-crop 'Hass' avocado trees (Table 6.4) (Lovatt, 2011a). Maximum return bloom was attained on the non-fruiting shoots of the 'off'-crop trees, for which there was no localized effect of fruit, and the crop load effect of fruit was minimal. These data also confirm the greater contribution of summer plus fall shoots versus spring shoots of 'off'-crop trees to return bloom the following year. The combined localized and whole-tree effect of fruit was strongest for buds on Year 1 spring shoots that set fruit on 'on'-crop trees, where the combined localized effect of fruit set on a shoot and the whole-tree effect of crop load were observed at return bloom in Year 2. Comparison of fruiting and non-fruiting shoots on 'off'-crop trees provided evidence that 'Hass' fruit exert a strong effect on the return bloom localized to shoots that set fruit. In contrast, the whole-tree effect of fruit that was evident for non-fruiting shoots of 'on'-crop trees showed a less inhibitory effect on flowering. Thus, for avocado, the localized effect of a single fruit set on a shoot is more inhibitory to flowering than the whole-tree effect of the 'on' crop. These data, and similar data not presented herein, document that for trees setting an 'on' crop, return bloom and yield were highly dependent on the number of shoots that did not set fruit, the minority of shoots of an 'on'-crop tree, fruiting shoots being the majority. This is an important fact that reinforces that mitigation of alternate bearing requires increasing the number of non-fruiting shoots using cultural practices that reduce crop load at the proper time through flower or fruit removal by hand, chemical thinning or pruning. Reducing the number of fruit per tree must be achieved sufficiently early to promote the growth of the vegetative shoots that contribute the majority of inflorescences to return bloom.

Negative effects of the 'on' crop of fruit are cumulative

Classic fruit removal experiments designed to identify when the 'on' crop was exerting its negative effect on the contribution of spring-, summer- and fall-flush vegetative shoots to return bloom of the 'Hass' avocado in California demonstrated that the negative effect of the 'on' crop on floral intensity at return bloom the following year was cumulative over time (Lovatt, 2011a). The reduction in the number of inflorescences at return bloom was the sum of the negative effects of the 'on' crop on each of the three shoot flushes (Table 6.5). For this experiment, only the current setting fruit were present on the trees; the previous year's mature crop had been harvested before June. Young developing fruit were removed from different sets of 'on'-crop trees monthly from June through January. The number of floral and vegetative shoots and inactive buds at bloom the following spring was quantified. For Year 1 spring-flush shoots, fruit removal from June through September significantly increased the per cent bud break and the number of floral shoots produced at return bloom (Table 6.5). Progressively later fruit removal from October through December reduced both the percentage bud break and the floral intensity of Year 1 spring shoots to a number equal to and not significantly different from the contribution of Year 1 spring shoots of either 'off'- or 'on'-crop trees, respectively.

Removal of the 'on' crop in June and July increased the number of summer vegetative shoots that developed and the number of floral shoots produced by these shoots at return bloom. Progressively later fruit removal reduced the growth of summer shoots and, commensurately, their contribution to bloom the following spring. Later fruit removal reduced the growth of summer vegetative shoots and their contribution to the return bloom and resulted in the growth of only fall-flush vegetative shoots. A significant proportion of the apical buds of the fall-flush shoots in this study remained vegetative. With each additional month that the 'on' crop remained on the trees past December through February, there was an incremental decrease in floral shoot number at return bloom that paralleled a reduction in per cent bud break on Year 1 spring, summer and fall shoots compared to 'off'-crop control trees and 'on'-crop trees with all fruit removed prior to December (data not shown).

In summer, the 'on' crop clearly inhibits bud break of the terminal buds of vegetative shoots from which summer vegetative shoots develop, thereby reducing the number of nodes that bear buds that can become inflorescences at return bloom (Table 6.5). In addition, the 'on' crop of fruit inhibits spring bud break, further reducing floral intensity at return bloom the following spring. The number of buds that remained quiescent (inactive) through spring bloom was consistently significantly greater in the spring following the heavy 'on' crop compared to the spring following the light 'off' crop (Escobedo-Solórzano and Lovatt, 2014). Apical buds on 1-year-old non-fruiting spring shoots (shoots that did not set fruit the previous spring) excised from 'off'-crop trees in March had a greater percentage of bud break than lateral buds and produced predominantly floral shoots. In contrast, both apical and lateral buds on 1-year-old fruiting spring shoots (shoots that set fruit the previous year with fruit present through the following March) excised from 'on'-crop trees in spring (March) underwent limited bud break. PGR treatments (discussed later in this section) increased bud break and resulted in apical and lateral inflorescences on both non-fruiting and fruiting shoots of 'on'-crop trees. This confirmed that a portion of quiescent (inactive) buds of 'on'-crop trees remained viable floral buds. Consistent with these data, the 'on' bloom that sets the 'on' crop is earlier, commensurate with bud break being less inhibited in the presence of the 'off'-crop set the previous year, and the bloom occurs over a longer period than the 'off'-crop bloom; whereas the 'off' bloom, which follows the 'on' crop occurs later, consistent with greater inhibition of bud break, and is shorter in duration (Garner and Lovatt, 2008).

Leaving the Year 1 mature fruit of an 'on' crop on the trees through the spring and summer to October of Year 2 (the young setting Year 2 crop was removed in June from these trees to observe the effect of the mature fruit only) reduced the return bloom in Year 3 to a similar degree as young developing fruit of an 'on' crop (Table 6.6). The reduction in floral shoot number at return bloom was due to a decrease in the number of vegetative shoots that developed

Table 6.6. Effect of delayed harvest of the mature 'on' crop until October of Year 2 on the contribution of the Year 2 summer shoots to the return bloom in Year 3 compared to the effects of the current young developing 'on' and 'off' crops of the 'Hass' avocado. Reproduced from Lovatt (2011a).

	Floral shoots			Vegetative shoots	Inactive buds
	Total	Indeterminate	Determinate		
	Average no./8 shoots per tree				
'Off'-crop control	18.0 a	5.0 a	13.0 a	0.2 c	24.4 b
Mature 'on' crop	5.2 b	3.0 a	2.2 bc	3.2 b	26.8 ab
'On'-crop control	8.8 ab	4.0 a	4.8 abc	5.0 ab	36.2 a
P-value	0.0248	0.3344	0.0409	0.0016	0.0494

[z]Values in a vertical column followed by different letters are significantly different at specified *P*-values by Fisher's protected LSD test.

during the summer in both cases. The mature crop was harvested in mid-October; there was no effect on fall vegetative shoot growth or the contribution of the fall-flush shoots to the return bloom. Thus, delaying the harvest of a heavy 'on' crop through summer to attain greater fruit size or to wait for better market prices indeed exacerbates alternate bearing. These data reinforce the benefit of an early harvest prior to this critical period of avocado tree phenology and explain the relatively high yields obtained in semi-warm climates, since in these regions, harvest is usually completed 1 to 3 months before bloom (Salazar-García *et al.*, 2007c).

Effect of the 'on' crop on floral development

Two mechanisms by which the 'on' crop of fruit indirectly causes reduced floral intensity in the spring following an 'on'-crop year are (i) inhibition of bud break, which prevents summer and fall vegetative shoot growth, thereby reducing the number of nodes that can bear floral buds and contribute floral shoots at return bloom the following year; and (ii) inhibition of spring bud break, which further reduces the number of inflorescences at return bloom. The third mechanism is a more direct effect of the fruit of the 'on' crop in inhibiting floral development at the level of gene expression first demonstrated in 'Hass' avocado by Ziv *et al.* (2014).

In California, the annual phase transition of avocado occurred in the summer, at the end of July to the beginning of August, with two to four secondary-axis inflorescence meristems evident by 30 August, consistent with the commitment of the PAM to floral development (Salazar-García *et al.*, 1998) (Figs. 6.4 and 6.9). From August through October, secondary-axis meristems of the inflorescence developed, with ten secondary-axis meristems present by 30 October. Despite the high temperatures (27°C–42°C) that occurred from August to October in the year of this research, reversion of apical buds to vegetative growth was not observed (Salazar-García *et al.*, 1998), consistent with apical buds of summer-flush shoots being irreversibly committed to flowering. Table 6.5 shows that progressively later removal of the 'on' crop through September resulted in fewer summer vegetative shoots. However, these shoots consistently contributed two to three floral shoots per summer shoot at return bloom. In addition, there was a significant increase in the number of floral shoots contributed by spring shoots to return bloom over this period compared to both 'on'- and 'off'-crop trees. These results suggest that fruit do not inhibit the phase transition of the terminal buds on spring or summer shoots. In addition, inhibition of phase transition of the terminal bud would have resulted in a concomitant increase in vegetative shoot number at return bloom, which did not occur (Table 6.5). The low number of floral shoots produced per summer shoot instead suggests the possibility that the 'on' crop inhibits phase transition of some axillary buds, which become determined at a later time than the apical bud. Consistent with this interpretation, the number of determinate inflorescences was reduced by 50% when fruit removal was delayed to September.

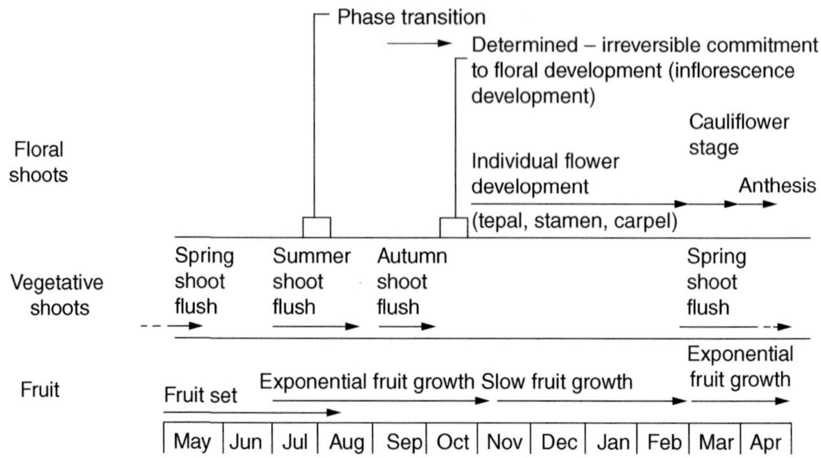

Fig. 6.9. 'Hass' avocado tree phenology in California. (After Salazar-García, 1998, with permission.)

By 30 November, development of terminal flowers on secondary and tertiary-inflorescence axes was initiated, but by 30 December, only the oldest (basal) secondary axis had cymes with fully formed flowers. By 30 January, the reproductive structures (gynoecium and stamen) were developing in the oldest flowers (Salazar-García *et al.*, 1998). The data in Table 6.5 document the dramatic reduction in floral shoot number that occurred when the 'on' crop was not removed until December. This comparison suggests the possibility that the 'on' crop might inhibit subsequent development of the more apical secondary axes and individual flowers, which could explain the presence of nonviable buds on Year 1 spring and summer shoots that abscise at bloom following the 'on' crop (Davenport, 1986).

For fall-flush vegetative shoots of 'Hass', delaying fruit removal from October through January reduced the floral shoot number and increased the vegetative shoot number at spring bloom (Table 6.5), suggesting the likelihood that phase transition was inhibited for late-developing fall-flush vegetative shoots. Whether this was due to the presence of the 'on' crop or climate was not resolved in that study, but 'on'-crop trees produced more vegetative shoots at return bloom than 'off'-crop trees, consistent with the negative effect of the 'on' crop on floral development.

The canopy of the avocado tree is comprised of vegetative shoots of varied developmental ages that render floral development asynchronous, a condition that typifies the interactions between climate and the presence of fruiting and non-fruiting shoots (Schaffer *et al.*, 1991). Despite this, the results of fruit removal experiments are useful for determining when the 'on' crop affects the return bloom to gain insight into possible mechanisms underlying the effect and to develop mitigation strategies.

Effect of the 'on' crop on floral development at the level of gene transcription

In a study in Israel using monthly fruit removal from 'on'-crop 'Hass' avocado trees, Ziv *et al.* (2014) determined that the 'on' crop significantly reduced flowering from October through November independent of its effect in reducing summer or fall vegetative shoot growth. Delaying removal of the 'on' crop until October or November progressively decreased the number of terminal buds on summer vegetative shoots that produced inflorescences and increased the development of vegetative shoots. In contrast, 100% of the terminal buds of the 'off'-crop control trees produced inflorescences and 100% of the terminal buds of the 'on'-crop control trees produced vegetative shoots. During this critical period, for 'off'-crop control trees, Ziv *et al.* (2014) quantified a significant increase in the accumulation of *FT* transcripts in leaves, which returned to a basal level in early December with the subsequent upregulation of bud floral meristem identity genes, *AP1* and *LFY*, in October and November, respectively. As reported earlier in this chapter, the accumulation of *AP1* and *LFY* transcripts to sufficient levels is essential for bud determinacy, ICF in avocado. At this time, buds of the 'off'-crop control trees had developed secondary- and tertiary-inflorescence meristems, consistent with the commitment of the buds to floral development. For the 'on'-crop control trees, leaf *FT* expression and bud *AP1* and *LFY* expression remained significantly lower than that in buds of the 'off'-crop control trees. The authors proposed that in avocado, the protein product of *FT* produced in leaves might serve as a phloem-mobile florigen (flowering) signal that controls the transition from vegetative growth to floral development, which appears to occur in late fall to early winter in the central district area of Israel (32°12′N, 34°58′E). The results of Ziv *et al.* (2014) identified the critical first step in the regulation of floral development at the level of gene expression in avocado in general, and in response to crop load in particular. The identification of a leaf-expressed *FT* as a key gene regulating floral induction provides a tool to investigate the myriad hypotheses on the putative roles played by hormones and carbohydrates in perpetuating the 'on'/'off' cycles in alternate-bearing avocado orchards and to further aid in the development of reliable strategies for mitigating alternate bearing.

Relationship among bud hormone concentrations, crop load and floral intensity at return bloom

Gibberellin

In woody perennial species, GAs appear to play an inhibitory role in floral induction (Ebert and Bangerth, 1981; Guardiola *et al.*, 1982; Monselise and Goldschmidt, 1982; Wilkie *et al.*, 2008; Tang and Lovatt, 2019). It has been proposed for avocado that the 'on' crop results in high concentrations of fruit- or seed-produced GAs that inhibit floral initiation and flowering the following year, but no studies have quantified the export of GAs from avocado fruit or the accumulation of GAs at the time of annual phase transition in avocado (Paz-Vega, 1997).

Exogenous application of GA_3 has been shown to inhibit floral development in favour of vegetative shoot growth (Guardiola *et al.*, 1982; Monselise and Goldschmidt, 1982). To have this effect, GA_3 must be applied before the bud is determined (committed to floral development); GA_3 applied after this stage of development does not inhibit floral development in woody perennials (Guardiola *et al.*, 1982; Lord and Eckard, 1987; Salazar-García and Lovatt, 1998). GA_3 applied to 'Hass' avocado trees in California prior to ICF stimulated vegetative shoot growth at the expense of floral shoot development (Salazar-García and Lovatt, 1998, 2000). If the secondary axes of the floral shoot had differentiated at the time of GA_3 application, the floral shoot developed precociously compared to floral shoots on the untreated branches; even partially developed inflorescences (<10 secondary axes) underwent bud break. Application of GA_3 in November reduced the inflorescence number and increased the vegetative shoot number of apical buds primarily on fall shoots (Salazar-García and Lovatt, 1998). Stimulation of vegetative shoot growth at the apex of indeterminate floral shoots by GA_3 inhibited bud break of axillary buds. Progressively later applications of GA_3 in December, January or February, when the proportion of fully formed inflorescences was greater, had no effect on the number of floral shoots produced by summer- or fall-flush shoots.

In apple, GA_1 and GA_4 accumulate in the buds of 'on'-crop trees at the time of floral bud development with a concomitant reduction in return bloom (Kittikorn *et al.*, 2010). However, it must be determined for avocado whether GA_1 and GA_4 (or other GA produced by avocado) have the same effect and whether the GA is exported from the fruit at the proper time to influence phase transition or subsequent floral development. Thus, the putative role of GA in inhibiting floral development during the 'on'-crop year of avocado remains to be demonstrated.

Auxin, cytokinin and abscisic acid

The question of whether the inhibition of summer vegetative shoot growth that occurs during the 'on'-crop year is due to hormones was investigated. The effects of 'off' and 'on' crops and fruit removal from 'on'-crop 'Hass' trees on endogenous hormone concentrations of the five apical buds (terminal bud plus distal four buds) of shoots were quantified (B. Eyheraguibel and C.J. Lovatt, 2007, unpublished results). As in previous experiments, June fruit removal from the 'on'-crop trees increased the number of summer vegetative shoots to a value above that of both 'on'- and 'off'-crop control trees, with a concomitant increase in floral shoot number at return bloom that was greater than that of the 'on'-crop trees and equal to that of the 'off'-crop trees. In response to removal of the 'on' crop in June, there was a significant increase in the concentration of the CK IPA to a value greater than that of the 'on'- and 'off'-crop trees by August. In addition, the concentration of IAA was significantly reduced at this time. Bud concentrations of ABA were not influenced by crop load or June fruit removal. The ratio of auxin to CK was significantly greater in buds from the 'on'-crop trees than 'off'-crop trees. Trees with the 'on' crop removed in June had the lowest auxin to CK ratio. Taken together, these results suggest that bud break of the buds producing the summer vegetative shoot flush is inhibited in the 'on'-crop trees by a high auxin to CK ratio typical of apical dominance (endodormancy), but more accurately described as paradormancy due to the putative role of fruit. Concurrent analysis of roots indicated that crop load and fruit removal had no effect on root hormone homeostasis during the summer months. The results suggest that IAA is exported from the fruit and accumulates in the buds apical to the fruit. Although a controversial idea, it is consistent with the conclusion of Cowan *et al.* (2005) that IAA export from avocado fruit is essential to maintain fruit growth. In contrast, IAA accumulation in apical buds could be due to reduced polar transport from the shoot apex (Bangerth, 2009) and, thus, be an indirect effect of the fruit. Studies with the *Arabidopsis* auxin transport mutant *pin1* provided evidence that normal polar transport of IAA was not critical to transitioning from vegetative to reproductive growth; *pin1* mutants produced inflorescences. Normal polar transport of IAA was, however, essential for the formation of flower primordia within the inflorescence (Cheng and Zhao, 2007). IAA also regulated the number and identity of the floral organs that developed.

The five apical buds from the shoots of the 'on'-crop 'Hass' trees, with all fruit removed each month from June through October, had high IPA concentrations and low IAA and ABA concentrations by February relative to the 'on'-crop trees that did not have their fruit removed until November, December or January ($P < 0.0001$; B. Eyheraguibel and C.J. Lovatt, 2007 unpublished results). In spring, the low IPA and high IAA and ABA concentrations in buds from the 'on'-crop trees with fruit removed after October corresponded with the reduction in floral shoot number for these trees ($P = 0.0002$) compared to the 'off'-crop trees and 'on'-crop trees with fruit removed prior to October.

In our early research, we did not pay attention to whether we were analysing fruiting or non-fruiting shoots; thus, results obtained for hormone concentrations in buds from 'off'- and 'on'-crop trees were frequently not significantly different, whereas changes in bud hormone concentrations obtained in response to fruit removal were clear (Lovatt, 2006). However, taken together, the results provide evidence that bud concentrations of IPA decreased while those of IAA and ABA increased, and hence the ratios of IAA and ABA to IPA increased in the apical and axillary buds in November through February of the 'on'-crop year in parallel with the decrease in the number of buds that underwent bud break in spring.

Both inhibition of bud break of summer shoots by a high ratio of IAA to IPA and spring bud break by high ratios of IAA and ABA to IPA during the 'on'-crop year have previously been demonstrated for *Citrus reticulata* (Verreynne, 2005). In citrus, as in avocado, concurrent analysis of roots indicated that crop load and fruit removal had no effect on root hormone homeostasis.

To test the possibility that these changes in bud hormone homeostasis were related to inhibition of the spring bud break, 'on'-crop 'Hass' avocado trees were trunk-injected with the auxin transport inhibitor tri-iodobenzoic acid (TIBA) combined with a CK in January, when the maturing 'on' crop was still on the trees (Year 2). The treatment increased the total number of floral shoots that developed at return bloom, with the greatest contribution due to an increase in determinate floral shoots borne on summer-flush shoots, most notably on fruiting shoots ($P = 0.09$) (Lovatt, 2011b). In subsequent experiments, trunk-injecting the 'on'-crop 'Hass' avocado trees in July and again in January with TIBA and a CK increased the number of determinate floral shoots borne on summer-flush shoots and total floral shoot number at return bloom to a value intermediate to 'off'- and 'on'-crop trees (Lovatt, 2011b).

Strategies to increase the spring floral bud break of 'Hass' avocado to increase floral intensity hold promise for increasing fruit set, yield and grower income. Such strategies would be especially important during the spring following a heavy 'on' crop when trees flower poorly. Both apical and lateral buds on 1-year-old fruiting spring shoots (shoots that set fruit the previous year with fruit present through bloom in California) excised from 'on'-crop trees in spring (March) underwent minimal bud break, and PGRs had no effect on the lateral bud break even when the apical bud was removed from the shoot (Escobedo-Solórzano and Lovatt, 2014). In contrast, apical buds on 1-year-old non-fruiting spring shoots (shoots that did not set fruit the previous spring) excised from 'off'-crop trees had a greater per cent bud break than lateral buds and produced predominantly floral shoots. When the apical bud was removed, the CK, 6-benzyladenine (BA), increased the lateral bud break of both floral and vegetative buds. In a commercial orchard, removal of the apical bud before the end of March increased the bud break of lateral vegetative buds on 1-year-old fruiting and non-fruiting shoots of 'on'- and 'off'-crop 'Hass' avocado trees, respectively. Early spring (February) foliar applications of BA significantly increased the bud break of apical buds, which were floral, but not lateral buds, of non-fruiting shoots on 'off'-crop trees; BA had no effect on the apical or lateral bud break on fruiting shoots of 'on'-crop trees. In spring, 'Hass' avocado floral buds are paradormant due to the inhibition of both apical and lateral buds caused by mature fruit on fruiting shoots of 'on'-crop trees and also due to the apical dominance of lateral buds. A January trunk injection of a proprietary product to increase bud CK concentration combined with the auxin transport inhibitor TIBA significantly increased the bud break of apical and lateral floral buds on both non-fruiting and fruiting shoots of 'on'-crop trees.

In reviews of the role of hormones in floral induction, evidence pointed to CKs among all the hormones as having a promotive role in floral induction in woody perennials, with IAA and GA having inhibitory effects on floral induction (Hanke *et al.*, 2007; Bangerth, 2009). Apical dominance is the classic example of the interaction between IAA and CK in regulating development. In apical dominance, IAA negatively regulates CK biosynthesis at the node of the shoot (point of attachment and vascular connection of the leaf and axillary bud to the shoot) by controlling the expression level of the gene *adenosine phosphate-isopentenyl transferase* (*IPT*), which encodes the key regulatory enzyme in CK biosynthesis (Tanaka *et al.*, 2006). This interaction might similarly play a regulatory role in flowering. Application of CKs, with or without an auxin transport inhibitor, to 'on'-crop trees increased flowering in avocado (Escobedo-Solórzano and Lovatt, 2014), citrus, olive and pistachio (Fichtner *et al.*, 2021) by increasing summer vegetative shoot growth, promoting spring bud break and reducing floral bud abscission.

Relationship between carbohydrate availability and the effect of the 'on' crop on floral intensity at return bloom

The starch concentration of a tissue is an excellent barometer of carbohydrate availability. Starch concentrations of vegetative tissues have repeatedly been demonstrated to be greater at the end of the 'off'-crop year and lower at the end of the 'on'-crop year for deciduous and evergreen tree crop species (reviewed in Paz-Vega, 1997). The high oil content of the avocado fruit provides a logical explanation for the depletion of carbohydrate reserves during the production of an 'on' crop. Whiley and Wolstenholme (1990) reported starch accumulation in fall through winter, with the highest concentration at bloom. They recorded a decrease in starch concentration during flowering, fruit set and fruit growth, with the lowest concentration in summer. Given that no differences in starch concentrations during the summer of 'on'- or 'off'-crop years were observed but large differences in the amount of summer vegetative shoot growth are consistently reported and known to impact the intensity of return bloom, it is clear that additional factors must be involved in perpetuating the alternating 'on'-/'off'-yield cycles in avocado. This conclusion is supported by the results of glucose and starch analyses of buds from 'off'- and 'on'-crop trees and 'on'-crop trees with fruit removed from June through January (Lovatt, 2006). The glucose concentrations in the five apical buds from non-fruiting shoots of 'off'-crop trees and fruiting shoots from 'on'-crop trees were not significantly different from August through February, providing further evidence that carbohydrates are not likely a factor limiting summer vegetative shoot growth. Removing the 'on' crop monthly from June through November significantly decreased bud glucose concentrations to a progressively greater degree through February to values below that of both the 'on'- and 'off'-crop trees. This is in direct contrast to the fact that 'on'-crop trees with fruit removed from June through October produced more floral shoots at return bloom than 'on'-crop control trees (Table 6.5). By February, bud starch concentrations were significantly lower for 'on'-crop trees than for trees in all other treatments (Lovatt, 2006). 'On'-crop trees with fruit removed in July through December had bud starch concentrations equal to 'off'-crop trees, but floral intensity at return bloom among the trees in these treatments was significantly different (Table 6.5). Buds from trees with fruit removed in June and January had equal starch concentrations that were significantly lower than 'off'-crop trees but greater than 'on'-crop trees. It is noteworthy that despite low bud starch concentrations, the 'on'-crop trees with fruit removed in June flowered profusely (equal to 'off' trees) at return bloom, whereas trees with fruit removed in January with equivalent starch concentrations flowered poorly (equal to 'on'-crop trees) (Table 6.5). There were no significant differences in the concentrations of glucose or starch in roots from August through February among the 'off'- and 'on'-crop trees and 'on'-crop trees with fruit removed from June through January, with the single exception that root starch concentrations were significantly lower for 'on'-crop trees than 'off'-crop trees in December. Thus, in avocado, the role that carbohydrate availability plays in regulating floral intensity during alternate bearing remains elusive.

In other plant species, during the transition from vegetative to reproductive development, sucrose levels rise rapidly but transiently in the phloem and shoot apices (Cho *et al.*, 2018). For several plant species, the addition of exogenous sucrose or glucose promotes flowering. Sucrose acts as a signal in the leaf phloem for the upregulation of *FT* expression and generation of the phloem-mobile FT-protein, which activates downstream floral gene expression in the bud. In *Arabidopsis thaliana*, trehalose-6-phosphate serves as a signal to induce flowering in the SAM to promote floral development downstream from *FT*. This role for trehalose-6-phosphate has not yet been reported in other plant species. However, Gould *et al.* (2019) demonstrated that crop load regulated the delivery of carbohydrates to the 'Hass' avocado flower. Differences in sugar delivery rate to the flower paralleled differences in the expression of the trehalose phosphate synthesis (*TPS1*) gene. The results suggested that flowers required accumulated sugars for successful development and that trehalose-6-phosphate concentration regulated the opening of the avocado flower in the female stage, independent of carbohydrate and boron concentrations in the flower. In light of recent research results, sugar signalling might be an important factor regulating floral development that should be investigated further in alternate-bearing 'Hass' avocado trees.

Pruning

Pruning is a valuable tool for reducing crop load to mitigate the negative effects of a heavy 'on' crop when alternate bearing is initiated and to maintain uniform annual yields of CVS fruit. Additionally, pruning is an orchard management practice essential for avocado production. Pruning to manage alternate bearing can be integrated with standard pruning practices necessary for keeping orchard rows open to facilitate tree management, for example, canopy-applied pesticides, fertilizers or PGRs, controlling tree height to support harvest, raising tree skirts to allow cold air to drain through the orchard and prevent sprinklers from wetting low-hanging fruit, and opening the canopy for better light penetration – no light, no flowers, no fruit (Kour *et el.*, 2018). Care should be taken when pruning not to remove an excessive number of vegetative shoots critical to a good return bloom. Thus, it is important to know the contribution made by each flush of vegetative shoots to return bloom for a given cultivar in a specific growing area. The time of pruning is critical to ensure that the terminal and axillary buds on the vegetative shoots that regrow in response to pruning have sufficient time to mature and transition from vegetative to reproductive development by the following spring bloom. Flower pruning in 'Hass' during the 'on'-crop year is an additional pruning technique for mitigating alternate bearing. Removal of 20–30% of the terminal panicles can be done before full bloom or from full bloom to 10% fruit set. Flowering pruning reduces ABI and also increases fruit size (Roe and Morudu, 2000). All pruning techniques require the growers to take a close look at the stage of tree phenology and the 'on'/'off' status of the trees in their orchard. Over-pruning, rather than evening out alternate bearing, can result in simply shifting 'on'-crop trees to 'off'-crop trees.

In regions where a single bloom occurs and commercial harvest of mature (non-overcrowded) trees occurs about 2–3 months before bloom, light pruning to stimulate the production of new shoots should be done immediately after harvest. However, pruning at this time is handicapped by not being able to determine to what degree shoots that will flower in the current season are being removed relative to those that just produced fruit and will not flower until the following year. Non-fruiting vegetative shoots that developed the previous year will flower producing a regular (not excessive) bloom, whereas new shoots that develop after pruning will not flower until the following year. When fruit are on the tree through bloom, size picking and light selective pruning help to obtain a regular bloom. However, under these conditions, there is a strong tendency to not prune shoots with fruit approaching harvest in favour of pruning non-fruiting shoots, which are the source of the approaching spring's inflorescences. Delaying pruning until fruit have attained legal maturity and marketable size prior to full bloom is advantageous. In this case, early spring pruning is efficacious because of visibly being able to see shoots bearing mature fruit that will not contribute inflorescences to the current spring's bloom and can be pruned to flower the following year and the current year's flowering shoots. Pruning should be used to achieve a balance between shoots flowering in the current year, especially in an 'on'-bloom year, and those that will flower the following year. Pruning should increase the complexity of the canopy with regard to shoot age and potential fruiting points.

There are several pruning styles or techniques for young, mature or old (overcrowded) orchards. However, growers do not always utilize pruning due to the potential loss of crop and income that usually results. Lack of bloom during the pruning year or the next year is a common cause of reduced yield. However, applying the proper pruning strategy at the optimal time to reduce the 'on' bloom or 'on' crop can help to minimize crop losses due to lack of bloom. This problem was addressed for olive. Pruning to mitigate alternate bearing and maintain uniform annual yields of CVS fruit was best when imposed 28 days after full bloom to one side of 'on'-bloom 'Manzanillo' olive trees and then the other side biennially (E.J. Fitchner and C.J. Lovatt, 2023, unpublished results). Over a 4-year period starting with an 'on' crop, for untreated 'on'-crop control trees, ABI for total yield was 0.94 and ABI for yield of CVS fruit was 0.80. Pruning 28 days after full bloom to one side of the tree and then the other side *annually* reduced the ABI for total yield and yield of CVS fruit to 0.72 and 0.61, respectively. In contrast, pruning 28 days after full bloom to one side of the tree and then the other side *biennially* reduced the ABI for total yield and yield of CVS fruit to 0.47 and 0.37, respectively. There were no differences in 4-year cumulative total yields among treatments, whereas cumulative yields of CVS fruit were 40% greater for trees pruned biennially than trees pruned annually for the same period ($P < 0.01$). It is not essential to prune the entire tree every year, especially if fruiting and non-fruiting shoots remain in balance, unless pruning is required for other aspects of orchard management.

In older overcrowded orchards, the practice of stumping and topworking with 'Hass' has worked to effectively rejuvenate the trees and get them back into flowering and production. A good description and a proposed timetable for the stump-topwork rejuvenation method are described by Hofshi *et al.* (2010). The best time to stump or cut back selected branches and limbs of giant, overcrowded trees is after harvest of an 'on'-crop year. This will allow the trees to produce a crop in the next cropping year.

In intensive high-density avocado orchards, a combination of selective side and top pruning combined with plant growth retardants results in regular bloom and good yields (Gardiazábal *et al.*, 2011). When the use of plant growth retardants is not an option, it is recommended that limb or branch pruning be done after buds have reached ICF. This will ensure that limbs and branches that were not pruned will flower in the same season. Alternate bearing in high-density orchards can also be managed by pruning opposing quadrants of the tree every other year.

As a tool for reducing flower or fruit number in an 'on'-crop year, appropriate pruning and at the proper time is an effective strategy for reestablishing the adequate balance between fruiting and non-fruiting shoots in the canopy of avocado trees to mitigate alternate bearing. Additionally, pruning should be used regularly to maintain yield of CVS fruit. It is not always necessary to prune the entire tree every year if fruiting and non-fruiting shoots remain in balance and fruit size is good. Selective pruning of tree parts can be used for orchard management, particularly opening the canopy to increase light penetration and maintain tree size.

Compared to apple and other deciduous tree crops, pruning in avocado production is relatively new. Similarly, use of PGRs in avocado production continues to be limited relative to deciduous tree crops. As knowledge of the regulation and timing of avocado floral development continues to advance with the implementation of new molecular techniques, additional research fine-tuning the science of pruning avocado trees and developing PGR strategies to control vegetative regrowth should be conducted to improve the management of larger trees in older established orchards and the maintenance of tree size in high-density plantings. This will also help mitigate the frequent reoccurrence of alternate bearing driven by erratic temperatures associated with climate change and maximize annual total yields of CVS fruit.

Conclusions

The reproductive plasticity of avocado trees is being repeatedly challenged by climate changes in avocado-producing areas of the world. Significant advances in understanding the regulation of the avocado floral development pathway and the effects of climate are increasingly possible through the utilization of high-throughput molecular approaches quantifying biological molecules at a resolution that previously was not possible. These technologies are being developed and implemented across the broad fields of genomics, metagenomics, epigenomics, transcriptomics, proteomics, metabolomics and phenomics. 'Multi-omics' research provides a comprehensive understanding of the biological system under study and has been used successfully to elucidate growth, yield, senescence and response to biotic and abiotic stress in important crop plants (Yang *et al.*, 2021). It holds great promise for elucidating the complex regulatory roles of plant hormones in developmental processes. Additionally, RNA deep sequencing enables the identification of epigenetic events, including changes in DNA methylation and quantification of RNA expression, making it possible to delineate the effects of the environment on the genome (Goldman and Domschke, 2014) and thus, to understand interactions among gene/s, the developmental stage of the plant, and its environment. Further, genome-editing approaches, such as CRISPR/Cas9 and next-generation versions that make a qualitative change in the genome, enable the determination of the functions of specific genes. These technologies make it possible to engineer a plant's genome without introducing 'foreign' genes from another species (nontransgenic) to improve yield, fruit size and quality, disease and pest tolerance, and heat and drought tolerance as examples.

New molecular techniques could contribute to the generation of transgenic lines of commercially important avocado scion cultivars with fast-flowering and fruit-bearing capabilities (shortened juvenile phase) that enable researchers to more rapidly examine phenotypic changes associated with commercially important traits related to flowering, fruit size and quality, resistance to pests and diseases, and abiotic stress tolerance through the use of gene stacking technology (Orbović *et al.*, 2021). The 'FasTrack' breeding system for plum (*Prunus domestica*) utilizes plants overexpressing the *Populus tremuloides FLOWERING LOCUS T* (*FT*) floral timing gene to shorten the juvenile phase and accelerate flowering and fruiting (Petri *et al.* 2018).

Whether a successful analogous avocado system can be developed to speed up the process of breeding new improved avocado cultivars remains to be tested.

The promise of these new technologies will only be realized through further understanding of avocado tree phenology, floral development, pollination, fertilization (syngamy), fruit retention and fruit growth using classic field and laboratory research approaches. The results of these investigations establish when key developmental events occur in the phenology of the avocado tree and the factors that influence them. This information is fundamental to the future direction and subsequent utilization of the results of the new molecular technologies to develop new cultural management strategies, especially in the context of global climate change.

In the past two decades, significant progress has been made in advancing basic knowledge of avocado reproductive biology. The foundational avocado phenology model of Whiley *et al.* (1988) stimulated similar research in other avocado-producing countries. Subsequently, avocado tree architecture was deciphered, revealing the importance of sylleptic versus proleptic shoots to flowering and productivity (Thorp and Sedgley, 1993a, b). Additionally, the contribution made by each of the annual vegetative shoot flushes to floral intensity the following spring is now known for several avocado-growing regions. The seminal research of Salazar-García *et al.* (1998) established that July–August (northern hemisphere) is the time of phase transition for summer-flush vegetative shoots, rather than winter. This information, coupled with the subsequent identification of when floral buds are determined (committed to floral development), including the macro- and microscopic anatomical characteristics associated with irreversible commitment to flowering (Salazar-García *et al.*, 1999) and taken together with the demonstration of the roles played by *LFY* and *API* in bud determinacy (Ziv *et al.*, 2014; Acosta-Rangel *et al.*, 2021), and leaf-expressed *FT* in floral induction in avocado (Ziv *et al.*, 2014), will prove invaluable for elucidating the hormonal control of floral development and the impact of climate. Collectively, information developed over the past two decades provides an invaluable set of tools to not only researchers but also growers. Further, adaptation of this information to different climatic zones will continue to prove instructive in how and when to manipulate vegetative shoot growth for maximum bloom or to shift the time of bloom to increase fruit set or change fruit maturity date. At a minimum, this information will help growers optimize pruning and other cultural practices and avoid applying such practices at a time that would reduce flowering and yield.

Hormonal regulation of flowering, fruit set and fruit development is an area of avocado reproductive biology where significant advances will logically be made through the use of new molecular genetic tools, more sensitive methods for quantifying plant hormone concentrations, and integrative multi-omics approaches. Results reported herein suggest that hormonal control of flowering and fruit development in avocado will be integral to identifying the physiological and genetic mechanisms by which fruit number (crop load) one year regulates flowering the next and to the development of new PGR strategies to increase yield, fruit size and quality, and mitigate alternate bearing. There are currently excellent PGR products commercially registered for use on other evergreen woody perennial tree crops that could help solve avocado production problems. Whereas avocado acreage is too limited in some countries to warrant registration of a PGR product for use on avocado at the local level, globally, for a major international company, avocado would appear a reasonable market that could be developed through collaborative efforts of avocado researchers worldwide.

Most research has been performed using the global industry standard cultivar, Hass. Future studies should include other cultivars to optimize avocado productivity in avocado-growing areas of the world experiencing increasingly volatile climates. There are an increasing number of 'Hass' relatives and 'Hass'-like cultivars, any one of which might prove valuable for expanding avocado production into a new area or preserving a local avocado industry affected by climate change.

Acknowledgments

The authors thank J. Dixon, New Zealand; Z. Jalata, Ethiopia; S. Newett, Australia; and A. Rebolledo-Roa, Colombia, for their generous contributions.

Dedication

This chapter is dedicated to the memory of our colleague and friend Dr Samuel Salazar-García, a world leader in avocado research, a major contributor to our understanding of avocado reproductive biology, and a pioneer in the use of PGRs in avocado production.

References

Acosta-Rangel, A., Li, R., Mauk, P., Santiago, L. and Lovatt, C.J. (2021) Effects of temperature, soil moisture and light intensity on the temporal pattern of floral gene expression and flowering of avocado buds (*Persea americana* cv. Hass). *Scientia Horticulturae* 280:109940. doi:10.1016/j.scienta.2021.109940.

Adato, I. (1990) Effects of paclobutrazol on avocado (*Persea americana* Mill.) cv. Fuerte. *Scientia Horticulturae* 45, 105–115.

Adato, I. and Gazit, S. (1977a) Role of ethylene in avocado fruit development and ripening. I. Fruit drop. *Journal of Experimental Botany* 28, 636–643.

Adato, I. and Gazit, S. (1977b) Role of ethylene in avocado fruit development and ripening. II. Ethylene production and respiration by harvested fruits. *Journal of Experimental Botany* 28, 644–649.

Afik, O., Dag, A., Kerem, Z. and Shafir, S. (2006a) Analyses of avocado (*Persea americana*) nectar properties and their perception by honey bees (*Apis mellifera*). *Journal of Chemical Ecology* 32(9), 1949–1963.

Afik, O., Dag, A. and Shafir, S. (2006b) The effect of avocado (*Persea americana*) nectar composition on its attractiveness to honey bees. *Apidologie* 37, 317–325.

Ahsan, M.U., Hayward, A., Irihimovitch, V., Fletcher, S., Tanurdzic, M. *et al.* (2019) Juvenility and vegetative phase transition in tropical/subtropical tree crops. *Frontiers in Plant Science* 10:729. doi:10.3389/fpls.2019.00729.

Alcaraz, M.L. and Hormaza, J.I. (2009) Selection of potential pollinizers for 'Hass' avocado based on flowering time and male-female overlapping. *Scientia Horticulturae* 121(3), 267–271.

Alcaraz, M.L. and Hormaza, J.I. (2011) Influence of physical distance between cultivars on yield, outcrossing rate and selective fruit drop in avocado (*Persea americana*, Lauraceae). *Annals of Applied Biology* 158, 354–361.

Alcaraz, M.L. and Hormaza, J.I. (2014) Optimization of controlled pollinations in avocado (*Persea americana*, Lauraceae). *Scientia Horticulturae* 180, 79–85.

Alcaraz, M.L. and Hormaza, J.I. (2021) Fruit set in avocado: pollen limitation, pollen load size, and selective fruit abortion. *Agronomy* 11(8), 1603.

Alcaraz, M.L., Montserrat, M. and Hormaza, J.I. (2011) In vitro pollen germination in avocado (*Persea americana* Mill.): optimization of the method and effect of temperature. *Scientia Horticulturae* 130, 152–156.

Alcaraz, M.L., Thorp, T.G. and Hormaza, J.I. (2013a) Phenological growth stages of avocado (*Persea americana*) according to the BBCH scale. *Scientia Horticulturae* 164, 434–439.

Alcaraz, M.L., Hormaza, J.I. and Rodrigo, J. (2013b) Pistil starch reserves at anthesis correlate with final flower fate in avocado (*Persea americana*). *PLoS One* 8(10): e78467.

Ashworth, V.E.T.M., Chen, H., Calderón-Vázquez, C.L., Arpaia, M.L., Kuhn, D.N. *et al.* (2019) Quantitative trait locus analysis in avocado: the challenge of a slow-maturing horticultural tree crop. *Journal of the American Society for Horticultural Science* 144(5), 352–362.

Bangerth, F. (2000) Abscission and thinning of young fruit and their regulation by plant hormones and bioregulators. *Plant Growth Regulation* 31, 43–59.

Bangerth, K.F. (2009) Floral induction in mature, perennial angiosperm fruit trees: similarities and discrepancies with annual/biennial plants and the involvement of plant hormones. *Scientia Horticulturae* 122, 153–163.

Benlloch, R., Berbel, A., Serrano-Mislata, A. and Madueno, F. (2007) Floral initiation and inflorescence architecture: a comparative view. *Annals of Botany* 100, 659–676. doi:10.1093/aob/mcml46.

Bergh, B.O. (1975) Avocados. In: Janick, J. and Moore, J.N. (eds) *Advances in Fruit Breeding*. Purdue University Press, West Lafayette, Indiana, pp. 541–567.

Bergh, B.O. (1986) *Persea americana.* In: Halevy, H.A. (ed.) *Handbook of Flowering*. Vol. 5. CRC Press, Boca Raton, Florida, pp. 253–268.

Bergh, B.O. and Lahav, E. (1996) Avocados. In: Janick, J. and Moore, J.N. (eds) *Fruit Breeding*. Vol. 1. *Tree and Tropical Fruits*. Wiley, West Lafayette, Indiana, pp. 113–166.

Bergh, B.O., Garber, M.J. and Gustafson, C.D. (1966) The effect of adjacent trees of other avocado varieties on Fuerte fruit-set. *Proceedings of the American Society for Horticultural Science* 89, 167–174.

Bernier, G. (1988) The control of floral evocation and morphogenesis. *Annual Review of Plant Physiology and Plant Molecular Biology* 39, 175–219.

Bernier, G., Kinet, J.M. and Sachs, R.M. (1981a) *The Physiology of Flowering*. Vol. 1. CRC Press, Boca Raton, Florida.

Bernier, G., Kinet, J.M. and Sachs, R.M. (1981b) *The Physiology of Flowering*. Vol. 2. CRC Press, Boca Raton, Florida.

Bernier, G., Havelange, A., Houssa, C., Petitjean, A. and Lejeune, P. (1993) Physiological signals that induce flowering. *Plant Cell* 5, 1147–1155.

Bertling, I. and Bower, J.P. (2005) Sugars as energy sources – is there a link to avocado fruit quality? *South African Avocado Growers' Association Yearbook* 28, 24–27.

Bertling, I. and Bower, J.P. (2006) Avocado sugars during early fruit development. *South African Avocado Growers' Association Yearbook* 29, 38–39.

Bertling, I., Tesfay, S.Z. and Bower, J.P. (2007) Antioxidants in 'Hass' avocado. *South African Avocado Growers' Association Yearbook* 30, 17–19.

Biran, D. (1979) Fruitlet abscission and spring growth retardation – their influence on avocado productivity. MSc. thesis, The Hebrew University of Jerusalem, Israel. (In Hebrew.)

Blanke, M.M. and Lovatt, C.J. (1993) Anatomy and transpiration of the avocado inflorescence. *Annals of Botany* 71, 543–547. doi:10.1006/anbo.1993.1070.

Blanke, M.M. and Lovatt, C.J. (1998) Determinate versus indeterminate inflorescences of the 'Hass' avocado. In *Proceedings of the III World Avocado Congress*. World Avocado Congress, Tel Aviv, Israel, pp. 33–36.

Blazquez, M.A., Ferrandiz, C., Madueno, F. and Parcy, F. (2006) How floral meristems are built. *Plant Molecular Biology* 60, 855–870.

Blumenfeld, A. and Gazit, S. (1970a) Cytokinin activity in avocado seeds during fruit development. *Plant Physiology* 46, 331–333.

Blumenfeld, A. and Gazit, S. (1970b) The role of the seedcoat in avocado fruit development (growth and maturation). *California Avocado Society Yearbook* 54, 100–104.

Blumenfeld, A. and Gazit, S. (1972) Gibberellin-like activity in the developing avocado fruit. *Physiologia Plantarum* 27, 116–120.

Blumenfeld, A. and Gazit, S. (1974) Development of seeded and seedless avocado fruits. *Journal of the American Society for Horticultural Science* 99, 442–448.

Boldingh, H.L., Alcaraz, M.L., Thorp, T.G., Minchin, P.E.H., Gould, N. and Hormaza, J.I. (2016) Carbohydrate and boron content of styles of 'Hass' avocado (*Persea americana* Mill.) flowers at anthesis can affect final fruit set. *Scientia Horticulturae* 198, 125–131. doi:10.1016/j.scienta.2015.11.011.

Borrone, J.W., Olano, C.T., Kuhn, D.N., Brown, J.S., Schnell, R.J. and Violi, H.A. (2008) Outcrossing in Florida avocados as measured using microsatellite markers. *Journal of the American Society for Horticultural Science* 133(2), 255–261.

Bower, J.P. and Cutting, J.G. (1988) Avocado fruit development and ripening physiology. In: Janick, J. (ed.) *Horticultural Reviews*. Vol. 10. Timber Press, Portland, Oregon, pp. 229–271.

Bower, J.P. and Cutting, J. (1992) The effect of selective pruning on yield and fruit quality in 'Hass' avocado. *Acta Horticulturae* 296, 55–58.

Bower, J.P., Lovatt, C.J., Cutting, J. and Blanke, M.M. (1990) Interaction of plant growth regulator and carbohydrate in flowering and fruit set. *Acta Horticulturae* 275, 425–434.

Bowman, J.L., Smyth, D.R., Meyerowitz, E.M. (1991) Genetic interactions among floral homeotic shoots by removal of terminal buds early in the flower bud induction period. *Proceedings of the Florida State Horticultural Society* 124, 60–64.

Buchholz, A. (1986) Carbohydrate partitioning between fruitlets and young vegetative growth as a possible factor involved with fruitlet abscission in avocado. MSc. thesis. The Hebrew University of Jerusalem, Israel.

Buttrose, M.S. and Alexander, D.M. (1978) Promotion of floral initiation in 'Fuerte' avocado by low temperature and short day length. *Scientia Horticulturae* 8, 213–217.

Buzgo, M., Chanderbali, A.S., Kim, S., Zheng, Z., Oppenheimer, D.G., Soltis, P.S. and Soltis E.D. (2007) Floral development morphology of *Persea americana* (avocado, Lauraceae): the oddities of male organ identity. *International Journal of Plant Science* 168(3), 261–284.

Cameron, S.H., Mueller, R.T. and Wallace, A. (1952) Nutrient composition and seasonal losses of avocado trees. *California Avocado Society Yearbook* 37, 201–209.

Can-Alonzo, C., Quezada-Euán, J.J.G., Xiu-Ancona, P., Moo-Valle, H., Valdovinos-Nuñez, G.R. and Medina-Peralta, S. (2005) Pollination of 'criollo' avocados (*Persea americana*) and the behavior of associated bees in subtropical Mexico. *Journal of Apicultural Research* 44(1), 3–8.
Chaikiattiyos, S., Menzel, C.M. and Rasmussen, T.S. (1994) Floral induction in tropical fruit trees: effects of temperature and water supply. *Journal of Horticultural Sciences* 69, 397–415.
Chanderbali, A.S., Kim, S., Buzgo, M., Zheng, Z., Oppenheimer, D.G.,Soltis, P.S. and Soltis, E.D. (2006) Genetic footprints of stamen ancestors guide perianth evolution in *Persea* (Lauraceae). *International Journal of Plant Science* 167, 1075–1089. doi:10.1086/507586.
Chanderbali, A.S., Albert, V.A., Leebens-Mack, J., Naomi, S.A., Soltis, D.E. and Soltis, P.S. (2009) Transcriptional signatures of ancient floral development genetics in avocado (*Persea americana*; Lauraceae). *Proceedings of the National Academy of Sciences of the United States of America* 106(22), 8929–8934.
Chanderbali, A.S., Yoo, M.Y. and Zahn, L.M. (2010) Conservation and canalization of gene expression during angiosperm diversification accompany the origin and evolution of the flower. *Proceedings of the National Academy of Sciences of the United States of America* 107, 22570–22575.
Chanderbali, A.S., Berger, B.A., Howarth, D.G., Soltis, P.S. and Soltis, D.E. (2016) Evolving ideas on the origin and evolution of flowers: new perspectives in the genomic era. *Genetics* 202, 1255–1265.
Chen, H., Ashworth, V.E.T.M., Xu, S. and Clegg, M.T. (2007) Quantitative genetic analysis of growth rate in avocado. *Journal of the American Society for Horticultural Science* 132(5), 691–696.
Chen, Q.X., Pan, D.M. and Lu, L.X. (2010) Effect of temperature on flower-bud induction of longan in phytotron. *Acta Horticulturae* 863, 279–284.
Cheng, Y. and Zhao, Y. (2007) A role for auxin in flower development. *Journal of Integrative Plant Biology* 49(1), 99–104.
Chhun, T., Aya, K., Asano, K., Yamamoto, E., Morinaka, Y. *et al.* (2007) Gibberellin regulates pollen viability and pollen tube growth in rice. *The Plant Cell* 19, 3876–3888.
Chica, E.J. and Albrigo, L.G. (2013a) Changes in *CsFT* transcript abundance at the onset of low-temperature floral induction in sweet orange. *Journal of the American Society for Horticultural Science* 13, 184–189. DOI: 10.21273/JASHS.138.3.184.
Chica, E.J. and Albrigo, L.G. (2013b) Expression of flower promoting genes in sweet orange during floral inductive water deficits. *Journal of the American Society for Horticultural Science* 138, 88–94. doi:10.21273/JASHS.138.2.88.
Cho, L.H., Pasriga, R., Yoon, J., Jeon, J.S. and An, G. (2018) Roles of sugars in controlling flowering time. *Journal of Plant Biology* 61, 121–130.
Church, C.G. and Chace, E.M. (1922) *Some changes in the composition of* California *avocados during growth*. United States Department of Agriculture *Bulletin* 1073., Washington, D.C. https://www.biodiversitylibrary.org/bibliography/108927 (accessed 19 October 2024).
Coen, E.S. and Meyerowitz, E.M. (1991) The war of the whorls: genetic interactions controlling flower development. *Nature* 353, 31–37.
Continella, G., Tribulato, E. and D'Aquino, S. (1992) Observations on flower behaviour and fruit set of avocado in Sicily. *Acta Horticulturae* 296, 59–68. doi:10.17660/ActaHortic.1992.296.7.
Cossio-Vargas, L.E., Salazar-García, S., González-Durán, I.J.L. and Medina-Torres, R. (2007) Modelos de predicción de la determinación irreversible a la floración en los aguacates 'Booth-8' y 'Choquette'. In *Proceedings of the VI World Avocado Congress*. World Avocado Congress, Viña del mar, Chile, pp. 1–11.
Cossio-Vargas, L.E., Salazar-García, S. and Medina-Torres, R. (2008a) Desarrollo floral de los aguacates 'Choquette' y 'Booth-8' en clima cálido. Parte I. Desarrollo vegetativo y floral. *Agricultura Técnica en Mexico* 34, 43–49.
Cossio-Vargas, L.E., Salazar-García, S., González-Durán, I.J.L. and Medina-Torres, R. (2008b) Fenología del aguacate 'Hass' en el clima semicálido de Nayarit, Mexico. *Revista Chapingo Serie Horticultura* 14, 325–330.
Cowan, A. (2004) Metabolic control of avocado fruit growth: 3-hydroxy-3-methylglutaryl coenzyme a reductase, active oxygen species and the role of C7 sugars. *South African Journal of Botany* 70, 75–82.
Cowan, A.K., Moore-Gordon, C.S., Bertling, I. and Wolstenholme, B.N. (1997) Metabolic control of avocado fruit growth. *Plant Physiology* 114, 511–518.
Cowan, A.K., Taylor, N.J. and Van Staden, J. (2005) Hormone homeostasis and induction of the small-fruit phenotype in 'Hass' avocado. *Plant Growth Regulation* 45(1), 11–19.
Crane, J. and Nelson, M. (1971) The unusual mechanism of alternate bearing in the pistachio. *HortScience* 6, 489–490.

Cummings, K. and Schroeder, C.A. (1942) Anatomy of the avocado fruit. *California avocado Society Yearbook* 27, 56–64.

Cutting, J. (1993) The cytokinin complex as related to small fruit in 'Hass' avocado. *Acta Horticulturae* 329, 147–149.

Cutting, J. and Bower, J.P. (1990) Relationship between auxin transport and calcium allocation in vegetative and reproductive flushes in avocado. *Acta Horticulturae* 275, 469–475.

Cutting, J., Lishman, A.W., Hofman, P.J. and Wolstenholme, B.N. (1986) Plant growth substance trends in developing avocado fruit as determined by radioimmunoassay. *Acta Horticulturae* 175, 285–290. doi:10.17660/ActaHortic.1986.175.42.

Dahan, Y., Rosenfeld, R., Zadiranov, V. and Irihimovitch, V. (2010) A proposed conserved role for an avocado *fw2.2-like* gene as a negative regulator of fruit cell division. *Planta* 232(3), 663–676. doi:10.1007/s00425-010-1200-3.

D'Asaro, A., Reig, C., Martínez-Fuentes, A., Mesejo, C., Farina, V. and Agusti, M. (2021) Hormonal and carbohydrate control of fruit set in avocado 'Lamb Hass'. A question of the type of inflorescence? *Scientia Horticulturae* 282:110046.

Davenport, J.B. and Ellis, S.C. (1959) Chemical changes during growth and storage of the avocado fruit. *Australian Journal of Biological Sciences* 12, 445–454.

Davenport, T.L. (1982) Avocado growth and development. *Proceedings of the Florida State Horticultural Society* 95, 92–96.

Davenport, T.L (1986) Avocado flowering. *Horticultural Reviews* 8, 257–289.

Davenport, T.L. (2019) Cross- vs. self-pollination in 'Hass' avocados growing in coastal and inland orchards of southern California. *Scientia Horticulturae* 246, 307–316.

Davenport, T.L. and Manners, M.M. (1982) Nucellar senescence and ethylene production as they relate to avocado fruitlet abscission. *Journal of Experimental Botany* 33, 815–825.

Davenport, T.L., Parnitzki, P., Fricke, S. and Hughes, M.S. (1994) Evidence and significance of self-pollination of avocados in Florida. *Journal of the American Society for Horticultural Science* 119, 1200–1207.

Davenport, T.L., Ying, Z. and Schnell, R.J. (2007) Use of simple sequence repeats (SSR) to determine incidence and effectiveness of self-and cross-pollinated avocado fruit in Southern California, Year 3 of 3. In *Proceedings of the California Avocado Research Symposium 2006*. University of California, Riverside, California, pp. 64–80.

Davies, W.J. and Zhang, J. (1991) Root signals and the regulation of growth and development of plants in drying soil. *Annual Review of Plant Physiology and Plant Molecular Biology* 42, 55–76.

Davis, T.D., Steffens, G.L. and Sankhla, N. (1988) Triazole plant growth regulators. *Horticultural Reviews* 10, 63–105.

Degani, C., Goldring, A., Gazit, S. and Lavi, U. (1986) Genetic selection during the abscission of avocado fruitlets. *HortScience* 21, 1187–1188.

Degani, C., Goldring, A. and Gazit, S. (1989) Pollen parent effect on outcrossing rate in 'Hass' and 'Fuerte' avocado plots during fruit development. *Journal of the American Society for Horticultural Science* 114, 106–111.

Degani, C., Goldring, A., Adato, I., El-Batsri, R. and Gazit, S. (1990) Pollen parent effect on outcrossing rate, yield, and fruit characteristics of 'Fuerte' avocado. *HortScience* 25, 471–473.

Degani, C., El-Batsri, R. and Gazit, S. (1997) Outcrossing rate, yield, and selective fruit abscission in 'Ettinger' and 'Ardith' avocado plots. *Journal of the American Society for Horticultural Science* 122, 813–817.

Demirkol, A. (2002) Phenological and pomological characteristics and yields of some avocado cultivars grown in Antalya ecological conditions. *Anadolu* 12, 49–64. (In Turkish.)

Dixon, J., Cotterell, C., Hofstee, B. and Elmsly, T.A. (2008) 'Hass' avocado tree phenology 2004–2009 in the western Bay of Plenty. *New Zealand Avocado Growers' Association Annual Research Report* 8, 35–57.

Dymond, K., Celis-Diez, J., Potts, S., Howlett, B., Willcox, B. and Garratt, M. (2021) The role of insect pollinators in avocado production: a global review. *Journal of Applied Entomology* 145, 369–383.

Eaks, I.L. (1990) Change in the fatty acid composition of avocado fruit during ontogeny, cold storage and ripening. *Acta Horticulturae* 269, 141–145.

Ebert, A. and Bangerth, F. (1981) Relation between the concentrations of diffusible and extractable gibberellin-like substances and the alternate bearing behavior in apple as affected by chemical fruit thinning. *Scientia Horticulturae* 15, 45–52.

Escobedo-Solórzano, V. and Lovatt, C.J. (2014) Forcing spring bud break in 'Hass' avocado. *Acta Horticulturae* 1042, 207–216. doi:10.17660/ActaHortic.2014.1042.25.

Falcão, M., Paraluppi, N.D., Clement, C.R., Kerr, W.E. and Silva, M.F. (2001) Phenology and yield of avocado (*Persea americana* Mill.) in Central Amazonia. *Acta Amazonica* 31, 3–9. (In Portuguese.)

Fichtner, E.J., Chao, Y.Y., Ferguson, L., Verreynne, J.S., Tang, L. and Lovatt, C.J. (2021) Repeating cycles of ON and OFF yields in alternate bearing olive, pistachio and citrus trees – different mechanisms, common solutions. *Acta Horticulturae* 1315, 1–10. doi:10.17660/ActaHortic.2021.1315.1.

Finazzo, S.F., Davenport, T.L. and Schaffer, B. (1994) Partitioning of photoassimilates in avocado (*Persea americana* Mill.) during flowering and fruit set. *Tree Physiology* 14, 153–164.

Free, J.B. (1970) *Insect Pollination of Crops*. Academic Press, London.

Gardiazábal I.F., Mena, V.F., Magdahl, S.C., Adriazola, C.C. and Torres B.J. (2011) Nuevos sistemas de poda en árboles adultos de paltos (*Persea americana* Mill.) cv. Hass. In *Proceeding of the VII World Avocado Congress*. World Avocado Congress, Cairns, Queensland, Australia, pp. 295–302.

Garner, L.C. (2004) Characterization and manipulation of flower and fruit abscission of 'Hass' avocado (*Persea americana* Mill.). PhD dissertation, University of California, Riverside, California.

Garner, L.C., Ashworth, V.E.T.M., Clegg, M.T. and Lovatt, C.J. (2008) The impact of outcrossing on yields of 'Hass' avocado. *Journal of the American Society for Horticultural Science* 133(5), 648–652.

Garner, L.C., Klein, G., Zheng, Y., Khuong, T. and Lovatt, C.J. (2011) Response of evergreen perennial tree crops to gibberellic acid is crop load-dependent. II. GA_3 increases yield and fruit size of 'Hass' avocado only in the on-crop year of an alternate bearing orchard. *Scientia Horticulturae* 130, 753–761.

Garner, L.C. and Lovatt, C.J. (2008) The relationship between flower and fruit abscission and alternate bearing of 'Hass' avocado. *Journal of the American Society for Horticultural Science* 133(1), 3–10.

Garner, L.C. and Lovatt, C.J. (2016) Physiological factors affecting flower and fruit abscission of 'Hass' avocado. *Scientia Horticulturae* 199, 32–40.

Gazit, S. (1977) Pollination and fruit set of avocado. In: *Proceedings of the First International Tropical Fruit Short Course: The Avocado*. University of Florida, Gainesville, Florida, pp. 88–92.

Gazit, S. and Blumenfeld, A. (1972) Inhibitor and auxin activity in the avocado fruit. *Physiologia Plantarum* 27, 77–82.

Gazit, S. and Gafni, E. (1986) Effect of hand pollination with different pollen donors on initial fruit-set in avocado. *Israel Agrisearch* 1, 3–17. (In Hebrew.)

Goldman, D. and Domschke, K. (2014) Making sense of deep sequencing. *International Journal of Neuropsychopharmacology* 17, 1717–1725.

Goldring, A., Gazit, S. and Degani, C. (1987) Isozyme analysis of mature avocado embryos to determine outcrossing rate in a 'Hass' plot. *Journal of the American Society for Horticultural Science* 112, 389–392.

Gonzalez, C., Zheng, Y. and Lovatt, C. (2010) Properly timed foliar fertilization can and should result in a yield benefit and net increase in grower income. *Acta Horticulturae* 868, 273–286.

Gould, N., Minchina, P.E.H., Boldingh, H., Gould, E.M. and Thorp, G. (2019) Biochemical and molecular analyses of potential markers for evaluating the competition for resources between mature fruit and open flowers in 'Hass' avocado trees. *New Zealand Journal of Crop and Horticultural Science* 47(3), 143–154. doi:10.1080/01140671.2019.1606020.

Guardiola, J.L., Monerri, C. and Agusti, M. (1982) The inhibitory effect of gibberellic acid on flowering in Citrus. *Physiologia Plantarum* 55, 136–142.

Hanke, M.V., Flachowsky, H., Peil, A. and Hättasch, C. (2007) No flower no fruit – Genetic potentials to trigger flowering in fruit trees. *Genes, Genomes and Genomics* 1(1), 1–20.

Hapuarachchi, N.S., Kämper, W., Wallace, H.M., Bai, S.H., Ogbourne, S.M., Nichols, J. and Trueman, S.J. (2022) Boron effects on fruit set, yield, quality and paternity of Hass avocado. *Agronomy* 12(6): 1479.

Hoddle, M.S., Hofshi, R. and Arpaia, M.L. (2010) Are thrips important pollinators of avocados in California? Research Report-2010, California Avocado Commission. Available at: https://www.californiaavocadogrowers.com/research-library/are-thrips-important-pollinators-avocados-california (accessed 19 October 2024).

Hodgson, R.W. (1947) The California avocado industry. University of California, Agriculture Extension Service, Circular 43, Berkeley, California, pp. 49–52.

Hofshi, R., Tapia, M. and Arpaia, M.L. (2010) Stump and topwork – a technique for rejuvenating mature avocado trees. *California Avocado Society Yearbook* 93, 51–71.

Hong, Y. and Jackson, S. (2015) Floral induction and flower formation – the role and potential applications of miRNAs. *Plant Biotechnology Journal* 13, 282–292.

Hu, J., Mitchum, M.G., Barnaby, N., Ayele, B.T., Ogawa, M. *et al.* (2008) Potential sites of bioactive gibberellin production during reproductive growth in *Arabidopsis*. *The Plant Cell* 20, 320–336.

Illsley-Granich, C., Brokaw, R. and Ochoa-Ascencio, S. (2011) Hass Carmen®, a precocious flowering avocado tree. In *Proceedings of the VII World Avocado Congress*. World Avocado Congress, Cairns, Queensland, Australia, pp. 569–574.

Inoue, H. and Takahashi, B. (1990) Studies on the bearing behavior and yield composition of the avocado tree. *Journal of the Japanese Society for Horticultural Science* 59, 487–501. (In Japanese.)

Ish-Am, G., Barrientos-Priego, A.F., Castañeda-Vildózola, A. and Gazit, S. (1999) Avocado (*Persea americana* Mill.) pollinators in its region of origin. *Revista Chapingo Serie Horticultura* 5, 137–143.

Ish-Am, G. and Eisikowitch, D. (1992) New insight into avocado flowering in relation to its pollination. *California Avocado Society Yearbook* 75, 125–137.

Ish-Am, G. and Eisikowitch, D. (1993) The behaviour of honey bees (*Apis mellifera*) visiting avocado (*Persea americana*) flowers and their contribution to its pollination. *Journal of Apicultural Research* 32, 175–186.

Ish-Am, G. and Eisikowitch, D. (1998) Low attractiveness of avocado (*Persea americana* Mill.) flowers to honeybees (*Apis mellifera* L.) limits fruit set in Israel. *Journal of Horticultural Science and Biotechnology* 73(2), 195–204.

Ish-Am, G. and Lahav, E. (2011) Evidence for a major role of honeybees (*Apis mellifera*) rather than wind during avocado (*Persea americana* Mill.) pollination. *Journal of Horticultural Science and Biotechnology* 86(6), 589–594.

Jaganath, I.B. (1993) Effect of boron and/or nitrogen on pollen tube growth, ovule viability and yield in 'Hass' avocado. MSc thesis, University of California, Riverside, California.

Jaganath, I. and Lovatt, C.J. (1998) Efficacy studies on prebloom canopy applications of boron and/or urea to 'Hass' avocado. In *Proceedings of the III World Avocado Congress*. World Avocado Congress, Tel Aviv, Israel, pp. 181–184.

Johannsmeier, M.F. and Morudu, T.M. (1999) Some factors affecting pollination and yield of Hass avocados. *South African Avocado Growers' Association Yearbook* 22, 22–25.

Jonkers, H. (1979) Biennial bearing in apple and pear: a literature survey. *Scientia Horticulturae* 11, 303–317.

Kaiser, C. and Wolstenholme, B.N. (1993) Aspects of late hung 'Hass' avocado (*Persea americana* Mill.) fruit in the Natal Midlands. II. Whole tree starch cycling. *South African Avocado Growers' Association Yearbook* 16, 46–55.

Kalmer, D. and Lahav, E. (1976) Water requirement of the avocado tree in the Western Galilee (1968–1974). *Pamphlet* 157. Agricultural Research Organization, Volcani Center, Division of Scientific Publications, Bet Dagan, Israel.

Kämper, W., Ogbourne, S.M., Hawkes, D. and Trueman, S.J. (2021) SNP markers reveal relationships between fruit paternity, fruit quality and distance from a cross-pollen source in avocado orchards. *Scientific Reports* 11:20043. doi:10.1038/s41598-021-99394-7.

Kittikorn, M., Shiraishi, N., Okawa, K., Ohara, H., Yokoyama, M. *et al.* (2010) Effect of fruit load on 9,10-ketol-octadecadienoic acid (KODA), GA and jasmonic acid concentrations in apple buds. *Scientia Horticulturae* 124, 225–230.

Kobayashi, M., Lin, J., Davis, J., Francis, L. and Clegg, M.T. (2000) Quantitative analysis of avocado outcrossing and yield in California using RAPD markers. *Scientia Horticulturae* 86, 135–149.

Köhne, S. (1989) Spring growth of avocado trees as influenced by pruning and the use of growth regulators to stabilize yield and improve fruit quality. PhD dissertation, Hohenheim University, Stuttgart, Germany.

Köhne, J.S. and Kremer-Köhne, S. (1987) Vegetative growth and fruit retention in avocado as affected by a new plant growth regulator: paclobutrazol. *South African Avocado Growers' Association Yearbook* 10, 64–65.

Köhne, J.S. and Kremer-Köhne, S. (1990) Results of a high-density avocado planting. *South African Avocado Growers' Association Yearbook* 13, 31–32.

Kour, D., Bakshi, P., Wali, V.K., Sharma, N., Sharma, A. and Iqbal, M. (2018) Alternate bearing in olive – a review. *International Journal of Current Microbiology and Applied Sciences* 7(9), 2281–2297.

Krizek, B.A. and Fletcher, J.C. (2005) Molecular mechanisms of flower development: an armchair guide. *Nature Reviews Genetics* 6, 688–698.

Kushad, M.M., Yelenosky, G. and Knight, R. (1988) Interrelationship of polyamine and ethylene biosynthesis during avocado fruit development and ripening. *Plant Physiology* 87, 463–467.

Lahav, E. and Kalmer, D. (1977) Water requirements of avocado in Israel. II. Influence on yield, fruit growth and oil content. *Australian Journal of Agricultural Research* 28, 869–877.

Lahav, E. and Zamet, D. (1975) Flower, fruitlet and fruit abscission in avocado trees. *Alon Hanotea* 29, 556–562. (In Hebrew.)

Lee, S.K. (1981) A review and background of the avocado maturity standard. *California Avocado Society Yearbook* 65, 101–109.

Lee, T.D. (1984) Patterns of fruit maturation: a gametophyte competition hypothesis. *American Naturalist* 123, 427–432.

Lesley, J.W. and Bringhurst, R.S. (1951) Environmental conditions affecting pollination of avocados. *California Avocado Society Yearbook* 36, 169–173.

Levin, A. (1981) Factors affecting inflorescence and vegetative development regulation in avocado (*Persea americana* Mill.). MSc thesis, The Hebrew University of Jerusalem, Israel. (In Hebrew.)

Liu, X., Robinson, P.W., Madore, M., Witney, G. and Arpaia, M.L. (1999a) 'Hass' avocado carbohydrate fluctuations. I. Growth and phenology. *Journal of the American Society for Horticultural Science* 124, 671–675.

Liu, X., Robinson, P.W., Madore, M., Witney, G. and Arpaia, M.L. (1999b) 'Hass' avocado carbohydrate fluctuations. II. Fruit growth and ripening. *Journal of the American Society for Horticultural Science* 124, 676–681.

Liu, X., Sievert, J., Arpaia, M.L. and Madore, M.A. (2002) Postulated physiological roles of the seven-carbon sugars, mannoheptulose, and perseitol in avocado. *Journal of the American Society for Horticultural Science* 127(1), 108–114.

Looney, N.E. (1998) Plant bioregulators in fruit production: an overview and outlook. *Journal of the Korean Society for Horticultural Sciences* 39, 125–128.

Lord, E.M. and Eckard, K.J. (1987) Shoot development in *Citrus sinensis* L. (Washington navel orange). II. Alteration of developmental fate of flowering shoots after GA_3 treatment. *Botanical Gazette* 148, 17–22.

Loupassaki, M.H., Vasilakakis, M. and Androulakis, I.I. (1995) The time of flowering of avocado and the female and male opening of flowers in Crete. *Advances in Horticultural Sciences* 9, 37–42.

Loupassaki, M.H., Vasilakakis, M. and Androulakis, I.I. (1997) Effect of pre-incubation humidity and temperature treatment on the in vitro germination of avocado pollen grains. *Euphytica* 94, 247–251.

Lovatt, C.J. (1994) Improving fruit set and yield of 'Hass' avocado with a spring application of boron and/or urea to the bloom. *California Avocado Society Yearbook* 78, 167–173.

Lovatt, C.J. (1997) Pollination biology and fruit set in avocado. In: Cutting, J. (ed.) *Proceedings of the Australian Avocado Growers' Federation and the New Zealand Avocado Growers' Association Conference*. Australian Avocado Growers' Federation and the New Zealand Avocado Growers' Association, Rotorua, New Zealand, pp. 98–105.

Lovatt, C.J. (2001) Properly timed soil-applied nitrogen fertilizer increases yield and fruit size of 'Hass' avocado. *Journal of the American Society for Horticultural Science* 126, 555–559.

Lovatt, C.J. (2006) Eliminating alternate bearing of the 'Hass' avocado. Research Report-2006, California Avocado Commission. Available at: https://www.californiaavocadogrowers.com/sites/default/files/Eliminate-alternate-bearing-in-avocados-2006.pdf (accessed 19 October 2024).

Lovatt, C.J. (2010) Developing field strategies to correct alternate bearing (II). Research Report-2010, California Avocado Commission. Available at: http://www.californiaavocadogrowers.com/assets/Uploads/Growers-Site/Production-Research-Progress-Reports/2010-End-of-Year-Reports/65304-Alternate-Bearing-2010-End-of-Year.pdf (accessed 5 August 2023).

Lovatt, C.J. (2011a) Alternate bearing of 'Hass' avocado. *California Avocado Society Yearbook* 93, 125–140.

Lovatt, C.J. (2011b) *Developing field strategies to correct alternate bearing (II).* Research Report-2011, California Avocado Commission. Available at: https://www.californiaavocadogrowers.com/sites/default/files/Developing-field-strategies-to-correct-alternate-bearing-Lovatt-Year-End-Report-2011_0.pdf (accessed 5 August 2023).

Lovatt, C.J. (2022) ProGibb LV Plus® at the cauliflower stage of avocado inflorescence development. *The Avocado Quarterly* Winter 2021/2022, 21–30.

Lovatt, C.J., Zheng, Y. and Hake, K.D. (1988) Demonstration of a change in nitrogen metabolism influencing flower initiation in Citrus. *Israel Journal of Botany* 37(2–4), 181–188.

Lovatt, C.J., Bertling, I. and Blanke, M.M. (1994) Comparison of determinate *vs.* indeterminate inflorescences to determine the roles of PGRs, carbohydrates, nitrogen, and other nutrients in fruit set of the 'Hass' avocado. *California Avocado Society Yearbook* 78, 183–186.

Lovatt, C.J., Daoudi, H. and Ferguson, L. (2006) Efficacy of foliar-applied cytokinins and nitrogen to increase floral bud retention and to reduce alternate bearing of pistachio. *Acta Horticulturae* 727, 353–364. doi:10.17660/ActaHortic.2006.727.42.

Lu, M., Chang, T., Jong, T. and Chen, I. (2004) Observations on flowering phenology of main avocado (*Persea americana* Mill.) cultivars in Taiwan. *Journal of the Chinese Society for Horticultural Science* 50, 141–156.

Ma, H. (1994) The unfolding drama of flower development: recent results from genetic and molecular analyses. *Genes & Development* 8, 745–756.

Malerbo-Souza, D.T., Toledo, V.A.A., Silva, S.R. and Sousa, F.F. (2000) Pollination in avocado flowers (*Persea americana* Mill.). *Acta Scientiarum* 22(4), 937–941. (In Portuguese.)

McDaniel, C.N. (1994) Photoperiodic induction, evocation and floral initiation. In: Greyson, R.L. (ed.) *The Development of Flowers*. Oxford University Press, New York, pp. 25–43.

Mickelbart, M.V., Robinson, P.W., Witney, G. and Arpaia, M.L. (2012) 'Hass' avocado tree growth on four rootstocks in California. *I. Yield and flowering. Scientia Horticulturae* 143, 184–188.

Monselise, S.P. and Goldschmidt, E.E. (1982) Alternate bearing in fruit trees. *Horticultural Reviews* 4, 128–173.

Moore-Gordon, C.S., Cowan, A.K., Bertling, I., Botha, C.E.J. and Cross, R.H.M. (1998) Symplastic solute transport and avocado fruit development: a decline in cytokinin/ABA ratio to appearance of the Hass small fruit variant. *Plant and Cell Physiology* 39, 1027–1038.

Mougheith, M.G. and Abdel-Hamid, M.F. (1978) Physiological studies on fruits of some avocado cultivars II. Chemical changes in fruit fleshy portion. *Annals of Agricultural Science, Moshtohor* 10, 171–185.

Mougheith, M.G., Abdel-Hamid, M.F. and El-Ashram, M. (1978) Physiological studies on fruits of some avocado cultivars. I. Fruit growth and development. *Annals of Agricultural Science, Moshtohor* 10, 147–169.

Müller-Xing, R., Clarenz, O., Pokorny, L., Goodrich, J. and Schubert, D. (2014) Polycomb-group proteins and *FLOWERING LOCUS T* maintain commitment to flowering in *Arabidopsis thaliana*. *Plant Cell* 26, 2457–2471.

Muñoz, A.E., Plantegenest, M., Amouroux, P. and Zaviezo, T. (2021) Native flower strips increase visitation by non-bee insects to avocado flowers and promote yield. *Basic and Applied Ecology* 56, 369–378.

Muñoz-Fambuena, N., Mesejo, C., González-Mas, M., Iglesias, D., Primo-Millo, E. *et al.* (2012) Gibberellic acid reduces flowering intensity in sweet orange [*Citrus sinensis* (L.) Osbeck] by repressing *CiFT* gene expression. *Journal of Plant Growth Regulation* 31, 529–536.

Muñoz Perez, R.B., Jankiewicz, L.S. and Acosta-Zamudio, C. (1988) Growth and abscission of avocado fruits (*Persea americana* Mill.) cv. Fuerte. *Acta Agrobotanica* 41, 47–59. doi:10.5586/aa.1988.006.

Nakajima, Y., Susanto, S., Hasegawa, K. and Agriculture, F. (1992) Influence of water stress in autumn on flower induction and fruiting in young pomelo trees (*Citrus grandis* [L.] Osbeck). *Journal of the Japanese Society for Horticultural Science* 62(1), 15–20.

Nevin, J.M. and Lovatt, C.J. (1989) Changes in starch and ammonia metabolism during low temperature stress-induced flowering in Hass avocado – a preliminary report. *South African Avocado Growers' Association Yearbook* 12, 21–25.

Núñez-Elisea, R. and Davenport, T.L. (1994) Flowering of mango trees in containers as influenced by seasonal temperature and water stress. *Scientia Horticulturae* 58, 57–66. doi:10.1016/0304-4238(94)90127-9.

Okello, E.N., Amugune, N.O., Mukiama, T.K. and Lattorff, H.M.G. (2021) Abundance and community composition of flower visiting insects of avocado (*Persea americana* Mill) in the East African region. *International Journal of Tropical Insect Science* 41, 2821–2827.

Orbović, V., Ravanfar, S.A., Acanda, Y., Narvaez, J., Merritt, B.A., Levy, A. and Lovatt, C.J. (2021) Stress-inducible *Arabidopsis thaliana RD29A* promoter constitutively drives *Citrus sinensis APETALA1* and *LEAFY* expression and precocious flowering in transgenic *Citrus* spp. *Transgenic Research* 30, 1–13.

Papademetriou, M.K. (1974) Pollen tube growth in avocados (*Persea americana* Mill.). *California Avocado Society Yearbook* 58, 99–102.

Papademetriou, M.K. (1975) Percentage fruit set in avocados (*Persea americana* Mill.). *California Avocado Society Yearbook* 59, 135–142.

Papademetriou, M.K. (1976) Some aspects of the flower behavior, pollination and fruit set of avocado (*Persea americana* Mill.) in Trinidad. *California Avocado Society Yearbook* 60, 106–152.

Parcy, F., Bomblies, K. and Weigel, D. (2002) Interaction of *LEAFY*, *AGAMOUS* and *TERMINAL FLOWER1* in maintaining floral meristem identity in *Arabidopsis*. *Development* 129, 2519–2527.

Pattemore, D.E., Buxton, M.N., Cutting, B.T., McBrydie, H.M., Goodwin, R.M. and Dag, A. (2018) Low overnight temperatures delay 'Hass' avocado (*Persea americana*) female flower opening, leading to nocturnal flowering. *Journal of Pollination Ecology* 23, 127–135.

Paz-Vega, S. (1997) Alternate bearing in the avocado (*Persea americana* Mill.). *California Avocado Society Yearbook* 81, 117–148.

Pearce, S.C. and Dobersek-Urbanc, S. (1967) The measurements of irregularity in growth and cropping. *Journal of Horticultural Sciences* 42:2305.

Peña, J.F. and Carabalí, A. (2018) Effect of honey bee (*Apis mellifera* L.) density on pollination and fruit set of avocado (*Persea americana* Mill.) cv. Hass. *Journal of Apicultural Science* 62, 5–14.

Pérez-Balam, J., Quezada-Euán, J.J.G., Alfaro-Bates, R., Medina, S., McKendrick, L., Soro, A. and Paxton, R.J. (2012) The contribution of honeybees, flies and wasps to avocado (*Persea americana*) pollination in southern Mexico. *Journal of Pollination Ecology* 8, 42–47.

Peterson, P.A. (1955) Avocado flower pollination and fruit set. *California Avocado Society Yearbook* 39, 163–169.

Petri, C., Alburquerque, N., Faize, M., Scorza, R. and Dardick, C. (2018) Current achievements and future directions in genetic engineering of European plum (*Prunus domestica* L.). *Transgenic Research* 27, 225–240.

Piper, R.B. and Gardner, F.E. (1943) Comparative seasonal development of avocado fruits and seeds. *Proceedings of the Florida State Horticultural Society* 56, 122–124.

Reece, P.C. (1942) Differentiation of avocado blossom buds in Florida. *Botanical Gazette* 104, 323–328.

Richings, E.W., Cripps, R.F. and Cowan, A.K. (2000) Factors affecting 'Hass' avocado fruit size: carbohydrate, abscisic acid and isoprenoid metabolism in normal and phenotypically small fruit. *Physiologia Plantarum* 109(1), 81–89.

Ritonga, F.N. and Chen, S. (2020) Physiological and molecular mechanism involved in cold stress tolerance in plants. *Plants* 9(590), 1–13. doi:10.3390/plants9050560.

Robinson, P.W., Mickelbart, M.V., Liu, X., Adams, C., Witney, G. and Arpaia, M.L. (2002) Development of a phenological model of avocado tree growth in California. *Acta Horticulturae* 575, 859–864.

Rocha-Arroyo, J.L., Salazar-García, S. and Bárcenas-Ortega, A.E. (2010) Determinación irreversible a la floración del aguacate 'Hass' en Michoacán. *Revista Méxicana de Ciencias Agrícolas* 1(4), 469–478.

Rocha-Arroyo, J.L., Salazar-García, S., Bárcenas-Ortega, A.E., González-Durán, I.J.L. and Cossio-Vargas, L.E. (2011a) Fenología del aguacate 'Hass' en Michoacán. *Revista Méxicana de Ciencias Agrícolas* 2(3), 303–316.

Rocha-Arroyo, J.L., Salazar-García, S., Bárcenas-Ortega, A.E., González-Durán, J.L. and Medina-Torres, R. (2011b) Crecimientos vegetativo y reproductivo del aguacate 'Hass' en diversos climas de Michoacán, Mexico. In: *Proceedings of the VII World Avocado Congress*. World Avocado Congress, Cairns, Queensland, Australia, pp. 463–445.

Roe, D.J. and Morudu, T.M. (2000) Hass avocado yields as affected by dwarfing rootstocks and flower pruning. *South African Avocado Growers' Association Yearbook* 23, 30–32.

Rossouw, T., Robbertse, P.J., Kremer-Köhne, S. and Köhne, J.S. (2000) Effect of gibberellic acid treatments on flowering of avocado. *South African Avocado Growers' Association Yearbook* 23, 43–45.

Sablowski, R. (2007) Flowering and determinacy in *Arabidopsis*. *Journal of Experimental Biology* 58(5), 899–907. doi:10.1093/jxb/erm002.

Sagwe, R.N., Peters, M.K., Dubois, T., Steffan-Dewenter, I. and Lattorff, H.M.G. (2022) Pollinator efficiency of avocado (*Persea americana*) flower insect visitors. *Ecological Solutions and Evidence* 3(4): e12178. doi:10.1002/2688-8319.12178.

Sakai, W.S. and Nagao, M.A. (1985) Fruit growth and abscission in *Macadamia integrifolia*. *Physiologia Plantarum* 64, 455–460.

Salakpetch, S., Turner, D.W. and Dell, B. (1990) The flowering of carambola (*Averrhoa carambola* L.) is more strongly influenced by cultivar and water stress than by diurnal temperature variation and photoperiod. *Scientia Horticulturae* 43, 83–94.

Salazar-García, S. and Lovatt, C.J. (1998) GA_3 application alters flowering phenology of 'Hass' avocado. *Journal of the American Society for Horticultural Science* 123, 791–797.

Salazar-García, S. and Lovatt, C.J. (1999) Winter trunk injections of gibberellic acid altered the fate of 'Hass' avocado buds: effects on inflorescence type, number and rate of development. *Journal of Horticultural Science and Biotechnology* 74, 69–73.

Salazar-García, S. and Lovatt, C.J. (2000) Use of GA_3 to manipulate flowering and yield of 'Hass' avocado. *Journal of the American Society for Horticultural Science* 125, 25–30.

Salazar-García, S., Lord, E.M. and Lovatt, C.J. (1998) Inflorescence and flower development of the 'Hass' avocado (*Persea americana* Mill.) during 'on' and 'off' crop years. *Journal of the American Society for Horticultural Science* 123, 537–544.

Salazar-García, S., Lord, E.M. and Lovatt, C.J. (1999) Inflorescence development of the 'Hass' avocado: commitment to flowering. *Journal of the American Society for Horticultural Science* 124, 478–482.

Salazar-García, S., Zamora-Cuevas, L. and Vega-López, R.J. (2005) Update on the avocado industry of Michoacán, Mexico. *California Avocado Society Yearbook* 87, 31–44.

Salazar-García, S., Cossio-Vargas, L.E., Lovatt, C.J., González-Durán, I.J.L. and Pérez Barraza, M.H. (2006) Crop load affects vegetative growth flushes and shoot age influences irreversible commitment to flowering of 'Hass' avocado. *HortScience* 41(7), 1541–1546.

Salazar-García, S., Cossio-Vargas, L.E., González-Durán, I.J.L. and Lovatt, C.J. (2007a) Desarrollo floral del aguacate 'Hass' en clima semicálido. I. Influencia de la carga del fruto y edad de los brotes. *Revista Chapingo Serie Horticultura* 13, 87–92.

Salazar-García, S., Cossio-Vargas, L.E., González-Durán, I.J.L. and Lovatt, C.J. (2007b) Desarrollo floral del aguacate 'Hass' en clima semicálido. II. Generación y validación de modelos de predicción del desarrollo floral. *Revista Chapingo Serie Horticultura* 13, 93–101.

Salazar-García, S., Cossio-Vargas, L.E., González-Durán, I.J.L. and Lovatt, C.J. (2007c) Foliar-applied GA_3 advances fruit maturity and allows off-season harvest of 'Hass' avocado. *HortScience* 42, 257–261.

Salazar-García, S., Cossio-Vargas, L.E. and González-Durán, I.J.L. (2008) Desarrollo floral de los aguacates 'Choquette' y 'Booth-8' en clima cálido. II. Determinación irreversible a la floración. *Agricultura Técnica en Mexico* 34, 51–56.

Salazar-García, S., Cossio-Vargas, L.E. and González-Durán, I.J.L. (2009) Validación de modelos de predicción del desarrollo floral del aguacate 'Hass' desarrollados para Nayarit, en varios climas de Michoacán. *Revista Chapingo Serie Horticultura* 15(3), 281–288.

Sánchez-González, E.I., Gutiérrez-Díez, A. and Mayek-Pérez, N. (2019) Outcrossing rate and genetic variability in Mexican race avocado. *Journal of the American Society for Horticultural Science* 145, 53–59.

Schaffer, B., Whiley, A.W. and Kohli, R.R. (1991) Effects of leaf age on gas exchange characteristics of avocado (*Persea americana* Mill.). *Scientia Horticulturae* 48, 21–28. doi:10.1016/0304-4238(91)90149-S.

Schnell, R.J., Tondo, C.L., Brown, J.S., Kuhn, D.N., Ayala-Silva, T. *et al.* (2009) Outcrossing between 'Bacon' pollinizers and adjacent 'Hass' avocado trees and the description of two new lethal mutants. *HortScience* 44(6), 1522–1526.

Scholefield, P.B. (1982) A scanning electron microscope study of flowers of avocado, litchi, macadamia and mango. *Scientia Horticulturae* 16(3), 263–272.

Scholefield, P.B., Sedgley, M. and Alexander, D.M. (1985) Carbohydrate cycling in relation to shoot growth, floral initiation and development and yield in the avocado. *Scientia Horticulturae* 25, 99–110.

Schroeder, C.A. (1942) Pollen germination in the avocado. *Proceedings of the American Society for Horticultural Science* 41, 181–182.

Schroeder, C.A. (1944) The avocado inflorescence. *California Avocado Society Yearbook* 28, 39–40.

Schroeder, C.A. (1951) Flower bud development in the avocado. *California Avocado Society Yearbook* 35, 159–163.

Schroeder, C.A. (1952) Floral development, sporogenesis, and embryology in the avocado, *Persea Americana*. *Botanical Gazette* 113, 270–278.

Schroeder, C.A. (1953) Growth and development of the 'Fuerte' avocado fruit. *Proceedings of the American Society for Horticultural Science* 68, 253–258.

Sedgley, M. (1976) Control by the embryosac over pollen tube growth in the style of the avocado (*Persea americana* Mill.). *New Phytologist* 77, 149–152.

Sedgley, M. (1977a) The effect of temperature on floral behaviour, pollen tube growth and fruit set in the avocado. *Journal of Horticultural Sciences* 52, 135–141.

Sedgley, M. (1977b) Reduced pollen tube growth and the presence of callose in the pistil of the male floral stage of the avocado. *Scientia Horticulturae* 7, 27–36.

Sedgley, M. (1979a) Inter-varietal pollen tube growth and ovule penetration in the avocado. *Euphytica* 28, 25–35.

Sedgley, M. (1979b) Light microscope study of pollen tube growth, fertilization and early embryo and endosperm development in the avocado varieties Fuerte and Hass. *Annals of Botany* 44, 353–359.

Sedgley, M. (1979c) Structural changes in the pollinated and unpollinated avocado stigma and style. *Journal of Cell Science* 38, 49–60.

Sedgley, M. (1980) Anatomical investigation of abscised avocado flowers and fruitlets. *Annals of Botany* 46, 771–777.

Sedgley, M. and Annells, C.M. (1981) Flowering and fruit-set response to temperature in the avocado cultivar 'Hass'. *Scientia Horticulturae* 14, 27–33.

Sedgley, M. and Grant, W.J.R. (1983) Effect of low temperatures during flowering on floral cycle and pollen tube growth in nine avocado cultivars. *Scientia Horticulturae* 18, 207–213.

Sharma, K.D. and Nayyar, H. (2016) Regulatory networks in pollen development under cold stress. *Frontiers in Plant Science* 7:402. doi:10.3389/fpls.2016.00402.

Shen, J., Xiao, Q., Qiu, H., Chen, C. and Chen, H. (2016) Integrative effect of drought and low temperature on litchi (*Litchi chinensis* Sonn.) floral initiation revealed by dynamic genome-wide transcriptome analysis. *Scientific Reports* 6:32005. doi:10.1038/srep32005.

Shoval, S. (1987) Pollination rate and pollen tube growth of avocado, in relation to yield. MSc Thesis, The Hebrew University of Jerusalem, Israel. (In Hebrew.)

Singh, U.R. (1960) Studies in the fruit drop of mango (*Mangifera indica* L.). II. Nature and extent of fruit drop. *Horticultural Advance* 4, 142–154.

Singh, D.P., Jermakow, A.M. and Swain, S.M. (2002) Gibberellins are required for seed development and pollen tube growth in Arabidopsis. *The Plant Cell* 14, 3133–3147.

Sitrit, Y., Blumenfeld, A. and Riov, J. (1987) Ethylene biosynthesis in tissues of young and mature avocado fruits. *Physiologia Plantarum* 69, 504–510.

Slabbert, M.J. (1981) Flower and fruit drop. *South African Avocado Growers' Association Yearbook* 4, 89–91.

Solares, E., Morales-Cruz, A., Figueroa Baldera, R., Focht, E., Ashworth, V.E.T.M. *et al.* (2023) Insights into the domestication of avocado and potential genetic contributors to heterodichogamy. *G3 Genes|Genomes|Genetics* 13(2): jkac323. doi:10.1093/g3journal/jkac323.

Soltis, D.E., Chanderbali, A.S., Kim, S., Buzgo, M. and Soltis, P.S. (2007a) The ABC model and its applicability to basal angiosperms. *Annals of Botany* 100, 155–163. doi:10.1093/aob/mcm117.

Soltis, D.E., Ma, H., Frohlich, M.W., Soltis, P.S., Albert, V.A. *et al.* (2007b) The floral genome: an evolutionary history of gene duplication and shifting patterns of gene expression. *Trends in Plant Science* 12, 358–367. doi:10.1016/j.tplants.2007.06.012.

Southwick, S.M. and Davenport, T.L. (1986) Characterization of water stress and low temperature effects on flower induction in Citrus. *Plant Physiology* 81, 26–29. doi:10.1104/pp.81.1.26.

Stahl, P., Mirom, Y.L., Stern, R.A. and Goldway, M. (2019) Comparing 'Iriet' and 'Ettinger' avocado cultivars as pollinators of 'Hass' using SNPs for paternal identification. *Scientia Horticulturae* 248, 50–57.

Stephenson, A.G. and Winsor, J.A. (1986) *Lotus corniculatus* regulates offspring quality through selective fruit abortion. *International Journal of Organic Evolution* 40, 453–458.

Stern, R.A., Adato, I., Goren, M., Eisenstein, D. and Gazit, S. (1993) Effects of autumnal water stress on litchi flowering and yield in Israel. *Scientia Horticulturae* 54, 295–302. doi:10.1016/0304-4238(93)90108-3.

Stern, R.A., Kigel, J., Tomer, E. and Gazit, S. (1995) 'Mauritius' lychee fruit development and reduced abscission after treatment with the auxin 2,4,5-TP. *Journal of the American Society for Horticultural Science* 120, 65–70.

Stern, R.A., Rozen, A., Eshed, R., Zviran, T., Sisai, I. *et al.* (2021) Bumblebees (*Bombus terrestris*) improve 'Hass' avocado (*Persea americana*) pollination. *Plants* 10(7): 1372.

Steyn, E.M.A., Robbertse, P.J. and Smith, D. (1993) An anatomical study of ovary-to-cuke development in consistently low-producing trees of the 'Fuerte' avocado (*Persea americana* Mill.) with special reference to seed abortion. *Sexual Plant Reproduction* 6, 87–97.

Stout, A.B. (1923) A study in cross-pollination of avocados in Southern California. *California Avocado Association Annual Report* 7, 29–45.

Stout, A.B. (1933) The pollination of avocados. *University of Florida Agricultural Experiment Station Bulletin* 257, 4–44.

Sulaiman, Z., Collins, G., Witherspoon, J. and Sedgley, M. (2004) Identification of pollen donors for the avocado cultivar Gwen in a mixed orchard by isozyme analysis. *Journal of Horticultural Science and Biotechnology* 79(4), 571–575.

Symons, P.R.R. and Wolstenholme, B.N. (1989) Responses of potted Hass avocado plants to paclobutrazol drenches. *South African Avocado Growers' Association Yearbook* 12, 40–42.

Takeno, K. (2016) Stress-induced flowering: the third category of flowering response. *Journal of Experimental Botany* 67(17), 4925–4934.

Tanaka, M., Takei, K., Kojima, M., Sakakibara, H. and Mori, H. (2006) Auxin controls local cytokinin biosynthesis in the nodal stem in apical dominance. *The Plant Journal* 45, 1028–1036.

Tang, L. and Lovatt, C.J. (2019) Effects of low temperature and gibberellic acid on floral gene expression and floral determinacy in 'Washington' navel orange (*Citrus sinensis* L. Osbeck). *Scientia Horticulturae* 243, 92–100.

Tang, L. and Lovatt, C.J. (2022) Effects of water-deficit stress and gibberellic acid on floral gene expression and floral determinacy in 'Washington' navel orange. *Journal of American Horticultural Science* 147(4), 183–195.

Taylor, N.J. and Cowan, A.K. (2001) Plant hormone homeostasis and the control of avocado fruit size. *Plant Growth Regulation* 35, 247–255.

Taylor, N.J. and Cowan, A.K. (2004) Xanthine dehydrogenase and aldehyde oxidase impact plant hormone homeostasis and affect fruit size in 'Hass' avocado. *Journal of Plant Research* 117(2), 121–130.

Thomas, T.V., Eicker, A. and Robbertse, P.J. (1994) Possible role of fungi in negatively affecting fruit-set in avocados. *South African Journal of Botany* 60, 251–256.

Thorp, T.G. and Sedgley, M. (1993a) Architecture analysis of tree form in a range of avocado cultivars. *Scientia Horticulturae* 53, 85–98.

Thorp, T.G. and Sedgley, M. (1993b) Manipulation of shoot growth patterns in relation to early fruit set in 'Hass' avocado (*Persea americana*). *Scientia Horticulturae* 56, 147–156.

Thorp, T.G., Aspinall, D. and Sedgley, M. (1993) Influence of shoot age on floral development and early fruit set in avocado (*Persea americana* Mill.) cv. Hass. *Journal of Horticultural Sciences* 68, 645–651.

Thorp, T.G., Aspinall, D. and Sedgley, M. (1994) Preformation of node number in vegetative and reproductive proleptic shoot modules of *Persea* (Lauraceae). *Annals of Botany* 73, 13–22.

Tingwa, P.O. and Young, R.E. (1975) Studies on the inhibition of ripening in attached avocado (*Persea americana* Mill.) fruits. *Journal of the American Society for Horticultural Science* 100, 447–449.

Tomer, E. and Gazit, S. (1979) Early stages in avocado (*Persea americana* Mill.) fruit development: anatomical aspects. *Botanical Gazette* 140, 304–309.

Tomer, E., Gazit, S. and Eisenstein D. (1980) Seedless fruit in 'Fuerte' and 'Ettinger' avocado. *Journal of the American Society for Horticultural Science* 105, 341–346.

Torres, A.M. and Bergh, B.O. (1978) Isozymes as indicators of out-crossing among 'Pinkerton' seedlings. *California Avocado Society Yearbook* 62, 103–110.

Valmayor, R.V. (1967) Cellular development of the avocado from blossom to maturity. *The Philippine Agriculturist* 50, 907–976.

Van der Walt, M., Davie, S.J. and Smith, D.G. (1993) Carbohydrate and other studies on alternate bearing Fuerte and Hass avocado trees. *South African Avocado Growers' Association Yearbook* 16, 82–85.

Verreynne, J.S. (2005) The mechanism and underlying physiology perpetuating alternate bearing in 'Pixie' mandarin (*Citrus reticulata* Blanco). PhD dissertation, University of California, Riverside, California.

Violi, H.A., Brown, J.S., Tondo, C.L., Borrone, J.W. and Schnell, R.J. (2009) Microsatellite markers reveal low breeding system efficacy and pollen contamination can limit production of full-sib avocado progeny. *Scientia Horticulturae* 120(3), 360–366.

Vithanage, V. (1990) The role of the European honeybee (*Apis mellifera* L.) in avocado pollination. *Journal of Horticultural Sciences* 65, 81–86.

Vrecenar-Gadus, M. and Ellstrand, N.C. (1985) The effect of planting design on out-crossing rate and yield in the 'Hass' avocado. *Scientia Horticulturae* 27, 215–221.

Whiley, A.W. (1990) CO_2 assimilation of developing fruiting shoots of cv. Hass avocado (*Persea americana* Mill). *South African Avocado Growers' Association Yearbook* 13, 28–30.

Whiley, A.W. and Schaffer, B. (1994) Avocado. In: Schaffer, B. and Anderson, P.C. (eds) *Handbook of Environmental Physiology of Fruit Crops: Volume* II. *Sub-tropical and Tropical Crops*. CRC Press, Boca Raton, Florida, pp. 3–35.

Whiley, A.W., Saranah, J., Cull, B. and Pegg, K. (1988) Manage avocado growth cycle for productivity gains. *Queensland Agricultural Journal* 114, 29–36.

Whiley, A.W., Köhne, J.S., Arpaia, M.L. and Bender, G.S. (1990) Future prospects with new avocado cultivars and elite rootstocks. *South African Avocado Growers' Association Yearbook* 13, 16–20.

Whiley, A., Saranah, J. and Rasmussen, T. (1992) Effect of the time of harvest on fruit size and yield on trunk starch concentration. In *Proceedings of the II World Avocado Congress*. World Avocado Congress, Orange, California, pp. 155–159.

Whiley, A.W., Rasmussen, T.S., Saranah, J.B. and Wolstenholme, B.N. (1996a) Delayed harvest effects on yield, fruit size and starch cycling in avocado (*Persea americana* Mill.) in two subtropical environments. I. The early-maturing cv. Fuerte. *Scientia Horticulturae* 66, 23–34.

Whiley, A.W., Rasmussen, T.S., Saranah, J.B. and Wolstenholme, B.N. (1996b) Delayed harvest effects on yield, fruit size and starch cycling in avocado (*Persea americana* Mill.) in subtropical environments. II. The late-maturing cv. Hass. *Scientia Horticulturae* 66, 35–49.

Whiley, A.W. and Wolstenholme, B. (1990) Carbohydrate management in avocado trees for increased production. *South African Avocado Growers' Association Yearbook* 13, 25–27.

Wilkie, J.D., Sedgley, M. and Olesen, T. (2008) Regulation of floral initiation in horticultural trees. *Journal of Experimental Botany* 59(12), 3215–3228.

Williams, M.W. and Edgerton, L.J. (1981) Fruit thinning of apples and pears with chemicals. *United States Department of Agriculture Information Bulletin* 289, 1–22.

Winer, L. and Apelbaum, A. (1986) Involvement of polyamines in the development and ripening of avocado fruits. *Journal of Plant Physiology* 126, 223–233.

Winslow, M.M. and Enderud, J. (1955) Flowering behavior and yields of some avocado varieties at Riverside. *California Avocado Society Yearbook* 39, 133–135.
Wolstenholme, B.N. (1986) Energy costs of fruiting as a yield-limiting factor with special reference to the avocado. *Acta Horticulturae* 175, 121–125.
Wolstenholme, B.N. (1990) Resource allocation and vegetative-reproductive competition: opportunities for manipulation in evergreen fruit trees. *Acta Horticulturae* 275, 451–459.
Wolstenholme, B.N., Whiley, A.W. and Saranah, J.B. (1990) Manipulating vegetative: reproductive growth in avocado (*Persea americana* Mill.) with paclobutrazol foliar sprays. *Scientia Horticulturae* 41, 315–327.
Yang, Y., Saand, M.A., Huang, L., Abdelaal, W.B., Zhang, J. *et al.* (2021) Applications of multi-omics technologies for crop improvement. *Frontiers in Plant Science* 12, 563953. doi:10.3389/fpls.2021.563953.
Yeshitela, T., Robbertse, P.J. and Stassen, P.J.C. (2004) Paclobutrazol suppressed vegetative growth and improved yield as well as fruit quality of 'Tommy Atkins' mango (*Mangifera indica*) in Ethiopia. *New Zealand Journal of Crop and Horticultural Science* 32(3), 281–293. doi:10.1080/01140671.2004.9514307.
Ying, Z., Davenport, T.L., Faber, B., Zhang, T., Schnell, R.J. and Tondo, C.L. (2009a) Re-evaluation of the roles of honeybees and wind on pollination in avocado. *Journal of Horticultural Science and Biotechnology* 84(3), 255–260.
Ying, Z., Davenport, T.L., Zhang, T., Schnell, R.J. and Tondo, C.L. (2009b) Selection of highly informative microsatellite markers to identify pollen donors in 'Hass' avocado orchards. *Plant Molecular Biology Reporter* 27(3), 374–380.
Young, P.J. (1992) Inherent influence of rootstock race on avocado fruit maturity. In *Proceedings of the II World Avocado Congress*. World Avocado Congress, Orange, California, pp. 149–154.
Zilkah, S. and David, I. (1993) Uptake and translocation of paclobutrazol through different canopy organs and roots of avocado (*Persea americana* Mill.). *Acta Horticulturae* 329, 98–100.
Zilkah, S., Klein, I., Feigenbaum, S. and Weinbaum, S.A. (1987) Translocation of foliar-applied urea^{15}N to reproductive and vegetative sinks of avocado and its effect on initial fruit set. *Journal of the American Society for Horticultural Science* 112, 1061–1065.
Ziv, D., Zviran, T., Zezak, O., Samach, A. and Irihimovitch, V. (2014) Expression profiling of *FLOWERING LOCUS T*-like gene in alternate bearing 'Hass' avocado trees suggests a role for *PaFT* in avocado flower induction. *PLoS One* 9(1): e110613. doi:10.1371/journal.pone.0110613.

7 Ecophysiology

B. Schaffer[1]*, A.W. Whiley[2], P.M. Gil[3], M.V. Mickelbart[4] and R. Castillo-Argaez[5]
[1]*Tropical Research and Education Center, University of Florida, Homestead, Florida;* [2]*Sunshine Horticultural Services Pty Ltd, Buderim, Queensland, Australia;* [3]*Departamento de Fruticultura y Enología, Facultad de Agronomía e Ingeniería Forestal, Pontificia Universidad Católica de Chile, Chile;* [4]*Department of Botany & Plant Pathology and Department of Horticulture & Landscape Architecture, Purdue University, West Lafayette, Indiana;* [5]*Department of Agronomy, University of Florida, Gainesville, Florida*

Abstract

This chapter discusses physiological, growth and yield responses of avocado trees to abiotic (environmental) factors including light, temperature, humidity, drought, flooding, wind, salinity, elevated atmospheric CO_2 and air pollution. Also discussed are the effects of two important plant diseases, Phytophthora root rot and laurel wilt, on whole-plant physiology of avocado. Methods to reduce or alleviate plant stress due to these factors are discussed, where information on mitigation exits.

Introduction

Environmental factors influence all aspects of plant growth and development and determine the extent to which genotypic potential is achieved. In natural woody plant communities, irradiance and water stress are environmental factors that usually have the greatest impact on plant growth and development (van Schaik *et al.*, 1993; Whiley, 1994). However, domesticated fruit tree species are also influenced by temperature (Sedgley and Grant, 1983; Whiley and Winston, 1987; Issarakraisila and Considine, 1994; Whiley, 1994), since the production of these crops has often been extended into locations with temperature extremes beyond their evolutionary limits. In fact, the temperature range often defines where a particular fruit species is commercially successful (Schaffer *et al.*, 1994). The impact of human population growth and the associated environmental consequences on agricultural production is of increasing concern (Vörösmarty *et al.*, 2000; FAO, 2008; Koning *et al.*, 2008). In many communities, competition for limited domestic versus agricultural water resources is increasing (Strzepek and Boehlert, 2010). The long-term effects of global warming on crop productivity are unknown, and there has been limited research relative to other crop species on the effects of increased atmospheric carbon dioxide (CO_2) concentrations on subtropical and tropical fruit crops (Schaffer *et al.*, 1997, 1999). Other environmental concerns include increasing atmospheric pollution, for example ozone and acid rain (Eissenstat *et al.*, 1991a,b), and chloride toxicity as a result of saltwater intrusion or accumulation from fertilizer use (Mass, 1994; Khondoker *et al.*, 2023). Determining the impacts of environmental variables on crop production will become increasingly important as the land available for agriculture becomes more limited, and crop production is

*Corresponding author: bas56@ufl.edu

DOI: 10.1079/9781800621824.0007

expanded or displaced into environments with higher levels of abiotic stress. An understanding of physiological and growth responses of fruit crops to the environment is crucial to minimize the deleterious impact of suboptimal conditions and to develop management strategies for maximum productivity (Schaffer and Andersen, 1994).

Avocado (*Persea americana* Mill.) is grown primarily in three distinct climatic zones: cool, semi-arid with winter-dominant rainfall (e.g. California, Chile, Israel, Peru and Spain); humid, subtropical with summer-dominant rainfall (e.g. eastern Australia, Mexico and South Africa); and tropical or semitropical with summer-dominant rainfall (e.g. Brazil, Columbia, Florida and Indonesia). *P. americana* is divided into three ecological races that are given varietal status within the species: *P. americana* var. *drymifolia* (Mexican race); *P. americana* var. *guatemalensis* (Guatemalan race); and *P. americana* var. *americana* (West Indian race) (Bergh *et al.*, 1973; Scora and Bergh, 1990; Ashworth and Clegg, 2003; Schnell *et al.*, 2003) (see Chanderbali *et al.*, Chapter 3, this volume). There are distinct differences among races in relation to their adaptability to environmental conditions. Hybridization has occurred freely among the races, resulting in increased genetic diversity and environmental plasticity of the species (Whiley and Schaffer, 1994).

Climatic shifts due to global climate change are expected to have a significant effect on crop production and food security worldwide (IPCC, 2022). Like most crops, future distribution and production of avocado are expected to be significantly affected by temperature and precipitation changes resulting from global warming (Ramírez-Gil *et al.*, 2019; Grüter *et al.*, 2022; Fernandes da Silva *et al.*, 2023). In a recent study, using modelling to predict climate change scenarios on current and future suitability of avocado production, the authors concluded that there will be both increases and decreases in suitable areas for avocado production. They suggested that increases in areas that become more suited for production will be due to higher minimum temperatures, whereas both increases and decreases in suitable avocado production areas will be due to changes in precipitation caused by higher or lower potentials for drought or flooding, depending on location (Grüter *et al.*, 2022). Understanding the physiological responses of avocado trees to abiotic and biotic stresses will be essential for mitigating responses of this important fruit crop species to changing environmental and edaphic conditions in the face of predicted global climate shifts. This chapter focuses on the ecophysiology of avocado and summarizes the impact of environmental variables, including light, temperature, water, salinity, carbon dioxide, pollution and major diseases on physiology, growth and development of avocado trees.

Leaf Age and Developmental Physiology

As leaves transition through various developmental stages from emergence to senescence and from sink to source, their carbon contribution to the tree changes, as does their response to environmental fluctuations (Dickson *et al.*, 2000). Avocado shoot growth is episodic, with leaves of different developmental stages often present among and within different cohorts of shoot flushes at a given point in time. Avocado leaf longevity ranges from 12 to 18 months (Whiley and Schaffer, 1994; Wolstenholme and Whiley, 1999); thus, within this period, trees completely renew their canopy. To understand the overall tree response to the environment, it is important to consider leaf and shoot developmental stages.

The time from budbreak to full leaf expansion in field-grown avocado trees is approximately 30 days, and leaves attain positive net CO_2 assimilation (also referred to as net photosynthesis) values slightly before or at full leaf expansion (Whiley, 1994; Liu *et al.*, 2002a). The sum of the work done to date suggests that individual avocado leaves become net producers of carbon 20–24 days after budbreak. However, maximum net CO_2 assimilation is reached between 40 and 70 days after budbreak (Whiley, 1990, 1994; Schaffer *et al.*, 1991; Liu *et al.*, 2002a). Dark respiration decreases until approximately 40 days after budbreak, at which point it plateaus (Schaffer *et al.*, 1991).

Mature avocado leaves have stomata only on the abaxial side of the leaf (Freeland, 1948; Mickelbart *et al.*, 2000), and gas exchange effectively does not occur on the adaxial leaf surface (Mickelbart *et al.*, 2000). In Australia, Whiley (1994) observed a maximum net CO_2 assimilation rate of 23 μmol CO_2 m^{-2} s^{-1} in leaves of field-grown avocado trees, while Mwelase *et al.* (2022) measured a maximum net CO_2 assimilation rate of 25.2 μmol CO_2 m^{-2} s^{-1} in field-grown avocado trees in South Africa. These net CO_2 assimilation rates are significantly higher than those reported by other researchers for container-grown trees (≈7 μmol CO_2 m^{-2} s^{-1}) (Bower *et al.*, 1978; Kimelmann, 1979; Scholefield *et al.*, 1980; Mickelbart and Arpaia, 2002; Yin *et al.*, 2022). Schaffer *et al.* (1987, 1991) found that net CO_2 assimilation of avocado trees in an orchard in southern Florida (7–10 μmol CO_2 m^{-2} s^{-1}) was similar to that previously reported for container-grown trees. In southern Florida, the rhizosphere of the hard, gravelly loam soil mimics that of containers, resulting in severe root restriction of mature trees, which can inhibit net CO_2 assimilation (Crane *et al.*, 1994; Schaffer *et al.*, 1994). In a later study with container- and field-grown 'Hass' trees in Australia, Whiley *et al.* (1999) reported that the maximum net CO_2 assimilation rate of container-grown trees (5.2 μmol CO_2 m^{-2} s^{-1}) was only 27% of that of field-grown trees (19.0 μmol CO_2 m^{-2} s^{-1}). The leaf starch content measured in these trees at the time of net CO_2 assimilation measurements was 28.0 ± 2.6 and 17.4 ± 1.4 mg g dry weight^{-1} in container- and field-grown trees, respectively. Root restriction, either due to high bulk-density soils or container limitations, likely results in reduced root sink strength, leading to accumulation of starch in avocado leaves, which inhibits net CO_2 assimilation (Fig. 7.1).

Avocado cultivars vary in their maximum net CO_2 assimilation. In the same orchard in California, net CO_2 assimilation of 'Lamb Hass' was higher than that of 'Hass' in autumn and summer, but not spring (Liu *et al.*, 2002a). The difference was correlated with the higher leaf chlorophyll concentration in 'Lamb Hass' and may account for inherent variances in productivity between cultivars. When net CO_2 assimilation was expressed on a per unit chlorophyll basis, cultivar differences did not exist. In an orchard in southern Florida, mature leaves of 'Booth 8' had higher net CO_2 assimilation than 'Peterson', although the chlorophyll concentrations in leaves were similar (Schaffer *et al.*, 1991). Avocado rootstocks have a large effect on leaves of the scion (Mickelbart *et al.*, 2007a), which could presumably lead to differences in net CO_2 assimilation. Some studies have demonstrated differences in leaf gas exchange among scions on different rootstocks (Ploetz *et al.*, 1989; Castro *et al.*, 2009), while in other studies, scion differences were not observed (Mickelbart and Arpaia, 2002). Given the wide range of effects of various rootstocks on scion physiology, this is not surprising.

Avocado shoot growth typically occurs in two-yearly flushes in subtropical Australia (Cull,

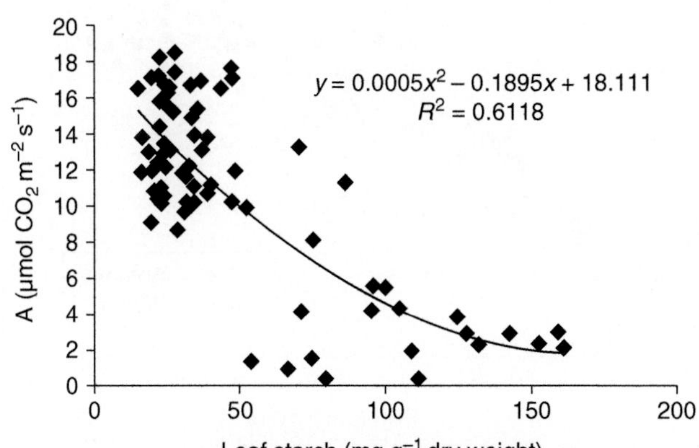

Fig. 7.1. Correlation between leaf starch concentration and net CO_2 assimilation (*A*) of container-grown 'Hass' avocado trees. One-year-old trees were grown in 56.8-l containers under a constant temperature of 28°C. Data were collected from multiple leaves on five trees. (M.V. Mickelbart, 2004, unpublished data).

1986; Whiley *et al.*, 1988b), South Africa (Wolstenholme, 2001), New Zealand (Dixon *et al.*, 2006), Mexico (Salazar-García *et al.*, 2018) and California (Robinson *et al.*, 2002), between two and four flushes in Florida (Venning and Lincoln, 1958; Ploetz *et al.*, 1993) and as many as six flushes in lowland tropical regions such as Trinidad (Gregoriou, 1982). The number of shoot flushes produced in a season varies among cultivars (Davenport, 1982) and growing locations within a region (Thorp *et al.*, 1995). Leaves within and among these various flushes vary in their physiology (Liu *et al.*, 2002a).

During their transition from net carbon importers (sinks) to net carbon exporters (sources), avocado leaves may compete with developing fruit on the same shoot for available photoassimilates (Biran, 1979; Buchholz, 1986; Cutting and Bower, 1990; Wolstenholme, 1990). The epidermis of a developing avocado fruit contains chlorophyll and has functioning stomata (Cran and Possingham, 1973; Blanke, 1991; Whiley *et al.*, 1992; Blanke and Whiley, 1995). During early ontogeny, fruit have a positive net CO_2 assimilation rate in adequate light and thus contribute slightly to their own carbon requirements for growth (Whiley *et al.*, 1992). As fruit advance in their development, their net CO_2 assimilation rate becomes increasingly less than CO_2 evolution at night (dark respiration), and the relative contribution to their own carbon requirement becomes negligible (Whiley *et al.*, 1992; Whiley and Schaffer, 1994).

Light

The environmental factor exerting perhaps the greatest influence on net CO_2 assimilation in an orchard is the availability of incident light. Quantifying the extent to which light affects net CO_2 assimilation and water loss in avocado leaves may lead to science-based management strategies to maximize yields and minimize environmental stress (Heath *et al.*, 2007a). To quantify photosynthetic responses to light level, irradiance is generally expressed as photosynthetic photon flux (PPF), also referred to as photosynthetically active radiation (PAR). This represents the number of photons (or quanta) in the spectral wavelength (400–700 nm) of light used for net CO_2 assimilation. In full sun at mid-day in summer, at low to mid-latitudes, irradiance is about 2000 μmol m^{-2} s^{-1}. A light response curve (Fig. 7.2) quantifies the relationship between light level (PPF) and net CO_2 assimilation of an individual leaf.

The photosynthetic light saturation point is the irradiance at which increasing the light level does not result in an increase in net CO_2 assimilation. There are conflicting reports in the literature about the effects of light on net photosynthesis

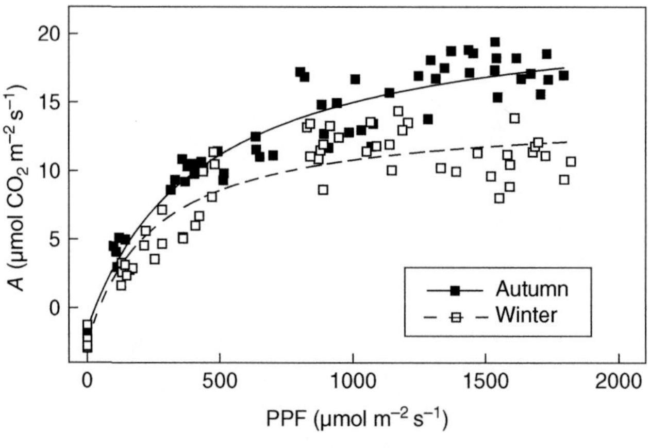

Fig. 7.2. Net CO_2 assimilation (*A*) response of field-grown 'Hass' avocado trees to varying PPF. The regression line for *A* of field-grown trees in summer is represented by $y = 22.08((-30.67 + x)/(427.43 + x))$, $r^2 = 0.94$; and that for field-grown trees in winter is represented by $y = 14.17((-46.72 + x)/(250.05 + x))$, $r^2 = 0.86$. (Redrawn from Whiley *et al.*, 1999, with permission).

of avocado trees. The reported light saturation point for leaves of container-grown avocado trees is generally low: approximately 500 μmol m^{-2} s^{-1} (about 25% of that of full sunlight) for 'Fuerte' (Scholefield *et al.*, 1980) and 'Hass' (M.V. Mickelbart *et al.*, West Lafayette, IN, USA, 2003, personal communication). Bower *et al.* (1978) measured a light saturation point of 660 μmol m^{-2} s^{-1} (or about 33% of full sunlight) of whole canopies of small container-grown 'Edranol' trees. The higher light saturation point of individual leaves (as opposed to a whole canopy) was most likely due to a portion of canopy leaves being shaded, thus requiring a higher irradiance to saturate all leaves (Whiley, 1994; Whiley and Schaffer, 1994). In an orchard in Queensland, Australia, Whiley (1994) observed that the light saturation point for photosynthesis of mature 'Hass' leaves was much higher (1110 μmol m^{-2} s^{-1}) than that of container-grown trees.

The light compensation point (the irradiance level at which net photosynthesis is 0) for avocado leaves has been reported to be 63 μmol m^{-2} s^{-1} for container-grown 'Fuerte' trees (Scholefield *et al.*, 1980) and observed to be 12 μmol m^{-2} s^{-1} for container-grown 'Hass' trees (M.V. Mickelbart, West Lafayette, IN, USA, 2003, personal communication). The light compensation point for field-grown 'Hass' trees in Australia was 10 μmol m^{-2} s^{-1} (Whiley, 1994; Whiley and Schaffer, 1994) (Fig. 7.2). All of these values are considerably less than the 200–120 μmol m^{-2} s^{-1} found 1–1.5 m inside an avocado canopy on clear days (Whiley *et al.*, 1992). These relatively low light compensation points undoubtedly reflect the environmental conditions under which avocado trees have evolved. In their natural rainforest habitat, avocado trees grow like understorey trees (Whiley and Schaffer, 1994). The bias to vegetative growth (Wolstenholme, 1990), with an extensive leaf plasticity response to light (Whiley, 1994), undoubtedly helps maximize photosynthesis, thereby assisting with competition among other tree species in their indigenous environment.

The initial slope of the linear portion of the photosynthetic light response curve is referred to as the apparent maximum quantum yield. Quantum yield is the efficiency of light utilization for photosynthesis, expressed as the number of moles of CO_2 fixed per mole of photons (or quanta) absorbed by the leaf (Singsaas *et al.*, 2001). The estimate of apparent maximum quantum yield is based on incident light, whereas true quantum yield takes into account the portion of incident light reflected or transmitted from the leaf surface. The first report of apparent maximum quantum yield for avocado was 0.013 μmol CO_2 $μmol^{-1}$ quanta from measurements of container-grown 'Fuerte' trees under laboratory conditions (Scholefield *et al.*, 1980). This is well below the average value of 0.08 defined for C_3 plants (Singsaas *et al.*, 2001) and is likely due to the previously mentioned feedback inhibition of net CO_2 assimilation in container-grown trees, where maximum net CO_2 assimilation of leaves is reduced due to a restricted root sink. Later research by Whiley *et al.* (1999) determined that the apparent maximum quantum yield of field-grown 'Hass' avocado trees was 0.06 μmol CO_2 $μmol^{-1}$. They were also able to show that quantum yield in avocado is significantly reduced by sink limitations of net CO_2 assimilation when trees were grown under conditions of root restriction (0.021 μmol CO_2 $μmol^{-1}$). Leaf age can also affect apparent maximum quantum yield. Leaves from three successive flushes (from oldest to youngest) of container-grown trees had mean apparent maximum quantum yield values of 0.046, 0.063 and 0.071 once the newest flush was fully expanded (M.V. Mickelbart, West Lafayette, IN, USA, 2004, personal communication).

There have been limited reports on the effects of light on growth of avocado trees. In a greenhouse study, reducing the irradiance from ≈1350 to ≈725 μmol m^{-2} s^{-1} did not reduce biomass accumulation of 'Fuerte' avocado trees in containers (Chirachint and Turner, 1988). However, the lower light level was above the light saturation point for net CO_2 assimilation of container-grown trees, thereby accounting for the lack of growth differences. Although light response curves for net CO_2 assimilation of avocado leaves have been developed, the effects of incident light on avocado growth and development need to be integrated over time to provide a clear understanding of the relationship between light absorption and tree productivity.

Full canopies of orchard crops only absorb 65%–70% of the available radiant energy, thereby limiting production potential (Jackson, 1980). Verheij (1968) reported that selective pruning of crowded apple trees resulted in production benefits with an associated reduction in canopy light absorption from 80% to 70%. Wilkie *et al.* (2019) found that mature (prior to crowding) avocado canopies in low-density planted 'Hass' orchards intercepted 80%–84% of the light available to the orchard surface, but penetration into the canopy was low in most situations. In most temperate orchard crops, methods such as selective pruning allow for maximization of light absorption by the canopy, and pruning methods that enhance light absorption in avocado canopies can result in increased yields (Westling *et al.*, 2021). Selective pruning combined with strategic use of triazole growth regulators resulted in yields ranging from 30 to 43 tonnes per hectare in blocks of a high-density planted 'Hass' orchard in New Zealand (Hermosilla-Madrid, Kaitaia, NZ, 2022, personal communication; Hermosilla-Madrid and Whiley, 2023). It is likely that this orchard, planted at a density of 814 trees per hectare and maintained at a height of ≈2.5 m, had consistently improved light interception, with fruiting sites closely coupled to well-lit leaves. Other factors such as resource distribution between vegetative and reproductive growth due to triazole use (Wolstenholme *et al.*, 1990) may have also assisted in setting and maturing crops of this tonnage. Wilkie *et al.* (2019) provided information on the relationship between light interception and yield in low-density-planted 'Hass' avocado orchards. Similar studies on the dynamics of light interception and distribution in high-density orchards and their impact on yield would be a useful contribution to the management of annual yield fluctuations, considering the trend toward production with smaller trees, which makes harvesting a ground-based operation in the absence of mechanical platforms (A.W. Whiley, Queensland, Australia, 2023, personal communication).

Avocado canopies consist of relatively short-lived leaves (Whiley, 1994) that are exposed to varying light conditions throughout their lifespan. Because of the sequential flushing of avocado shoots, most leaves will develop in full sun and then transition into deep shade following new shoot growth, while other leaves develop and remain in shaded portions of the canopy. Leaves that develop in shade have lower chlorophyll a/b ratios and lower carotenoid concentrations (Förster *et al.*, 2009). Shade leaves of avocado have lower photosynthetic rates, which are not due to differences in stomatal density (Mickelbart *et al.*, 2000) but due to reduced stomatal conductance at lower light levels (Sterne *et al.*, 1977).

Leaves in avocado canopies must maintain a balance between optimizing photosynthetic efficiency and photo-protection in changing light environments. Leaves that are in shade are often exposed to transient sunflecks, created by gaps in the canopy. These sunflecks are short periods (seconds to minutes) of intense light that may be ten times the light level to which the leaves are acclimated (M.V. Mickelbart and R. Heath, Riverside, CA, USA, 2000, personal communication). The ability to adjust to these changing light levels is important on two levels: (i) to maintain carbon fixation, which allows for vegetative and reproductive growth; and (ii) to avoid photoinhibition caused by sudden increases in light intensity. Measurements of avocado leaf response to simulated sunflecks revealed that the half-time for leaves to reach maximum photosynthetic rates when exposed to a sunfleck was approximately 120 seconds (Fig. 7.3). Avocado leaves can, therefore, respond very quickly to changes in light; however, this response is also dependent on the stomatal conductance at the time the leaf is exposed to the sunfleck (M.V. Mickelbart, West Lafayette, IN, USA, 2003, personal communication).

Xanthophyll pigments in leaves function for both energy transfer to photosynthetic reaction centres and energy dissipation and detoxification of reactive oxygen species (ROS). The βcarotene-derived violaxanthin cycle is ubiquitous among plants, but the α-carotene-derived lutein epoxide (Lx) cycle is unique to specific genera, including avocado (Esteban *et al.*, 2007). Upon exposure to light, de-epoxidation of Lx to lutein (L) occurs quickly, while the reverse reaction upon exposure to shade happens very slowly in avocado (Förster *et al.*, 2009). Concentrations of Lx are higher in older than in younger avocado leaves (Esteban *et al.*, 2008), which may enhance the light-harvesting efficiency of shade leaves (Matsubara *et al.*, 2004). Furthermore, the accumulated Lx found in shaded avocado leaves likely allows for a photo-convertible pool of L, which provides photo-protection against sudden increases in light (Esteban *et al.*, 2008). Taken together, these studies suggest that avocado canopies are very efficient at utilizing

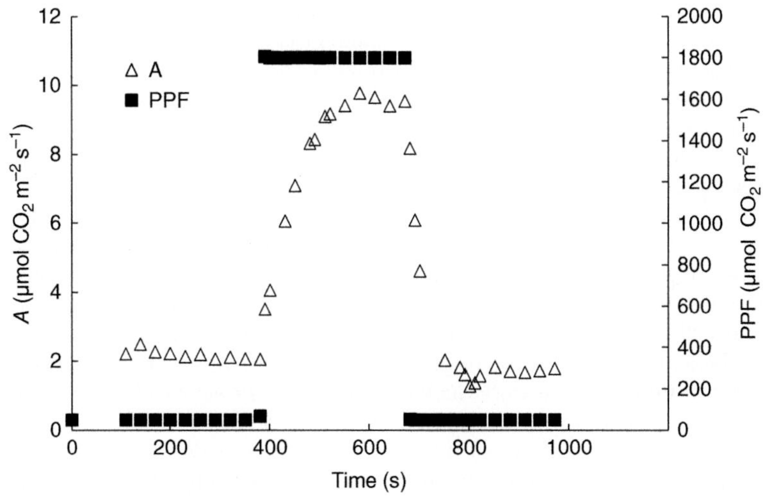

Fig. 7.3. Representative simulated sunfleck response curve of leaves of container-grown 'Hass' avocado trees. Leaves were adapted to a PPF of 50 μmol m^{-2} s^{-1} and then exposed to a PPF of 1500 μmol m^{-2} s^{-1} for 5 minutes while net CO_2 assimilation (A) was measured. First-order kinetics of the response were calculated to determine time to half-maximum A ($t_{1/2}$ = 120.9 s), A at $t_{1/2}$ ($A_{1/2}$ = 7.61 μmol m^{-2} s^{-1}) and maximum A (A_{max} = 9.56 μmol m^{-2} s^{-1}). Leaves of two subsequent flushes were not significantly different for any of the calculated parameters. One-year-old trees were grown in 56.8-l containers under a constant temperature of 28°C. (Data were collected from multiple leaves on three trees. M.V. Mickelbart, 2004, unpublished data).

the low light levels present within the canopy to maintain carbon assimilation and that avocado leaves have mechanisms that allow them to respond quickly to a changing light environment. This capacity is no doubt a reflection of the environment that nurtured the evolutionary development of this species. However, photoinhibition of overwintered leaves is a feature of cold, sunny and dry subtropical winters, when 'surplus' irradiance is not fully utilized in growth. Additionally, a recent study on leaves of young 'Hass' trees indicated that the combination of high temperature and high light can have severe negative effects on the integrity and activity of photosystem II (PS II), higher than that of either stress alone. Therefore, under field conditions, where high temperatures and high light intensity can occur simultaneously, there is potential for substantial crop damage (Shapira *et al.*, 2021).

Temperature

The three ecological races of avocado differ in their response to temperature (Krezdorn, 1970; Whiley and Schaffer, 1994; Soares *et al.*, 2002; see also Ramírez-Gil, Valbuena-Goana and Wolstenholme, Chapter 5, this volume). The Mexican race, which is native to cool, subtropical mountainous regions, is considered to be the most cold tolerant of the three races. However, there are differences in cold tolerance among Mexican race cultivars. Krezdorn (1970) evaluated the performance of several Mexican race cultivars in central and northern Florida during a 9-year period (1961–1970), where temperatures dropped below −6.7°C in 5 of the 9 years. A range of coldtolerance was observed among Mexican race cultivars, with 'Gainesville' and 'Mexicola' having the highest cold tolerance of the cultivars evaluated. The Guatemalan race cultivars are reported to have intermediate cold tolerance, whereas West Indian race cultivars are the most susceptible to chilling or freeze damage. During a 1917 freeze in California, temperatures of −6.1°C resulted in damage to only young leaves of some Mexican race cultivars, whereas the buds of all Guatemalan cultivars were killed (Webber, 1918). In another observation, West Indian race cultivars were the most sensitive to cold temperatures and were severely damaged at temperatures of −2.8°C (Webber, 1918). During the 1990 freeze in California, trees of the Mexican race cultivar 'Bacon' and the Mexican × Guatemalan hybrid 'Zutano' sustained minimal damage, whereas trees of the Mexican × Guatemalan hybrid cv. 'Hass' were severely damaged or killed following 14 consecutive nights at or below −2.8°C (Witney and Arpaia, 1991). In southern Florida, West Indian race cultivars were more severely damaged than Guatemalan race cultivars when temperatures fell to ≈−4°C for 12 h (Campbell *et al.*, 1977). Another example of a cold-tolerant cultivar is 'Negra de la Cruz', cultivated in Chile, a Mexican × Guatemalan hybrid. Cold tolerance to −5°C has been reported for this cultivar and although it has a strong alternate-bearing tendency and low fruit yield, it can be grown in areas with a high risk of frost (Lemus *et al.*, 2005).

Avocado leaves exposed to a gradual reduction in temperature have the ability to adapt, thereby increasing cold and freezing tolerance, as demonstrated by laboratory hardening experiments (Scorza and Wiltbank, 1975) and seasonal changes in tolerance (McKellar *et al.*, 1992), although the mechanism by which this occurs is unknown. When cold-hardened avocado trees in containers were exposed to sub-zero temperatures for 1 h, seedlings of the Mexican race cultivar 'Gainesville' had a lethal freezing point of −7.8°C compared with −6.7°C for two other Mexican cultivars 'Mexicola' and 'Topa Topa'; West Indian seedlings suffered lethal damage when the temperature reached −5.6°C (Scorza and Wiltbank, 1976). Below-freezing temperatures did not damage the wood of Mexican race cultivars as it did for cultivars of the other races, although some premature leaf senescence did occur in Mexican race cultivars (Scorza and Wiltbank, 1976). McKellar *et al.* (1992) observed a range of freezing tolerance of three cultivars, from −5.1°C to −9.3°C. In that study, there were no differences among the races studied (Mexican, Mexican × Guatemalan and Guatemalan × West Indian), but the lethal freezing temperature changed in response to different locations and seasonal temperatures.

The effects of temperature on avocado growth and development differ somewhat among cultivars. For example, growth and dry matter accumulation of grafted 'Fuerte' and 'Hass' avocado trees were maximized when air temperatures were held between 21°C/14°C and 33°C/26°C (day/night) (Lahav and Trochoulias, 1982). However, growth and dry matter accumulation in 'Hass' were less affected by low- and high-temperature extremes than 'Fuerte', which had reduced growth at temperatures below 25°C/18°C and above 29°C/22°C (day/night). The greater temperature adaptability of 'Hass' is likely one of the reasons for its commercial viability under more diverse climatic conditions. The accumulation of root dry matter is generally highest between 18°C and 28°C and is reduced when soil temperatures are below 13°C or above 32°C (Yusof *et al.*, 1969; Whiley *et al.*, 1990).

When fruit development occurs at lower mean temperatures, fruit generally become more elongated, with the pedicel end lengthening in pyriform fruit and round fruit becoming more oval (A.W. Whiley, Queensland, Australia, 2023, personal communication). Fruit are also generally more sensitive to chilling than other parts of the tree and may drop without substantial damage to the foliage. For example, mature 'Hass' fruit may abscise following exposure to temperatures less than −4°C, whereas long-term exposure to sub-zero temperatures (°C) can have a negative effect on postharvest physiology, with fruit ripening causing darkened areas within the flesh (A.W. Whiley, Queensland, Australia, 2023, personal communication).

There is some evidence that mean temperatures during fruit development affect final fruit size in some cultivars. In subtropical Australia, 'Hass' fruit are approximately 17% smaller when grown in warm coastal environments compared with those grown in cooler highland environments (Whiley *et al.*, 1996b). A similar trend was noted for 'Hass' in South Africa, with smaller fruit produced in warmer northern production areas compared with the cooler, highland regions of KwaZulu-Natal (B.N. Wolstenholme, Pietermaritzburg, South Africa, 1998, personal communication). Also, 'Hass' fruit grown in southern Florida, where mean max./min. temperatures are 31.5°C/24.4°C during the 4 months preceding maturity, are much smaller compared to those grown under cooler conditions in California (C. Balerdi, Homestead, Florida, 1999, personal communication). Dahan *et al.* (2010) identified several genes controlling avocado fruit cell division and ultimately fruit size. However, the regulation of these genes by environmental factors such as temperature has not been investigated. High ambient temperatures during growth and development of avocado fruit can affect postharvest ripening physiology, which is further discussed by Defilippi *et al.*, Chapter 15, this volume.

Avocado has a unique flowering behaviour described as 'complementary, synchronous dichogamy' with genotypes classified as either Group 'A' or 'B' based on flower behaviour during anthesis. The dichogamous cycle is temperature sensitive, with Group 'B' genotypes more susceptible to temperature extremes than Group 'A' genotypes (see Hormaza *et al.*, Chapter 6, this volume).

Photoassimilation and quantum yield

From several studies of avocado trees in containers, there is consensus that the optimal leaf temperature range for maximum net CO_2 assimilation is 20°C–30°C (Bower *et al.*, 1978; Kimelmann, 1979; Scholefield *et al.*, 1980). These temperature optima for net CO_2 assimilation generally coincide with temperatures that are optimal for whole-plant growth rates (Lahav and Trochoulias, 1982). At temperatures outside of this range, net CO_2 assimilation is reduced.

In an avocado orchard in Queensland, Australia, Whiley *et al.* (1999) reported that the maximum net CO_2 assimilation of 'Hass' leaves decreased from 19.0 μmol CO_2 m^{-2} s^{-1} during autumn, when minimum daily temperatures were ≥14°C, to 10.9 μmol CO_2 m^{-2} s^{-1} in winter when minimum temperatures were ≤10°C (Fig. 7.2). There was also an 18% reduction in the irradiance required for light saturation of net CO_2 assimilation in winter (from 1270 to 1040 μmol m^{-2} s^{-1}) with a corresponding 67% increase in the irradiance to reach the light compensation point for net CO_2 assimilation (PPF from 30 to 50 μmol m^{-2} s^{-1}). In studies with field-grown 'Hass' trees, Whiley (1994) also observed that temperatures below 10°C during winter significantly reduced apparent maximum quantum yield of leaves from 0.055 to 0.034 μCO_2 $μmol^{-1}$ quanta (Fig. 7.4). These results are consistent with those for other subtropical and tropical species, where it has been shown that short-term exposure to temperatures below 10°C can cause chilling injury, resulting in inhibition of photosynthesis (Taylor and Rowley, 1971).

Photoinhibition

Chlorophyll fluorescence is often measured as an indicator of plant stress (Krause, 1988). A discussion of the complex reactions involved in chlorophyll fluorescence is beyond the scope of this chapter, and the interested reader is referred

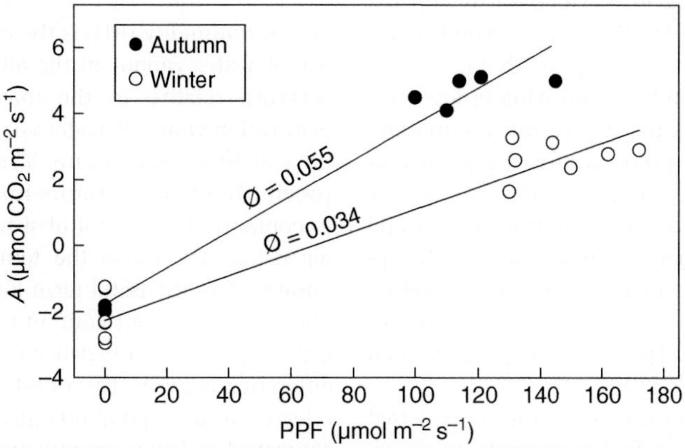

Fig. 7.4. Quantum yield (Φ) of leaves of field-grown 'Hass' trees measured in the autumn or winter. Data points represent values of single leaves, and CO_2 partial pressure was 330 μmol mol^{-1} at the time of measurement. Φ differs significantly ($P < 0.01$) based on a T-test comparing the slopes. (From Whiley, 1994, with permission).

to the review by Baker (2008). At high light levels, chilling can result in photoinhibitory damage to PS II (Powles, 1984; Smillie *et al.*, 1988), which can be quantified by measuring a decrease in the maximum potential quantum efficiency of PS II (the ratio of variable to maximum chlorophyll fluorescence; F_v/F_m) (Björkman, 1987; Demmig and Björkman, 1987).

Avocado leaves are susceptible to physiological damage following prolonged exposure to low temperatures, as experienced during winter. For example, in an avocado orchard in Queensland, Australia, minimum temperatures of 8°C in July resulted in visible signs (leaf yellowing) of chlorophyll photo-oxidation of some leaves (Whiley, 1994). Photo-oxidation often develops after long-term exposure to non-freezing, low temperatures when irradiance (PPF) is high (Taylor and Rowley, 1971; Taylor *et al.*, 1974; Powles, 1984; Robinson, 1993). Chung *et al.* (2022) reported degradation of chlorophyll a, but not chlorophyll b, in 'Hass' avocado leaves after short-term exposure to low temperatures. Whiley (1994) reported that the F_v/F_m of avocado was 0.79–0.81 when minimum temperatures in an orchard were above 12.9°C. However, when minimum temperatures dropped below 10°C, the mean F_v/F_m was 0.41, indicating cold-induced damage to PS II. The relative level of frost-induced damage to PS II is greater in the cold-sensitive 'Hass' than in the relatively cold-tolerant 'Ettinger' (Weil *et al.*, 2022), suggesting a genetic basis for cold tolerance in avocado (Soares *et al.*, 2002), which is likely related to the environmental diversity of evolutionary centres for the species. It appears that protective shading of avocado leaves growing at low temperatures is not due to temperature modulation, but rather due to protection of the photosynthetic apparatus from a combination of low temperatures and high irradiance (Kondo and Honsho, 2018; Chernoivanov *et al.*, 2022). Application of sodium hydrosulfide prior to a chilling event also may act to protect the photosynthetic apparatus and prevent damage to avocado leaves (Joshi *et al.*, 2020).

High temperatures can result in decreased productivity in avocado. For example, in an orchard in Riverside, California, there was a linear decline in net CO_2 assimilation of 'Hass' trees when temperatures increased above approximately 30°C (Liu *et al.*, 2002a), which was likely due to increased stomatal and mesophyll resistances at high temperatures (Kimelmann, 1979). Extreme temperatures exceeding 50°C can result in severe reductions in net CO_2 assimilation and reduced F_v/F_m, ultimately resulting in irreversible damage (Shapira *et al.*, 2021). There is evidence that avocado leaves contain a heat acclimation mechanism to prevent photo-oxidative damage to PS II over a moderate increase in temperature (from 21°C to 35°C). Havaux and Lannoye (1987) used chlorophyll fluorescence measurements to determine the distribution of absorbed light energy between photosystem I (PS I) and PS II. As the temperature increased, there was a shift in some of the energy absorbed from PS II to PS I. Since PS II is more sensitive to heat than PS I, it was concluded that the adjusted light distribution in favour of PS II provides a protective mechanism, limiting over-excitation and subsequent photo-oxidative damage to PS II. However, irreversible heat damage to PS II is typical in avocado leaves exposed to extremely high temperatures (Havaux *et al.*, 1987; Shapira *et al.*, 2021). Alon *et al.* (2022) demonstrated that overhead shading of avocado canopies could reduce temperatures and increase photosynthesis under high-temperature conditions, despite a significant reduction in light levels.

Humidity

Relative humidity (RH) is the ratio of the quantity of water vapour in the air at a given temperature relative to the maximum quantity (saturation value) of water vapour the air could hold at that temperature. The interaction between RH and temperature makes it impossible to compare the amount of water vapour in the air using RH unless the temperature is also known. A more useful term for relating atmospheric water vapour content to plant responses is the vapour-pressure deficit (VPD), which is the difference between the moisture content (also referred to as vapour pressure) of air and the saturated moisture content (i.e. the maximum amount of moisture the air can hold) of air; VPD takes into account the effect of temperature on the quantity of water the air can hold (Anderson, 1936).

Avocado fruit production in subtropical, semi-arid or Mediterranean climates can average 8–12 t ha^{-1}, whereas good growers in warm and humid subtropical climates can average 12–16 t ha^{-1} (Wolstenholme and Whiley, 1999). These yield differences may partially be the result of differences in tree physiology under humid subtropical conditions compared with drier Mediterranean climates. In semi-arid avocado production areas, such as Israel, high VPD has been shown to be detrimental to avocado flowering, fruit set and yield (Tomer, 1977; Levin, 1981; Argaman, 1983; Gafni, 1984; Lomas, 1988, 1992). Studies in Israel have demonstrated that the effects of adverse climatic conditions on canopy water status during flowering and fruit set significantly impact the development of avocado trees (Levinson and Adato, 1991; Lomas, 1992; Lomas and Zamet, 1994). Periods of high air temperature accompanied by low RH (high VPD) cause excessive abscission of leaves, flowers and fruit, thus reducing potential yield (Tomer, 1977; Argaman, 1983). Under these high-VPD conditions, water supplied by irrigation was insufficient to offset canopy water stress and fruit drop (Honing and Lavee, 1989). More recently, average yields consistently exceeding 18 t ha^{-1} have been reported from high-density-planted orchards from the cool, semi-arid Mediterranean region of Chile and the cool, humid subtropics of New Zealand (Gardiazábal and Mena, 2011; Hermosilla-Madrid and Whiley, 2023). Climatically, these locations have very different VPD profiles, indicating other environmental and tree physiology factors probably have greater influence on yield under these production systems.

Stomatal conductance and photoassimilation

Avocado leaves are responsive to changes in VPD, and stomatal conductance of leaves has been observed to decrease as VPD increases (Sterne *et al.*, 1977; Scholefield *et al.*, 1980; Whiley *et al.*, 1988a; Heath *et al.*, 2007b; P.M. Gil, Santiago, Chile, 2006–07, personal communication). Whiley *et al.* (1988a) found that stomatal conductance of avocado leaves was highest (425 mmol m^{-2} s^{-1}) shortly after sunrise (0700 h) when VPD was lowest (−0.4 kPa) for the day, but by 1100 h, it had fallen to 105 mmol m^{-2} s^{-1} when VPD had reached −1.3 kPa (Fig. 7.5). Several studies on the effects of VPD on avocado trees have shown a direct relationship between reductions in net CO_2 assimilation and reduced stomatal conductance resulting

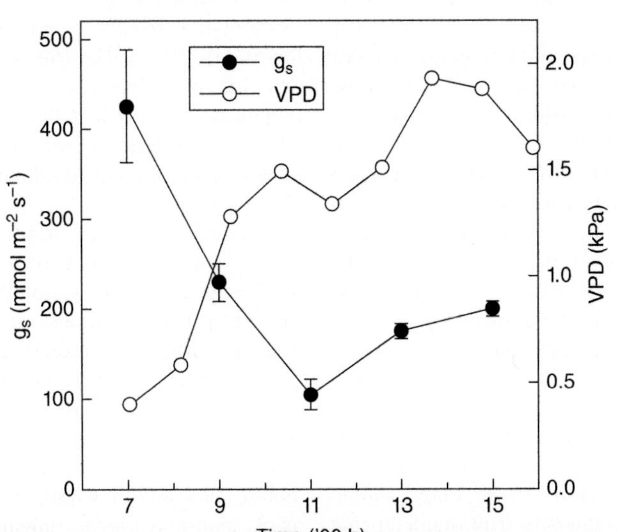

Fig. 7.5. Changes in stomatal conductance (g_s) of 'Fuerte' avocado as related to VPD. (From Whiley *et al.*, 1988a, with permission).

from increased VPD (Bower, 1978; Ramadasan, 1980; Ploetz and Schaffer, 1989; Schaffer and Ploetz, 1989; Sharon, 1999; Gil, 2008) or reduced RH (Takayama *et al.*, 2013). Bower *et al.* (1978) observed a 50% decrease in net CO_2 assimilation of container-grown 'Edranol' avocado trees due to a 50% reduction in stomatal conductance resulting from increased VPD.

Management

Options for modifying the VPD in fruit tree canopies are limited. However, there are some management strategies that can be adopted to modify the VPD during critical hours. Among these is intermittent spraying or misting of water at high pressure, which has been shown to reduce leaf temperature and improve stomatal conductance in 'Hass' avocado trees (Hermoso *et al.*, 2003). In Israel, overhead irrigation reduced the air temperature in an orchard by 4°C–7°C and increased RH by 6%–27% (Lomas and Mandel, 1973). Also, in Israel, intermittent application of water above avocado tree canopies during heatwaves significantly reduced air temperatures by 10°C–12°C and increased RH by 16%–18% (Aharoni *et al.*, 1968). In that study, increased fruit production in some orchards was attributed to reduced VPD.

While avocado is able to withstand some seasonal variations in water supply, successful cropping is related to adequate water availability throughout the reproductive cycle. Water requirements increase during flowering, with inflorescences contributing to increased water loss from the canopy (Whiley *et al.*, 1988a). There is sufficient evidence to conclude that the reproductive organs of avocado trees have higher rates of water loss per unit area than the vegetative components. For example, Whiley *et al.* (1988a) reported that the diurnal water deficit of inflorescences is greater than that of adjacent mature leaves. They concluded that inflorescences were either less efficient in containing water loss, or there is preferential water movement to adjacent leaves as water deficits develop. Schroeder and Wieland (1956) found significant diurnal fluctuations in fruit diameter that were correlated with leaf transpiration. They reported that fruit typically reached their maximum diameter at 0800 h, after which time fruit diameter contracted throughout the day until 1500 h, when expansion occurred. Reducing leaf transpiration by misting the canopy with water stopped fruit shrinkage, and within a short time, fruit diameter increased. Diurnal fluctuations in fruit diameter were proportionally greater in young developing fruit than in mature fruit. This is likely related to the status of water control by the canopy and fruit as well as varying environmental conditions at different stages of fruit ontogeny (Schroeder and Wieland, 1956; Whiley and Schaffer, 1994).

Drought

The future supply of fresh water for agriculture is threatened because demand is increasing, and climate change predictions suggest greater disruption from weather events such as floods, drought, changes in annual precipitation patterns, depletion of water storage and soil moisture and increased salinization (Mahato *et al.*, 2022). With respect to current global avocado production, the areas most threatened by increased drought, diminishing water resources and salinization are the semi-arid Mediterranean climatic zones of southern Europe, south-western North and South America and southern Australia (Grafton and Hussy, 2011; Cholet *et al.*, 2022), which together currently produce ca. 50% of the world's annual avocado crop (Anonymous, 2022). The effects are already being felt in some Mediterranean climatic zones, where lack of sufficient water resources for agriculture threatens the commercial viability of avocado production (Balikian and Genskow, 2022).

The evolution of avocado in the cool, montane cloud forests of Central America, where high summer/autumn rainfall but relatively dry winter/spring conditions occur, has equipped trees to withstand seasonal variations in water supply (Wolstenholme and Whiley, 1999). Subsequent commercialization of the avocado outside its indigenous range has resulted in successful production across a diverse range of environments, from warm, humid subtropical to cool, semi-arid Mediterranean climates. There are anatomical features that enable avocados to tolerate semi-arid environments, provided sufficient soil moisture is maintained to meet growth requirements (Whiley and Schaffer, 1994). The adaxial and abaxial surfaces of mature leaves have waxy cuticles, which reduce cuticular water loss (Whiley *et al.*, 1988a; Blanke and Lovatt, 1993; Mickelbart *et al.*, 2000). In addition, the abaxial surface of young leaves (before the leaf cuticle is fully formed) and the flower sepals and petals are densely pubescent (Whiley *et al.*, 1988a; Blanke and Lovatt, 1993), creating an effective boundary layer, which minimizes water loss. Also, fruit have a waxy film covering the surface of the exocarp that reduces water loss (Cummings and Schroeder, 1942; Blanke and Bower, 1990). Kofidis *et al.* (2004) reported a decrease in stomatal density in response to water stress in 'Fuerte', but not 'Hass' leaves, suggesting different genotypic responses, while Neuhaus *et al.* (2009) measured a reduction in leaf conductance in response to drought. Leaf xylem embolism is a major factor in leaf necrosis under water-limiting conditions. There appears to be large variation in the stem water potential of avocado leaves within the canopy at which embolisms occur, which may act to buffer the entire canopy from hydraulic failure in the face of drought (Cardoso *et al.*, 2020). Furthermore, the capacity to minimize embolism development may explain the ability of avocado trees to recover from severe water stress (Martinez-Ferri *et al.*, 2019).

While dryland production of avocados is practiced in some countries, its success largely depends on the quantity and distribution of annual rainfall, the level of atmospheric stress and the water storage features of the soil. The fleshy, non-suberized roots of trees that colonize the top 300–450 mm of soil (Whiley, 1994) are prone to desiccation when the rhizosphere dries out during extended periods of drought. However, Levinson and Adato (1991) reported dense feeder root production under prolonged but stable, low-moisture conditions, which aided adaptation to drier soils. The annual replacement of leaves provides some protection during drought periods through accumulation of litter beneath trees, which reduces evaporative losses from the soil. This can be supplemented with suitable mulching materials such as barley straw or composted pine bark, resulting in improved tree performance in some areas (Gregoriou and Kumar, 1984; Moore-Gordon *et al.*, 1996), but different strategies may be needed for regions with year-round rainfall and strong alternate-bearing cycles (Dixon *et al.*, 2007).

Soil moisture

Water stress caused by periods of soil moisture depletion can severely inhibit physiological processes and thus productivity of avocado, particularly if drought stress occurs when trees are actively growing. Monitoring soil water (Kiggundu *et al.*, 2012; Beyá-Marshall *et al.*, 2022), trunk diameter (Assouline *et al.*, 2021) and/or trunk sap flow (Fernández *et al.*, 2007) improves the efficiency of irrigation management and is a useful tool to manage limited water resources, since production is dependent on adequate soil water (Zuazo *et al.*, 2021), especially in high-yielding years (Neuhaus *et al.*, 2009).

Measurement of the pressure potential in the xylem of plants using a Scholander pressure chamber (or similar device) has become the predominant method for determining plant water status in the fields of plant physiology and ecophysiology. Under controlled pressure, the xylem fluid of the plant sample placed in the pressure chamber is forced out through xylem tissue and can be viewed at the cut end of a protruding stem or petiole, the point when the pressure is recorded. Sterne *et al.* (1977) determined that in avocado, pressure potential measured with a pressure chamber provides a good estimate of plant water potential, as is the case with many other plant species. Over the past 40 years that have been covered by discussions in this chapter, the nomenclature of the measurement has changed due to subtle modifications in the procedures used to determine the pressure potential. For uniformity and avoidance of confusion, this discussion refers to stem water potential measured in MPa for all citations reporting data on water deficits in avocado trees.

For avocado trees growing in cool, semi-arid environments, mid-day values of stem water potential (≈ -1.4 MPa) are more negative than those found in humid environments (≈ -1.0 MPa). This is undoubtedly the result of higher VPDs found in drier climates (Sterne *et al.*, 1977; Bower *et al.*, 1978; Scholefield *et al.*, 1980; Wolstenholme, 1990). From a study in Greece during the dry season (May–October), Chartzoulakis *et al.* (2002) observed that the trunk diameter was 34% and 39% greater for non-stressed (irrigated when the soil matric potential reached -30 kPa) 'Hass' and 'Fuerte' avocado trees, respectively, compared with water-stressed trees (irrigated when the soil matric potential reached -500 kPa) after a 6-month period. Also, moderate water stress resulted in 57% and 69% less leaf area and 63% and 80% less total plant dry weight for 'Fuerte' and 'Hass', respectively, compared with non-stressed trees. After the stress period, water-stressed trees of each cultivar had smaller leaves and lower specific leaf weights than non-stressed trees. Additionally, prolonged water stress resulted in reduced fibrous root weight compared with non-stressed trees of each cultivar, with more visible damage to the root system of 'Hass' than 'Fuerte'. Almost all histological components of the mesophyll and the entire lamina were thinner in the leaves of water-stressed trees.

Several studies have shown reductions in stomatal conductance, stem water potential, trunk contraction, and/or net CO_2 assimilation as a result of water stress in avocado trees (Sterne *et al.*, 1977; Bower, 1978; Ramadasan, 1980; Scholefield *et al.*, 1980; Whiley *et al.*, 1988a; Barrientos Pliego and Rodríguez, 1998; Chartzoulakis *et al.*, 2002; Ferreyra *et al.*, 2009). From these studies, the time it takes to evoke stress responses varies widely (from 4 to 30 or more days) and is related to edaphic conditions as well as climatic factors such as temperature and VPD. Sterne *et al.* (1977) observed a reduction in soil matric potential to −80 kPa or lower at all depths measured after withholding water from 'Bacon' avocado trees for 20 days. However, at that time, the stem water potential of stressed plants during day and night was similar to that of non-stressed (irrigated) trees. After 26 days, the stem water potential was somewhat lower during the day in stressed trees, but at night the stem water potential was between −0.1 and −0.2 MPa, similar to that of non-stressed trees. After withholding water for 30 days, transpiration and stem water potential were significantly lower in stressed trees during the day and night. The reduction of transpiration with reduced stem water potential suggests that stomata were beginning to close at a stem water potential between −1.2 and −1.5 MPa (Sterne *et al.*, 1977). During the dry season (May–October) in Greece, Chartzoulakis *et al.* (2002) observed that water stress (irrigating when the soil matric potential reached −500 kPa) resulted in a decline in pre-dawn stem water potential, net CO_2 assimilation and stomatal conductance of 'Hass' and 'Fuerte' avocado trees after withholding irrigation for 12 days. Although significant decreases in the pre-dawn stem water potential of water-stressed trees were observed after 12 days of withholding irrigation, reductions in net CO_2 assimilation and stomatal conductance were observed after 5 and 4 days of imposing water stress, respectively. After irrigating water-stressed trees, pre-dawn stem water potential, net CO_2 assimilation and stomatal conductance of water-stressed trees returned to initial (non-stressed) values within 2 days.

More recent irrigation studies in Chile have provided additional and complementary information on responses of avocados to water stress. In the first study carried out at a site with a cool, dry Mediterranean climate (with maximum daily VPD ranging from 1.15 to 3.79 kPa during the experimental period), 3-year-old 'Hass' avocado trees, planted on a northern exposed slope, were subjected to periods of water stress (irrigation withheld for 10 days) during summer. Water-stressed trees had an average stem water potential of −0.86 MPa after 5 days and −1.1 MPa after 10 days, whereas irrigated trees had a stem water potential of −0.55 and −0.53 MPa after 5 and 10 days, respectively. The average stomatal conductance after 10 days was 131.1 and 199.5 mmol m^{-2} s^{-1} in stressed and irrigated trees, respectively (P.M. Gil *et al.*, Santiago, Chile, 2010, personal communication). In a second Chilean study where trees were grown in a cool, humid (marine influence) Mediterranean climate (VPD during the experimental period ranged from 0.2 kPa on cloudy days to 3.6 kPa on sunny days), stem water potential was lower, and maximum daily trunk shrinkage was higher in non-irrigated trees after 4 days of withholding irrigation from 'Hass' trees grafted onto 'Mexicola' seedling rootstock (Celedón *et al.*, 2012). Withholding irrigation for 13 days resulted in a stem water potential as low as −0.9 MPa and mid-day trunk shrinkage of 290 μm, whereas the stem water potential of irrigated trees remained between −0.45 and −0.60 MPa, and mid-day trunk shrinkage remained between 35 and 75 μm, depending on VPD. These results clearly show reductions in stem water potential and stomatal conductance and increased trunk shrinkage in avocado trees with progressive development of water stress when irrigation was withheld.

Osmotic adjustment

Osmotic adjustment is a net increase in solute content of cells, which is independent of volume changes that result from loss of water (Hsiao *et al.*, 1976; Taiz and Zeiger, 2010). In a South African study, Bower (1985) observed a lower osmotic potential in 'Fuerte' avocado trees exposed to prolonged water stress but was unconvinced that active osmotic adjustment had a significant role in turgor maintenance. Chartzoulakis *et al.* (2002) reported a decrease in osmotic potential 12 days after 'Fuerte' and 'Hass' avocado trees were subjected to water stress (irrigating when soil matric potential reached −500 kPa) compared to non-stressed trees (irrigating when soil matric potential reached −30 kPa). Osmotic adjustment contributed to the maintenance of positive turgor in water-stressed trees, where it was calculated that the net increase of solutes in cells was 40.6% in 'Fuerte' and 29.4% in 'Hass'; initial solute values were attained 2 days after reirrigating. The lower osmotic potential in water-stressed trees was attributed primarily to dehydration and only partly to active accumulation of solutes in stressed trees. Cell-wall elasticity in leaves of stressed trees also decreased during the stress period, and tissue elasticity was determined to be the predominant physiological adaptation mechanism of avocado trees to drought stress.

Photoassimilation

Net CO_2 assimilation of a crop canopy will decline under water stress due to stomatal closure and the detrimental effects on chloroplast processes (Beadle *et al.*, 1985). In avocado trees, stomatal conductance begins to decline when stem water potential falls to approximately −0.4 MPa and continues to decrease until stomatal closure occurs at a stem water potential of −1.0 to −1.2 MPa (Sterne *et al.*, 1977; Bower, 1978; Scholefield *et al.*, 1980; Whiley *et al.*, 1988a). Because net CO_2 assimilation in avocado is highly correlated with stomatal conductance over a wide range of stomatal conductance values (Chartzoulakis *et al.*, 2002; Liu *et al.*, 2002a; Heath *et al.*, 2007a), reductions in net CO_2 assimilation occur soon after water stress develops.

Chartzoulakis *et al.* (2002) found a moderate, positive linear relationship between net CO_2 assimilation and intercellular CO_2 concentration for 'Hass' and 'Fuerte' trees. However, leaves of 'Hass' trees recovering from water stress had much higher net CO_2 assimilation than leaves of stressed trees with the same intercellular CO_2 concentration. In that study, net CO_2 assimilation was higher for 'Hass' than 'Fuerte' when both cultivars were at the same level of water stress. Two days after reirrigating, net CO_2 assimilation and stomatal conductance of drought-stressed 'Fuerte' trees returned to pre-stress levels, whereas net CO_2 assimilation and stomatal conductance of water-stressed 'Hass' trees, each remained 20% lower than those of the non-stressed 'Hass' trees.

Histological and molecular responses to different levels of water stress were recently reported for clonally propagated, non-grafted 'Dusa' avocado trees. In an experiment with potted trees in which plants were deficit-irrigated until soil water content reached 50% (mild water stress) or 25% (severe water stress) of field capacity and were then rewatered, stem water potential, net CO_2 assimilation, transpiration and stomatal conductance significantly decreased under both water stress treatments compared with the control treatment, where soil water was maintained at field capacity. However, after rewatering, stressed plants showed a rapid recovery of most physiological variables measured. A cDNA microarray analysis to identify possible genes related to these responses showed a wide transcriptome response associated with a higher degree of water stress, and functional enrichment of differentially expressed genes revealed the richness of common sequences associated with water stress, such as genes related to abscisic acid (ABA) synthesis, redox homeostasis, activation of antioxidants, osmoprotection and cell-wall organization (Guillermo *et al.*, 2021).

Reproductive function

Plant water status often affects the reproductive behaviour of fruit-bearing trees, since it may influence floral induction, cause premature flower abortion and reduce fruit retention and growth, among other responses. Whiley *et al.* (1988a) reported that floral structures of avocado significantly contribute to water loss during flowering, with the canopy surface available for transpiration losses increasing by ca. 90% over this period. The increased water demand in avocado orchards during flowering has been recorded by water-smart technology (real-time soil moisture probes and trunk-measuring dendrometers) (A.W. Whiley, Queensland, Australia, 2023, personal communication) indicating that increased irrigation is required during this period. A study of 3.5-year-old, container-grown 'Hass' avocado trees showed that low soil moisture for 6 weeks (8% to 12% volumetric water content) resulted in a noon stem water potential of ≤-2.0 MPa and visible stress symptoms, which were associated with a failure of floral induction (Acosta-Rangel *et al.*, 2021). Silber *et al.* (2012) demonstrated that water stress during the early period of fruit growth increased fruitlet abscission and thus reduced yields. In addition, Silber *et al.* (2013) found that daily water uptake was 40% lower in fruiting compared to nonfruiting 'Hass' trees. Also, stomatal conductance and net CO_2 assimilation were significantly higher in fruiting trees compared with nonfruiting trees, probably as a result of a higher leaf carbohydrate concentration in the nonfruiting trees, which may be a trigger for stomatal closure (Paul and Pellny, 2003). Consequently, fruit load is an important consideration when estimating crop water requirements.

Water stress during critical stages of fruit ontogeny has been linked to fruit disorders. For example, ring-neck, a disorder of avocados in which corky lesions develop at the abscission sites on the fruit stalk, has been attributed to water deficit in trees during fruit development (Whiley *et al.*, 1986; Whiley and Schaffer, 1994). The fruit stalk lesions impede the movement of resources into affected fruit, resulting in the premature death of the seed testa and smaller fruit size. 'Hass' fruit from water-stressed trees may also be more elongated than fruit from well-watered trees (P.J. Hofman, Queensland, Australia, 2000, personal communication). In field studies, trees were rewatered when the soil matric potential at 300 mm below the soil surface reached either −20 (well-watered) or −70 kPa (water-stressed) and continued until the soil matric potential at 600 mm below soil level reached 0 kPa. The differences in fruit shape began to show about 100 days after fruit set and were greater at fruit maturity. The reason for the changed fruit shape is not known but probably relates to differences in cell division between treatments, as this continues throughout fruit growth.

Bower *et al.* (1989) reported increased polyphenol oxidase activity in fruit from trees, where the soil matric potential measured at 300 mm below soil surface was allowed to fall to −80 kPa before rewetting, compared with fruit grown where the soil matric potential at 300 mm did not fall below −55 kPa. High polyphenol oxidase concentration in fruit at maturity may cause flesh discoloration when they are ripening. Fruit from the more frequently irrigated treatment also had higher flesh calcium concentrations during the first 16 weeks of growth. High fruit calcium concentrations improve membrane structure and reduce physiological fruit disorders and postharvest diseases in avocado (Bangerth, 1979; Chaplin and Scott, 1980; Ferguson, 1984).

Flooding and Poor Soil Aeration

Avocado is considered to be a flood-sensitive species, with rapid tree responses occurring when soils become waterlogged or flooded (Schaffer *et al.*, 1992). The roots are particularly sensitive to anaerobic conditions caused by flooding. Stolzy *et al.* (1967) reported that oxygen diffusion rates of less than 0.18 μg cm^{-1} min^{-1} resulted in 44%–100% mortality of 'Mexicola' roots that were healthy prior to flooding. Reduced soil oxygen availability due to flooding severely affects root functions and results in resistance to water movement, dehydration, low productivity and tree death (Schaffer *et al.*, 1992, 2006; Gil *et al.*, 2012). Most locations where avocados are grown do not experience regular flooding. However, free surface water in avocado orchards has been observed in southern Florida, where a combination of a rising water table and heavy rain from tropical storms results in flooding (Crane *et al.*, 1997). Where avocados are grown in the humid subtropics of eastern Australia and South Africa, exceptionally high summer rainfall events can result in the saturation of soil profiles, creating anaerobic conditions, leading to deterioration in tree performance (Pegg *et al.*, 1982).

For avocado trees, short periods of flooding usually result in the inhibition of leaf expansion, a reduction in root and shoot growth, root necrosis and moderate-to-severe leaf abscission (Stolzy *et al.*, 1967; Ploetz and Schaffer, 1987, 1989; Schaffer and Ploetz, 1989; Schaffer *et al.*, 1992). Some reports have documented a very rapid death of trees when saturation of soil profiles occurs in avocado orchards for short periods (Pegg *et al.*, 1982). In southern Florida, after Tropical Storm Dennis in 1981, avocado trees that appeared to have the largest canopy volume often died as a result of standing in water for 3–5 days (S. Goldweber, Homestead, USA, 1998, personal communication). Since the rhizosphere of the hard, calcareous soils in southern Florida severely restricts root growth (Schaffer *et al.*, 1994), it is likely that under flooded conditions, the root volume was insufficient to meet the oxygen demands of roots and water demands of trees with large canopies, thereby resulting in tree mortality.

Some subtropical and tropical fruit tree species have evolved anatomical or morphological adaptations to increase plant survival under conditions of low soil oxygen. These include the development of adventitious roots for increased oxygen absorption (e.g. mango; Larson *et al.*, 1993), development of aerenchyma tissue in the stem for increased internal oxygen transport under hypoxic conditions (e.g. *Annona* species; Nuñez-Elisea *et al.*, 1999) and the development of hypertrophic (swollen) stem lenticels that function to increase oxygen absorption and for the excretion of potentially toxic metabolites resulting from anaerobic metabolism in the roots (e.g. mango; Larson *et al.*, 1991a, 1991b, 1993). Although none of these anatomical or morphological adaptations have been reported for orchard-grown avocado trees, hypertrophic stem lenticels have been observed on 'Edranol' avocado trees (Fig. 7.6) that were flooded in containers (R. Hofshi, Fallbrook, USA, 2009, personal communication).

Fig. 7.6. Stem of flooded 'Velvick' avocado seedling tree showing hypertrophied (swollen) stem lenticels. (Photograph courtesy of R. Hofshi).

Photoassimilation and stomatal conductance

Early measurable physiological responses of avocado trees to flooding include reductions in transpiration, stomatal conductance and net CO_2 assimilation (Ploetz and Schaffer, 1989, Schaffer *et al.*, 1992, 2006, Gil *et al.*, 2007, Reeksting *et al.*, 2014; Lin *et al.*, 2022) as well as increased photoinhibition measured by a reduction in F_v/F_m (Reeksting *et al.*, 2014; Lin *et al.*, 2022).

In an organic soil with a high water-holding capacity, 5 days of continuous flooding almost totally inhibited transpiration in avocado (Ploetz and Schaffer, 1989). Schaffer *et al.* (1992) concluded that the reduction of transpiration caused by flooding is most likely the result of reduced stomatal conductance rather than a hydraulic effect, since flooding did not significantly decrease stem water potential. In a later study with 1-year-old 'Beta' and 'Hass' avocado scions grafted onto 'Waldin' seedling rootstock, continuous flooding for 11 days caused a significant reduction in transpiration and stomatal conductance after 10 days, a reduction in net CO_2 assimilation 12 days after flooding began, but no differences in stem water potential during the evaluation period. 'Beta' was more susceptible to flooding than 'Hass', with significant reductions in stomatal conductance, transpiration and net CO_2 assimilation 8 days after the flooding treatment was initiated. However, 14 days after commencing the 11-day flooding period, stem water potential was significantly lower than that of non-flooded trees, indicating a delayed response to hydraulic conductivity. There was no correlation between stem water potential and stomatal conductance or the number of open stomata (Gil *et al.*, 2007). Lin *et al.* (2022) reported that following the cessation of flooding, the recovery of stomatal conductance varied across genotypes, indicating differences in tolerance to hypoxic soil conditions.

Net CO_2 assimilation of combined flooded and non-flooded 'Simmonds' avocado trees on 'Waldin' seedling rootstock was positively correlated with stomatal conductance, with values for flooded trees generally grouped at the bottom of the regression line (Fig. 7.7; Ploetz and Schaffer, 1991), indicating a strong relationship between reduced net CO_2 assimilation and reduced stomatal conductance due to flooding. Whiley and Schaffer (1994) postulated that stomatal closure in avocado trees exposed to soil flooding may be due to a build-up of CO_2 in the leaf, which leads to stomatal closure (Raschke, 1975; Mansfield *et al.*, 1990). This was based on studies where flooding-induced reductions of stomatal conductance in avocado trees were observed concomitantly with decreases in net CO_2 assimilation and increased intercellular CO_2 concentrations in leaves (Schaffer *et al.*, 1992). Although the sequence of flooding-induced reductions in stomatal conductance and net CO_2 assimilation has not been elucidated in avocado trees, reductions in net CO_2 assimilation may occur before those in stomatal conductance, resulting in an increased intercellular CO_2 concentration. However, this theory assumes that stomatal conductance decreases after reductions in net CO_2 assimilation, a sequence that lacks experimental verification (Schaffer *et al.*, 1992; Gil *et al.*, 2009). Schaffer *et al.* (1992) cite studies with several fruit crops, indicating that photosynthetic

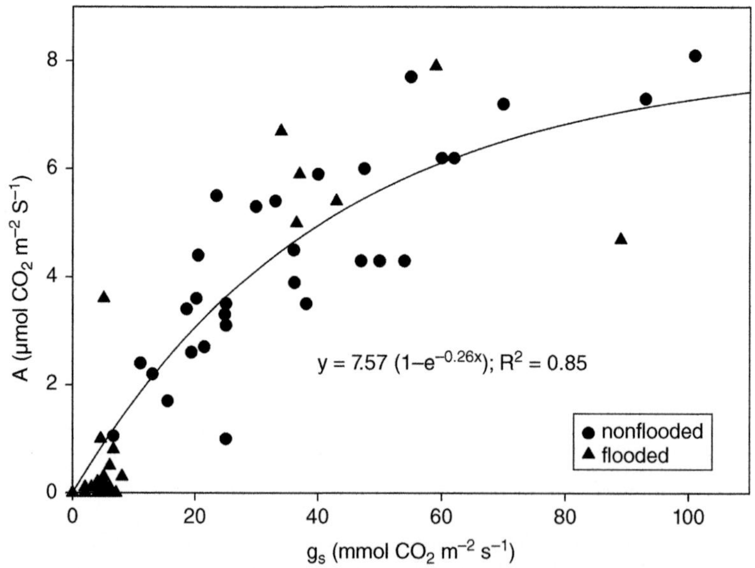

Fig. 7.7. Net CO_2 assimilation (*A*) and stomatal conductance of CO_2 (g_s) for flooded and non-flooded 'Simmonds' avocado scions grafted to 'Waldin' seedling rootstock. Each symbol represents a single plant. (Redrawn from Ploetz and Schaffer, 1991, with permission).

reductions from flooding may be due to non-stomatal factors, such as biochemical changes associated with photosynthetic reactions, which are further discussed in this section under 'Biochemical changes within trees'.

The reduction of stomatal conductance in avocado, resulting from soil flooding, is exacerbated when VPD is high (B. Schaffer, Homestead, USA, 2007, personal communication), presumably due to the lack of sufficient root absorption to offset the high evaporative demand during periods of high VPD. Thus, in addition to the direct effects of reduced stomatal conductance due to stomatal closure, VPD also exerts a significant influence on the response of avocado trees to environmental stresses.

Soils

Different soil types have a wide range of water-holding capacities and their response to anaerobic conditions after flooding or heavy rainfall events. In organic soils with a high water-holding capacity, net CO_2 assimilation of avocado leaves declined shortly after trees were flooded (Schaffer *et al.*, 1992, 2006; Whiley and Schaffer, 1994). In contrast, net CO_2 assimilation of trees with healthy roots growing in the porous calcareous soils of southern Florida did not decrease as quickly when flooding occurred (Ploetz and Schaffer, 1989). The difference in flood tolerance for trees grown in organic compared to a rocky, calcareous soil is likely related to higher O_2 content in the root zone of the latter due to its porous nature.

In addition to flooding, heavy soil texture, high bulk soil density or poor irrigation management combined with low soil macroporosity can result in poor soil aeration. Similar to flooded soils, these situations often result in physiological damage, reductions in growth and yield and even mortality of avocado trees. The soil water-to-air ratio varies with soil type and is related to soil physical properties (Gil, 2008). Ferreyra *et al.* (2008) reported that low soil air content (5%–18%) resulted in reduced stomatal conductance of 'Hass' trees grafted onto 'Mexicola' seedling rootstock. The same authors established that soil air content lower than 17% restricts the oxygen diffusion rate (ODR) to less than 0.2 μg cm^{-2} min^{-1} and that soil macroporosity values were correlated with both soil O_2 and CO_2 content. In a study by Gil (2008), 'Hass' avocado trees grafted onto 'Mexicola' seedling rootstock were grown in containers in five different soils with soil water-to-air ratios of 1.7, 1.3, 0.6, 0.4 or 0.3. Soil water content was maintained near field capacity for 2 years. There was a good correlation between the soil water-to-air ratio and the ODR in the soil (Fig. 7.8). Stomatal conductance, transpiration, stem water potential, and net CO_2 assimilation of avocado trees were generally higher in

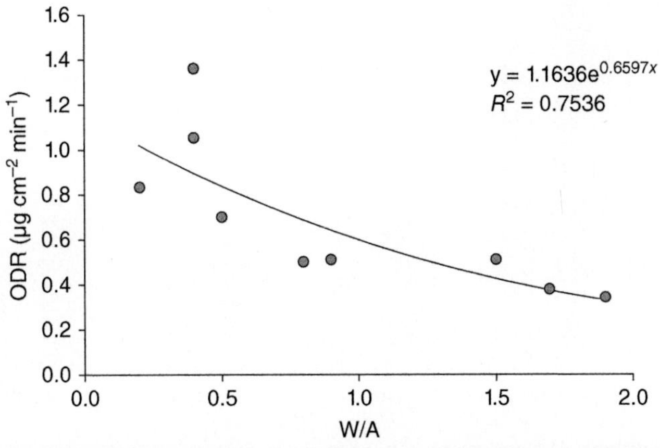

Fig. 7.8. Relationship between the soil water-to-air ratio (W/A) and the oxygen diffusion rate (ODR) in the soils from avocado production areas in Chile. Symbols represent mean W/A and ODR values from each of five soils during two seasons (2005/06 and 2006/07). (From Gil, 2008, with permission).

soils with low water-to-air ratios than in soils with high water-to-air ratios. Gil *et al.* (2012) also reported that 'Hass' avocado trees grown in soils with low water-to-air ratios (0.4 and 0.3) had higher biomass, leaf area and root density, and longer leaf retention compared with trees grown in soils with higher water-to-air ratios (1.7, 1.3 and 0.6) during 2 years of tree growth and development (Fig. 7.9).

Biochemical changes within trees

Reactive oxygen species (ROS) are oxygen-containing radicals that are capable of independent existence with one or more unpaired electrons. They are signal molecules and regulators of stress responses but can be harmful to plants, causing tissue damage and membrane destabilization. When generation of ROS exceeds the plant's ability to neutralize or eliminate them, a stress condition results termed 'oxidative stress' (Shankar and Mehendale, 2014; Huang *et al.*, 2019). Plants can counter oxidative stress by producing antioxidants and protective osmolytes. The activities of ROS, antixoxidants, and osmolytes can be used to determine the tolerance of fruit trees to flooding (McGee *et al.*, 2021). Yin *et al.* (2023) found that the relative flood tolerance of avocado cultivars is partially related to the activity of ROS and antioxidants in leaf and root tissues. Doupis *et al.* (2017) reported that waterlogging of 'Hass' and 'Fuerte' trees on 'Zutano' seedling rootstock resulted in less oxidative stress to 'Hass' compared with 'Fuerte'. However, they could not correlate reduced oxidative stress in waterlogged 'Hass' to the activity of the four antioxidant enzymes they analysed, suggesting that there was some other mechanism involved in protection of 'Hass' from root hypoxia-induced stress.

Characterizing ROS and antioxidant activities in response to flooding among the three horticultural races of avocado provides insight into their inherent mechanisms for overcoming oxidative stress due to root-zone hypoxia. In a study by Yin *et al.* (2023) comparing physiological and biochemical responses and survival to flooding among clonally propagated non-grafted trees of each race, potted trees of the Guatemalan race were more sensitive to short-term (2 days) flooding compared with trees of Mexican or West Indian races. Flooded Guatemalan race trees had significantly lower net CO_2 assimilation, stomatal conductance and transpiration compared with non-flooded trees during the

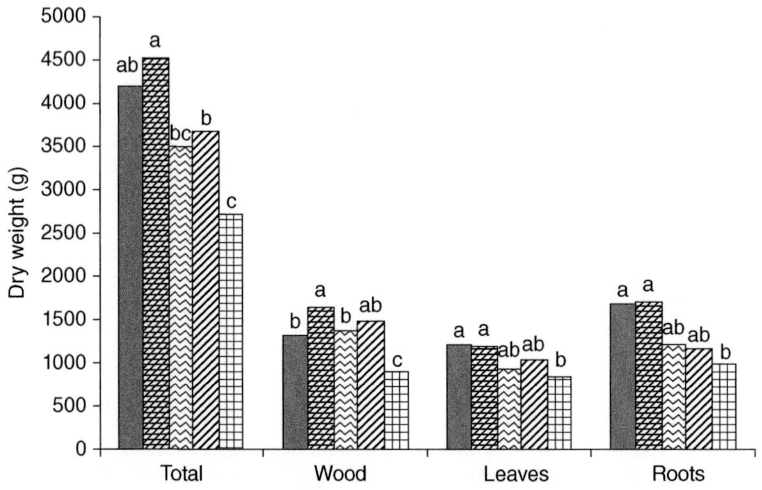

Fig. 7.9. Tissue dry weights of 'Hass' avocado trees grown in soils with different water-to-air ratios (W/A). Different letters indicate significant differences among means ($p \leq 0.1$). The W/A was 1.7, 1.3, 0.6, 0.4 and 0.3 for treatments T1, T2, T3, T4 and T5, respectively; $n = 5$. (From Gil, 2008, with permission).

flooding period and for several days after flooding was discontinued. However, there was generally no difference in leaf gas exchange between flooded and non-flooded trees of Mexican or West Indian races. Also, the percentage of plants that survived flooding was higher for the Mexican and West Indian cultivars compared with Guatemalan cultivars. In Guatemalan cultivars, there was less partitioning of sugars, particularly mannoheptulose, to the roots of flooded than non-flooded trees compared with Mexican or West Indian cultivars. In that study, principal component analysis showed distinct separation of flooded trees among races based on ROS and antioxidant profiles (Fig. 7.10). The authors concluded that differences in partitioning of sugars and activities of ROS and antioxidants among races in response to flooding may explain the greater flooding sensitivity of non-grafted Guatemalan trees compared with Mexican or West Indian trees. In another study with 'Simmonds', a West Indian scion, grafted onto clonally propagated West Indian ('Waldin'), Mexican ('Duke 7') or Guatemalan ('Reed') rootstocks, trees on the West Indian rootstock were more flood-tolerant than trees on the Guatemalan or Mexican rootstock (Yin *et al.*, 2023). In that study, flooding reduced leaf gas exchange and leaf dry weight of trees on the Mexican or Guatemalan rootstocks but not on the West Indian rootstock. All trees on the West Indian rootstock survived flooding, whereas 87% of trees on the Mexican rootstock, and 50% of the trees on the Guatemalan rootstock died by the end of the study.

Reduced leaf gas exchange may be associated with the reduction of aquaporin expression (membrane proteins that aid in water transport into cells), and although little evidence exists for aquaporins in avocado trees, aquaporin genes were found to be down-regulated in avocado roots exposed to flooding, indicating reduced water uptake under flooded conditions (Reeksting *et al.*, 2016).

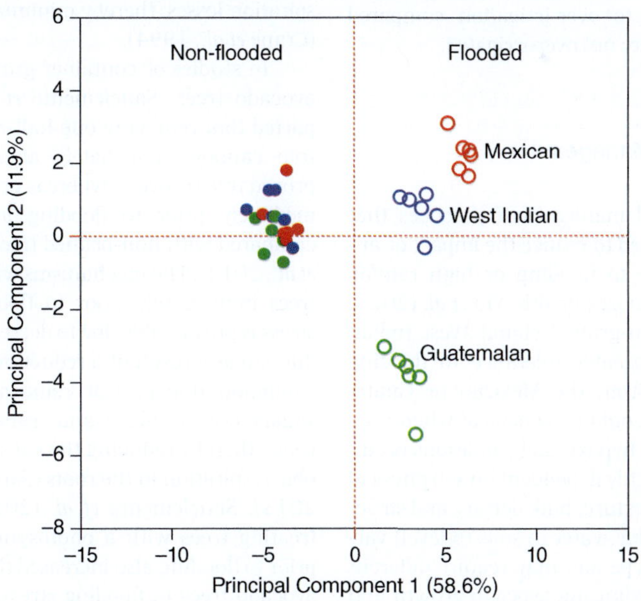

Fig. 7.10. Principal component analysis scatterplot showing clustering of races and flooding treatments based on antioxidant activity, reactive oxygen species, lipid peroxidation and osmolytes in leaves and roots of clonally produced, non-grafted avocado trees. Filled symbols indicate non-flooded trees and hollow symbols indicate flooded trees. Red symbols denote Mexican race ('Duke 7'), green symbols denote Guatemalan race ('Reed') and blue symbols denote West Indian race ('Waldin') trees. Principal Component 1 (PC1) accounted for 58.6% of the variation among flooding treatments and PC2 accounted for 11.9% of the variation among races. (From Yin *et al.*, 2022, with permission).

Nutrients

Waterlogging and the associated reduction in soil oxygen content reduced the nitrogen, phosphorus, potassium, magnesium, zinc, copper and boron concentrations in the leaves of plants compared with those trees grown under non-flooded or non-waterlogged conditions (Labanauskas *et al.*, 1968). In contrast, the manganese and iron concentrations in the leaves of partially waterlogged trees increased (Stolzy *et al.*, 1967; Labanauskas *et al.*, 1968), which was likely due to their increased solubility under these conditions (Stolzy *et al.*, 1967). Flooding results in more reduced forms of manganese and iron, which are readily available for uptake by plants (Ponnamperuma, 1972; Larson *et al.*, 1991b). In a more recent study, root waterlogging reduced the leaf and stem contents of most nutrients in 'Hass' and 'Fuerte' avocado trees, but the reductions were more severe in 'Fuerte' compared with 'Hass' (Tzatzani *et al.*, 2020). 'Fuerte' leaves had 63.4% less boron and up to 80% less magnesium, calcium and nitrogen when soil was waterlogged due to over-irrigation compared with trees that were not over-irrigated.

Management

There are several management practices that can be implemented to reduce the impact of anaerobic soils, due to flooding or high rainfall events, on avocado tree health. Yin *et al.* (2023) observed that non-grafted clonal West Indian rootstocks have greater tolerance to flooding than rootstocks from the Mexican or Guatemalan races and could be beneficial when used where short-term hypoxic soil conditions occur. Soil aeration is highly dependent on soil physical features such as texture, bulk density and structure. Plant-available water in soils (p-level) varies between soil type and may require different tolerances for maximizing avocado growth and production. However, studies by Ferreyra (2017) to determine the p-level for different soils reported that in sandy and clay loam soils, the p-level can reach 60% without affecting avocado water status, yield or fruit size. With the rapid advancement of electronic monitoring devices, soil water content can be effectively managed through irrigation scheduling based on real-time data from farm-smart instrumentation that measures soil water potential, sap flow and changes in trunk diameter (Siddiqi *et al.*, 2021). This water-smart technology is increasingly being used by fruit tree growers to maximize water use efficiency and crop production from their orchards. Preplant mounding along the rows of the orchard floor is another practical method of improving soil aeration, thus protecting avocado tree health when waterlogging conditions from major rainfall events occur (Grunennvaldt, 2022).

Bayabil *et al.* (2022) stated that the removal of one-third of avocado tree canopies once flooding has receded will assist in the recovery of trees. This practice has long been successfully implemented in the wet subtropics of Australia and New Zealand following damaging rainfall events (A.W. Whiley, Queensland, Australia, 2023, personal communication). Due to extensive water-related root damage, or partially uprooted trees, reduction of tree volume allows easier stabilization of the trees, reducing transpiration losses, thereby minimizing desiccation (Crane *et al.*, 1994).

In studies of container-grown 'Choquette' avocado trees, Sanclemente *et al.* (2014) reported that removing one-half to two-thirds of tree canopy immediately after flooding improved tree recovery, whereas pruning trees immediately prior to flooding increased stress compared with non-pruned trees (Sanclemente *et al.*, 2013). The mechanisms whereby pruning trees immediately prior to flooding increases stress is presumably due to decreased sugar production as a result of a reduction of net CO_2 assimilation due to leaf removal. Thus, fewer sugars are produced and translocated to the roots, thereby reducing the substrate for anaerobic respiration in the roots (Sanclemente *et al.*, 2013). Sanclemente *et al.* (2013) found that treating trees with a photosynthetic inhibitor prior to flooding also increased the sensitivity of avocado trees to flooding stress, similar to the effects of leaf removal prior to flooding. Leaf removal prior to flooding also decreased the content of the C7 sugars, mannoheptulose and perseitol, in the roots, supporting the idea of susceptibility to flooding being related to carbohydrate production in the leaves and transport to the roots.

Recent results by Yin *et al.* (2023) indicated that 'Simmonds' avocado trees had greater flood tolerance when grafted onto a clonal West Indian race rootstock ('Waldin') compared with the same cultivar grafted onto a clonally propagated Guatemalan rootstock. However, in earlier studies, trees with the same West Indian seedling rootstock were very sensitive to flooding stress (Ploetz and Schaffer, 1989). Therefore, there appears to be a difference in flood tolerance between clonally propagated and seedling trees, which should be further investigated. Since rootstocks from the West Indian race are also associated with higher salinity tolerance, further research is warranted to provide genetic solutions to meet the challenges of changes in global water quality (Khondoker *et al.*, 2023) and predicted increased frequency of high rainfall events (Robinson *et al.*, 2021).

Chemical priming (exposure to a mild stress to increase tolerance to a more severe future stress) of avocado trees may be another mechanism to reduce flooding stress. Yin *et al.* (2022) applied a soil drench of the brassinosteroid, 24-epibrassinolide, to potted 'Reed' and 'Monroe' trees prior to flooding. The results showed that flooding reduced leaf gas exchange and F_v/F_m in trees in a non-treated control, but there was little effect of flooding on these variables in trees treated with 24-epibrassinolide. Also, a higher percentage of flooded trees treated with 24-epibrassinolide survived compared to untreated, flooded trees or those treated with 28-homobrassinolide, another brassinosteroid.

Wind

Transpiration and evaporation (together termed evapotranspiration or ET) are dependent on energy supply, vapour-pressure gradient and wind speed (Allen *et al.*, 1988). Wind greatly affects plant transpiration, because the thickness of the boundary layer is determined primarily by wind speed. When air surrounding the leaf is very still, the layer of saturated air on the surface of the leaf (the boundary layer) may be sufficiently thick to cause primary resistance to water vapour loss from the leaf. Increases in stomatal aperture under such conditions have little effect on the transpiration rate. When wind velocity is high, moving air reduces the thickness of the boundary layer at the leaf surface, reducing the resistance of this layer. Under such conditions, stomatal resistance will largely control water loss from leaves (Taiz and Zeiger, 2010).

There is very little quantitative information in the literature about the direct effect of wind on avocado tree physiology. In a study of the effects of wind on tree growth and water loss in potted young 'Puebla' avocado seedling trees, Haas (1939) suggested that exposure to wind reduced transpiration, presumably due to partial closing of stomata. Wind velocities in that study were variable and of low intensity, reaching only 2.4 km h^{-1}. In a greenhouse study of potted 'Hass' avocado trees, wind at speeds of 6 mph (9.6 km h^{-1}) increased stem sap flow (and thus transpiration) by 31% (R. Heath, M.L. Arpaia and M.V. Mickelbart, Riverside, USA, 2003, personal communication).

There are no quantitative reports of the effect of wind on growth, development and production of avocado trees. However, anecdotal reports in the literature suggest that wind velocity affects growth, particularly for young trees, because high wind velocities may result in tree bending, structural deformation, shading and bud death (Lemus *et al.*, 2005). Young orchards are the most susceptible to damage because for a number of years after planting, there is no self-sheltering to reduce wind speed. High wind speeds coupled with low air temperatures (below 12°C–14°C) have been reported to significantly decrease fruit set of avocados, although specific wind speeds were not reported (Rosenberg, 1987). Excessive wind speed can result in mechanical damage to limbs and loss of buds, flowers and fruit (Gardiazábal, 2004) and can also affect fruit quality due to wounding of the exocarp (Gardiazábal, 2008). Intermittently, significant skin-ridging occurs on New Zealand fruit, which is generally industry-wide, suggesting an environmental event, with strong wind during a particular phase of fruit development being the most likely agent (A.W. Whiley, Queensland, Australia, 2023, personal communication). During flowering, wind speeds of 10 km h^{-1} or higher have been reported to impede the flying ability of honeybees, thereby reducing pollination.

Wind damage can be reduced by establishing natural or artificial windbreaks, which consist of fast-growing tree species that do not interfere with the crop and are not hosts for avocado insects or diseases (Gardiazábal, 2004; Lemus *et al.*, 2005). Due to the strong marine influence where avocados can be grown in New Zealand, shelter-belt provision is practically mandatory for a successful outcome (Fig. 7.11; Hermosilla-Madrid and Whiley, 2023).

Salinity

Salinity is generally associated with avocado production in semi-arid, Mediterranean climates (e.g. California, Chile, Israel, Mexico, Peru and Spain), where the individual or combined effects of poor water quality and low annual rainfall can lead to high salt concentrations in soils. Salinity is seldom a problem in humid subtropical climates such as eastern Australia, New Zealand or South Africa, where substantial summer rainfall reduces salt accumulation in soil, although deteriorating water quality with lower rainfall has recently led to salinity issues in some areas of eastern Australia (A.W. Whiley, Queensland, Australia, 2023, personal communication). Development of saline soils is mostly due to poor-quality irrigation water. Since dissolved salts conduct electricity, salinity is quantified by the electrical conductivity (EC) of the irrigation water or soil solution, and the measurement taken is usually expressed in deciSiemens per metre (dS m^{-1}). The EC of irrigation water (EC_{iw}) is generally lower than that of soil water (EC_{sw}) due to ET from the soil, which increases the salt:water ratio (Crowley, 2008; Kourgialas and Dokoa, 2021).

Fig. 7.11. *Pinus radiata* shelter belts in an avocado orchard near Kaitaia, New Zealand, which is in close proximity to the Tasman Sea and Pacific Ocean.

Avocado is very sensitive to high salinity compared with most other crops (Haas, 1929; Ayers *et al.*, 1951; Kadman, 1963, 1964; Patel *et al.*, 1975; Downton, 1978; Chirachint and Turner, 1988; Bernstein *et al.*, 2004) and is grown in many regions of the world where irrigation water has an $EC_{iw} \geq 0.75$ dS m^{-1} and a chloride content ≥100 mg/l (Crowley, 2008). Significant efforts are being made to identify the physiological bases of salt tolerance and management strategies for growing avocados under saline conditions, which is critical to maintain production in many parts of the world (Reints *et al.*, 2020; Kourgialas and Dokou, 2021).

Among the salts causing physiological and production problems in avocado orchards, NaCl and carbonates are the most damaging. The harm caused by these salts is due to the single effect of sodium (Na^+) or chloride (Cl^-) when acting as cations or anions after dissolving in water. Carbonates can cause physiological damage and thus production problems in avocado orchards due to alkaline stress and reduction in the availability of nutrients such as iron and zinc, which become insoluble in high-pH soils. Calcium carbonate ($CaCO_3$) and bicarbonate (HCO_3^-) are among the more common carbonates found in avocado production areas (Ferreyra *et al.*, 2008; Granja and Covarruvias, 2018).

Toxicity symptoms of Na^+ are expressed as necrotic spots near the margin or interior of the leaf lamina (Ayers *et al.*, 1951), although Bingham and Nelson (1970) reported severe leaf necrosis and defoliation in avocado trees from a sand-culture study when leaf Na^+ concentrations exceeded 0.8%, which infers symptoms similar to Cl^- toxicity (see below). Necrosis is reduced in shaded leaves, even if there is no difference in leaf Na^+ concentration between exposed and shaded leaves (Chirachint and Turner, 1988). Sodium can accumulate in the roots and woody tissue of the tree (Kadman, 1964; Downton, 1978; Mickelbart *et al.*, 2007b), with accumulation in roots resulting in reduced root growth (Kadman, 1964; Alvarez-Acosta *et al.*,

2018). Tolerance of avocado to high levels of soil Na^+ has been attributed to the existence of a physiological root barrier, the effectiveness of which determines the Na^+ susceptibility of a rootstock (Kadman, 1964).

In contrast, Cl^- is not sequestered in the roots and woody organs (Fenn *et al.*, 1968; Mickelbart *et al.*, 2007b; Celis *et al.*, 2018) but moves rapidly with the transpiration stream and accumulates in leaf tissue (Ayers *et al.*, 1951; Patel *et al.*, 1975). The result of Cl^- accumulation in the leaves is tip burn that progresses basipetally and sometimes along the leaf margin (Kadman, 1963, 1964). Symptoms are typically seen in 'Hass' when the leaf Cl^- concentration reaches 0.25%–0.3%, and moderate to severe defoliation occurs during flowering when leaf chloride concentrations exceed ca. 0.5% (Crowley, 2008; A.W. Whiley, Queensland, Australia, 2023, personal communication). However, avocado leaf necrosis as a result of root-zone salinity may not simply be due to the amount of Cl^- accumulated in the leaves. Mickelbart and Arpaia (2002) reported that leaf Na^+ concentration was correlated with the level of leaf necrosis, even though symptoms observed were typical of Cl^- injury. They suggested that elevated leaf concentrations of Na^+ together with high Cl^- content in leaves may intensify the visual symptoms of Cl^- injury in 'Hass' avocado leaves. Celis *et al.* (2018) and Lazare *et al.* (2021b) concluded that damage to avocados growing under saline conditions appears to be primarily related to Cl^- accumulation in the leaf tissues. With conflicting reports, it remains unclear if leaf Na^+ or Cl^- is the primary reason for observed leaf necrosis, but the ultimate expression of leaf damage is likely the net effect of genetics and environmental factors that include the salinity composition of the irrigation water.

Kadman (1963) studied the effects of salinity on seedling avocado trees representing discrete populations drawn from the three horticultural races. He concluded that there were two possible mechanisms for tolerance to high levels of salinity. The first was less uptake by roots and subsequent transfer to leaves of Cl^-, and the second was a greater tolerance to higher leaf Cl^- concentrations before the associated onset of leaf necrosis. In later studies, it was concluded that root exclusion of soil Cl^-, or restriction of Cl^- from leaves, despite high root concentrations, appears to be the mechanism of tolerance shown by Guatemalan and West Indian race cultivars when used as rootstocks, thereby allowing shoot growth to be maintained under saline conditions (Embleton *et al.*, 1961; Wiesman, 1995, Lazare *et al.*, 2021b). In another study, root Cl^- concentrations were similar among three different Mexican race rootstocks grafted onto 'Hass' when the EC_{iw} was ≥3 dS m^{-1}, whereas scion Cl^- concentrations varied across rootstocks and were correlated with leaf necrosis (Mickelbart *et al.*, 2007b).

Salinity reduces growth of avocado trees (Bingham *et al.*, 1968; Kalmar and Lahav, 1977; Bernstein *et al.*, 2001), primarily through lower growth rates during flushing (Mickelbart and Arpaia, 1995), and reduces fruit yield (Lahav and Kalmar, 1977; Steinhardt *et al.*, 1995; Oster *et al.*, 2007; Celis *et al.*, 2018; Acosta-Rangel *et al.*, 2019). Oster *et al.* (2007) reported that avocado yields began to decline in 'Hass' trees grafted onto rootstocks of mostly Mexican race origin growing in California once irrigation water salinity reached an EC_{iw} of 0.57 dS m^{-1}. Conversely, Kourgialas and Dokoa (2021) found that fruit yields of 'Hass' trees in Israel declined when the EC_{iw} was ≥0.75 dS m^{-1}. In the latter case, trees were grafted onto rootstocks of predominantly West Indian race origin, thus difference in rootstock genotype between the two studies is likely the reason for differences in EC thresholds. Mickelbart *et al.* (2007b) reported that rootstocks also modify the concentrations of other nutrients that accumulate in the scion in response to salinity. Analyses of the older leaves indicated that, in general, Mg and Ca contents were reduced with increased salinity while K content increased, but there were variations among the three rootstocks tested. Under severe conditions, salinity can threaten tree survival (Celis *et al.*, 2018); flowering is advanced and more prolific, being a stress response when trees are exposed to moderate salinity, but is retarded at higher salinity levels (Downton, 1978; A.W. Whiley, Queensland, Australia, 2023, personal communication). The recommended chloride concentration thresholds in irrigation water vary broadly according to the race of the rootstock, with suggested ranges of 120–150 mg l^{-1} for Mexican rootstocks, 200–250 mg l^{-1} for Guatemalan rootstocks and 350–500 mg l^{-1} for West Indian rootstocks (Kourgialas and Dokou, 2021).

Effect of salinity on tree physiology

Although avocado trees grown in saline soils typically have reduced transpiration (Bingham *et al.*, 1968; Acosta-Rangel *et al.*, 2019) and leaf water potential (Salazar-García and Larque-Saavedra, 1985; M.V. Mickelbart and M.L. Arpaia, Riverside, USA, 1995, personal communication), leaf water content is usually not reduced (Bingham *et al.*, 1968), except at high salinity levels (Mickelbart and Arpaia, 2002). In fact, exposure to salinity results in increased leaf succulence (Downton, 1978; Salazar-García *et al.*, 1987; M.V. Mickelbart and M.L. Arpaia, Riverside, USA, 1995, personal communication). Osmotic adjustment occurs in response to salinity in recently developed avocado leaves, but only at high (>3.0 dS m^{-1}) salinity (Marler and Zozor, 1996; M.V. Mickelbart and M.L. Arpaia, Riverside, USA, 1995, personal communication).

Net CO_2 assimilation in avocado leaves is reduced at relatively low salinity concentrations (Mickelbart and Arpaia, 2002; Castro *et al.*, 2009). This reduction is correlated with a reduction in leaf chlorophyll concentration (Mickelbart and Arpaia, 2002; Musyimi *et al.*, 2007), which is a commonly observed consequence of high soil salinity (Musyimi *et al.*, 2008; Alvarez-Acosta *et al.*, 2018). At least part of the reduction in net CO_2 assimilation with increased salinity appears to be due to reduced stomatal conductance (Downton, 1978; Salazar-García and Larque-Saavedra, 1985; Musyimi *et al.*, 2007), although possible non-stomatal constraints to CO_2 assimilation from increased salinity have not been quantified. Reductions in net CO_2 assimilation may be related to high accumulation of toxic ions, as it is typically not lower in plants grown with high soil salinity until visible partial necrosis occurs (Mickelbart and Arpaia, 2002; Acosta-Rangel *et al.*, 2019).

When avocado trees were put under alkaline stress, greater reductions in net CO_2 assimilation, stomatal conductance and transpiration were observed compared to salinity stress. Also, greater oxidative stress was observed for alkaline than saline conditions, since the alpha-tocopherol (an antioxidant that protects plants from oxidative stress) concentration was higher in trees exposed to alkaline stress. In addition, greater damage to the photosynthetic apparatus was observed for avocado trees under alkaline stress compared with trees under salt stress (Alvarez-Acosta *et al.*, 2018).

Rootstocks

In general, Mexican race rootstocks are considered to be the most sensitive to saline conditions, Guatemalan race rootstocks are intermediate, and West Indian race rootstocks the most tolerant of saline conditions (Haas, 1950; Embleton *et al.*, 1961; Kadman, 1963, 1964; Bergh, 1975; Kadman and Ben-Ya'acov, 1976; Downton, 1978). However, there is significant genetic variability in salt tolerance among avocado rootstocks both within and among the three horticultural races that can be commercially utilized depending on the severity of the saline conditions (Alvarez-Acosta *et al.*, 2018; Celis *et al.*, 2018; Acosta-Rangel *et al.*, 2019; Lazare *et al.*, 2021a,b). Although the Mexican race cultivars are regarded as the most sensitive to salinity, there is also significant genetic variability for rootstock-mediated salinity tolerance within this group (Mickelbart and Arpaia, 2002; Castro *et al.*, 2009). Alvarez-Acosta *et al.* (2018) exposed six clonal rootstocks (two Mexican race, three West Indian race and one Guatemalan × West Indian hybrid) to salinity (NaCl) and alkaline ($NaHCO_3$) stresses for 5 months. They found that both stresses reduced aerial biomass, while alkaline stress reduced root biomass. The West Indian genotypes had a higher root biomass than those of Mexican race origin, indicating a greater tolerance of the West Indian race rootstocks to alkaline conditions.

Cultivar responses to soil salinity are highly dependent on the rootstock. Lazare *et al.* (2021a) reported that 'Hass' was less affected in terms of physiology, anatomy and development when grafted onto the West Indian rootstocks VC 159, VC 140 or VC 152 and exposed to high Cl^- concentrations compared with other rootstocks. In a second study, Lazare *et al.* (2021b) found that despite high susceptibility to salinity-induced damage, physiological and metabolic mechanisms of salt tolerance are indeed present in avocado. They suggested that mutual rootstock–scion relationships enable marked tolerance to salinity stress. Although salt ions accumulate mainly in roots, the primary metabolic response of a relatively salt-tolerant avocado tree to salinity was observed in the scion rather than at the root level. In that study, 'Hass' were grafted onto VC152 (least salt-sensitive) or VC840 (most salt-sensitive) rootstock, and it was observed that the rootstock VC840 did not restrict the transport of irrigation solution components to the scion, leading to salt accumulation in the trunk and leaves. The second rootstock, VC152, functioned selectively, moderating the movement of toxic ions to the scion organs by accumulating them in the roots. Despite the restricted flow of toxic ions, the scion grafted onto VC152 acquired the standard level of essential minerals without being exposed to excessive salt concentrations, and no salt damage or nutrient deficiency was detected.

Management

Salinity tolerance of crops is influenced by the interrelationships among plant, soil and environmental factors; hence, successful management strategies for production of a species under these conditions may vary across regions. While production of avocados under saline conditions is challenging, there are a number of management strategies available that, if implemented, can improve the likelihood of success.

Genetic resistance to salinity within the avocado genotype is well documented, both among and within the three horticultural races, and research continues to look for rootstock material that provides commercial resistance together with productivity (Oster and Arpaia, 1992; Bernstein *et al.*, 2001, 2004; Mickelbart and Arpaia, 2002; Celis *et al.*, 2018). Most research assessments for salinity tolerance in avocado rootstocks have focused on shoot growth and leaf responses, but minimizing leaf damage under salt stress has not prevented negative effects on avocado productivity (Ben-Ya'acov *et al.*, 1979; Ben-Ya'acov and Michelson, 1995). Bernstein *et al.* (2004) reported that root growth of avocado can be a significant casualty under moderate saline conditions. Seedlings of the West Indian race rootstock 'Degania 117' were grown in a complete nutrient solution containing 15 mM NaCl, which did not affect the rate of leaf emergence. Eight days following exposure to NaCl, there was a 10% reduction in leaf biomass compared to control trees, but a 43% reduction in root elongation together with a 33% decrease in volumetric growth rate of roots. The results of Bernstein *et al.* (2004) challenge the use of leaf damage as a criterion for determining salinity tolerance in avocado, since other tree responses may have a significantly greater impact on resistance and productivity.

Irrigation management is a principal tool for reducing the effects of salinity on avocado growth and production (Kourgialas and Dokou, 2021). Approximately 80% of avocado roots that take up water and nutrients are located in the top 20–30 cm of the soil, with the majority of roots occurring in the top 10 cm. Therefore, this upper soil zone is the most critical for salinity management (Whiley, 1994; Crowley, 2008), as the concentrations of Na^+ and Cl^- ions can exceed that of irrigation water under poor water management. Regular deep leaching can be effective in limiting the effects of salinity, and this is best managed by the installation of soil moisture-monitoring smart sensors at different depths that provide real-time information of water movement through the soil profile.

The first commercial magnetic device for treating poor water quality was patented in Belgium (Vermeiren, 1958), but magnetic water treatment (MWT) has remained a controversial process for the treatment of industrial and commercial water (Surendran *et al.*, 2016). More recently, there has been increased interest in the use of MWT for the management of salinity in agricultural and horticultural crops. This is a non-polluting, ecologically sustainable treatment where magnets are placed around a primary irrigation line directing water through a strong magnetic field. However, the success of the treatment depends on the characteristics of the magnetic fields generated from the equipment installed, such as polarity, intensity, exposure time and magnet type (Surendran *et al.*, 2016). MWT does not change the chemical parameters of water; however, it changes physical characteristics, and, according to published results, the magnetic fields change surface tension, viscosity, zeta-potential, solubility and diffusion of the treated water (Gang *et al.*, 2012). Based on a study with soils from an avocado orchard in California, Zlotopolski (2017) reported that after leaching 90-cm-long columns of soil daily for 2 months with untreated well water (control) and MTW well water ($Na^+ = 150.8\ mg\ l^{-1}$,

Cl^- = 286.3 mg l^{-1} and EC_{iw} = 191 mS m^{-1}), Na^+ concentrations in the control treatment at 30 cm were 15% higher, and at 60 cm were 21% higher than those in the MTW treatment, respectively. Similarly, Cl^- concentrations in the control column at 60 cm were 18% higher, and at 60 cm were 30% higher than those in the MWT column, respectively. Conversely, at the depth of 90 cm, the EC_{sw} value was higher in the column irrigated with the MTW leaching solution, suggesting that MTW had displaced a greater content of Na^+ and Cl^- ions below the root zone than the untreated water. Calcium and magnesium concentrations were not significantly different among the three sampled depths in the columns. The use of MWT on seasonally variable saline irrigation water (Na^+ = 98–120 mg l^{-1}, Cl^- = 180–250 mg l^{-1} and EC_{iw} = 0.65–1.25) used for avocado production in south-eastern Queensland, Australia, has mostly eliminated what was previously severe leaf damage and improved crop yields (A.W. Whiley, Queensland, Australia, 2023, personal communication). Within 2 years after installation of MWT, soil (≤30 cm depth) and mature leaf Cl^- concentrations were reduced from 348 to 193 mg.kg^{-1} and 0.65% to 0.35%, respectively.

Other management approaches, such as shading (Downton, 1978), the application of nitrate nitrogen (Bar and Kafkafi, 1992; Bar *et al.*, 1997) and overhead irrigation (Lazare *et al.*, 2022), can reduce the uptake and accumulation of Na^+ and/or Cl^- in avocado leaves. More recently, seaweed extract has been shown to alleviate salt-stress symptoms in potted avocado trees but only in the early growth stages, likely by contributing additional essential elements (Bonomelli *et al.*, 2018). Also, halotolerant soil bacteria are associated with avocados growing in saline soils (Nadeem *et al.*, 2012) and may mitigate the effects of saline irrigation water on avocado nursery stock (Barra *et al.*, 2017).

Atmospheric CO_2 Concentration

The CO_2 concentration of the Earth's atmosphere has been increasing rapidly over the last 35 years and continues to rise, primarily due to burning of fossil fuels. Currently, the median atmospheric CO_2 concentration of the Earth is ≈420 μmol CO_2 mol^{-1} (NOAA, 2022). The response of plants to increased atmospheric CO_2 is species-dependent and subject to regulation by other environmental factors, since the impact of elevated CO_2 and warming on growth responses is largely determined via shifts in CO_2 assimilation, photorespiration and respiration, which include both carbon sources and sinks. In a comprehensive review of many plant species grown at multiple temperatures, focusing on net CO_2 assimilation and leaf dark respiration rates (A_{growth} and R_{growth}, respectively), measured at optimum growth temperatures, Dusenge *et al.* (2019) reported that the ratio of A_{growth} to R_{growth} is generally homeostatic across most species and temperatures. However, they found that trees showed the highest response to elevated CO_2 concentrations (a 47% increase in maximum CO_2 assimilation) but over time acclimate to elevated CO_2 concentrations, resulting in lower photosynthetic stimulation. Thus, with predictions of future concomitant increases in ambient CO_2 concentrations and mean temperatures, it is uncertain whether rises in crop productivity due to atmospheric elevation of CO_2 will be sustained (Dusenge *et al.*, 2019).

Photoassimilation responses to enhanced CO_2 concentrations

In a study of 'Hass' avocado trees, Whiley *et al.* (1999) reported that the photosynthetic rate of leaves increased asymptotically with short-term rises in ambient CO_2 concentration until maximum carboxylation was reached at 1350–1470 μmol CO_2 μmol^{-1} (Fig. 7.12). The maximum saturated CO_2 assimilation for autumn-measured leaves was 50.6 μmol m^{-2} s^{-1} but was lower in winter with a maximum of 34.9 μmol m^{-2} s^{-1}. During both autumn and winter, there was a linear relationship between ambient CO_2 concentration and intercellular partial pressure of CO_2, implying that diffusion through the mesophyll tissue was never limiting net CO_2 assimilation. The reduction in net CO_2 assimilation of winter leaves was likely due to either chlorophyll inhibition as minimum daily temperatures were ≤10°C, or sink-related as starch levels in winter-measured leaves were 35% higher than in autumn-measured leaves. During winter, growth of avocado trees is relatively quiescent with starch accumulating in major organs (Whiley *et al.*, 1996a,b). Other non-structural carbohydrates

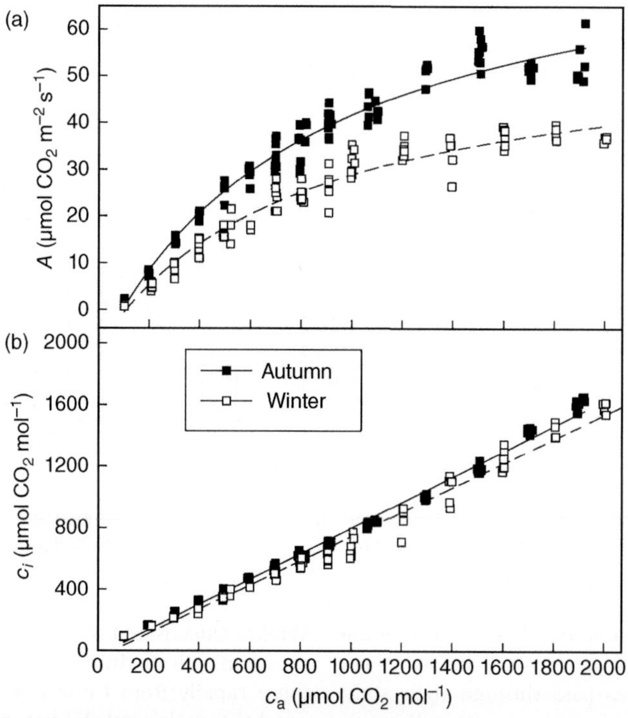

Fig. 7.12. Net CO_2 assimilation (A) and intercellular partial pressure of CO_2 (c_i) responses of container and field-grown 'Hass' avocado trees to varying ambient CO_2 partial pressures (c_a). The regression line for A of field-grown trees in summer (a) is represented by $y = 85.49\ ((-96.72 + x)/(852.75 + x))$, $r^2 = 0.96$; and field-grown trees in winter by $y = 58.54\ ((-110.10 + x)/(805.77 + x))$, $r^2 = 0.96$. The regression line for Ci (b) of field-grown trees in summer is represented by $y = 0.83x - 33.53$, $r^2 = 0.99$; and field-grown trees in winter by $y = 0.793x - 49.057$, $r^2 = 0.98$. (From Whiley *et al.*, 1999, with permission).

found in avocado trees such as fructose, glucose, D-mannoheptulose and perseitol, which are the primary sugars of photosynthetic CO_2 fixation in avocado (Liu *et al.*, 1999a, b; 2002b), may also accumulate in leaves as a result of sink restriction. The accumulation of photosynthetic end-products in leaves, particularly hexoses, may repress genes that code for rubisco, resulting in end-product inhibition of net CO_2 assimilation (Stitt, 1991; Koch, 1996; Drake *et al.*, 1997).

Carbon partitioning in elevated CO_2 atmospheres

While previous studies provided evidence that CO_2 enrichment of the atmosphere leads to a substantial increase in plant biomass in C_3 plants through increased photoassimilation (Downton *et al.*, 1987; Allen *et al.*, 1988; Idso and Idso, 1994; Poorter *et al.*, 1997; Schaffer *et al.*, 1997, 1999), beneficial consequences from this significant environmental change will only be realized if partitioning of increased biomass favours increases in the harvest index, either through genetic selection or manipulation of the crop (Wolstenholme *et al.*, 1990; Whiley *et al.*, 1991).

Schaffer *et al.* (1999) reported on dry matter partitioning in 'Hass' avocado trees grown at either 350 or 600 μmol CO_2 mol^{-1} for 6 months (Fig. 7.13). There was a significantly greater amount of dry matter partitioned to old leaves in trees growing at 350 μmol CO_2 mol^{-1} than those at 600 μmol CO_2 mol^{-1}. This may have been due to a greater investment in new branches and leaves by trees growing at 600 μmol CO_2 mol^{-1}.

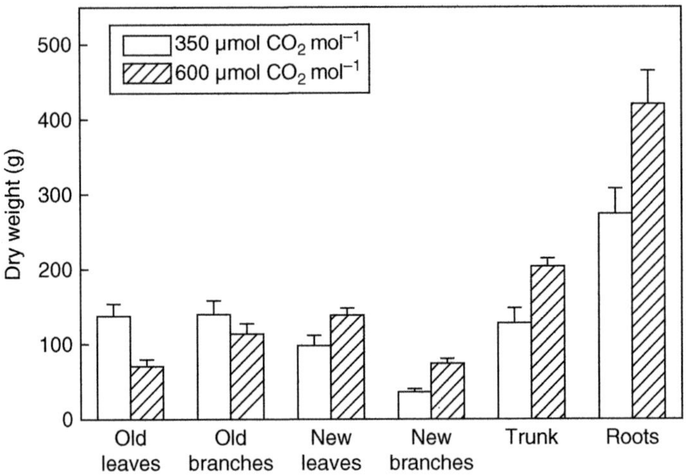

Fig. 7.13. Partitioning of dry matter in avocado trees grown for 6 months in atmospheres of either 350 or 600 µmol CO_2 mol^{-1}. Columns represent means (n = 6) ± se. (Redrawn from Schaffer *et al.*,1999, with permission).

Whiley (1994) observed that with renewal spring shoot growth in avocado, older leaves are shed as new leaves pass through their sink/source transition, but older leaves will be retained if new growth is delayed or weak. There was also a greater allocation of dry matter to trunks of avocado trees growing at 600 µmol CO_2 mol^{-1} when compared with trees growing at 350 µmol CO_2 mol^{-1} (Schaffer *et al.*, 1999). The trunks of avocado are an important storage organ for surplus photoassimilates (Whiley *et al.*, 1996a,b), and it is likely that higher dry matter accumulated in trunks of trees growing at 600 µmol CO_2 mol^{-1} was composed of starch and/or other non-structural carbohydrates. The roots of avocado trees growing at 600 µmol CO_2 mol^{-1} also had significantly increased dry matter allocation. Increased partitioning of dry matter to roots has also been reported in other studies with container-grown crops exposed to CO_2 enrichment (Pearson and Brooks, 1995; Schaffer *et al.*, 1997, 1999). Presumably, when conditions are not sink-limited, such as in an orchard, increased water and nutrient uptake resulting from increased root mass would eventually encourage greater assimilate partitioning to above-ground organs, as observed for citrus (Idso *et al.*, 1991a,b).

With respect to fruit retention, avocado trees grown at 350 µmol CO_2 mol^{-1} held fewer fruit than those at 600 µmol CO_2 mol^{-1} (A.W. Whiley, Queensland, Australia, 1999, personal communication). Initially, fruit loss occurred more rapidly from those trees grown at 600 µmol CO_2 mol^{-1}, but 25 days after the completion of anthesis, fruit loss at this atmospheric CO_2 concentration levelled off, whereas fruit continued to drop from trees grown at 350 µmol CO_2 mol^{-1} (Fig. 7.14). Forty-five days after anthesis, there were more fruit retained on the trees grown at 600 µmol CO_2 mol^{-1} compared with trees grown at 350 µmol CO_2 mol^{-1}. The higher fruit loss in the first 25 days following anthesis from trees grown at 600 µmol CO_2 mol^{-1} may be due to increased partitioning of photoassimilates into renewal shoot growth, which occurs concurrently with early fruit development and hence competes for resources (Biran, 1979; Wolstenholme *et al.*, 1990). While that study did not take fruit through to maturity, there is a direct relationship between fruit retained 40–50 days after anthesis and final yield (Wolstenholme *et al.*, 1990). Thus, it is likely that increased atmospheric CO_2 concentrations will provide some benefit to productivity of avocado.

Air Pollution

The most common gaseous air pollutants that can cause injury to horticultural crops are ozone (O_3), sulfur dioxide (SO_2), nitrogen oxides (NO_x),

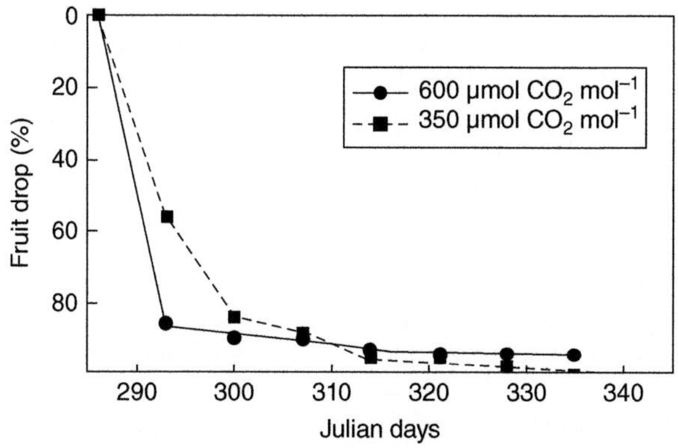

Fig. 7.14. Effect of enhanced CO_2 concentration on fruit drop during the first 45 days following anthesis from avocado trees growing in atmospheres of either 350 or 600 µmol CO_2 mol^{-1}. Fruit drop is expressed as the percentage of the total fruit set on trees immediately following anthesis. Data points are mean values from six trees ± se. (A.W. Whiley and C. Searle, Nambour, Australia, 1996, unpublished data).

hydrogen fluoride (HF) and peroxyacetyl nitrate (PAN) (Ormrod, 1986). Additionally, the conversion of SO_2 and NO_3 into strong acids in the atmosphere results in acid rain, which can cause significant crop damage (Likens *et al.*, 1976; Ormrod, 1986). Of these pollutants, the effects of ozone, sulfur dioxide and simulated acid rain on avocado have been investigated.

Ozone, formed by photochemical reactions of nitrogen oxides and volatile organic compounds on warm, sunny days, is a major widespread air pollutant in the lower troposphere. Ozone enters plant tissues primarily through the stomata, affecting metabolic processes and causing tissue damage that inhibits plant growth and yield (Ormrod and Hale, 1994). Exposing 'Zutano' avocado seedlings to smog generated from ozone (3.5 ml min^{-1}) and 1-*N*-hexene vapour (30 ml min^{-1}) for 7 h each day for 10 days resulted in phytotoxic damage to the underside of newly expanded leaves, leaf tip burn and marginal leaf necrosis (Taylor *et al.*, 1958). Taylor (1958) observed that 8 weeks of exposure to ozone smog resulted in a 58% reduction in dry matter accumulation. Similarly, Eissenstat *et al.* (1991b) observed that exposing 'Simmonds' and 'Pancho' avocado trees (both West Indian race cultivars) to 75 and 108 nl l^{-1} of ozone for 32 weeks reduced growth by 20% and 61%, respectively. Reductions in dry matter accumulation in plants due to ozone exposure have been attributed to decreased net CO_2 assimilation, because ozone exposure disrupts chloroplast metabolism (Ormrod and Hale, 1994).

There is some evidence that tolerance to ozone or sulfur dioxide differs among ecological races. For example, when hybrids with varying percentages of Mexican heritage were exposed to 300 nl l^{-1} of ozone (16 h) or 1000 nl l^{-1} of sulfur dioxide (48 h), severe leaf injury occurred to Mexican × Guatemalan hybrids that had 0%–15% Mexican heritage, whereas minimal leaf damage occurred to hybrids that had 50% or 100% Mexican heritage. Thus, it was suggested that Mexican cultivars may have a higher resistance to air pollution compared to cultivars of predominantly Guatemalan origin (Olszyk *et al.*, 1987).

Elevated ozone concentrations and acid rain can inhibit freeze resistance in woody plants (Fincher *et al.*, 1989; Eissenstat *et al.*, 1991a,b; Eamus and Murray, 1992; Eamus, 1993). Eissenstat *et al.* (1991a,b) studied the interactions between ozone and acid rain exposure and freeze resistance of 'Simmonds' and 'Pancho' avocados. Plants exposed to ozone concentrations of two to three times that of ambient (30 or 39 nl l^{-1}) for 12 or 24 h each day for 16 or 32 weeks had reduced freezing resistance, as evidenced by increased electrolyte leakage from leaf discs. Acid rain (pH = 3.3) increased the detrimental effects of ozone on electrolyte leakage in avocado at sub-zero temperatures. However, acid rain (pH = 3.3–5.5) alone had little detrimental effect on growth of avocado.

Avocado is relatively tolerant to air pollution compared to other plant species (Fincher *et al.*, 1989; Schultze, 1989). Nevertheless, phytotoxicities will occur if the exposure is sufficiently severe. Since ozone and acid rain can negatively affect growth and freeze resistance of avocado, and genotypic variance has been reported, there may be opportunities to reduce the damaging effects of air pollution through selection and breeding programmes. However, this may not be a viable long-term solution, since concentrations of pollutants in the atmosphere continue to rise. The best method to control the injurious effects of air pollution on crops, including avocado, is likely to be through elimination or reduction of the source of pollution (Ormrod, 1986).

Ecophysiological Responses to Pathogens

In addition to abiotic factors influencing plant performance, biotic agents such as pathogens can drastically affect plant physiology and anatomy. There has been considerable research on ecophysiology related to two avocado diseases, Phytophthora root rot and laurel wilt. This section will discuss these diseases in response to the ecophysiology of infected avocado trees.

Phytophthora root rot

The most important and widely distributed avocado disease in countries where avocados are produced is Phytophthora root rot caused by the oomycete *Phytophthora cinnamomi* Rands (see Dann and Gazis, Chapter 13, this volume for a detailed discussion of this disease). Once infection occurs, the pathogen downregulates the host's genes and metabolites, which leads to a hormonal imbalance in the host (Van den Berg *et al.*, 2021), causing a reduction of cytokinins (Cahill *et al.*, 1986). This triggers upregulation of stress-responsive genes and hormones, including an increase in ABA in leaves, resulting in stomatal closure and a decrease in net CO_2 assimilation (Davison, 2011; Ha *et al.*, 2012). Ploetz and Schaffer (1987), using non-grafted seedlings of the West Indian cultivars, 'Waldin' and 'Simmonds', and the West Indian × Guatemalan cultivar, 'Lula', showed that early responses of susceptible avocado trees to *P. cinnamomi* include a reduction in net CO_2 assimilation and stomatal conductance. Similar responses were observed in 'Hass' trees after inoculation with *P. cinnamomi*, where net CO_2 assimilation, stomatal conductance and transpiration decreased, and water use efficiency increased when 10% to 15% of the roots were infected with the pathogen (Ramírez-Gil and Morales-Osorio, 2020). Symptom severity was related to the percentage of root necrosis and pathogen density. Feeding and growth of *P. cinnamomi* mycelium on the extensive network of non-suberized roots exacerbates the decrease of water transport and nutrient uptake. As a consequence, profuse leaf chlorosis, defoliation and wilting were observed when more than 25% of the roots were colonized by the pathogen (Schaffer and Ploetz, 1989; Oßwald *et al.*, 2014; Ramírez-Gil and Morales-Osorio, 2020).

Recent studies have shown that Phytophthora root rot resistance is attributed to the triggering of defence-related genes in the host, which is time-sensitive; tolerant and resistant cultivars respond faster than susceptible cultivars (Van den Berg *et al.*, 2018b). Some of the defence mechanisms associated with avocado resistance to *P. cinnamomi* are rapid callose and lignin deposition at the infection site to prevent entry of the pathogen and to protect the healthy cells, secretion of root metabolites that inhibit pathogen growth, regeneration of roots, formation of tyloses (outgrowth of the parenchymal cells that expand into xylem vessels) to limit the spread of the pathogen through the xylem, production of polyphenolic compounds that prevent growth of the pathogen, isolation of infected tissue by creating necrophylactic periderm, and production of proteinase inhibitors to counter-attack the catalytic pathogen enzymes (Phillips *et al.*, 1987, 1991; Nicholson and Hammerschmidt, 1992; Sánchez-Pérez *et al.*, 2009; De Micco *et al.*, 2016; Van den Berg *et al.*, 2018a; Van den Berg *et al.*, 2021).

Net CO_2 assimilation, stomatal conductance and transpiration of 'Dusa', a Phytophthora-tolerant rootstock, inoculated with *P. cinnamomi*, remained similar to that of non-inoculated trees, even 1 month after inoculation (Reeksting *et al.*, 2014). Also, the photochemical machinery was not affected by the disease as indicated by no difference in PS II efficiency, photochemical quenching and non-photochemical quenching between infected and control plants. However, 'Duke 7', another Phytophthora-tolerant rootstock, developed visual symptoms, with net CO_2 assimilation, stomatal conductance, transpiration, PS II efficiency and photochemical quenching reduced by about one-half in inoculated trees compared with non-inoculated trees (Reeksting *et al.*, 2014).

Resistance to *P. cinnamomi* varies greatly among avocado cultivars (Backer *et al.*, 2022). In a study that compared three non-grafted, cloned rootstock cultivars for Phytophthora root rot susceptibility, 'Dusa' was the most tolerant, followed by R0.06 and R0.12, the latter being susceptible to the disease (Fick *et al.*, 2022). The R0.12 rootstock exhibited a decline in plant photochemistry with a significant reduction in PS II efficiency, accompanied by a decrease in photochemical quenching and an increase in non-photochemical quenching, which decreases leaf gas exchange (Reeksting *et al.*, 2014).

Pathogenesis as a result of *P. cinnamomi* infection is influenced by abiotic factors. Environmental conditions and edaphic conditions, such as soil water content and temperature, can exacerbate plant susceptibility (Trivedi *et al.*, 2022). Several authors have reported that under flooded conditions, avocado trees are more susceptible to Phytophthora root rot.

Interactions between flooding and Phytophthora

There are additive or synergistic effects of flooding and Phytophthora root rot on stomatal conductance and net CO_2 assimilation of avocado that vary with soil type (Ploetz and Schaffer, 1987, 1988, 1989; Schaffer and Ploetz, 1987; Schaffer *et al.*, 1992; Whiley and Schaffer, 1994). In southern Florida, shortly after tropical storm Dennis in 1981, the shoot:root ratios of trees may have been further increased by a reduction in root volume caused by Phytophthora root rot, which was found in most orchards. In the unique calcareous soils of southern Florida, Phytophthora root rot only causes severe damage to avocados under flooded conditions (Ploetz and Schaffer, 1987, 1989; Schaffer and Ploetz, 1989; Schaffer *et al.*, 1992). In a study of avocado trees grown in containers with soil collected from an avocado orchard in southern Florida, non-flooded trees were able to sustain up to 50% root necrosis without a significant decrease in net CO_2 assimilation. Even as much as 90% root necrosis from Phytophthora root rot only resulted in a 65% decrease in net CO_2 assimilation (Schaffer and Ploetz, 1989). However, under flooded conditions, as little as 20% root necrosis from *P. cinnamomi* resulted in almost a complete inhibition of photoassimilation (Fig. 7.15).

Fischer *et al.* (2020) examined the effect of *P. cinnamomi* and flooding in non-grafted seedlings

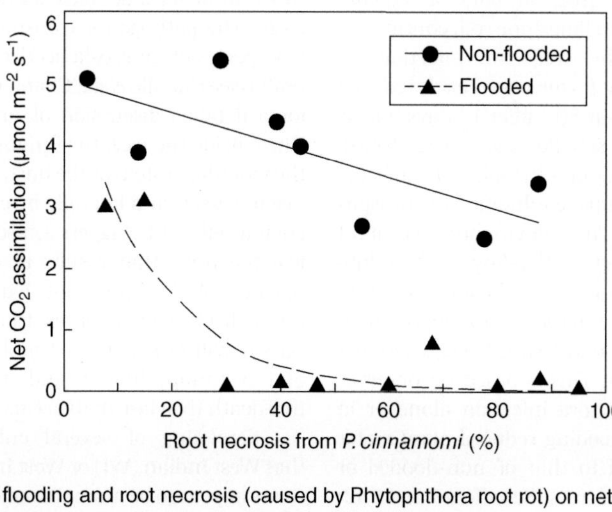

Fig. 7.15. Effect of flooding and root necrosis (caused by Phytophthora root rot) on net CO_2 assimilation of avocado trees. (Redrawn from Schaffer and Ploetz, 1989, with permission).

of 'Hass' and 'Margarida' avocados. They found that the interaction between the flooding period (12 or 24 h) and inoculum concentration (0%, 0.1% or 1%) was related to a decrease in leaf number, the percentage of diseased roots and severity of infection by the pathogen. However, flooding appeared to have a greater impact than *P. cinnamomi* infection, because there was no relationship between the inoculum concentration and plant damage once the trees were flooded (Fischer *et al.*, 2020). A study of Phytophthora root rot and flooding interactions of two Phytophthora-tolerant rootstocks, non-grafted 'Dusa' and 'Duke 7' found that plants under combined flooding and Phytophthora stress showed more symptoms than either flooding or Phytophthora stress alone, and net CO_2 assimilation decreased nearly to zero, while stomatal conductance and transpiration were drastically reduced in plants under both stresses compared to trees with no infection that were not flooded or trees with Phytophthora root rot only (Reeksting *et al.*, 2014). Similar results were observed by Farrow *et al.* (2011), where leaf gas exchange in 'Duke 7' and 'R0.09' (rootstocks with some tolerance to Phytophthora root rot) was drastically decreased due to the additive or synergistic effects of flooding and Phytophthora root rot. The authors attributed the decline in net CO_2 assimilation to a combination of stomatal limitation and biochemical failure, since there was a build-up of internal partial pressure of CO_2 in the leaves.

For 'Simmonds' trees on 'Lula' or 'Waldin' seedling rootstocks in limestone soil, continuous flooding of healthy trees or infection of non-flooded trees by *P. cinnamomi* each reduced transpiration by about 50% after 14 days (Ploetz and Schaffer, 1987; Schaffer *et al.*, 1992, 2006). In addition, flooding of Phytophthora-infected trees resulted in complete inhibition of transpiration after 14 days. Thus, Phytophthora root rot exacerbates the effects of flooding on the inhibition of transpiration. Reduced transpiration in avocado trees due to Phytophthora root rot, with or without flooding, may be due to decreased hydraulic conductivity. This is based on observations that Phytophthora infection alone or in conjunction with flooding reduced plant water potential compared to that of non-flooded or flooded, non-infected trees (Schaffer *et al.*, 1992; Whiley and Schaffer, 1994). Whiley and Schaffer (1994) proposed that reduced transpiration in flooded, non-diseased plants is due to reduced stomatal conductance, whereas reductions in transpiration of flooded, diseased trees are the result of decreased hydraulic conductivity.

Interactions between water deficit and Phytophthora

Plant water status strongly influences plant growth and biomass production through its effect on cell expansion (Beadle *et al.*, 1985). Conversely, factors that inhibit root expansion, such as drought, are likely to reduce stem water potential, resulting in decreased biomass production. Phytophthora root rot also lowers the stem water potential in avocado trees (Sterne *et al.*, 1978; Whiley *et al.*, 1986). In trees with severe root rot, the stem water potential is similar to trees under severe water stress, even when soil moisture is adequate. In field studies, reduced stem water potential of Phytophthora-infected avocado trees was correlated with a sixfold decrease in transpiration (Sterne *et al.*, 1977, 1978).

Laurel wilt

Laurel wilt, caused by the fungal pathogen *Harringtonia lauricola* (T.C. Harr., Fraedrich & Aghayeva) Z.W. de Beer and M. Procter, previously called *Raffaelea lauricola* (de Beer *et al.*, 2022), is a relatively recent biotic threat to avocado production (see Dann and Gazis, Chapter 13, this volume for a detailed discussion of this disease). The pathogen is transmitted locally by a few species of ambrosia beetles or through root grafts (see Carrillo *et al.*, Chapter 14, this volume for a detailed discussion of ambrosia beetles). Once inside the tree, the fungus moves through the vascular system of the host, vertically via the xylem vessels and laterally by colonizing parenchyma cells. This triggers a cascade of biochemical reactions that results in a hypersensitive reaction of the host plant. External symptoms range from wilted leaves that remain green (green wilting), chlorosis and leaf abscission, and eventually branch and stem dieback and tree death (Navia-Urrutia *et al.*, 2022).

Screening of several cultivars indicated that West Indian (WI) or West Indian hybrid cultivars appear to be more susceptible to laurel wilt than Mexican (M) or Guatemalan (G) cultivars

(Ploetz *et al.*, 2011). In a study of 'Russell' (WI), 'Brogdon' (M × G × WI) and 'Marcus Pumpkin' (G) scions grafted onto 'Waldin' (WI) seedling rootstock, Ploetz *et al.* (2015) observed that cultivars with a higher xylem sap flow rate prior to inoculation with *H. lauricola* developed more severe symptoms than cultivars with lower sap flow rates. It was hypothesized that laurel wilt susceptibility is related to the speed of movement of fungal conidia in the xylem, which increases with the rate of xylem sap flow. In a study by Beier *et al.* (2021), WI cultivars had wider diameter vessels and thus higher xylem-specific potential hydraulic conductivity compared with Guatemalan or Mexican race cultivars. However, that study only examined a few cultivars of each race. In subsequent studies, Castillo-Argaez *et al.* (2021a) found that there is considerable variation in vessel size and density within cultivars of the same race. In another study, Castillo-Argaez *et al.* (2022) determined that laurel wilt susceptibility is related to xylem anatomy; plants with a high density of xylem vessels and wider vessels are more susceptible to laurel wilt than plants with fewer, narrower vessels. Similarly, studies on other vascular wilt diseases of woody plants, such as Dutch elm disease and Esca disease of grapevines, have demonstrated that plants with a high density of wider xylem vessels are more susceptible to infection than those with less dense, narrower vessels due to the faster transport of the pathogen and its metabolites in the xylem (Solla and Gil, 2002; Pouzoulet *et al.*, 2014, 2017; Venturas *et al.*, 2014).

Physiological and anatomical responses of avocado trees infected with *H. lauricola* start with xylem embolism and the formation of tyloses and gels, causing impairment of water transport and internal symptoms (sapwood discoloration). Water stress due to drought or infection by a vascular wilt pathogen can substantially increase tension in the xylem (to over -100 MPa), and as a result, air is introduced into the xylem conduits, forming embolisms and increased resistance to water flow (Tyree *et al.*, 1994; Pockman and Sperry, 1997; Pérez-Donoso, 2007). Plant vulnerability to cavitation is related to xylem anatomical characteristics such as xylem vessel size (Tyree and Sperry, 1989). Wider vessels have higher hydraulic conductivity and sap flow rates compared to narrow vessels but are more susceptible to cavitation (Ewers and Fisher, 1989; Tyree and Sperry, 1989; Pockman and Sperry, 2000; Carlquist, 2001; Tyree and Zimmermann, 2002; Sperry *et al.*, 2006). In 'Simmonds' avocado, which is highly susceptible to laurel wilt, 3 days after inoculation with *H. lauricola*, 14% of the xylem was embolized and 1% of the vessels were occluded; after 1 week, internal symptoms appeared, and the xylem lost about 27% of its functionality (Inch and Ploetz, 2012; Inch *et al.*, 2012). As the disease progresses, plants enter a state of induced drought that causes stomatal closure, which leads to a reduction of leaf gas exchange, further decreasing plant hydraulic conductivity and the rate of sap flow in the xylem. Usually, this occurs during the second or third week after inoculation, when 20–40% of the vessels are occluded (Inch *et al.*, 2012). During this period, net CO_2 assimilation, stomatal conductance, transpiration, water use efficiency and sap flow in inoculated plants are significantly lower compared with non-inoculated plants; the xylem sap flow rate can decrease by nearly 50% and hydraulic conductivity can decrease by 50%–70% (Ploetz *et al.*, 2015, Castillo-Argaez *et al.*, 2021a,b). In a study comparing avocado cultivars of different races, 'Waldin' (West Indian), 'Nabal' (Guatemalan), and a hybrid cultivar 'Zutano' (Mexican × Guatemalan), all inoculated plants exhibited an initial decrease in net CO_2 assimilation about 4 days after inoculation. However, continuous differences between the control and inoculated plants occurred 13 days after inoculation (Fig. 7.16). This coincided with the emergence of external symptoms, such as green wilting of young leaves on terminal branches, as well as an initial decrease of sap flow and a reduction in F_v/F_m (Castillo-Argaez *et al.*, 2020a). The decrease in F_v/F_m occurred simultaneously with an increase in intracellular CO_2 concentration, indicating an inhibition of CO_2 fixation, photoinhibition and damage to PS II. Based on these results and leaf carbon isotope ratios of laurel wilt-infected and control trees, it was determined that the reduction in net CO_2 assimilation in infected plants was due to a reduction in plant metabolism and not solely due to stomatal closure (Castillo-Argaez *et al.*, 2020a). Twenty days after inoculation, net CO_2 assimilation, stomatal conductance and xylem sap flow in infected plants decreased significantly when approximately only 20%–30% of the total canopy

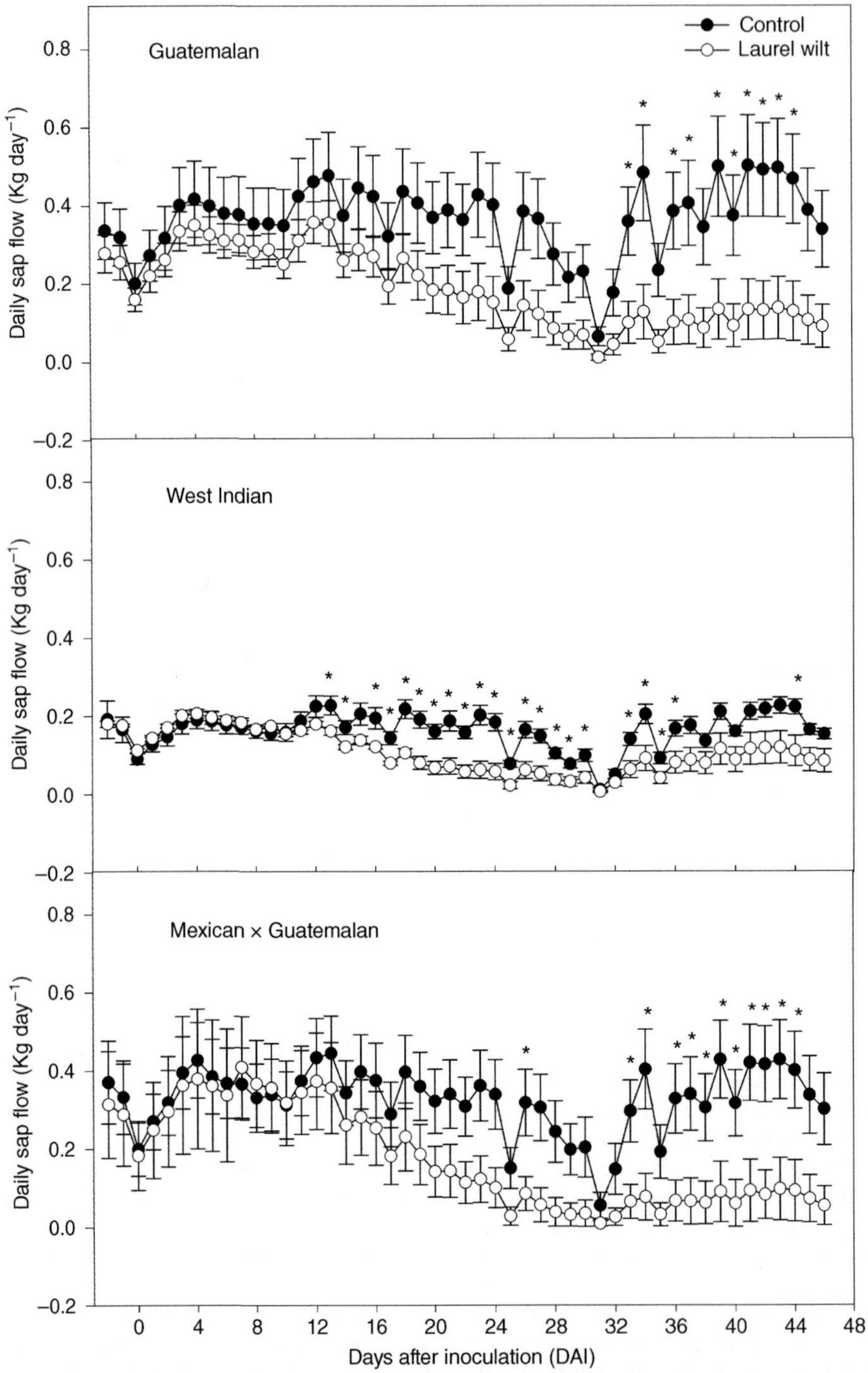

Fig 7.16. Daily sap flow rate of avocado cultivars of the Guatemalan, West Indian and Mexican × Guatemalan races in laurel wilt-inoculated and non-inoculated (control) treatments. Symbols represent means and error bars represent standard errors. Asterisks indicate significant differences between treatments within the same cultivar. (From Castillo-Argaez *et al.*, 2022, with permission).

exhibited laurel wilt symptoms. By 46 days after inoculation, hydraulic conductance of inoculated plants was reduced by two-thirds compared to the control plants (Castillo-Argaez *et al.*, 2020a).

Although the progression of anatomical and physiological responses to laurel wilt is the same in all avocado cultivars, the timing and severity among and within cultivars can be highly variable. For example, highly susceptible cultivars, such as 'Simmonds', 'Waldin', 'Russell' (all West Indian), 'Zutano' (Mexican × Guatemalan) and 'Lula' (West Indian × Guatemalan), exhibited green wilting followed by browning of the leaves. Other cultivars, such as 'Reed' (Guatemalan) and 'Hass' (Guatemalan × Mexican), showed a mixture of green wilting and browning at the early stages of infection. However, non-grafted clonal Mexican and Mexican hybrid cultivars 'Duke 7', 'Dusa' and 'Toro Canyon', with abundant narrow vessels, were less susceptible to laurel wilt compared with the other cultivars tested, with less than 20% of the total canopy affected and no significant decrease in xylem sap flow in the infected trees.

Grafting is another factor that impacts susceptibility to laurel wilt. The graft union modifies xylem anatomical attributes, thereby affecting the resistance to water flow (Olmstead *et al.*, 2006). A study comparing scion/rootstock combinations from the different avocado races found that the scion/rootstock combination had a significant effect on laurel wilt disease severity and sap flow (Castillo-Argaez *et al.*, 2021a). The highly susceptible 'Simmonds' (West Indian) scion developed more symptoms when grafted onto clonally propagated 'Reed' (Guatemalan) or 'Waldin' (West Indian) rootstocks, than when grafted onto clonally propagated 'Duke 7' (Mexican) rootstock. When the rootstock was either Guatemalan ('Reed') or Mexican ('Duke 7'), trees with West Indian scions showed the most symptoms, followed by trees with Guatemalan race scions, and the Mexican race scions showed the fewest symptoms (Castillo-Argaez *et al.*, 2021a). When clonally propagated 'Reed' rootstock was grafted with either 'Waldin', 'Reed' or 'Duke 7' scions, *H. lauricola*-inoculated trees with West Indian or Guatemalan scions had a significantly lower net CO_2 assimilation, stomatal conductance and F_v/F_m compared to control trees that were inoculated with distilled water. However, for inoculated trees with the Mexican ('Duke 7') scion, there was no difference in these physiological variables between inoculated and control trees (Fig. 7.17). Also, in that study, xylem sap flow in non-inoculated trees was higher in the West Indian and Guatemalan scions compared to the Mexican scion, supporting the concept that laurel wilt susceptibility among avocado races or cultivars may be related to the inherent rate of xylem sap flow.

Preliminary studies have shown that 'Hass' (Guatemalan × Mexican) is highly susceptible to laurel wilt. However, severity varies depending on the rootstock. Inoculated 'Hass' trees grafted onto 'Dusa' (Mexican × Guatemalan) rootstock had a disease severity index of 10 (91–100% of the canopy was wilted, based on Navia-Urrutia *et al.*, 2022 scale). However, when 'Hass' was grafted onto 'Duke 7' rootstock, disease severity was reduced by half, with only 50–60% of the canopy showing symptoms (Castillo-Argaez *et al.*, 2020b). Compared to other cultivars, 'Duke 7' had narrower vessels (~50 μm diameter) and lower vessel density (Castillo-Argaez *et al.*, 2018). Thus, trees with this cultivar, either as the scion or rootstock, developed fewer symptoms, presumably due to decreased sap flow as a result of less-dense, narrower vessels.

Conclusions

Physiological responses of avocado trees to environmental factors reflect the natural evolution of the species in the highland and lowland rainforests of Central America (Whiley, 1994; Wolstenholme and Whiley, 1999). The vegetatively biased growth and plasticity of light response (e.g. shade tolerance but high light saturation point for CO_2 assimilation and rapid response to sunflecks) provide avocado trees with an adaptive advantage in their indigenous habitat, allowing them to colonize small gaps when the forest canopy is disturbed (Whiley, 1994). As a consequence of their predominantly mountain rainforest evolution (see Galindo Tovar and Bost, Chapter 2, this volume), avocado trees have an inefficient, shallow root system with a high oxygen requirement. Thus, water uptake is relatively poor, and trees are intolerant of short periods of waterlogging (Whiley and Schaffer,

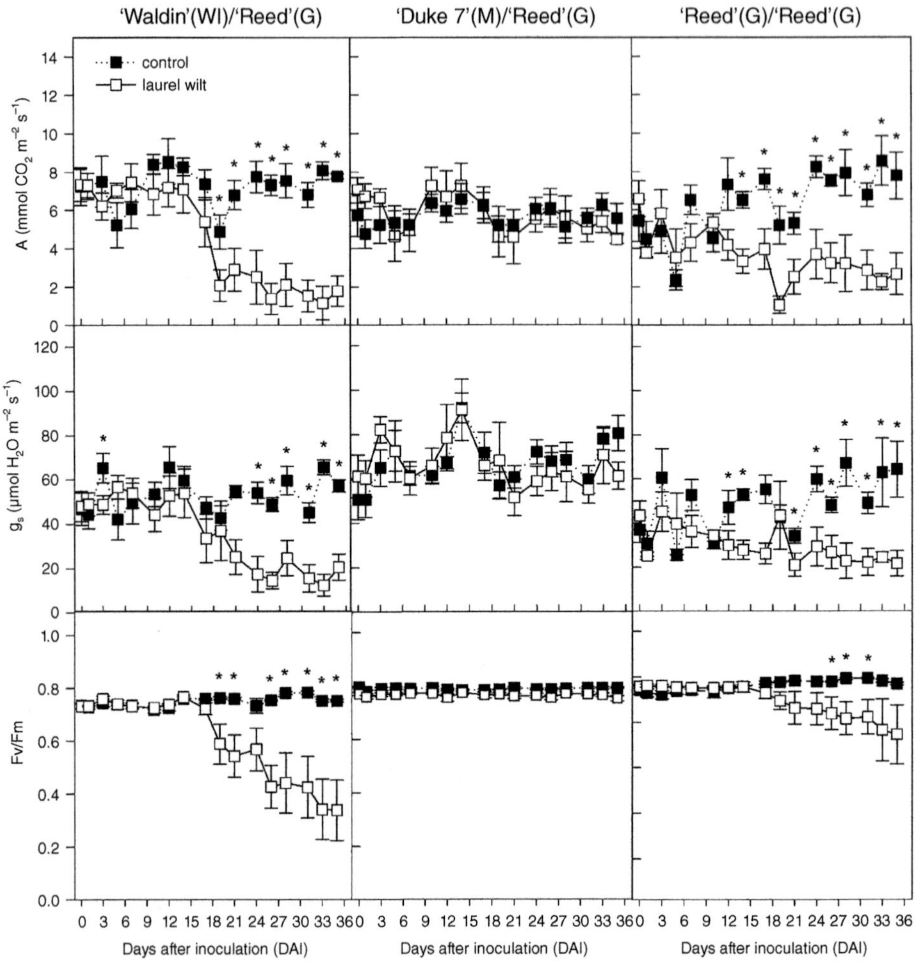

Fig. 7.17. Net CO_2 assimilation (*A*), stomatal conductance (g_s) and the ratio of variable to maximum chlorophyll fluorescence (F_v/F_m) of 'Waldin' (West Indian), 'Reed' (Guatemalan) and 'Duke 7'(Mexican) avocado scions grafted onto 'Reed' (Guatemalan) avocado rootstock inoculated with laurel wilt and a non-inoculated (control) treatment. An asterisk indicates a significant difference between the control and laurel wilt treatment. (From Castillo-Argaez *et al.*, 2021a, with permission).

1994). Also, physiological performance and thus potential production and growth are highly limited not only by waterlogging or flooding but also by high water-to-air ratios, which depend not only on rain or irrigation management, but also on soil physical characteristics such as texture, structure and bulk density (Gil, 2008; see also Ramírez-Gil, Valbuena-Gaona and Wolstenholme, Chapter 5, this volume).

Within its production zone, avocado exhibits its adaptive plasticity to several environmental parameters, which can be related to the three ecological races. Differences exist among Mexican, Guatemalan and West Indian race cultivars in tolerances to temperature, salinity and air pollution. Thus, genetic potential exists to improve the species tolerance to environmental stresses through rootstock and cultivar selection and breeding, since hybridization occurs freely among ecological races. However, expanding the database of physiological × environmental responses of avocado would provide an improved framework for selecting parental material.

There has been a significant amount of research on cultural management practices for avocado (see Mena *et al.*, Chapter 13, this volume). However, as is generally the case for subtropical and tropical fruit tree species compared with temperate species (Schaffer and Andersen, 1994), relatively little research has been conducted on environmental physiology of avocado. For example, light interception and utilization within avocado canopies has only recently been studied in some detail (Wilke *et al.*, 2019) but not sufficiently quantified across changing orchard designs (see Mena *et al.*, Chapter 13, this volume). Seasonal water use requirements of the crop relative to soil and plant water status have not been adequately defined for different edaphic and climatic conditions. Also, there are limited data on the effects of atmospheric CO_2 enrichment and air pollution on growth and physiology of avocado. Phytophthora root rot and laurel wilt disease are two potentially lethal diseases of avocado trees that negatively impact critical physiological functions within the tree. Research over the past 10 years has contributed to a greater understanding of the modes of infection of these diseases and identified the presence of genetic resistance for commercial exploitation. Through evolutionary pressures, *P. americana* has a wide genetic diversity with respect to environmental responses embedded in its three horticultural races. A greater understanding of how to harness the cross-race physiological and growth responses to environmental variables would undoubtedly lead to improved cultural management practices for the crop and possibly allow production to expand into areas which are now considered marginal for commercial use. This is becoming increasingly important, as areas of avocado production have increased dramatically in the past 10 years, particularly in South America. Also, a better understanding of the physiological responses of avocado trees to environmental and biotic factors will help in planting and managing orchards in the face of increasing climatic shifts as a result of global warming.

References

Acosta-Rangel, A.M., Li, R., Celis, N., Suarez, D.L., Santiago, L.S., Arpaia, M.L. and Mauk, P.A. (2019) The physiological response of 'Hass' avocado to salinity as influenced by rootstock. *Scientia Horticulturae* 256(3), 108629.

Acosta-Rangel, A.M., Li, R., Mauk, P.A., Santiago, L.S. and Lovatt, C.J. (2021) Effects of temperature, soil moisture and light intensity on the temporal pattern of floral gene expression and flowering of avocado buds (*Persea americana* cv. Hass). *Scientia Horticulturae* 280(9), 109940.

Aharoni, Y., Schiffmann-Nadel, M. and Zauberman, G. (1968) Effect of gradually decreasing temperatures and polythene wraps on the ripening and respiration of avocado fruit. *Israel Journal of Agricultural Research* 18, 77–82.

Allen, L.H., Vu, J.C.V., Valle, R.R., Boote, K.J. and Jones, P.H. (1988) Non-structural carbohydrates and nitrogen of soybean under carbon dioxide enrichment. *Crop Science* 28, 84–94.

Alon, E., Shapira, O., Azoulay-Shemer, T. and Rubinovich, L. (2022) Shading nets reduce canopy temperature and improve photosynthetic performance in 'Pinkerton' avocado trees during extreme heat events. *Agronomy* 12:1360.

Alvarez-Acosta, C., Marrero-Dominguez, A., Gallo-Llobet, L. and Gonzalez-Rodriguez, A.M. (2018) Physiological response of selected avocados (*Persea americana*) subjected to NaCl and $NaHCO_3$ stress. *Scientia Horticulturae* 237, 81–88.

Anderson, D.B. (1936) Relative humidity of vapor pressure deficit. *Ecology* 17, 277–282.

Anonymous (2022) Avocados area harvested 2022. Food and Agriculture Organization Statistics Database. FAOSTAT. Available at: http://faostat.fao.org/site/567/default.aspx#ancor (accessed 15 October 2024).

Argaman, E. (1983) Effect of temperature and pollen source on fertilization, fruit-set and abscission in avocado (*Persea americana* Mill.). MSc. thesis, The Hebrew University of Jerusalem, Rehovot, Israel, p. 143.

Ashworth, V.E.T. and Clegg, M.T. (2003) Microsatellite markers in avocado (*Persea americana* Mill.): genealogical relationships among cultivated avocado genotypes. *Journal of Heredity* 94, 407–415.

Assouline, S., Hochberg, U. and Silber, A. (2021) The impact of tree phenology on the response of irrigated avocado: the hysteretic nature of the maximum trunk daily shrinkage. *Agricultural Water Management* 256:107104.

Ayers, A.D., Aldrich, D.G. and Coony, J.J. (1951) Sodium and chloride injury of Fuerte leaves. *California Avocado Society Yearbook* 36, 174–178.

Backer, R., Engelbrecht, J. and van den Berg, N. (2022) Differing responses to *Phytophthora cinnamomi* infection in susceptible and partially resistant *Persea americana* (Mill.) rootstocks: a case for the role of receptor-like kinases and apoplastic proteases. *Frontiers in Plant Science* 13, 928176.

Baker, N.R. (2008) Chlorophyll fluorescence: a probe of photosynthesis *in vivo*. *Annual Review of Plant Biology* 59, 89–113.

Balikian, R. and Genskow, K. (2022) Losing aguacate: what if water costs kill avocado farming in San Diego County? *Case Studies in the Environment* 6:1.

Bangerth, F. (1979) Calcium-related physiological disorders of plants. *Annual Review of Phytopathology* 17, 97–122.

Bar, Y. and Kafkafi, U. (1992) Nitrate-induced iron-deficiency chlorosis in avocado (*Persea americana* Mill.) rootstocks and its prevention by chloride. *Journal of Plant Nutrition* 15, 1739–1746.

Bar, Y., Apelbaum, A., Kafkaki, U. and Goren, R. (1997) Relationship between chloride and nitrate and its effect on growth and mineral composition of avocado and citrus plants. *Journal of Plant Nutrition* 20, 715–733.

Barra, P.J., Inostroza, N.G., Mora, M.L., Crowley, D.E. and Jorquera, M.A. (2017) Bacterial consortia inoculation mitigates the water shortage and salt stress in an avocado (*Persea americana* Mill.) nursery. *Applied Soil Ecology* 111, 39–47.

Barrientos Pliego, A.F. and Rodríguez, J.L. (1998) Respuesta de plantas de aguacate cv. Hass bajo condiciones de sequía. *Revista Chapingo Serie Horticultura* 2, 191–198.

Bayabil, H.K., Li, Y., Crane, J.H., Schaffer, B., Smyth, A.R. *et al.* (2022) Saltwater intrusion and flooding: risks to south Florida's agriculture and potential management practices. EDIS Publication AE572, University of Florida, Gainesville, Florida, 7 pp.

Beadle, C.L., Ludlow, M.M. and Honeysett, J.L. (1985) Water relations. In: Coombs, J., Hall, D.O., Long, S.P. and Scurlock, J.M.O. (eds) *Techniques in Bioproductivity and Photosynthesis*. Pergamon Press, Oxford, UK, pp. 50–61.

Beier, G.L., Lund, C.D., Held, B.W., Ploetz, R.C., Konkol, J.L. and Blanchette, R.A. (2021) Variation in xylem characteristics of botanical races of *Persea americana* and their potential influence on susceptibility to the pathogen *Raffaelea lauricola*. *Tropical Plant Pathology* 46, 232–239.

Ben-Ya'acov, A. and Michelson, E. (1995) Avocado Rootstocks. *Horticultural Reviews* 17, 381–429.

Ben-Ya'acov, A., Michaelson, E. and Sela, I. (1979) *Field research for selection of avocado rootstocks and scion in commercial plantations: report of the first decade of research, 1968–1978*. The Division of Subtropical Horticulture, Agricultural Research Organization, Bet Dagan, Israel. (In Hebrew.)

Bergh, B.O. (1975) Avocado. In: Janick, J. and Moore, J.N. (eds) *Advances in Fruit Breeding*. Purdue University Press, West Lafayette, Indiana, pp. 541–567.

Bergh, B.O., Scora, R.W. and Storey, W.B. (1973) A comparison of leaf terpenes in *Persea* subgenus *Persea*. *Botanical Gazette* 134, 130–134.

Bernstein, N., Meiri, A. and Zilberstaine, M. (2001) Root growth of avocado is more sensitive to salinity than shoot growth. *Journal of the American Society for Horticultural Science* 129, 188–192.

Bernstein, N., Ioffe, M. and Zilberstaine, M. (2004) Salt-stress effects on avocado rootstock growth. I. Establishing criteria for determination of shoot growth sensitivity to the stress. *Plant and Soil* 233, 1–11.

Beyá-Marshall, V., Arcos, E., Seguel, O., Galleguillos, M. and Kremer, C. (2022) Optimal irrigation management for avocado (cv. 'Hass') trees by monitoring soil water content and plant water status. *Agricultural Water Management* 271, 107794.

Bingham, F.T. and Nelson, C.O. (1970) The effects of sodium on mature avocado trees. *California Avocado Society Yearbook* 54, 75–78.

Bingham, F.T., Fenn, L.B. and Oertli, J.J. (1968) Sandculture study of chloride toxicity to mature avocado trees. *Proceedings of the Soil Science Society of America* 32, 249–252.

Biran, D. (1979) Fruitlet abscission and spring growth retardation – their influence on avocado productivity. MSc thesis, Hebrew University of Jerusalem, Rehovot, Israel. (In Hebrew with English summary.)

Björkman, O. (1987) Low temperature chlorophyll fluorescence in leaves and its relationship to photon yield of photosynthesis photoinhibition. In: Kyle, D.J., Osmond, C.B. and Arntzen, C.J. (eds) *Topics in Photoinhibition*. Vol. 9. Elsevier Press, Amsterdam, pp. 123–144.

Blanke, M.M. (1991) Photosynthesis of avocado fruit. In: Lovatt, C.J., Holthe, P.A. and Arpaia, M.L. (eds) *Proceedings of the Second World Avocado Congress*. Vol. 1. University of California, Riverside, California, pp. 179–190.

Blanke, M.M. and Bower, J.P. (1990) Surface features of the avocado fruit. *Tropical Agriculture* 67, 379–383.

Blanke, M.M. and Lovatt, C.J. (1993) Anatomy and transpiration of the avocado inflorescence. *Annals of Botany* 71, 543–547.

Blanke, M.M. and Whiley, A.W. (1995) Bioenergetics, respiration cost and water relations of developing avocado fruit. *Journal of Plant Physiology* 145, 87–92.

Bonomelli, C., Celis, V., Lombardi, G. and Martiz, J. (2018) Salt stress effects on avocado (*Persea americana* Mill.) plants with and without seaweed extract (*Ascophyllum nodosum*) application. *Agronomy* 8:64.

Bower, J.P. (1978) The effects of shade and water relations in the avocado cv. Edranol. *South African Avocado Growers' Association Research Report* 2, 59–61.

Bower, J.P. (1985) Some aspects of water relations on avocado *Persea americana* (Mill.) tree and fruit physiology. PhD thesis, Department of Horticultural Science, University of Natal, Pietermaritzburg, South Africa.

Bower, J.P., Wolstenholme, B.N. and de Jager, J.M. (1978) Incoming solar radiation and internal water status as stress factors in avocado, *Persea americana* Mill. cv. Edranol. *Crop Production* 7, 129–133.

Bower, J.P., Cutting, J.G.M. and Wolstenholme, B.N. (1989) Effect of pre- and post-harvest water stress on the potential for fruit quality defects in avocado. *South African Journal of Plant and Soil* 6, 219–222.

Buchholz, A. (1986) Young vegetative growth as a possible factor involved with fruitlet abscission in avocado. MSc thesis, Hebrew University of Jerusalem, Rehovot, Israel. (In Hebrew with English summary.)

Cahill, D.M., Weste, G.M. and Grant, B.R. (1986) Changes in cytokinin concentrations in xylem extrudate following infection of *Eucalyptus marginata* Donn ex Sm with *Phytophthora cinnamomi* Rands. *Plant Physiology* 81, 1103–1109.

Campbell, C.W., Knight, R.J. and Zareski, N.L. (1977) Freeze damage to tropical fruits in southern Florida in 1977. *Proceedings of the Florida State Horticultural Society* 90, 254–257.

Cardoso, A.A., Batz, T.A. and McAdam, S.A.M. (2020) Xylem embolism resistance determines leaf mortality during drought in *Persea americana*. *Plant Physiology* 182, 547–554.

Carlquist, S. (2001) *Comparative Wood Anatomy*. Springer, Berlin, Germany.

Castillo-Argaez, R., Schaffer, B., Ploetz, R.C., Vargas, A.I., Konkol, J., Fernandez, R. and Vazquez, A. (2018) Susceptibility to laurel wilt related to physiology of three clonal avocado rootstocks. *HortScience* 53:S58.

Castillo-Argaez, R., Schaffer, B., Vazquez, A. and Sternberg, L.D. (2020a) Leaf gas exchange and stable carbon isotope composition of redbay and avocado trees in response to laurel wilt or drought stress. *Environmental and Experimental Botany* 171, 103948.

Castillo-Argaez, R., Vargas, A.I. and Schaffer, B. (2020b) Response of 'Hass' avocado trees on different clonal rootstocks to laurel wilt disease. *HortScience* 55:S320.

Castillo-Argaez, R., Konkol, J.L., Vargas, A.I., Ploetz, R.C. and Schaffer, B. (2021a) Disease severity and ecophysiology of rootstock/scion combinations of different avocado (*Persea americana* Mill.) genotypes in response to laurel wilt. *Scientia Horticulturae* 287, 110250.

Castillo-Argaez, R., Vazquez, A., Konkol, J.L., Vargas, A.I., Ploetz, R.C., Etxeberria, E. and Schaffer, B. (2021b) Sap flow, xylem anatomy and photosynthetic variables of three *Persea* species in response to laurel wilt. *Tree Physiology* 41(6), 1004–1018.

Castillo-Argaez, R., Vazquez, A., Konkol, J.L., Vargas, A.I., Ploetz, R.C. and Schaffer, B. (2022) Laurel wilt susceptibility of three avocado (*Persea americana* Mill.) ecotypes in relation to xylem anatomy, sap flow and leaf gas exchange. *Trees* 36, 1649–1663.

Castro, M., Iturrieta, R. and Fassio, C. (2009) Rootstock effect on the tolerance of avocado plants cv. Hass to NaCl stress. *Chilean Journal of Agricultural Research* 69, 316–324.

Celedón, J.M., Gil, P.M., Ferreyra, R., Maldonado, P. and Barrera, C. (2012) Sensitivity and variability of two plant water stress indicators: exploring criteria for choosing a plant monitoring method for avocado irrigation management. *Chilean Journal of Agricultural Research* 72, 379–387.

Celis, N., Suarez, D.L., Wu, L., Li, R., Arpaia, M.L. and Mauk, P. (2018) Salt tolerance and growth of 13 avocado rootstocks related best to chloride uptake. *HortScience* 53(12), 1737–1747.

Chaplin, G.R. and Scott, K.J. (1980) Association of chilling injury susceptibility of stored avocados. *HortScience* 15, 514–515.

Chartzoulakis, K., Patakas, A., Kofidis, G., Bosabalidis, A. and Nastou, A. (2002) Water stress affects leaf anatomy, gas exchange, water relations and growth of two avocado cultivars. *Scientia Horticulturae* 95, 39–50.

Chernoivanov, S., Neuberger, I., Levy, S., Szenes, N. and Rubinovich, L. (2022) Covering young 'Reed' avocado trees with shading nets during winter alleviates cold stress and promotes vegetative growth. *European Journal of Horticultural Science* 87:1.

Chirachint, W. and Turner, D.W. (1988) Shade reduces the foliar symptoms of 'Fuerte' avocado affected by salt, without significantly changing the concentration of sodium and chloride in the leaves. *Scientia Horticulturae* 36, 1–15.

Cholet, C., Houle, D., Sylvain, J., Doyon, F. and Maheu, A. (2022) Climate Change increases the severity and duration of soil water stress in the temperate forest of eastern North America. *Frontiers in Forest and Global Change* 5:879382

Chung, S.W., Rho, H., Lim, C.K., Jeon, M.K., Kim, S., Jang, Y.J. and An, H.J. (2022) Photosynthetic response and oxidative activity of 'Hass' avocado cultivar treated with short-term low temperature. *Scientific Reports* 12:11593.

Cran, D.G. and Possingham, J.V. (1973) The fine structure of avocado plastids. *Annals of Botany* 37, 993–997.

Crane, J.H., Balerdi, C., Campbell, R. and Goldweber, S. (1994) Managing fruit orchards to minimize hurricane damage. *HortTechnology* 4, 21–27.

Crane, J.H., Balerdi, C.F., Lamberts, M., Hull, D. and Olczyk, T. (1997) Flood damage of agricultural crops in south Dade County as a result of Tropical Storm Gordon. *Proceedings of the Florida State Horticultural Society* 110, 152–155.

Crowley, D. (2008) Salinity management in avocado orchards. *California Avocado Society 2008 Yearbook* 91, 83–104.

Cull, B.W. (1986) A phenological cycling approach to tree crop productivity research. *Acta Horticulturae* 175, 151–156.

Cummings, K. and Schroeder, C.A. (1942) Anatomy of the avocado fruit. *California Avocado Society Yearbook* 1942, 56–64.

Cutting, J.G.M. and Bower, J.P. (1990) Spring vegetative flush removal: the effect on yield, size, fruit mineral composition and quality. *South African Avocado Growers' Association Yearbook* 13, 33–34.

Dahan, Y., Rosenfeld, R., Zadiranov, V. and Irihimovitch, V. (2010) A proposed conserved role for an avocado *fw2.2-like* gene as a negative regulator of fruit cell division. *Planta* 232, 663–676.

Davenport, T.L. (1982) Avocado growth and development. *Proceedings of the Florida State Horticultural Society* 95, 92–96.

Davison, E.M. (2011) How do *Phytophthora* spp. de Bary kill trees? *New Zealand Journal of Forestry Science* 41S, S25–S37.

de Beer, Z.W., Proctor, M., Wingfield, M.J., Marincowitz, S. and Wuong, T.A. (2022) Generic boundaries in the *Ophiostomatales* reconsidered and revised. *Studies in Mycology* 101, 57–120.

De Micco, V., Balzano, A., Wheeler, E.A. and Baas, P. (2016) Tyloses and gums: a review of structure, function and occurrence of vessel occlusions. *IAWA Journal* 37, 186–205.

Demmig, B. and Björkman, O. (1987) Comparison of the effect of excessive light on chlorophyll fluorescence (77k) and photon yield in leaves of higher plants. *Planta* 171, 171–184.

Dickson, R.E., Tomlinson, P.T. and Isebrands, J.G. (2000) Allocation of current photosynthate and changes in tissue dry weight within northern red oak seedlings: individual leaf and flush carbon contribution during episodic growth. *Canadian Journal of Forest Research* 30, 1296–1307.

Dixon, J., Elmsly, T.A., Dixon, E.M., Fields, F.P., Greenwood, A.C. and Smith, D.B. (2006) 'Hass' avocado tree phenology 2004–2006 in the Western Bay of Plenty. *New Zealand Avocado Growers' Association Annual Research Report* 6, 1–12.

Dixon, J., Elmsly, T.A., Fields, F.P., Smith, D.B., Mandemaker, A.J., Greenwood, A.C. and Pak, H.A. (2007) 'Hass' avocado tree phenology 2004–2006 in the Western Bay of Plenty. *New Zealand Avocado Growers' Association Annual Research Report* 7, 49–72.

Doupis, G., Kavroulakis, N., Psarras, G. and Papadakis, I.E. (2017) Growth, photosynthetic performance and antioxidative response of 'Hass' and 'Fuerte' avocado (*Persea americana* Mill.) plants grown under high soil moisture. *Photosynthetica* 55, 655–663.

Downton, W.J.S. (1978) Growth and flowering in salt-stressed avocado. *Australian Journal of Agricultural Research* 29, 523–534.

Downton, W.J.S., Grant, W.J.R. and Loveys, B.R. (1987) Carbon dioxide enrichment increases yield of Valencia orange. *Australian Journal of Plant Physiology* 14, 493–501.

Drake, B.G., Gonzàles-Meler, M.A. and Long, S.P. (1997) More efficient plants: a consequence of rising atmospheric CO_2. *Annual Review of Plant Physiology and Molecular Biology* 48, 609–639.

Dusenge, M.E., Duarte, A.G. and Way, D.A. (2019) Plant carbon metabolism and climate change: elevated CO_2 and temperature impacts on photosynthesis, photorespiration and respiration. *New Phytologist* 221(1), 32–49.

Eamus, D. (1993) Assimilation and stomatal conductance responses of red spruce to midwinter frosts and the constituent ions of acid mist. *Tree Physiology* 13, 145–155.

Eamus, D. and Murray, M.B. (1992) The impact of constituent ions of acid mist on assimilation and stomatal conductance of Norway spruce prior to and post midwinter freeze. *Environmental Pollution* 72, 23–44.

Eissenstat, D.M., Syvertsen, J.P., Dean, T.J., Johnson, J.D. and Yelenosky, G. (1991a) Interaction of simulated acid rain with ozone on freeze resistance, growth, and mineral nutrition in citrus and avocado. *Journal of the American Society for Horticultural Science* 116, 838–845.

Eissenstat, D.M., Syvertsen, J.P., Dean, T.J., Yelenosky, G. and Johnson, J.D. (1991b) Sensitivity of frost resistance and growth in citrus and avocado to chronic ozone exposure. *New Phytologist* 118, 139–146.

Embleton, T.W., Matsumura, M., Storey, W.B. and Garber, J.J. (1961) Chloride and avocado rootstocks. *California Avocado Society Yearbook* 45, 110–115.

Esteban, R., Jiménez, E.T., Jiménez, M.S., Morales, D., Hormaetxe, K., Becerril, J.M. and García-Plazaola, J.I. (2007) Dynamics of violaxanthin and lutein epoxide xanthophyll cycles in Lauraceae tree species under field conditions. *Tree Physiology* 27, 1407–1414.

Esteban, R., Jiménez, M.S., Morales, D., Jiménez, E.T., Hormaetxe, K., Becerril, J.M., Osmond, B. and García-Plazaola, J.I. (2008) Short- and long-term modulation of the lutein epoxide and violaxanthin cycles in two species of the Lauraceae: sweet bay laurel (*Laurus noblis* L.) and avocado (*Persea americana* Mill.). *Plant Biology* 10, 288–297.

Ewers, F.W. and Fisher, J.B. (1989) Variation in vessel length and diameter in stems of six tropical and subtropical lianas. *American Journal of Botany* 76, 1452–1459.

FAO (2008) *The State of Food Insecurity in the World*. Food and Agriculture Organization of the United Nations, Rome, Italy.

Farrow, B.J., Taylor, N. and Van den Berg, N. (2011) The physiological response of two tolerant *Persea americana* rootstocks to *Phytophthora cinnamomi* and flooding. In: *Proceedings of the VIII World Avocado Congress*. Cairns, Australia, p. 9.

Fenn, L.B., Bingham, F.T. and Oertli, J.J. (1968) On the mechanism of chloride toxicity. *California Avocado Society Yearbook* 52, 113–116.

Ferguson, I.B. (1984) Calcium in plant senescence and fruit ripening. *Plant Cell and Environment* 7, 477–489.

Fernández, J.E., Green, S.R., Caspari, H.W., Diaz-Espejo, A. and Cuevas, M.V. (2007) The use of sap flow measurements for scheduling irrigation in olive, apple, Asian pear and grapevines. *Plant and Soil* 305(1), 91–94.

Fernandes da Silva, D., Villa, F. and da Silva, G.J. (2023) Climate change implications on cultivation of avocado (*Persea americana* Mill.). In: Abul-Soad, A.A. and Al-Khayri, J.M. (eds) *Cultivation for Climate Change Resilience*. Vol. 1. CRC Press, Boca Raton, Florida, pp. 164–190.

Ferreyra, R. (2017) *Manejo de agua y suelo en aguacate y su influencia sobre la pudrición de raíces*. Memorias del V Congreso Latinoamericano del Aguacate. Ciudad Guzmán, México.

Ferreyra, R., Sellés, G., Maldonado, P., Celedón, J., Gil, P. and Torres, A. (2008) Soil air content effects on the water status of avocado trees. *Acta Horticulturae* 792, 291–296.

Ferreyra, R., Celedón, J., Gil, P.M., Maldonado, P., Barrera, C. and Sellés, G. (2009) Sensitivity and variability of stem water potential (SWP) and maximum daily trunk shrinkage as water stress indicators in avocado (*Persea americana* Mill.). In *Proceedings of the 8th Fruit, Nut and Vegetable Production Engineering Symposium*. Frutic Chile, Concepción, Chile, pp. 131–139.

Fick, A., Swart, V., Backer, R., Bombarely, A., Engelbrecht, J. and van den Berg, N. (2022) Partially resistant avocado rootstock Dusa® shows prolonged upregulation of nucleotide binding-leucine rich repeat genes in responses to *Phytophthora cinnamomi* infection. *Frontiers of Plant Science* 13, 793644. doi:10.3389/fpls.2022.793644.

Fincher, J., Cumming, J.R., Alscher, R.G., Rubin, G. and Weinstein, L. (1989) Long-term ozone exposure affects winter hardiness of red spruce (*Picea rubens* Sarg.) seedlings. *New Phytologist* 113, 85–96.

Fischer, I.H., Costa Sampaio, A., Marques de Almeida Bertani, R., Leite Gonçalves, B.H. and Feichtenberger, E. (2020) Aggressiveness of *Phytophthora cinnamomi* in avocado seedlings and effect of pathogen inoculum concentration and substrate flooding periods on root rot and development of the plants. *Revista Brasileira de Fruticultura* 42(6): e-352. doi: 10.1590/0100-29452020352.

Förster, B., Osmond, C.B. and Pogson, B.J. (2009) *De novo* synthesis and degradation of Lx and V cycle pigments during shade and sun acclimation in avocado leaves. *Plant Physiology* 149, 1179–1195.

Freeland, R.O. (1948) Photosynthesis in relation to stomatal frequency and distribution. *Plant Physiology* 23, 595–600.

Gafni, E. (1984) Effect of extreme temperature regimes and different pollinizers on the fertilization and fruit set processes in avocado. MSc thesis, Hebrew University of Jerusalem, Rehovot, Israel. (In Hebrew.)

Gang, N., St-Pierre, L.S. and Persinger, M.A. (2012) Water dynamics following treatment by one hour 0.16 T static magnetic fields depend on exposure volume. *Water* 3, 122–131.

Gardiazábal, F. (2004) Factores agronómicos a considerar en la implantación de un huerto de paltos. 2° *Seminario Internacional de Paltos*. Sociedad Gardiazabal y Magdahl Ltda., Quillota, Chile.

Gardiazábal, F. (2008) Palto y cítricos. Generalidades del cultivo. In: Ripa, R. and Larral, P. (eds) *Manejo de Plagas en Paltos y Cítricos*. Instituto de Investigaciones Agropecuarias, INIA, Ministerio de Agricultura, Centro Regional de Investigación La Cruz, La Cruz, Chile, p. 399.

Gardiazábal, F. and Mena, F. (2011) The avocado industry in Chile and its evolution. *California Avocado Society Yearbook* 94, 52–68.

Gil, P.M. (2008) Study of avocado (*Persea americana* Mill.) physiological responses to different soil water-to-air ratios and the transmission of root-to-shoot electrical signals in response of soil water deficit and root hypoxia. PhD thesis, Facultad de Agronomía e Ingeniería Forestal, Pontificia Universidad Católica de Chile, Santiago, Chile.

Gil, P.M., Schaffer, B., Gutiérrez, S.M. and Li, C. (2007) Effect of waterlogging on plant water status, leaf gas exchange and biomass of avocado. In: *Proceedings of the Sixth World Avocado Congress*. Ediciones Universitarias de la Universidad Catolica de Valparaiso, Valparaiso, Chile. http://faostat.fao.org/site/567/default.aspx#ancor (accessed 15 October 2024).

Gil, P.M., Gurovich, L., Schaffer, B., García, N. and Iturriaga, R. (2009) Electrical signaling, stomatal conductance, ABA and ethylene content in avocado trees in response to root hypoxia. *Plant Signaling and Behavior* 4, 100–108.

Gil, P.M., Bonomelli, C., Schaffer, B., Ferreyra, R. and Gentina, R. (2012) Effect of soil water-to-air ratio on biomass and mineral nutrition of avocado trees. *Journal of Soil Science and Plant Nutrition* 12, 609–630.

Grafton, R.Q. and Hussy, K. (2011) *Water Resources Planning and Management*. Cambridge University Press, Cambridge, UK.

Granja, F. and Covarrubias, J.I. (2018) Evaluation of acidifying nitrogen fertilizers in avocado trees with iron deficiency symptoms. *Journal of Soil Science and Plant Nutrition* 18, 157–172.

Gregoriou, C. (1982) Some aspects of shoot and root growth of avocado under lowland tropical conditions. *California Avocado Society Yearbook* 66, 127–144.

Gregoriou, C. and Kumar, D.R. (1984) Effects of mulching on shoot and root growth of avocado (*Persea americana* Mill.) and mango (*Mangifera indica* L.). *Journal of Horticultural Science* 59, 109–117.

Grunennvaldt, R. (2022) Avocado soil health literature review. Avocados Australia – Best Practice Resources. Available at: https://avocado.org.au/wp-content/uploads/2022/11/Soil-health-literature-review_NOV-22.pdf (accessed 15 October 2024).

Grüter, R., Trachsel, T., Laube, P. and Jaisli, I. (2022) Expected global suitability of coffee, cashew and avocado due to climate change. *PLOS One* 17:e0261976.

Guillermo, M.-O., Adela, Z., Antonio, M., Olivier, N.A., van den Berg, N. *et al.* (2021) Physiological and molecular responses of 'Dusa' avocado rootstock to water stress: insights for drought adaptation. *Plants* 10:2077.

Ha, S., Vankova, R., Yamaguchi-Shinozaki, K., Shinozaki, K. and Tran, L.S.P. (2012) Cytokinins: metabolism and function in plant adaptation to environmental stresses. *Trends in Plant Science* 17, 172–179.

Haas, A.R.C. (1929) Composition of avocado trees in relation to chlorosis and tip burn. *Botanical Gazette* 37, 422–430.

Haas, A.R.C. (1939) Growth and transpiration in avocado seedlings as affected by artificial winds of low intensity. *California Avocado Society Yearbook* 24, 92–96.

Haas, A.R.C. (1950) Effect of sodium chloride on Mexican, Guatemalan and West Indian avocado seedlings. *California Avocado Society Yearbook* 35, 153–160.

Havaux, M. and Lannoye, R. (1987) Reversible effects of moderately elevated temperature on the distribution of excitation energy between the two photosystems of photosynthesis in intact avocado leaves. *Photosynthesis Research* 14, 147–158.

Havaux, M., Canaani, O. and Malkin, S. (1987) Inhibition of photosynthetic activities under slow water stress measured *in vivo* by the photoacoustic method. *Physiologia Plantarum* 70, 503–510.

Heath, R.L., Arpaia, M.L. and Mickelbart, M.V. (2007a) Assimilation productivity from canopy to fruit as determined by avocado tree physiology. 2007 California Avocado Commission Production Research Report, California Avocado Commision, p. 17. Available at: https://www.californiaavocadogrowers.com/sites/default/files/Assimilation-Productivity-from-Canopy-to-Fruit-as-Determined-by-Avocado-Tree-Physiology-2007.pdf (accessed 15 October 2024).

Heath, R.L., Mickelbart, M.V., Arpaia, M.L., Fassio, C. and Miller, R. (2007b) Water relations of avocado in response to changes in relative humidity. *HortScience* 41:1058.

Hermoso, J.M., Soria, J.T., Cortés, S. and Farré, J.M. (2003) Aplicación intermitente de agua sobre el follaje del aguacate Hass. In: *Proceedings of the Fifth World Avocado Congress*. Consejería de Agricultura y Pesca, Malaga, Spain, pp. 295–300.

Hermosilla-Madrid, C. and Whiley, A.W. (2023) High-density (HD) avocado production in New Zealand. In: *Proceedings of the Tenth World Avocado Congress*. Auckland, New Zealand. Available at: https://industry.nzavocado.co.nz/wp-content/uploads/2023/04/1415-Claudia-Hermosilla-.pdf (accessed 25 September 2024).

Honing, Y. and Lavee, Z. (1989) About irrigation of avocado. *Water and Irrigation* 270, 17–18.

Hsiao, T.C., Acevedo, E., Fereres, E. and Henderson, D.W. (1976) Stress metabolism. Water stress, growth and osmotic adjustment. *Philosophical Transactions of the Royal Society of London, Series B* 273, 479–500.

Huang, H., Ullah, F., Zhou, D.X., Yi, M. and Zhao, Y. (2019) Mechanisms of ROS regulation of plant development and stress responses. *Frontiers in Plant Science* 10:800.

Idso, K.E. and Idso, S.B. (1994) Plant responses to atmospheric CO_2 enrichment in the face of environmental constraints: a review of the past 10 years' research. *Agriculture and Forest Meteorology* 69, 153–203.

Idso, K.E., Kimball, B.A. and Allen, S.G. (1991a) CO_2 enrichment of sour orange trees: 2.5 years into a long-term experiment. *Plant Cell and Environment* 14, 351–353.

Idso, K.E., Kimball, B.A. and Allen, S.G. (1991b) Net photosynthesis of sour orange trees: 2.5 years into a long-term experiment. *Agriculture and Forest Meteorology* 54, 95–101.

Inch, S.A. and Ploetz, R.C. (2012) Impact of laurel wilt, caused by *Raffaelea lauricola,* on xylem function in avocado, *Persea americana*. *Forest Pathology* 42, 239–245.

Inch, S., Ploetz, R., Held, B. and Blanchette, R. (2012) Histological and anatomical responses in avocado, *Persea americana*, induced by the vascular wilt pathogen, *Raffaelea lauricola*. *Botany* 90, 627–635.

IPCC (2022) Climate change 2022: impacts, adaptation, and vulnerability. In: Pörtner, H.-O., Roberts, D.C., Tignor, M., Poloczanska, E.S., Mintenbeck, K. *et al.* (eds) *Contribution of Working Group II to the Sixth Assessment Report of the Intergovernmental Panel on Climate Change*. Cambridge University Press, Cambridge, UK, p. 3056. doi:10.1017/9781009325844.

Issarakraisila, M. and Considine, J.A. (1994) Effects of temperature on microsporogenesis and pollen viability in mango cv. Kensington. *Annals of Botany* 73, 231–240.

Jackson, J.E. (1980) Light interception and utilization by orchard systems. *Horticultural Reviews* 2, 208–287.

Joshi, N.C., Yadav, D., Ratner, K., Kamara, I., Aviv-Sharon, E., Irihimovitch, V. and Charuvi, D. (2020) Sodium hydrosulfide priming improves the response of photosynthesis to overnight frost and day high light in avocado (*Persea americana* Mill, cv. 'Hass'). *Physiologia Plantarum* 168, 394–405.

Kadman, A. (1963) The uptake and accumulation of chloride in avocado leaves and the tolerance of avocado seedlings under saline conditions. *Proceedings of the American Society for Horticultural Science* 83, 280–286.

Kadman, A. (1964) The uptake and accumulation of sodium in avocado seedlings. *Proceedings of the American Society for Horticultural Science* 85, 179–182.

Kadman, A. and Ben-Ya'acov, A. (1976) Selection of avocado rootstocks for saline conditions. *Acta Horticulturae* 57, 189–197.

Kalmar, D. and Lahav, E. (1977) Water requirements of avocado in Israel. I. Tree and soil parameters. *Australian Journal of Agricultural Research* 28, 869–877.

Khondoker, M., Mandel, S., Gurav, R. and Hwang, S. (2023) Freshwater shortage, salinity increase, and global food production: a need for sustainable irrigation water desalination – a scoping review. *Earth* 4, 223–240.

Kiggundu, N., Migliaccio, K.W., Schaffer, B. and Crane, J.H. (2012) Water savings, nutrient leaching, and fruit yield in a young avocado orchard as affected by irrigation and nutrient management. *Irrigation Science* 30, 275–286.

Kimelmann, R. (1979) The influence of different temperatures and light intensities on gas exchange of avocado leaves (*Persea americana* Mill.). MSc. thesis, Hebrew University of Jerusalem, Rehovot, Israel. (In Hebrew with English summary.)

Koch, K.E. (1996) Carbohydrate-modulated gene expression in plants. *Annual Review of Plant Physiology and Molecular Biology* 47, 509–540.

Kofidis, G., Bosabalidis, A.M. and Chartzoulakis, K. (2004) Leaf anatomical alterations induced by drought stress in two avocado cultivars. *Journal of Biological Research* 1, 115–120.

Kondo, T. and Honsho, C. (2018) Effect of shading on low-temperature damage in 'Bacon' avocado (*Persea americana*). *Tropical Agriculture and Development* 62, 132–135.

Koning, N.B., Van Ittersum, M.K., Becx, G.A., Van Boekel, M.A.J.S., Brandenburg, W.A. *et al.* (2008) Long-term global availability of food: continued abundance or new scarcity? *Wageningen Journal of Life Sciences* 55, 229–292.

Kourgialas, N.N. and Dokou, Z. (2021) Water management and salinity adaptation approaches of avocado trees: a review for hot-summer Mediterranean climate. *Agricultural Water Management* 252, 106923.

Krause, G.H. (1988) Photoinhibition of photosynthesis. An evaluation of damaging and protective mechanisms. *Physiologia Plantarum* 74, 566–574.

Krezdorn, A.H. (1970) Evaluation of cold-hardy avocados in Florida. *Proceedings of the Florida State Horticultural Society* 83, 382–386.

Labanauskas, C.K., Stolzy, L.H., Zentmyer, G.A. and Szuskiewicz, T.E. (1968) Influence of soil oxygen and soil water on the accumulation of nutrients in avocado seedlings (*Persea americana* Mill.). *Plant and Soil* 29, 391–406.

Lahav, E. and Kalmar, D. (1977) Water requirements of avocado in Israel. II. Influence on yield, fruit growth and oil content. *Australian Journal of Agricultural Research* 28, 869–877.

Lahav, E. and Trochoulias, T. (1982) The effect of temperature on growth and dry matter production of avocado plants. *Australian Journal of Agricultural Research* 33, 549–558.

Larson, K.D., Davies, F.S. and Schaffer, B. (1991a) Floodwater temperature and stem lenticel hypertrophy of *Mangifera indica* L. *American Journal of Botany* 78, 1397–1403.

Larson, K.D., Graetz, D.A. and Schaffer, B. (1991b) Flood-induced chemical transformations in calcareous agricultural soils of South Florida. *Soil Science* 152, 33–40.

Larson, K.D., Schaffer, B. and Davies, F.S. (1993) Floodwater oxygen content, lenticel hypertrophy, and ethylene evolution from mango trees. *Journal of Experimental Botany* 44, 665–671.

Lazare, S., Cohen, Y., Goldshtein, E., Yermiyahu, U., Ben-Gal, A. and Dag, A. (2021a) Rootstock-dependent response of Hass avocado to salt stress. *Plants* 10:1672.

Lazare, S., Yasuor, H., Yermiyahu, U., Kuhalskaya, A., Brotman, Y., Ben-Gal., A. and Dag, A. (2021b) It takes two: reciprocal scion rootstock relationships enable salt tolerance in 'Hass' avocado. *Plant Science* 312:111048.

Lazare, S., Vitoshkin, H., Alchanatis, V., Reshef, G., Ziv, D., Simenski, E. and Dag, A. (2022) Canopy-cooling systems applied on avocado trees to mitigate heatwave damages. *Scientific Reports* 12:12563.

Lemus, G., Ferreyra, R., Gil, P., Maldonado, P., Toledo, C., Barrera, C. and Celedón, J. (2005) *El Cultivo del Palto*. INIA Handbook No. 129. Instituto de Investigaciones Agropecuaria, La Cruz, Chile, 75 pp.

Levin, A. (1981) Factors affecting inflorescence and vegetative development regulation in avocado. MSc. thesis, Hebrew University of Jerusalem, Rehovot, Israel.

Levinson, B. and Adato, I. (1991) Influence of reduced rates of water and fertilizer application using daily intermittent drip irrigation on the water requirements, root development and responses of avocado trees (cv. Fuerte). *Journal of Horticultural Science* 66, 449–463.

Likens, G.E., Johnson, N.M., Galloway, J.N. and Bormann, F.H. (1976) Acid precipitation: strong and weak acids. *Science* 194, 643–645.

Lin, S.Y., Chen, P.A. and Zhuang, B.W. (2022) The stomatal conductance and Fv/Fm as indicators of stress tolerance of avocado seedlings under short-term waterlogging. *Agronomy* 12:1084.

Liu, X., Robinson, P.W., Madore, M.A., Witney, G.W. and Arpaia, M.L. (1999a) 'Hass' avocado carbohydrate fluctuations. I. Growth and phenology. *Journal of the American Society for Horticultural Science* 124, 671–675.

Liu, X., Robinson, P.W., Madore, M.A., Witney, G.W. and Arpaia, M.L. (1999b) 'Hass' avocado carbohydrate fluctuations. II. Fruit growth and ripening. *Journal of the American Society for Horticultural Science* 124, 676–681.

Liu, X., Mickelbart, M.V., Robinson, P.W., Hofshi, R. and Arpaia, M.L. (2002a) Photosynthetic characteristics of avocado leaves. *Acta Horticulturae* 575, 865–874.

Liu, X., Sievert, J., Arpaia, M.L. and Madore, M.A. (2002b) Postulated roles of seven-carbon sugars, mannoheptulose, and perseitol in avocado. *Journal of the American Society for Horticultural Science* 127, 108–114

Lomas, J. (1988) An agrometeorological model for assessing the effect of heat stress during the flowering and early fruit set on avocado yields. *Journal of the American Society for Horticultural Science* 113, 172–176.

Lomas, J. (1992) Analysis of the effect of heat stress during flowering on the yield of avocado under Mediterranean climatic conditions. *Agricultural and Forest Meteorology* 59, 207–216.

Lomas, J. and Mandel, M. (1973) The quantitative effects of two methods of sprinkler irrigation on the microclimate of a mature avocado plantation. *Journal of Agricultural Meteorology* 12, 35–48.

Lomas, J. and Zamet, D. (1994) Long-term analysis and modelling of agroclimatic effects on national avocado yields in Israel. *Agricultural and Forest Meteorology* 71, 315–336.

Mahato, A., Upadhyay, S. and Sharma, D. (2022) Climate change and its impact on agriculture. *Plant Archives* 22(1), 64–69.

Mansfield, T.A., Hetherington, A.M. and Atkinson, C.J. (1990) Some current aspects of stomatal physiology. *Annual Review of Plant Physiology and Plant Molecular Biology* 41, 55–75.

Marler, T.E. and Zozor, Y. (1996) Osmotic potential and photosynthetic characteristics of 'Waldin' avocado seedlings as influenced by mild salinity. In: *Proceedings of the International Conference on Tropical Fruits*. Malaysian Agricultural Research and Development Institute (MARDI), Kuala Lumpur, Malaysia, pp. 89–95.

Martinez-Ferri, E., Moreno-Ortega, G. and Prieto, C.P. (2019) Mild water stress-induced priming enhance tolerance to *Rosellinia necatrix* in susceptible avocado rootstocks. *BMC Plant Biology* 19:458.

Mass, E.V. (1994) Tolerance and responses of tropical crops to salinity stress. In: Davenport, T.L. and Harrington, H.M. (eds) *Plant Stress in the Tropical Environment*. Proceedings of a meeting held at Kailua-Kona, Hawaii. U.S. Department of Agriculture, Cooperative States Research Service.

Matsubara, S., Naumann, M., Martin, R., Nichol, C., Rascher, U. *et al.* (2004) Slowly reversible de-epoxidation of lutein-epoxide in deep shade leaves of a tropical tree legume may 'lock-in' lutein-based photoprotection during acclimation to strong light. *Journal of Experimental Botany* 56, 461–468.

McGee, T., Shahid, M.A., Beckman, T.G., Chaparro, J.X., Schaffer, B. and Sarkhosh, A. (2021) Physiological and biochemical characterization of six *Prunus* rootstocks in response to flooding. *Environmental and Experimental Botany* 183, 104368.

McKellar, M.A., Buchanan, D.W., Ingram, D.L. and Campbell, C.W. (1992) Freezing tolerance of avocado leaves. *HortScience* 27, 341–343.

Mickelbart, M.V. and Arpaia, M.L. (1995) Effects of salinity on growth and photosynthesis of 'Hass' avocado on three rootstocks. *1995 California Avocado Research Symposium Proceedings* 127, 649–655.

Mickelbart, M.V. and Arpaia, M.L. (2002) Rootstock influences changes in ion concentrations, growth, and photosynthesis of 'Hass' avocado in response to salinity. *Journal of the American Society for Horticultural Science* 127, 649–655.

Mickelbart, M.V., Miller, R., Parry, S., Arpaia, M.L. and Heath, R. (2000) Avocado leaf surface morphology. *California Avocado Society Yearbook* 84, 139–150.

Mickelbart, M.V., Bender, G.S., Witney, G.W., Adams, C. and Arpaia, M.L. (2007a) Effects of clonal rootstocks on 'Hass' avocado yield components, alternate bearing, and nutrition. *Journal of Horticultural Science and Biotechnology* 82, 460–466.

Mickelbart, M.V., Melser, S. and Arpaia, M.L. (2007b) Salinity-induced changes in ion concentrations of 'Hass' avocado trees on three rootstocks. *Journal of Plant Nutrition* 127, 649–655.

Moore-Gordon, C.S., Wolstenholme, B.N. and Levin, J. (1996) Effect of composted pinebark mulching on *Persea americana* Mill. cv. Hass fruit growth and yield in a cool subtropical environment. *Journal of the South African Society for Horticultural Science* 6, 23–26.

Musyimi, D.M., Netondo, G.W. and Ouma, G. (2007) Effects of salinity on gas exchange and nutrient uptake in avocados. *Journal of Biological Sciences* 7, 496–505.

Musyimi, D.M., Netondo, G.W. and Ouma, G. (2008) Salinity and temperature effects on CO_2 assimilation in leaves of avocados. *American Journal of Plant Physiology* 3, 40–49.

Mwelase, S., Mditshwa, A., Magwaza, L.S., Tesfay, S.Z. and Mashilo, J. (2022) Relationship between photosynthetic efficiency and maturity of 'Lamb Hass' avocado fruit. *Acta Horticulturae* 1349, 85–94.

Nadeem, S.M., Shaharoona, B, Arshad, M. and Crowley, D.E. (2012) Population density and functional diversity of plant growth promoting rhizobacteria associated with avocado trees in saline soils. *Applied Soil Ecology* 62, 147–154.

Navia-Urrutia, M., Sánchez-Pinzón, L., Parra, P.P. and Gazis, R. (2022) A diagnostic guide for laurel wilt disease in avocado. *Plant Health Progress* 23, 345–354.

Neuhaus, A., Turner, D.W., Colmer, T.D. and Blight, A. (2009) Drying half of the root-zone from mid-fruit growth to maturity in 'Hass' avocado (*Persea americana* Mill.) trees for one season reduced fruit production in two years. *Scientia Horticulturae* 120, 437–442.

Nicholson, R.L. and Hammerschmidt, R. (1992) Phenolic compounds and their role in disease resistance. *Annual Review of Phytopathology* 30, 369–389.

NOAA (2022) Trends in carbon dioxide (CO_2). United States Department of Commerce, National Oceanic and Atmospheric Administration, Earth System Research Laboratory, Global Monitoring Division. Available at: http://www.esrl.noaa.gov/gmd/ccgg/trends/ (accessed 26 April 2023).

Nuñez-Elisea, R., Schaffer, B., Fisher, J., Colls, A.M. and Crane, J.H. (1999) Influence of flooding on net CO_2 assimilation, growth, and stem anatomy of *Annona* species. *Annals of Botany* 74, 771–780.

Olmstead, M.A., Lang, N.S., Lang, G.A., Ewers, F.W. and Owens, S.A. (2006) Examining the vascular pathway of sweet cherries grafted onto dwarfing rootstocks. *HortScience* 41, 674–679

Olszyk, D.M., Bergh, B.O., Kats, G. and Whitsell, R.H. (1987) Relative sensitivity of avocado varieties to photochemical smog (ozone) and sulfur dioxide. *California Avocado Society Yearbook* 71, 235–240.

Ormrod, D.P. (1986) Gaseous air pollution and horticultural crop production. *Horticultural Reviews* 8, 1–42.

Ormrod, D.P. and Hale, B.A. (1994) Physiological responses of plants and crops to ozone stress. In: Pessarakli, M. (ed.) *Handbook of Plant and Crop Physiology*. Marcel Dekker, New York, pp. 753–760.

Oster, J.D. and Arpaia, M.L. (1992) Effects of salinity and applied water on yields of 'Hass' avocado on Mexican seedling rootstock. In: *Proceedings of the 2nd World Avocado Congress*. University of California, Riverside, Riverside, California, pp. 209–214.

Oster, J.D., Stottlmyer, D.E. and Arpaia, M.L. (2007) Salinity and water effects on 'Hass' avocado yields. *Journal of the American Society for Horticultural Science* 132, 253–261.

Oßwald, W., Fleischmann, F., Rigling, D., Coelho, A.C., Cravador, A. *et al.* (2014) Strategies of attack and defence in woody plant–*Phytophthora* interactions. *Forest Pathology* 44, 169–190.

Patel, P.M., Wallace, A. and Mueller, R.T. (1975) Salt tolerance of Huntalas compared with other avocado rootstocks. *California Avocado Society Yearbook* 59, 78–85.

Paul, M.J. and Pellny, T.K. (2003) Carbon metabolite feedback regulation of leaf photosynthesis and development. *Journal of Experimental Botany* 54, 539–547.

Pearson, M. and Brooks, M.L. (1995) Influence of elevated CO_2 on growth and age-related changes in leaf gas exchange. *Journal of Experimental Botany* 46, 1651–1659.

Pegg, K.G., Forsberg, L.I. and Whiley, A.W. (1982) Avocado root rot. *Queensland Agricultural Journal* 108, 162–168.

Pérez-Donoso, A.G. (2007) *Xylella fastidiosa* infection and ethylene exposure result in xylem and water movement disruption in grapevine shoots. *Plant Physiology* 143, 1024–1036.

Phillips, D., Grant, B.R. and Weste, G. (1987) Histological changes in the roots of an avocado cultivar, Duke 7, infected with *Phytophthora cinnamomi. Phytopathology* 77, 691–698.

Phillips, D., Weste, G. and Hinch, J.M. (1991) Resistance to *Phytophthora cinnamomi* in callus tissue derived from three avocado cultivars. *Canadian Journal of Botany* 69, 2026–2032.

Ploetz, R.C. and Schaffer, B. (1987) Effects of flooding and Phytophthora root rot on photosynthetic characteristics of avocado. *Proceedings of the Florida State Horticultural Society* 100, 290–294.

Ploetz, R.C. and Schaffer, B. (1988) Damage thresholds for flooded and nonflooded avocado with Phytophthora root rot. *Phytopathology* 78:1544.

Ploetz, R.C. and Schaffer, B. (1989) Effects of flooding and Phytophthora root rot on net gas exchange and growth of avocado. *Phytopathology* 79, 204–208.

Ploetz, R.C. and Schaffer, B. (1991) Effects of flooding and Phytophthora root rot on net gas exchange of avocado in Dade County, Florida. In: *Proceedings of the Second World Avocado Congress,* Orange, California, pp. 111–117.

Ploetz, R.C., Ramos, J.L., Parrado, J.L., Schaffer, B. and Lara, S.P. (1989) Performance of clonal avocado rootstocks in Dade County, Florida. *Proceedings of the Florida State Horticultural Society* 102, 234–236.

Ploetz, R.C., Peña, J.E., Smith, J.A., Dreaden, T.J., Crane, J.H., Schubert, T. and Dixon, W. (2011) Laurel wilt, caused by *Raffaelea lauricola*, is confirmed in Miami-Dade County, center of Florida's commercial avocado production. *Plant Disease* 95, 1589–1589.

Ploetz, R.C., Ramos, J.L. and Parrado, J.L. (1993) Periodicity of shoot and root growth in grafted avocado. *Tropical Agriculture* 70, 248–251.

Ploetz, R.C., Schaffer, B., Vargas, A.I., Konkol, J.L., Salvatierra, J. and Wideman, R. (2015) Impact of laurel wilt, caused by *Raffaelea lauricola*, on leaf gas exchange and xylem sap flow in avocado, *Persea americana*. *Phytopathology* 105, 433–440.

Pockman, W.T. and Sperry, J.S. (1997) Freezing-induced xylem cavitation and the northern limit of *Larrea tridentata*. *Oecologia* 109, 19–27.

Pockman, W.T. and Sperry, J.S. (2000) Vulnerability to xylem cavitation and the distribution of Sonoran Desert vegetation. *American Journal of Botany* 87, 1287–1299.

Ponnamperuma, F.N. (1972) The chemistry of submerged soils. *Advances in Agronomy* 24, 29–96.

Poorter, H., van Berkel, Y., Baxter, R., Den Hertog, J., Dukstra, P. *et al.* (1997) The effect of elevated CO_2 on the chemical composition and construction costs of leaves of 27 C_3 species. *Plant Cell and Environment* 20, 472–482.

Pouzoulet, J., Pivovaroff, A.L., Santiago, L.S. and Rolshausen, P.E. (2014) Can vessel dimension explain tolerance toward fungal vascular wilt diseases in woody plants? Lessons from Dutch elm disease and esca disease in grapevine. *Frontiers in Plant Science* 5:253.

Pouzoulet, J., Scudiero, E., Schiavon, M. and Rolshausen, P.E. (2017) Xylem vessel diameter affects the compartmentalization of the vascular pathogen *Phaeomoniella chlamydospora* in grapevine. *Frontiers in Plant Science* 8:1442.

Powles, S.B. (1984) Photoinhibition of photosynthesis induced by visible light. *Annual Review of Plant Physiology* 35, 15–44.

Ramadasan, A. (1980) Gas exchange in the avocado leaves under water stress and recovery. *California Avocado Society Yearbook* 64, 147–151.

Ramírez-Gil, J.G., Cobos, M.E., Jiménez-García, D., Morales-Osorio, J.G. and Peterson, A.T. (2019) Current and potential future distributions of Hass avocados in the face of climate change across the Americas. *Crop Pasture Science* 70, 694–708.

Ramírez-Gil, J.G. and Morales-Osorio, J.G. (2020) Development and validation of severity scales of avocado wilt complex caused by *Phytophthora cinnamomi*, *Verticillium dahliae* and hypoxia-anoxia disorder and their physiological responses in avocado plants. *Agronomía Colombiana* 38, 85–100.

Raschke, K. (1975) Stomatal action. *Annual Review of Plant Physiology* 26, 309–340.

Reeksting, B.J., Taylor, N.J. and Van den Berg, N. (2014) Flooding and *Phytophthora cinnamomi*: effects on photosynthesis and chlorophyll fluorescence in shoots of non-grafted *Persea americana* (Mill.) rootstocks differing in tolerance to Phytophthora root rot. *South African Journal of Botany* 95, 40–53.

Reeksting, B.J., Olivier N.A. and van den Berg, N. (2016) Transcriptome responses of an ungrafted Phytophthora root rot tolerant avocado (*Persea americana*) rootstock to flooding and *Phytophthora cinnamomi*. *BMC Plant Biology* 16:205.

Reints, J., Dinar, A. and Crowley, D. (2020) Dealing with water scarcity and salinity: adoption of water efficient technologies and management practices by California avocado growers. *Sustainability* 12:3555.

Robinson, A., Lehmann, J., Barriopedio, D., Rahmstorf, S. and Coumou, D. (2021) Increasing heat and rainfall extremes now far outside the historical climate. *Climate and Atmospheric Science* 4(1), 45.

Robinson, J.C. (1993) Climate-induced problems of growing bananas in the subtropics. In: Robinson, J.C. (ed.) *Handbook of Banana Growing in South Africa*. Agricultural Research Council, Nelspruit, South Africa, pp. 11–17.

Robinson, P.W., Mickelbart, M.V., Liu, X., Adams, C., Witney, G. and Arpaia, M.L. (2002) Development of a phenological model of avocado tree growth in California. *Acta Horticulturae* 575, 859–864.

Rosenberg, G. (1987) *El Palto*. Pontificia Universidad Católica de Chile, Santiago, Chile, p. 200.

Salazar-García, S. and Larque-Saavedra, A. (1985) Effect of progressive soil salinity on the leaf water potential and stomatal conductance in avocado (*Persea americana* Mill.). *California Avocado Society Yearbook* 72, 101–104.

Salazar-García, S., Cortes, J.I., Alcaldo, S. and Zarate, G.P. (1987) Salinity injuries in 'Fuerte' avocado trees in Atlixco, Puebla. *Agrociencia* 68, 115–134.

Salazar-Garcia, S., Ibarra-Estrada, M.E. and Gonzalez-Valdivia, J. (2018) Phenology of 'Mendez' avocado in southern Jalisco, Mexico. *Agrosciencia* 52, 991–1003.

Sánchez-Pérez, J.D., Jaimes-Lara, M., Salgado-Garciglia, R. and López-Meza, J.E. (2009) Root extracts from Mexican avocado (*Persea americana* var. drymifolia) inhibit the mycelial growth of the oomycete *Phytophthora cinnamomi. European Journal of Plant Pathology* 124, 595–601.

Sanclemente, M.A., Schaffer, B., Gil, P.M., Davies, F.S. and Crane, J.H. (2013) Leaf removal before flooding influences recovery of avocado (*Persea americana* Mill.) trees from flooding stress. *Scientia Horticulturae* 150, 154–163.

Sanclemente, M.A., Schaffer, B., Gil, P.M., Vargas, A.I. and Davies, F.S. (2014) Pruning after flooding hastens recovery of flood-stressed avocado (*Persea americana* Mill.) trees. *Scientia Horticulturae* 169, 27–35.

Schaffer, B. and Andersen, P.C. (1994) Introduction. In: Schaffer, B. and Andersen, P.C. (eds) *Handbook of Environmental Physiology of Fruit Crops*. Vol. 2: *Subtropical and Tropical Crops*. CRC Press, Boca Raton, Florida, pp. 165–197.

Schaffer, B. and Ploetz, R.C. (1987) Effects of Phytophthora root rot and flooding on net gas exchange of potted avocado seedlings. *HortScience* 22:1141.

Schaffer, B. and Ploetz, R.C. (1989) Net gas exchange as a damage indicator for Phytophthora root rot of flooded and nonflooded avocado. *HortScience* 24, 653–655.

Schaffer, B., Ramos, L. and Lara, S.P. (1987) Effect of fruit removal on net gas exchange of avocado leaves. *HortScience* 22, 925–927.

Schaffer, B., Whiley, A.W. and Kholi, R.R. (1991) Effects of leaf age on gas exchange characteristics of avocado (*Persea americana* Mill.). *Scientia Horticulturae* 48, 21–28.

Schaffer, B., Andersen, P.C. and Ploetz, R.C. (1992) Responses of fruit crops to flooding. *Horticultural Reviews* 13, 257–313.

Schaffer, B., Whiley, A.W. and Crane, J.H. (1994) Mango. In: Schaffer, B. and Andersen, P.C. (eds) *Handbook of Environmental Physiology of Fruit Crops*. Vol. 2: *Subtropical and Tropical Crops*. CRC Press, Boca Raton, Florida, pp. 165–197.

Schaffer, B., Whiley, A.W., Searle, C. and Nissen, R. (1997) Leaf gas exchange, dry matter partitioning, and mineral element concentration in mango (*Mangifera indica* L.) as influenced by elevated atmospheric carbon dioxide and root restriction. *Journal of the American Society for Horticultural Science* 122, 849–855.

Schaffer, B., Whiley, A.W. and Searle, C. (1999) Atmospheric CO_2 enrichment, root restriction, photosynthesis, and dry-matter partitioning in subtropical and tropical crops. *HortScience* 34, 1033–1037.

Schaffer, B., Davies, F.S. and Crane, J.H. (2006) Responses of subtropical and tropical fruit trees to flooding in calcareous soil. *HortScience* 41, 549–555.

Schnell, R.J., Brown, J.S., Olano, C.T., Power, E.J., Krol, C.A., Kuhn, D.N. and Motamayor, J.C. (2003) Evaluation of avocado germplasm using microsatellite markers. *Journal of the American Society for Horticultural Science* 128, 881–889.

Scholefield, P.B., Walcott, J.J., Kriedemann, P.E. and Ramadasan, A. (1980) Some environmental effects on photosynthesis and water relations of avocado leaves. *California Avocado Society Yearbook* 64, 93–105.

Schroeder, C.A. and Wieland, P.A. (1956) Diurnal fluctuation in size in various parts of the avocado tree and fruit. *Proceedings of the American Society for Horticultural Science* 68, 253–258.

Schultze, E.D. (1989) Air pollution and forest decline in a spruce (*Picea abies*) forest. *Science* 244, 776–783.

Scora, R.W. and Bergh, B. (1990) The origins and taxonomy of avocado (*Persea americana* Mill.) Lauraceae. *Acta Horticulturae* 275, 387–394.

Scorza, R. and Wiltbank, W.J. (1975) Evaluation of avocado cold hardiness. *Proceedings of the Florida State Horticultural Society* 88, 496–499.

Scorza, R. and Wiltbank, W.J. (1976) Measurement of avocado cold hardiness. *HortScience* 11, 267–268.

Sedgley, M. and Grant, J.R. (1983) Effect of low temperatures during flowering on floral cycle and pollen tube growth in nine avocado cultivars. *Scientia Horticulturae* 18, 207–213.

Shankar, K. and Mehendale, H.M. (2014) Oxidative Stress. In: Wexler P. (ed.) *Encylopedia of Toxicology*. 3rd edn. Academic Press; Oxford, UK, pp. 735–737.

Shapira, O., Chernoivanov, S., Neuberger, I., Levy, S. and Rubinovich, L. (2021) Physiological characterization of young 'Hass' avocado plant leaves following exposure to high temperatures and low light intensity. *Plants* 10:1562.

Sharon, Y. (1999) Aspects of the water economy of Hass avocado trees (*Persea americana*, cv. Hass). I. Plant water status and gas exchange. *South African Avocado Growers' Association Yearbook* 22, 106–109.

Siddiqi, S.A., Al-Mulla, Y.A., McCann, I., AbuRumman, G, Makram Belhaj, M. *et al.* (2021) Smart monitoring, sap-flow, stem-psychrometer and soil-moisture measurements tools for precision irrigation and water saving of date palm. *International Journal of Agriculture and Biology* 26(5), 568–578.

Silber, A., Israeli, Y., Levi, M., Keinan, A., Shapira, O. *et al.* (2012) Response of 'Hass' avocado trees to irrigation management and root constraint. *Agricultural Water Management* 104, 95–103.

Silber, A., Israeli, Y., Levi, M., Keinan, A., Chudi, G. *et al.* (2013) The roles of fruit sink in the regulation of gas exchange and water uptake: a case of study for avocado. *Agricultural Water Management* 116, 21–28.

Singsaas, E.L., Ort, D.R. and DeLucia, E.H. (2001) Variation in measured values of photosynthetic quantum yield in ecophysiological studies. *Oecologia* 128, 15–23.

Smillie, R.M., Hetherington, S.E., He, J. and Nott, R. (1988) Photoinhibition at chilling temperatures. *Australian Journal of Plant Physiology* 15, 207–222.

Soares, N.B., Pedro Junior, M.J., Teixera, L.A.J. and de Castro, J.L. (2002) Tolerance to low temperature of avocado (*Persea americana*) cultivars. *Revista Brasileira de Fruticultura* 24, 721–723.

Solla, A. and Gil, L. (2002) Xylem vessel diameter as a factor in resistance of *Ulmus minor* to *Ophiostoma novo-ulmi*. *Forest Pathology* 32, 123–134.

Sperry, J.S., Hacke, U.G. and Pittermann, J. (2006) Size and function in conifer tracheids and angiosperm vessels. *American Journal of Botany* 93, 1490–1500.

Steinhardt, R., Kalmar, D., Miari, A. and Lahav, E. (1995) Salinity damage to avocado trees on Mexican rootstock. B. Damage to yield and its economic significance. *Alon Hanotea* 49, 374–382. (In Hebrew.)

Sterne, R.E., Kaufmann, M.R. and Zentmyer, G.A. (1977) Environmental effects on transpiration and leaf water potential in avocado. *Physiologia Plantarum* 41, 1–6.

Sterne, R.E., Kaufmann, M.R. and Zentmyer, G.A. (1978) Effect of Phytophthora root rot on water relations of avocado: interpretation with a water transport model. *Phytopathology* 68, 595–602.

Stitt, M. (1991) Rising CO_2 levels and their potential significance for carbon flow in photosynthetic cells. *Plant Cell and Environment* 14, 741–762.

Stolzy, L.H., Zentmyer, G.A., Klotz, L.J. and Labanauskas, C.K. (1967) Oxygen diffusion, water, and *Phytophthora cinnamomi* in root decay and nutrition of avocados. *Proceedings of the American Society for Horticultural Science* 90, 67–76.

Strzepek, K. and Boehlert, B. (2010) Competition for the food system. *Philosophical Transactions of the Royal Society* 365, 2927–2940.

Surendran, U.P., Sandeep, O. and Joseph, E.T. (2016) The impacts of magnetic treatment of irrigation water on plant, water and soil characteristics. *Agricultural Water Management* 178, 21–29.

Taiz, L. and Zeiger, E. (2010) *Plant Physiology*. 5th edn. Sinauer Associates, Sinauer, New York.

Takayama, K., King, D., Robinson, S.A. and Osmond, B. (2013) Integrating transient heterogeneity of non-photochemical quenching in shade-grown heterobaric leaves of avocado (*Persea americana* L.): responses to CO_2 concentration, stomatal occlusion, dehydration and relative humidity. *Plant & Cell Physiology* 54, 1852–1866.

Taylor, A.O. and Rowley, J.A. (1971) Plants under climatic stress. I. Low temperature, high light effects on photosynthesis. *Plant Physiology* 47, 713–718.

Taylor, A.O., Slack, C.R. and McPherson, H.G. (1974) Plants under climatic stress. VI. Chilling and light effects on photosynthetic enzymes of sorghum and maize. *Plant Physiology* 54, 696–701.

Taylor, O.C. (1958) Air pollution with relation to agronomic crops. IV. Plant growth suppressed by exposure of air-borne oxidants (smog). *Agronomy Journal* 50, 556–558.

Taylor, O.C., Cardiff, E.A., Mersereau, J.D. and Middleton, J.T. (1958) Effect of air-borne reaction products of ozone and 1-N-hexene vapour (synthetic smog) on growth of avocado seedlings. *Proceedings of the American Society for Horticultural Science* 71, 320–325.

Thorp, T.G., Anderson, P. and Camilleri, M. (1995) Avocado tree growth cycles – a quantitative model. In: *Proceedings of the Third World Avocado Congress*, Tel Aviv, Israel, pp. 76–79. Available at: http://www.avocadosource.com/temp/OLD%20WAC%20III/WAC3_p076.htm (accessed 15 October 2024).

Tomer, E. (1977) The effect of girdling on flowering, fruit setting and abscission in avocado trees. PhD thesis, The Hebrew University of Jerusalem, Israel.

Trivedi, P., Batista, B.D., Bazany, K.E. and Singh, B.K. (2022) Plant–microbiome interactions under a changing world: responses, consequences and perspectives. *New Phytologist* 234, 1951–1959.

Tyree, M.T. and Sperry, J.S. (1989) Vulnerability of xylem to cavitation and embolism. *Annual Review of Plant Biology* 40, 19–36.

Tyree, M.T., Kolb, K.J., Rood, S.B. and Patiño, S. (1994) Vulnerability to drought-induced cavitation of riparian cottonwoods in Alberta: a possible factor in the decline of the ecosystem? *Tree Physiology* 14, 455–466.

Tyree, M.T. and Zimmermann, M.H. (2002) Xylem Structure and the Ascent of Sap. 2nd ed. Springer, Berlin, Germany.

Tzatzani, T., Kavroulakis, N., Doupis, G., Psarras, G. and Papadakis, L. (2020) Nutritional status of 'Hass' and 'Fuerte' avocado (*Persea americana* Mill.) plants subjected to high soil moisture. *Journal of Plant Nutrition* 43, 327–334.

Van den Berg, N., Christie, J.B., Aveling, T.A.S. and Engelbrecht, J. (2018a) Callose and β-1, 3-glucanase inhibit *Phytophthora cinnamomi* in a resistant avocado rootstock. *Plant Pathology* 67, 1150–1160.

Van den Berg, N., Mahomed, W., Olivier, N.A., Swart, V. and Crampton, B.G. (2018b) Transcriptome analysis of an incompatible *Persea americana-Phytophthora cinnamomi* interaction reveals the involvement of SA-and JA-pathways in a successful defense response. *PLOS One* 13:e0205705.

Van den Berg, N., Swart, V., Backer, R., Fick, A., Wienk, R., Engelbrecht, J. and Prabhu, S.A. (2021) Advances in understanding defense mechanisms in *Persea americana* against *Phytophthora cinnamomi*. *Frontiers in Plant Science* 12, 636339.

Van Schaik, C.P., Terborgh, J.W. and Wright, S.J. (1993) The phenology of tropical forests. Adaptive significance and consequences for primary consumers. *Annual Review of Ecology and Systematics* 42, 353–377.

Venning, F.D. and Lincoln, F.B. (1958) Developmental morphology of the vegetative axis of avocado (*Persea americana* L.) and its significance to spacing, pruning practices, and yields of the grove. *Proceedings of the Florida State Horticultural Society* 71, 350–356.

Venturas, M., López, R., Martín, J.A., Gascó, A. and Gil, L. (2014) Heritability of *Ulmus minor* resistance to Dutch elm disease and its relationship to vessel size, but not to xylem vulnerability to drought. *Plant Pathology* 63, 500–509.

Verheij, E.W.M. (1968) Yield-density relationships in apple; results of a planting system experiment in Hungary. *Instititute of Tuinb. Technology. Wageningen* 37:31.

Vermeiren, T. (1958) Magnetic treatment of liquids for scale and corrosion prevention. *Anti-Corrosion Methods and Materials* 5(7), 215–219.

Vörösmarty, C.J., Green, P., Salisbury, J. and Lammers, R.B. (2000) Global water resources: vulnerability from climate change and population growth. *Science* 289, 284–288.

Webber, H.I. (1918) Cold resistance of the avocado. *California Avocado Association Annual Report* 1917, 49–50.

Weil, A., Rubinovich, L., Tchernov, D. and Liran, O. (2022) Comparative study between the photosynthetic parameters of two avocado (*Persea americana*) cultivars reveals natural variation in light reactions in response to frost stress. *Agronomy* 12:1129.

Westling, F., Underwood, J. and Bryson, M. (2021) A procedure for automated tree pruning suggestion using LiDAR scans of fruit trees. *Computers and Electronics in Agriculture* 187, 106274.

Whiley, A.W. (1990) CO_2 assimilation of developing fruiting shoots of cv. Hass avocado (*Persea americana* Mill.) – a preliminary report. *South African Avocado Growers' Association Yearbook* 13, 28–30.

Whiley, A.W. (1994) Ecophysiological studies and tree manipulation for maximisation of yield potential in avocado (*Persea americana* Mill.). PhD thesis, University of Natal, Pietermaritzburg, South Africa.

Whiley, A.W. and Schaffer, B. (1994) Avocado. In: Schaffer, B. and Andersen, P.C. (eds) *Handbook of Environmental Physiology of Fruit Crops*. Vol. 2: *Subtropical and Tropical Crops*. CRC Press, Boca Raton, Florida, pp. 165–197.

Whiley, A.W. and Winston, E.C. (1987) Effect of temperature at flowering on varietal productivity in some avocado growing areas in Australia. *South African Avocado Growers' Association Yearbook* 10, 45–47.

Whiley, A.W., Pegg, K.G., Saranah, J.B. and Forsberg, L.I. (1986) The control of Phytophthora root rot of avocado with fungicides and the effect of this disease on water relations, yield and ring-neck. *Australian Journal of Experimental Agriculture* 26, 249–253.

Whiley, A.W., Chapman, K.R. and Saranah, J.B. (1988a) Water loss by floral structures of avocado (*Persea americana* Mill.) cv. Fuerte during flowering. *Australian Journal of Agricultural Research* 39, 457–467.

Whiley, A.W., Saranah, J.B., Cull, B.W. and Pegg, K.G. (1988b) Manage avocado tree growth cycles for productivity gains. *Queensland Agriculture Journal* 114, 29–36.

Whiley, A.W., Wolstenholme, B.N., Saranah, J.B. and Anderson, P.A. (1990) Effect of root temperatures on growth of two avocado rootstock cultivars. *Acta Horticulturae* 275, 153–160.

Whiley, A.W., Saranah, J.B., Wolstenholme, B.N. and Rasmussen, T.S. (1991) Use of paclobutrazol sprays at mid-anthesis on fruit size and yield of avocado (*Persea americana* Mill. cv. Hass). *Journal of Horticultural Science* 66, 593–600.

Whiley, A.W., Schaffer, B. and Lara, S.P. (1992) Carbon dioxide exchange of developing avocado (*Persea americana* Mill.) fruit. *Tree Physiology* 11, 85–94.

Whiley, A.W., Rasmussen, T.S., Saranah, J.B. and Wolstenholme, B.N. (1996a) Delayed harvest effects on yield, fruit size and starch cycling in avocado (*Persea americana* Mill.) in two subtropical environments. I. The early-maturing cv. Fuerte. *Scientia Horticulturae* 66, 23–34.

Whiley, A.W., Rasmussen, T.S., Saranah, J.B. and Wolstenholme, B.N. (1996b) Delayed harvest effects on yield, fruit size and starch cycling in avocado (*Persea americana* Mill.) in two subtropical environments. II. The late-maturing cv. Hass. *Scientia Horticulturae* 66, 35–49.

Whiley, A.W., Searle, C., Schaffer, B. and Wolstenholme, B.N. (1999) Cool orchard temperatures or growing trees in containers can inhibit leaf gas exchange of avocado and mango. *Journal of the American Society for Horticultural Science* 124, 46–51.

Wiesman, Z. (1995) Rootstock and nitrate involvement in 'Ettinger' avocado response to chloride stress. *Scientia Horticulturae* 62, 33–43.

Wilkie, J.D., Conway, J., Griffin, J. and Toegel, H. (2019) Relationships between canopy size, light interception and productivity in conventional avocado planting systems. *Journal of Horticultural Science and Biotechnology* 94, 481–487.

Witney, G. and Arpaia, M.L. (1991) Tree recovery after the December 1990 freeze. *California Avocado Society Yearbook* 75, 63–70.

Wolstenholme, B.N. (1990) Resource allocation and vegetative-reproductive competition: opportunities for manipulation in evergreen fruit trees. *Acta Horticulturae* 275, 451–460.

Wolstenholme, B.N. (2001) Understanding the avocado tree – introductory ecophysiology. In: De Villiers, E.A. (ed.) *The Cultivation of Avocado*. Institute for Tropical and Subtropical Crops, Nelspruit, South Africa, pp. 45–61.

Wolstenholme, B.N and Whiley, A.W. (1999) Ecophysiology of the avocado (*Persea americana* Mill.) tree as a basis for pre-harvest management. *Revista Chapingo Serie Horticultura* 5, 77–88.

Wolstenholme, B.N., Whiley, A.W. and Saranah, J.B. (1990) Manipulating vegetative: reproductive growth in avocado (*Persea americana* Mill.) with paclobutrazol foliar sprays. *Scientia Horticulturae* 41, 315–327.

Yin, M.H., Gutierrez-Rodriguez, E.A., Vargas, A.I. and Schaffer, B. (2022) Chemical priming with brassinosteroids to mitigate responses of avocado (*Persea americana*) trees to flooding stress. *Horticulturae* 8:1115.

Yin, M.H., Vargas, A.I., Fuentealba, C., Shahid, M.A., Bassil, E. and Schaffer, B. (2023) Differences in physiological and biochemical responses to short-term flooding among the three avocado (*Persea americana* Mill.) races. *Plant Physiology and Biochemistry* 196, 925–939.

Yusof, I.M., Buchman, D.W. and Gerber, J.F. (1969) The response of avocado and mango to soil temperature. *Journal of the American Society for Horticultural Science* 94, 619–621.

Zlotopolski, V. (2017) The impact of magnetic water treatment on salt distribution in a large unsaturated soil column. *International Soil and Water Conservation Research* 5, 253–257.

Zuazo, V.H.D., Lipan, L., Rodriguez, B.C., Sendra, E., Tarifa, D.F. *et al.* (2021) Impact of deficit irrigation on fruit yield and lipid profile of terraced avocado orchards. *Agronomy for Sustainable Development* 41:69.

8 Cultivars and Rootstocks

J.H. Crane[1,*], P. Manosalva[2], B.A. Faber[3], A.F. Barrientos-Priego[4], E.D. Focht[5], M.L. Arpaia[5], G.S. Bender[6] and C.F. Balerdi[7]

[1]Tropical Research and Education Center, University of Florida, Homestead, USA; [2]Department of Microbiology and Plant Pathology, University of California, Riverside, USA; [3]University of California, Cooperative Extension Service, Ventura County, USA; [4]Departamento de Fitotecnia, Universidad Autónoma Chapingo, México; [5]Department of Botany and Plant Sciences, University of California, Riverside, USA; [6]University of California, Cooperative Extension Service, San Diego County, USA; [7]University of Florida/Miami-Dade County Cooperative Extension Service, Homestead, USA

Abstract

This chapter updates and expands the information and progress in selecting new scions and rootstocks for commercial, subtropical and tropical avocado production. The information on recently released and new scions is important, as the industry seeks to offer superior cultivars for specific markets, tolerance to adverse environmental conditions (e.g. salinity and flooding), greater resistance to foliar diseases and rootstock appropriate for high-density production systems. Much progress has been made in developing Phytophthora root rot disease (PRR)- and salinity-tolerant rootstocks, and many are described in this chapter. Tolerance to the lethal laurel wilt pathogen (*Harringtonia lauricola*) is the next major challenge facing avocado production worldwide. While significant progress has been made in understanding the pathogen, the ambrosia beetle vectors and the pathogen scion–rootstock interactions, much more in the areas of genomics and metabolomics of this plant patho-system is needed; that effort is only just beginning.

Avocado Selection and Dispersal

Avocado cultivar and selection programmes started during the mid-20th century whereas informal selection has occurred for thousands of years. Early cultivar and rootstock efforts were primarily searches for and/or introductions of germplasm and seeds of potentially beneficial seedlings and selections from Latin America to the USA (i.e. California and Florida). The benefits from breeding and selection programmes in California (USA), Israel, South Africa, México and others are apparent in new 'Hass'-like cultivars (e.g. 'Lamb Hass' and 'Maluma') and numerous rootstocks selected to address diseases (mainly PRR but also *Collectotrichum* spp.), poorly drained soils, soil and/or irrigation water salinity, and fruit production. Today, with the acceleration in cost-effective genetic mapping tools, the completion of several avocado genomes along with phenotypic data, and the use of metabolomics to bridge phenotypic, genetic and physiological data, the worldwide avocado industry can look forward to more efficient avocado cultivar and

*Correspondence: jhcr@ufl.edu

DOI: 10.1079/9781800621824.0008

rootstock breeding (Ben-Ya'acov and Michelson, 1995; Barrone *et al.*, 2009; Boza *et al.*, 2018; Ge *et al.*, 2019; Kuhn *et al.*, 2019a,b; Rendón-Anaya *et al.*, 2019; Reyes-Herrera *et al.*, 2020; Kämper *et al.*, 2021; Nath *et al.*, 2022; Beyer *et al.*, 2022; Tamayo-Ramos *et al.*, 2022; Anonymous, 2022; Núñez-Lillo *et al.*, 2024).

Prior to the arrival of humans in the Americas, wild ancestors of avocado were consumed and dispersed by Pleistocene megafauna (0.01–1.6 million years ago), such as the giant ground sloths (Megatheridae, Mylodontidae and Megalonychidae) and mastodon-like ancestors (gomphotheres), which roamed western North America and Mesoamerica (Janzen and Martin, 1982; Barlow, 2000; Galindo-Tovar and Bost, Chapter 2, this volume). These megafaunas may have been attracted to the anise aroma of this fruit, which had a high oil content. With the cooling climactic changes in North America at the close of the Pleistocene era, and the loss of most of the natural megafauna involved in seed dispersal of *Persea* species, the arrival of humans in the Americas (>15,000 years ago) became an important mechanism for the dispersal and ecotypic morphological and physiological differences of avocado adapted to diverse environments (Barlow, 2000; Galindo-Tovar *et al.*, 2008; Meltzer, 2021).

Archaeological evidence suggests that avocado was one of the first domesticated tree species in the neotropics (Smith, 1966, 1969; Galindo-Tovar *et al.*, 2008). Avocado seeds excavated and analysed from the Coxcatlán Cave in the Tehuacán-Cuicatlán Valley, Puebla, México, ranged in the time period from 7000 to 200 BC (Smith, 1966). Over this period of time, there was an increase in seed size (and, by implication, fruit size) and a change from seeds with concave to convex bases (and, by implication, fruit shape from round to more ovoid or ellipsoid), clearly demonstrating human selection for larger fruit size over time. Furthermore, the archaeological evidence of indigenous tree species throughout the valley indicates that the Tehuacán-Cuicatlán Valley had a xeric (dry) climate for at least 12,000 years (Smith, 1966). Avocado trees in the valley were generally restricted to areas of human habitat with sources of water nearby (e.g. streams). These findings suggest that some of these seeds originated from the barrancas (gorges) found in the Sierra de Zongolica on the eastern side of the Tehuacán-Cuicatlán Valley and were purposefully planted. Additionally, commerce of food, fibre and stone objects among the ancient indigenous peoples throughout México, Central America and northern South America has been well-documented (Gama-Campillo and Gomez-Pompa, 1992; Coe, 2005; Coe and Koontz, 2008; Galindo-Tovar *et al.*, 2008; Chen *et al.*, 2009; see Galindo-Tovar and Bost, Chapter 2, this volume for more details).

As primitive avocado germplasm was moved southward from México to Central America and eventually to Perú, human preferences and environmental adaptations to local habitats, via natural and human plant selection, probably occurred multiple times (Smith, 1966, 1969; Barrientos-Priego, 2000; Sahagún and Temprano, 2000; Ge *et al.*, 2009; Chen *et al.*, 2009). The evidence for a common ancestor, and the gradual divergence of the primitive avocado into three or more ecotypes come from the botanical, archaeological and linguistic evidence that avocado germplasm was taken to diverse environments (e.g. cool uplands and lowland tropical areas) first by megafauna and then by indigenous peoples (Gama-Campillo and Gomez-Pompa, 1992; Galindo-Tovar *et al.*, 2008). Archaeological evidence documents the change in fruit size and shape over time (Smith, 1966; Janzen and Martin, 1982; Storey *et al.*, 1986; Boza *et al.*, 2018). Early human intervention and selection included development of forestry practices that established and favoured specific perennial fruit species such as land clearing, deliberate planting of seeds from superior fruit in suitable or new habitats (e.g. near water sources and at lower elevation), near home sites and/or seasonal hunting and gathering routes, and the elimination of adjacent tree species that competed for light and/or water (Casas, 2001).

Avocado is a highly variable species, and this plasticity is evident by the three recognized ecotypes, the archaeological evidence of pre-Colombian selection from 5000 to 8000 BC (Smith, 1966, 1969; Galindo-Tovar *et al.*, 2008; Chen *et al.*, 2009; Berdugo-Cely *et al.*, 2023), the fact that all three ecotypes easily hybridize, and that trees of mixed-ecotype origin are found in native habitats and germplasm repositories in parts of México, Central America and northern South America (Popenoe and Williams, 1947;

Schnell *et al.*, 2003). Historically, three botanical subspecies or ecotypes (races) of *P. americana* Mill. were recognized: Mexican (M) (*P. americana* var. *drymifolia* Schlecht. Et Cham. Blake), Guatemalan (G) (*P. americana* var. *guatemalensis* L. Wms.) and West Indian (WI) (Antillean) (*P. americana* Mill. var. *americana*) (Popenoe, 1935; Bergh and Ellstrand *et al.*, 1986; Lavi *et al.*, 1991; Mhameed *et al.*, 1997; Schnell *et al.*, 2003; Alcaraz and Hormaza, 2007), all of which have the same somatic chromosome number (n = 24) and easily hybridize (Scora *et al.*, 2002). This classification into ecotypes was based on their distinct botanical, phenological and physiological characteristics and climatic and edaphic adaptations (Barrientos-Priego *et al.*, 1992; Ge *et al.*, 2019). However, more recent genetic analyses will undoubtedly result in taxonomic changes (Schnell *et al.*, 2003; Borrone *et al.*, 2007; Boza *et al.*, 2018; Ge *et al.*, 2019; Kuhn *et al.*, 2019a and 2019b; Berdugo-Cely *et al.*, 2023) (for taxonomic clarification, see the Taxonomy section of Chanderbali *et al.*, Chapter 3, this volume).

The avocado ecotypes are based on plant morphology, climatic tolerances and geographic location (Popenoe, 1935; Scora *et al.*, 2002; Chanderbali *et al.*, 2013; Pliego-Alfaro *et al.*, 2020). In this chapter, the ecotype identification of the cultivars and rootstocks will be based on the historical classification of the three ecotypes (races), but, where appropriate, cultivar identification will be based on recent genetic analysis (Boza *et al.*, 2018; Ge *et al.*, 2019; Kämper *et al.*, 2021). Typically, the major cultivars grown commercially are hybrids among the three ecotypes: 'Hass' (G × M), 'Lamb Hass' (G × M), 'Fuerte' (G × M), 'Semil 34' (G × WI) and 'Monroe' (G × WI); although this is not always the case (e.g. 'Simmonds' (WI), 'Ettinger' (M) and 'Reed' (G)).

The first recorded clonal propagation (grafting) of avocado was in 1901 by George Cellon in Florida (Fairchild, 1945). This opened the possibility of rapidly propagating superior selections and developing an avocado industry based on known cultivars with superior plant characteristics and fruit quality. Beginning in the early decades of the 1900s, M, G and WI avocado seeds and some plants from México and Central America were introduced into the USA (Florida and California) by government, university and private interests (Fairchild, 1945; Storey *et al.*, 1986; Campbell and Knight, 1998; Knight and Campbell, 1999). This diverse germplasm freely hybridized, resulting in the rapid selection of hybrids with unique and useful characteristics such as superior taste, early or very late fruit maturation, adaptation to local climatic and edaphic conditions, prolonged on-tree fruit storage capability, increased crop yields, prolonged postharvest storage and shelf life, and better pest and disease resistance. Eventually, some of these selected lines became named cultivars. Meanwhile, exploration for clonal and seedling material and selection of superior seedlings in California was ongoing since the early 1900s, resulting in the selection of 'Fuerte', 'Zutano' and 'Hass' (Platt, 1975, 1976; Bergh, 1992). Breeding and selection programmes in California, México, Israel, South Africa and Australia have resulted in some new cultivars and rootstocks which are grown commercially (e.g. Bergh and Martin, 1988; Ben-Ya'acov and Michelson, 1995; Ben-Ya'acov *et al.*, 1995b; Witney and Martin, 1995; Ben-Ya'acov *et al.*, 2003; Köhne, 2004; Kremer-Köhne and Köhne, 2001, 2003, 2004, 2007; Kremer-Köhne *et al.*, 2011; Castro *et al.*, 2015). However, cultivars derived from selected seedlings remain the dominant avocados grown commercially throughout the world. Rootstock selection for PRR resistance, on the other hand, has been mainly the result of formal breeding programmes. However, in the near future, tolerance and/or resistance to the laurel wilt (LW) pathogen (*H. lauricola*) may become a major scion and rootstock disease-tolerance (resistance) attribute sought (Inch *et al.*, 2012; Ploetz *et al.*, 2012; Crane *et al.*, 2015; Ploetz *et al.*, 2015; Everett and Seibert, 2018; Castillo-Argaez *et al.*, 2020, 2021, 2022). To date, no LW pathogen-tolerant scions or rootstocks have been detected (Table 8.1) (Crane *et al.*, 2015).

The cultivars and rootstocks described below are broken into two main ecological groups, those adapted to subtropical climates and those best adapted to tropical lowland and warm-to-hot subtropical climates. However, in many geographically tropical production areas, elevation allows the production of tropical types in the lowlands and subtropical types in the highlands. Some of the cultivars described may be of historical importance, some continue to appear in the commercial markets, and some

Table 8.1. Mature[z] and young container-grown avocado cultivars killed by the LW pathogen, *H. lauricola*.[y]

Mature orchard trees		Mature orchard trees[z]	
Cultivar name	Scion ecotype[x]	Cultivar name	Scion ecotype[x]
Arue	WI	Miguel	G-WI
Bernecker	WI	Monroe	G-WI
Day	WI	Nadir	G-WI
Donnie	WI	Nesbitt	G-WI
Dupuis	WI	Tonnage	G-WI
Hardee	WI	Tower-2	G-WI
Peterson	WI	Wheeling	G-WI
Pollock	WI	Brogdon	G-M-WI
Russell	WI	Marcus Pumpkin	G
Simmonds	WI	Winter Mexican	G-M
Waldin	WI	Toni	Nd
Beta	G-WI	Jim Lapeck	Nd
Booth 7	G-WI	Greenhouse inoculated trees[y]	
Booth 8	G-WI	Catalina	WI
Brooks Late	G-WI	Trapp	WI
Buck II	Nd	Hass[w]	G-M
Choquette	G-WI	Ettinger	G-M
Hall	G-WI	Pinkerton	G-M
Loretta	G-WI	Bacon	G
Lula	G-WI	Reed	G

z, Ecotype: WI=West Indian; G=Guatemalan; M=Mexican; Nd=not determined.
y, Young container-grown trees artificially inoculated with *H. lauricola*.
x, Trees inoculated via natural means, i.e. ambrosia beetle infestation and/or root–graft transmission among adjacent avocado trees and on seedling 'Waldin' or 'Lula' rootstock.
w, 7-year-old 'Hass' on 'Waldin' seedling rootstock.

may be used as pollinizers and/or windbreaks, though their fruit may not be widely sold. Cultivar and rootstock descriptions and attributes are in alphabetical order to facilitate the reader's ability to locate the information upon future reference.

Cultivars and Rootstocks of Subtropical Climates

The cultivars primarily grown in subtropical climates are derived from G and M ecotypes and possess the physiological attributes for these areas, primarily adaptations to cooler temperatures compared with the WI ecotype cultivars. Exploration of México and Central America by horticulturists and the collection and importation of clonal and seedling germplasm into production areas outside of their native habitat (e.g. the USA (California and Florida), Israel, México, South America and Asia) have led to the selection of numerous cultivars and rootstocks. Most of the international avocado trade in the western hemisphere and Europe is based on selections made in California during the 20th century. There are currently several new cultivars that have been released or are currently being evaluated worldwide (Kremer-Köhne and Köhne, 2001; Kremer-Köhne and Mokgalabone, 2003; Arpaia, 2004; Köhne, 2004; Kremer-Köhne and Köhne, 2001, 2004, 2007; Bruwer, 2006; Bruwer and Mokgalabone, 2006; Kruger *et al.*, 2020). Some of these are from México ('Hass Carmen'), South Africa ('Grace' and 'Maluma'), the USA ('Lamb Hass', 'GEM', 'Harvest' and 'Sir-Prize' among others) and Australia ('Llanos Hass' and 'Turner Hass'). The Israeli breeding programme has released 'Iriet', 'Eden', 'Galil', 'Arad' and 'Lavi' (Lahav *et al.*, 2005; Regev *et al.*, 2005). Most of these cultivars have traits in common with 'Hass', primarily peel colour, but some of their other properties are quite different, and they may have superior performance to 'Hass' in some production areas. Further experience and evaluation in different environments and with varying commercial practices are needed to determine how well they will eventually perform.

Cultivars

'Bacon' (California)

A G × M hybrid that is seldom planted today. It originated in 1928 in Buena Park, California. The trees are tall and upright with a pointed crown; the leaves emit an anise aroma when crushed; red flecking is observed on the wood of new shoots; flowers belong to Group B. The fruit is ovate, medium to large in size, weighing 170–510 g; the peel is thin, green and glossy with a leathery texture; the seed is large. The fruit is early-maturing with very pale yellow-green pulp and about 61% pulp recovery (Fig. 8.1). A precocious cultivar with consistent production and higher yields compared to 'Fuerte' in colder areas; cold tolerance is widely reported from California, Australia, Italy and Corsica; frost tolerance is down to −4.4°C; thus, production is suited to colder regions. Susceptible to insect attack, extremely susceptible to anthracnose, unsuitable for humid subtropical areas; the peel is susceptible to wind scarring in severe cases, especially with overmature fruit splitting to expose the seed. 'Bacon' is a successful pollinizer of 'Hass'. The fruit has poor storage life and is susceptible to internal pulp discolouration (Pehrson, 1957; Vogel, 1971; Platt, 1975, 1976; Alexander, 1978; Sarooshi *et al.*, 1979; Banks, 1980; Alexander, 1983a; Chalker, 1985; Calabrese *et al.*, 1998; Gardiazabal and Gandolfo, 1998; Ge *et al.*, 2019).

Fig. 8.1. 'Bacon'.

'Edranol' (California)

A G or a G hybrid that originated in 1930 in Vista, California, from an open-pollinated 'Lyon' seedling. The growth habit is upright and vigorous; flower Group B. The fruit is pyriform in shape, necky, medium to large, weighing 255–500 g; the peel is dark green, of medium thickness, corky texture, medium gloss; seed size is small to medium. The fruit matures mid-season with a buttery-yellow pulp, about 77% pulp recovery, a good nutty flavour and good pulp quality. Trees are slow to begin production and have an irregular but heavy production habit. It is sensitive to climatic extremes and withstands temperatures as low as −2°C. The fruit drops from the tree when mature in warm weather. It is susceptible to zinc deficiency, grey pulp, peel russeting and anthracnose. 'Edranol' is a successful pollinizer of 'Hass' and is widely used as a pollinizer in Chile for this purpose. It has been used as a 'nurse' seedling for clonal propagation in South Africa (see Whiley *et al.*, Chapter 9, this volume; Alexander, 1978; Banks, 1980; Rowell and Durand, 1982; Piccone and Whiley, 1986a; Durand, 1990; Whiley and Wolstenholme, 1995; Gardiazabal and Gandolfo, 1998; Blight *et al.*, 2000).

'Ettinger' (Israel)

A G × M hybrid selected from a seedling of 'Fuerte' in Kefar Malal, Israel, in 1947. The growth is upright with a strong central leader. Leaves emit a weak anise aroma when crushed; flowering Group B; the fruit is pyriform in shape, medium to large, weighing 170–570 g; the peel is bright green, very thin with a slightly rough surface; the seed is large, sometimes sits loose in the cavity, the seed coat adheres to the pulp; early-maturing with a clear light cream to yellow-coloured pulp, soft and melting texture; pulp recovery is similar to that of 'Fuerte'; fruit has a short on-tree life (the peel cracks) and does not store as well as 'Fuerte' or 'Hass'. This cultivar is more sensitive to chilling injury during storage than 'Hass' or 'Fuerte'. It has moderate-to-heavy production and is the earliest-maturing export cultivar in Israel, harvested well before 'Fuerte'. The tree is more freeze-resistant than 'Fuerte'. Severe problems can occur with peel cracking, anthracnose and the fruit spotting bug in Queensland, Australia, and stem-end rot in South Africa. 'Ettinger' is an outstanding pollinizer, consistently increasing 'Hass' yields in Israel. In 2018, it accounted for 18% of avocado production in Israel, much less than 'Hass' and 'Hass'-like production (>50%) (Storey and Bergh, 1963; Vakis, 1982; Degani *et al.*, 1989; Durand, 1990; Blumenfeld and Koussevitzky, 1995; Newett and Whiley, 1996; UCR, 2012a; Boza *et al.*, 2018; Imbert, 2018; Ge *et al.*, 2019).

'Eugenin' (Chile)

The 'Eugenin' is a 'Hass'-like variety that originated in the commercial orchard of Javier Eugenin in Chile in 1997. It is also known as 'Andes 3'. The 'Eugenin' is internationally protected by U.S. Patent (Anonymous, 2015) and Plant Breeders' Rights (PBR) and is exclusively managed by Avotopia LLC, a company based in California, USA. Recent genetic analysis suggests that 'Eugenin' may be a somatic mutant of 'Hass' (Solares, Focht, Arpaia, personal communication). The tree growth habit and appearance are very similar to 'Hass'. It has similar susceptibility to *Persea* mite like 'Hass'. The fruit peel colouration, texture and overall appearance is like that of 'Hass' except that the 'Eugenin' fruit is consistently larger; this is the primary distinguishing difference between 'Hass' and 'Eugenin'. The fruit size of 'Eugenin' is 278 g compared to 240 g for 'Hass' in a 4-year comparative study. The fruit maturity is similar to that of 'Hass', although data show that flowering may be slightly earlier than 'Hass' (Arpaia, personal communication).

'Flavia' (Chile)

The 'Flavia' is a 'Hass'-like variety that originated in a planting of Alfredo Shiappacasse Macchiavello in Chile in 1998. It is also known as 'Andes 4' or 'Shiappacasse'. The 'Flavia' is internationally protected by U.S. Patent (Anonymous, 2016) and PBR and is exclusively managed by Avotopia LLC, a company based in California, USA. Recent genetic analysis suggests that 'Flavia' may be a somatic mutant of 'Hass' (Solares, Focht, Arpaia, personal communication). The tree growth habit and appearance are very similar to 'Hass'. It has similar susceptibility to *Persea* mite (*Oligonychus perseae*) as 'Hass'. The primary distinguishing differences between 'Hass' and 'Flavia' are fruit size and fruit maturation. The fruit of 'Flavia' is typically larger than that of 'Hass', with an average fruit size (4-year average) of 291 g compared to 240 g for 'Hass' and may have a slightly more pronounced neck under some conditions. 'Flavia' fruit reach maturity 4–5 weeks earlier than 'Hass'.

'Fuerte' (México)

A G × M hybrid collected by Carl Schmidt in 1911 from Atlixco, Puebla, México, from the home of Alejandro Le Blanc. It survived the extremely cold weather in California in January 1913 and consequently was named 'Fuerte', Spanish for 'strong' or 'hardy'. Several strains (variants) of 'Fuerte' exist. A vigorous cultivar; trees are large with spreading crowns; leaves emit an anise aroma when crushed, red flecking is observed on the wood of new shoots; flower Group B. The fruit is pyriform in shape, with a distinct neck but variable, ranging from elongated with a long narrow neck, to dumpy with a short broad neck, medium-to-large size, weighing 170–500 g; the peel is thin, green, and has a medium gloss, supple leathery texture with a pimpled surface (Fig. 8.2). Seed size is medium to large, conical with a pointed apex. The fruit is early-maturing with a pale-yellow pulp, with 75%–77% pulp recovery, is of excellent quality with some nutty aftertaste; good on-tree storage but a short shelf life when ripe. 'Fuerte' is slow to reach production, has low yield in cooler climates, a marked tendency for erratic cropping, and sensitivity to low temperatures (the frost tolerance limit is −2.8°C) during flowering and fruit set (Gazit and Degani, 2002), all of which are reasons for the gradual loss of its popularity and replacement of this cultivar by 'Hass' in California. The fruit is susceptible to anthracnose, stem-end rot and insect attack, which can cause severe pre- and postharvest losses (especially in summer-rainfall climates). The tree has tolerance to *Persea* mite. Long-term storage is limited by the development of internal physiological disorders. When cultivated under low soil calcium availability, it may develop a disorder termed ‘pulp spot’. Formerly the world’s leading commercial cultivar and the first-generation standard of excellence, it retains its importance in low-humidity environments, where disease and insect pressure is low. ‘Fuerte’ accounts for about 20% of South Africa avocado production and less than 3% of Mexican production and has been largely replaced by ‘Hass’ and ‘Hass’-like cultivars. ‘Fuerte’ comprises 3% of the crop in Israel and very small percentages in Australia, México, California and New Zealand (Hodgson, 1945; Kellogg, 1971; Platt, 1975; Alexander, 1978; Biran, 1979; Banks, 1980; Hodgkin, 1980; Chalker, 1983; Muirhead, 1983; Ledger and Barker, 1995a; du Plessis *et al.*, 1995; Hoddle and Kerguelen, 1999; Imbert, 2019; Anonymous, 2000a; Osoyo, 2020; Ge *et al.*, 2019).

‘GEM’ (California)

‘GEM’ is a ‘Hass’-like cultivar from the University of California breeding programme, with better fruit quality and later fruit maturity compared to ‘Hass’. Its optimum season in California, where it originated, is spring through late summer. The trees are compact and thus suited to higher planting densities compared to ‘Hass’. The cultivar has a U.S. Plant Patent (issued in 2003, under the name ‘3-29-5’; Anonymous, 2003b) and PBR, and, from 2011, is managed under licence by Westfalia Estate in South Africa (van Rooyen, 2011; Anonymous, 2012b). The PBR is active in several countries including México (until 2029). The tree is somewhat vigorous and vase-like, similar to its maternal parent, ‘Gwen’, with an upright growth habit and is very precocious, often flowering and setting fruit within 1 year of planting. Generally, fruit is borne inside the canopy. ‘GEM’ has an A-type flowering habit. Trees are less alternate-bearing compared to ‘Hass’. The fruit is larger than ‘Hass’ (average, 235 g) and ellipsoid-shaped without shoulders and has a slightly pebbled, black peel (Fig. 8.3). The buttery pulp is yellow-green in colour with a nutty flavour. The seed is comparable to ‘Hass’ or slightly larger and broadly ovate-shaped. Fruit yields are similar to or greater than those of ‘Hass’, with a significantly reduced alternate-bearing tendency (Anonymous, 2012b). The variety has a postharvest storage potential similar to ‘Hass’, with 28 to 40 days possible. ‘GEM’ accounted for only 0.93% of the avocado volume in California (CAC, 2023) but is gaining in popularity with California growers due to its horticultural characteristics, in particular in the San Joaquin Valley, where it is believed to be slightly more cold-/heat-tolerant than ‘Hass’ and bears fruit of good size. It accounts for approximately 10% of the new plantings in South Africa (Imbert, 2019).

Fig. 8.2. ‘Fuerte’.

Fig. 8.3. ‘GEM’.

'Gwen' (California)

'Gwen' is an open-pollinated seedling of 'Thille' released in 1982 by B.O. Bergh at the University of California, Riverside, and patented in 1984 (Clegg *et al.*, 2007; Chen *et al.*, 2009). 'Thille' is an open-pollinated seedling of 'Hass' (G × M hybrid). 'Gwen' is an M × G hybrid with mostly G fruit characteristics. The trees possess an umbrella-like rounded canopy; flower Group A; fruit are oval-shaped, with a green peel, slightly smaller than 'Hass' (100–250 g) with a rich, nutty flavour. The peel is moderately thick with a texture more finely pebbled than 'Hass', dull green when ripe; seed size is medium to small (Fig. 8.4). When compared to 'Hass', 'Gwen' is a very precocious, higher-yielding, smaller, less vigorous tree under California conditions. It tends to be less prone to alternate-bearing than 'Hass'. The fruit set is improved when 'Gwen' is interplanted with cultivars in flower Group B. The fruit harvest season is slightly later than it is for 'Hass'. Fruit production in young trees can be so high that trees may need to be sprayed with whitewash to prevent sunburn of the fruit, but once the trees are established, the variety hides its fruit within its dense canopy. 'Gwen' is about as cold-hardy as 'Hass', down to −3.6°C. The tree is very susceptible to damage by *Persea* mite, which may cause extensive leaf drop (Bergh and Martin, 1988; Hoddle and Kerguelen, 1999; Anonymous, 2011e). Due to the current market preference for black-skin fruit like 'Hass', the 'Gwen' is not grown to a large extent in California (B. Faber, UC, *personal communication*).

Fig. 8.4. 'Gwen'.

'Hass' (California)

At present, 'Hass' is overwhelmingly the most important avocado cultivar in the world, having replaced 'Fuerte' as the 'standard of excellence' in subtropical areas by the 1970s. It accounts for about 90% of the export trade. 'Hass' is grown primarily in cool subtropical regions and at higher elevations in some tropical areas (e.g. Dominican Republic and Colombia). It is a G × M hybrid from a chance seedling of unknown parentage selected in the early 1920s by Mr Rudolph Hass at La Habra Heights, California, for its better pulp quality, higher yield and later maturity than 'Fuerte'. The cultivar was patented in 1935. It is a medium-to-large tree, with a semi-upright growth habit, almost as broad as it is tall with a rounded crown, flower Group A. The fruit is ovate, although fruit shape is dependent on the growing conditions (warm growing conditions result in a more pyriform fruit), small to medium in size, weighing 140–400 g (average 250–350 g); the peel is medium to thick and leathery with a coarse, corky texture and a rough and pebbled surface; pebbling is largely absent when grown at high altitudes (e.g. Mexican highlands). The fruit is dark green on the tree, although it may become purplish black on the tree at the end of its harvest season (Fig. 8.5). The seed is medium-sized and round. The fruit mature from mid to late season; have a creamy yellow pulp; with 66%–70% pulp recovery; are of excellent quality with a rich and nutty flavour; with typical oil content of 18%–20% and good on-tree storage ability through to late summer of the second growing season in cooler climates, for example California and New Zealand, but may become mealy and rancid if harvested too late. 'Hass' is precocious and produces regular, heavy crops but late on-tree storage of fruit will accentuate alternate-bearing (see Hormaza *et al.*, Chapter 6 and Whiley *et al.*, Chapter 12, this volume). Flowering and fruit set are less sensitive to cold temperatures than 'Fuerte' and 'Ettinger', but 'Hass' leaves are more cold-sensitive (damaged below −1.1°C). The medium-to-thick fruit peel provides some

Fig. 8.5. 'Hass'.

tolerance to pests and diseases, but leaves are susceptible to *Persea* mite and fruit to avocado thrips (*Scirtothrips perseae*). This cultivar tends to bear large numbers of phenotypically undersized fruit (<200 g), especially in warmer, more stressful climates, and the percentage of small fruit increases as trees age or health deteriorates. Trees grown in cooler climates produce larger fruit. The tree is often slow to produce a new leaf cover in spring, when a large percentage of determinate flower shoots occur, resulting in sunburnt limbs and fruit. Late-summer pruning often removes fruiting wood for the following spring, resulting in poor flowering and fruit set. The variety has excellent postharvest storage potential when properly handled and is successfully shipped internationally to distant markets. Under controlled-atmosphere storage, the fruit can be successfully shipped and marketed up to 30 to 42 days after harvest (M.L. Arpaia, UC-Riverside, *personal communication*). These features contribute to the popularity of 'Hass' when compared with most other cultivars, along with the change in peel colour from green to black, which provides an easy index for ripeness and masks minor rind imperfections. During the 2020–2021 season, 'Hass' accounted for 96.7% of the California avocado volume and 70.5% of the Australian volume (Avocados Australia, 2023; CAC, 2023). 'Hass' accounted for 97% of the Mexican avocado production (Osoyo, 2020). In the last 50 years, 'Hass' has become the most dominant cultivar grown in subtropical climates in 2023 and probably accounted for 80% of the worldwide avocado production (Anonymous, 1974; Platt, 1975; Alexander, 1978; Banks, 1980; Alexander, 1983a; Chalker, 1983; Whiley, 1983; Bergh and Ellstrand, 1986; Piccone and Whiley, 1986a; Ish-Am and Eisikowitch, 1989; Witney *et al.*, 1990; Köhne and Schutte, 1991; Whiley, 1991; Blumenfeld and Koussevitzky, 1995; Kremer-Köhne and Köhne, 1995; Story *et al.*, 1996; Hoddle and Kerguelen, 1999; Handwerk, 2017; Ge *et al.*, 2019; Rendón-Anaya *et al.*, 2019; Anonymous, 2021a). The high regard in which 'Hass' is held in subtropical growing areas and by consumers is reflected in the increasing number of 'Hass'-like selections that have been released and are described elsewhere in this chapter. Some of these include 'Hass Carmen', 'GEM', 'Harvest', 'Lamb Hass', 'Lavi', 'Luna University of California-Riverside' and 'Maluma'. These are either chance seedlings or mutations found in 'Hass' orchards, with some characteristic(s) that confer an economic advantage and are regarded as an improvement on 'Hass' deficiencies or involve 'Hass' as a parent in conventional breeding programmes.

'Hass Carmen' (México)

This variety, called 'Hass Carmen' or 'Mendez Hass', is a 'Hass' look-alike, is harvested earlier than 'Hass', with good, consistent bearing and a medium-sized (200–285 g) fruit with postharvest quality comparable to that of 'Hass'. It was selected in México by Carlos Mendez Vega and was first known as 'Mendez 1'. 'Hass Carmen' is internationally protected by U.S. Patent and PBR and is exclusively managed by Westfalia Estate in sub-Saharan Africa and the Middle East (Illsley-Granich *et al.*, 2011; van Rooyen, 2011). Recent genetic analysis suggests that 'Hass Carmen' may be a somatic mutant of 'Hass' (Kämper *et al.*, 2021). The fruit is very 'Hass'-like, but the tree differs in terms of the occurrence of a heavy 'flor loca' or off-bloom in the late summer and fall. This results in a second fruit set in both warm and cool subtropical conditions. The growth habit of the tree differs from 'Hass' in that the summer/fall vegetative wood is stiff and largely non-flexible and this growth can appear to be 'candelabra'-like, since the apical meristem may be inhibited, and growth is dominated by growth of subtending proleptic buds. Depending on the environment in which it is grown, the tree may bloom 1–7 months earlier than 'Hass', and fruit reach physiological maturity in 9–10 months (earlier than 'Hass') (Anonymous, 2012a; Illsley-Granich *et al.*, 2011). The tree is moderately vigorous and more cold-tolerant than 'Hass'. Phenotypically, the fruit is slightly smaller than 'Hass', but fruit quality is very similar to that of 'Hass' (Fig. 8.6). Since 2015–2016, 'Hass Carmen' has made up about 15% of the new 'Hass'-like plantings in South Africa (Imbert, 2019), and in México is the second most cultivated avocado; however, in the country production statistics, it is considered as 'Hass' (A.F. Barrientos-Priego, Universidad Autónoma Chapingo, México, *personal communication*).

'Harvest' (California)

The 'Harvest' is an open-pollinated seedling of 'Gwen' selected from the University of California breeding programme with larger fruit and greater productivity compared to 'Hass' (Anonymous, 2012c; Kremer-Köhne and Mokgalabone, 2004), and the U.S. Patent expired in 2022 (Anonymous, 2003a). The PBR is active with Australia until 2036. It has also been known as 'N4(-)5'. The tree has an open and spreading growth habit and is about as vigorous as 'Hass'. The tree is more open than 'Hass', and this can lead to some sunburning of the fruit. 'Harvest' has a Group A flowering habit. The trees bear fruit on the inside of the canopy and have an alternate-bearing habit similar to 'Hass'. The fruit are highly spheroid-shaped, with an average weight of 204 g; the peel is slightly pebbled and generally black when ripe; the pulp is yellow-green and of good quality when fully ripe (Fig. 8.7) (Bruwer, 2006). Postharvest quality is moderate, and the storage life, without controlled atmosphere storage, is limited to about 21 days (M.L. Arpaia, UC-Riverside, *personal communication*).

'Holiday' (California)

A G-type fruit, also known as 'XX3', selected from the University of California breeding programme in 2001 (UCR, 2012a). The tree is dwarf with a compact, spreading growth habit. The variety is less cold-hardy compared to 'Hass'. 'Holiday' has a Group A flowering habit. The fruit is obovate-shaped and weighs 510–680 g. The peel is slightly pebbled and green; the pulp is yellow-green and of good quality. The seed is medium-sized (Fig. 8.8). It is largely grown as a backyard variety.

'Lamb Hass' (California)

'Lamb Hass' is a seedling of 'Gwen' propagated by B.O. Bergh. It was planted at the Lamb family ranch in Camarillo, California, and patented in 1996 by the University of California, Riverside (Anonymous, 1996b). The variety is still protected in the

Fig. 8.6. 'Hass Carmen'.

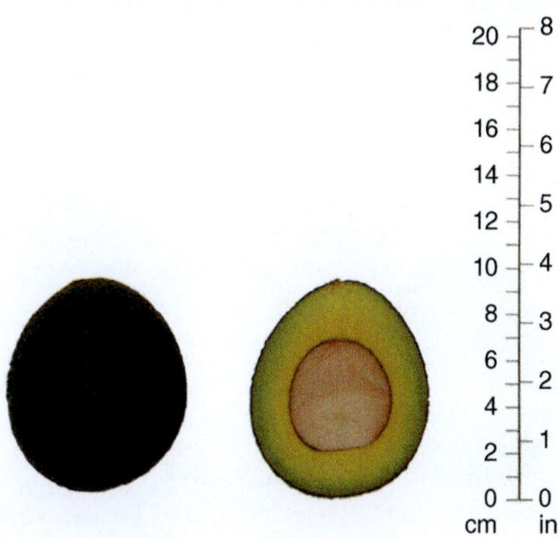

Fig. 8.7. 'Harvest'.

European Union until 2039 and licensed there to Eurosemillas SA, a company based in Spain. 'Lamb Hass' is very precocious, bearing fruit on the inside of the canopy, flower Group A. The fruit is pyriform in shape with a distinctive flat shoulder. The fruit is larger than 'Hass' (235–405 g) and matures several months later than 'Hass' in cool climates; the peel is finely pebbled, medium-thick and brittle; the fruit turns dark purple (black) upon ripening. The seed is small and round; the pulp is yellow in colour and of good to very good quality (Fig. 8.9). If the fruit is harvested before proper maturity, the ripe fruit will tend to be stringy. The fruit is less susceptible to wind damage than 'Hass', as it is held on the inside of the tree with a thick pedicel. 'Lamb Hass' has an upright canopy, which lends itself to high-density plantings, is less susceptible to the *Persea* mite, and has a similar or less resistance to cold than 'Hass'. The variety has a postharvest storage potential similar to 'Hass'. 'Lamb Hass' is increasing in popularity with small plantings being made in several countries; for example, 4% of nursery sales in 2009–2010 in South Africa were 'Lamb Hass' (Retief, 2011). 'Lamb Hass' accounted for 3.9% of California production during the 2020–2021 season (Anonymous, 2000a,b; CAC, 2023; Kremer-Köhne and Köhne, 2001; Dixon *et al.*, 2008; Anonymous, 2011f).

'Lavi' (Israel)

This variety is an open-pollinated seedling of 'Hass' selected in 1994/1995. The fruit has a 'Hass'-like form but is larger (260–400 g) and was patented in 2006 (Regev *et al.*, 2005). The tree is small (~6 m, non-pruned) with dense foliage. 'Lavi' has a Group A flowering habit, produces well (~20 t/ha), and has less tendency towards alternate-bearing than 'Hass'. The peel is glossy, pebbly, stony, breakable and thicker than 'Hass'. The seed is smaller than that of 'Hass'. The harvest season is like 'Hass'. Preliminary results indicate that 'Lavi' fruit can be stored for 3 weeks at 5°C and 95% relative humidity (RH) with little decay following ripening.

'Luna UCR' (California)

'Luna UCR' is a hybrid, previously referred to as 'Marvel', and patented under the name 'BL516' (Anonymous, 2023a). It is a seedling of 'Gwen' and is from the University of California, Riverside, scion breeding programme. The cultivar has U.S. Patent and PBR, and from 2023 has been managed under an exclusive international licence by Eurosemillas SA, a company based in Spain. Fruit is similar to 'Hass', with an average fruit size of 170–250 g, peels easily, pear- to oval-shaped, attractive pulp, peel green or green and black when hard, black when soft, and the seed is medium in size. The fruit typically has a distinct small indentation at the base (Fig. 8.10). The fruit maturity season in California is mid-spring to late summer. The postharvest storage potential is comparable to that of 'Hass'. The tree has a narrow, upright growth habit with drooping branches with the fruit borne on the tree interior, which provides protection from the

Fig. 8.8. 'Holiday'.

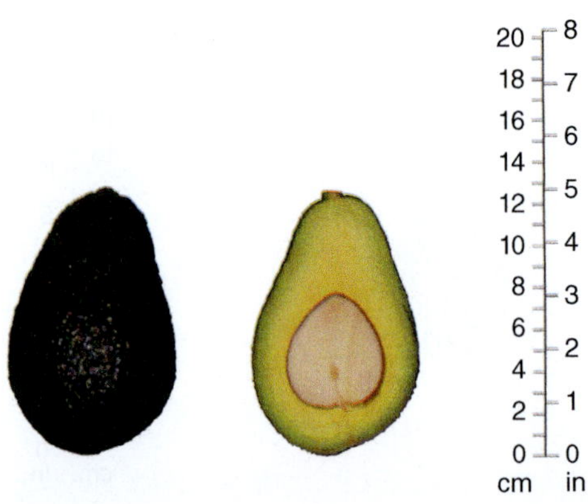

Fig. 8.9. 'Lamb Hass'.

Fig. 8.10. 'Luna UCR'.

wind and sun. The tree is approximately half the size of a 'Hass' tree, making it a good candidate for high-density plantings. Based on productivity per tree, it has a high yield efficiency, given its smaller stature. It has flower type B, so it can be used as a pollinizer variety for 'GEM' or 'Lamb Hass' under high density. Alternate-bearing is similar to 'Gwen', 'GEM' and 'Lamb Hass'. The variety has about the same cold and heat tolerance as 'Hass' but has higher tolerance to *Persea* mite (M.L. Arpaia and E.D. Focht, UC-Riverside, *personal communication*, 2023).

'Maluma' (South Africa)

'Maluma' is a G × M 'Hass'-like cultivar released in 2007 by Allesbeste Nursery in Tzaneen, South Africa, and patented (Anonymous, 2010; Ernst, 2011; Ernst *et al.*, 2015). The tree is precocious, with an upright central leader and lateral branching, suited to central leader pruning in high-density plantings. The tree is less vigorous than 'Hass' and may be managed in high-density plantings without growth retardants. 'Maluma' has an A-type flowering habit. The fruit has a shiny, semi-rough, pebbly and leathery peel, which turns purple-black upon ripening. It weighs 150–400 g and is pyriform in shape. The seed is obovate with a flat base. The smooth pulp is yellow with light-green colour near the rind and has an excellent, rich and tasty flavour (Ernst, 2011; Ernst and Ernst, 2011) (Fig. 8.11). 'Maluma' accounts for about 15% of the new avocado plantings in South Africa since 2015–2016 (Imbert, 2019).

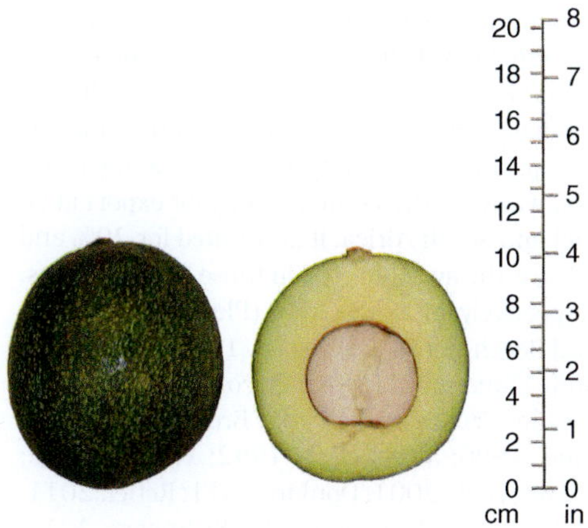

Fig. 8.11. 'Reed'.

'Pinkerton' (California)

A G × M hybrid that originated around 1959 as a 'Rincon' seedling on the property of John W. Pinkerton in Ventura, California, and thought to be a 'Rincon' × 'Hass' hybrid; patented in 1975. The tree is semidwarf, moderately spreading with a growth rate similar to that of 'Hass'; flower Group A. The fruit is pyriform in shape but can be excessively 'necky' in warm climates; fruit shape can be improved with uniconazole sprays, where permitted. Fruit is medium to large, weighing 230–425 g; the fruit retains its size when grown in hot, subtropical climates; the peel has medium thickness, is dark green, prominently pebbled, leathery and tough, easy to peel. The seed size is relatively small. It is a mid-season fruit with an attractive cream-coloured pulp, with up to 82% pulp recovery, and a rich and nutty flavour with typical oil content of 18%–25%. The fruit ripens more slowly than most varieties, thus requiring postharvest protocols that are different from other mainstream cultivars. 'Pinkerton' is very precocious with consistent, heavy production and low alternate-bearing. It has cold tolerance similar to that of 'Hass' and 'Reed', tolerating temperatures as low as −2°C. Extended flowering (from autumn to spring) results in wide variation of maturity at harvest. 'Pinkerton' is relatively resistant to anthracnose, but the fruit can have a high percentage of internal fruit disorders, including uneven ripening. Fruit-quality problems are more likely in mesic, non-stressful climates with fertile soils, and are associated with high nitrogen and low calcium in the fruit pulp (van Rooyen and Bower, 2005; Defilippi *et al.*, Chapter 15, this volume). Guidelines are now available to reduce 'Pinkerton' postharvest quality problems (Kruger and Magwaza, 2010). Grown mainly for export in Israel and South Africa, it accounted for 20% and 12% of the avocado crop in those two countries, respectively, in 2018–2019 (Platt, 1976; Torres and Bergh, 1978; Offer *et al.*, 1985; Blumenfeld and Elimelech, 1986; Piccone and Whiley, 1986b; Prusky *et al.*, 1988; Brokaw, 1989; Durand, 1990; Sippel *et al.*, 1992; Whiley, 1992; Newett *et al.*, 2001; Donkin, 2011; Retief, 2011; UCR, 2012a; Boza *et al.*, 2018; Imbert, 2018, 2019; Ge *et al.*, 2019).

'Reed' (California)

A G type that originated on the property of James Reed in Carlsbad, California, in about 1948; registered in 1960, the patent expired in 1977. It is probably an 'Anaheim' × 'Nabal' hybrid (both G-type cultivars). The tree is slender, distinctly upright with downward-hanging branches that protect the fruit from sunburn, flower Group A. The fruit is round, medium to large, weighing 270–680 g (average 400–500 g); the peel is medium to thick, green, corky, slightly pebbled and easy to peel (Fig. 8.11). The seed is medium to large in size and rounded. 'Reed' has a late-season maturity with an attractive, cream or pale-to-buttery, yellow-coloured pulp. The oil content is typically 19%–20%. It has a pulp recovery of 71%–72% and a rich and nutty flavour. The pulp surface of the cut fruit does not darken. The fruit can hang on trees in warm, subtropical climates through the summer, although the oil content increases, and biennial bearing is aggravated when this happens. 'Reed' is very precocious, consistently producing heavy crops even when the trees are older. It has cold tolerance down to −1.1°C. In California conditions, it flowers later than other cultivars. It has some resistance to *Persea* mite. The tree is used in ultra-high-density orchards in California due to its upright growth and consistently high yields (see Whiley *et al.*, Chapter 12, this volume). In 2018, it accounted for 8% of the production in Israel, 1% in California, and is an important minor variety in Florida due to its late maturity, high pulp quality and ability to crop in semitropical conditions (Platt, 1975, 1976; Alexander, 1978; Banks, 1980; Sive and Resnizky, 1989; Hoddle and Kerguelen, 1999; Anonymous, 2000a; Newett *et al.*, 2001; UCR, 2012a; Imbert, 2018; Ge *et al.*, 2019).

'Ryan' (syn. 'Summer Fuerte') (California)

An M × G hybrid selected by Edward R. Ryan in about 1927 in Whittier, California. The tree is spreading and medium to large in size; flower Group B. The fruit are pyriform in shape, medium sized, weighing 225–420 g; the peel has medium thickness and a green, leathery texture with a pimpled surface; the seed is large. The fruit has late-season maturity with golden-yellow pulp, about 68% pulp recovery, bland flavour, sometimes mealy and poor-to-medium pulp quality. 'Ryan' has unreliable production in cool climates but heavy and consistent production in the warm subtropics of South Africa. This cultivar has problems with uneven ripening. The fruit has an attractive shape but bruises easily and is susceptible to vascular browning and fruit diseases. The seed germinates in the fruit with advanced maturity (vivipary). The tree is very frost-tolerant. In 2019, it accounted for 8% of South African production, but it is not a commercially significant cultivar in any other country. The South African production of this mediocre cultivar is due solely to its very late maturity in cool growing areas, especially for the summer market when avocados fetch high prices (Anonymous, 1946; Alexander, 1978; Banks, 1980; Chalker, 1983; Smith, 1984; Durand, 1990; UCR, 2012a; Donkin, 2011; Imbert, 2019).

'Sharwil' (Australia)

Predominantly G with some M genes selected in 1951 by Sir Frank Sharpe at Redland Bay, Queensland, with unknown parentage. The name 'Sharwil' is an amalgamation of Sharpe and Wilson (J.C. Wilson was the first propagator). The trees are large and rounded in shape, with a broad crown, extremely vigorous, forming a large dome and are more upright than 'Fuerte'.

New leaves are distinctively red but turn green when fully developed; red flecking is observed on the wood of new shoots; flower Group B. The fruit are pyriform to ovate in shape, medium-sized, weighing 245–475 g; the peel is medium to thick, green with medium gloss and corky with a wrinkled surface. The seeds are small and conical. The fruit has mid-season maturity, with the pulp being buttery to golden-yellow in colour. Pulp recovery is about 81% with excellent quality, a rich nutty flavour and a high palatability over the full maturity range (21%–30% dry matter) (Fig. 8.12). The fruit stores well both on the tree and postharvest. This cultivar has the potential to set heavy crops, but flowering and fruit set are sensitive to cool temperatures; thus, reliable production only occurs in warm subtropical climates. 'Sharwil' has good tolerance to anthracnose and insect pests, presumably due to a medium-to-thick peel. It is among the most susceptible cultivars to boron deficiency. Growth is vigorous, but the large limbs are brittle, with protection from strong winds required. In 2022, 'Sharwil' was the principal commercial cultivar in Hawaii, but production is limited elsewhere (except in localized warm areas) due to erratic cropping (Sharpe, 1965; Alexander, 1978; Whiley and Brown, 1980; Van Velsen, 1983; Whiley, 1983; Winston, 1983; Whiley, 1987; Durand, 1990; Whiley *et al.*, 1996; UCR, 2012a; J.H. Crane, UF-IFAS, 2022, *personal communication*).

Fig. 8.12. 'Sharwil'.

'Shepard' (California)

Predominantly Mexican, from an open-pollinated seedling of 'Edranol' selected in 1952 by Stanley L. Shepard in Carpinteria, California (Anonymous, 2021a; Whiley and Lavers, 2021). 'Shepard' is a semidwarf tree with a compact growth habit, suited for medium to high-density plantings. Spring vegetative growth tends to be early and can be at an advanced stage by the time of anthesis, flower Group B. The flowering cycle of 'Shepard' (i.e. female and male opening sequence) is more sensitive (easily disrupted) due to cool temperatures compared to 'Hass' (Whiley and Lavers, 2021). The fruit is pyriform in shape and small, weighing 155–285 g; the green peel has medium thickness and is pebbly; the seed size is large. The fruit is early-maturing (before 'Hass') with a buttery-yellow pulp, about 69% pulp recovery; the pulp does not discolour readily when cut; it is of excellent eating quality. The fruit is susceptible to postharvest chilling injury, with current information suggesting that it should be stored at temperatures 2°C–3°C higher than that of 'Hass'; fruit softens rapidly and thus has a short shelf life. This cultivar is precocious, with reliable crop yields when grown in warm subtropical climates, but it is not suited to cooler regions due to sensitivity of flowering and fruit set to low temperatures. Small fruit and erratic flowering can be significant production problems. Early harvesting reduces the risk of biennial bearing. Fruit is less susceptible to disease and insect damage compared to other early-maturing cultivars. 'Shepard' matures 3–6 weeks ahead of 'Fuerte'. 'Shepard' never attained significant popularity in California. However, it is an important cultivar in the warmer, lower-lying areas of the Atherton Tableland (c.17°S latitude) of northern Queensland, Australia, due to its very early (late summer) maturity and strong marketing promotion. In Australia, Tolga is the southern limit for production, and Walkamin is considered the ideal location for 'Shepard' production. However, south of the Bundaberg/Childers area in central Queensland, market advantage is progressively lost, and yields decline (Anonymous, 1956, 2023c; Whiley, 1987, 1992; Whiley and Lavers, 2021; Kernot, 1995; Ledger and Barker, 1995a,b; Rudge and Story, 1995; I, Kernot, Mareeba, Australia, 2001, *personal communication*; Newett *et al.*, 2001).

'Shepard' accounts for about 25% of the Australian production (Avocados Australia, 2023).

'SirPrize' (California)

The 'SirPrize', also known as 'Sir-Prize', is an M × G hybrid from the University of California, Riverside, avocado variety breeding programme. It was released in 1996 and was patented within the USA (Anonymous, 1996a). The tree is a seedling of 'HX48,' which is a seedling of 'Hass'. The tree is somewhat upright but spreading in nature, with semi-weeping foliage hiding the fruit within the tree canopy, flower Group B. The fruit matures early, and the harvest season extends from late fall into late winter. The fruit is typically green on the tree, although late-season fruit may appear black. When ripe, fruit colour is purple to purple-black. Although the peel is thinner than 'Hass', it peels easily. The seed size is small to medium. Once cut, the flesh darkens at a slower rate than many varieties, notably 'Hass'. Production is erratic, and the tree may take 4 to 5 years to come into production, after which production is somewhat regular. Limited field studies indicate that it is a good pollinizer for 'Hass'. Although appreciated for its high eating quality and non-browning characteristics, the tree is not commonly used as a pollinizer variety primarily due to its irregular bearing habit.

'Wurtz' (syns. 'Wertz', 'Minicado', 'Littlecado') (California)

An M × G hybrid from a chance seedling, introduced by Mr. Roy Wertz to Encinitas, California, in 1948. The tree is semidwarf with sprawling habit, a densely matted, weeping foliage, flower Group A. The fruit is pyriform to ovate in shape, medium to large, weighing 175–560 g; the peel varies from thick in California to thin in the subtropics; is glossy green, leathery and pimpled. The seed is large and conical. The fruit matures mid- to late season in Australia. The pulp has a light cream to pale yellow colour; with 65%–69% pulp recovery; fruit quality is questionable with a slightly astringent flavour when first mature. It is a precocious cultivar with heavy production in warm subtropical climates but becomes biennial if fruit remains on the tree for long periods. 'Wurtz' has similar cold-hardiness to 'Fuerte'. It is prone to attack by a large range of insects and susceptible to anthracnose, its dense canopy making pest and disease control difficult. The semidwarf tree size and late fruit maturity account for its popularity in some production centres. 'Wurtz' is a very minor cultivar grown in Israel and Australia. (Alexander, 1978; Banks, 1980; Alexander, 1983a; Chalker, 1983; Whiley, 1983; Winston, 1983; Prusky *et al.*, 1988; Waite, 1988; Newett and Whiley, 1996; Newett *et al.*, 2001; UCR, 2012a, Imbert, 2018; Avocados Australia, 2023).

'Zutano' (California)

G × M hybrid, selected by W.L. Ruitt at Fallbrook, California, in 1926. The trees have upright growth; red flecking on the wood of new shoots; flower Group B. The fruit are ovate to pyriform in shape, medium in size, weighing 200–400 g; the peel is thin, light green, smooth, glossy and leathery with waxy bumps. The seed is medium to large in size and has a conical shape. The fruit is early-maturing with a pale whitish-yellow pulp; about 65% pulp recovery; has a poor-to-fair-quality pulp with a watery texture; poor in flavour with, at times, an unpleasant aftertaste; the fruit has a short tree life once it matures. 'Zutano' produces consistent and heavy crops and is predominantly grown in colder regions due to frost tolerance (down to −3.3°C). The peel blemishes readily and is susceptible to peel cracking, lenticel corking, anthracnose and insect attack. This cultivar is very susceptible to PRR and often exhibits scion overgrowth when used as a rootstock (Belisle *et al.*, 2019).

Rootstocks

Avocado rootstocks have a profound effect on avocado growth, development, production potential and disease tolerance (e.g. Ben-Ya'acov, 1992; Ben-Ya'acov and Michelson, 1995; Kremer-Köhne and Köhne, 2007; Menzel and Lagadec, 2014; Villiers and Ernst, 2015; Lazare *et al.*, 2020). Currently, the major attributes sought in avocado rootstocks are resistance to *Phytophthora cinnamomi* (e.g. Menge *et al.*, 2012; Engelbrecht and van den Berg, 2013; Sánchez-González *et al.*, 2019a, 2019b), salinity tolerance (e.g. Ben-Ya'acov and Michelson, 1995; Castro *et al.*, 2015; Celis *et al.*, 2018; Acosta-Rangel *et al.*, 2019; Lazare *et al.*, 2021a,b), adaptability to drought conditions (e.g. Moreno-Ortega *et al.*, 2021) and smaller trees with high sustainable yields of quality fruit (Whiley, 1992; Menzel and Lagadec, 2014; Dann, 2016; Acosta-Rangel *et al.*, 2019). Rootstocks can significantly influence the incidence of anthracnose on fruit (e.g. Willingham *et al.*, 2001, 2006; Coates *et al.*, 2011; Hernández *et al.*, 2023), internal fruit quality (i.e. dry matter and oil content) (Herrera-González *et al.*, 2013; Reyes-Herrera *et al.*, 2020) and the incidence of internal fruit disorders (Smith, 1993; Whiley, 2001). Others have reported rootstock effects on leaf mineral content (e.g. Acosta-Rangel *et al.*, 2019; Lazare *et al.*, 2020, 2021b; Tamayo-Vélez *et al.*, 2022). Mickelbart *et al.* (2007b) reported significant differences in nutrient uptake and yield of 'Hass' as influenced by different clonal rootstocks.

The first reported rootstock selection programme for avocados was productivity-based, beginning in the 1940s by F.F. Halma in California (Halma and White, 1951). However, when *P. cinnamomi* was discovered as a devastating root pathogen of avocados in California in 1942, Dr George A. Zentmyer, at the University of California, Riverside, changed the prime focus of rootstock selection to disease resistance (Zentmyer, 1957, 1963, 1978; Ben-Ya'acov and Michelson, 1995). The propagation of clonal rootstocks was greatly facilitated by the development of the 'nurse-seed' etiolation technique by Frolich and Platt (1972) and perfected by Brokaw (1987). (see Whiley *et al.*, Chapter 9, this volume). To the present time, the development of PRR-resistant rootstocks has been of major interest in all countries where this disease is a significant problem. However, despite the ongoing efforts, no rootstocks exhibiting complete resistance to *P. cinnamomi* are commercially available. By definition, resistance traits reduce the harm caused by the disease by preventing infection or limiting the pathogen growth, while tolerance traits do not inhibit infection but instead reduce or offset its negative fitness consequences by reducing host mortality or restoring the reproductive capacity of infected hosts. Here, we will use the term resistance rather than tolerance if the pathogen population is reduced after infection (Horns and Hood, 2012). The first widely planted PRR-tolerant rootstock was 'Duke 7', and since that discovery, rootstock programmes have been ongoing in California, Israel, México, Australia, South Africa (Zentmyer, 1963; Ben-Ya'acov and Michelson, 1995; Menge, 2001; Wolstenholme, 2003; Kremer-Köhne and Köhne, 2007; Arpaia and Smith, 2009–2010; Sánchez-González *et al.*, 2019a, b), and more recently Chile (Castro *et al.*, 2003; Fassio *et al.*, 2021) and Colombia (Henao *et al.*, 2017).

With the genome of avocado and advanced molecular techniques, the understanding of resistance, tolerance and susceptibility to PRR promises to improve and accelerate the selection of superior PRR-resistant rootstocks (e.g. Engelbrecht and van den Berg, 2013; Berg *et al.*, 2018; Rendón-Anaya *et al.*, 2019; Berg *et al.*, 2021; Kämper *et al.*, 2021; Backer *et al.*, 2022). After more than 80 years, rootstock development and selection have intensified and resulted in numerous rootstocks with varying attributes selected to address the biotic and/or abiotic constraints of local and regional production areas (e.g. México, Australia, California, South Africa, Israel, Chile). Today, some of the commercially available PRR-partially resistant/-tolerant rootstocks include 'Duke 7', 'Dusa'®, 'Bounty', 'Toro Canyon', 'Zentmyer', 'Steddom', 'Miriam (VC801)', 'Leola', 'Zerala', 'Velvick' and 'Borchard'. Some of these rootstocks, such as 'Dusa'®, 'Toro Canyon', 'Miriam' and 'Steddom', also exhibited some levels of salinity tolerance (Menge *et al.*, 2012; Celis *et al.*, 2018; Lazare *et al.*, 2021a).

Historically, the search and selection for salinity-tolerant rootstocks were second in importance after PRR tolerance (Haas, 1950). Saline soils and water have long plagued avocado production in Israel and have now become a major production constraint in California, Chile and Spain, to name a few. Long-term testing and selection for salinity-tolerant rootstocks has resulted in a better understanding of the physiological parameters that confer salinity tolerance (Ben-Ya'acov and Michelson, 1995; Bernstein *et al.*, 2001; Castro *et al.*, 2009; Mickelbart and Arpaia, 2002; Crowley and Arpaia, 2002; Celis *et al.*, 2018; Lazare *et al.*, 2021a,b; Cohen *et al.*, 2023; see Schaffer *et al.*, Chapter 7 of this volume). The rootstock appears to play a dominant role in salinity tolerance (Celis *et al.*, 2018; Acosta-Rangel *et al.*, 2019; Bernstein *et al.*, 2004; Mickelbart and Arpaia, 2002), although most investigations tested only 'Hass' as the scion. Most of the salinity-tolerant rootstocks that have been released come from the breeding or selection programmes of Israel (e.g. 'Tami' and VC152), California (e.g. PP40 and PP35) and South Africa (e.g. 'Zerala' and R0.18) (Mauk *et al.*, 2017; Celis *et al.*, 2018; Acosta-Rangel *et al.*, 2019; Lazare *et al.*, 2021a,b). However, additional genetic and metabolomic investigations into the complex scion–rootstock disease relationships and long-term field trials under various climatic conditions are needed.

Poor water quality (salinity) and calcareous soils plague avocado production in drier parts of the world and where sea-level rise has contaminated fresh water sources. Oppenheimer (1947) identified salinity tolerance in trees having WI ecotype in their genetic composition growing in Israel, and Ben-Ya'acov (1992) selected high-producing trees growing under saline conditions and recovered the rootstocks. Approximately 50 accessions containing different percentages of the WI ecotype in their genetic composition were selected as salinity-tolerant rootstocks for use by the Israeli industry (Ben-Ya'acov and Michelson, 1995; Ben-Ya'acov and Zilberstaine, 1999; Ben-Ya'acov *et al.*, 2003; Lazare *et al.*, 2021a). More recently, a range of WI, M and hybrid rootstocks have been evaluated for iron chlorosis and other problems, because avocado trees are grown in alkaline soils in Florida, Spain, Israel and Cyprus and, to a lesser degree, in California; salt stress can be reduced through the selection of tolerant rootstocks. WI rootstocks were also found to be highly resistant to lime-induced iron chlorosis when trees were grown in calcareous soils, while those of the G ecotype or G × M hybrids were determined to be highly sensitive (Kadman, 1985; Ploetz *et al.*, 1989; Ben-Ya'acov and Michelson, 1995).

With high-density orchards becoming increasingly desirable for economic returns, dwarfing rootstocks with root rot tolerance and other properties will be increasingly sought-after. 'Ashdot', 'Colin V-33' and 'Maoz' are examples of size-reducing rootstocks. The combination of size-reducing rootstocks along with smaller scions, such as 'GEM' or 'Luna UCR',® may lead to more high-density orchards.

The following descriptions of the named rootstocks are in alphabetical order, with the country where they were selected or developed in parentheses.

'Ashdot' series (Israel)

Predominantly WI, the 'Ashdot' series originated in the 'Ashdot' orchard in Israel. Trees in this series are used as seedling rootstocks, and several selections have performed well for different scions, for example, 'Ashdot 17' grafted with a 'Fuerte' scion gave the highest cumulative yield (t ha^{-1}) and 'Ashdot 27' grafted onto 'Ettinger' resulted in the highest yield from smaller trees (kg m^{-3} of tree canopy). Both 'Ashdot 17' and 'Ashdot 27' were the recommended rootstocks for all cultivars grown in Israel. 'Ashdot 17' gave high cumulative yields in the Jordan Valley of Israel when grafted with 'Pinkerton'. Lagadec (2011) compared the seedling 'Ashdot' with 21 other seedling rootstocks and 11 clonal rootstocks grafted onto 'Hass' over a 7-year period. 'Hass' on 'Ashdot' produced the smallest trees but had the highest yield efficiency (kg fruit per m^3 canopy volume). However, whereas 'Hass' on 'Ashdot' may be suitable for high-density plantings, it was observed to be more susceptible to environmental stresses. 'Ashdot 17' was recommended for new orchards where calcareous soils and/or salinity are present (Ben-Ya'acov *et al.*, 1995a,b).

'Barr Duke' (California)

This pure M ecotype rootstock is a seedling of 'Duke 6', selected and released in 1989 because of its superior performance in an area with severe PRR infestation in California. Trees grafted onto cloned 'Barr Duke' showed the best performance under high-PRR incidence at that time, and the *Phytophthora cinnamomi* resistance was comparable with those exhibited by 'Thomas'; however, 'Barr Duke' was shown to be susceptible to *Phytophthora citricola* and sensitive to saline conditions. In South Africa, the yield efficiency of 'Hass' on 'Barr Duke' was superior to that of 'Hass' on 'Duke 7' (Coffey and Guillemet, 1987b; Gabor *et al.*, 1990; Menge *et al.*, 1992; Oster and

Arpaia, 1992; Tsao *et al.*, 1992; Roe *et al.*, 1999). 'Barr Duke' does not tolerate saline soil and irrigation water (Ben-Ya'acov and Michelson 1995; Crowley and Arpaia, 2002).

'Ben-Ya'acov1 (syn. VC66)' (Israel)

The WI vegetative clone (VC) hybrid ecotype, 'Ben-Ya'acov1'(VC66), was identified in an orchard in Ma'agan Michael, Israel (Anonymous, 2019a) and is property of the Agricultural Research Organization of the State of Israel. This rootstock was commercially released in California in 2018 based on field data collected in Israel, indicating a lower tendency to alternate bear and some level of tolerance to *P. cinnamomi* and salinity. This rootstock is moderately tolerant to salinity and its salinity tolerance is better than that of 'Dusa'® VC207 ('Day') and 'Tami' (VC801) (Lazare *et al.*, 2021a). In addition, Haberman *et al.* (2020) reported that this rootstock was highly susceptible to verticillium wilt caused by *Verticillium dahliae* and exhibited low productivity in fields infected with this pathogen. In addition, 'Hass' grafted on to 'Ben-Ya'acov1' had similar levels of productivity as 'Hass' on 'Ashdot 17', 'Degania 117' and 'Degania 113, 'Fairchild', and other VCs evaluated in fields that were not affected by *V. dahliae* (Cohen *et al.*, 2023). Currently, this rootstock is being tested for its field performance under several growing conditions in California by Brokaw Nursery LLC.

'Borys' (México)

'Borys' is a WI ecotype selected under saline conditions in Zempoala, Veracruz, México (Karp and Gasic, 2022). Trees with 'Hass' scions are vigorous, produce good yields, display early flowering and high salinity and drought tolerance.

'Borchard' (California)

This M race rootstock is described as one of the most susceptible to *P. cinnamomi* but is resistant to *P. citricola*, which causes stem (trunk) and crown rot damage and is a significant disease of avocado trees in California. This rootstock performs better than 'Duke 7' in saline conditions and is a successful special-purpose rootstock that does well in alkaline soils, where avocados on other rootstocks show intense chlorosis. In a California trial with 'Hass' on ten clonal rootstocks, trees on 'Borchard' and 'Duke 7' had the highest yield, with 'Borchard' having the most growth. However, PRR and salinity were not problems at that site (Oster *et al.*, 1985; Brokaw, 1986; Gabor *et al.*, 1990; Menge *et al.*, 1992; El-Hamalawi *et al.*, 1994; Ben-Ya'acov and Michelson, 1995; Mickelbart *et al.*, 2007a).

'Bounty' (South Africa)

'Bounty' is an M × G rootstock selected as a survivor tree at the Bounty Farm in Kiepersol, South Africa (ANFIC, 2023; Riversun, 2023). This rootstock is patented by the Agricultural Research Council Institute for Tropical and Subtropical Crops in Nelspruit and by the Fruit Farm Group. This rootstock was selected for its high tolerance to PRR, outperforming 'Dusa'® under heavily infested soils, and its PRR tolerance appears to be similar to that of the 'Duke 7' rootstock. In addition, 'Bounty' exhibited tolerance to heavy soils and does not suffer from waterlogging like 'Dusa'®. 'Maluma' grafted on the 'Bounty' rootstock has a smaller, more open tree compared to all other rootstocks, for example, 'Dusa.®' It is advised not to plant this rootstock under good soil conditions, because the trees will be too vigorous. A salinity-tolerant phenotype is unknown for this rootstock, so it needs to be extensively tested under conditions of high soil and water salinity.

'Bravo' (México)

'Bravo' is a WI ecotype selected in Nayarit, México (Karp and Gasic, 2022). 'Hass' trees grafted onto 'Bravo' produce good yields, exhibit medium vigour and show resistance to PRR and salinity and drought tolerance.

'Campeón' (México)

A WI ecotype, 'Campeón', was selected in Nayarit, México, and tested under drought conditions

(Karp and Gasic, 2022). Trees grafted onto 'Campeón' are vigorous with a spreading growth habit. Trees grafted with the 'Hass' scion may yield 60% more fruit than 'Hass' on local rootstocks. The trees display high drought tolerance and PRR resistance.

'Colin V-33' (México)

This rootstock is predominantly an M type selected in México from a seedling of 'Fuerte', characterized by its dwarf stature in México (Ben-Ya'acov and Michelson, 1995; Useche-Carrillo *et al.*, 2021). In México, 'Colin V-33' was used successfully as a dwarfing inter-stock for 'Fuerte' scions. These three-part trees consisted of an M ecotype seedling rootstock, 'Colin V-33' inter-stock and 'Fuerte' scion. The low vigour of 'Colin V-33' has been attributed to a high density of vessel elements. However, in South Africa, 'Colin V-33' had very little effect on reducing tree vigour and improving yields when used as an inter-stock for 'Hass' scions on 'Duke 7' rootstocks, possibly due to effective control of boron deficiency (Barrientos-Priego *et al.*, 1987; Sanchez-Salvador, 1989; Vega, 1989; Roe and Morudu, 2000).

'Day (syn. VC207)' (Florida)

This WI hybrid cultivar was selected in Florida. The fruit is small (8–16 oz.), pear-shaped with an excellent rich and nutty taste. The skin is green in colour and dull. The tree is very productive and cold-hardy. In California, 'Day' is used as a rootstock since it is moderately resistant to the current populations of *P. cinnamomi* in California, and its resistance levels are comparable to those exhibited by 'Dusa'® (Manosalva, *personal communication*; Douhan, 2011). In addition to PRR resistance, 'Day' also exhibited moderate resistance to Verticillium wilt, exhibiting good productivity in highly infested fields in Israel (Haberman *et al.*, 2020). 'Day' exhibited salinity tolerance under field conditions in California and Israel, and this tolerance is better than that of 'Dusa'® (Lazare *et al.* 2021a).

'Duke 9' (California)

This rootstock was selected in California following gamma irradiation of the 'Duke' budwood. It had a semidwarf stature when grown as a non-grafted tree in the absence of PRR. In California, trees grafted onto clonal 'Duke 9' rootstocks had high levels of resistance to *P. cinnamomi* (similar to 'Barr Duke' and 'Dusa'®) (Ben-Ya'acov and Michelson, 1995; Manosalva, *personal communication*). However, in the absence of PRR, trees grafted onto 'Duke 9' had poor yield (kg fruit m^{-3} of canopy) compared with trees on other rootstocks. In South Africa, field evaluations under subtropical conditions showed that 'Hass' trees grafted onto clonal 'Duke 9' rootstocks are comparable with 'Hass' grafted onto clonal 'Duke 7' rootstocks regarding good productivity (Coffey, 1987; Gabor *et al.*, 1990; Arpaia *et al.*, 1992; Menge *et al.*, 1992; Roe *et al.*, 1997; Mickelbart *et al.*, 2007b). 'Duke 9' does not tolerate high-salinity soil and irrigation water conditions (Ben-Ya'acov and Michelson, 1995). However, field evaluations in California indicated that this rootstock exhibited levels of salinity tolerance similar to those of the 'Duke 7' rootstock and better than those of 'Thomas', which is highly susceptible to salinity (Bender *et al.*, 2010; Douhan, 2011).

'Degania' series (Israel)

The Degania series is predominantly West Indian ecotypes. 'Degania 112', 'Degania 115' and 'Degania 117' were superior-yielding rootstocks when grafted with the 'Fuerte' scion, the former two having a dwarfing effect. For 'Ettinger' trees planted in heavy calcareous soils with high salinity, 'Degania 113', 'Degania 115', 'Degania 117' and 'Peker 2' are the preferred rootstocks. 'Degania 62' and 'Degania 112' are no longer used as rootstocks. 'Degania 62' has been reported to exhibit good levels of salinity tolerance, better than 'Dusa'® 'Tami (VC801)', 'Day (VC207)' and 'Ben-Ya'acov1' (VC66) (Lazare *et al.*, 2021a). 'Degania 112' grafted with 'Fuerte' and 'Ettinger' scions was found to be particularly sensitive to Verticillium wilt (Ben-Ya'acov and Frenkel, 1974; Ben-Ya'acov, 1995a; E. Lahav, Bet Dagan, Israel, 2011, *personal communication*). In contrast,

Haberman *et al.* (2020) reported clonal 'Degania 189' to be highly tolerant to Verticillium wilt.

'Duke 7' (California)

Seedlings of the cold-resistant M cultivars 'Duke', 'Duke 6' and 'Duke 7' were selected by George Zentmyer for their moderate field resistance to *P. cinnamomi*. 'Duke 7' was selected in the 1950s and released commercially as a clonal rootstock in the mid-1970s. 'Duke 7' is intolerant to waterlogging and should only be used on deep, well-drained soils and does not appear to be suitable for replant situations in California. There is accumulating evidence that 'Duke 7' is a productive rootstock for 'Hass'. However, experience in South Africa and Australia has shown that 'Duke 7' is susceptible to trunk canker caused by *P. cinnamomi*. A 10-year trial (Mickelbart *et al.*, 2007b) showed that 'Hass' on 'G755B' and 'G755C' had the lowest alternate-bearing index, with 'Duke 7' intermediate but with significantly higher mean production. Sánchez-González *et al.* (2019b) compared container-grown clonal 'Duke 7' and seedlings of ten named M-ecotype avocados for tolerance and susceptibility to PRR (PRR) and 48 h of flooding. After 11 weeks, the seedlings of 'Todo el Año' were rated tolerant to PRR whereas 'Duke 7' was rated highly susceptible. The salinity tolerance of 'Duke 7' is not clear and may be due to differences in experimental methods and/or scion–rootstock combinations (e.g. Oster *et al.*, 1985; Oster and Arpaia, 1992). As the second decade of the 2000s begins, the preeminence of 'Duke 7' has decreased as the evaluation and development of new rootstocks has accelerated (Arpaia *et al.*, 1992; Köhne, 1992; Kremer-Köhne and Köhne, 1992; Whiley, 1992; Roe *et al.*, 1998; Arpaia, 1999; Mickelbart *et al.*, 2007a,b; Smith *et al.*, 2011; Mickelbart *et al.*, 2012; Acosta-Rangel *et al.*, 2019; Mauk *et al.*, 2017; Ge *et al.*, 2019; Sánchez-González *et al.*, 2019a; Hernández *et al.*, 2023).

'Dusa'® (formerly 'Merensky 2') (South Africa)

A G × M hybrid, 'Dusa'® was selected from a seedling of unknown origin at Westfalia Estate, South Africa (J.S. Köhne, Tzaneen, South Africa, 2011, *personal communication*; Ashworth and Clegg, 2003; Ge *et al.*, 2019). Data collected in South Africa, California, Australia and Chile over many years have shown that, under certain conditions, 'Hass' trees grafted onto 'Dusa'® are significantly more tolerant of PRR and more productive than 'Hass' on 'Duke 7' (Roe *et al.*, 1997; Menge, 1998, 2002; Kremer-Köhne and Mukhumo, 2003; Rose, 2003; Smith *et al.*, 2011). The partial PRR resistance of 'Dusa'® has been shown to be reduced by more virulent isolates of *P. cinnamomi* collected in Southern California (Belisle *et al.*, 2019). 'Dusa'® is purported to have good tolerance to cold winters (Kremer-Khöhne and Khöhne, 2007). More recently, 'Dusa'® was rated as one of two rootstocks with tolerance to the soil-borne *V. dahliae* (Haberman *et al.*, 2020). 'Dusa'® was rated as tolerant (Martínez-Ferri *et al.*, 2019) and, in a different study, susceptible to *Rosellinia necatrix* (white root rot) (Zumaquero *et al.*, 2019). The tolerance of 'Dusa'® to saline conditions is not clear, with some field comparisons rating it tolerant (Crowley and Arpaia, 2002; Crowley, 2004; Mauk *et al.*, 2017; Celis *et al.*, 2018) and others not tolerant (Lazare *et al.*, 2021a). 'Dusa'® is highly susceptible to heavy soils and waterlogging conditions and performs poorly when planted in poorly drained soils (Spann, 2008). Since 2002, 'Dusa'® has been made available to avocado growers in several countries and continues to gain popularity. Because of its good productivity, salinity tolerance and PRR resistance 'Dusa'® is still a popular rootstock and still the most widely planted rootstock in California; however, other rootstocks are becoming popular in California, including 'Tami (VC801)', 'Zerala'® and 'Steddom'.

'Galindo' (México)

This M-ecotype rootstock was identified by the Instituto Nacional de Investigaciones Forestales, Agrícolas y Pecuarias (INIFAP) as an escape tree in a highly infested *P. cinnamomi* orchard in México State, México (Karp and Gasic, 2022). Field testing indicated that 'Galindo' is highly resistant to *P. cinnamomi* and 'Hass' on 'Galindo' is moderately vigorous with good yields.

'Jefe' (México)

This WI ecotype rootstock was discovered in Nayarit, México (Karp and Gasic, 2022). Field testing determined that 'Jefe' is highly resistant to *P. cinnamomi* and is salinity- and drought-tolerant. 'Hass' trees grafted onto 'Jefe' are vigorous with high yields.

'Latas' (formerly 'Merensky 1') (South Africa)

'Latas' was selected at the Westfalia Estate, South Africa, where it was growing in seasonally waterlogged soil conditions (Roe *et al.*, 1995). This M × G hybrid clonal rootstock is thought to be a 'Fuerte' seedling (Haberman *et al.*, 2020). In South African conditions, 'Hass' grafted onto 'Latas' out-produced 'Hass' grafted onto 'Duke 7' in the first two cropping seasons by ≥29%. In a 7-year field evaluation comparing 'Hass' grafted onto 14 rootstocks, production of 'Hass' on 'Latas' was generally lower than 'Hass' grafted onto 'Dusa'® 'Velvick', or 'Duke 7' (Guirado *et al.*, 2015). This rootstock has good tolerance to PRR (Roe *et al.*, 1998; Menge *et al.*, 2012; UCR, 2012b). In South Africa (J.S. Köhne, Tzaneen, South Africa, 2011, *personal communication*) and California (Douhan, 2009), 'Hass' grafted onto 'Latas' has continued to produce well in PRR-infested soils for many years but has not been marketed as a commercial rootstock due to the success of 'Dusa,®'. In the short term, 'Hass' grafted onto 'Latas' or VC801 had the lowest chloride leaf accumulation of 15 different rootstocks (Crowley and Arpaia, 2002).

'Leola'® (formerly 'Merensky 6') (South Africa)

This rootstock was selected in a breeding block by the Westfalia Estate in South Africa from their *P. cinnamomi* screening programme (Karp and Gasic, 2022). The rootstock is patented (Anonymous, 2020b). 'Leola'® was selected for its partial resistance to *P. cinnamomi* and was commercially released in California in 2020. Its PRR resistance against the current *P. cinnamomi* population in California is comparable to the moderate resistance of 'Dusa'® and 'Zerala'® under greenhouse conditions (P. Manosalva, UC-Riverside, *personal communication*). This rootstock exhibited competitive yields when grafted onto 'Hass' or 'GEM' compared to other commercially available rootstocks in California (Brokaw Nursery LLC, Ventura, California, personal communication). The salinity tolerance of 'Leola'® is unknown. Currently, this rootstock is being tested for its field performance under several growing conditions in California by Brokaw Nursery LLC.

'Mexicola' (California)

This M race seedling rootstock was once commonly used as a seedling rootstock in California and is the main seedling rootstock in Chile (Castro *et al.*, 2003), where cold tolerance is important. The fruit is obovate, small (113–184 g) with a black, smooth peel (Anonymous, 2012d; Fig. 8.13). This rootstock is highly susceptible to PRR and leaf burn under saline conditions, and high levels of sodium are accumulated in the leaves. As a rootstock, 'Mexicola' had the highest cumulative yield over 10 years when grafted

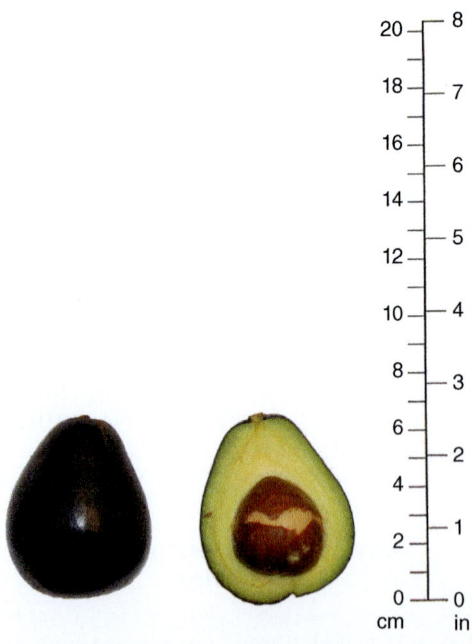

Fig. 8.13. 'Mexicola'.

with 'Fuerte' in Cyprus (Haas, 1952; Kadman, 1962; Zentmyer, 1972; Allen, 1977; Whiley, 1986; Gregoriou, 1992; Boza *et al.*, 2018). 'Hass' fruit grafted onto seedling 'Mexicola' or clonal 'Duke 7' had similar long-term postharvest handling characteristics under regular and controlled atmospheric storage conditions (Hernández *et al.*, 2023).

'Maoz' (syn. 'VC 43') (Israel)

Although not in use in Israel anymore, this WI rootstock is one of the most tolerant rootstocks to calcareous soils in Israel and is also highly tolerant to saline irrigation water. In controlled studies, seedlings showed no leaf scorch or other damage when irrigated with solutions containing 800 mg kg^{-1} of chloride. 'Maoz' also shows some dwarfing effect on 'Ettinger', 'Fuerte' and 'Hass'. However, the yield from 'Ettinger' on this rootstock was unacceptable. In the Canary Islands, 'Maoz' seedlings produced a high proportion of PRR-tolerant lines (Kadman and Ben-Ya'acov, 1980; Kadman, 1985; Whiley, 1990; Ben-Ya'acov *et al.*, 1995a; E. Lahav, Bet Dagan, Israel, 2011, *personal communication*).

'Martin Grande' (California) ('G755' series)

'Martin Grande' is a composite of three individual seedlings: 'G755A', 'G755B' and 'G755C', which are hybrids of coyo (*P. schiedeana*) and *avocado* (*P. americana*), collected as wild-collected coyo fruit from the Cobá, Guatemalan market, in September 1975 and evaluated for PRR tolerance in California. The compatibility of 'Martin Grande' with various important commercial scions such as 'Hass', 'Gwen', 'Pinkerton' and 'Fuerte' has been demonstrated. The clones of all three seedlings have similar resistance to PRR and are significantly more resistant than 'Duke 7'. 'Martin Grande' will tolerate higher salinity compared to 'Duke 7' and is vigorous and shows tolerance to winter cold in California coastal conditions. Unfortunately, due to pronounced vigour, trees are relatively non-productive and have strong alternate-bearing when grafted with 'Hass'. 'Martin Grande' is also susceptible to stem, crown and root infection by *P. mengei* as well as *Phomopsis perseae*. 'Martin Grande' clonal rootstocks are no longer recommended in California, even in PRR-infested conditions (Mickelbart *et al.*, 2007b), and have also lost favour in South Africa (Coffey and Guillemet, 1987b; Coffey *et al.*, 1988; Ellstrand *et al.*, 1986; Whiley, 1990; Van der Merwe and Kotzé, 1991; Arpaia *et al.*, 1992; Tsao *et al.*, 1992; Arpaia, 1993; Martin, 1993; Mickelbart *et al.*, 2007b).

'Miriam (syn. VC218)' (Israel)

A WI hybrid, 'Miriam' was selected from an orchard in Givat Haim, Israel (Anonymous, 2019b). This rootstock was commercially released in California in 2018 and is property of the Israel Agricultural Research Organization of the State of Israel. The commercial release of 'Miriam' was based on field data from Israel, indicating its partial resistance to PRR and tolerance to drought, alkaline soil and salinity. Before its commercial release, this rootstock was grafted with 'Hass' scion and tested under several growing conditions in California by the University of California Riverside (UCR) Rootstock Breeding Program, indicating good field performance under PRR and salinity conditions (Douhan, 2009, 2011; Douhan *et al.*, 2011). The partial resistance of this rootstock is similar to that of 'Dusa'® against the current *P. cinnamomi* populations in California (P. Manosalva, UC-Riverside, *personal communication*). Trees grafted onto 'Miriam' are of medium vigour and possess a semi-upright growth habit. Currently, this rootstock is being re-evaluated for its field performance under several growing conditions in California by Brokaw Nursery LLC.

'Procado' (Australia)

'Procado' (SHSR-04) was selected during the early 2000s as a surviving tree in a highly *P. cinnamomi*-infested orchard in South Kolan, Australia (Whiley, 2020; Whiley and Dann, 2021). Patented in 2021, 'Procado' is also trademarked (Anonymous, 2021b). The tree is a G hybrid with a semi-upright growth habit and has moderate vigour and type-A flower habit. However, 'Hass' and 'Shepard' grafted onto 'Procado' are precocious and have low-to-moderate vigour. Results from rootstock field trials of grafted 'Hass' and 'Shepard' planted in orchards with high-PRR incidence reported superior *P. cinnamomi* tolerance of 'Procado' to many current rootstocks such as 'Latas', 'Velvick' (seedling and clonal), 'Duke 7', 'Barr Duke', 'Zutano' (seedling), 'Thomas' and 'Reed'. In some cases, 'Dusa'® (clonal) and 'Velvick' (seedling) performed as well. Results from the second year of a field comparison among BW2 (seedling) and clonal 'Dusa'®, 'Procado' and 'Velvick' grafted with 'Hass' indicated that 'Hass' grafted onto 'Procado' had a smaller canopy volume than all other scion–rootstock combinations but the second highest yield efficiency after 'Hass' grafted onto BW2 (Whiley, 2023). The yield of 25-month-old 'Hass' on 'Procado' was much greater than 'Hass' grafted onto 'Dusa'® or 'Velvick', whereas trees on 'Procado' had the smallest canopy volume compared to trees on 'Dusa'® or 'Velvick' (A. Whiley, *personal communication*). 'Procado' may be a suitable rootstock for high-density plantings and provides partial resistance to *P. cinnamomi*. To date, 'Procado' has only been adequately tested under Australian conditions.

'Rudo' (México)

This M ecotype rootstock was identified in México State and tested in a *P. cinnamomi*-infested orchard and under drought conditions (Karp and Gasic, 2022). Field testing indicated that trees grafted onto 'Rudo' are resistant to *P. cinnamomi* and moderately tolerant to drought conditions. The trees are medium-sized with good yields.

'Salazar' (México)

This WI ecotype rootstock was identified by the INIFAP, Veracruz State and tested under drought conditions (Karp and Gasic, 2022). Field testing indicated that 'Hass' grafted onto 'Salazar' is highly drought-tolerant and a vigorous tree with 26.5% more yield than local rootstocks.

'Steddom' (California)

'Steddom' is an M × G seedling from 'Toro Canyon' that was commercially released in 2014 by the UCR Avocado Rootstock Breeding Program in California. 'Steddom' has been tested in California as selection 'PP24' since 1999 (Menge *et al.*, 2012) and is patented (Anonymous, 2014c). 'Steddom' was selected because of its high partial resistance to *P. cinnamomi* under greenhouse and field conditions when grafted with 'Hass' scion (Douhan *et al.*, 2011; Menge *et al.*, 2012). Once grafted, the trees are relatively slow-growing but heavy-yielding (Douhan, 2009). Therefore, it appears to have a high yield/canopy volume ratio and may be an excellent choice for high-density or hedgerow avocado plantings. 'Steddom' exhibited moderate tolerance to salinity, comparable with 'Dusa'® (Menge *et al.*, 2012; Mauk *et al.*, 2017; Celis *et al.*, 2018). This rootstock is one of the best performers when compared with other commercial rootstocks, including 'Duke 7', 'Dusa'®, 'Zerala'™, 'Leola'™, 'Uzi', or 'Zentmyer', grafted with different scions, including 'Hass', 'Carmen', 'GEM', 'Lamb Hass', or 'Reed', regarding yield and yield efficiency (2015–2020) in a rootstock × scion trial in Ventura, California (M.L. Arpaia, UC-Riverside, *personal communication*). Due to its PRR resistance, salinity tolerance, small stature, low vigour and good yield and yield efficiency, 'Steddom' is gaining popularity among California growers. This rootstock variety has only been adequately tested under California conditions thus far; however, 'Steddom' is being tested grafted onto different scions under high-salinity conditions in Texas.

'Tami' (syn. VC801) (Israel)

A WI hybrid rootstock, 'Tami' was an escape growing in Givat Haim, Israel (Anonymous, 2018a). This rootstock was commercially released in California in 2018 and is property of the Agricultural Research Organization of the State of Israel. This rootstock was selected in Israel as a high-yielding rootstock growing under high-salinity conditions. It was imported into California by the UCR rootstock breeding programme, where it was tested for many years under different growing conditions. The early field testing of 'Tami' grafted with 'Hass' scion in California indicated good performance under PRR and high-salinity conditions in terms of tree health and yield (Douhan, 2011, Crowley and Arpaia, 2002). 'Tami' is reportedly moderately resistant to *P. cinnamomi* in field evaluations. The partial resistance of this rootstock is similar to that of 'Dusa'® against the current *P. cinnamomi* populations in California (P. Manosalva, UC-Riverside, *personal communication*). The tolerance to salinity of 'Tami' is comparable to 'Day' and better than 'Dusa'® (Lazare *et al.*, 2021a). In addition, this rootstock exhibited moderate susceptibility to Verticillium wilt, and it was less productive than 'Dusa'® under Verticillium wilt disease pressure (Haberman *et al.*, 2020). Currently, this rootstock is being re-evaluated for its field performance under several growing conditions in California by Brokaw Nursery LLC.

'Thomas' (California)

This is an M-ectotype clonal rootstock recovered in 1979 from an escape 'Fuerte' tree growing in a PRR-infested area in Escondido, California. 'Thomas' was commercially released by the UCR Rootstock Breeding Program in 1989 for its high resistance to *P. cinnamomi* and was initially an impressive rootstock. In California, the yield efficiency (fruit per cubic metre of canopy) of 'Hass' on either 'Duke 7' or 'Thomas', 5 years after planting, was significantly higher than that of 'Hass' on nine other rootstocks. However, over time, it became clear that 'Thomas' was also highly sensitive to salinity and did not yield very well in many locations in California. In addition, this rootstock is highly susceptible to the current *P. cinnamomi* populations in California. 'Thomas' is also susceptible to stem, crown and root infection by *P. mengei*. These shortcomings have resulted in this rootstock's popularity declining in California. Moreover, in South Africa, 'Hass' trees grafted onto 'Thomas' only produced acceptable yield at one site, and it is not a recommended rootstock for that country (Coffey and Guillemet, 1987a,b; Gabor *et al.*, 1990; Arpaia *et al.*, 1992; Menge *et al.*, 1992; Tsao *et al.*, 1992; Oster and Arpaia, 1992; Roe *et al.*, 1997; Crowley and Arpaia, 2001–2002; Mickelbart and Arpaia, 2002; Mickelbart *et al.*, 2007b).

'Topa Topa' (California)

This M-ectotype seedling rootstock was originally selected on the Topa Topa Ranch in Ojai, California. 'Topa Topa' was the most commonly used rootstock in California during the years of rapid expansion of the avocado industry, as trees produced heavy crops, and seedlings were vigorous, relatively uniform and easy to graft. However, 'Topa Topa' is highly susceptible to *P. cinnamomi* and *P. mengei* and has poor tolerance to salinity (Henao *et al.*, 2017). In a 10-year California trial, 'Hass' scions on 'Topa Topa' had a high alternate-bearing index, yield and canopy efficiency, but this was in a PRR-free orchard without salinity problems (Mickelbart, 2007b). Today, PRR sensitivity rules out this rootstock, except where cold tolerance is paramount and PRR-free soils occur. A degree of scion overgrowth also leads to root starvation, aggravating alternate-bearing and PRR susceptibility in this rootstock (Zentmyer, 1972; Zentmyer *et al.*, 1976; Coffey and Guillemet, 1987a; Gabor *et al.*, 1990; Tsao *et al.*, 1992; Whiley, 1992; Ben-Ya'acov *et al.*, 1995a; Whiley and Whiley, 2011).

'Topara'™ (KB1) (Perú)

This rootstock is a selection originating from a vigorous productive tree in the Topara region of Perú. This rootstock is patented and was commercially released in California in 2018 as KB1 (Anonymous, 2018b). It is susceptible to *P. cinnamomi* but induces high vigour when grafted with the 'Hass' scion (Belisle *et al.*, 2019). A 'Topara'™ salinity-tolerant phenotype is unknown; however, it is believed to harbour some level of salinity tolerance because of its origins in the sandy Peruvian coast.

'Toro Canyon' (California)

An M × G rootstock cloned from a survivor tree in a *P. cinnamomi*-infested area in California. Resistance to *P. cinnamomi* is described as comparable to that of 'Martin Grande', although studies in Spain found that 'Toro Canyon' did not have resistance to *P. cinnamomi*. This rootstock has a high degree of tolerance to *P. citricola*. In an evaluation of 'Hass' productivity on ten avocado rootstocks in the absence of *P. cinnamomi*, 'Toro Canyon' produced an intermediate yield compared with the other nine rootstocks, but induced pronounced alternate-bearing (Coffey, 1987; Arpaia *et al.*, 1992; Oster and Arpaia, 1992; Tsao *et al.*, 1992; Gallo-Llobet and Siverio, 1995; Mickelbart *et al.*, 2007b). 'Toro Canyon' exhibited some levels of salinity tolerance, which is why this rootstock continues to be one of the top-selling rootstocks in California (Brokaw Nursery LLC, personal communication).

'Uzi' (California)

The 'Uzi' M rootstock was selected from 'G6' and has been tested in California as selection 'PP14' since 1999 (Menge *et al.*, 2012). 'Uzi' is patented and named after Uzi Afek, a scientist from the Volcani Center in Bet Dagan, Israel (Anonymous, 2014b). This rootstock was commercially released in 2014 by the UCR Avocado Rootstock Breeding Program in California for its high *P. cinnamomi* resistance (Douhan *et al.*, 2011; Menge *et al.*, 2012). 'Uzi' is an extremely vigorous and fast-growing rootstock that, when grafted with the 'Hass' scion, can support tree growth to a height of 4 m in 2 years, and its yields are generally high and consistent (Douhan, 2009). 'Uzi' is not salt-tolerant and exhibits leaf-tip burn due to salt damage (Menge *et al.*, 2012; Mauk *et al.*, 2017; Celis *et al.*, 2018). This rootstock has only been adequately tested under California conditions thus far and is a good rootstock for planting under high-PRR conditions with no salinity problems.

'Velvick' (Australia)

This rootstock is a WI × G-race hybrid from a seedling selected in Queensland, Australia (Rendón-Anaya *et al.*, 2019; Kämper *et al.*, 2021). A vigorous rootstock that performs well in PRR-infested soils and has been described as an outstanding rootstock for subtropical Australia. 'Velvick' is intermediate in its resistance to *P. citricola*. In a South African field experiment evaluating PRR tolerance of 'Hass' trees grafted onto eight different rootstocks including 'Duke 7', 'Latas' ('Merensky 1') and 'Dua'®, 'Velvick' was rated in the highest group with respect to tree health 2 and 3 years after planting (Lavers, 1989; El-Hamalawi *et al.*, 1994; Kremer-Köhne and Duvenhage, 2000; A.W. Whiley, Nambour, Australia, 2001, *personal communication*). Willingham *et al.* (2001, 2006) reported that the 'Velvick' rootstock can significantly reduce the incidence of anthracnose fruit rots compared to the M race rootstocks. 'Velvick' is used overwhelmingly as a seedling rootstock in Australia, where expensive clonal rootstocks do not appear to consistently increase yield or orchard uniformity, at least in new plantings. In previous field trials, selected and proven seedling rootstocks, including 'Velvick' seedlings, often out-yielded equivalent clonal rootstocks over five cropping cycles. It was important, however, that the 'Velvick' seed source came from isolated (self-pollinated) trees. The choice of seedling versus clonal 'Velvick' depended on the local circumstances. Previously, Whiley and Whiley (2011) in Australia concluded that proven seedling rootstocks are recommended for new plantings, and clonal rootstocks with good tolerance to PRR are preferred for replant sites.

'Zentmyer' (California)

'Zentmyer', an M rootstock, has been tested in California as selection 'PP4' since 1996 and was selected from 'Thomas' (Menge *et al.*, 2012). 'Zentmyer' is patented and named after George A. Zentmyer, who was a plant pathologist at the University of California, Riverside, and a pioneer for the avocado rootstock breeding programmes and a collector of avocado germplasm worldwide (Anonymous, 2014a; Douhan *et al.*, 2011; Menge *et al.*, 2012). 'Zentmyer' is an extremely vigorous, highly durable clonal rootstock that is highly resistant to *P. cinnamomi*, exhibiting better resistance than 'Dusa'® and other commercially available rootstocks (P. Manosalva, UC-Riverside, *personal communication*). 'Hass' on 'Zentmyer' performed well compared to 'Hass' on 'Dusa'®, 'Steddom' or 'Uzi'; however, this rootstock is severely damaged by salt and is not recommended for locations where saline conditions are a problem (Douhan, 2009; Crowley and Arpaia,

2002; Mauk *et al.*, 2017). The 'Zentmyer' rootstock, when grafted with 'Hass' scion, often does not yield well under non-root-rot conditions in comparison with other new rootstocks such as 'Steddom' and 'Uzi' (Menge *et al.*, 2012). Therefore, it is recommended that the 'Zentmyer' rootstock be used in replant situations where root rot-infested soils are a problem but where salinity is not an issue. This rootstock has only been adequately tested under California conditions thus far.

'Zerala'™ (formerly 'Merensky 5') (South Africa)

This rootstock was discovered during the late 1980s in a 'Fuerte' orchard highly infested with *P. cinnamomi*, and subsequent testing of clonal plants verified its resistance to *P. cinnamomi* (Karp and Gasic, 2022; Brokaw, 2024). 'Zerala'™ is patented and was selected for its partial resistance to *P. cinnamomi*. It was commercially released in California in 2020 (Anonymous, 2020a). 'Zerala'™ exhibited similar levels of salinity tolerance as 'Dusa'® and other UCR-advanced rootstocks, such as PP40 (Celis *et al.*, 2018). 'Zerala'™ rootstocks often support larger canopy volumes. It is recommended to be planted on berms, hillsides, or well-drained soils, since it is susceptible to waterlogging and low-oxygen conditions (Brokaw Nursery LLC, personal communication). Currently, field testing of this rootstock grafted with the 'Hass' scion under different growing conditions in California had comparable productivity to other commercially available rootstocks, including 'Dusa'® and 'Toro Canyon' (Brokaw Nursery LLC, personal communication).

'Zutano' (California)

This G × M hybrid was initially noted for its tolerance to saline conditions and cold. It was extensively used as a seedling rootstock in Sunraysia, Australia, where the water quality is marginal for avocados. 'Zutano' was also the main rootstock used in New Zealand and is sometimes used as a 'nurse' seed when cloning rootstocks in California and South Africa (Alexander, 1983b; R. Hofshi, Fallbrook, California, 2001, *personal communication*). Seedlings result in variable trees with low tolerance to PRR and saline soil conditions (Crowley and Arpaia, 2002). 'Zutano' is a successful pollinizer for 'Hass'. 'Zutano' played a role in the development of the California industry due primarily to its cold tolerance; however its use as a seedling rootstock has decreased, because it is highly susceptible to the current *P. cinnamomi* populations in California (Belisle *et al.*, 2019) and often exhibits scion overgrowth when used as a rootstock. Despite these disadvantages, 'Zutano' is used as seedling rootstock in other cold production areas, including New Zealand and Chile (Frolich and Platt, 1972; Platt, 1975; Alexander, 1978; Banks, 1980; Alexander, 1983a; Arpaia, 1987; Faber *et al.*, 1995; Gardiazabal and Gandolfo, 1998; Anonymous, 2000a; Blight *et al.*, 2000; UCR, 2012a; Ge *et al.*, 2019).

Cultivars and Rootstocks of Tropical Climates

WI and WI hybrids are grown commercially and in subsistence agriculture in Africa (e.g. Cameroon, Kenya, Central African Republic and Democratic Republic of Congo), Asia (e.g. China, Indonesia, Philippines, Taiwan and Thailand), Central and South America (e.g. Perú, Brazil, Colombia, Costa Rica, Guatemala and Panama), the Caribbean (e.g. Cuba, Dominican Republic, Bahamas, Haiti, Jamaica, Puerto Rico and St. Lucia) and parts of North America (e.g. Veracruz, Chiapas, Sinaloa, Campeche, Yucatan and Quintana Roo in México; Florida, Puerto Rico and Hawaii in the USA; Anonymous, 2021a).

WI-type avocados were introduced into Florida from Cuba sometime before 1835 and were found naturalized in southern Florida by 1850 (Knight and Campbell, 1999). During the early 1900s, the U.S. Department of Agriculture introduced G and hybrid G × WI avocados, which subsequently crossed with the local seedling WI types. Soon afterward (1920s), W.J. Krome and W. Booth established private seedling avocado collections of G, WI and G × WI hybrids, and superior hybrids were subsequently selected, named and vegetatively propagated. Hybrid (mainly G × WI) and WI seedling selections by Florida residents and producers have continued throughout the 20th and 21st centuries and resulted in several hundred named cultivars. Since the early 1930s, many of these Florida selections have been introduced to numerous tropical countries throughout the world (e.g. Kenya/Africa, Colombia/South America, Costa Rica/Central America, Dominican Republic/Caribbean and Thailand/Asia) and formed the basis for their industries and/or their own cultivar selections (Campbell and Malo, 1976; Knight and Campbell, 1999; Papademetriou, 2000). Indeed, most countries have now selected their own cultivars from seedling selections, and many people have seedling trees in the home landscape as a source of the fruit. Most fruit from WI and G × WI cultivars are consumed locally by the countries that produce them, the Dominican Republic being an exception. Because of the importance and wide use of the Florida selections in tropical avocado production throughout the world and the Dominican Republic's large tropical avocado export industry, this section will focus on the Florida cultivars with some selections grown in the Dominican Republic.

The Florida Avocado Administrative Committee recognizes 24 major and 37 minor cultivars (Anonymous, 2023b). However, most of the production is derived from 17 major and ten minor cultivars. The Florida avocado season extends from mid-May to March, with July to December being the peak production period. Based on recent production, the top cultivars include 'Bernecker', 'Beta', 'Choquette', 'Donnie', 'Simmonds', 'Hall', 'Loretta', 'Lula', 'Miguel', 'Monroe', 'Nadir', 'Nesbitt' and 'Tower 2'. A brief description of the origin and characteristics of the major and selected minor cultivars is provided. The major cultivars grown in the Dominican Republic include 'Semil 34', 'Lula', 'Pollock', 'Melendez' and 'Carla'. Numerous cultivars are grown in Asia, and some of the production in Asian countries (e.g. Indonesia and Myanmar) is from unnamed seedling trees. The cultivar and rootstock descriptions and attributes are in alphabetical order to facilitate the reader's ability to locate the information upon future reference.

Cultivars

'Arue' (Florida)

'Arue' is a minor early WI ecotype that originated in the Society Islands and was introduced by the US Department of Agriculture, Agricultural Research Service (USDA-ARS). The fruit is pyriform in shape; medium to large in size, 80–105 mm in diameter and weighs 400–620 g; the peel is green with many pronounced yellow lenticels, smooth; pulp is of fair quality; the fruit oil content is 4%–6%. The seed is medium-sized and sits loose in the cavity (Fig. 8.14). The fruit has an early harvest season (late May to early June in Florida). The flowers belong to Group A. The tree has low cold tolerance. The safest low storage temperature is 13°C. The fruit is of fair quality with numerous green-yellow lenticels on the peel and fair postharvest handling (Anonymous, 2011c; Campbell, 1979; Chen *et al.*, 2009; Hatton *et al.*, 1963; 1964; 1965; Knight, Jr, 1971; Malo *et al.*, 1977; Popenoe, 1962; Rhodes *et al.*, 1971; C.F. Balerdi, Homestead, Florida, 2011, *personal communication*).

'Bernecker' (Florida)

A major WI cultivar, probably a seedling of 'Waldin'; flower Group A. The fruit is ellipsoid-spheroid, medium to large in size, weighing 397–623 g, typically measuring 82–86 mm in diameter but may be larger; the peel is light green to green and smooth (Fig. 8.15); the seed is medium-sized and sits tightly in the cavity. The fruit is harvested early in the season, and the quality of the pulp is good to excellent . The fruit oil content is unknown. The tree has low cold tolerance. The recommended postharvest storage temperature is 13°C. Defects include susceptibility to *Cercospora* spot (Hatton *et al.*, 1965;

Fig. 8.14. 'Arue'.

W.H. Krome, Homestead, Florida, 1999, *personal communication*; Schnell *et al.*, 2003; Kuhn *et al.*, 2019).

'Beta' (Florida)

A major WI × G hybrid from a 'Waldin' seedling selected at the home orchard of William H. Krome in Homestead, Florida, flower Group B. The fruit is elliptical, medium to large in size, 84–89 mm in diameter, and weighs 397–680 g. The peel of 'Beta' is green and smooth. The seed is medium in size and sits tightly in the cavity (Fig. 8.16). The fruit is harvested mid-season, and the quality of the pulp is fair; fruit oil content is unknown. The tree has moderate cold tolerance. The recommended postharvest storage temperature is 4°C–10°C. Postharvest 1-MCP treatments extended the shelf life by 75% without a noticeable decrease in fruit quality (Pereira *et al.*, 2013). Trees tend to overbear, and limb breakage and sunburn of outer canopy fruit can be problematic; however, with good cultural practices, trees usually recover (Hatton *et al.*, 1965; Campbell, 1979; Rouse and Knight, 1991; Crane *et al.*, 1996; W.H. Krome, Homestead, Florida, 1999, *personal communication*; Pereira *et al.*, 2013; Boza *et al.*, 2018; Ge *et al.*, 2019; FDACS-DFV, 2023; J.H. Crane, UF-IFAS, *personal communication*).

Fig. 8.15. 'Bernecker'.

Fig. 8.16. 'Beta'.

'Booth 7' (Florida)

'Booth 7' is a major G × WI hybrid cultivar from a seedling of unknown parentage originating at the orchard of W. Booth in Homestead, Florida. It first fruited in 1927, in flower Group B. The fruit is spheroid-obovate with a rounded fruit apex and level pedicel insertion, small to medium in size, weighing 283–567 g, and is 79–100 mm in diameter; the peel is bright green, slightly pebbled, glossy, thick and woody. The seed is medium in size and sits tightly in the cavity. Fruit is harvested mid-season to the beginning part of the late season; good-quality pulp; oil content is 7%–13%. The tree has moderate cold tolerance. The recommended postharvest fruit storage temperature is 12°C–13°C. Defects include overbearing, and the fruit is moderately susceptible to scab (Hatton *et al.*, 1963, 1964, 1965; Ruehle, 1963; Knight, 1971; Malo *et al.*, 1977; Crane *et al.*, 1996).

'Booth 8' (Florida)

'Booth 8' is a major G × WI hybrid cultivar from a seedling of unknown parentage originating at the orchard of W. Booth in Homestead, Florida. It first fruited in 1927; flower Group B. The fruit is oblong-obovate with a rounded apex and an asymmetrical pedicel insertion; it is small to medium-large in size, weighs 283–680 g and measures 78–100 mm in diameter; the peel is medium green, dull and slightly rough. The seed is medium in size and sits tightly in the cavity. The fruit is harvested from mid- to late season and yields pulp of fair to good quality; the oil content is 7%–13.1%. The tree has moderate cold tolerance. The recommended postharvest fruit storage temperature 4°C. Defects include overbearing, limb breakage, small fruit size and moderate susceptibility to scab (Ruehle, 1963; Hatton *et al.*, 1964, 1965; Knight, 1971; Malo *et al.*, 1977; Crane *et al.*, 1996).

'Carla' (Dominican Republic)

A patented G × WI hybrid from a seedling selection by C. Pimentel and C. Antonio in the Ocoa River Valley, Dominican Republic, in 2006 (Anonymous, 2006). The flower belongs to Group B. The fruit is obovate-spheroid with a round fruit apex and a somewhat flattened fruit bottom; asymmetrical pedicel insertion; occasionally fruit have a raised ridge (chimera) running along a side from the stem downward; medium to large in size, weigh 425–1243 g (920 g average) and measure 110–130 mm in diameter; the peel is green to light green and smooth to slightly rough. The seed is medium to large in size and sits tightly in the cavity. The fruit is harvested in the late season and yields excellent-quality pulp, a high percentage of which is edible; the fruit oil content is 8%–13%. The recommended postharvest storage temperature is 7.2°C–7.5°C (Anonymous, 2006). The cold tolerance of the tree is unknown. The trees tend to overbear with some limb breakage and sunburn of the outer canopy fruit. This is a major late-season cultivar grown in the Dominican Republic (Pérez, 2017; J.H. Crane, UF-IFAS, 2023, *personal observation*).

'Choquette' (Florida)

A major G × WI hybrid cultivar from a seedling of unknown origin selected at the property of R. D. Choquette in Miami, Florida. It first fruited in 1934 and belongs to flower Group A. The fruit are elliptical, flattened obliquely towards the apex on one side, with a central pedicel insertion; medium to large in size, weighing 510–1133 g, and measuring 95–111 mm in diameter but may be larger. The peel is light to dark green, nearly smooth with some undulations, glossy and somewhat leathery (Fig. 8.17). The seed is medium to large and fairly sits tightly in the cavity. The fruit is harvested mid- to late season and yields pulp of good to excellent quality; the oil content is 8%–13%. The tree has moderately high cold tolerance. The recommended postharvest fruit storage temperature is 4°C–10°C. Defects include some susceptibility to *Cercospora* spot (Ruehle, 1963; Hatton *et al.*, 1964, 1965; Knight, 1971; Malo *et al.*, 1977; Crane *et al.*, 1996; Schnell *et al.*, 2003; Boza *et al.*, 2018; Ge *et al.*, 2019).

'Donnie' (Florida)

A major WI cultivar; probably a seedling of 'Fuchs' or 'Waldin' originating at the Donaldson orchard in Homestead, Florida. The flowers belong to Group A. The fruit is oblong to elongated and pyriform in shape; medium to large, weighing 397–680 g and typically measuring 82–84 mm in diameter but may be larger; the peel is green and smooth (Fig. 8.18). The seed is medium in size, sits sits tightly in the cavity when first matures but then becomes loose. Fruit harvested in the early part of the season may not have

Fig. 8.17. 'Choquette'.

Fig. 8.18. 'Donnie'.

good-quality pulp but if picked later in the season yield good-quality pulp; the fruit oil content is unknown. The tree has low cold tolerance. The recommended postharvest fruit storage temperature is 13°C. The fruit have a relatively short postharvest storage life (Hatton *et al.*, 1965; Crane *et al.*, 1996; C.F. Balerdi, Homestead, Florida, 1999, *personal communication*; Ge *et al.*, 2019).

'Dr. Dupuis' (syn. 'Dupuis No. 2' and 'Dupuis') (Florida)

'Dr. Dupuis' is a minor cultivar, WI type from a seedling originating at the Dupuis orchard in Hialeah, Florida. Flowers belong to Group A. The fruit is obovate to pyriform in shape, medium to large, weighing 340–680 g and measuring 79–87 mm in diameter; the peel is green and nearly smooth. The seed is medium in size and sits tightly in the cavity. The fruit is harvested early in the season and yields pulp of excellent quality; the exact fruit oil content is unknown, but low. The tree has low cold tolerance. The recommended postharvest fruit storage temperature is 13°C. Defects include irregular bearing and difficulty in controlling tree size (Hatton *et al.*, 1965; Hatton and Reeder, 1965; Campbell, 1979; Crane *et al.*, 1996; W.H. Krome, Homestead, Florida, 1999, *personal communication*; Schnell *et al.*, 2003). 'Dr. Dupuis' is grown on a small scale in the Dominican Republic (Pérez, 2017).

'Hall' (Florida)

A major G × WI hybrid cultivar from a seedling of unknown parentage selected at the property of W. Hall in Miami, Florida, that first fruited in 1937. Flowers belong to Group B. The fruit is pyriform in shape with a rounded base; pedicel insertion is very asymmetrical; fruit are medium to large in size, weighing 510–850 g and typically measuring 89–98 mm in diameter but may be larger; the peel is nearly smooth, dark green and moderately thick. The seed is medium to large in size and sits tightly in the cavity. The fruit is harvested late in the season and has a good-quality pulp; the oil content is 10%–16%. The tree has good cold tolerance. The recommended postharvest fruit storage temperature is 4°C–10°C. Defects include susceptibility to scab, *Cercospora* spot, anthracnose and to limb breakage (Ruehle *et al.*, 1949; Ruehle, 1963; Hatton *et al.*, 1964, 1965; Malo *et al.*, 1977; Crane *et al.*, 1996; Boza *et al.*, 2018).

'Hardee' (Florida)

'Hardee' is a minor cultivar, a WI type from a seedling of unknown parentage selected at the property of Capt. Thomas Hardee in Coconut Grove, Florida, in about 1908. The flower belongs to Group B. The fruit is oblong to elongated and pyriform in shape, medium to large, weighing 340–680 g and typically measuring 75–90 mm in diameter but may be larger; the peel is pale yellow, developing a red colour at maturity and smooth. The seed is medium in size and sits loose in the cavity, sometimes accompanied by water. The fruit is harvested early in the season and yields pulp of fair quality; the oil content is 4%–6%. The tree has low cold tolerance. The recommended postharvest storage temperature is 13°C. Defects include water in the seed cavity and poor postharvest handling characteristics (Wolfe *et al.*, 1946; Hatton *et al.*, 1964, 1965; Malo *et al.*, 1977; Campbell, 1979; Crane *et al.*, 1996).

'Loretta' (Florida)

A major cultivar of G × WI parentage from a seedling found at the Munz orchard in Homestead, Florida. Flowers belong to Group B. The fruit is elliptical and elongated, medium to large in size, weighing 567–1020 g and measuring 100–106 mm in diameter; the peel is dark green, smooth to slightly pebbly (Fig. 8.19). The seed is very small and sits tightly in the cavity. The fruit is harvested mid-season and yields pulp of fair to good quality; the fruit oil content is unknown. The tree has low-to-moderate cold tolerance. The recommended postharvest fruit storage temperature is 4°C–10°C. The fruit is moderately susceptible to scab and *Cercospora* spot (Hatton *et al.*, 1965; Crane *et al.*, 1996; W.H. Krome, Homestead, Florida, 1999, *personal communication*; Ge *et al.*, 2019).

'Lula' (Florida)

'Lula' is a minor cultivar, a G × WI hybrid from a seedling of 'Taft' selected at the property of G.B. Cellon in Miami, Florida, and named after his wife. It first fruited in 1919; flower belongs to Group A. The fruit is pyriform in shape with a rounded apex, level pedicel insertion; medium to large in size, weighing 397–680 g and measuring 81–105 mm in diameter; the peel is green to dark green and slightly rough. The seed is large and sits tightly in the cavity (Fig. 8.20). The fruit is harvested late in the season and yields pulp of good to excellent quality (slightly sweet), about 65% of which is edible; the fruit oil content is 6.0%–15%. The tree has high cold tolerance. The recommended postharvest fruit storage temperature is 4°C. Defects include fruit being highly susceptible to scab and difficulty in controlling tree size. It is grown for fruit production, as a seedling rootstock in Florida and grown commercially in the Dominican Republic for fresh fruit (Beach, 1920; Toy, 1929; Stahl, 1931; Fairchild, 1945; Ruehle, 1963; Hatton *et al.*, 1964, 1965; Knight, 1971; Malo *et al.*, 1977; Crane *et al.*, 1996; Schnell *et al.*, 2003; Ge *et al.*, 2019).

Fig. 8.19. 'Loretta'.

'Miguel' (Florida)

A major WI cultivar from a seedling selected at the Kendall Foods' Krystow orchard. The flowers belong to Group B. The fruit is elliptical, medium to large in size, weighing 510–850 g and typically measuring 92–97 mm in diameter, but may be larger; the peel is dark green and smooth. The seed is medium in size and sits tightly in the cavity (Fig. 8.21). The fruit is harvested early to mid-season and yields pulp of good to excellent quality; the fruit oil content is unknown. The tree has low cold tolerance. The recommended postharvest fruit storage temperature is 4°C–10°C. Defects include overbearing, limb breakage and fruit being susceptible to *Cercospora* spot (Hatton *et al.*, 1965; Campbell, 1979; Crane *et al.*, 1996; C.F. Balerdi, Homestead, Florida, 1999, *personal communication*; Boza *et al.*, 2018; Ge *et al.*, 2019).

'Monroe' (Florida)

A major G × WI hybrid cultivar from a seedling of unknown parentage originating at the orchard of J. J. L. Phillips in Homestead, Florida.

Fig. 8.20. 'Lula'.

Fig. 8.21. 'Miguel'.

It first fruited in 1935; flower belongs to Group B. The fruit are elliptical, flattened obliquely towards the apex on one side, with asymmetrical pedicel insertion; medium to large in size, weighing 453–1133 g and measuring 92–111 mm in diameter; the peel is dark green, glossy and slightly rough. The seed is medium to large in size and sits tightly in the cavity (Fig. 8.22). It is harvested late in the season and yields pulp of good to excellent quality; the oil content is 6.1%–14%. The tree has moderate cold tolerance. The safest fruit storage temperature is 4°C–10°C. Defects include fruit being moderately susceptible to scab, *Cercospora* spot and anthracnose; fruit drop; and limb breakage (Ruehle *et al.*, 1949; Ruehle, 1963; Hatton *et al.*, 1964, 1965; Knight, 1971; Crane *et al.*, 1996; Schnell *et al.*, 2003).

'Nadir' (Florida)

A minor cultivar, presumably a G × WI hybrid (possibly WI race only) from a seedling selected at the Lipchurch orchard in Homestead, Florida. The flower belongs to Group A. The fruit is pyriform in shape, small to medium-large in size, weighing 283–623 g, and typically measuring 57–86 mm in diameter but may be larger; the peel is green and smooth. The seed is medium in size and sits tightly in the cavity. The fruit is harvested early in the season and yields good-quality pulp; the fruit oil content is 6%–7%. The tree has very low cold tolerance. The recommended postharvest storage temperature is 4°C–10°C. Defects include irregular bearing, overbearing and limb breakage (Hatton *et al.*, 1963, 1965; Knight, 1971; Rhodes *et al.*, 1971; Malo *et al.*, 1977; Crane *et al.*, 1996; W.H. Krome, Homestead, Florida, 1999, *personal communication*).

Fig. 8.22. 'Monroe'.

'Nesbitt' (Florida)

A major G × WI hybrid cultivar from a seedling tree found at Mrs Nesbitt's home in Homestead, Florida. The flower belongs to Group A. The fruit is elliptical-pyriform in shape, medium to large in size, weighing 397–737 g and measuring 81–95 mm in diameter; the peel is green and smooth with pronounced lenticels. The seed is medium in size and sits tightly in the cavity. The fruit is harvested early in the season and yields good-quality pulp; the fruit oil content is unknown. The tree has low cold tolerance. The recommended postharvest fruit storage temperature is 4°C–10°C. Other than low cold tolerance, the cultivar has no major defects (Hatton *et al.*, 1965; Malo *et al.*, 1977; Crane *et al.*, 1996; W.H. Krome, Homestead, Florida, 1999, *personal communication*).

'Pollock' (Florida)

'Pollock' is a minor cultivar of WI background from a seedling of unknown parentage selected at the property of H.S. Pollock in Homestead, Florida. It first fruited in about 1896. Flowers belong to Group B. The fruit is oblong to pyriform in shape with a rounded base and an asymmetrical pedicel insertion; medium to large in size, weighing 397–1416 g and measuring 84–100 mm in diameter, although it may be larger; the peel is green to dark green, smooth, thin and glossy. The seed is medium to large in size and often sits loose in the cavity (Fig. 8.23). The fruit is harvested early in the season and yields excellent-quality pulp; 78% of the pulp is edible; the fruit oil content is 4%–8%. The tree has low cold tolerance. The recommended postharvest fruit storage temperature is 13°C. Its major defect is poor fruit production (Toy, 1929; Stahl, 1931; Ruehle *et al.*, 1949; Campbell, 1979; Hatton *et al.*, 1963, 1964, 1965; Ruehle, 1963; Malo *et al.*, 1977; Crane *et al.*, 1996; Schnell *et al.*, 2003; Boza *et al.*, 2018). 'Pollock' is grown commercially on a very small scale in Florida but is

Fig. 8.23. 'Pollock'.

Fig. 8.24. 'Simmonds'.

moderately important in the Dominican Republic (sometimes under the name 'Poll') (Pérez, 2017).

'Simmonds' (Florida)

A major WI cultivar from a 'Pollock' seedling selected at the United States Department of Agriculture, Subtropical Horticulture Research Station in Miami, Florida. It first fruited in 1913 The flower belongs to Group A. The fruit is obovate, flattened obliquely towards the apex on one side, with asymmetrical pedicel insertion; medium to large in size, weighing 453–963 g and measuring 78–98 mm in diameter; the peel is light green to green, smooth and glossy. The seed is medium in size and sits tightly in the cavity (Fig 8.24). The fruit is harvested early in the season and yields pulp of good to excellent quality; about 76% of the pulp is edible; and the oil content is 3.3%–5.0%. The recommended postharvest storage temperature is 13°C. The tree has low cold tolerance. Other defects include lack of vigour and excessive fruit drop (Toy, 1929; Stahl, 1931; Ruehle *et al.*, 1949; Ruehle, 1963; Hatton *et al.*, 1964, 1965; Knight, 1971; Malo *et al.*, 1977; Crane *et al.*, 1996; Schnell *et al.*, 2003; Ge *et al.*, 2019; Boza *et al.*, 2018).

'Tower 2' (Florida)

A major WI cultivar, probably from a seedling of 'Waldin'. The flower belongs to Group B. The fruit is ellipsoid-spheroid in shape, small to medium in size, weighing 340–565 g and typically measuring 82–86 mm in diameter but may be larger; the peel is green and smooth. The seed is medium in size and sits tightly in the cavity. The fruit is harvested mid-season and yields pulp of good to excellent quality; the fruit oil content is unknown. The tree has low cold tolerance. The recommended postharvest storage temperature is 4°C–10°C (Hatton *et al.*, 1965; Crane *et al.*, 1996; W.H. Krome, Homestead, Florida, 1999, *personal communication*; Schnell *et al.*, 2003).

'Semil 34' (Dominican Republic)

A major WI × G hybrid cultivar from a seedling selected at Finca El Semil, Villalba, Puerto Rico, in 1947. The flower belongs to Group A. The fruit is obovate/pyriform in shape and flattened obliquely toward the apex on one side; with asymmetrical pedicel insertion; medium to large in size, weighing 396–700 g and typically measuring 84–92 mm in diameter, but may be larger; the peel is green, slightly bumpy to smooth and glossy. The seed is medium in size and sits tightly in the cavity (Fig. 8.25). The fruit is harvested mid-season and yields pulp of good to excellent quality; about 78% of the pulp is edible; the fruit oil content is 7.5%–15.0%. The recommended postharvest fruit storage temperature is 13°C (Abram, 1960; Jeanteur, 1970; Knight, 1971; Ben-Ya'acov *et al.*, 2003; Schnell *et al.*, 2003; Ayala-Silva *et al.*, 2007; Cuevas, 2007; Boza *et al.*, 2018). 'Semil 34' is the major tropical cultivar grown in and exported from the Dominican Republic (Cuevas, 2007; Pérez, 2017).

Fig. 8.25. 'Semil 34'.

Fig. 8.26. 'Wheeling'.

'Tonnage' (Florida)

'Tonnage' is a minor cultivar. A G × WI hybrid from a seedling of 'Taylor' selected at the orchard of D.M. Roberts in Homestead, Florida. It first fruited in 1921, with flower belonging to Group B. The fruit is pyriform in shape with a rounded base; the pedicel insertion is very asymmetrical; fruit is small to medium in size, weighing 340–567 g and measuring 76–90 mm in diameter, but may be larger; the peel is pebbled, dark green, glossy and thick. The seed is medium in size and fairly sits tightly in the cavity. The fruit is harvested mid-season and yields fair-quality pulp with oil content of 8%–10%. The tree has high cold tolerance. The recommended postharvest fruit storage temperature is 4°C–10°C. 'Tonnage' has fair fruit quality, an upright tree growth habit and the fruit is moderately susceptible to scab (Toy, 1933; Ruehle *et al.*, 1949; Ruehle, 1963; Hatton *et al.*, 1964, 1965; Knight, 1971; Campbell, 1979; Crane *et al.*, 1996; Schnell *et al.*, 2003).

'Wheeling' (Florida)

'Wheeling' is a minor cultivar of purported G × WI origin patented by J.R. Brooks, Brooks Tropicals, Homestead, Florida (Anonymous, 2007). The tree has a B-type flowering habit. The fruit is ovate with a smooth-textured, thick, hard-shelled exocarp, dark green with numerous yellowish-green lenticels (Fig. 8.26). The fruit weight is -340–453 g and it averages 114 mm in length. The seed is medium sized; the pulp is green near the peel and yellow near the seed. The fruit is harvested late in the season and yields excellent-quality pulp, and the oil content is not known. The tree has good cold tolerance, and fruit are tolerant of *Cercospora purpurea* and *Colletotrichum gloeosporoides*.

Rootstocks

A significant amount of research on the characteristics of all three avocado races and hybrids for use as rootstocks for saline soil and water conditions and tolerance to PRR has occurred since the late 1930s (Ben-Ya'acov, 1995a; Ben-Ya'acov and Michelson, 1995). Previous research in Israel identified superior salinity tolerance of some WI and G × WI hybrids (Ben-Ya'acov and Kadman, 1976). Rootstocks of WI origin used in Israel included 'Maoz', 'Shiller No. 10', 'Degania 117', 'VC 6', 'VC 28', 'VC 51', 'VC 66' and 'Zrifin 99' (Ben-Ya'acov, 1995b). More recent releases include 'Ben-Ya'acov1' (Israel), 'Zerala' (South Africa) and 'Steddom' (Menge *et al.*, 2012; Celis *et al.*, 2018; Anonymous, 2019a; Karp and Gasic, 2022).

In the Caribbean region and Brazil, mostly WI seedlings are used for rootstocks (Ben-Ya'acov and Michelson, 1995; Moraes *et al.*, 2022). WI seedling rootstocks are also used in parts of the Canary Islands because of their salinity tolerance (Galan-Sauco and Fernandez-Galvan, 1985). WI seedling rootstocks were investigated and used in Texas because of their superior salinity tolerance (Cooper, 1951) and in Florida because of their tolerance to alkaline soils and better iron absorption compared to seedling rootstocks of G or M origin (Ruehle, 1963).

Early experience in Florida showed that seeds of 'Lula' (G × WI) and 'Waldin' (WI) produced a high percentage of vigorous, relatively uniform seedling rootstocks that appeared well-adapted to the rocky, highly calcareous soils of south Miami-Dade County (Ruehle, 1963). Thus, seedlings of these two cultivars (especially 'Waldin') have continued to be used commercially in Florida. 'Lula' is recommended as a salinity-tolerant rootstock in Texas (J.L. Jifon, Texas A&M, *personal communication*). 'Lula' is also a minor fresh-market cultivar in Florida.

Historically, there has been limited rootstock research in Florida. This was due to the general satisfaction with the performance of 'Waldin' and 'Lula' seedling rootstocks and that PRR is not a serious disease of avocados in Florida unless flooding occurs (Ploetz and Schaffer, 1989; Schaffer and Ploetz, 1989; Ploetz *et al.*, 2002; Schaffer *et al.*, 2006). Thus, until recently, the impetus for a systematic rootstock programme was lacking. More recently, water table management in south Florida has led to higher groundwater levels and increased flooding frequency and duration. This, along with the introduction and destruction caused by the LW pathogen (*H. lauricola*), has resulted in renewed interest in developing PRR- and LW-tolerant or resistant rootstocks (Ploetz *et al.*, 2002; J.H. Crane, UF-IFAS, personal observation).

Early rootstock investigations in Florida compared the yields of 9-year-old 'Booth 7' (G × WI) and 'Lula' grafted onto seedling rootstocks of 'Waldin' or 'Lula' (Harkness and Lincoln, 1955). Over a 5-year period, 'Booth 7' yields were not significantly affected by rootstock source whereas 'Lula' grafted onto 'Lula' seedling rootstock yielded significantly more fruit than 'Lula' on 'Waldin' seedling rootstock. However, when these same trees were 15–18 years old, yields for 'Booth 7' and 'Lula' on 'Waldin' seedling rootstock were consistently greater than those of 'Booth 7' or 'Lula' on 'Lula' seedling rootstock (Malo and Harkness, 1965). Hence, the limited history of these trees suggests that the performance of scion/rootstock combinations may change as trees age. Krezdorn (1973) reported that rootstocks of G and/or M origin did not improve the cold-hardiness of WI scions.

A PRR rootstock program in Florida screened thousands of seedlings from 51 accessions of diverse ecotype backgrounds (Ploetz *et al.*, 2002; Violi *et al.*, 2006). Far more seedlings of WI (15.7%) and G × WI (14.6%) background were tolerant compared to seedlings of other pedigrees (3.1%). 'Choquette' (G × WI) was grafted onto six clonal PRR-tolerant rootstocks from California (i.e. 'Bo Borchard', 'Duke 7', Parida', 'Thomas', Toro Canyon', G755b, G755c and G1033), and the hybrids were evaluated for their tolerance to alkaline soil conditions in south Florida (Ploetz *et al.*, 1989). 'Choquette' grafted onto 'Parida' was significantly more iron chlorotic than 'Choquette' on all other rootstocks growing in alkaline soil, indicating that tolerance to high soil pH was an important attribute for production in areas with alkaline soils (Ploetz *et al.*, 1989). Furthermore, inheritance of PRR tolerance is most likely multigenic, and 0.45 broad-sense heritability of this trait suggests significant progress could be made by screening additional progeny of the PRR-tolerant accessions. Data mining of publicly available expressed sequence tags from the Plant Genome Network (PGN, http:/pgn.cornell.edu/) resulted in an additional 70 informative microsatellite loci from 24 WI avocado accessions that will be useful in marker-assisted development of PRR-tolerant rootstocks (Borrone *et al.*, 2007).

More recently, the need to identify the susceptibility and tolerance mechanisms and to understand the genetics of avocado scions and rootstocks to the lethal LW pathogen, *H. lauricola* (*Hl*), has become critical for the survival of commercial avocado production in Florida. This is because LW kills mature trees 4–8 weeks after inoculation with the pathogen. The pathogen is vectored at least four ambrosia beetle (AB) species (Carrillo *et al.*, 2012, 2014; Cruz *et al.*, 2021; see Dann and Gazis, Chapter 13 and Carrillo *et al.*, Chapter 14, this volume) and moves through root grafts of LW-affected adjacent avocado trees (Crane *et al.*, 2024; Ploetz *et al.*, 2017). In addition, to avocado, LW affects native trees in the Lauraceae and multiple AB species have been spreading the pathogen through natural and urban areas. Currently, LW has been detected in 12 south-eastern USA. states and is moving west toward California and México, threatening native lauraceous plants and avocado production (Kendra *et al.*, 2023; Crane *et al.*, 2024).

An investigation to identify tolerance or resistance to the lethal LW , *H. lauricola* (*Hl*), on 24 young container-grown avocado cultivars of diverse genetic backgrounds (i.e. G, G × M, G × M × WI, G × WI and WI) grafted onto seedling rootstock suggested LW disease severity of WI ecotypes was greater than G × M, G × M × WI and G × M cultivars (Ploetz *et al.*, 2012). The disease severity of G × M (e.g. 'Hass' and 'Winter Mexican') and G (e.g. 'Bacon' and 'Reed') cultivars progressed more slowly within 2 months of *Hl* inoculation compared to WI (e.g. 'Simmonds' and 'Donnie') and G × WI (e.g. 'Hall' and 'Monroe') types. However, mature trees of 34 commercial avocado cultivars of diverse genetic backgrounds have been documented to succumb to LW (Table 8.1).

The effect of *Hl* on avocado physiology and xylem anatomy determined *Hl* induced significant tylose formation in vessel elements which blocks water movement (Inch *et al.*, 2012; Ploetz *et al.*, 2015; Beier *et al.*, 2021). As the amount of tylose formation increased, sap flow and gas exchange were reduced, and external and internal plant symptoms increased. Xylem vessel element diameter is greatest for WI ecotypes compared to the M, G and G × M hybrid types and the rate of internal and external LW symptoms increased with xylem vessel element diameter (Castillo-Argaez *et al.*, 2020, 2022; Inch *et al.*, 2012; Ploetz *et al.*, 2012; Ploetz *et al.*, 2015; Beier *et al.*, 2021).

LW significantly reduced carbon assimilation and gas exchange of non-grafted clonal 'Zutano' (M × G) and 'Nabal' (G) more rapidly than 'Waldin' (WI) trees, and the diameter of xylem vessel elements of 'Zutano' and 'Nabal' was significantly greater than 'Waldin' (Castillo-Argaez *et al.*, 2021; See Schaffer *et al.*, Chapter 7, this volume). The sap flow of *Hl*-inoculated 'Zutano' and 'Nabal' was reduced more than that of 'Waldin'. A comparison of multiple scion–rootstock combinations demonstrated that susceptibility to LW is affected by scion, rootstock, scion/rootstock combination and rootstock pedigree (i.e. open-pollinated seedling vs. clonal) (Castillo-Argaez *et al.*, 2020, 2021). For example, 'Duke 7'/ 'Duke 7' (scion/rootstock) was much more susceptible to LW than 'Duke 7'/'Reed' and sap flow rate and symptom development of 'Duke 7'/'Reed' and 'Reed'/'Reed' was lower and slower, respectively, compared to 'Waldin'/'Reed'. These results indicate rootstock and scion can affect xylem anatomy, physiological parameters and tolerance to a xylem-mobile pathogen. Understanding the physiological basis for susceptibility to *Hl* continues, and a project to screen for and identify the parts of the avocado genome that influence LW susceptibility of a wide diversity of avocado germplasm is under way (R. Gazis, UF-IFAS, *personal communication*). At this time, no LW-resistant rootstocks have been identified.

'Waldin' (Florida)

A WI type from a seedling of unknown parentage selected at the property of B.A. Waldin in Homestead, Florida. It first fruited in 1913; flower belongs to Group A. Fruit are frequently produced in clusters; the fruit is ellipsoid-spheroid, flattened obliquely toward the apex on one side, has a central pedicel insertion, medium to large in size, weighing 397–680 g and typically measuring 87–102 mm in diameter, but may be larger; the peel is pale green to greenish yellow, smooth and leathery. The seed is medium to large and sits fairly tightly in the cavity. The seed is used to propagate the rootstock. Fruit is harvested mid-season, yielding good to excellent-quality pulp; about 64% of the pulp is edible; the oil content is 6%–12%. The tree has low cold tolerance. The recommended postharvest storage temperature is 13°C. Seedlings are vigorous and easily grafted or budded. Defects include large seed, susceptibility to *Cercospora* spot and tree susceptibility to Zn, Fe and Mn deficiency (Goldberg, 1921; Stahl, 1931; Ruehle *et al.*, 1949; Hatton *et al.*, 1963, 1964, 1965; Ruehle, 1963; Knight, 1971; Malo *et al.*, 1977; Crane *et al.*, 1996; Schnell *et al.*, 2003; Boza *et al.*, 2018). This is a minor fruit cultivar in Florida and is grown primarily as a source of seedling rootstock. 'Waldin' is also used as a nurse seed for the clonal propagation of PRR-tolerant rootstocks (see Whiley *et al.*, Chapter 9 of this volume for detained description of clonal propagation of avocado).

Conclusions

Avocado trees have been mostly vegetatively propagated, mainly onto seedling rootstocks, for 130 years. 'Fuerte' was selected over 100 years ago and became the first-generation cultivar 'standard of excellence' for California and the world's subtropical growing areas. Its decline in the 1950s was due to low and irregular yield in California and postharvest fruit problems. The black-skinned 'Hass', which replaced it, partially remedied these problems and is now, by far, the world's most widely grown and exported single cultivar. Like hundreds of other named cultivars, most of which were not widely grown, 'Fuerte' and 'Hass' were selections from chance seedlings.

The need for a wider range of elite cultivars led to dedicated breeding programmes, involving the screening of thousands of seedlings from both purposeful crosses and open-pollinated seeds. These programmes were also aimed at selecting rootstocks with tolerance to PRR (California, and later South Africa, Australia and others) and to soil salinity and calcareous soils (i.e. Israel; see Hormaza *et al.*, Chapter 4, this volume). 'Duke 7' was the first clonal PRR-tolerant rootstock released by the California program and has been widely used there and in South Africa for over 40 years. The tropical lowland avocado industries have been mainly served by cultivars developed for markets that are familiar with large avocados and by local selections made in countries located in the warm subtropics and lowland tropics.

The difficulties of conventional avocado breeding are underscored by the successive dominance for over 120 years of only two cultivars in the subtropics and a range of seedling selections in the tropics. Several 'Hass'-like selections released in the past 10 to 20 years extend the 'Hass' season with good-quality larger fruit and have gained some wider acceptance, for example, 'Lamb Hass', 'GEM' and 'Hass Carmen'. There has also been good progress in rootstock breeding and research. The 'Dusa'® clonal rootstock from South Africa, plus several clonal rootstocks from the California program provide both good PRR tolerance as well as impart good scion yield and fruit quality and tolerance of some soil stress factors. 'Provacado' rootstock from Australia appears promising, but only testing in various edaphic and environmental conditions in the presence of PRR will indicate its suitability.

Both scion and rootstock breeding are expected to benefit the industry in the second and third decades of the 2000s from advances in genomics and biotechnology. The world's subtropical and tropical (lowland) industries will be best served in the future by a range of three to five elite, proven cultivars and a similar number of proven rootstocks, probably both clonal and seedling. Novel cultivars will cater to changing tastes and different markets. A true and lasting PRR resistance, plus desirable horticultural attributes for varying conditions, remains a challenge for avocado breeders and researchers. A new and equally important challenge is to identify and select for LW-resistant scions and rootstocks, a challenge that Florida and México are working on.

References

Abram, P.R. (1960) El cultivo del aguacate en Puerto Rico. *Revista de Agricultura de Puerto Rico* 47, 68–74.

Acosta-Rangel, A.M., Rui, L., Celis, N., Suarez, D.L., Santiago, L.S., Arpaia, M.L. and Mauk, P.A. (2019) The physiological response of 'Hass' avocado to salinity as influenced by rootstock. *Scientia Horticulturae* 256(3), 108629. doi:10.1016/j.scienta.2019.108629.

Alcaraz, M.L. and Hormaza, J.I. (2007) Molecular characterization and genetic diversity in an avocado collection of cultivars and local Spanish genotypes using SSRs. *Hereditas* 144, 244–253.

Alexander, D.McE. (1978) *Some Avocado Varieties for Australia*. Commonwealth Scientific and Industrial Research Organisation, Australia.

Alexander, D.McE. (1983a) Avocado performance in Sunraysia. In: Allen, R.N. (ed.) *Proceedings of the 2nd Australian Avocado Research Workshop*. Agriculture Research Centre, Wollongbar, Australia, pp. 41–43.

Alexander, D.McE. (1983b) Some effects of frost damage, scions and rootstocks of avocados in Sunraysia. In: Allen, R.N. (ed.) *Proceedings of the 2nd Australian Avocado Research Workshop*. Agriculture Research Centre, Wollongbar, Australia, pp. 17–22.

Allen, R.N. (1977) Virus indexing. In: *Proceedings of the Australian Avocado Research Workshop*, Lamington National Park, Queensland, Australia, pp. 36–40.

ANFIC (2023) Australian Nurserymen's Fruit Improvement Company. Queensland, Australia. Available at: https://anfic.com.au/variety/bounty-avocado-rootstock/ (accessed 27 November 2023).

Anonymous (1946) Checklist of avocado varieties. *California Avocado Society Yearbook* 32:49.

Anonymous (1956) Avocado varieties registered, January 1, 1955 to June 1, 1956. *California Avocado Society Yearbook* 40:27.

Anonymous (1974) The 'Hass' mother tree. *California Avocado Society Yearbook* 57, 16–17.

Anonymous (1996a) Avocado tree named 'Sir-Prize'. U.S. Plant Patent USPP 9,709 (Issued 26 November 1996). Available at: https://ppubs.uspto.gov/pubwebapp/static/pages/ppubsbasic.html (accessed 4 March 2024).

Anonymous (1996b) Avocado tree named 'Lamb/Hass'. U.S. Plant Patent USPP 9,753 (Issued 17 December 1996). Available at: https://ppubs.uspto.gov/pubwebapp/static/pages/ppubsbasic.html (accessed 4 March 2024).

Anonymous (2000a) California Avocado Commission, Santa Ana, California.

Anonymous (2000b) Panel on the 'Lamb Hass' avocado variety. *California Avocado Society Yearbook* 84, 39–57.

Anonymous (2003a) Avocado tree named 'Harvest'. U.S. Plant Patent USPP 14,238 P3 (Issued 14 October 2003). Available at: https://ppubs.uspto.gov/pubwebapp/static/pages/ppubsbasic.html (accessed 4 March 2024).

Anonymous (2003b) Avocado tree named '3-29-5'. U.S. Plant Patent USPP 14,239 P3 (Issued 14 October 2003). Available at: https://ppubs.uspto.gov/pubwebapp/static/pages/ppubsbasic.html (accessed 4 March 2024).

Anonymous (2006) Avocado tree named 'Carla'. *United States Patent and Trademark Office: Department of Commerce*. Available at: USPP16594P3 - Avocado tree named 'Carla' (Issued 30 May 2006).

Anonymous (2007) Avocado tree named 'Wheeling'. U.S. Plant Patent USPP US 2007/0044183 P1, U.S. Plant Patent Application Publication (Issued 22 February, 2007).

Anonymous (2010) Avocado tree named 'Maluma'. U.S. Plant Patent USPP 21,099 P3 (Issued 29 June 2010). Available at: https://ppubs.uspto.gov/pubwebapp/static/pages/ppubsbasic.html (accessed 4 March 2024).

Anonymous (2011c) Avocados grown in south Florida and imported avocados: revision of the maturity requirements for fresh avocados. *Federal Register* 64(190), 53181–53186.

Anonymous (2011e) Gwen avocado. U.S. Plant Patent USPP 5,298 (Issued 23 October 1984). Available at: http://www.google.com/patents?id=XKgAAAEBAJ&printsec=abstract&source=gbs_overview_r&cad=0#v= onepage&q&f=false (accessed 9 August 2011).

Anonymous (2011f) Lamb Hass avocado. U.S. Plant Patent USPP 9,753 (Issued 17 December 1996). Available at: http://www.google.com/patents?id=iGwqAAAAEBAJ&pg=PA1&dq=Lamb+Hass+avocado&hl=en&ei =4oBCTsPcCcny0gHU2ti9CQ&sa=X&oi=book_result&ct=result&resnum=1&ved=0CCkQ6aEwAA#v=onepage&q&f=false (accessed 9 August 2011).

Anonymous (2012a) Carmen avocado. U.S. Plant Patent USPP 11,173 (Issued 4 January 2000). Available at: http://www.google.com/patents?id=ZQUBAAAAEBAJ&pg= PA1&dq=U.S.+patent+'Mendez+1'+avocado&hl=en&sa=X&ei=neQJT6m_K4mltwe4ncjQBg&ved=0CDIQ6aEwAA#v=onepage&q=U.S.%20patent%20'Mendez%201'%20avocado&f=false (accessed 8 January 2012).

Anonymous (2012b) GEM avocado. U.S. Plant Patent USPP14239 (Issued 14 October 2003). Available at: http://www.google.com/patents?id=MGCXAAAAEBAJ&pg=PA1&dq=GEM+avocado&hl=en&sa=X&ei=otsJT7CuH9ShtweynlDQBg&ved=0CDMQ6aEwAA#v=onepage&q=GEM%20avocado&f=false (accessed 8 January 2012).

Anonymous (2012c) Harvest avocado. U.S. Plant Patent USPP 14,238 P3 (Issued 14 October 2003). Available at: http://www.google.com/patents/USPP14238?printsec=abstract&dq=Harvest+avocado&ei=wh0LT_aWMIr4tgf3r7GoBw&source=gbs_overview_r&cad=0#v=onepage&q=Harvest%20avocado&f=false (accessed 9 January 2012).

Anonymous (2012d) Mexicola avocado. *Avocado Information, Avocado Varieties, University of California, Agriculture and Natural Resources.* Available at: http://ucavo.ucr.edu/AvocadoVarieties/VarietyFrame.html#Anchor-47857 (accessed 9 January 2012).

Anonymous (2014a) Avocado rootstock named 'Zentmyer'. U.S. Plant Patent USPP 24,258 P3 (Issued 25 February 2014). Available at: https://ppubs.uspto.gov/pubwebapp/static/pages/ppubsbasic.html (accessed 4 March 2024).

Anonymous (2014b) Avocado variety named 'Uzi'. U.S. Plant Patent USPP 24,278 P3 (Issued 4 March 2014). Available at: https://ppubs.uspto.gov/pubwebapp/static/pages/ppubsbasic.html (accessed 4 March 2024).

Anonymous (2014c) Avocado rootstock named 'Steddom'. U.S. Plant Patent USPP 24,279 P3 (Issued 4 March 2014). Available at: https://ppubs.uspto.gov/pubwebapp/static/pages/ppubsbasic.html (accessed 4 March 2024).

Anonymous (2015) Avocado tree named 'Eugenin'. U.S. Plant Patent USPP 25,635 P3 (Issued 23 June 2015). Available at: https://ppubs.uspto.gov/pubwebapp/static/pages/ppubsbasic.html (accessed 4 March 2024).

Anonymous (2016) Avocado tree named 'Flavia'. U.S. Plant Patent USPP 26,798 P3 (Issued 7 June 2016). Available at: https://ppubs.uspto.gov/pubwebapp/static/pages/ppubsbasic.html (accessed 4 March 2024).

Anonymous (2018a) Avocado rootstock named 'KB1'. U.S. Plant Patent USPP 29,767 P3 (Issued 23 October 2018). Available at: https://ppubs.uspto.gov/pubwebapp/static/pages/ppubsbasic.html (accessed 4 March 2024).

Anonymous (2018b) Tami avocado. U.S. Plant Patent US 2018/0103567 P1 (Issued 12 April 2018). Available at: https://ppubs.uspto.gov/pubwebapp/static/pages/ppubsbasic.html.

Anonymous (2019a) Ben-Ya'acov1 avocado. U.S. Plant Patent US PP30,447 P3 (Issued 30 April 2019). Available at: https://ppubs.uspto.gov/pubwebapp/static/pages/ppubsbasic.html.

Anonymous (2019b) Miriam avocado. U.S. Plant Patent US PP30,353 P3 (Issued 9 April 2019). Available at: https://ppubs.uspto.gov/pubwebapp/static/pages/ppubsbasic.html.

Anonymous (2020a) Avocado rootstock named 'Merensky 5'. U.S. Plant Patent USPP 32,464 P2 (Issued 17 November 2020). Available at: https://ppubs.uspto.gov/pubwebapp/static/pages/ppubsbasic.html (accessed 4 March 2024).

Anonymous (2020b) Avocado rootstock named 'Merensky 6'. U.S. Plant Patent USPP 32,465 P2 (Issued 17 November 2020). Available at: https://ppubs.uspto.gov/pubwebapp/static/pages/ppubsbasic.html (accessed 4 March 2024).

Anonymous (2021a) Avocados area harvested 2021. Food and Agriculture Organization Statistics Database. FAOSTAT. Available at: http://faostat.fao.org/faostat/en/#data/QCL (accessed 11 March 2023).

Anonymous (2021b) Avocado tree named 'SHSR-04. U.S. Plant Patent USPP 33,355 P3 (Issued 17 August 2021). Available at: https://ppubs.uspto.gov/pubwebapp/static/pages/ppubsbasic.html (accessed 4 March 2024).

Anonymous (2022) Researchers map Hass avocado genome. Available at: https://phys.org/news/2022-11-hass-avocado-genome.html (accessed 23 February 2023).

Anonymous (2023a) Avocado tree named 'BL516'. U.S. Plant Patent USPP 35,444 P2 (Issued 31 October 2023). Available at: https://ppubs.uspto.gov/pubwebapp/static/pages/ppubsbasic.html (accessed 4 March 2024).

Anonymous (2023b) *Florida Avocado Administrative Committee Shipping Schedule 2023–2024.* Federal-State Marketing Order 915 and Federal Marketing Agreement 121, Homestead, Florida.

Anonymous (2023c) Shepard avocado. Avocado Variety Database, UC Riverside, Avocado Information. Available at: https://avocado.ucr.edu/ (accessed 3 July 2023).

Arpaia, M.L. (1987) United States avocado production. *South African Avocado Growers' Association Yearbook* 10, 88–89.

Arpaia, M.L. (1993) Avocado clonal rootstock production trial. *California Avocado Research Symposium*, Riverside, California, pp. 17–23.

Arpaia, M.L. (1999) Plant breeding and genetics panel summary. In: Arpaia, M.L. and Hofshi, R. (eds) *Proceedings of Avocado Brainstorming '99*. California Avocado Commission and the University of California, Riverside, California, pp. 7–8.

Arpaia, M.L. (2004) Enhancement of avocado productivity. Plant improvement: selection and evaluation of improved varieties and rootstocks. *Proceedings of the California Avocado Research Symposium*. University of California and California Avocado Commission, Riverside, California, pp. 9–23.

Arpaia, M.L., Bender, G.S. and Witney, G.W. (1992) Avocado clonal rootstock production trial. In: Lovatt, C., Holthe, P.A. and Arpaia, M.L. (eds) *Proceedings of the Second World Avocado Congress*. University of California, Riverside, California, pp. 305–310.

Arpaia, M.L. and Smith, H. (2009–10) Enhancement of avocado productivity: plant improvement – selection and evaluation of improved varieties and rootstocks. California Avocado Commission, Breeding and Genetics, 2009–2010 Production Research Report. 9s p.

Ashworth, V.E.T.M. and Clegg, M.T. (2003) Microsatellite markers in avocado (*Persea americana* Mill.): genealogical relationships among cultivated avocado genotypes. *Journal of Heredity* 94, 407–415.

Avocados Australia (2023) Facts at a glance 2020/21 for the Australian avocado industry (accessed 20 March 23) [2020-21_AAL-Facts-at-a-glance3.pdf (avocado.org.au)].

Ayala-Silva, T., Schnell, R. and Winterstein, M. (2007) Phenotypic characterization of commercial and rootstock avocado varieties. *Proceedings of the VI World Avocado Congress,* Viña Del Mar, Chile. Available at: http://www.avocadosource.com/WAC6/en/Extenso/1a-29.pdf (accessed 2 August 2011).

Backer, R., Englebrecht, J. and Berg van den, N. (2022) Differing responses to *Phytophthora cinnamomic* infection in susceptible and partially resistant *Persea americana* (Mill.) rootstocks: a case for the role of receptor-like kinases and apoplastic proteases. *Frontiers in Plant Science* 12, 928176. doi:10.3389/fpls.2022.928176.

Banks, A. (1980) *Keys and Descriptions for Identifying the Fruit and Trees of Thirteen Avocado Varieties Grown in the North Moreton Region*. Queensland Department of Primary Industries, Nambour, Queensland, Australia.

Barlow, C. (2000) *The Ghosts of Evolution.* Perseus Books Group, Jackson, Tennessee, pp. 7–12, 51–69.

Barrientos-Priego, A.F., Lopez-Jimenez, A. and Sanchez-Colin, S. (1987) Effect of cv. Colin V-33 as an interstock on avocado (*Persea americana* Mill.) growth, cv. Fuerte. *South African Avocado Growers' Association Yearbook* 10, 62–64.

Barrientos-Priego, A.F., Ben-Ya'acov, A., Borys, M.W., Bufler, G., de la Cruz, E. and Lopez, L.L. (1992) *Descriptors for Avocado (Persea* sp.*)*. International Board for Plant Genetic Resources, Rome.

Barrientos-Priego, A.F., López-López, A.F. and López-López, L. (2000) Historia y Genetica del Aguacate. In: Téliz, D. (ed) *El aguacate y su manejo integrado*. Ediciones Mundi Prensa, México, D.F., pp. 33–51.

Barrone, J.W., Brown, J.S., Tondo, C.L., Mauro-Herrera, M., Kuhn, D.N. *et al.* (2009) An EST-SSR-based linkage map for Persea americana Mill. (avocado). *Tree Genet Genomes* 5, 553–560.

Beach, J.B. (1920) Guatemalan and Mexican avocados fruiting in Florida. *Proceedings of the Florida State Horticultural Society* 33, 60–65.

Beier, G.L., Lund, C.D., Held, B.W., Ploetz, R.C., Konkol, J.L. and Blanchette, R.A. (2021) Variation in xylem characteristics of botanical races of *Persea americana* and their potential influence on susceptibility to the pathogen *Raffaelea lauricola*. *Tropical Plant Pathology* 46, 232–239.

Belisle, R.J., McKee, B., Hao, W., Crowley, M., Arpaia, M.L. *et al.* (2019) Phenotypic characterization of genetically distinct *Phytophthora cinnamomi* isolates from avocado. *Phytopathology* 109, 384–394. doi:10.1094/PHYTO-09-17-0326-R.

Bender, G.S., Crowley, D. and Arpaia, M.L. (2010) The search for salinity tolerance in avocado: an update on a frozen rootstock trial. *Topics in Subtropics* 8(3), 5–7.

Ben-Ya'acov, A. (1992) Recommended rootstocks for new planting of avocado – 'Ettinger' cultivar. *Alon Hanotea* 46, 919–926. (In Hebrew.)

Ben-Ya'acov, A. (1995a) A sum-up of experiments with seedling avocado rootstock and scion sources. III. Selection of West Indian seedling rootstocks for the 'Ettinger' cultivar. *Alon Hanotea* 49, 574–577.

Ben-Ya'acov, A. (1995b) A comparison between Mexican and West Indian rootstocks for the Ettinger cultivar at Kvutzat Shiller, Israel. *California Avocado Society Yearbook* 79, 165–169.

Ben-Ya'acov, A. and Frenkel, M. (1974) Sensitivity of avocado rootstocks to Verticillium wilt. *California Avocado Society Yearbook* 57, 111–113.

Ben-Ya'acov, A. and Kadman, A. (1976) Selection of avocado rootstocks for saline conditions. *Acta Horticulturae* 57, 1189–1197.

Ben-Ya'acov, A. and Michelson, E. (1995) Avocado rootstocks. *Horticultural Reviews* 17, 381–429.

Ben-Ya'acov, A. and Zilberstaine, M. (1999) Clonal avocado (*Persea americana* Mill.) rootstocks in Israel. *Revista Chapingo Serie Horticultura Num. Especial* V, 39–42.

Ben-Ya'acov, A., Michelson, E., Sela, I. and the avocado growers team at Kibbutz Ha'Ogen (1995a) A field study of seeded avocado rootstocks and scion sources: final report on experiments planted as commercial orchards. X. Experiments at Kibbutz Ha'Ogen, planted in 1977. *Alon Hanotea* 49, 506–511.

Ben-Ya'acov, A., Eshel, I., Shefer, S., Michelson, E. and the avocado growers of Sha'ar ha'Golan (1995b) Rootstocks experiments for 'Pinkerton' avocado cultivar, at Kibbutz Sha'ar ha'Golan (intermediate report). *Alon Hanotea* 49, 234–239.

Ben-Ya'acov, A., Zilberstaine, M., Goren, M. and Tomer, E. (2003) The Israeli avocado germplasm bank: where and why the items had been collected. *Proceedings of the V World Avocado Congress*, Malaga, Spain, pp. 13–19.

Berdugo-Cely, J.A., Cortés, A.J., López-Hernádez, F., Delgadillo-Durán, Cerón-Souza, I., Reyes-Herrera, P.H., Navas-Arboleda, A.A. and Yockteng, R. (2023) Pliestocene-dated genomic divergence of avocado trees supports cryptic diversity in the Colombian germplasm. *Tree Genetics and Genomes* 19:42. doi:10.1007/s11295-023-01616-8.

Berg van den, N., Christie, J.B., Aveling, A.S. and Englebrecht, J. (2018) Callose and β-1, 3-glacanase inhibit Phytophthora cinnamomic in a resistant avocado rootstock. *Plant Pathology* 67, 1150–1160.

Berg van den, N., Swart, V., Backer, R., Fick, A., Wienk, R., Englebrecht, J. and Prabhu, S.A. (2021) Advances in understanding defense mechanisms in *Persea americana* against *Phytophthora cinnamomic*. *Frontiers in Plant Science* 12, 636339. doi:10.3389/fpls.2021.636339.

Bergh, B.O. (1992) The origin, nature and genetic improvement of avocado. *California Avocado Society Yearbook* 76, 61–75.

Bergh, B.O. and Ellstrand, N. (1986) Taxonomy of the avocado. *California Avocado Society Yearbook* 70, 135–145.

Bergh, B. and Martin, G. (1988) The Gwen avocado. *California Avocado Society Yearbook* 72, 195–207.

Bernstein, N., Ioffe, M. and Zilberstaine, M. (2001) Salt-stress effects on avocado rootstock growth. I. Establishing criteria for determination of growth sensitivity to the stress. *Plant and Soil* 233, 1–11.

Bernstein, N., Meiri, A., and Zilberstaine, M. (2004) Root growth of avocado is more sensitive to salinity than shoot growth. *Journal of the American Society for Horticultural Science* 129, 188–192.

Beyer, C.P., Barrientos-Sanhueza, C., Ponce, E., Pedreschi, R., Cuneo, I.F. and Alvaro, J.E. (2022) Differential hydraulic properties and primary metabolism in fine root of avocado trees rootstocks. *Plants* 11, 1059. doi:10.3390/plants11081059.

Biran, D. (1979) Fruitlet abscission and spring growth retardation – their influence on avocado productivity. M.Sc. thesis, The Hebrew University of Jerusalem, Rehovot, Israel.

Blight, A., Newett, S.D.E., Searle, C. and Waite, G. (2000) *AV98013, Technology Exchange at the 4th International Avocado Congress in México, October 1999*. Horticultural Research and Development Corporation, 150 pp.

Blumenfeld, A. and Elimelech, M. (1986) New avocado cultivars. III. 'Pinkerton'. *Alon Hanotea* 40, 777–780. (In Hebrew.)

Blumenfeld, A. and Koussevitzky, S. (1995) Determining freezing sensitivity of avocado. In: *Proceedings of the III World Avocado Congress,* Tel Aviv, Israel (Conference abstracts), p. 20.

Borrone, J.W., Schnell, R.J., Violi, H.A. and Ploetz, R.C. (2007) Seventy microsatellite markers from *Persea americana* Miller (avocado) expressed sequence tags. *Molecular Ecology Notes* 7, 439–444.

Boza, E.J., Tondo, C.L., Ledesma, N., Campbell, R.J., Bost, J., Schnell, R.J. and Gutiérrez. (2018) Genetic differentiation, races, and interracial admixture in avocado (*Persea americana* Mill.), and *Persea* spp. evaluated using SSR markers. *Genet Resources Crop Evolution* 65, 195–1215.

Brokaw, W.H. (1986) Selecting rootstock. *California Avocado Society Yearbook* 70, 111–114.

Brokaw, W.H. (1987) Avocado clonal rootstock propagation. *Combined Proceedings of the International Plant Propagation Society* 37, 97–103.

Brokaw, W.H. (1989) Introducing new cultivars: needs vs. risks. *California Avocado Society Yearbook* 73, 121–126.

Brokaw, W.H. (2024) 'Zerala' (Merensky 5) fact sheet. Brokaw Nursery LLC. Available at: https://www.brokawnursery.com/ (accessed March 2024).

Bruwer, A.T. (2006) Evaluation of the Hass-like avocado cultivars Harvest, GEM and Grace. *South African Avocado Growers' Association Yearbook* 30, 32–33.

Bruwer, A.T. and Mokgalabone, M.L. (2006) Evaluation of the Hass-like avocado cultivars Harvest, GEM and Grace. *South African Avocado Growers' Association Yearbook* 29:41.

Calabrese, F., Azzena, M., Barone, F., Continella, G., D'Ascanio, R. *et al.* (1998) Research on avocado in Italy. In: *Proceedings of the III World Avocado Congress*, Tel Aviv, Israel, pp. 457–460.

CAC (California Avocado Commission). (2023) California Avocado Commission Pounds and Dollars by Variety. Jan. 4, 2022 (accessed 20 March 2023) [Pounds and Dollars By Variety I California Avocado Commission (californiaavocadogrowers.com)].

Campbell, C.W. (1979) Characteristics of Florida avocado cultivars. *Homestead Agricultural Research and Education Center Research Report SB79-5*.

Campbell, C.W. and Knight, R.J., Jr. (1998) The Kromes, pioneers in the tropical fruit industry of Florida. *Proceedings of the Florida State Horticultural Society* 111, 194–195.

Campbell, C.W. and Malo, S.E. (1976) A survey of avocado cultivars. In: Sauls, J.W., Phillips, R.L. and Jackson, L.K. (eds) *Proceedings of the First International Tropical Fruit Short Course: The Avocado*. University of Florida Cooperative Extension Service, IFAS, Gainesville, Florida, pp. 20–24.

Carrillo, D., Duncan, R.E., and Peña, J.E. (2012) Ambrosia beetles (Coleoptera: Curculionidae: Scolytinae) that breed in avocado wood in Florida. *The Florida Entomologist* 95(3), 573579.

Carrillo, D., Duncan, R.E. Ploetz, J.N., Campbell, A.F., Ploetz, R.C. and Peña, J.E. (2014) Lateral transfer of a phytopathogenic symbiont among native and exotic ambrosia beetles. *Plant Pathology* 63, 54–62. doi:10.1111/ppa.12073.

Casas, A. (2001) Silvicultura y domensticación de plantas en Mesoamérica. In: Aguilar, B.R., Domínguez, S.R. Nieto, J.C. and Alfaro, M.A.M. (eds) *Plantas, Cultura y Sociedad: Estudio Sobre la Relación Entre Seres Humanos y Plantas en los Arboles del Siglo XXI*. Universidad Autónoma Metropolitana, Iztapalapa, México. pp. 123–157.

Castillo-Argaez, R., Vazquez, A., Konkol, J.L., Vargas, A.I., Ploetz, R.C., Etxeberria, E. and Schaffer, B. (2020) Sap flow, xylem anatomy and photosynthetic variables of three Persea species in response to laurel wilt. *Tree Physiology* 41, 1004–1018.

Castillo-Argaez, R., Konkol, J.L., Vargas, A.I., Ploetz, R.C. and Schaffer, B. (2021) Disease severity and ecophysiology of rootstock/scion combinations of different avocado (*Persea americana* Mill.) genotypes in response to laurel wilt. *Scientia Horticulturae* 287, 110250.

Castillo-Argaez, R., Vazquez, A., Konkol, J.L., Vargas, A.I., Ploetz, R.C. and Schaffer, B. (2022) Laurel wilt susceptibility of three avocado (*Persea americana* Mill.) ecotypes in relation to xylem anatomy, sap flow and leaf gas exchange. *Trees* 36, 1649–1663.

Castro, M., Cautin, R., Fassio, C. and Darrouy, N. (2003) Introduction, selection and propagation program for avocado rootstocks and cultivars in Chile. In: *Proceedings of the V World Avocado Congress.* Malaga, Spain (Conference abstracts), pp. 120–121.

Castro, M., Iturrieta, R. and Fassio, C. (2009) Rootstock effect on the tolerance of avocado plants cv. Hass to NaCl stress. *Chilean Journal of Agricultural Research* 69(3), 316324.

Castro, M., Fassio, C., Cautin, R. and Ampuero, J. (2015) UCV7, an avocado rootstock tolerant to salinity. *Revista Fitotecnia Mexicana* 38(1), 85–92.

Celis, N., Suarez, D.L., Wu, L., Li, R., Arpaia, M.L. and Mauk, P. (2018) Salt tolerance and growth of 13 avocado rootstocks related best to chloride uptake. *HortScience* 53(12), 1737–1745.

Chalker, F.C. (1983) Avocado varieties in northern NSW. In: Allen, R.N. (ed.) *Proceedings of the 2nd Australian Avocado Research Workshop*. Agriculture Research Centre, Wollongbar, Australia, pp. 35–39.

Chalker, F.C. (1985) *Avocado growing*. Agfacts – NSW Department of Agriculture, Sydney. No. H6. 1. 8 pp.

Chanderbali, A.S., Soltis, D.E., Soltis, P.S., and Wolstenholme, B.N. (2013) *Chapter X. Taxonomy and Botany*. In: *The Avocado: Botany, Production and Uses*, 2nd Edn. CABI International, New York, NY. Pages 268–300.

Chen, H., Morrell, P.L., Vanessa, E.T.M., Ashworth, M.C. and Clegg, M.T. (2009) Tracing the geographic origins of major avocado cultivars. *Journal of Heredity* 200, 56–65.

Clegg, M.T., Ashworth, V.E.T.M. and Chen, H. (2007) *Assessing the Genetic Determination of Valuable Avocado Traits Using Microsatellite (SSR) Markers and Quantitative Trail Locus (QTL) Analysis*. 2007 Production and Research Report, California Avocado Commission, Breeding, Varieties and Genetics, p. 5.

Coates, L.M., Dann, E.K., Shuey, L.S., Smith, L.A., Dean, J.R. *et al.*(2011) Effects of rootstocks on avocado fruit quality – assessment of postharvest disease, major cations and biochemical traits. *Proceedings of the VII World Avocado Congress,* Cairns, Australia. 9 pp. Available at: http://worldavocadocongress2011.com/userfiles/file/Louise%20Shuey%200900-0920.pdf (accessed 17 February 2012).

Coe, M.D. (2005) *The Maya.* Thames and Hudson, Inc., New York USA.

Coe, M.D. and Koontz, R. (2008) *México: from the Olmecs to the Aztecs*. Thames and Hudson, Inc., New York USA.

Coffey, M.D. (1987) Current avocado rootstocks. *Queensland Fruit and Vegetable News* 58:25.

Coffey, M.D. and Guillemet, F.B. (1987a) Profiles of UCR clonal rootstocks. *California Avocado Society Yearbook* 71, 169–171.

Coffey, M.D. and Guillemet, F.B. (1987b) Avocado rootstocks. *California Avocado Society Yearbook* 71, 173–179.

Coffey, M.D., Guillemet, F.B., Schieber, E. and Zentmyer, G.A. (1988) *Persea schiedeana* and Martin Grande – the period from 1920 to 1975. *California Avocado Society Yearbook* 72, 107–120.

Cohen, H., Bar-Noy, Y., Irihimovitch, V. and Rubinovich, L. (2023) Effects of seedling and clonal West Indian rootstocks irrigated with recycled water on 'Hass' avocado yield, fruit weight and alternate bearing. *New Zealand Journal of Crop and Horticultural Science* 51(1), 39–51.

Cooper, W.C. (1951) Salt tolerance of avocado on various rootstocks. *Texas Avocado Society Yearbook* 1951, 24–28.

Crane, J.H., Balerdi, C.F. and Campbell, C.W. (1996) The Avocado. *Florida Cooperative Extension Service Circular 1034*. IFAS, University of Florida, Gainesville, Florida.

Crane, J.H., Evans, E.A., Carrillo, D., Ploetz, R.C. and Palmateer, A.J. (2015) The potential for laurel wilt to threaten avocado production is real. *Proceedings of the VIII Congreso Mundial de la Palta*, Lima, Perú 13–18 September, Actas:9–16.

Crane, J.H., Carrillo, D., Evans, E.A., Gazis, R., Schaffer, B., Ballen, F., and Wasielewski, J. (2024) Recommendations for control and mitigation of laurel wilt and ambrosia beetle vectors in commercial avocado groves in Florida. Horticultural Sciences Department, UF/IFAS Extension. https://edis.ifas.ufl.edu/publication/HS1360 (8 pages).

Crowley, D. (2004) Salinity tolerance in avocado. *Proceedings of the California Avocado Research Symposium,* October 30, 2004, University of California, Riverside and California Avocado Commission, Irvine, California, pp. 108–111.

Crowley, D. and Arpaia, M.L. (2002) *Rootstock Selections for Improved Salinity Tolerance of Avocado*. California Avocado Summit, pp. 83–86.

Cruz, L.F., Menocal, O., Kendra, P.E. and Carrillo, D. (2021) Phoretic and internal transport of Raffaelea lauricola by different species of ambrosia beetle associated with avocado trees. *Symbiosis* 84, 151–161.

Cuevas, M.J. (2007) Non-destructive methods and optimum harvesting time of 'Semil 34' avocado (*Persea americana* Mill.) in the Dominican Republic. *Proceedings of the VI World Avocado Congress,* Viña Del Mar, Chile. Available at: http://www.avocadosource.com/WAC6/en/Extenso/4a-164.pdf (accessed 1 August 2011).

Dann, E. (2016) Avocado rootstock assessment and improvement-interim, final report. Horticulture Innovation Australia Limited, Sydney, Australia. 195 p.

Degani, C., Goldring, A. and Gazit, S. (1989) Pollen parent effect on outcrossing rate in 'Hass' and 'Fuerte' avocado plots during fruit development. *Journal of the American Society for Horticultural Science* 114, 106–111.

Dixon, J., Cotterell, C., Hofstee, B. and Elmsly, T.A. (2008) University of California avocado cultivars 'Lamb Hass' and 'GEM' maturity and fruit quality results from New Zealand evaluation trials. *New Zealand Avocado Growers' Association Annual Research Report* 8, 15–26.

Donkin, D. (2011) Overview of the South African avocado industry. Abstracts, *VII World Avocado Congress*, Cairns, Australia. Available at: http://worldavocadocongress2011.com/?PageID=34 (accessed 12 November 2011).

Douhan, G.W. (2009) Current and future perspectives regarding avocado rootstock breeding in USC. *California Avocado Society Yearbook* 92, 87–111.

Douhan, G.W. (2011) Screening and evaluation of new rootstocks with resistance to Phytophthora cinnamomic. 2011 Final Production Research Report, California Avocado Commission, Breeding and Genetics. Avocadosource.com (accessed 26 April 2024).

Douhan, G.W., Fuller, E., McKee, B. and Pond, E. (2011) Genetic diversity analysis of avocado (*Persea americana* Miller) rootstocks selected under greenhouse conditions for tolerance to phytophthora root rot caused by *Phytophthora cinnamomi. Euphytica* 182, 209–217.

Du Plessis, S.F., Koen, T.J. and Abercrombie, R.A. (1995) Nutritional requirements of the Fuerte avocado: a summary of 21 years of research in South Africa. *Proceedings of the III World Avocado Congress,* Tel Aviv, Israel, pp. 160–165.

Durand, B.J. (1990) Avokado cultivars. Navorsingsinstituut vir Sitrus en Subtropiese Vrugte, Nelspruit. Uitgegee deur die Departement van Landbou-Ontwikkeling en verkrygbaar van die Direktoraat Landbouinligting, Privaatsak X144, Pretoria 0001. (In Afrikaans).

El-Hamalawi, Z.A., Menge, J.A. and Guillemet, F.B. (1994) Comparison of resistance to *Phytophthora citricola* in nineteen avocado rootstocks under greenhouse conditions. *California Avocado Society Yearbook* 78, 121–129.

Ellstrand, C., Lee, J.M.E., Bergh, B.O., Coffey, M.D. and Zentmyer, G.A. (1986) Isozymes confirm hybrid parentage for 'G 755' selections. *California Avocado Society Yearbook* 70, 199–203.

Engelbrecht, J. and van den Berg, N. (2013) Expression of defence-related genes against *Phytophthora cinnamomia* in five avocado rootstocks. *South African Journal of Science* 109(11/12), Art. #2012-0058, 8. doi:10.1590/ sajs.2013/20120058.

Ernst, A.A. (2011) Interaction of storage, ethylene and ethylene inhibitors on post harvest quality of 'Maluma'. *Proceedings of the VII World Avocado Congress,* Cairns, Australia. Available at: http://worldavocadocongress2011.com/userfiles/file/Andre%20Ernst%201030-1050.pdf (accessed 16 February 2012).

Ernst, Z.R. and Ernst, A.A. (2011) High density cultivation: a case study of central leader pruning with Maluma. *Proceedings of the VII World Avocado Congress,* Cairns, Australia. Available at: https://www.avocadosource.com/wac7/wac7_toc.htm (accessed 16 February 2012).

Ernst, A.A., Ernst, Z.R. and Ernst, E.D. (2015) 'Maluma': establishing a new generation avocado cultivar commercially. *Actas – Proceedings VIII Congreso Mundial de la Palta 2015*, Lima, Perú. pp. 53–59.

Everett, K.R. and Siebert, B. (2018) Exotic plant disease threats to the New Zealand avocado industry and climatic suitability: a review. *New Zealand Plant Protection* 71, 25–38.

Faber, B., Downer, J., Menge, J. and Ohr, H. (1995) Mulch effects on avocado root rot. In: *Proceedings of the III World Avocado Congress,* Tel Aviv, Israel (Conference abstracts), p. 49.

Fairchild, D. (1945) Personal recollections of George B. Cellon, horticultural pioneer of south Florida. *Proceedings of the Florida State Horticultural Society* 58, 205–209.

Fassio, C., Castro, M. and Cruz, R. (2021) Factores que afectan la unión injerto/portainjerto en la propagación clonal de *Persea americana* Mill. *Proceedings VI Congreso Latinoamericano- del Aguacate*, La Antigua, Guatemala, pp.7–11.

FDACS-DFV (2023) Florida Avocado Varieties. Florida Department of Agriculture and Consumer Services, Division of Fruit and Vegetables (accessed 25 March 23) [Florida Avocado Varieties (usda.gov)].

Frolich, E.F. and Platt, R.G. (1972) Use of the etiolation technique in rooting avocado cuttings. *California Avocado Society Yearbook* 55, 97–109.

Gabor, B.K., Guillemet, F.B. and Coffey, M.D. (1990) Comparison of field resistance to *Phytophthora cinnamomi* in twelve avocado rootstocks. *HortScience* 25, 1655–1656.

Galan-Sauco, V. and Fernandez-Galvan, D. (1985) Avocado culture in the Canary Islands. *California Avocado Society Yearbook* 69, 73–79.

Galindo-Tovar, M.E., Ogata-Aguilar, N. and Arzate-Ferαndez, A. (2008) Some aspects of avocado (*Persea americana* Mill.) diversity and domestication in Mesoamerica. *Genetic Resources and Crop Evolution* 55, 441–450.

Gallo-Llobet, I. and Siverio, F. (1995) Evaluation of West Indian avocado seedling tolerance-resistance to *Phytophthora cinnamomi* Rands compared with clonal resistant rootstocks 'Duke 7', 'Thomas' and 'Toro Canyon', under field conditions. In: *Proceedings of the III World Avocado Congress,* Tel Aviv, Israel (Conference abstracts), p. 56.

Gama-Campillo, L. and Gomez-Pompa, A. (1992) An ethnoecological approach for the study of *Persea*: a case study in the Maya area. In: *Proceedings of the Second World Avocado Congress*, Riverside, California, pp. 1–17.

Gardiazabal, F.J. and Gandolfo, S. (1998) A study of self-pollination and cross-pollination in avocado (*Persea americana* Mill.) cv. Hass of different varieties. In: *Proceedings of the III World Avocado Congress*, Tel Aviv, Israel, pp. 52–56.

Gazit, S. and Degani, C. (2002) Reproductive biology. In: Whiley, A.W., Schaffer, B. and Wolstenholme, B.N. (eds) *The Avocado: Botany, Production and Uses*. CAB International, Wallingford, UK, pp. 101–133.

Ge, Y., Zhang, T., Wu, B., Tan, L., Ma, F. *et al.* (2019) Genome-wide assessment of avocado germplasm determined from specific length amplified fragment sequencing and transcriptomes: population structure, genetic diversity, identification, and application of race-specific markers. *Genes* 10:215. doi:10.3390/genes10030215.

Goldberg, E.R. (1921) Avocado culture in south Dade County. *Proceedings of the Florida State Horticultural Society* 34, 81–85.

Gregoriou, C. (1992) Yield growth and fruit quality of 'Fuerte' and 'Ettinger' cultivars of avocado on four rootstocks in Cyprus. *California Avocado Society Yearbook* 76, 159–164.

Guirado, E., Hermoso, J.M. and Farré, J.M. (2015) Comparación de portainjertos de aguacate Hass, observaciones durante los primeros 7 años. *Actas, Proceedings VIII Congreso Mundial de la Palta*, Lima, Perú. pp.75–81.

Haas, A.R.C. (1950) Effect of sodium chloride on Mexican, Guatemalan, and West Indian avocado seedlings. *California Avocado Society Yearbook* 34, 153–160.

Haas, A.R.C. (1952) Sodium effects on avocado rootstocks. *California Avocado Society Yearbook* 37, 159–166.

Haberman, A., Leah, T., Lazare, S., Hazanovsky, M., Lebiush, S. *et al.* (2020) Management of verticillium wilt of avocado using tolerant rootstocks. *Plants* 9:531. doi:10.3390/plants9040531.

Halma, F.F. and White, F.A. (1951) Relative susceptibility of avocado rootstocks to chlorosis. *California Avocado Society Yearbook* 35, 153–156.

Handwerk, B. (2017) Holy guacamole: how the Hass avocado conquered the world. *Smithsonian Magazine*. Available at: https://www.smithsonianmag.com/science-nature/holy-guacamole-how-hass-avocado-conquered-world-180964250/#:%7E:text=Per%20capita%20consumption%20of%20avocados,single%20variety%3A%20the%20ubiquitous%20Hass (accessed 23 February 23).

Harkness, R.W. and Lincoln, F.B. (1955) Avocado culture studies, State Project 276. In: *Florida Agriculture Experiment Station Annual Report 1955* p. 306.

Hatton, T.T. Jr. and Reeder, W.F. (1965) Maturity of minor varieties of Florida avocados, 1964–65. *Proceedings of the Florida State Horticultural Society* 78, 327–330.

Hatton, T.T. Jr., Harding, P.L., Reeder, W.F., Yeatman, J.N. and Krome, W.H. (1963) *Fruit Weights and Corresponding Diameters for Florida Avocados*. AMS-515. United States Department of Agriculture/Agricultural Marketing Service Market Quality Research Division, Washington, DC.

Hatton, T.T. Jr., Harding, P.L. and Reeder, W.F. (1964) *Seasonal Changes in Florida Avocados*. United States Department of Agriculture Technical Bulletin 1310, US Government Printing Office, Washington, DC.

Hatton, T.T. Jr., Reeder, W.F. and Campbell, C.W. (1965) *Ripening and Storage of Florida Avocados*. United States Department of Agriculture-Agricultural Research Service Marketing Research Report No. 697, Washington, DC.

Henao, E.R., Arana, A.C., Valencia, A.L.E. and Florez, J.E.M. (2017) Evaluation of tolerance to *Phytopthora cinnamomi* Rands in avocado (*Persea americana* Miller.) germplasm. *Acta Agronómica* 66(1), 128–138.

Hernández, I., Molina, V., Fuentealba, C., Alvaro, J.W., Defilippi, B.G. and Pedreschi, R. (2023) Do rootstocks influence global fruit quality, postharvest performance and metabolite profiles of *Persea americana* cv. Hass? *Horticulturae* 9:184. doi:10.3390/horticulturae9020184.

Herrera-González, J.A., Salazar-García, S., Gutiérrez-Martínez, P. and González-Durán, I.J.L (2013) Postharvest performance of 'Hass' avocado fruit is influenced by rootstock. *Revista Mexicana de Ciencias Agricolas* 4(1), 19–32.

Hoddle, M.S. and Kerguelen, V. (1999) Determining the susceptibility of avocado cultivars to feeding by the Persea mite, *Oligonychus persea* (Acari: Tetranychidae). *Revista Chapingo Serie Horticultura Núm Especial* V, 301–308.

Hodgkin, G.B. (1980) A pilgrimage (1929) to the parent 'Fuerte' tree. *California Avocado Society Yearbook* 64, 63–65.

Hodgson, R.W. (1945) Suggestive evidence of the existence of strains in the 'Fuerte' avocado variety. *California Avocado Society Yearbook* 30, 24–26.

Horns, F. and Hood, M. (2012) The evolution of disease resistance and tolerance in spatially structured populations. *Ecology Evolution* 2(7), 1705–1711.

Illsley-Granich, C., Brokaw, R. and Ochoa-Ascencio, S. (2011) Hass Carmen®, a precocious flowering avocado tree. *Proceedings of the VII World Avocado Congress,* Cairns, Australia. Available at: http://avocadosource.com/ (accessed 1 August 2012).

Imbert, E. (2018) Producer Country File: the Avocado in Israel. *FRuiTROP* 259, 74–79.

Imbert, E. (2019) Producer Country File: the Avocado in South Africa. *FRuiTROP* 262, 39–49.

Inch, S., Ploetz, R., Held, B. and Blanchette, R. (2012) Histological and anatomical responses in avocado, *Persea americana*, induced by the vascular wilt pathogen, *Raffaelea lauricola*. *Botany* 90, 627–235.

Ish-Am, G. and Eisikowitch, D. (1989) The influence of temperature on flower morphology and flowering phenology in the avocado cultivars Fuerte, Ettinger and Hass. *Alon Hanotea* 43, 747–766. (In Hebrew.)

Janzen, D.H. and Martin, P.S. (1982) Neotropical anachronisms: the fruits the gomphotheres ate. *Science* 215(4528), 19–27.

Jeanteur, P. (1970) Quelques caractéristiques de la culture de l'avocatier en Floride et á Porto-Rico. *Fruits* 25, 817–823.

Kadman, A. (1962) Selection of avocado rootstocks, tolerant to saline irrigation water, and their vegetative propagation. PhD thesis, The Hebrew University of Jerusalem, Rehovot, Israel.

Kadman, A. (1985) Selection of avocado and mango rootstocks for calcareous soils. *Acta Horticulturae* 158, 63–67.

Kadman, A. and Ben-Ya'acov, A. (1980) Maoz avocado rootstock selection. *HortScience* 15:207.

Kämper, W., Trueman, S.J., Cooke, J., Kasinadhuni, N., Brunton, A.J. and Ogbourne, S.M. (2021) Single-nucleotide polymorphisms that uniquely identify cultivars of avocado (*Persea americana*). *Applications in Plant Sciences* 9(6): e1140. doi:10.1002/aps3.11440.

Karp, D. and Gasic, K. (2022) Register of New Fruit and Nut Cultivars List 51. *HortScience* 57, 1175–1233.

Kellogg, G. (1971) The 'Fuerte' avocado in Yorba Linda. *California Avocado Society Yearbook* 54, 47–48.

Kendra, P.E., Guillén, L., Tabanca, N., Montgomery, W.S., Schnell, E.Q., Deyrup, M.A. and Cloonan, K.R. (2023) Risk assessment of Hass avocado and Mexican Lauraceae for attach by redbay ambrosia beetle (Coleoptera: Curculionidae: Scolytinae). *Agricultural and Forest Entomology* 1–18. doi:10.1111/afe.12551.

Kernot, I. (1995) Selling 'Shepard'. In: *Proceedings of the Australian Avocado Growers' Federation Conference '95 'The Way Ahead'*, Fremantle, Australia, pp. 103–107.

Knight, R.J. Jr. (1971) Flowering behavior (A and B classification) of avocado cultivars. *Proceedings of the Tropical Region of the American Society for Horticultural Science* 15, 14–18.

Knight, R.J. Jr. and Campbell, C.W. (1999) Florida's contribution to the world avocado industry. *Proceedings of the Florida State Horticultural Society* 112, 233–236.

Köhne, J.S. (1992) Field evaluation of 'Hass' avocado grown on 'Duke 7', 'G6' and 'G7555C' rootstocks. In: Lovatt, C., Holthe, P.A. and Arpaia, M.L. (eds) *Proceedings of the Second World Avocado Congress*. University of California, Riverside, California, pp. 301–303.

Köhne, J.S. (2004) Avocado tree named 'Merensky 2'. U.S. Plant Patent PP15,309 P2. U.S. Patent Office, Washington, D.C.

Köhne, J.S. and Schutte, J.H. (1991) Increasing 'Hass' fruit size. *South African Avocado Growers' Association Yearbook* 14:38.

Kremer-Köhne, S. and Duvenhage, J.A. (2000) Field testing of new avocado rootstocks for tolerance to root rot. *South African Avocado Growers' Association Yearbook* 23, 70–71.

Kremer-Köhne, S. and Köhne, J.S. (2007) 25 years of avocado rootstock development in South Africa. Proc. VI World Avocado Congress (Actas VI Congreso Mundial del Aguacate, Chile). ISBN No 978-956-17-0413-8.

Kremer-Köhne, S. and Köhne, J.S. (1992) Yield and fruit quality of 'Fuerte' and 'Hass' on clonal rootstocks. *South African Avocado Growers' Association Yearbook* 15:69.

Kremer-Köhne, S. and Köhne, J.S. (1995) Approaches to solving the 'Hass' small fruit problem. *South African Avocado Growers' Association Yearbook* 18, 59–60.

Kremer-Köhne, S. and Köhne, J.S. (2001) Lamb Hass evaluation at Westfalia Estate – final report. *South African Avocado Growers' Association Yearbook* 24, 67–68.

Kremer-Köhne, S. and Mokgalabone, M.L. (2003) Evaluation of new 'Hass'-like avocado cultivars in South Africa. *Proceedings of the V World Avocado Congress,* Malaga, Spain, pp. 129–133.

Kremer-Köhne, S. and Mokgalabone, M.L. (2004) New Hass-like avocado cultivars at Merensky Technological Services: 2003 progress report on the evaluation of Harvest and GEM in 2003. *South African Avocado Growers' Association Yearbook* 27, 17–18.

Kremer-Köhne, S. and Mukumo, M. L. (2003) Breeding and field evaluation of new rootstocks for increased Hass yields and resistance to root rot in South Africa. *Proceedings of the V World Avocado Congress,* Malaga, Spain, pp. 555–560.

Kremer-Köhne, S., van Rooyen, Z. and Köhne, J.S. (2011) Avocado rootstock evaluation techniques over the last 30 years. *Proceedings of the VII World Avocado Congress,* Cairns, Australia. Available at: http://worldavocadocongress2011.com/userfiles/file/Sylvie%20Kremer-K%C3%B6hne%200940-1000.pdf (accessed 16 February 2012).

Krezdorn, A.H. (1973) Influence of rootstock on cold hardiness of avocado. *Proceedings of the Florida State Horticultural Society* 86, 346–348.

Kruger, F. and Magwaza, L. (2010) Avocado fruit mineral content norms made easy. *South African Avocado Growers' Association, Avoinfo* 170:18.

Kruger, F.J., Volschenk, G.O. and Volschenk, L.C. (2020) Effects of cultivar, season and altitude on South African avocado fruit's soluble solid contents and early season quality. *African Avocado Growers' Association Yearbook* 43, 88–91.

Kuhn, D.N., Groh, A., Rahaman, J., Freeman, B., Arpaia, M.L., Van den Berg, N., Abeysekara, N., Manosalva, P. and Chambers, A.H. (2019a) Creation of an avocado unambiguous genotype SNP database for germplasm curation and as an aid to breeders. *Tree Genetics and Genomes* 15:71. doi:10.1007/s11295-019-1374-1.

Kuhn, D., Livingstone, D. III, Richards, J., Manosalva, P., Van den Berg, N. and Chambers, A. (2019b) Application of genomic tools to avocado (*Persea americana*) breeding: SNP discovery for genotyping and germplasm characterization. *Scientia Horticulturae* 246, 1–11.

Lagadec, D.L. (2011) Field evaluation of superior avocado rootstocks with 'Hass' as scions. *Proceedings VII World Avocado Congress 2011 (Actas VII Congreso Mundial del Aguacate 2011)*, Cairns, Australia. 10 p.

Lahav, E., Israeli, T., Regev, I., Ardity, H., Lavi, U. *et al.* (2005) 'Arad' – a new avocado cultivar. *HortScience* 40:488.

Lavers, D. (1989) Report from Australia – an update. *California Avocado Society Yearbook* 73, 127–131.

Lavi, U., Hillel, J., Vainstein, A., Lahav, E. and Sharon, D. (1991) Application of DNA fingerprints for identification and genetic analysis of avocado. *Journal of the American Society for Horticultural Science* 116, 1078–1081.

Lazare, S., Haberman, A., Yermiyahu, U., Erel, R., Simenski, E. and Dag, A. (2020) Avocado rootstock influences scion leaf mineral content. *Archive of Agronomy and Soil Science* 66(10), 1399–1409.

Lazare, S., Cohen, Y., Goldshtein, E., Yermiyahu, U., Ben-Gal, A. and Dag, A. (2021a) Rootstock-dependent response of Hass avocado to salt stress. *Plants* 2021, 10, 1672. doi:10.3390/81672.

Lazare, S., Yasuor, H., Yermihyahu, U., Kuhalskaya, A., Brotman, Y., Ben-Gal, A. and Dag, A. (2021b) It takes two: reciprocal scion-rootstock relationships enable salt tolerance in 'Hass' avocado. *Plant Science* 312, 111048. doi:10.1016/j.plantsci.2021.111048.

Ledger, S. and Barker, L. (1995a) Shelf life varies with variety. *Talking Avocados* 6(1):27.

Ledger, S. and Barker, L. (1995b) Black avocados – the inside story. In: *Proceedings of the Australian Avocado Growers' Federation Conference '95, 'The Way Ahead'*, Fremantle, Australia, pp. 71–76.

Malo, S.E. and Harkness, R.W. (1965) Avocado culture studies, State Project 276. In: *Florida Agriculture Experiment Stations Annual Report 1965*, pp. 371–372.

Malo, S.E., Orth, P.G. and Brooks, N.P. (1977) Effects of the 1977 freeze on avocados and limes in south Florida. *Proceedings of the Florida State Horticultural Society* 90, 247–151.

Martin, G.E. (1993) Report from the variety/nursery section. *California Avocado Society Yearbook* 77, 39–41.

Martínez-Ferri, E., Moreno-Ortega, G., van den Berg, N. and Pliego, C. (2019) Mild water stress-induced priming enhances tolerance to *Rosellinia necatrix* in susceptible avocado rootstocks. *BMC Plant Biology* 19:458, doi:10.1186/s12870-019-2016-3.

Mauk, P., Li, R., Arpaia, M.L., Suarez, D., Celis, N. and Rangel, A.A. (2017) The challenge of salinity: hope for the future with new avocado rootstocks. *From the Grove, Spring*, 24–27.

Meltzer, D.J. (2021) *First Peoples in a New World: Populating Ice Age America*. United Kingdom: Cambridge University Press, Cambridge, UK.

Menge, J.A. (1998) Screening and evaluation of new rootstocks with resistance to *Phytophthora cinnamomi*. In: *Proceedings of the California Avocado Research Symposium*, pp. 41–43.

Menge, J.A. (2001) Non-fungicidal control strategies for *Phytophthora cinnamomi* root rot of avocado. In: *Proceedings of the Australian and New Zealand Avocado Growers' Conference 'Vision 2020'.* Australian Avocado Growers' Federation, Brisbane, Australia, pp. 1–10.

Menge, J.A. (2002) Screening and evaluation of new rootstocks with resistance to *Phytophthora cinnamomi. Proceedings of the California Avocado Research Symposium 2002*, pp. 55–59.

Menge, J.A., Guillemet, F.B., Campbell, S., Johnson, E. and Pond, E. (1992) The performance of rootstocks tolerant to root rot caused by *Phytophthora cinnamomi* under field conditions in Southern California. In: Lovatt, C., Holthe, P.A. and Arpaia, M.L. (eds) *Proceedings of the Second World Avocado Congress*. University of California, Riverside, California, pp. 53–59.

Menge, J.A., Douhan, G.W., McKee, B., Pond, E., Bender, G.S. and Faber, B. (2012) Three new avocado rootstocks tolerant to *Phytophthora cinnamomi;* 'Zentmyer' 'Uzi', and 'Steddom'. *HortScience* 47, 1191–1194.

Menzel, C.M. and Lagadec, M.D. Le (2014) Increasing the productivity of avocado orchards using high-density plantings: a review. *Scientia Horticulturae* 177, 21–36.

Mhameed, S., Sharon, D., Kaufman, D., Lahav, E., Hillel, J., Degani, C. and Lavi, U. (1997) Genetic relationships within avocado (*Persea americana* Mill.) cultivars and between *Persea* species. *Theoretical and Applied Genetics* 94, 279–286.

Mickelbart, M.V and Arpaia, M.L. (2002) Rootstock influences changes in ion concentrations, growth, and photosynthesis of 'Hass' avocado trees in response to salinity. *Journal of the American Society for Horticultural Science* 127, 649–655.

Mickelbart, M.V, Mesleer, S. and Arpaia, M.L. (2007a) Salinity-induced changes in ion concentration of 'Hass' avocado trees on three rootstocks. *Journal of Plant Nutrition* 30, 105–122.

Mickelbart, M.V., Bender, G.S., Witney, G.W., Adams, C. and Arpaia, M.L. (2007b) Effects of clonal rootstocks on 'Hass' avocado yield components, alternate-bearing, and nutrition. *Journal of Horticultural Science and Biotechology* 82(3), 460–466.

Mickelbart, M.V., Robinson, P.W., Witney, G. and Arpaia, M.L. (2012) 'Hass' avocado tree growth on four rootstocks in California. Yield and flowering. *Scientia Horticulturae* 143, 184–188.

Moraes, A.F.G., Micheletti, L.B., Santoro, M.B, dos Santos N.T., Avilés, T.C., and da Silva, S.R. (2022) Horticultural performance of 'Hass' avocado grafted onto seedling and clonal rootstocks under tropical wet-dry climate conditions. *Scientia Horticulturae* 302 (2022). doi:10.1016/j.scienta.2022.111155.

Moreno-Ortega, G., Zumaquero, A., Matas, A., Nicholas, A.O., van den Berg, N. *et al.* (2021) Physiological and molecular responses of 'Dusa®' avocado rootstock to water stress: insights for drought adaptation. *Plants* 10:2077. doi:10.3390/plants10102077.

Muirhead, I.F. (1983) Progress in postharvest disease control in avocados. In: Allen, R.N. (ed.) *Proceedings of the 2nd Australian Avocado Research Workshop*. Agriculture Research Centre, Wollongbar, Australia, pp. 89–93.

Nath, O., Fletcher, S.J., Shaw, L.M., Masouleh, A.K., Furtado, A., Henry, R.J. and Mitter, N. (2022) A haplotype resolved chromosomal level avocado genome allows analysis of novel avocado genes. *Horticultural Research* 9: uhac157. doi:10.1093/hr/uhac157.

Newett, S.D.E. and Whiley, A.W. (1996) Avocado production in Israel. *Talking Avocados* 7(1), 20–26.

Newett, S.D.E., Whiley, A.W., Dirou, J., Hofman, P.J., Ireland, G. *et al.* (2001) Growing the crop. In: Vock, N. (ed.) *Avocado Information Kit*. Agrilink Series, Queensland Department of Primary Industries, Brisbane, Australia, pp. 3–54.

Núñez-Lillo, G., Ponce, E., Beyer, C.P., Álvaro, Meneses, C. and Pedreschi, R. (2024) A first omics data integration approach in Hass avocados to evaluate rootstock-scion interactions: from aerial and root plant growth to fruit development. *Plants* 13:603. doi:10.3390/plants13050603.

Offer, R., Elimelech, M. and Blumenfeld, A. (1985) Harvesting dates for 'Pinkerton'. *Alon Hanotea* 40, 77–82. (In Hebrew.)

Oppenheimer. (1947) The acclimatisation of new tropical and subtropical fruit trees in Palestine. *Bulletin of the Agricultural Research Station Rehovot* 44 (cited by Ben-Ya'acov and Michelson, 1995).

Osoyo, A. (2020) Avocado Annual Report Number MX2020-0049. USDA-Foreign Agricultural Service, Global Agricultural Information Network. Sept. 10, 2020 (accessed 20 March 2023) [DownloadReportByFileName (usda.gov)].

Oster, J.D. and Arpaia, M.L. (1992) 'Hass' avocado response to salinity as influenced by clonal rootstocks. In: Lovatt, C., Holthe, P.A. and Arpaia, M.L. (eds) *Proceedings of the Second World Avocado Congress*. University of California, Riverside, California, pp. 209–214.

Oster, J.D., Brokaw, R., Strohman, R.A. and Tracy, J.E. (1985) The influence of salinity and rootstock on avocado seedling growth – progress report. *California Avocado Society Yearbook* 69, 105–110.

Papademetriou, M.K. (2000) *Avocado Production in Asia and the Pacific.* FAO Regional Office for Asia and the Pacific, Bangkok, Thailand, pp. 1–72.

Pehrson, J.E. (1957) Bringing home the 'Bacon'. *California Avocado Society Yearbook* 41, 87–89.

Pereira, M.E.C., Sargent, S.A., Sims, C.A., Huber, D.J., Moretti, C.L. and Crane, J.H. (2013) Aqueous 1-methylcyclopropene extends longevity and does not affect sensory acceptability of Guatemalan-West Indian hybrid avocado. *HortTechnology* 23(4), 468–473.

Pérez, H.G. (2017) The avocado crop: great potential for the Dominican Republic. Instituto Interamericano de Cooperación para la Agricultura (accessed 23 January 2023) [https://www.iica.int/es/prensa/noticias/avocado-crop-great- potential-dominican-republic].

Piccone, M.F. and Whiley, A.W. (1986a) Varieties. Avocado Fact Sheet. Queensland Department of Primary Industries, Australia, Horticulture Branch. RQN 86015 5 pp.

Piccone, M.F. and Whiley, A.W. (1986b) Don't rush into planting 'Pinkerton'! *Queensland Fruit and Vegetable News* 57(4), 26–31.

Platt, R.G. (1975) Climatic zones of avocado maturity in California. *California Avocado Society Yearbook* 58, 49–53.

Platt, R.G. (1976) Avocado varieties recently registered with the California Avocado Society. *California Avocado Society Yearbook* 59, 41–51.

Pliego-Alfaro, F., Palomo-Ríos, E., Mercado, J.A., Pliego, C., Barceló-Muñoz, A. *et al.* (2020) 12.1 *Persea americana* Avocado. In: Litz, R.E., Pliego-Alfaro, F. and Hormaza, J.I. (eds) *Biotechnology of Fruit and Nut Crops*. CAB International, Boston, MA, pp. 258–281.

Ploetz, R.C., Konkol, J.L., Pérez-Martínez, J.M. and Fernandez, R. (2017) Management of laurel wilt of avocado, caused by *Raffaelea lauricola*. *European Journal of Plant Pathology*. doi:10.1007/s10658-017-1173-1.

Ploetz, R.C., Pérez-Martínez, J.M., Smith, J.A., Hughes, M., Dreaden, T.J., Inch, S.A. and Fu, Y. (2012) Responses of avocado to laurel wilt, caused by *Raffaelea lauricola*. *Plant Pathology* 61:801–808.

Ploetz, R.C., Ramos, J.L., Parrado, J.L, Schaffer, B. and Lara, S.P. (1989) Performance of clonal avocado rootstocks in Dade County, Florida. *Proceedings of the Florida State Horticultural Society* 102, 234–236.

Ploetz, R.C. and Schaffer, B. (1989) Effects of flooding and Phytophthora root rot on net gas exchange and growth of avocado. *Phytopathology* 79, 204–208.

Ploetz, R.C., Schaffer, B, Vargas, A.I., Konkol, J.L., Salvatierra, J. and Wideman, R. (2015) Impact of laurel wilt, caused by *Raffaelea lauricola*, on leaf gas exchange and xylem sap flow in avocado, *Persea americana*. *Phytopathology* 105(4), 433–440.

Ploetz, R.C., Schnell, R.J. and Haynes, J. (2002) Variable response of open-pollinated seedling progeny of avocado to Phytophthora root rot. *Phytoparasitica* 30, 1–7.

Popenoe, J. (1962) Summer avocado varieties. *Proceedings of the Florida State Horticultural Society* 75, 358–360.

Popenoe, W. (1935) Origin of the cultivated races of avocados. *California Avocado Society Yearbook* 19, 184–194.

Popenoe, W. and Williams, L.O. (1947) The expedition to México of October 1947. *California Avocado Society Yearbook* 32, 22–28.

Prusky, D., Kobiler, I. and Jacoby, B. (1988) Involvement of epicatechin in cultivar susceptibility of avocado fruits to *Colletotrichum gloeosporioides* after harvest. *Journal of Phytopathology* 123, 140–146.

Regev, I., Ardity, H., Israeli, T., Lahav, E., Lavi, U. *et al.* (2005) 'Lavi' – a new avocado cultivar. *HortScience* 40:489.

Rendón-Anaya, M., Ibarra-Laclette, E., Méndez-Bravo, A., Lan, T., Zheng, C. *et al.* (2019) The avocado genome informs deep angiosperm phylogeny, highlights introgressive hybridization, and reveals pathogen-influence gene space adaptation. *Proceedings of the National Academy of Science*, USA 116(34), 17081–17089

Retief, W. (2011) Suid-Afrikaanse avocado boomverkope. Avoinfo 177, 22–23 (In Afrikaans).

Reyes-Herrera, P., Muñoz-Baena, L., Velásquez-Zapata, V., Patiño, L., Delgado-Paz, O.A. *et al.* (2020) Inheritance of rootstock effects in avocado (Persea americana Mill.) cv. Hass. *Frontiers in Plant Science* 11, 555071. doi:10.3389/fpls.2020.555071.

Rhodes, A.M., Malo, S.E., Campbell, C.W. and Carmer, S.G. (1971) A numerical taxonomic study of the avocado (*Persea americana* Mill.). *Journal of the American Society for Horticultural Science* 96, 391–395.

Riversun (2023) Avocado rootstock characteristics. Available at: https://www.riversun.co.nz/ (accessed 27 November 2023).

Roe, D.J. and Morudu, T.M. (2000) 'Hass' avocado yields as affected by dwarfing rootstocks and flower pruning. *South African Avocado Growers' Association Yearbook* 23, 30–32.

Roe, D.J., Kremer-Köhne, S. and Köhne, J.S. (1995) Local and imported avocado rootstocks in South Africa. *Proceedings of the III World Avocado Congress*, Tel Aviv, Israel, pp. 132–139.

Roe, D.J., Kremer-Köhne, S. and Köhne, J.S. (1997) Avocado rootstock and cultivar evaluation at Merensky Technological Services: a progress report. *South African Avocado Growers' Association Yearbook* 20, 36–38.

Roe, D.J., Morudu, T.M. and Köhne, J.S. (1998) Commercial avocado rootstock evaluation at Merensky Technological Services: Further progress in 1997. *South African Avocado Growers' Association Yearbook* 21, 3–5.

Roe, D.J., Morudu, T.M. and Köhne, J.S. (1999) Performance of commercially grown 'Hass' avocado on clonal rootstocks at Westfalia Estate, South Africa. *Revista Chapingo Serie Horticultura Núm Especial* V, 35–38.

Rose, L.S. (2003) Horticultural characteristics of 'Hass' avocado on commercial clonal and seedling rootstocks in California. *Proceedings of the V World Avocado Congress,* Malaga, Spain, pp. 171–175.

Rouse, R.E. and Knight, Jr., R.J. (1991) Evaluation and observations of avocado cultivars for subtropical climates. *Proceedings of the Florida State Horticultural Society* 104, 24–27.

Rowell, A.W.G. and Durand, B.J. (1982) Avocado fruit quality studies in the Nelspruit area from 1977–1981. *South African Avocado Growers' Association Yearbook* 5, 28–29.

Rudge, T. and Story, A. (1995) Study of handling practices for avocados: at wholesale and retail level, June – August 1994. *Proceedings of the Australian Avocado Growers' Federation Conference '95, 'The Way Ahead'*, Fremantle, Australia, pp. 61–69.

Ruehle, G.D. (1963) *The Florida Avocado Industry*. Bulletin 602, University of Florida, Agriculture Experiment Station, Gainesville, Florida.

Ruehle, G.D., Wolfe, H.S., Toy, L.R. and Stahl, A.L. (1949) *Avocado Production in Florida.* Agriculture Extension Service Bulletin 141, University of Florida, Gainesville, Florida.

Sahagún, B. d. and Temprano (2000) *Historia general de las Cosas de la Nueva Espana*. Dastin, Madrid.

Sánchez-González, E.I., Barrientos-Priego, A.F., Ochoa-Ascensio, S., Gutiérrez-Soto, J.G., Olivares-Sáenz, E. and Gutiérrez-Díez, A. (2019a) Assessment of resistance to *Phytophthora cinnamomi* Rands in Mexican race avocado genotypes by electrical conductivity. *Journal of Animal and Plant Sciences* 39(2), 6426–6424.

Sánchez-González, E.I., Gutiérrez-Soto, J.G., Olivares-Sáenz, E., Gutérriez-Díez, A., Barrientos-Priego, A.F. and Ochoa-Ascencio, S. (2019b) Screening progenies of Mexican race avocado genotypes for resistance to *Phytophthora cinnamomi* Rands. *HortScience* 54(5), 809–813.

Sanchez-Salvador, C. (1989) CICTAMEX, its origin and perspectives. *California Avocado Society Yearbook* 73, 141–144.

Sarooshi, R.A. Blundell, D.R. and Peasley, D.L. (1979) Blemishes and abnormalities of avocado fruit. *Agricultural Gazette of New South Wales* 90(2), 18–20.

Schaffer, B. and Ploetz, R.C. (1989) Net gas exchange as a damage indicator for Phytophthora root rot of flooded and nonflooded avocado. *HortScience* 24, 653–655.

Schaffer, B., Davies, F.S. and Crane, J.H. (2006) Responses of subtropical and tropical fruit trees to flooding in calcareous soil. *HortScience* 41, 549–555.

Schnell, R.J., Brown, J.S., Olano, C.T., Power, E.J., Krol, C.A., Kuhn, D.N. and Motamayor, J.C. (2003) Evaluation of avocado germplasm using microsatellite markers. *Journal of the American Society for Horticultural Science* 128, 881–889.

Scora, R.W., Wolstenholme, B.N. and Lavi, U. (2002) Taxonomy and botany. In: Whiley A.W., Schaffer, B. and Wolstenholme, B.N. (eds) *The Avocado: Botany, Production and Uses*. CAB International, New York, pp. 15–37.

Sharpe, F.V. (1965) The 'Sharwil' avocado comes from the Redland Bay area. *Queensland Fruit and Vegetable News,* p. 471.

Sippel, A.D., Conradie, W. and Claassens, N.J.F. (1992) Growth rates of 'Pinkerton' avocado fruit. *South African Avocado Growers' Association Yearbook* 15, 72–74.

Sive, A. and Resnizky, D. (1989) CA storage as a solution to a number of problems in the avocado industry. *Alon Hanotea* 44(1), 39–46.

Smith, C.E. Jr. (1966) Archaeological evidence for selection in avocado. *Economic Botany* 20, 169–175.

Smith, C.E. Jr. (1969) Additional notes on pre-conquest avocados in México. *Economic Botany* 23, 135–140.

Smith, D.G. (1993) Horticultural performance of imported cultivars and rootstocks in the north-eastern Transvaal. *South African Avocado Growers' Association Yearbook* 16, 28–30.

Smith, J.H.E. (1984) Internal quality analysis of commercial avocado cultivars continuously through the 1983 season. *South African Avocado Growers' Association Yearbook* 7:44.

Smith, L.A., Dann, E.K., Pegg, K.G., Whiley, A.W., Giblin, F.R., Doogan, V. and Kopittke, R. (2011) Field assessment of avocado rootstock selections for resistance to Phytophthora root rot. *Australasian Plant Pathology* 40, 39–47.

Stahl, A.L. (1931) Composition of Florida avocados. *Proceedings of the Florida State Horticultural Society* 44, 81–86.

Storey, W.B. and Bergh, B.O. (1963) Some interesting early-maturing avocados. *California Avocado Society Yearbook* 46, 87–101.

Storey, W.B., Bergh, B.O. and Zentmyer, G.A. (1986) The origin, indigenous range and dissemination of the avocado. *California Avocado Society Yearbook* 70, 127–133.

Story, A., Fuss, A. and Rudge, T. (1996) Survey of internal quality of avocados in Brisbane retail shops. *Talking Avocados* 7(1), 10–12.

Tamayo-Ramos, D.I., Salazar-González, J.A., Casson, S.A., and Urrea-López. (2022) Old and new horizons on *Persea americana* transformation techniques and applications. *Plant Cell, Tissue and Organ Culture* 150, 253–266.

Tamayo-Vélez, A. de J., Bernal-Estrada, J.A., Díaz-Díez, C.A., Cano-Gallego, L.E. and Córdoba-Gaona, O. de J. (2022) Effect of rootstock/scion compatibility on fruit and foliar nutrient composition in avocado (*Persea americana* Mill.) cv. Hass in Colombia. *Colombiana de Ciencias Hortíicolas* 16(3): e14833. doi:10.17584/rcch.2022v16i3.14833.

Torres, A.M. and Bergh, B.O. (1978) Isozymes as indicators of outcrossing among 'Pinkerton' seedlings. *California Avocado Society Yearbook* 62, 103–110.

Toy, L.R. (1929) *Avocados in Florida.* Florida Department of Agriculture Bulletin 24, Tallahassee, and University of Florida, Gainesville, Florida.

Toy, L.H. (1933) The search for new and promising varieties of avocados. *Proceedings of the Florida State Horticultural Society* 46, 133–138.

Tsao, P.H., Alizadeh, A., Kraesan, E., Fang, J.G., Menge, J. and Guillemet, F.B. (1992) Susceptibility to *Phytophthora citricola* of certain avocado rootstock cultivars known to be tolerant to *P. cinnamomi*. In: Lovatt, C., Holthe, P.A. and Arpaia, M.L. (eds) *Proceedings of the Second World Avocado Congress*. University of California, Riverside, California, pp. 89–94.

UCR (2012a) Avocado varieties. Avocado Information, University of California, Riverside, California. Available at: http://www.ucavo.ucr.edu/AvocadoVarieties/AvocadoVarieties.html (accessed 16 February 2012).

UCR (2012b) Avocado rootstocks. Avocado information, University of California, Riverside, California. Available at: www.ucavo.ucr.edu/Rootstocks/UCRootstocks.html (accessed 16 February 2012).

Useche-Carrillo, N.V., Barrientos-Priego, A.F., Núñez-Colín, C.A., Campos-Rojas, E. and Ayala-Arreola, J. (2021) Stem and leaf anatomical and physiological characteristics of 'Colín V-33' avocado seedlings. *Advances in Horticultural Science* 35(2), 183–194.

Vakis, N.J. (1982) Storage behaviour of 'Ettinger', 'Fuerte' and 'Hass' avocados grown on Mexican rootstock in Cyprus. *Journal of Horticultural Science* 57, 221–226.

Van der Merwe, M.de V. and Kotzé, J.M. (1991) *Phomopsis perseae* the cause of trunk canker of Martin Grande avocado rootstocks in South Africa. *South African Avocado Growers' Association Yearbook* 14, 85–86.

Van Rooyen, Z. (2011) New developments in horticultural research at Westfalia, South Africa. *Proceedings of the VII World Avocado Congress,* Cairns, Australia. Available at: http://worldavocadocongress2011.com/userfiles/file/Zelda%20Van%20Rooyen%200940-1000.pdf (accessed 16 February 2012).

Van Rooyen, Z. and Bower, J.P. (2005) The role of fruit mineral composition on fruit softness and mesocarp discolouration in 'Pinkerton' avocado (*Persea Americana* Mill.). *Journal of Horticultural Science and Biotechnology* 80, 793–799.

Van Velsen, R. (1983) Avocado varieties – Riverland. In: Allen, R.N. (ed.) *Proceedings of the Second Australian Avocado Research Workshop*. Agriculture Research Centre, Wollongbar, Australia p. 45.

Vega, D.E.S. (1989) Propagation in vitro of rootstocks of avocado. *California Avocado Society Yearbook* 73, 149–151.

Villiers, A.I. De. and Ernst A.A. (2015). Actas – Proceedings VIII Congreso Mundial de la Palta, Lima, Perú. pp. 40–45.

Violi, H.A., Borrone, J.W., Ploetz, P., Schnell, R.J. and Olano, C.T. (2006) Development of phytophthora root rot-resistant avocado rootstocks for the Caribbean. *Proceedings of the Florida State Horticultural Society* 119, 41–44.

Vogel, R. (1971) The behaviour of avocado-tree varieties under cold conditions at the Corsican Citrus Research Station. *Fruits* 26, 769–772, 801.

Waite, G.K. (1988) Avocado pests the past, the present and the future. In: *Proceedings of the Australia Avocado Grower's Biennial Conference 'Avocados Towards 2000',* Caloundra, Australia, pp. 64–67.

Whiley, A.W. (1983) Performance of some avocado cultivars in South East Queensland. In: Allen, R.N. (ed.) *Proceedings of the 2nd Australian Avocado Research Workshop*. Agriculture Research Centre, Wollongbar, Australia, pp. 31–34.

Whiley, A.W. (1986) Avocados in the west. *Queensland Fruit and Vegetable News* 57, 19–24.

Whiley, A.W. (1987) The Australian avocado industry. *South African Avocado Growers' Association Yearbook* 10, 28–30.

Whiley, A.W. (1990) Avocados across two continents and an archipelago. In: *Proceedings of the Australian Avocado Growers' Federation Conference, 'Profit Through Quality'*, Gold Coast, Australia, pp. 45–50.

Whiley, A.W. (1991) *'Hass' – past, present and future. Talking Avocados*, 8th edn. November/December, pp. 5–6.

Whiley, A.W. (1992) Avocado varieties and rootstocks – a review. In: *Proceedings of the Avocado Research Workshop*. Queensland Department of Primary Industries, Brisbane, Australia, pp. 15–20.

Whiley, A.W. (2001) Adoption of field practices to assist in expanding avocado markets. In: *Proceedings of the Australian and New Zealand Avocado Growers' Conference 'Vision 2020'.* Australian Avocado Growers' Federation, Brisbane, Australia, Session 1/1, 22 p.

Whiley, T. (2020) New rootstock released for use by the Australian avocado industry. *Talking Avocados* 31(3), 60–61.

Whiley, T. (2023) Procado® - a rootstock for the future. Procado® Newsletter from Commercial Nursery Partners – May 2023 4 pages.

Whiley, A.W. and Brown, B.I. (1980) Avocado maturity times in Queensland. *Queensland Fruit and Vegetable Newsletter* 51(7), 242–243.

Whiley, T. and Dann, L. (2021) Procado®: a new Australian avocado rootstock. *Talking Avocados* 32(1), 57–64.

Whiley, T. and Lavers, D. (2021) Finding a home for Shepard – the Australian story. *Talking Avocados* 32(1), 31–36.

Whiley, A.W. and Whiley, D.G. (2011) Australian rootstock programme update. *Proceedings of the VII World Avocado Congress,* Cairns, Australia (abstract). Available at: http://www.worldavocadocongress2011.com/userfiles/file/Anthony%20Whiley%201050-1110.pdf (accessed 16 February 2012).

Whiley, A.W. and Wolstenholme, B.N. (1995) Growing avocado fruit to travel. In: *Proceedings of the Australian Avocado Growers' Fereration Conference '95. 'The Way Ahead',* Fremantle, Australia, pp. 147–156.

Whiley, A.W., Smith, T.E., Wolstenholme, B.N. and Saranah, J.B. (1996) Boron nutrition of avocados. *South African Avocado Growers' Association Yearbook* 19, 1–7.

Willingham, S.L., Pegg, K.G., Cooke, A.W., Coates, L.M., Langdon, P.W.B. and Dean, J.R. (2001) Rootstock influences postharvest anthracnose development in 'Hass' avocado. *Australian Journal of Agricultural Research* 52, 1017–1022.

Willingham, S.L., Pegg, K.G., Anderson, J.M., Cooke, A.W., Dean, J.R., Giblin, F.R. and Coates, L.M. (2006) Effect of rootstocks and nitrogen fertilizer on postharvest anthracnose development in Hass avocado. *Australasian Plant Pathology* 35, 619–629.

Winston, E.C. (1983) Avocado varieties – far north Queensland. In: Allen, R.N. (ed.) *Proceedings of the 2nd Australian Avocado Research Workshop*. Agriculture Research Centre, Wollongbar, Australia, pp. 49–53.

Witney, G. and Martin, G. (1995) Taking the California avocado breeding program into the next century. *Proceedings of the III World Avocado Congress*, Tel Aviv, Israel, pp.114–118.

Witney, G.W., Hofman, P.J. and Wolstenholme, B.N. (1990) Effect of cultivar, tree vigour and fruit position on calcium accumulation in avocado fruits. *Scientia Horticulturae* 44, 269–278.

Wolfe, H.S., Toy, L.R. and Stahl, A.L. (1946) *Avocado Production in Florida.* University of Florida Cooperative Extension Service Bulletin 129, Gainesville, Florida.

Wolstenholme, B.N. (2003) Avocado rootstocks: what do we know; are we doing enough research? *South African Avocado Growers' Association Yearbook* 26, 106–112.

Zentmyer, G.A. (1957) The search for resistant rootstocks in Latin America. *California Avocado Society Yearbook* 41, 101–106.

Zentmyer, G.A. (1963) The Duke avocado. *California Avocado Society Yearbook* 47, 28–36.

Zentmyer, G.A. (1972) Avocado root rot (*Phytophthora cinnamomi*). *California Avocado Society Yearbook* 55, 29–36.

Zentmyer, G.A. (1978) Origin of root rot resistant rootstocks. *California Avocado Society Yearbook* 62, 87–89.

Zentmyer, G.A., Guillemet, F.B., Goodall, G., Gustafson, C.D. and Lee, B.W. (1976) Avocado rootstocks resistant to *Phytophthora cinnamomi. California Avocado Society Yearbook* 60, 165–167.

Zumaquero, A., Martínez-Ferri, E., Matas, A.J., Reeksting, B., Olivier, N.A., Pliego-Alfaro, F. *et al.* (2019) *Rosellinia necatrix* infection induces differential gene expression between tolerant and susceptible avocado, rootstocks. *PLOS One* 14(2): e0212359. doi:10.1371/journal.pone.0212359.

9 Propagation

A.W. Whiley
Sunshine Horticultural Services, Queensland, Australia

Abstract

Propagation reviews the changes and progress made with avocado from when trees were planted as self-rooted seedlings to the most recent development of *in vitro* rootstock propagation. The chapter also reviews nursery production protocols and includes grafting techniques used for scion cultivars in the nursery and field when top-working older orchards. Also discussed are the two types of rootstocks originating from different propagation techniques and when either should be used, since this issue remains contested.

Introduction

Avocados were propagated in Mexico and Central America as self-rooted seedlings for thousands of years by the Amerindian tribes, who valued the fruit as a food source. The seeds of fruit with superior qualities were often collected from the forest and planted closer to tribal villages, thus providing a crude genetic improvement process. Wider distribution (as seeds) from their indigenous areas occurred post-European colonization of the Americas by the new settlers, presumably recognizing their value as a food source (Popenoe, 1920). Propagation for commercial production in California began in the late 1800s, with seeds imported from Mexico and Central America by botanical explorers and hobbyists. However, the genetic variability of seedling trees often resulted in extended periods before the first crop set and fruit quality was unreliable. Vegetatively propagated (budded/grafted) trees only occurred after the discovery and introduction of 'Fuerte' into California in 1911 (Platt and Frolich, 1965; Whitsell *et al.*, 1989; Ben-Ya'acov and Michelson, 1995). The benefits from grafting scions of superior lines to seedling rootstocks, which maintained the attributes of precocity, yield and fruit quality, were quickly recognized by the fledgling industry. Initially, citrus propagation techniques were adopted, where buds from parent trees were inserted into T-shaped cuts in the bark of seedling rootstocks. These trees were field grown and at the time of sale, 2–3 years from planting seed, were dug up and the root ball wrapped in burlap (Whitsell *et al.*, 1989). Around 1950, Walter Beck, a nurseryman in Fallbrook, California, pioneered a new propagation technique of tip-grafting

*Corresponding author: tonywhiley43@gmail.com

DOI: 10.1079/9781800621824.0009

container-grown seedlings produced in a warm greenhouse. A modification of tip grafting, also known as cleft grafting, became the dominant method of propagation in California, and its use has spread to most other avocado-growing countries. Due to occasional, damaging sub-zero winter temperatures, the Californian industry migrated to Mexican race seedling rootstocks due to their greater cold tolerance (Wolstenholme, 2003), and because of their leading status in avocado production, some other countries later followed; for example, Spain and South Africa. During the 1970s, the Australian industry briefly converted to the Mexican race rootstock 'Topa Topa'. However, following disastrous results due to high Phytophthora root rot susceptibility, the industry reverted to home-grown rootstock lines, predominantly of Guatemalan and West Indian race origin (A.W. Whiley, SHS, Buderim, QLD, Australia, 2024, personal observation) introduced through the 1800s trans-Pacific shipping trade.

About 1970, California nurseries began vegetative propagation of avocado rootstocks in response to the release of Phytophthora root rot-tolerant lines by the University of California, since clonal production was the only method of retaining genetically acquired disease resistance. Frolich and Platt (1972) developed and described the 'nurse-seed/etiolation' technique for the clonal propagation of avocado rootstocks, which was modified by Brokaw (1987a), who reported that 1977 marked the first year of significant commercial production of cloned rootstocks in California. In 1977, Brokaw Nursery, Ventura, CA, produced 4000 'Hass', 'Pinkerton' and 'Bacon' trees grafted onto clonal 'Duke 7' rootstock, which increased to 500,000 trees propagated on cloned rootstocks within 10 years. The growth in clonal rootstock production has been driven by other genetically acquired attributes such as improved salinity (Mickelbart and Arpaia, 2002; Celis *et al.*, 2018; Lazare *et al.*, 2021), high pH tolerance (Ben-Ya'acov and Michelson, 1995) and improved precocity from lower-vigour rootstocks (Fig. 9.1) (Brokaw, 1987b; A.W. Whiley, SHS, Buderim, QLD, Australia, 2024, personal observation).

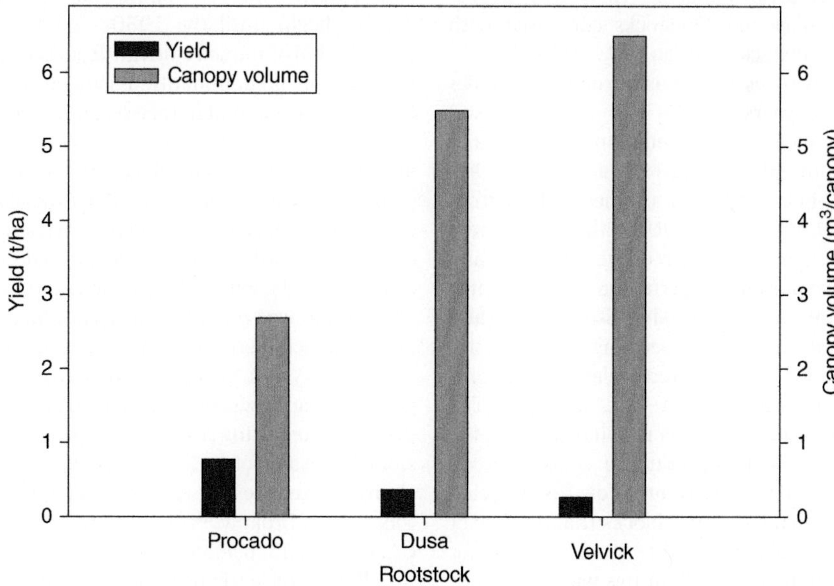

Fig. 9.1. First crop yield and canopy size of 'Hass' grafted onto three clonally propagated rootstocks growing in subtropical, central Queensland, Australia, 2 years after planting. Trees are spaced at 7.5 × 3 m (443 trees ha^{-1}). Data are mean values from 40 trees of each rootstock/scion combination. 'Hass' grafted onto 'Procado®' are smaller trees with increased precocity compared with 'Hass' grafted onto 'Dusa' or 'Velvick'. 'Dusa' and 'Procado®' have commercially useful partial resistance to Phytophthora root rot (Smith *et al.*, 2011). (Figure supplied by A.W. Whiley, Sunshine Horticultural Services, Queensland, Australia)

The commercialization of avocado in California and the associated development of propagation technology provided a lead to other countries producing avocados. In Israel, commercial orchards were planted from 1954, with a rapid expansion in area from 1960 (Ben-Ya'acov and Michelson, 1995). Samish and Gur (1962) reported that autumn budding of 1-year-old seedlings, mostly of the Mexican race but some West Indian race hybrids, was initially the most favoured propagation method. However, the high mortality of buds (about 50%) subsequently led to the adoption of tip-grafting techniques similar to those used in California. Me and Arzate-Fernández (2010) have reported 'Lowland' as descriptively more accurate than the 'West Indian' race, but due to entrenchment in the literature, this chapter will continue to use West Indian nomenclature. Ben-Ya'acov *et al.* (1992) concluded that developed avocado industries were shifting to clonal rootstocks, as trees show good vegetative development and precocity and predominantly outperformed trees grafted onto seedling rootstock. However, more recent Israeli research has not supported the higher performance of 'Hass' trees propagated onto cloned rootstocks compared with seedling rootstock (Cohena *et al.*, 2023).

Following comprehensive rootstock studies over many years, Ben-Ya'acov and Michelson (1995) made recommendations that significantly consolidated rootstock and scion numbers used in Israel. Rootstock lines declined from 600 in 1970 to 30 in 1990. Mother blocks were established to supply seeds of the best types, and only this material was permitted for use by nurseries. Ben-Ya'acov and Michelson (1995) also noted that West Indian seedling lines used for propagation in Israel were relatively uniform compared with Mexican race seedlings. The source of scion wood was rationalized with the elimination of non-productive scion sources. Certified scion sources are now used, being mainly grown in mother blocks that are free of avocado sunblotch viroid (ASBv). A Centre for Coordination of New Plantings was established to choose the best material for each grower according to local conditions (Ben-Ya'acov and Michelson, 1995).

In South Africa, Le Roux (1940) reported that the budding and grafting methods initially practiced for field-grown seedlings gave poor results, leading to a scarcity of nursery trees. As a result of the investigations done by W. Carter, of the Subtropical Horticultural Research Station in Nelspruit, South Africa, a modified form of patch budding was described whereby the wood of the bud shield was carefully removed before T-budding. This method was found to be superior to the methods previously used. From the mid-1950s, budding was replaced by grafting using container-grown seedling rootstock trees. Van der Meulen (1952) adapted the tip-(cleft-) grafting technique used in California, but a heated greenhouse was not needed under South African conditions. South Africa was one of the first countries to adopt the Brokaw Nursery modified etiolation technique for clonal propagation of rootstocks, mainly due to the severity of Phytophthora root rot caused by the pathogen *Phytophthora cinnamomi* Rands (Durand, 1990; see Dann and Gazis, Chapter 13, this volume).

The first known mention of avocados in the Canary Islands, Spain, was made in 1867, where it was discovered growing as an ornamental tree (Galán Saúco and Fernandez Galvan, 1985). However, commercial production did not begin until the 1950s, following the introduction of named cultivars from California. Propagation predominantly occurred on West Indian and Guatemalan race seedlings, presumably from local trees grown from earlier introductions into the archipelago prior to records being kept. During the late 1970s, avocado production developed around Malaga on the south-eastern seaboard of mainland Spain (Galán Saúco *et al.*, 1981), where the Californian rootstock 'Topa Topa' was predominantly used as a seedling rootstock (Pérez Jiménez *et al.*, 2005). Over the past 40 years, Spain has grown to be the largest European producer of avocados (Anonymous, 2024a), integrating a significant percentage of cloned 'Dusa' rootstocks into its expanding orchards to reduce the effects of Phytophthora root rot (R. Brokaw, Brokaw Nursery, Ventura, CA, USA, 2023, personal communication).

European settlement of Australia during the mid-1850s included ocean traffic from the Americas, resulting in the arrival of avocado seed and the establishment of a genetically diverse population of seedling trees along the north-eastern seaboard. This included material from the three botanical cultivars or horticultural races described by Scora *et al.* (2002). Whiley *et al.* (2007) reported that the Australian avocado industry was still largely using seedling rootstocks selected from the progeny of these early introductions. There has been a relatively slow uptake of trees grafted onto cloned rootstocks that provide superior tolerance to Phytophthora root rot; for example, 'Dusa' and more recently, 'Procado®' (A.W. Whiley, Buderim, QLD, Australia, 2023, personal observation), and to date, these rootstocks account for only a small percentage of total trees planted each year.

The first avocados grown in New Zealand were in the 1920s from seeds that were introduced and distributed by the Department of Agriculture to interested growers (Schroeder, 1965). Later introductions of Californian cultivars were made and grown in the Gisborne area, including 'Zutano', which later became the main source of seedling rootstocks as the industry developed during the 1980s–1990s. 'Zutano' seedlings remained the favoured rootstocks for commercial use until about 2015, when an increasingly strong preference for cloned rootstocks emerged. In 2023, it was estimated that about 90% of the nursery trees are produced on cloned rootstocks, apportioned between 'Astro' (55%), a rootstock with good Phytophthora root rot resistance, recently developed in New Zealand; 'Bounty' (25%); 'Dusa' (15%) and 'Latas' (5%) (S. Wade, Lynwood Nursery, Whangarei, New Zealand, 2023, personal communication).

Currently, the Chilean avocado industry predominantly relies on seedling rootstocks (mainly 'Mexicola' and 'Zutano'), with an increasing use of 'Velvick' in high-density, replant situations. To a much lesser extent, clonal (vegetatively propagated) rootstocks (mainly 'Dusa' and 'Duke 7') are being propagated and grafted with 'Hass', in part for their increased uniformity and improved ability to adapt to stress factors, both pathological and edaphic (Hernández *et al.*, 2023). Other countries of the neotropical region, where production over the past 10–20 years has rapidly expanded, have planted 'Hass' orchards mostly grafted onto open-pollinated, seedling rootstocks from selected 'criollo' trees (Cañas-Hernández *et al.*, 2022; C. Fassio, Quillota, Chile, 2023, personal communication), which are mainly of the Mexican and West Indian horticultural races (Corrales-García and Méndez-Zúñiga, 2020; G. Aldunate, Asesorias Agricolas Los Alerces, Quillota, Chile, 2024, personal communication). However, more recently, in some areas, the availability and uptake of cloned rootstock have occurred (C. Fassio, Quillota, Chile, 2023, personal communication).

This chapter reviews the current knowledge on rootstock type and the various propagation techniques developed for avocado and describes the more common procedures currently being used for production of nursery trees and field grafting (top-working). For a more detailed discussion of *in vitro* propagation of avocado, see Pliego-Alfaro *et al.*, Chapter 10, this volume.

Rootstocks

There has been an exponential growth in global avocado production in the past 10 years, rising from 4.41 million MT in 2012 to 8.98 million MT in 2022, an increase of 104% (source: Anonymous, 2024b). Alarmingly, the majority of this growth has been demand-driven, rather than by technical innovation that generates improved efficiencies in orchard production, and does not take into account production from the millions of trees planted in the last 2–3 years.

Despite the unprecedented growth in global production, avocado trees grafted onto seedling rootstock remain the substantial majority of nursery trees grown due to their availability and ease of propagation, making them clear winners for mass nursery production (Wolstenholme, 2003). Genetic improvement in fruit tree crops is slow and unpopular, with funding fatigue generally occurring long before objectives are achieved. This is unfortunate, since advances made in crop production with genetic material are enduring, and progress can be lost during the intervening time between funding cycles, which are often considerably protracted. Continuity in such programmes is the key to success, and 20-year time frames are the norm for advances in genetic improvement with fruit tree crops.

Scion cultivars grafted onto rootstocks are the foundation of modern orchards, with the latter providing tree anchorage, supplying scion cultivars with water, auxins and cytokinins, and may also act as a seasonal repository for surplus metabolites (Wolstenholme, 2003; Hayat *et al.*, 2019). Additionally, rootstocks can either enhance or diminish scion yield by minimizing soil stresses (Ben-Ya'acov and Michelson, 1995; Smith *et al.*, 2011; Menge *et al.*, 2012; Celis *et al.*, 2018; Castillo-Argaez *et al.*, 2021; Lazare, *et al.*, 2021); contribute to anatomical/physiological graft incompatibility with scion cultivars (Thomas, 1997; Atkinson *et al.*, 2003; Whiley, 2003; Castillo-Argaez *et al.*, 2021); affect tree size through scion/cultivar vigour (Palmer, 1988; Ben-Ya'acov and Michelson, 1995; Webster, 2002; Atkinson and Else, 2004; Whiley and Dann, 2021; Hayat *et al.*, 2022) and reduce susceptibility to postharvest anthracnose fruit lesions through modification of the mineral nutrient content of fruit (Willingham *et al.*, 2001; Marques *et al.*, 2003; Whiley *et al.*, 2013; Lazare *et al.*, 2020). This section broadly reviews current knowledge on avocado rootstocks and discusses the choices available to astute orchardists.

Graft compatibility

Seedling populations of avocado (*Persea americana* Mill.) used commercially for rootstocks can be highly diverse, being drawn from three botanical varieties, *P. americana* var. *drymifolia*, *P. americana* var. *guatemalensis* and *P. americana* var. *americana* (Scora *et al.*, 2002; Castro *et al.*, 2011; Whiley *et al.*, 2013), which are both pollen and graft compatible. Variation in climatic and edaphic responses among seedlings of the botanical varieties is anchored in their origin (Ben-Ya'acov and Michelson, 1995) and expressed through varietal lines, or selectively in the numerous hybrids, most being serendipitously created through anthropogenic intervention (Popenoe, 1920; Crane *et al.*, 2013).

Fruit tree graft compatibility is defined as 'the union of a root system (rootstock) and apex (scion cultivar) of two plants, so that vascular continuity is established resulting in extended survival and proper function of a composite plant', and is usually found in graft partners that are phylogenetically close (Adhikari *et al.*, 2022). The three horticultural races (botanical varieties) of avocado have varying levels of graft compatibility, with most rootstock/scion cohorts forming a surviving union. However, significant numbers of rootstock/scion unions would be considered horticulturally unviable due to various forms of anatomical and physiological incompatibility that prematurely result in the decline and death of trees (Ben-Ya'acov and Michelson, 1995). Graft incompatibility in fruit tree crops is due to cellular physiological intolerance caused by metabolic, developmental, genetic and/or anatomical and physiological differences between stock and scion and are discussed in the reviews of Andrews and Serrano Marquez (1993), Nawaz *et al.* (2016) and Habibi *et al.* (2022). Within avocado, the anatomical and physiological causes behind incompatibility responses differ widely between rootstock/scion combinations, resulting in visual differences both at and on either side of graft unions (Fig. 9.2). However, there is an absence of science-tested data available on their effect on long-term scion productivity. Based on morphological observations, there is evidence of 'translocation incompatibility', which causes reduced scion performance in many fruit tree species, where interruption of metabolic (particularly carbohydrates) flow from scion to rootstock occurs (Adhikari *et al.*, 2022). 'Translocation incompatibility' in fruit tree and vine crops has been associated with starch accumulation above the union and its absence below (Moing *et al.*, 1987); phloem degeneration; normal vascular continuity at the union with overgrowth of the scion (Fig. 9.3) (Mosse, 1962; Pereira *et al.*, 2014), where distinct differences in bark thickness and texture between scion and rootstock can occur in avocados (Fig. 9.2) (A.W. Whiley, SHS, Buderim, QLD, Australia, 2024, personal observation), reduced tree vigour 1–2 years after planting; and some avocado rootstock/scion combinations produce yellowed leaves at full expansion, with premature senescence during winter (A.W. Whiley, SHS, Buderim, QLD, Australia, 2024, personal observation). The occurrence and potential effects of 'translocation incompatibility' in relation to avocado are further discussed.

With advances in genomic analysis, it is now possible to separate the Mexican, Guatemalan and West Indian races into distinctly separate subpopulations (Boza *et al.*, 2018), with general consensus that *P. americana* var. *drymifolia* (Mexican) is inherently more distant from *P. americana* var. *guatemalensis* (Guatemalan) and *P. americana* var. *americana* (West Indian), than the latter two are from each other. These genetic differences may explain the relatively wide occurrence of differing levels of incompatibility

Fig. 9.2. Different types of graft unions and bark pattern changes on commercial rootstocks grafted onto 'Hass' and 'Dusa'. The (a) union is with cloned 'Procado'® (G), where the rootstock and scion diameters are equal, but bark texture changes are visible on either side of the graft; the (b) union is with cloned 'Dusa' (Mexican × Guatemalan), showing a slight scion overgrowth and distinct change in bark pattern; the (c) union is with cloned 'Velvick' (West Indian × Guatemalan), showing a moderate scion overgrowth and distinct change in bark pattern; the (d) union is with seedling 'BW2' (Guatemalan) and shows moderate scion overgrowth; the (e) union is between 'Dusa' and an unknown rootstock, showing a strong rootstock overgrowth, which results in weak scion growth. (Photographs supplied by A.W. Whiley, Sunshine Horticultural Services, Qld, Australia)

between avocado rootstock/scion combinations, which is likely due to sufficient genome separation when scion cultivars are grafted onto some rootstocks. The probable graft 'translocation incompatibility' between Guatemalan scion cultivars ('Lyon' and 'Murrieta') and Mexican seedling rootstock was reported by Schroeder and Frolich (1955), where resultant plants were frequently dwarfed or short-lived. However, when the scion cultivars were grafted onto Guatemalan seedling rootstocks, the resultant combinations were successful and trees grew normally.

The New Zealand avocado industry was founded on seedling rootstocks of the hybrid cultivar 'Zutano' (Cutting, 2002), with an attributed 43% Mexican and 56% Guatemalan parentage (Chen *et al.*, 2009). When grafted 'Hass' (42% Mexican and 58% Guatemalan, Chen *et al.*, 2009; ≈60% Mexican and 40% Guatemalan, Rendón-Anaya *et al.*, 2019; ≈25% Mexican and 75% Guatemalan, Solares *et al.*, 2023), trees with significant scion overgrowth at graft unions commonly occurred across orchards (Fig. 9.4), most likely induced by poor phloem translocation between scion and rootstock (Mosse, 1962; Pereira *et al.*, 2014). In this situation, trees with significant scion/rootstock overgrowth developed strong alternate-bearing cycles, compared to trees with minimal difference between rootstock/scion diameters at the graft union (J.G.M. Cutting, NZAGA, Tauranga, New Zealand, 2002, personal communication). Interestingly, the percentage of trees showing significant scion overgrowth varied considerably among orchards, with some ≤15% and others ≥30% affected (Whiley and Anderson, 2002). This mixed response within the 'Zutano' seedling rootstocks when

Fig. 9.3 Effects of scion overgrowth on container-grown, two-year-old avocado trees. (a) illustrates an extreme scion overgrowth of a cloned 'SS3–1' rootstock (West Indian sourced from Canary Islands) grafted onto 'Shepard' (Mexican × Guatemalan), with (b) the resultant poorly developed root system. (c) illustrates a small rootstock overgrowth, with (d) the resultant healthy, dense root system when 'Shepard' (Mexican × Guatemalan) is grafted onto seedling 'Velvick' (West Indian × Guatemalan). (Photographs supplied by A.W. Whiley, Sunshine Horticultural Services, Qld, Australia)

grafted with 'Hass' may be caused by the intensity of outcrossing during flowering, influenced by the parent 'Zutano' trees' proximity to other cultivars, thus changing the magnitude and composition of seedling heterozygosity (Whiley, 2003). The Chilean avocado industry has similar scion overgrowth incompatibility issues, with its most-used seedling rootstock 'Mexicola' (99% Mexican, Chen *et al.*, 2009) grafted with 'Hass' (Fig. 9.4), which may contribute to reduced yield and alternate-bearing cropping patterns in affected trees (Mena *et al.*, 2015). However, the picture is not clear, as not all scion overgrowths result in negative consequences on production. Le Lagadec (2010) reported the highest cumulative yield from 22 seedling rootstock/scion ('Hass') combinations, measured over 5 consecutive years, was from 'BW2', which has moderate scion overgrowth of the rootstock (Fig. 9.2d). In many instances, seedling rootstocks grafted to 'Hass' or other mainstream cultivars, consistently form normal graft unions, which become nearly invisible within 3–4 years of field planting (A.W. Whiley, SHS, Buderim, QLD, Australia,

Fig. 9.4. Scion overgrowth of rootstock which intermittently occurs with 'Hass' grafted onto (a) seedling 'Zutano' rootstocks in New Zealand, and (b) 'Mexicola' rootstocks in Chile. (Photographs supplied by A.W. Whiley, Sunshine Horticultural Services, Qld, Australia)

2024, personal observation). Alternatively, some graft unions produce slight rootstock overgrowths, which in the case of 'Shepard' (Whiley and Lavers, 2021) grafted to 'Velvick', reduces tree vigour and increases crop yield compared with a range of other seedling rootstocks grafted to 'Shepard' (Le Lagadec, 2010; Whiley *et al.*, 2013).

Cano-Gallego *et al.* (2023) compared the growth and respiration of 'Hass' fruit grafted with seedling rootstocks, showing no visible graft incompatibility to fruit from 'Hass' grafted onto seedlings showing significant scion overgrowth of rootstocks. From the results, it was concluded that there was no significant difference between fruit from either rootstock/scion combination.

Productivity – Seedling versus Cloned Rootstocks

Flowering in avocado trees is heterodichogamous (Solares *et al.*, 2023), which favours outcrossing, resulting in seedling populations that are highly heterozygous (Lahav and Lavi, 2009). This feature is an evolutionary trait for competitive fitness and survival in forest communities (Barrett, 2010; Tito de Morais *et al.*, 2019). The use of seedling avocados from the same genetic source for rootstocks introduces genetic diversity, and potentially large performance differences among grafted trees, in both growth and production when measured over a number of years. However, this may not necessarily disadvantage the use of seedlings compared to cloned rootstocks, when assessing sustainable production over the life of an orchard. Currently, the choice of clones is mostly restricted to those selected for Phytophthora root rot tolerance (Smith *et al.*, 2011; Menge *et al.*, 2012) and salinity (Mickelbart and Arpaia, 2002; Lazare *et al.*, 2021; Cohena *et al.*, 2023). In situations where neither of these soil stress factors is an issue, seedling lines that induce high yields when grafted with 'Hass' may be a better choice. However, the selection of seedling lines, particularly with respect to their compatibility with scion cultivars, is an important but largely overlooked factor when choosing propagation material, which often reverts to availability rather than best practice. A number of case studies are presented below, where data from multiple years of production from seedling and clonal rootstocks grafted with 'Hass' have been compiled.

AUSTRALIA 1. Whiley *et al.* (2013) studied orchard performance of a number of rootstocks drawn from the three botanical races, grafted with either 'Hass' or 'Shepard'. Rootstocks were propagated either from seed, or the parent tree was cloned using the microcloning technique described by Ernst (1999). The six experimental sites covered a diverse range of edaphic and climatic environments, between latitudes 17°S and 34°S. The research was commissioned to answer three basic questions: (i) Due to the wide genetic diversity of potential rootstock material in Australia, drawn from the three horticultural races, viz. Mexican, Guatemalan and West Indian, is there any specific genetic group, when grafted with 'Hass' or 'Shepard', that will consistently improve yield? (ii) Will different rootstocks be more suited for each production region across the country, considering significant differences in edaphic and climatic conditions? (iii) Will the use of cloned rootstocks enhance production over the seedling rootstocks currently used by providing greater genetic uniformity? 'Shepard' was only planted at the two warm subtropical sites (Childers and Walkamin in central and North Queensland, respectively) where this cultivar is commercially grown (Whiley and Lavers, 2021).

The selection of seed for this study was rigorous, with material sourced from isolated maternal trees that were unlikely to be outcrossed with other cultivars (Whiley and Whiley, 2011). This limited genetic diversity to that contained within the maternal parent trees. Seed and cloning material was sourced from the same tree, and nursery seedlings were carefully inspected, with off-types removed before grafting. Experimental sites were planted as randomized blocks, with ten replications of each treatment, and all experimental sites were planted in new ground for avocado production. One exception was the Walkamin site in North Queensland, where the land was fallowed for 2 years with a succession of cover crops following the removal of a 20-year-old avocado orchard. For the duration of the study, Phytophthora root rot was not a production issue at any of the research locations. The experiments were mostly planted in 2005 with yield data collected consecutively over 5–6 years.

Enhanced statistical analysis was applied to the data of Whiley *et al.* (2013) and the results reported by Dann (2016). The trees on 'Velvick' seedling rootstock grafted with 'Hass' were, overall, the highest yielding across all sites and significantly greater than their counterparts cloned from the maternal parent tree. However, production from trees on seedling 'Velvick' was not dominant at all sites where the yield from other rootstock/scion combinations were randomly higher. The reduced performance of trees grafted onto 'Velvick' clones may be due to scion overgrowth of the rootstock (Fig. 2c), which is consistently present when the parent 'Velvick' tree is cloned, but rarely in 'Hass' or 'Shepard' when grafted onto 'Velvick' seedlings. Overall yields were significantly greater from the Guatemalan × West Indian hybrid and Guatemalan rootstocks, while generally, Mexican race rootstocks and their hybrids were in the lower yielding quartile. 'Hass' grafted onto clonal 'Zutano' was the highest yielding of the cloned rootstocks growing at Pemberton, WA, and Walkamin, Queensland, and no scion overgrowth of the cloned rootstock was observed. The results also showed that 'Hass' or 'Shepard' grafted onto clonally produced rootstocks were not necessarily more uniform than the same scion cultivars grafted onto seedlings in terms of production-related criteria, particularly as they aged. Mostly, there was no statistically significant difference in yield between 'Hass' grafted onto seedling and clonally propagated rootstocks. However, there was an overall effect of propagation type on yield efficiency (measured as kg of fruit per m^3), with trees grown on seedling rootstocks having higher mean yield efficiencies than those grown on cloned rootstocks.

The results from this study challenge the expectation that greater genetic uniformity of cloned rootstocks will increase orchard productivity when compared with seedling rootstocks. While the use of some cloned rootstock lines increased yield, there were only 3 of the 26 matched rootstock pairs where production from clones was significantly higher than trees grown on their paired seedling rootstocks. In seven of the seedling/cloned rootstock pairs, production from seedling rootstocks was significantly greater than their paired clones, and 16 combinations were not statistically different. Additionally, in a separate rootstock study planted in central Queensland, with 22 seedling and 11 cloned rootstocks of Guatemalan and West Indian origin grafted to 'Hass', Le Lagadec (2010) reported that overall the cumulative yields from trees grown on cloned rootstock were less than the yields from trees grafted onto their paired seedling rootstock. Within the cloned group of rootstocks, the cumulative yield from 'Hass' grafted onto 'Duke 7' was in the bottom quartile of the group. Whiley *et al.* (2013) also reported significant differences in postharvest fruit quality. There was less disease development in fruit from trees grafted onto Guatemalan/West Indian race rootstocks compared with those grown on Mexican race rootstocks, which has important commercial implications for fruit grown in humid, subtropical climates, which favour fruit pathogens. These results corroborate the findings of Willingham *et al.* (2001) and Marques *et al.* (2003). Production diversity within some seedling rootstock groups, particularly those from West Indian/Guatemalan heritage, was higher than expected. Within some seedling lines, individual trees grafted with 'Hass' produced 130%–200% higher yield than the mean of the other trees in their replication (Whiley *et al.*, 2013). Interestingly, the yield from these higher-producing trees, either increased annually or between some high-cropping years, remained static over the 6 consecutive years that data were collected. Lower-yielding trees on the same rootstock within the replication developed different intensities of alternate bearing. This diversity is lost when the maternal parent tree is used as a cloned rootstock, and in the majority of cases in this study, the random recombination of genes in seedling populations has, overall, given similar or, in a few cases, statistically significantly higher yield as a group, compared with their cloned parent where genetic performance is fixed (Whiley *et al.*, 2013).

AUSTRALIA 2. Smith *et al.* (2011) and Dann *et al.* (2013) evaluated eight different seedling and cloned rootstocks grafted with 'Hass' for resistance to Phytophthora root rot. The evaluation block was a replant site with a high level of *P. cinnamomi* infestation, and tree health and yield data were collected from 2009 to 2013, inclusive. Rootstocks included 'A8' (seedling G), 'A10' (seedling M), 'Reed' (seedling G), 'Velvick^{A}' (seedling WI/G hybrid), 'Velvick^{L}' (seedling WI/G hybrid), 'Dusa' (cloned M/G hybrid), 'Latas' (race unknown) and 'Velvick' (cloned WI/G hybrid). Prior observations suggested performance differences when 'Hass' was grafted onto two sources of 'Velvick' seedlings from genetically identical parent trees that were growing at different locations: 'Velvick^{A}', in a block of mixed rootstocks grown for seed, with the opportunity to outcross to other rootstock cultivars; and 'Velvick^{L}', at a location where there was no opportunity for cross-pollination from other cultivars. The study was concluded after 2013 due to exceptionally wet summer conditions in 2011 and 2013, which resulted in the death of a number of trees. 'Velvick^{L}' grafted with 'Hass' produced the highest cumulative yield for the 5 years of data collection, which was significantly higher than 'Reed' and 'Velvick^{A}' but not significantly different from the other rootstocks (Fig. 9.5). With respect to Phytophthora root rot decline, at the completion of the study, 'Dusa' and 'Velvick^{L}' trees were the healthiest, and both significantly healthier than 'Hass' grafted onto 'Velvick^{A}' (Dann *et al.*, 2013). Due to funding constraints, molecular studies were not carried out to explore genetic differences between the two seedling 'Velvick' lines. However, it is hypothesized that seeds that have resulted from outcrossing may have changed field performance characteristics compared to rootstocks of the same cultivar grown from self-pollinated seeds, since significant differences were found between seedling 'Velvick' rootstock from separate sources of parent 'Velvick' trees. These results also support the hypothesis that the 'Zutano' seed produced from outcrossing may result in rootstocks that cause significant variation in the yield profiles of 'Hass' due to indiscriminate scion overgrowth in some New Zealand orchards (Fig. 9.4).

ISRAEL. The results of a recent research comparing the productivity of 'Hass' trees grafted onto seedling and clonal West Indian rootstocks have been reported by Cohena *et al.* (2023). In their research, four seedling and five clonal rootstocks, selected for proven salt tolerance, since irrigation was with recycled water (chloride content was between 200 and 260 $mg\ l^{-1}$), were grafted with 'Hass' and separate experiments planted at two locations. Yield data were collected for 6 consecutive years, and the

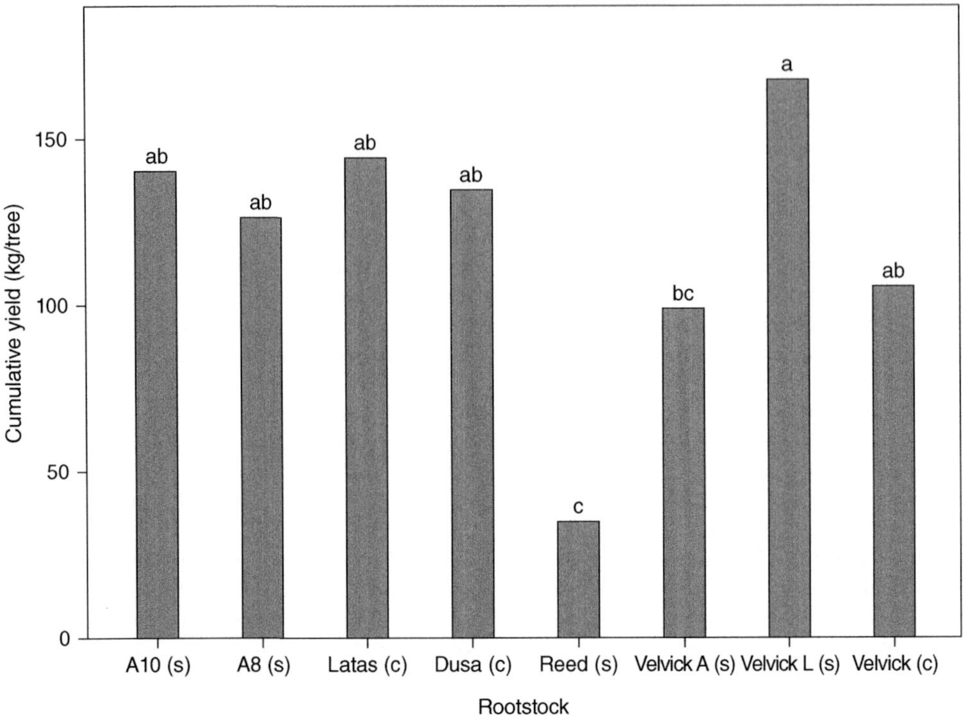

Fig. 9.5. Mean cumulative yield (kg) for 'Hass' trees grafted onto different rootstocks planted in 2006 at Childers, QLD, Australia, in a replant site with natural, high Phytophthora root rot pressure. Means are calculated from ten trees for each rootstock/scion combination with the exception of cloned 'Velvick', which had nine trees. Columns with different letters are significantly different ($P \leq 0.05$). (The figure is redrawn from Dann *et al.*, 2013, with permission)

alternate-bearing index (Hoblyn *et al.*, 1937) was calculated annually for each rootstock/scion combination. Cohena *et al.* (2023) reported that significantly different yields were recorded between rootstocks in some individual years; however, there was no significant difference in the 6-year cumulative yield between any of the rootstock/scion combinations. With respect to the alternate-bearing index, Cohena *et al.* (2023) found that, at both experimental sites, there was no significant difference in the multi-year mean index values between the nine rootstock/scion combinations, although trees with cloned rootstock consistently had lower index values. They suggested that this may reflect greater genetic uniformity in trees grafted onto cloned rootstocks. In conclusion, based on the results from this research, Cohena *et al.* (2023) recommended that Israeli growers should choose trees grafted onto seedling rootstocks, rather than clones from the group of material evaluated, as this approach offers cost savings at planting (trees are cheaper when grafted onto seedlings).

NEW ZEALAND. A comparison in production between 'Hass' grafted onto cloned 'Dusa' and seedling 'Zutano' rootstocks was made in an orchard near Kaitaia, in the Far North of New Zealand, using farm block yield data collected annually for 13 consecutive years. Trees on both rootstocks were planted in 2008, in adjacent blocks on deep, well-drained soils derived from ancient sand dunes. The total number of trees in the 'Dusa' block was 2300, while the 'Zutano' block had 2495 trees; both were planted at high density (814 trees ha^{-1}). Irrigation to complement the 1337 mm mean annual rainfall was delivered via under-tree sprinklers, which were also used for fertigation. Soil probes provided data for irrigation scheduling, and soil and leaf analyses were taken annually to support nutritional programmes. Tree canopies were managed through strategic pruning and fertigated applications of paclobutrazol (Payback®), which is registered for soil application in New Zealand. The blocks had interspersed pollinizer cultivars (4% of the trees), and honeybees (11 hives ha^{-1}) were introduced each year to cover the flowering period (Hermosilla-Madrid and Whiley, 2023). Despite the high-rainfall conditions, the presence of Phytophthora root rot in these blocks remained low and was managed with phosphonate fungicides when required. At harvest, the bins of fruit from each block were weighed and values recorded. The first crop was harvested in 2010, 2 years after planting, and thereafter, harvests were made annually through to 2022. The orchard is at a latitude (35°S) that delays fruit maturity, and, for a short time, trees may simultaneously carry both fully mature and newly set fruit, which, as trees mature, induces stronger alternate bearing (see Whiley *et al.*, Chapter 12, this volume). Scion overgrowth of the rootstock in the seedling 'Zutano' block was estimated at 15%–20% (C. Hermosilla-Madrid, Awanui, New Zealand, 2024, personal communication). Cumulative production over the first 7 years slightly favoured 'Hass' grafted onto the seedling 'Zutano', with a mean annual yield of 6.3 t ha^{-1} compared to 5.0 t ha^{-1} for the 'Hass'/'Dusa' block. During the following 6 years, 'Hass' trees grafted onto 'Dusa' rootstocks were slightly more productive, with a mean annual yield of 17.2 t ha^{-1}, compared to 15.8 t ha^{-1} for the 'Hass'/'Zutano' block. A similar trend was observed by Whiley *et al.* (2013), with seedling rootstocks grafted with 'Hass' outcropping their cloned equivalents in the first few years of orchard establishment but with trees grafted onto cloned rootstocks eventually producing similar crops to those on seedling rootstocks. At Kaitaia, the mean annual yield over the 13 years of cropping was 10.7 t ha^{-1} for 'Hass' grafted onto seedling 'Zutano', and 10.6 t ha^{-1} for 'Hass' grafted onto cloned 'Dusa'. Crop load peaked in both blocks in 2019, with yield reaching 27.2 and 23.8 t ha^{-1} for 'Dusa' and seedling 'Zutano', respectively. However, subsequent production has declined across both rootstock lines due to orchard crowding and implementation of more severe pruning interventions. Interpretation of this long term yield data suggests that in this low Phytophthora root rot environment, there has been no advantage in using cloned 'Dusa' as a rootstock and considering the significant difference in tree costs at planting, for economic value, the 'Hass'/'Zutano' combination has been a better choice.

SOUTH AFRICA. De Villiers and Ernst (2015) compared production from 'Hass' grafted onto seedling and cloned 'Duke 7'. Rootstock treatments were planted alternatively along rows at a density of 204 trees ha^{-1}. They reported that 'Hass' on clonal 'Duke 7' rootstock outperformed 'Hass' on seedling 'Duke 7' rootstock, over the 6-year period (2010–2015) of data collection. The mean annual yield for 'Hass'/clonal 'Duke 7' trees was 67 kg $tree^{-1}$ compared to 35 kg $tree^{-1}$ for 'Hass'/seedling 'Duke 7' trees. There are large differences between the two types of 'Duke 7' rootstocks, and this result has not been reproduced in Australia. In the research by Whiley *et al.* (2013), yield comparisons between 'Hass' grafted onto both seedling and cloned 'Duke 7' across multiple research sites have either shown no significant difference or, in one instance, were higher from 'Shepard' grafted onto seedling 'Duke 7'. However, there is a large age difference between the trees from which de Villiers and Ernst (2015) collected their data and the age of the trees in the Australian research, which may explain the divergence between the results. In South Africa, data were collected from 23-year-old trees (planted in 1987), whereas in Australia yield data were collected from trees in their first 5–6 years of production. The cloned 'Duke 7' parent has some Phytophthora root rot tolerance (Gabor *et al.*, 1990), and its seedlings are very susceptible to the root disease. It is likely that there was a disproportionate level of root infection between rootstock types in the 23-year-old orchard, which reflects in yield differences, since visually the trees on cloned 'Duke 7' were, overall, healthier than those on the seedling rootstock (de Villiers and Ernst, 2015). The experimental orchards in Australia were mostly planted on new ground, and Phytophthora root infection was not an issue for the duration of the research (Whiley *et al.*, 2013).

Discussion

The most likely reasons for the relatively slow uptake of cloned rootstocks outside of California, Spain and South Africa are that more recently developed industries have comparatively low Phytophthora root rot infections; the current availability of clones is primarily restricted to 'single-factor' rootstocks (VC801 is an exception, being selected in Israel for its high tolerance to salinity, *P. cinnamomi* and *Verticillium dahliae*; Duman *et al*., 2020); and their significantly higher cost due to more complex propagation requirements compared with trees grafted onto seedlings. Most clonally produced rootstocks have been selected for tolerance/resistance to Phytophthora root rot, for example, 'Duke 7', 'Dusa', 'Latas', 'Leola', 'Steddom', 'Uzi' and 'Zentmyer' (Menge *et al*., 2012; Maxwell, 2023); salinity tolerance, for example, VC140, VC152, VC159 and 'Zerala' (Lazare *et al*., 2021; Maxwell, 2023); and hypoxic stress that periodically occurs in heavy soils, for example, 'Bounty' (B.N. Wolstenholme, Pietermarizburg, Natal, South Africa, 2024, personal communication).

Host plant mechanisms that prevent or reduce attack from pathogens fall into two broad classes: resistance, which limits pathogen growth and infection; and tolerance, which does not limit infection, but instead reduces or offsets the negative consequences of infection (Horns and Hood, 2012). Commercially useful pathogen defence in avocado has been identified by van den Berg *et al*. (2021) from quantitative polymerase chain reaction (PCR) studies that showed reduced amounts of *P. cinnamomi* DNA in 'Dusa' roots following *in vitro* exposure to the pathogen. They suggested that partial resistance was a more accurate descriptor compared to tolerance to Phytophthora in this rootstock. The currently available Phytophthora-resistant/-tolerant rootstocks still require fungicidal support for commercial success where soils are heavily infested with *P. cinnamomi* (Smith *et al*., 2011), otherwise longevity declines, albeit at a slower rate than in rootstocks that are highly susceptible.

Limited resistance in host (avocado) and excessive aggressiveness in the pathogen (*P. cinnamomi* Rand) reflect the lack of prior coevolution (Brasier, 2001) and usually result in substantial production losses where environments favour the pathogen. Since the evolutionary theatres of *P. cinnamomi* Rands and *P. americana* Mill, were geographically separated, the discovery of 'stand-alone' resistance is unlikely in the foreseeable future, and complete disease resistance is more likely to be attained through molecular techniques than from breeding or 'escape' tree recovery. In the interim, cloned rootstocks with identified Phytophthora root rot tolerance/partial resistance should be utilized along with other management strategies where the presence of Phytophthora infection is a threat to production and longevity of orchards, viz. replant orchards where residual infection propagules remain high (see Dann and Gazit, Chapter 13, this volume).

The data reviewed above suggest that, from the currently available rootstocks, long-term commercial benefits from cloned Phytophthora-resistant rootstocks may only be realized where orchards are planted in the presence of moderate-to-high Phytophthora root rot pressure. Low-cost alternatives, such as high-quality seedling lines, should be considered for continued use in situations where Phytophthora root infection is a low risk, for example, planting new avocado ground with well-drained soil. The variable genetic composition of seedling rootstocks may be an influencing factor in their performance, and Whiley *et al*. (2013) recommended that seed production for nursery trees should be in isolation, restricting outcrossing opportunities.

Two recently released avocado rootstocks, 'Astro' (probably a 'Zutano' seedling), in New Zealand, and 'Procado'® (an unknown Guatemalan race seedling), in Australia, have shown improved survival compared to 'Dusa', when cloned and grafted with 'Hass' and grown in soils infested with *P. cinnamomi* (Smith *et al*., 2011; C.J. Partridge, Tauranga, New Zealand, 2023, personal communication). However, longer-term evaluations over a range of edaphic and climatic conditions are required to validate these results. As these two Phytophthora-tolerant rootstocks have been identified outside of the gene pool with majority Mexican race inherence, there are indications that they may provide additional contributions to the next generation of Phytophthora-resistant rootstocks, such as improved tolerance to salinity and hypoxic soil conditions, which are nominally found within the Guatemalan race.

Compared to pip, stone and citrus fruit, the development of avocados as a modern, sophisticated crop has junior status, as only a little over 100 years (which roughly equates to two generations of growers) have elapsed since superior cultivars were first selected and grafted to produce homogenous volumes of fruit. Currently, the most widely grown cultivar, 'Hass', was commercialized approximately 90 years ago (Hass, 1935), and despite the emergence of many new cultivars, it remains firmly entrenched as the bastion of the global industry. Likewise, rootstock improvement has been incrementally slow to change, with selection for resistance to root diseases and salinity tolerance prioritized. With the rapid development of commercial avocado production in Colombia, planted on heavy clay soils subject to periodic hypoxic conditions due to high-rainfall events (Ramírez Gil, 2018), identification of suitable rootstocks for local conditions has begun (Reyes-Herrera *et al.*, 2020) using molecular techniques to screen local seedling avocado populations. Predominantly, Mexican race rootstocks were selected for use in California due to their superior cold tolerance. South Africa turned to California for crop technology and began importing Mexican race rootstocks that included the root rot-tolerant 'Duke 7'. Today, Mexican race cultivars dominate the rootstock gene pool largely excluding Guatemalan and West Indian race cultivars, which are likely to deliver higher yields when grafted with 'Hass'. Avocado production in Chile is largely based on seedling 'Mexicola' rootstock, but a comparative evaluation with seedling 'Velvick' rootstock, across 8 years of recorded production, found that trees grafted onto 'Mexicola' had a mean annual yield of 14.2 t ha^{-1} compared to 22.0 t ha^{-1} from 'Velvick' trees grafted with 'Hass'. These data support the results of Whiley *et al.* (2013), who reported that 'Hass' trees grafted onto rootstocks of Guatemalan and West Indian origin had and overall higher cumulative yield than those grafted onto Mexican race rootstocks. Trees grafted onto seedling 'Velvick' are now being used in high-density replant sites, since they have been successful on the heavy, clay soils of Chile and are more affordable than trees grafted onto cloned rootstocks (F. Mena, GAMA, Quillota, Chile, 2024, personal communication). Apart from the development of high-density orchards in Chile (see Whiley *et al.*, Chapter 12 this volume), mean productivity (t ha^{-1}) languishes at low levels in most avocado-producing countries. Recent research has identified individual seedling rootstocks grafted with 'Hass' producing 75% to 200% greater yields than their siblings from data recorded over a 4–6-year production cycle (Whiley *et al.*, 2013; Mena *et al.*, 2015). Additionally, in these high-yielding individuals, alternate bearing was either reduced or eliminated. These results indicate the importance of seedling rootstock populations as a genetic resource for productivity improvement. With their continued erosion through loss of evolutionary habitat and replacement with vegetatively propagated rootstocks, exploitation of the current seedling resource should be a global priority while opportunities for improvement exist. The replacement of seedling rootstock populations comes with an evolutionary price, namely reducing both the genetic base and biodiversity, which endangers, and in extreme cases, impedes further plant improvement within crops (Harris *et al.*, 2002; Janick, 2005).

Production of Disease-free Nursery Trees

A reliable source of true-to-type, disease-free planting material is essential for the continued success of cropping systems. While avocado trees can be infected by many diseases, the two most likely to threaten orchard productivity by dissemination from nursery trees are Phytophthora root rot and ASBv, an *in planta* viroid. *P. cinnamomi* is easily spread with the sale of nursery trees planted in infected potting media or through contamination in the nursery due to poor hygiene standards, while ASBv can be transmitted through using infected seed and/or scion wood when propagating nursery trees (see Dann and Gazis, Chapter 13, this volume). In many instances, saleable nursery trees will show no visual symptoms of either of these diseases even though they may be infected. Hence, if selection of propagation material and nursery hygiene are not sufficiently stringent, nursery trees may become a potential infection source of ASBv and/or Phytophthora root rot in newly established orchards. In relatively recent times, *Calonectria ilicicola*, causing black root rot, has been identified as a severe root pathogen of avocado nursery trees in south-eastern Queensland, Australia (Dann *et al.*, 2011, 2012; Prabhakaran and Dann, 2022). Black root rot may be lethal in newly planted trees, adding significant costs to orchard establishment. However, once trees increase in size and become established, they are no longer threatened by *C. ilicicola*. Whitsell *et al.* (1989) provided a list of recommendations to assist nurseries in excluding root diseases and ASBv from avocado trees in California (Table 9.1), which are complemented by the recommendations of Toerien (1977) and Claassens (2001). These are essentially the procedures followed by California nurseries if they have entered into a certification programme monitored by the California Department of Agriculture. This programme has been supported by studies carried out by the University of California since the 1950s. To enter the certification programme, all mother trees (seed and scion) need to be indexed to be proved free from ASBv (Wallace and Drake, 1972).

The severity of Phytophthora root rot in the humid subtropics of Australia and South Africa is arguably greater than that found in the

Table 9.1. Recommended protocol to follow for the production of disease-free nursery trees (Toerien, 1977; Whitsell *et al.*, 1989; Claassens, 2001).

1. Use an isolated, dust-free site free of *P. cinnamomi* and without runoff water entering the nursery from adjacent areas.
2. Fence the nursery so that all incoming and outgoing traffic can be controlled with preferably only one entrance. Restrict the movement of people and vehicles.
3. Provide a fungicidal dip, protected from rain, at the nursery entrance, which caters to both wheel and foot traffic. Copper sulfate, copper oxychloride, sodium hypochlorite and quaternary ammonium compounds are effective and commonly used fungicides.
4. Disinfect approach and internal roads weekly, as well as nursery benches (structures) and floors (walkways), with a copper fungicide.
5. Nursery tools may not leave the nursery area and should frequently be disinfected.
6. Each entry point must have a handwashing facility provided with a cleaning agent.
7. All water used in the nursery must be clean and pathogen-free (regularly tested). Deep well/borehole water is the best, but if not available, the water needs to be disinfected (from *P. cinnamomi* spores) through filtration and chlorination.
8. Keep hose nozzles off floors and feet off benches.
9. Place plant bags or pots (containers) on permanent and sturdy structures at least 200 mm in height (600 mm high in high-rainfall areas), designed to minimize cross-contamination of containers during irrigation or rainfall.
10. Containers must have adequate drainage holes and be filled with a suitable growing medium (high air-filled porosity and acceptable water-holding capacity).
11. Regularly disinfect all nursery equipment including transport for shipping trees.
12. Either fumigate all nursery potting media with methyl bromide (which is currently being phased out throughout the world due to environmental concerns) or steam pasteurization using aerated steam. Bio-fumigation methods are not suited for preparing potting media, to eradicate Phytophthora and other fungal species, for the production of avocado trees.
13. Fumigate/sterilize outdoor areas if plants are to be placed on the ground. Elevated bench placement is strongly recommended for production in high-rainfall areas (<800 mm per annum).
14. Heat-treat all seeds collected for propagation in water at 50°C for 30 min.
15. Only collect propagation material from registered sources indexed (using the real-time PCR (RT-PCR) technique) to be free of ASBv, with a known origin and production (this practice should be followed with seed/rootstock and scion material). Only collect material from shoulder height and higher to avoid fungal contamination by rain-splash.
16. Remove all shade trees, shrubs and weeds within or in close proximity to the nursery area. No outside plant should enter the nursery area.
17. Regularly cull and discard weak-looking trees. Maintain a low disease pressure through practicing high phytosanitary standards. Regularly scout for diseases and insect pests.
18. Do not keep trees too long in the nursery, thereby preventing them from becoming root-bound and creating a possible disease risk.
19. Nursery staff must be well trained and acquainted with all phytosanitary procedures that should be rigidly adhered to on a daily basis.

semi-arid Mediterranean climates, where avocados are grown, with the potential for more rapid spread of the disease. Consequently, the industries of these two countries have introduced nursery production schemes that provide growers with trees certified free of Phytophthora root rot and ASBv. In Australia, the Avocado Nursery Voluntary Accreditation Scheme (ANVAS) was started by the Australian Avocado Growers' Federation (AAGF) in 1978. The scheme evolved over many years, meeting the requirements of the industry, and in 2018 underwent a major upgrade with incorporation into the Nursery Industry Accreditation Scheme, Australia (NIASA). Entry into the scheme remains voluntary, but to attain ANVAS status, an avocado nursery must now be accredited with NIASA and be found at audit to have implemented the procedures detailed in Appendix 13 (Avocado Nursery Stock Specification) of their operational manual to a satisfactory standard. The basic requirements of ANVAS have been retained with improvements to their implementation where technology allows.

Sourcing propagation material from true-to-type trees that are free of ASBv is mandatory, with trees used for seed and bud-stick collection regularly indexed using a real-time PCR (RT-PCR) technique (see Dann and Gazis, Chapter 13, this volume). The scheme remains largely modelled on the procedures used by the University of California system for the production of pathogen-free, container-grown plants (Baker, 1957) and includes the recommendations of Whitsell *et al.* (1989) with modifications to suit the local conditions. For example, all water used in the nursery, other than from rainwater storage or an approved deep well/borehole, must be treated with chlorine, wherein a residual concentration of ≥2 μg ml^{-1} of chloride for 30 min is achieved (note: chlorine is ineffective as a decontaminant if the pH is above 8.0); all trees must be grown on steel-mesh benches at least 600 mm above ground level, with concrete or bitumen pathways, and the remaining floor area must be covered with 50 mm of clean, coarse gravel. To ensure freedom from *P. cinnamomi* and other root pathogens, potting media may be sourced from accredited suppliers of disease-free products and must be stored in a manner that avoids contamination. However, if the nursery obtains materials for the production of potting media, it must be pasteurized using aerated steam to raise the temperature to 60°C for 30 min (Allan *et al.*, 1981). After 30 min at 60°C, steam should be turned off, but air should continue to be blown through the media to assist in rapid cooling so that surviving beneficial organisms are not overexposed to high temperatures. Systemic fungicidal drenches (e.g. metalaxyl or phosphonate) to temporarily suppress populations of *P. cinnamomi* and reduce root damage in nursery stock are not permitted until trees leave the nursery, as these products will not entirely eliminate pathogens and may mask the detection of the disease in test samples.

Independent inspections are carried out twice yearly, with soil and plant samples collected to evaluate their disease-free status. Nurseries that gain accreditation under the scheme are entitled to use the words 'ANVAS Nursery' and the registered logo in advertising/promotional material and are listed as approved suppliers of nursery stock in the Avocados Australia quarterly journal '*Talking Avocados*', which is distributed to all Australian growers. Accredited nurseries are also listed on the Avocados Australia Limited website as recommended sources of disease-free trees. Following introduction and implementation of this scheme, there was a major improvement in survival rates with field establishment of nursery-produced trees (Pegg, 1992).

The South African Avocado Plant Improvement Scheme (APIS) was first proposed in 1983 as a result of the formation of the Avocado Nurserymen's Association (ANA) and is a combined venture of ANA and the South African Avocado Growers' Association (SAAGA) (Partridge, 1984). It was gradually implemented over time (Partridge, 1986) and became fully operational by 1989. APIS is similar to ANVAS, with voluntary participation, and in 2010, there were six member nurseries. Together with the Avocado Nursery Improvement Scheme (ANIS), the broad aim is to raise the productivity of avocado production by ensuring high-quality nursery trees (both from a horticultural and phytosanitary point of view) are supplied to avocado growers. Cultivar evaluation blocks of all SAAGA-approved commercial cultivars and rootstocks serve as source material for participating nurseries and have been established in all major production areas from the selected mother trees.

Korsten *et al.* (1987) conducted a general screening (dot-blotch method) of the South African mother trees to eliminate the sources of ASBv infection. This was followed by indexing the trees for ASBv, using the RT-PCR technique (Luttig and Manicom, 1999), after which the trees were re-indexed every 3 years. Nursery protocols (ANA, 2000; Claassens, 2001) are like those outlined by Whitsell *et al.* (1989) and used by ANVAS and are monitored twice yearly by independent nursery inspections. Participating nurseries must conform to minimum standards, which include the use of pathogen-free irrigation water and growing media and good sanitation and propagation techniques without the use of suppressive fungicides such as metalaxyl and/or phosphonate. Through APIS and ANIS, nurseries are graded as 0–5 stars on a points system. Tree certification is provided by successful nurseries, and growers can request labelling on all certified trees. Approved nurseries can also display the ANA logo on their letterhead. APIS-certified nurseries maintain a database of all certified propagation material, along with the sales of certified trees, and provide buyers with additional certification for following consignments.

Sexual/seed propagation

The avocado fruit is a botanical berry with a mono-embryonic seed (Storey, 1974; Oliveira *et al.*, 2010). The seed is dominated by two large, nutrient-dense cotyledons tightly enclosing the embryo axis. The dry, vascularized seed-coat is pachychalazal in origin, as is typical of the Lauraceae family (Heo *et al.*, 1998). The large seed size is presumed to be an evolutionary strategy, reflecting competition in the Mesoamerican native forest habitat. Wolstenholme and Whiley (1999) noted that avocado fruit and seeds have characteristics supportive of Janzen and Martin's (1982) extinct megafaunal dispersal syndrome, a theme elaborated by Barlow (2000).

As nursery seedlings, the West Indian ecotypes are noticeably less variable than the other races. There is evidence that this carries through to grafted tree performance. In Israel, the 'Fuerte' yield was found to be more uniform on West Indian race rootstocks, as compared with an often more variable set on the standard Mexican race stocks, using seeds from a single tree of each race. 'Maoz' seedlings (a West Indian race line), according to Kadman, are exceptionally uniform and so greatly reduce the need for asexual propagation in Israel (Bergh, 1975). The Australian rootstock improvement project, 2002–2013 (Whiley *et al.*, 2013), used seedling rootstocks from the three horticultural races, where in the nursery it was observed that seedlings from the West Indian and Guatemalan races (or their hybrids) were comparatively uniform compared with those of Mexican race origin ('Barr Duke', 'Duke 7') (A.W. Whiley, SHS, Buderim, QLD, Australia, 2013, personal observation). Each seed line was sourced from a sole tree. Seedling populations of Mexican race origin required significant pre-grafting rogueing to remove stunted and deformed plants compared with other seedling lines selected for the experiments.

Seeds should be harvested directly from the tree when mature, avoiding contact with the ground. However, if fallen fruit are selected or there is doubt surrounding the fruit collection process, then seed should be treated with hot water at 50°C for 30 min to eliminate any infection by *P. cinnamomi* (Zentmyer *et al.*, 1958; Ohr *et al.*, 1991). Temperature control must be precise, as seeds will lose viability above 52°C. After hot-water treatment, seeds should immediately be placed in clean cold water to cool and then planted in a pasteurized potting mix. Care should be taken to ensure mother trees are not infected with ASBv, as this disease is highly seed-transmissible.

Seed storage

The avocado seed is described as 'physiologically recalcitrant' (Wolstenholme and Whiley, 1999) and requires careful management once extracted from the fruit. If not required in the short term, seeds may be extracted from fruit and cool-stored, where they can be successfully held for several months (Halma and Frolich, 1949; Spalding *et al.*, 1976; Whitsell *et al.*, 1989). In Florida, the storage of seed from cultivars that mature at different times permits planned planting, thereby assisting in nursery management (Spalding *et al.*, 1976). Seed storage is useful in California where an unreliable source of seed, due to biennial bearing, disrupts nursery production (Paulus *et al.*, 1976). In subtropical climates such as Australia and South Africa, where propagation is carried out in spring and autumn, seed storage is commonly practiced, as it increases the flexibility of nursery management.

Whitsell *et al.* (1989) reported that if immature seeds are cool stored, there may be a gradual decline in their viability. Successful temperatures used for cool storage range between 4.5°C and 8.9°C (Halma and Frolich, 1949; Paulus *et al.*, 1976; Spalding *et al.*, 1976; Bergh, 1988). The common problems encountered during storage are seed (embryo) desiccation and decay (Paulus *et al.*, 1976; Spalding *et al.*, 1976). In cool-storage (8.9°C), studies with 'Topa Topa', 'Yama' and 'Henry Select' seeds, Paulus *et al.* (1976) found that acceptable storage and germination were achieved; they were sealed in 1.5-mm-thick polythene bags for 12 months. Although the seeds were treated with one of a number of different fungicides, there was no clear evidence that these treatments were beneficial, because untreated seeds in sealed polythene bags were also disease free at the end of storage. Additionally, seed storage in perforated polythene bags or wood shavings was unsatisfactory. Supporting evidence for successful storage of avocado seeds was reported by Spalding *et al.* (1976), although a shorter storage period was tested. They found 100% germination of 'Lula' seeds after storage in sealed 1.5-mm-thick polythene bags at 4.4°C for up to 5 months. Seed viability was markedly reduced (down to 8%) when seeds were stored for 5 months in perforated polythene or plastic mesh bags. More recently, Gálvez-Cendegui *et al.* (2017) reported on subsequent growth of different-sized 'Esther' seeds following storage in polypropylene tarp bags for 10, 30 and 100 days in a refrigerated chamber at 5°C ± 1°C and 80% ± 5% relative humidity. They found that there was no effect on seed size but storage from 30 to 100 days increased the size (length and diameter of growth) of the seedlings compared with growth from 10 days of storage. In a seed viability study, Makhoul and Attaf (2020) reported the highest viability (87.5%) was attained when avocado seeds were planted directly following extraction from fruit, while after storing at 5°C ± 2°C for 60 days, germination fell to 4.2%. In this study, no detail was given on whether the seeds were protected from desiccation during storage.

In summary, avocado seeds can be successfully stored, for up to 12 months if precautions are taken against desiccation, and temperatures are lowered, although there may be differences in viability among avocado cultivars in how long their seed may be held prior to planting. Only sound, fully mature seeds from healthy trees should be stored, as these are more likely to have sufficient carbohydrate and mineral nutrient levels to sustain the embryo through the storage period. If required, hot-water disinfestation of seeds should not be undertaken until they have been removed from cool storage. Seeds should be sealed in polythene bags (about 1.5 mm thick) and pretreatment with a fungicide may be beneficial, for example, copper oxychloride. The recommended storage temperatures range from 4.4°C (5 months' storage) to 9°C (12 months' storage).

Seed germination and growth

Seeds are not only a reservoir of growth regulators for the development of fruit but also for cytokinins, which influence the germination and growth of the plantlets (Blumenfeld and Gazit, 1970). However, avocado seeds often germinate slowly and irregularly, which can be due to either their postharvest treatment or the seedling line chosen (Platt and Frolich, 1965; Leal *et al.*, 1976). Seeds should be planted soon after removal from fruit or stored in a manner that prevents dehydration, as a high mortality will result if the recalcitrant seeds are allowed to dry out (Storey *et al.*, 1986; Whitsell *et al.*, 1989). Seed germination is hypogeal (Oliveira *et al.*, 2010).

Several studies have been completed on avocado seed germination, which have examined the effects of seedcoat removal and/or cutting sections off cotyledons (collectively termed scarification), and the use of plant growth regulators (PGRs) to improve germination (Eggers, 1942; Halma and Frolich, 1949; Hodgson, 1951; Kadman, 1963; Burns *et al.*, 1965; Leal *et al.*, 1976; Sauls and Campbell, 1980; Bergh, 1988; Whiley and Whiley, 2005). From these studies, it has been consistently reported that removal of the seedcoat will significantly increase both the speed and percentage of germination, particularly for seed that has been cool stored. The firmly attached seedcoats of seeds extracted from soften fruit can be removed by wetting the seeds and allowing them to dry in the sun (Hartmann *et al.*, 2011). However, to overcome the difficulty of removing the seedcoat, seed extraction from hard, unripe fruit is recommended. By applying sufficient force to cut through to the cotyledons, the seed in hard fruit comes away cleanly from the seedcoat, and the cotyledons are lightly scored, thereby applying scarification on two fronts during the seed extraction process and enhancing the germination rate (Whiley and Whiley, 2005; Adjei *et al.*, 2011). An additional pretreatment consisting of cutting a cross-sectional slice from the base, taking care not to damage the embryo, and a similar section from the top of the seed was shown to improve the germination of an 'Indio' line that was coolstored at 4.5°C for 11 days (Bergh, 1988). Cutting lateral slices from the sides of seeds further increased germination. However, not all seed scarification studies were able to improve seed germination, possibly because of excessive mechanical damage and/or failure to control fungal rots (Kadman, 1963; Sauls and Campbell, 1980).

Burns *et al.* (1965) reported improved germination of 'Duke' seeds following the removal of the seedcoat, trimming both ends of the seed and soaking in gibberellic acid (GA). In later studies, Leal *et al.* (1976) found that soaking freshly harvested seeds for 24 h in 500 $\mu g\ ml^{-1}$ GA overcame the inhibitory effects of the seedcoat to some extent. However, germination was significantly greater when the seed-coat was removed, and the seed was treated the same way with GA. Their research was carried out with seeds from several cultivars, and they found, on average, a 22% increase in germination after soaking the seeds for 24 h in 500 $\mu g\ ml^{-1}$ GA for 7 weeks following seedcoat removal. In contrast, when working with 'Waldin' seeds, Sauls and Campbell (1980) reported that soaking seeds for 24 h, in either 250 or 500 $\mu g\ ml^{-1}$ GA, initially increased the rate of germination, but after 34 days had no practical effect on total germination. In a more recent study, Arabi and Bertling (2021) reported that soaking 'Fuerte' seed for 30 min in a 2.5% solution of *Moringa oleifera* leaf extract gave a small improvement in the rate of germination when seed was extracted from fruit that had recently reached maturity.

From the number of studies reported, there is little doubt that the removal of the seedcoat will improve the rate and uniformity of germination of avocado seeds, particularly when they have been cool stored. The mechanism involved is not well-understood, but prevention of desiccation is certainly involved. Perumal (1961) reported that ethyl alcohol extracts of avocado seedcoats contained apparent auxin–inhibitor complexes that markedly reduced tomato seed germination and growth of avocado stems. Increased germination following treatment of seeds with intact seedcoats supports the presence of a biochemical inhibitor mechanism (Leal *et al.*, 1976). Conversely, the mechanism may be mechanical, as seedcoats in many species act as a physical barrier to water and gas exchange and must be disrupted before germination takes place. Observations of improved germination when seedcoats were cracked were reported by Bergh (1988) and appear to support a mechanical barrier mechanism. It is also conceivable that both mechanisms have some control over seed germination, with pretreatment of seeds being particularly effective following cool storage. Scarification of cool stored seeds by either seedcoat removal and/or cutting basal/top slices from the seed is practiced in Australia, California, Cyprus, Florida, Mexico, Puerto Rico and South Africa (Hume, 1951; Malan and van der Meulen, 1954; Johnston and Frolich, 1956; Kadman, 1963; Ruehle, 1963; Alexander, 1977; Gregoriou *et al.*, 1984; Borys *et al.*, 1985; Whitsell *et al.*, 1989). Seed germination and seedling growth are determined by the length of the cut made into the seed and seed size. Medium (3–4 cm in diameter) and large (>4 cm in diameter) seeds germinated faster than small (<3 cm in diameter) seeds, which could be attributed to their proportional food reserves (Adjei *et al.*, 2011). Scarification is not considered necessary when the seed is planted immediately after removal from the fruit (Johnston and Frolich, 1956). However, Shindre *et al.* (2022) reported a decline in seed germination commensurate with the time held before planting after the seedcoat was removed.

In studying the effect of seed maturity on germination, Traub and Auchter (1933) achieved 95% success with very immature seeds of 'Collins' (Guatemalan race), 'Winslowson' (Guatemalan × West Indian race) and 'Mayapan' (Guatemalan race) collected from fruit blown from the trees in Orlando, Florida, during the September 1933 hurricane. These seeds produced excellent plants. Even the immature seeds from the late-maturing cultivar 'Collins' germinated successfully. The survival rate of seedlings, however, is affected by the removal of the cotyledons. Avocado seedlings shorter than 8 cm did not survive more than 35 days after cotyledon removal, as the seedling depends on the cotyledon reserves during the early stage of development. Seedlings taller than 13 cm survived longer than the evaluation period; however, with an unequal development in stem diameter, stem length and number of leaves (Barrientos-Priego *et al.*, 1995). Traub and Auchter (1933) also demonstrated a technique whereby the number of plants per seed can be increased through fractional embryos created by cutting the cotyledon in such a manner that the meristematic tissue (embryo) is split in half with each cut. Successful sprouting is possible with a quarter of meristematic tissue.

At the other extreme, when overmature fruit are stored on trees for a long time, it is not unusual for some seeds to start germination (emergence of the taproot) while fruit are still on the trees. This phenomenon is called vivipary and occurs more frequently in very mature fruit that is still hanging during warm, wet weather. It is an indication that the avocado seed is primed for immediate germination if conditions are favourable. It is perhaps pertinent here to invoke Janzen and Martin's (1982) gomphotheres (and other extinct megafaunal dispersers) in indigenous habitats, swallowing ancestral avocado fruit whole, with the slippery seedcoat (of the slightly bitter/poisonous seed) facilitating fast passage through the gut. The intact seed, perhaps scarified by molars and stomach juices would then be deposited at a distance from the parent tree in a fertile pile of dung on a moist, shady montane forest floor protected against desiccation. Fast germination would have followed, at least for some seeds (Wolstenholme and Whiley, 1999; Barlow, 2000).

In most localities, seeds can be planted in the spring without supplementary heating, but if planted in the autumn/winter period, heating will be required at temperate latitudes. If the nursery environment is maintained at 23°C–25°C, germination occurs within a month (Alexander, 1977). In the warmer areas of California, seeds may be planted outside during the autumn/winter period in a seedbed with a plastic hot cap placed over each seed. However, most nurseries now germinate the seeds in a hothouse (S.L. Maddock, Fallbrook, California, 2011, personal communication).

Seeds are planted with the large, flat end downwards (i.e. as the fruit hung on the tree) in a pathogen-free germination medium in seedbeds, leaving the top of the seed flush with the soil surface (Johnston and Frolich, 1956; Platt and Frolich, 1965). The germinating seed requires good aeration; hence, it is important that they are maintained in a well-drained substrate. Vermiculite, perlite or fresh coarse sawdust (from *Pinus elliotii* or *Pinus radiata*) are suitable materials for use in seedbeds, providing moisture retention and aeration (Anderson, 1977; Whitsell *et al.*, 1989). Seeds are transplanted into nursery bags at the first signs of growth. As seed technology and propagation procedures (particularly rootstock cloning) have advanced, it has become increasingly more common to sow directly into nursery bags, reducing the handling and associated costs (Whitsell *et al.*, 1989; Ernst, 1999). In California, this procedure involves direct planting into 230 × 60 mm polythene nursery bags with perforated bases. The nursery bags are placed into wooden or plastic boxes, usually about 270 plants m^{-2}, which provide greater efficiency in handling. The seedlings can be graded for size, with weak or off-type plants discarded. Thus, batches of sorted seedlings can all be grafted at the same time (Whitsell *et al.*, 1989). A similar process is used in South Africa for the production of clonal rootstocks where seeds are directly sown into 400 × 75 mm polythene nursery sleeves that have been folded back along their length so that they are initially 200 mm tall. Each bag has 12 holes to facilitate drainage (D.J. Roe, Tzaneen, South Africa, 2000, personal communication).

Initially, it is not necessary to fertilize seedlings, as cotyledons provide adequate nutrition for the young plant (Platt and Frolich, 1965; Chalker and Robinson, 1969). By the time the second set of leaves fully expand, some nurseries begin to apply a weak liquid fertilizer (5–2–2, N-P-K) with minor elements (S.L. Maddock, Fallbrook, California, 2011, personal communication), although care must be taken, as seedlings are easily burnt by soluble fertilizers (Chalker and Robinson, 1969). Slow-release fertilizers such as Osmocote® and Nutricote® can also be applied to support growth, which considerably reduces the risk of seedling damage (A.W. Whiley, SHS, Buderim, QLD, Australia, 2024, personal observation).

The media used in nursery containers vary according to the availability and cost of materials. For example, the University of California, Riverside, uses steam-sterilized UC No. 2 mix containing 2:1:1 sand, peat moss, and nitrogenated redwood compost, plus added fertilizers (Whitsell *et al.*, 1989). A variation of this mix used in Australia is 1:1 coarse river sand and peat moss (Anderson, 1977) or, more recently, 1:1 composted pine bark and coarse river sand mix (A.W. Whiley, SHS, Buderim, QLD, Australia, 2024, personal observation). Some nurseries in San Diego, California, use a 1.5:1 mix of sponge rock (perlite) and peat moss, incorporating a wetting agent to promote water penetration (S.L. Maddock, Fallbrook, California, 2011, personal communication). In South Africa, commonly used potting media are 100% composted milled pine bark, or a 1:1:1 mix of composted pine bark, composted sugarcane filter press, and coarse river sand (N. Claassens, Tzaneen, South Africa, 2000, personal communication). Filter press is an organic by-product of the sugarcane milling process.

For optimum growth of seedlings, the greenhouse temperature should be maintained at about 27°C with high humidity, although some believe that cooler night temperatures (13°C) produce sturdier plants (Teague, 1966; Whitsell *et al.*, 1989). When seedlings are 300–450 mm in height, usually 2–4 months after planting, they are ready to graft (Teague, 1966; Chalker and Robinson, 1969; Anderson, 1977; Whitsell *et al.*, 1989). In California, grafted seedlings are grown in greenhouses for 6 weeks, then repotted and moved outdoors. At Maddock Nursery, they have found that grafted trees do best if they are repotted when the leaves are about one-quarter to one-half fully expanded. Immediately after moving outdoors, trees are sprayed with a weak water-based white paint (2:1 water:paint) to reduce sunburn (S.L. Maddock, Fallbrook, California, 2011, personal communication). If seeds have been directly sown in small nursery bags, they are usually repotted into larger containers before or shortly after grafting (Anderson, 1977; Whitsell *et al.*, 1989). In Australia and South Africa, where nursery trees are grown on elevated benches, the final container is a 400 × 125 mm polythene nursery bag with a perforated base (Anderson, 1977). However, in California, it is usually a 400 × 125 mm bottomless tar paper or polythene container (Brokaw, 1977; Koch, 1983). Production of a saleable, grafted tree on a seedling rootstock in California normally takes 16–18 months (Teague, 1966), although, under ideal conditions, some trees may be ready to field plant 10–12 months after the seed is planted (S.L. Maddock, Fallbrook, California, 2011, personal communication). In contrast, nursery production of similar trees under the more favourable conditions of subtropical Australia and South Africa results in a saleable product within 8–9 months of sowing the seed.

Containerized production of commercial avocado trees has become a universal practice, with production from seed to sale of nursery trees taking place in either polythene nursery bags or rigid plastic pots. The challenge facing nursery growers is to economically produce an avocado tree with the potential to develop the architectural, engineering and biological features approximating those that develop from a seed that germinates and grows *in situ* in its natural environment. Poor nursery practices, or on-farm retention in containers for long periods after delivery can promote multiple root system defects that may lead to catastrophic failure within the early years after planting. The most common root defects that develop with containerized production are 'kinked' and 'circling' roots. 'Kinked' roots can be the source of major structural weakness at the trunk/root interface as the tree grows in the field, making it susceptible to windthrow, and in severe cases, the trunk can snap off at the root interface (Ford, 1996; Dann *et al.*, 2019). 'Kinked' roots normally form after incorrect transplanting operations and are less likely to occur when seed is directly sown and grown in containers. 'Circling' roots develop when the tree becomes too large for the solid container and root growth circles around the walls. In more severe cases, after field-planting trees with 'circling' roots, trunks and/or other roots can be girdled, as they grow radially and lignify, restricting the flow of water and metabolites through the root-crown area. This can lead to significant tree stress and poor soil anchorage with increased susceptibility to windthrow (Whitcomb, 1988). Dann *et al.* (2019)

reported on a significant event in 3-year-old fruiting avocado trees in WA, Australia, following orchards being buffeted by a period of gusty winds. Hundreds of young trees were lodged (poor anchorage) or, in more severe cases, snapped off at the trunk/root interface, where in some cases growth distortions were, observed. Evidence of severe root 'circling' was present (Fig. 9.6).

The development of air-pruning devices to correct root growth deformities in container nursery production has been available for in excess of 20 years. Recently, Flemings Nurseries Qld, a subtropical and tropical fruit tree nursery in S.E. Queensland, has introduced 4.5 l Pioneer® air-pruning pots to avocado production. The pot consists of a mesh basket insert that holds the potting media and developing tree and a solid outer holder. The mesh basket provides superior drainage to other pots/bags in common use and keeps the potting media well aerated, which encourages root development throughout the whole volume of the container. Roots are able to grow through the mesh where they are air pruned, preventing the development of circling roots (Fig. 9.7).

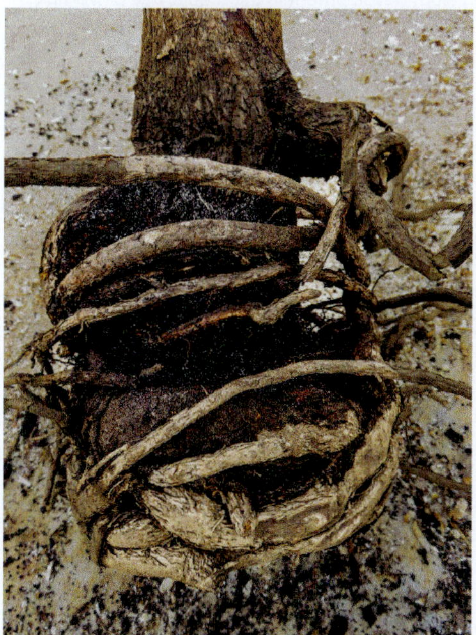

Fig. 9.6. Severe root 'circling' on a 3-year-old avocado tree growing in WA, Australia. (Photograph supplied by E.K. Dann, Hort Health, Brisbane, Australia)

Since Pioneer® pots are relatively expensive, immediately before customer dispatch, trees are removed from the mesh basket and placed into solid, plastic containers of similar size for shipment. A research project comparing tree establishment, growth and production between trees produced in polythene planter bags and Pioneer® air-pruning pots is in progress with early results indicating superior growth from trees produced in air-pruning pots compared with conventional nursery bags. (A.W. Whiley, Buderim, QLD, Australia, 2024, personal observation).

Asexual/vegetative propagation

Vegetative propagation is primarily used to perpetuate the unique genetic characteristics found in a rootstock or cultivar that make it valuable in a production system. Considerable effort has gone into research on rooting avocado cuttings as a simple means of harnessing the genetic traits of a material selected for its potential to improve tree performance, for example, rootstocks with tolerance to Phytophthora root rot or salinity, and which confer desirable horticultural attributes to the scion. However, it is quite difficult to vegetatively propagate avocado rootstocks from physiologically mature wood with known beneficial characteristics (Ben-Ya'acov and Michelson, 1995).

Stem cuttings

Physiologically juvenile stem cuttings from young avocado seedlings root quite readily (Eggers and Halma, 1937; Gillespie, 1956, 1957), while only a few cultivars can be successfully rooted from physiologically mature stem cuttings (Eggers and Halma, 1936; Haas, 1937a,b; Wallace and North, 1957). However, Gillespie (1956) demonstrated, with a 4-year-old Mexican-type seedling, that a residual physiologically juvenile condition exists in the latent axillary buds on the basal portion of the trunk. Haas (1937a) and Ernst (1978) were unsuccessful in demonstrating the movement of root-promoting substances, inducing root

Fig. 9.7. Avocado production in Pioneer® air-pruning pots at Flemings Nurseries Qld, Nambour, Australia. (a) Root development in the air-pruning mesh basket. (b) Root development in the media ball of 'Velvick' seedling grafted onto 'Hass'. (Photographs supplied by L. Gillies, Flemings Nurseries Qld, Nambour, Australia)

formation from a piece of seedling root grafted to the basal part of a leafy-twig cutting. In spite of the limited successes in rooting physiologically mature stem cuttings, it was generally accepted that vegetative propagation of avocado is difficult, and green stem cuttings do not root sufficiently well for commercial propagation in nurseries (Halma, 1953; Kadman and Ben-Ya'acov, 1965; Ben-Ya'acov and Michelson, 1995).

The reasons for difficulties in rooting stem cuttings of most avocado cultivars are not understood. Kadman (1976) provided evidence for the role of physiological ageing in avocado seedlings. Cuttings of the Mexican race seedling 'Northrop 28/5', aged 6, 8, 10 or 12 months from seed, were rooted under intermittent mist. The rooting percentage declined from 100% in cuttings from 6-month-old seedlings to 30% of cuttings from 12-month-old seedlings. In subsequent experiments, Kadman and Ben-Ya'acov (1965) and Reuveni and Raviv (1981) concluded that the Mexican race cultivars generally root quicker and more successfully than Guatemalan or West Indian race cultivars, with the latter being the most difficult to root. However, in a comparative trial, Ernst (1984) found that rooting commenced after 60 days with 'Fuerte' (Guatemalan × Mexican race) and after up to 240 days with 'Duke 7' (Mexican race), with the rooting percentage of the latter being significantly lower. Ernst (1981, 1984) and Ernst and Holtzhausen (1987) reported that fungicide treatment of cuttings combined with either indole-3-butyric acid (IBA) or indole-3-acetic acid (IAA) at 5–15 g l^{-1} (quick dip), and a well-aerated medium with bottom heat at 27°C–30°C, promoted rooting of the 'Fuerte' and 'Duke 7' stem cuttings under intermittent mist. Because of the deleterious effect of bottom heat at 32°C and above, Leal and Krezdorn (1964) found IBA treatments inconclusive. Although root development from stem tissue was observed, it is more likely to occur from callus tissue. Callus formation at the bases of the greenwood cuttings was associated with rooting (Leal and Krezdorn, 1964; Ernst, 1981, 1984; Ernst and Holtzhausen, 1987). This is in contrast with the findings of Kadman and Ben-Ya'acov (1965), who reported that roots only penetrate through callus tissue and that no connection between callus formation and the rooting of cuttings has been confirmed. The rooting of terminal cuttings was more successful than that of subterminal or basal cuttings.

Gomez *et al.* (1971) suggested that the configuration of the perivascular sclerenchyma ring might contribute to the difficulty of rooting, by acting as a barrier to root emergence in young stems. They also suggested it may explain why air layers of Mexican race cultivars are rooted best and West Indian race cultivars poorest (Gomez *et al.*, 1971). In later studies, Ernst (1984) found no anatomical explanation for rooting differences between etiolated and greenwood avocado cuttings and confirmed that sclerenchyma does not interfere with root development, which is initiated in the vascular cambium. Ernst (1984) and Ernst and Holtzhausen (1987) observed the differentiation of vascular tissue within the callus at the base of semi-hardwood adult greenwood avocado cuttings, during the first 120 days under intermittent mist. Root initiation took place in the vascular cambium after 60 days and developed through the scattered sclerenchyma cells without any difficulty. It was externally visible after 90 days and confirmed the importance of callus formation as an important precursor to rooting. Most of the rooting took place (became visible) between 120 and 240 days. This also explains why the days to rooting of greenwood cuttings are substantially longer than those of juvenile and etiolated shoots. As a result of these findings, Ernst and Holtzhausen (1987) suggested that to achieve a higher rooting percentage, the treatment of cuttings (callus) with an auxin should be considered after callus formation and subsequent vascular differentiation (approximately 60 days). In contrast to IBA, IAA tends to stimulate callus formation. In such an event, the initial treatment with IAA followed by the retreatment of the cutting's callus with IBA (after 30–60 days) is recommended. Since the stimulatory effect of auxins on root formation does not take place unless the tissues also contain an appropriate concentration of cytokinins (Mastalerz, 1977), Ernst and Holtzhausen (1987) suggested considering the use of cytokinins in combination with auxins for callus treatment.

The importance of leaves and their retention for rooting avocado cuttings has been demonstrated (Young, 1961; Leal and Krezdorn, 1964; Kadman and Ben-Ya'acov, 1965; Ernst, 1978, 1981, 1984; Reuveni and Raviv, 1981; Raviv and Reuveni, 1984a; Ernst and Holtzhausen, 1987). Ernst (1984) and Ernst and Holtzhausen (1987) reported that during rooting, dieback at the base of the cutting initially preceded defoliation, after which defoliation became the first sign of cutting deterioration. Reuveni and Raviv (1981) reported a direct correlation between leaf retention and the rooting of avocado cuttings. They found that starch accumulation at the base of leafy cuttings was correlated with rooting capability, but there was no relationship between rooting and leaf mineral nutrient content, except for a negative relationship between manganese (Mn) and rooting. In later studies, Raviv and Reuveni (1984a) found that leaf senescence and abscission layer formation were affected at different rates, alone and in combination, by auxins and cytokinins. The optimal combination significantly delayed defoliation with improved rooting. The cytokinin 6-benzylaminopurine (BA) and the auxin naphthaleneacetic acid (NAA) together had a synergistic effect on leaf retention in 'Fuchs 20'. It has been suggested that each rootstock cultivar requires different levels of PGRs in order to delay leaf senescence or petiole abscission. Raviv and Reuveni (1984b) reported a direct correlation between the concentration of a root-promoting substance in leaves and the rooting ability of avocado cuttings. It was established that the rooting promoter was non-auxinic and that it acts at different stages during root development than IAA (Raviv *et al.*, 1986).

The use of PGRs to improve rooting of hardwood cuttings has been widely reported. IAA, a naturally occurring auxin, was demonstrated by investigators in 1935 as having a practical use in stimulating root formation of cuttings susceptible to IAA-oxidase (IAA-oxidase) and, therefore, is most commonly used commercially (Hartmann *et al.*, 2011). Different synthetic auxins belonging to the indole, naphthalene, phenoxy and benzo groups are, to a greater or lesser extent, successfully used in the rooting of cuttings. At the same concentration, phytotoxicity and other aftereffects increase in the order from indole, naphthalene, phenoxy and benzo groups (Leopold and Kriedemann, 1975). Auxins are not readily dissolved in water and must be dissolved in a solvent (ethanol DMSO) or a base (1N NaOH) before being quickly added to water. Potassium salts of IBA and NAA (potassium IBA (KIBA) and potassium NAA (KNAA)) are auxin formulations that dissolve easily in water and are available commercially (Hartmann *et al.*, 2011).

With avocado, Kadman and Ben-Ya'acov (1965) and Gustafson and Kadman (1969) found that, in most cases, standard PGR treatments did not affect the success of rooting green cuttings but only improved the quality (development) of the root system. This is in contrast with the findings of Ernst (1978), who reported an interaction between rooting temperature (bottom heat) and IBA at 10 g l^{-1}, resulting in a significant increase in rooting from 18°C to 26°C. In preliminary studies comparing IAA, NAA, and KIBA as quick dips or 24 h soaks, highly variable rooting was reported, with KIBA producing the best results. Follow-up research suggested that the optimal KIBA concentration was between 10 and 100 mg l^{-1} (24 h soak). Within this range, the concentration was cultivar/race sensitive, that is, the West Indian race cultivar 'Fuchs 20' responded best to a concentration of 10 mg l^{-1} while the Mexican race cultivar 'GA–13' responded more favourably to 100 mg l^{-1} of KIBA (Gustafson and Kadman, 1969; Kadman and Gustafson, 1971). About 50% to 100% ethanol, acetone or methylated spirits can be successfully used as carriers for non-water-soluble auxins such as IBA, IAA and NAA. Under artificially aerated (4.9 l dm^{-3} oxygen) medium conditions, treatment with a 10 g kg^{-1} IBA in talcum gave 75% rooting of 'Fuerte' cuttings compared to 43% in the absence of IBA, underlining the importance of sufficient oxygen for respiration at the cutting bases (Ernst, 1984). Cuttings taken in late autumn or winter are rooted better under intermittent mist than summer cuttings (Kadman and Gustafson, 1971).

Attempts have been made to rejuvenate physiologically mature avocado wood by grafting scions onto juvenile rootstocks *in vitro* (Pliego-Alfaro and Murashige, 1987) or through gibberellin injection into trees from which cuttings were taken (Cutting and van Vuuren, 1988). This has been largely unsuccessful. Pliego-Alfaro (1988) reported that 100% of juvenile microcuttings (20 mm long) can be rooted if successively cultured *in vitro* between two rooting media. García-Gómez *et al.* (1994) attempted to clarify the role of auxin in root formation of juvenile avocado microcuttings. They found that untreated microcuttings showed no IAA increase throughout rooting, whereas at the base of IBA-treated cuttings, the IAA levels increased twofold in the first 6 days. Additionally, levels of the conjugate indole-3-acetyl-aspartic acid (IAAsp) increased during the first part of the rooting process, that is, activation of cambium cells (days 0–3) and division of cambium cell derivatives (days 3–6). They concluded that the formation of IAAsp could be a mechanism for detoxifying excess levels of free IAA.

More recently, Turkinje Nursery, located in North Queensland (latitude 17°S, altitude about 430 m above sea level), has commercialized the production of some avocado rootstocks through minicuttings. After preparation (details are not provided due to 'commercial in confidence'), the cuttings are placed under intermittent misting in a greenhouse facility equipped with evapotranspiration cooling that is able to maintain close-to-summer ambient day temperatures (Fig. 9.8).

'Velvick', thought to be a West Indian/Guatemalan hybrid, for many years, has been a popular rootstock in Australia but is prone to alternate bearing, making seed supply difficult to manage from year to year. Using minitip cuttings from 'Velvick' seedlings at the time of grafting (5–6 months after planting seed) and harvesting later from the axillary bud growth on grafted seedlings can significantly increase rootstock numbers, since rooting of this physiological juvenile material is about 90% (Fig. 9.9a). The total time from rooting 'Velvick' seedling cuttings to the sale of a grafted tree is about 15 months. Genetically, the diversity of 'Velvick' cutting rootstocks resembles that of the seedling population and has been accepted by growers ordering seedling 'Velvick' rootstocks.

Physiologically mature cuttings of 'Dusa' are also successfully rooted using a similar process with minitip cuttings from rootstock plants collected at the time of grafting. Shoots from auxiliary buds, which grow once the scion cultivar has been grafted, are also harvested for propagation of 'Dusa' cuttings. About 63% of 'Dusa' cuttings placed under mist successfully root (Fig. 9.9b). The total time taken from placing cuttings under mist until the sale of a grafted tree ready for field planting is approximately 18 months, which, under North Queensland conditions, is a similar production time to using the nurse-seed/etiolation technique. However, the cutting method has the advantage of no requirement for etiolation, and only one graft is made for the scion cultivar. Taking into account local labour costs, the cutting technology has been the preferred methodology for producing nursery trees grown on cloned 'Dusa' rootstock.

It is clear that more research is required on the fundamental mechanism of rooting physiologically mature stem cuttings of avocado. In spite of some exceptions through various innovative techniques, commercially cost-effective rooting of avocado cuttings remains challenging (Gillespie, 1952; Hendry and van Staden, 1982; Salazar-García and Borys, 1983; Barrientos-Priego *et al.*, 1986; Alves-de Oliviera *et al.*, 1999; Da Silveira *et al.*, 2004).

Fig. 9.8. A bench of avocado cuttings in a propagation facility at Turkinje Nursery, North Queensland, Australia.

Fig. 9.9. (a) Recently rooted mini-tip cutting at Turkinje Nursery, North Queensland, Australia; and (b) root mass on 'Dusa'-cloned rootstock approximately 6 months after placing cuttings in the mist propagation house. (Photographs supplied by A.W. Whiley, Sunshine Horticultural Services, Qld, Australia)

Etiolated cuttings

Etiolation is the development of plants or plant parts in the absence of light, resulting in characteristics such as small unexpanded leaves, elongated shoots, and lack of chlorophyll, which result in a yellowish or whitish colour of the treated growth. The ease of rooting etiolated shoots was reported as early as 1937 and has proved a useful technique for the propagation of difficult-to-root species (Gardner, 1937; Knight and Witt, 1937; Hartmann *et al.*, 2011). Frolich (1951) first reported the success of etiolation in rooting physiologically mature Guatemalan race cuttings. A method was described whereby the basal portion of an avocado shoot, growing from a cutback tree, was allowed to develop in darkness and high humidity while the apical portion was allowed to develop leaves under normal conditions. Before severing the etiolated shoot from the parent plant for rooting in a humidity chamber with bottom heat, it was found necessary to have a minimum of two, but preferably three to five fully expanded leaves. Rooting was normally achieved 4–8 weeks after detaching from the parent plant (Frolich, 1951, 1961).

Further research with etiolation methodology led to the development of a double-grafting ('nurse' seed) technique suitable for the commercial production of 'cloned' trees (Frolich and Platt, 1972). Large seeds (selected as a reservoir of carbohydrates to impart vigour) were grown in 1 l containers to provide a temporary 'nurse' root system. Each seedling was grafted as soon as possible (to conserve seed carbohydrates) with a scion from the chosen rootstock cultivar, just above the surface of the potting medium. The grafts (scions) were allowed to grow in light until the unions were well established. Shoots of the scion were then cut back to near their base, and plants were placed in a dark chamber as soon as the buds had again started growth. Without light, growth was elongated, and new shoots lacked chlorophyll. When new shoots reached 200–250 mm in length, plants were removed from the chamber and each enclosed in a cylinder 150 mm high filled with a moist rooting medium. Shoot tips continued to elongate, regaining chlorophyll and turning green again in sunlight. After the new leaves matured, shoots, including the etiolated stem base, were severed from the 'nurse' seedling and placed in an open medium in a humidity chamber with bottom heat (minimum 24°C) to complete rooting. Shoots can also be rooted by girdling (ring-barking) below the etiolated base before re-enclosing them with the 150 mm cylinder. The application of auxin had no beneficial effect. The rooted cuttings were transplanted into 100 mm peat pots filled with soil. After the plants were established and hardened off, they were transplanted to soil-filled 11 l containers. When well established and after the required size was reached, they were grafted with the scion cultivar retaining several leaves below the graft and grown until ready for field planting. Leaves were retained on the rooted cutting (rootstock) until the new graft had produced several of its own leaves.

Reuveni and Goren (1982) described a variation of the 'Frolich and Platt' technique, where a side branch of a grafted mother plant (of the selected rootstock) was grown in normal light during etiolation of the offspring from the main branch. The bases of the etiolated shoots were covered with a black paste, allowing further growth to continue in light. After leaf expansion and hardening off, the cuttings were severed from the mother plant and rooted in a chamber under intermittent mist. This method allowed for repeated and extended usage of the mother plant.

Barrientos-Priego *et al.* (1986) reported a modification to the 'Frolich and Platt' method for the production of etiolated cuttings. They cut back 11-month-old grafted plants and placed them in a dark chamber with high humidity until new regrowth reached 150 mm in length (18–24 days). Several treatments were applied to the shoots that included etiolation alone; etiolation plus a 3 mm-wide cincture at the base of shoots; etiolation plus auxin application (10 g l^{-1} KIBA or 0.3 g l^{-1} NAA) to the base of shoots; or various combinations of the above. After treatment, a black plastic collar was placed around the shoots and filled with a sterile potting mix; hence, cuttings were not removed from the mother plant. An evaluation 36 days after treatment showed the highest rooting percentage (about 90%) was from shoots that were etiolated, cinctured and auxin treated. Shoots that were etiolated and cinctured had about 60% rooting, those only etiolated had about 24% rooting, and shoots that were etiolated and auxin-treated had about 17% rooting. Non-etiolated shoots did not develop any roots despite cincturing and/or auxin applications. The experiment clearly showed that etiolation was a prerequisite for rooting, while the addition of cincturing or cincturing plus auxin treatments significantly improved the percentage of rooting.

Ernst (1978) and Ernst and Holtzhausen (1978) forced growth from lateral buds on 'nurse' branches, placed horizontally under 50 mm of vermiculite in a cold frame, to produce etiolated shoots suitable for rooting. As soon as the shoots grew into the light, leaf development with consequential photosynthesis took place. The etiolated cuttings were severed from the 'nurse' branches and successfully rooted under intermittent mist. In contrast to the findings of Frolich and Platt (1972), Ernst (1978) found a synergistic reaction between etiolation and auxin treatment (3 g kg^{-1} IBA dissolved in 50% ethanol). Ernst (1978) reported that it was possible to produce a rooted cutting with less effort, about 4–8 weeks sooner than with the 'Frolich and Platt' method. The highest percentage of etiolated shoots was produced in the shortest time at 25°C. The disadvantage of the technique is the large quantity of branches needed for commercial quantities of etiolated shoots. Mohammed and Sorhaindo (1984) reported a 93% sprout promotion of 'nurse' cuttings by treating them with 100 mg l^{-1} GA (GA_3). Multiple shoot development was induced in the 'nurse' cutting with a 50 mg l^{-1} BA treatment. According to Ernst (1984), this is the only known technique whereby etiolated cuttings are independently produced in the absence of a rooted mother or 'nurse' plant, consequently eliminating the risk of virus transmission from one genotype to another.

Moll and Wood (1980) in South Africa developed a system that deviated from the technique described by Frolich and Platt (1972), to produce multiple etiolated cuttings from each graft established on a 'nurse' seedling. They reported that during etiolation of the selected rootstock scion grafted onto a 'nurse' seedling, new growth is allowed to reach 300 mm in length before removal from the plant. The etiolated shoot is then cut into 50 mm lengths, each containing at least one node with a leaf, and placed into a rooting medium of 1:1:1 peat moss, perlite, and vermiculite in a mist bed with bottom heat set at 27°C. The mist bed was regulated to a 5 s spray every 60 s. Under these conditions, cuttings were rooted in 14–40 days with an 80–90% success rate. The original 'nurse' plant can be placed back into the dark chamber and 'cropped' 4–6 times, producing 15–25 cuttings. However, this technique can only be applied successfully to rootstock cultivars having some leaf development during etiolation. It has found some commercial acceptance by the South African nursery industry.

Whiley and Whiley (2005) reported that during clonal propagation, the type of wound applied to the etiolated stem prior to KIBA treatment can determine the success and quality of rooting. Easy-to-root rootstocks require minimum wounding, while a full 360° scrape of the etiolated stem where the root-promoting auxin is applied is usually required for those that root with greater difficulty. The 360° scrape wound prior to auxin treatment gave the most consistent results with 'Velvick', a difficult-to-root Guatemalan × West Indian hybrid cultivar, which develops excessive wound callus, with the standard stem tissue wound cuts resulting in very sparse or zero rooting.

More recently, Duman *et al.* (2020) have reported that a short period of de-etiolation following etiolation of rootstock material increased rooting success in the difficult-to-root VC801 avocado rootstock, which is in demand in Israel for its high tolerance to salinity and the soil pathogens *P. cinnamomi* and *Verticillium dahliae*. The scions of VC801 were grafted onto 'Reed' seedlings, and when growth from the scion commenced, plants were placed in the dark and etiolated. A number of these plants were then placed in light for 1 week, after which the etiolated growth regreened. Leafy cuttings of the etiolated and regreened shoots were then treated at their base with 60 mg l−1 of KIBA, then incubated for 24 h, before transferring to a rooting facility with intermittent mist for 1–2 months. Several batches of the cuttings were treated this way, and the rooting success of de-etiolated cuttings varied between 5% and 40%, while no rooting occurred from any of the etiolated-only cuttings. Additional de-etiolation studies were carried out on VC804 and VC66 rootstock lines, and rooting in etiolated and de-etiolated cuttings was 33%–40%, %–58%, 20%–25% and 40%–75%, respectively.

A number of physiological parameters were measured on etiolated and de-etiolated cuttings to define any differences that may have occurred during pretreatment. Duman *et al.* (2020) reported that short de-etiolation provides an advantage for adventitious root formation in shoots of the clonal avocado rootstock VC801 (as well as VC804 and VC66), which was removed from the nurse seedling on which it grew. They reported that a short de-etiolation period leads to changes in chlorophyll content, cell-wall composition and hormone profiles, which compensate for the loss of support from the nurse seedling, while at the same time, the advantages provided by etiolation are not completely eliminated. Their results suggest that the combined changes in de-etiolated cuttings lead to increased activation of the cambium, as detected by greater cambial cell numbers, and fine-tuned optimization of the conditions for adventitious root formation in de-etiolated cuttings. Interestingly, increased cambial cell numbers were accompanied by a corresponding increase in xylem cells in the de-etiolated cuttings. Duman *et al.* (2020) also observed that beyond rooting compatibility, de-etiolation may have also made these cuttings less sensitive to pathogens during their time in the rooting facility, which may have contributed to their higher survival, thus enabling better rooting percentages.

Commercial rootstock cloning using the 'nurse-seed/etiolation' system

Brokaw Nursery in Ventura County, California, modified and improved the cloning technique pioneered by Frolich and Platt (1972) sufficiently to have a US patent granted on 22 March 1977 (Patent no. 4012866). This method (Fig. 9.10) streamlined the commercial propagation of

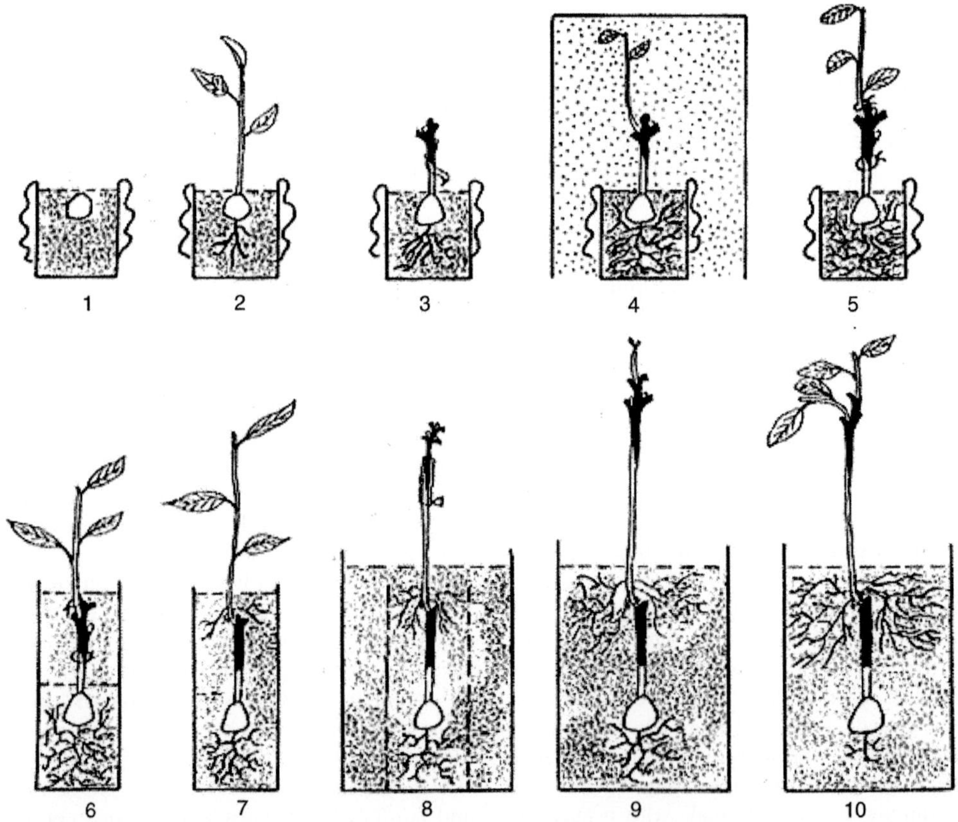

Fig. 9.10. The Frolich and Platt 'nurse'-seed technique as modified by Brokaw Nursery for the rooting of clonal avocado rootstocks (Claassens, 2001).

clonal trees. Subsequently, other nurseries adopted the 'Brokaw' technique under a sublicense agreement until the patent expired in 1994. In the 'Brokaw' method, a large 'nurse' seed is planted about one-third from the bottom of a 300 × 70 mm polythene sleeve where the top has been folded down on the bottom half of the sleeve. When sufficient growth has occurred, the seedling is grafted close to the soil surface with a scion of the desired rootstock. A single strong bud is retained when the graft begins to grow, and the plant is placed in darkness. The plant is removed from the dark chamber once 300–400 mm of new etiolated growth has occurred, and a metal or plastic ring is clamped near the base of the shoot. The bag is then extended to its full length and filled with moist potting medium. The ring gradually constricts and eventually kills the 'nurse' seedling, usually within 1 year, leaving the rooted scion as the new clonal rootstock.

The scion cultivar can be grafted onto the etiolated stem immediately after the girdling ring is applied and the bag has been extended, or it can be grafted later after roots have formed from the stem. The fruiting scion is grafted at a height of 200–250 mm so that, in the case of graft failure, there is sufficient room to regraft (Brokaw, 1975, 1987a). Profuse suckering in field trees was observed in South Africa, when the 'nurse' seedling failed to separate and/or die. Poor clonal root formation also resulted in stunted, uneven, and unstable tree growth in commercial orchards (Ernst, 1999; de Villiers and Ernst, 2007).

A further modification to the Brokaw Nursery technique was described by Ernst (1999) and developed by Allesbeste Nursery in Tzaneen, South Africa, in 1990, to combat rising nursery costs and the requirement for economically priced clones for use in high-density plantings (Fig. 9.11). The production of microclones, by encouraging two or more shoots to develop during etiolation through removal of apical dominance, and their subsequent treatment as individual plants by encasing the etiolated base of each shoot in a 55 ml polythene microcontainer filled with a rooting medium, has resulted in greater production from the original 'nurse' seed (Fig. 9.12). To assist in rooting, IBA (7 g l^{-1}) is applied to a wound made at the base of each etiolated shoot before it is enclosed in the microcontainer. Plants are returned to shaded sunlight to permit photosynthesis and grafted at approximately 300 mm above the base of the shoot, to the selected scion cultivar as soon as sufficient growth above the microcontainer has occurred.

When new growth on the fruiting scions has reached 50 mm in length, the microclones are cut from the grafted 'nurse' seedling (approximately 4–6 weeks after grafting the scion cultivar) (Fig. 9.13a) and hardened off in a greenhouse (Fig. 9.13b). The whole propagation process from 'nurse' seed to hardened microclone takes 8–10 months and a further 4–6 months to grow a tree suitable for field planting (Ernst, 1999).

The 'Allesbeste' microcloning technique has advantages over the 'Brokaw' method in that several clones can be produced from one 'nurse' seed; there is complete separation of the microclone and the 'nurse'-seedling root systems (buffer zone) (Fig. 9.13a(i)), so that rooting of the microclone can be definitively established, and small microclones can be transported at low cost to other nurseries (domestic or international), where they can be grown further for sale (Ernst, 1999). According to Hofshi (1996a), this method is rapid, produces a large number of healthy roots, and does not destroy the 'nurse' seed. Since the microclone relies fully on its own roots, and the technique allows for frequent inspection and culling, it is possible to deliver trees with a guaranteed clonal root system of the highest quality and with less failure in the field (de Villiers and Ernst, 2007). Through field observations, de Villiers and Ernst (2011) concluded that tree uniformity among clonal rootstock trees requires a reliable rooting technique. Poor field uniformity with clonal rootstocks is the direct result of uneven root distribution, poor root quality and quantity and has no genetic explanation. It is, therefore, important that the technique used should make provision for surveys to inspect root development and quality at different stages of development in the nursery.

Hofshi (1996a,b) reported a modification of the 'Allesbeste' microcloning procedure. After removing the 'nurse' plant from the dark chamber, the etiolated shoot (approximately 400 mm in length) is vertically scored on two sides, about 75 mm above the graft, and the area is treated with 10 g l^{-1} IBA. A long bamboo stake is then embedded in the container and

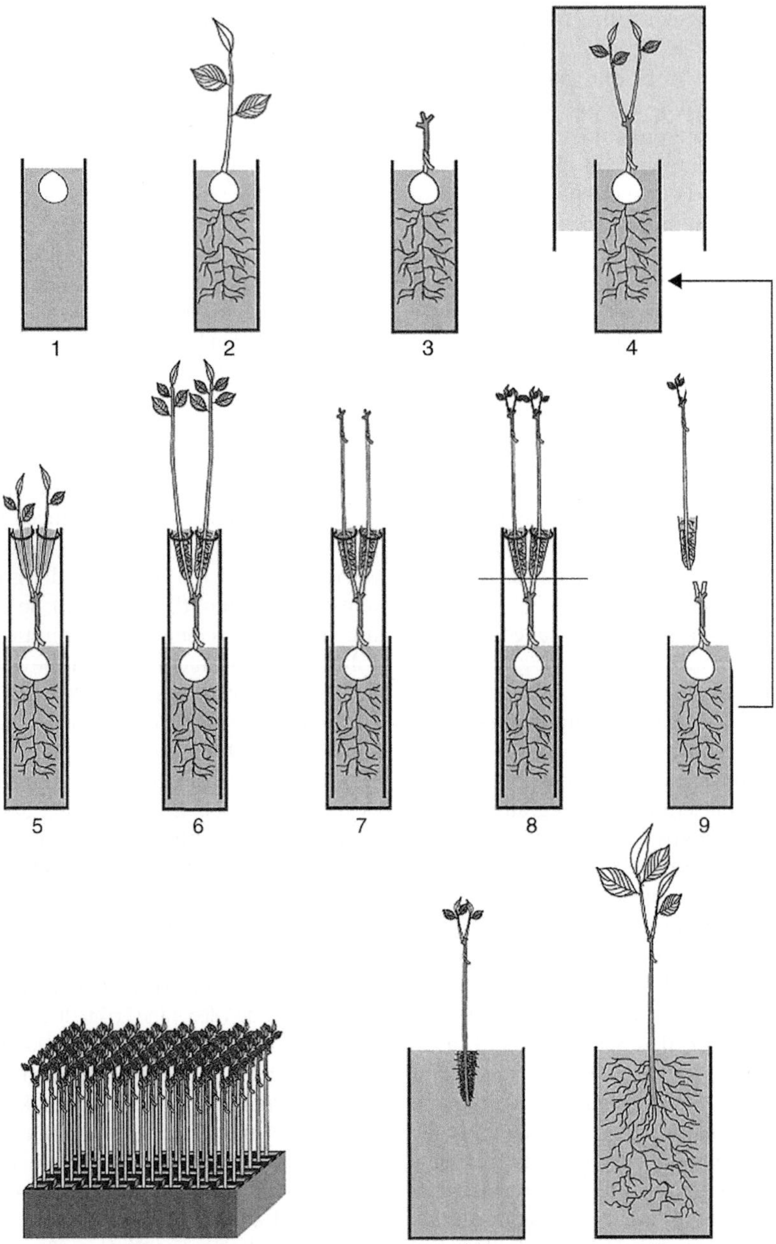

Fig. 9.11. The 'Allesbeste' microcloning technique, a modification of the 'Brokaw' method, for cloning avocado rootstocks (Ernst, 1999).

placed alongside the etiolated shoot. A sliced, clear plastic cup (240 ml) is taped around the stem with the scored area about one-quarter of the distance below the rim of the cup. The cup is filled with sterile rooting medium and taped to a stake for support (Fig. 9.14a). If the shoot is long enough, a second cup is placed in the same manner, about 200 mm above the first

Fig. 9.12. Rootstock etiolation using the Allesbeste Nursery microcloning technique. Tip pruning of the established rootstock scion produces two rooted plants from each nurse seed. If sufficient reserves remain in the seed, a third plant can be produced. (Photograph supplied by M.L. Arpaia, UCR, California, USA)

cup. The clear cup allows visual inspection of root development so that progress can easily be monitored. Under optimal conditions, the cup is usually filled with roots within 4 weeks (Fig. 9.14b). When sufficient top growth on the rootstock has occurred, the scion cultivar is grafted onto the newly cloned rootstock, which is removed from the 'nurse' seedling once sufficient leaves (3–4 fully expanded) have grown on the scion to support a positive carbon balance. The new tree, still in the clear cup, is placed under mist in the greenhouse for further growth. Subsequently, it is replanted into a large nursery bag, transferred outside, and grown to the required size for field planting. Following detachment of the rooted, grafted plant from the 'nurse' seedling, a second cloned rootstock can be produced. However, in this case, the cloned rootstock may not be grafted to the scion cultivar, until after removal from the 'nurse' seedling. This latter stage depends on the residual vigour of the 'nurse' seedling and the prevailing climatic conditions (time of year). Following propagation of the second clonal rootstock, the 'nurse' seedling can be used as a standard seedling rootstock.

Commercial rootstock cloning through tissue culture

While successful commercially viable techniques have been developed for the mass production of clonal rootstocks (Frolich and Platt, 1972; Brokaw, 1975; Hofshi, 1996a,b; Ernst, 1999), the ultimate goal has been mass production of rootstock material through tissue culture. Advantages of *in vitro* propagation include the elimination of seed dependency for propagation; disease-free rootstocks produced from the aseptic culture environment; large numbers of rootstocks can be propagated at any time regardless of season; selection of mother plants with proven, field performance results in the replication of genetically uniform rootstocks with identical attributes; and elimination of double grafting used in the 'nurse'-seed cloning technique, which saves significant time and cost. Some of the earliest *in vitro* studies with avocado are attributed to Schroeder *et al.* (1962) and Schroeder (1971, 1977, 1980), where various avocado tissues were placed on modified Nitsch or MS medium, resulting in callus production, and in some cases adventitious root growth (Schroeder

Fig. 9.13. (a) The severing of the rooted and grafted 'Allesbeste' microclone from the grafted 'nurse' seedling (approximately 4–6 weeks after grafting the fruiting scion). The buffer zone (i) provides separation between the microclone and nurse-seed root systems. (b) Hardening off in the microliner area of a greenhouse (approximately 6–8 weeks after severing).

et al., 1962; Schroeder 1980). However, histological studies were not conducted to determine if the roots were derived from the callus or tissues of the original explant. With woody species like avocado, it is less difficult to produce roots from physiologically juvenile material compared to mature (post-flowering) shoots. Pliego-Alfaro (1988) reported 100% rooting success *in vitro* with physiologically juvenile avocado microcuttings, but about 30 years have passed before a small select group of *in vitro*-produced rootstocks from physiologically mature material have been offered at commercial volumes (O. Markovich, Rahan Meristem, Rosh HaNikra, Israel, 2023, personal communication). Progress has been slow, as protocols for the *in vitro* propagation of avocado are highly cultivar-specific and, once developed, require optimization to increase proliferation and rooting success (Hiti-Bandaralage *et al.*, 2017). Successful shoot-tip culture provides commercial outcomes, since a single shoot tip can produce 500 plants or more within 8–12 months (Hiti-Bandaralage *et al.*, 2017). The development of successful *in vitro* protocols appears easier for Mexican race cultivars and more challenging for West Indian race cultivars (O. Markovich, Rahan Meristem, Rosh HaNikra, Israel, 2023, personal communication; L. Gilles, Flemings Nursery, Nambour, QLD, Nambour, Australia, 2024, personal communication), which is consistent with reports on other methodologies for rooting avocado cuttings. The discussion on *in vitro* propagation in this chapter is about commercial outcomes, and for details on methodology and protocols, see Pliego-Alfaro *et al.*, Chapter 10, this volume.

AUSTRALIA. Anderson Horticulture is the longest-operating avocado nursery in Australia and has an exclusive license to propagate avocado rootstock using tissue culture protocols developed by the University of Queensland. While test plots of 'Reed' rootstock sourced from tissue

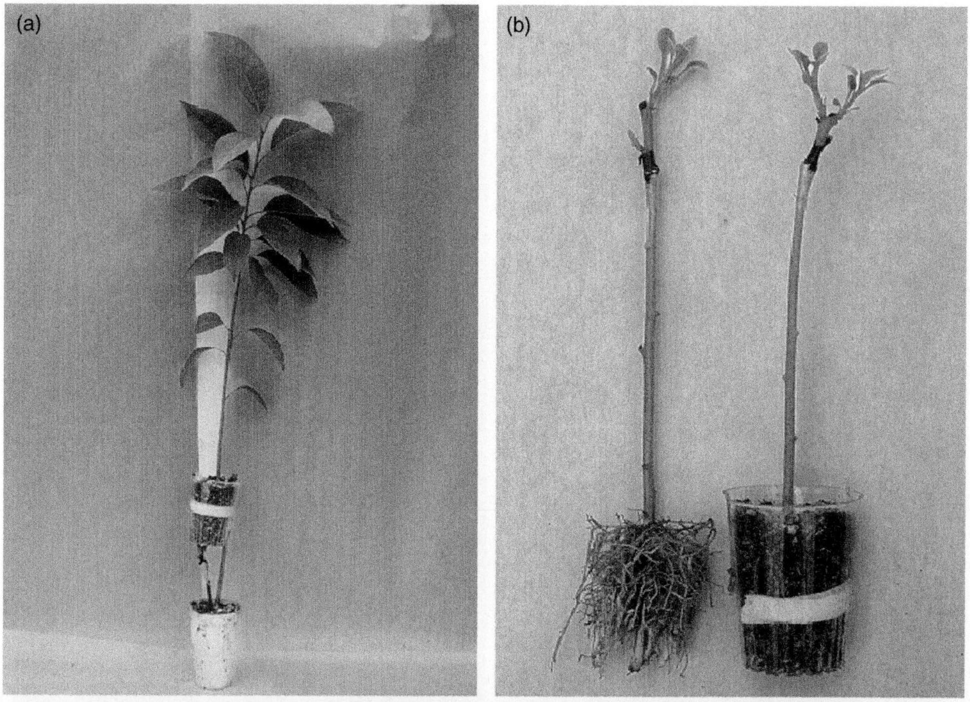

Fig. 9.14. (a) The rootstock cloning of Hofshi (1996a,b) with application of a 240 ml clear plastic cup filled with a rooting medium placed on a pretreated portion of the etiolated shoot to induce rooting.
(b) Root development in the cup after 4–6 weeks and the subsequent grafting onto a scion cultivar and detachment from the 'nurse' seed. (Photographs supplied by M.L. Arpaia, UCR, California, USA)

culture and grafted to 'Hass' have been planted and preliminary data collected (Best *et al.*, 2021), to date, there has been no commercial offer from Anderson Horticulture of trees grafted onto tissue-cultured rootstock. It should be noted that in the research of Smith *et al.* (2011) and Dann *et al.* (2013), 'Reed' seedlings used as a rootstock had high susceptibility to Phytophthora root rot. While there are no reports of the tolerance of cloned rootstocks produced from the parent 'Reed' tree, caution is required with its use, particularly in replant blocks, until its susceptibility has been defined.

Flemings Nursery Qld specializes in subtropical and tropical fruit tree crops and is a subsidiary of Flemings Nurseries, one of the largest nursery businesses in Australia. Flemings Nursery Qld has successfully developed avocado tissue culture technology (Fig. 9.15) in-house at their Queensland facility and now offers grafted trees on multiple rootstock choices that include 'Zutano' (Mexican × Guatemalan hybrid), 'BW2' (Guatemalan), 'Velvick' (West Indian × Guatemalan hybrid), 'Ashdot' (West Indian) and 'BW78' (West Indian × Guatemalan hybrid), which are in ascending order of difficulty in developing optimized commercial protocols.

ISRAEL. Rahan Meristem has tissue culture operations in Israel, Australia, Colombia, Costa Rica, Ecuador and the Philippines and is the world's largest producer of micropropagated bananas. Rahan Meristem has been researching tissue culture protocols for *in vitro* propagation of avocado rootstocks for a number of years, developing successful procedures for 'VC65', 'VC256' and 'VC804' (Markovich, 2023), which are 'elite' Israeli rootstocks with known salinity and Phytophthora root rot tolerance. The material required to establish the initial cultures of these rootstocks was sourced from physiologically mature mother plants, thus preserving the identified genetic attributes of each cultivar (O. Markovich, Rahan Meristem, Rosh HaNikra, Israel, 2023, personal communication). Additionally, Rahan

Fig. 9.15. Commercial avocado rootstock tissue culture production at Flemings Nursery Qld. (a) Aseptic culture room where rootstocks are grown and multiplied prior to potting up and on-grown to grafting size. (b) A bench of avocado rootstocks sourced from tissue culture and near-ready for grafting. (Photographs supplied by Leanne Gillies, Flemings Nursery Qld, Australia)

Meristem offers *in vitro* propagated rootstocks produced from selected seedlings of 'Ashdot 17', 'Degania 62', 'Degania 117', 'Fairchild' and 'Shiller' (Fig. 9.16). Of these, the former four belong to the West Indian race, while 'Shiller' is Mexican. The West Indian group of rootstocks promote high productivity with medium/high vigour and provide tolerance to heavy soils, and medium/high tolerance to salinity (Haskelberg, 2021). About 40% of Rahan Meristem's nursery production of avocado trees are grafted onto rootstocks propagated through tissue culture while the remaining 60% are on seedling or cloned rootstocks produced via the 'nurse'-seed technique. From 2020, Rahan Meristem has planted extensive field trials in 12 sites across Israel using their tissue-cultured rootstocks grafted with 'Hass', 'Pinkerton' and 'Ettinger' to validate the performance of *in vitro*-grown rootstocks grafted with commercial scion cultivars (Markovich, 2023).

SPAIN. Grajera Nurseries is one of the larger EU Mediterranean producers of avocado trees supplying local demand and exporting trees to nearby countries. To meet the demand for trees, Grajera Nurseries began in-house research to develop a commercially scaled methodology for the *in vitro* propagation of avocado rootstock using material from the Israeli cultivar 'Degania' due to its tolerance of clay, alkaline soils and salinity (García Romero, 2024). They now offer trees grafted onto *in vitro*-produced rootstocks.

Differences in root morphology between seedling and cloned 'Duke 7'

Fassio *et al.* (2016) reported on the differences in root morphology and biomass allocation in two-year-old, field-grown 'Duke 7' trees that were either grown from seedlings or clonally propagated.

Fig. 9.16. Tissue culture procedures at Rahan Meristem. (a) Microplants in rooting media. (b) Successfully rooted microplants. (c) Developing root system on tissue culture rootstock ex-culture. (Photographs supplied by O. Markovich, Rahan Meristem, Israel)

A subset of trees propagated by both methods was not grafted, while a second subset was grafted with 'Hass'. One year after field planting, trees were excavated and their root systems examined. Propagation methods and grafting affected how roots developed. Non-grafted 'Duke 7' seedlings grew a main taproot with narrow-angled lateral roots, which penetrated deep into the soil, increasing aboveground biomass. Contrary to this, non-grafted cloned 'Duke 7' produced a 'crown' of roots from a relatively short stem and was shallow-rooted. The roots had wider angles at branching than are found on seedlings and more fine roots that overall increased root length and density (total length of roots per unit of soil volume). Grafting resulted in similar root architecture for seedling and cloned rootstocks and reduced root and shoot biomass and the volume of soil explored compared with non-grafted trees. The root systems of cloned trees also had greater abundance of larger-diameter roots with unbranched growth when compared to roots on seedling trees.

Additional results on root branching and anatomy in non-grafted 'Duke 7', field-grown seedlings and cloned trees were reported by Fassio *et al.* (2020). Clonal trees produced third-order roots with larger stele expanses than seedling trees. Additionally, in cloned trees, there was more frequent generation of 'pioneer' roots, which have increased diameter and proportionally greater cortex area than other roots. Hence, cloned trees had more extensive root branching, which increased the volume of soil that could be exploited for water and mineral nutrients. The increased expanse of the stele in cloned roots may be due to greater exposure to shade because of propagation requirements during the first 12 months of growth, since reduced sunlight has increased the stele component in roots of other species (Wahl *et al.*, 2001).

Of additional interest from this study is the reduction in root and shoot biomass that grafting made to both rootstocks, irrespective of the propagation method. Both root systems were shallow, and grafted clonal trees had a higher root-to-shoot ratio than grafted seedlings. These changes brought through grafting are most likely due to the disruption of photoassimilate exchange between the two genetically different components of the grafted tree. As discussed under 'Graft compatibility', it is likely that different genetic combinations of scion and rootstock affect outcomes on root/shoot balances that impact overall tree performance. Finding more closely aligned genetic combinations when grafting may lead to significant benefits in productivity.

Scion Propagation (Nursery)

Many different grafting techniques have been developed to accommodate the multiplication of commercial cultivars under either nursery or field conditions. This effort has resulted in numerous choices, from which techniques appropriate for the circumstances can be selected and are discussed below.

Rooted scions

By means of the 'Allesbeste' microcloning technique (Ernst, 1999), the scion cultivar 'Hass' was cloned as a rootstock and included together with other rootstock selections in a field assessment trial by Smith *et al.* (2011) to test for resistance to Phytophthora root rot. The cloned 'Hass' rootstocks were not grafted and were, therefore, planted as rooted cuttings. 'Hass' was among the rootstocks consistently in the healthiest group, under high Phytophthora root rot disease pressure for the length of the assessments (4 years after planting). However, continued observations of the non-grafted 'Hass' trees revealed a rapid decline in health, which was likely associated with increasing crop load in the presence of the aggressive soil pathogen, *P. cinnamomi* (A.W. Whiley, SHS, Buderim, QLD, Australia, 2024, personal observation). This unexpected early performance of the non-grafted clonal 'Hass' trees was suspected to be due to the absence of a graft union, with no potential graft union incompatibility, resulting in a more vigorous root system that was able to overcome the damage caused by Phytophthora root rot. However, with stronger fruiting sinks developing with tree age, it is likely that root repair/replacement was not able to outgrow the disease. There was no apparent genetic-based resistance to Phytophthora root rot in 'Hass' roots.

Budding and grafting (nursery)

Nursery tree propagation of fruiting scions has evolved considerably since the recognition of superior scion cultivars and the requirement for vegetative propagation to retain their genetic lines. Procedures for nursery propagation are discussed below.

Selection of scion wood

Selection of good propagation material is one of the most important tasks for the propagator. Suitable buds are found on mature terminal growth that is quiescent. Terminals carrying plump, dormant buds with a dark green leaf colour in late spring through summer and autumn are selected. Scion wood with extra-large rounded buds during early winter through to spring are likely to be floral and may not produce vegetative growth. Other buds to avoid are those that are slim and elongated with a small leafy growth at the end, as these are not mature (Whitsell *et al.*, 1989).

For spring propagation, scion wood should be chosen in late autumn to early winter before floral growth begins as due to an extended period of quiescence, this material contains relatively high concentrations of carbohydrates and can be successfully stored longer; in southern California, scion wood is usually best obtained in December or January. Buds on new growth with fully expanded, mature leaves should be selected for autumn propagation (Platt and Frolich, 1965). In southern hemisphere subtropical climates, spring shoot growth matures in November and can be used for propagation into the summer (A.W. Whiley, SHS, Buderim, QLD, Australia, 2024, personal observation). Prolific quantities of scion wood suitable for grafting are generally found on young trees or older trees that have been cut back to force strong new growth. However, if the growth has been too vigorous, the scion wood may have a central soft pith that contributes to drying of the wood. Scion wood for grafting should be firm and not flexible; the latter indicates immaturity. It is sometimes helpful to tip-prune flexible, immature wood a week before cutting scion material, as this will usually hasten the maturity of the basal buds (Whitsell *et al.*, 1989). Scion wood that has brownish bark, internal browning in the wood or abscised buds is not suitable for propagation.

When selecting scions for tip grafting, sticks cut 50–75 mm in length and 5–6 mm in diameter normally produce the best results. Leaves about 6 mm away from the buds should be removed, and the scions immediately wrapped in moist burlap or placed in a polythene bag with a damp cloth or paper towel, to reduce water loss (Platt and Frolich, 1965; Whitsell *et al.*, 1989). Each scion should have at least one good bud toward the terminal end and preferably no axillary buds in the basal portion to cause irregularities when matching cambium during grafting (Platt and Frolich, 1965; Whitsell *et al.*, 1989). Scions should be selected from trees that are not water-stressed and are best cut in the early morning to avoid heat and wind, which can cause desiccation as the material is collected.

When cutting large quantities of scion wood in subtropical climates, the use of a portable cool-storage facility (an insulated cold box) will prevent grafting sticks from overheating, since day temperatures can commonly exceed 30°C (A.W. Whiley, SHS, Buderim, QLD, Australia, 2024, personal observation). Larger scions used for top-working trees in the field are usually cut 150–200 mm in length and 9–18 mm in diameter. The ideal scion will have three to four good buds on each stick.

Care must be taken to select scions from trees free of ASBv. Mother trees can be symptomless carriers of the viroid, and transmission with contaminated cutting tools has been demonstrated (see Dann and Gazis, Chapter 13, this volume).

Storage and protection of scion wood

Scion wood for cleft grafting (described below) is normally used soon after it is removed from the tree. The wood can be stored but is best used within 3 weeks if stored at 4–5°C, although wood cut during late autumn to early winter can be successfully stored for 2 months. The larger scion wood described above can be stored for 2–3 months if sealed in polythene bags and kept at 5°C. Storage life can be extended by treating the wood with a mild fungicide, taking care that it is surface-dry before storage. Because of respiration, with the resultant release of heat and moisture, no more than 50 scion sticks should be stored in a bag at one time (Whitsell *et al.*, 1989). During the grafting operation, grafting wood should be covered with a damp cloth to prevent desiccation. Scions stored in polythene bags should be kept out of sunlight, as heat will quickly reach lethal temperatures.

Budding

Budding with a 'shield' bud is the oldest method of asexual propagation used in commercial avocado nurseries. It is still used when scion wood is scarce or expensive, but its use is limited to times when bark slips easily from the rootstock. The success rate is not as high as with cleft grafting.

With a single stroke, a bud with a small underlying slice of wood (shield) is cut from the scion stick using a sharp knife, beginning 10 mm below the bud and ending at the same distance above. The cut is made with a large enough shield to be firm. For thicker rootstocks, a larger bud shield should be used to avoid the bud being engulfed in callus tissue (Whitsell *et al.*, 1989).

The rootstock to be budded should have a diameter of at least 12 mm (about pencil thickness). A vertical cut about 12 mm long is made through the bark with a horizontal cut at the top of the slit to complete a 'T'. The bud shield is then inserted under the vertical flaps, sliding it in from the top (Fig. 9.17). The bud is wrapped with a rubber strip or a length of polyvinyl grafting tape, leaving the bud exposed. Recently, Parafilm® has become popular for wrapping buds tightly to the stem. In this case, the bud is covered by a 25 × 40 mm patch of Parafilm®, which seems to be advantageous, as it keeps the temperature and humidity higher than ambient around the bud. Rubber strips and Parafilm® eventually deteriorate and allow the bud to expand, but polyvinyl grafting tape has to be cut off 3–4 weeks after budding otherwise strangulation may occur. The bud can be forced by pruning the rootstock immediately above the

Fig. 9.17. 'Shield' or 'T' budding sometimes used for propagating avocados. (Illustrated by G.H. Manglallan.)

bud either at the time of budding or 2 weeks later. The new shoot from the bud will need to be tied to a stake until it has firmly established.

Grafting

CLEFT GRAFTING. Cleft grafting, also known as tip grafting, is the most commonly used propagation method in avocado nurseries today. The method is rapid, and the success rate is high, particularly when both sides of the scion are matched with the cambium of similar-diameter rootstocks (Whitsell *et al.*, 1989).

Small scions, about 5–8 mm in diameter, are suitable for cleft grafting nursery trees. The base of the scion is shaped into a wedge by making opposite tapered cuts 25–50 mm long. A rootstock is selected with a matching diameter of about 200–300 mm above soil level and is pruned at this point. The rootstock stub is split down the centre, making the cut a little longer than the scion wedge. The scion is then inserted into the cleft with the cambium matched on both sides (Fig. 9.18). If the scion is too small, a successful graft can be made with the cambium aligned to match one side. If too large, the scion can be tapered on one side as long as the cambium matches the other side of the graft (Whitsell *et al.*, 1989). The union is then wrapped with a rubber strip about 8 mm wide and 150 mm long, or with polyvinyl grafting tape or Parafilm®. If cut, the tip of the scion is sealed with an asphalt emulsion or white acrylic paint to prevent desiccation. Asphalt emulsion should not be used near the graft union, as it may damage cambium tissues. The scion should begin to grow within 3–4 weeks, depending on ambient temperatures (Whitsell *et al.*, 1989). When winter grafting in the subtropics, scion growth may not occur for 8–10 weeks, so long as the scions have been covered with small polythene bags to prevent desiccation (A.W. Whiley, SHS, Buderim, QLD, Australia, 2024, personal observation).

SPLICE (WHIP GRAFTING). Before cleft grafting became popular, splice (or whip grafting) was often used to graft nursery seedlings. The method is a little more difficult than cleft grafting because precise long cuts have to be made, and the cambium of scion and rootstock must be carefully matched and held together while being securely taped.

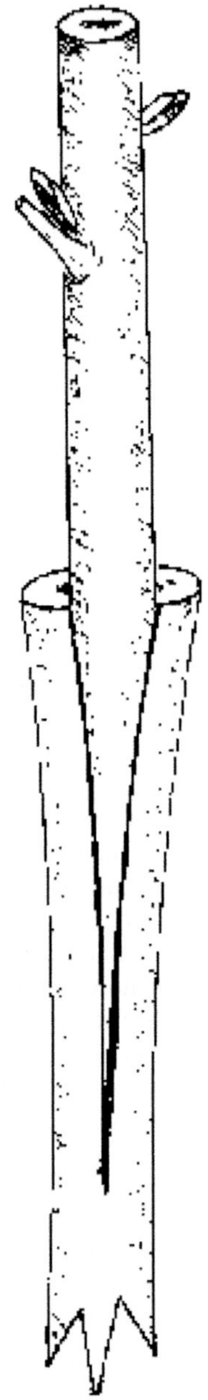

Fig. 9.18. Cleft graft commonly used when propagating avocado trees in the nursery. (Illustrated by G.H. Manglallan.)

To make a splice graft, a sloping, clean cut about 60 mm long is made on a rootstock about 6–10 mm in diameter. The cut should be made 200–300 mm above soil level, completely severing the top of the rootstock. A similar-diameter scion is cut to match the rootstock, and the two cut surfaces are bound together with a rubber strip, polyvinyl grafting tape or Parafilm® (Fig. 9.19). The tip of the scion is treated as previously described to reduce desiccation. Larger or smaller scions can be used if the cambium is matched on one side, but tight wrapping may tend to centre the unions and cause the cambium alignment to shift.

A modification of this method is the whip- and- tongue graft. An extra, vertical cleft cut is made down the middle of the scion and rootstock to permit a solid attachment (Fig. 9.20). This allows both hands to stay free while binding the scion to the rootstock and is a good procedure for a beginner to use (Whitsell *et al.*, 1989).

Aftercare of grafts (scion)

In warmer climates, young grafted plants may suffer desiccation before a physical union between the scion and rootstock occurs. Some nurseries use polythene tents over the benches to increase humidity for 1–2 weeks following grafting. Alternatively, small polythene bags are pulled on and tied below the union and removed once bud growth on scions is seen. Under hot conditions, the greenhouse should be painted with whitewash to reduce sunburn. Some nurseries use small white paper bags placed over each plant to provide some protection from direct sunlight or, alternatively, 40% shade cloth is erected. It is important to maintain optimum soil moisture at all times. Greenhouse-produced trees can be placed outside under 40–50% shade once strong scion growth begins. In California, plants are kept in shade for at least 2 weeks and, thereafter, placed in full sun to be hardened off.

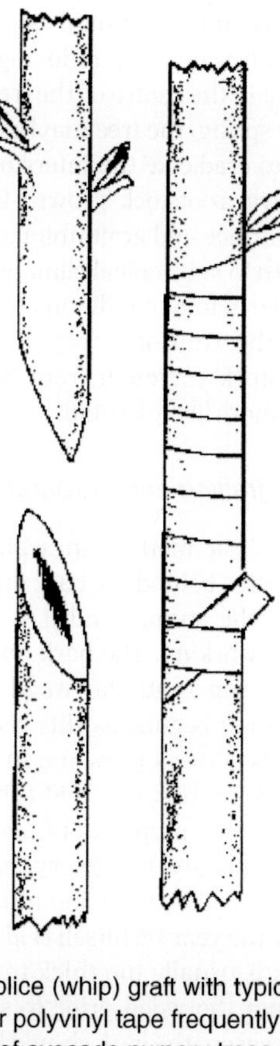

Fig. 9.19. Splice (whip) graft with typical rubber band wrap or polyvinyl tape frequently used for the propagation of avocado nursery trees. (Illustrated by G.H. Manglallan.)

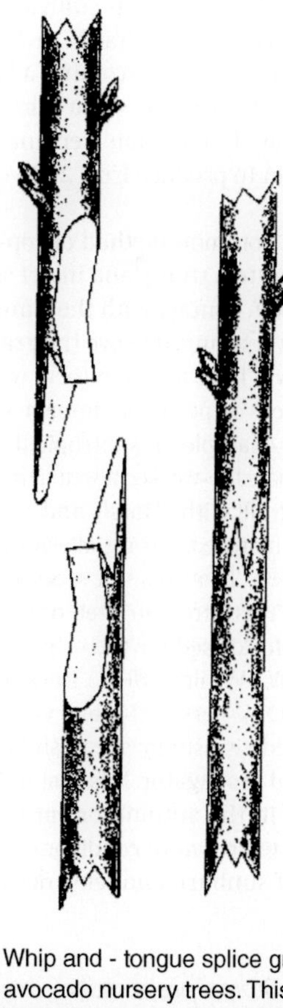

Fig. 9.20. Whip and - tongue splice graft used to propagate avocado nursery trees. This technique is relatively slow but gives a secure match of rootstock and scion and thus is helpful to use when first developing grafting skills. (Illustrated by G.H. Manglallan.)

After 2 weeks in the sun, they can be repotted into larger containers and grown more for sale (Whitsell *et al.*, 1989). In warm subtropical conditions in Australia and South Africa, nursery plants are normally produced entirely outdoors on benches in 40% shade. Trees are hardened off in full sunlight by some nurseries, a few weeks prior to sale.

Top-working propagation (field)

Avocado trees are commonly top-worked in the field if a pollinizer is added to improve fruit set or a change in cultivars is required to meet market requirements. Cultivars belonging to opposite 'flower groups' (see Hormaza *et al.*, Chapter 6, this volume) that can provide cross-pollination to improve yields of the main cultivar are commonly introduced into orchards in Mediterranean climates through top-working a branch or selected trees. Healthy trees are fairly easy to top-work, grow faster than replant nursery trees, and begin to produce fruit 2–3 years after grafting

The most common method of top-working is to cut the tree to a stump and insert scions beneath the bark in contact with the cambium tissue. If the tree is stumped above the graft union, the new tree will have an interstock, which may either negatively or positively impact scion performance. For example, in subtropical Australia 'Fuerte' trees were top-worked about a metre above the original graft with 'Hass', and as the new scion cultivar matured, strong alternate-bearing cycles developed. This was not seen in an adjacent block of 'Hass' trees grafted onto the same seedling rootstock used for the original 'Fuerte' planting (A.W. Whiley, SHS, Buderim, QLD, Australia, 2024, personal observation). Top-working avocado is usually successful if done by an experienced propagator. It is best to top-work avocado trees in the autumn or early spring to avoid extreme temperature conditions as in summer the risk of sunburn and scion desiccation is much greater.

Stump preparation

Older avocado orchards that are being stumped for cultivar conversion may have a level of Phytophthora rot root infection present, which should be suppressed with fungicides (metalaxyl and/or phosphonic acid if they are registered for use), treating trees about a month before beginning operations. When trees are stumped, there is major root loss, so it is preferable to cut trees back coming out of winter when soil temperatures are increasing, since this provides maximum opportunity for root regeneration and regrowth of the tree over the summer months. In California, trees are stumped at a height of about 600 mm with a chainsaw, preferably where the bark is smooth. The freshly cut surface and stump are immediately treated with a 1:1 mixture of white acrylic paint and water in order to protect the bark and cambium from sunburn; oil- or mineral-based paints should not be used because they are phytotoxic. Just before grafting, a second smooth horizontal cut is made. Heavy white paper or the freshly cut block of wood can be placed back on top of the cut surface to protect the cambium until the grafting operation is started. The second cut should be on a slight angle to drain off water, thus reducing the risk of rots developing in the centre of the tree. Alternatively, in early spring, the tree maybe stumped at ground level to eradicate the interstock and encourage renewal rootstock growth from below both the soil surface and graft union. Due to vigorous regrowth in subtropical climates, there is a tendency under windy conditions for regrowth to blow out of the trunk once they start cropping, whereas rootstock regrowth from beneath the soil surface is more likely to hold.

Bark grafting and preparation

Bark grafting is a method in which tapered scions are inserted behind the bark into the cambium layer of the stump, and it is the easiest method of top-working. The bark on the stump should readily slip from the wood before the grafting operation begins, as this is a sign that the cambium is actively growing. In California, this usually occurs between mid-February and April (spring). Some propagators have designed special wedges that are used to pry the cambium from the wood to allow insertion of the scion at other times of the year (Whitsell *et al.*, 1989).

The bark is usually too thick to allow easy separation from the wood prior to scion insertion. For this reason, the bark should be trimmed down to about 6 mm thickness at the points to be grafted. A knife or curved two-handled pruning saw can be used to reduce bark thickness (Fig. 9.21). The bark is slit vertically to the length of the tapered cuts on the scion using a heavy-bladed sharp knife. About two to four vertical cuts are made in the bark, evenly spaced around the perimeter of the cut stump (Fig. 9.22). Just before the insertion of scions, the bark flaps are gently peeled back with the knife. If the bark does not slip easily, it can be peeled back with a special wedge knife inserted into the cambium.

Scion selection and placement

Suitable scions for field grafting are 10–18 mm in diameter and are cut to 150–200 mm in length. The portion of the scion that projects above the cut stump should have two to three buds and should not be higher than 75 mm above the stump to reduce desiccation. A scion for bark grafting is prepared by making a long, tapered cut, usually 60–75 mm long, on the side in contact with the cambium, and 12 mm long on the side facing the bark. It has been suggested that longer cuts (75–125 mm) on both sides give additional cambium contact, permitting a stronger union, but more skill is required to make good, longer cuts (Whitsell *et al.*, 1989). Rough, uneven, or concave cuts are a common source of graft failure.

The scion is inserted from the top of the stump, either in the centre of the two bark flaps or under just one bark flap (Fig. 9.22). If inserted in the centre, the edges of the outer part of the scion are tapered to allow a snug fit; if the scion is inserted under one bark flap, only one side is tapered to allow a snug fit. Four scions will usually be inserted into the larger stumps while smaller stumps (150 mm in diameter) can be grafted with three scions, and 100-mm-diameter stumps grafted with two scions. The extra scions will ensure against graft failure and will promote faster healing of the trunk surface due to the extra cambial growth around the perimeter of the cut surface. The best scion should be placed on the side of the tree facing the prevailing wind, as the trunk will brace the new growth as it leans into the tree (Whitsell *et al.*, 1989).

Graft protection

When top-working in the field, extra precautions are required to protect scions until they become well established. The scions are bound tightly to the stump with 25-mm-wide polyvinyl tape. Correct binding begins at the top of the stump and continues in a downward spiral, finishing with a half-hitch knot just below the

Fig. 9.21. Top-working avocado stumps in the field through bark grafting. A heavy knife or saw is used to reduce the bark thickness prior to inserting the scion. (Illustrated by G.H. Manglallan.)

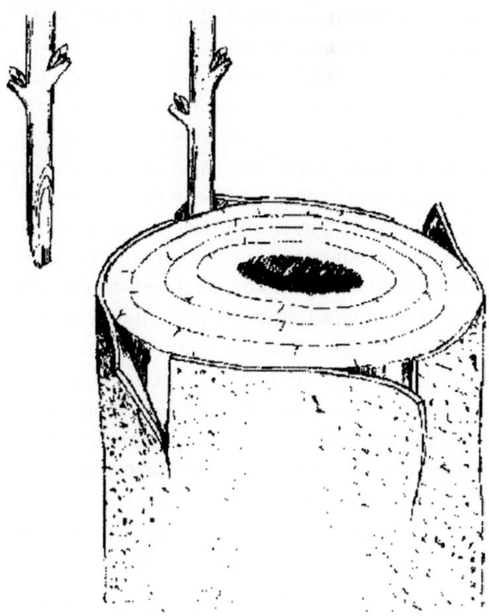

Fig. 9.22. Top-working avocado stumps in the field through bark grafting. Scions are inserted under bark flaps cut at intervals around the stump. (Illustrated by G.H. Manglallan.)

vertical cuts. Tension should be kept on the tape during the process, and the spirals of tape should overlap to reduce desiccation. The cut surfaces on the scions should be sealed with full-strength asphalt emulsion or white acrylic paint to prevent loss of moisture. The cut stump surface can be painted with diluted asphalt emulsion or acrylic paint. Full-strength emulsion on the cut stump reduces aeration, traps moisture and has led to increased trunk rot caused by *Poria* spp. It is important to prevent the sealant from draining into the cracks between the wood and the bark where scions are inserted, as this will prevent cambial growth (Whitsell *et al.*, 1989).

Black asphalt emulsion can heat up rapidly in the full sun and, when used, should be covered with heavy, white butcher's paper (or more weather-resilient material in humid climates) wrapped around the stump twice to form a cylinder. For added protection, the black asphalt surfaces should be painted with white acrylic paint. Sometimes, thin bamboo stakes are tied to the stump near each graft to support the paper cylinder and the young grafts. During cold weather, the paper can be stapled shut to reduce frost damage to the tender growth. As the shoots begin to grow, the paper cap can be opened up to allow some light to enter but still provide some shade protection for the scions.

Grafts can be dislodged, most often by wind or feral animals, and should be supported until well established. New growth should be staked after emerging through the paper cylinder. This is usually done by nailing a 50 × 50 mm stake, 1.2–1.8 m long, to the stump next to each of the grafts. The binding material should be wide enough not to cut into shoots and strong enough to withstand wind gusts. Shoots should be tied with a 75–100 mm space from the stake to allow room for growth. As new growth gets larger, it should be tied 3–4 more times for additional support. Ties should be periodically loosened and re-tied to prevent constriction. The stakes should be kept on the tree until the base of the scion is strong enough to withstand strong winds. Support maybe required for 2–3 years after grafting.

Developing the leader

The strongest graft should be chosen to regrow the new tree. In a windy area, it is common to choose a graft on the windward side of the tree. The other grafts will compete, and thus these should be pruned back but not removed, as they provide insurance if the main graft is lost. After the first year, the unwanted grafts can be removed, with the top of the stump cut at a downward slope from the new leader (Whitsell *et al.*, 1989). This will minimize the risk of water retention on top of the stump, which can lead to the development of rots.

If the leader is growing too upright, the top can be pruned to encourage side branching. It is common to let 'Hass' grow to its natural spreading shape. Constant sucker removal, both from the stump and roots, is very important during the recovery process. Suckers from the stump will often overgrow the young scion if not removed early.

Saw-kerf (notch graft)

The saw-kerf or notch graft can be made at any time of year, as bark does not need to slip from the wood. However, the technique requires more skill and larger scion wood than that required for bark grafting. For this method, a single vertical radial cut is made with the saw in the stump for each scion. The cut should extend inward 50 mm on the cut surface of the stump and down the side about 75 mm.

The scion will usually have a diameter between 120 mm and 250 mm and is cut to make a wedge 60–75 mm long, with the side of the scion that is facing out thicker than the inward-facing side. The saw cut is widened with a chisel to fit the scion (Fig. 9.23). The scion is tapped into the cut, thick edge outward, with the cambium of the stump matching the cambium of the scion (Fig. 9.24). It should be noted that to match the cambiums, the bark of the scion will be slightly recessed compared with the thicker bark of the stump. The scions do not need to be tied or nailed. Some propagators wrap the top of the stump with tape to protect the upper bark of the stump. The scions should not project more than 75 mm above the top of the stump to reduce desiccation. All cut surfaces should be painted with sealant.

Sucker (coppice shoot) grafting

Sucker grafting became popular in San Diego in the 1980s, when many growers were top-working

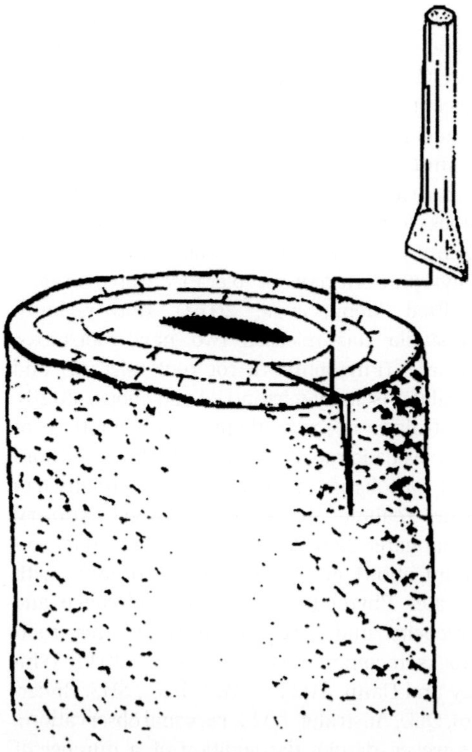

Fig. 9.23. Top-working avocados in the field using the saw-kerf method. A wedge is used to pry open the cut. (Illustrated by G.H. Manglallan.)

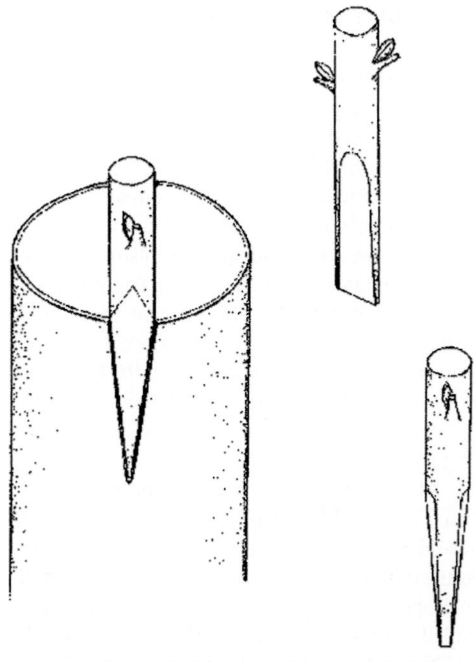

Fig. 9.24. Top-working avocados in the field using the saw-kerf method. The scion is shaped and tapped into the cut, thick edge outward, with the cambium of the stump matching the cambium of the scion. (Illustrated by G.H. Manglallan.)

older cultivars to 'Gwen'. Due to the demand for scion wood, most of the available material was small. Sucker grafting allowed the use of thinner scion wood, required less skill and could be done over a longer period of the year than bark grafting. This technique demonstrated that sucker grafting is a viable alternative to other top-working techniques when a new cultivar becomes popular.

For sucker grafting, the tree is cut down to a 400–600-mm-high stump. This short stump encourages the growth of suckers (coppice shoots) from dormant buds at the base of trees. Trees are usually cut down immediately after harvest (or at the end of winter) with the intention of obtaining well-grown suckers, which, in warmer climates, can be grafted prior to the following winter (Whitsell *et al.*, 1989; A.W. Whiley, SHS, Buderim, QLD, Australia, 2024, personal observation). The stump is treated with white acrylic paint mixed with a copper fungicide (if needed and allowed under local pesticide registration) immediately after cutting to reduce sunburn and disease.

Three suckers are generally selected around the tree, and the others are removed. Suckers chosen for grafting should be at least 500 mm high. Side branches should be trimmed from the sucker to a height of about 300 mm. Suitable leaves should be retained to provide shade for the scion, and exposed stems should be painted white unless grafted immediately. Grafting is usually done using either a cleft or splice graft near the top of the sucker. Since suckers are usually thicker than the available scion wood, care should be taken to match the cambium on at least one side of the graft. A sealant should be applied to exposed cut surfaces and the graft wrapped with polyvinyl tape. Some propagators have tried completely wrapping the scion/stock with Parafilm®, instead of the traditional polyvinyl tape and asphalt emulsion, and reported superior results with reduced time for grafting and a greater percentage of scions growing. Similar protection should be given to new growth, as previously mentioned for field grafting.

Care of the top-worked tree

Water use is low when trees have a reduced canopy, and most water loss from soil will be due to evaporation and/or weed growth. It is important that irrigation be reduced following top-working so that roots will not experience anoxic or hypoxic conditions in saturated soil. The most effective way to maintain adequate moisture in the soil is to irrigate when the soil matric potential falls to about 10 kPa. This can be monitored with a soil moisture probe or tensiometer. Some growers will not water at all during the first year after grafting, but the growth of the new grafts is usually slowed during summer without some irrigation.

The top-worked trees will not need to be fertilized during the first year. After that, fertilizer should be applied as for a young developing tree. Pests such as snails, crows, deer, rabbits, small antelopes and wallabies will need to be controlled, as they will feed on the young growth and, if not checked, may kill the top-worked tree.

Conclusions

In the second half of the 20th century, new propagation techniques were developed for the relatively young avocado industry. The evolution of procedures for vegetatively propagating rootstocks became essential to capture improved tolerance to Phytophthora root rot and salinity, as increasing soil disease pressure and declining water quality occurred in some important production regions. Significant advances from the Frolich and Platt (1972) 'nurse-seed/etiolation' method were made, which enable commercial-scale production of the vegetatively propagated rootstocks. In 2002, Westfalia, South Africa, commercialized 'Dusa' ('Merensky 2'), with partial resistance to Phytophthora rot root, which over 20 years has developed into the world's most successful cloned rootstock. Not long after, 'Bounty' (thought to be of Guatemalan race origin), recommended as a cloned rootstock for use on heavy soils where short periods of hypoxic conditions may develop, was commercialized in South Africa and is now available in Australia, New Zealand and South America (A.W. Whiley, SHS, Buderim, QLD, Australia, 2024, personal observation; S. Wade, Lynwood Nursery, Whangarei, New Zealand, 2023, personal communication; M. Buhl, Avotopia, De Luz, CA, USA, 2024, personal communication). Recent research (Derebe *et al.*, 2023) concluded that Guatemalan race rootstocks were more tolerant to hypoxic conditions than those of Mexican race, albeit from a small sample of rootstocks tested. In California, three new rootstocks, 'Zentmyer', 'Uzi' and 'Steddom', with improved Phytophthora root rot tolerance were commercialized (Menge *et al.*, 2012). More recently, Westfalia has released two new rootstocks, Leola™ (Phytophthora rot root tolerant) and Zerala™ (salinity-tolerant) (Anonymous, 2020). Australasia has recently made contributions of new avocado rootstocks, with the Phytophthora root rot-tolerant 'Astro' being commercialized in New Zealand (S. Wade, Lynwood Nursery, Whangarei, New Zealand, 2023, personal communication), and in Australia, 'Procado®', with greater Phytophthora root rot tolerance and precocity, and lower vigour than other rootstocks in common use (Smith *et al.*, 2011; Whiley and Dann, 2021; A.W. Whiley, SHS, Buderim, QLD, Australia, 2024, personal observation). However, despite the addition of a number of Phytophthora root rot-tolerant/-resistant rootstocks over the past 20 years, full resistance to *P. cinnamomi* without molecular intervention is unlikely due to the host/disease history being noncompliant with coevolutionary principles for the development of host resistance (Brasier, 2001).

Recently, successful *in vitro* propagation at a commercial level has occurred, with nurseries in Australia, Israel and Spain now offering trees grafted to a selection of rootstocks propagated in tissue culture. However, the requirement for cultivar-specific, *in vitro* protocols, and varying degrees of vegetative propagation difficulty between botanical varieties (Eggers and Halma, 1937; O. Markovich, Rahan Meristem, Kibbutz Rosh Hanikra, Israel, 2023, personal communication), suggests progress in the foreseeable future will be slow in adding new rootstock cultivars, particularly those from physiologically mature sources with proven, commercially valuable attributes, due to much greater difficulty in establishing successful *in vitro* cultures from this material. However, future success from the increasing numbers of tissue culture facilities working globally, to develop commercially viable *in vitro* protocols for avocado, should add to the diversity of the cloned material available.

In the years ahead, tissue culture propagation of avocado rootstocks is set to revolutionize the production of avocado trees, since one explant can produce in excess of 500 cloned units in 1 year (Hiti-Bandaralage *et al.*, 2017). Production is highly industrialized and reliable, as there is no dependence on seed supply, which is variable from year to year, and propagation is independent of weather conditions, since it occurs under highly controlled, enclosed environments. However, rootstock material chosen for clonal production must have proven genetic superiority over seedling lines to command the higher price asked, and not be implemented to exclusively improve management of product flow and nursery outcomes. Real value to growers purchasing *in vitro*-produced rootstocks, in most cases, will be achieved with rootstock cultivars originating from physiologically mature parent material, with scientifically proven attributes encompassing higher productivity (Whiley *et al.*, 2013; Mena *et al.*, 2015), improved disease (root and fruit) and salinity resistance/tolerance (Ben-Ya'acov and Zilberstaine, 1999; Willingham *et al.*, 2001; Whiley *et al.*, 2007, 2013; Smith *et al.*, 2011; Menge *et al.*, 2012; Celis *et al.*, 2018; Lazare *et al.*, 2021), and greater adaption to soil conditions such as alkalinity and hypoxia (Ben-Ya'acov *et al.*, 1992; Derebe *et al.*, 2023).

Most currently available clones from 'nurse-seed/etiolation' propagation have been selected for 'single-factor' attributes (Cañas-Gutiérrez *et al.*, 2022) and may not outperform the yield of seedling rootstocks when grown outside the biotic or abiotic stress conditions they were chosen to address (Dann *et al.*, 2013; Cohena *et al.*, 2023; C. Hermosilla-Madrid, Awanui, New Zealand, 2024, personal communication). Risk management and financial implications are primary considerations when planting long-term fruit tree crops. Both aspects should receive careful consideration when choosing between seedling or cloned rootstocks regarding their suitability for new orchards or replanting old avocado land. The integrity of both rootstock sources should be well known before committing to a purchase.

With avocados, we are dealing with the complexity of genetically divergent material, from three botanical races, which readily hybridize. Interactions between scions and rootstocks are very complex, but increasingly, research has attempted to uncover the processes involved, including graft-induced phenotypic changes that can occur. Productivity gains in avocado may come from grafting scions to rootstocks that are genetically more closely aligned, providing greater cohesion in compatibility across graft unions; for example, production of 'Hass' seems notably higher when grafted onto rootstocks of Guatemalan and West Indian parentage in the absence of biotic and abiotic stresses. Identification of rootstocks from seedling populations that impart consistently higher production should be prioritized, as it has been for Phytophthora and salinity tolerance, since this offers the best long-term solution to raising yield performance. Recent insights in grafting research hint at the presence of complex scion–rootstock communications. Movement of water and nutrients between plant components and long-distance signalling through auxins, cytokinins, GA, proteins, metabolites and messenger RNAs to coordinate successful vascular connections have been reported (Wang *et al.*, 2017; Lu *et al.*, 2020).

Reyes-Herrera *et al.* (2020) and Cañas-Hernández *et al.* (2022) studied the inheritance of seedling avocado rootstock effects on scion ('Hass') characteristics drawn from the Colombian 'criolla' seedling population. The genetic screening used 13 microsatellite markers in simple sequence repeats (SSRs) that were related to phenotypic traits, including fruit yield and quality. Both studies identified the heritable characteristics passed on to the scion cultivar, within the open-pollinated seedling rootstocks used. Advances in tissue culture and molecular biology are closer to the possibility of genetic modification, potentially introducing disease resistance into 'elite' rootstocks selected for producing consistently high yields in scion cultivars.

References

Adhikari, P.B., Xu, Q. and Notaguchi, M. (2022) Compatible graft establishment in fruit trees and its potential markers. *Agronomy* 12(8), 1981. Available at: doi:10.3390/agronomy12081981 (accessed 3 October 2023).

Adjei, P.Y., Banful, B.K. and Idun, I.A. (2011) Seed size and seed cut-length's effect on germination behaviour and seedling growth of avocado (*Persea americana* Mill.). *International Journal of Agricultural Research* 6(3), 299–305.

Alexander, D.McE. (1977) Propagation. In: *Proceedings of the Australian Avocado Research Workshop*. New South Wales Department of Agriculture, Wollongbar, Australia, pp. 21–25.

Allan, P., Lamb, D. and Chalton, D. (1981) Sterilization and pasteurization of soil mixes. *South African Avocado Growers' Association Yearbook* 4, 124–127.

Alves-de Oliviera, A., Carlos-Koller, O. and Villegas-Monter, Á. (1999) Propagación vegetative de aguacate selección 153 (*Persea* spp.) por acodo en contenedor. *Revista Chapingo Serie Horticultura Núm* Especial V, 221–225. (In Spanish with English summary.)

ANA (2000) *Avokado Plantverbeteringskema (APVS).* South African Avocado Growers' Association, Tzaneen, South Africa, 13 pp. (In Afrikaans.)

Anderson, G. (1977) Nursery programmes and techniques. In: *Proceedings of the Australian Avocado Research Workshop*. New South Wales Department of Agriculture, Woolongbar, Australia, pp. 26–30.

Andrews, P.K. and Serrano Marquez, C. (1993) Graft Incompatibility. In: Janick, J. (ed.) *Horticultural Reviews*. Vol. 15. Wiley-Interscience, Hoboken, NJ, pp. 183–218.

Anonymous (2020) Westfalia launched two new co-owned avocado rootstocks. Westfalia Fruit. Available at: https://www.westfaliafruit.com/westfalia-launches-two-new-co-owned-avocado-rootstocks/#:~:-text=Trees%20of%20these%20two%20new,as%20confirmed%20in%20recent%20trials (accessed 11 March 2024).

Anonymous (2024a) The European market potential for avocados. Available at: https://www.cbi.eu/market-information/fresh-fruit-vegetables/avocados/market-potential#:~:text=Spain%2C%20Europe's%20largest%20 avocado%2Dproducing,the%20role%20of%20international%20distributor (accessed 6 February 2024).

Anonymous (2024b) Global avocado production 2022, by country. Statista. Available at: https://www.statista.com (accessed 6 February 2024).

Arabi, M.A. and Bertling, I. (2021) *Moringa oleifera* leaf extract affects avocado (*Persea americana*) seed phenolics, sugars and starch as well as germination. *Acta Horticulturae* 1306, 185–192.

Atkinson, C.J. and Else, M.A. (2004) Enhancing Harvest Index in temperate fruit tree crops through the use of dwarfing rootstocks. In: Bekele, F., End, M.J. and Eskes, A.B. (eds) *Proceedings of the International Workshop on Cocoa Breeding for Improved Production Systems,* October 19–21, 2003, INGENIC and the Ghana Cocoa Board, Accra, Ghana, pp. 118–131.

Atkinson, C.J., Else, M.A., Taylor, L. and Dover, C.J. (2003) Root and stem hydraulic conductivity as determinants of growth potential in grafted trees of apple (*Malus pumila* Mill.). *Journal of Experimental Botany* 54(385), 1221–1229. Available at: doi:10.1093/jxb/erg132 (accessed 11 December 2023).

Baker, K.F. (ed.) (1957) *The UC System for Producing Healthy Container-grown Plants*. California Agricultural Experiment Station Manual 23, University of California, Riverside, California, 332 pp.

Barlow, C. (2000) *The Ghosts of Evolution: Nonsensical Fruit, Missing Partners, and Other Ecological Anachronisms*. Basic Books, New York.

Barrett, S.C. (2010) Darwin's legacy: the forms, function and sexual diversity of flowers. *Philosophical Transactions the Royal Society London, Series B, Biological Sciences* 365(1539), 351–368. Available at: doi:10.1098/rstb.2009.0212 (accessed 24 August 2023).

Barrientos-Priego, A.F., Borys, M.W. and Barrientos-Perez, F. (1986) Rooting of avocado cuttings (*Persea americana* Mill.) cvs. Fuerte and Colin V-33. *California Avocado Society Yearbook* 70, 157–163.

Barrientos-Priego, A.F., Morales-Nieto, M.J., Almaguer-Vargas, G., Borys, M.W., Martinez-Damián, T. and Barrientos-Pérez, F. (1995) Response of cotyledons detachments over the development of nursery seedlings of avocado (*Persea americana* Mill.). In: *Proceedings of the III World Avocado Congress*. Tel Aviv, Israel, pp. 211–216.

Ben-Ya'acov, A. and Michelson, E. (1995) Avocado rootstocks. *Horticultural Reviews* 17, 381–429.

Ben-Ya'acov, A. and Zilberstaine, M. (1999) Clonal avocado (*Persea americana* Mill.) rootstocks in Israel. *Revista Chapingo Serie Horticultura Núm* Especial V, 39–42.

Ben-Ya'acov, A., Michelson, E., Zilberstaine, M., Barkan, Z. and Sela, I. (1992) Selection of clonal avocado rootstocks in Israel for high productivity under different soil conditions. In: Lovatt, C.J., Holthe, P.A. and Arpaia, M.L. (eds) *Proceedings of the II World Avocado Congress*. Riverside, California, pp. 521–526.

Bergh, B.O. (1975) Avocado research in Israel. *California Avocado Society Yearbook* 58, 103–126.

Bergh, B.O. (1988) The effect of pre-treatments on avocado germination. *California Avocado Society Yearbook* 72, 215–221.

Best, T., Thorne, H., Axtens, J., Gleeson, M. and Hiti-Bandaralage, J. *et al.* (2021) *Tissue Culture for Australian Avocados: Project Information Pack*. Central Queensland University, Brisbane, Australia. Available at: https://qaafi.uq.edu.au/files/72009/Tissue%20Culture%20for%20Australian%20Avocados%20-%20PIP.pdf (accessed 6 March 2024).

Blumenfeld, A. and Gazit, S. (1970) Cytokinin activity in avocado seeds during fruit development. *Plant Physiology* 46, 331–333.

Borys, M.W., Barrera-Guerra, J.L. and Luna-López, J. (1985) Some root characteristics of avocado (*Persea americana* Mill.) seedlings of the West Indian and Guatemalan races. *California Avocado Society Yearbook* 69, 111–222.

Boza, E.J., Tondo, C.L, Ledesma, N., Campbell R.J., Bost, J. *et al.* (2018) Genetic differentiation, races and interracial admixture in avocado (*Persea americana* Mill.), and *Persea* spp. evaluated using SSR markers. *Genetic Resources and Crop Evolution* 65, 1195–1215. Available at: doi:10.1007/s10722-018-0608-7 (accessed 30 April 2024).

Brasier, C.M. (2001) Rapid evolution of introduced plant pathogens via interspecific hybridization: hybridization is leading to rapid evolution of Dutch elm disease and other fungal plant pathogens. *BioScience* 51(2), 123–133. Available at: doi:10.1641/0006-3568(2001)051[0123:REOIPP]2.0.CO;2 accessed 12 March 2024).

Brokaw, W.H. (1975) Root rot resistant avocado clonal rootstocks. *Plant Propagation* 21(4), 7–8.

Brokaw, W.H. (1977) Subtropical fruit tree production: avocado as a case study. *Proceedings of the International Plant Propagators' Society* 27, 113–121.

Brokaw, W.H. (1987a) Avocado clonal rootstock propagation. *Proceedings of the International Plant Propagators' Society* 37, 97–103.

Brokaw, W.H. (1987b) Field experiences with clonal rootstocks. *South African Avocado Growers' Association Yearbook* 10, 34–36.

Burns, R.M., Mircetich, S.M., Coggins, C.W. Jr and Zentmyer, G.A. (1965) Gibberellin increases size of Duke avocado seedlings. *California Avocado Society Yearbook* 50, 118–120.

Cañas-Gutiérrez, G.P., Sepulveda-Ortega, S., López-Hernández, F., Navas-Arboleda, A.A. and Cortés, A.J. (2022) Inheritance of yield components and morphological traits in avocado cv. Hass from "Criollo" elite trees via half-sib seedling rootstocks. *Frontiers in Plant Science.* Available at: doi: 10.3389/fpls.2022.843099 (accessed 24 April 2024).

Cañas-Hernández, G.P., Sepulveda-Ortega, López-Heranádez, F., Navas-Arboleda, A.A. and Cortés, A.J. (2022) Inheritance of yield components and morphological traits in avocado cv. Hass from 'Criollo' elite trees via half-sib seedling rootstocks. *Frontiers of Plant Science* 13. Available at: doi:10.3389/fpls.2022.843099 (accessed 2 April 2024).

Cano-Gallego, L.E., Estrada, J.A.B., Arredondo, J.D.H., Correo-Londono, G.A. and Cordoba-Gaona, O. de J. (2023) Graft compatibility effects on cv. Hass avocado fruit growth. *Ciencia e Agrotecnologia* 47:e014022.

Castillo-Argaez, R., Konkol, J.L., Vargas, A.I., Ploetz, R.C. and Schaffer, B. (2021) Disease severity and ecophysiology of rootstock/scion combinations of different avocado (*Persea americana* Mill.) genotypes in response to laurel wilt. *Scientia Horticulturae* 287(9), 110250. Available at: doi:10.1016/j.scienta.2021.110250 (accessed 3 February 2024).

Castro, M., Fassio, C., Darrouy, N., Aedo, M. and Jorquera, L. (2011) Different alternatives for producing sprouts of avocado (Persea americana) rootstocks. *Ciencia e Investigación Agrarian* 38(2), 219–226. Available at: doi:10.4067/S0718-16202011000200007 (accessed 4 February 2024).

Celis, N., Donald, L., Suarez, D.L., Wu, L., Li, R., Arpaia, M.L. and Mauk, P. (2018) Salt tolerance and growth of 13 avocado rootstocks related best to chloride uptake. *HortScience* 53(12), 1737–1745. Available at: doi:10.21273/HORTSCI13198-18 (accessed 28 April 2024).

Chalker, F.C. and Robinson, P.W. (1969) Propagating avocados. *Agricultural Gazette of New South Wales* 80, 400–405.

Chen, H., Morrell, P., Ashworth, V.E.T.M., de la Cruz, M. and Clegg, M. (2009) Tracing the geographic origins of major avocado cultivars. *Journal of Heredity* 100(1), 56–65. Available at: doi:10.1093/jhered/esn068 (accessed 19 March 2024).

Claassens, N.J.F. (2001) The avocado nursery. In: De Villiers, E.A. (ed.) *The Cultivation of Avocado*. Institute for Tropical and Subtropical Crops, Nelspruit, South Africa, pp. 91–97.

Cohena, H., Bar-Noya, Y., Irihimovitchband, V. and Rubinovich, L. (2023) Effects of seedling and clonal West Indian rootstocks irrigated with recycled water on 'Hass' avocado yield, fruit weight and alternate bearing. *New Zealand Journal of Crop and Horticultural Science* 51(1), 39–51. Available at: doi:10.1080/01140671.2022.2098779 (accessed 12 March 2024).

Corrales-Garcia, J.E. and Méndez-Zúñiga, S.M. (2020) Criollo avocado of Mexican race (*Persea americana* var. *drymifolia*): an underutilized species in horticulture. *Horticulture International Journal* 4(6), 221–223. Available at: doi:10.15406/hij.2020.04.00186 (accessed 11 February 2024).

Crane, J.H., Douhan, G., Faber, B.A., Arpaia, M.L., Bender, G.S. *et al.* (2013) Cultivars and rootstocks. In: Schaffer, B.A., Wolstenholme, B.N. and Whiley, A.W. (eds) *The Avocado: Botany, Production and Uses*. 2nd edn. CAB International, Wallingford, UK, pp. 200–233.

Cutting, J. (2002) New avocado cultivars and rootstocks – what's best for the New Zealand avocado industry. Avoscene. Available at: http://www.avocadosource.com/Journals/AvoScene/CuttingEdge_Sept_2002.PDF (accessed 22 March 2024).

Cutting, J.G.M. and van Vuuren, S.P. (1988) Rooting leafy non-etiolated avocado cuttings from gibberellin-injected trees. *Scientia Horticulturae* 37, 171–176.

Dann, E.K. (2016) *Avocado Rootstock Assessment and Improvement – Interim (AV13018)*. Horticulture Innovation Australia Ltd, Sydney, New South Wales, Australia.

Dann, E., Bransgrove, K., Newett, S. and Thomas, G. (2019) Lodging of avocado trees. *Talking Avocados* 29(4), 40–45.

Dann, E.K., Forsberg, L.I., Cooke, A.W., Pegg, K.G., Shivas, R.G. and Tan, Y.P. (2011) The 'Cylindro' complex of avocado root pathogens. In: *Proceedings of the VII World Avocado Congress*. Cairns, Queensland, Australia, pp. 64–70.

Dann, E.K., Cooke, A.W., Forsberg, L.I., Pegg, K.G., Tan, Y.P. and Shivas, R. G. (2012) Pathogenicity studies in avocado with three nectriaceous fungi, *Calonectria ilicicola*, *Gliocladiopsis* sp. and *Ilyonectria liriodendri*. *Plant Pathology* 61, 896–902.

Dann, E.K., Whiley, T., Pegg, K., Dean, J., Chandra, K. and Smith, L. (2013) Evaluation of fruit yields and tree health of Hass trees grafted to 8 rootstocks in high Phytophthora root rot conditions at Childers, Qld. *Talking Avocados* 24(3), 34–37.

Da Silveira, S.V., De Souza, P.V.D. and Koller, O.C. (2004) Propagação vegetativa de abacteiro por estaquia. *Revista Brasileira de Fruticultura* 26(1), 191–192. (In Portuguese with English summary.)

Derebe, A.D., Dema, M.W. and Roro, A.G. (2023) Impact of waterlogging stress on grafted avocado (Persea americana) seedlings growth and physiological performance. *Cogent Food & Agriculture* 9(1). Available at: doi:10.1080/23311932.2023.2261837 (accessed 29 January 2024).

De Villiers, A.I. and Ernst, A.A. (2007) Practical value of the Allesbeste micro-cloning technique. In: *Proceedings of the VI World Avocado Congress, Viña Del Mar, Chile*. Available at: http://www.avocadosource.com/wac6/en/extenso/1a-5.pdf (accessed 4 October 2012).

De Villiers, A.I. and Ernst, A.A. (2011) Tree uniformity: determining factors – a field observation. In: *Proceedings of the VII World Avocado Congress*. Cairns, Queensland, Australia, pp. 267–270.

De Villiers, A.I. and Ernst, A.A. (2015) Avocado rootstock research: principles and practices. In: *Proceedings of the VIII World Avocado Congress*. Lima, Peru. Available at: https://www.avocadosource.com/WAC8/Section_02/DeVilliersAI2015.pdf (accessed 12 February 2024).

Duman, Z., Hadas-Brandwein, G., Eliyahu, A., Belausov, E., Abu-Abied, M. *et al.* (2020) Short de-etiolation increases the rooting of VC801 avocado rootstock. *Plants* 9(11), 1481. Available at: doi:10.3390/plants9111481 (accessed 12 January 2024).

Durand, B.J. (1990) Clonal propagation of avocado rootstocks. *Farming in South Africa Pamphlet Series, Avocados*, D.5.

Eggers, E.R. (1942) Effect of the removal of the seed coats on avocado seed germination. *California Avocado Society Yearbook* 27, 41–43.

Eggers, E.R. and Halma, F.F. (1936) Propagating the avocado by means of stem cuttings. *California Avocado Association Yearbook* 21, 63–66.

Eggers, E.R. and Halma, F.F. (1937) Rooting avocado cuttings. *California Avocado Association Yearbook* 22, 121–125.

Ernst, A.A. (1978) Beworteling van avocado *(Persea americana* Mill.) steggies. MSc thesis, University of Pretoria, Pretoria, South Africa. (In Afrikaans.)

Ernst, A.A. (1981) Die invloed van mediumdeurlugting, swamdoder en indoolbottersuur op die beworteling van avokadosteggies. *South African Avocado Growers' Association Yearbook* 4, 121–123. (In Afrikaans with English summary.)

Ernst, A.A. (1984) Beworteling van avokado *(Persea americana* Mill.) steggies. DSc thesis, University of Pretoria, Pretoria, South Africa. (In Afrikaans.)

Ernst, A.A. (1999) Micro cloning: a multiple cloning technique for avocados using micro containers. *Revista Chapingo Serie Horticultura Núm* Especial V, 217–220.

Ernst, A.A. and Holtzhausen, L.C. (1978) New promising techniques for rooting difficult-to-root avocado (*Persea americana* Mill.) cuttings. *Citrus and Subtropical Fruit Journal* 532, 6–10.

Ernst, A.A. and Holtzhausen, L.C. (1987) Callus development – a possible aid in rooting avocado cuttings. *South African Avocado Growers' Association Yearbook* 10, 39–41.

Fassio, C., Cautin, R., Perez-Donoso, A., Bonomelli, C. and Castro, M. (2016) Propagation techniques and grafting modify the morphological traits of roots and biomass allocation in avocado trees. *HortTechnology* 26(1), 63–69.

Fassio, C., Cautin, R., Perez-Donoso, A.G., Castro, M. and Bonomelli, C. (2020) Comparative branching order and root anatomy of clonal and seed-grown avocado trees (*Persea americana* Mill.). *International Journal of Agriculture and Natural Resources* 47(2), 134–144.

Ford, D. (1996) Landscapers beware of poorly grown trees. *Australian Horticulture* 2, 98–101.

Frolich, E.F. (1951) Rooting Guatemalan avocado cuttings. *California Avocado Society Yearbook* 36, 136–138.

Frolich, E.F. (1961) Etiolation and rooting of cuttings. In: *Proceedings of the 11th Annual Meeting Plant Propagators' Society*, pp. 277–283.

Frolich, E.F. and Platt R.G. (1972) Use of the etiolation technique in rooting avocado cuttings. *California Avocado Society Yearbook* 55, 97–109.

Gabor, B.K., Guillemet, F.B. and Coffey, M.D. (1990) Comparison of field resistance to Phytophthora cinnamomi in twelve avocado rootstock. *HortScience* 25(12), 1655–1656.

Galán Saúco, V. and Fernandez Galvan, D. (1985) Avocado culture on the Canary Islands. *California Avocado Society Yearbook* 69, 73–80.

Galán Saúco, V., Mansito Perez, P. and Cabellero Ruano, M. (1981) Horticulture in the Canary Islands. *HortScience* 16(2), 148–150.

Gálvez-Cendegui, L., Peñaloza, P., Oyanedel, E. and Castro, M. (2017) Storage, size and vigor of 'Ester' avocado seeds (*Persea americana* Mill.). *Ciencia e Investigación Agraria* 44(1), 94–99.

García-Gómez, M.L., Sánchez-Romero, C., Barceló-Munoz, A., Heredia, A. and Pliego-Alfaro, F. (1994) Levels of endogenous indole-3-acetic acid and indole-3-acetyl-aspartic acid during adventitious rooting in avocado microcuttings. *Journal of Experimental Botany* 45(6), 865–870.

García Romero, J.A. (2024) We are the first in the world to get avocado plants grafted on clonal rootstocks from *in vitro* culture. Fresh Plaza. Available at: https://www.freshplaza.com/north-america/article/9363915/we-are-the-first-in-the-world-to-get-avocado-plants-grafted-on-clonal-rootstocks-from-in-vitro-culture/ (accessed 7 March 2024).

Gardner, F.E. (1937) Etiolation as a method of rooting apple variety stem cuttings. *Proceedings of the American Society for Horticultural Science* 34, 323–329.

Gillespie, H.L. (1952) Asexual reproduction of rootstock from heavy-producing Fuerte avocado trees. *California Avocado Society Yearbook* 37, 97–102.

Gillespie, H.L. (1956) Preliminary investigation of 'residual juvenility' in avocado seedling stems. *California Avocado Society Yearbook* 40, 132–134.

Gillespie, H.L. (1957) Stem-rooting clones by means of 'juvenile growth phase' leafy-stem cuttings. *California Avocado Society Yearbook* 41, 94–96.

Gomez, R.E., Soule, J. and Malo, S.E. (1971) Avocado air layers – a study of seven varieties during a year's cycle. *Proceedings of the American Society for Horticultural Science, Tropical Region* 15, 113–120.

Gregoriou, C., Papademetriou, M. and Christofides, L. (1984) Propagation of avocado plants in Cyprus (comparison between budding and grafting of four avocado varieties). *California Avocado Society Yearbook* 68, 121–126.

Gustafson, C.D. and Kadman, A. (1969) Effect of some plant hormones on the rooting capacity of avocado cuttings. *California Avocado Society Yearbook* 53, 97–100.

Haas, A.R.C. (1937a) Propagation of the Fuerte avocado by means of leafy twig cuttings. *California Avocado Association Yearbook* 22, 126–130.

Haas, A.R.C. (1937b) Further progress in the rooting of Fuerte avocado cuttings. *California Avocado Association Yearbook* 22, 130–132.

Habibi, F., Liu, T., Folta, K. and Sarkhosh, A. (2022) Physiological, biochemical, and molecular aspects of grafting in fruit trees. *Horticulture Research* 9:uhac032. Available at: doi:10.1093/hr/uhac032 (accessed 21 November 2023).

Halma, F.F. (1953) Avocado rootstock experiments – a 10-year report. *California Avocado Society Yearbook* 38, 79–86.

Halma, F.F. and Frolich, E.F. (1949) Storing avocado seeds and hastening germination. *California Avocado Society Yearbook* 34, 136–138.

Harris, D.R., Robinson, J.P. and Juniper, B.E. (2002) Genetic clues to the origin of the apple. *Trends in Genetics* 18, 426–430.

Hartmann, H.T., Kester, D.E., Davies, F.T. and Geneve, R.L. (2011) *Hartmann & Kester's Plant Propagation Principles and Practices*. 8th edn. Prentice-Hall, Upper Saddle River, New Jersey, pp. 373–381, 749.

Haskelberg, U. (2021) Israeli avocados in 2021. Available at: https://canacado.com/israel-avocados/ (accessed 8 March 2024).

Hass, R.G. (1935) Plant Patent 139. United States Patent Office. Available at: https://www.avocadosource.com/links/hasspatent_1935.pdf (accessed 12 April 2024).

Hayat, F., Li, J., Iqbal, S., Peng, Y. and Hong, L. *et al.* (2022) A mini review of citrus rootstocks and their role in high-density orchards. *Plants* 11(21), 2876. Available at: doi:10.3390/plants11212876 (accessed 16 January 2024).

Hayat, F., Qiu, C., Xu, X., Wang, Y., Wu, T. *et al.* (2019) Rootstocks influence morphological and biochemical changes in young 'Red Fuji' apple plants. *International Journal of Agriculture and Biology* 21, 1097–1105. Available at: doi:10.17957/IJAB/15.0999 (accessed 13 December 2023).

Hendry, N.S. and van Staden, J. (1982) Attempts at propagation of avocado using various techniques. *South African Avocado Growers' Association Yearbook* 5, 71–73.

Heo, K., van der Werf, H. and Tobe, H. (1998) Embryology and relationships of Lauraceae (Laurales). *Botanical Journal of the Linnaean Society* 126, 295–322.

Hermosilla-Madrid, C. and Whiley, A.W. (2023) High-density (HD) avocado production in New Zealand. In: *Proceedings of the Tenth World Avocado Congress*. Auckland, New Zealand. Available at: https://industry.nzavocado.co.nz (accessed 9 October 2023).

Hernández, I., Molina, V., Fuentealba, C., Alvaro, J.E., Defilippi, B and Pedreschi, R. (2023) Do rootstocks Influence global fruit quality, postharvest performance and metabolite profiles of *Persea americana* cv. Hass? *Horticulturae* 9(2), 184. Available at: doi:10.3390/horticulturae9020184 (accessed 3 October 2023).

Hiti-Bandaralage, J.C.A., Hayward, A. and Mitter, N. (2017) Micropropagation of avocado (*Persea americana* Mill.). *American Journal of Plant Sciences* 8(11), 2898–2921.

Hoblyn, T.N., Grubb, N.H., Painter, A.C. and Wates, B.L. (1937) Studies in biennial bearing. *I. Journal of Pomology and Horticultural Science* 14, 39–76. Available at: doi:10.1080/03683621.1937.11513464 (accessed 11 September 2023).

Hodgson, R.W. (1951) Report of long-term research on the avocado – studies on alternate bearing, varieties, rootstock, propagation and storage by Division of Subtropical Horticulture. *California Agriculture* 5, 8–11.

Hofshi, R. (1996a) Clone your own avocado at home. *Subtropical Fruit News* 4(2), 4–6.

Hofshi, R. (1996b) Experiments with cloning avocado rootstocks. *California Avocado Society Yearbook* 80, 103–108.

Horns, F. and Hood, M.E. (2012) The evolution of disease resistance and tolerance in spatially structured populations. *Ecology and Evolution* 2(7), 1705–1711. Available at: doi:10.1002/ece3.290 (accessed 15 September 2023).

Hume, E.P. (1951) Growing Avocados in Puerto Rico. Puerto Rico (Mayaquez) Federal Experiment Station (U.S. Department of Agriculture) *Circular* 33, 93 pp.

Janick, J. (2005) The origin of fruits, fruit growing and fruit breeding. *Plant Breeding Reviews* 25, 255–320.

Janzen, D.H. and Martin, P.S. (1982) Neotropical anachronisms: the fruits the gomphotheres ate. *Science* 215, 19–27.

Johnston, J.C. and Frolich, E.F. (1956) Avocado propagation. *California Avocado Society Yearbook* 40, 89–98.

Kadman, A. (1963) Germination experiments with avocado seeds. *California Avocado Society Yearbook* 47, 58–60.

Kadman, A. (1976) Effect of the age of juvenile stage avocado seedlings on the rooting capacity of their cuttings. *California Avocado Society Yearbook* 59, 58–60.

Kadman, A. and Ben-Ya'acov, A. (1965) A review of experiments on some factors influencing the rooting of avocado cuttings. *California Avocado Society Yearbook* 49, 67–72.

Kadman, A. and Gustafson, C.D. (1971) The use of potassium salt of indole butyric acid (KIBA) in rooting avocado cuttings. *California Avocado Society Yearbook* 54, 96–99.

Knight, R.C. and Witt, A.W. (1937) The propagation of fruit tree stocks by stem cuttings. *Journal of Pomology* 6, 47–60.

Koch, F.D. (1983) Nursery practices. In: Thomas, P.H. (ed.) *Avocado Growers' Handbook*. Bonsall Publications, Bonsall, California, pp. 56–59.

Korsten, L., Bar-Joseph, M. and Kotze, J.M. (1987) Monitoring avocado sunblotch disease in South Africa. *South African Avocado Growers' Association Yearbook* 10, 128–129.

Lahav, E. and Lavi, U. (2009) Avocado genetics and breeding. In: Mohan Jain, S. and Priyadarshan, P.M. (eds) *Breeding Plantation Tree Crops: Tropical Species*. Springer Science, New York.

Lazare, S., Cohen, Y., Goldshtein, E., Yermiyahu, U., Ben-Gal, A. and Dag, A. (2021) Rootstock-dependant response of Hass avocado to salt stress. *Plants* 10(8), 1672. Available at: doi:10.3390/plants10081672 (accessed 12 January 2024).

Lazare, S., Haberman, A., Yermiyahu, U., Erel, R., Simenski, E. and Dag, A. (2020) Avocado rootstock influences scion leaf mineral content. *Archives of Agronomy and Soil Science* 66(10), 1399–1409. Available at: doi:10.1080/03650340.2019.1672163 (accessed 5 November 2023).

Leal, F.J. and Krezdorn, A.H. (1964) Rooting avocado cuttings. *Proceedings of the Florida State Horticultural Society* 77, 358–362.

Leal, F.J., Krezdorn, A.H. and Marte, R.J. (1976) The influence of gibberellic acid on the germination of avocado seeds. *Proceedings of the Florida State Horticultural Society* 89, 258–261.

Le Lagadec, D. (2010) Five years data collection for clonal and seedling rootstocks with 'Hass' as a scion. *Talking Avocados* 21(4), 36–40.

Le Roux, J.C. (1940) *The Avocado in South Africa*. South African Department of Agriculture and Forestry Horticultural Series No. 1, p. 15

Leopold, A.C. and Kriedemann, P.E. (1975) *Plant Growth and Development*. 2nd edn. McGraw-Hill, London.

Lu, X., Liu, W., Wang, T., Zhang, J., Li, X. and Zhang, W. (2020) Systemic long distance signaling and communication between rootstock and scion in grafted vegetables. *Frontiers of Plant Science* 11, 460. Available at: doi:10.3389/fpls.2020.00460 (accessed 3 October 2023.

Luttig, M. and Manicom, B.Q. (1999) Application of a highly sensitive avocado sunblotch viroid indexing method. *South African Avocado Growers' Association Yearbook* 22, 55–60.

Makhoul, G. and Attaf, W.A. (2020) Effect of storage time on germination of avocado seeds and seedlings quality. *Tishreen University Journal for Research and Scientific Studies – Biological Sciences Series* 42(4), 19–24.

Malan, E.F. and van der Meulen, A. (1954) Propagation of avocados. *Farming in South Africa* 29, 499–502.

Markovich, O. (2023) Avocado rootstock propagation via tissue culture. In: *Proceedings of the WAC*. Auckland, New Zealand. Available at: https://industry.nzavocado.co.nz/world-avocado-congress-nz-2023/presentations/ (accessed 7 March 2024).

Marques, J.R., Hofman, P.J. and Wearing, A.H. (2003) Rootstocks influence 'Hass' avocado fruit quality and fruit minerals. *The Journal of Horticultural Science and Biotechnology* 78(5), 673–679.

Mastalerz, J.W. (1977) *The Greenhouse Environment*. John Wiley & Sons, New York.

Maxwell, M. (2023) Westfalia showcases avocado advances. Eurofruit. Available at: https://www.fruitnet.com/eurofruit/westfalia-showcases-avocado-advances/248322.article (accessed 15 March 2024).

Me, G.-T. and Arzate-Fernández, A.M. (2010) West Indian avocado: where did it originate? *Phyton (B Aires)* 79(1), 203–207. Available at: doi:10.32604/phyton.2010.79.203 (accessed 13 January 2024).

Mena, F., Gardiazabal, F., Torres, J. and Pinto, A. (2015) Comportamiento poblacional de 200 árboles de Paltos (*Persea americana* Mill.) cv Hass cultivados en Panquehue, Chile. In: *Proceedings of the VIII World Avocado Congress*. Lima, Perú. Available at: https://avocadosource.com/wac8/section_04/menafrancisco2015.pdf (accessed 6 October 2024).

Menge, J.A., Douhan, G.W., McKee, B., Pond, E., Bender, G.S. and Faber, B. (2012) Three new avocado rootstock cultivars tolerant to Phytophthora root rot: 'Zentmyer', 'Uzi', and 'Steddom'. *HortScience* 47(8) 1191–1194.

Mickelbart, M.V. and Arpaia, M.L. (2002) Rootstock influences changes in ion concentrations, growth, and photosynthesis of 'Hass' avocado trees in response to salinity. *Journal of the American Society of Horticultural Science* 127(4), 649–655.

Mohammed, S. and Sorhaindo, C.A. (1984) Production and rooting of etiolated cuttings of West Indian and hybrid avocado. *Tropical Agriculture* 61(3), 200–204.

Moing, A., Salesses, G. and Saglio, P.H. (1987) Growth and the composition and transport of carbohydrate in compatible and incompatible peach/plum grafts. *Tree Physiology* 3, 345–354.

Moll, J.N. and Wood, R. (1980) An efficient method for producing rooted avocado cuttings. *Subtropica* 1, 9–12.

Mosse, B. (1962) *Graft-incompatibility in Fruit Trees with Particular Reference to its Underlying Causes*, Technical. Commonwealth Agricultural Bureaux, Farnham Royal, UK.

Nawaz, M.A., Imtiaz, M., Kong, Q., Cheng, F., Ahmed, W. *et al.* (2016) Grafting: a technique to modify ion accumulation in horticultural crops. *Frontiers in Plant Science* 21(7), 1457. Available at: doi:10.3389/fpls.2016.01457 (accessed 7 December 2024).

Ohr, H.D., Zentmeyer, G.A. and Menge, J.A. (1991) *Avocado root rot*. University of California Cooperative Extension Publication, Riverside, CA. p. 2440.

Oliveira, I.V. de M., Costa, R.S., Môro, F.V., Martins, A.B.G. and da Silva, R.R.S. (2010) Caracterização morfológica do fruto, da semente e desenvolvimento pós-seminal do abacateiro. *Comunicata Scientiae* 1(1), 69–73. (In Portuguese.)

Palmer, J.W. (1988) Annual dry matter production and partitioning over the first 5 years of a bed system of Crispin/M.27 apple trees at four spacings. *Journal of Applied Ecology* 25, 569–578.

Partridge, C.J. (1984) Plant improvement scheme – long-term planning. *South African Avocado Growers' Association Yearbook* 7:23.

Partridge, C.J. (1986) The avocado plant improvement scheme. *South African Avocado Growers' Association Yearbook* 9, 15–16.

Paulus, A.O., Nelson, J. and Zentmyer, G.A. (1976) Avocado seed treatment and storage trial. *California Avocado Society Yearbook* 59, 70–71.

Pegg, K.G. (1992) Avocado research – achievements in disease control. In: *Proceedings of the 3rd Avocado Research Workshop*. Queensland Department of Primary Industries, Brisbane, Australia, pp. 1–5.

Pereira, I.D.S., Fachinello, J.C., Antunes Corrêa, L.E., Campos, A. and Diniz Pina, A. (2014) Incompatibilidade de enxertia em Prunus graft incompatibility in Prunus. *Ciência Rural* 44(9), 1519–1526. Available at: doi:10.1590/0103-8478cr20131074 (accessed 13 August 2023).

Pérez Jiménez, R.M., Zea Bonilla, T. and López Herrera, C.J. (2005) Avocado root rots in Andalucía: a review. *South African Avocado Growers' Association Yearbook* 28, 43–46.

Perumal, A. (1961) Some characteristics of growth-inhibiting material in the alcohol-soluble-extract from avocado seed coats. MSc thesis, Oklahoma State University, Stillwater, Oklahoma (cited by Bergh, 1988).

Platt, R.G. and Frolich, E.F. (1965) Propagation of avocados. In: *California Agricultural Experiment Station Extension Service, Circular 531*. University of California, Riverside CA. 19 pp.

Pliego-Alfaro, F. (1988) Development of an *in-vitro* rooting bioassay using juvenile-phase stem cuttings of *Persea americana* Mill. *Journal of Horticultural Science* 63, 295–301.

Pliego-Alfaro, F. and Murashige, T. (1987) Possible rejuvenation of adult avocado by graftage onto juvenile rootstocks *in vitro*. *HortScience* 22, 1321–1324.

Popenoe, W. (1920) The avocado. In: *Manual of Tropical and Subtropical Fruit*. Hafner Press, New York. Available at: https://ia801605.us.archive.org/31/items/manualoftropical00poperich/manualoftropical00poperich.pdf (accessed 8 April 2024).

Prabhakaran, A.D. and Dann, E.K. (2022) Evaluation of fungicide soil drench treatments to manage black root rot disease of avocado. *Plant Disease* 106, 2026–2030.

Ramírez Gil, J.G. (2018) Avocado wilt complex disease, implications and management in Colombia. *Revista Facaltad Nacional de Agronomía Medellín* 71(2), 8525–8541. Available at: doi:10,15446/rfna.v71n2.66465 (accessed 18 April 2024).

Raviv, M. and Reuveni, O. (1984a) Mode of leaf shedding from avocado cuttings and the effect of its delay on rooting. *HortScience* 19, 529–531.

Raviv, M. and Reuveni, O. (1984b) Endogenous content of leaf substance(s) associated with rooting ability of avocado cuttings. *Journal of the American Society for Horticultural Science* 109, 284–287.

Raviv, M., Reuveni, O. and Goldschmidt, E.E. (1986) Evidence for the presence of a native, non-auxinic rooting promoter in avocado (*Persea americana* Mill.). *Plant Growth Regulation* 4, 95–102.

Rendón-Anaya, M., Ibarra-Laclette, E., Méndez-Bravo, A., Lan, T., Zheng, C. *et al.* (2019) The avocado genome informs deep angiosperm phylogeny, highlights introgressive hybridization, and reveals pathogen-influenced gene space adaptation. In: *Proceedings of the National Academy of Sciences of the United States of America* 116, 17081–17089. Available at: doi:10.1073/pnas.1822129116 (accessed 3 March 2024).

Reuveni, O. and Goren, M. (1982) Production and rooting of base-etiolated avocado cuttings. In: *International Society of Horticultural Science. XXI International Horticultural Congress* Abstract No. 1348.
Reuveni, O. and Raviv, M. (1981) Importance of leaf retention to rooting of avocado cuttings. *Journal of the American Society for Horticultural Science* 106, 127–130.
Reyes-Herrera, P.H., Muñoz-Baena, L., Velásquez-Zapata, V., Patiño, L., Delgado-Paz, O.A. *et al.* (2020) Inheritance of rootstock effects in avocado (*Persea americana* Mill.) cv. Hass. *Frontiers of Plant Science* 11:555071. Available at: doi:10.3389/fpls.2020.555071 (accessed 14 January 2024).
Ruehle, G.D. (1963) *The Florida Avocado Industry*. Florida Agriculture Experiment Station Bulletin. Gainesville, Florida. 602.
Salazar-García, S. and Borys, M.W. (1983) Clonal propagation of the avocado through 'franqueamiento'. *California Avocado Society Yearbook* 67, 69–72.
Samish, R.M. and Gur, A. (1962) Experiments with budding avocado. *Proceedings of the American Society for Horticultural Science* 81, 94–201.
Sauls, J.W. and Campbell, C.W. (1980) Avocado seed germination studies. *Proceedings of the Florida State Horticultural Society* 93, 153–154.
Schroeder, C.A. (1965) Avocado and other subtropical fruits of New Zealand. *California Avocado Society Yearbook* 49, 29–34.
Schroeder, C.A. (1971) The response of avocado pericarp tissue to temperature and light *in vitro*. *California Avocado Society Yearbook* 54, 85–89.
Schroeder, C.A. (1977) Longevity of plant tissue cultures. *California Avocado Society Yearbook* 56, 138–141.
Schroeder, C.A. (1980) Avocado tissue *in vitro*. *California Avocado Society Yearbook* 64, 139–141.
Schroeder, C.A. and Frolich, E.F. (1955). Avocado rootstock studies. *California Agriculture*, 11–12.
Schroeder, C.A., Kay, E. and Davis, L.H. (1962) Totipotency of cells from fruit pericarp tissue *in vitro*. *Science* 138, 595–596.
Scora, R.W., Wolstenholme, B.N. and Lavi, U. (2002). Taxonomy and botany. In: Whiley, A.W., Schaffer, B. and Wolstenholme, B.N. (eds) *The Avocado Botany Production and Uses*. CAB International, Wallingford, UK, pp. 15–37.
Shindre, P., Yallesh Kumar, H.S., Narayana, S. Mavarkar, S., Ganapathi, M. and Chaitanya, H.S. (2002) Effect of viability on seed germination, growth and graft-take in avocado (*Persea americana* Mill.) under hill zone of Karnataka. *Biological Forum* 14(4), 803–806.
Smith, L.A., Dann, E.K., Pegg, K.G., Whiley, A.W., Giblin, F.R., Doogan, V. and Kopittke, R. (2011) Field assessment of avocado rootstock selections for resistance to Phytophthora root rot. *Australasian Plant Pathology* 40, 39–47.
Solares, E., Morales-Cruz, A., Balderas, R.F., Focht, E., Ashworth, V.E.T.M. *et al.* (2023) Insights into the domestication of avocado and potential genetic contributors to heterodichogamy. *G3 Genes|Genomes|Genetics* 13(2), jkac323. Available at: doi:10.1093/g3journal/jkac323 (accessed 12 October 2023).
Spalding, D.H., Knight, R.J. and Reeder, W.F. (1976) Storage of avocado seed. *Proceedings of the Florida State Horticultural Society* 89, 257–258.
Storey, W.B. (1974) What kind of fruit is the avocado? *California Avocado Society Yearbook* 57, 70–71.
Storey, W.B., Bergh, B.O. and Zentmyer, G.A. (1986) The origin, indigenous range and dissemination of the avocado. *California Avocado Society Yearbook* 70, 127–133.
Teague, C.P. (1966) Avocado tip-grafting. *Proceedings of the International Plant Propagators' Society* 16, 50–51.
Thomas, G. (1997) Rootstock influence on yield of 'Hass' avocado. In: *Proceedings of the Australian Avocado Growers' Federation and the New Zealand Avocado Growers' Association Conference '97, 'Searching for Quality'*. Rotorua, New Zealand, pp. 138–146.
Tito de Morais, C., Kettle, C.J., Philipson, C.D., Maycock, C.R. and Burslem, D.F.R.P. *et al.* (2019) Exploring the role of genetic diversity and relatedness in tree seedling growth and mortality: a multispecies study in a Bornean rainforest. *Journal of Ecology* 108, 1174–1185. Available at: doi:10.1111/1365-2745.13331 (accessed 3 March 2024).
Toerien, J.C. (1977) Production of avocado trees in the nursery. *South African Avocado Growers' Association Proceedings of the Technical Committee* 1, 21–22.
Traub, H.P. and Auchter, E.C. (1933) Propagation experiments with avocado, mango and papaya. *Proceedings of the American Society for Horticultural Science* 30, 382–386.
Van den Berg, N., Swart, V., Backer, R., Fick, A., Wienk, R., Engelbrecht, J. and Prabhu, S.A. (2021) Advances in understanding defense mechanisms in *Persea americana* against *Phytophthora cinnamomi*. *Frontiers in Plant Science* 12:636339. Available at: https://doi.org/10.3389/fpls.2021.636339 (accessed 23 October 2024).

Van der Meulen, A. (1952) The modified Forkert method of budding. *Farming in South Africa* 27, 289–290.

Wahl, S., Ryser, P. and Edwards, P.J. (2001) Phenotypic plasticity of grass root anatomy in response to light intensity and nutrient supply. *Annuals of Botany* 88, 1071–1078.

Wallace, A. and North, C.P. (1957) The Scott avocado. *California Avocado Society Yearbook* 41, 82–84.

Wallace, J.M. and Drake, R.J. (1972) Report on the programme for production of avocado nursery trees free of sunblotch. *California Avocado Society Yearbook* 55, 120–123.

Wang, J., Jiang, L. and Wu, R. (2017) Plant grafting: how genetic exchange promotes vascular reconnection. *New Phytologist* 214, 56–65.

Webster, T. (2002) Dwarfing rootstocks: past, present and future. *The Compact Fruit Tree* 35, 67–72.

Whiley, A.W. (2003) Rootstock incompatibility – does it affect tree performance? *Talking Avocados* 14(2), 13–15.

Whiley, A. and Anderson, G. (2002) *Study tour of South Africa, New Zealand, the UK and the USA (Also Incorporating Chile)*. Milestone Report for AV01007 – Rootstock Improvement for the Avocado Industry. Horticultural Australia Ltd, Sydney, New South Wales, Australia, pp. 1–32. Available at: https://www.avocado.org.au/wp-content/uploads/2016/12/AV01007-Study-Tour-of-South-Africa-NZ-UK-USA-and-Chile.pdf (accessed 18 March 2024).

Whiley, A.W. and Whiley, D.G. (2005) Rootstock improvement for the Australian avocado industry – a preliminary report. In: *New Zealand and Australia Avocado Growers' Conference '05*. Tauranga, New Zealand 10 pp.

Whiley, A.W. and Whiley, D.G. (2011) Australian rootstock program update 2011. In: *Proceedings of the VII World Avocado Congress*. Cairns, Queensland, Australia, 5–9 September 2011 (abstract). Available at: https://www.avocadosource.com/wac7/Section_14/WhileyAnthony2011.pdf (accessed 4 October 2012).

Whiley, A., Giblin, F., Pegg, K. and Whiley, D. (2007) Preliminary results from avocado rootstock research in Australia. In: *Proceedings of the VI World Avocado Congress*. Viña Del Mar, Chile. Available at: https://www.avocadosource.com/wac6/en/extenso/1a-7.pdf (accessed 4 October 2012).

Whiley, A.W., Coates, L.M., Hofman, P.J., Dann, E.K., Pegg, K.G. *et al.* (2013) *Rootstock Improvement for the Australian Avocado industry – Phase 3*. Horticulture Australia Ltd Project No. AV08000 pp. 154. Available at: https://www.avocado.org.au/wp-content/uploads/2016/12/AV08000-Rootstock-Improvement-for-the-Australian-Avocado-Industry-Phase-3.pdf (accessed 8 February 2024).

Whiley, T. and Dann. L. (2021) Procado: a new Australian avocado rootstock. *Talking Avocados* 32(1), 57–64.

Whiley, T. and Lavers, D. (2021) Finding a home for Shepard: the Australian story. *Talking Avocados* 32(1), 31–36.

Whitcomb, C.E. (1998) *Plant Production in Containers*. Revised edn. Lacebark Publications, Stillwater, Oklahoma.

Whitsell, R.H., Martin, G.E., Bergh, B.O., Lypps, A.V. and Brokaw, W.H. (1989) *Propagating Avocados: Principles and Techniques of Nursery and Field Grafting*. University of California, Division of Agriculture and Natural Resources Publication 21461, Oakland, California, 30 pp.

Willingham, S.A., Pegg, K.G., Cooke, A.W., Coates, L.M., Langdon, P.W.B. and Dean, J.R. (2001) Rootstock influences postharvest anthracnose development in 'Hass' avocado. *Australian Journal of Agricultural Research* 52, 1017–1022.

Wolstenholme, B.N. (2003) Avocado rootstocks: what do we know; are we doing enough research? *South African Avocado Growers' Association Yearbook* 26,106–112.

Wolstenholme, B.N. and Whiley, A.W. (1999) Ecophysiology of the avocado (*Persea americana* Mill.) tree as a basis for pre-harvest management. *Revista Chapingo Serie Horticultura Núm* Especial V, 77–88.

Young, L.B. (1961) Vegetative propagation in avocado by means of marcotage and the rooting of cuttings. *California Avocado Society Yearbook* 45, 63–66.

Zentmyer, G.A., Thorn, W.A., Paulus, O.A. and Burns, R.M. (1958) Hot water treatment of avocado seed. *California Avocado Society Yearbook* 42, 108–110.

10 Biotechnology

F. Pliego-Alfaro[1,*], J.I. Hormaza[1], C. Pliego[2], R. Folgado[3], A. Barceló-Muñoz[2], R. López-Gómez[4], J.A. Mercado[1], I. Narváez[1], E. Palomo-Ríos[1] and R.E. Litz[5]

[1]Instituto de Hortofruticultura Subtropical y Mediterránea 'La Mayora', Universidad de Málaga-Consejo Superior de Investigaciones Científicas (IHSM La Mayora-UMA-CSIC), Spain; [2]IFAPA, Centro de Churriana (Málaga), Spain; [3]The Huntington Library, Art Museum and Botanical Gardens, San Marino, California, USA; [4]Laboratorio de Fisiología Molecular de Plantas, Universidad Michoacana de San Nicolás de Hidalgo, Mexico; [5]Tropical Research & Education Center, University of Florida, USA

Abstract

This chapter reviews biotechnological approaches that could be useful for developing avocado rootstocks with enhanced resistance to biotic and abiotic stresses. This chapter also discusses the use of molecular approaches to understand the mechanisms involved in gene regulation and expression necessary to obtain transgenic plants of horticultural value.

Avocado breeding is slow due to heterozygosity and the long juvenile period of this species (Pliego-Alfaro and Bergh, 1992); however, the variability of desirable horticultural traits is almost limitless. Breeding programmes for scion selections have focused on improved fruit yield and quality, whereas for rootstock selections, the focus has been on enhanced resistance to biotic and abiotic stresses and/or reduced tree size. The use of seedlings as avocado rootstocks has been linked to a lack of uniformity in orchards with respect to production, vigour and/or tolerance of soil-borne pathogens or adverse soil conditions (Ben-Ya'acov and Michelson, 1995). Phytophthora root rot (PRR), caused by *Phytophthora cinnamomi* Rands, is a very serious problem in avocado orchards, and much effort has focused on developing rootstocks that are tolerant to this pathogen (Bergh *et al.*, 1976; Kellam and Coffey, 1985; Köhne, 1992; Kremer-Köhne and Mukhumo, 2003; Kremer-Köhne *et al.*, 2011). Attempts have also been made to identify germplasm with tolerance to saline and calcareous soils (Kadman and Ben-Ya'acov, 1980; Kadman, 1985). Rootstock selections can be vegetatively propagated by the 'nurse' seed/etiolation method (Frolich, 1951; Frolich and Platt, 1972; Brokaw, 1988; see Chapter 9, Propagation, in this volume), a relatively expensive and time-consuming procedure.

Biotechnological approaches could be useful for developing avocado rootstocks because (i) rootstock selections could be micropropagated at theoretically low cost; (ii) identification of somaclonal variants with increased tolerance of PRR (van den Bulk, 1991) would be possible, especially since resistance to *P. cinnamomi* at the cellular level appears to be similar to that at the whole-plant level (Phillips *et al.*, 1991); and (iii) disease resistance could be enhanced by transformation with various classes of genes,

*Correspondence: ferpliego@uma.es

 The Avocado: Botany, Production and Uses (eds D. Carrillo, B. Schaffer, A.W. Whiley and B.N. Wolstenholme)
DOI: 10.1079/9781800621824.0010

for example defensins, genes for pathogenesis-related proteins, and ribonucleases. In addition, recovery of cultivars with fruit showing extended shelf life and on-the-tree storage is feasible. Finally, embryo rescue by *in vitro* culture could be a very useful tool to recover genotypes from selected crosses in conventional breeding programmes.

Molecular tools are important in taxonomic and systematic studies (Chanderbali *et al.*, 2001; Rendón-Anaya *et al.*, 2019; Berdugo-Cely *et al.*, 2023). In addition, many genes involved in the regulation of specific horticultural traits, for example, fruit ripening (Smith *et al.*, 1988; Botella, 2002; Ibarra-Laclette *et al.*, 2015) or resistance to fungal diseases (Broglie *et al.*, 1991; Stuiver, 2006; Backer *et al.*, 2015; Zumaquero *et al.*, 2019), have been identified. Transformation with these genes will allow specific changes to be made in the genome while maintaining the genetic integrity of the clone (Schuerman and Dandekar, 1993). Understanding the mechanisms involved in gene regulation and expression is necessary to obtain transgenic plants of horticultural value (Jorgensen, 1993). The number of fruit and nut crops regenerated by cell cultures has increased considerably in the last decade (Litz *et al.*, 2020). In avocado, different genes have been introduced into embryogenic cells, generally derived from immature embryos, and some regenerants have been obtained; however, more efforts are needed to improve the conversion rate of somatic embryos.

In Vitro Morphogenesis

Avocado plants have been recovered *in vitro* via shoot culture and somatic embryogenesis. The former approach has great utility as an alternative method for vegetatively propagating plant material, particularly new rootstock selections, when a limited number of copies are needed. Hiti-Bandaralage *et al.* (2017) described problems encountered in developing methodologies that could be used commercially; however, a protocol has been developed using meristematic tissue as starting material (Hiti-Bandaralage *et al.*, 2019). Somatic embryogenesis, involving regeneration from single cells, is critical for genetic manipulation using genetic transformation and *in vitro* mutation induction.

Shoot Culture

Multiplication of plants by stimulating axillary bud proliferation is the most reliable micropropagation method for maintaining the genetic integrity of elite selections (George, 1993). Cultures are initiated from explants with intact meristems, and newly formed shoots are used as propagules. In woody species such as avocado, morphogenic competence is generally greater in juvenile than in adult material (Favre and Juncker, 1987; Pliego-Alfaro and Murashige, 1987; Arrillaga *et al.*, 1991). Micropropagation of many woody horticulturally important species has been addressed using juvenile material in order to establish guidelines for the multiplication of explants of adult origin.

Juvenile material

Experiments involving shoot tips or nodal stem segments of avocado seedlings and related species have demonstrated the importance of salt formulation of the basal medium for survival and shoot proliferation. Murashige and Skoog (1962) (MS) formulation causes scaly leaves (González-Rosas and Salazar-García, 1984; González-Rosas *et al.*, 1985), leaf tip burn (Witjaksono, 1991), shoot apical necrosis (Pliego-Alfaro, 1981; Witjaksono, 1991; Nhut *et al.*, 2008) and leaf abscission (Pliego-Alfaro, 1981; Witjaksono, 1991), all of which are similar to ammonium toxicity symptoms in avocado (Lovatt, 1987). Satisfactory shoot proliferation and survival can be obtained when the plant growth medium contains lower salt concentrations than MS medium. Nel *et al.* (1983) and Campos and Pais (1996) used MS formulation with half-strength macroelements to obtain shoot elongation and limited proliferation from nodal explants of *Persea indica*. Half-strength MS salts have also been used to promote axillary bud growth (Vega-Solórzano, 1989; Biasi *et al.*, 1994). Diluted MS nutrient solution has also been recommended for avocado by Barceló-Muñoz *et al.* (1990), while Cooper (1987) preferred Woody Plant Medium (WPM) (Lloyd and McCown, 1980). A 100% survival rate has been obtained on MS medium containing 20 mM KNO_3 without NH_4NO_3 (Witjaksono, 1991;

Witjaksono *et al.*, 1999a), whereas concentrations of KNO_3 > 20 mM resulted in less growth. Optimum growth has been achieved on a modified MS medium containing 20 mM KNO_3 and 10 mM NH_4NO_3 (Witjaksono *et al.*, 1999a). Nhut *et al.* (2008) found that 2% peptone added to basal MS medium stimulated shoot formation and elongation.

Benzyladenine (BA) in the range 4.44–8.88 μM is the most suitable cytokinin for proliferation of juvenile shoots of *P. indica* (Nel *et al.*, 1983) and *Persea americana* (Nel *et al.*, 1983; Cooper, 1987; Barceló-Muñoz *et al.*, 1990; Witjaksono *et al.*, 1999a). BA is superior to thidiazuron (TDZ) since the latter stimulates bud proliferation without shoot elongation from explants (Mohamed-Yasseen, 1993; Barringer *et al.*, 1996). Since the presence of BA during successive subcultures can cause miniaturization of shoots, an elongation phase, for example, subculture in lower BA concentrations in liquid media (Barceló-Muñoz *et al.*, 1990) or in combination with auxin (Cooper, 1987), has been used prior to rooting. According to Barceló-Muñoz *et al.* (1990), to avoid hyperhydricity, propagules should not be maintained for more than 2 weeks in a liquid medium. Leaves of hyperhydric shoots are thicker than those on normal shoots, showing an accumulation of carbohydrates and a decrease in peroxidase activity (Barceló-Muñoz, 1995). Hyperhydric shoots cannot survive *ex vitro*.

Kane *et al.* (1989) multiplied seedling-derived *Persea palustris* shoots in liquid MS medium, and plantlets were acclimatized. De la Viña *et al.* (2001) studied the effects of medium texture, sucrose concentration and solid versus double phase in combination with irradiance (35–85 μmol m^{-2} s^{-1}) on avocado shoot quality during proliferation. Culture in double-phase medium caused hyperhydricity of microcuttings, deformed stomata and poor development of epicuticular waxes. Increasing the irradiance level decreased the leaf chlorophyll and carotenoid contents but did not affect hyperhydricity. De la Viña *et al.* (1999) reported that avocado plantlets growing in high atmospheric CO_2 (100 Pa), high irradiance (85 μmol m^{-2} s^{-1}) and 14.61 mM sucrose had increased ribulose bisphosphate carboxylase (Rubisco) activity. Photosynthetic activity and growth rate were enhanced when compared with plantlets in the presence of 87.64 mM sucrose; however, their survival during acclimatization (70%) was not improved compared with plants grown on high sucrose. Witjaksono *et al.* (1999a) observed that proliferating avocado shoot cultures and plantlets were photosynthetically active on a medium supplemented with 30 g l^{-1} sucrose and 4.44 μM BA with a 16 h photoperiod (120–150 μmol m^{-2} s^{-1}). The net CO_2 assimilation rates of shoots and plantlets in an ambient CO_2 environment were 17 ± 2 and 31 ± 7 μmol CO_2 m^{-2} s^{-1}, respectively. When measured at ambient atmospheric CO_2 concentration, the net CO_2 assimilation rates for shoots and plantlets were higher for plants grown in ambient CO_2 than for those grown in a CO_2-enriched environment, indicating reduced photosynthetic efficiency of shoots and plantlets grown in an enriched atmospheric CO_2 environment. The net CO_2 assimilation rates of *in vitro*–grown plantlets were comparable to those of seedlings *ex vitro*.

In vitro shoots of avocado and related species can be rooted in media containing auxin. Examples include 36.9–49 μM indole-3-butyric acid (IBA) with 50% rooting (González-Rosas and Salazar-García, 1984), 4.9–9.8 μM IBA with rooting frequency of 30% (Barringer *et al.*, 1996); 9.8 μM IBA with 65% rooting of *P. indica* shoots in liquid medium with a filter paper bridge (Nel *et al.*, 1983), 9.8 μM IBA with 13.9 μM kinetin to root shoots of *Persea schiedeana* (González-Rosas *et al.*, 1985) or 16 μM naphthalene acetic acid (NAA) alone (Cooper, 1987) or in the presence of 2% peptone (Nhut *et al.*, 2008) for a rooting frequency of 100%.

Pliego-Alfaro (1988) found that two-step rooting resulted in 100% rooting frequency of shoot tips derived from *in vitro* seedlings. This procedure includes an induction step in which shoots are cultured on a medium containing 111 μM IBA for 3 days, followed by a development step in which the shoots are transferred to 33% strength MS medium with 1 mg l^{-1} activated charcoal but without auxin. Witjaksono *et al.* (1999a) also obtained high rooting frequency using this procedure with seedling-derived shoots, although Biasi *et al.* (1994) obtained only 45% rooting in shoots of 'Ouro Verde'. Variants of the induction step, including a pulse with 4.92 μM IBA (Cooper, 1987; Barceló-Muñoz *et al.*, 1990), can enhance rooting. A quick dip for 1–2 s in 4.9–24.6 μM IBA resulted in 100% rooting of *P. indica* shoots (Campos and Pais, 1996). In some cases, the auxin exposure should be carried out in darkness (de la Viña *et al.*, 1996). Activated charcoal can promote rooting when used in the auxin-free phase (Pliego-Alfaro, 1988; Barceló-Muñoz *et al.*, 1990). Successful rooting *ex vitro* has also been reported by Cooper (1987): a quick immersion in a solution of 16 mM NAA followed by transplanting the shoots into a fine pumice-peat substrate. A quick dip for 5 min in 245 μM IBA followed by transfer of shoots into plug trays with Vegro Klay Mix A resulted in 100% rooting of *P. palustris* (Kane *et al.*, 1989).

Using juvenile shoots, Premkumar *et al.* (2002, 2003) studied the effect of sucrose in the culture medium on different physiological parameters of rooting and acclimatization. Changes in sucrose levels did not affect chlorophyll or carotenoid contents; however, increasing the sucrose concentration caused a decrease in Rubisco during rooting that was even more pronounced during the acclimatization phase, suggesting a possible role of Rubisco as a source of reduced nitrogen (Premkumar *et al.*, 2002). Lower sucrose levels observed in the leaves were linked to increased starch content, indicating improved sucrose metabolism and starch biosynthesis during acclimatization (Premkumar *et al.*, 2003).

Generally, survival of juvenile avocado material during acclimatization has been acceptable (ca. 80%) (Cooper, 1987). However, inoculation with mycorrhizal fungi, such as *Glomus fasciculatum* or *Glomus deserticola*, can improve root and shoot growth, enhance the shoot/root ratio and increase the content of N, P and K, enabling plants to tolerate the stress caused by transplanting (Azcón-Aguilar *et al.*, 1992; Vidal *et al.*, 1992; de la Viña *et al.*, 1996).

Adult material

Early attempts to micropropagate adult phase shoots were generally unsuccessful. Harty (1985) used modified MS salts containing reduced ammonium and elevated nitrate concentration (Dixon and Fuller, 1976) and obtained 90% survival and sixfold multiplication at 9-week intervals, even though tissue necrosis was not completely eliminated and shoots failed to root. Cooper (1987) indicated that etiolation of mother plants or incubation of explants for 24 h under continuous light enhanced bud break; however, proliferation of shoots did not occur. Zirari and Lionakis (1994) used material derived from etiolated cuttings for culture establishment but demonstrated that etiolated explants did not have positive effects on 'Topa-Topa', 'Fuerte' or 'Hass', although it did improve sprouting of buds from 'Duke 7'. Attempts to root microshoots were unsuccessful due to severe shoot necrosis following transfer to rooting medium. An etiolation pretreatment of mother plants has been reported to improve culture establishment of 'Hass' explants (Restrepo-Osorio *et al.*, 2018). Hiti-Bandaralage *et al.* (2015), using nodal sections of 'Velvick', reported a synergistic action between meta-topolin and gibberellic acid (both at 0.1 mg l^{-1}), resulting in well-developed shoots of good quality; however, subsequent proliferation of this material was not reported.

Zulfiqar *et al.* (2009) working with 'Fuerte' obtained positive results with modified MS medium (75% macro and micro salts) and 1 mg l^{-1} BA during the proliferation phase (4.8 shoots per explant) or 1 mg l^{-1} IBA for rooting (53% rooting). Generally, successful micropropagation from adult trees requires selection of explants that have some degree of tissue juvenility, as a result of either hand pruning or partial rejuvenation by grafting adult tissue onto juvenile seedlings.

Pliego-Alfaro *et al.* (1987) established cultures of two rootstocks, 'IV-8', a highly productive 'Topa-Topa' seedling, and 'GA-13', a selection with tolerance to saline and calcareous soil conditions (Kadman and Ben-Ya'acov, 1980). Explants were from actively growing shoots. The 'IV-8' rootstock tree was forced to sprout after heavy pruning, while adult 'GA-13' scions were grafted onto seedling rootstocks and forced to sprout in the greenhouse. With both selections, reduced MS macroelement concentrations and 1.3–4.44 μM BA enhanced bud sprouting and shoot growth.

Schall (1987) used relatively high BA dosages (22 μM) to stimulate proliferation of adult 'Fuerte' explants derived from 2- to 4-year-old grafted trees; however, after several subcultures, significant oxidation and browning caused death of the propagules. Capote *et al.* (2000)

induced bud sprouting and shoot growth of 'Duke', using a double-phase medium with 8.88 μM BA and 5.77 μM gibberellic acid (GA_3); however, growth decreased with subculturing.

Although apical necrosis is a problem during shoot proliferation, adult shoots of 'GA-13' could be maintained for more than 1 year on medium with 4.44 μM BA with weekly subcultures. Apical necrosis was overcome in 'IV-8' rootstock using the double-phase medium (Pliego-Alfaro *et al.*, 1987); however, after several subcultures, severe hyperhydricity occurred. The combined use of liquid medium on a rotary shaker (2 weeks) and double-phase medium (6 weeks) lowered hyperhydricity to 20% in proliferating 'IV-8' shoots. Replacement of MS salts with B5 medium (Gamborg *et al.*, 1968) resulted in more vigorous and elongated shoots with no leaf necrosis, which could be continuously maintained in active proliferation (Fig. 10.1) with a threefold multiplication rate (Barceló-Muñoz *et al.*, 1999). Hiti-Bandaralage *et al.* (2019) have emphasized the importance of meristematic tissue for culture establishment; these authors were able to obtain up to 500 clonal shoots from a single shoot tip of 'Reed', establishing the basis of a commercial micropropagation protocol for avocado. Hiti-Bandaralage *et al.* (2020) found no differences in miRNA156 and miRNA172 levels, involved in juvenile-to-adult phase change, between mature shoot tips and micropropagated material of 'Velvick', following acclimatization in greenhouse conditions. These authors concluded that *in vitro* culture does not induce rejuvenation in adult material; however, a transient increase in miRNA156 was observed during the elongation phase, which could be due to the hormonal components of the culture medium.

Fig. 10.1. Adult avocado shoots of the 'IV-8' rootstock at the proliferation stage in double-phase medium.

Although rooting capacity of adult avocado shoots is very low and does not increase with subculturing (Pliego-Alfaro *et al.*, 1987), restoration of rooting competence can be achieved by grafting adult buds onto *in vitro* germinated seedlings (Pliego-Alfaro and Murashige, 1987; Barceló-Muñoz, 1995) or after successive severe prunings of mother plants (Barceló-Muñoz *et al.*, 1999). With 'GA-13', the percentage of rooting (about 90%) was similar to that of juvenile material after 16 successive *in vitro* grafts. The restored rooting capacity appeared to be a stable trait since it was constant for nine subcultures (Barceló-Muñoz, 1995). Working with 'IV-8' rootstock, Barceló-Muñoz *et al.* (1999) obtained > 90% rooting after successive pruning over several years.

Rooting is affected by medium texture (> 70% in liquid medium vs 30% in semi-solid medium) and by the salt formulation of the medium in which the shoots proliferated prior to rooting. Shoots that proliferated in half-strength MS medium rooted significantly less (74%) than those in B5 medium (90%) (Barceló-Muñoz *et al.* 1999). Histological analysis of two genotypes, differing in their rooting capacity, revealed a positive correlation between root formation and the presence of a continuous fascicular cambium ('Reed'), while abundance of phloem fibres ('Velvick') could have a negative effect on rooting (Hiti-Bandaralage *et al.*, 2022). After optimization of culture and environmental conditions, up to 100% rooting has been reported for 'Reed' (Hiti-Bandaralage *et al.*, 2019).

Acclimatization of rooted adult shoots has occurred in a peat-moss substrate with 70% survival after 8 weeks (Fig. 10.2) (Barceló-Muñoz *et al.*, 1999). More than 96% plantlet survival following acclimatization was observed by Hiti-Bandaralage *et al.* (2018) with 'Reed', using either of two methods: (i) 2–3-day root soaking followed by a 4-week dome-cover period for plantlets transplanted in soil; and (ii) 4-week in-jar hardening (including 2 weeks under dome cover and 2 weeks open dome) followed by transplanting to soil.

An alternative procedure for propagating adult avocado material involves the use of nodal sections and is useful when a limited number of copies of specific genotypes are required. Barceló-Muñoz *et al.* (2001) micropropagated seven seedling rootstocks showing the partial tolerance of *Rosellinia necatrix*, while Cortés-Rodríguez *et al.* (2011) micropropagated adult Mexican race trees. The procedure involved an establishment phase in B5 medium supplemented with 0.3 mg l^{-1} BA (Barceló-Muñoz *et al.*, 2001) or in MS medium supplemented with 0.5 mg l^{-1} BA and 0.1 mg l^{-1} IBA (Cortés-Rodríguez *et al.*, 2011). After 6–8 weeks, axillary buds produce shoots that can be excised, rooted *in vitro* and acclimatized. Nodal sections with basal buds can be recultured in fresh medium for continuous shoot production. To enhance rooting capacity, Barceló-Muñoz *et al.* (2001) recommended the use of successive pruning of recalcitrant genotypes. Using this approach and following the protocol of Barceló-Muñoz *et al.* (1999), *in vitro* rooting rates in the 40–100% range could be obtained (Table 10.1).

Fig. 10.2. Micropropagated adult avocado shoots of the 'IV-8' rootstock at the acclimatization phase.

The protocol developed by Hiti-Bandaralage *et al.* (2019) with 'Reed' based on the use of meristematic tissues (shoot tips) appears to be commercially viable, although given the great variability found in avocado, extension to other genotypes will require optimization of medium and culture conditions for each genotype. Thorne *et al.* (2022) evaluated the industry's acceptance of micropropagated material and reported a favourable reception by Australian growers.

Embryo Culture

The avocado fruitlet has a high abscission rate. Therefore, very few fruit from controlled crosses remain on the tree, which is a serious obstacle in avocado breeding programmes (Alcaraz and Hormaza, 2014). *In vitro* culture of immature embryos allows plant recovery from interesting crosses that would otherwise be lost (Sharma *et al.*, 1996). The first attempt to rescue avocado embryos was described by Skene and Barlass (1983), who cultured immature embryos from abscised avocado fruit to recover genotypes obtained after hybridization. They used a liquid medium with diluted MS macroelements and 2.2 μM BA (M1); embryos needed to be at least 6 weeks old to survive, and ca. 30–50% of shoots formed roots. Therefore, shoots had to be grafted on greenhouse-grown seedlings to assure survival.

Perán-Quesada *et al.* (2005) studied the anatomical differentiation and storage product accumulation during the development of avocado zygotic embryos, and results were correlated with *in vitro* germination capacity. Histodifferentiation was completed 100 days after pollination (DAP) in 16–18-mm-long embryos, while maturation

Table 10.1. *In vitro* rooting percentages of avocado seedling rootstocks after successive prunings at ground level. Shoots were rooted according to the protocol of Barceló-Muñoz *et al.* (1999).

	Number of successive prunings at ground level		
Genotype	1	2	3
BG70	100% rooting		
BG194	98% rooting		
BG81	72% rooting		
BG71	No response	90% rooting	
BG181	No response	100% rooting	
BG183	No response	No response	100% rooting
BG189	No response	40% rooting	48% rooting

started 125 DAP with embryos 24–26 mm in length, and was linked to accumulation of starch granules and protein bodies. In previous studies, Sánchez-Romero *et al.* (2002) had shown that storage proteins began to accumulate in smaller embryos (7–8 mm in length), although most protein was found at the end of the growth period. Moreover, they indicated that storage proteins represent 83% of the total proteins in avocado embryos. When examining the germination capacity *in vitro*, in relation to the developmental stage, Perán-Quesada *et al.* (2005) observed that embryos 64–95 DAP showed a > 20% rooting capacity, which increased to > 60% between 125 and 204 DAP. Germination rates > 90% were observed with 305 DAP embryos linked to low moisture content and complete physiological maturity. Pliego-Alfaro (1988) also obtained nearly 100% plant recovery from excised embryo axes with a piece of cotyledon from mature avocado seeds on MS basal medium. Investigating the different germination requirements of the embryos at different developmental stages, Sánchez-Romero *et al.* (2007) found that very immature embryos (5 mm long) germinated better in liquid M1 medium, while 15-mm-long embryos responded better on semi-solid M1 medium, although cotyledon removal was necessary. These authors indicated that cotyledons may be involved in avocado embryo dormancy. *In vitro* germination of immature avocado embryos can be improved following a partial desiccation process (Sánchez-Romero *et al.*, 2003; Márquez-Martín *et al.*, 2007, 2017).

To increase germination capacity of very immature embryos, Márquez-Martín *et al.* (2009) developed a system whereby < 10-mm-long avocado embryos underwent an *in vitro* maturation phase. This included culture on B5 medium supplemented with Jensen's amino acids (Jensen, 1977), 88 mM sucrose and 6 g l^{-1} agar, prior to germination on M1 medium. They obtained a 65% germination rate and healthy and vigorous plants. Histological studies revealed that at the end of the *in vitro* maturation period, histo-differentiation had been achieved, and starch granules and proteins were abundant. This technique could be used as a reliable method for plant recovery in controlled crosses in avocado breeding programmes (Pliego-Alfaro and Litz, 2007).

Callus Cultures

In early studies, avocado callus was established from different explants, although there was no evidence of plant regeneration. Culture of cotyledons, etiolated stem sections, peduncles and leaf petioles on Nitsch medium (Nitsch, 1951; Schroeder, 1977), stem sections on MS medium with 1.3 μM BA and 5.7 μM indole-3-acetic acid (IAA) (Blickle *et al.*, 1986) or 10 μM BA and NAA (Phillips *et al.*, 1991) as well as stem and leaf sections on Anderson medium (Anderson, 1975) with 1 μM 2,4 dichlorophenoxyacetic acid (2,4-D) (Young, 1983) resulted in callus proliferation (Schroeder, 1977; Young, 1983) in complete darkness (Schroeder, 1971) or under low light (Young, 1983). Blumenfeld and Gazit (1971) compared the *in vitro* responses of cotyledon and mesocarp-derived calluses and reported that cotyledon callus synthesized its own cytokinin and could be grown on a medium without this hormone.

Growth requirements for callus proliferation from seedling-derived stem explants have been evaluated extensively for avocado and other *Persea* species, that is, *P. indica*, *P. borbonia* and *P. nubigena* (Aaouine, 1986). Generally satisfactory callus growth for all species was obtained on a medium consisting of half concentration of MS major salts except for $MgSO_4$, MS minor salts, vitamins with 1.4 μM 2,4-D and 4.9 μM 2-isopentenyladenine (2iP) (Aaouine, 1986). Cell suspensions could be established with 0.5 mg l^{-1} picloram from callus derived from 'Hass' avocado peel initiated on MS medium supplemented with 0.1 mg l^{-1} picloram (Prusky *et al.*, 1996).

Adventitious root development was observed after several subcultures of fruit pericarp-derived callus on Nitsch medium supplemented with 59 μM IAA (Schroeder *et al.*, 1962) or at the base of etiolated stem sections which had formed callus on Nitsch or MS medium with 29.4–147 μM 2iP (Schroeder, 1980). No histological studies were conducted to determine if the roots were derived from the callus or tissues of the original explant.

Adventitious shoots have been induced in the presence of 4.4 μM BA and 2.5 μM IBA. Explants were etiolated internodal sections from *in vitro* germinated 'Lula' avocado seedlings. The regeneration rate was about 6% with 1–2 regenerated shoots per explant (R. Perán-Quesada, IFAPA-Centro de Churriana, Málaga, Spain, 2001, personal communication).

Somatic Embryogenesis

Embryogenic avocado cultures have been induced in the presence of picloram (Pliego-Alfaro, 1981; Mooney and van Staden, 1987; Pliego-Alfaro and Murashige, 1988; Raviv *et al.*, 1998; Witjaksono and Litz, 1999a) from zygotic embryo explants of various sizes and developmental stages, ranging from globular (0.1 mm) to heart and cotyledonary (1.2 mm) (Mooney and van Staden, 1987; Pliego-Alfaro and Murashige, 1988; Witjaksono and Litz, 1999a). Raviv *et al.* (1998) induced embryogenic cultures from cotyledon segments excised from 7 to 10 mm embryos, while Suarez (2003) and Suarez *et al.* (2006) used nucellar explants. Witjaksono and Litz (1999a) demonstrated that an induction basal medium consisting of B5 major salts with MS minor salts and organic addenda was superior to MS medium. This medium was also utilized by López-Encina *et al.* (2014) to induce embryogenic cultures of four genotypes. Using small fruitlets (≤ 0.3 cm) and explanting the bisected ovules that contain zygotic embryos at pre-globular to globular stages, induction frequencies of about 25% can be achieved (Witjaksono, University of Florida, Tropical Research and Education Center, Homestead, Florida, 1997, personal communication).

Embryogenic avocado cultures consist of proembryonic masses (PEMs) and early stages of hyperhydric somatic embryos and first appear 8–25 days after explanting on induction medium. Although early reports described the newly initiated cultures as being embryogenic callus, it is clear in retrospect that they are typical embryogenic cultures that consist of proembryonic cells and PEMs. This is confirmed by the high-magnification photographs of all published reports and the histological preparations of Mooney and van Staden (1987), and later confirmed by Witjaksono and Litz (1999a,b) and Witjaksono *et al.* (1999b).

Embryogenic cultures have been maintained on induction medium based on MS basal formulation (MSP) (Pliego-Alfaro and Murashige, 1988) or on Dixon and Fuller medium supplemented with 3 mg l^{-1} each of isopentenyladenosine and benzothiazole-2-oxyacetic acid (Mooney and van Staden, 1987). The medium and long-term maintenance of embryogenic avocado cultures on semi-solid medium and in suspension was described in detail by Witjaksono and Litz (1999a). They observed that the proliferation of PEMs is optimum on a semi-solid MSP induction medium formulation. Maintenance on semi-solid medium requires subculture at 3–5 week intervals, and only the smallest PEMs are inoculated on to fresh medium.

Embryogenic culture maintenance is optimal in liquid MS medium that has been modified to contain 12 mg l^{-1} NH_4NO_3 and 30.3 mg l^{-1} KNO_3 (MS3:1P), which results in the highest culture fresh weight increase compared with other medium formulations (Witjaksono and Litz, 1999b); MS3:1P is supplemented with 30–50 g l^{-1} sucrose, 100 mg l^{-1} myo-inositol, 4 mg l^{-1} thiamine HCl and 0.1 mg l^{-1} picloram (Witjaksono and Litz, 1999a,b). Embryogenic suspension cultures are initiated by inoculating 0.5–1.0 g PEMs into 40 ml or 80 ml liquid MS3:1P medium in 125 ml or 250 ml Erlenmeyer flasks, respectively, maintained on a rotary shaker at about 100–125 rpm and subcultured at 2-week intervals. Newly initiated SE-type suspension cultures (see below) must be sieved, and only < 0.8 mm PEMs are used as inoculum for subculturing.

Witjaksono and Litz (1999a) recognized two types of embryogenic avocado cultures: (i) genotypes that proliferate as PEMs in the presence of auxin (PEM type); and (ii) genotypes in which heart and later stages of somatic embryos can develop in the presence of auxin (SE type). The latter response (SE type) appears to be more common. The description of SE and PEM types of embryogenic cultures has been confirmed by Márquez-Martín *et al.* (2012), who demonstrated that maturation depended not only on culture type but also on inoculum size and time in suspension. In the case of SE type, an initial inoculum of 0.4 g and 9 days of culture in 40 ml of MSP medium were optimal conditions for appearance of white-opaque embryos following transfer to maturation medium, while in the case of PEM-type cultures, an inoculum density of 4 g and 14 days culture were required. Somatic embryos and PEM-type cultures also differ in their protein profile (Guzmán-García *et al.*, 2013b). Witjaksono and Litz (1999a) demonstrated that the time required for loss of embryogenic potential is genotype-dependent, varying from 3 months for 'Yon' to >2 years for 'Esther'. Quintero-Jimenez *et al.* (2020) reported that embryogenic cultures of 'San Miguel' lost regeneration potential after 6 months on maintenance medium. Loss of embryogenic potential is associated with increasing disorganization of PEMs. Maintenance cultures on semi-solid medium are incubated in darkness at 25°C, whereas suspension cultures are incubated in semi-darkness.

Early reports indicated that somatic embryos develop following the transfer of embryogenic cultures onto semi-solid medium without picloram (Mooney and van Staden, 1987; Pliego-Alfaro and Murashige, 1988). The absence of picloram, however, is not a prerequisite for somatic embryo development, as somatic embryos of the SE-type cultures can develop in the presence of picloram (Witjaksono and Litz, 1999a). Raviv *et al.* (1998) also observed that cotyledonary-stage somatic embryos can develop on semi-solid proliferation medium supplemented with 9.04 μM 2,4-D and 2.22 μM BA. These somatic embryos are generally hyperhydrous and cannot develop to maturity. Somatic embryo development in avocado is genotype-dependent; SE-type cultures produce somatic embryos readily, whereas PEM-type cultures are less efficient in this respect (Witjaksono and Litz, 1999a,b, 2002; Márquez-Martín *et al.*, 2012). Witjaksono and Litz (1999b) demonstrated that various factors could affect the development of good-quality somatic embryos and recommended the use of the MS formulation, while better results have been reported by Perán-Quesada *et al.* (2004) following the use of B5 major salts. According to Perán-Quesada *et al.* (2004), MS formulation favours the formation of somatic embryos at early development stages, while maturity is enhanced by B5 major salts. Increasing the gelling agent and sucrose concentrations affects both the size and number of somatic embryos. The optimum response for recovery of high-quality somatic embryos occurs on medium supplemented with 6–7 g l^{-1} gellan gum and 90 g l^{-1} sucrose; however, this concentration of sucrose also suppresses somatic embryo development, and further embryo growth requires 30 g l^{-1} sucrose. Better results have been obtained using 10 g l^{-1} agar (Sigma A-1296) in comparison with 6–7 g l^{-1} gellan gum (Márquez-Martín *et al.*, 2011). High gellan gum concentrations together with high sucrose content in the maturation medium are effective for reducing the occurrence of hyperhydricity. Perán-Quesada *et al.* (2004) also observed a positive effect of sugar on embryo maturation, but Márquez-Martín (2007) indicated that the combined effects of high agar concentration and high sugar decreased embryo quality; embryos turned dark beige and their surface became irregular. Supplementing maturation medium with 20% (v/v) filter-sterilized coconut water has also been shown to improve the quality of recovered somatic embryos (Witjaksono and Litz, 2002; Perán-Quesada *et al.*, 2004; López-Encina *et al.*, 2014). Palomo-Ríos *et al.* (2013) combined the use of high agar and high sugar concentrations with 20% (v/v) coconut water, obtaining embryos with lower necrosis signals than those maturated without coconut water. Somatic embryos mature in darkness at 25°C. The effect of abscisic acid (ABA) has also been studied in avocado somatic embryo maturation, but its role is unclear; ABA seems to increase the number of globular translucent embryos (Perán-Quesada *et al.*, 2004). Desiccation of immature zygotic embryos of avocado under high relative humidity improved maturation and plant recovery (Márquez-Martín *et al.*, 2017); this could have important implications for controlling the maturation of avocado somatic embryos, although desiccation had a negative effect on germination of zygotic embryos maturated *in vitro*. An alternative approach was carried out by Palomo-Ríos *et al.* (2013) using cellulose acetate semi-permeable membranes as prematuration treatment to reduce water availability to the somatic embryos. Following this treatment, the number of matured embryos and their quality increased, showing less callus formation, fewer brown or necrotic areas and a more uniform white-opaque appearance. ABA content and water potential of these embryos were lower than those maturated without the membrane. The embryos obtained after this culture were small, and a two-step procedure involving maturation medium (MS with B5 major salts and 10 g l^{-1} agar) followed by transfer to a medium with a higher sucrose content (45 g l^{-1}) and 20% (v/v) coconut water was required to achieve a suitable size for germination.

Well-developed, opaque and mature somatic embryos (≥ 0.5 cm diameter) (Fig. 10.3) are transferred individually to semi-solid germination medium, which is similar to maintenance medium, but without picloram, and supplemented with 4.44 μM BA and 2.89 μM gibberellic acid (GA_3) (Witjaksono and Litz, 1999b). After 4–6 months on germination medium, only a few somatic embryos develop shoots or roots only or are bipolar. Shoot development from somatic embryos is generally ≤5% (Mooney and van Staden, 1987; Pliego-Alfaro and Murashige, 1988; Witjaksono and Litz, 1999b). Other reports, including Raviv *et al.* (1998), Witjaksono *et al.* (2009) and Avenido *et al.* (2009), indicated that 11%, 20% and 26.9%, respectively, of somatic embryos were able to form shoots. López-Encina *et al.* (2014) obtained up to 35% shoot sprouting following a two-step regeneration system which involves a culture period in liquid germination medium, supplemented with glutamine followed by transfer to solid medium of the same composition. A similar approach was followed by Palomo-Ríos *et al.* (2013) using somatic embryos prematurated over semi-permeable membranes, for example, culture of somatic embryos for 3 days in the liquid germination medium of Witjaksono and Litz (1999b), followed by the subsequent transfer to a solid medium of the same composition during three recultures of 5 weeks each (Fig. 10.4). This resulted in 40% germination. Very few somatic embryos (ca. 1%) developed both a shoot and a root. In fact, most mature avocado somatic embryos are not bipolar (Mooney and van Staden, 1987; Pliego-Alfaro and Murashige, 1988; Witjaksono and Litz, 1999b). Pliego-Alfaro and Murashige (1988) ascribed this anomaly of avocado somatic embryo development to failure of either the apical or root meristem to become organized.

To compensate for the low recovery of shoots from somatic embryos, Witjaksono *et al.* (1999a) and Perán-Quesada *et al.* (2004) demonstrated that shoot tips and nodal sections from somatic embryo shoots can be micropropagated (see above). This procedure can be utilized for producing a large number of shoots from the relatively few somatic embryos derived from protoplasts and embryogenic cultures, respectively (Witjaksono *et al.*, 1998, 1999b). Avocado shoots derived from somatic embryos can be individually rooted according to the two-step procedure of Pliego-Alfaro (1988). Individual shoots (1.5–2 cm long with 1–3 leaf primordia and non-expanded leaves) are pulsed for 3 days on medium supplemented with 122.6 μM IBA before subculture on basal medium in GA7 vessels. Culture conditions included a 16 h photoperiod (100–120 μmol m^{-2} s^{-1} provided by cool white fluorescent bulbs) at 25°C (Witjaksono

Fig. 10.3. Well-developed, white-opaque, mature avocado somatic embryos.

Fig. 10.4. Sprouted somatic embryo after culture in liquid germination medium for 3 days (Witjaksono and Litz, 1999a) followed by transfer to solid medium of the same composition during three recultures of 5 weeks each.

and Litz, 1999b); however, Perán-Quesada *et al.* (2004) recommend the use of a lower IBA concentration (4.92 μM) during the 3-day pulse to avoid excessive callus formation at the shoot base.

Raharjo and Litz (2005) and Avenido *et al.* (2009) demonstrated that somatic embryo shoots could be micrografted onto seedling rootstocks. The former obtained success rates of 59–100% with micrografts, depending on the scion genotype, and reported success rates of 52–76% when somatic embryo shoots were grafted *ex vitro* onto seedling rootstocks. Raharjo *et al.* (2008) successfully air-layered somatic embryo-derived shoots that had been grafted onto seedling rootstocks, thereby restoring the rooting ability of the avocado scion clones.

Palomo-Ríos *et al.* (2017) reported that plant recovery efficiency declines when transgenic somatic embryos are used. To increase the number of recovered shoots, white-opaque mature somatic embryos showing small sprouts and buds failing to emerge were cultured in MS medium supplemented with 4.44 μM BA and 4.55 μM TDZ after partial excision of a cotyledon. More than 50% of somatic embryos formed new shoots, although they were weak and small. These shoots were elongated in MS medium with 4.44 μM BA and subsequently micrografted onto *in vitro* germinated seedlings to induce vigorous growth prior to rooting and acclimatization. Micrografting success rate was in the range of 60–80%.

Protoplast Isolation, Culture and Plant Regeneration

Protoplasts are cells that have been deprived of their cell walls. Protoplasts have been used in breeding to obtain somatic hybrids for transferring useful traits from wild relatives to agronomic crops (Louzada *et al.*, 1993; Kumari *et al.*, 2020). Recently, protoplast technology has become crucial for applying CRISPR/Cas-mediated gene editing (Yue *et al.*, 2021). However, protoplast regeneration is difficult to achieve and may require development of genotype-specific protocols (Yue *et al.*, 2021); moreover, regeneration from single cells is frequently associated with a high frequency of somaclonal variation (Roest and Gilissen, 1993; Fossi *et al.*, 2019; Eeckhaut *et al.*, 2020).

Avocado protoplasts have been isolated from non-morphogenic callus for studying sunblotch viroid replication (Blickle *et al.*, 1986) and from fruit mesocarp tissue to study fruit ripening (Percival *et al.*, 1991). Witjaksono *et al.* (1998) demonstrated that protoplasts isolated from embryogenic cultures underwent somatic embryo development under appropriate conditions. Protoplasts are isolated by incubating 0.8–1.2 g of 8–14-day-old suspension cultures in 60 × 15 mm Petri dishes in a mixture of 1.5 ml enzyme solution and 2.5 ml avocado protoplast culture medium. The enzyme solution consisted of filter-sterilized 1% cellulase Onozuka RS, 1% Macerase R 10, 0.2% Pectolyase Y-23, 0.7 M mannitol and salts (Grosser and Gmitter, 1990). The protoplast culture medium (MS-8P) is a combination of avocado maintenance medium without picloram, in which NH_4NO_3 has been replaced by 3.1 g l^{-1} glutamine, and Kao and Michayluk (1975) organic addenda as modified by Grosser and Gmitter (1990). Medium osmolarity is maintained with 0.15 M sucrose and 0–0.55 M mannitol as needed. Protoplasts are purified following filtration through a 45-μm mesh stainless-steel screen and centrifuged at 100 rpm for 5 min. Following the removal of supernatants, protoplast pellets are further purified by gradient centrifugation, according to Grosser and Gmitter (1990). Embryogenic protoplasts ranged in size from 20 to 40 μm. Consistently high yields of > 3×10^6 protoplasts g^{-1} can be obtained from embryogenic cultures maintained in liquid medium for > 8 months (Witjaksono, University of Florida, Tropical Research and Education Center, Homestead, Florida, 2000, personal communication).

Avocado protoplasts can be cultured either in liquid (Witjaksono *et al.*, 1998) or in agarose-solidified medium (Witjaksono *et al.*, 1999b); however, the former is a more efficient and simpler procedure for recovery of somatic embryos. Protoplasts develop as microcalluses and PEMs in plant growth medium. The recovery of PEMs from protoplasts in liquid medium is dependent on medium osmolarity, source of nitrogen, plating density and interaction of osmolarity and nitrogen source. PEMs have been recovered in medium with 0.4 M osmolarity, while only microcalluses develop in medium with 0.6 M osmolarity. With 0.4 M medium, the replacement of NH_4NO_3 of MS formulation with 3.1 g l^{-1}

glutamine increases the frequency of occurrence of PEMs. A plating density of 0.8×10^5 ml^{-1} also results in more PEMs than a higher plating density of 1.6×10^5 ml^{-1}. Therefore, the optimum conditions for recovery of PEMs from protoplasts are provided by 0.4 M $MS^{-}8P$ with a protoplast density of 0.8×10^5 ml^{-1}.

Protoplasts are cultured in 2–3 ml liquid medium distributed in 60 × 15 mm sterile plastic dishes sealed with Nescofilm® and maintained in darkness at 25°C. Under these conditions, 5% of the protoplasts underwent the first division after 5 days of culture. The plating efficiency after 12 days was 25%, and PEMs were visible 14 days after plating. The number and mass of protoplast-derived PEMs depend on subculture age, dilution rate and their interaction. Protoplast-derived PEMs develop as somatic embryos on maturation medium, and mature somatic embryos have germinated, with a low frequency of conversion.

Protoplast isolation, culture and regeneration are genotype- and culture-age-dependent (Witjaksono, 1997). Cultures consisting of highly organized PEMs and somatic embryos yield only a few protoplasts that do not divide. Old cultures consisting of disorganized PEMs yield high numbers of protoplasts that form microcalluses without somatic embryo development. Cultures that have been maintained in liquid medium for at least 6 months are usually good sources for regenerable protoplasts.

A goal of avocado protoplast technology has been to create somatic hybrids between avocado and sexually and graft-incompatible *Persea* species that are resistant to PRR, for example, *P. borbonia*, *Persea cinerascens* and *Persea pachypoda*. Witjaksono (1997) addressed this objective by fusing protoplasts derived from embryogenic cultures with leaf mesophyll protoplasts from *in vitro*-grown *Persea* spp. plantlets, and with protoplasts isolated from non-morphogenic callus of the *Persea* spp. Somatic hybridization occurred between avocado and *P. cinerascens*, and somatic embryos having the rose colour of *P. cinerascens* were recovered (Fig. 10.5). However, the frequency of recovery was low, and the apical meristem failed to organize on the somatic hybrid embryos (Witjaksono, University of Florida, Tropical Research and Education Center, Homestead, Florida, 2000, personal communication).

Fig. 10.5. Somatic embryos of putative somatic hybrid of *P. americana* + *P. cinerascens*.

Induced Mutations

The exploitation of induced mutations for horticultural crop improvement has had a long and successful history (Joint FAO/IAEA, 2010), although it has largely been superseded by genetic transformation. According to the Joint FAO/IAEA Mutant Variety and Genetic Stock Database, there are currently ca. 3100 mutant varieties. If current animosity to genetically modified crops prevails in the EU and elsewhere, the directed *in vitro* selection for a specific trait of mutated embryogenic cultures can be a very efficient process for avocado improvement.

The selection 'D9', which shows some promise as a rootstock with moderate resistance to PRR, was derived from an irradiated 'Duke' seedling. De la Cruz *et al.* (1993, 1995, 1998) and Sánchez-Colín *et al.* (1990) described the irradiation of 'Hass' budwood and observed variability with respect to fruit set, flowering time, plant height and so on. Fuentes *et al.* (2009) irradiated zygotic embryos of 'Duke 7' at an LD_{50}

of ca. 25 Gy and micropropagated the resulting plants in an effort to recover material with tolerance to saline conditions. Screening for salt tolerance occurred at an LD_{20} of 157 mM NaCl. Coto *et al.* (2014) described a similar protocol for the recovery of salt-tolerant 'Duke 7', 'Catalina' and 'Hass' and reported preliminary studies to select for disease resistance by challenging tissues with the culture filtrate of *Phytophthora palmivora.*

Witjaksono and Litz (2004) irradiated embryogenic cultures of 'Fuerte' and 'T362' and determined an LD_{50} for both genotypes at 35 Gy. Later studies by Witjaksono *et al.* (2009) and Avenido *et al.* (2009) involved the irradiation of embryogenic cultures of 'Semil' and Indonesian dooryard trees followed by regeneration of plants. The work plan called for selection *in vitro* for resistance to the culture filtrate of *P. cinnamomi* and regeneration of putatively PRR-resistant plants. Evidence that resistance to PRR operates at the cellular level was provided by Phillips *et al.* (1991). These authors established callus from different avocado cultivars that differed in their susceptibility to PRR, for example, 'Topa-Topa' (susceptible), 'Duke- 7' (fairly resistant) and 'Martin Grande' (moderately resistant). *In vitro* responses following inoculation with *P. cinnamomi* were quite similar to those observed in the field. Moreover, although no clear differences in infection rate could be detected between the two tolerant cultivars, in both cases a hypersensitive type of reaction was observed; for example, rapid necrosis occurred in cells surrounding the inoculated area.

Working with the soil-borne pathogen *R. necatrix*, Moreno-Pérez *et al.* (2023) were able to select an embryogenic line of 'Anaheim' capable of surviving in the presence of 80% fungal culture filtrate (FCF). Following exposure of cells from this line and those of the susceptible control to 40% FCF, genes related with defence (lignin biosynthesis, pathogenesis-related proteins, proteinase inhibitors, etc.) showed higher expression in the resistant line in comparison to the susceptible control, demonstrating the feasibility of using cell cultures in avocado-pathogen interaction studies.

Micrografting and Nucellar Culture for Viroid Elimination

Avocado sunblotch viroid (ASBVd), a member of the Avsunviroidae family, causes sunblotch disease, a serious production problem of avocado in many areas. There is no resistance to the disease in the genus, and the viroid is spread by pollen, root grafting, mechanical injury and so on. ASBVd replicates in chloroplasts, unlike other pathogenic microorganisms. Viroids in the Pospiviroidae family replicate in the nucleus and can be eliminated by meristem culture (Durán-Vila *et al.*, 1988), micrografting (Navarro *et al.*, 1975) or nucellar culture (Bitters *et al.*, 1972). Suarez *et al.* (2005) micrografted excised meristems of ASBVd-infected 'Donaldson' and 'Vero Beach' on *in vitro* seedling rootstocks of ASBVd-free 'Simmonds' and 'Wilson Popenoe'. All scion plants that were recovered were assayed for the presence of ASBVd by RT-PCR; meristem culture was unsuccessful for eliminating the viroid. In a parallel study, Suarez *et al.* (2006) induced embryogenic cultures from the nucellus of ASBVd-infected 'Vero Beach'. The somatic embryo–derived plants were also assayed for the presence of the viroid by RT-PCR. All regenerants were ASBVd positive.

Genetic Transformation

Genetic transformation has the potential for addressing plant breeding objectives that are difficult to achieve by conventional methods. Advances in avocado transformation have been recently reviewed by Tamayo-Ramos *et al.* (2022). Primary goals have included: (i) enhanced resistance to diseases caused by oomycetes (*P. cinnamomi*), viroids (avocado sunblotch) and fungi (*R. necatrix*); and (ii) controlled ripening to improve shelf life by blocking ethylene biosynthesis. Transient transformation protocols can also be utilized for functional genomic studies.

a) *Agrobacterium rhizogenes*

Avocado transformation with *A. rhizogenes* resulted in the recovery of chimeric plants having a radicular system composed of transgenic and non-transgenic roots supporting a non-transformed aerial part. These composite plants are useful for functional genomic studies to understand avocado interaction with *P. cinnamomi*. Prabhu *et al.* (2017) obtained composite plants following inoculation of the shoot bases of *ex vitro* germinated seedlings with *Agrobacterium rhizogenes* ARqua1 strain, which included pBINUbiGUSint (GUS, β-glucuronidase) transformation vector. To enhance root formation, application of a rooting hormone was required at the inoculation site prior to air layering. Transgenic roots were highly branched, although the number of transformed roots on each plant was low.

b) *Agrobacterium tumefaciens*

The genetic transformation of avocado has been based on the well-defined protocol for regeneration from embryogenic suspension cultures. The growth of embryogenic suspensions of the PEM type can be suppressed by as much as 50% by 50 mg l^{-1} kanamycin sulfate, whereas 50% growth suppression on semi-solid medium requires 100 mg l^{-1} kanamycin sulfate (Cruz-Hernández *et al.*, 1998). Complete suppression of growth of embryogenic cultures occurs on semi-solid medium containing 200 mg l^{-1} kanamycin sulfate. In SE-type cultures derived from zygotic embryos of 'Duke 7', 50 mg l^{-1} kanamycin sulfate was sufficient for growth suppression >50% on semi-solid medium (Palomo-Ríos *et al.*, 2012). Moreover, when isolated somatic embryos were cultured on semi-solid selection medium, 12.5 g l^{-1} kanamycin impaired the growth of new embryos.

Cruz-Hernández *et al.* (1998) utilized a two-step selection procedure for recovery of genetically transformed embryogenic avocado cultures (PEM type) that were resistant to kanamycin and that expressed the *uidA* (β-glucuronidase) gene. Embryogenic cultures on semi-solid maintenance medium were gently abraded with a soft camel hairbrush and then inoculated with log phase acetosyringone-activated disarmed *Agrobacterium tumefaciens* strain 9749 ASE2, which harboured a cointegrate vector pMON9749 containing a selectable kanamycin sulfate-resistant marker gene (*nptII*) and the reporter gene *uidA*, both of which were driven by the CaMV35S constitutive promoter. The PEMs were co-cultured with *A. tumefaciens* in liquid maintenance medium for 3 days at 100 rpm. *Agrobacterium* was eliminated by incubating the cultures in maintenance medium supplemented with 200 mg l^{-1} cefotaxime and 50 mg l^{-1} kanamycin sulfate. An initial selection for antibiotic resistance in liquid maintenance medium containing 50 mg l^{-1} kanamycin sulfate for 2–4 months was followed by a second, more intensive selection with 100 mg l^{-1} kanamycin sulfate for 2 months. The cultures were finally transferred to maintenance medium containing 200 mg l^{-1} kanamycin sulfate. Somatic embryo development occurred on maturation medium without kanamycin sulfate, followed by subculture onto maturation medium containing 200 mg l^{-1} kanamycin sulfate. Transformed somatic embryos were recovered that stained positively for *uidA* in the X-GLUC reaction (Jefferson, 1987), and the integration of *npt*II and *uidA* into the avocado genome was confirmed by PCR and Southern hybridization. Transgenic plants were not regenerated in that study.

Inhibition of 'Hass' embryogenic suspension culture growth by phosphinothricin (the active ingredient of Basta® or Finale®) herbicide was reported by Raharjo *et al.* (2008), who observed that 3 mg l^{-1} phosphinothricin inhibited the growth of embryogenic cultures, and was utilized at this rate for selection of cultures transformed with the pGPTV-BPDF1.2. The construct contained the reporter gene *uidA*, the *bar* gene that confers resistance to phosphinothricin and the antifungal *pdf1.2* defensin gene, all of which were controlled by the CaMV35S constitutive promoter. A large fraction of embryogenic 'Hass' suspension cultures in their logarithmic growth phase were collected on sterile filtration fabric (1000 μm pore size), and ca. 300 mg was abraded with a sterile brush on sterile filter paper. The tissue was then co-cultured with log phase acetosyringone-activated *A. tumefaciens* strain EHA 105 with pGPTV-BPDF1.2 in 50 ml of liquid medium for 3 days. The PEMs were then transferred to liquid maintenance medium supplemented with 200 mg l^{-1} cefotaxime and 500 mg l^{-1} carbenicillin and after 2 weeks were subcultured into fresh liquid maintenance medium with 200 mg l^{-1} cefotaxime, 500 mg l^{-1} carbenicillin and 3 mg l^{-1} phosphinothricin for 3 months. Embryogenic cultures were selected continuously with phosphinothricin and stained periodically for *uidA* until they were apparently completely transformed. Transformed embryogenic cultures were transferred to somatic embryo development medium that contained 3 mg l^{-1} phosphinothricin. The shoots from somatic embryos transformed with *pdf1.2* were excised and grafted *in vitro* on to decapitated 'Peterson' seedling rootstocks in the nursery (Raharjo and Litz, 2005), and transformed scion shoots were later air-layered, and self-rooted plants were recovered (Raharjo *et al.*, 2008). 'Hass' plants were resistant to foliar application of the herbicide Finale® (Fig. 10.6).

A protocol for genotypes that show poor proliferation in liquid media was developed by Palomo-Ríos *et al.* (2012). These authors inoculated isolated 1–2-mm globular somatic embryos, derived from a zygotic embryo of 'Duke 7', with an $O.D._{600nm}$ 0.5 *A. tumefaciens* AGL1 strain harbouring the plasmid pBINUbiGUSint for 20 min followed by a 2-day coculture. This binary vector contains *nptII* and *uidA* as marker genes under the control of the nopaline synthase promoter and the promoter of the ubi-1 gene of maize polyubiquitin, respectively. *Agrobacterium*-inoculated embryos were cultured on solid selection medium supplemented with 50 mg l^{-1} kanamycin and 250 mg l^{-1} timentin for 2 months. Subsequently, kanamycin concentration in the selection medium was increased up to 100 mg l^{-1} for 2 additional months; 12% of inoculated embryos proliferated in this selection medium (a transformation efficiency of 6%). Two of these lines were cultured in liquid selection medium for 1 month, tolerating up to 400 mg l^{-1} kanamycin. Transgenic somatic embryos and plants expressing the GUS gene could be recovered from five independent embryogenic lines; in addition, *nptII* gene insertion was proved by amplification by PCR and Southern blot analysis. Three *A. tumefaciens* strains were tested in this study (LBA4404, EHA105 and AGL1), with AGL1 showing the highest transformation efficiency. A study of the usefulness of fluorescent proteins as visual markers during the transformation of avocado somatic embryos was reported by Palomo-Ríos *et al.* (2017). Embryogenic lines expressing the green (GFP) or red (DsRed) fluorescent proteins under the control of the constitutive promoter CaMV35S, with pK7FNF2 (double *gfp*), pK7RNR2 (*gfp* and *uidA*) and pK7S*NF2 (double *DsRed*) could be obtained. After 2 months of culture in solid selection medium, all kanamycin-resistant somatic embryos showed a level of fluorescence significantly higher than control embryos, using the three vectors. Red fluorescence was higher and more easily detected than green fluorescence throughout the selection process as well as in mature somatic embryos and regenerated plants.

Fig. 10.6. 'Hass' transformed with reporter gene *uidA*, the *bar* gene and the antifungal *pdf1.2* defensin gene showed resistance to the herbicide Finale®.

Ahmed *et al.* (2012) conducted a study in which six *Agrobacterium* strains were compared for their efficacy following the infection of embryonic shoots derived from 'Hass'. Strain GV3850/pB1121 provided optimum results. Assays were confirmed by the X-gluc (GUS) reaction and by PCR.

Disease resistance

NPR1 is a key regulatory gene of the salicylic acid (SA)-mediated systemic acquired resistance in *Arabidopsis* (Cao et al., 1997). Studies in different species have shown that constitutive expression of *AtNPR1* in transgenic plants induces a higher expression of pathogen-related genes and also resistance to a wide range of different pathogens (Cao *et al.*, 1998; Lin *et al.*, 2004; Wally *et al.*, 2009). Palomo-Ríos *et al.* (2010, 2011) obtained four transgenic avocado lines containing the *NPR1* gene from *Arabidopsis thaliana*, and plants have been recovered (Fig. 10.7). The lines showed different expression levels of the *NPR1* gene; moreover, they differed *in vitro*. Attempts to propagate these lines to test their resistance to the fungus *R. necatrix* are in progress. In an attempt to control PRR, Mitter *et al.* (2011) designed hairpin RNA constructs targeting *P. cinnamomi* essential genes; proof of concept was carried out using *A. thaliana* transformants as model system.

Sunblotch elimination

The expression of double-stranded RNA-specific ribonucleases in transgenic plants has been an

Fig. 10.7. Avocado plants transformed with *At*NPR1 gene growing in a confined greenhouse.

efficient approach to enhance plant resistance to viroids and also to some viruses in potato (Sano *et al.*, 1997), tobacco (Ishida *et al.*, 2002) and chrysanthemum (Ogawa *et al.*, 2005). In a strategy for controlling avocado sunblotch disease caused by the ASBVd, embryogenic 'Y381' cultures have been transformed with the *pac*1 ribonuclease gene from yeast *Schizosaccharomyces pombe* in a construct containing a transit peptide (Litz *et al.*, 2010). Preliminary results based on RT-PCR analysis indicate that the ASBVd was eliminated in most of the transformed lines.

Shelf life

Avocado is a climacteric fruit; therefore, control of ethylene biosynthesis can extend fruit shelf life (Meyer and Terry, 2010; Zhang *et al.*, 2011). ACC synthase, a key enzyme in ethylene biosynthesis, has been cloned in several species (Bapat *et al.*, 2010). Tomato plants transformed with antisense ACC synthase show ca. 99% inhibition of ethylene biosynthesis compared with non-transformed plants (Klee *et al.*, 1991). Exelisis (formerly Agritope, Beaverton, Oregon, 1997, personal communication) utilized S-adenosylmethionine hydrolase (SAMase) to convert SAM into a non-toxic by-product that is recycled within the plant cell, so that SAM cannot be converted to ACC. Bacterial ACC deaminase causes ethylene synthesis to be reduced by 90–97% in tomato (Klee *et al.*, 1991; Klee, 1993).

Raharjo *et al.* (2003) and Litz *et al.* (2007a,b) described the transformation of avocado embryogenic cultures with SAMase under the influence of an avocado fruit-ripening promoter in order to block ethylene biosynthesis in 'Suardia' (West Indian race) avocados with the goal of controlling fruit ripening and promoting on-the-tree fruit storage. These transgenic did not flower when observed in a screened enclosure (R.E. Litz, University of Florida, Tropical Research and Education Center, Homestead, Florida, 2012).

a) *Biolistic*

A transient transformation protocol through biolistic has been described for avocado embryogenic cultures (Chaparro-Pulido *et al.*, 2014). Six cm target distance, 900 psi helium pressure and 1.8 μg DNA per shoot were determined as optimal parameters. The expression of *uidA* with four constitutive promoters and a trichome-specific promoter was tested, showing the highest GUS expression with the sunflower polyubiquitin and the trichome-specific promoters. GFP and DsRed fluorescent signals were also tested. The developed protocol proved to be a suitable tool to test promoters, marker genes and transient expression studies.

b) *In planta transformation*

Salazar-Gonzalez *et al.* (2023) described the transient *Agrobacterium*-mediated genetic transformation of avocado leaves with the *Agrobacterium tumefaciens* strain LBA4404 and the *RUBY* reporter. Success was dependent on: (i) the stage of development of leaves; and (ii) microwounding together with exposure to 250 μM jasmonic acid (JA) prior to agroinfiltration. The authors speculate that this procedure could be used to explore gene function and metabolic pathways in avocado leaves.

In Vitro Conservation

In the region of origin of avocado, a large pool of genetic diversity within the genus *Persea* and particularly within the species *P. americana* is present (Lahav and Lavi, 2002; Boza *et al.*,

2018; Barrientos-Priego, 2019). This diversity is being threatened by biotic and abiotic stress factors affecting tropical and subtropical forests. Hence, besides the *in situ* conservation of natural habitats, the maintenance and increase of existing germplasm collections are urgently needed. Avocado field collections are the conventional way to conserve its genetic resources, but this is hampered by high costs of land and labour associated with orchard management; moreover, field collections are highly vulnerable to natural disasters, pests and diseases. Conventional seed banking, the other traditional *ex situ* conservation method, is not possible for avocado because it has recalcitrant seeds that do not tolerate desiccation and cold storage. Moreover, there is a need for long-term storage of selected embryogenic lines to ensure a continuous supply of material. Thus, for avocado, *in vitro* conservation represents a valuable tool for diversity conservation, thereby securing germplasm for future use (Pence *et al.*, 2020) and contributing to breeding programmes.

Low-temperature storage

Mexican genotypes of avocado are more cold-tolerant than Guatemalan or West Indian types, and this has also been observed *in vitro*. Vidales-Fernández *et al.* (2011) preserved axillary buds of Mexican genotypes ('Criollo') for 270 days at 5°C in a medium with low ionic strength (0.25–0.5X MS salts); however, for Guatemalan × West Indian hybrids, Valladares (2005) found a temperature of 12.5°C to be more suitable. Vidoy-Mercado *et al.* (2008) observed that 17°C is adequate for the conservation of juvenile material of the Guatemalan genotype 'Anaheim' as well as adult material of PC4 genotype (a selection of unknown origin and tolerant to *R. necatrix*). Molecular analysis using 15 SSR loci indicated that this material was genetically stable after a year in storage. Raya (2004) maintained embryogenic cultures for 150 days at 15°C on a medium with half-strength MS mineral elements, 30 g l^{-1} sucrose and 30 g l^{-1} mannitol.

Cryopreservation

Cryopreservation, involving storage of materials at ultra-low temperatures (−196°C in liquid nitrogen and below −150°C in the vapour phase), is an option for conservation of germplasm collections of different species (Reed, 2008; Wang *et al.*, 2021). This technology would allow the long-term conservation of avocado genetic resources for backing up field collections and the maintenance of embryogenic cultures.

Efendi and Litz (2003) described cryogenic storage of embryogenic avocado cultures and the recovery of somatic embryos. Two procedures were described: (i) stepwise cooling (−1°C min^{-1} from room temperature to −80°C) followed by rapid cooling to −196°C; and (ii) vitrification or rapid cooling from 25°C to −196°C.

i. Stepwise cooling: Embryogenic cultures of avocado were transferred into a cryoprotectant mixture of 5% DMSO and 5% glycerol with either 0.13 M or 1.0 M sucrose (Efendi, 2003). Approximately 200 mg of air-dried embryogenic cultures were mixed with 1.0 ml of sterile cryoprotectant solution in 1.2 ml Corning® cryogenic vials and maintained on ice for 30 min. The vials were inserted into Nalgene® 'Mr. Frosty' containers with 250 ml isopropanol. The containers were placed in a low-temperature (–80°C) freezer for 2 h for controlled cooling at the rate of −1°C min^{-1}. The vials were then removed and inserted into aluminium cryo-cane holders and subsequently immersed directly into liquid nitrogen (−196°C) in a Bio-Cane 20 Dewar container.

ii. Vitrification: Embryogenic cultures (approximately 200 mg) were suspended in 1.0 ml sterile loading solution (2.0 M glycerol and 0.4 M sucrose) in 1.2 ml cryovials for 15 min (Efendi, 2003). The loading solution was then replaced with 1.0 ml plant vitrification solution number 2 (PVS2) (Sakai *et al.*, 1990) and maintained on ice. PVS2 consisted of 15% DMSO, 30% glycerol and 15% ethylene glycol. The vitrification solution was decanted after 60 min and the cryovials were inserted into cryosleeves and plunged into liquid nitrogen in a Bio-Cane 20 Dewar container. The cryovials were removed from liquid nitrogen and immersed for 5 min in a 40°C water bath (Efendi, 2003). The embryogenic cultures were decanted on to sterile semi-solid MSP medium in Petri dishes. Cryopreserved embryogenic cultures proliferated on MSP medium and as suspension cultures in MS3:1P liquid medium. Somatic embryos were recovered and developed to maturity.

Guzmán-García *et al.* (2013a) compared three cryopreservation protocols based on slow cooling or vitrification. The droplet-vitrification method developed by Panis *et al.* (2005) appeared to be optimal for avocado, with recovery rates of 77.78–100%. The genotype showed an important effect, and the best results were obtained when embryogenic cultures containing somatic embryos at early developmental stages were used. O'Brien *et al.* (2016) compared four different cultivars and focused on the sucrose concentration of the loading solution (0.2, 0.4, 0.6 and 0.8 M), temperature of vitrification solution and incubation time. Optimal sucrose concentration was found to be genotype-dependent; PVS2 ice-cooled was the preferred temperature for all avocado cultivars while 60 min exposure provided maximum viability.

Early attempts for cold storage of *in vitro* avocado buds of Mexican genotypes were reported by Vargas (2008), who tested viability after storage at −20°C for 1 h. Buds had been dissected from shoots cultured in MS medium supplemented with brassinolide (0–10 mg l^{-1}); prior to cold storage they were air desiccated for 60 min and incubated in plant vitrification solution 4 (PVS4, liquid solution containing 35% glycerol, 20% ethylene glycol and 0.6 M sucrose) for 30 min. Brassinolide enhanced bud sprouting after cold storage. O'Brien *et al.* (2018, 2020b) determined that ascorbic acid at 100 and 250 mg l^{-1} reduced browning of avocado shoot tips after being treated with vitrification solutions. O'Brien *et al.* (2020a) exposed *in vitro* shoots of 'Velvick' (West Indian × Guatemalan) and 'Reed' (Guatemalan) to 0.3 M sucrose and 10°C for 2 weeks before shoot tip dissection. Results showed that the pretreatment effect was genotype dependent. The cold pretreatment improved regrowth of 'Reed' (cold-tolerant but sensitive to osmotic stress), while 'Velvick' responded better to the high sucrose (cold-sensitive but tolerant to salinity). O'Brien *et al.* (2021) described the cryopreservation of *in vitro* meristems using droplet vitrification (Panis *et al.*, 2005): (i) pretreatment of shoots on high sucrose or exposed to chilling temperature; (ii) dehydration by incubating the shoot tips in loading solution for 20 min and ice-cooled vitrification solution for 0–40 min; (iii) exposure to liquid nitrogen; and (iv) rewarming in unloading solution for 20 min. Modified VSL solution (Suzuki *et al.*, 2008) resulted in recovery of better-quality shoots than PVS2 (Sakai *et al.*, 1990).

Although genotypes might differ in their responses to pretreatments and more investigations are needed to understand the tolerance to cryopreservation, this research opens the way for setting up a cryobank of *Persea* spp.

Molecular Biology

The avocado is a diploid (2n=2x=24) species with a haploid genome size of 896 Mb (Arumuganathan and Earle, 1991). Initial molecular biology studies were focused on the generation and application of various molecular markers (for genotype fingerprinting, differentiation among main races, diversity and phylogenetic studies, linkage map construction and screening for traits of interest) and the identification of avocado genes and expressed sequence tags (ESTs). More recently, advances in genomics and other omics are making a qualitative change for utilizing genetic diversity in this species.

The number of linkage maps constructed to date is still low in avocado. The first avocado genetic linkage map was developed with 50 SSRs, 23 minisatellite DNA fingerprint (DFPs) and 17 RAPD markers, using the progeny of a cross between 'Pinkerton' × 'Ettinger', covering 352.6 cM (Sharon *et al.*, 1997). Another linkage map was developed using SSR makers and 715 F1 individuals (456 from 'Tonnage' × 'Simmonds' and 259 from the reciprocal cross 'Simmonds' × 'Tonnage') (Borrone *et al.*, 2009), covering 1087.4 cM. Ashworth *et al.* (2019) reported a linkage map from a 'Gwen' × 'Fuerte' progeny that was constructed using SSRs and SNPs covering 1044.7 cM. In this case, several molecular markers located in linkage group 10 showed significant association with flowering type. In addition, markers linked to QTLs associated with fruit levels of β-sitosterol (on linkage group 1) and vitamin E (α-tocopherol) (on group 3) were found.

Gene Isolation

Avocado is a member of the Lauraceae family in the order Laurales, that, together with the Canellales, Piperales and Magnoliales, are included in the early divergent angiosperm clade magnoliids, representing a basal lineage among flowering plants (Chase *et al.*, 2016, see Chapter 3, Chanderbali *et al.*, in this volume). In this sense, comparative genomic studies have been carried out to provide insight into the evolution of flowering traits in angiosperms. A large-scale transcriptomic project, the Floral Genome Project, collected nearly 10,000 EST sequences from early developmental stages of flowers of 'Hass' (Albert *et al.*, 2005). This information has been useful for designing flower microarrays and performing comparative gene expression analysis with flowers of different angiosperms (Chanderbali *et al.*, 2009). Custom microarrays targeting 6068 genes collected from *Persea* floral buds by the Floral Genome Project (Albert *et al.*, 2005) identified 4797 significantly different expressed genes ($P < 0.05$) among eight sampled tissues: inflorescence buds, premeiotic floral buds, outer tepals, inner tepals, stamens (including staminodes), carpels, initiating fruit and leaves (Chanderbali *et al.*, 2009). Two homologues of SEPALLATA MADS box, PEamAGL9.1 and PEamAGL9.2, were identified in a phylogenetic study of this floral organ identity gene family among divergent angiosperm taxa (Zahn *et al.*, 2005). A study that focused on flower initiation allowed the isolation of FLOWERING LOCUS T-like gene from 'Hass' (*Pa*FT) (Ziv *et al.*, 2014). The ectopic expression of this gene induced early flowering in *Arabidopsis*. The expression of *Pa*FT in leaves, as well as the floral identity genes *Pa*AP1 and *Pa*LFY in buds, was upregulated in plants with fruits removed, suggesting that fruit load might reduce flowering by repressing the *Pa*FT gene.

Several avocado genes involved in fruit development and maturation have been cloned and sequenced; most of them have been isolated from the mesocarp of 'Hass' fruit, for example, cellulase (Cass *et al.*, 1990), cytochrome P-450 (Bozak *et al.*, 1990), polygalacturonase (Kutsunai *et al.*, 1993) or ACC oxidase (McGarvey *et al.*, 1992). Owino *et al.* (2002) reported the isolation and cloning of two ACC synthase genes (*PA-ACS1*, *PA-ACS2*) and a putative ethylene response sensor (*PA-ERS1*) expressed during fruit ripening and wounding and also in response to 1-MCP and propylene treatments.

Two putative transcription factors, the ethylene response factor (ERF) and the auxin-repressed protein-like factor, have also been identified in ESTs of avocado fruits. These two genes have been reported during the ripening of fruits of different species, and it has been suggested that they could act as crossroads between ethylene and auxin during fruit ripening and development (Reddy and Poovaiah, 1990; Wang *et al.*, 2007; Trainotti *et al.*, 2007; Chaabouni *et al.*, 2009). The B12D-like protein gene is another messenger found in mesocarp; this is a protein of unknown function that has been isolated from senescing sweet potato (Huanga *et al.*, 2001). The transcriptomes of vegetative tissues and fruits at three different ripening stages, preclimacteric, climacteric and post-climacteric, were analysed in avocado genotypes of the Mexican race (Ibarra-Laclette *et al.*, 2015). A set of 83,650 unigenes was generated, from which 81% were annotated. An expression map of these unigenes allowed the identification of organ-specific genes as well as differentially expressed genes during fruit development and ripening. Expression of 382 unigenes was increased in fruit compared to flowers, among them, metallothioneins, chitinases and pectate lyases were included. During fruit ripening, 1230 unigenes were differentially expressed, including ACC synthase, ACC oxidase, ethylene response factor ERF110, ethylene receptor EIN4, NOR transcription factor, thaumatin-like protein and several chitinases. The expression patterns of genes implicated in fatty acid metabolism during fruit ripening were also analysed using this assembled transcriptome.

Regarding fruit size, four distinct avocado cell proliferation-related genes were isolated by Dahan *et al.* (2010) and their expression in normal and small fruit of 'Hass' was compared. Three cDNAs (*Pa*CYCA1, *Pa*CYCB1 and *Pa*PCNA), encoding two mitotic cyclins and a proliferating cell nuclear antigen (PCNA), were isolated from young normal fruit, and transcripts were shown to accumulate predominantly in mitotically active organs, including young fruitlets, leaves and roots. Accumulation of transcripts of these genes decreased earlier in small fruit compared to average-size fruit, concomitant with an earlier arrest of cell division. A fourth full-length cDNA (*Pa*fw2.2-like), which encodes an FW2.2 (fruit-weight)-like protein that has been proposed as a putative negative regulator for cell division in tomato fruit, was isolated from small fruit. The accumulation of the mitotic cyclins and PCNA transcripts gradually decreased in normal fruit, whereas levels in small fruit had already decreased at earlier stages of fruit development. In contrast, mRNA levels of *Pa*fw2.2-like showed higher accumulation in small fruit compared with average-size fruit at all growth stages examined, concomitant with an earlier arrest of fruit cell division activity. Sabag *et al.* (2013) isolated four cDNAs encoding a D-type cyclin (*Pa*CYCD1), two cyclin-dependent kinases (CDKs) and an inhibitor of CDKs (ICKs/KRP). The expression of *Pa*KRP was enhanced in phenotypically small fruits associated with an earlier decline of CDKs and *Pa*CYCD1 levels as well as with an accumulation of ABA.

García-Rojas *et al.* (2012) constructed a suppression subtractive hybridization (SSH) cDNA library to identify genes involved in the lower quality of late harvested fruit. A total of 117 differentially expressed genes involved in different biochemical pathways were identified. Polygalacturonase, putative NAD^+-dependent sorbitol dehydrogenase and three putative acyl-CoA synthase genes were upregulated in late-harvested fruit. Two genes encoding different subunits of the acetyl-CoA carboxylase complex (ACCase), which might be involved in internal browning induced by low-temperature storage, were also identified in this SSH library (Gudenschwager *et al.*, 2013).

Avocado fruit accumulate high levels of oils, predominantly oleic acid, in the mesocarp. Kilaru *et al.* (2015) performed a comparative transcriptome analysis of selected lipid metabolic pathways in 'Hass' fruit, in oil palm and in rapeseed and castor bean. The results showed similar oil biosynthesis patterns in the different species, although some unique features were detected in avocado, that is, the higher contribution of long-chain acyl-CoA synthetase (LACS) orthologs encoding isoforms associated with the endoplasmic reticulum instead of genes encoding plastidial LACS that are more common in the other oil-producing species.

Avocado is a climacteric fruit exhibiting a burst in ethylene biosynthesis during ripening, which usually occurs after harvest since the fruit does not ripen while still attached to the tree (Chanderbali *et al.*, 2008). Avocado fruit softening during the ripening process is mainly due to the degeneration of parenchymatic cell walls mediated by cell-wall hydrolases. Starrett and Laties (1993) analysed ethylene and wound-induced genes in the mesocarp of ripening 'Hass' fruit. Dopico *et al.* (1993) cloned and characterized genes that showed differential expression during cold storage of 'Hass' fruit. Some of those genes showed homology with cysteine proteinase inhibitor (oryzacystatin), polygalacturonase, thaumatin-like protein and endochitinase. Ripening-related polygalacturonase (Kanellis *et al.*, 1991; Dopico *et al.*, 1993), cellulase (Christoffersen *et al.*, 1984), β-galactosidase (Tateishi *et al.*, 2007) and α-L-arabinofuranosidase genes (Kamiyoshihara *et al.*, 2018) have been cloned from avocado fruit.

Storage of avocado fruit at low temperatures induces chilling injury disorders. In a study of mesocarp discolouration, a fragment of the ethylene receptor *Pa*ETR was amplified from seedless fruit of 'Arad' (Hershkovitz *et al.*, 2009a). An evaluation of the expression of genes related to ethylene production during chilling stress by quantitative real-time PCR (qRT-PCR) was performed by Hershkovitz *et al.* (2009b). *Pa*ETR expression decreased gradually from the base of the seed to the blossom end, suggesting a role of ethylene in the development of chilling injury symptoms.

Avocado fruits are also a source of genes related to plant defence, for example, Kimura *et al.* (1995) determined the amino acid sequence of cystatin (a proteinase inhibitor) from fruit. Identification and cloning of an endochitinase gene (Prs a 1) have also been reported. This protein, when expressed in yeast, showed inhibition of *Fusarium oxysporum*; however, it is a potent allergen, raising concern about using chitinases to enhance fungal tolerance in transgenic crops (Sowka *et al.*, 1998). Two cysteine-rich antimicrobial peptide (AMPs) genes have been isolated from pulp and seed of Mexican race avocado accessions. One defensin, *Pa*Def, abundant in mesocarp tissue, showed antimicrobial activity following heterologous expression in bovine endothelial cell line BVE-E6E7 (Guzmán-Rodríguez *et al.*, 2013a). A snakin gene (*Pa*Sn), abundantly expressed in avocado seeds, was identified (Guzmán-Rodríguez *et al.*, 2013b). Conditioned media in which transfected endothelial cells BVE-E6E7 line had been grown showed antimicrobial activity, indicating that this protein might have a protective role in avocado seeds. Three metallothioneins (*Pa*MT2a,

*Pa*MT2b, *Pa*MT3) and a bLHLH (*Pa*bLHLHS1) transcription factor, possibly involved in conferring tolerance to drought stress in avocado, have also been identified through transcriptomic studies (Suárez-Rodríguez *et al.*, 2017).

Studies have provided understanding of avocado interactions with different pathogens. Working with *Colletotrichum gloeosporioides*, the causal agent of avocado anthracnose, Madi *et al.* (2003) found that several factors, for example, inoculation with the fungus, exposure to ethylene and wounding, enhanced expression of the avocado homologue avfad9 encoding Δ9-stearoyl-ACP desaturase; this change in expression is associated with an increase in an antifungal diene (AFD) and the AFD increment is linked to enhanced fruit resistance to this fungus. Wang *et al.* (2004) demonstrated a temporal relationship among transcripts from the avfad12 gene family, accumulation of the AFD precursor linoleic acid, AFD content and decrease of fungal growth. To accomplish a successful cDNA synthesis, Djami-Tchatchou and Straker (2012) developed a CTAB-based extraction procedure to obtain high-quality RNA from avocado fruit skin and flesh; they constructed cDNA libraries enriched in differentially expressed genes from harvested and unharvested fruit infected with *C. gloeosporioides*. Djami-Tchatchou *et al.* (2012) inoculated harvested and unharvested fruit with this fungus and identified several categories of genes, for example, defence, transcription factors, metabolism/photosynthesis, protein synthesis, oxidative burst, stress response, transport and others of unknown function. As expected, expression of general defence genes was higher in unharvested fruit. Djami-Tchatchou *et al.* (2013) searched for specific genes involved in the avocado response to this pathogen. Differences in the expression of several genes, for example, salicylic acid binding protein 2 (SABP2), ethylene-responsive element binding protein (EREBP), catalase, endochitinase, endo β-1,4 glucanase and pathogenesis-related proteins (PR5 and PR6), were found between infected and uninfected fruit; moreover, the occurrence of a modulatory type of response could indicate a compatible-type interaction. This study represents an early step to understanding resistance of avocado fruit to *C. gloeosporioides*; however, further investigations will be required to elucidate the signalling pathways and the specific role of SA and ethylene in this process.

Attempts to reduce anthracnose incidence in harvested avocado fruit were carried out by Bill *et al.* (2016) using thyme oil fumigation. Treated fruits showed a lower disease incidence than in untreated controls, although there were differences between genotypes; less fruit damage positively correlated with increased β-1,3 glucanase and chitinase activities. Further attempts to improve control of anthracnose through the use of chitosan were carried out by Xoca-Orozco *et al.* (2017). This molecule acts as an elicitor inducing multiple metabolic pathways, some of which are involved in fungal growth inhibition, improving defence and reducing disease damage. However, further studies are needed to elucidate the specific function of genes, up- and downregulated, in the chitosan-avocado-*Colletotrichum* interaction system.

Understanding the biochemical and molecular basis of avocado responses to the causal agent of root rot, the hemibiotrophic oomycete *P. cinnamomi* is of key importance when searching for resistant rootstocks. Using a proteomic approach, Acosta-Muñiz *et al.* (2012) identified differentially expressed proteins in avocado following inoculation with *P. cinnamomi*. A glycine-rich RNA-binding protein and a small heat-shock protein were downregulated, while 16 proteins were upregulated with five of them only found on infected roots. Differentially expressed proteins showed homology to proteins such as isoflavone reductase, glutathione S-transferase, cinnamyl alcohol dehydrogenase or cysteine synthase, among others. Proteinase inhibitors have also been shown to be produced by avocado roots in response to extracellular proteases released by the oomycete. Colonization by the pathogen was inhibited following root pretreatment with these inhibitors (Encino-López *et al.*, 2011).

In a comparative study between susceptible and resistant rootstocks, van den Berg *et al.* (2018a) found increased β-1,3-glucanase activity 6 h after pathogen inoculation (hpi) in resistant rootstocks, while an increase in catalase occurred at 24 hpi. The response in susceptible rootstocks was associated with lignin deposition. According to these authors, resistance to the pathogen is based on a multilayered response.

Using a mycelial suspension to inoculate avocado roots, Mahomed and van den Berg (2011) identified general (metallothionein and thaumatin-like, among others) and specific (thaumatin, PR-10 and LRR resistance) defence genes. Based on transcriptome data, they hypothesized that the response is probably polygenic in nature. In a later study, responses of five avocado rootstocks, differing in their susceptibility to this pathogen, were evaluated (Engelbrecht and van den Berg, 2013). Differences in transcript abundance were only found for phenylalanine ammonia-lyase and lipoxygenase, indicating that they could be involved in tolerance of this pathogen, while defence-related genes, such as PR-5 and endochitinase, were expressed in all rootstocks. Partial resistance to the oomycete probably requires interactions of many genes at different levels.

Flooding conditions favour infections by *P. cinnamomi*. Using massive sequencing, Reeksting *et al.* (2014) confirmed this observation at the gene level since an increased expression of flood-responsive genes, for example, sucrose synthase, non-symbiotic haemoglobin and pyruvate decarboxylase, could be detected in infected plants under flooding conditions compared to non-infected flooded plants and infected and non-infected non-flooded plants. Flooding could also cause a slight increase in endochitinase expression. Further studies on the avocado flooding–*P. cinnamomi* interactions revealed that flooding causes large changes in gene expression of both infected and non-infected control plants. Probably, this huge impact on plant metabolism masks plant responses to the oomycete. Many of the repressed genes under flooding were defence-related, which could explain the enhanced susceptibility to the pathogen under these conditions. The observed reduction in stomatal conductance and subsequent decrease of water uptake under flooding could be linked to a drastic downregulation of aquaporins (Reeksting *et al.*, 2016). Trying to identify the role played by SA and JA in the interaction of this hemibiotrophic oomycete with a resistant rootstock, van den Berg *et al.* (2018b) reported the appearance of transcripts associated with SA in the early response (6 hpi) in which auxin, reactive oxygen species and Ca2+ signalling also played a role. By contrast, genes related to JA response appeared at 24 hpi. At this time, other processes, such as phenolics formation and cell-wall strengthening, also occurred. At 18 hpi, both SA and JA defence signalling occurred. According to van den Berg *et al.* (2021), in resistant rootstocks, there is an earlier regulation followed by a more complete suppression of the SA route, allowing a more rapid initiation of the JA pathway. The *NPR1* gene (*NONEXPRESSOR OF PATHOGENESIS-RELATED GENES 1*) and NPR1-like genes family are known to play a key role in plant defence (Backer *et al.*, 2019). Trying to shed light on the role played by NPR1-like genes in SA signalling and SA-JA crosstalk following avocado infection with this *P. cinnamomi*, Backer *et al.* (2015) identified five NPR1-like genes. Expression analysis in avocado roots following treatments with SA, JA or *P. cinnamomi* infection provided evidence that three of them could be related to plant defence, while the other two seemed to be involved in plant development. They also indicated that PR-1 expression linked to the SA pathway is induced earlier (6 hpi) in resistant rootstocks in comparison to susceptible material, following infection with the pathogen; in addition, a drastic downregulation of PR-1 occurred at 24 hpi in this material.

Nucleotide-binding leucine-rich repeat (NLR) proteins are part of the effector-triggered immune (ETI) responses in plant–pathogen interactions; in a recent study, Fick *et al.* (2022) reported the expression of 84 *Pa*NLR at 24 hpi in an avocado rootstock with partial resistance to *P. cinnamomi* (Dusa®), while in susceptible material (RO.12) NLRs expression occurred at 6 hpi; the more prolonged ETI response could explain the partial resistance of the rootstock. A thorough review of the interaction of avocado–*P. cinnamomi* can be found in van den Berg *et al.* (2021).

Another soil-borne pathogen, the causal agent of white root rot (WRR) in avocado is the necrotrophic fungus *R. necatrix*. Zumaquero *et al.* (2019) carried out a comparative study between rootstocks differing in their response to the fungus, highlighting the role of protease inhibitors in WRR tolerance. Genes related to salt and osmotic stress are upregulated, and these authors concluded that degree of tolerance could be associated with capacity to withstand osmotic imbalance.

In the last decade, Fusarium dieback, a beetle-vectored disease, has become a serious threat to avocado orchards. In this case, the complex *Fusarium* spp. ambrosia beetle is the causal agent of the disease. In an attempt to decipher the molecular basis of this interaction, Pérez-Torres *et al.* (2021) carried out a transcriptome analysis following infection of avocado stems with *Fusarium kuroshium*, a symbiont of the Kuroshio shot hole borer. Unigenes could be gathered into six groups and most important relevant functions were linked to photosynthesis, stress responses or water deprivation, among others. Four NLR genes involved with immunity and hypersensitivity responses were identified. They emphasized the need for similar experiments using whole plants to validate the results obtained in this study and continue the exploration for new genes involved in resistance or tolerance to this disease.

Regarding the interaction between avocado and the viroid causing sunblotch (ASBVd), López-Rivera *et al.* (2017) found changes in expression of PR-5 and PR-6 genes in fruit infected by the viroid, while no changes were observed in *Pa*NPR1 and EREBP genes. PR-5 expression was higher in fruit from trees with no symptoms. It appears that both genes may play a role in a coordinated response to the viroid.

Genomics

New technologies and strategies for manipulating the genome have revolutionized the study of genetics in many different crops. The sequencing of cDNA clones has been an important part of large-scale sequencing projects. The goal of these efforts is to determine the nucleotide sequence from the highest number of expressed genes in an organism.

Next-generation sequencing approaches in avocado have so far mostly been focused on transcriptome analyses and SNPs development. These investigations have been mainly applied to study disease resistance, plant development, stress tolerance and genetic diversity (Ibarra-Laclette *et al.*, 2015; Kilaru *et al.*, 2015; Ge *et al.*, 2019a,b,c; Kuhn *et al.*, 2019a,b; Rubinstein *et al.*, 2019; Talavera *et al.*, 2019; Vergara-Pulgar *et al.*, 2019; Chabikwa *et al.*, 2020; Kämper *et al.*, 2021; Pérez-Torres *et al.*, 2021; Fick *et al.*, 2022; Hernández *et al.*, 2022). Several avocado nuclear genome sequences have been published. Rendón-Anaya *et al.* (2019) developed genomic sequence drafts of 'Hass' as well as of a wild accession of the Mexican race. The 'Hass' assembly consisted of 8135 contigs of which 915 scaffolds were anchored to a genetic map and the resulting genome covered ca. half (46.2%) of the estimated avocado genome. Furthermore, ten genomes representing the three avocado subspecies or races, as well as 'Hass'-related cultivars and *P. schiedeana*, were re-sequenced helping to better understand the genetic diversity and the evolutionary history of avocado. Another 'Hass' genome reference has been generated by Sharma *et al.* (2021), resulting in an assembly of 788 Mb representing ca. 88% of the avocado genome size; however, this assembly was not anchored into scaffolds. A chromosome-level genome assembly for 'Hass' has been published (Nath *et al.*, 2022). The total assembly was 913 Mb with a contig N50 of 84 Mb. Contigs assigned to the 12 chromosomes represented 874 Mb with 62.6% repeat elements. Annotation of protein coding sequences identified 48,915 avocado genes, of which 39,207 could correspond to ascribed functions. Additional work sequencing 'Gwen' (Solares *et al.*, 2022) resulted in an N50 of 3.37 Mb and organized in 12 pseudo-chromosomes with a total of 49,450 genes. In a recent study aimed at characterizing genomic diversity in Colombian avocados, Berdugo-Cely *et al.* (2023) sequenced 144 accessions from the Colombian National GENEBANK and 240 accessions from local commercial orchards in the Colombian north-western Andean region. Merging resulting reads with sequences of reference genotypes allowed the discovery of 4931 SNPs, while the analysis of population structure and phylogenetic diversity revealed the existence, in addition to the three well-recognized races (Mexican, Guatemalan and West Indian), of a new Colombian group closely related to the West Indian race. According to the authors, these findings represent a source of allelic novelty that could be very useful in avocado breeding programmes. All these genomic advances need adequate bioinformatic platforms to host avocado omics data. A recent example is Avocadobase (https://www.avocado.uma.es), a web portal dedicated to avocado genomics, which provides multiple interactive bioinformatics tools, sequences and annotations to analyse, visualize and download omics data related to avocado.

Conclusions

Biotechnology could be an important tool in avocado breeding. Micropropagation of shoots via axillary branching is being used in breeding programmes for fungal resistance to generate a limited number of copies of specific selections. Recently, use of stem tips as explants has allowed mass cloning of some genotypes, although application of the technology to other selections will require optimization for each genotype. Significant improvements have been achieved in embryo rescue techniques, thereby allowing the efficient recovery of genotypes from selected crosses.

Establishment of embryogenic lines from immature zygotic embryos is now routine; however, efforts are needed to improve embryogenic culture induction in explants of adult origin, such as nucellus. Nucellar explants are promising sources of embryogenic cultures, although the embryogenic response is probably highly genotype-dependent. Methods for transforming avocado embryogenic cultures are currently available and, in some cases, plants have been recovered from transgenic embryos, although the methodology to convert transformed somatic embryos into plants needs to be improved.

Cryopreservation could be useful for preserving shoot tips and selected embryogenic cell lines, and avoiding the loss of embryogenic competence, which occurs with time. Although significant advances have been made, efforts are needed to improve these techniques, in order to use *in vitro* material as backup for avocado germplasm collections and to facilitate international movement of important genetic resources, particularly from the great genetic repositories of the USA, Israel, Mexico and Australia.

The number of genes with identified function in avocado is still small. However, recent advances in high-throughput sequencing allow a qualitative change in the development of genomics and other omic technologies that are becoming increasingly cheaper and faster. As a consequence, new information at the gene level will be available for use in biotechnological breeding in the near future. The availability of more powerful gene-editing tools, such as those based on CRISPR/Cas, together with the increasing number of genes with known function, will surely result in a wider use of biotechnological approaches in avocado breeding in the future; however, currently available protoplast regeneration protocols will need to be optimized for efficient application of these technologies.

Acknowledgements

Support provided by research projects AGL2017-83368-C2-1-R (PLAN NACIONAL I+D+I, MEC, Spain) and PID2022-141851OB-I00 (MCIN/AEI/10.13039/501100011033 and "ERDF A way of making Europe" by the European Union), the California Avocado Commission and the USDA TSTAR Special Grants program (USA), is gratefully acknowledged.

References

Aaouine, M. (1986) Tissue culture investigation of graft incompatibility among *Persea* species. PhD thesis, University of California, Riverside, California.

Acosta-Muñiz, C.H., Escobar-Tovar, L., Valdes-Rodriguez, S., Fernández-Pavia, S., Arias-Saucedo, J.L., Espindola Barquera, M.C. and Gómez Lim, M.A. (2012) Identification of avocado (*Persea americana*) root proteins induced by infection with the oomycete *Phytophthora cinnamomi* using a proteomic approach. *Physiologia Plantarum* 144, 59–72.

Albert, V.A., Soltis, D.E., Carlson, J.E., Farmerie, W.G., Wall, P.K., *et al.* (2005) Floral gene resources from basal angiosperms for comparative genomics research. *BMC Plant Biology* 5, 1–15.

Alcaraz, M.L. and Hormaza, J.I. (2014) Optimization of controlled pollinations in avocado (*Persea americana*, Lauraceae). *Scientia Horticulturae* 180, 79–85.

Ahmed, M.F., Kantharajah, A.S. and Holford, P. (2012) Genetic transformation studies on avocado cultivar 'Hass' (*Persea americana*). *American Journal of Plant Sciences* 3, 1225–1231.

Anderson, W.C. (1975) Propagation of rhododendrons by tissue culture. Part I. Development of a culture medium for multiplication of shoots. *International Plant Propagators' Society Combined Proceedings* 25, 129–135.

Arrillaga, I., Marzo, T. and Segura, J. (1991) Micropropagation of juvenile and adult *Sorbus domestica* L. *Plant Cell, Tissue and Organ Culture* 27, 341–348.

Arumuganathan, K. and Earle, E.D. (1991) Nuclear DNA content of some important plant species. *Plant Molecular Biology Reporter/ISPMB* 9, 208–218.

Ashworth, V.E., Chen, H., Calderón-Vázquez, C.L., Arpaia, M.L., Kuhn, D.N., *et al.* (2019) Quantitative trait locus analysis in avocado: the challenge of a slow-maturing horticultural tree crop. *Journal of the American Society for Horticultural Science* 144, 352–362.

Avenido, R.A., Galvez, H.F., Dimaculangan, J.G., Welgas, J.L., Frankie, R.B. and Damasco, O.P. (2009) Somatic embryogenesis and embryo culture coupled with gamma irradiation for generating avocado (*Persea americana* Miller) mutants in the Philippines. In: *Induced Mutation in Tropical Fruit Trees, IAEA-TECDOC-1615*. International Atomic Energy Agency, Vienna, pp. 47–69.

Azcón-Aguilar, C., Barceló, A., Vidal, M.T. and de la Viña, G. (1992) Further studies on the influence of mycorrhizae on growth and development of micropropagated avocado plants. *Agronomie* 12, 837–840.

Backer, R., Mahomed, W., Reeksting, B.J., Engelbrecht, J., Ibarra-Laclette, E. and van den Berg, N. (2015) Phylogenetic and expression analysis of the *NPR1*-like gene family from *Persea americana* (Mill.). *Frontiers in Plant Science* 6:300.

Backer, R., Naidoo, S. and van den Berg, N. (2019) The NONEXPRESSOR OF PATHOGENESIS-RELATED GENES 1 (NPR1) and related family: mechanistic insights in plant disease resistance. *Frontiers in Plant Science* 10:102.

Bapat, V.A., Trivedi, P.K., Ghosh, A., Sane, V.A., Ganapathi, T.R. and Nath, P. (2010) Ripening of fleshy fruit: molecular insight and the role of ethylene. *Biotechnology Advances* 258, 94–107.

Barceló-Muñoz, A. (1995) Micropropagación de aguacate (*Persea americana* Mill.). Aspectos histológicos del proceso de vitrificación. PhD thesis, Facultad de Ciencias, Universidad de Málaga, Malaga, Spain. (In Spanish.)

Barceló-Muñoz, A., Pliego-Alfaro, F. and Barea, J.M. (1990) Micropropagación de aguacate (*Persea americana* Mill.) en fase juvenil. *Actas de Horticultura* 1, 503–506 (In Spanish).

Barceló-Muñoz, A., López-Encina, C., Simón-Pérez, E. and Pliego-Alfaro, F. (1999) Micropropagation of adult avocado. *Plant Cell, Tissue and Organ Culture* 58, 11–17.

Barceló-Muñoz, A., Zea-Bonilla, T., Pérez-Jiménez, R., López-Herrera, C. and Pliego-Alfaro, F. (2001) Selección y micropropagación de portainjertos de aguacate tolerantes a la podredumbre blanca. In: *Proceedings of the IV Congreso Ibérico de Ciencias Hortícolas*, Cáceres, Spain. Libro de Resúmenes, p. 194. (In Spanish.)

Barrientos-Priego, A.F. (2019) Recursos Genéticos del Aguacate: Perspectivas y Futuro. In: *Proceeedings of the IX World Avocado Congress*, Medellín, Colombia. Available at: http://avocadosource.com/WAC9/01-BarrientosAlejandro2019.pdf (accessed 15 July 2023). (In Spanish.)

Barringer, S.A., Yasseen, M.Y. and Splittstoesser, W.E. (1996) *In vitro* multiplication and plantlet establishment of avocado. *In vitro Cell Developmental Biology – Plant* 32, 119–121.

Ben-Ya'acov, A. and Michelson, E. (1995) Avocado rootstocks. *Horticultural Reviews* 17, 381–429.

Berdugo-Cely, J.A., Cortés, A.J., Lopez-Hernandez, F., Delgadillo-Duran, P., Ceron-Souza, I., *et al.* (2023) Pleistocene-dated genomic divergence of avocado trees supports cryptic diversity in the Colombian germplasm. *Tree Genetics & Genomes* 19:42.

Bergh, B.O., Zentmyer, G.A., Whitsell, R.M., Boswell, S.B. and Storey, W.B. (1976) Avocado rootstock breeding, especially in relation to *Phytophthora*. *Acta Horticulturae* 57, 237–240.

Biasi, L.A., Koller, O.C. and Kampf, A.N. (1994) Micropropagation of avocado 'Ouro-Verde' through tissue culture of nodal segments. *Pesquisa Agropecuaria Brasileira* 29, 1051–1058.

Bill, M., Sivakumar, D., Beukes, M. and Korsten, L. (2016) Expression of pathogenesis-related (PR) genes in avocados fumigated with thyme oil vapors and control of anthracnose. *Food Chemistry* 194, 938–943.

Bitters, W.P., Murashige, T., Rangan, T.S. and Nauer, E. (1972) Investigation on establishing virus free citrus plants through citrus culture. In: *Proceedings of the International Organization of Citrus Virologists*. University of Florida Press, Gainesville, Florida, pp. 267–271.

Blickle, W., Mühlbach, H.P. and Sänger, H.L. (1986) Conditions for the isolation of protoplasts from callus cultures of avocado (*Persea americana* Mill.). In: *Proceedings of the VI International Congress of Plant Tissue and Cell Culture*, Minneapolis, p. 384 (Abstract).

Blumenfeld, A. and Gazit, S. (1971) Growth of avocado fruit callus and its relation to exogenous and endogenous cytokinins. *Physiologia Plantarum* 25, 369–371.

Borrone, J.W., Brown, J.S., Tondo, C.L., Mauro-Herrera, M., Kuhn, D.N., *et al.* (2009) An EST-SSR-based linkage map for *Persea americana* Mill. (avocado). *Tree Genetics & Genomes* 5, 553–560.

Botella, J.R. (2002) Genes selected for their role in modifying post-harvest life. In: Valpuesta, V. (ed) *Fruit and Vegetable Biotechnology*. CRC Press LLC, Florida, USA, pp. 136–151.

Boza, E.J., Tondo, C.L., Ledesma, N., Campbell, R.J., Bost, J., Schnell, R.J. and Gutiérrez, O.A. (2018) Genetic differentiation, races and interracial admixture in avocado (*Persea americana* Mill.), and *Persea* spp. evaluated using SSR markers. *Genetic Resources and Crop Evolution* 65, 1195–1215.

Bozak, K.R., Yu, H., Sirevag, R. and Christoffersen, R.E. (1990) Sequence analysis of ripening-related cytochrome P-450 cDNAs from avocado fruit. *Proceedings of the National Academy of Science USA* 87, 3904–3908.

Broglie, K., Chet, J., Holliday, M., Cressman, R., Biddle, P., *et al.* (1991) Transgenic plants with enhanced resistance to the fungal pathogen *Rhizoctonia solani*. *Science* 254, 1194–1197.

Brokaw, W.H. (1988) Avocado clonal rootstock propagation. *Combined Proceedings of the International Plant Propagators' Society* 37, 97–103.

Campos, P.S. and Pais, M.S.S. (1996) *In vitro* micropropagation of the macaronesian evergreen tree *Persea indica* (L.) K. Spreng. *In Vitro Cellular and Developmental Biology–Plant* 32, 184–189.

Cao, H., Glazebrook, J., Clarke, J.D., Volko, S. and Dong, X.N. (1997) The *Arabidopsis* NPR1 gene that controls systemic acquired resistance encodes a novel protein containing ankyrin repeats. *Cell* 88, 57–63.

Cao, H., Li, X. and Dong, X.N. (1998) Generation of broad-spectrum disease resistance by overexpression of an essential regulatory gene in systemic acquired resistance. *Proceedings of the National Academy of Science* 95, 6531–6536.

Capote, M., Rodríguez, N.N. and Blanco, M. (2000) *In vitro* propagation of avocado. *Tropical Fruits Newsletter* 36/37, 3–7.

Cass, L.G., Kirven, K.A. and Christoffersen, R.E. (1990) Isolation and characterization of a cellulose gene family member expressed during avocado fruit ripening. *Molecular Gene Genetics* 223, 76–86.

Chaabouni, S., Jones, B., Delalande, C., Wang, H., Zhenggou, L., *et al.* (2009) Sl-IAA3, a tomato Aux/IAA at the crossroads of auxin and ethylene signaling involved in differential growth. *Journal of Experimental Botany* 60, 1349–1362.

Chabikwa, T.G., Barbier, F.F., Tanurdzic, M. and Beveridge, C.A. (2020) *De novo* transcriptome assembly and annotation for gene discovery in avocado, macadamia and mango. *Scientific Data* 7:9.

Chanderbali, A.S., van der Werff, H. and Renner, S.S. (2001) Phylogeny and historical biogeography of Lauraceae: evidence from the chloroplast and nuclear genomes. *Annals of the Missouri Botanical Garden* 88, 104–134.

Chanderbali, A.S., Albert, V.A., Ashworth, V.E.T.M., Clegg, M.T., Litz, R.E., Soltis, D.E. and Soltis, P.S. (2008) *Persea americana* (avocado) bringing ancient flowers to fruit in the genomic era. *BioEssays* 30, 386–396.

Chanderbali, A.S., Albert, V.A., Leebens-Mackd, J., Altmane N.S., Soltis, D.E. and Soltis, P. (2009) Transcriptional signatures of ancient floral developmental genetics in avocado (*Persea americana*; Lauraceae). *Proceedings of the National Academy of Sciences* 106, 8929–8934.

Chaparro-Pulido, C.A., Montiel, M.M., Palomo-Ríos, E., Mercado, J.A. and Pliego-Alfaro, F. (2014) Development of an efficient transient transformation protocol for avocado (*Persea americana* Mill.) embryogenic callus. *In Vitro Cellular and Developmental Biology-Plant* 50, 292–298.

Chase, M.W., Christenhusz, M.J.M., Fay, M.F., Byng, J.W., Judd, W.S., *et al.* (2016) An update of the Angiosperm Phylogeny Group classification for the orders and families of flowering plants: APG IV. *Botanical Journal of the Linnean Society* 181, 1–20.

Christoffersen, R.E., Tucker, M.L. and Laties, G.G. (1984) Cellulase gene expression in ripening avocado fruit: the accumulation of cellulase mRNA and protein as demonstrated by cDNA hybridization and immunodetection. *Plant Molecular Biology* 3, 385–391.

Cooper, P.A. (1987) Advances in micropropagation of avocado (*Persea americana* Mill.). *Acta Horticulturae* 212, 571–575.

Cortés-Rodríguez, M.A., López-Gómez, R., Martínez-Pacheco, M.M., Suárez-Rodríguez, L.M., Hernández-García, A., *et al.* (2011) *In vitro* propagation of Mexican race avocado (*Persea americana* Mill. var. *drymifolia*). *Acta Horticulturae* 923, 47–52.

Coto, O., Rodriguez, N.N., Fuentes, J.L., Alvarez, A., Machado, M., *et al.* (2014) Mutation-based breeding of avocado in Cuba: state of the art. In: Tomiekova, N.B., Kozgar, M.J. and Wani, M.R. (eds) *Mutagenesis: Exploring Genetic Diversity of Crops*. Wageningen Academic Publishers, Wageningen, The Netherlands, pp. 265–282.

Cruz-Hernández, A., Witjaksono, Litz, R.E. and Gómez-Lim, M.A. (1998) *Agrobacterium tumefaciens* mediated transformation of embryogenic avocado cultures and regeneration of somatic embryos. *Plant Cell Reports* 17, 493–503.

Dahan, Y., Rosenfeld, R., Zadiranov, V. and Irihimovitch, V. (2010) A proposed conserved role for an avocado *fw2.2* like gene as a negative regulator of fruit cell division. *Planta* 232, 663–676.

De la Cruz, E., Hernandez, M., Rubi, M. and Saavedra, C. (1993) Radiosensibility study on Hass avocado for breeding purposes. In: *Proceedings of Salvador Sánchez Colín Foundation*. CICTAMEX, Coatepec Harinas, pp. 121–128.

De la Cruz, E., Rubi, M. and Falcon, B. (1995) Efecto de la radiación gamma sobre el desarrollo vegetativo y reproductivo de aguacate cv. Hass. In: *Proceedings of Salvador Sánchez Colín Foundation*. CICTAMEX, Coatepec Harinas, pp. 55–60. (In Spanish.)

De la Cruz, E., Rubi, M., Sandoval, L. and García-Andrade, J.M. (1998) Architecture variability on Hass avocado subjected to ^{60}Co gamma irradiation. In: *Proceedings of the 17th Phytogenetic Congress*, Acapulco, p. 76.

De la Viña, G., Azcón-Aguilar, C., Azcón, R., Barea, J.M., Barceló, A. and Pliego-Alfaro, F. (1996) Effects of arbuscular mycorrhizas in the acclimatization of micropropagated avocado plants. In: Azcón-Aguilar, C. and Barea, J.M. (eds) *Mycorrhizas in Integrated Systems (From Genes to Plant Development)*. EUR Report 16728-EN. Directorate-General XII Science; Research and Development, Brussels, pp. 491–494.

De la Viña, G., Pliego-Alfaro, F., Driscoll, S.P., Mitchell, V.J., Parry, M.A. and Lawlor, D.W. (1999) Effects of CO_2 and sugars on photosynthesis and composition of avocado leaves grown *in vitro*. *Plant Physiology and Biochemistry* 37, 587–595.

De la Viña, G., Barceló-Muñoz, A. and Pliego-Alfaro, F. (2001) Effect of culture media and irradiance level on growth and morphology of *Persea americana* Mill. microcuttings. *Plant Cell, Tissue and Organ Culture* 65, 229–237.

Dixon, R.A. and Fuller, K.W. (1976) Effects of synthetic auxin levels on phaseollin production and phenylalanine ammonia lyase (PAL) activity in tissue cultures of *Phaseolus vulgaris* L. *Physiological Plant Pathology* 11, 287–292.

Djami-Tchatchou, A.T. and Straker C.J. (2012) The isolation of high quality RNA from the fruit of avocado (*Persea americana* Mill.). *South African Journal of Botany* 78, 44–46.

Djami-Tchatchou, A.T., Straker C.J. and Allie, F. (2012) 454 Sequencing for the identification of genes differentially expressed in avocado fruit (cv. Fuerte) infected by *Colletotrichum gloeosporioides. Journal of Phytopathology* 160, 449–460.

Djami-Tchatchou, A.T., Allie, F. and Straker, C.J. (2013) Expression of defence-related genes in avocado fruit (cv. Fuerte) infected with *Colletotrichum gloeosporioides. South African Journal of Botany* 86, 92–100.

Dopico, B., Lowe, A.L., Wilson, I.D., Merodio, C. and Grierson, D. (1993) Cloning and characterization of avocado fruit mRNAs and their expression during ripening and low-temperature storage. *Plant Molecular Biology* 21, 437–449.

Durán-Vila, N., Juárez, J. and Arregui, J.M. (1988) Production of viroid-free grapevines by shoot tip culture. *American Journal of Ecological Viticulture* 39, 217–220.

Eeckhaut, T., van Houtven, W., Bruznican, S., Leus, L. and van Huylenbroeck, J. (2020) Somaclonal variation in *Chrysanthemum morifolium* protoplast regenerants. *Frontiers in Plant Science* 11, 607171.

Efendi, D. (2003) Transformation and cryopreservation of embryogenic cultures of avocado *(Persea americana* Mill.) cultures. PhD thesis, University of Florida, Gainesville, Florida.

Efendi, D. and Litz, R.E. (2003) Cryoconservation of avocado. In: *Proceedings of the V World Avocado Congress,* Vol II, Málaga, pp. 111–114.

Encino-López, J., Castro-Mercado, E., Rangel-Sánchez, G. and García-Pineda, E. (2011) Interaction of extracellular proteinases of *Phytophthora cinnamomi* with proteinase inhibitors produced by avocado root. *Journal of Plant Pathology and Microbiology* 2, 4.

Engelbrecht, J. and van den Berg, N. (2013) Expression of defence-related genes against *Phytophthora cinnamomi* in five avocado rootstocks. *South African Journal of Science* 109, 1–8.

Favre, J.M. and Juncker, B. (1987) *In vitro* growth of buds taken from seedlings and adult plant material in *Quercus robur* L. *Plant Cell, Tissue and Organ Culture* 8, 49–60.

Fick, A., Swar, V., Backer, R., Bombarely, A., Engelbrecht, J. and van den Berg, N. (2022) Partially resistant avocado rootstock Dusa® shows prolonged upregulation of nucleotide binding-Leucine rich repeat genes in response to *Phytophthora cinnamomi* infection. *Frontiers in Plant Science* 13, 793644.

Fossi, M., Amundson, K., Kuppu, S., Britt, A. and Comai, L. (2019) Regeneration of *Solanum tuberosum* plants from protoplasts induces widespread genome instability. *Plant Physiology* 180, 78–86.

Frolich, E.F. (1951) Rooting Guatemalan avocado cuttings. *California Avocado Society Yearbook* 36, 136–138.

Frolich, E.F. and Platt, R.G. (1972) Use of the etiolation technique in rooting avocado cuttings. *California Avocado Society Yearbook* 55, 97–109.

Fuentes, J.L., Santiago, L., Rodríguez, N.N., Coto-Arbelo, O., Alvarez, A., *et al.* (2009) Combining zygotic embryo culture and mutation induction to improve salinity tolerance in avocado (*Persea americana* Mill). In: *Induced Mutation in Tropical Fruit Trees, IAEA-TECDOC-1615*. International Atomic Energy Agency, Vienna, pp. 71–82.

Gamborg, O.L., Miller, R.A. and Ojima, K. (1968) Plant cell cultures. I. Nutrient requirements of suspension cultures of soybean root cells. *Experimental Cell Research* 50, 151–158.

García-Rojas, M., Gudenschwager, O., Defilippi, B.G. and González-Agüero, M. (2012) Identification of genes possibly related to loss of quality in late-season 'Hass' avocados in Chile. *Postharvest Biology and Technology* 73, 1–7.

Ge, Y., Cheng, Z., Si, X., Ma, W., Tan, L., *et al.* (2019a) Transcriptome profiling provides insight into the genes in carotenoid biosynthesis during the mesocarp and seed developmental stages of avocado (*Persea americana*). *International Journal of Molecular Sciences* 20, 4117.

Ge, Y., Tan, L., Wu, B., Wang, T., Zhang, T., *et al.* (2019b) Transcriptome sequencing of different avocado ecotypes: de novo transcriptome assembly, annotation, identification and validation of EST-SSR markers. *Forests* 10:411.

Ge, Y., Zhang, T., Wu, B., Tan, B.L., Ma, F., *et al.* (2019c) Genome-wide assessment of avocado germplasm determined from specific length amplified fragment sequencing and transcriptomes: population structure, genetic diversity, identification, and application of race-specific markers. *Genes* 10:215.

George, E.F. (1993) *Plant Propagation by Tissue Culture. I. The Technology*. Exegetics, Edington, UK.

González-Rosas, H. and Salazar-García, S. (1984) Root induction and vegetative development from avocado plantules (*Persea americana* Mill.). *California Avocado Society Yearbook* 68, 167–169.

González-Rosas, H., Salazar-García, S. and Valdivia, V.V. (1985) Propagation *in vitro* of Chinini (*Persea schiedeana* Nees). *California Avocado Society Yearbook* 69, 125–131.

Grosser, J.W. and Gmitter, F.G. Jr (1990) Protoplast fusion and citrus improvement. *Plant Breeding Reviews* 8, 339–374.

Gudenschwager, O., García-Rojas, M., Defilippi, B.G. and González-Agüero, M. (2013) Identification and characterization of two putative genes encoding acetyl-coenzyme A carboxylase subunits that are possibly associated with internal browning during cold storage of 'Hass' avocados (*Persea americana* Mill.). *Postharvest Biology and Technology* 84, 74–80.

Guzmán-García, E., Bradai, F. and Sanchez-Romero, C. (2013a) Cryopreservation of avocado embryogenic cultures using the droplet-vitrification method. *Acta Physiologia Plantarum* 35, 183–193.

Guzmán-García, E., Sanchez-Romero, C. Panis, B. and Carpentier, S.C. (2013b) The use of 2D-DIGE to understand the regeneration of somatic embryos in avocado. *Proteomics* 13, 3498–3507.

Guzmán-Rodríguez, J., Ibarra-Laclette, E., Suárez-Rodríguez, L.M., Herrera-Estrella, L., Ochoa-Zarzosa A., *et al.* (2013a) Analysis of expressed sequence tags (ESTs) of avocado seed (*Persea americana* var. *drymifolia*) reveals abundant expression of the antimicrobial peptide snakin gene. *Plant Physiology and Biochemistry* 70, 318–324.

Guzmán-Rodríguez, J.J., López-Gómez, R., Ochoa-Zarzosa, A., Suárez-Rodríguez, L.M., Salgado-Garciglia, R., *et al.* (2013b) Antibacterial activity of defensin (*Pa*Def) from avocado fruit (*Persea americana* var. *drymifolia*) expressed in endothelial cells against *Escherichia coli* and *Staphylococcus aureus*. *BioMed Research International* 2013, 986273.

Harty, P.A. (1985) Propagation of avocado by tissue culture: development of a culture medium for multiplication of shoots. *South African Avocado Growers' Association Yearbook* 8, 70–71.

Hernández, I., Uarrota, V., Fuentealba, C., Paredes, D., Defilippi, B.G., *et al.* (2022) Transcriptome and hormone analyses reveals differences in physiological age of 'Hass' avocado fruit. *Postharvest Biology and Technology* 185, 111806.

Hershkovitz, V., Friedman, H., Goldschmidt, E.E. and Pesis, E. (2009a) The role of the embryo and ethylene in avocado fruit mesocarp discoloration. *Journal of Experimental Botany* 60, 791–799.

Hershkovitz, V., Friedman, H., Goldschmidt, E.E., Feygenberg, O. and Pesis, E. (2009b) Induction of ethylene in avocado fruit response to chilling stress on tree. *Journal of Plant Physiology* 166, 1855–1862.

Hiti-Bandaralage, J.C.A., Hayward, A., O'Brien, C. and Mitter, N. (2015) Gibberellin and cytokinin in synergy for a rapid nodal multiplication system of avocado. In: *Proceedings of the VIII World Avocado Congress*, Lima. Available at: https://www.avocadosource.com/WAC8/Section_02/HitiBandaralageJCA2015.pdf (accessed 15 September 2023).

Hiti-Bandaralage, J.C.A., Hayward, A. and Mitter, N. (2017) Micropropagation of avocado (*Persea americana* Mill.). *American Journal of Plant Science* 8, 2898.

Hiti-Bandaralage, J.C.A., Hayward, A., O'Brien, C., Beveridge, C. and Mitter, N. (2018) Acclimatization of micropropagated mature avocado. *Acta Horticulturae* 1224, 13–20.

Hiti-Bandaralage, J.C.A., Hayward, A., O'Brien, C., Gleeson, M., Nak, W. and Mitter, N. (2019) Advances in avocado propagation for the sustainable supply of planting materials. In: Yahia E.M. (ed) *Achieving Sustainable Cultivation of Tropical Fruit*. Burleigh Dodds Science Publishing, Cambridge, pp. 215–238.

Hiti-Bandaralage, J.C.A., Hayward, A., O'Brien, C., Ahsan, U., Gleeson, M., Xue, Y. and Mitter, N. (2020) Phase change related microRNA profiles in the plant regeneration process of avocado through shoot-tip culture. *Annals of Advances Agricultural Sciences* 4, 9–17.

Hiti-Bandaralage, J.C.A., Hayward, A. and Mitter, N. (2022) Structural disparity of avocado rootstocks in vitro for rooting and acclimation success. *International Journal of Plant Biology* 13, 426–442.

Huanga, Y.J., Tob, K.Y., Yapc, M.N., Chianga, W.J., Suenc, D.F. and Grace Chenc, S.C. (2001) Cloning and characterization of leaf senescence up-regulated genes in sweet potato. *Physiologia Plantarum* 113, 384–391.

Ibarra-Laclette, E., Méndez-Bravo, A., Pérez-Torres, C.A., Albert, V.A., Mockaitis, K., *et al*. (2015) Deep sequencing of the Mexican avocado transcriptome, an ancient angiosperm with a high content of fatty acids. *BMC Genomics* 16:599.

Ishida, I., Tukahara, M., Yoshioka, M., Ogawa, T., Kakitani, M. and Toguri, T. (2002) Production of anti-virus, viroid plants by genetic manipulations. *Pest Management Science* 58, 1132–1136.

Jefferson, R.A. (1987) Assaying chimeric genes in plants: the GUS gene fusion system. *Plant Molecular Biology* 5, 387–405.

Jensen, C.J. (1977) Monoploid production by chromosome elimination. In: Reinert, J. and Bajaj, Y.P.S. (eds) *Applied and Fundamental Aspects of Plant Cell, Tissue and Organ Culture*. Springer, Berlin, pp. 299–340.

Joint FAO/IAEA. (2010) Mutant Variety and Genetic Stock Database. Available at: http://mvgs.iaea.org/ (accessed 11 October 2011).

Jorgensen, R. (1993) The modification of horticultural plant phenotypes by direct gene transfer. *Scientia Horticulturae* 55, 1–4.

Kadman, A. (1985) Selection of avocado and mango rootstocks for calcareous soils. *Acta Horticulturae* 158, 63–67.

Kadman, A. and Ben-Ya'acov, A. (1980) 'GA-13' avocado rootstock selection. *HortScience* 15:206.

Kamiyoshihara, Y., Mizuno, S., Azuma, M., Miyohashi, F., Yoshida, M., *et al*. (2018) Opposite accumulation patterns of two glycoside hydrolase family 3 α-L-arabinofuranosidase proteins in avocado fruit during ripening. *The Horticulture Journal* 87, 430–435.

Kämper, W., Trueman, S.J., Cooke, J., Kasinadhuni, N., Bruton, A.J. and Ogbourne, S.M. (2021) Single-nucleotide polymorphisms that uniquely identify cultivars of avocado (*Persea americana*). *Applications in Plant Sciences* 9: e11440.

Kane, M.E., Sheehan, T.J., Dehgan, B. and Philman, N.L. (1989) *In vitro* propagation of Florida native plants: *Styrax americana* and *Persea palustris*. *Proceedings of the Florida State Horticultural Society* 102, 264–268.

Kanellis, A.K., Solomos, T. and Roubelakis-Angelakis, K.A. (1991) Suppression of cellulase and polygalacturonase and induction of alcohol dehydrogenase isoenzymes in avocado fruit mesocarp subjected to oxygen stress. *Plant Physiology* 96, 269–274.

Kao, K.N. and Michayluk, M.R. (1975) Nutritional requirements for growth of *Vicia hajstana* cells and protoplasts at a very low population density in liquid media. *Planta* 126, 105–110.

Kellam, M.K. and Coffey, M.D. (1985) Quantitative comparison of the resistance to *Phytophthora* root rot in three avocado rootstocks. *Phytopathology* 75, 230–234.

Kilaru, A., Cao, X., Dabbs, P.B., Sung, H.J., Rahman, M.M., *et al*. (2015) Oil biosynthesis in a basal angiosperm: transcriptome analysis of *Persea americana* mesocarp. *BMC Plant Biology* 15:203.

Kimura, M., Ikeda, T., Fukumoto, D., Yamasaki, N. and Yonekura, M. (1995) Primary structure of a cysteine proteinase inhibitor from the fruit of avocado (*Persea americana* Mill.). *Bioscience, Biotechnology, and Biochemistry* 59, 2328–2329.

Klee, H.J. (1993) Ripening physiology of fruit from transgenic tomato (*Lycopersicon esculentum*) plants with reduced ethylene synthesis. *Plant Physiology* 102, 911–916.

Klee, H.J., Hayford, M.B., Kretzmer, K.A., Barry, G.F. and Kishore, G.M. (1991) Control of ethylene synthesis by expression of bacterial enzyme in transgenic tomato. *Plant Cell* 3, 1187–1193.

Köhne, J.S. (1992) Field evaluation of 'Hass' avocado grown on Duke 7, G6 and G755C rootstocks. In: Lovatt, C., Holthe, P.A. and Arpaia, M.L. (eds) *Proceedings of the Second World Avocado Congress,* Vol. I. University of California, Riverside, California, pp. 301–303.

Kremer-Köhne, S. and Mukhumo, M.L. (2003) Breeding and field evaluation of new rootstocks for increased 'Hass' yields and resistance to root rot in South Africa. In: *Proceedings of the V World Avocado Congress,* Vol. II, Málaga, Spain, pp. 555–560.

Kremer-Köhne, S., van Rooyen, Z. and Köhne, J.S. (2011) Avocado rootstock evaluation techniques over the last 30 years. In: *Proceedings of the VII World Avocado Congress*, Cairns, Australia. Available at: http://www.avocadosource.com/wac7/Section_14/KremerKohneS2011.pdf (accessed 2 February 2012).

Kuhn, D.N., Groh, A., Rahaman, J., Freeman, B., Arpaia, M.L., *et al.* (2019a) Creation of an avocado unambiguous genotype SNP database for germplasm curation and as an aid to breeders. *Tree Genetics & Genomes* 15:71.

Kuhn, D.N., Livingstone D.S., III, Richards, J.H., Manosalva, P., van den Berg, N. and Chambers, A.H. (2019b) Application of genomic tools to avocado (*Persea americana*) breeding: SNP discovery for genotyping and germplasm characterization. *Scientia Horticulturae* 246, 1–11.

Kumari, P., Singh, K.P., Kumar, S. and Yadava, D.K. (2020) Development of a yellow-seeded stable allohexaploid Brassica through inter-generic somatic hybridization with a high degree of fertility and resistance to *Sclerotinia sclerotiorum. Frontiers in Plant Science* 11, 575591.

Kutsunai, S.Y., Lin, A.C., Percival, F.W., Laties, G.G. and Christoffersen, R.E. (1993) Ripening-related polygalacturonase cDNA from avocado. *Plant Physiology* 163, 289–290.

Lahav, E. and Lavi, U. (2002) Genetics and classical breeding. In: Whiley, A.W., Schaffer, B. and Wolstenholme, B.N. (eds) *The Avocado: Botany, Production and Uses*. CAB International, Wallingford, UK, pp. 39–69.

Lin, W.C., Lu, C.F., Wu, J.W., Cheng, M.L., Lin, Y.M., *et al.* (2004) Transgenic tomato plants expressing the *Arabidopsis* NPR1 gene display enhanced resistance to a spectrum of fungal and bacterial diseases. *Transgenic Research* 13, 567–581.

Litz, R.E., Efendi, D., Raharjo, S.H.T., Padilla, G. and Moon, P.A. (2007a) Genetic transformation of avocado with S-adenosylmethionine hydrolase (samase) and evaluation of transformants after three years. In: *Proceedings of the VI World Avocado Congress,* Viña del Mar. Available at: https://www.avocadosource.com/wac6/en/Extenso/1c-24.pdf (accessed 15 July 2023).

Litz, R.E., Raharjo, S.H.T., Efendi, D., Witjaksono and Gomez-Lim, M.A. (2007b) Transformation of avocado with anti-fungal protein and SAM hydrolase genes. *Acta Horticulturae* 738, 447–450.

Litz, R.E., Perea-Arango, I. and Suárez Padron, I. (2010) A comparison of biotech methods for controlling infections by the avocado sunblotch viroid (ASBVD). In: *Proceedings of the 28th International Horticultural Congress,* Vol. II. Lisbon, Portugal, p. 280 (Abstract).

Litz, R.E., Pliego-Alfaro, F. and Hormaza, I. (2020) *Biotechnology of Fruit and Nut Crops*. 2nd edn. CAB International, Wallingford, UK.

Lloyd, G.B. and McCown, B.H. (1980) Commercially feasible micropropagation of mountain laurel, *Kalmia latifolia,* by use of shoot-tip culture. *International Plant Propagators' Society Combined Proceedings* 30, 421–427.

López-Encina, C.L., Parisi, A., O'Brien, C. and Mitter, N. (2014) Enhancing somatic embryogenesis in avocado (*Persea americana* Mill.) using a two-step culture system and including glutamine in the culture medium. *Scientia Horticulturae* 165, 44–50.

López-Rivera, L.A., Ramírez-Ramírez, I., González-Hernández, V.A., Cruz-Huerta, N. and Téliz-Ortiz, D. (2017) Differential gene expression of avocado defense genes in response to avocado sunblotch viroid infection. *Revista Mexicana de Fitopatología* 36, 151–161.

Louzada, E.S., Grosser, J.W. and Gmitter, F.G., Jr (1993) Intergeneric somatic hybridization of sexually incompatible parents: *Citrus sinensis* and *Atalantia ceylanica*. *Plant Cell Reports* 12, 687–690.

Lovatt, C.J. (1987) Stress. *California Avocado Society Yearbook* 71, 251–255.

Madi, L., Wang, X., Kobiler, E., Lichter, A. and Prusky, D. (2003) Stress on avocado fruits regulates Δ^9desaturase stearoyl ACP desaturase, expression, fatty acid composition, antifungal diene level and resistance to *Colletotrichum gloeosporioides* attack. *Physiological and Molecular Plant Pathology* 62, 277–283.

Mahomed, W. and van den Berg, N. (2011) EST sequencing and gene expression profiling of defence related genes from *Persea americana* infected with *Phytophthora cinnamomi*. *BMC Plant Biology* 11:167.

Márquez-Martín, B., Barceló-Muñoz, A., Pliego-Alfaro, F. and Sánchez-Romero, C. (2007) Efecto de la maduración y desecación *in vitro* sobre la conversión en planta de embriones inmaduros de aguacate. *Actas de Horticultura* 48, 93–96. (In Spanish.)

Márquez-Martín, B., Guzmán-García, E., Barceló-Muñoz, A., Pliego-Alfaro, F. and Sánchez-Romero, C. (2009) Effects of an *in vitro* maturation treatment on plant recovery from avocado zygotic embryos. *Scientia Horticulturae* 122, 532–539.

Márquez-Martín, B., Sesmero, R., Quesada, M.A., Pliego-Alfaro, F. and Sánchez-Romero, C. (2011) Water relations in culture media influence maturation of avocado somatic embryos. *Journal of Plant Physiology* 168, 2028–2034.

Márquez-Martín, B., Barceló-Muñoz, A., Pliego-Alfaro, F. and Sánchez-Romero, C. (2012) Somatic embryogenesis and plant regeneration in avocado (*Persea americana*, Mill.): influence of embryogenic culture type. *Journal of Plant Biochemistry and Biotechnology* 21, 180–188.

Márquez-Martín, B., Pliego-Alfaro, F. and Sánchez-Romero, C. (2017) Improvement of plant recovery from avocado zygotic embryos by desiccation under high relative humidity conditions. *Scientia Horticulturae* 222, 169–174.

McGarvey, D.J., Sirevag, R. and Christoffersen, R.E. (1992) Ripening-related genes from avocado fruit. *Plant Physiology* 8, 554–559.

Meyer, M.D. and Terry, L.A. (2010) Fatty acid and sugar composition of avocado, cv. Hass, in response to treatment with an ethylene scavenger or 1-methylcyclopropene to extend storage life. *Food Chemistry* 121, 1203–1210.

Mitter, N., Parisi, A., Mitchell, R., O'Brien, C., Dietzgen, R.G., Bailey, A. and Niblett, C.L. (2011) RNA silencing based resistance to *Phytophthora cinnamomi*. In: *Proceedings of the VII World Avocado Congress*, Cairns, Australia. Available at: http://www.avocadosource.com (accessed 2 February 2012).

Mohamed-Yasseen, Y. (1993) *In vitro* propagation of avocado (*Persea americana* Mill.). *California Avocado Society Yearbook* 77, 107–111.

Mooney, P.A. and van Staden, J. (1987) Induction of embryogenesis in callus from immature embryos of *Persea americana* Mill. *Canadian Journal of Botany* 65, 622–626.

Moreno-Pérez, A., Zumaquero, A., Martínez-Ferri, E., López-Herrera, C., Pliego-Alfaro, F., Palomo-Ríos, E. and Pliego, C. (2023) A comparative transcriptome analysis of avocado embryogenic lines susceptible or resistant to *Rosellinia necatrix* exudate. *Agronomy* 13, 1354.

Murashige, T. and Skoog, F. (1962) A revised medium for rapid growth and bioassays with tobacco tissue cultures. *Physiologia Plantarum* 15, 473–497.

Nath, O., Fletcher, S.J., Hayward, A., Shaw, L.M., Masouleh, A.K., *et al.* (2022) A haplotype resolved chromosomal level avocado genome allows analysis of novel avocado genes. *Horticulture Research* 9: uhac157.

Navarro, L., Roistacher, C.N. and Murashige, T. (1975) Improvement of shoot-tip grafting *in vitro* for virus-free citrus. *Journal of the American Society for Horticultural Science* 100, 471–479.

Nel, D.A., Kotzé, J.M. and Snyman, C.P. (1983) Progress in tissue culture of avocado. *South African Avocado Growers' Association Yearbook* 6, 90–91.

Nhut, D.T., Thi, N.N., Khiet, B.L.T. and Luan, V.Q. (2008) Peptone stimulates *in vitro* shoot and root regeneration of avocado (*Persea americana* Mill.). *Scientia Horticulturae* 115, 124–128.

Nitsch, J. (1951) Growth and development *in vitro* of excised ovaries. *American Journal of Botany* 38, 566–567.

O'Brien, C., Constantin, M., Walia, A., Yiing, J.L.Y. and Mitter, N. (2016) Cryopreservation of somatic embryos for avocado germplasm conservation. *Scientia Horticulturae* 211, 328–335.

O'Brien, C., Hiti-Bandaralage, J.C.A., Hayward, A. and Mitter, N. (2018) Avocado (*Persea americana* Mill). In: Jain, S.M. and Gupta, P. (eds) *Step Wise Protocols for Somatic Embryogenesis of Important Woody Plants*. *Forestry Sciences*, Vol. 85. Springer, Cham, Switzerland, pp. 305–328.

O'Brien C., Hiti-Bandaralage, J.C.A., Folgado, R., Lahmeyer, S., Hayward, A., Folsom, J. and Mitter, N. (2020a) A method to increase regrowth of vitrified shoot tips of avocado (*Persea americana* Mill.): first critical step in developing a cryopreservation protocol. *Scientia Horticulturae* 266, 109305.

O'Brien, C., Hiti-Bandaralage, J.C.A., Folgado, R., Lahmeyer, S., Hayward, A. and Mitter, N. (2020b) Developing a cryopreservation protocol for avocado (*Persea americana* Mill.) apical shoot tips using different antioxidants. *Acta Horticulturae* 1285, 15–22.

O'Brien, C.O., Hiti-Bandaralage, J.C.A., Folgado, R., Lahmeyer, S., Hayward, A., Folsom, J. and Mitter, N. (2021) First report on cryopreservation of mature shoot tips of two avocado *(Persea americana* Mill.) rootstocks. *Plant Cell Tissue and Organ Culture* 144, 103–113.

Ogawa, T., Toguri, T., Kudoh, H., Okamura, M., Momma, T., *et al.* (2005) Double-stranded RNA-specific ribonuclease confers tolerance against Chrysanthemum stunt viroid and tomato spotted wilt virus in transgenic chrysanthemum plants. *Breeding Science* 55, 49–55.

Owino, W.O., Nakano, R., Kubo, Y. and Inaba, A. (2002) Differential regulation of genes encoding ethylene biosynthesis enzymes and ethylene response sensor ortholog during ripening and in response to wounding in avocados. *Journal of the American Society for Horticultural Science* 127, 520–527.

Palomo-Ríos, E., Pliego, C., Muñoz-Blanco, J., Mercado, J.A. and Pliego-Alfaro, F. (2010) *Agrobacterium*-mediated transformation of avocado (*Persea americana* Mill.) with *At*NPR1 gene. In: *Proceedings of the 28th International Horticultural Congress,* Vol. II, Lisbon, p. 286 (Abstract).

Palomo-Ríos, E., Barceló, M., Barceló-Muñoz, A., Pliego, C., Blanco-Portales, R., *et al.* (2011) *Agrobacterium*-mediated transformation to induce tolerance to avocado white root rot. In: *Proceedings of the VII World Avocado Congress*, Cairns, Australia. Available at: http://www.avocadosource.com (accessed 2 February 2012). (In Spanish.)

Palomo-Ríos, E., Barceló-Muñoz, A., Mercado, J.A. and Pliego-Alfaro, F. (2012) Evaluation of key factors influencing *Agrobacterium*-mediated transformation of somatic embryos of avocado (*Persea americana*, Mill). *Plant Cell Tissue and Organ Culture* 109, 201–211.

Palomo Rios, E, Pérez, C., Mercado, J.A. and Pliego-Alfaro, F. (2013) Enhancing frequency of regeneration of somatic embryos of avocado (*Persea americana* Mill.) using semi-permeable cellulose acetate membranes. *Plant Cell Tissue and Organ Culture* 115, 199–207.

Palomo-Ríos, E., Cerezo, S., Mercado, J.A. and Pliego-Alfaro, F. (2017) Agrobacterium-mediated transformation of avocado (*Persea americana* Mill.) somatic embryos with fluorescent marker genes and optimization of transgenic plant recovery. *Plant Cell Tissue and Organ Culture* 128, 447–455.

Panis, B., Piette, B. and Swennen, R. (2005) Droplet vitrification of apical meristems: a cryopreservation protocol applicable to all Musaceae. *Plant Science* 168, 45–55.

Pence, V.C., Ballesteros, D., Walters, C., Reed, B.M., Philpott, M., *et al.* (2020) Cryobiotechnologies: tools for expanding long-term ex situ conservation to all plant species. *Biological Conservation* 250, 108736.

Perán-Quesada, R., Sánchez-Romero, C., Barceló-Muñoz, A. and Pliego-Alfaro, F. (2004) Factors affecting maturation of avocado somatic embryos. *Scientia Horticulturae* 102, 61–73.

Perán-Quesada, R., Sánchez-Romero, C., Pliego-Alfaro, F. and Barceló-Muñoz, A. (2005) Histological aspects of avocado embryo development and effect of developmental stages on germination. *Seed Science Research* 15, 125–132.

Percival, F.W., Cass, L.G., Bozak, C. and Christoffersen, R. (1991) Avocado fruit protoplasts: a cellular model system for ripening studies. *Plant Cell Reports* 10, 512–516.

Pérez-Torres, C.A., Ibarra-Laclette, E., Hernández-Domínguez, E.E., Rodríguez-Haas, B., Pérez-Lira, A.J., *et al.* (2021) Molecular evidence of the avocado defense response to *Fusarium kuroshium* infection: a deep transcriptome analysis using RNA-Seq. *Peer Journal* 9: e11215.

Phillips, D., Weste, G. and Hinch, J.M. (1991) Resistance to *Phytophthora cinnamomi* in callus derived from three avocado cultivars. *Canadian Journal of Botany* 69, 2026–2032.

Pliego-Alfaro, F. (1981) A morphogenetic study of the avocado (*Persea americana* Mill.) *in vitro*. I. Development of a rooting bioassay and its application to studying restoration by grafting of rooting competence in adult shoots. II. Somatic embryogenesis in callus. PhD thesis, University of California, Riverside, California.

Pliego-Alfaro, F. (1988) Development of an *in vitro* rooting bioassay using juvenile-phase stem cuttings of *Persea americana* Mill. *Journal of Horticultural Science* 63, 295–301.

Pliego-Alfaro, F. and Bergh, B.O. (1992) Avocado. In: Hammerschlag, F.A. and Litz, R.E. (eds) *Biotechnology of Perennial Fruit Crops*. CAB International, Wallingford, UK, pp. 323–333.

Pliego-Alfaro, F. and Litz, R.E. (2007) Biotechnology and avocado improvement. *California Avocado Society Yearbook* 90, 149–167.

Pliego-Alfaro, F., López-Encina, C. and Barceló-Muñoz, A. (1987) Propagation of avocado rootstocks by tissue culture. *South African Avocado Growers' Yearbook* 10, 27–28.

Pliego-Alfaro, F. and Murashige, T. (1987) Possible rejuvenation of adult avocado by graftage onto juvenile rootstocks *in vitro*. *HortScience* 22, 1321–1324.

Pliego-Alfaro, F. and Murashige, T. (1988) Somatic embryogenesis in avocado (*Persea americana* Mill.) *in vitro*. *Plant Cell, Tissue and Organ Culture* 12, 61–66.

Prabhu, S.A., Ndlovu, B., Engelbrecht, J. and van den Berg, N. (2017) Generation of composite *Persea americana* (Mill.) (avocado) plants: a proof of concept study. *PLOS ONE* 12: e0185896.

Premkumar, A., Barceló-Muñoz, A., Pliego-Alfaro, F., Quesada, M.A. and Mercado, J.A. (2002) Influences of exogenous sucrose on juvenile avocado during *in vitro* cultivation and subsequent *ex vitro* acclimatization. *Trees* 16, 569–575.

Premkumar, A., Barceló-Muñoz, A., Quesada, M.A., Mercado, J.A. and Pliego-Alfaro, F. (2003) Influence of sucrose concentration on *in vitro* rooting, growth, endogenous sugars and *ex vitro* survival of juvenile avocado. *Journal of Horticultural Science & Biotechnology* 78, 46–50.

Prusky, D., Hamdan, H., Ardi, R. and Keen, T. (1996) Induction of biosynthesis of epicatechin in avocado suspension cells treated with an enriched CO_2 atmosphere. *Physiological and Molecular Plant Pathology* 48, 171–178.

Quintero-Jimenez, A., Heredia-Garcia, E., Aguirre-Mancilla, C.L., Raya-Perez, J.C., Ramirez-Pimentel, J.G. and Iturriaga, G. (2020) Embryogenesis and organogenesis of Mexican creole avocado. *Revista Mexicana de Ciencias Agrícolas* 11:7.

Raharjo, S.H.T. and Litz, R.E. (2005) *In vitro* and *in vivo* grafting for somatic embryo rescue and plant recovery in avocado (*Persea americana* Mill.). *Plant Cell Tissue and Organ Culture* 82, 1–9.

Raharjo, S., Witjaksono, Efendi, D., Suárez, I., Gasca Vazquez, I., Gomez-Lim, M.A. and Litz, R. (2003) Genetic transformation of avocado. In: *Proceedings of the V World Avocado Congress (Actas V Congreso Mundial del Aguacate),* Vol I, Málaga, Spain, pp. 115–118.

Raharjo, S.H.T., Witjaksono, Gómez-Lim, M.A., Padilla G. and Litz, R.E. (2008) Recovery of avocado (*Persea americana* Mill.) plants transformed with the antifungal plant defensin gene pdf1.2. *In Vitro Cellular and Developmental Biology – Plant* 44, 254–262.

Raviv, A., Avenido, R.A., Tisalona, L.F., Damasco, O.P., Mendoza, E.M.T., Pinkas, Y. and Zilkah, S. (1998) Callus and somatic embryogenesis of *Persea* species. *Plant Tissue Culture and Biotechnology* 4, 196–206.

Raya, M.Y.A. (2004) Efecto de la temperatura, sales minerales y osmorreguladores sobre la conservación *in vitro* de masas proembriogénicas y embriones somáticos de aguacate (*Persea americana* Mill.) var. Hass. Tesis de Licenciatura. Universidad Michoacana de San Nicolás de Hidalgo, Michoacán. (In Spanish.)

Reddy, A.S. and Poovaiah, B.W. (1990) Molecular cloning and sequencing of a cDNA for an auxin-repressed mRNA: correlation between fruit growth and repression of the auxin-regulated gene. *Plant Molecular Biology* 14, 127–136.

Reed, B.M. (2008) *Plant Cryopreservation: A Practical Guide*. Springer-Business Media LLC, New York.

Reeksting, B.J., Coetzer, N., Mahomed, W., Engelbrecht, J. and van den Berg, N. (2014) *De Novo* sequencing, assembly, and analysis of the root transcriptome of *Persea americana* (Mill.) in response to *Phytophthora cinnamomi* and flooding. *PLOS ONE* 9: e86399.

Reeksting, B.J., Olivier, N.A. and van den Berg, N. (2016) Transcriptome responses of an ungrafted Phytophthora root rot tolerant avocado (*Persea americana*) rootstock to flooding and *Phytophthora cinnamomi. BMC Plant Biology* 16: 205.

Rendón-Anaya, M., Ibarra-Laclette, E., Méndez-Bravo, A., Lan, T., Zheng, C., *et al.* (2019) The avocado genome informs deep angiosperm phylogeny, highlights introgressive hybridization, and reveals pathogen-influenced gene space adaptation. *Proceedings of the National Academy of Sciences* 116, 17081–17089.

Restrepo-Osorio, C., Gómez-Velasquez, F.A., Gil-Correal, A., Torres-Bonilla, J.M. and Urrea-Trujillo, A.I. (2018) *In vitro* propagation of avocado (*Persea americana* Mill.) cv. Hass through morphogenesis. *Acta Agronomica* 67, 160–167.

Roest, S. and Gilissen, L.J.W. (1993) Regeneration from protoplasts, a supplementary literature review. *Acta Botanica Neerlandica* 41, 1–23.

Rubinstein, M., Eshed, R., Rozen, A., Zviran, T., Kuhn, D.N. *et al.* (2019) Genetic diversity of avocado (*Persea americana* Mill.) germplasm using pooled sequencing. *BMC Genomics* 20, 1–10.

Sabag, M., Ari, G.B., Zviran, T., Biton, I., Goren, M. *et al.* (2013) PaKRP, a cyclin-dependent kinase inhibitor from avocado, may facilitate exit from the cell cycle during fruit growth. *Plant Science* 213, 18–29.

Sakai, A., Kobayashi, S. and Oiyama, I. (1990) Cryopreservation of nucellar cells of navel orange (*Citrus sinensis* Osb. *var. brasiliensis* Tanaka) by vitrification. *Plant Cell Reports* 9, 30–33.

Salazar-Gonzalez, J.A., Castro-Medina, M., Bernardino-Rivera, L.E. Martinez-Terrazas, E., Casson, S.A. and Urrea-Lopez, R. (2023) In planta transient transformation of avocado (*Persea americana*) by vacuum agroinfiltration of aerial plant parts. *Plant Cell Tissue and Organ Culture* 152, 635–646.

Sánchez-Colín, S., Rubi, M. and Sosa, R. (1990) Variability induction by irradiation of avocado (*Persea americana* Mill.) scions. In: *Proceedings of the Salvador Sánchez Colín Foundation*. CICTAMEX, Coatepec Harinas, pp. 41–48.

Sánchez-Romero, C., Perán-Quesada, R., Barceló-Muñoz, A. and Pliego-Alfaro, F. (2002) Variations in storage protein and carbohydrate levels during development of avocado zygotic embryos. *Plant Physiology and Biochemistry* 40, 1043–1049.

Sánchez-Romero, C., Perán-Quesada, R., Márquez-Martín, B., Barceló-Muñoz, A. and Pliego-Alfaro, F. (2003) Efecto de la desecación parcial sobre la germinación de embriones zigóticos inmaduros de aguacate. In: *Proceedings of the V World Avocado Congress*, Vol. 1, Málaga, Spain, pp. 83–87. (In Spanish.)

Sánchez-Romero, C., Perán-Quesada, R., Márquez-Martín, B., Barceló-Muñoz, A. and Pliego-Alfaro, F. (2007) *In vitro* rescue of immature avocado (*Persea americana* Mill.) embryos. *Scientia Horticulturae* 111, 365–370.

Sano, T., Nagayama, A., Ogawa, T., Ishida, I. and Okada, Y. (1997) Transgenic potato expressing a double-stranded RNA-specific ribonuclease is resistant to potato spindle tuber viroid. *Nature Biotechnology* 15, 1290–1294.

Schall, S. (1987) La multiplication de l'avocatier (*Persea americana* Mill., cv. Fuerte) par microbouturage *in vitro*. *Fruits* 42, 171–176. (In French.)

Schroeder, C.A. (1971) The response of avocado pericarp tissue to temperature and light *in vitro*. *California Avocado Society Yearbook* 54, 85–89.

Schroeder, C.A. (1977) Longevity of plant tissue cultures. *California Avocado Society Yearbook* 56, 138–141.

Schroeder, C.A. (1980) Avocado tissue *in vitro*. *California Avocado Society Yearbook* 64, 139–141.

Schroeder, C.A., Kay, E. and Davis, L.H. (1962) Totipotency of cells from fruit pericarp tissue *in vitro*. *Science* 138, 595–596.

Schuerman, P.L. and Dandekar, A.M. (1993) Transformation of temperate woody crops: progress and potentials. *Scientia Horticulturae* 55, 101–124.

Sharma, D.R., Kaur, R. and Kumar, K. (1996) Embryo rescue in plants. A review. *Euphytica* 89, 325–337.

Sharma, P., Al-Dossary, O., Alsubaie, B., Al-Mssallem, I., Nath, O. *et al.* (2021) Improvements in the sequencing and assembly of plant genomes. *Gigabyte* 2021, 1–10.

Sharon, D., Cregan, P.B., Mhameed, S., Kusharska, M., Hillel, J., Lahav, E. and Lavi, U. (1997) An integrated genetic linkage map of avocado. *Theoretical and Applied Genetics* 95, 911–921.

Skene, K.G.M. and Barlass, M. (1983) *In vitro* culture of abscissed immature avocado embryos. *Annals of Botany* 52, 667–672.

Smith, C.J.S., Watson, C.F., Ray, J., Bird, C.R., Morris, P.C., Schuch, W. and Grierson, D. (1988) Antisense RNA inhibition of polygalacturonase gene expression in transgenic tomatoes. *Nature* 334, 724–726.

Solares, E., Morales-Cruz, A., Figueroa-Balderas, R., Focht, E., Ashworth, V.E.T.M., *et al.* (2022) Insights into the domestication of avocado and potential genetic contributors to heterodichogamy. *G3 Genes|Genomes|Genetics*, 13(2), jkac323.

Sowka, S., Hesieh, L.S., Krebitz, M., Akasawa, A., Martin, B.M., *et al.* (1998) Identification and cloning of Prs a 1, a 32-kDa endochitinase and major allergen of avocado, and its expression in the yeast *Pichia pastoris*. *The Journal of Biological Chemistry* 43, 28091–28097.

Starrett, D.A. and Laties, G.G. (1993) Ethylene and wound-induced gene expression in the preclimacteric phase of ripening avocado fruit and mesocarp discs. *Plant Physiology* 103, 227–234.

Stuiver, M. (2006) Engineering fungal resistance in crops. In: Halford, N. (ed.) *Plant Biotechnology*. Wiley, Toronto, Canada, pp. 225–239.

Suarez, I.E. (2003) Avocado sunblotch viroid (ASBVd) in cell and tissue cultures of avocado (*Persea americana* Mill.). PhD thesis, University of Florida, Gainesville, Florida.

Suarez, I.E., Schnell, R.A., Kuhn, D.A. and Litz, R.E. (2005) Micrografting of ASBVd-infected avocado (*Persea americana* Mill.) plants. *Plant Cell Tissue and Organ Culture* 80, 179–185.

Suarez, I.E., Schnell, R.A., Kuhn, D.A. and Litz, R.E. (2006) Recovery of avocado plants (*Persea americana* Mill.) from embryogenic nucellar cultures from an avocado sunblotch viroid-infected tree. *Plant Cell Tissue and Organ Culture* 84, 27–37.

Suárez-Rodríguez, L.M., Sánchez-Albarrán, F., León-Corona, H. and López-Gómez, R. (2017) Transcriptome (ESTs) of avocado "native" mexicano early seed development shows abundance of regulatory, antioxidant and defense genes. In: Jimenez-Lopez, J.C. (ed.) *Advances in Seed Biology*. IntechOpen, London, pp. 45–63.

Suzuki, M., Tandon, P., Ishikawa, M. and Toyomasu, T. (2008) Development of a new vitrification solution, VSL, and its application to the cryopreservation of gentian axillary buds. *Plant Biotechnology Reports* 2, 123–131.

Talavera, A., Soorni, A., Bombarely, A., Matas, A.J. and Hormaza, J.I. (2019) Genome-wide SNP discovery and genomic characterization in avocado (*Persea americana* Mill.). *Scientific Reports* 9, 1–13.

Tamayo-Ramos, D.I., Salazar-González, J.A., Casson, S.A. and Urrea-López, R. (2022) Old and new horizons on *Persea americana* transformation techniques and applications. *Plant Cell, Tissue and Organ Culture* 150, 253–266.

Tateishi, A., Shiba, H., Ogihara, J., Isobe, K., Nomura, K., Watanabe, K. and Inoue, K. (2007) Differential expression and ethylene regulation of β-galactosidase genes and isozymes isolated from avocado (*Persea americana* Mill.). *Postharvest Biology and Technology* 45, 56–65.

Thorne, H.B., Axtens, J., and Best, T. (2022) Perceptual factors influencing the adoption of innovative tissue culture technology by the Australian avocado industry. *Agriculture* 12, 1288.

Trainotti, L., Tadiello, A. and Casadoro G. (2007) The involvement of auxin in the ripening of climacteric fruits comes of age: the hormone plays a role of its own and has an intense interplay with ethylene in ripening peaches. *Journal of Experimental Botany* 58, 3299–3308.

Valladares, R.S. (2005) Conservación *in vitro* de yemas axilares micropropagadas de aguacate (*Persea americana* Mill.). Tesis de Licenciatura. Universidad Michoacana de San Nicolás de Hidalgo, Michoacán. (In Spanish.)

Van den Berg, N., Christie, J.B., Aveling, T.A.S. and Engelbrecht, J. (2018a) Callose and β-1,3-glucanase inhibit *Phytophthora cinnamomi* in a resistant avocado rootstock. *Plant Pathology* 67, 1150–1160.

Van den Berg, N., Mahomed, W., Olivier, N.A., Swart, V. and Crampton, B.G. (2018b) Transcriptome analysis of an incompatible *Persea americana-Phytophthora cinnamomi* interaction reveals the involvement of SA-and JA-pathways in a successful defense response. *PLOS ONE* 13, e0205705.

Van den Berg, N., Swart, V., Backer, R., Fick, A., Wienk, R., Engelbrecht, J. and Prabhu, S.A. (2021) Advances in understanding defence mechanisms in *Persea americana* against *Phytophthora cinnamomi*. *Frontiers in Plant Science* 12, 636339.

Van den Bulk, R.W. (1991) Application of cell and tissue culture and *in vitro* selection for disease resistance breeding – a review. *Euphytica* 56, 269–285.

Vargas, V.M. (2008) Efecto fisiológico de brasinoesteroides y crioprotectores sobre yemas axilares de aguacate criollo producidas *in vitro*. PhD thesis, Universidad Autónoma de Nayarit. (In Spanish.)

Vega-Solórzano, D.E. (1989) Propagation *in vitro* of rootstocks of avocado. *California Avocado Society Yearbook* 73, 149–151.

Vergara-Pulgar, C., Rothkegel, K., González-Agüero, M., Pedreschi, R., Campos-Vargas, R., Defilippi, B.G. and Meneses, C. (2019) De novo assembly of *Persea americana* cv. 'Hass' transcriptome during fruit development. *BMC Genomics* 20, 1–14.

Vidal, M.T., Azcón-Aguilar, C., Barea, J.M. and Pliego-Alfaro, F. (1992) Mycorrhizal inoculation enhances growth and development of micropropagated plants of avocado. *HortScience* 27, 785–787.

Vidales-Fernández, I., Larios-Guzmán, A., Tapia-Vargas, L.M., Guillén-Andrade, H. and Villaseñor-Ramírez, F. (2011) Avocado germplasm cryopreservation. In: *Proceedings of the VII World Avocado Congress*, Cairns, Australia. Available at: http://www.avocadosource.com/wac7/Section_14/VidalesFernandezI2011b.pdf (accessed 2 February 2012). (In Spanish.)

Vidoy-Mercado, I., Imbroda-Solano, I., Hormaza, J.I., Pliego-Alfaro, F. and Barceló-Muñoz, A. (2008) Frigoconservación *in vitro* de germoplasma de aguacate. *Actas de Horticultura* 51, 71–72. (In Spanish.)

Wally, O., Jayaraj, J. and Punja, Z.K. (2009) Broad-spectrum disease resistance to necrotrophic and biotrophic pathogens in transgenic carrots (*Daucus carota* L.) expressing an *Arabidopsis* NPR1 gene. *Planta* 231, 131–141.

Wang, A., Tan, D., Takahashi, A., Li, T.Z. and Harada, T. (2007) MdERFs, two ethylene-response factors involved in fruit ripening. *Journal of Experimental Botany* 58, 3743–3748.

Wang, M.R., Lambardi, M., Engelmann, F., Pathirana, R., Panis, B., Volk, G.M. and Wang, Q.-C. (2021) Advances in cryopreservation of in vitro-derived propagules: technologies and explant sources. *Plant Cell Tissue and Organ Culture* 144, 7–20.

Wang, X., Beno-Moualem, D., Kobiler, I., Leikin-Frenkel, A., Lichter, A. and Prusky, D. (2004) Expression of Δ^{12} fatty acid desaturase during the induced accumulation of the antifungal diene in avocado fruits. *Molecular Plant Pathology* 5, 575–585.

Witjaksono. (1991) The tissue culture medium of 'Pinkerton' avocado (*Persea americana* Mill.). In: *Proceedings of the Seminar and National Congress on Biology X. Indonesian Biological Society and IUC Life Science*, IPB, Bogor, pp. 411–417. (In Indonesian with English abstract.)

Witjaksono. (1997) Development of protocols for avocado tissue culture: somatic embryogenesis, protoplast culture, shoot proliferation and protoplast fusion. PhD thesis, University of Florida, Gainesville, Florida.

Witjaksono and Litz, R.E. (1999a) Induction and growth characteristics of embryogenic avocado (*Persea americana* Mill.) cultures. *Plant Cell, Tissue and Organ Culture* 58, 19–29.

Witjaksono and Litz, R.E. (1999b) Maturation and germination of avocado (*Persea americana* Mill.) somatic embryos. *Plant Cell, Tissue and Organ Culture* 58, 141–148.

Witjaksono and Litz, R.E. (2002) Somatic embryogenesis of avocado (*Persea americana* Mill.) and its application for plant improvement. *Acta Horticulturae* 575, 133–138.

Witjaksono and Litz, R.E. (2004) Effect of gamma irradiation on embryogenic avocado cultures and somatic embryo development. *Plant Cell, Tissue and Organ Culture* 77, 139–147.

Witjaksono, Litz, R.E. and Grosser, J.W. (1998) Isolation, culture and regeneration of avocado (*Persea americana* Mill.) protoplasts. *Plant Cell Reports* 18, 235–242.

Witjaksono, Schaffer, B., Colls, A., Litz, R.E. and Moon, P.A. (1999a) Avocado shoot culture, plantlet development and net CO_2 assimilation in an ambient and enhanced CO_2 environment. *In Vitro Cellular and Developmental Biology – Plant* 35, 238–244.

Witjaksono, Litz, R.E. and Pliego-Alfaro, F. (1999b) Somatic embryogenesis in avocado (*Persea americana* Mill.). In: Jain, S.M., Gupta, P.K. and Newton, R.J. (eds) *Somatic Embryogenesis in Woody Plants,* Vol. 5. Kluwer Academic Publishers, Dordrecht, The Netherlands, pp. 197–214.

Witjaksono, Nugraheni, U.K., Hoesen, D.H. and Litz, R.E. (2009) Regeneration from irradiated avocado (*Persea americana* Mill.) embryogenic cultures. In: *Induced Mutation in Tropical Fruit Trees, IAEA-TEC-DOC-1615.* International Atomic Energy Agency, Vienna, pp. 83–90.

Xoca-Orozco, L.A., Cuellar-Torres, E.A., González-Morales, S., Gutiérrez-Martínez, P., López-García, U., *et al.* (2017) Transcriptomic analysis of avocado Hass (*Persea americana* Mill) in the interaction system fruit-chitosan-*Colletotrichum*. *Frontiers in Plant Science* 8:956.

Young, M.J. (1983) Avocado callus and bud culture. *Proceedings of the Florida State Horticultural Society* 96, 181–182.

Yue, J.-J., Yuan, J.-L., Wu, F.-H., Yuan, Y.-H., Cheng, Q.-W., Hsu, C.-T. and Lin, C.-S. (2021) Protoplasts: from isolation to CRISPR/Cas genome editing application. *Frontiers in Genome Editing* 3, 717017.

Zahn, L.M., Kong, H., Leebens-Mack, J.H., Kim, S., Soltis, P.M., *et al.* (2005) The evolution of the *SEPALLATA* subfamily of MAD-Box genes: a preangiosperm origin with multiple duplications throughout angiosperm history. *Genetics* 169, 2209–2223.

Zhang, Z.K., Huber, D.J. and Rao, J.P. (2011) Ripening delay of mid-climacteric avocado fruit in response to elevated doses of 1-methylcyclopropene and hypoxia-mediated reduction in internal ethylene concentration. *Postharvest Biology and Technology* 60, 83–91.

Zirari, A. and Lionakis, S.M. (1994) Effect of cultivar, explant type, etiolation pretreatment and the age of plant material on the *in vitro* regeneration ability of avocado (*Persea americana*). *Acta Horticulturae* 365, 69–76.

Ziv, D., Zviran, T., Zezak, O., Samach, A. and Irihimovitch, V. (2014) Expression profiling of FLOWERING LOCUS T-Like gene in alternate bearing 'Hass' avocado trees suggests a role for *Pa*FT in avocado flower induction. *PLOS ONE* 9, e110613.

Zulfiqar, B., Abbasi, N.A., Ahmad, T. and Hafiz, I.A. (2009) Effect of explant sources and different concentrations of plant growth regulators on *in vitro* shoot proliferation and rooting of avocado (*Persea americana* Mill.) cv. Fuerte. *Pakistan Journal of Botany* 41, 2333–2346.

Zumaquero, A., Martínez-Ferri, E., Matas, A.J., Reeksting, B., Olivier, N.A., *et al.* (2019) *Rosellinia necatrix* infection induces differential gene expression between tolerant and susceptible avocado rootstocks. *PLOS ONE* 14: e0212359.

11 Water and Nutrient Management

P.M. Gil[1], T. Cantuarias-Avilés[2], F. Meza[1], E. Lahav[3] and A.W. Whiley[4]
[1]*Facultad de Agronomía y Sistemas Naturales, Pontificia Universidad Católica de Chile, Chile;* [2]*Luiz de Queiroz College of Agriculture, University of São Paulo, Brazil;* [3]*Institute of Horticulture, ARO, The Volcani Center, Israel;* [4]*Sunshine Horticultural Services, Queensland, Australia*

Abstract

This chapter discusses water and nutrient management in the context of the physiological and productive needs of avocado crops in different places and production systems. It also discusses the industry's current challenges driven by climate change and market trends, as well as the traditional and new water and nutrient management technologies.

Modern avocado production faces challenges centred around sustainable production and reduced environmental impacts, and is driven by public scrutiny and emerging agricultural regulations. Climate change substantially threatens global agricultural productivity, jeopardizing food security and the livelihoods of stakeholders across the food chain, avocados being no exception. Yield reductions are expected due to rising temperatures and diminished precipitation, with reductions possibly reaching up to 40% (Lobell *et al.*, 2006).

Avocado evolved in subtropical and tropical climates in central and northern South America and characterized by wet summer and dry winter cycles. Commercial avocado orchards are predominantly situated in regions with either a dry summer Mediterranean or wet summer subtropical/tropical climate, where irrigation is a pivotal factor influencing tree growth, productivity and fruit quality. This chapter explores the water requirements and emerging irrigation technologies specific to avocado.

In recent decades, avocado nutrient management has also undergone significant transformations, emphasizing the importance of soil biology for tree nutrition and health. The escalating adoption of biofertilizers, organic fertilizers and tree row mulching in avocado orchards across diverse edaphoclimatic environments aims to meet consumer demand for healthier produce while mitigating pollution from overfertilization. This chapter explores current avocado nutrient management based on nutritional diagnoses throughout the tree's phenological cycle.

Climate Change and Water Requirements

Climate change is expected to cause changes in the suitability of some areas for avocado production (see Ramírez-Gil *et al.*, Chapter 5, this volume for a detailed discussion of climate change related to avocado production). While some areas may lose potential productivity due to increasing temperatures and a higher frequency of heatwaves, other areas currently constrained

 The Avocado: Botany, Production and Uses (eds D. Carrillo, B. Schaffer, A.W. Whiley and B.N. Wolstenholme)
DOI: 10.1079/9781800621824.0011

by low temperatures and frost may become attractive in the coming decades (Ramírez-Gil *et al.*, 2019; Parker *et al.*, 2021; Grüter *et al.*, 2022). Areas suitable for avocado cultivation are expected to expand globally, although areas with potential high productivity will likely decrease. Similar to other perennial fruit trees, production will likely expand northward in the northern hemisphere and southward in the southern hemisphere (Meza *et al.*, 2023).

Howden *et al.* (2005) presented a complete summary of climatic factors affecting avocado growth and development. The most important factors are frost frequency and intensity; extreme heat (usually temperatures greater than 35°C); intense precipitation that can cause flooding, root asphyxia and favour fungal diseases; and changes in precipitation that affect the soil-water balance and increase water requirements. Table 11.1 summarizes projected changes in some of these factors in major avocado-producing regions/countries according to the Intergovernmental Panel on Climate Change (IPCC WG1) atlas (https://interactive-atlas.ipcc.ch/).

Almost all regions are expected to have fewer frost days, particularly radiative frost, due to increasing temperature, which should positively impact average crop productivity. However, as air temperatures increase, extremely hot days and increased solar radiation are expected to cause crop heat stress and sunburn, likely overcoming the marginal gains generated by reduced frost. Climate change is also expected to affect precipitation patterns. Regions with less seasonal precipitation will rely more on supplemental irrigation, whereas areas with greater precipitation will likely experience more frequent waterlogging or flooding. Regardless of the trend in total rainfall, climate change increases the frequency of intense precipitation events (see the maximum daily precipitation column in Table 11.1). In addition, physiological plant responses are expected with higher atmospheric CO_2 concentrations due to climate change. For instance, atmospheric CO_2 enrichment positively impacts carbon assimilation and biomass accumulation and produces partial stomatal closure, thereby increasing water use efficiency in the short term.

Implementing measures to minimize adverse impacts and/or capitalize on emerging favourable conditions may be needed in most avocado-producing regions. Supplemental irrigation and adequate water management practices are two top priorities. For extremely hot temperatures, likely responses include adjustments in irrigation scheduling to satisfy increasing water requirements and implementing soil management practices that increase water storage

Table 11.1. Projected changes in key climate indices for avocado production from the IPCC WG1 Interactive Atlas. Results correspond to the average of global circulation model runs under shared socio-economic pathway 5-8.5 (SSP5-8.5) scenario achieving a 2°C increase in global temperature (Iturbide *et al.*, 2021; Gutiérrez *et al.*, 2021)

Region of production	IPCC WG I Interactive Atlas reference region	Frost days	Change in median: Days above 35°C	Change in median: Total precipitation (%)	Maximum daily precipitation (%)
USA	Western North America	–18	12	3.2	8.9
Mexico	Northern Central America	–5.3	17.6	–4.9	6.3
Central America	Southern Central America	0	8.4	–4.4	2.2
Chile–Perú	North-western South America	4.3	1.6	–11.4	–0.7
Colombia–Venezuela	Northern South America	0	44.1	–6.6	7
Brazil	North-eastern South America	0	24.6	–2.7	9.7
Kenya	North-eastern Africa	0	37.5	11.2	21.9
South Africa	Eastern Southern Africa	–1.8	15.8	–3.6	9.8
Australia–New Zealand	Eastern Australia	–1.7	11.7	–1.0	5.7
Israel	Mediterranean	–8.8	16.2	–9.8	4.7

capacity. Relocation of crops to cooler microclimates or using evaporative cooling irrigation systems and shading nets may be required in warmer environments.

Ecophysiological and Production Considerations for Irrigation Management

Much of the formal knowledge regarding avocado irrigation has been developed in Mediterranean environments, where irrigation is deemed essential for production. In contrast, limited experimentation has been conducted in humid areas with summer rainfall where irrigation is often considered supplementary. Nevertheless, principles of water use established in Mediterranean regions can be adapted to other environments. For example, in the Andean region of Colombia, with a positive water balance during most of the production cycle, 'Hass' avocado orchards are traditionally managed without irrigation due to uniform rainfall distribution. However, a study by Grajales *et al.* (2018) determined that rain may not consistently meet the water requirements of the crop, and irregular rainfall often results in water deficit during the flowering and fruit-filling stages (June-August, South Hemisphere). When trees were subjected to three irrigation treatments based on reference evapotranspiration (ETo) equivalent to 50%, 75% or 100% ETo compared with non-irrigated (rainfed) trees, the best performance was observed at 75% ETo. Other researchers corroborated the significance of irrigation in enhancing avocado production in Tropical or Subtropical regions. In Colombia, Dorado *et al.* (2017) demonstrated a 23% increase in production when drip irrigation was used in a 'Hass' orchard. A study by Erazo-Mesa *et al.* (2021) revealed that 99.8% of Colombia's current and potential 'Hass' production zones require irrigation for at least 1 month annually. Similarly, Cantuarias *et al.* (2019) reported an 18.2% increase in fruit yield in an irrigated compared with a rainfed 'Hass' orchard in Brazil. Despite the demonstrated benefits of effective irrigation management, many avocado orchards in tropical countries are not irrigated (Erazo-Mesa *et al.*, 2021), although some growers in different areas of Colombia and Florida (USA) are already using irrigation to complement water needs during the dry season (Fig. 11.1).

In Mediterranean environments, high evaporative demand, often compounded by soil salinity and water scarcity in the summer months, is the main challenge growers face in

Fig. 11.1. Sprinkler system for complementary irrigation in a West Indian avocado orchard, subtropical southern Florida, USA.

maintaining optimal soil moisture levels for sustained avocado cultivation. Additionally, salinity management requires good quality water, exacerbating the need for increased irrigation in arid conditions (Ferreyra *et al.*, 2012).

Regardless of the production area, inappropriate irrigation, particularly in clay or loam clay soils, can lead to root hypoxia (Ferreyra *et al.*, 2012). In addition, moderate-to-extreme rainfall events can lead to temporary waterlogging, adversely affecting the delicate root system of avocado trees. Impaired oxygen availability in waterlogged or poorly drained soils hampers root function and can harm overall tree health and productivity. In humid environments with summer rainfall, this issue is aggravated by Phytophthora root rot caused by *Phytophthora cinnamomi* (see Schaffer *et al.*, Chapter 7 and Dann and Gazis, Chapter 13 of this volume). The disease develops rapidly in very wet conditions, characterized by a soil matric tension of −10 kPa or higher (Sterne *et al.*, 1977b). Phytophthora root rot may also be enhanced on slopes where water movement over or under the soil surface facilitates the spread of the *P. cinnamomi* (Zentmyer and Richards, 1952; Valoras *et al.*, 1964).

Physiological Aspects of Water Use

Elevated temperatures (>30°C) reduce net CO_2 assimilation and increase internal CO_2 concentration, leading to stomatal closure, which results in heat stress and productive loss. Studies have shown rapid stomatal closure and changes in avocado leaf anatomy at leaf water potentials below −0.9 MPa (Bower *et al.*, 1977; hi *et al.*, 2002; Roets *et al.*, 2012). Sterne *et al.* (1977a) found declining leaf gas exchange at stem water potentials (SWPs) of −0.4 MPa and stomatal closure at −1.2 MPa in avocado in response to drought.

Avocado stomata are located on the abaxial leaf surface. Stomatal densities ranging from 390 to 510 stomata mm^{-2} have been observed in 'Fuerte' avocado (Blanke, 1992). Avocado fruit also have stomata, with recently set fruit having 50–70 stomata mm^{-2}, whereas mature fruit have 5 stomata mm^{-2}. Despite the low stomatal density of individual fruit, the overall count per fruit can reach as high as 20,000–30,000 stomata (Blanke, 1992). Therefore, plants with high fruit loads can consume more water than those with lower yields (Silber *et al.*, 2015). Torres (2017) found a positive correlation between avocado fruit load and stomatal conductance, a condition already demonstrated in olives (Bustan *et al.*, 2016) and apples (Atay *et al.*, 2021).

Avocado is categorized as an 'isohydric' species, meaning that stomata close and stomatal conductance is reduced in response to declining soil-water potential or dry atmospheric conditions. Studies indicate that avocado may transition from isohydric behaviour to anisohydric behaviour (where drought does not induce rapid stomatal closure) based on crop load (Naor, 1998; Silber *et al.* 2012; Kaneko *et al.*, 2024). This transition is associated with dynamic stomatal sensitivity to varying environmental conditions (Zhang *et al.*, 2012). Isohydric stomatal regulation is a drought avoidance strategy often found in humid regions, whereas anisohydric stomatal regulation occurs in arid regions (Hu *et al.*, 2023). There is currently a lack of information regarding isohydric versus anisohydric responses of avocado to drought, making further research on this subject a compelling avenue for exploration.

Understanding critical irrigation periods is essential for maximizing production and/or efficiently managing limited water resources. Avocado trees, although evergreen, undergo an annual cycle encompassing quiescence, budbreak, flowering, leaf and shoot flushing, root growth, fruit growth, fruit drop and harvest. Aligning tree phenology with climatic conditions to optimize water application has been considered, particularly in Australia, where avocados thrive in tropical, subtropical and Mediterranean environments (Anonymous, 2009). In all cases, adequate irrigation is imperative during flowering and early fruit development. Water management during flowering is critical for fruit set, given the increased water losses from flowers (Whiley *et al.*, 1988a). Withholding water from young 'Hass' avocado trees led to leaves wilting before the inflorescences did; fruit set did not occur during the dry period, but resumed upon rewatering, although fruit abscised shortly thereafter (Neuhaus *et al.*, 2007). The rapid fruit growth phase marks another critical irrigation period, where effective irrigation management reduces fruit drop and enhances final fruit size (Bower, 1985a; Whiley *et al.*, 1988b; Wolstenholme *et al.*, 1990). Lahav and Kalmar (1983) cautioned that in summer, with accelerated fruit growth and heightened environmental demand, soil-water content depletes rapidly, necessitating increased irrigation frequency.

In Mediterranean climates, winter irrigation is generally unnecessary unless extreme drought conditions prevail. Limited information exists on the relationship between avocado yield and winter irrigation, but periods of drought and elevated temperatures can impede fruit growth, particularly in cultivars with later harvests such as 'Hass' (Winer *et al.*, 2007). Also, water deficit in autumn, a practice used for inducing flowering in lemons, has proven ineffective for avocados (Lahav and Kalmar, 1983; Bower, 1985a). When winter rainfall is insufficient to saturate the soil to water capacity, a prolonged irrigation event in winter is recommended to maintain optimal moisture levels for the first peak of root growth, promoting vegetative and reproductive development throughout the season (Ferreyra and Sellés, 2012, 2013).

Another critical irrigation period corresponds to the root growth flush, typically occurring twice in Mediterranean climates, once before spring shoot growth and again at the end of autumn. Given that roots are shallow and lack root hairs, maintaining adequate soil moisture content is crucial during root growth. A favourable water-to-air ratio in the soil (0.3–0.6) ensures water availability for transpiration and tree growth, along with oxygen for optimal physiological functioning of the roots (Gil, 2008; Gil *et al.*, 2012).

The transpiration rate in avocado trees is influenced by water supply, atmospheric water demand, root development and performance, and morphological and physiological factors. Evaporation from the soil coupled with water vapour released through stomata (transpiration) collectively contributes to crop evapotranspiration (ETc). Environmental demand for water can vary, ranging from 2 to 10 mm day^{-1}, contingent on factors such as air vapour pressure deficit (VPD), sunlight and wind speed (Allen *et al.*, 1998). Understanding the dynamics of supply, capacity and demand is pivotal for effective irrigation management for avocado trees.

The choice of avocado cultivar significantly influences water use. Recent research by Beyer *et al.* (2021) highlighted differences in acrial and root growth and water use efficiency among 'Hass' avocado trees on different rootstocks (clonal 'Dusa' and seedling 'Mexícola'). These differences were attributed to differences in root hydraulic properties, with fine roots of 'Mexícola' seedling rootstocks exhibiting a greater water uptake than 'Dusa'. Additionally, Fassio *et al.* (2009) found that 'Hass' avocado grafted onto 'Duke 7' (clonal) rootstock had 29% higher sap flow rate and transpiration than 'Hass' grafted onto 'Toro Canyon' (seedling) rootstock. Intergrafting can also modify stomatal density in avocado leaves. For example, Ayala-Arreola *et al.* (2010) reported increased stomatal density and smaller stomata size for 'Hass' and 'Fuerte' scions grafted onto Guatemalan seedling rootstock when intergrafted with 'Colín V-33' interstsock, enabling quicker responses to unfavourable conditions.

Water Requirements for Growth, Productivity and Fruit Quality

As observed for many plant species, there is a high positive correlation between water input and avocado yield (Fig. 11.2). Even when other crop management practices are executed effectively, achieving maximum yield is unattainable if the required water is unavailable. Effective irrigation management necessitates consideration of several factors, including crop water consumption, soil-water holding capacity, irrigation system characteristics, and the specific behaviour and requirements of the irrigated species (Allen *et al.*, 1998).

Avocado trees typically take 3–6 years to achieve full ground cover and maximum ETc, depending on tree spacing, management practices and location. In Mediterranean climates, daily water applications in mid-summer increase from 4–8 l tree^{-1} in the first year to 80–150 l tree^{-1} in the fourth year. Water consumption of mature avocado trees is also influenced by weather conditions, leaf area, phenological stage and physiological behaviour. A study in South Africa, employing the energy balance principle and Eddy Covariance technology, determined the mean daily water use for the season to be 2.69 mm day^{-1}, with summer and winter mean daily water use of 3.98 and 1.64 mm day^{-1}, respectively (Mazhawu *et al.*, 2018). That study estimated the crop coefficient (Kc) for avocado and demonstrated a good correlation

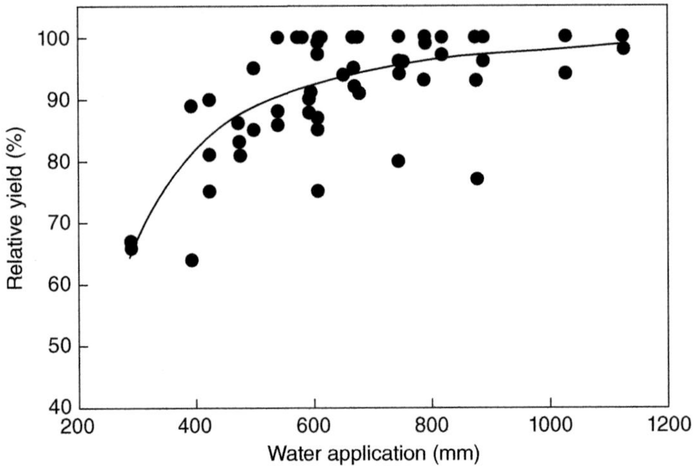

Fig. 11.2. The relationship between seasonal water application (mm year^{-1}) and relative yield of three cultivars ('Ettinger', 'Fuerte' and 'Hass') combined. Data were obtained from ten experiments on the coastal plain of Israel. From Steinhardt (1991) with permission.

between ETc and reference evapotranspiration (ETo), indicating the potential for using ETc and ETo to manage irrigation in avocado orchards.

Recent reports from the subtropical Spanish coast with a Mediterranean climate indicated irrigation requirements ranging from 530 to 700 mm water ha^{-1} year^{-1}. In the central zone of Chile, mature orchards require 700–900 mm water ha^{-1} year^{-1} (Ferreyra and Sellés, 2007). In the coastal plain of Israel, optimum yield was attained using a summer irrigation requirement (ETc) of 650 mm ha^{-1} year^{-1} in addition to the 500–600 mm supplied by winter rains (Steinhardt, 1991). In a mixed orchard of 'Ettinger', 'Fuerte' and 'Hass', when water application ranged between 300 and 600 mm ha^{-1} year^{-1}, each additional 100 mm of water added 1.57 ± 0.30 t ha^{-1} of fruit, but when water application ranged between 600 and 850 mm, the fruit added by each 100 mm increment was only 0.33 ± 0.09 t ha^{-1} (Fig. 11.2; Steinhardt, 1991).

Shoot growth in avocado is intricately linked to water availability. A long-term irrigation experiment in Israel inducing water deficit through varying irrigation intervals revealed significant impacts on tree growth (Kalmar and Lahav, 1977). Trees irrigated at 7-day intervals were larger and exhibited increased trunk growth than those irrigated at 28-day intervals. The response to the irrigation regime was most pronounced for 'Ettinger', followed by 'Hass' and 'Fuerte'. Water scarcity, as observed in California, restricted the growth of 'Hass' trees, resulting in fewer flushes per season, and reduced tree height, trunk circumference and magnitude of diurnal changes in trunk diameter (Kalmar and Lahav, 1977; Meyer *et al.*, 1990).

Water plays a critical role in controlling the growth of young, non-fruiting trees and mature, fruiting trees. Excessive vegetative growth induced by frequent irrigation can diminish yields in mature trees, primarily due to shading and the need for aggressive hedging to prevent overcrowding between rows. A well-planned irrigation programme can effectively adjust the balance between growth and productivity. Current recommendations favour using growth retardants (where governmental regulations permit) to restrict growth in mature trees, while accelerating young trees through water management (see Whiley *et al.*, Chapter 12 of this volume).

Silber *et al.* (2012) highlighted that fruit yield tends to be lower for 'Hass' avocado compared with other commercial fruit crops due to high abscission rates of flowers and fruit. It has been proposed that carbohydrate stress during the rapid fruit growth and oil accumulation period, coinciding with high temperatures and high evaporative demand, are significant factors contributing to fruit abscission.

Limited information is available about how irrigation management influences alternate bearing (AB) and the magnitude of yield fluctuation between 'on' and 'off' crop years. An experiment was conducted in Chile's southern production area (Peumo, Cachapoal Valley) to assess the impact of different irrigation rates on AB. The treatments involved applying 25%, 50%, 75% and 100% of the reference evapotranspiration (ETo). Trees subjected to 25% and 50% ETo in the 'off' year had only 22% of the average yield of the two 'on' years, whereas trees with 75% and 100% ETo in the "off" year had 42% of the average yield of the two 'on' years. Thus, adequate water supply for avocados is important even during 'off' years (Holzapfel *et al.*, 2017).

Avocado fruit size, peel colour, oil content and internal and external damage play a crucial role in the market, influencing prices and demand. Irrigation practices influence fruit size, and in most cases, shorter intervals between irrigation events contribute to increased fruit size (Lahav and Kalmar, 1977; Adato and Levinson, 1988; Meyer *et al.*, 1990; Michelakis *et al.*, 1993). Fruit oil concentration tends to increase over time and is correlated with yield and fruit size. Although irrigation influences fruit size by accelerating the rate of fruit growth, evaluating the direct impact of irrigation on oil concentration is challenging. In uniformly sized fruit, shorter irrigation intervals and increased water quantities have been associated with higher oil concentrations (Lahav and Kalmar, 1977). For example, irrigation at 7-day intervals (889 mm water year^{-1}) resulted in significantly higher fruit oil concentration (10.0%) compared with irrigation at 28-day intervals (594 mm water year^{-1}), which resulted in fruit with a 9.5% oil concentration (Lahav and Kalmar, 1977). Given that oil concentration is an indicator of harvest readiness, it is plausible to advance maturity by irrigating more frequently to enhance oil concentration, accompanied by an increase in fruit size. However, in addition to irrigation frequency, the quantity of water is a primary factor affecting both yield and fruit quality. Studies in Spain demonstrated that reductions of 40% and 60% of the estimated irrigation needs during the growing season led to a 30% decrease in yield and a smaller proportion of large fruit (Moreno *et al.*, 2019).

Despite evidence supporting the positive effects of higher irrigation frequency on avocado fruit size and content, there is compelling evidence suggesting potential negative impacts on root development, particularly when the soil-water/air ratio capacity is not taken into account, and frequent watering leads to over-irrigation and root hypoxia (Ferreyra *et al.*, 2007). This not only affects fruit yield, size and oil content but can also impact orchard longevity (Lahav and Kalmar, 1983; Menge and Marais, 2000; Ferreyra *et al.*, 2007; Gil *et al.*, 2007). Avocado roots require soils with more than 30% air capacity and poorly aerated soils with less than 1% O_2 inhibit root growth (Menge and Marais, 2000; Ferreyra *et al.*, 2007). Soil air content below 17% restricts oxygen diffusion rates to less than 0.2 μg cm^{-2} min^{-1}, when avocado roots cease growing (Valoras *et al.*, 1964; Ferreyra *et al.*, 2007). In addition, avocado trees may experience significant root damage in soils with O_2 diffusion rates below 0.17 μg cm^{-2} min^{-1} (Valoras *et al.*, 1964; Stolzy *et al.*, 1967). Effective irrigation management goes beyond supplying water equivalent to crop water consumption; it requires providing water at appropriate intervals, considering soil physical features influencing water holding and aeration capacity. Soil-water holding capacity, influenced by soil texture and bulk density, is an important factor to consider when determining irrigation frequency to ensure water availability without causing root asphyxiation (Ferreyra *et al.*, 2007; Gil *et al.*, 2012). In an 'Ettinger' avocado orchard in Israel, short irrigation cycles were applied during the spring flowering and fruit set periods, resulting in excessive soil moisture, reduced soil aeration and a cooled soil, which can potentially harm roots (Lahav and Kalmar, 1983). In studies in Chile (Gil *et al.*, 2012) and Crete (Tzatzani *et al.*, 2022), extended exposure to excessive soil moisture affected the mineral element concentrations in 'Hass' (Gil *et al.*, 2012) and 'Fuerte' (Tzatzani *et al.*, 2022) avocado trees, with 'Fuerte' being more sensitive than 'Hass' to waterlogging (Tzatzani *et al.*, 2022). In addition, Gil *et al.* (2012) found in Chile that in 'Hass', total contents of nitrogen (N), phosphorus (P), potassium (K), calcium (Ca), magnesium (Mg), carbon (C), nitrogen (N) and boron (B) in roots and whole plants were highest in soils with lower water-to-air ratios (i.e. sandy and sandy loam soils) maintained at field capacity, compared with soils with a higher water-to-air ratio (i.e. clay soils). These findings underscore the significant impact of soil water-to-air ratio on the growth and mineral nutrition of avocado trees, highlighting its relevance in avocado site selection and the formulation of nutrition and irrigation management strategies.

Water Consumption and Total Water Footprint

Avocado production has stirred global controversy due to its water requirements; avocado is often accused of being one of the highest consumers of irrigation water among fruit-bearing trees (Sommaruga and Eldridge, 2021). This has prompted growers and industry stakeholders to intensify efforts to improve water management practices and certify the best production methods for optimizing water use. However, it is important to challenge this notion as somewhat misconceived. In fact, avocado's water requirement, ranging between 600 and 1200 mm year^{-1}, is lower than that of many other fruit crops, including cherry, citrus, walnut, almond and plum (Steduto *et al.*, 2012). Water requirements vary greatly depending on factors such as geographical location, climatic conditions, the efficiency of irrigation methods and irrigation management practices. For instance, Moreno *et al.* (2019) observed that on the Spanish coast, adult commercial avocado orchards typically receive irrigation ranging from 530 to 700 mm water ha^{-1} year^{-1}, presenting a contrast with other regions such as the central zone of Chile, where mature orchards demand 700–900 mm water ha^{-1} year^{-1} (Ferreyra and Sellés, 2007), or certain production areas in Perú, where the avocado crop utilizes an average of 1694 mm water ha^{-1} year^{-1} (Apaza *et al.*, 2019). Verbist *et al.* (2020) highlighted the widespread adoption of modernized irrigation systems among avocado producers in Chile, qualifying water resource use by avocado as efficient compared with other crops. The estimated consumption of 898 mm water ha^{-1} year^{-1} aligns with the findings reported by Ferreyra and Pérez (2013).

Today, one of the indicators of orchard water use is the total water footprint, which considers the sum of the green (rainfall), blue (water through irrigation) and grey (contaminated) water footprints. The concept of water footprint takes into account the total amount of water utilized per kilogram of product produced (litres or cubic metres of water per kilogram or tonne of product) throughout the entire production chain up to the consumer (Hoekstra, 2008). This encompasses the water required for production and the wastewater resulting from processing, representing the contaminated water generated during production. The blue footprint, involving water lost by evapotranspiration and water integrated into the product, signifies fresh water, surface water and/or groundwater used by a crop. The green water footprint measures water utilization from rainfall stored in the soil and transpired by plants, excluding rainwater recharging aquifers or lost via surface runoff. The grey water footprint assesses the extent of freshwater contamination during production, surpassing irrigation water quality standards due to fertilizers or pesticides. It quantifies the volume of water required to dilute discharged contaminants to meet environmental water quality standards. The total water footprint of a product amalgamates these components, expressed as water volume per unit mass of product obtained.

For avocado, the total water footprint ranges between 480 (Ferreyra and Pérez, 2013) and 2300 l kg^{-1} (Mekonnen and Hoekstra, 2011), with a reported global average of 750–849 m^3 t^{-1} (Tapia and Bravo, 2009; Mekonnen and Hoekstra, 2011; Holmqvist, 2020). The observed blue water footprint (ground and surface water consumed) for avocado is between 250 and 1000 l kg^{-1}. According to those data, the total water footprint for avocado is lower than that of other fruit crops such as bananas (790 l kg^{-1}), apples (822 l kg^{-1}), walnuts and almonds (4000 l kg^{-1}) and much lower than that of agronomic crops such as wheat (1000 l kg^{-1}) and rice (2500 l kg^{-1}) (Mekonnen and Hoekstra, 2010; Donoso and Franco, 2013; Donoso *et al.*, 2016). However, the growing demand for avocado has induced a significant expansion of cultivated areas, sometimes impacting natural resources in areas with a high agricultural water footprint (Novoa *et al.*, 2018). This scenario obligates the industry to be more proactive, avant-garde and strict with the sustainable use of water, incorporating high-efficiency irrigation technologies.

Irrigation Systems

Irrigated avocado orchards commonly adopt localized irrigation systems, with dripper, microsprinkler and jet irrigation being the most prevalent choices due to their high efficiency and uniformity. Research by Salgado and Cautín (2008) demonstrated that drip irrigation led to approximately 25% more root production compared with microsprinklers, emphasizing the influence of the irrigation system on root density. When selecting an irrigation system for avocados, careful consideration of root allocation and the necessary wetted soil area is essential for achieving optimum development and fruit yields. For localized irrigation systems such as drippers or microsprinklers, it is recommended to cover 30–50% of the soil. For drippers, this equals two or three dripper lines per row, depending on planting density. Flood or overhead irrigation with sprinkler systems is discouraged due to their low efficiency, high water consumption and the potential for runoff, erosion or excessive soil-water content. These factors can exacerbate the impact of Phytophthora root rot on avocado (Ploetz and Schaffer, 1989).

Irrigation Scheduling

Sustainable water management hinges on various factors, including orchard water availability, addressing water quality limitations, utilizing efficient irrigation systems and infrastructure (such as drip or microsprinkler systems), ensuring proper maintenance and calibration of the irrigation system and monitoring soil and/or plant water status (Gil and Barrera, 2018). Among these factors, scheduling irrigation operations is often overlooked but holds significant potential to improve water use.

Irrigation scheduling involves determining how much water to apply and when. Scheduling takes into account 'irrigation agronomy', which considers crop-specific aspects such as soil and climatic characteristics that determine proper water management for production. Decisions about how much water to apply and when to irrigate should be based on estimates or measurements of crop water consumption, soil-water holding capacity, irrigation system characteristics and adjusting the irrigation budget by accounting for the water status of the soil and trees (water status monitoring) (Allen *et al.*, 1998). Generally, the crop water requirement is determined by considering the FAO 56 Model (Allen *et al.*, 1998),

$$\mathrm{ETc} = \mathrm{Kc} \times \mathrm{ETo} \qquad \text{(Eqn 1)}$$

ETo can be obtained from local portable weather stations or from agrometeorological networks, available in different countries and production areas (i.e. CIMIS in California, RAN in Chile, RIA in Spain). Irrigation should also be based on the crop evapotranspiration coefficients (Kc) commonly obtained from literature. The Kc can be estimated from vegetative covering and/or status indicators such as the normalized difference vegetation index (NDVI, Rouse *et al.*, 1973) or the soil-adjusted vegetation index (SAVI, Huete, 1988) generated from remote sensing (RS), which have been used to predict Kc at field and regional scales (Bausch, 1995; Duchemin *et al.*, 2006; Gontia and Tiwari, 2010). Irrigation intervals are determined by the rate of soil-water depletion considering crop water consumption.

The Kc is affected by local situations and varies from one location to another. Table 11.2 shows Kc values for several avocado production countries. In the table, the values for Israel are for a standard orchard in the coastal plain, based on many observations and irrigation experiments. The Californian data also cover irrigation during the winter if required (Meyer *et al.*, 1990). The peak irrigation season is from June to July in California and from August to October in Israel. The Kc may vary according to tree vigour, leaf nutrition and soil albedo, among other factors (Allen, 1998). The reference Kc is generally for mature orchards and cannot be used in recently established orchards (Allen *et al.*, 1998; Fereres *et al.* 2012). The reported Kc can differ even if obtained with the same crop and management procedures. For example, in the central zone of Chile, Kc values of 0.5–0.6 in winter, 0.6–0.7 in spring and near one in summer seem to satisfy the needs of the avocado crop (Cardemil, 1999). However, research in the same region but with trees planted on slopes showed that using 77% of ETo with a Kc of 0.77 for 11 months resulted in the best yield and fruit quality (Gil *et al.*, 2011). Nevertheless, the most commonly used Kc for avocado in Chile is 0.72 for all months of the year (Gardiazábal *et al.*,

Table 11.2. Crop coefficients (Kc) based on Class A pan evaporation for irrigation of avocado in the Mediterranean climates of California (Meyer *et al.*, 1990), the coastal plain of Israel (Anonymous, 2001 cited by Kourgialas and Dokou, 2021), the central zone of Chile (Gardiazábal *et al.*, 2003), the south of Spain (Rodríguez Pleguezuelo *et al.*, 2018) and South Africa (Kruger *et al.*, 2001)

Month	California	Israel	Chile	Spain	South Africa
January	0.35	–	0.72	–	0.83–0.99
February	0.40	–	0.72	–	0.73–0.85
March	0.45	–	0.72	–	0.56–0.68
April	0.45	0.35–0.4	0.72	0.55–0.6	0.49–0.53
May	0.5	0.45–0.5	0.72	0.57–0.66	0.44–0.46
June	0.55	0.55–0.6	0.72	0.57–0.66	0.36–0.34
July	0.55	0.6	0.72	0.57–0.66	0.42–0.43
August	0.5	0.6–0.65	0.72	0.57–0.66	0.45–0.49
September	0.45	0.6–0.65	0.72	0.57–0.66	0.41–0.50
October	0.45	0.6–0.65	0.72	0.55–0.6	0.55–0.66
November	0.45	0.5	0.72	–	0.72–0.82
December	0.40	–	0.72	–	0.89–1.07

2003). It is common for growers to increase irrigation in proportion to crop load, which can be done by modifying the Kc. For example, trees yielding 18 t ha^{-1} or more receive 15% more water than those yielding less (Anonymous, 2001 cited by Kourgialas and Dokou, 2021). However, increasing irrigation for high-yielding trees increases fruit size to a certain point, above which factors other than water supply become limiting. In addition, the Kc can be changed and incorporated into the water balance equation (irrigation = ET + percolation + difference in soil-water content – rain precipitation) to calculate the amount of water required.

When an orchard has not reached its complete ground coverage by the canopy, has been severely pruned or the trees are subject to shade, canopy coverage of the ground should be taken into account to adjust the water requirements according to the tree size. In these cases the equation for estimating ETc is:

$$\text{ETc} = \text{ETo Kc Kr} \quad \text{(Eqn 2)}$$

where Kr is an empirical coefficient relating the ET of an orchard with incomplete cover to that of a mature orchard, related to the horizontal projection of the tree shade (ground cover). Cover factors (Kr) have not been reported for avocado, but for fruit-bearing trees Kr can be estimated using the following equation:

$$\text{Kr} = -0.00012\,(\text{G\%})2 + 0.0226(\text{G\%}) \quad \text{(Eqn 3)}$$

where G% corresponds to the ground cover by the trees (Fereres *et al.*, 2012).

In recent years, agriculture has shifted from decision making based on limited samples to more precise management, referred to as site-specific management (Pinter *et al.*, 2003; Jones and Vaughan, 2010). Different methods have been developed to estimate biophysical variables such as leaf surface temperature and evapotranspiration (Jones and Vaughan, 2010), as well as Kc using RS data from satellites (i.e. Sentinel, Lansat and Modis) and unmanned aerial vehicles (drones), demonstrating variability both temporally and spatially. Many of these methods are based on the dual crop coefficient method of the Food and Agricultural Organization of the United Nations (FAO), Irrigation and Drainage Paper No. 56 (FAO-56) (Allen *et al.*, 1998), which intended to improve daily simulation of ETc by considering separately the contribution of evaporation from soil. The dual method utilizes basal crop coefficients (Kcb) representing ET from crops having a dry soil surface. It separately predicts evaporation from bare soil (Ke) based on the water balance of the soil surface layer.

As previously mentioned, NDVI and SAVI have been highly correlated with crop canopy cover for various crops in semi-arid areas and can be converted into Kcb values by empirical relationships. For many fruit tree species such as grapevines (Kamble *et al.*, 2013), apple (Marsal *et al.*, 2013; Odi-Lara *et al.*, 2016), olive (Pôças *et al.*, 2015) and citrus (Ippolito *et al.*, 2023), the correlation between NDVI and/or SAVI and Kcb is high and thus can be used for ET estimation. Avocado Kc and its components, Kcb and Ke, have been determined using RS technologies and vegetation indices such as NDVI (Rivero, 2020). However, there is still much to explore, as many models do not match avocado requirements and often overestimate ET values. Riveros (2020) evaluated Kc maps to enhance avocado irrigation in Peumo, Chile. Using Sentinel 2 satellite data and *in situ* measurements, Kc, Kcb and Ke were generated, representing the spatio-temporal variability of the orchard under study. The results were highly correlated with simulations using the HYDRUS 1D model, a public domain model of water flow and solute transport in variably saturated porous media. Temporal data indicated an average potential crop coefficient for avocado of 0.78. These findings underscore the feasibility of managing irrigation with local crop coefficient data, combined with a method for detecting soil-water availability and generating management zones associated with Kc and its components.

Taylor *et al.* (2021) compared three models for estimating canopy ET in South African avocado orchards. They studied two orchards with intermediate canopy size and one mature orchard. Transpiration coefficients were consistent across orchards, with variations due to canopy cover. Monthly average coefficients differed from previous avocado curves, providing better seasonal water use estimates. Allen's and Pereira's parameters led to overestimations. The Penman-Monteith equation with Jarvis-based estimates worked well for intermediate orchards, while adjustments improved results for mature orchards. The study emphasized the need for accurate methods for adjusting maximum canopy conductance (gc max). Biweekly estimates offered reliable water use data for all orchards.

Irrigation scheduling is very important for good water management, but it is often not considered or estimated incorrectly. Calculation of the irrigation programme or schedule involves determining the amount of water to replenish for the crop and the frequency of application. Water replenishment considers the crop's water consumption (ETc), the inherent inefficiencies of the irrigation system and whether additional water is needed for leaching salts. The ETc depends on meteorological conditions as well as crop vigour and developmental stage. Conversely, the irrigation frequency considers the soil's water storage capacity and the crop's water consumption rate. The soil's water storage capacity is related to physical characteristics of the soil, such as rock content and rooting depth, as well as the portion of the soil wetted by the irrigation system. Additionally, it involves an irrigation criterion determined by the producer, known as the depletion factor. This factor dictates the proportion of the maximum water storage capacity allowed to deplete before re-irrigation.

Computing the irrigation interval is an important practice for avocado orchards, due to the high impact of water and air availability on tree development and productivity. Concepts such as total available water (TAW) and readily available water (RAW) in the soil should be taken into account for sustainable water management (Allen *et al.*, 1998). TAW is the theoretical total amount of water a crop can extract from its root zone; its magnitude depends on the soil type and rooting depth. The magnitude of TAW, which corresponds to the moisture range between field capacity and permanent wilting point, depends largely on soil texture (Table 11.3). TAW can be calculated from the difference between the water content at field capacity and wilting point, and considering the rooting depth as follows:

$$\mathrm{TAW} = (\Theta\mathrm{FC} - \Theta\mathrm{PWP}) \times \mathrm{Zr} \qquad \text{(Eqn 4)}$$

where TAW is the total available soil water in the root zone [mm], ΘFC is the volumetric water content at field capacity [m^3 m^{-3}], ΘPWP is the volumetric water content at the permanent wilting point [m^3 m^{-3}] and Zr is the rooting depth [mm] (Allen *et al.*, 1998).

When adding other soil components that determine the total water holding capacity of soil, and thus the 'tank capacity', the stoniness and the wetted surface of the soil by the irrigation system can be included. The equation for TAW is then modified as follows:

$$\mathrm{TAW} = (\Theta\mathrm{FC} - \Theta\mathrm{PWP}) \times \mathrm{Zr} \times (1 - \mathrm{s}) \times \%\mathrm{WSS} \qquad \text{(Eqn 5)}$$

Table 11.3. Field capacity and permanent wilting point for different soil textures. Adapted from Santa Olalla and De Juan (1993)

Soil texture	Field capacity (volumetric water content, cm^3)	Permanent wilting point (volumetric water content, cm^3)	Total available water (TAW, cm^3)
Sandy	0.15	0.07	0.08
Loam sandy	0.21	0.09	0.12
Loam	0.31	0.14	0.17
Loam clay	0.36	0.18	0.18
Clay sandy	0.4	0.2	0.2
Clay	0.44	0.21	0.23

where s is the soil stoniness (0–1) and WSS is the wetted soil surface by the irrigation system (0.3–0.5 for localized irrigation) (Ferreyra and Sellés, 2013).

An avocado orchard should not reach the permanent wilting point because, at that point, crop water uptake is highly reduced and plants suffer from irreversible water stress. When the soil-water content drops below a threshold value, soil water can no longer be transported quickly enough towards the roots to respond to the transpirational demand and the crop begins to experience water stress. Thus, to avoid water stress in the orchard, irrigation management should consider a fraction of TAW so that the crop can extract water from the root zone without suffering. This concept is called the readily available soil water (RAW):

$$RAW = p\, TAW \quad \text{(Eqn 6)}$$

where RAW is the readily available soil water in the root zone (mm), and p is the average fraction of total available soil water (TAW) that can be depleted from the root zone before water stress (0–1), also called the 'depletion factor' (Allen *et al.*, 1998).

Avocado trees develop relatively shallow root systems compared with other fruit trees. The maximum rooting depth in deep, well-drained soils is 1.2–1.5 m, with the main root system (70–80%) in the 0–60 cm zone. However, roots can be much deeper in well-drained sandy soils or under restricted water supply. In Western Galilee, Israel, with a 7-day irrigation interval, 14% of the total roots of 'Ettinger' and 'Fuerte' were found in the deep clay soil layers (60–120 cm), but irrigating once each month almost doubled the proportion of roots in the 60–120 cm zone. A strong correlation was observed between the number of roots and water uptake (Kalmar and Lahav, 1976). In Crete, Michelakis *et al.* (1993) reported a similar relationship between water use and avocado roots, concluding that irrigation should be directed to the 0–60 cm zone where most roots occur. Reducing the amount of water applied or increasing the irrigation interval will generally induce a deeper root system. However, this may not occur in compacted soils where aeration is poor.

Erazo-Mesa (2022a) assessed the potential of the soil surface water content (SSWC) for scheduling avocado irrigation. The variable SSWC was simulated in three 'Hass' avocado orchards using Hydrus-1D software. The study, carried out in three 2 ha plots in Colombia, identified soil types and characteristics. Portable weather stations and soil-water potential monitoring stations were employed for monitoring climate and soil-water dynamics. Mathematical models (Richards equation) were used for estimating surface water dynamics. Results showed considerable variation in soil-water content in the upper 15 cm of the soil during the validation period, while water accumulation and redistribution were slower at the 15–60 cm depth range. The study concluded that soil-water content at a 5–10 cm soil depth can be used for crop irrigation scheduling. Additionally, water availability in the upper 15 cm soil depth significantly influenced the total water required.

Lahav and Kalmar (1977) compared irrigation intervals (7, 14 and 28 days) for 'Ettinger', 'Fuerte' and 'Hass' grown on clay soil in Israel, and only the 28-day interval reduced yield (≈13%). The normal irrigation frequencies for avocado in hot climates are every 7–12 days for under-tree sprinkling, 2–7 days for microsprinklers and 1–3 days for drip irrigation. In very hot weather or on sandy soils, several irrigations per day are required because the soil next to the roots dries as water flows from the roots to the leaves to meet the high demand of the atmosphere. In this case, frequent irrigation allows the water lost from soil adjacent to roots to be replenished, and water further away from the roots may then move to it, according to the demand by trees. Silber *et al.* (2012) studied the responses of 'Hass' avocado trees in lysimeters – filled with coarse turf and perlite substrate – to three irrigation frequencies. They compared pulsed drip irrigation, applied for 10–20 min every 30 min throughout the day, with daily night-to-morning irrigation and alternate-day irrigation. The results showed that pulsed irrigation significantly outperformed night-time irrigation, which surpassed the alternate-day treatment regarding trunk growth and yield.

Irrigation frequency, defined by the depletion factor (p) indicating the fraction of water that can be extracted from the root zone before inducing water stress, varies with soil type and plant species. For avocados, the P value typically ranges from 0.4 to 0.7. Allen *et al.* (2006) specified a P value of 0.7 for avocados, while Ferreyra *et al.* (2011) found that a P value of 0.6 in sandy loam soil did not affect yield, fruit size or plant water status. Ferreyra *et al.* (2012) determined that allowing a P value of 0.5 or 0.6 between irrigation cycles did not compromise yield or fruit size compared to more frequent irrigation. Optimal irrigation frequencies, which deplete approximately 25–30% of the tree's water reservoir (P = 0.25–0.3), are generally suitable for most soils, striking a balance between adequate water and oxygen supply to the root system. When using drip irrigation, it is essential to leach excess salts from the potential root zone and ensure adequate soil moisture, especially in shallow, coarse-textured soils.

The water holding capacity in the root zone is mostly a function of the soil properties and the effective rooting depth of the trees. Avocado roots can grow very shallow, including the zone where leaf litter is accumulated or organic mulch is applied. In estimating the soil-water holding capacity of the root zone, the layer of mulch is usually neglected. The mulch can absorb most of the water with little flow to the soil and roots beneath. In addition, the mulch may or may not contain roots. Applying water below the mulch, such as with drip irrigation, avoids this problem.

Irrigation for Managing Soil Salinity

Salinity is a primary factor influencing avocado productivity worldwide, especially in Mediterranean regions where reduced precipitation correlates with elevated levels of dissolved salts in irrigation water (Crowley, 2008; Berkessa, 2020; Kourgialas and Dokou, 2021). While irrigation contributes to soil salinization by introducing water with dissolved ions, rainfed orchards benefit from nearly pure water, facilitating effective salt leaching from the soil. Sodium (Na) and chloride (Cl) ions are among the most important salts, exhibiting mobility within the soil and capable of leaching below the root zone if irrigation exceeds the soil-water holding capacity.

Avocado is sensitive to salinity requiring high amounts of irrigation for leaching salts from the soil (Schaffer and Whiley, 2002; Shalhevet, 2006; see Schaffer *et al.*, Chapter 7 of this volume). Trees exhibit high sensitivity to chloride accumulation leading to leaf tip burning and adverse effects on physiological processes, resulting in diminished yields due to restricted water uptake by the roots (Ayers, 1951; Marsh *et al.*, 1978; Shalhevet, 2006). The symptoms of salinity damage in avocados vary based on the tree developmental stage, leaf age and nutritional status (Bingham and Fenn, 1966; Bar *et al.*, 1987b, 1991, see Schaffer *et al.*, Chapter 7 of this volume). Chloride toxicity symptoms appear as marginal necrosis, mainly affecting the tips of older leaves (Fig. 11.3). While leaf scorching serves as an initial warning sign for growers, root growth is more sensitive to Cl toxicity (Bernstein *et al.*, 2004). Decreased fine root numbers can hinder nutrient and water uptake before visible foliage symptoms emerge. Elevated salt concentrations in the soil solution can lead to iron deficiency chlorosis, osmotic drought, toxicity from high Na and Cl retention and nutritional imbalances (Castro *et al.*, 2009; Musyimi *et al.*, 2007; Alvarez-Acosta *et al.*, 2018). Furthermore, salt accumulation in the soil solution diminishes soil-water potential,

Fig. 11.3. A commercial avocado orchard in Maipo Valley, central zone of Chile with 'Hass' trees on 'Mexícola' seedling rootstock. Pictures from A. Link (with permission).

hindering root growth, water and nutrient uptake, thereby reducing plant growth and crop yields (see Schaffer *et al.*, Chapter 7 of this volume).

Soil salinity is influenced by water quantity and irrigation methods. Strategies for salinity management focus on preventing salt accumulation in the root zone to levels that restrict root water uptake and minimize salinity's detrimental effects on avocado trees for optimal crop growth. Sandy soils with good drainage are easier to flush than clay soils. Microsprinkler irrigation results in uniform salt concentration at the same depth, with a gradual increase in concentration as depth increases. Drip irrigation typically results in lower salt concentrations at the water application point and increasing salt concentration radiating away from it.

Salinity is often determined by measuring the electrical conductivity (EC) of the soil or irrigation water. Avocado trees struggle to extract water from the soil when the EC reaches 4 dS m^{-1} (Crowley, 2008). Water with an EC below 0.5 dS m^{-1} is considered high quality for agriculture, although values between 0 and 3 dS m^{-1} are commonly used for avocado irrigation. Water uptake by 'Hass' avocados on Mexican seedling rootstocks becomes restricted when the EC in the root zone reaches 4.4 dS m^{-1}, equivalent to a soil EC of 2.2 dS m^{-1}. This threshold aligns with the findings of Bingham and Richards (1958) and Kalmar and Lahav (1977), who also identified limiting soil EC levels of 2 dS m^{-1} for avocado trees.

Avocado fruit yield begins to decline with irrigation water EC above 0.75 dS m^{-1} and Cl concentrations exceeding 100 ppm. It is generally advised to maintain a leaching fraction (LF) of 10–20% to keep soil EC values below 2.0 dS m^{-1} and Cl concentrations under 100 ppm (Mickelbart *et al.*, 2007; Crowley, 2008).

Water Cl concentration thresholds vary by cultivar based on salinity tolerance. These thresholds are 120–150 ppm for Mexican, 140–250 ppm for Guatemalan and 177–350 ppm for West Indian varieties, according to reports by Maas (1984), Bingham and Richards (1958), Kalmar and Lahav (1977), Oster *et al.* (2007) and A. Link (2024, personal communication). These thresholds are crucial considerations for commercial variety selection, rootstock selection and leaching salt management (Table 11.4). Bernstein *et al.* (2004) found that the threshold NaCl concentration causing root and shoot growth reductions ranged from 5 to 15 mM. A 15 mM NaCl concentration led to a 10% decrease in leaf biomass production, a 43% reduction in the root elongation rate and a 33% decline in the root volumetric growth rate. This 5 mM threshold corresponds to approximately 200 ppm Cl, a level easily reached in saturated soils after two irrigation cycles without leaching or after one cycle as soil water decreases to 50% of its available water-holding capacity.

Leaf chloride analysis has been reported as a diagnostic tool for identifying salt-sensitive rootstocks, suggesting that the threshold for salinity damage to avocado trees grafted onto Mexican and West Indian race rootstocks is estimated to be 70–100 and 300–400 mg kg^{-1} leaf Cl, respectively (Steinhardt *et al.*, 1995a,c). Notably, an average yield loss of 12% is observed with a mere 35 mg kg^{-1} increase in Cl concentration (Steinhardt *et al.*, 1995b).

Field studies showed that avocado trees lose their ability to extract water when the total dissolved salt concentrations in the soil surpass an EC of 4 $dS \cdot m^{-1}$, even under saturated soil conditions. This limitation mainly stems from the osmotic effects of salts that govern the initial water transport into the root tissue. Oster *et al.* (2007) identified an EC threshold for avocado of 0.57 $dS \cdot m^{-1}$, leading to a 65% yield reduction for each $dS \cdot m^{-1}$ increase beyond this level. Research from Israel estimated a 12% yield reduction for every milliequivalent of Cl in the irrigation water (1 meq l^{-1} Cl = 35.5 ppm). Observations from the Maipo River Valley in Chile demonstrated that salt toxicity, specifically Cl, can decrease productivity in mature orchards by 10–15 tonnes ha^{-1}, especially in 'Hass' grafted on to 'Mexícola' rootstock (A. Link, 2023, personal communication).

To estimate the leaching requirement (LR) accurately, the EC of the irrigation water and the crop's tolerance to soil salinity must be determined. Water salinity can be assessed through laboratory analysis, while salinity tolerance should be derived from the appropriate rootstock tolerance values. Table 11.4 shows general chloride and EC tolerance values for avocado rootstocks of each race.

The LF and LR are used interchangeably. Both terms refer to an extra portion of irrigation water that must pass through the root zone to control salts at a specified level. The necessary LR can be calculated using the equation proposed by Rhoades (1974) and Rhoades and Merrill (1976):

$$\mathrm{LR} = \mathrm{ECw}/[5(\mathrm{ECe}) - \mathrm{ECw}] \qquad \text{(Eqn 7)}$$

where LR is the minimum LR necessary to manage salts within the crop's tolerance level, ECw is the salinity of the applied irrigation water in dS m^{-1}, and ECe is the average soil salinity tolerated by the crop as measured in a soil saturation extract (if soil EC is lower than tolerance, then the actual soil EC is used).

Soil salinity must be taken into account under situations of water scarcity. Water scarcity may require using recycled water for irrigation (Kourgialas, 2021), thereby increasing salinity of the irrigation water. Recent findings indicate that prolonged irrigation (over 7 years) with treated wastewater in avocado orchards can have adverse effects on soil quality. In clay soils, extended use of recycled wastewater for irrigation leads to soil salinity, reduced aeration and disruption of soil aggregates and structure (Assouline *et al.*, 2016; Nemera *et al.*, 2021). A decade-long study of avocado orchards in Israel revealed a 20–40% reduction in yield when irrigated with reclaimed wastewater compared to fresh water (Assouline and Narkis, 2013). Nemera *et al.* (2021) suggested several salinity mitigation techniques for avocado orchards in

Table 11.4. Common salinity chloride and electrical conductivity (EC) threshold values for yield limitations according to avocado rootstocks races tolerance

Rootstock (race)	Chloride threshold in irrigation water (ppm)	EC threshold (dS m^{-1}) in soil saturation extract	EC threshold (dS m^{-1}) in irrigation water
West Indian	177–350	2.5	2.0
Guatemalan	140–250	2.0	1.5
Mexican	120–150	2.2	0.6

Adapted from Maas (1984), Bingham and Richards (1958), Kalmar and Lahav (1977), Oster *et al.* (2007) and A. Link (2024, personal communication).

water-scarce regions in Mediterranean climates. One recommendation is to blend fresh water and treated wastewater in a 1:1 ratio to reduce damage caused by high salinity. Another mitigation strategy for lowering salt quantities in recycled water is using osmotic reversal treatments already practiced in the Maipo Valley, Chile. Other potential methods for mitigating salinity stress in avocado trees are discussed by Schaffer *et al.* in Chapter 7 of this volume.

Calculations for Efficient Irrigation Taking Soil Salinity into Account

The main factors for irrigation scheduling are irrigation time determination (T) and frequency or watering interval (IF). These can be calculated by using the previously described equations as follows:

$$T(\text{h/day}) = \text{TWR}(\text{mm/day}) / \text{AR}(\text{mm/h}) \qquad \text{(Eqn 8)}$$

where T is the time the irrigation system needs to run (h/day), TWR is the total crop water requirement (mm/day) and AR is the application rate of the system (mm/h).

$$\text{TWR}(\text{mm/day}) = \left[\text{ETc}(\text{mm/day}) \times (1+\text{LF})\right]/\text{Ef} \qquad \text{(Eqn 9)}$$

where ETc is the crop water requirement or crop water use (see Eqn 1), Ef is the efficiency of the irrigation system (0.9 for drip irrigation, 0.85 for microsprinkler) and LF is the leaching fraction requirements (see Eqn 6).

$$\text{AR}(\text{mm/day}) = \left[\text{No emitters/plant} \times Q(\text{L/h})\right] / \text{plant area}(\text{m}^2) \qquad \text{(Eqn 10)}$$

where No emitters/plant is the number of drippers or microsprinklers/plant, Q is the emitter flow (l/h) and plant area is the area occupied by each plant according to the planting layout.

$$\text{WI}(\text{days}) = \text{RAW}(\text{mm})/\text{ETc}(\text{mm/day}) \qquad \text{(Eqn 11)}$$

where WI is the watering interval (days), RAW is the readily available soil water in the root zone [mm] (see Eqn 5) and ETc is the crop water requirement (see Eqn 1). If the watering interval of the system is over 1 day, then

$$T(\text{mm/irrigation event}) = [\text{RAW}(\text{mm/day}) \times (1+\text{LR})]/\text{AR} \qquad \text{(Eqn 12)}$$

Irrigation computing, although simple, may be overwhelming for growers. Today, technology makes this operation much easier through different irrigation scheduling decision support systems (DSS). Examples of DSS software available for avocado irrigation include the irrigation scheduling calculator (Hofshi and Hofshi, 2007), AquaCrop (Steduto *et al.*, 2009) and smartphone applications, such as SmartIrrigation Avocado (Migliaccio *et al.*, 2016), available for Florida, USA, the IrrisatSms approach (Hornbuckle *et al.*, 2009) available for Australia, Irrimaps (www.agrisat.es) available for Spain and the RiegaBien App (Gil *et al.*, 2021) available for Chile. While those are useful tools to help growers manage irrigation, field knowledge and experience cannot be overlooked.

Example of irrigation scheduling determination taking soil EC into account

Below are calculations for a weekly irrigation schedule in January for a mature 'Hass' avocado orchard on 'Mexícola' seeding rootstock in a Mediterranean climate in the Southern Hemisphere (summer), with no rainfall input, tree spacing of 5 m between rows and 3 m within rows, using drip irrigation with two lines per row and drippers with a flow rate of 4 l h^{-1} spaced 50 cm apart. A 90% irrigation efficiency is assumed. The ETo is 38.5 mm $week^{-1}$ (5.5 mm day^{-1}), the Kc commonly used in Chile for January = 0.72. The calculations are for a loam sandy soil with root development up to a 60 cm depth, a wetted soil percentage of 50%, containing 5% rocks and a water depletion factor of 40%. The soil has an EC of 1.25 dS m^{-1}, and the water has an EC of 0.75 dS m^{-1}. Based on the aforementioned variables, the irrigation schedule is calculated as follows:

Irrigation efficiency:	0.9	90% efficiency	
Planting spacing:	15.0	5 × 3 m spacing	
AR	3.2	AR (mm day^{-1}) = [N° emitters/plant × Q (L h^{-1})]/ plant area (m^2)	(Eqn 10)
ETc (mm/day)	4.0	ETc = Kc × ETo	(Eqn 1)
Leaching fraction	0.14	LF = ECw/ [5 (ECe) – ECw]	(Eqn 7)
TWR	5.1	TWR (mm day^{-1}) = [ETc (mm day^{-1}) × (1+LR)]/ Ef	(Eqn 9)
T (h/day)	1.6	T (h day^{-1})= TWR (mm day^{-1})/ AR (mm h^{-1})	(Eqn 8)
TAW (mm)	34.2	TAW = ($\Theta_{FC} - \Theta_{PWP}$) × Z_r× (1–s) × %WSS	(Eqn 5)
RAW (mm)	13.7	RAW = p TAW	(Eqn 6)
WI (days)	3.14	WI (days) = RAW (mm)/ ETc (mm day^{-1})	(Eqn 11)
Irrigation time (h/day)	1.6		
Wetting interval days (days)	3.34 ≈ 3		
Total irrigation time for 3 days (h/irrigation event)	4.8 (1.6 hr/day)		(Eqn 12)

Influence of Rootstock on Salinity Tolerance

While salt leaching via irrigation is often an effective salinity management strategy (see salinity section of Schaffer *et al.*, Chapter 7 of this volume for other saltmitigation strategies), the most sustainable and cost-effective means of mitigating salt toxicity is using tolerant rootstocks (see Schaffer *et al.*, Chapter 7 of this volume and Ben-Ya'acov and Michelson, 1995). West Indian race avocado rootstocks are more tolerant of high soil salinity than Mexican or Guatemalan race cultivars (Ben-Ya'acov, 1992; Ben-Yaácov and Michelson, 1995). In Israel, Lazare *et al.* (2021) investigated the relative salinity tolerance of 'Hass' avocado trees on 20 different rootstocks by adding NaCl to the fertigation solution to reach an EC of 1.4–1.5 (corresponding to 200–300 mg Cl L^{-1}) and comparing physiological growth and yield responses among rootstocks. They ranked VC 159, VC 140 and VC 152 as the most salttolerant and VC 840, 'Dusa', VC802 as the most salt sensitive of the rootstocks tested. Other research by Link and Alegría (2019) in the Maipo Valley of Chile evaluated the effects of high Cl in the irrigation water on 'Hass' trees on different rootstocks over 3 years. Results consistently showed that seedling West Indian rootstocks were the most tolerant of high soil salinity, followed by 'Nabal' (Guatemalan race) rootstock. In contrast, 'Mexícola' (Mexican race) seeding rootstock exhibited the least salinity tolerance, whereas 'Duke 7' showed some salinity tolerance. In that study, 'Hass' grafted on to West Indian rootstock yielded an average of 21.6 kg m^{-2} compared with 9.98 kg m^{-2} for 'Hass' on 'Mexícola'.

Studies in California by Celis *et al.* (2018) revealed significant variability in response to irrigation with saline water of 'Hass' avocado on different clonal rootstocks. Rootstocks with lower leaf Cl concentrations, such as R0.05, PP40, R0 and 'Dusa', were more salttolerant, with higher yields, larger trunk diameters and increased survival rates under saline conditions than other rootstocks tested, highlighting the importance of leaf Cl analysis for identifying salt-sensitive rootstocks. A study in central Chile of 'Hass' trees on clonal 'Duke 7', or seedling 'Nabal', 'Degania 117', 'Mexícola' or 'Zutano' rootstocks (in 55l containers) compared vegetative growth and plant nutrient contents between trees treated with 30 mg l^{-1} of Cl and trees treated with 200 mg l^{-1} of Cl. In the high salinity treatment, trees on 'Nabal' rootstock had the highest carbon assimilation rate and root Cl concentrations, but low leaf Cl concentrations indicated promising salt tolerance for this rootstock due to its ability to retain Cl in the roots, thus reducing translocation to the leaves (Castro *et al.*, 2009). Crowley (2008) proposed multiple strategies for salinity management, including tolerant rootstock selection, ensuring adequate water quality, optimizing irrigation timing, monitoring EC of the irrigation water and soil and using sufficient water to leach salts below the root zone. Based on experiences with farmers in California with various crops and soils, LFs of 0.30 are achievable for sand and loam-textured soils. An LF of 0.30 is projected to result in maximum yields with an EC of 0.5 dS m^{-1}. Thus, if maximum potential yields are the goal, EC should be maintained at 0.6 dS m^{-1} or less for 'Hass' avocado trees on Mexican seedling rootstock.

Water Status Monitoring

An important aspect of efficient irrigation management is to continuously monitor the water status of the soil and/or plants. Given that much of the scheduling data are theoretical, these assessments are crucial for making necessary adjustments to the irrigation schedule. Soil-water measurements provide reliable information on soil-water status in mature orchards where water is applied to the entire soil surface. However, soils may exhibit textural and structural heterogeneity that cannot be detected with point measurements using the existing *in situ* methods for soil-water moisture monitoring (Araya Vargas *et al.*, 2021). Additionally, under drip or microsprinklers, where only part of the soil in the orchard is irrigated, soil-water measurements provide limited quantitative information due to the partial and uneven distribution of water in the soil. Despite the inability of a single sensor to provide information about soil-water status across the entire root zone, continuous or periodic monitoring of soil-water potential at fixed positions (distances and soil depths relative to the emitters) is recommended to adjust irrigation schedules. When using monitoring systems, the selection of the monitoring point and proper analysis of the information are very important for irrigation management (Tolomio and Raffaele, 2020).

In situ soil-water data can be collected gravimetrically or by periodic or continuous monitoring with tensiometers, neutron probes or capacitance systems such as time domain reflectometry (TDR) or frequency domain reflectometry (FDR) (Marsh and Gustafson, 1958; du Plessis, 1991; Zekri *et al.*, 1999; Núñez-Elisea *et al.*, 2001), and by subjective water monitoring by trench observation (Ferreyra and Sellés, 2013). In Mediterranean countries (i.e. Chile, Israel, California, Perú and Spain), soil-based sensors, such as capacitance probes, are widely used and are considered the most developed methods (Carr, 2013). Typically, avocado growers using soil-based sensors focus on an irrigation depth of up to 60 cm, where over 70% of roots are concentrated (Lahav and Whiley, 2002). Tensiometers are typically buried to a depth of 30 cm and trigger irrigation when soil matric potential reaches between −20 and −50 kPa (du Plessis, 1991). There are some situations where soil-water status is monitored only in the top soil layers, for example, in calcareous soils in Florida, soil-water measurements with tensiometers were highly correlated with soil moisture data from capacitance sensors at 10 cm below the soil surface. However, there was a poor correlation at 30 cm below the soil surface in these soils due to the larger soil particles, resulting in poor contact between the tensiometers and the soil (Núñez-Elisea *et al.*, 2001).

Other monitoring systems rely on plant physiological indicators. Fruit and trunk diameter changes (Kalmar and Lahav, 1977; Celedón *et al.*, 2012), leaf thickness (Schroeder and Wieland, 1956; Sharon, 1999), leaf and SWP (Bower *et al.*, 1977; Sterne *et al.*, 1977a; Ferreyra *et al.*, 2007), the xylem sap flow rate (Cantuarias-Avilés, 1995) and stem-imbedded microtensiometers (Lakso *et al.*, 2022) have been used to control or adjust irrigation scheduling for avocados with varying degrees of success.

Among all the plant indicators, monitoring trunk diameter is perhaps the simplest and most likely to be used commercially for irrigation scheduling. Determination of the daily maximum trunk contraction in relation to a baseline reference was a useful tool for scheduling irrigation in avocado orchards (Gurovich *et al.*, 2006; Winer and Zachs, 2007; Celedón *et al.*, 2012). SWP can also be used for irrigation scheduling. In well-watered avocado trees, SWP ranges from −0.5 to −0.6 MPa on a typical summer day, serving as reference thresholds (Ferreyra *et al.*, 2007). Mild water deficits induce mid-day SWP values between −0.6 and −1.0 MPa, whereas values below −1.0 MPa, reaching even −2.0 to −3.0 MPa, indicate more severe water deficits (Naor,

1998). Celedón *et al.* (2012) found that measurements of the maximum trunk contraction were a more sensitive plant water stress indication than SWP and, therefore, a more useful plant-based monitoring method for irrigation scheduling in avocado orchards.

RS and digital agriculture (DA) offer promising solutions for better water management. DA, also known as smart agriculture or agriculture 4.0, integrates various technologies such as sensors, robotics and artificial intelligence (AI) (Karmas *et al.*, 2016; Kamilaris *et al.*, 2017). However, implementing DA faces challenges, requiring reliable information and user-friendly solutions (Ramírez-Gil *et al.*, 2018; Rose and Chilvers, 2018; Rijswijk *et al.*, 2019). Digital tools are already available for irrigation scheduling in avocado and other crops, many of them using information from agroclimatic networking and NDVI or SAVI satellite information for estimating crop water consumption. Other applications can complement *in situ* information, such as soil moisture values, by using radar images, which are accessible and have extensive coverage compared with conventional *in situ* technologies. However, DA tools have not yet been widely adopted globally for irrigation scheduling in avocado orchards (Erazo-Mesa *et al.* 2022b).

Strategies for Irrigation Water Savings

Deficit irrigation (DI) refers to applying water during critical growth stages, but withholding or reducing the amount of water applied outside of these growth stages. Regulated deficit irrigation (RDI) is an effective approach for enhancing water use efficiency (da Silva *et al.*, 2018; González-Dugo *et al.*, 2018; Romero-Trigueros *et al.*, 2020). Despite being introduced more than 40 years ago, RDI's implementation methods, scope of application and detailed mechanisms remain poorly understood (Ju *et al.*, 2019). Initially, RDI methods were simple, involving targeted water deficit during critical phenological periods to conserve water without compromising plant growth (Ju *et al.*, 2019). This approach considers differential water requirements and the varying effects of water deficits at different growth stages (Ju *et al.*, 2019). Implementation of RDI has resulted in water savings for various fruit crops such as citrus, grapes, blueberries, pears, mangoes and peaches, but its implementation has been limited in avocado because it requires precise water control, and investments in drip irrigation and water measurement equipment (Ju *et al.*, 2019; Yang *et al.*, 2022).

Much of the experimental evidence suggests that RDI is not recommended for avocado irrigation due to the heightened sensitivity of commercial yields to water deficits during the majority of the irrigation season (Holzapfel *et al.*, 2017; Silber *et al.*, 2018; Moreno *et al.*, 2019; Ferreyra *et al.*, 2012; Durán Zuazo, 2021). In contrast to RDI, where water is limited or withheld at specific phenological stages, sustained deficit irrigation (SDI) is an irrigation strategy whereby the amount of water applied is limited throughout the crop cycle. In a study in Andalucia, Spain (Durán Zuazo *et al.*, 2021), mature 'Hass' avocado trees on 'Lula' rootstock were subjected to 33%, 50% or 75% SDI compared with a control treatment receiving full irrigation (100% ETc). Water stress negatively impacted vegetative growth, flowering intensity, fruit set and yield, with severely reduced irrigation treatments (33% SDI) having negative effects on fruit size and weight. Significant differences in fruit characteristics were observed among different SDI treatments, accompanied by decreased leaf water potential and stomatal conductance, indicating physiological stress. The 75% SDI treatment had relatively milder effects on tree physiology and yield. Another study on the coast of Spain showed that water stress via DI based on 40–60% of ETC during the growing season reduced avocado yield by 30% (Moreno *et al.* 2019).

Crop Management for Reducing Evapotranspiration

A management strategy for reducing water use in fruit crops, with minor reduction of yields, is based on reducing evapotranspiration. Although some growers are already using available alternatives, such as soil mulching, or plastic or mesh covers above the canopy, little information has been reported about the effect on water saving. Ferreyra *et al.* (2018a,b) evaluated plastic cover above the canopy and plastic soil mulching as water-saving strategies for avocado orchards. The plastic cover was intended to change micrometeorological conditions by reducing the amount of radiation and wind, and therefore, the VPD (Fig. 11.4), whereas mulching was intended to reduce evaporation from the soil, increasing water availability for the plant. In both treatments, 15–20% reductions of ET were observed (Ferreyra *et al.*, 2018a, b; Antúnez *et al.*, 2019).

Other water-saving strategies include the application of soil silicates, which resulted in 50% water reduction over two growing seasons in a 'Hass" orchard, with minimum effect on yield (Cea *et al.*, 2019). Kaolin and biostimulants based on algae extract have been effectively used to increase water use efficiency of other fruit-bearing species such as apples, citrus, grapes and walnuts (Brito *et al.*, 2019). However, little information is available about these methods for avocado. New irrigation technologies can also increase water use efficiency. Buried irrigation and low flow irrigation have water use efficiencies higher than 95% and have been adopted in other fruit and annual crops but not yet in avocado (Tong *et al.*, 2022).

Fig. 11.4. A 'Hass' avocado tree under plastic cover and energy flux balance device for actual ET estimation.

Microclimate Management for Heat Stress Mitigation and Frost Protection

Irrigation is often used to protect orchards from two main climatic hazards: (i) low temperatures and light frosts; and (ii) high temperatures and low humidity during flowering and fruit set and expansion.

Water management for cold and frost

Many of the major avocado-growing areas worldwide suffer periodically from severe frost. The susceptibility of avocado to frost depends on the dormancy status of the trees at the time of sub-zero temperatures. Inactive or quiescent trees are much less sensitive to frost. However, frost can damage flower buds in dormant trees, such that the following spring flush is strongly vegetative with a few weak flowers (A. McCarthy, Bunbury, Australia, 2011, personal communication). Reducing water during autumn can enhance cold hardiness by avoiding late growth flushes that are extremely sensitive to frost.

Irrigation is regarded as one of the most economical methods for frost protection. In general, trees in dry topsoil are more susceptible to frost damage than trees in wet soil because moist soil can store and conduct heat more than dry soil. Thus, irrigation can increase the ability of soil to accumulate heat during the day and radiate it back to the canopy at night. With soil surface irrigation, water increases orchard temperatures mainly by releasing heat from the water during application. With overhead irrigation at night, when temperatures are below zero, latent heat is released to the surrounding environment as water freezes (heat of fusion) (80 cal g^{-1}). Overhead irrigation is the main method of frost protection in Florida; however, high-volume under-tree irrigation is preferred since overhead irrigation can lead to limb breakage due to the weight of ice that accumulates on dense foliage (J.H. Crane, Homestead, Florida, 2001, personal communication). Overhead irrigation has also been used in new avocado production areas in Chile's central coastal zone, with good results (A. Link, 2023, personal communication).

Mini-sprinklers can protect the lower limbs of a tree during a radiation freeze, typically with a strong temperature inversion. The dense canopy of avocado trees helps retain the heat released by the water when passing from liquid to ice (80 cal gr^{-1}). However, during an advective freeze with high winds and low dew point, evaporative cooling may cause more damage to wet than to dry plant parts. Under such freezing conditions, damage will be prevented only if large volumes of water are applied. The emitters should have at least 4 mm h^{-1} discharge. Irrigation must continue, even if a thin layer of ice is formed for 2 h after sunrise or until temperatures rise above 2–3°C (Fig. 11.5).

Water management to mitigate heat damage

Water application to mitigate heat damage during flowering and fruit set is critical in avocado production. The tree may lose its crop if hot and dry winds occur during the fruit development period (Miller *et al.*, 1963). In Israel, temperatures can reach 45°C with very low relative humidity (RH) over several consecutive days (and nights) during flowering. These conditions may cause shoot damage, leaf burn and, in severe instances, total leaf shedding. Frequent irrigation can reduce heat damage, and daily pulses are preferable. Overhead irrigation provides better protection than under-canopy sprinkling, and both are preferred to drip irrigation for reducing heat damage. Overhead irrigation wets the canopy and induces a higher evaporation rate, reducing leaf temperatures. Overhead irrigation reduces air temperature by 4–7°C and increases RH by 6–27% (Miller *et al.*, 1963; Lomas and Mandel, 1973).

Fig. 11.5. A 'Hass' avocado with water sprinkler system for frost protection in the central coastal zone of Chile (Picture by A. Link, 2023, with permission).

In many cases, because of the high water pressure required for overhead irrigation, the whole orchard cannot be simultaneously irrigated. Drip irrigation may then be preferred, since it uses lower water volumes and will protect the entire orchard. An efficient way to overcome hot and dry spring winds (variously called khamsin, Santa Ana, berg or Terral winds) is to have an over-the-canopy fogging system operated separately from the regular irrigation system. However, the cost prevents wide-scale use, although it significantly reduces temperature and increases RH.

To reduce crop damage, an additional 50% of the budgeted amount of water should be applied the day before an expected heatwave. The threshold for operating the irrigation system in Israel is 38°C and/or 40% RH. A 2-h pulse interval should be applied (Anonymous, 2001). Overhead sprinkling may increase the hazard of Na and Cl damage to the leaves and should be implemented only with low salinity water.

Nutrition

The ecophysiology of the avocado tree is the product of its evolutionary history, where its centre of diversity encompassed the highland neotropical forests of Mexico, Guatemala and Honduras (Kopp, 1966). Competitive selection has resulted in ecophysiological features in *Persea americana* Mill., which most likely identify it with climax species (Swaine and Whitmore, 1988; Whitmore, 1990) that colonize small gaps when the forest canopy is disturbed. Successful capture of canopy space is dependent on the presence of viable seed or seedlings, the latter with the capacity to survive for long periods in low light regimes (Denslow, 1987); high plasticity of light response with leaves tolerant to shade but able to utilize full sunlight, which is converted into fast growth when opportunities arise (Pearcy and Francheschi, 1986); and rapid leaf redundancy with effective recovery of mineral nutrients during senescence, since shading (inducing reduced function) occurs quickly from successive cohorts of leaves (Chabot and Hicks, 1982). These features have been identified in avocado by Cameron *et al.* (1952), Whiley (1994), Wolstenholme and Whiley (1999), Barlow (2000), Salazar-García *et al.* (2007) and Wolstenholme (2002). The avocado tree is generally described as an evergreen species since non-stressed trees retain a year-round, leafy canopy. However, both phenologically and physiologically, they behave more like deciduous fruit crops and fit the description of a small group of species classified as 'winter-green' (Chabot and Hicks, 1982). Unlike other evergreen fruit crop species, where leaf longevity ranges from 3 to 5 years, longevity of avocado leaves is 9–10 months with total canopy replacement every 12 months (Whiley, 1994). This has significant implications for carbon assimilation capacity, and nutrient accumulation, retention, storage and recovery with these physiological functions more closely aligned to deciduous trees. Additionally, stored carbohydrates in trunks and branches can reach concentrations similar to deciduous species (Scholefield *et al.*, 1985) and are remobilized to support inflorescence growth and fruit set.

Soils supporting the neotropical forests of Mexico, Guatemala and Honduras are generally considered highly weathered and infertile having low base saturation and acidic pH (Fujii *et al.*, 2018). However, accumulation of senesced leaves and other organic debris on the forest floor provides a source of nutrition (Fujii *et al.*, 2018), which is continually recycled by maintaining forest biomass. The root system of avocado has adapted to exploit this source of relative mineral abundance, producing a dense network of non-suberized, fleshy, feeder roots that populate the top 200–300 mm of the soil profile (Whiley, 1994), and become colonized with arbuscular mycorrhizae (AM). The hypha of AM and root hairs are the main absorption structures for water and nutrients in mycorrhizal trees (Bárcenas *et al.*, 2007; Fujii *et al.*, 2018), greatly increasing the root surface area available for water and mineral uptake. However, avocado roots do not develop root hairs (Burgis and Wolfe, 1945; Salazar-García, 2002) and are reliant on AM for increased efficiency of water and nutrient absorption. The deeper penetrating suberized roots provide anchorage for trees and contribute relatively little to water and nutrition requirements. From rhizotron studies in subtropical Australia, Whiley (1994) reported that root growth followed a bimodal cycle with peak growth activity during the spring months, generally concomitant with new spring shoot growth, with the second flush of root growth starting in late summer, and peaking during winter just prior to anthesis. The relatively shallow distribution of avocado feeder roots in soils highlights the need to use fertilizer practices in commercial orchards that place and maintain nutrients within this zone. Movement and loss of minerals due to differences in soil retention and leaching pressure will vary considerably within and between climatic zones, viz. subtropical versus Mediterranean, and fertilizer strategies will require a level of customization to optimize benefits from applied nutrients. This chapter describes deficiency and toxicity symptoms, reviews the effects of mineral nutrition on crop yield and fruit quality, explores the potential management of nutritional requirements via pheno-physiological relationships and discusses application timing for some critical nutrients.

Nutrient dynamics in avocado orchards

Nutrient levels within an avocado tree need to be sufficient to meet specific physiological requirements during each phenological stage of growth. By knowing the timing and length of growth periods for the relative organs (fruit, shoot and root), it is possible to schedule applications of fertilizers, mulch and soil amendments to support tree requirements in a timely manner (Whiley *et al.*, 1988b; Whiley 1994; Wolstenholme 2004; da Silva *et al.*, 2017). For example, flowering occurs over a relatively short time with a high demand for both mineral and carbon resources (Whiley, 1994). These are drawn from sources within the tree, for example, leaves, trunks, branches and roots, that have accumulated during the quiescent period following summer/autumn, and can be enhanced through appropriate management practices (Wolstenholme and Whiley 1997, 1999; Lovatt, 2013b).

The main nutrient pools available to orchard-grown trees are soil minerals and organic matter, mulch both applied and from falling leaves, and so on, vegetation growth on the orchard floor that includes cover crops and/or weeds, and minerals accumulated within the trees, all of which are in a fluid state due to their interconnectivity. Inputs and losses of mineral nutrients from the whole orchard system also need consideration. Additional gains can come from irrigation water, rainfall (nitrification via lightening), atmospheric dust and fertilizer and soil amendment applications. The main mineral losses are in fruit at harvest, leaching of soluble elements below the root zone and/or through surface runoff, fixation in the soil and ammonia volatilization of surface-applied ammonium-based fertilizer, including manures. Some soil-fixed nutrients, such as phosphorus and zinc, may be slowly released over time, depending on soil pH.

Nutrients move between the pools with those in trees recycled through abscission of leaves, twigs, flowers and fruit that fall to the ground and decompose, thus releasing minerals into the soil. These may be re-absorbed by tree roots or other vegetation and retained within the orchard system. Transient pools of nutrients within trees, especially nitrogen (N), phosphorus (P) and potassium (K), may be recovered from ageing organs and transferred to new growth, thereby held within trees. Cameron *et al.* (1952) reported that about 50% of N, 57% of P, 25% of K and 33% of sulfur (S) present in leaves returned to 'Fuerte' trees in California, prior to abscission. However, only 4% of calcium (Ca) and 10% of magnesium (Mg) were captured before leaf fall. They estimated the annual nutrient transfer to the soil from abscised leaves and flowers was 50 kg of N, 6 kg of P, 38 kg of K, 86 kg of Ca and 27 kg of Mg ha^{-1}, based on a planting density of 156 trees ha^{-1}. In Israel, Lahav and Zamet (1999) reported 26 kg of N, 4 kg of P, 31 kg of K, 11 kg of Ca and 9 kg of Mg ha^{-1}, respectively, removed by abscised buds, flowers, fruitlets and fruit from 'Fuerte'. The transfer of micronutrients through abscission was relatively low with more iron (Fe) (1.4 kg ha^{-1}) than zinc (Zn) (0.2 kg ha^{-1}) returning to the soil. The two studies are not directly comparable since Lahav and Zamet (1999) did not include minerals from abscised leaves in their results. However, from calculations made between the two sets of data, leaves appear to contribute about 50% of N, 33% of P, 18% of K, 87% Ca and 66% of Mg of the total minerals lost from trees in senesced organs.

Thunderstorms are considered a significant source of nitrogen oxides and contribute around 500 million tonnes of nitrogen fertilizer to the planet every year (Tie *et al.*, 2022). However, distribution is irregular and in subtropical climates, storms are generally seasonal occurring during the spring and summer months. The contribution to N nutrition from storms in most cases is relatively small, difficult to measure and often prematurely lost to deep leaching by heavy rainfall events associated with thunderstorm activity. In intensively cultivated areas, groundwater used for irrigation may be contaminated with nitrates (Singh and Craswell, 2021); thus, N from depth is transferred back to the root zone. In other situations, recycled urban water is used for irrigation (Tal, 2016) and carries elemental contaminants that add to the nutrient supply of orchards. In such cases, mineral analysis of the water is recommended so that the quantity of nutrients can be accounted for when fertilizer application rates are determined.

Mineral nutrient losses from orchards, caused by leaching and runoff, can be considerable but are difficult to quantify. Three important nutrients for avocados, which are prone to leaching from the soil during high rainfall, are N, Ca and B. Under these conditions, fertilization rates are typically raised by up to 20% above calculated requirement, but caution is required with B fertilization, especially on light sandy soils, since this element can easily reach toxic levels. As N fertilizing comprises about 45% of total fertilizer applied to avocado orchards, its application during wet weather may result in reduced uptake and environmental pollution through surface runoff and deep leaching. Kiggundu *et al.* (2012) assessed nitrate leaching under several irrigation regimes applied to young trees in Florida. They found that under greatly varying treatments in the amount of irrigation water applied, the volume of irrigation water used had greater contribution to leaching than the amount of fertilizer applied. In a 12-year-old 'Hass' orchard in Michoacán (Mexico) fertilized with 220 kg N ha^{-1} year^{-1}, LFs of 25–35 ppm of N-NO_3 to a 90 cm depth were observed in soil irrigated every 8 days with under-tree sprinklers, which increased to 80–100 ppm of N-NO_3 when trees were irrigated by a hose discharging close to the trunk, every 18–21 days (Tapia *et al.*, 2007). Other factors contributing to high nutrient losses in the soil are poor health of the root system (i.e. Phytophthora root rot, root hypoxia and translocation incompatibility: see Schaffer *et al.*, Chapter 7, this volume, and Whiley, Chapter 9, this volume), low cation exchange capacity of the soil and over-fertilization.

Distribution of nutrients within the tree

The first report on total mineral distribution in avocado trees, including roots and fruit, was by Marchal and Bertin (1980) for 'Lula' adult trees grown in Martinique (Table 11.5). The leaves, twigs and branches had similar concentrations of N, P, K, Ca, Mg, Na and Mn. The branches contained 78% of the tree chloride (Cl) and 62% of the zinc (Zn), while the roots were high in Fe (57% of the total). In South Africa, similar distribution values were observed for Ca and K in 'Fuerte' and 'Hass' trees (Witney *et al.*, 1990a). Ca and K were equally distributed between leaves and branches, while almost 50% of the Mg was found in the leaves. The roots had about 10% Ca, 15% Mg and 16–23% K, while fruit were very low in Ca and Mg. Stassen *et al.* (1997) reported similar results for N, Ca and Mg in 'Hass' fruit, but leaves had lower P (9%) and K (8%). Some of these anomalies in nutrient distribution within trees may be related to differences between cultivars, rootstocks and soil types, viz. 'Lula' (West Indian) in Martinique versus 'Hass' (58% Guatemalan × 42% Mexican; Chen *et al.*, 2009) in South Africa. In 17-year-old 'Hass' trees grafted on to 'Zriffin' (West Indian) rootstock growing in the central coastal region of Perú, Tomassini-Vidal and Pino-Huasacca (2022) reported that N was more abundant as nitrate than ammonium in fruit and leaves. Ca and Mg in leaves were higher than in fruit and P, K, Ca and Mg concentrations decreased as the fruit grew.

Nitrogen distribution in avocados was studied in detail by Lovatt (1996), who reported differences from previously published data. Proportionally, less N was found in roots (≈4%) and leaves (≈18%) and more in branches (≈50%). Differences may have been enhanced by the high yield on trees (100 kg tree^{-1}) that resulted in 26% of total tree N being removed in fruit at harvest. In Israel, Zilkah *et al.* (1999) used ^{15}N to study distribution in container-grown avocado trees. Leaves contained 10–25% of the total dry matter throughout the year and held 30–40% of the trees' N, significantly more than in the trunk, branches or roots. These results are similar to those reported by Marchal and Bertin (1980) and may reflect N distribution in trees without fruit or

Table 11.5. Mineral distribution (g) in a 7-year-old 'Lula' avocado in Martinique (Marchal and Bertin, 1980)

Nutrient	Leaves and twigs	Branches	Trunk and dead wood	Roots	Fruit[a]	Total	Proportion removed by fruit (%)
N	706	567	76	195	89	1633	5.5
P	59	71	9	29	11	179	6.1
K	523	795	107	210	145	1780	8.1
Ca	439	344	65	85	4	937	0.4
Mg	96	62	10	23	6	197	3.0
Cl	49	270	13	13	2	347	0.6
Na	1.7	2.7	0.8	2.8	0.3	8.3	3.6
Fe	5.3	55.0	8.6	92.9	0.5	162.3	0.3
Mn	6.5	4.3	0.6	2.2	0.04	13.64	0.3
Zn	1.1	4.8	0.5	1.2	0.2	7.8	2.6

Based on a medium crop of 32 kg tree^{-1}.

with a small crop (32 kg on 7-year-old trees), thus in the relative absence of a strong sink.

Nutrient removal by the crop

Some orchard managers use estimates of the nutrients removed in the harvested crop as a basis for calculating fertilizer rates to apply in the orchards, since the concentration of nutrients in fruit dry matter is less affected by nutrient supply (primary sinks) than in other plant tissues. The percentage of nutrients in fruit is small in relation to the total amount in trees. For instance, the percentage of K (8%) in the fruit of 'Lula' is slightly greater than the percentage of P or N (6%) while Mg, Na or Zn in fruit each represent only 3% of the tree total (Table 11.6). Wolstenholme (1991) compared these data with the nutrients removed by 10 t ha^{-1} of oranges (*Citrus* sp.) and concluded that avocados are 'nutrient cheap' placing low demands on the mineral resources of soils, compared with oranges. For example, 10 t ha^{-1} of avocados would remove one-twelfth of the Ca, one-quarter of the P and Zn, one-third of the K and one-half of the N compared to a 10 t ha^{-1} crop of oranges. A similar conclusion was reached by Marchal and Bertin (1980), who compared the amounts of nutrients in avocados with that of citrus, pineapple and bananas. Aside from comparatively low mineral concentrations in the fruit, avocado trees also produce lower yields than oranges because of their high oil concentration (oil synthesis requires more energy than carbohydrates, which are the main 'energy' constituent of citrus fruits).

The low amount of nutrients removed by fruit justifies the comparatively small amount of fertilizer (NPK) applied to avocados under most conditions. Furthermore, nutrient hoarding and recycling in healthy trees appears well developed due to survival mechanisms evolved during the avocados eco-evolutionary history (Fujii *et al.*, 2018). The fertilizer rationale developed by Huett and Dirou (2000) calculates nutrient loss through crop removal, and then takes account of losses due to leaching, soil fixation and runoff. This usually predicts a quantity of nutrients that contribute less than current fertilizer recommendations. The balance is accounted for by the addition of the amount of nutrients estimated to support the new cohorts of seasonal shoot growth that sustains fruit retention and development, and finally corroborated through soils and leaf mineral analysis to ensure targets were met.

Nutrient loss through crop removal has been determined for 'Hass' in Venezuela (Avilan *et al.*, 1979), Australia (Dirou and Huett, 2001), Israel (Lahav and Kadman, 1980), USA (California) (Rosecrance *et al.*, 2012), Mexico (Salazar-García and Lazcano-Ferrat, 2001) and Colombia (Tamayo-Vélez *et al.*, 2018; Rebolledo-Roa and Burbano-Diaz, 2023; Table 11.6). From the data reported, the mean quantity of N removed at harvest in 'Hass' fruit was 2.95 kg t^{-1} with a range of 2.2–3.8 kg t^{-1}. Scaling this up to a commercial level, it represents ≈30 kg of N for a 10 t ha^{-1} crop and ≈53 kg for an 18 t ha^{-1} crop, which roughly matches the country median and higher

Table 11.6. Estimates of nutrients (kg t^{-1}) removed by one tonne of avocado fruit from different cultivars

Author	Country	Cultivar	N	P	K	S	Ca	Mg
Dirou and Huett (2001)	Australia	Hass	3.8	0.66	5.7	0.5	0.48	0.55
Salazar-García and Lazcano-Ferrat (2001)	Mexico	Hass	2.6	1.03	4.7	0.35	0.08	0.30
		Choquette	1.5	0.69	3.0	0.19	0.09	0.16
		Hall	1.5	0.50	3.0	0.18	0.07	0.17
		Booth-8	1.8	0.58	2.7	0.22	0.10	0.22
Stassen *et al.* (2010)	South Africa	Hass	2.4	0.60	5.3		0.10	0.30
Rosecrance *et al.* (2012)	USA (CA)	Hass	2.2	0.40	3.0	0.50	0.10	0.30
Snijder, B. (2015)	South Africa	Maluma	5.3					
Silber *et al.* (2018)	Israel	Hass	4.0	0.83	8.0			
Tamayo-Vélez *et al.* (2018)	Colombia	Hass	3.3	0.48	4.0		0.31	
Rebolledo-Roa and Burbano-Diaz (2023)	Colombia	Hass	2.6	0.36	4.1	0.25	0.14	0.24

sustainable crops attained globally. Potassium had the highest loss of macronutrients through fruit removal from 'Hass' trees with a mean quantity of 4.85 kg t^{-1} across the eight data sets reported. For 10 and 18 t ha^{-1} crops, K removal was 48.5 and 87.3 kg ha^{-1}, respectively. In contrast, the mean N and K content in fruit of the West Indian cultivars 'Choquette', 'Hall' and 'Booth 8' was 1.6 and 2.9 kg t^{-1}, respectively, which is 54% (N) and 60% (K) less than the content in 'Hass' fruit.

Ideally, fertilizers should be managed to minimize waste but supply sufficient nutrients to meet expected productivity. To achieve this nutrition, programmes should allow for the replacement of minerals removed with fruit at harvest along with other losses from the orchard. These include depletion through leaf senescence, premature fruit and residual flower drop, leaching to deep drainage and sequestration in annual growth. At best, only estimates can be made, but outcomes can be improved through strategic soil and leaf mineral analyses, and correctly timing nutrient applications to meet the changing annual requirements of tree and fruit growth. Improved efficiency in nutrient management will contribute to higher sustainable yield and more robust fruit, thus meeting economic and environmental objectives.

Rootstocks as accumulators and filters of nutrients

Rootstock/scion relationships are of practical importance, since commercial production is based on selected cultivars grafted to seedling or clonal rootstocks. However, there is little information available on the physiological compatibility of avocado rootstocks and scions in common use. Coetzer *et al.* (1994) and Robbertse *et al.* (1998) observed an initial accumulation of B in the roots of 'Hass' avocados grafted on to 'Duke 7' rootstocks. This suggests a level of incompatibility between the rootstock and scion, altering the transport of nutrients to the scion (see Whiley, Chapter 9, this volume).

Roots also produce auxins and cytokinins, which influence the growth and development of branches, shoots and leaves. There is an increasing body of evidence reporting the effects of avocado rootstocks on tree nutrition. For example, Haas (1950a,b), Embleton *et al.* (1962) and Ben-Ya'acov *et al.* (1992) found higher leaf Ca concentrations in trees grafted to Guatemalan compared with Mexican rootstocks. In contrast, trees grafted to Mexican rootstocks had higher leaf K concentrations than those on Guatemalan rootstocks (Haas, 1950b). In Israel, Oppenheimer *et al.* (1961) reported rootstock effects on the distribution of several nutrients in trees. For example, 'Fuerte' trees grafted to Mexican rootstocks had higher leaf N concentrations than those on Guatemalan rootstocks. Additionally, trees on West Indian rootstocks had higher leaf Zn (Oppenheimer *et al.*, 1961) and Mn (Lahav *et al.*, 1993) concentrations than those on Mexican rootstocks. Whiley *et al.* (1996) reported higher leaf B concentrations in 'Hass' trees grafted on 'Velvick' (West Indian × Guatemalan hybrid) compared with those on 'Duke 7' (Mexican) rootstock (Fig. 11.6) and found when B supply is limited, trees on Mexican rootstocks have stunted growth and low yield compared with trees on 'Velvick' rootstocks. Bard (1997) reported higher leaf B concentrations in 'Hass' trees grafted to 'Edranol' (Guatemalan) rootstock compared with 'Hass' grafted to 'Duke 7' (Mexican).

Australian research has also found considerable variation in fruit Ca concentrations among individual 'Hass' trees on seedling rootstocks, suggesting an effect of rootstock or rootstock/scion combination on uptake or transfer of Ca within trees (Vuthapanich, 1998). In south-east Queensland, in a uniformly managed orchard of 'Hass' on seedling Guatemalan rootstocks of unknown origin, the rootstock contributed to a twofold variation in Ca concentration in the pulp of fruit, between trees (Hofman *et al.*, 2002). Whiley *et al.* (2013a) and Dann *et al.* (2016) reported 'Hass' grafted to rootstocks from the Guatemalan and West Indian gene pool produced fruit with lower N:Ca ratios and improved postharvest quality than those grown on Mexican race rootstocks. From a study in Colombia on 'Hass' grafted to seedling rootstocks of Antillean origin with different levels of physical rootstock/scion compatibility, Tamayo-Vélez *et al.* (2022) reported that except for Zn, the scion/rootstock physical compatibility at the graft union did not affect the nutritional content of the peel, pulp, seedcoat or seed of avocado fruit. They also found no

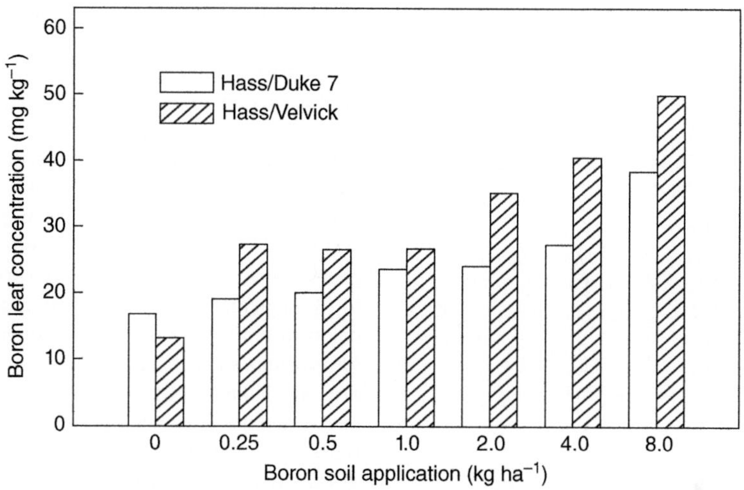

Fig. 11.6. The effect of soil boron (B) application on leaf B concentrations of 'Hass' grafted on to 'Duke 7' (Mexican) or 'Velvick' (West Indian × Guatemalan hybrid) rootstocks. From Whiley *et al.* (1996), with permission.

difference between treatments on leaf mineral nutrient content. While these results differ from others reported, it could be due to insufficient variability in the seedling rootstock population since they were obtained from fruits of commercial clone with uniform characteristics.

In semi-arid California and Israel, West Indian race rootstocks are more resistant to salinity than Mexican and Guatemalan rootstocks (Embleton *et al.*, 1962; Kadman, 1963b). Mickelbart and Arpaia (2002) reported the relative tolerance of rootstocks to salinity is due primarily to their ability to exclude Na and Cl from the scion. Overall, Mexican race rootstocks tend to comparatively increase the concentrations of N, K, Na and Cl in leaves; West Indian rootstocks increase the concentrations of P, Mg and the trace elements Mn, Fe and Zn, while Guatemalan rootstocks increase the concentrations of Ca, Mg and B (Table 11.7).

While there are numerous reports of rootstock effects on tree nutrition, there are relatively few data available that identify phenotypic or physiological parameters that may be responsible for genotypic differences. In general terms, the place and number of lateral roots that occur along a primary root, and their diameter and angle to the primary root, determine the plant's capacity to explore soil. Consequently, this will affect the availability of mineral nutrients and water for plant growth. Borys *et al.* (1985) and Borys (1986, 1989) studied the architecture and growth of seedling rootstock populations of West Indian, Guatemalan and Mexican races and found significant differences between populations within races as well as between races. They reported that West Indian seedlings had a larger root system than Mexican and Guatemalan seedlings, with significantly more first-, second- and third-order laterals. There was a direct correlation between the root system and the size of the aboveground growth, with West Indian seedlings producing larger trees. This may be attributed partly to a root system more capable of exploiting the soil environment. More recently, Fassio *et al.* (2016, 2020) reported different rootstock propagation methods (seedling vs clones) and grafting changed avocado root architecture, which may affect nutrient uptake as envisaged above (see Whiley, Chapter 9, this volume).

In studies of K uptake by avocado roots, Zilberstaine *et al.* (1992) found differences between root sectors of excised primary and lateral roots from Mexican and West Indian trees. Potassium uptake occurred along the entire root segment, although it was the highest through the unsuberized (white) roots and declined with distance

Table 11.7. The relative effect of rootstock on the scion leaf nutrient concentration (H, high; M, medium; L, low; –, no data)

	Rootstock		
Nutrient	Mexican	West Indian	Guatemalan
N	H	L	L
P	M	H	L
K	H	M	L
Ca	L	L	H
Mg	L	H	H
Na	H	L	L
Cl	H	L	L
Mn	L	H	–
Fe	M	H	L
Zn	L	H	–
B	L	–	H

Based on data from Haas (1950a,b), Oppenheimer *et al.* (1961), Embleton *et al.* (1962), Zilberstaine *et al.* (1992), Ben-Ya'acov *et al.* (1992), Lahav *et al.* (1993), Whiley *et al.* (1996) and Bard (1997).

from the apex. Since most of the root is suberized (brown), it was estimated that most of the K was absorbed by suberized roots. Roots from Mexican trees took up about 13% more K than those from West Indian trees, while the thinner second-order laterals were more efficient than the thicker first-order laterals and primary roots. The differences between roots in K uptake were directly related to the air volume fraction within the roots (Zilberstaine *et al.*, 1992). It is likely the relationship between root type and segment and the efficiency of K uptake is similar for other mineral nutrients, and may partly explain differences between rootstocks and races. However, it is known that grafting rootstocks with different cultivars changes the size of the root system along with its horizontal and vertical extension (Wallace *et al.*, 1955; Whiley, 1994; Fassio *et al.*, 2016, 2020); hence, the synergism between rootstock/scion is an important factor in determining whole-tree performance.

Nutrition and mycorrhizae

Arbuscular mycorrhizal fungi (AMF) are found in all soils that support plant growth (Huey *et al.*, 2020) and form symbiotic associations when colonizing plant roots that improve uptake of soil nutrients, particularly phosphorus (Peterson *et al.*, 1984). Ginsburg and Avizohar-Hershenson (1965) were the first to document mycorrhizal associations with avocado roots in Israeli orchards, and in Australia, Dann found 65% intracellular self-colonization of the roots of field-grown avocado trees by mycorrhizae species (E.K. Dann, QAAFI, Brisbane, Australia, 2014, personal communication). Bárcenas *et al.* (2007) also reported colonization of avocado roots by several species of AM from orchard trees in Michoacán, Mexico. It is believed that AMF colonization of avocado roots is most beneficial for avocado, since it greatly increases the surface area available for nutrient and water absorption of a system that evolved without developing root hairs (Burgis and Wolfe, 1945; Salazar-García, 2002). Menge *et al.* (1978) reported the benefits of mycorrhizal fungi (AMF) on growth and reduced transplant loss of avocado in California. Menge *et al.* (1980) later found that in the presence of mycorrhizae, seedlings of 'Topa' increased absorption of N, P, Cu and Zn, resulting in significantly improved growth compared with seedlings grown in sterile soil without mycorrhizae. Fertilizing with phosphorus did not alter leaf P concentration of non-mycorrhizal seedlings, but increased leaf P in seedlings inoculated with AMF. In supporting research, Violi *et al.* (2007) demonstrated that mycorrhizal fungi colonize avocado roots and increase the uptake of P, Zn and Fe. Vidal *et al.* (1992) found that growth and uptake of N, P and K were improved when avocado plantlets were transferred to *ex vitro* conditions in the presence of mycorrhizae. This result has become relevant since avocado trees grafted to tissue-culture grown rootstocks in recent years are being offered for sale (see Whiley, Chapter 9, this volume) and post *in vitro* inoculation with mycorrhizae may provide commercial benefits with improved growth rates. Da Silveira *et al.* (2002) evaluated the influence of six AMF strains on inoculated avocado trees in the nursery and after field planting. Results varied according to AMF species, with *Scutellospora heterogama*, *Acaulospora scrobiculata*, *Glomus etunicatum* and *Glomus clarum* being the most effective in colonizing avocado roots and improving vegetative growth. In studies with micropropagated avocado plants, under greenhouse conditions, Montoya and Osorio (2009) investigated the growth of avocado seedlings inoculated with AMF in response to different P levels in the soil solution. AMF-inoculated plants had faster growth rates at a lower soil P content, suggesting that avocado may benefit from mycorrhizal inoculation when soil P levels are low. In a Mexican study, 'creole' avocado seedlings were inoculated with *Glomus fasciculatum*, *Glomus constrictum*, *Glomus tortuosum*, *Glomus geosporum*, *Glomus mosseae*, *Glomus Cubense* and *A. scrobiculata*. The AMF increased the seedling growth rate and produced better-quality nursery trees (Castro *et al.*, 2013).

Balderas-Alba *et al.* (2019) found that applying organic and synthetic fertilizers to trees had negative effects on the intra-radical fungal structures and decreased arbuscular mycorrhizal colonization. In a study that evaluated the interaction between AMF inoculation and chicken manure application on the growth of seedling avocado rootstocks, Carreón-Abud *et al.* (2015) observed negative effects from high chicken manure rates on seedling growth and health. In a study where 'Zutano' seedling rootstocks growing in saline soil were inoculated with *Glomeromycota* mycorrhizal fungi, Solórzano-Acosta *et al.* (2023) reported improved growth and reduced concentrations of Cl and Na in the roots and leaves.

Results from the above-mentioned studies overwhelmingly show benefits in tree performance when avocado roots are colonized by AMF. However, all results have come from controlled experiments with seedling nursery trees. There is a dearth of information on responses from inoculating field-established avocado trees, and since natural colonization occurs, it is unlikely commercial benefits would be gained from additional inoculation. An exception may be with the introduction of additional AMF species to saline soils to reduce the uptake of Cl. However, commercial opportunities with AMF inoculation are likely with the treatment of nursery trees established in pasteurized soils since AMF species are destroyed by the elevated temperatures of this treatment (Vicca *et al.* 2009).

Primary and secondary nutrients

Nitrogen (N)

Nitrogen is an important nutrient for growth and production in avocado trees. Nitrogen deficiency restricts shoot growth and deficiency symptoms include small pale green to yellowish leaves and premature leaf shedding (Fig. 11.7). In cases of acute N deficiency, leaf veins turn yellow.

Avocado nitrogen (N) management is a complex issue requiring site-specific and season-specific adjustment, due to large variations between climatically distinct subtropical to temperate regions and the range of soil types used for avocado production. Additionally, avocado cultivars have discrete requirements for N, and such differences must be considered when programming N fertilization. Nitrogen management strategies currently applied in commercial avocado production include consideration of (i) rates and timing of N application; (ii) dynamic contributions of compost and mulch to short-, medium- and long-term N supply; (iii) slow-release inorganic fertilizers; and (iv) fertilizers with nitrification inhibitors (Newett *et al.*, 2018). Various nitrogenous compounds are used to fertilize avocado, including urea, ammonium nitrate, calcium nitrate, ammonium sulfate, ammonium di-phosphate, potassium nitrate and compound solid and liquid NPK fertilizers. Fertilizers formulations should be selected according to soil pH and management practices, including the presence or absence of irrigation.

Fig. 11.7. Nitrogen deficiency: small, narrow, pale green to yellowish leaves that become stiff and slightly inwardly rolling; growth is stunted with reduced crop; leaf N = 0.9%. Photograph from Weir and Cresswell (1995), with permission of NSW Agriculture.

Avocados grown in many parts of the world are periodically exposed to low temperatures that may cause chilling damage to trees. From nutrition studies with N, P and K in Florida, Lynch and Goldweber (1956) concluded that high rates of leaf N prior to winter reduced damage to avocado trees that were exposed to freezing temperatures. Similarly, Lahav *et al.* (1987) reported a linear relationship between frost damage and the amount of N applied prior to winter; the more N applied, the lower the frost damage (Fig. 11.8). A relationship between the N status of trees and freeze damage was also reported from California following the 1990 freeze. In a survey conducted immediately following a freeze event, Witney and Arpaia (1991) reported that trees showing N deficiency were more severely affected by cold than healthy trees. In laboratory and orchard studies (Zilkah *et al.*, 1996), three foliar applications of 2% urea raised leaf N concentration by 26% and these leaves had increased freezing hardiness and delayed senescence. These results indicate that a useful level of chill protection is induced in avocado through maintaining N leaf concentrations at the upper recommended concentrations. In some cases, it may be appropriate to apply 2% urea sprays to trees before freeze conditions are anticipated.

Nitrogen as a fertilizer is available in nitrate and ammonium formulations. Zilkah *et al.* (1999) observed that when ^{15}N isotope was applied to container-grown avocado trees, nitrate N was taken up at a significantly higher rate throughout the year than ammonium-N. These results were not repeated under field conditions. Nitrate N is subjected to rapid loss from the soil, since it is not adsorbed by the soil cation exchange complex. There is some evidence to support the use of ammonium-N in the presence of Phytophthora root rot as Lee and Zentmyer (1982) found that ammonium-N suppressed sporangial development and the release of zoospores. However, high concentrations of ammonia are toxic to avocado feeder roots (Pegg, 2010).

Nitrogen supplied to trees from the mineralization of soil organic matter can be significant and should be considered when developing fertilizer application programmes (Wolstenholme, 2004). Mineralization of soil organic matter is governed by factors, including soil type, soil pH, the C:N ratio of the organic material, temperature, moisture content and aeration. Theoretically, it can range from about 75 kg N ha^{-1} year^{-1} in soils with an organic matter content of less than 2%, to about 150 kg N ha^{-1} year^{-1} in soils with an organic matter content of more than 4%; however, only a proportion of this potential N is mineralized

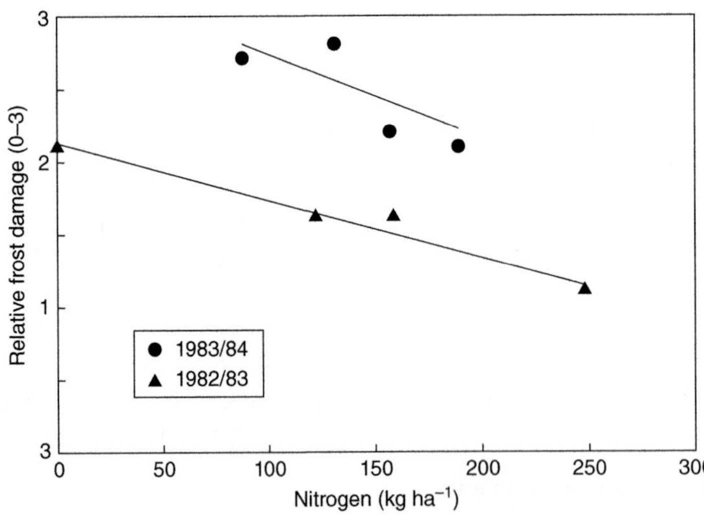

Fig. 11.8. The relationship between nitrogen (N) applied and relative frost damage (0 = none; 3 = total leaf burn) of 'Hass' avocado trees in two winter seasons. The regression line for 1983/84 is represented by $y = 3.317–0.0058x$; $r^2 = 0.67$ and for 1982/83 $y = 2.129–0.0039x$; $r^2 = 0.95$. From Lahav *et al.* (1987), with permission.

each year. Guidelines need to be developed to estimate the amount of N made available through mineralization each year in different situations (Newett *et al.*, 2018). In South Africa, research on fruit quality problems in avocados from trees growing in soils with a high organic N content, the supply of N from mineralization of organic matter was sufficient to cause quality problems, even with no addition of N fertilizer (Snijder *et al.*, 2003). The use of liquid organic soil amendments (Gardiazábal, 2023) or organic fertilizers (Tzatzani *et al.*, 2022) may partially substitute or complement a conventional inorganic fertilization programme.

Nitrification inhibitors slow the conversion of N from the ammonium to nitrate and can reduce the amount of N leached beyond the root zone. In Chile, 'Hass' avocado trees planted at 6 × 4 m were fertilized over a 4-year period with three annual applications of ammonium sulfate containing DMPP nitrification inhibitor (Entec® Solub 21) and compared with the use of urea with equivalent N rates. Trees fertilized with DMPP nitrification inhibitor had larger trunk diameters and tree growth and could be fertilized with 25% less N than those receiving conventional fertilizers (Gardiazábal *et al.*, 2003). More recently, Torres *et al.* (2023) found significant increase in fruit yield over a 3-year period by applying an extra application of a DMPP nitrification inhibitor 3 weeks after spring N fertilization.

Phosphorus (P)

Symptoms of phosphorous deficiency are rare in avocado orchards. From avocado trees grown in sand, Furr *et al.* (1946) described deficiency symptoms of P as decreased vegetative growth, small round brownish-green and burnt leaves and early leaf shedding and branch dieback. Novoa *et al.* (2018) described leaf deficiency symptoms as a diffuse dull leaf chlorosis in mature leaves that turned into irregular necrotic spots. Under P deficiency, avocado leaves developed a leathery texture, while preserving a green dull leaf colour. No general chlorosis was observed in young P-deficient leaves. Phosphorus deficiency inhibits shoot and bud growth, leading to reduced plant growth (Novoa *et al.*, 2018). Fruit of P-deficient trees have a floury pulp texture and are more sensitive to low storage temperatures; fruit are smaller with malformed seed (Ataucusi, 2015). Frequent application of phosphorous in compound fertilizer blends to avocados may cause excess soil levels of P resulting in induced Zn deficiency (Newett *et al.*, 2018), especially if soil Zn content is low and pH is high.

The most common P fertilizers are rock phosphate (only in acidic soils), phosphoric acid (H_3PO_4), which is commonly used to remove precipitated carbonates from microirrigation lines, superphosphate and mono-ammonium phosphate (MAP). The annual application will depend on soil and leaf concentrations determined by analysis. Solid P compounds should be applied to the wetted soil area in two applications (spring and autumn) combined with two to three irrigations. Phosphorus fertilizers should not be applied if leaf P concentration exceeds 0.14%

Potassium (K)

Potassium is unique among the essential plant nutrients in that it does not become part of complex organic molecules. Rather, K remains in ionic form within the plant and acts as a regulator of many essential processes, including the opening and closing of stomates, regulating enzymes and controlling cellular pH. Potassium is taken up by plants as K^+ ions dissolved in the soil solution (Spann, 2019). Avocado has a relatively high requirement for K, which occurs in greater quantity in mature fruit than N (Table 11.6). Characteristic symptoms of K deficiency are small and narrow leaves with brownish-red, irregular necrotic spots that develop on older leaves and subsequently coalesce over the entire leaf blade between the main veins (Fig. 11.9). With severe deficiency, shoot growth is thin and some dieback occurs (Furr *et al.*, 1946). Charpentier *et al.* (1967) reported that deficiency symptoms appear first on the base and petiole of leaves and then advance through the central and secondary veins towards the tip.

Potassium chloride is the most cost-effective form of K applied to avocado orchards. However, since avocado is sensitive to chloride, in areas of suspected salinity, potassium sulfate or potassium nitrate is recommended. The latter also supplies N in a nitrate form. The recommended annual application is 170–250 kg ha^{-1} K, depending on leaf analyses and soil characteristics. When leaf K concentration is above 1.2%, no K fertilizer should be applied.

Fig. 11.9. Potassium deficiency symptoms: large, irregular necrotic spots develop on older leaves in the interveinal areas, usually following the development of interveinal chlorosis. Leaf K = 0.14%. Photograph from Weir and Cresswell (1995) with permission of NSW Agriculture.

Fig. 11.10. Magnesium deficiency symptoms: interveinal chlorosis progresses inward from the end margins and from the tip towards the leaf base. Develops first on older leaves. Leaf Mg = 0.14%. Photograph from Weir and Cresswell (1995) with permission of NSW Agriculture.

Calcium (Ca)

Calcium is an essential element in plant cell wall and membrane structure, providing mechanical strength to fruit tissues and conferring protection against cell-wall-degrading enzymes (Wehr *et al.*, 2004; Hocking *et al.*, 2016). Additionally, fluctuations in Ca concentrations in compartmentalized plant cell solutions act as a signal to regulate metabolism, for example, to trigger plant defence processes against biotic (e.g. fungal pathogens) and environmental (e.g. heat) stress (White and Broadley, 2003). Calcium uptake by plants is a passive process, driven largely by transpiration and root pressure (Hocking *et al.*, 2016). The rate of Ca uptake is greatest at the root tip (Clarkson, 1984), while increasing soil acidity limits Ca availability and root uptake (White, 2012).

Foliar Ca deficiency symptoms have been documented from sand cultures but to our knowledge have not been identified in field-grown trees. Symptoms include scorched tips and small-sized leaves (Furr *et al.*, 1946). Calcium deficiency symptoms are sometimes similar to those of *P. cinnamomi* damage, because they both cause root systems to collapse and disintegrate (Groot Obbink and Alexander, 1976). However, with Ca deficiency, roots regenerate, whereas in plants affected by Phytophthora root rot, root disintegration is followed by loss of leaf turgor, and a faster dieback in the aerial portion of infected trees.

Calcium is supplied to avocado orchards in superphosphate and mixed fertilizers, or as gypsum to supply Ca without changing pH, or in the carbonate form as agricultural lime or dolomite. The latter are usually applied to adjust soil pH and/or reduce toxic Al or Mn concentrations, rather than to increase the supply of Ca as a plant nutrient. The amount of lime applied depends on the change in soil pH required, the soil Al concentration and the buffering capacity of the soil. A common rate for avocados in acid soils is 3–6 t ha^{-1} every 3–5 years. Maintenance applications of 1–3 t ha^{-1} of gypsum are recommended for fruiting orchards on acid soils to achieve the mild fungistatic effect on root pathogens and increase postharvest fruit quality (Hofman *et al.*, 2005).

Magnesium (Mg)

Magnesium is a significant component of the chlorophyll molecule, which is essential for photosynthesis, and acts as an activator of numerous plant enzymes important for normal plant growth. Magnesium is mobile within plants, and deficiency symptoms appear first in older leaves (Spann, 2019). Deficiency symptoms of Mg were reported by Furr *et al.* (1946) and Robinson (1961) as interveinal chlorosis of older leaves (Fig. 11.10).

Magnesium deficiency can be corrected through the application of magnesium sulfate

($MgSO_4$), which is highly soluble and most easily applied by fertigation. Dolomitic limestone ($CaCO_3 \cdot MgCO_3$) is also a source of Mg (and Ca), as well as a liming material to ameliorate soil acidity.

Chloride (Cl)

The avocado is particularly susceptible to Cl toxicity (Celis *et al.*, 2018). Symptoms of Cl toxicity are leaf burn of tips and margins aggravated during the summer and autumn to the point of premature leaf shedding (Fig. 11.11). There is a distinct correlation between the extent of the burn and the Cl concentration in the leaves. Chloride concentration in normal leaves is 0.2–0.3%, compared with 0.35–1.48% in leaves with tip and marginal burn (Cooper and Gorton, 1950). In analyses carried out with leaves of various avocado rootstocks, >2% Cl was found in severely scorched leaves of Mexican race rootstock (Kadman, 1963b). Hot, dry winds are known to intensify leaf burn, and Kadman (1970) reported a positive correlation between increased transpiration and higher rates of absorption and accumulation of Cl in avocado leaves.

In a Californian field experiment that evaluated the salt tolerance of 'Hass' on 13 different clonal rootstocks, Celis *et al.* (2018) found that the more tolerant rootstocks, with larger trunk diameter, and a survival rate after 23 months of saline treatment, were characterized as Cl^- ion excluders, based on leaf and root Cl^- ion accumulation. In contrast, there was no correlation between Na^+ accumulation and trunk diameter. Leaf Cl^- analysis proved to be a useful method to identify salt-sensitive rootstocks rapidly and serves as a diagnostic tool for salt damage. For additional information on salt stress and moderation in avocados, see Schaffer *et al.*, Chapter 9, this volume.

Fig. 11.11. Chloride toxicity symptoms: marginal and interveinal necrosis that usually begins at the tips and progresses back along the leaf margins. Symptoms become more severe as leaves age. Leaf Cl = 0.78%.

Sodium (Na)

Avocado trees are highly susceptible to excess Na, which is expressed by interveinal necrotic spots in leaves, which spread as the Na concentration increases (Fig. 11.12). Over time, leaf buds are affected, and young and mature branches completely desiccate (Kadman, 1964). Irrigation water with sodium absorption ratios >3 are considered sodic, and soils with exchangeable Na above 4–5% are considered hazardous for avocados. Leaf sodium levels of ≈0.4–0.5% are indicative of leaf injury and levels above 1% result in acute Na toxicity (Bingham and Nelson, 1970).

In saline-resistant rootstocks (West Indian race), translocation of Na from roots to the canopy was reduced, whereas salinesusceptible rootstocks (Mexican) translocated appreciable quantities of Na from the roots to the leaves (Kadman, 1964). Leaves on trees, grafted to sensitive rootstocks, are severely affected by excess Na, and in Kadman's study contained ≥1.0% Na

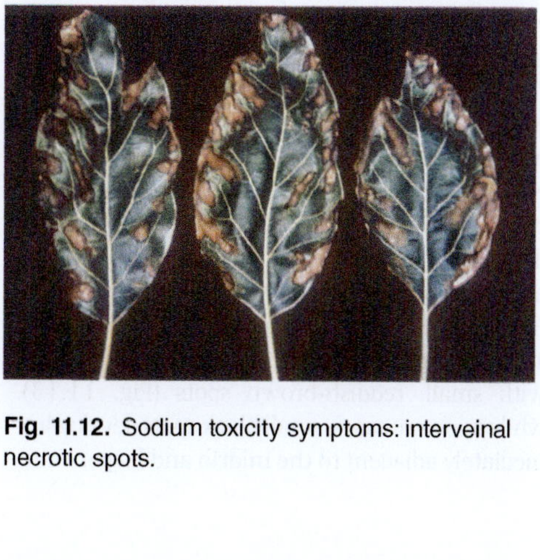

Fig. 11.12. Sodium toxicity symptoms: interveinal necrotic spots.

versus 0.02–0.1% Na found in normal (non-Na-resistant) trees (Kadman, 1964).

Sulfur (S)

Sulfur is taken up from the soil in the form of sulfate ($SO4^{2-}$). In plants, sulfur is a component of amino acids and is necessary for protein synthesis. Like nitrate, sulfate is also leachable from soils and deficiencies can occur in most soil types. However, in low rainfall areas like California, sulfate can form precipitates in soil as gypsum (calcium sulfate). Sulfur is immobile within plants and deficiency symptoms first appear on new leaves. Typical symptoms include yellowing of new leaves, smaller leaf size and slow growth (Barnard *et al.*, 1991). In many respects, S deficiency symptoms are similar to those of N deficiency, except that young leaves are more sensitive than older leaves, in contrast to N deficiency where all the leaves are affected. There are no data on the effect of S on yield.

Sulfur is usually supplied to avocados by the soil, water or fertilizers such as ammonium sulfate, potassium sulfate or superphosphate. Where deficiency conditions occur, S should be applied separately. Gypsum ($CaSO_4 \cdot 2H_2O$) is a good source of S. After application, oxidation of S-containing fertilizers forms H_2SO_4; hence, S-containing fertilizers can be used to reduce the pH of alkaline soils (Granja and Covarrubias, 2018).

Manganese (Mn)

Manganese plays an important role in plants where it is required for chloroplast formation, photosynthesis, nitrogen metabolism and the synthesis of some enzymes. Manganese deficiency occurs mainly due to heavy soil liming (high pH) and in coarse-textured sandy soils. Symptoms include interveinal chlorosis of leaves, which become pale green. In extreme cases, the whole leaf becomes a dull yellowish green (Furr *et al.*, 1946; Barnard *et al.*, 1991). Decreasing Mn concentrations were thought to contribute to the development of mesocarp discolouration of 'Pinkerton' avocados (Van Rooyen and Bower, 2005). However, Mn is more commonly toxic to avocado. Toxicity symptoms appear as interveinal chlorosis with small, reddish-brown spots (Fig. 11.13), while in severe cases, small black spots develop immediately adjacent to the midrib and larger veins of leaves (Tracy, 1983; Weir and Cresswell, 1995). An inverse relationship was found between leaf Mn concentrations and productivity of 'Hass' trees (Lahav *et al.*, 1993). Higher soil acidity increased Mn concentrations of avocado leaves up to 2000–2500 mg kg^{-1} (Fernandez Falcon *et al.*, 1984). High Mn concentrations in leaves were negatively correlated with rooting capability of avocado cuttings (Reuveni and Raviv, 1981). Toxic Mn concentrations in leaves can be reversed by decreasing available Mn in the soil through increasing pH and improving soil drainage. Injection of a lime solution into the soil resulted in rapid recovery of trees suffering Mn toxicity in south-western Australia (A. McCarthy, Bunbury, Australia, 2011, personal communication).

Fig. 11.13. Manganese toxicity symptoms: interveinal chlorosis with small reddish-brown spots more advanced on older leaves. Leaf Mn = 942 mg kg^{-1}. Photograph from Weir and Cresswell (1995) with permission of NSW Agriculture.

Iron (Fe)

Iron is required for the synthesis of chlorophyll and in reception and transfer of energy in photosynthesis and respiration. It acts on oxide-reduction reactions such as the reduction of nitrates. Iron is solubilized in acidic soils and precipitates in alkaline soils, becoming increasingly deficient in plants with rising soil pH. Iron deficiency is characterized by interveinal yellowing in the young leaves with narrow dark-green strips along the veins (Fig. 11.14). With progressive deficiency, leaves of the summer flush become smaller, more delicate and while young are pale yellow to white. Later, severely chlorotic leaves show scorch symptoms at their tips and along margins and in extreme cases abscise. With severe deficiency, desiccation of shoots follows, and fruit change from dark to lighter green. Iron deficiency is normally not a problem for avocado unless trees are grown on calcareous soils with a pH greater than 7.0 or in acid soils with high available P or Mn that suppress Fe availability.

Zinc (Zn)

Zinc is an essential microelement in the formation of auxins, and the synthesis of nucleic acids, proteins and the transformation of sugars into starch. It has a direct effect on fruit set in relation to pollen formation and viability, subsequent fruit growth and ripening. Zinc also assists plants develop immune responses to repel soil pathogens. The first report of Zn deficiency in avocados was from California, where symptoms were described as leaf mottling between veins caused by impaired chlorophyll synthesis (Fig. 11.15a), reduced leaf size and shoot growth retardation that caused shortened internode length on branches, which gave a very distinctive rosette or 'feather-duster' appearance (Fig. 11.15b). Other symptoms of Zn deficiency include reduced fruit size that becomes round in shape (Fig. 11.15c), occasionally with a reddish skin appearance (Coit, 1928; Parker, 1936; Wallihan *et al.*, 1958). Symptoms are expressed in leaves and fruit when leaf concentrations fall below 15 mg kg^{-1} (Wallihan *et al.*, 1958). Advanced Zn deficiency reduces yield and if not corrected, a severe decline in tree health occurs (Coit, 1928; Ruehle, 1940; Gustafson, 1973; Lee, 1973).

Fig. 11.14. Iron deficiency symptoms: interveinal yellowing and narrow dark-green strips along the veins. Symptoms develop first in younger leaves. Leaf Fe = 30 mg kg^{-1}.

Zinc deficiency in avocado is widespread and occurs in most countries where this crop is grown and is considered an important limiting factor in fruit quality and tree health (Crowley *et al.*, 1996). It is commonly found where calcareous soils occur (California, Florida, Israel, Western Australia and Chile), in which availability is limited by its extremely low solubility at high soil pH. Nonetheless, Zn deficiency also occurs on acidic sandy soils with low total Zn content, and in warm, semi-arid regions that typically have low soil organic matter, as well as in highly leached, acidic sandy soils in subtropical and south-western Australia, Brazil and South Africa. Strategies for correction of Zn deficiency depend largely on the soil pH, and the amount of organic matter available to form metal complexes (Labanauskas *et al.*, 1959; Srivastava *et al.*, 1981). Other factors that influence Zn deficiency are high levels of N and P fertilizers (Labanauskas *et al.*, 1959) and seasonal fluctuations in leaf micronutrient content (Bingham, 1961; Labanauskas *et al.*, 1961) that may be related to irrigation and climatic factors affecting root growth

Fig. 11.15. Zinc deficiency symptoms: (a) leaf mottled between veins with reduced size, leaf Zn = 16 mg kg^{-1}; (b) shoot symptoms when internode length is shortened producing a feather-duster appearance. In severe cases, leaves are arranged in a rosette; (c) fruit are smaller and round (LHS) and can develop a reddish colour compared with normal fruit (RHS).

and nutrient uptake (Crowley *et al.*, 1996). Zinc deficiency can occur during spring when soils are too cold for uptake (Newett *et al.*, 2018). Zinc deficiency has been correlated with pulp spot in 'Fuerte' in South Africa, where fruit without pulp spot had a higher Zn concentration than affected fruit (Vorster and Bezuidenhout, 1988). It was suggested that in fruit with higher Zn concentration, increased Ca was released from various chelating and complexing agents and moved into the fruit. Higher amounts of mobile or active Ca in fruit reduce physiological disorders (Bower and Cutting, 1988).

Copper (Cu)

Copper is one of the essential trace elements that play a significant role in regulating many physiological and biochemical processes in plants (Kohatsu *et al.*, 2021). The first symptoms of Cu deficiency are dark-green foliage (Barnard *et al.*, 1991). Later, terminal growth is affected showing shortened internodes and dieback of growing terminals. Fungicide treatments containing Cu can supply sufficient amounts of this nutrient since it is either washed to the orchard floor during rainfall events or added to the soil through leaf senescence. However, Cu is also reported to accumulate in toxic quantities in Florida soils due to multi-application of fungicidal sprays. High concentrations of Cu have been found on the exterior surfaces of 'Lula' avocado roots (Malo, 1976). Build-up of Cu residues in orchard soils after years of repeated fungicidal sprays is of growing concern in several countries and has been shown to have detrimental effects on beneficial soil microorganisms.

Recently, copper-based nanoparticles (CuNPs) have generated considerable interest for agricultural purposes due to their antimicrobial properties and because they can improve plant yield and quality compared to conventional micronutrient formulations. In a study with CuNPs applied by different methods to avocado nursery trees in Mexico, López-Luna *et al.* (2023) found that foliar spraying improved the uptake and translocation of CuNPs in plant tissues and significantly stimulated shoot growth 60 days after application.

Boron (B)

Boron deficiency is one of the most common nutrient disorders of avocados worldwide, particularly in soils with low organic matter content and in the acidic soils of subtropical regions of Australia and South Africa (Whiley *et al.*, 1996; Bard and Wolstenholme, 1998), with symptoms developing when leaf concentrations fall below 25 mg kg^{-1} (Whiley *et al.*, 1996). Reasons for low B content in soils are due to the high mobility of the borate ion, considered second only to that of the nitrate ion under irrigation or natural leaching events.

One of the early symptoms of B deficiency is inhibition of root growth and development. Also, B-deficient roots are more susceptible to soil pathogens due to the weakening of cell walls. Foliar B deficiency symptoms include yellowing and a shot-holed effect in leaves (Fig. 11.16a), particularly those of spring-grown leaves when the demand for B is high. Shoot symptoms include arrested growth, swelling of the nodes (Fig. 11.16b), gradual death of the apical and axillary growing points that cause a horizontal rather than vertical growth due to loss of apical 'dominance', and in severe cases bark splitting, which may develop into trunk

Fig. 11.16. Boron deficiency symptoms: (a) irregular-sized holes with pale green halos in leaves more common in spring growth; (b) swollen nodes with light-coloured lesions on small branches; (c) corky bark lesions on branches and trunks that can develop into cankers (leaf B = 8 mg kg^{-1}).

cankers (Fig. 11.16c) and limb dieback. Additional B deficiency symptoms include shortening of internodes, dry buds, low pollen viability and premature abscission of flowers. Some fruit on B-deficient trees may show considerable growth distortions. In some instances, fruit are 'bumpy', while others are sickle shaped (Fig. 11.16d) with corresponding angled fruit stalk insertion. In severely affected fruit, a lesion develops on the indented side and may penetrate through the flesh to the seed, often showing dead tissue on the cotyledons. Trees on Mexican rootstocks, for example, 'Mexícola', 'Topa Topa' and 'Duke 7', are more susceptible to B deficiency than those on Guatemalan/West Indian rootstocks (Fig. 11.1). The Australian cultivar 'Sharwil' is highly sensitive to B deficiency and a good indicator tree in the orchard for visual expression of B deficiency. Care is required when correcting B deficiency since toxicity results when mature leaf concentrations exceed 100 mg kg^{-1}. Symptoms of toxicity are marginal leaf burn with small interveinal necrotic spots and premature leaf fall.

Boron is most commonly found in soils as boric acid $[B(OH)_3]$ and is a key micronutrient in avocado nutrition where a deficiency can have a major disruption on growth and production, causing irregular-sized holes on leaves, swollen nodes on small branches and corky bark lesions that can develop into cankers on branches and trunk. Its general role in plants is providing integrity of cell walls and the plasma membrane, carbohydrate and protein metabolism, the synthesis of ribonucleic acid, pollen tube growth, root elongation and meristem cell division, along with other functions (Pereira *et al.*, 2021).

There appear to be some benefits from using strategically targeted foliar B applications to improve reproductive function. Avocado studies in South Africa (Robbertse *et al.*, 1989, 1992), Australia (Smith *et al.*, 1997) and California (Jaganath and Lovatt, 1998) have shown that pollen viability, ovule fertilization and fruit set can be improved by targeting newly emerged inflorescences with foliar B sprays. The benefits from these applications were greatest in B-deficient trees and in years when climatic conditions during flowering were less conducive to fruit set. More recently, Abdel-Karim *et al.* (2015) validated the effect of combined Zn and B foliar sprays on 18-year-old 'Fuerte' avocado trees growing in Egypt during two successive seasons. It was reported that three foliar applications at the beginning of flower bud induction (October), at bud burst (January) and at anthesis (April) with 1 g l^{-1} Zn, as zinc sulfate monohydrate (35% Zn), and 1 g l^{-1} B, as borax (17.5% B), had a synergistic effect in improving pollen germination, fruit set and yield, with larger fruit size and increased leaf chlorophyll content.

In South Africa, Robbertse and Coetzer (1990) and Robbertse *et al.* (1992) have encouraged foliar application of boron (as Solubor®) at flowering and fruit set to enhance tissue concentrations, thereby improving pollination, ovule fertilization and fruit set. Coetzer *et al.* (1993) demonstrated that foliar-applied ^{10}B was absorbed by mature leaves and transported to adjacent developing inflorescences. However, the dynamics of B re-mobilization are complex and rely on the accumulation of a surplus to growth requirements inside the cell vacuole (Eaton, 1944). Hence, when young expanding avocado leaves were sprayed with ^{10}B, some of the material was absorbed but not translocated and was probably bound by the developing cells (Coetzer *et al.*, 1993). There is some evidence that B is more likely to be in short supply during 'on' crop years when foliar sprays alone may be insufficient and timely soil application more effective. Bard and Wolstenholme (1998) surveyed South African avocado orchards where B nutrition was managed through foliar applications. They reported that although leaf concentrations of B were above the threshold (25 mg kg^{-1}; Whiley *et al.*, 1996) where symptoms are usually expressed, visual deficiency symptoms persisted in all orchards visited. They concluded that foliar B applications to avocados were ineffective in correcting chronic deficiencies, and the B sprays likely inflated analytical results through surface contamination from leaf samples.

Most studies on boron have focused on improving fruit set and yield with little focus on avocado fruit quality. Soil-applied B has been shown to increase B levels in the skin and the results of Hapuarachchi *et al.* (2022) reported soil-applied B at recommended rates increased fruit flesh concentrations by 55% higher than fruit from untreated trees. Unfortunately, postharvest quality was not determined in this study since higher B flesh concentrations in 'Hass' fruit have been linked with lower incidence and severity of diffuse flesh discolouration and vascular browning, the absence of a postharvest flesh browning disorder (endocarp black spot) near the stem end of the seed cavity and lower severity of body rots (Newett *et al.*, 2021).

Molybdenum (Mo) and Nickel (Ni)

The absence of Mo (Brusca and Haas, 1955) or Ni (Haas, 1952) from nutrient solutions reduced the growth of avocado seedlings grown in sand culture. Concentrations of 6.5–26.0 µM Mo in the nutrient solution promoted the growth of 'Harman' (Mexican race) seedlings, while concentrations beyond this range reduced growth. 'Fuerte' grafted on to 'Harman' rootstocks showed similar responses to Mo. However, Haas and Brusca (1957) observed translucent spots on full-sized leaves of potted avocado trees grown without Mo. As the leaves aged, they abscised prematurely. The symptoms were only expressed on plants under glasshouse conditions. Transferring plants outdoors did not reverse the symptoms on affected leaves but symptoms did not appear on new leaves when they reached full size. Slater *et al.* (1975) recorded Ni concentrations of 0.0–22.0 mg/kg dry weight in the edible pulp of 'Fuerte' and 'Hass' fruit grown in California.

Molybdenum is an essential plant trace element and has a number of roles including being implicated with N metabolism and as a cofactor for several essential enzymes. Cowan *et al.* (2001) reported that molybdenum participates as an enzymatic cofactor in the metabolic control of avocado fruit growth and final fruit size. Deficiency symptoms can appear the same as those of N deficiency, due to its role in N metabolism. Molybdenum is required in very small amounts that are likely endemic in most orchards in sufficient quantities for normal growth. An additional source is as a contaminant in commonly used fertilizers. It is the only nutrient where availability increases with pH, and adequate levels are generally present in alkaline soils but can be deficiency in some acid soils. There is an inverse relationship between molybdenum and manganese, so soils high in available manganese (e.g. acid soils with high water content) can be deficient in Mo, especially if organic matter levels are low. In addition, sulfate and molybdate ions compete strongly for root uptake from soils, so sulfate-based fertilizers, including gypsum, will suppress Mo uptake. Since Mo is required in such small amounts and is highly phloem mobile, foliar sprays are likely to be successful in addressing deficiencies (Newett *et al.*, 2018).

Interactions between nutrients

A large number of minerals are naturally available in soils; however, only 17 are known to be essential to support the physiological and structural requirements for plant growth. While there is a large body of published work detailing deficiency and toxicity limits of essential elements in plants, current knowledge on interactions between multiple nutrient minerals suggest a synergism that affects uptake, transport and/or assimilation between each other (Kumar *et al.*, 2021). The importance of the nutritional balance in avocado trees is found in rootstock/scion interactions, the correlation between a number of nutrients and fruit quality, the interpretation of leaf analyses and the effects of salinity on leaf nutrient concentrations (Table 11.8). This section reviews known intra-plant interactions with nutrient minerals in avocado, particularly in relation to increasing the supply on one nutrient to the detriment of others.

Increasing N application to avocado trees increased leaf P concentration in leaves (Haas, 1947; Lynch *et al.*, 1954; Embleton *et al.*, 1958b). Lahav *et al.* (1990) also reported that N applications (calcium nitrate or ammonium nitrate) enhanced the translocation of P from old 'Fuerte' and 'Hass' leaves into new growth in spring. Embleton *et al.* (1958b) found that applying P to 'Fuerte' trees in California increased leaf Ca, Mg and Cl but reduced leaf K concentration. Using the same trees as Embleton, Labanauskas *et al.* (1958a) reported that increased P fertilization increased the Mn concentration but reduced Zn, Cu and B concentrations in leaves. Where the highest amounts of P were applied, Zn reduction was severe enough to induce deficiency symptoms. Similar results were recorded by Lahav (1984), who increased leaf P concentration with organic manure. The antagonistic effect resulted in significant reductions in B, Zn and Mn concentrations and trees developed Fe

Table 11.8. Effect of N, P and K applications and salinity on the changes on concentrations of other nutrients in avocado leaves

	Nutrient applied			
Nutrient affected	N	P	K	Salinity
N	+	+	0/+	–
P	+	+	0	0–
K	±	–	0/?	+
Ca	±	+	–	
Mg	±	+	–	–/?
Na	+			
Cl	–	+	+	
Zn	±	–	–	+
Mn	+	±	+	–
Fe	+	–/?		
Cu	±	–	+	+
B	–	–	–	

+, increase; –, decrease; 0, no effect; ±, variable results; ?, non-significant effect. Based on data from Ruehle (1940); Haas (1947); Bingham and Martin (1956); Embleton *et al.* (1958a,b); Labanauskas *et al.* (1958a,b); Lahav *et al.* (1976, 1990); Lahav (1984); Limaco Sing and McNeil (1991); Zilberstaine *et al.* (1992); Jamie *et al.* (1992); Bar and Kafkafi (1992); Steinhardt *et al.* (1995a); Arpaia *et al.* (1996) and Loupassaki (1998).

chlorosis. In a study of the relationship between Cl, NO_3 and Fe deficiency in Mexican and West Indian rootstocks, an increase in the concentration of NO_3 in the nutrient solution led to a decrease in Cl accumulation in all plant parts in rootstocks of both races (Bar and Kafkafi, 1992). High nitrate supply increased Fe chlorosis. Nitrate assimilation leads to increased pH around the root, due to the efflux of HCO_3^- or OH^- (Kirkby, 1969). The increased pH in close proximity to roots is likely to have reduced the availability of Fe (Nye, 1986) through lower Fe solubility and its precipitation on the outer surface of the roots (Bar and Kafkafi, 1992). A significant increase in Fe chlorosis due to increased N application was found in three avocado cultivars. An increase in N from 80 to 640 kg ha^{-1} increased the Fe chlorosis rating sixfold in 'Ettinger', sevenfold in 'Ardith' and tenfold in 'Hass' (Lahav, 1999).

Similar results with avocados growing on calcareous soils were reported by Ruehle (1940) with respect to Zn deficiency. The high soil pH reduced the availability of Zn and increased N fertilization increased the expression of Zn deficiency symptoms. The antagonistic interaction between N and Zn nutrition in avocados may be due to the enhancement of P uptake under higher N regimes, as there is a well-documented, negative interaction between P and Zn in avocado (Bingham and Martin, 1956; Labanauskas *et al.*, 1958a; Lahav, 1984; Lahav *et al.*, 1990). However, it is likely that the reduction in the leaf concentration of other nutrients is due to limited soil availability and the dilution of leaf concentrations with increased growth in trees promoted by increased N applications.

Embleton *et al.* (1958b) found that ammonium sulfate reduced the concentration of Ca and Mg in leaves. However, applications of calcium nitrate or ammonium nitrate increased leaf Ca (Embleton *et al.*, 1958a; Lahav *et al.*, 1990) and Mg concentrations (Lahav *et al.*, 1990). The differences between these studies may be related to the different N sources or rootstock/scion combinations. Increased N application also increased leaf Mn and Fe and reduced B and Zn concentrations (Ruehle, 1940; Labanauskas *et al.*, 1958a; Lahav *et al.*, 1990; Loupassaki, 1998).

Fertilizing with K reduced the concentration of Ca and Mg in leaves of 'Hass' trees in California (Embleton *et al.*, 1958b) and increased the Cu and Mn leaf concentrations (Labanauskas *et al.*, 1958a). However, Lahav *et al.* (1976) reported that K fertilizers applied to 'Fuerte' and 'Hass' trees in Israel had no effect on other leaf nutrient concentrations with the exception of Zn, which was significantly reduced in 'Fuerte' leaves over a 3-year leaf sampling programme. The effect of K on 'Hass' was not consistent over time. A similar K/Zn relationship has been reported by Limaco Sing and McNeil (1991).

In a 21-year-old 'Fuerte' avocado orchard in Mexico, Aguilar-Melchor *et al.* (1997) found that N and P applied simultaneously increased yield, even though N and P reduced the content of leaf Zn. Both leaf Ca and Mg were higher in trees that received phosphorus application. Embleton *et al.* (1958a,b) reported that N fertilization had no effect on leaf K concentration. In contrast, Lynch *et al.* (1954), Lahav *et al.* (1990) and Loupassaki (1998) reported that N applications decreased leaf K concentration.

There have been many studies on mineral nutrient content in fruit and their effect on postharvest physiological disorders, disease resistance and quality, which are addressed in Defilippi *et al.*, Chapter 15, this volume.

Fertilizer practices

Fertilizers are generally applied to crops by soil or foliar applications. Farmyard manures were the first reported method of applying nutrients to avocado trees (Popenoe, 1939). However, the sustenance of trees by the addition of manures has largely been discontinued on industrial-scale orchards due to the high costs of transport and application of these materials. Continued use of manures is still widely practied by organic growers who generally have smaller production units and command a premium return for their product. However, while there is an increasing interest in organic production, the topic will not be discussed in detail in this chapter. In recent years, there has been a move to augment orchard floor leaf trash with organic mulches to improve the general root environment through increasing soil carbon content, thereby adding to the pool of available nutrients, and promoting populations of beneficial microorganisms that assist with Phytophthora root rot management (Pegg and Giblin, 2008; Pegg, 2010; Grunennvaldt, 2022) (see Whiley *et al.*, Chapter 12, this volume for a discussion on mulching).

Comparative use of inorganic and organic fertilizers on avocados trees has been reported from different countries (Goodall *et al.*, 1965; Ben-Ya'acov, 1972; Gustafson, 1973; Abou Aziz *et al.*, 1975; Lahav and Kadman 1980; Lahav, 1984; Wolstenholme *et al.*, 1998; Salazar-García *et al.*, 2009; da Silva *et al.*, 2017; Eman *et al.*, 2022). Other studies have reported improvements in tree growth, yield and uptake of micronutrients from applying a combination of organic and inorganic fertilizers (Elmetwalli, 2019; Campos-Mariscal *et al.*, 2020; Hazra and Santosa, 2022). Benefits from fertilizing will largely depend on source, application time and amount applied, and on tree water status at application. In Brazil, da Silva *et al.* (2017) recommended scheduling the application of fertilizers, mulch and soil amendments during the root growth flushes, as well as foliar sprays during the shoot flushes, which can be defined by determining the extent of each phenological stage in the orchard. The well-known shallow and spreading root systems of avocado trees should be considered to obtain maximum efficiency when applying fertilizers to orchards (Lahav and Kadman, 1980; Whiley, 1994).

Application of foliar nutrients has developed in horticultural crops to quickly supply nutrients, particularly when nutrient symptoms develop. Investigations with foliar-applied nutrients on avocado have had mixed results as to their effectiveness in correcting nutrient deficiencies (Newett, 2005). However, when new cohorts of leaves are expanding, in some instances foliar application can be effective (Robbertse *et al.*, 1989, 1992; Bender, 1997; Smith *et al.*, 1997b; Jaganath and Lovatt, 1998).

Synthetic fertilizer products are widely used in the production of avocados and are available in a solid form, which can be either slow release or highly soluble. These are either regularly broadcast on to the orchard floor or dissolved and applied through fertigation. Some of the less soluble mineral nutrients such as Ca and K are now available in a liquid thiosulfate formulation designed specifically for fertigation application.

Fertilizer rates and application

Avocado orchards are generally not cultivated once trees are planted, and major soil improvements including pH adjustment, correction of mineral nutrient deficiencies, incorporation of organic matter (cover crops, manures) and drainage improvements (mounding/ridging) are generally implemented prior to planting. In the summer-wet subtropics, inter-row cover crops of a biofumigant sorghum (Fumigator™) (Duff and Firrell, 2021) are recommended for the first 2 years after planting to increase soil organic matter and reduce populations of soil diseases. Under most conditions, mineral nutrients are added on an annual basis to maintain tree health and yield. For efficient uptake, fertilizer should be applied to the zone of highest feeder root activity, which is between the trunk and drip line of the tree, the area normally wetted by sprinkler or dripper irrigation. In rainfed orchards in subtropical environments, avocado trees are fertilized with soil applications during the wet season, from mid-spring to the end of the summer. In sandy soils with low fertility and cation exchange capacity, nutrient application should be split into multiple applications (weekly during peak growing periods) to match the more immediate needs of the crop. Where there is a danger of nutrient fixation or a need to create a zone of availability for better uptake, a concentrated band along the drip line can be effective for enhancing nutrient supply, for example, soil zinc applications (Crowley *et al.*, 1996). On the krasnozem soils of eastern Australia, P is bound to soil complexes, and banding of this nutrient is more effective than broadcasting (Newett *et al.*, 2018).

Soil applications of solid fertilizers have a comparative higher cost for labour. Alternatively, the application of plant nutrients to avocado trees via fertigation is cost-effective and an efficient method for delivering soluble fertilizers in accurate amounts to the precise locality of roots, thus potentially improving production and decreasing water and mineral percolation to soil layers below the root zone compared with conventional fertilizer applications (Abbasi *et al.*, 2012). Controlling soil mineral content at the various phenological stages of tree growth and development provides a means for regulating nutrient uptake at any given time. The success of the high-density avocado orchards in the semi-arid climate of Chile is largely due to the control of tree growth by limiting the root zone through mounding with a reduced wetted area, and applying plant growth regulators and nutrients through fertigation (Gardiazábal *et al.*, 2015).

Most commercially available foliar nutrient formulations are not cost-effective compared with soil-applied products and should only be used when benefits have been reliably demonstrated. Uptake of foliar nutrients applied to avocados can be minimal. This appears partly due to the presence of waxy cuticles on both the adaxial and abaxial surfaces of mature leaves (Whiley *et al.*, 1988a), where stomata are only present on the abaxial side (Whiley *et al.*, 1988a; Mickelbart *et al.*, 2000). These waxy surfaces are effective not only in controlling water loss from mature leaves but also reduce uptake of foliar-applied nutrients. Adding wetting agents and using highly soluble and nanoscale fertilizer formulations improve the efficiency of foliar fertilization. However, to be effective, foliar-applied nutrients must be phloem mobile to re-translocate to other plant tissues. Nutrients identified as phloem mobile are N, P, K and S. Partially phloem-mobile microelements are Zn, Fe, Mg, Mo and B. Calcium is not phloem mobile, hence is ineffective as a foliar application for avocado. Lovatt (2013a) reported the usefulness of foliar-applied nutrients on avocado trees during spring when soils are wet and cold. Under these conditions, nutrient uptake is restricted and additionally compromised in the presence of salinity and where soil chemistry reduces nutrient availability.

Nitrogen. Recommended amounts of N range from 10 kg ha^{-1} in the first year to 120–250 kg ha^{-1} from the sixth year onwards. However, the rate should be adjusted in accordance with leaf tissue concentration, cultivar, the potential of soil organic matter mineralization and yield. In N-deficient trees, higher rates of N can be applied in the short term. However, excessive N application is most likely to leach and contaminate groundwater (Lovatt, 2001). Cultivars with high productivity require higher N applications to ensure adequate leaf area to support fruit growth, especially in an ‘on’ crop year. Nitrogen application should be reduced in an ‘off’ year so that vegetative growth is not overstimulated. Usually, cultivars of Guatemalan origin contain more N in their tissues and require more nitrogenous fertilizer than Mexican cultivars (Lahav and Kadman, 1980).

Abou Aziz *et al.* (1975) evaluated the effect of soil- and foliar-applied N fertilizers on yield and fruit size of ‘Fuerte’ avocados. A foliar N-spray (500 g N tree^{-1} split over four applications from spring to mid-summer) significantly increased yield and fruit size compared with four applications to the soil. In Israel, Klein and Zilkah (1986) reported a 65–85% uptake of urea within 2–5 days of dipping ‘Fuerte’ leaves into urea solutions containing surfactants. Phytotoxicity occurred in leaves and flowers when urea concentrations exceeded 2% and 0.5–1%, respectively. In further studies, Zilkah *et al.* (1987) found that about 27% of labelled ^{15}N (applied as foliar urea) was translocated to the developing inflorescence. The N from the foliar-applied urea also moved to developing fruit and new spring leaves where, 14 days after application, it made up 14% (fruit) and 11% (leaves) of the total N found in these organs. Uptake and translocation of N into developing fruit were greater when the spring growth was left unpruned. Zilkah *et al.* (1987) also reported that N from labelled urea was taken up and basipetally translocated from new spring leaves to developing fruit.

In subtropical South Africa (Koen and du Plessis, 1992), N was applied in three equal applications to ‘Fuerte’ in winter, summer and autumn over four consecutive seasons, which produced a curvilinear yield response to increasing N, indicating an inflection point where excessive N reduces yield. Lahav and Kalmar (1987) increased spring applications of N to 'Fuerte' and 'Ettinger' trees in Israel but there was no effect on yield or leaf N concentration. They suggested that low soil temperatures in spring reduced N uptake. In contrast, another field experiment in Israel found that an efficient N fertilization programme must start early in the spring and end late in autumn (I. Adato, Upper Galilee, Israel, 1991, personal communication).

Additional N (17% of the annual application) applied at anthesis/fruit set to 'Hass' in California increased the cumulative 4-year yield by 31%, and when the same amount was applied in summer, yield increased by 41% (Lovatt, 2001). The increase in yield was mainly due to larger fruit size. There was no consistent effect of treatments on leaf N concentration over the 4 years of the study, and hence, no relationship could be established between leaf N concentration and yield (Lovatt, 2001). From earlier research, Embleton *et al.* (1968) concluded that the time of N application was more important than the amount of N with respect to increasing yield and fruit size of 'Hass' grown in California.

Lynch *et al.* (1954) concluded that among N, P and K, nitrogen has the greatest effect on yield of West Indian cultivars in Florida. In later studies, high rates of N applied to 'Tonnage' avocado trees in Florida resulted in light cropping (Young and Koo, 1977). Nitrogen fertilization increased yield from 'Fuerte' trees in Egypt, Greece and South Africa (Abou Aziz *et al.*, 1975; Koen and du Plessis, 1992; Loupassaki, 1998). Yields of 'Hass' were also increased in Israel after an autumn application of N and P (Winer *et al.*, 1998) and in California following an increased application of N in autumn (Lovatt, 2001). In Israel, under particularly harsh climatic conditions during flowering, Silber *et al.* (2018) concluded that continuous fertigation (NPK + trace elements) over the entire growth period gave the best results on yield and fruit quality, and emphasized the importance of fertilizing the tree prior to flowering and to match demand for water and nutrients during the flowering and fruit-set processes. In a study conducted in California on nitrogen fertilization strategies to increase 'Hass' fruit yield and quality, Salvo and Lovatt (2016) found that summer vegetative shoots contribute 60–70% of the total inflorescences in the subsequent flowering, suggesting that N fertilizer should be applied to the soil during the summer, at an adequate rate to mitigate June drop and support competing growth processes of the fruit and summer vegetative shoots. The above-mentioned studies demonstrate the importance of the timing of soil N application and explain the effect of extra soil N at full bloom through meeting tree requirements at this critical stage of production.

Jaganath and Lovatt (1998) reported that the treatment of 'Hass' trees with a 1% foliar application of urea during early inflorescence development (cauliflower stage; Fig. 11.17) increased ovule viability as well as the number of pollen tubes that reached the ovule. The treatment also increased the 3-year cumulative yield by 23%. Similar applications of urea to flowers in New Zealand over three seasons did not significantly affect fruit set (Dixon *et al.*, 2007), presumably due to the high nitrogen mineralization in andosol soils. Foliar N sprays with low biuret urea can be used to quickly correct N deficiency or assist tree recovery from disease (Lovatt, 1999; Dixon *et al.*, 2005).

Phosphorus. Lynch *et al.* (1954) concluded that soil-applied P had a negligible effect on leaf concentrations and yield of West Indian cultivars growing in the unique calcareous soils of southern Florida, and Malo (1976) reported

Fig. 11.17. Cauliflower stage of avocado flowering. Photograph supplied by A.W. Whiley, Sunshine Horticultural Services, Queensland, Australia.

that there was no effect on productivity after 15 years without P application when avocados were grown on these soils. However, it is likely that the soil was well supplied with P prior to planting (Li, 2009), since in the calcareous soils of Florida, P accumulates to very high concentrations. Phosphorus soil applications increased P concentration in leaves of 'Fuerte' trees in California but did not affect yield or fruit quality (Embleton *et al.*, 1958b). Koen and du Plessis (1992) concluded that P applications had little effect on yield of 'Fuerte' in South Africa, since there was only 1 year out of 5 when a significant yield increase was recorded following soil treatment. Avocado roots are readily colonized by AMF and respond favourably to mycorrhizal inoculation (Bárcenas *et al.*, 2007; Bañuelos *et al.*, 2013), which is known to assist the uptake of P. This association may explain the low requirements for phosphate fertilizers in avocado orchards (see Nutrition and mycorrhizae section, this chapter).

Potassium. Embleton and Jones (1964) reported that from a 12-year study in California, increasing leaf K concentrations from 0.9% to 1.3% had no effect on yield. Similar results were obtained from a 6-year study with 'Fuerte' in South Africa (Koen and du Plessis, 1992). Lynch and Goldweber (1956) found an interaction between N, P and K in West Indian cultivars in Florida. They reported N at 80 kg ha^{-1} $year^{-1}$ gave low yields and tree growth with poor fruit quality at all K and P levels. At 160 kg ha^{-1} $year^{-1}$ of N, yield was not appreciably affected by variations in K and P, but the lowest (no K) and highest amounts of K (300 kg ha^{-1} $year^{-1}$) were detrimental to high yields. When K was kept at a high level, increasing P from 0 to 156 kg ha^{-1} $year^{-1}$ depressed yield. Potassium application increased yield when a high rate of N (240 kg ha^{-1} $year^{-1}$) and a low rate of P (52 kg ha^{-1} $year^{-1}$) were applied to trees. In that study, there was no effect of K on fruit size.

Lahav *et al.* (1976) researched the effect of applied K on tree growth, yield and fruit quality of 'Ettinger', 'Fuerte', 'Hass' and 'Nabal' avocados growing in heavy soil in Israel over a 7-year period. Three rates of K, applied as potassium nitrate at 150, 300 and 450 kg ha^{-1} and potassium sulfate at 300 kg ha^{-1}, were compared with a zero K (control). All treatments were supplemented with 160 kg ha^{-1} of N, equivalent to the amount received by the 450 kg ha^{-1} treatment of potassium nitrate. The highest rate of K produced a small increase in 'Fuerte' leaf content but had no effect on K leaf concentration of 'Hass'. The highest rate of KNO_3 (450 kg K ha^{-1}) significantly increased the height of 'Hass' trees by 20% and 'Fuerte' trees by 28% over 7 years. However, there was no effect of any of the treatments on fruit yield during the entire experiment. Potassium at the highest rates consistently increase the fruit size of 'Fuerte' and 'Nabal'.

In Mexico, annual applications of 1 kg K_2O $tree^{-1}$ over a 5-year period increased cumulative fruit yield by 44% (Aguilera and Salazar, 1996), while in rainfed 'Hass' avocado orchards in Mexico, Salazar-Garcia *et al.* (2009) reported that potassium fertilization, applied at 2.5 kg K_2O $tree^{-1}$, resulted in a 63% increase in the production of large-sized fruit. In another Mexican study, Guerrero-Polanco *et al.* (2018) assessed the effects of three applications of KNO_3 on fruit production and canopy volume of 'Hass' trees, under rainfed and irrigated conditions. It was concluded that fertilization with KNO_3 significantly increased fruit yield (kg $tree^{-1}$) by 59% and 73% under both water regimes, respectively. It is likely that the greater effect in irrigated orchards was due to the increased efficiency of mineral nutrient uptake with the availability of continuous soil moisture (Guerrero-Polanco *et al.*, 2018). Also, under irrigated conditions, a positive correlation was observed between leaf N and K concentrations and fruit yield. Research in subtropical Ecuador evaluated two levels of nitrogen and two levels of potassium applied by fertigation on 'Hass' avocado growth (Viera *et al.*, 2021). It was concluded that the application rate of nitrogen influenced absorption of P and other nutrients, and the interaction of both nutrients applied in rates of 200 kg ha^{-1} of N and 250 kg ha^{-1} of K positively affected vegetative growth of the trees studied.

In California, Lovatt (2013b) found that that potassium phosphite (0N–12.2P–21.6K) sprayed at 6.5 L ha^{-1} at the cauliflower stage of flowering, significantly increased the 3-year cumulative yield of commercially valuable-sized 'Hass' fruit without reducing total yield. Limaco Sing and McNeil (1991) investigated the effectiveness of foliar applications of KNO_3 on 'Hass' avocados growing on a serpentine-derived soil with high Mg content in California. Trees were sprayed to runoff with a 3.6% solution of KNO_3. Treatments were applied to new growth at different stages of leaf development, which were either at half or full leaf expansion, or 1 month after full leaf expansion. In addition, different combinations of treatment times were applied. The two-spray application at half leaf expansion and 1 month after full expansion and the treatment of three sprays at each treatment time significantly increase leaf K content compared to unsprayed control. Additionally, the three-spray treatment significantly increased leaf Mn concentration and all K treatments significantly increase leaf Zn content. Limaco Sing and McNeil (1991) also reported that banded, soil-applied K_2SO_4 was more cost-effective than two to three sprays of KNO_3 in supplying tree K requirements under the experimental conditions.

In summary, the response from soil-applied K fertilizers to avocado was minimal at experimental sites in semi-arid, Mediterranean climates. Increased yield and greater fruit size from K applications more often occurred when research was conducted in subtropical climates with summer or year-round rainfall, where leaching losses of K from soils are more likely to occur. The opportunity for increasing 'Hass' fruit size by K application warrants further investigation over a wider range of soils and climatic conditions, as size is a significant problem for 'Hass' grown in warm, subtropical climates. Increased K fertilization has improved fruit size in 'Valencia' oranges grown in similar environments (Erner *et al.*, 1993).

Calcium. Some studies have reported avocado fruit yield increases in response to soil-applied Ca. In eastern Australia, seven soil management treatments that included calcareous amendments and organic matter to 'Fuerte' grafted on to 'Fuerte' seedling rootstock on an acid, red basaltic clay loam were evaluated over a 6-year period (Trochoulias *et al.*, 1986; Broadbent *et al.*, 1989). Annual dolomite applications of 5 t ha^{-1} increased the exchangeable Ca and Mg concentrations in soil to a depth of 30 cm and raised the soil pH from 5.1 to 6.1, but had no effect on leaf nutrient concentrations. In contrast, annual gypsum applications of 10 t ha^{-1} increased the exchangeable Ca concentrations in soil but reduced exchangeable Mg to a depth of 50 cm and had no effect on soil pH. Gypsum increased leaf Ca concentrations and yield by 40% more than the dolomite treatment over the 6-year study (Broadbent *et al.*, 1989). There appears to be no direct nutritional effect from gypsum that would account for the increase in yield, but the mild fungistatic properties of gypsum on *P. cinnamomi* may have accounted for improved tree performance, as root rot was present at this site. Gypsum added to soils in avocado orchards in California reduced sporangial production and size, and zoospore production of *P. cinnamomi*, thereby reducing the disease pressure on trees (Messenger *et al.*, 2000).

Soil applications of dolomitic lime, calcium silicate and gypsum consistently increased yield of 'Edranol' avocados during a 3-year study on acid soils in South Africa (du Plessis and Koen, 1987). However, except for 1 year, calcium hydroxide had no effect on yield. In that study, treatments were applied annually from 1979 to 1982 and were selected to give comparable neutralizing values of Ca concentrations (Table 11.9). High applications of all Ca products decreased yield. The trees were growing on a red clayey soil (red latosol) with a low pH (4.6–4.8) and high extractable Al (90–135 mg kg^{-1}). Dolomitic lime, calcium silicate and calcium hydroxide increased soil pH and reduced extractable Al in the topsoil, but there was little effect in the subsoil (30–60 cm). Gypsum was the most effective product to increase the Ca concentration of the subsoil with Ca concentrations 2.7 times higher compared with dolomitic lime or calcium silicate. There was little effect of Ca liming treatments on either leaf or fruit nutrient concentrations, with K and Ca being the only nutrients to change significantly compared with untreated trees. Barnard (1989) found that soil-applied gypsum displaced K from both the top and subsoil of acidic avocado soils in South Africa, and du Plessis and Koen (1987) found that calcium hydroxide applied to soil reduced leaf and fruit K concentrations (Table 11.9).

Studies with soil-applied Ca liming products in Australia have produced similar results in soil chemistry to those in South Africa. In both countries, there was no correlation between soil pH and yield, and dolomitic products increased soil Ca concentrations to a depth of 15–30 cm, but had no effect at the subsoil level. In contrast, soil-applied gypsum in both countries increased yield and soil Ca concentrations to a depth of 50–60 cm. While gypsum was the most effective

Table 11.9. Effects of different calcium sources applied to 'Edranol' trees in South Africa, on chemical soil composition, leaf and fruit potassium concentration, and fruit yield. From du Plessis and Koen (1987), with permission

Treatments[a]							
Ca source	Application rate ($t\ ha^{-1}\ year^{-1}$)	Topsoil pH	Al ($mg\ kg^{-1}$)	Subsoil Ca ($mg\ kg^{-1}$)	Leaf K (%)	Fruit K (%)	Yield ($kg\ tree^{-1}$)
Control	0	5.16c	93ab	38b	0.86a	1.69a	44.3c
Dolomitic lime	4.7	5.86a	34c	100b	0.79a	1.55a	141.4a
Ca silicate	5.0	5.71ab	57bc	100b	0.75ab	1.35ab	138.2a
Ca hydroxide	3.7	5.38bc	68bc	100b	0.60b	1.13b	62.7c
Gypsum	9.2	4.70d	123a	269a	0.72ab	1.55a	102.8b
LSD ($P \leq 0.05$)		0.4	50	69	0.15	0.32	31.7
CV (%)		5.4	46.4	39.8	14.6	17.2	29.4

[a]Three levels of each Ca source were applied but the data presented are of the middle level and for one year (1983). Values in columns with different letters are significantly different ($P \leq 0.05$).

treatment tested for increasing Ca concentrations in the soil, it also displaced K and Mg from the top and subsoil. Therefore, the continued use of gypsum may be detrimental to tree health and fruit quality, unless K and Mg are also supplied.

Magnesium. There have been few studies on the effect of varying Mg supply on avocado yield and those reported have invariably assessed its effect on fruit quality. For example, in a 3-year study, Kremer-Köhne *et al.* (1993) investigated the effect of soil-applied N, K and Mg on 'Fuerte' trees. Five applications of 2 kg $tree^{-1}$ of $MgSO_4$ (10 kg $tree^{-1}$ total) were applied to adult trees from spring to mid-summer each year. In the second year of the study, Mg significantly increased yield but the 3-year cumulative yield was not significantly different from that of untreated trees. Mg application did not affect leaf or fruit Mg concentrations, uptake of N and K or physiological fruit disorders.

Most Californian soils on which avocados are grown have a reasonable supply of Mg and deficiency symptoms are rare. In soils with naturally high Mg content, sufficient calcium must be available to compensate for competitive uptake between cations (Spann, 2019). Magnesium fertilization has been recommended for use on the calcareous soils of Florida with formulas of N-K-Mg in a 1:0.8:0.3 ratio (Malo, 1966) and in South Africa at 0.5 t ha^{-1} of $MgSO_4$ (Koen and Langenegger, 1971). Soil type and leaching potential influence the annual of quantity of Mg fertilizer required to maintain avocado leaf concentrations within the optimum range (0.25–0.8%) defined by Embleton and Jones (1964). When grown on the clay loam krasnozems of eastern Australia (annual rainfall from 1000 to 3000 mm $year^{-1}$) an annual application of 10–40 kg ha^{-1} $MgSO_4$ is sufficient to prevent deficiency, while for production on the deep sands in the Far North of New Zealand (annual rainfall of 1350 mm), 50–60 kg ha^{-1} of $MgSO^4$ are fertigated during the months of active growth to maintain leaf Mg levels (A.W. Whiley, SHS, Queensland, Australia, 2024, personal communication).

Zinc. Zinc deficiency may be corrected by application of various Zn products applied to soil (Gustafson, 1957; Wallihan *et al.*, 1958; Embleton and Wallihan, 1966), or as foliar sprays (Parker, 1936; Ruehle, 1940; Wallihan *et al.*, 1958; Goodall *et al.*, 1965; Lee, 1973). In California, foliar sprays of Zn compounds onto expanded spring-grown leaves were reported to be effective (Lee, 1973). However, former results from Israel challenged the benefit of foliar applications for correcting zinc deficiency (Ben-Ya'acov, 1972; Lahav and Zamet, 1977). In foliar application studies using ^{65}Zn on West Indian and Mexican seedlings, Kadman and Cohen (1977) reported that after daily treatment for 1 week, there was negligible penetration into the leaf by the radio-labelled Zn. Some penetration occurred from Zn sprays when leaves on new growth were about 2/3 of full expansion. They concluded that there was little value in correcting Zn deficiency of avocado trees using foliar sprays. The application of zinc sulfate or zinc chelate

(Zn-EDTA) to the soil, either banded in the irrigation zone or dissolved in the irrigation water, was shown by Crowley *et al.* (1996) to be more reliable and effective for correcting plant deficiencies than foliar applications, even in high pH soils. Usually, $ZnSO_4$ is applied at 200 kg ha^{-1}. Because Zn movement through the soil profile is slow, a single application can be effective for 3 years.

In a 4-year experiment when foliar zinc sulfate was applied at the cauliflower stage of flowering with emerging vegetative growth, Salazar-García *et al.* (2008) reported that spraying with 1.46 g Zn·lt water^{-1} was ineffective in correcting Zn deficiency, with no effect on yield or fruit size. However, two annual soil applications of zinc sulfate (35.5% Zn) at 0.75 kg tree^{-1} increased yield and fruit size and normalised fruit shape. Foliar applications of zinc sulfate, oxide and chelate formulations (Crowley *et al.*, 1996) and foliar-applied organic acids plus ferrous sulfate solutions (Crane *et al.*, 2007b) have been effective in increasing leaf Zn and Fe levels, respectively, when sprayed on to avocado leaves.

Boron. Carefully controlled soil applications are more efficient for remedying chronic B deficiency in the summer-wet subtropics where soils are naturally deficient (Whiley *et al.*, 1996). This is confirmed by the fact that unrecognized B deficiency existed for decades in avocado orchards in the summer-wet subtropics of Australia and South Africa, despite applications of foliar sprays at flowering, fruit set and during the autumn months. Although some B translocation can occur in phloem, possibly as a B-perseitol complex (Minchin *et al.*, 2011), it is obviously insufficient to meet the B requirements of the whole tree in B-deficient soils, especially roots.

Thorp *et al.* (2009) investigated the transport of B from leaves to inflorescences with the view of determining the role of B in AB. Thorp *et al.* (2009) girdled branches of 'off' and 'on' bearing trees and increased the starch concentration in the leaves by twofold to threefold, especially in 'off' bearing trees. Girdling reduced the concentration of B in leaves by more than 42% and increased B in flowers by a similar proportion, irrespective of the fruit load. Thorp *et al.* (2011) suggest that B transport to avocado flowers is regulated by carbohydrate supply, and therefore by crop load with fruit competing strongly for B.

Smith *et al.* (1997a) reported that when 'Hass' trees with marginal leaf B concentrations of 15–29 mg kg^{-1} growing on krasnozems (red/brown clay loams) in Australia were treated with soil-applied borax, mean fruit size at maturity was increased by about 13–16%. In similar studies on oxisols in South Africa, Bard and Wolstenholme (1998) found that soil-applied borax increased fruit size of 'Hass' by about 4%. More recently, Hapuarachchi *et al.* (2022) reported a pre-flowering soil application of B at 15 g tree^{-1} on 5-year-old 'Hass' in a Queensland orchard increased fruit mass by 5%, fruit diameter by 2%, flesh mass by 9%, flesh B concentration by 55% and the relative abundance of unsaturated fatty acids by 1% compared with control trees. However, applying B at double the recommended rate (30 g tree^{-1}) significantly reduced fruit yield.

Boron deficiency is most likely to occur in high-rainfall areas with acidic soils or those with naturally high pH. Excessive B can be toxic, and care should be taken not to exceed recommended leaf concentrations. However, the avocado is particularly poor in responding to foliar and soil-applied B, tolerating soil applications up to ten times greater than those that cause severe toxicity in other crops, for example, citrus, macadamia, mango and papaya (Whiley *et al.*, 1988c). This has been attributed to the absence of root hairs (reduced absorption from the soil) and the accumulation of high B concentrations in roots (poor translocation from roots to aerial parts of the tree) (Coetzer *et al.*, 1993, 1994). Soil B applications need to be carefully managed, as soil B concentrations can easily build up to toxic levels and optimal soil boron concentrations vary greatly depending on soil texture (Newett *et al.*, 2021). Boron toxicity can be problematic in semi-arid areas due to either a high concentration of B in soil and/or irrigation water.

Boron can be applied to the soil by broadcasting ulexite (sodium calcium borate) or borax (sodium borate, Solubor®) under trees or through fertigation with boric acid or sodium borate. When the mature leaf B concentration is less than 20 mg kg^{-1}, B at 1–2 g m^{-2} of tree canopy should be applied three times annually (spring, summer and autumn) or more frequently on sandy soils with correspondingly lower rates. Careful monitoring of leaf B concentrations should be undertaken as rootstocks, soil type (especially organic matter content) and rainfall/irrigation practices (leaching) will all affect availability and uptake (Whiley *et al.*, 1996; Bard, 1997). Adjustments to the base application rate can be made once a response profile has been developed.

There is evidence that when B is deficient in avocados, spraying with Bortrac™ (15% B and 6.5% N) or similar products at the tight cauliflower stage of flowering and 10 days later, may increase fruit set due to the improved viability of pollen (A.W. Whiley, SHS, Queensland, Australia, 2024, personal communication). Developing floral structures have no waxy cuticle (Whiley *et al.*, 1988a) and readily absorb B, which is sufficiently translocated acropetally to be effective (Robbertse *et al.*, 1989, 1992; Bender, 1997; Smith *et al.*, 1997a; Jaganath and Lovatt, 1998). In a Mexican rainfed 'Hass' avocado orchard with boron deficiency symptoms, Cossio-Vergas *et al.* (2009) reported a foliar spray at 2 kg ha^{-1} of B applied at early fruit set improved fruit size, while single spray at 1 kg ha^{-1} of B at fruit set, followed by a second spray a month later, increased yield and fruit size. In Spain, Torres *et al.* (2003) observed that foliar applications of B at flowering, gave higher B concentrations in autumn-sampled leaves. Smith *et al.* (1997a) concluded that foliar B sprays at flowering only induced a response when leaf B was deficient (<30 ppm). In New Zealand, Dixon *et al.* (2006) found no effect from foliar B sprays when trees had sufficient B content. Three foliar applications of 1.2 g l^{-1} B applied as boric acid given to 2.5-year-old pot-grown 'Hass' trees in Mexico, increased the cell-wall thickness of buds and leaf amino acid concentration (González-Gervacio *et al.*, 2019).

Iron. Iron deficiency can occur during spring if soils are too cold for uptake (Newett *et al.*, 2018). Iron deficiency may be remedied in calcareous soils by applying Fe chelate or by reducing pH and improving drainage in soils with high available Mn. Soil application of Fe chelate (Fe EDDHA) generally controls Fe deficiency, with trees responding within 2–3 weeks (Kadman, 1963a). Soil treatments may last for 2 years, but continuous application of 1–2 mg kg^{-1} of Fe EDDHA through fertigation is recommended. The usual rate of soil application is 15 g $tree^{-1}$ in the first year of growth to 150 g $tree^{-1}$ from the sixth year onwards. In the calcareous soils of southern Florida, the common standard practice for preventing Fe deficiency in avocado trees is periodic soil drenches of Fe EDDHA (Crane *et al.*, 2007b). Foliar applications of different Fe formulations (Kadman and Lahav, 1972, 1982) produced variable results, probably due to poor absorption and translocation through avocado leaves. A unique method of direct pressure injection was tested in Israel (Kadman and Lahav, 1982). Injection of 0.5% to 1% FeEDDHA or 2% $FeSO_4$ solution into the trunk or main branches of severely chlorotic 'Edranol' and 'Wurtz' trees, produced a very rapid greening of leaves, giving a significant increase in the Fe and chlorophyll concentration of leaves and total recovery within 7–10 days. Iron content in 'Hass' avocado pulp has been utilized to predict susceptibility of fruit to chilling injury on a seasonal basis, since ≥60 ppm Fe in fruit pulp by February significantly reduced chilling injury in 'Hass' and 'Fuerte' avocados (Magwaza *et al.*, 2008).

Newett (2018) surveyed the fertilization programmes of nine different avocado-growing regions. An updated version of that survey is presented in Table 11.10.

Fertilization frequency and timing

Nutrient uptake is a continuous process and is strongly correlated to soil mineral concentrations subject to availability. Other factors influencing avocado mineral uptake from soil are rootstock (Haas, 1950a,b; Oppenheimer *et al.*, 1961; Embleton *et al.*, 1962; Ben-Ya'acov *et al.*, 1992; Lahav *et al.*, 1993; Whiley *et al.*, 1996), soil moisture content, soil temperature and root diseases (Whiley *et al.*, 1987). Frequency of application varies widely, being dependent on soil type and climatic conditions at each location. Frequent recurrent applications are especially required when production is on sandy soils with low CEC and fertility and at those locations with high rainfall: both scenarios will potentially have excessive leaching losses. Yates *et al.* (1992) reported that splitting applications of granular fertilizer in avocado orchards to eight per year reduced nutrient leaching as opposed to applying fertilizer twice a year.

Silber *et al.* (2012, 2018) conducted lysimeter and field experiments to investigate the critical periodic demand for water and nutrients in 'Hass' avocado. They concluded that

Table 11.10. Fertilization rates in nine avocado-growing regions in the world. Adapted from Newett (2018).

Growing region	Main fertilizer rates
Australia[1]	In orchards with 278 trees ha^{-1}, 80% of canopy covering of total orchard floor and optimum foliar nutrient levels: 110 kg N and 80 kg K (96 kg K_2O) ha^{-1} $year^{-1}$, with phosphorus ranging from 0 to 100 kg P (0–229 kg P_2O_5) ha^{-1} $year^{-1}$.
Brazil[2]	For expected yields of 10 to 25 t ha^{-1} under rainfed conditions: 60–120 kg N ha^{-1} $year^{-1}$ when leaf N is below 2%; 9-53 kg P (20–121 kg P_2O_5) ha^{-1} $year^{-1}$ and 28–140 kg K (34–169 kg K_2O) ha^{-1} $year^{-1}$.
California[3,4]	For expected yields of 10 to 20 t ha^{-1}: 72–100 kg N, 65–130 kg K_2O ha^{-1} $year^{-1}$ and 2,2 kg ha^{-1} P_2O_5 for each 1000 kg of fruit harvested. P and K are not applied if foliar levels are higher than 0.14% and 1.2%, respectively.
Florida[5]	In 3-year-old 'Simmonds' avocado orchards in a very gravelly limestone soil: 156 kg N, 25 kg P, 146 kg K and 58 kg Mg ha^{-1} $year^{-1}$.
Chile[6,7]	For avocados planted in densities of 416–833 trees ha^{-1} and producing 18–25 t ha^{-1} $year^{-1}$: 180–250 kg N, 0–80 kg P and 0–50 kg K_2O ha^{-1} $year^{-1}$. N is reduced to 120–170 kg N ha^{-1} $year^{-1}$ by adding DMPP nitrification inhibitor. For ultra-high-density orchards (1000+ trees ha^{-1}): 200–250 N ha^{-1} $year^{-1}$. N is reduced to 0–80 kg N ha^{-1} $year^{-1}$ by adding DMPP nitrification inhibitor. Potassium is applied only if leaf levels are <0.50% K. If foliar Zn levels are between 40–70 ppm, then zinc is soil-applied at a rate of 51 kg Zn (243 kg zinc sulfate) ha^{-1} $year^{-1}$, while for leaf Zn levels lower than 40 ppm zinc is soil-applied at a rate of 66 kg Zn (314 kg zinc sulfate) ha^{-1} $year^{-1}$.
Israel[8]	50 kg ha^{-1} of N at the beginning of March and later on 200–250 kg ha^{-1} of N and K from April to October. P is not commonly applied during the growing season and only added in October–November, in form of phosphoric acid (20–30 kg P ha^{-1}) for cleaning the irrigation system and preventing emitter clogging.
New Zealand[9]	In high-yielding 'Hass' orchards planted on wide spacing: 280 kg N, 36 kg P and 165 kg K ha^{-1} $year^{-1}$.
Peru[10]	In orchards in alluvial loamy soils and yielding about 18–20 t ha^{-1}: 180–200 kg N, 44–48 kg P (100–110 Kg P_2O_5) and 208–332 kg K (250–400 kg K_2O) ha^{-1} $year^{-1}$. In orchards with sandy soils and yielding about 16–17 t ha^{-1}: 240–260 kg N, 60–100 kg P (137–230 kg P_2O_5) and 300 kg K (361 kg K_2O) ha^{-1} $year^{-1}$. In orchards with sandy soils and yielding about 20–32 t ha^{-1}: 200–380 kg N, 0–76 kg P and 160–373 kg K ha^{-1} $year^{-1}$.
South Africa[11]	For orchards in clay soils and yielding about 20 t ha^{-1}: 90–110 kg N, 20 kg P and 140–160 kg K ha^{-1} $year^{-1}$.

[1,9,10,11]Newett (2018); [2] T. Cantuarias-Avilés, USP, *Piracicaba,* 2024, personal communication; [3]California Fertilization Guidelines for Avocados, UC Davis (2024). Available at: http://geisseler.ucdavis.edu/Guidelines/Avocado.html (accessed 3 June 2024); [4]Crowley (2011) Verbal presentation. Available at: https://www.avocadosource.com/papers/research_articles/crowley david2011.pdf (accessed 3 June 2024); [5]Kiggundu *et al.* (2012); [6]F. González, Belloto Agro Ltda., São Paulo, Brazil, 2024, personal communication; [7]Francisco Gardiazábal, GAMA, Quillota, Chile, 2024 personal communication; [8]Noy (2006).

unavailability of water and/or nutrients may lead to a malfunction of the embryo or seed, causing fruitlet drop and emphasized the importance of fertilizing the tree prior to flowering to match demand for water and nutrients during the flowering and fruit set stages. In California, Rosecrance *et al.* (2012) found that N fertilization in spring increased fruit size, yield and reduced the severity of AB compared with trees receiving N at any other time of the year, except during flowering. The rate of N and P accumulation in the fruit continued at moderate levels as the fruit matured.

Pheno-physiology modelling (Whiley, 1994) establishes the relationship between visible seasonal changes in plant growth and the associated molecular, cellular and biochemical (mostly invisible) changes that are simultaneously occurring. Whiley *et al.* (1988b) and Whiley (1994) propose pheno-physiology modelling as a tool to predict the timing of fertilizer application (particularly N and Ca) for avocado. Leaves are the primary organs of photosynthesis and take place in chloroplasts found in leaf cells, where absorbed atmospheric CO_2 is synthesized into glucose, which provides metabolic energy and downstream carbon-based products for the production of biomass. At different phenological stages of growth, excess carbohydrate (starch) is diverted to storage organs from where it can be remobilized to support growth requirements. Mineral nutrients play an integral role in the photosynthetic process and must be present in leaves in adequate quantities to facilitate optimal plant performance. The photosynthetic capacity of leaves is directly correlated to nitrogen content (DeJong and Doyle, 1985; Syvertsen, 1987) as the proteins of the Calvin cycle and thylakoids (chlorophyll) contain the majority of leaf nitrogen (Evans, 1989; Bojović and Marković, 2009). As an adaption to its evolutionary past, the photosynthetic rate of avocado leaves exhibits high plasticity responding to shade and full sun, as well as leaf N content. The maximum CO_2 assimilation rate of mature leaves on orchard-grown avocado trees is ≥18 $\mu mol{\cdot}m^{-2}{\cdot}s^{-1}$ measured across mild temperate, Mediterranean and subtropical climates (Whiley *et al.*, 1999; Liu *et al.*, 2002; Schaffer and Whiley, 2003; Mandemaker, 2007). Similar to deciduous species (Ye *et al.*, 2020), there is a relatively low carbon investment in leaves with maximum photoassimilation capacity reached within 40–50 days after bud break.

Avocado is a 'winter-green' species, annually replacing its leaves while continually retaining leaf coverage. In subtropical Queensland, Australia, Whiley (1994) studied the dynamics of summer-grown leaves and total spring-grown shoot CO_2 assimilation, and the associated fluctuations in leaf N and chlorophyll content. The research particularly focused on the period covering the reproductive processes of flower bud swelling, anthesis, fruit set and growth through to the transition from sink to source of spring shoot growth. The leaf nitrogen concentration remained relatively stable from mid-autumn through to mid-winter, followed by a sharp decline during inflorescence growth. Leaf N concentrations showed some recovery during anthesis but declined during fruit set and spring shoot growth. Leaf chlorophyll content and the rate of CO_2 assimilation approximately mirrored leaf N concentrations. The significant changes in leaf N concentration were related to critical stages of tree phenology, with a significant fall coinciding with inflorescence development. It is likely that re-mobilization of N occurred to support growth of the proximal reproductive sink. A further decline in leaf N occurred during spring shoot growth which ended in senescence and leaf fall approximately at the time of the transition of spring growth from sink to source. The decline in leaf N concomitant with fruit set and shoot growth has similarly been reported in citrus where it was concluded that young vegetative flushes draw nitrogen from reserves in old leaves (Erner, 1988).

To test the hypothesis that overwintered leaves contribute to fruit set and retention in 'Hass' avocados, three 2.5 m-high trees were defoliated by hand in early winter when flower buds began to swell. The treatment was carried out for 2 consecutive years and there was no significant difference in counts of newly set fruit on individual indeterminate shoots at the end of anthesis (mean fruit counts on defoliated and control fruiting shoots were 20.9 ± 2.0 and 22.5 ± 2.0, respectively). However, at harvest the mean cumulative weight of fruit was 6.7 ± 0.7 and 30.6 ± 0.9 for defoliated and control trees, respectively. It was concluded that the presence of functional overwintered leaves, through to the sink/source transition of new spring shoot growth (about 40 days after budbreak), was essential for fruit retention (Whiley, 1994), since most fruit dropped between setting and the sink-to-source transition of spring-grown shoots.

Excessive spring applications of N may invigorate indeterminate shoot growth and increase drop of newly set fruit. In another experiment, Whiley (1994) soil-applied spring N- and foliar-sprayed trees with paclobutrazol at peak anthesis to reduce indeterminate shoot growth, thereby favouring resource allocation to fruit (Wolstenholme *et al.*, 1990), and in a second treatment, only foliar-sprayed paclobutrazol. Both treatments increased leaf N and CO_2 assimilation of overwintered leaves during spring compared with control. Over 2 consecutive years, the soil-applied N plus foliar-sprayed paclobutrazol treatment significantly increased cumulative yield compared to control trees. To maximize CO_2 assimilation during periods of competing sinks, the timing of N application was adjusted to apply 20% of annual N during the last month of winter and the first month of spring (peak inflorescence growth and anthesis for this location); 30% of annual N applied in the last

2 months of autumn (trees are quiescent and 'Hass' fruit have reached maturity and are no longer strong sinks), and the remaining 50% of annual N applied evenly over the remaining spring and summer months. Under this N regime, sustainable production of 18+ t ha^{-1} in 'Hass' orchards has been achieved (Hermosilla-Madrid and Whiley, 2023).

In field studies with timing and amount of soil-applied N in California, Lovatt (2001) reported increased yield in 'Hass' avocado from applying additional N in autumn and again in spring during anthesis while retaining smaller applications during the latter part of spring and summer. Sperling *et al.* (2022) reported on annual fluctuations in mineral nutrient demand from lysimeter studies with avocado in Israel, determined by the difference between nutrient inputs and contents of the drainage discharge. They found the highest demand for N was during the winter months, which is concomitant with floral development, with lower quantities of N taken up over the summer months. There is consistency between results of the research of Whiley (1994), Lovatt (2001) and Sperling *et al.* (2022) despite different methodologies being used to determine seasonal N requirements, which gives confidence in programming N applications appropriate to tree requirements at different stages of the phenological cycle.

Apart from being a critical structural mineral nutrient for plants, there are many reports on the importance of calcium content in avocado fruit with respect to internal physiological disorders and postharvest quality (Bower, 1985b; Hofman *et al.*, 2002; Marques *et al.*, 2003). Witney *et al.* (1990b) reported on the accumulation of Ca in 'Hass' fruit from fruit set through to maturity, growing in a subtropical climate in South Africa. They found a rapid increase in the Ca concentration (mg kg^{-1} DM) of fruit of over the first 6 weeks of growth followed by a swift decline through to about the 12th week after fruit set, when the reduction in Ca concentration plateaued, through to fruit maturity. Calcium uptake and intra-plant transport occur mainly through mass flow in transpiration streams and polar indoleacetic (IAA) acid transport, which is greatest in organs with high physiological activity (Witney *et al.*, 1990b). The rapid accumulation of Ca during the first 6 weeks if growth is likely due to relatively higher transpiration flow through small fruitlets, which have a high surface-area-to-volume ratio compared with older fruit. Also, Blanke (1992) reported a significant reduction in stomate density on the skin of avocado fruit as they grew, which would be consistent with reduced transpiration flow with age. Research in Australia found a direct relationship between the Ca concentration in fruit flesh at harvest and postharvest quality, with high fruit Ca concentration at 6 weeks resulting in higher Ca concentration at maturity directly correlated to improved postharvest quality outcomes (P.J. Hofman, DPI, Nambour, Australia, 1996, personal communication).

While Ca and avocado fruit physiology-related research has provided strong, direct correlations between Ca and postharvest quality, the agronomic challenge lies in increasing fruit Ca concentrations during the first 6 weeks of growth, which provides a narrow window of opportunity to load fruit with this nutrient. Foliar Ca applications are ineffective, since Ca is not phloem translocated and the only pathway for this nutrient into fruit is through root uptake. Placement of Ca within the rhizosphere by the time ovule fertilization (fruit set) occurs is critical to achieve maximum flesh concentration at fruit maturity. Cation interactions need to be considered since K may preferentially be taken up to Ca in soils with low CEC (Johansen *et al.*, 1968). Fortunately, maximum K demand in avocado is during the fruit sizing phase over summer (Lovatt, 2001; Sperling *et al.*, 2022), thus meeting fruit requirements for these two cations is not conflicted.

Boron deficiency in avocado most often occurs on highly leached, acid soils typical of production areas in Australia and South Africa. Soil application is the most effective method of correcting B deficiency and maintaining optimum intra-plant concentrations (Bard and Wolstenholme, 1998). Boron is a trace element; while it is only required in small amount for normal growth in avocado, exceeding maximum leaf concentrations (100 mg kg^{-1}) will lead to toxicity starting with leaf burn followed by premature defoliation. When taken up by avocado trees, B accumulates in leaves over time thus the highest concentration is in the oldest leaves. To prevent toxicity developing B should be applied monthly in small amounts through fertigation (depending on soil type between 2 and 10 kg ha^{-1} $annum^{-1}$ of boric acid). Boron should not be applied to the soil while trees are quiescent, since this increases the risk of toxicity developing. If mature, summer-grown leaf levels exceed 60 mg kg^{-1}, further soil applications should be substantially reduced until analysis verifies the direction in which tissue concentrations are moving (Whiley *et al.*, 1996).

Organic Fertilization Practices

The global organic fruit and vegetable markets' compound annual growth rate over the 2018–2032 period is estimated to be between 4.78% and 8.25% (Market Reports World, 2023; Zion Market Research, 2022). In 2013, organic avocado production represented ≈8% of the total avocado cultivated area worldwide, with 86% of the area under organic management in Mexico (Granatstein *et al.*, 2013). By 2018, it was estimated that the Mexican area under organic production had increased by 260% (Delfín *et al.*, 2018). Due to its eco-evolutionary history in neotropical forests, the avocado is well adapted to extracting nutrients from decaying organic litter accumulated on the forest floor. The use of organic manures as the primary source of crop nutrition provides opportunities to access premium markets with health-conscious, affluent customers. The use of organic manures has proven effective in rehabilitating depleted soils following long-term use of chemical fertilizers (Bender *et al.*, 1992; Abd-Rabou, 2006; Sotomayor *et al.*, 2019; Infante-Jimenéz *et al.*, 2020; Tzatzani *et al.*, 2022; Joseph *et al.*, 2020; Mena *et al.*, 2023; Mohale *et al.*, 2023). However, the sole reliance on organic manures for avocado production does have disadvantages since the supply of manure nutrients is dependent on biotic factors, which tend to deliver a constant supply of nutrients making it more difficult to control vegetative growth and meet changing seasonal requirements (see *Fertilizer frequency and timing*, this chapter). Improved orchard sustainability for mainstream production in the longer term is more likely to be achieved if mulches and manures are integrated with strategic applications of synthetic fetilizers and biofertilizers.

Biofertilization

The use of biofertilization as a companion or alternative strategy to crop production has gained some momentum over the past 10 years with the greatest uptake in Europe and Latin America (Daniel *et al.*, 2022). However, a high proportion of biofertilizers marketed have not been subjected to independent scientific scrutiny, resulting in poor products with little or no impact on soil fertility and improved yield and quality. Understandably, there is still an element of scepticism in some circles as to the real benefits that may be accrued by their use. By definition, a biofertilizer is a substance that contains living microorganisms which, when applied to seeds, plants or soil, colonize the rhizosphere and/or roots and promote growth by increasing the supply or availability of primary nutrients to the host plant.

There have been some studies that have shown positive effects on avocado growth after the application of biofertilizers. In Egypt, 15-year-old 'Fuerte' avocado trees were treated with soil applications of effective microorganisms (EM) and seaweed extract (SW) during two seasons. Six combined soil applications of EM+SW at high rates significantly increased foliar N, P, K, Fe and Zn concentrations and fruit yield, weight, size and fruit quality (El-Shamma *et al.*, 2017). In a Brazilian study over 2 consecutive years, twice yearly soil applications of rock dust and a liquid biofertilizer formulated with 50% humic acid + 25% fulvic acid, were made to rainfed, 8-year-old 'Hass' avocado trees. The treatment improved soil carbon content and biological activity and increased root growth fruit yield and size (T. Cantuarias-Avilés, Piracicaba, Brazil, 2024, personal communication). Finally, Viera *et al.* (2023) reported a combined application of conventional chemical fertigation and soil-applied microorganisms had no effect on yield of young 'Hass' trees, but significantly increased fruit firmness and nutritional value of fruit pulp.

For maintenance of high-producing orchards, it is likely that biofertilizers will serve as support products for conventional synthetic fertilizers where increased biological activity may allow reductions in baseline use of chemical fertilizers. As stand-alone products, they are unlikely to have the flexibility to strategically increase nutrient supply in sufficient quantities at critical times of tree demand as driven by rapidly changing phenological conditions (see Fertilization frequency and timing, this chapter).

Methods for monitoring and estimating nutrient requirements

The analysis of soil, leaves and fruit to determine nutrient concentrations at critical stages of development is a valuable tool supporting decisions on the management of crop nutrition, which have both yield and fruit quality consequences (Joyce *et al.*, 2022). The measurement of leaf nutrient status is commonly used across the global avocado industry and is supported by predetermined nutrient standards (Embleton and Jones, 1964; Lahav and Aycicegi-Lowengart, 2003). While leaf analyses are mostly performed annually in autumn, soils are generally analysed every 2–4 years. For accurate nutritional diagnosis of commercial avocado orchards, sampling of both soil and leaves in autumn, at the end of the fertilization season, provides more timely information on the soil nutrient profile and availability of essential minerals to the trees. This allows for early intervention should potential deficient or excessive mineral concentrations be revealed. Additionally, the interpretation of soil analysis should consider differences in the optimum ranges of nutrients as defined for different soil textures.

Leaf mineral concentrations are a measure of their nutrient status, which influences their physiological efficiency thus defining potential tree performance. Leaf analysis is a useful tool for developing a strategy for fertilizing an avocado orchard. For consistent accuracy, mature healthy leaves are sampled from non-fruiting terminals, using either mature spring or autumn-grown leaves. The defined optimum mineral levels are used for both spring and autumn-grown leaves. Leaf nutrient thresholds were developed empirically in various studies that demonstrated the relationship between leaf mineral concentration and fertilization (mainly N) (Koen and du Plessis, 1992; Lovatt, 1996, 2001; Lovatt and Witney, 2001; Salvo and Lovatt, 2016). Ideally, a relationship between leaf nutrient content and yield is established (Lazare *et al.*, 2021), and from this, the critical concentration (or range) can be determined for each essential nutrient. If leaf nutrient content falls below the minimum concentration, a response to increased supply can be expected. Since many factors, apart from nutrient supply, affect nutrient concentrations in leaves, and many factors other than nutrition affect yield, the relationship between nutrient leaf content and yield is variable but robust enough to be a useful guide for fertilizer practice. Leaf analysis can indicate whether or not a nutrient should be applied, but it does not indicate the quantity to apply. Careful attention to factors that influence avocado leaf nutrient concentration is required before meaningful interpretations of analytical results can be made. Standardized sampling techniques provide the basis for reliable interpretation of leaf analysis data, while annual leaf analyses allow fine tuning of fertilizer management by detection of trends in mineral composition and early warning of incipient deficiencies and toxicities.

Not all organs on a tree respond in the same way to increases in nutrient concentrations in the root zone. The response of leaves to increased N applications differs from that of the fruit stalk (pedicel) and the inflorescence. Of these three organs, the leaf is the most buffered one (Razeto and Salgado, 2004). The association between the nutrient concentration in a plant organ and yield is empirically determined, and for comparisons to be made it is important to use a standard tissue. Currently, leaves are used, but this does not negate the fact that good correlations may be found with other tissues. In Australia, N:Ca ratios in fruit flesh or skin are being commercially used to adjust nutrition to reduce postharvest physiological disorders and predict postharvest quality (see DeFillipi *et al.*, Chapter 15, this volume).

Other alternative methods for monitoring nutrient content include chemical analysis of the fruit peduncle (Razeto and Salgado, 2004) and fruit pulp (Razeto and Palacios, 2007). Campisi-Pinto *et al.* (2017) reported that analyses of nutrient concentrations in inflorescence tissues at the cauliflower stage may serve as a good predictor for yield. Since fruit rather than leaves are the main sink in avocado trees, Silber *et al.* (2018) suggested analysing fruit pulp as an effective tool for assessing nutrient requirements of avocado orchards, and using leaf and/or inflorescence analyses for control and surveillance of the fertilization programme. Salazar-Reque *et al.* (2023) studied different vegetation indices (VIs) extracted from aerial RGB (red, green and blue) images acquired in different flights, and proposed a modified green-red vegetation index (MGRVI) to be used as a correlation variable in studies with RGB sensors, in order to monitor the nutritional and water status of 'Hass' avocado orchards.

Standardizing leaf age and season for leaf analysis samples

Leaf nutrient analyses are laborious and time-consuming, and this may cause delays between foliar sampling and fertilizer application (Dimkpa *et al.*, 2017; Prado and Rozane, 2020). Salazar-García *et al.* (2015) proposed a mathematical model to determine the correct time for leaf sampling in avocado trees. There are several reports on the effect of leaf age and season of sampling on nutrient concentrations in avocado leaves (Embleton *et al.*, 1958b; Bingham, 1961; Labanauskas *et al.*, 1961; Wutscher and Maxwell, 1975; Koo and Young, 1977; Bar *et al.*, 1987a; Lahav *et al.*, 1990; Koen and du Plessis, 1992), which have been interpreted in Fig. 11.18.

Nitrogen concentrations in spring-grown leaves initially increase for the first 3–4 months, briefly plateau and then decline. The rate of decline accelerates during flowering when leaves are about 10 months old. During autumn/winter, the N concentration of summer-grown leaves is significantly higher than that of spring-grown leaves (Lahav *et al.*, 1990). Phosphorus and K are at their highest concentration in young leaves and gradually decline with leaf age. There is a rapid increase in Mg and Cl concentrations in spring-grown leaves in the 3 months following full leaf expansion, but thereafter the concentration tends to plateau. Zinc and Cu concentrations decrease while the less mobile Mn and Fe increase with leaf age. Boron leaf concentrations increase with age but significantly decline about the time of flowering. This is consistent with the pattern found in avocado leaves in California (Bingham, 1961) and in Queensland, Australia (Whiley and Schaffer, 1994), suggesting that re-mobilization of B may occur during periods of high sink demand at flowering, since it is required for successful pollination and rapid cell division in fruit.

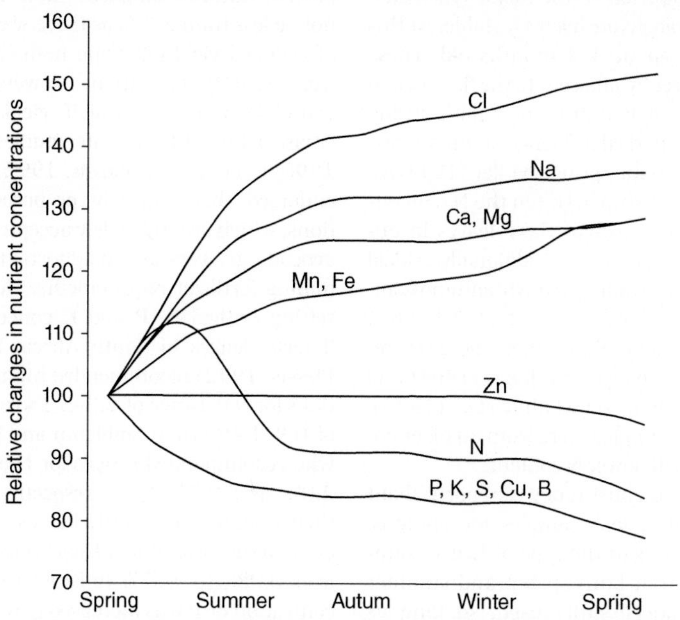

Fig. 11.18. Relative changes in nutrient concentrations of spring-flush leaves, according to leaf age and season. Based on Embleton *et al.* (1958b); Bingham (1961); Labanauskas *et al.* (1961); Oppenheimer *et al.* (1961); Wutscher and Maxwell (1975); Koo and Young (1977); Bar *et al.* (1987a); Lahav *et al.* (1990) and Koen and du Plessis (1992).

In setting critical leaf nutrient concentrations for avocado, Embleton and Jones (1964) as well as Koen and du Plessis (1991) chose the youngest, fully expanded mature leaves from non-fruiting, spring-grown shoots that were not flushing. Leaves (about 7 months old) were sampled in early autumn due to the relative stability of N, P and K concentrations and the correlation between leaf N concentration and yield. In Israel, Lahav *et al.* (1990) reaffirmed autumn sampling of spring-grown leaves as the most appropriate time and leaf stage for nutrient analysis, since the concentration of most mineral nutrients is relatively stable at that time. They also set the critical N concentration in Israel at ≤1.6% for 'Fuerte' and 1.8% for 'Hass'.

In humid, subtropical Australia, nutrient analysis of leaves sampled in late autumn is also recommended as a tool for managing tree nutrition. However, instead of using spring-grown leaves as the sample unit, the recommendation is to sample fully expanded leaves from non-fruiting, summer-grown shoots. Leaves should not be collected until all shoot growth on trees has finished but before flower buds begin to swell, since during this period N (the major yield determinant) concentrations are relatively stable. At this time leaves are about 3–4 months old. These leaves will be directly adjacent to the flowers and setting fruit and their nutritional health during this period has a marked influence on the success of fruit retention in the spring. Whiley (1994) established a relationship between the N concentration of summer-grown 'Hass' leaves in autumn and fruit set and yield. Additionally, critical N concentrations for summer-grown, autumn-sampled 'Hass' leaves have been set at 2.4–2.6%, which is substantially higher than the 2.0% recommended by Embleton and Jones (1964) and the 1.8% recommended by Lahav *et al.* (1990), albeit the leaves sampled were from an older cohort of growth with lower N content.

While there is consistency in reports about the time to collect leaf samples for analysis, there are differences in the type of leaf to sample. It appears that both spring and summer leaves can be successfully used, so long as there is consistency in selecting material of the same physiological age for analysis, and the relationship between leaf age and nutrient concentration is considered when developing recommendations. It is not surprising that there is a difference between critical leaf concentrations developed for the different avocado-growing environments, since average yields in California and Israel are lower than in Australia and South Africa (Wolstenholme, 1987), where higher leaf N concentrations for 'Fuerte' and 'Hass' are recommended. Leaves should be sampled only from healthy trees. Trees showing leaf burn, chlorosis or suffering from any other problem should be sampled separately to healthy trees. Sampling procedures were given by Lahav and Kadman (1980) and Marchal (1987).

Determining nutrient-critical leaf nutrient concentrations

In California, Embleton *et al.* (1959) established a significant curvilinear relationship between the N concentration in spring-grown, autumn-sampled leaves and yield of 'Fuerte' trees on Mexican seedling rootstocks. A leaf N concentration of 1.8% gave the highest yield, which declined significantly either above or below this concentration. Embleton *et al.* (1968) later recommended that the N concentration of spring-grown, autumn-sampled 'Hass' leaves should not be less than 2.0%, as there was no evidence of reduced yield following high N (≈1800 g N tree^{-1} year^{-1}) applications. However, such rates would be uneconomical if yield was not increased (Table 11.11). Later studies (Lahav *et al.*, 1990; Koen and du Plessis, 1992) have largely endorsed these initially proposed concentrations, which are still widely accepted and used by avocado growers as a diagnostic tool for determining fertilizer requirements. However, when setting critical N, P and K concentrations for 'Fuerte' leaves in South Africa, Koen and du Plessis (1992) recommended higher concentrations for N (a range of 2.0–2.3%) and K (a range of 0.9–1.4%) than Embleton and Jones (1964), who recommended ranges for N and K of 1.6–2.0% and 0.75–2.0%, respectively. Following their studies, Koen and du Plessis (1992) also increased the lower critical leaf concentration of P from 0.08% to 0.17%. In Israel, the critical concentration of P was increased to 0.14% and K to 1.0%. The differences in critical N concentrations may relate to either differences in the physiological age and location of the leaf tissue sampled for nutrient analysis, sampling time, the environmental conditions in which the crop

Table 11.11. Standards of avocado leaf analysis for diagnosing the nutrient status of mature trees

Nutrient	Deficient (less than)	Commercial range	Excess (more than)
N (%)	1.60	1.6–2.8	3.0
P (%)	0.14	0.14–0.25	0.3
K (%)	0.9	0.9–2.0	3.0
Ca (%)	0.50	1.0–3.0	4.0
Mg (%)	0.15	0.25–0.80	1.0
S (%)	0.05	0.20–0.60	1.0
Mn (mg kg^{-1})	10–15	30–500	1000
Fe (mg kg^{-1})	20–40	50–200	?
Zn (mg kg^{-1})	10–20	40–80	100
B (mg kg^{-1})	10–20	40–60	100
Cu (mg kg^{-1})	2–3	5–15	25
Mo (mg kg^{-1})	0.01	0.05–1.0	?
Cl (%)	?	–	0.25–0.50
Na (%)	?	–	0.25–0.50

is grown (semi-arid Mediterranean vs humid subtropical), or average crop load. A higher concentration for P is also supported by Tracy (1985), who reported that 'Hass' avocados in California were P deficient when leaf P concentrations in spring-grown, early autumn-sampled leaves were less than 0.10%.

In California, new optimum nutrient levels for hardened summer-flush leaves sampled in autumn were proposed by Crowley *et al.* (2016) and Spann (2016), who surveyed data from production areas in Southern California in order to model the relationships between leaf nutrient concentrations and the yields. Newett *et al.* (2018) reviewed the new optimum ranges and compared them with the levels still in use in Australia. Crowley's upper limit (0.15%) for his suggested new optimum range for leaf P compares closely with Lahav *et al.* (2013) who state that no phosphorus fertilizer should be applied when leaf phosphorus is above 0.14%. Similarly, for K, Lahav *et al.* (2013) recommend not to apply K fertilizer to the soil if leaf potassium is above 1.2%, while Crowley *et al.* (2016) suggested that the upper limit for leaf K should be lowered from 2.0% to 0.9%. These authors also reported that excessive applications of N and K to the soil increase nutrient levels in the trees beyond the optimum values, causing nutrient imbalances between N and K that may be implicated with AB and losses in production.

Nutrient concentrations (on a dry weight basis) were determined from expanded mature leaves from non-fruiting and non-flushing terminals on the spring shoot growth. The leaves were normally 5–7 months old. Data are based on Embleton and Jones (1964), Embleton *et al.* (1968), Lahav and Kadman (1980), Reuter and Robinson (1986) and Whiley *et al.* (1996).

Factors affecting leaf mineral concentrations

A number of factors, independent of nutrient supply, influence the concentration of minerals in leaves. These are position of leaves on the tree, rainfall and irrigation, crop load ('on' or 'off' year), girdling, cultivars, rootstocks, diseases, nutrient and pesticide residues on leaves. The significance of each of these factors was assessed in several countries and the magnitude of their effects on leaf nutrient concentrations is related to local conditions. For additional information, see Lahav and Whiley (2002) and Newett *et al.* (2018).

According to Lovatt (2013a,b), the multiple factors that influence uptake of soil-applied nutrients make it difficult to predict when and how much is taken up over a given period of time. Moreover, the amount of fertilizer being leached past the root zone with each irrigation or rain event is unknown. Hence, it is difficult to know if the tree's nutritional needs are being met by soil-applied fertilization during periods of high nutrient demand, especially if they occur when existing soil conditions restrict nutrient uptake by roots (low soil temperatures in winter or periods of soil hypoxia).

In Australia, Dann *et al.* (2016) reported on the effect of rootstock on fruit mineral content and the susceptibility of fruit flesh body rots. Rootstock significantly affected marketability of fruit in 58% of the fruit samples assessed at different regions, and 'Hass' fruit with higher postharvest quality were harvested from trees grafted to Guatemalan and West Indian rootstocks, e.g. 'Velvick'. Fruit quality was frequently poor from trees grafted to Mexican race rootstocks regardless of growing location. Fruit from rootstocks with superior fruit quality were often associated with lower N and higher Ca skin concentrations. The N:Ca ratios in the skin of unripe avocado fruit provide the useful indicators of potential postharvest disease in ripe fruit, and these nutrients need to be managed accordingly (see *Fertilizer frequency and timing* section, this chapter).

In a young, high-density avocado orchard in Israel, Lazare *et al.* (2021) found large variations in leaf nutrient content in 'Hass' and 'Ettinger' trees across 20 rootstocks, with different genetic/racial origins (West Indian or Mexican), as well as between clonal or seedling rootstocks of the same cultivar. In both cultivars, leaf mineral composition was analysed and found to be correlated with crop load.

Crop load is an important factor driving the uptake of nutrients by avocado trees. The greater the number of fruit set, the greater the uptake of nutrients (Rosecrance *et al.*, 2012). Therefore, fertilization programmes in commercial orchards are commonly adjusted once fruit set is known (T. Cantuarias-Avilés, Piracicaba, Brazil, 2024, personal communication). In Brazil, Cazetta *et al.* (2023) found significant differences in leaf nutrient content among the 'Fortuna', 'Geada' and 'Quintal' local tropical cultivars (West Indian × Guatemalan hybrids), which showed large seasonal variations of foliar nutrient content, which is consistent with other studies.

Salinity

Avocado is widely recognized as a salt-sensitive species with injury commonly associated with excessive concentrations of Na and Cl (Haas, 1929; Ayers *et al.*, 1951; Kadman, 1963b, 1964; Downton, 1978; Chirachint and Turner, 1988, Crowley, 2008; Berkessa, 2020; Kourgialas and Dokou, 2021). Toxicity symptoms of salinity injury to avocado vary, depending on the stage of tree development, leaf age and nutrition (Bingham and Fenn, 1966; Bar *et al.*, 1987b, 1991). High salt concentrations in irrigation water cause toxicity due to excessive chloride and sodium retention in leaves, and nutritional imbalances of other minerals (Musyimi *et al.*, 2007; Castro *et al.*, 2009; Alvarez-Acosta *et al.*, 2018). Salt accumulation in the soil solution decreases soil-water potential, reducing root growth and water and nutrient uptake, and subsequently diminishes plant growth and avocado yields. As a direct consequence of these effects, avocado trees produce increased amounts of ethylene that cause root growth inhibition and initiation of leaf senescence in a feedback loop that finally leads to plant death (Bar *et al.*, 1987b; Barra *et al.*, 2017).

Differences in the severity of damage between production areas suggest that microclimatic factors may be important in the expression of toxicity symptoms (Chirachint and Turner, 1988). The most commonly observed symptoms are tip and marginal leaf burn, eventually resulting in premature leaf abscission (Ayers, 1950; Embleton *et al.*, 1962; Bingham and Fenn, 1966). The threshold of salinity damage to avocado trees grafted onto Mexican and West Indian race rootstocks is estimated to be 70–100 and 300–400 mg kg^{-1} leaf Cl, respectively (Steinhardt *et al.*, 1995a,c). An average yield loss of 12% resulted from an increase of 35 mg kg^{-1} in Cl above the identified threshold (Steinhardt *et al.*, 1995b). In addition, Downton (1978) observed that high NaCl concentrations (20 mM) in the potting medium reduced the numbers of floral buds, flowers per bud, inflorescences per bud and flowers per inflorescence of 'Fuerte' scions on Mexican and Guatemalan rootstocks.

The effect of salinity on leaf mineral content is summarized in Table 11.12. A number of studies demonstrated that increased nitrate N applications to avocado trees reduce leaf Cl concentration and damage to the leaves (Haas, 1928, 1929; Haas and Brusca, 1955; Embleton *et al.*, 1958a; Moore and Richards, 1958). In potted Mexican and West Indian race seedlings, Bar *et al.* (1987b, 1991) reported that when trees were supplied with Cl at 2, 4, 8 or 16 mM, they developed toxicity symptoms that were most severe at 16 mM of Cl. However, when nitrate at 16 mM was added to the potting medium, no Cl damage was observed as the plants preferentially took up nitrate. They proposed that where Cl is a problem, the use of potassium nitrate (or ammonium nitrate) as a fertilizer may be a practical and simple way to reduce salinity damage to trees.

Crowley (2008) draws attention to the need to balance increased water use efficiency with

Table 11.12. Effect of N, P, K applications and salinity on the changes on concentrations of other nutrients in avocado leaves.

Nutrient affected	Nutrient applied N	P	K	Salinity
N	+	+	0/+	–
P	+	+	0	0–
K	±	–	0/?	+
Ca	±	+	–	
Mg	±	+	–	–/?
Na	+			
Cl	–	+	+	
Zn	±	–	–	+
Mn	+	±	+	–
Fe	+	–/?		
Cu	±	–	+	+
B	–	–	–	

+, increase; –, decrease; 0, no effect; ±, variable results; ?, non-significant effect. Data are based on Ruehle (1940); Haas (1947); Bingham and Martin (1956); Embleton *et al.* (1958a,b); Labanauskas *et al.* (1958a,b); Lahav *et al.* (1976, 1990); Lahav (1984); Limaco Sing and McNeil (1991); Zilberstaine *et al.* (1992); Jamie *et al.* (1992); Bar and Kafkafi (1992); Steinhardt *et al.* (1995a); Arpaia *et al.* (1996) and Loupassaki (1998).

the management of salinity. Strategies suggested are the selection of rootstocks that tolerate saline soils, EC monitoring of irrigation water and soil, timing of irrigation and the amount of water needed to leach salts below the root-zone. For further discussion on salinity, see Schaffer *et al.*, Chapter 7, this volume.

Lime-induced chlorosis

Avocado is one of the most susceptible fruit tree crops to iron (Fe) deficiency, with chlorosis symptoms appearing at soil pH above 7.0. These are usually alkaline soils with high $CaCO_3$ content (Manthey and Crowley, 1997) and can be aggravated by irrigation if aeration is limited (Wallihan *et al.*, 1968). Initially, alkaline soil reduces the availability of Fe to trees. Under wet soil conditions, air is displaced out of soil pores and root respiration depletes oxygen and increases CO_2 concentrations in the soil. The resulting increase in HCO^{-3} leads to an immobilization of Fe and chlorosis develops (Zude-Sasse and Schaffer, 2000). In poorly aerated soils, Fe deficiency is often overcome by reducing the amount of water applied or increasing the intervals between irrigation. It is generally necessary to apply Fe in a chelated form in calcareous soils to control lime-induced chlorosis (J.H. Crane, Homestead, Florida, 2024, personal communication). Controlling iron chlorosis with synthetic Fe chelates is costly, and their application can have adverse environmental impacts due to the risks of leaching and potential groundwater contamination (Rombolà and Tagliavini, 2006).

In a commercial 'Hass' avocado orchard planted in a calcareous soil in Chile, Granja and Covarrubias (2018) investigated the effectiveness of different acidifying nitrogen and nitrification inhibitors to control Fe chlorosis. They found that fertilization with a combination of ammonium sulfate and DMPP or urea soil application can be an effective and sustainable alternative to synthetic Fe chelates, for preventing Fe deficiency in avocado. Alternatively, Kadman and Lahav (1982) were able to control lime-induced chlorosis in commercial avocado orchards with soil applications of FeEDDHA iron chelate and by direct pressure injections of a 0.5–1% FeEDDHA solution into the trunk or main branches of severely chlorotic mature trees. The latter resulted in rapid re-greening of leaves and full recovery within 7–10 days following injection. However, foliar applications of several Fe formulations (including FeEDDHA) did not improve Fe chlorosis, most likely due to poor leaf absorption translocation to other parts of the tree (Kadman and Lahav, 1982).

Conclusions

The evolutionary history of avocado has shaped the species for adaption to cultivation over a wide range of edaphic and climatic conditions. However, there are three key features that place limits on the success of avocado as a commercial crop: the high sensitivity of avocado's shallow roots to hypoxia, which can develop in heavy clay soils, with rising water tables and excessive rainfall events; the high sensitivity of avocado to saline conditions; and its predisposition to prioritize vegetative growth when abundant nitrogenous nutrients are present (an adaptive feature for rainforest survival). Both water and nutrient management have been discussed at length in this chapter and offer insights on past and developing commercial strategies required for sustainable production of this crop.

Compared to other fruit tree crops, the overall productivity of avocado (tonnes per hectare) globally remains low, even when taking into account the biological cost of producing an energy-rich fruit with high calorific value. There are some exceptions where management has succeeded in consistently producing crops near to the estimated genetic potential of the species (Wolstenholme, 1987).

In most environments where avocado is grown, access to water for irrigation is essential, either year-round in semi-arid regions or to supplement rainfall in the moist, temperate and subtropical areas of production. Climate change, inducing periods of lack or excess of water, and water salinization, challenges avocado production in different parts of the world. Therefore, the avocado industry must incorporate new knowledge and technologies to improve irrigation management. Salination of irrigation water is an ongoing concern in semi-arid regions where changing weather patterns and urban competition are placing greater demands on a limited supply. To some extent this is being resolved through the development of rootstocks with improved salinity tolerance, and the gradual adoption of plant-based, smart water monitoring devices (sap flow meters, dendrometers and soil moisture probes) that result in more efficient water use. Additionally, effective use of soil moisture probes in orchards reduces the likelihood of overwatering and leaching nutrients past the active root zone into deep water drainage where pollution of groundwater is a likely outcome.

Knowledge of the role of rootstocks on nutrient uptake and scion transfer is slowly improving. Further research is required to identify genetic combinations that outperform their peers. In addition, large populations of avocado trees that are grafted to seedling rootstocks remain a valuable resource for future genetic improvement. Nitrogen is arguably the key mineral nutrient that influences avocado production since the band of optimal leaf content is narrow and yield markedly falls if the tree is not maintained within the identified range. A combination of both earlier and new research in this chapter has identified with confidence critical periods in the annual phenological changes, when to apply nitrogen to support tree demand without driving excessive vegetative growth. Similarly, application times to meet tree and fruit demand have been identified for other key nutrients (Ca and K). Refinement of the original critical leaf nutrient concentrations defined by Embleton and Jones (1964) through ongoing research is continuing, contributing to more efficient use of fertilizers.

With growing pressures being made on preserving our planet's environment, agricultural activities are coming under increasing scrutiny with legislation in many countries placing restrictions on outflows of contaminants from food production activities. The adoption of dripper and microsprinkler irrigation installed with fertigation capability improves efficiency of both water and nutrient use. Mulching orchard floors supplementing the organic input from leaf fall is becoming more common, conserving soil moisture and stabilizing soil nutrients, while improving general soil health and tree performance. New nutrition compounds are emerging in the form of biofertilizers, which in some cases improve tree health and yield, although not all products have been proven through rigorous scientific scrutiny. There remain further opportunities to improve avocado production through research with outcomes that may further increase the efficiency of limited water and nutrient resources.

References

Abbasi, F., Rezaee, H.T., Jolaini, M. and Alizadeh, H.A. (2012) Evaluation of fertigation in different soils and furrow irrigation regimes. *Irrigation and Drainage* 61, 533–541.

Abdel-Karim, H.A., Nehad, M.A., El-Rouby, K.M. and Roshdy, K.A. (2015) Effect of foliar application of boron and zinc on fruit set, yield and some fruit characteristics of Fuerte avocado. *Research Journal of Pharmaceutical, Biological and Chemical Sciences* 6, 443–449.

Abd-Rabou, F.A. (2006) Effect of microbien, phosphorine and effective microorganisms (EM) as bio-stimulants on growth of avocado and mango seedlings. *Egyptian Journal of Applied Science* 21, 673–693.

Abou Aziz, A.B., Desouki, I. and El-Tanahy, M.M. (1975) Effect of nitrogen fertilizer on yield and fruit oil content of avocado. *Scientia Horticulturae* 3, 89–94.

Adato, I. and Levinson, B. (1988) Influence of daily intermittent drip irrigation on avocado (cv Fuerte) fruit yield and trunk growth. *Journal of Horticultural Science* 63, 675–685.

Aguilar-Melchor, J.J., López-Jiménez, A., Cortes-Flores, J.I., Samano-Guevara, R.A. and Martínez-Garza, A. (1997) Efecto de la fertilización orgánica e inorgánica en el crecimiento, producción y estado nutrimental del aguacate. Memoria Fundación Salvador Sánchez Colin, CICTAMEX S.C., Coatepec Harinas, Mexico.

Aguilera, L. and Salazar, S. (1996) Efecto del Nitrógeno, Fósforo y Potasio en el rendimiento y tamaño del fruto de aguacate. INIFAP, CIRPAC, Campo Experimental Uruapan, Michoacán, México.

Allen, R.G., Pereira, L.S. and Raes, D. (1998a) *Crop Evapotranspiration-Guidelines for Computing Crop Water Requirements.* FAO Irrigation and Drainage Paper 56. FAO, Rome, 300(9): D05109.

Allen, R.G., Pereira L.S., Raes D., Smith M. (1998b) Crop evapotranspiration. Guidelines for computing crop water requirements. FAO Irrigation and Drainage Paper 56, 1–333.

Allen, R.G., Pereira, L.S. and Raes, D. and Smith, M. (2006) *Crop Evapotranspiration - Guidelines for Computing Crop Water Requirements* - FAO Irrigation and Drainage Paper 56. FAO, Rome, Italy.

Alvarez-Acosta, C., Marrero-Dominguez, A., Gallo-Llobet, L. and Gonzalez-Rodriguez, A.M. (2018) Physiological response of selected avocados (*Persea americana*) subjected to NaCl and $NaHCO_3$ stress. *Scientia Horticulturae* 237, 81–88. doi:10.1016/j.scienta.2018.04.010.

Anonymous (2001) Avocado Ecofresh Protocol. Based on AG 9000 Specifications and Eurep GAP Requirements. *Agrexco, The Israeli Avocado Association and the Ministry of Agriculture, Israel.* (In Hebrew.)

Anonymous (2009) Guidelines for managing under limited water resources – avocados. Avocados Australia, Woolloongabba, Qld, Australia. Available at: http://www.avocado.org.au (accessed 21 May 2024).

Antúnez, A., Rojas, V., Gil, P.M., Cea, D. and Ferreyra, R. (2019) Improving water productivity of avocado (*Persea americana* Mill. var. Hass) by using a reinforced plastic cover over the canopy at Petorca Valley, Region of Valparaíso, Chile. In: *Proceedings of the IX International Symposium on Irrigation of Horticultural Crops, 17–20 Junio 2019, Matera, Italia*.

Apaza, W., Quiroz, P. and Julca-Otiniano, A. (2019) Characterisation of avocado and asparagus farms in the Chavimochic irrigation project in La Libertad, Peru. *Peruvian Journal of Agronomy*3(3), 91–103. 10.21704/pja.v3i3.1342

Araya Vargas, J., Gil, P., Meza, F., Yáñez, G., Menanno, G., *et al.* (2021) Soil electrical resistivity monitoring as a practical tool for evaluating irrigation systems efficiency at the orchard scale: a case study in a vineyard in Central Chile. *Irrigation Science* 39(1–2), 1–21.

Arpaia, M.L., Meyer, J.L., Witney, G.W., Bender, G.S., Stottlemeyer, D.S. and Robinson, P.R. (1996) The Cashin Creek nitrogen fertilizer trial – what did we learn? *California Avocado Society Yearbook* 80, 85–98.

Assouline, S., Narkis, K., Gherabli, R. and Sposito, G. (2016) Combined effect of sodicity and organic matter on soil properties under long-term irrigation with treated wastewater. *Vadose Zone Journal* 15(4), 1–10.

Assouline, S. and Narkis, K. (2013) Effects of long-term irrigation with treated wastewater on the root zone environment. *Vadose Zone Journal* 12(2), 1, vzj2012.0216–10. doi:10.2136/vzj2012.0216.

Ataucusi, S. (2015) Manejo técnico del cultivo del palto. Primera Edición – Cáritas del Perú. PRA-Buenaventura. Lima, Perú. 41 p. Available at: https://www.slideshare.net/slideshow/manual-palta-74682172/74682172 (accessed 23 May 2024).

Atay, E., Crété, X., Loubet, D. and Lauri, P. E. (2021) Effects of different crop loads on physiological, yield and fruit quality of 'Joya™' apple trees: high crop load decreases maximum daily trunk diameter and does not affect stem water potential. *International Journal of Fruit Science* 21(1), 955–969. doi:10.1080/15538362.2021.1951922.

Avilan, R.L., Chirinos, A. and Figueroa, M. (1979) Quantification of some minerals extracted from the soil by an avocado crop. *Proceedings of the Tropical Region of the American Society for Horticultural Science* 27, 108–113.

Ayala-Arreola, J., Barrientos-Priego, A.F., Colinas-León, M.T., Sahagún-Castellanos, J. and Reyes-Alemán, J.C. (2010) Relaciones injerto-interinjerto y características anatómicas y fisiológicas de la hoja de cuatro genotipos de aguacate. *Revista Chapingo. Serie Horticultura* 16(2), 147–154.

Ayers, A.D. (1950) Salt tolerance of avocado trees grown in culture solutions. *California Avocado Society Yearbook* 35, 139–148.

Ayers, A.D., Aldrich, D.G. and Coony, J.J. (1951) Sodium and chloride injury of Fuerte leaves. *California Avocado Society Yearbook* 36, 174–178.

Balderas-Alba, A., Luna-Esquivel, G. and Vega-Frutis, R. (2019) Arbuscular mycorrhizal colonization in avocado orchards with two different farm management practices. *Interciencia* 44, 649–652. Available at: https://www.redalyc.org/journal/339/33962174007/html/ (accessed 21 May 2024).

Bañuelos, J., Trejo, D., Lara, L., Gavito, M. and Carreón, Y. (2013) Effects of seven different mycorrhizal innoculum in *Persea americana* in sterile and non-sterile soil. *Tropical and Subtropical Agroecosystems* 16, 423–429.

Bar, Y. and Kafkafi, U. (1992) Nitrate-induced iron-deficiency chlorosis in avocado (*Persea americana* Mill.) rootstocks and its prevention by chloride. *Journal of Plant Nutrition* 15, 1739–1746.

Bar, Y., Lahav, E. and Kalmar, D. (1987a) Seasonal changes in nitrogen concentration in avocado leaves associated with leaf age and fertilizer regime. *South African Avocado Growers' Association Yearbook* 10, 57–58.

Bar, Y., Kafkafi, U. and Lahav, E. (1987b) Nitrate nutrition as a tool to reduce chloride toxicity in avocado. *South African Avocado Growers' Association Yearbook* 10, 47–48.

Bar, Y., Kafkafi, U. and Lahav, E. (1991) Reducing chloride toxicity in avocado by nitrate. In: Lovatt, C., Holthe, P.A. and Arpaia, M.L. (eds) *Proceedings of the Second World Avocado Congress*. University of California, Riverside, California, 1:373.

Bárcenas, A., Almaraz, C., Reyes, L., Varela, L., Lara, B., *et al.* (2007) Diversity of arbuscular mycorrhizal fungi on avocado orchards from Michoacan. In: *Proceedings of the VI World Avocado Congress*, Viña del Mar, Chile. Available at: https://www.researchgate.net/publication/241385598_Diversity_of_arbuscular_mycorrhizal_fungi_on_Avocado_orchards_from_Michoacan (accessed 1 June 2024).

Bard, Z.J. (1997) Soil boron applications for the control of boron deficiency in the avocado (*Persea americana* Mill.). MSc thesis, University of Natal, Pietermaritzburg, South Africa.

Bard, Z.J. and Wolstenholme, B.N. (1997) Soil boron application for the control of boron deficiency in KwaZulu-Natal avocado orchards. *South African Avocado Growers' Association Yearbook* 20, 13–15.

Bard, Z.J. and Wolstenholme, B.N. (1998) Soil boron application for control of boron deficiency in the avocado in the Kwa-Zulu-Natal midlands. *South African Avocado Growers' Association Yearbook* 21, 22–25.

Barlow, C. (2000) *The Ghosts of Evolution: Nonsensical Fruit, Missing Partners, and Other Ecological Anachronisms*. Basic Books, New York.

Barnard, R.O. (1989) Cation distribution during soil profile amelioration with lime and gypsum. *South African Avocado Growers' Association Yearbook* 12, 43–47.

Barnard, R.O., Cillie, G.E.B. and Kotzé, J.M. (1991) Deficiency symptoms in avocados. *South African Avocado Growers' Association Yearbook* 14, 67–71.

Barra, P.J., Inostroza, N.G., Mora, M.L., Crowley, D.E. and Jorquera, M.A. (2017) Bacterial consortia inoculation mitigates the water shortage and salt stress in an avocado (*Persea americana* Mill.) nursery. *Applied Soil Ecology* 111, 39–47. doi:10.1016/j.apsoil.2016.02.014.

Bausch, W.C. (1995) Remote sensing of crop coefficients for improving the irrigation scheduling of corn. *Agricultural Water Management* 27, 55–68.

Bender, G. (1997) Bloom sprays may improve fruit set. *California Grower* 21(5), 17–18.

Bender, G.S., Casale, W.L. and Rahimian, M. (1992) Use of worm-composted sludge as a soil amendment for avocados in phytophthora-infested soil. In: Lovatt, C., Holthe, P.A. and Arpaia, M.L. (eds) *Proceedings of the Second World Avocado Congress*. University of California, Riverside, California, 1, pp. 143.

Ben-Ya'acov, A. (1972) Effect of potassium nitrate, zinc sulfate and magnesium sulfate on avocado trees growing in light soil. In: *Preliminary Report No. 732*. Agricultural Research Organisation, Bet Dagan, Israel. (In Hebrew with English summary.)

Ben-Ya'acov, A. (1992) Recommended rootstocks for new planting of avocado-"Ettinger" cultivar (In Hebrew). *Alon Hanotea* 46, 919–926.

Ben-Ya'acov A. and Michelson, E. (1995) Avocado rootstocks. In: Janick, J. (ed.) *Horticultural Reviews* 17, 381–429. John Wiley and Sons, Inc. New York, NY.

Ben-Ya'acov, A., Michelson, E. and Sela, I. (1992) Field study of avocado seeded rootstocks and scion sources: final report of experiments planted as commercial orchards 1. Investigation of rootstocks, scion and manure at Bene-Deror Fuerte orchard planted in 1969. *Alon Hanotea* 46, 769–782. (In Hebrew with English summary.)

Berkessa, A.J. (2020) Salinity and avocado production, a review. *International Journal of Forestry and Horticulture* 6, 32–38. doi:10.20431/2454-9487.0601004.

Bernstein, N., Meiri, A. and Zilberstaine M. (2004) Root growth of avocado is more sensitive to salinity than shoot growth. *Journal of the American Society for Horticultural Science* 129, 188–192. doi:10.21273/JASHS.129.2.0188.

Beyer, C., Cuneo, I., Alvaro, J.E. and Pedreschi, R. (2021) Evaluation of aerial and root plant growth behavior, water and nutrient use efficiency and carbohydrate dynamics for Hass avocado grown in a soilless and protected growing system. *Scientia Horticulturae* 277, 109830. doi:10.1016/j.scienta.2020.109830.

Bingham, F.T. (1961) Seasonal trends in nutrient composition of Hass avocado leaves. *Proceedings of the American Society for Horticultural Science* 78, 149–160.

Bingham, F.T. and Fenn, L.B. (1966) Chloride injury to Hass avocado trees: a sand culture experiment. *California Avocado Society Yearbook* 50, 99–106.

Bingham, F.T. and Martin, J.P. (1956) Effects of soil phosphorus on growth and minor element nutrition of citrus. *Proceedings of the American Society for Soil Science* 20, 382–385.

Bingham, F.T. and Nelson, C.O. (1970) The effects of sodium on mature avocado trees. *California Avocado Society Yearbook* 54, 75–78.

Bingham, F.T. and Richards, S.J. (1958) Effects of irrigation treatments and rates of nitrogen fertilization on young avocado trees. III. Changes in soil chemical properties. *Proceedings of the American Society for Horticultural Science* 71, 305–309.

Blanke, M.M. (1992) Photosynthesis of avocado fruit. In: *Proceedings of the Second World Avocado Congress*, Anaheim, USA, pp. 179–189.

Bojović, B. and Marković, A. (2009) Correlation between nitrogen and chlorophyll content in wheat (*Triticum aestivum* L.). *Kragujevac Journal of Science* 31, 69–74.

Borys, M.W. (1986) Root/shoot relations and some root characteristics in seedlings of avocado and Chinini. *California Avocado Society Yearbook* 70, 175–198.

Borys, M.W. (1989) Potential of various root forms identified in avocado (*Persea americana* Mill.) and *Persea schiedeana* Nees. in Mexico. *Revista Chapingo* 13–16, 23–30. (In Spanish.)

Borys, M.W., Barrera-Guerra, J.L. and Luna-Lopez, J. (1985) Some root characteristics of avocado (*Persea americana* Mill.) seedlings of the West Indian and Guatemalan races. *California Avocado Society Yearbook* 69, 111–122.

Bower, J.P. (1985a) Some aspects of water relations on avocado (*Persea americana* Mill.) tree and fruit physiology. PhD thesis, University of Natal, Pietermaritzburg, South Africa.

Bower, J.P. (1985b) The calcium accumulation pattern in avocado fruit as influenced by long-term irrigation regime. *South African Avocado Growers' Association Yearbook* 8, 97–99.

Bower, J.P. and Cutting, J.G.M. (1988) Avocado fruit development and ripening physiology. *Horticultural Reviews* 10, 229–261.

Bower, J.P., Wolstenholme, B.N. and de Jager, J.M. (1977) Incoming solar radiation and internal water status as stress factors in avocado *Persea americana* (Mill.) cv. Edranol. In: *Proceedings of the Technical Committee, South African Avocado Growers' Association*, pp. 35–40.

Brito, C., Dinis, L.T., Moutinho-Pereira, J. and Correia, C. (2019) Kaolin, an emerging tool to alleviate the effects of abiotic stresses on crop performance. *Scientia Horticulturae* 50, 310–316. doi:10.1016/j.scienta.2019.02.070.

Broadbent, P., Trochoulias, T., Baigent, D.R., Abbott, T.S. and Dettmann, E.B. (1989) Effect of soil management on avocados in a krasnozem soil. *Scientia Horticulturae* 38, 87–104.

Brusca, J.N. and Haas, A.R.C. (1955) Effect of molybdenum on avocado trees. *California Avocado Society Yearbook* 39, 175–178.

Burgis, D.S. and Wolfe, H.S. (1945) Do avocado roots develop root-hairs. *Proceedings of the Florida State Horticultural Society* 58, 197–198.

Bustan, A., Dag, A., Yermiyahu, U., Erel, R., Presnov, E., *et al.* (2016) Fruit load governs transpiration of olive trees. *Tree Physiology* 36, 380–391. doi:10.1093/treephys/tpv138.

Cameron, S.H., Mueller, R.T. and Wallace, A. (1952) Nutrient composition and seasonal losses of avocado trees. *California Avocado Society Yearbook* 36, 201–209.

Campisi-Pinto, S., Zheng, Y., Rolshausen, P.E., Crowley, D.E., Faber, B., *et al.* (2017) Optimal nutrient concentration ranges of 'Hass' avocado cauliflower stage inflorescences-potential diagnostic tool to optimize tree nutrient status and increase yield. *HortScience* 52, 1707–1715.

Campos-Mariscal, J.L., Álvarez-Sánchez, M.E., Maldonado-Torres, R. and Almaguer-Vargas, G. (2020) Application of organic fertilizers in yield and root development in avocado cultivation. *Revista Mexicana de Ciencias Agrícolas* 11, 263–274.

Cantuarias, A.T., Silva, S.R., Figueiredo, A.S., Brogio, B.A., Baptista, G.E. and Micheletti, B.L. (2019) Water status and productivity of 'Hass' avocado trees in response to supplemental irrigation during winter. *Pesquisa Agropecuária Brasileira* 54: e00237, 1–7.

Cantuarias-Aviles, T.E. (1995) Transpiration rate and water status of a mature avocado orchard as related to wetted soil volume. MSc thesis, Hebrew University of Jerusalem, Rehovot, Israel.

Cardemil, G. (1999) Aproximación a los requerimientos hídricos del palto (*Persea americana* Mill) cv. Hass, para la provincia de Quillota. Taller de titulación, Pontificia Universidad Católica de Valparaíso.

Carr, M.K.V. (2013) The water relations and irrigation requirements of avocado (*Persea americana* Mill.): a review. *Experimental Agriculture* 49(2), 256–278. doi:10.1017/S0014479712001317.

Carreón-Abud, Y., Vega-Fraga, M. and Gavito, M.E. (2015) Interaction of arbuscular mycorrhizal inoculants and chicken manure in avocado rootstock production. *Journal of Soil Science and Plant Nutrition* 15, 867–881. doi:10.4067/S0718-95162015005000060.

Castro, A.E., Chávez, B.A.T., Garcia, S.P.A., Reyes, R.L. and Bárcenas, O.A.E. (2013) Effect of mycorrhizal inoculants in the development of Mexican landrace avocado rootstocks. *Tropical and Subtropical Agroecosystems* 16, 407–413.

Castro, M., Iturrieta, R. and Fassio, C. (2009) Rootstock effect on the tolerance of cv. Hass avocado plants to NaCl stress. *Chilean Journal of Agricultural Research* 69, 316–324. doi:10.4067/s0718-58392009000300003.

Cazetta, J.O., Zanata, R.M.P. and Barbosa, P.W.G. (2023) Seasonal nutrient content variation in avocado leaves of 'Geada', 'Fortuna' and 'Quintal' genotypes. *Revista Brasileira de Fruticultura* 45: e-425. doi:10.1590/0100-29452023425.

Cea, D., Gil P., Bonomelli C. and Martiz, J. (2019) Evaluación de la aplicación de silicatos como medida de mitigación en palto (*Persea americana* Mill) cv. Hass bajo estrés hídrico. In: *Proceeedings of the IX World Avocado Congress*, Medellín, Colombia.

Celedón, J.M., Gil, P.M., Ferreyra, R., Maldonado, P. and Barrera, C. (2012) Sensitivity and variability of two plant water stress indicators: exploring criteria for choosing a plant monitoring method for avocado irrigation management. *Chilean Journal of Agricultural Research* 72(3), 379–387. doi:10.4067/S0718-58392012000300012.

Celis, N., Suarez, D.L., Wu, L., Li, R., Arpaia, M.L. and Mauk, P. (2018) Salt tolerance and growth of 13 avocado rootstocks related best to chloride uptake. *HortScience* 53, 1737–1745. doi:10.21273/HORTSCI13198-18.

Chabot, B.E. and Hicks, D.J. (1982) The ecology of leaf life spans. *Annual Reviews of Ecology Systems* 13, 229–259.

Charpentier, J.M., Martin Prével, P. and Lacoeuilhe, J.I. (1967) Etude des carences minérales chez l'avocatier. *Fruits* 22, 213–233. (In French.)

Chartzoulakisa, K., Patakasb, A., Kofidisc, G., Bosabalidisc, A. and Nastoub, A. (2002) Water stress affects leaf anatomy, gas exchange, water relations and growth of two avocado cultivars. *Scientia Horticulturae* 95, 39–50.

Chen, H., Morrell, P.L., Ashworth, V.E., de la Cruz, M. and Clegg, M.T. (2009) Tracing the geographic origins of major avocado cultivars. *Journal of Heredity* 100(1), 56–65. doi:10.1093/jhered/esn068.

Chirachint, W. and Turner, D.W. (1988) Shade reduces the foliar symptoms of 'Fuerte' avocado affected by salt, without significantly changing the concentration of Na, K or Cl in the leaves. *Scientia Horticulturae* 36, 1–15.

Clarkson, D.T. (1984) Calcium transport between tissues and its distribution in the plant. *Plant, Cell & Environment* 7, 449–456.

Coetzer, L.A., Robbertse, P.J. and Janse van Vuuren, B.P.H. (1993) The role of boron in avocados: theory, practice and reality. *South African Avocado Growers' Association Yearbook* 16, 2–4.

Coetzer, L.A., Robbertse, P.J., Barnard, R.O. and Tomer, E. (1994) Uptake and transport of boron in avocado seedlings. *South African Avocado Growers' Association Yearbook* 17, 95–98.

Coit, J.E. (1928) Pests and diseases of the avocado. *California Avocado Growers' Association Yearbook* 1928, 18–21.

Cooper, W.C. and Gorton, B.S. (1950) Relation of leaf composition to leaf burn of avocados and other subtropical fruits. *Texas Avocado Society Yearbook* 3, 32–38.

Cossio-Vergas, L.E., Salazar-Garcia, S. and Gonzáles-Durán, I.J.L. (2009) Respuesta del aguacate 'Hass' a la fertilización con boro en huertos sin riego. *III Congreso Latinoamericano del Aguacate*, Medellín, Colombia, pp. 4–17.

Cowan, A.K., Cripps, R.F., Richings, E.W. and Taylor, N.J. (2001) Fruit size: towards an understanding of the metabolic control of fruit growth using avocado as a model system. *Physiologia Plantarum* 111, 127–136. doi:10.1034/j.1399-3054.2001.1110201.x.

Crane, J.H., Balerdi, C.F. and MaGuire, I. (2007a) Avocado growing in the Florida home landscape. In: *University of Florida, IFAS Cooperative Extension Service Circular 1034*, Gainesville, Florida pp. 14.

Crane, J.H., Schaffer, B., Li, Y., Evans, E., Montas, W. and Li, C. (2007b) Effect of foliarly-applied acids and ferrous sulfate on iron nutrition of avocado trees. In: *Proceedings of the VI World Avocado Congress*, Viña Del Mar, Chile, pp. 1–13.

Crowley, D.E. (2008) Salinity management in avocado orchards. *California Avocado Society Yearbook* 91, 83–104.

Crowley, D.E. (2011) Fertilizers programs for avocado production. Index Fresh Seminar Series. Available at: https://www.indexfresh.com/wp-content/uploads/2011/02/seminar-presentations_fertilization-mulching-1.pdf (accessed 21 May 2024).

Crowley, D.E. Campisi, S., Lovatt, C., Rolshausen, P., Arpaia, M., Dinar, A. and Escalera, J. (2016) Decision support tools for management of avocado nutrition and chloride toxicity, final report. Available at: https://www.californiaavocadogrowers.com/sites/default/files/DST-Final-Report-Crowley-2016_1.pdf (accessed 16 October 2024).

Crowley, D.E., Smith, W., Faber, B. and Manthey, J.A. (1996) Zinc fertilization of avocado trees. *HortScience* 31, 224–229.

da Silva, S.R., Cantuarias-Avilés, T.E., Chiavelli, B., Martins, M.A. and Oliveira, M.S. (2017) Phenological models for implementing management practices in rain-fed avocado orchards. *Pesquisa Agropecuária Tropical* 47, 321–327. doi:10.1590/1983-40632016v4747140.

da Silva, J.R., Rodrigues, W.P., Ferreira, L.S., Bernado, W.D., Paixao, J.S., *et al.* (2018) Deficit irrigation and transparent plastic covers can save water and improve grapevine cultivation in the tropics. *Agricultural Water Management* 202, 66–80. doi:10.1016/j.agwat.2018.02.013.

da Silveira, S.V., Souza, P.V.D. de and Koller, O.C. (2002) Efeito de fungos micorrízicos arbusculares no desenvolvimento do abacateiro. *Pesquisa Agropecuária Brasileira* 37, 1597–1604.

Daniel, A.I., Fadaka, A.O., Gokul, A., Bakare, O.O., Aina, O. *et al.* (2022) Biofertilizer: the future of food security and food safety. *Microorganisms* 10(6), 1220. doi:10.3390/microorganisms10061220.

Dann, E.K., Coates, L.M., Pegg, K.G., Dean, J.R., Cooke, A.W. *et al.* (2016) Rootstock selection, nitrogen and calcium influence postharvest disease in avocado. *Acta Horticulturae* 1120, 391–397.

DeJong, T.M. and Doyle, J.F. (1985) Seasonal relationships between leaf nitrogen content (photosynthetic capacity) and leaf canopy light exposure in peach (*Prunus persica*). *Plant, Cell & Environment* 8(9), 701–706.

Delfín, V., Bonales, J. and Rocha, C.C. (2018) Competitividad internacional del aguacate orgánico en Michoacán, un estudio basado en la metodología Partial Least Squares Odette. *Revista Nicolaita de Estudios Económicos* 12, 119–139.

Denslow, J.S. (1987) Tropical rainforest gaps and tree species diversity. *Annual Review of Ecology and Systematics* 18, 431–451.

Dimkpa, C., Bindraban, P.S., McLean, J., Gatere, L., Singh, U. and Hellums, D. (2017) Methods for rapid testing of plant and soil nutrients. In: Lichtfouse, E. (ed) *Sustainable Agriculture Reviews. Vol.*25. Springer, New York, NY, USA, pp. 1–43.

Dirou, J. and Huett, D. (2001) Crop nutrient replacement for avocado. *Talking Avocados* 12(1), 25–27.

Dixon, J., Elmsly, T.A., Dixon, E.M., Mandemaker, A. and Pak, H.A. (2007) Factors influencing fruit set of Hass avocados in New Zealand. In: *Proceedings of the VI World Avocado Congress*, Viña Del Mar, Chile, pp. 1–11.

Dixon, J., Smith, D.B., Elmsly, T.A. and Fields, F.P. (2005) The impact of foliar applications of nitrogen and boron on 'Hass' avocado fruitset in 2004. *New Zealand Avocado Growers' Association Annual Research Report* 5, 27–34.

Dixon, J., Elmsly, T.A. and Dixon, E.M. (2006) Seasonal variations in leaf mineral content. *New Zealand Avocado Growers' Association Annual Research Report* 6, 21–34.

Donoso, G. and Franco, G. (2013) La huella hídrica agrícola de Chile. Agronomía y Forestal n°47:10–15. Facultad de Agronomía e Ingeniería Forestal. Pontificia Universidad Católica de Chile, Santiago.

Donoso, G., Franco, E.G. and Blanco, J.L. (2016) Water footprints and irrigated agricultural sustainability: the case of Chile. *International Journal of Water Resources and Development* 32, 738–748.

Dorado, D., Grajales, L. and Rebolledo, A. (2017) Requerimientos hídricos del cultivo de aguacate (*Persea americana*) variedad Hass en zonas productoras de Colombia. *Corporación Colombiana de Investigación Agropecuaria (Corpoica)*. doi:10.21930/agrosavia.manual.7402322.

Downton, W.J.S. (1978) Growth and flowering in salt-stressed avocado. *Australian Journal of Agricultural Research* 29, 523–534.

Duchemin, B., Hadria, R., Erraki, S., Boulet, G., Maisongrande, P., *et al.* (2006) Monitoring wheat phenology and irrigation in Central Morocco: on the use of relationships between evapotranspiration, crops coefficients, leaf area index and remotely-sensed vegetation indices. *Agricultural Water Management* 79, 1–27.

Du Plessis, S.F. (1991) Factors important for optimal irrigation scheduling of avocado orchards. *South African Avocado Growers' Association Yearbook* 14, 91–93.

Du Plessis, S.F. and Koen, T.J. (1987) Comparison of different calcium sources on avocado production. *South African Avocado Growers' Association Yearbook* 10, 49–51.

Duff, J.D. and Firrell, M.C. (2021) Biofumigation: a cover crop option 12 months of the year to manage three soilborne pathogens ailing the Australian vegetable industry. *Global Journal of Agricultural Innovation Research and Development* 8, 104–116. doi:10.15377/2409-9813.2021.08.8.

Durán-Zuazo, V.H.D., Lipan, L., Sendra, E., Tarifa, D.F., Nems, A. *et al.* (2021) Impact of deficit irrigation on fruit yield and lipid profile of terraced avocado orchards. *Agronomy for Sustainable Development* 41:69. doi:10.1007/s13593-021-00731.

Eaton, F.M. (1944) Deficiency, toxicity and accumulation of boron in plants. *Journal of Agricultural Research* 69, 237–277.

Elmetwalli, A. (2019) Influence of irrigation method and fertilization type on avocado yield and quality. *Misr Journal of Agricultural Engineering* 36, 141–156.

El-Shamma, M.S., Mona, E.M., Helal, M.A., Maksoud, F., Khalil, H. and Mansour, E.M. (2017) Effect of some bio-stimulants on nutritional status, yield and fruit quality of avocados. *Middle East Journal of Agriculture Research* 6, 692–699.

Eman, A.A., El-Moniem, A., Thabet, A.Y.I., Abdelaziz, A.M.R.A., Baiea, M.H.M. and Amin, O.A. (2022) Reducing chemical fertilizers partially by using natural alternative sources of organic fertilizers and its impact on 'Hass' avocado trees. *Egyptian Journal of Chemistry* 65, 1255–1264. Available at: https://www.researchgate.net/publication/360865992_Reducing_chemical_fertilizers_partially_by_using_natural_alternative_sources_of_organic_fertilizers_and_its_impact_on_Hass_avocado_trees (accessed 3 June 2024).

Embleton, T.W. and Jones, W.W. (1964) Avocado nutrition in California. *Proceedings of the Florida State Horticultural Society* 77, 401–405.

Embleton, T.W. and Jones, W.W. (1966) Avocado and mango nutrition. In: Childers, N.F. (ed.) *Temperate to Tropical Fruit Nutrition*. Horticultural Publications, Rutgers – The State University, New Jersey, pp. 51–76.

Embleton, T.W. and Wallihan, E.F. (1966) Soil applications of zinc for avocados. *California Avocado Society Yearbook* 50, 87–93.

Embleton, T.W., Garber, M.J., Jones, W.W. and Richards, S.J. (1958a) Effects of irrigation treatments and rates of nitrogen fertilization on young Hass avocado trees. IV. Macronutrient content of leaves. *Proceedings of the American Society for Horticultural Science* 71, 310–314.

Embleton, T.W., Jones, W.W. and Garber, M.J. (1959) Curvilinear relationship between leaf nitrogen and yield of Fuerte avocados. *Proceedings of the American Society for Horticultural Science* 74, 378–382.

Embleton, T.W., Jones, W.W., Garber, M.J. and Boswell, S.B. (1968) Nitrogen fertilization of the Hass avocado. *California Avocado Society Yearbook* 52, 131–134.

Embleton, T.W., Jones, W.W., Kirkpatrick, J.D. and Gregory-Allen, D.G. (1958b) Influence of sampling date, season and fertilization on macronutrients in Fuerte avocado leaves. *Proceedings of the American Society for Horticultural Science* 72, 309–320.

Embleton, T.W., Matsumura, M., Storey, W.B. and Garber, M.J. (1962) Chloride and other elements in avocado leaves as influenced by rootstock. *Proceedings of the American Society for Horticultural Science* 80, 230–236.

Erazo-Mesa, E., Ramírez-Gil, J.G. and Sánchez, A.E. (2021) Avocado cv. Hass needs water irrigation in tropical precipitation regime: evidence from Colombia. *Water* 13(14), 1942. doi:10.3390/w13141942.

Erazo-Mesa, E., Echeverri-Sánchez, A.E. and Ramírez-Gil, J.G. (2022b) Advances in Hass avocado irrigation scheduling under digital agriculture approach. *Revista Colombiana de Ciencias Horticolas* 16:1.

Erazo-Mesa, E., Hincapié Gómez, E. and Echeverri-Sánchez, A. (2022a) Surface soil water content as an indicator of Hass avocado irrigation scheduling. *Agricultural Water Management* 273, 107864. doi:10.1016/j.agwat.2022.107864.

Erner, Y. (1988) Effects of girdling on the differentiation of inflorescence types and fruit set in 'Shamouti' orange trees. *Israeli Journal of Botany* 37, 173–180.

Erner, Y., Kaplan, Y., Artzi, B. and Hamu, M. (1993) Increasing citrus fruit size using auxins and potassium. *Acta Horticulturae* 329, 112–119.

Evans J.R. (1989) Photosynthesis and nitrogen relationships in leaves of C_3 plants. *Oecologia* 78(1), 9–19.

Fassio, C., Cautin, R., Perez-Donoso, A., Bonomelli, C. and Castro, M. (2016) Propagation techniques and grafting modify the morphological traits of roots and biomass allocation in avocado trees. *HortTechnology* 26(1), 63–69.

Fassio, C., Cautin, R., Perez-Donoso, A.G., Castro, M. and Bonomelli, C. (2020) Comparative branching order and root anatomy of clonal and seed-grown avocado trees (*Persea americana* Mill.). *International Journal of Agriculture and Natural Resources* 47(2), 134–144.

Fassio, C., Heath, R., Arpaia, M.L. and Castro, M. (2009) Sap flow in 'Hass' avocado trees on two clonal rootstocks in relation to xylem anatomy. *Scientia Horticulturae* 120, 8–13.

Fereres, E., Goldhamer, D. and Sadras, V. (2012) Yield response to water of fruit trees and vines. In: Steduto, P., Hsiao, T., Fereres, E. and Raes, D. (eds) *Crop Yield Response to Water*, Rome, Italy. pp. 246–295.

Fernandez Falcon, M., Fox, R.L. and Trujillo, E.E. (1984) Interactions of soil pH, nutrients and moisture on Phytophthora root rot of avocado. *Plant and Soil* 81, 165–176.

Ferreyra, R. and Sellés, G. (2007) Manejo del riego y suelo en palto [en línea]. Santiago: Boletín INIA - Instituto de Investigaciones Agropecuarias. no. 160.

Ferreyra, R., Sellés, G. and Gil, P. (2011) Asfixia radicular en huertos de paltos: Manejos del riego y suelo. La Cruz, Chile: Boletín INIA-Instituto de Investigaciones Agropecuarias. no. 231.

Ferreyra, R. and Sellés, G. (2012) Avocado, In: Steduto, P., Hsio, T., Fereres, E. and Raes, D. (eds) Crop yield response to water. FAO irrigation and Drainage paper 66, Rome, Italy, pp. 414–432.

Ferreyra, R. and Pérez, A. (2013) Huella hídrica en palta. En Osorio U., Alfonso (ed.) Determinación de la huella del agua y estrategias de manejo de recursos hídricos. La Serena, Chile: Serie Actas - Instituto de Investigaciones Agropecuarias. N° 50.

Ferreyra, R. and Sellés, G. (2013) Manual de riego para especies frutales: uso eficiente del agua de riego y estrategias para enfrentar periodos de escasez [en línea]. La Cruz, Chile: Boletín INIA - Instituto de Investigaciones Agropecuarias. no. 278. Available at: https://hdl.handle.net/20.500.14001/7681 (accessed 29 Nov 2024).

Ferreyra, R., Antúnez, A., Rojas, V., Cea, D. and Gil, P.M. (2018a) Evaluation of an avocado orchard producing under plastic greenhouse as a strategy to reduce water consumption for irrigation. In: *Proceedings of the Avocado Brainstorming 2018*, Tzaneen, South Africa.

Ferreyra, R., Antúnez, A., Rojas, V., Cea, D. and Gil, P.M. (2018b) Effect of plastic mulching on soil moisture and evaporation in avocado orchards. In: *Proceedings of the Avocado Brainstorming 2018*, Tzaneen, South Africa.

Ferreyra, R., Sellés, G. and Fereres, E. (2012) Fruit trees and vines: avocado. In: Steduto, P., Hsiao, T.C., Fereres E. and Raes D. (eds) *FAO Irrigation and Drainage Paper 66*. Food and Agricultural Organization of the United Nations, Rome, Italy, p. 500.

Ferreyra, R., Sellés, G., Maldonado, P., Celedón, J. and Gil, P.M. (2007) Effects of environment, leaf characteristics and measuring methodology on stem water potential in avocado trees (*Persea americana* Mill.). *Agricultura Técnica (Chile)* 67(2), 182–188.

Fujii, K., Shibata, M., Kitajima, K., Ichie, T., Kitayama, K. and Turner, B.L. (2018) Plant–soil interactions maintain biodiversity and functions of tropical forest ecosystems. *Ecological Research* 33, 149–160. doi:10.1007/s11284-017-1511-y.

Furr, J.R., Reece, P.C. and Gardner, F.E. (1946) Symptoms exhibited by avocado trees grown in outdoor sand cultures deprived of various mineral nutrients. *Proceedings of the Florida State Horticulture Society* 59, 138–145.

Gardiazábal, F., Magdahl, C., Mena, F. and Wilhelmy C. (2003) Determinación del coeficiente de cultivo (Kc) para paltos cv. Hass en Chile. In: *Proceedings of the V World Avocado Congress (Actas V Congreso Mundial del Aguacate)*, pp. 329–334.

Gardiazábal, F. (2023) Efecto del Bioaminol e Serenade en la productividad de huertos de paltos Hass en Chile. In: *Proceedings of the 10 World Avocado Congress*, New Zealand.

Gardiazabal, F., Mena, F., Torres, J. and Pinto A. (2015) Efecto de la aplicación de Cultar® al suelo, sobre la productividad y desarrollo de paltos (Persea americana Mill.) Cv Hass. In: *Proceedings of the VIII Worlds Avocado Congress,* Lima, Perú.

Gil, P.M. (2008) Study of avocado (*Persea americana* Mill.) physiological responses to different soil water-to-air ratios and the transmission of root-to-shoot electrical signals in response of soil water deficit and root hypoxia. PhD thesis, Pontificia Universidad Catolica de Chile, Santiago, Chile.

Gil, P.M., Schaffer, B., Gutiérrez, S.M. and Li, C. (2007) Effect of waterlogging on plant water status, leaf gas exchange and biomass of avocado. In: *Proceedings of the VI World Avocado Congress, 12–16 November 2007, Viña del Mar, Chile*.

Gil, P.M., Ferreyra, R., Karlezi, D., Troncoso, C., Saavedra, J., Robledo, P. and Defilippi, B. (2011) Determining the effect of different amounts of water on the physiology, yield and postharvest of avocado cv. Hass. In: *Proceedings of the VII World Avocado Congress 2011 (Actas VII Congreso Mundial del Aguacate 2011)*.

Gil, P.M. and Barrera M.C. (2018) *El rol del regante ante la escasez hídrica* [en línea]. *Agronomia y Forestal* UC. (no. 54), 10–14.
Gil, P.M., Bonomelli, C., Schaffer, B., Ferreyra, R. and Gentina, C. (2012) Effect of soil water-to-air ratio on biomass and mineral nutrition of avocado trees. *Journal of Soil Science and Plant Nutrition* 12(3), 609–630.
Gil, P.M., Odi, M., Cea, D., Meza, F., Balbontín, C., Arias, M.C., Kleiman, I. and Barrientos, A. (2021) Manual de uso para aplicación telefónica Coquimbo RiegaBien Versión 1.0. Pontificia Universidad Católica de Chile.
Ginsburg, O. and Avizohar-Hershenson, Z. (1965) Observations on vesicular-arbuscular mycorrhiza associated with avocado roots in Israel. *Transactions of the British Mycological Society* 48, 101–104.
Gontia, N.K. and Tiwari, K.N. (2010) Estimation of crop coefficient and evapotranspiration of wheat (Triticum estivum) in an irrigation command using remote sensing and GIS. *Water Resources Management* 24, 1399–1414.
Gonzalez-Dugo, V., Ruz, C., Testi, L., Orgaz, F. and Fereres, E. (2018) The impact of deficit irrigation on transpiration and yield of mandarin and late oranges. *Irrigation Science* 36(4–5), 227–239.
González-Gervacio, C., Castillo-González, A.M., Gómez-Merino, F.C. and Trejo-Téllez, L.I. (2019) Foliar boron fertilization in 'Hass' avocado increases amino acid concentrations and cell wall thickness. *Interciencia* 44, 362–368.
Goodall, G.E., Embleton, T.W. and Platt, R.G. (1965) Avocado fertilization. *University of California Cooperative Extension Bulletin 2024*. University of California, California p. 6.
Grajales, L., Dorado, D. and Rebolledo-Roa, A. (2018) Assessment of efficient water management indicators in Hass avocado (Persea americana mill.) production in Colombia. In: *Proceedings of the 21st World Congress of Soil Science (21WCSS), 12–17 August 2018, Rio de Janeiro*.
Granatstein, D., Kirby, E. and Willer, H. (2013) Global area and trends of organic fruit production. *Acta Horticulturae* 1001, 383–394. doi:10.17660/ActaHortic.2013.1001.45.
Granja, F. and Covarrubias, J.I. (2018) Evaluation of acidifying nitrogen fertilizers in avocado trees with iron deficiency symptoms. *Journal of Soil Science and Plant Nutrition* 18, 157–172. doi:10.4067/S0718-95162018005000702.
Groot Obbink, J. and Alexander, D.McE. (1976) Calcium deficiency and *Phytophthora cinnamomi* symptoms in young avocado seedlings. *Journal of the Australian Institute of Agricultural Science* 42, 131–134.
Grunennvaldt, R. (2022) Avocado soil health literature review. Queensland Department of Agriculture and Fisheries, Hort Innovation and Avocado Fund, 54 p. Available at: https://avocado.org.au/wp-content/uploads/2022/11/Soil-health-literature-review_NOV-22.pdf (accessed 21 May 2024).
Grüter, R., Trachsel, T., Laube, P. and Jaisli, I. (2022) Expected global suitability of coffee, cashew and avocado due to climate change. *PLOS ONE* 17(1), e0261976. doi:10.1371/journal.pone.0261976.
Guerrero-Polanco, F., Alejo-Santiago, G., Sánchez Hernández, R., Bugarín-Montoya, R., Aburto-González, C. and Isiordia-Aquino, N. (2018) Fertilization with potassium nitrate and fruit production in 'Hass' avocado (*Persea americana* Mill.). *Acta Agronómica* 67, 425–430. doi:10.15446/acag.v67n3.68858.
Gurovich, L.A., Ton, Y. and Vergara, L.M. (2006) Irrigation scheduling of avocado using phytomonitoring techniques. *Ciencia e Investigación Agraria* 33, 117–124.
Gustafson, C.D. (1957) Zinc soil treatment on avocados. *California Citrograph* 42, 225–227.
Gustafson, C.D. (1973) Fertilization of avocado trees. *California Avocado Society Yearbook* 56, 51–53.
Gutiérrez, J.M., Jones, R.G., Narisma, G.T., Alves, L.M., Amjad, M. *et al.* (2021) Atlas. In: *Climate Change 2021: The Physical Science Basis*. Contribution of Working Group I to the Sixth Assessment Report of the Intergovernmental Panel on Climate Change [V. Masson-Delmotte, P. Zhai, A. Pirani, S.L. Connors, C. Péan, S. Berger, N. Caud, Y. Chen, L. Goldfarb, M.I. Gomis, M. Huang, K. Leitzell, E. Lonnoy, J.B.R. Matthews, T.K. Maycock, T. Waterfield, O. Yelekçi, R. Yu, and B. Zhou (eds)]. Cambridge University Press. Interactive Atlas. Available at: http://interactive-atlas.ipcc.ch/ (accessed 21 May 2024).
Haas, A.R.C. (1928) Relation of chloride content to tipburn of avocado leaves. *California Avocado Association Yearbook*, 27.
Haas, A.R.C. (1929) Composition of avocado seedlings in relation to chlorosis and tipburn. *Botanical Gazette* 87, 422–430.
Haas, A.R.C. (1947) Nitrogen effects on avocado seedlings in soil cultures. *California Avocado Society Yearbook* 32, 51–54.
Haas, A.R.C. (1950a) Effect of sodium chloride on Mexican, Guatemalan and West Indian avocado seedlings. *California Avocado Society Yearbook* 35, 153–160.

Haas, A.R.C. (1950b) Rootstock influence on the composition of scion avocado leaves. *California Avocado Society Yearbook* 35, 149–152.

Haas, A.R.C. (1952) Importance of inorganic elements in the growth of avocado trees. *California Avocado Society Yearbook* 37, 187–195.

Haas, A.R.C. and Brusca, J.N. (1955) Chloride toxicity in avocados. *California Agriculture* 9, 12–14.

Haas, A.R.C. and Brusca, J.N. (1957) Leaf malady of avocado trees. *California Agriculture* 11, 14–15.

Hapuarachchi, N.S., Kämper, W., Wallace, H.M., Hosseini Bai, S., Ogbourne, S.M., Nichols, J. and Trueman, S.J. (2022) Boron effects on fruit set, yield, quality and paternity of Hass avocado. *Agronomy* 12(6), 1479. doi:10.3390/agronomy12061479.

Hazra, F. and Santosa, A. (2022) Evaluation of the use of biofertilizer on the growth of avocado plants (*Persea americana* Mill.) at Superavo orchard, Subang. *Journal of Soil Science and Environment* 24, 14–19. (In Bahasa Indonesia.) doi:10.29244/jitl.24.1.14-19.

Hermosilla-Madrid, C. and Whiley, A.W. (2023) High-density (HD) avocado production in New Zealand. In: *Proceedings of the Tenth World Avocado Congress, Auckland, New Zealand.* Available at: https://industry.nzavocado.co.nz (accessed 12 October 2023).

Hocking, B., Tyerman, S.D., Burton, R.A. and Gilliham, M. (2016) Fruit calcium: transport and physiology. *Frontiers in Plant Science* 7, 569.

Hoekstra, A.Y. (2008) The water footprint of food. In: Förare, J. (ed.) *Water and Food,* The Swedish Research Council for Environment. Agricultural Sciences and Spatial Planning, Stockholm, pp. 49–60.

Hofman, P.J., Marques, J.R., Searle, C., Stubbings, B. and Moody, P. (2005) Improving avocado fruit quality through tree nutrition - present knowledge. In: *Proceedings of the New Zealand and Australia Avocado Grower's Conference*, Tauranga, New Zealand. Available at: https://www.avocadosource.com/Journals/AUSNZ/AUSNZ_2005/HofmanPeter2005.pdf (accessed 21 May 2024).

Hofman, P.J., Vuthapanich, S., Whiley, A.W., Klieber, A. and Simons, D.H. (2002) Tree yield and fruit minerals concentrations influence 'Hass' avocado fruit quality. *Scientia Horticulturae* 92, 113–123. doi:10.1016/S0304-4238(01)00286-2.

Hofshi, R. and Hofshi, S. (2007) Irrigation scheduling calculator. Tools. In: www. Avocadosource.com. The Hofshi Foundation. Available at: http://avocadosource.com/.

Holmqvist, J.Å. (2020) Sustainable water footprint of avocado imported to Norway from Chile: Norwegian retailer's information and influence on water management. Master's thesis, Norwegian University of Life Sciences, Ås, Norway.

Holzapfel, E., de Souza, J.A., Jara, J. *et al.* (2017) Respuestas de la producción de aguacate a la variación en los niveles de riego. *Irrigation Science* 35, 205–215. doi:10.1007/s00271-017-0533-0.

Hornbuckle, J., Car, N.J., Christen, E.W., Stein, T.M. and Williamson, B. (2009) IrriSatSMS Irrigation water management by satellite and SMS – a utilisation framework, CSIRO and CRC for Irrigation Futures, CSIRO Land and Water Science Report No 04/09.

Howden, M., Newett, S. and Deuter, P. (2005) Climate change: risks and opportunities for the avocado industry. In: *Proceedings of the New Zealand and Australian Avocado Grower's Conference* Tauranga, New Zealand, pp. 1–28.

Hu, Y., Sun, Z., Zeng, Y., Ouyang, S., Chen, L. *et al.* (2023) Tree-level stomatal regulation is more closely related to xylem hydraulic traits than to leaf photosynthetic traits across diverse tree species. *Agricultural and Forest Meteorology* 329, 109291. doi:10.1016/j.agrformet.2022.109291.

Huete, A.R. (1988) A soil-adjusted vegetation index (SAVI). *Remote Sensing of Environment* 25(3), 295–309.

Huett, D.O. and Dirou, J.F. (2000) An evaluation of the rationale for fertiliser management of tropical fruit crops. *Australian Journal of Experimental Agriculture* 40, 1137–1143. doi:10.1071/EA00047.

Huey, C.J., Gopinath, S.C.B., Uda, M.N.A., Zulhaimi, H.I., Jaafar, M.N., Kasim, F.H. and Yaakub, A.R.W. (2020) Mycorrhiza: a natural resource assists plant growth under varied soil conditions. *3 Biotech* 10(5), 204. doi:10.1007/s13205-020-02188-3.

Infante Jiménez, Z., Ortega Gómez, P. and Coutiño Puchuli, A.E. (2020) Las biofábricas y su relación con el desarrollo sostenible en Michoacán, México. In: *Factores críticos y estratégicos en la interacción territorial desafíos actuales y escenarios futuros*. Universidad Nacional Autónoma de México y Asociación Mexicana de Ciencias para el Desarrollo Regional A.C, Coeditores, Ciudad de México, pp. 401–418. Available at: https://ru.iiec.unam.mx/5122/1/5-049-Infante-Ortega-Coutino.pdf (accessed 21 May 2024).

Ippolito, M., De Caro, D., Ciraolo, G., Minacapilli, M. and Provenzano, G. (2023) Estimating crop coefficients and actual evapotranspiration in citrus orchards with sporadic cover weeds based on ground and remote sensing data. *Irrigation Science* 41, 5–22. doi:10.1007/s00271-022-00829-4.

Iturbide, M., Fernández, J., Gutiérrez, J.M., Bedia, J., Cimadevilla, E., *et al.* (2021) Repository supporting the implementation of FAIR principles in the IPCC-WG1 Atlas. Zenodo. Available at: https://github.com/IPCC-WG1/Atlas (accessed 21 May 2024).

Jaganath, I. and Lovatt, C.J. (1998) Efficacy studies on pre-bloom canopy applications of boron and/or urea to 'Hass' avocados in California. In: *Proceedings of the 3rd World Avocado Congress*, Tel Aviv, Israel, pp. 181–184.

Jamie, S., Subires, M.J., Soria, J.T. and Aguilar, A. (1992) Interaction of K-B in avocado (*Persea americana* Mill.) culture. *Acta Horticulturae* 296, 75–81.

Johansen, C., Edwards, D. and Loneragan' J. (1968) Interactions between potassium and calcium in their absorption by intact barley plants. I. Effects of potassium on calcium absorption. *Plant Physiology* 43, 1717–1721. doi:10.1104/pp.43.10.1717.

Jones, H.G. and Vaughan, R.A. (2010) *Remote Sensing of Vegetation*. Oxford University Press, New York.

Joseph, S., Pow, D., Dawson, K., Rust, J., Munroe, P., *et al.* (2020) Biochar increases soil organic carbon, avocado yields and economic return over 4 years of cultivation. *Science of the Total Environment* 724, 138153. doi:10.1016/j.scitotenv.2020.138153.

Joyce, D., Perkins, M., Newett, S., Coates, L. and Irvine Brown, S. (2022) Mineral nutrition for avocado fruit quality – a brief review and recommendations. *Acta Horticulturae* 1333, 285–290. doi:10.17660/ActaHortic.2022.1333.37.

Ju, Y.L., Yang, B.H., He, S., Tu, T.Y., Min, Z., Fang, Y.L. and Sun, X.Y. (2019) Anthocyanin accumulation and biosynthesis are modulated by regulated deficit irrigation in Cabernet Sauvignon (*Vitis Vinifera* L.) grapes and wines. *Plant Physiology and Biochemistry* 135, 469–479. doi:10.1016/j.plaphy.2018.11.013.

Kadman, A. (1963a) Soil treatments with iron chelates to cure chlorotic avocado trees in Israel. *California Avocado Society Yearbook* 46, 73–75.

Kadman, A. (1963b) The uptake and accumulation of chloride in avocado leaves and the tolerance of avocado seedlings under saline conditions. *Proceedings of the American Society for Horticultural Science* 83, 280–286.

Kadman, A. (1964) The uptake and accumulation of sodium in avocado seedlings. *Proceedings of the American Society for Horticultural Science* 85, 179–182.

Kadman, A. (1970) Factors influencing intake and accumulation of chlorine and sodium in avocado plants. In: *The Division of Subtropical Horticulture, The Volcani Institute of Agricultural Research* 1960–1969, pp. 33–40.

Kadman, A. and Cohen, A. (1977) Experiments with zinc applications to avocado trees. *California Avocado Society Yearbook* 61, 81–85.

Kadman, A. and Lahav, E. (1972) Experiments with various treatments to cure chlorotic avocado trees. *California Avocado Society Yearbook* 55, 176–178.

Kadman, A. and Lahav, E. (1982) Experiments to correct iron deficiency in avocado trees. *Journal of Plant Nutrition* 5, 961–966.

Kalmar, D. and Lahav, E. (1976) Water requirements of avocado in the Western Galilee. a) The effect of different irrigation treatments on water consumption, salt content in the soil and root distribution. *Alon Hanotea* 30, 629–643. (In Hebrew.)

Kalmar, D. and Lahav, E. (1977) Water requirements of avocado in Israel. 1. Tree and soil parameters. *Australian Journal of Agricultural Research* 28, 859–868.

Kamble, B., Kilic, A. and Hubbard, K. (2013) Estimating crop coefficients using remote sensing-based vegetation index. *Remote Sensing* 5(4), 1588–1602. doi:10.3390/rs5041588.

Kamilaris, A., Kartakoullis, A. and Prenafeta Boldú, F. (2017) A review on the practice of big sata analysis in agriculture. *Computers and Electronics in Agriculture* 143. doi:10.1016/j.compag.2017.09.037.

Kaneko, T., Gould, N., Campbell, D. and Clearwater, M.J. (2024) Isohydric stomatal behaviour alters fruit vascular flows and minimizes fruit size reductions in drought-stressed 'Hass' avocado (*Persea americana* Mill.). *Annals of Botany* 133(7), 969–982. doi:10.1093/aob/mcae024.

Karmas, A., Tzotsos, A. and Karantzalos, K. (2016) Geospatial big data for environmental and agricultural applications. In: Yu, S. and Guo, S. (eds) *Big Data Concepts, Theories, and Applications*. Springer, Cham. https://doi.org/10.1007/978-3-319-27763-9_10.

Kiggundu, N., Migliaccio, K.W., Schaffer, B., Li, Y. and Crane, J.H. (2012) Water savings, nutrient leaching, and fruit yield in a young avocado orchard as affected by irrigation and nutrient management. *Irrigation Science* 30, 275–286.

Kirkby, E.A. (1969) Ion uptake and ionic balance in plants in relation to the form of nitrogen nutrition. In: Rorison, I.H. (ed.) *Ecological Aspects of Mineral Nutrition of Plants*. British Ecological Society, Blackwell Scientific Publications, Oxford, UK, pp. 215–235.

Klein, I. and Zilkah, S. (1986) Urea retention and uptake by avocado and apple leaves. *Journal of Plant Nutrition* 9, 1415–1425.

Koen, T.J. and du Plessis, S.F. (1992) Optimal leaf analysis norms for avocado (cv. Fuerte). In: Lovatt, C., Holthe, P.A. and Arpaia, M.L. (eds) *Proceedings of the Second World Avocado Congress*. University of California, Riverside, California, 1, 289–299.

Koen, T.J. and Langenegger, W. (1971) Fertilizing the avocado. *Queensland Fruit and Vegetable News* 437, pp. 451–452.

Kohatsu, M.Y., Pelegrino, M.T., Monteiro, L.R., Freire, B.M., Pereira R.M. *et al.* (2021) Comparison of foliar spray and soil irrigation of biogenic CuO nanoparticles (NPs) on elemental uptake and accumulation in lettuce. *Environmental Science and Pollution Research* 28, 16350–16367. doi:10.1007/s11356-020-12169-x.

Koo, C.J. and Young, T.W. (1977) Effects of age, position, and fruiting status on mineral composition of 'Tonnage' avocado leaves. *Journal of the American Society for Horticultural Science* 102, 311–313.

Kopp, L.E. (1966) A taxonomic revision of the genus *Persea* in the Western Hemisphere. (*Persea-Lauraceae*), *Memoirs N.Y. Botanic Gardens* 14, 1–120.

Kourgialas, N.N. and Dokou, Z.(2021) Water management and salinity adaptation approaches of Avocado trees: A review for hot-summer Mediterranean climate. *Agricultural Water Management* 252, 106923. doi:10.1016/j.agwat.2021.106923.

Kourgialas, N.N. (2021) A critical review of water resources in Greece: The key role of agricultural adaptation to climate-water effects. *Science Of The Total Environment* 775, 145854. doi:10.1016/j.scitotenv.2021.145857.

Kremer-Köhne, S., Köhne, J.S. and Schutte, J.M. (1993) Effect of potassium, magnesium and nitrogen soil applications on Fuerte avocado fruit quality. *South African Avocado Growers' Association Yearbook* 16, 33–36.

Kruger, F.J., du Plessis, M.H., Kritzinger, M., Malumane, R., Penter, M., Snijder, B. and Claassens, N. (2001) Updating Pinkerton export parameters and evaluation of new and upgraded avocado postharvest applications. *South African Avocado Growers' Association Yearbook* 24, 49–51.

Kumar, S., Kumar, S. and Mohapatra, T. (2021) Interaction between macro- and micronutrients in plants. *Frontiers of Science* 12, 665583. doi:10.3389/fpls.2021.665583.

Labanauskas, C.K., Embleton, T.W., Garber, M.J. and Richards, S.J. (1958b) Effects of irrigation treatments and rates of nitrogen fertilization on young Hass avocado trees. V. Micronutrient content of leaves. *Proceedings of the American Society for Horticultural Science* 71, 315–319.

Labanauskas, C.K., Embleton, T.W. and Jones, W.W. (1958a) Influence of soil applications of nitrogen, phosphate, potash, dolomite and manure on the micronutrient content of avocado leaves. *Proceedings of the Amerkican Society for Horticultural Science* 71, 285–291.

Labanauskas, C.K., Embleton, iT.W. and Jones, W.W. (1959) Fertilizer effects on micronutrient nutrition in avocado. *California Avocado Society Yearbook* 43, 96–99.

Labanauskas, C.K., Embleton, T.W., Jones, W.W. and Garber, M.J. (1961) Seasonal changes in concentrations of zinc, copper, boron, manganese and iron in Fuerte avocado leaves. *Proceedings of the American Society for Horticultural Science* 77, 173–179.

Lahav, E. (1984) The effect of organic manure on avocado trees in Israel. In: *Proceedings of the VI International Colloquium for the Optimization of Plant Nutrition*, Montpellier, France. *Actes Proceedings* 3, pp. 1051–1062.

Lahav, E. (1999) Avocado nutrition. *Subtropical Fruit News* 7(1), 5–7.

Lahav, E. and Aycicegi-Lowengart, A. (2003) Avocado mineral nutrition the water-nutrients relationship. In: *Proceedings of the 5th World Avocado Congress*, Granada, Málaga, Spain, pp. 349–357.

Lahav, E. and Kadman, A. (1980) Avocado fertilization. *Bulletin of the International Potash Institute No. 6*. Worblaufen-Bern, Switzerland.

Lahav, E. and Kalmar, D. (1977) Water requirements of avocado in Israel. 2. Influence on yield, fruit growth and oil content. *Australian Journal of Agricultural Research* 28, 869–877.

Lahav, E. and Kalmar, D. (1983) Determination of irrigation regimes for an avocado plantation in spring and autumn. *Australian Journal of Agricultural Research* 34, 717–724.

Lahav, E. and Kalmar, D. (1987) The use of early spring nitrogen applications to avocados. *Alon Hanotea* 41, 699–701. (In Hebrew.)

Lahav, E. and Whiley A.W. (2002) Irrigation and mineral nutrition. In: Whiley, A.W., Schaffer, B. and Wolstenholme, B.N. (eds) *The Avocado: Botany, Production and Uses*. CAB International, Wallingford, UK, pp. 259–298.

Lahav, E. and Zamet, D. (1977) Zinc supply to avocado trees in heavy soil. *Alon Hanotea* 4, 269–273. (In Hebrew.)

Lahav, E. and Zamet, D. (1999) Mineral losses of avocado tree resulting from abscission of flowers, fruitlets and fruits. *Revista Chapingo Serie Horticultura Número Especial* V, 101–102.

Lahav, E., Bar, Y. and Kalmar, D. (1990) Effect of nitrogenous fertilization on the annual variations in nutrients in avocado leaves. *Communications in Soil Science and Plant Analysis* 21, 1353–1365.

Lahav, E., Bareket, M. and Zamet, D. (1976) Potassium fertilizer experiment with avocado trees on heavy soils. *California Avocado Society Yearbook* 60, 181–186.

Lahav, E., Kalmar, D. and Bar, Y. (1987) Nitrogen fertilization, a guarantee for relative resistance of avocado trees to frost. *Journal of Plant Nutrition* 10, 1859–1868.

Lahav, E., Steinhardt, R. and Kalmar, D. (1993) Effect of salinity on the nutritional level of the avocado. In: Fragoso, M.A.C. and Beusichen, M.L. (eds) *Optimization of Plant Nutrition*. Kluwer Academic Publishers, Netherlands, pp. 593–596.

Lahav, E., Whiley, A.W. and Turner, D.W. (2013) Irrigation and mineral nutrition. In: Schaffer, B., Wolstenholme, B. N. and Whiley, A.W. (eds) The Avocado: Botany, Production and Uses, 2nd edn. CABI, Wallingford, UK, pp. 301–341.

Lakso, A.N., Santiago, M. and Stroock, A.D. (2022) Monitoring stem water potential with an embedded microtensiometer to inform irrigation scheduling in fruit crops. *Horticulturae* 8(12), 1207. doi:10.3390/horticulturae8121207.

Lazare, S., Cohen, Y., Goldshtein, E., Yermiyahu, U., Ben-Gal, A. and Dag, A. (2021) Rootstock-dependent response of Hass avocado to salt stress. *Plants (Basel)* 10(8), 1672. doi:10.3390/plants10081672.

Lee, B.S. and Zentmyer, G.A. (1982) Influence of calcium nitrate and ammonium sulfate on Phytophthora root rot of *Persea indica*. *Phytopathology* 72, 1558–1564.

Lee, B.W. (1973) The efficacy of aerial applications of zinc to avocado trees. *California Avocado Society Yearbook* 56, 121–123.

Li, Y. (2009) *Calcareous soils in Miami-Dade County*. Soil and Water Science Department, Florida Cooperative Extension Service, IFAS, University of Florida. Fact Sheet SL 183.

Limaco Sing, J. and McNeil, R.J. (1991) The effectiveness of foliar potassium nitrate sprays on the 'Hass' avocado (*Persea americana* Mill.). In: Lovatt, C., Holthe, P.A. and Arpaia, M.L. (eds) *Proceedings of the Second World Avocado Congress*. University of California, Riverside, California, 1, 337–342.

Link, A. and Alegría, N. (2019) Alternativas a portainjertos de semilla Mexicola y Nabal bajo condiciones de alta salinidad en Chile. IX World Avocado Congress, Medellín, Colombia, 24–26 September 2019.

Liu, X., Mickelbart, M.V., Robinson, P.W., Holshi, R. and Arpais, M.L. (2002) Photosynthetic characteristics of avocado leaves. *International Symposium of Tropical & Subtropical Fruits*, pp. 856–874.

Lobell, D.B., Field, C.B., Cahill, K.N. and Bonfils, C. (2006) Impacts of future climate change on California perennial crop yields: model projections with climate and crop uncertainties. *Agricultural Meteorology* 141, 208–218.

Lomas, J. and Mandel, M. (1973) The quantitative effects of two methods of sprinkler irrigation on the microclimate of a mature avocado plantation. *Agricultural Meteorology* 12, 35–48.

López-Luna, J., Nopal-Hormiga, Y., López-Sánchez, L., Mtz-Enriquez, A.I. and Pariona, N. (2023) Effect of methods application of copper nanoparticles in the growth of avocado plants. *Science of the Total Environment* 880, 163341. doi:10.1016/j.scitotenv.2023.163341.

Loupassaki, M.H. (1998) The effect of nitrogen fertilizer on the growth, yield and mineral contents of leaves of the avocado cv Fuerte. In: *Proceedings of the 3rd World Avocado Congress*, Tel Aviv, Israel, pp. 168–171.

Lovatt, C.J. (1996) Nitrogen allocation within the 'Hass' avocado. *California Avocado Society Yearbook* 80, 75–83.

Lovatt, C.J. (1999) Timing citrus and avocado foliar nutrient applications to increase fruit set and size. *HortTechnology* 9, 607–612.

Lovatt, C.J. (2001) Properly timed soil-applied nitrogen fertilizer increases yield and fruit size of 'Hass' avocado. *Journal of the American Society for Horticultural Science* 126, 555–559.

Lovatt, C.J. (2013a) 'Hass' avocado nutrition research in California. A summary of basic information to assist growers in managing their orchards. *California Avocado Society* 96, 74–105.

Lovatt, C.J. (2013b) Properly timing foliar-applied fertilisers increases efficiency: a review and update on timing foliar nutrient applications to citrus and avocado. *HortTechnology* 23, 536–541.

Lovatt, C.J. and Witney, G. (2001) Nitrogen fertilization of the 'Hass' avocado in California. *AvoResearch* 1, 1–5.

Lynch, S.J. and Goldweber, S. (1956) Some effects of nitrogen, phosphorus and potassium fertilization on the yield and tree growth of avocados. *Proceedings of the Florida State Horticultural Society* 69, 289–292.

Lynch, S.J., Goldweber, S. and Rich, C.E. (1954) Some effects of nitrogen, phosphorus and potassium on the yield, tree growth, and leaf analysis of avocados. *Proceedings of the Florida State Horticultural Society* 67, 220–224.

Maas, E.V. (1984) Salt tolerance of plants. In: Christie, R.B. (ed.) *The Handbook of Plant Science in Agriculture*. CRC Press, Boca Raton (Florida).

Magwaza, L.S., Lemmer, D., Ntandane, J. and Kruger, F.J. (2008) Effect of N and Fe fertilizer treatments on the incidence of chilling injury in 'Fuerte' and 'Hass' avocados. *South African Avocado Growers' Association Yearbook* 31, 6–8.

Malo, S.E. (1966) Mineral nutrition of avocados and other tropical fruits in South Florida. In: *Proceedings of the Caribbean Region, American Society for Horticultural Science, XIV Annual Meeting, El Salvador*, pp. 101–107.

Malo, S.E. (1976) Mineral nutrition of avocados. In: Sauls, J.W., Phillips, R.L. and Jackson, L.K. (eds) *Proceedings of the First International Tropical Fruit Short Course. The Avocado*. University of Florida, Gainesville, Florida, pp. 42–46.

Mandemaker, A.J. (2007) Review: photosynthesis of avocado. *New Zealand's Avocado Growers Association Annual Research Report* 7, 1–10.

Manthey, J.A. and Crowley, D.E. (1997) Leaf and root responses to iron deficiency in avocado. *Journal of Plant Nutrition* 20, 683–693.

Marchal, J. (1987) Avocado. In: Martin Prevel, P., Gagnard, J. and Gautier, P. (eds) *Plant Analysis as a Guide to the Nutrient Requirements of Temperate and Tropical Crops*. Lavoisier Publishing, New York, pp. 366–376.

Marchal, J. and Bertin, Y. (1980) Contenu en elements mineraux des organes de l'avocatier 'Lula' et relations avec la fumure. *Fruits* 35, 139–149. (In French.)

Market Reports World (2023) Global organic avocado market 2024–2032: size, projections of growth and share. Available at: https://www.marketreportsworld.com/agriculture-industry-market-research-reports. (accessed 22 May 2024)

Marques, J.R., Hofman, P.J. and Wearing, A.H. (2003) Rootstocks influence 'Hass' avocado fruit quality and fruit minerals. *Journal of Horticultural Science and Biotechnology* 78, 673–679.

Marsal, J., Girona, J., Casadesus, J., Lopez Velasco, G. and Stockle, C. (2013) Crop coefficient (Kc) for apple: comparison between measurements by a weighing lysimeter and prediction by CropSyst. *Irrigation Science* 31, 455–463. doi:10.1007/s00271-012-0323-7.

Marsh, A.W. and Gustafson, C.D. (1958) Orchard irrigation. *California Avocado Society Yearbook* 42, 30–33.

Marsh, A., Branson, R., Davis, S. and Gustafson, D. (1978) Irrigating new avocado orchards. *California Agriculture* 32(5), 19–20.

Mazhawu, E., Clulow, A., Savage, M.J. and Taylor, N.J. (2018) Water use of avocado orchards – year 1. *South African Avocado Growers' Association Yearbook* 41, 37–41.

Mena, F., Torres, J., Zulueta, C., Pinto, M.A. and Gardiazábal, F. (2023) Nitrogen distribution and its effect on productivity in avocado trees var. Hass in Chile. In: *Proceedings of the 10th World Avocado Congress*, New Zealand. Available at: https://industry.nzavocado.co.nz/wp-content/uploads/2023/05/10th-World-Avocado-Congress-all-full-papers.pdf (accessed 22 May 2024).

Menge, J.A. and Marais, L.J. (2000) Soil environmental factors and their relationship of avocado root rot. *Subtropical Fruit News* 8, 1–11.

Menge, J.A., Davis, R.M., Johnson, E.L.V. and Zentmyer, G.A. (1978) Mycorrhizal fungi increase growth and reduce transplant injury in avocados. *California Agriculture* 32, 6–7.

Menge, J.A., LaRue, J., Labanauskas, C.K. and Johnson, E.L.V. (1980) The effect of two mycorrhizal fungi upon growth and nutrition of avocado seedlings grown with six fertilizer treatments. *Journal of the American Society for Horticultural Science* 105, 400–404.

Mekonnen, M.M. and Hoekstra, A.Y. (2011) The green, blue and grey water footprint of crops and derived crop products. *Hydrology and Earth System Sciences* 15, 1577–1600.

Mekonnen, M.M. and Hoekstra, A.Y. (2010) The green, blue and grey water footprint of crops and derived crop products. In: *Value of Water Research Report Series, Nr. 47 (Appendix II)*. UNESCO-IHE, Delft, The Netherlands.

Messenger, B.J., Menge, J.A. and Pond, E. (2000) Effects of gypsum on zoospores and sporangia of *Phytophthora cinnamomi* in field soil. *Plant Disease* 84, 617–621.

Meyer, J.L., Arpaia, M.L., Yates, M.V., Takele, E., Bender, G. and Witney, G. (1990) Irrigation and fertilization management of avocado. *California Avocado Society Yearbook* 74, 71–83.

Meza, F., Darbyshire, R., Farrell, A., Lakso, A., Lawson, J. *et al.* (2023) Assessing temperature-based adaptation limits to climate change of temperate perennial fruit crops. *Global Change Biology* 29(9), 2557–2571. doi:10.1111/gcb.16601.

Michelakis, N., Vougioucalou, E. and Clapaki, G. (1993) Water use, wetted soil volume, root distribution and yield of avocado under drip irrigation. *Agricultural Water Management* 24, 119–131.

Mickelbart, M.V. and Arpaia, M.L. (2002) Rootstock influences changes in ion concentrations, growth, and photosynthesis of 'Hass' avocado trees in response to salinity. *Journal of the American Society for Horticultural Science* 127, 649–655.

Mickelbart, M.V., Melser, S. and Arpaia, M.L. (2007) Salinity-induced changes in ion concentrations of 'Hass' avocado trees on three rootstocks. *Journal of Plant Nutrition* 30, 105–122.

Mickelbart, M.V., Miller, R., Parry, S., Arpaia, M.L. and Heath, R. (2000) Avocado leaf surface morphology. *California Avocado Society Yearbook* 84, 139–150.

Migliaccio, K.W., Morgan, K.T., Vellidis, G., Zotarelli, L., Fraisse, C. *et al.* (2016) Smartphone apps for irrigation scheduling. *American Society of Agricultural and Biological Engineers* 59(1), 291–301. doi:10.13031/trans.59.11158.

Miller, M.P., Turrell, F.M. and Austin, S.W. (1963) Cooling avocado trees by sprinkling. *California Avocado Society Yearbook* 47, 102–105.

Minchin, P.E.H., Thorp, G.T., Boldingh, H., Cooney, J.M., Negm, F.B. *et al.* (2011) Phloem transport of boron within avocado trees. In: *Proceedings of the VII World Avocado Congress*, Cairns, Australia. Available at: https://www.avocadosource.com/wac7/Section_09/MinchinPeter2011.pdf (accessed 22 May 2024).

Mohale, P.M., Manyevere, A., Parwada, C. and Zerizghy, M.G. (2023) Effect of eucalyptus wood-based compost application rates on soil chemical properties in semi-organic avocado plantations, Limpopo province, South Africa. *PLOS ONE* 18, e0265728. doi:10.1371/journal.pone.0265728.

Montoya, B. and Osorio, W. (2009) Mycorrhizal dependency of avocado at different levels of soil solution phosphorus. *Suelos Ecuatoriales* 39, 143–147.

Moore, P.W. and Richards, S.J. (1958) Effects of irrigation treatments and rates of nitrogen fertilization on young Hass avocado trees. II. Relation to tipburn, tree sunburn, shoot dieback, leaf scorch, leaf colour, leaf size, tree vigour and leaf moisture deficits. *Proceedings of the American Society for Horticultural Science* 71, 298–303.

Moreno, O.G., Pliego, C., Sarmiento, D., Barceló, A., and Martínez-Ferri, E. (2019) Yield and fruit quality of avocado trees under different regimes of water supply in the subtropical coast of Spain. *Agricultural Water Management* 221, 192–201. doi:10.1016/j.agwat.2019.05.001.

Musyimi, D.M., Netondo, G.W. and Ouma, G. (2007) Effects of salinity on gas exchange and nutrients uptake in avocados. *Journal of Biological Science* 7, 496–505. doi:10.3923/jbs.2007.496.505.

Naor, A. (1998) Relationships between leaf and stem water potential and stomatal conductance in three field grown woody species. *Journal of Horticultural Sciences* 73, 431–436.

Nemera, D.B., Bar-Tal, A., Levy, G.J., Tarchitzky, J., Rog, I., Klein, T. and Cohen, S. (2021) Mitigating negative effects of long-term treated wastewater irrigation: leaf gas exchange and water use efficiency response of avocado trees (*Persea americana* Mill.). *Agricultural Water Management* 256, 107126. doi:10.1016/j.agwat.2021.107126.

Neuhaus, A., Turner, D.W., Colmer, T.D., Kuo, J. and Eastham, J. (2007) Drying half the root-zone of potted avocado (*Persea americana* Mill., cv. Hass) trees avoids the symptoms of water deficit that occur under complete root-zone drying. *Journal of Horticultural Science and Biotechnology* 82, 679–689.

Newett, S. (2005) Little evidence to support the use of foliar applied nutrients in avocado. *Topics in Subtropics Newsletter* 3, 1–5. Available at: https://cesandiego.ucanr.edu/newsletters/July_-_September_200526160.pdf. Accessed 22 May 2024.

Newett, S. (2018) *Achieving more consistent yields of quality fruits in the Australian Avocado Industry*. Final report AV14000. Horticulture Innovation Australia Ltd., Sydney.

Newett, S., Perkins, M., Coates, L., Irvine-Brown, S. and Joyce, D. (2021) Review of pre-harvest mineral nutrition for post-harvest quality, *Talking Avocados* 31, 54–57.

Newett, S., Rigden, P. and Carr, B. (2018) Avocado plant nutrition review Nambour: Department of Agriculture and Fisheries, 2018. 49p. Available at: https://avocado.org.au/wp-content/uploads/2018/08/AVOCADO-PLANT-NUTRITION-REVIEW-2018_cover.pdf (accessed 22 May 2024).

Novoa, M., Miranda, D. and Melgarejo, L. (2018) Efecto de las deficiencias y excesos de fósforo, potasio y boro en la fisiología y el crecimiento de plantas de aguacate (*Persea americana* cv. 'Hass'). *Revista Colombiana de Ciencias Hortícolas* 12, 293–307.

Noy, M. (2006) *Recommendations for Growing of Avocado*. Extension Service, Ministry of Agriculture, Israel.

Núñez-Elisea, R., Schaffer, B., Zekri, M., O'Hair, S.K. and Crane, J.H. (2001) *In situ* soil-water characteristic curves for tropical fruit orchards in trenched calcareous soil. *HortTechnology* 11, 65–69.

Nye, P.H. (1986) Acid-base changes in the rhizosphere. *Advances in Plant Nutrition* 2, 129–153.

Odi-Lara, M., Campos, I., Neale, C.M.U., Ortega-Farías, S., Poblete-Echeverría, C., Balbontín, C. and Calera, A. (2016) Estimating evapotranspiration of an apple orchard using a remote sensing-based soil water balance. *Remote Sensing* 8, 253. doi:10.3390/rs8030253.

Oppenheimer, C., Kadman, A. and Radziszewska, F. (1961) Some observations on the nutritional status of avocado trees in Israel. *California Avocado Society Yearbook* 45, 81–85.

Oster, J.D., Stottlmyer, D.E. and Arpaia, M.L. (2007) Salinity and water effects on 'Hass' avocado yields. *Journal of the American Society for Horticultural Science* 132(2), 253–261.

Parker, E.R. (1936) Mottle-leaf and sun-blotch disease control. *California Avocado Association Yearbook* 21, 149–151.

Parker, L., Pathak, T. and Ostoja, S. (2021) Climate change reduces frost exposure for high-value California orchard crops. *Science of the Total Environment* 762(1), 143971. doi:10.1016/j.scitotenv.2020.143971.

Pearcy, R.W. and Francheschi, V. (1986) Photosynthetic characteristics of C_3 and C_4 tree species grown in high light and low light environments. *Photosynthetic Research* 9, 317–331.

Pegg, K.G. and Giblin, F. (2008) Principles of Phytophthora root rot management in established orchards. *Talking Avocados* 19(1), 36–38.

Pegg, K.G. (2010) Pathology challenges in avocado: Phytophthora management. In: *Proceedings of the Talk to Southern Produce Avocado Conference, June 2010, New Zealand*.

Pereira, G., Siqueira, J.A., Silva, W., Cardoso, F., Nunes-Nesi, A. and Araújo, W. (2021) Boron: more than an essential element for land plants? *Frontiers in Plant Science* 11. doi:10.3389/fpls.2020.610307.

Peterson, R.L., Piche, T. and Plenchette, C. (1984) Mychorrhizae and their potential use in the agricultural and forestry industries. *Biotechnical Advances* 2, 101–120.

Pinter, P.J.J., Hatfield, J.L., Schepers, J.S., Barnes, E.M., Moran, M.S., Daughtry, C.S.T. and Upchurch, D.R. (2003) Remote sensing for crop management. *Photogrammetric Engineering and Remote Sensing* 69, 647–664. doi:10.14358/PERS.69.6.647.

Ploetz, R. and Schaffer, B. (1989) Effects of flooding and Phytophthora root rot on net gas exchange and growth of avocado. *Phytopathology* 79, 204–208.

Pôças, I., Paço, T.A., Paredes, P., Cunha, M. and Pereira, L.S. (2015) Estimation of actual crop coefficients using remotely sensed vegetation indices and soil water balance modelled data. *Remote Sensing* 7, 2373–2400. doi:10.3390/rs70302373.

Popenoe, W. (1939) The avocado. In: *Manual of Tropical and Subtropical Fruits*. Macmillan, New York, pp. 9–78.

Prado, R.M. and Rozane, D.E. (2020) Leaf analysis as diagnostic tool for balanced fertilization in tropical fruits. In: Srivastava, A.K. and Hu, C. (eds) *Fruit Crops: Diagnosis and Management of Nutrient Constraints*. Elsevier, Amsterdam, The Netherlands, pp. 131–141.

Ramírez-Gil, J.G., Morales, J.G. and Peterson, A.T. (2018) Potential geography and productivity of 'Hass' avocado crops in Colombia estimated by ecological niche modeling. *Scientia Horticulturae* 237, 287–295. doi:10.1016/j. scienta.2018.04.021.

Ramírez-Gil, J.G., Cobos, M.E., Jiménez-García, D., Morales-Osorio, J.G. and Townsend, T. (2019) Current and potential future distributions of Hass avocados in the face of climate change across the Americas. *Crop and Pasture Science* 70(8), 694–708. doi:10.1071/CP19094.

Razeto, B. and Salgado, J. (2004) The inflorescence and fruit peduncle as indicators of nitrogen status of the avocado tree. *HortScience* 39, 1173–1174.

Razeto, B. and Palacios, J. (2007) Fruit analysis as an alternative to leaf analysis for diagnosing iron status of avocado tree Viña Del Mar, Chile. In: *Proceedings of the VI World Avocado Congress*, Viña Del Mar, Chile.

Rebolledo-Roa, A. and Burbano-Diaz, R.A. (2023) Nutrient balance in Hass avocado trees as a tool to optimize crop fertilization management. *Agronomy* 13(8), 1956. doi:10.3390/agronomy13081956.

Reuter, D.J. and Robinson, J.B. (1986) *Plant Analysis: an Interpretation Manual*. Inkata Press, Melbourne.

Reuveni, O. and Raviv, M. (1981) Importance of leaf retention to rooting of avocado cuttings. *Journal of the American Society for Horticultural Science* 106, 127–130.

Rhoades, J. D. (1974) Drainage for salinity control. *Drainage For Agriculture* 17, 433–461.

Rhoades, J.D. and Merrill, S.D. (1976) Assessing the suitability for irrigation: theoretical and empirical approaches. *FAO Soil Bull* 31, 69–109.

Rijswijk, K., Klerkx, L. and Turner J.A. (2019) Digitalisation in the New Zealand agricultural knowledge and innovation system: initial understandings and emerging organisational responses to digital agriculture. *NJAS – Wageningen Journal of Life Sciences* 90–91, 100313. doi:10.1016/j. njas.2019.100313.

Rivero, L. (2020) Mapping of crop coefficient (kc) from remote information in avocado in Mediterranean Chile. Memoria de Título. Universidad de Chile Facultad de Ciencias Agronómicas. Available at: https://repositorio.uchile.cl/handle/2250/184208.

Robbertse, P.J. and Coetzer, L.A. (1990) Booropname deur avokadoblare. *South African Avocado Growers' Association Yearbook* 13, 37. (In Afrikaans with English summary.)

Robbertse, P.J., Coetzer, L.A. and Janse van Vuuren, B.P.H. (1992) Booropname by avokado. *South African Avocado Growers' Association Yearbook* 15, 89–93. (In Afrikaans with English summary.)

Robbertse, P.J., Coetzer, L.A., Slabbert, M. and Swart, N.G.N. (1989) Die invloed van blaar-en worteltoedienings van boor op vrugopbrengs by avokado. *South African Avocado Growers' Association Yearbook* 12, 74–75. (In Afrikaans.)

Robbertse, P.J., Coetzer, L.A., Tomer, E. and Barnard, R.O. (1998) Boron uptake by container-grown, rootstock avocado plants from different boron-containing media. In: *Proceedings of the 3rd World Avocado Congress*, Tel Aviv, Israel, pp. 185–188.

Robinson, J.B.D. (1961) A note on magnesium unbalance symptoms in avocado pear trees. *East African Agriculture and Forestry Journal* 27, 47–48.

Rodríguez Pleguezuelo, C.R., Martínez, J.R., García-Tejero, I.F., Gálvez Ruiz, B., Tarifa, D. and Durán Zuazo, V. (2018) Avocado (*Persea americana* Mill.) trends in water-saving strategies and production potential in a Mediterranean climate: the study case of Spain: a review. doi:10.1016/B978-0-12-813164-0.00014-4.

Roets, N.J.R., Schoeman, S.P., Cronje, R.B. and Murovhi, N.R. (2012) Calibrating avocado irrigation through xylem water potential measurements. Preliminary report. *South African Avocado Growers' Association Yearbook* 35, 79–83.

Rombolà, A.D. and Tagliavini, M. (2006) Iron nutrition of fruit tree crops. In: L. Barton and J. Abadía (eds) *Iron Nutrition in Plants and Rhizospheric Microorganisms*. Springer, Dordrecht, The Netherlands, pp. 61–83.

Romero-Trigueros, C., Cabanero, J.J.A., Tortosa, P.A.N., Gambin, J.M.B., Maestre-Valero, J.F. and Nicolas, E.N. (2020) Medium-long term effects of saline reclaimed water and regulated deficit irrigation on fruit quality of citrus. *Journal of the Science of Food and Agriculture* 100(3), 1350–1357. doi:10.1002/jsfa.10091.

Rose, D.C. and Chilvers J. (2018) Agriculture 4.0: broadening responsible innovation in an era of smart farming. *Frontiers in Sustainable Food Systems* 2, 87. doi: 10.3389/ fsufs.2018.00087.

Rosecrance, R., Faber, B. and Lovatt, C. (2012) Patterns of nutrient accumulation in 'Hass' avocado fruit. *Better Crops with Plant Food* 96, 12–21. Available at: http://www.ipni.net/publication/bettercrops.nsf/0/F2E59118B155B269062579AC005AE171/$FILE/Better%20Crops%202012-1%20p.12-13.pdf (accessed 8 May 2024).

Rouse, J.W., Haas, R.H., Schell, J.A. and Deering, D.W. (1973) Monitoring vegetation systems in the Great Plains with ERTS (Earth Resources Technology Satellite). In: *Proceedings of 3rd Earth Resources Technology Satellite Symposium, Greenbelt, 10–14 December*, SP-351, 309–317.

Ruehle, G.D. (1940) Zinc deficiency of the avocado. *Proceedings of the Florida State Horticultural Society* 53, 150–152.

Salazar-García, S. (2002) Nutrición del aguacate, principios y aplicaciones. *Instituto Nacional de Investigaciones Forestales*, Agrícolas y Pecuarias (INIFAP) en asociación con el Instituto de la Potasa y el Fósforo (INPOFOS), Querétaro, México. 165 p.

Salazar-García, S. and Lazcano-Ferrat, I. (2001) Identifying fruit mineral removal differences is four avocado cultivars. *Better Crops International* 15, 28–31.

Salazar-García, S., Cossio, V. and González, D. (2009) La fertilización de sitio específico mejoró la productividad del aguacate 'Hass' en huertos sin riego. *Agricultura Técnica en México* 35, 439–448.

Salazar-García, S., Cossio-Vargas, L. E. and Gonzáles-Durán, I.J.L. (2008) Chronic zinc deficiency correction in 'Hass' avocado. *Revista Chapingo Serie Horticultura* 14, 153–159.

Salazar-García, S., Cossio-Vargas, L.E. and González-Durán, I.J. (2007) Reciclamiento de nutrimentos por las hojas de aguacate 'Hass'. In: *Proceedings of the VI World Avocado Congress*, Viña Del Mar, Chile. (In Spanish with English summary.)

Salazar-García, S., González-Durán, I.J.L. and Ibarra-Estrada, M.E. (2015) Identification of the appropriate leaf sampling period for nutrient analysis in 'Hass' avocado. *HortScience* 50, 130–136.

Salazar-Reque, I., Arteaga, D., Mendoza, F., Rojas, M.E., Soto, J., Huaman, S. and Kemper, G. (2023) Differentiating nutritional and water statuses in Hass avocado plantations through a temporal analysis of vegetation indices computed from aerial RGB images. *Computers and Electronics in Agriculture* 213, 108246. doi:10.1016/j.compag.2023.108246.

Salgado, E. and Cautín, R. (2008) Avocado root distribution in fine and coarse-textured soils under drip and microsprinkler irrigation. *Agricultural Water Management* 95, 817–824.

Salvo, J.E. and Lovatt, C.J. (2016) Nitrogen fertilization strategies for the 'Hass' avocado that increase total yield without reducing fruit size. *HortTechnology* 26, 426–433.

Santa Olalla, M. and De Juan Valero, J.A. (1993) *Agronomía del riego*. Ediciones Mundi-Prensa, España, 732 p.

Schaffer, B. and Whiley, A.W. (2002) Environmental physiology. In: A.W. Whiley, B. Schaffer, B.N. Wolstenholme (eds) *Avocado: Botany, Production and Uses*. CABI Publishing, Wallingford, UK, pp. 135–160.

Schaffer, B. and Whiley, A.W. (2003) Environmental regulation of photosynthesis in avocado trees- a mini review. In: *Proceedings of the V World Avocado Congress (Actas del V Co9ngreso Mundial del Aguacate)*, pp. 335–342.

Scholefield, P.B., Sedgley, M. and Alexander, D.McE. (1985) Carbohydrate cycling in relation to shoot growth, floral initiation and development and yield in the avocado. *Scientia Horticulturae* 25, 99–110.

Schroeder, C.A. and Wieland, P.A. (1956) Diurnal fluctuation in size in various parts of the avocado tree and fruit. *Proceedings of the American Society for Horticultural Science* 68, 253–258.

Shalhevet, J. (2006) *Salinity and Water Management in Avocado*. Institute of Soils, Water, and Environmental Sciences, The Volcani Center, Bet-Dagan, Israel. Available at: https://citeseerx.ist.psu.edu/document?repid=rep1&type=pdf&doi=19723e9ebe3c2f8fd509f20f6854b39bfb51e4a5 (accessed 16 October 2024).

Sharon, Y. (1999) Aspects of water economy of Hass avocado trees (*Persea americana*, cv. Hass). 1. Plant water status and gas exchange. *South African Avocado Growers' Association Yearbook* 22, 106–109.

Silber, A., Israeli, Y., Levi, M., Keinan, A., Shapira, O., *et al.* (2012) Response of 'Hass' avocado trees to irrigation management and root constraint. *Agricultural Water Management* 104, 95–103.

Silber, A., Naor, A., Cohen, H., Bar-Noy, Y., Yechieli, N., *et al.* (2018) Avocado fertilization: matching the periodic demand for nutrients. *Scientia Horticulturae* 241, 231–240. doi:10.1016/j.scienta.2018.06.094.

Silber, A., Noy, M., Naor, A., Levi, M., Keinan, A., Chudi, G. and Assouline, S. (2015a) Actually close the avocado stomata during a daytime? In: *Proceedings of the VIII World Avocado Congress*, Lima, Perú.

Silber, A., Noy, M., Naor, A., Levi, M., Keinan, A., Chudi, G., & Assouline, S. (2015b) Actually close the avocado stomata during a daytime?. In VIII Congreso Mundial de la Palta 2015 Proceedings.

Singh, B. and Craswell, E. (2021) Fertilizers and nitrate pollution of surface and ground water: an increasingly pervasive global problem. *SN Applied Sciences* 3, 518. doi.org/10.1007/s42452-021-04521-8.

Slater, G.G., Shankman, S., Shepherd, J.S. and Alfin-Slater, R.B. (1975) Seasonal composition of California avocados. *Journal of Agricultural and Food Chemistry* 23, 468–474.

Smith, T.E., Stephenson, R.A., Asher, C.J. and Hetherington, S.E. (1997a) Boron deficiency of avocado. 1. Effects on pollen viability and fruit set. In: Bell, R.W. and Rerkasem, B. (eds) *Boron in Soils and Plants*. Kluwer Academic Publishers, Dordrecht, The Netherlands, pp. 131–134.

Smith, T.E., Asher, C.J., Stephenson, R.A. and Hetherington, S.E. (1997b) Boron deficiency of avocado. 2. Effects on fruit size and ripening. In: Bell, R.W. and Rerkasem, B. (eds) *Boron in Soils and Plants*. Kluwer Academic Publishers, Dordrecht, The Netherlands, pp. 135–137.

Snijder, B. (2015) Fertilizer needs for Maluma - a case study. In: *Proceedings of the VIII World Avocado Congress,* Lima, Perú.

Snijder, B., Mathumbu, J.M. and Kruger, F.J. (2003) Development of fruit maturity and mineral content norms for export avocado cultivars from different South African avocado growing regions. *South African Avocado Growers' Association Yearbook* 26, 51–54.

Solórzano-Acosta, R., Toro, M. and Zúñiga-Dávila, D. (2023) Plant growth promoting bacteria and arbuscular mycorrhizae improve the growth of *Persea americana* var. Zutano under salt stress conditions. *Journal of Fungi* 9, 233. doi:10.3390/jof9020233.

Sommaruga, R. and Eldridge, H.M. (2021) Avocado production: water footprint and socio-economic implications. *EuroChoices* 20, 48–53. doi:10.1111/1746-692X.12289.

Sotomayor, A., Gonzáles, A., Jin Cho, K., Villavicencio, A., Jackson, T. and Viera, W. (2019) Effect of the application of microorganisms on the nutrient absorption in avocado (*Persea americana* Mill.) seedlings. *Journal of the Korean Society of International Agriculture* 31, 17–24. doi:10.12719/KSIA.2019.31.1.17.

Spann, T. (2016) Final report issued giving tools for fertilization and salinity management. *From the Grove*, winter 2016, 26–29. Available at: https://www.californiaavocadogrowers.com/sites/default/files/documents/11-Final-Report-Issued-Giving-Tools-for-Fertilization-and-Salinity-Management-Winter-2016.pdf (accessed 21 May 2024).

Spann, T. (2019) Avocado fertilization: the macronutrients. *From the Grove,* 32–35. Available at: https://www.californiaavocadogrowers.com/sites/default/files/documents/11-Avocado-Fertilization-The-Macronutrients-Fall-2019.pdf (accessed 21 May 2024).

Sperling, O., Erel, R., Dag, A. and Yermiyahu, U. (2022) Mineral mass balances reveal the phenology of evergreen and deciduous tree crops' nutrient uptake. *Irrigation Science* 41, 649–661. doi:10.1007/s00271-022-00824-9.

Srivastava, O.P. and Sethi, B.C. (1981) Contribution of farmyard manure on the buildup of available Zn in an aridisol. *Communications in Soil Science and Plant Analysis* 12, 355–361.

Stassen, P., Kangueehi, G. and Woolridge, J. (2010) Macro and micronutrient requirements of fruit trees. *SA Fruit* Journal 9(2): 44–45.

Stassen, P.J.C., Janse van Vuuren, B.P.H. and Davie, S.J. (1997) Preliminary studies on macro-element utilization by Hass avocado trees. *South African Avocado Growers' Association Yearbook* 20, 68–73.

Steduto, P., Hsiao, T.C., Raes, D. and Fereres, E. (2009) AquaCrop-the FAO crop model to simulate yield response to water: I. Concepts and underlying principles. *Agronomy Journal* 101(3), 426–437.

Steduto, P., Hsiao, T.C., Fereres, E. and Raes, D. (2012) Crop yield response to water. *Irrigation and Drainage Paper Nr. 66*. FAO, Rome, Italy, 500 p.

Steinhardt, R. (1991) Estimating yield loss of avocado trees due to a cut in water application. *Hassadeh* 71, 1363–1366. (In Hebrew with English summary.)

Steinhardt, R., Kalmar, D. and Lahav, E. (1995a) Salinity damage to avocado trees on Mexican rootstock. A. Existing knowledge, methods of research and salinity effects on soil, growth and nutrient uptake. *Alon Hanotea* 49, 326–338. (In Hebrew with English summary.)

Steinhardt, R., Kalmar, D., Miari, A. and Lahav, E. (1995b) Salinity damage to avocado trees on Mexican rootstock. B. Damage to yield and its economic significance. *Alon Hanotea* 49, 374–382. (In Hebrew with English summary.)

Steinhardt, R., Kalmar, D., Miari, A. and Lahav, E. (1995c) VC65 rootstock inducing salinity resistance and productivity to Fuerte avocado. *Alon Hanotea* 49, 452–460. (In Hebrew with English summary.)

Sterne, R.E., Kaufmann, M.R. and Zentmyer, G.A. (1977a) Environmental effects on transpiration and leaf water potential in avocado. *Physiologia Plantarum* 41, 1–6.

Sterne, R.E., Zentmyer, G.A. and Kaufmann, M.R. (1977b) The influence of matric potential, soil texture and soil amendment on root disease caused by *Phytophthora cinnamomi*. *Phytopathology* 67, 1495–1500.

Stolzy, L.H., Zentmyer, G.A., Klotz, L.J. and Labanauskas, C.K. (1967) Oxygen diffusion, water and *Phytophthora cinnamomi* in root decay and nutrition of avocado. *Proceedings of the American Society for Horticultural Science* 90, 67–76.

Swaine, M.D. and Whitmore, T.C. (1988) On the definition of ecological species groups in tropical forests. *Plant Ecology* 75(1), 81–86. doi:10.1007/BF00044629.

Syvertsen, J.P. (1987) Nitrogen content and CO_2 assimilation characteristics of citrus leaves. *HortScience* 22(2), 289–291.

Tal, A. (2016) Rethinking the sustainability of Israel's irrigation practices in the Drylands. *Water Research* 90, 387–394. doi:10.1016/j.watres.2015.12.016.

Tamayo-Vélez, A., Bernal-Estrada, J.A. and Díaz-Díez, C.A. (2018) Composition and removal of nutrients by the harvested fruit of avocado cv. Hass in Antioquia'. *Revista Facultad Nacional de Agronomía Medellín*, 71, 8511–8516.

Tamayo-Vélez, A. de J., Bernal-Estrada, J.A, Díaz-Díez, C.A., Cano-Gallego, L.E. and Córdoba-Gaona, O. de J. (2022) Effect of rootstock/scion compatibility on fruit and foliar nutrient composition in avocado (*Persea americana* Mill.) cv. Hass in Colombia. *Revista Colombiana de Ciencias Hortícolas* 16, e14833. doi:10.17584/rcch.2022v16i3.14833.

Tapia, V.L.M. and Bravo, M.E. (2009) Consumo de agua. In *Impactos Ambientales Y Socioeconómicos Del Cambio De Uso Del Suelo Forestal A Huertos De Aguacate En Michoacán*; Bravo Espinoza, M., Ed.; Instituto Nacional de Investigaciones Forestales Agrícolas y Pecuarias (INIFAP): Mexico City, Mexico, 2009; pp. 25–27.

Tapia, L.M., Larios, A., Tapia, L., Anguiano, J. and Vidales, I. (2007) Nitrate leaching in two nutritional and water management systems of avocado in Michoacán. In: *Proceedings of the VI World Avocado Congress,* Viña Del Mar, Chile.

Taylor, N.J., Mazhawu, E., Clulow, A.D. and Savage, M.J. (2021) Water use of avocado orchards – year 4. *South African Avocado Growers' Association Yearbook* pp. 446–516.

Thorp, G., Boldingh, H., Barnett, A. and Minchin, P. (2009) Alternate bearing in 'Hass' avocado: possible role of carbohydrates and boron. In: *Proceedings of the 4th Australian and New Zealand Avocado Growers Conference*, Cairns, Australia, pp. 23–24.

Thorp, T.G., Barnett, A.M., Boldingh, H., Elmsly, T. and Minchin, P.E.H. (2011) Is boron transport to avocado flowers regulated by carbohydrate supply. In: *Proceedings of the VII World Avocado Congress,* Cairns, Australia (Abstract).

Tie X., Zhang R., Brasseur, G. and Lei, W. (2002) Global NOx production by lightning. *Journal of Atmospheric Chemistry* 43, 61–74. doi:10.1023/A:1016145719608.

Tolomio, M. and Raffaele, C. (2020) Dynamic crop models and remote sensing irrigation decision support systems: a review of water stress concepts for improved estimation of water requirements. *Remote Sensing* 12(23), 3945. doi:10.3390/rs12233945.

Tomassini-Vidal, L. and Pino-Huasacca, N. (2022) Macronutrient dynamics in avocado (*Persea americana* Mill.) 'Hass' under the central Peruvian coast conditions. *Peruvian Journal of Agronomy* 6, 53–77. doi:10.21704/pja.v6i1.1758.

Tong, X., Wu, P., Liu, X., Zhang, L., Zhou, W. and Wang, Z. (2022) A global meta-analysis of fruit tree yield and water use efficiency under deficit irrigation. *Agricultural Water Management* 260, 107321. doi:10.1016/j.agwat.2021.107321.

Torres, J. (2017) Efecto de la carga frutal sobre variables fisiológicas en palto (*Persea americana* Mill.) variedad Hass. Available at: https://repositorio.uchile.cl/handle/2250/151500. (accessed 19 May 2024).

Torres, J., Mena, F., Zulueta, C., Pinto, M.A. and Gardiazábal, F. (2023) Use of DMP - Nitrobrake®, to improve the urea fertilization efficiency in Chile. In: *Proceedings of the 10th World Avocado Congress*. Auckland, New Zealand. 1, 373. Available at: https://industry.nzavocado.co.nz/wp-content/uploads/2023/04/1630-Claudio -Zulueta.pdf (accessed 19 May 2024).

Torres, M.D., Hermoso, J.M. and Farré, J.M. (2003) Nutrición B, Zn y Cu del aguacate. Comparación de métodos de fertilización. In: *Proceedings of the V World Avocado Congress,* Málaga, Spain.

Tracy, J.E. (1983) Manganese toxicity in avocado (*Persea americana* Mill.). *California Avocado Society Yearbook* 67, 147–156.

Tracy, J.E. (1985) A preliminary report on phosphorus deficiency of Hass avocado. *California Avocado Society Yearbook* 69, 145–154.

Trochoulias, T., Broadbent, P. and Baigent, D.R. (1986) Response of avocado to calcareous and organic amendments. *Acta Horticulturae* 175, 179–181.

Tzatzani, T., Psarras, G., Scuderi, D., Kokolakis, E. and Papadakis, I.E. (2022) Effect of organic fertilizers on avocado trees (cvs. Fuerte, Hass, Lamb Hass) in Western Crete, a cool subtropical region. *Sustainability*, 14, 12221. doi:10.3390/su141912221.

Valoras, N., Letey, J., Stolzy, L.J. and Fraligh, E.F. (1964) The oxygen requirement for root growth of three avocado varieties. *Proceedings of the American Society for Horticultural Science* 85, 172–178.

Van Rooyen, Z. and Bower, J.P. (2005) The role of fruit mineral composition on fruit softness and mesocarp discolouration in 'Pinkerton' avocado (*Persea americana* Mill.). *Journal of Horticultural Science and Biotechnology* 80, 793–799.

Verbist, K., Rojas, P. and H. Maureira. (2020) A stress test for climate change impacts on water security. Case study from the limarí watershed in Chile. UNESCO, Paris, 45.

Vicca, S., Zavalloni, C., Fu, Y.S.H., Voets, L., Dupré de Boulois, H., *et al.* (2009) Arbuscular mycorrhizal fungi may mitigate the influence of a joint rise of temperature and atmospheric on soil respiration in grasslands. *International Journal of Ecology* 1, 10. doi:10.1155/2009/209768.

Vidal, M.T., Azcon-Aguilar, C., Barea, J.M. and Pliego-Alfaro, F. (1992) Mycorrhizal inoculation enhances growth and development of micropropagated plants of avocado. *HortScience* 27, 785–787.

Viera, W., Cartagena, Y., Toaquiza, J., Gaona, P., Viteri, P., Sotomayor, A. and Medina, L. (2021) Response of 'Hass' avocado under different nitrogen and potassium fertilizer regimes in subtropical Ecuador. *Acta Horticulturae* 1327, 175–180. doi:10.17660/ActaHortic.2021.1327.23.

Viera, W., Gaona, P., Samaniego, I., Sotomayor, A., Viteri, P., *et al.* (2023) Mineral content and phytochemical composition of avocado var. Hass grown using sustainable agriculture practices in Ecuador. *Plants* 12, 1791. doi:10.3390/plants12091791.

Violi, H.A., Treseder, K.K., Menge, J.A., Wright, S.F. and Lovatt, C.J. (2007) Density dependence and interspecific interactions between arbuscular mycorrhizal fungi mediated plant growth, glomalin production, and sporulation. *Canadian Journal of Botany* 85, 63–75. doi:10.1139/B06-151.

Vorster, L.L. and Bezuidenhout, J.J. (1988) Does zinc play a role in reducing pulp spot? *South African Avocado Growers' Association Yearbook* 11, 60.

Vuthapanich, S. (1998) Preharvest practices affecting postharvest quality and mineral composition of 'Hass' avocado fruit. PhD thesis, The University of Queensland, Brisbane, Australia.

Wallace, A., Shannon, L.M., North, C.P. and Mueller, R.T. (1955) Glasshouse studies on the salt tolerance and growth of Persea Floccosa as a rootstock. *California Avocado Society 1955 Yearbook* 39, 179–183.

Wallihan, E.F., Embleton, T.W. and Printy, W. (1958) Zinc deficiency in avocado. *California Avocado Society Yearbook* 42, 4–5.

Wallihan, E.F., Goodall, G.E. and Embleton, T.W. (1968) Iron chlorosis in relation to irrigation practices. *California Avocado Society Yearbook* 52, 120–124.

Wehr, J.B., Menzies, N.W. and Blamey, F.P.C. (2004) Inhibition of cell-wall autolysis and pectin degradation by cations. *Plant Physiology and Biochemistry* 42, 485–492.

Weir, R.G. and Cresswell, G.C. (1995) *Plant Nutrient Disorders 2. Tropical Fruit and Nut Crops*. Inkata Press, Melbourne, Australia. 101 pp.

Whiley, A.W. (1994) Ecophysiological studies and tree manipulation for maximisation of yield potential in avocado (*Persea americana* Mill.). PhD thesis, University of Natal, Pietermaritzburg, South Africa.

Whiley, A.W. and Schaffer, B. (1994) Avocado. In: Schaffer, B. and Anderson, P.C. (eds) *CRC Handbook of Environmental Physiology of Fruit Crops*. Vol. II. CRC Press Inc., Boca Raton, Florida, pp. 3–35.

Whiley, A.W., Pegg, K.G., Saranah, J.B. and Langdon, P.W. (1987) Influence of Phytophthora root rot on mineral nutrient concentrations in avocado leaves. *Australian Journal of Experimental Agriculture* 27, 173–177.

Whiley, A.W., Chapman, K.R. and Saranah, J.B. (1988a) Water loss by floral structures of avocado (*Persea americana* Mill.) cv. Fuerte during flowering. *Australian Journal of Agricultural Research* 39, 457–467.

Whiley, A.W., Saranah, J.B., Cull, B.W. and Pegg, K.G. (1988b) Manage avocado tree growth cycles for productivity gains. *Queensland Agricultural Journal* 114, 29–36.

Whiley, A.W., Winks, C.W., Stephenson, R.A. and Winston, E.C. (1988c) Boron nutrition of subtropical and tropical fruit and nut crops in Queensland. In: Winks, C.W. (ed.) *Maroochy Horticultural Research Station Report* 5, QDPI, Nambour, Queensland, pp. 103–104.

Whiley, A.W., Smith, T.E., Saranah, J.B. and Wolstenholme, B.N. (1996) Boron nutrition of avocados. *Talking Avocados*, 7, 12–15.

Whiley, A.W., Searle, C., Schaffer, B. and Wolstenholme, B.N. (1999) Cool orchard temperature or growing trees in containers can inhibit leaf gas exchange of avocado and mango. *Journal of the American Society of Horticultural Science* 124(1), 46–51.

Whiley, A.W., Coates, L., Hofman, P., Dann, E., Pegg, K., Marques, R., Dean, J., Shuey, L., Smith, L., Cooke, T., Stubbings, B., Smith, J. and Whiley, D. (2013a) *Rootstock Improvement for the Australian Avocado Industry - Phase 3. Final report AV08000.* Horticulture Australia Ltd., Sydney.

Whiley, A.W., Wolstenholme, B.N. and Faber, B. (2013b) Crop management. In: Schaffer, B. Wolstenholme, B.N. and Whiley, A.W. (eds) *The Avocado: Botany, Production and Uses*, 2nd edn. CABI, Wallingford, UK, pp. 342–379.

White, P. (2012) Ion uptake mechanisms of individual cells and roots: short-distance transport. In: Marschner, P. (ed.) *Marschner's Mineral Nutrition of Higher Plants*. Academic Press, London, pp. 7–47.

White, P.J. and Broadley, M.R. (2003) Calcium in Plants. *Annals of Botany* 92, 487–511.

Whitmore, T.C. (1990) Forest dynamics. In: *An Introduction to Tropical Rainforests*. Claredon Press, Oxford, pp. 99–132.

Winer, L. and Zachs, I. (2007) Daily trunk contraction in relation to a base line as an improved criterion for irrigation in avocado. In: *Proceedings of the 6th World Avocado Congress*, Viña Del Mar, Chile.

Winer, L., Reuveni, B., Bar, Y., Haas, J. and Zviebel, A. (1998) Influence of autumn fertilization with nitrogen and phosphorus on root activity and development in avocado. In: *Proceedings of the 3rd World Avocado Congress*, Tel Aviv, Israel, pp. 172–180.

Winer, L., Zachs, I. and Chudzik, P. (2007) Avocado fruit growth during winter: the importance of irrigation between rains. In: *Proceedings of the 6th World Avocado Congress*, Viña Del Mar, Chile.

Witney, G.W. and Arpaia, M.L. (1991) Tree recovery after the December 1990 freeze. *California Avocado Society Yearbook* 75, 63–70.

Witney, G.W., Hofman, P.J. and Wolstenholme, B.N. (1990a) Mineral distribution in avocado trees with reference to calcium cycling and fruit quality. *Scientia Horticulturae* 44, 279–291.

Witney, G.W., Hofman, P.J. and Wolstenholme, B.N. (1990b) Effect of cultivar, tree vigour and fruit position on calcium accumulation in avocado fruit. *Scientia Horticulturae* 44, 269–278.

Wolstenholme, B.N. (1987) Theoretical and applied aspects of avocado yield as affected by energy budgets and carbon partitioning. *South African Avocado Growers' Association Yearbook* 10, 58–61.

Wolstenholme, B.N. (1991) Making an avocado fruit: energy expensive but mineral cheap. *Avokad* 11, 8–9.

Wolstenholme, B.N. (2004) Nitrogen – the manipulator element: managing inputs and outputs in different environments. *South African Avocado Growers' Association Yearbook* 27, 45–61.

Wolstenholme, B.N. and Whiley, A.W. (1997) What do carbohydrate reserves tell us about avocado orchard management. *South African Avocado Growers' Association Yearbook* 20, 63–67.

Wolstenholme, B.N. and Whiley, A.W. (1999) Ecophysiology of the avocado (*Persea americana* Mill.) tree as a basis for pre-harvest management. *Revista Chapingo Serie Horticultura Núm Especial* V, 77–88.

Wolstenholme, B.N., Whiley, A.W. and Saranah, J.B. (1990) Manipulating vegetative: reproductive growth in avocado (*Persea americana* Mill.) with paclobutrazol foliar sprays. *Scientia Horticulturae* 41, 315–327.

Wolstenholme, B.N., Moore-Gordon, C.S. and Cowan, A.K. (1998) Mulching of avocado orchards: Quo vadis? *South African Avocado Growers' Association Yearbook* 21, 26–28.

Wolstenholme, B.N. (2002) Ecology: climate and the edaphic environment. In: Whiley, A.W., Schaffer, B. and Wolstenholme, B.N. (eds). *The Avocado: Botany, Production and Uses*, 1st edn, CABI, Oxfordshire, UK, pp. 71–99.

Wutscher, H.K. and Maxwell, N.P. (1975) Seasonal changes in 12 leaf nutrients of 'Lula' avocado with drip and flood irrigation. *HortScience* 10, 512–514.

Yang, B., Fu, P., Lu, J., Ma, F., Sun, X. and Fang, Y. (2022) Regulated deficit irrigation: an effective way to solve the shortage of agricultural water for horticulture. *Stress Biology* 2, 28.

Yates, M.V., Storrlemyer, D.E. and Meyer, J.L. (1992) Irrigation and fertiliser management to minimise nitrate leaching in avocado production. In: *Proceedings of the 2nd World Avocado Congress*, Orange, California, pp. 331–335.

Ye, X., Abe, S. and Zhang, S. (2020) Estimation and mapping of nitrogen content in apple trees at leaf and canopy levels using hyperspectral imaging. *Precision Agriculture* 21, 198–225.

Young, T.W. and Koo, R.C.J. (1977) Influence of soil and cultivar on mineral composition of avocado leaves in Florida. *Journal of the American Society for Horticultural Science* 102, 308–311.

Zhang, L., Niu, H., Wang, S., Zhu, X., Luo, C., Li, Y. and Zhao, X. (2012) Gene or environment? Species-specific control of stomatal density and length. *Ecology and Evolution* 2(5), 1065–1070. Available at: doi: 10.1002/ece3.233.

Zekri, M., Núñez-Elisea, R., Schaffer, B., O'Hair, S.K. and Crane, J.H. (1999) Multi-sensor capacitance probes for monitoring soil water dynamics in tropical fruit orchards in south Florida. *Proceedings of the Florida State Horticultural Society* 112, 178–181.

Zentmyer, G.A. and Richards, S.J. (1952) Pathogenicity of *Phytophthora cinnamomi* of avocado trees and the effect of irrigation on disease development. *Phytopathology* 42, 35–37.

Zilberstaine, M., Eshel, A. and Waisel, Y. (1992) Potassium uptake by avocado roots. In: Lovatt, C., Holthe, P.A. and Arpaia, M.L. (eds) *Proceedings of the Second World Avocado Congress*. University of California, Riverside, California, 1, 221–225.

Zilkah, S., David, I., Yeselson, Y. and Klein, I. (1999) Seasonal uptake and distribution of ^{15}N-labelled ammonia and nitrate in young avocado trees. *Alon Hanotea* 53, 416–425. (In Hebrew with English summary.)

Zilkah, S., Klein, I., Feigenbaum, S. and Weinbaum, S.A. (1987) Translocation of foliar-applied urea ^{15}N to reproductive and vegetative sinks of avocado and its effect on initial fruit set. *Journal of the American Society for Horticultural Science* 112, 1061–1065.

Zilkah, S., Wiesman, Z., Klein, I. and David, I. (1996) Foliar applied urea improves freezing protection in avocado and peach. *Scientia Horticulturae* 66, 85–92.

Zion Market Research (2022) Organic fruits and vegetables market size, share, growth report 2022-2030. Available at: www.zionmarketresearch.com /report/organic-fruits-vegetables-market (accessed 19 May 2024).

Zude-Sasse, M. and Schaffer, B. (2000) Influence of soil oxygen depletion on iron uptake and reduction in mango (*Mangifera indica* L.) roots. *Proceedings of the Florida State Horticultural Society* 113, 1–4.

12 Crop Management

A.W. Whiley[1], F.J. Gardiazabal[2], F.J. Mena[2,*], B.N. Wolstenholme[3] and J.H. Crane[4]
[1]*Sunshine Horticultural Services, Queensland, Australia;* [2]*GAMA, Quillota, Chile;* [3]*University of KwaZulu-Natal, Pietermaritzburg, South Africa;* [4]*Tropical Research and Education Center, University of Florida, Homestead, USA*

Abstract

Crop management discusses the use of avocado tree phenology models to assist with timing of management inputs, mulching and composting to improve tree health and performance, the application of plant bioregulators (primarily the triazoles and gibberellic acid) to modify tree responses, orchard design and canopy management as appropriate for differing climates and associated socio-economic conditions, sustainable production and improved regularity of cropping, including management techniques to reduce alternate bearing and the use of girdling/scoring to enhance productivity.

Introduction

Compared to citrus, pip and stone fruit crops, avocados have a relatively short period of commercial domestication with related intensive management. Among the most intractable horticultural problems of avocado orchards are almost universal low and erratic production. Average sustainable yield in many countries was given by Wolstenholme and Whiley (1998) and Gazit and Degani (2002) at less than 10 t ha^{-1}, with the best growers averaging 12–15 t ha^{-1} in the semi-arid winter rainfall subtropics, and 20–25 t ha^{-1} in the humid, summer rainfall subtropics. Wolstenholme (1987) estimated yield potential to be in excess of 30 t ha^{-1} with known cultivars, rootstocks and technology at the time. Progress has since been made with increasing average yields, especially in newer orchards using high-density plantings incorporating the use of plant growth bioregulators, strategic pruning and other techniques such as girdling (Gardiazabal *et al.*, 2015d). New crop management practices have and are being tested, aside from improvements in Phytophthora root rot control, which remains a major production-limiting factor worldwide (see Dann and Gazis, Chapter 13, this volume). This chapter discusses some of the significant management areas where the use of mulches and composts, improved orchard design, strategic use of plant bioregulants (PBRs) and techniques that reduce alternate bearing (AB) can be implemented to improve production performance. Irrigation and mineral nutrition are dealt with separately (see Gil *et al.*, Chapter 11, this volume).

Phenological Models

Phenology models describe a holistic and integrated way in which different growth events

Corresponding author: fmena@gama.cl

 The Avocado: Botany, Production and Uses (eds D. Carrillo, B. Schaffer, A.W. Whiley and B.N. Wolstenholme)
DOI: 10.1079/9781800621824.0012

occur in plants. In trees, these events are the visible expression of physiological changes driven by genetic × environmental interactions. Phenological modelling can be utilized as a powerful tool for managing orchards. It is particularly useful when dealing with fruit tree crops where current performance is affected by previous physiological events. Yield management involves altering resource (assimilate) allocation between reproductive and vegetative structures of the tree. Competing sinks can have a direct impact on productivity in avocado but timing of sink strength can be manipulated so that competition between reproductive and vegetative structures is lessened, thereby increasing potential yield (Wolstenholme *et al.*, 1990; Whiley *et al.*, 1991).

Chandler (1958) and Venning and Lincoln (1958) provided the first descriptions of phenological events in avocado, which were subsequently expanded by Kotzé (1979), Wolstenholme (1981) and Davenport (1982). However, the first detailed conceptual model of avocado phenology was published by Whiley *et al.* (1988) for the humid, warm subtropics of south-east Queensland, with management strategies (nutrition) linked to growth events (Fig. 12.1). With an idealized trunk starch concentration curve superimposed, the model has increased explanatory depth (Whiley *et al.*, 1995b). Under the relatively mesic conditions for which the model was developed, the seasonal cycle of starch reserves does not vary widely under good orchard management. The two shoot growth periods (spring and summer), alternating with early summer and autumn root flushes, do not vary greatly with season (Whiley and Wolstenholme, 1990). Consequently, yield is fairly regular from year to year in the absence of extreme weather events. However, phenology can differ widely among different environments. In more stressful semi-arid Mediterranean climates at higher latitudes, such as in California, Israel, Spain and Chile, there are three shoot flushes (an additional autumn flush). In the tropical Nayarit State of Mexico, Salazar-García *et al.* (2006) described four shoot flushes – a major late winter and three minor summer flushes, while in southern Florida, with a marine, warm/hot, subtropical climate with summer rainfall, avocado trees have up to four flushes distributed randomly from spring flowering through to late summer (J.H. Crane, University of Florida, USA, 2023, personal communication). All these flushes contribute flowering/fruiting sites for the next season, maintaining a balance between vegetative:reproductive growth. In most avocado-growing areas, deviations from the idealized phenological growth cycle are the norm, disrupted by abnormal weather

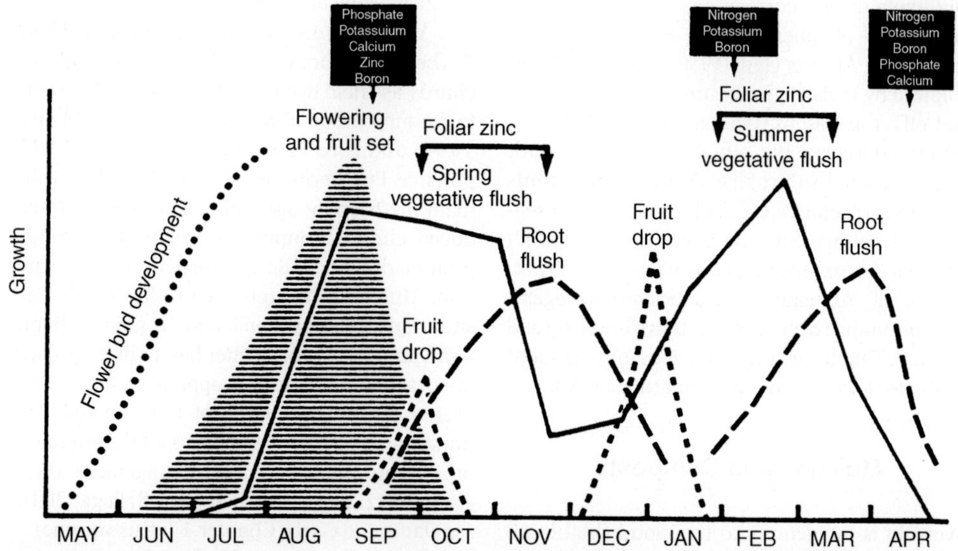

Fig. 12.1. Phenology model developed for 'Fuerte' avocado trees growing in a warm, subtropical climate at Nambour, south-east Queensland. Redrawn from Whiley *et al.* (1988).

events and seasonal variations in crop load, impacting the vegetative:reproductive balance, which can lead to AB, a major problem in many production areas.

The phenological model concept has been widely adopted, with modifications for local conditions, as a research and extension tool for more science-based timing of orchard management practices. For example, tree phenology has been successfully used to improve the efficiency of Phytophthora root rot control with trunk-injected or foliar-applied phosphonate fungicides. Timing of applications in relation to growth events (relative sink strength of roots in late autumn/early winter and late spring) maximizes root phosphonate concentrations, thereby enhancing protection (Whiley *et al.*, 1995a, 2001), although effective treatments can be made during summer to reinforce root protection during a period of high infection (E.K. Dann, Hort Health, Brisbane, Australia, 2024, personal communication).

More recently, phenological models have been developed to portray 'on' and 'off' AB cycles. The cool climate in the Bay of Plenty, New Zealand, requires 'Hass' fruit to hang on the tree for considerably longer than 12 months for acceptable horticultural maturity, complicating management and aggravating AB and irregular bearing. A circular two-season 'on'–'off' crop chart (Hardy *et al.*, 2008) diagrammatically explains the consequences of timing and magnitude of phenological events. A grower-oriented AB overview (Wolstenholme, 2009) followed by revised growth model charts for 'on' and 'off' crop seasons (Wolstenholme and Sheard, 2009) addressed the situation for the South African humid subtropics. Reproductive events dominate vegetative events in an 'on' crop season, with consequent greater drawdown of starch reserves and reduced vegetative (shoot and root) flushes. 'Off' crop seasons are characterized by vegetative dominance over reproduction (flowering and fruiting). The charts suggest obvious management strategies for reducing the amplitude of AB.

Mulches and Composts

Avocado is indigenous to the cloud (highland) and lowland forests of Mexico and Central America, where it has adapted to soils with abundant surface organic litter and high organic matter content. These provide a well-aerated substrate, rich in microorganisms and with a high water-holding capacity (see Ramirez *et al.*, Chapter 5, this volume). Under these conditions, a dense mat of 'feeder' (unsuberized, white) roots develops to exploit the supply of mineralized nutrients released from decaying vegetation and to take up water to meet tree requirements. The litter zone also provides a buffer between the air/soil interface, moderating the impact of atmospheric changes on the root environment, thereby protecting the fleshy roots from desiccation and large changes in temperature (Gregoriou and Rajkumar, 1984). Domestication of avocado has introduced this crop into a monoculture system, which has become reliant on chemical fertilizers and pesticides to reduce production costs. The benefit of maintaining high levels of organic matter in avocado soils to suppress *Phytophthora cinnamomi* (root rot) activity has been known for some time (Broadbent and Baker, 1974; Pegg and Whiley, 1987). Prior to the development of effective fungicides to control root rot in the late 1970s, mulching trees with organic substances to maintain tree health was widely practised in some countries (Pegg and Whiley, 1987). However, the availability of a low-cost and effective systemic fungicide (phosphorous acid) in the 1980s (Pegg *et al.*, 1985) reduced the reliance on organic additives, resulting in a decline in their widespread use in avocado orchards.

More recently, attention has switched back to the use of organic mulches in avocado orchards as their use provides additional benefits to the suppression of root rot (Turney and Menge, 1994) and is recognized as a more sustainable practice. Furthermore, sustainable and environmentally friendly agricultural systems with reduced chemical inputs are currently strongly promoted worldwide in horticultural production. Mulching has returned to favour in most situations; compost application is gaining popularity, and soil organic matter has assumed greater importance in natural cropping systems. Integrated disease management is promoted, and mulching is again regarded as a vital component of Phytophthora root rot management (Pegg *et al.*, 2002; Pegg and Giblin, 2008; Pegg, 2010; see Dann and Gazis, Chapter 13, this volume).

In a cool, mesic locality in subtropical South Africa, Moore-Gordon *et al.* (1996, 1997) and Wolstenholme *et al.* (1998) reported on the effects of mulching 'Hass' avocados with a 150 mm thick layer of composted pine bark. In summary, they found that over the 3 years of research, mulching increased mean fruit size by 6.6%, which gave significant economic benefits due to improved pack-outs, increased mean tree fruit numbers by 14.7% and increased yield by 22.6% (Table 12.1). The benefits of mulching on tree performance are likely due to improved root growth that was measured under mulched trees and the associated amelioration of stress. The reduced stress in mulched trees was apparent from their cooler canopies (about 3°C lower during summer), less photoinhibition of leaves during summer and autumn and a reduction in pedicel ring-neck (47% less) and premature seed-coat degeneration (39% less) (Wolstenholme *et al.*, 1998). The primary aim of that study was to reduce the proportion of phenotypically small 'Hass' fruit and to improve the understanding of avocado fruit growth physiology (Cowan *et al.*, 1997, 2005). Subsequently, Dahan *et al.* (2010) cloned distinct cell-proliferation genes from avocado and proposed a negative role for an fw2.2-like gene as a regulator of fruit cell division. They concluded that considering previous studies which argued that phenotypically small 'Hass' fruits might be linked to changes in sink strength and sugar content, and knowing that small fruit are characterized by decreased IAA content and elevated ABA levels (Cowan *et al.*, 1997, 2005; Moore-Gordon *et al.*, 1998; Richings *et al.*, 2000), limiting sucrose supply and/or changes in fruit hormone homeostasis might control the accelerated repression of transcription of the two mitotic cyclins and the three cDNAs termed PaCYCA1, PaCYCB1 and PaPCNA, found in Hass fruit.

Suitable material for mulching is still debated, and choice is often related to expense and local availability. Composted pine bark used in the South African study is usually an expensive material used more commonly in high-value, ornamental and protected horticulture. However, in the Moore-Gordon *et al.*'s (1996, 1997) experiment, the initial cost of pine bark was recouped from improved productivity within the first 2 years of application, with the product having an expected half-life of approximately 5 years (Moore-Gordon *et al.*, 1997). Wolstenholme *et al.* (1996, 1998) discussed the properties of suitable mulching material that ideally include a carbon:nitrogen (C:N) ratio between about 25:1 and 100:1 to avoid a serious soil nitrogen drawdown (N 'negative period') that can happen with the use of hardwood chips/sawdust (C:N ratios of 400–500:1). In areas with clay soils and heavy summer rains, the time of application (autumn) becomes an important decision as thick mulches composed of small particles can be too moisture-retentive, thus exacerbating root rot problems. Materials with low C:N ratios (e.g. poultry litter and cattle manure) have high nitrogen content

Table 12.1. Effect of pine bark mulching on 'Hass' avocado productivity in subtropical South Africa.

	Untreated*	Mulch	% increase
1993/94			
Mean fruit weight (g)	198.0	221.3	11.8**
Mean fruit number tree^{-1}	509	540	6.1
Yield (t ha^{-1})	21.2	23.8	18.5**
1994/95			
Mean fruit weight (g)	178.2	199.2	11.8**
Mean fruit number tree^{-1}	262	333	27.2**
Yield (t ha^{-1})	9.4	13.4	42.2**
1995/96			
Mean fruit weight (g)	216.1	220.4	2.0
Mean fruit number tree^{-1}	698	814	16.6**
Yield (t ha^{-1})	31.7	35.8	18.9**
Average of three seasons			
Mean fruit weight (g)	203.1	216.5	6.6**
Mean fruit number tree^{-1}	509	540	14.7**
Yield (t ha^{-1})	20.0	24.4	22.6**

*Data in columns are mean values of six trees. **denotes a significant ($P \leq 0.01$) increase in the response to mulching. Data are from Wolstenholme *et al.* (1998) with permission.

and thus are considered organic manures and usually act differently compared with true mulches, and in cases like chicken manure, parameters such as chloride and sodium content have to be evaluated before the use of these type of amendments. Mulching products commonly used in subtropical Australia include barley and sorghum straw, sugarcane tops, stalky grasses such as Rhodes grass (*Chloris gayana*), filter press (from sugarcane milling) and groundnut husks, although the latter two contribute a significant quantity of nitrogen that increases tree vigour. Pegg and Giblin (2008) also list aged hardwood chips, wheat straw, lucerne meal, cotton waste and chicken manure. Chipped avocado prunings are readily available and make excellent mulch. Increasingly, growers are making composts from locally available materials. Fire and frost are additional hazards that must be assessed before mulches are applied, as dry litter beneath trees can increase the risks of damage from both.

In a 3-year mulch trial in California, Faber *et al.* (2003) noted that avocado feeder roots grew out of a clay loam soil and were concentrated at the soil/mulch interface, being a zone with improved aeration and lower fungal pathogen numbers. Californian experience warns against mulches accumulating around the base of trees where they can encourage *Phytophthora citricola* crown infection.

In the very gravelly calcareous soils of southern Florida, mulching with a product primarily sourced from composted municipal waste and lawn clippings can increase water- and nutrient-holding capacity for tropical fruit trees (Schaffer, 1995), including avocado (B. Schaffer, University of Florida, USA, 2001, personal communication). This reduces potential leaching of agrochemicals into groundwater, which is of increasing concern in southern Florida where agricultural activities are adjacent to the environmentally sensitive natural areas of the Everglades and Biscayne National Parks (Schaffer, 1995). Until recently, application of composted sewage sludge called biosolids was a major source of avocado plant nutrients in commercial avocado orchards in southern Florida (J.H. Crane, University of Florida, USA, 2023, personal communication). The sludge was aerobically and anaerobically composted and tested to meet state and federal standards. However, food safety concerns have reduced its use, despite no factual data showing microbial contamination of fruit when used.

Crowley (2008) noted the recent interest in liquid organic amendments such as compost leachate or 'compost tea' and biologically enriched materials, such as organic fertilizers and root growth promoters for 'organically' produced avocados. There remains little scientific evidence for the superiority of most of these expensive products. Crowley reports that standard, lower cost organic mulches are 'transformed *in situ* by soil microorganisms to the same end-products that contribute to stable organic matter in the soil'. Soil inoculation with bacterial and fungal-based products has been previously discussed (Pegg and Giblin, 2008; Pegg, 2010). In recent years, a new group of bacterial soil inoculants have become available and are being exhaustively evaluated. Early results with some of these products in areas with low Phytophthora pressure (the central region of Chile) indicate that crop yield can be improved but a better understanding of the microbe composition in soil under different production conditions is required (F. Gardiazabal, GAMA, Quillota, Chile, 2023, personal communication).

It should be noted that in the absence of phosphonate and metalaxyl fungicides, organic amendments are insufficient to combat Phytophthora root rot, especially in high disease pressure situations. For example, organic orchards in Westfalia Estate at Tzaneen, South Africa, where there is very high Phytophthora root rot and fruit pathogen pressure, yields are only half that of conventionally managed orchards with first-grade pack-outs 30% lower. Under such conditions 'organic' avocado farming is difficult, expensive and probably unsustainable (D.G. Smith, Westfalia Estate, South Africa, 2011, personal communication). This reflects the parlous state of avocado industries in the humid subtropics up to the early 1980s, when rapid decline in 'organic' orchards was the norm as root–shoot ratios changed with season conditions and cropping cycles. For production under different environmental conditions, Faber (2007, 2008) has outlined the case for 'organic' orchards in California, where pest and disease pressure is lower. A review of the sustainability of organic agriculture in general is given by Raviv (2010). In conclusion, the practice of reinforcing natural leaf litter beneath avocado trees with organic mulching materials or compost is generally beneficial to tree and root health with improved performance under most situations, especially in stressful climates and where soils are low in organic matter. Careful choice of the material and its management is paramount to obtain the full benefits from this practice. The choice of mulch and its application will alter the irrigation and nutritional requirements of trees, and both should be carefully monitored with real-time, water-smart technology and leaf and soil analysis to ensure that correct balances are maintained. In some cases, mulching may reduce negative impacts on trees from soil salinity due to reduced spikes in salt concentrations as soil moisture fluctuations are moderated (Downer *et al.*, 2002). Mulching is also compatible with practices that promote orchard sustainability in an increasingly environmentally aware world, and, along with composting and application of organic manures, form the basis of 'organic' avocado production. However, care must be taken during production, harvest and subsequent handling to ensure that contaminated fruit is not marketed.

Plant Bioregulants

Phytohormones occur universally at low concentrations in all plants, where they promote or inhibit the growth of cells. In this role, they influence tissue differentiation and determine the reaction of plants to environmental cues such as gravity, temperature, light (day length) and water stress (Luckwill, 1980). PBRs are usually defined as synthetic compounds applied exogenously to modify plant growth. A PBR may be a related chemical that mimics hormone action or be identical to the naturally occurring hormone. However, other PBRs, such as the triazole group, are unrelated to hormones and either block the synthesis of a hormone or interfere with its translocation to achieve the required result. PBRs are extensively used in horticulture and are playing an increasingly important role in avocado production. Lovatt (2005) summarized research on PBRs in avocado production in California. Ethylene, a plant hormone extensively used in the postharvest management of avocado ripening and its action inhibitor 1 MCP, is discussed by Defilippi *et al.*, Chapter 15, this volume.

Auxins

Auxins have been shown to control shoot regrowth on avocado stumps following top-working trees and to improve the rooting of cuttings. Regrowth below the graft union when top-working avocado stumps to new cultivars will compete with new scion growth. Hand removal of watershoots is labour-intensive and expensive. Boswell *et al.* (1976) controlled regrowth by spraying stumps with either a 1% ethyl ester or a sodium salt formulation of naphthaleneacetic acid (NAA) in a 30% aqueous solution of a white acrylic paint when shoots were 10–80 mm long. Shoots wilted 2 h after treatment and were killed by both NAA formulations 45 days later, without any deleterious effect on the newly grafted scions that were shielded at the time of treatment. After subsequent research by Arpaia *et al.* (2007), NAA (Treehold®) was registered and commercialized in the USA as a 1.15% solution in latex paint for control of regrowth after stumping or pruning, thereby also facilitating tree size control and management of high-density plantings. Additionally, Treehold® is used to control regrowth along structural limbs following pruning.

The use of auxins to promote rooting of hardwood cuttings is widespread in propagation. The production of clonal avocado rootstocks is a common practice in California, Israel, New Zealand, Spain and South Africa and in more recent years is gaining traction in Australia, Chile and other South American countries. This propagation method requires rooting cuttings, even though it is usually done through the somewhat laborious 'nurse seedling' technique. Ernst (1981) reported on the success of using 1% indolebutyric acid (IBA) to increase the rooting of avocado cuttings placed in a mist bed with bottom heat at 30°C. However, when etiolated cuttings are rooted, as in the 'nurse seedling' or 'multiple cloning' techniques, the IBA concentration is reduced to 0.7% and applied to a small wound made at the base of the cutting in the etiolated region (Ernst, 1999). This treatment results in a high percentage of rooted clones. For additional discussion on propagation methods, see Whiley, Chapter 9, this volume.

Auxins are also being evaluated for their role in management of AB. The heavy 'on' crop causes correlative inhibition (auxin > cytokinin) in buds giving rise to reduced summer and autumn growth flushes, thereby reducing the number of inflorescence-bearing nodes the following spring. Foliar application of TIBA (tri-iodobenzoic acid, an auxin transport inhibitor) combined with BA (a cytokinin) in July (mid-summer) and January (mid-winter) increased return bloom, potentially increasing the 'off' crop. This research is ongoing (Lopez-Jiminez *et al.*, 2011) and is discussed in greater detail under AB (this chapter).

As part of a research project (Lovatt, 2005, 2010; Zheng *et al.*, 2011) on PBR use in avocado orchards, synthetic auxins such as 2,4 dichlorophenoxyacetic acid (2,4-D) and 3,5,6 trichloro-2-pyridyloxyacetic acid (3,5,6-TPA) were tested. Results show that over a 2-year period, there is no difference in total production but in the case of 2,4 D (111 g of acid equivalent ha^{1}) applied when fruit reached 17–20 mm in diameter, final size is increased. Ish-Am *et al.* (2007) showed there was no effect from the use of 2,4 D or 3,5,6 TPA (3 mg l^{-1}) on production or fruit size when applied during flowering in 'Ettinger' avocados.

Cytokinins and gibberellic acid

Cytokinins have been tested to increase fruit size in several research trials. Ish-Am *et al.* (2007) tried different PBRs on 'Ettinger' avocados. They found that 6 BA in concentrations of 30 and 40 mg l^{-1} can increase the production of larger fruit. Similarly, in California, with a full-bloom application of 25 mg l^{-1} of 6 BA, Zheng *et al.* (2011) reported increased numbers of larger fruit (> 270 g) in 'Hass'.

There are several applications for gibberellic acid (GA_3) in fruit crops and while its use with avocado is limited, some important commercial responses have been reported. For example, avocado seed germination was improved by soaking seeds for 24 h in a 500 μg ml^{-1} solution of GA_3 (see Whiley, Chapter 9, this volume). Salazar-García and Lovatt (1997) reported on the potential for manipulating flowering in avocado with GA_3. When 'Hass' trees were treated with foliar applications of GA_3 at 25 mg kg^{-1} in late autumn (November), winter (January) or spring (March) in an 'on' crop year, there was a trend for the treatments to increase yield in the following 'off' year. Untreated trees produced 18 kg of fruit, while those sprayed with GA_3 in either autumn, winter or spring produced 35, 27 and 34 kg $tree^{-1}$, respectively, although none were statistically different from untreated trees. The autumn GA_3 application resulted in increased numbers of small fruit (135–177 g) compared with untreated trees, while the spring application promoted larger fruit (213–269 g) (Salazar-García and Lovatt, 1997). Additionally, foliar applications of GA_3 at any of the three chosen times reduced the number of late-harvested fruit with black skins but had no other effect on fruit quality or maturity. It was concluded that there was potential for the use of foliar applications of GA_3 to increase yield and fruit size through the manipulation of flowering, with further studies recommended to develop reliable strategies for the management of this technology.

As part of a research programme on foliar-applied PBRs on 'Hass' yield in California (Zheng *et al.*, 2011), a 25mg l^{-1} GA_3 spray at the cauliflower stage of inflorescence development increased total yield as well as yield of commercially desirable large fruit (178–325 g) in the 'on' crop season. Application of GA_3 at the end of July (mid-summer) followed by prohexadione-Ca 30 days later increased total yield and large fruit (> 270 g) yield when crop load was both low and very high.

In South Africa, Rossouw *et al.* (2000) investigated the effect of single or multiple foliar sprays of GA_3 on container-grown trees of 'Fuerte', 'Hass' and 'Ryan' applied at either 50 or 250 mg l^{-1}. Spray applications were started in mid-summer (February) and finished in late autumn (May). In general, foliar applications of GA_3 at either concentration inhibited flowering, with the effect being greater with increased concentration and number of applications. They suggested that foliar applications of GA_3 could be used to inhibit flowering on trees used to supply nursery scion wood where selection of scions without floral bud differentiation is desirable. To achieve this, five applications of GA_3 at 250 mg l^{-1} were required for a satisfactory result. However, no studies were carried out to assess propagation results when scions from trees treated with GA_3 were used.

In similar research carried out in the central part of Chile (Torres, 2017), spraying in different months showed that mid-summer (January and February) foliar applications of 300 mg l^{-1} of GA_3, reduced subsequent spring flowering in 'Hass' avocados. December applications did not reduce flowering and later applications (March, April, May and June) showed no difference to the control. Flowering in this case was reduced between 56.3% and 66.7% for the January and February treatments, respectively. Histological analysis of treated buds showed that in December, there were only vegetative meristems in the buds, in January there were already primary floral meristems and from February on, secondary floral meristems were also present. These data differ those of from Ziv *et al.* (2014), who were able to find secondary floral meristems in late November (northern Hemisphere). Differences may be due to the type of material chosen to analyse. Torres (2017) worked with spring flush material and Ziv *et al.* (2014) worked with autumn flush material. Under central Chilean conditions, the autumn flush does not generally carry inflorescences for flowering in the same year.

Triazoles

The triazoles are a group of chemically related plant growth regulators that inhibit gibberellin biosynthesis when exogenously applied to plants (Davis *et al.*, 1988) and give predictable effects on vegetative growth of avocado trees (Köhne and Kremer-Köhne, 1987; Köhne, 1988; Adato, 1990; Wolstenholme *et al.*, 1990; Whiley *et al.*, 1991). Within this group, paclobutrazol (Cultar® or Payback®) and uniconazole (Magic® or Sunny®) have the same mode of action when applied to plants (Noguchi, 1987) and are registered for use on avocados in Australia, Chile, Israel, Perú, Colombia, New Zealand and South Africa. In studies with container-grown avocado trees, Köhne and Kremer-Köhne (1989) reported that foliar sprays of paclobutrazol and uniconazole both significantly reduced shoot extension, with the inhibitory effects greater for uniconazole compared to the same concentration of paclobutrazol. More recently, it has been observed that the effect on shoot extension from soil-applied paclobutrazol is much greater in 'Hass' than 'Shepard' when grown on the same seedling rootstock (BW2) in subtropical Queensland (A.W. Whiley, SHS, Queensland, Australia, 2023, personal communication). Thus, it is likely that sensitivity to treatment is dependent on the cultivar or cultivar/rootstock interactions.

Competition for resources between concurrently developing spring shoots and newly set fruit in avocado has been well documented. Biran (1979) and Cutting and Bower (1990) demonstrated that tip pruning spring shoots increases initial fruit retention but not necessarily final yield if insufficient leaf numbers are left to mature. Whiley (1994) found that the greatest production of fruit on vigorous 'Hass' trees in south-east Queensland came from functionally determinate inflorescences (those that do not produce shoots during spring), which also had larger fruit at maturity than those from functionally indeterminate inflorescences (those that produce a shoot from a terminal bud on the inflorescence, concurrently with fruit set and early growth). In Californian studies, Salazar-García and Lovatt (1998) reported that the greatest percentage of fruit from 'Hass' trees came from functionally indeterminate inflorescences due to their much greater representation in the tree. However, they also found that individual determinate inflorescences were three times more productive than indeterminate inflorescences.

With the availability of paclobutrazol (Cultar®) in the mid-1980s, an opportunity was seen to chemically retard vigorous spring shoot growth under invigorating environmental conditions, thereby reducing resource competition with developing fruit (Köhne and Kremer-Köhne, 1987; Adato, 1990; Wolstenholme *et al.*, 1990; Whiley *et al.*, 1991). Köhne and Kremer-Köhne (1987) reported that a 0.4% active ingredient (a.i.) foliar spray of paclobutrazol, applied to branches of 'Fuerte' trees in mid-bloom, reduced shoot extension growth by 43% and significantly increased fruit yield compared to untreated branches. Working with whole trees of 'Fuerte' and 'Hass', Wolstenholme *et al.* (1990) confirmed that mid-bloom foliar sprays of paclobutrazol applied at 0.25% and 0.5% a.i. reduced the length of spring shoots by about 40%, increased fruit retention during spring and increased the mean fruit size of 'Hass' by about 20%. They also showed that paclobutrazol affected redistribution of dry matter within the fruiting spring shoot, with a greater amount allocated to fruit and a reduction in allocation to leaves and stem of shoots on treated trees (Fig. 12.2). There was no significant difference in the total dry matter content between fruiting shoots from any of the treatments. In a series of experiments with 'Fuerte' in Israel, Adato (1990) was able to significantly increase yield with a single paclobutrazol spray of 0.8–1.0% a.i. immediately following anthesis (flower opening). Additionally, from a 3-year study with 'Hass' in Australia, Whiley *et al.* (1991) reported that a single mid-bloom foliar application of paclobutrazol at concentrations between 0.06% and 0.25% a.i. significantly increased mean fruit size and/or yield (Fig. 12.3). The 0.25% a.i. application of paclobutrazol increased mean fruit size by 14% but had little effect on yield. When a concentration of 0.125% a.i. was used, there was a significant increase in yield (40%) and fruit size (11%), while at 0.06% a.i., fruit yield was increased (43%) but not fruit size.

Some years following the commercial release of paclobutrazol, uniconazole was registered as Magic® (Israel) or Sunny® (Australia, Chile, Perú, Colombia, New Zealand and South Africa) and similarly has been used to increase yield and improve fruit size of avocados. In South Africa, 1-year studies showed that a single, mid-bloom foliar spray of uniconazole at 0.035–0.05% a.i. (0.7–1% Sunny®) increased yield and fruit size of 'Hass' (Erasmus and Brooks, 1998; Penter *et al.*, 2000). When 0.05% a.i. uniconazole mixed with an adjuvant (UP50, a urea/phosphate product at 2%) and sprayed on trees at a number of sites representing different environmental conditions, yield was increased by 7–36% (Erasmus and Brooks, 1998). Penter *et al.* (2000) reported a variable effect on yield when 'Hass' trees were treated with 0.035% a.i. uniconazole without adding an adjuvant. In two experiments, increases of 30% and 100% were recorded but in a third experiment there was no effect on yield. However, Penter *et al.* (2000) consistently found that 0.035% a.i. uniconazole (0.7% Sunny®) sprays increased mean fruit size by as much as 20%, which markedly improved the pack-out grades for export markets. They also found that mid-bloom 0.035% a.i. foliar sprays of uniconazole significantly increased the yield of pruned and non-pruned 'Fuerte' by 100% and 130%, respectively.

Results from a study in Australia with foliar-applied 0.025% or 0.05% a.i. uniconazole (0.5–1.0% Sunny®) on 'Hass' for two consecutive seasons showed that, while there was

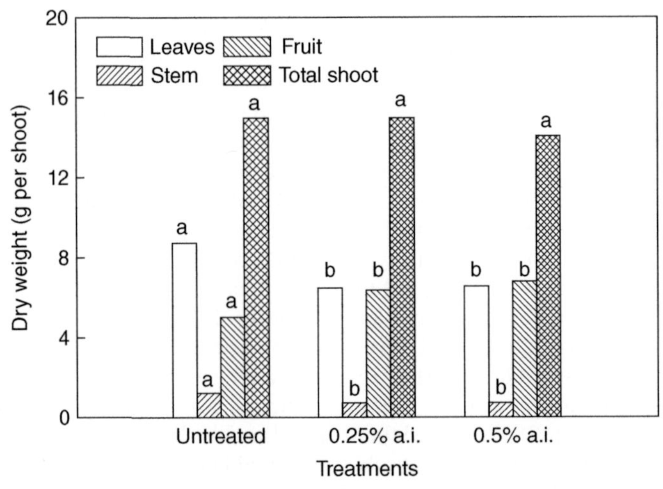

Fig. 12.2. Effect of mid-bloom sprays of paclobutrazol (as Payback®) on dry matter partitioning in fruiting spring shoots of 'Hass' avocado trees growing in south-east Queensland. Paclobutrazol was applied at 0.25 and 0.5% a.i. concentration to complete wetness of the foliage. Columns are mean data for 20 shoots from each of five trees. Different letters on columns for each organ across treatments indicate significant difference ($P \leq 0.05$). Drawn from the data of Wolstenholme *et al.* (1990).

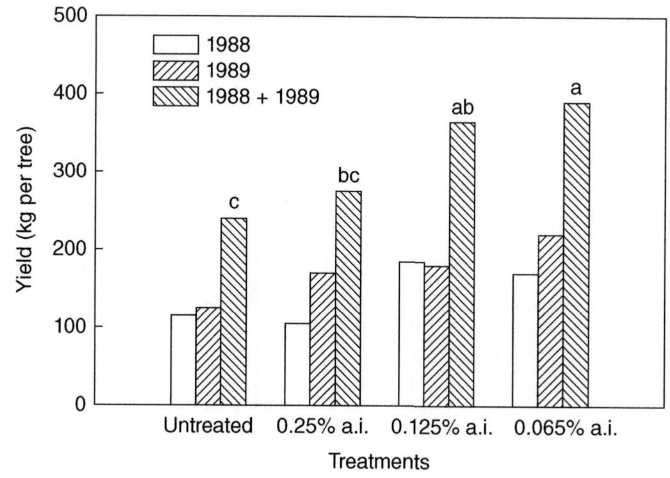

Fig. 12.3. Yield from 'Hass' avocado trees growing in south-east Queensland sprayed at mid-bloom with different concentrations of paclobutrazol (as Payback®). Columns are mean values of five trees. Significant differences were only recorded with the cumulative 1998 + 1999 yields and columns with different letters are significantly different ($P \leq 0.05$). Redrawn from the data of Whiley *et al.* (1991).

a general trend for cumulative yields to increase, they were not significantly different from those of untreated trees (Whiley, 2001). The greatest benefit from this treatment was the consistent increase in mean fruit size that improved the pack-out and return to growers. Similar to the data of Penter *et al.* (2000), there was considerable variation in responses to uniconazole between experimental sites in different locations. For example, when comparing the cropping patterns between north (Walkamin) and south-east (Glasshouse Mountains) Queensland, Australia, it was shown that at Walkamin, the yield increased across all treatments from 1999 to 2000, and for the most part, there was a consistent trend for increased cumulative yield from Sunny® treatments (Fig. 12.4). In contrast, at the Glasshouse Mountains site, the three highest producing Sunny® treatments in 1999 had a marked reduction in yield the following season with only a marginal increase in the cumulative yield from two of the Sunny® treatments. Harvest management was considerably different between the two sites with fruit at Walkamin picked in both seasons as soon as fruit reached physiological maturity (21% dry matter), while at the Glasshouse Mountains site, fruit was harvested at a more advanced stage of maturity (27% dry matter).

In summary, there is significant evidence from a number of studies to indicate that the use of either paclobutrazol or uniconazole as mid-bloom foliar sprays can improve avocado yield. The time of application is relatively critical, with most studies producing the best results when either paclobutrazol or uniconazole was applied at mid-bloom (Fig. 12.5). For sustainable results, fruit should be harvested as close to commercial maturity as possible. Most studies found improved yield in the season when either of these plant growth regulators was first used. Timely removal of fruit as close as possible to commercial maturity resulted in improved performance from trees that were successively treated, as additional recovery time between crops was required. As the use of these products optimizes tree performance, it is essential that management of light interception, nutrition, irrigation and Phytophthora root rot control is at the highest level.

Widespread use of paclobutrazol, and especially uniconazole which is less persistent in the tree, is a feature of the management of ultra-high-density avocado orchards on inland hillsides in central Chile. Tree height is limited to as little as 2 m by training and pruning, a non-invigorating climate and soil, root competition, extreme precocity and high yield, and dedicated use of triazoles – described in greater detail

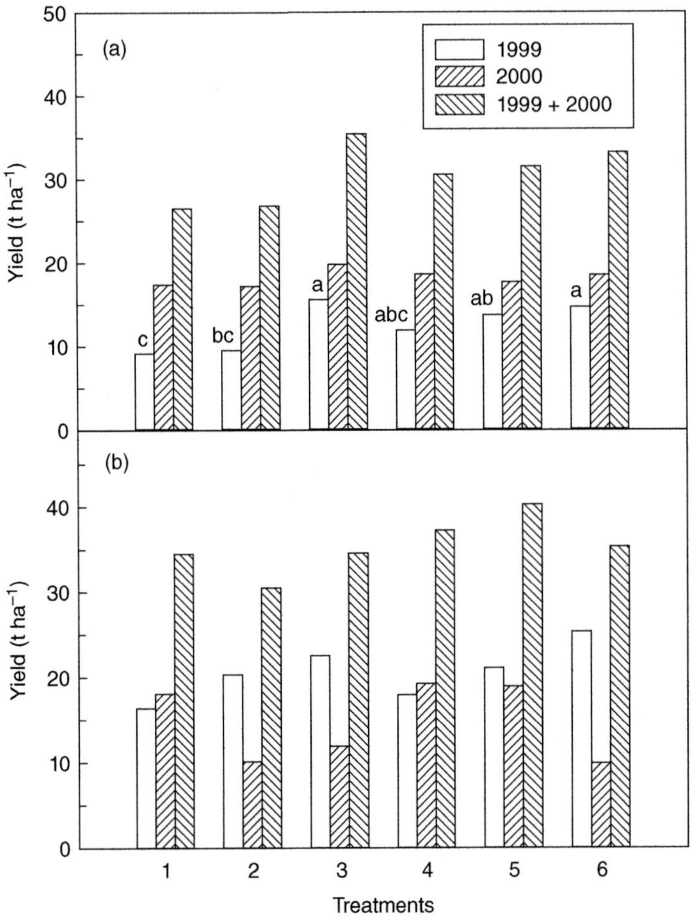

Fig. 12.4. Effect of foliar applications of Sunny® (uniconazole) on yield of 'Hass' avocado trees growing at (a) Walkamin and (b) Glasshouse Mountains in Queensland, Australia. The treatments used were: (i) untreated, (ii) 0.5% Sunny® + 1% UP50, (iii) 0.5% Sunny® + N, (iv) 1% Sunny® + 2% UP50, (v) 1% Sunny® + N and (vi) 1% Sunny® + 2% UP50 + N. Sunny®, and the respective adjuvants were applied at mid-bloom. Where applied, N was soil-applied as urea. Columns represent means values of six trees. Columns of any 1 year marked with different letters are significantly different ($P \leq 0.05$) as tested by ANOVA. Redrawn from the data of Whiley (2001).

under 'Bioregulator modification of alternate bearing' in this chapter. Triazoles are applied to help control tree vigour and increase yield (especially in lower crop seasons) and fruit size.

Foliar applications of the triazoles have been the norm in Chile, but soil applications have also increased productivity. Gardiazabal *et al.* (2015a,b,c) tried different times, rates, irrigation methods and soil types for paclobutrazol and uniconazole in the central area of Chile. Results showed a trend for soil-applied paclobutrazol and uniconazole to increase yield, albeit not always statistically significant. In soil-applied paclobutrazol studies in Chile, Gardiazabal *et al.* (2015c) concluded that root confinement (to ridges/mounds) and very localized irrigation systems (micro-sprinklers with stream deflectors or drippers) gave the best results, especially at higher annual rates (1 kg a.i. ha^{-1} in lower coverage irrigation systems to 1.5 kg a.i. ha^{-1} in fully covered wetted area). In the case of micro-sprinklers with stream deflectors (1 m wetted diameter) on sandy loam soils, an annual application of 1 kg a.i. ha^{-1} of paclobutrazol increased cumulative

Fig. 12.5. Flowering in avocado showing (a) the correct time to apply triazole foliar sprays and (b) when flowering becomes too advanced for triazole application. It should be noted that the stage of flowering will vary within and between trees. The application should be timed so that less than 15% of inflorescences have advanced to (b). Photographs supplied by A.W. Whiley, Sunshine Horticultural Services, Queensland, Australia.

yield by 120% over a 4-year period compared to the untreated control, with three out of 4 years of the study having similar or larger fruit size compared to fruit from untreated trees. In a similar study on clay loam soils, with soil-applied paclobutrazol using micro-sprinklers without stream deflectors (3.5 m wetted diameter), and annual treatment rates of 1.0 and 1.5 kg a.i. ha^{-1}, there was a trend for increased yield of 11% and 16%, respectively, though results were not statistically significant. When paclobutrazol was annually applied by drip irrigation on a sandy loam soil, the application rate of 1 kg a.i. ha^{-1} gave a 34% increase in final fruit numbers compared to the untreated trees.

Gardiazabal *et al.* (2015b) found that December-applied uniconazole at 0.4 kg a.i. ha^{-1} through the irrigation (drippers) increased yield by 61.8%, compared to untreated trees in central Chile. This was reflected in greater fruit numbers (60.1% more than control) with fruit marginally larger than those from untreated trees. Gardiazabal *et al.* (2015a) compared annual treatments of 0.1 kg a.i. ha^{-1} of soil-applied uniconazole at different times from October through to February and found that in the cool Mediterranean climatic conditions of central Chile, the third week of November produced the highest number of fruit and yield (kg) per tree. With respect to size, application in the third week of December produced the largest fruit but resulted in one of the lowest yields.

Reasons why soil-applied triazoles increase production in avocados are still not completely understood. Greater flowering intensity, which is observed at higher triazole application rates, may be a contributing factor (Gardiazabal *et al.*, 2015c), potentially providing more opportunities for fruit set. Alternatively, soil application of triazoles modifies tree architecture by producing more compact, willowy growth, resulting in smaller trees with darker green leaves (Fig. 12.6) compared with untreated plants, which has been correlated with higher chlorophyll and nitrogen content and higher photosynthetic rates (Davis *et al.*, 1988; Whiley, 1994). Additionally, reduced shoot growth may favour fruit sinks, thereby increasing retention while maintaining or promoting larger fruit.

It has been observed that when triazoles are used in areas where irrigation water has high salinity levels, leaf chloride symptoms are less severe compared to those expressed in untreated trees (F. Gardiazabal, GAMA, Quillota, Chile, 2023, personal communication). Mackay *et al.* (1990) and Kojima *et al.* (1996) found ABA concentrations increased in plants following triazole application. Bharath *et al.* (2021) reported that under drought, salinity or cold stress, ABA accumulation in plants causes partial stomatal closure to conserve water while upregulating genes to promote osmotic adjustment in leaves. Since uptake of chloride in soil water is through the xylem, resulting in accumulation in leaves

Fig. 12.6. Avocado trees showing the effect of triazole application on growth habit and leaf density in a high-density orchard in Chile. Photograph supplied by F. Mena, GAMA, Quillota, Chile.

that may reach toxic concentrations (≥0.3%), a triazole-induced increase in ABA resulting in partial stomatal closure with reduced inflows of chloride-carrying water into leaves should result in less leaf chloride accumulation and expression of toxicity symptoms. However, there have been no direct triazole/ABA studies with avocado published.

Gardiazabal *et al.* (2015b) reported there were no fruit residues at harvest following any of the soil-applied uniconazole treatments in avocado research carried out in Chile, and similar results were found in Australia following soil-applied paclobutrazol (A.W. Whiley, SHS, Buderim, QLD, Australia, 2023, personal communication). Triazoles are believed to be primarily transported acropetally through the xylem of most species (Davis *et al.*, 1988; Fletcher *et al.*, 2010; Desta and Amare, 2021) with little evidence of basipetal movement via phloem. Winkler and Knoche (2021) found that xylem inflows into developing European plum are significantly reduced during the second stage of growth when fruit were rapidly expanding, corresponding with increased phloem inflows supplying nutrients and sugars. Similar results were reported for a number of other fruit crops with sigmoidal growth curves as occurring in avocado (Schroeder, 1958; Blumenfeld and Gazit, 1974). Hence, it is likely that where soil-applied triazole treatments are used, particularly during the second rapid growth stage of avocado fruit, xylem inflows will be minimal being replaced with a significant increase in phloem inflows, hence the absence of triazole residues in reproductive organs (Davis *et al.*, 1988).

Growth control with triazole application is achieved by either foliar or soil applications. The length of control is strongly influenced by the application procedure (Barrett and Bartuska, 1982, with soil-applied triazole exerting growth control for a much longer period than foliar application (Köhne and Kremer-Köhne, 1992; F. Gardiazabal, GAMA, Quillota, Chile, 2023, personal communication). Lever (1986) proposed that a threshold concentration of paclobutrazol needs to be available in the shoot apex to maintain GA biosynthesis suppression. The vascular system behind the growing point may act as a reservoir if the paclobutrazol concentration remains relatively high. A sufficiently high concentration is more likely to be maintained in that region of the vascular system if continual supply is being translocated upwards from roots (soil application). Soil persistence of growth-retardant triazoles has been reported as 3–12 months (Davis *et al.*, 1988) depending on treatment rate and soil type. Residue analysis of soil in Chile, following treatment at 0.3 kg a.i. ha^{-1} of uniconazole, found no detectable traces after 6 months (F. Gardiazabal, GAMA, Quillota, Chile, 2023, personal communication). Little or no inhibitor reserve is available from triazole foliar or bud treatment (Lever, 1986).

Orchard Design and Canopy Management

The fundamental principle of cropping systems is to harvest light (energy) and convert it into economic yield, largely in the form of carbohydrate and oil- or fibre-based plant products. To enhance productivity, the design of any cropping system should optimize light interception over the life of the crop in a way that maximizes yield and yield stability. Due to access requirements, orchard trees usually only intercept about 70% of incoming solar radiation at full canopy, which may take several years to attain depending on tree spacing (Jackson, 1980). There are opportunities and challenges to reduce the time between planting and full canopy development in avocado, as well as to maintain a productive orchard once an effective full canopy has been attained. Orchard planting densities vary from about 90 to 1600 trees ha^{-1}, with trial plantings in Chile of up to 6400 trees ha^{-1}, which emphasize plasticity in orchard design, further aggravated by significant climate and soil differences between global production areas. Studies of the light regime in an avocado canopy (Hadari, 2004; Heath *et al.*, 2005) emphasize the importance of canopy light exposure optimization over the life of the orchard.

When establishing an orchard, the choice of planting density largely determines the time to full canopy development when light interception is maximized. However, selections for planting density and orchard layout are numerous for any given environmental/socioeconomic situation, and a choice from the potential possibilities should strike an appropriate balance between simplicity and complexity, to match the skills and tools available to manage and maintain the orchard. Where land is inexpensive and climate and soil conducive to high tree vigour, there is an opportunity for wide spacing within and between rows (10 × 8 to 12 × 9 m giving 92–120 trees ha^{-1}). This results in lower returns during the early life of the orchard and eventually larger trees with potentially poorer fruit quality and increased harvesting costs (Hofshi, 1999c). However, advantages include savings on nursery trees, eventual tree removal costs and generally less intense management inputs, for example, irrigation, fertilizers and pruning (Newett *et al.*, 2001).

From about 1994, Stassen and co-workers in South Africa investigated tree containment and shaping for avocado based on principles developed for deciduous fruit trees. They proposed that to meet the requirements of modern-day avocado production, at least 500 trees ha^{-1} should be planted and maintained as productive units for the life of the orchard (Stassen *et al.*, 1995, 1998). To achieve this, it was suggested that trees be planted using central leader training so that a conical-shaped hedgerow configuration is attained. To maintain a productive hedgerow, they recommended that trees should be pruned immediately following harvesting with further maintenance pruning during summer to ensure that light interception and penetration into the orchard are not compromised, thus keeping the lower canopy productive. However, while the principles proposed are horticulturally and physiologically sound, experience has dictated that for our most important cultivars, production at least in the mesic humid subtropics, where climate and soil can encourage very vigorous growth, is usually compromised by the intensive pruning regime required to control tree size and shape. Nevertheless, the research of Stassen *et al.* (1995, 1998) re-stimulated interest and debate on canopy management of avocado orchards, resulting in the development of new management techniques based on their concepts, especially in areas with less invigorating climates and soils. Some of these are further discussed herein.

Medium- to high-density orchards

The design of most contemporary avocado orchards has shifted to medium- and higher-density planting (6 × 4 to 9 × 7 m giving 159–416 trees ha^{-1}) where topography, cultivar and pruning/ training systems become key issues in determining tree spacing and orchard layout. More traditional spacings in the 1980s were 6 × 6, 7 × 7, 10 × 7 and 12 × 9 m (90–278 trees ha^{-1}). Close planting of avocado trees in semi-arid California was proposed by Platt *et al.* (undated) where a step-wise tree thinning (removal) strategy was recommended to combat crowding within the orchard. Based on a 6 × 4.8 m initial spacing (347 trees ha^{-1}), prior to canopies touching every second row of trees on the diagonal was removed across the orchard, with this repeated on the opposite diagonal to prevent subsequent crowding (Platt, 1976; Platt *et al.*, undated). The next stage of thinning was the removal of every second row across the block where retained trees were spaced at 12 × 9.6 m. A similar canopy management strategy has been used in subtropical Australia where 'Hass' trees are initially planted at 6 × 6 m (278 trees ha^{-1}), with timely tree removal of alternate rows on alternate diagonals to prevent crowding within the orchard (Newett *et al.*, 2001). A final removal of alternate rows is made leaving permanent trees spaced at 12 × 12 m. Once trees exceed 10 m in height they are stumped to about 1 m and allowed to regrow, beginning fruiting within 2–3 years. This latter process is repeated every 10–12 years, depending on soil and climatic conditions. Stumping orchards is carried out sequentially on a block-by-block basis to maintain positive cash flow.

Studies in humid, semi-tropical Florida with 34-year-old 'Lula' and 'Booth 8' orchards where trees were planted at 6 × 6 m found tree height reduction and thinning increased production and profitability within 4 years of treatment (Crane *et al.*, 1992). Reduction in tree height was particularly effective for 'Lula' when 12–18 m tall trees were topped to about 6 m. This resulted in yields of about 6.7 t ha^{-1} in non-thinned orchards and 7.3 t ha^{-1} where every second tree on the diagonal was removed. Most fruit were carried in the middle third of the canopy. A similar result was recorded for 'Booth 8' when 7.5–10.5 m tall trees were topped to 4.8–6 m and every second tree on the diagonal removed, with production being restored 3 years from pruning (Crane *et al.*, 1992). With this cultivar, the removal of every second tree, together with topping at either 4.8 or 6 m, was the most economically beneficial treatment to apply. As with 'Lula', the most productive cropping zone within the tree was the middle third of the canopy followed by the bottom third.

In subtropical South Africa, Köhne and Kremer-Köhne (1990, 1991) reported on a high-density orchard with 'Hass' planted at 5 × 2.5 m (800 trees ha^{-1}). Under these conditions, tree management was assisted by soil-applied paclobutrazol (see the 'Triazoles' section of this chapter), cincturing and tree removal (Köhne and Kremer-Köhne, 1990; Köhne, 1992a). Girdling was used strategically when 3.5 years after planting, the largest limbs of every second tree in the block were treated, and following harvest of the subsequent crop these trees were removed (see the 'Girdling for increased yield' section of this chapter). They concluded that high-density planting showed a positive return on investment 5 years after planting the orchard. In contrast, at a similar age, investment in the standard planting density orchard (400 trees ha^{-1}) had not been recovered. Nevertheless, such high-density plantings have not become the norm in this warm, humid subtropical area.

While there appears to be cash flow advantages in high-density planted orchards, there are some disadvantages that should be considered before adopting this strategy. First, the capital outlay on nursery trees grafted to clonal rootstocks is high, especially in areas where clonal rootstocks must be used to assist with Phytophthora root rot control. Some orchard designs, particularly those where a square grid layout is chosen, dictate that row orientation (orchard access) will change with each tree removal event. Therefore, topography should be suitable for access in several directions. Although now largely mechanized, tree removal is still an expensive operation and should be planned to fit in with cropping cycles. Finally, there is the disease risk consideration, particularly with respect to Phytophthora root rot and laurel wilt (*Harringtonia lauricola*), while Verticillium wilt and a Basidiomycete woodrotting fungus (*Phellinus noxius*) can be problematic in some situations (see Dann and Gazis, Chapter 13, this volume).

The laurel wilt pathogen can move rapidly among adjacent root-grafted avocado trees and time to root grafting is reduced the closer their spacing (Ploetz *et al.*, 2017a,b, Crane *et al.*, 2020). The grower needs to be confident that these diseases can be controlled; otherwise, loss of permanent trees can reduce the economic viability of this strategy.

High-density orchards

California

During the 1990s, high-density orchards started in California with the objectives of containing tree height to reduce harvest costs and improve worker safety since many orchards are planted on steep hillsides. In the early stages, cultivars with strong apical dominance (weak acrotony), resulting in a single dominant trunk (Thorp and Sedgley, 1993), were used to extend the productive life of high-density avocado orchards. Examples of such cultivars include 'Gwen', 'Lamb Hass' and 'Reed' which are also known to be high yielding (Fig. 12.7) (see Crane *et al.*, Chapter 8, this volume).

Small experimental orchards of 'Lamb Hass' and 'Reed' were planted in California where trees were spaced at 2.25 × 2.25 m (1973 trees ha^{-1}) (Hofshi, 1999a) (Fig. 12.8). The initial thought was that this type of orchard would be productive for 10 years and then replaced with new trees. It was found that training trees for high-density orchards should begin in the nursery where their future shape is reinforced through pruning (Hofshi, 1999b). Pruning is continued each year in the field, and once trees begin fruiting, it is mostly carried out after the following year's crop has been set. Field pruning was aimed at maintaining tree shape, height, light interception and orchard access while ensuring continued productivity. Pruning should also be implemented in such a way that crop load is maintained, as fruit plays a major role in reducing tree vigour and containing horizontal spread as their weight pulls branches into a more vertical plane. In fruiting orchards, tree height is maintained at a level that allows hand-harvesting fruit from the ground (Hofshi, 1999a). This reduces harvest costs, improves efficacy of size picking, benefits light interception and facilitates application of pesticides. Using these protocols, a 6-year-old, high-density 'Reed' orchard has produced 6.5, 26.0, 46.2 and 81.7 t ha^{-1} in the third to sixth years from planting, respectively (Hofshi, 1999a; R. Hofshi, Fallbrook, California, 2001, personal communication). Production in the fifth and sixth years from planting was two to three times higher than previously reported for avocado, indicating that high-density orchards can be profitable where markets exist for 'Reed', and nursery trees for planting can be purchased at viable prices (Hofshi, 1999b). Current limitations to this system of production are market acceptability of cultivars with architectural structure that suits close planting, for example, 'Gwen', 'Gem', 'Lamb Hass' and 'Reed', and the cost/return ratio of such production systems based on the length of the orchard cycle. With limited field data and extrapolations of future yield over the envisaged orchard life, Hofshi (1999a,b) has given some insight into the potential productivity of high-density avocado production under Californian conditions, where tree vigour is much less than in the humid, summer rainfall subtropics.

Fig. 12.7. Fruit on 'Lamb Hass' trees spaced at 2.25 × 2.25 m (1973 trees ha^{-1}) in a California orchard when trees were 2.5 years old. Photograph supplied by M.L. Arpaia, UCR, California, USA.

Chile

As the new millennium arrived and good and stable results were obtained with compact, upright varieties, the idea of high-density plantings was also seen as an opportunity for 'Hass' in California and in Chile. Early stages of these

Fig. 12.8. A high-density planting of 'Lamb Hass' in California. Trees are 2.5 years old and spaced at 2.25 × 2.25 m (1973 trees ha^{-1}). The stakes are 1.5 m tall. Photograph supplied by M.L. Arpaia, UCR, California, USA.

developments under cool, Mediterranean climatic conditions in central Chile started with plantings at 3 × 3 m, and from there moved into higher densities, where up to 6400 trees ha^{-1} were tried, initially with good results (Gardiazabal *et al.*, 2015d).

Such extreme departures from the accepted practice in most avocado-growing countries are made possible by several unique conditions in Chile, which favour high-density orchards. Many new plantings are on steep hillsides (Fig. 12.9) with lower land costs, larger orchard sizes, reduced frost risk, improved climatic conditions for fruit set compared with the colder river valleys and earlier fruit maturity. The cool Mediterranean climate with only 900–1200 heat units (base temperature 12.5°C) reduces tree vigour and growth. Rainfall is low and occurs in winter when soil temperatures are mostly too low for root growth, and summer irrigation is managed to further restrict root growth without sacrificing yield or fruit quality. Hillside soils are generally shallow (30–90 cm) and variable, often heavy clays with restricted drainage, necessitating not only ridging (mounding) but also assisting by reducing tree vigour (Gardiazabal *et al.*, 2011a). When 3 × 3 m orchards were first planted, Chilean avocado growers had already been successfully managing orchards with trees at either 6 × 3 m or 5 × 2 m. Consequently, the commercial risk of increasing tree density was negated by the option of row thinning if growth could not be controlled and tree crowding occurred, resulting in tree spacings that were already commercially successful.

Additional factors driving high-density planting in Chile are drought and water restrictions, decreasing land availability and increasing costs. Furthermore, orchard establishment costs are reduced by using seedlings rather than clonal rootstocks (see Whiley, Chapter 9, this volume) since the latter are not yet required as Phytophthora root rot overall remains at low levels. In contrast to rectangular tree spacing leading to hedgerows, hillside orchards are

Fig. 12.9. High-density (3 × 3 m; 1111 trees ha^{-1}) avocado orchards planted on hillsides in Chile. Trees are planted on small ridges to improve soil depth and drainage and to confine root systems to limit growth. Photograph supplied by F. Gardiazabal, GAMA, Quillota, Chile.

planted on the square (e.g. 3 × 3 m). This allows light interception on all sides of trees (especially on steep slopes); thus, individual tree rather than row management is necessary. Square planting also means that north–south row orientation, which is difficult on slopes with varying aspect, is not necessary, but it excludes orchard access for farm machinery. Consequently, good road design, along with comparatively cheap labour, is necessary to adequately service these steep orchards (Gardiazabal *et al.*, 2011a).

The first high-density orchards (1111 trees ha^{-1}) were planted in 2004 and showed very positive results with yields from the first two harvests significantly higher than usually obtained in the early years of traditionally planted 'Hass' orchards at lower tree densities (555 trees ha^{-1}). During the first year after planting, great importance was given to shaping trees but no major pruning was performed during the second and third years of growth. However, heavy pruning was required in 2007, which resulted in a significant loss of flowering with low production in 2008 (Table 12.2).

The collapse in the 2008 crop yield subsequently led to more focused pruning, with the removal of large lateral branches immediately after harvest to reduce the impact of shading. Growth management problems also led to the first Chilean trials studying soil-applied triazoles under high density conditions, which helped achieve higher production and easier maintenance of the tree canopy size. In 2006, in an orchard planted at nearby Panquehue, constant pruning was not performed with only one immediate postharvest pruning carried out in order to maintain light penetration. Also, soil-applied uniconazole at 0.1 l a.i. ha^{-1} was given on a yearly basis in January to help control tree size and increase flowering. Over time, this orchard became consistently high yielding (Table 12.3).

After exceeding average industry yields from orchards planted at 6 × 3 m and 5 × 2 m by successfully developing agronomic tools to manage trees at 3 × 3 m, further investigations were carried out to assess the commercial limits of tree spacing. Gardiazabal *et al.* (2015d) evaluated new orchards where planting density was increased up to 6400 trees ha^{-1}. Early results were promising with first harvests of up to 55 t ha^{-1}. However, these ultra-high-density orchards required intense management with fruit harvested immediately when maturity is reached, followed directly by pruning to avoid

Table 12.2. Annual production (average of 20.7 ha) of 'Hass' avocado grafted to seedling 'Mexicola' rootstocks in Llay Llay, Chile planted in August 2004 at a spacing of 3 × 3 m (1111 trees ha^{-1}) (Gardiazabal *et al.*, 2015d used with permission).

Year	2006	2007	2008	2009	2010	2011	2012	2013	2014
t ha^{-1}	9.3	17.7	4.4	39.5	19.1	31.5	16.9	42.5	18.0

Table 12.3. Annual production (average of 28.7 ha) of 'Hass' avocado grafted to seedling 'Mexicola' rootstocks at Panquehue, Chile planted in January 2006 at a spacing of 3 × 3 m (1111 trees ha^{-1}) (Gardiazabal *et al.*, 2015d, used with permission).

Year	2007	2008	2009	2010	2011	2012	2013	2014
t ha^{-1}	0.5	18.3	20.6	23.9	23.5	14.6	30.6	28.9

tree crowding. These specific timeframes are unworkable across medium-sized and large commercial orchards, and tree densities of this magnitude were never adopted by the Chilean avocado industry.

Currently, 2.5 × 2.5 m tree spacing (1600 trees ha^{-1}) is quite common in Chile using 'Hass' grafted to seedling rootstocks to reduce establishment costs. These orchards are precocious and remain productive if trees are maintained well-illuminated from the early developmental stages and onwards. Pruning is usually carried out twice yearly: the first directly after harvest, where excessive growth is removed (height maintenance) and light penetration is maintained (preservation of inter-row integrity); and the second in autumn–winter, removing vigorous watershoots that grew following postharvest pruning since they are physiologically juvenile and usually do not flower in spring. In areas where frost risk is present, the autumn–winter pruning is delayed until the risk of freeze damage has passed. In association with pruning, it is usual to apply annual mid-bloom, foliar triazoles to enhance fruit set, and soil-applied triazoles to help maintain tree size, reduce pruning operations, and enhance flowering to improve opportunities for fruit set.

While California and Chile have been at the forefront of the development of high-density avocado orchards, Australia, New Zealand and South Africa have investigated and, in some cases, have had partial success with commercialized high-density avocado production (A.W. Whiley, SHS, Buderim, QLD, Australia, personal communication; Hermosilla-Madrid and Whiley (2023); D.G. Smith, Tzaneen, South Africa, 2011, personal communication).

Australasia

Due to the relatively high costs of avocado production in Australia and New Zealand, the respective industries have traditionally planted low-intensity orchards (90–250 trees ha^{-1}) and relied on high levels of mechanization to contain labour costs. Invariably, orchards matured when trees reached 6–8 m in height, which required sophisticated equipment for pruning and harvesting. A large part of production in both countries is in areas that have mild to warm climates with moderate-to-high rainfall (1100–1350 mm annually), which promotes tree vigour during the summer/autumn months. Most of the research has focused on pruning strategies to prevent canopy-crowding that preserve production as orchards mature (Leonardi, 2001, 2005, 2008, 2009).

More recently, a study with 'Hass' grown at variable planting densities was concluded at a research site near Bundaberg, Queensland, Australia, which has a warm, humid, subtropical climate (Wilkie, 2021). Tree spacings selected were conventional low density at 9 × 5 m (222 trees ha^{-1}), medium density at 6 × 3 m (556 trees ha^{-1}) and high density at 4.5 × 2 m (1111 trees ha^{-1}). Trees in the medium- and high-density orchards were trained to central leader and those in the high-density configuration supported by vertical trellising. Two rootstocks were used to compare tree vigour ('Velvick', which is high vigour and in common industry use; and 'Ashdot', which is lower vigour with low industry adoption). Tree architecture and canopy management were mostly carried out through pruning using concepts similar to those applied in apple production. Plant growth regulators were only used as foliar applications to improve fruit set and retention. Wilkie (2021) reported that the medium- and high-density orchards produced higher yields/hectare from their first crop compared with the low-density orchard but thereafter failed to outyield the low-density orchard, which after 5 years significantly outperformed the higher-density options. 'Hass' trees grafted to 'Ashdot' were generally more productive than those grafted to 'Velvick'. A similar result with rootstocks was reported by Le Lagadec (2010) where 'Ashdot' grafted to 'Hass' had the greatest yield efficiency (kg m^{-3}) compared with 21 other rootstocks grafted to 'Hass'. Low vigour rootstocks such as 'Ashdot' are known to be highly susceptible to environmental/root disease stress and require optimum management to maintain long-term productivity (Ben-Ya'acov *et al.*, 1995; Le Lagadec, 2010), which may be pertinent to its general lack of use by Australian growers.

Commercial development of management techniques for higher-density planted orchards has been ongoing in the Bundaberg region for several years using planting densities ranging from 444 to 880 trees ha^{-1}. Management approaches include strategic use of pruning and soil-applied paclobutrazol (Payback®) to control tree size and enhance production (A.W. Whiley, SHS, Queensland, Australia, 2023, personal communication). Unlike Chile, planting configurations used allow access for equipment to apply pesticides and for harvesting. Orchards have been planted on two different soil types (a clay loam of volcanic origin vs a grey, sandy loam) with several varieties that include 'Hass', 'Gem', 'Maluma Hass' and 'Shepard' planted on a range of seedling rootstocks. The objective has been to produce commercial crops from trees restricted to 2–3 m in height so that harvest operations are ground based without the use of elevated platforms. Strategic pruning has been used for tree height and width control supporting growth suppression from soil-applied Payback®. Results to date clearly show differences in growth responses between varieties to soil-applied Payback®. 'Hass' and 'Maluma Hass' are far more responsive to Payback® than 'Shepard', while 'Gem' responds to a lesser extent than 'Hass' but greater than 'Shepard'. There are also clear differences in application rates required between the two different soil types. Work on this project is ongoing (A.W. Whiley, SHS, Queensland, Australia, 2023, personal communication).

The first blocks in a pioneering high-density orchard north of Kaitaia, in the Far North of New Zealand, were planted in 2006 on ancient sand dunes in a cool, year-round wet (1350 mm average annual rainfall), subtropical climate. Mostly 'Hass' grafted to seedling 'Zutano' rootstocks were planted but a small number of 'Hass' grafted to 'Dusa' rootstocks were included. The orchard was developed over an 11-year period, initially at 3.5 × 3.5 m spacings (812 trees ha^{-1}) with vehicle access left every eight rows of trees. However, the configuration was changed to 5 × 2.5 m (800 trees ha^{-1}) in later planted orchards (Fig. 12.10), which allowed vehicle access between rows. Due to the steep terrain in some areas of the orchard, together with restricted vehicle access, pesticides, foliar-applied Plant Growth Regulators (PGRs) and some nutrition are applied by helicopter.

In this invigorating climate, tree crowding soon became a management issue and initially growth control was attempted by selective, postharvest limb removal and April cincturing of one-third of each tree to induce heavier cropping. However, due to strong annual tree growth and labour costs, the programme was abandoned. Following the published results of Gardiazabal *et al.* (2011b), in relation to the successful management of high-density orchards in Chile with soil-applied uniconazole, selected blocks were treated with soil-applied paclobutrazol (registered for use in New Zealand), which was successful in controlling tree growth and increasing yield. Further management approaches that include postharvest strategic pruning to limit orchard crowding, the addition of pollinizer varieties at 11% of trees across blocks together with intensive bee management during flowering to ensure maximum opportunities for fruit set and a balanced nutrition programme have resulted in sustainable yields 70% higher than the New Zealand average. Average production from 26% of the orchard is ≥ 18 t ha^{-1}, with the highest yield reaching 43 t ha^{-1} for a single year (Hermosilla-Madrid and Whiley, 2023).

Results from both Australasian sites have shown that time of triazole soil application is critical and must be synchronized with tree phenology to get the maximum effect from the product, and the larger the trees become, the more difficult it becomes to control tree growth with PGRs. However, with lower tree densities

Fig. 12.10. A high-density (5 × 3 m, 667 trees ha^{-1}) 'Hass' orchard in New Zealand where tree growth has been retarded by soil-applied paclobutrazol. Irrigation is supplied via under-tree sprinklers, while elevated sprinklers are used for frost control. Windbreaks are necessary to reduce damage to trees and fruit. Photograph supplied by A.W. Whiley, Sunshine Horticultural Services, Queensland, Australia.

text needs to be justified than Chile is able to use, larger trees must be grown to achieve optimum light interception for equivalent production. To date, there has been no consistent formula developed to manage this technology, and decisions are reviewed and revised during each production cycle (A.W. Whiley, SHS, Buderim, QLD, Australia, 2023, personal communication).

South Africa

A small orchard of 'Hass' grafted to cloned 'Dusa' and planted at 3 × 3 m was established for evaluation purposes with a fall-back position to remove every second row if necessary (D.G. Smith, Tzaneen, South Africa, 2011, personal communication). After 2 years, due to crowding, 50% of trees were removed resulting in a 6 × 3 m planting, which in this subtropical location will likely result in future crowding, but nevertheless it was manageable at the time. In practical terms, in the humid subtropics, the 3 × 3 m planting presents significant challenges when growing 'Hass'. When planting more precocious and less vigorous cultivars (e.g. 3-29-5, 'Gem', 'Maluma Hass'), closer tree spacing may be an option. However, such closer spacing options would require a different approach to the application of pre-harvest fungicides than presently used in South Africa. Since a spray unit driving through every row of trees would be impractical, a 'fungicide canon' as used in some 'Hass' plantings in Chile, or by helicopter as used in parts of New Zealand, may be an option for closer spaced avocado orchards of low vigour cultivars (J.S. Khöne, Horticultural Advisory Services, South Africa, 2023, personal communication; A.W. Whiley, SHS, Buderim, QLD, Australia, 2024, personal communication).

The Future

One of the questions that remain unanswered is the lifespan of high-density avocado orchards. Probably, if disease pressure is low, they can be quite long lived. At least in Chile, the first planted, high-density 'Hass' orchards have achieved a longer lifespan than originally expected (10 years, Hofshi, 1999a) as they are already near 20 years old and still commercially viable. To date, the high-density orchards in New Zealand and Australia managed with soil-applied paclobutrazol are controllable for 6–8 years when crowding significantly impacts production and radical pruning is required. It is uncertain how effective soil-applied paclobutrazol will be on regrowth canopies if trees are retained at the original planting density or if selective tree removal will be required. The continued success of future high-density 'Hass' orchards likely relies on access to new, heavier-cropping, market-acceptable varieties grafted to lower-vigour (cloned) rootstocks with good Phytophthora tolerance that impart precocity and higher yields (see Crane *et al.*, Chapter 8, this volume), in combination with triazoles.

Fig. 12.11. Mechanical pruning of 'Hass' avocado trees to a hedgerow configuration. Tree spacing is 8 m between rows and 5 m within the row. Photograph supplied by A.W. Whiley, Sunshine Horticultural Services, Queensland, Australia.

Hedgerow orchards

Avocado orchards planted to hedgerow layouts without the necessity to remove trees remain more commonly accepted than high-density orchards and follow similar developments in the citrus and mango industries (Jackson, 1985; Crane *et al.*, 1997). North/south row orientation is favoured to maximize light interception, which becomes more important at higher latitudes. The hedgerow system mechanically pruned to a pyramidal or barn roof shape has become increasingly popular in Australia, Israel and South Africa (Lahav, 1999; Newett, 1999; Toerien, 1999; Leonardi, 2009) and is common in South Florida where the topography allows the passage of large equipment through orchards (Fig. 12.11). However, due to the peripheral fruiting habit of avocado and persistence of fruit on trees, pruning strategies that minimize damage to the current crop are required to maintain productivity and containing tree size.

Hedgerow orchards that are mechanically maintained are generally pruned at least twice each year, immediately following harvest (autumn/winter) and again in summer. Postharvest pruning establishes the shape of trees within which fruit are carried to the following season, allowing summer pruning of spring regrowth with minimal damage to fruit. The summer pruning is made beyond the winter cuts. In Australia, Chile, Perú, Israel, Colombia, New Zealand and South Africa, regrowth following summer pruning can be controlled using foliar-applied PGRs (usually paclobutrazol or uniconazole). New growth is treated when shoots are between 50 and 150 mm in length (Lahav, 1999; Leonardi, 2005, 2009). The timing of summer pruning is critical with respect to producing new shoots that are sufficiently mature to flower the next fruiting season and will differ between locations depending on tree phenology. In addition to growth restriction, treatment of summer regrowth with uniconazole has been found to significantly increase the number of terminal shoots on summer-pruned trees that subsequently flower the following spring (Leonardi, 2001).

On Westfalia Estate, South Africa, mechanical pruning of young hedgerow orchards is initially done in summer. The regrowth is sprayed with uniconazole to ensure that the flush is hardened off in February (late summer) to permit flowering the next spring (D.G. Smith, Westfalia Estate, South Africa, 2011, personal communication). Continued mechanical hedging will usually result in a solid wall of foliage, which makes light and spray penetration into the hedge difficult. This is overcome by 'window-pruning' the side walls of the hedge by strategic removal of large limbs, usually on an annual basis. To benefit from this style of canopy management, it is important that pruning is implemented at the correct time, particularly in relation to attaining sufficiently mature shoots for flower initiation at the end of summer, as maintaining productive trees is the most effective means of restricting shoot growth. Removal of fruit as soon as possible after reaching commercial maturity is also important for maintaining acceptable crop loads on trees (Whiley *et al.*, 1996a,b; Leonardi, 2009).

Insights into the management of hedgerow plantings in both the highly invigorating and somewhat drier as well as colder areas in subtropical South Africa have been provided (D.G. Smith, Westfalia Estate, South Africa, 2011, personal communication). In the Tzaneen area where excessive tree vigour occurs, new plantings of 'Hass' grafted to 'Dusa' rootstocks have been established at 8 × 4 m (313 trees ha^{-1}). Pruning is mechanized and implemented as infrequently as possible with fewer large cuts as these stimulate vigour. When trees reach 5–7 years of age, a more severe winter pruning is necessary. Experience has demonstrated that winter pruning both sides of every second hedgerow at the correct angle and height, about 1 m below the final height of 80% of row width and leaving a final inter-row access of 2.5 m, results in the best outcome. Once pruned, rows usually return to cropping two seasons later when the alternate rows left uncut are pruned. This procedure prevents rows from growing excessively tall and maintains light interception and cropping. Stumping of older blocks is only carried out for top-working to new scion varieties. For the hotter and drier Mooketsi area as well as Everdon Estate in the cool to cold mesic KwaZulu-Natal mist belt in South Africa, a 7 × 4 m (357 trees ha^{-1}) spacing is more appropriate for avocado production.

The average yield for the Tzaneen/Mooketsi hedgerow orchards, with a number of young orchards, plus 150 ha of 'organic' orchards (in total about 1000 ha) was 18 t ha^{-1} for 'Hass' and 21 t ha^{-1} for 'Fuerte' in 2010. Young 'Hass' orchards yield 20–25 t ha^{-1} in 'on' seasons and 13–18 t ha^{-1} in 'off' seasons. A young 'Hass' on clonal 'Dusa' orchard at the cool climate, Everdon Farm grown organically produced 17 t ha^{-1} in its fifth year, but average sustainable yields are only about 10 t ha^{-1} (D.G. Smith, Westfalia Estate, South Africa, 2011, personal communication).

Sustainable Production and Regularity of Cropping

Alternate (biennial) bearing (AB) is a widespread phenomenon that occurs in both deciduous and evergreen fruit trees and it is likely inherent in polycarpic perennial plants (Monselise and Goldschmidt, 1982). Evolutionary processes are thought to contribute to this trait, as competition in forest stands has increased the complexity of woody perennials to the extent that they have developed growth-cycle strategies that optimize competitiveness with other species (Dickson, 1991). The development of synchronous flowering in tree species increases the potential for cross-pollination, thereby improving the robustness of progeny (Opler *et al.*, 1976; Browning, 1977; Hormaza *et al.*, Chapter 6, this volume). In avocado, the opportunity for cross-pollination is further emphasized by a unique dichogamous mechanism during anthesis (Bergh, 1974). Another ecological benefit of synchronous flowering is the resultant heavy cropping episodes that provide an escape from seed predation (Janzen, 1971). AB is initiated by 'one-off' climatic or management factors that negatively impact production over 1 year through poor flowering or excessive fruit loss, for example, freeze damage, extreme temperatures during flowering and fruit set, excessively high winds, water deficits, nutritional disorders or poor pruning practices. The resultant poor crop load ('off' year) is typically followed by very high production ('on' year) and once the alternate cycle begins, it becomes entrenched through endogenous tree factors that ultimately affect the intensity of annual flowering (Salazar-García *et al.*, 1998). AB is not to be confused with irregular bearing which is driven by climatic events that impact negatively on crop load irrespective of flowering intensity (Whiley, 2013). Irregular bearing is more common in areas that are climatically marginal for avocado production and is usually due to low minimum temperature episodes during flowering, sometimes combined with consistent wet weather during anthesis that disrupt the dichogamous cycle and/or restricts pollination and ovule fertilization for most of the flowering event. Poor crops can occur several years in succession prior to favourable conditions prevailing that result in a heavy fruit set. The year following heavy fruit set is normally an 'off' year due to the establishment of an AB cycle, but a succession of poor crops can reoccur due to inclement climatic conditions. Irregular bearing can be partly mediated through interplanting with pollinizer varieties and modified cultural inputs to reduce tree vigour (Whiley, 2013).

Theories of alternate bearing

Flowering and fruiting are resource demanding, and the ability to store significant quantities of minerals and carbohydrates to support these events is well developed in avocado (Cameron and Borst, 1938; Cameron *et al.*, 1952; Scholefield *et al.*, 1985; Whiley *et al.*, 1995b). While mineral nutrient demands for growth and development of avocado fruit are relatively small per unit of fruit weight compared with other fruit crops (Wolstenholme, 1991), there is a substantial investment of photoassimilates required to produce large-seeded, oil-rich fruit (Wolstenholme, 1986, 1987). Indeed, Scholefield *et al.* (1985) clearly demonstrated in the semi-arid, cold growing area in southern Australia that following heavy cropping, there was a significant reduction in the pre-flowering accumulation of starch in the trunk of 'Fuerte' trees, with a subsequent poor crop. Thus, while it seems evident that there are eco-evolutionary factors predisposing avocado trees to AB, there are opportunities in crop management to reduce the intensity of the 'on'/'off' yield cycle, assuming that other environmental factors are not negatively impacting fruit set and Phytophthora root rot is under control.

A grower-oriented overview of AB in avocado (Wolstenholme, 2009) was accompanied by revised growth model charts for 'on' and 'off' years (Wolstenholme and Sheard, 2009) in typical moist, humid subtropical climates. Apart from the starch depletion theory, it has been hypothesized for apple trees that seed-produced gibberellins in 'on' seasons are exported and have an anti-flowering effect or compete with shoot apices for flowering hormones (Dennis and Nielsen, 1999). Certainly, in avocado, 'on' crops reduce flowering intensity the following spring (Salazar-García *et al.*, 1998, 1999) although not the percentage of fruit set in California conditions (Garner and Lovatt, 2008). Trials with early winter GA_3 sprays in California (Salazar-García and Lovatt, 1998, 2000) reduced 'on' season bloom and yield, but more research is needed. However, in apple (Schmidt *et al.*, 2009) and 'Pixie' mandarin (Verreyne and Lovatt, 2009), crop load appears to reduce floral intensity of return bloom by inhibiting bud break, before possible seed-produced gibberellin effects can be manifested. Something similar was shown in Israel by Ziv *et al.* (2014), in research that de-fruited 'Hass' avocado trees at different times after fruit set, from July through November and measured the effect on individual bud fate of summer flush shoots. As fruit removal is delayed, the percentage of apical buds flowering the next spring is reduced up to 100% in the non-de-fruited control. Reduction of apical bud flowering starts when all fruit is removed in September while fruit removal in early November only resulted in 25% of apical buds flowering. Fruit removal at the beginning of October allowed a certain degree of inflorescence formation until the eighth node, while fruit removal at the beginning of November limited the number of nodes capable of producing inflorescences to the first five nodes. Iturrieta (2017) reports that under California conditions, growth on fruit-bearing shoots is reduced in all vegetative flushes of the season. Some of these fruit-bearing shoots will produce flowers in the following year, but in a much-reduced number compared to non-fruit-bearing shoots. Sunlight exposure also affected the length of growing shoots, irrespective of their relation to fruit presence (Iturrieta, 2017).

Research with other fruit and nut crops has led to more comprehensive theories of flowering and its relationship to AB. Basic studies suggest that multi-level signals regulate the various phases of flowering. Floral induction is required for evocation, that is, irreversible commitment to the initiation of floral primordia in buds. The 'level one' floral signal is proposed to be a foliage-produced, phloem-translocated florigen (the long-postulated 'flowering hormone'), which triggers chromatin modification to start floral induction (Zeevaart, 2008; Shalit *et al.*, 2009). Endogenous phytohormones, including gibberellins (GAs), auxin (indole-acetic acid, 1AA), cytokinins and ethylene, act and interact as 'level two' long-distance floral signals regulating the initiation of flowering (Bangerth, 2006, 2009). A later, 'level three' signal involves carbohydrates, specifically sugars in the developmental phase of flowering (Koch, 1996). Wood (2011) combined these concepts into a proposed model of pistillate flowering in pecan. Pecan orchards show pronounced AB. Field evidence suggests that at the 'second level', a cytokinin–gibberellin balance in pistillate flower buds is partially modulated by auxin and ethylene. This stage seems potentially reactive to synthetic plant bioregulator control through the application of foliar sprays that promote subsequent spring flowering during an 'on' crop season and inhibition of subsequent spring flowering initiation during the previous 'off' crop season. Surprisingly, this model does not include the inhibitory hormone abscisic acid (ABA).

Ziv *et al.* (2014) found a pProtein (PaFT) in avocados that acts in a similar way as its relative FT (Flowering Locus T) in Arabidopsis that functions as a signal that moves from leaves to shoot meristem to mediate the floral transition. PaFT levels were higher in non-fruit-bearing buds on non-bearing shoots compared to those on fruit bearing. Nevertheless, the question of how fruit load modulates PaFT encoding genes is still open for interpretation. There is reasonable probability that all named factors (sugar availability, hormonal levels and gene expression) might interact to cause or reduce AB behaviour. Further investigation needs to be conducted to determine how these different factors interact and how they can be controlled to moderate AB. For further discussion on AB, see Hormaza *et al.*, Chapter 6, this volume.

Rootstock effects on alternate bearing

Rootstocks are known to affect AB of tree fruit crops. For example, in Israel, the regularbearing 'Shamouti' orange cultivar exhibits a degree of alternation when grafted to sour orange rootstock. The same rootstock is thought to be the cause of heavy AB when grafted to mandarin cultivars growing in the Mediterranean area (Monselise and Goldschmidt, 1982). Studies with 'Valencia' orange in Australia also found that rootstocks had a significant effect on alternate cropping. The intensity of alternation was greatest when 'Valencia' was grafted to 'Troyer' citrange and trifoliate orange and least when 'Emperor' mandarin and 'Valencia' orange stocks were used (El-Zeftawi and Thornton, 1975).

Studies in Israel with two Mexican race avocado rootstocks each grafted to 'Ettinger', 'Nabal' or 'Hass' suggested that there was no rootstock effect on AB, although there were marked differences between cultivars. 'Ettinger', an early maturing cultivar, had the least tendency for AB, while 'Nabal' and 'Hass', both late-maturing cultivars, showed strong alternation in yield (Monselise and Goldschmidt, 1982). However, a later study in Australia indicated that avocado rootstocks may have a profound effect on AB (Thomas, 1997). In a 'Hass' orchard in south-east Queensland, yield records maintained for individual trees over a 6-year period were examined. From a block of 86 trees, mean yield data were calculated for the three highest and three lowest yielding trees randomly located in the orchard. The trees were grafted on to seedling Guatemalan rootstocks and received the same level of management. The six selected trees were tested for sunblotch viroid using highly sensitive PCR technique (see Dann and Gazis, Chapter 13, this volume) and were found to be free of this disease. Alternation intensity (*I*) (or intensity of deviation in yield in successive years) was calculated by the method of Gur *et al.* (1969) where

$$I = \frac{1}{n-1}\left(\frac{a2-a1}{a2+a1}+\frac{a3-a2}{a3+a2}+\ldots+\frac{a(n-1)-an}{a(n-1)+an}\right)$$

n is the number of years; a1, a2,...$a_{(n-1)}$ is the yield in corresponding years. The calculation in brackets is done ignoring the signs of terms for each module calculation.

For high-yielding trees I = 0.15, and for low-yielding trees I = 0.69, where I = 0 represents regular cropping and I = 2 strong AB (Fig. 12.12). For each year of the study, the high-yielding trees significantly outcropped low-yielding trees, which displayed scion overgrowth and therefore a degree of root starvation. Although the specific identity of the rootstock material used in this block is unknown, it is likely that the trees were grafted onto seedlings from several maternal sources as was a common nursery practice at the time of orchard establishment. Hence, the genetic diversity that is represented in the trees used in that study clearly demonstrates the potential influence of avocado rootstocks on both total production and cropping patterns. Scion overgrowth as an indicator of enhanced AB has also been noted in New Zealand, Western Australia and Argentina (A.W. Whiley, SHS, Buderim, QLD, Australia, personal communication, 2001).

Mickelbart *et al.* (2007) conducted a 10-year study of 'Hass' avocado trees on ten clonal rootstocks in California, under soil conditions where Phytophthora root rot and salinity were not problematic. Trees on all rootstocks showed AB, ranging from a low of 0.33 on 'G755 C' to a

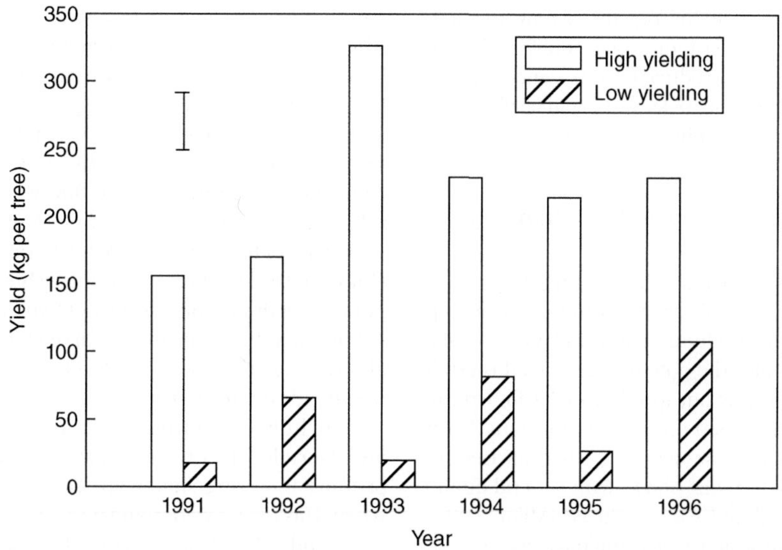

Fig. 12.12. Yield comparisons between consistently high- and low-yielding 'Hass' avocado trees grafted onto seedling Guatemalan rootstocks and growing in south-east Queensland. The vertical line represents the LSD at ($P \leq 0.05$) as determined by ANOVA. The alternation intensity (I) for high-yielding trees was 0.15 and for low-yielding trees was 0.69. Drawn from data of Thomas (1997), with permission.

high of 0.56 on 'Toro Canyon' and 0.55 on 'Topa Topa'. The 'G755' series had the lowest mean yields. 'Duke 7' with an *I* of 0.52 had the highest canopy efficiency (kg fruit m^{-3} canopy) and highest cumulative yield along with 'Borchard'. It is pertinent to point out that due to varying rootstock tolerances, the results would have differed in Phytophthora root rot–affected and saline soils.

Whiley *et al.* (2013) completed a 10-year study of avocado rootstocks grafted onto either 'Hass' or 'Shepard' planted across a diverse range of environments in Australia. The programme included rootstocks propagated either clonally or as seedlings. Following the analysis of 6 years of consecutive cropping data, there was strong evidence within the seedling populations of individual trees at three of the experimental sites having significantly higher cumulative yields (150–200% greater than the balance of the experimental population). These higher-yielding trees were distinguished by their consistent annual increase in production and were proposed for further commercial development, whereas the lower-yielding trees within the experimental population had already developed AB cropping features.

Mena *et al.* (2015a) recorded 4 years production from 193 trees of 'Hass' grafted on Mexicola seedling rootstock. The alternate bearing index (ABI) described by Lovatt (2003) was calculated each year for individual trees.

$$ABI = \frac{Production\ \text{'on'}\ year - Production\ \text{'off'}\ year}{Production\ \text{'on'}\ year + Production\ \text{'off'}\ year}$$

Values closer to 0 = lower AB, values closer to 1 = higher AB.

Over 70% of the trees showed an ABI higher than 0.5, an indication of moderate-to-high level of AB. The average production over 4 years from the 20 highest yielding trees was about 400% more than produced from the 20 lowest yielding trees. The 20 highest yielding trees also produced over 75% more fruit than the average of all trees, with an average ABI of 0.5. Additionally, most of the highest yielding trees produced fruit annually, while the lowest producing trees produced fruit 1 or 2 years out of the 4 years recorded, resulting in an average ABI of 0.9. The above-mentioned results from Australia and Chile suggest there are opportunities to select higher-yielding rootstocks from seedling populations that have been grafted onto 'Hass'. Clonal propagation of such rootstocks may result in overall more consistent and greater production than currently produced from genetically diverse seedling rootstocks and could be planted in areas of low Phytophthora pressure.

Flower pruning and fruit thinning

Pruning as a means of regulating crop load is widely used in fruit crops although more commonly with deciduous trees and vines (Mika, 1986). Usually, the potential crop load is established by pruning out wood excessive for fruiting requirements when the plant is dormant. Pruning studies have shown that when carried out during the dormant phase, fruit set and quality are improved and AB is reduced (Mika, 1986). Studies with selective pruning have been carried out on avocado to break or reduce strong AB patterns that occur with this crop, particularly with late-maturing cultivars growing in cool climates.

Farré *et al.* (1987) studied the effects of prebloom pruning on the cropping pattern and mean fruit size of 'Hass' trees growing in Spain that were in a strong AB cycle. Treatments included different pruning severity (light or heavy) and timing in relation to flowering phenology (early or late) that were applied only in the first year of the study when heavy flowering was indicated. They found that early, heavy pruning (selective removal of wood from the previous year's summer and spring growth at the beginning of March) significantly increased the cumulative yield for 2 successive years during which a strong 'on'/'off' cycle occurred (Table 12.4). This treatment also markedly reduced the severity of AB ($I = 0.30$ for early heavy pruning; $I = 1.19$ for the control). There was also a beneficial effect from late, heavy pruning (at the end of March coinciding with the first open flowers) that significantly increased cumulative yield compared with control trees and reduced AB (Table 12.4). However, the late, heavy pruning treatment was not as effective at increasing yield and reducing AB as early, heavy pruning. Farré *et al.* (1987) also reported there was no effect from any of the treatments on mean fruit size.

In humid, subtropical South Africa, Roe and Morudu (1999) investigated the effect of pruning flowering shoots on AB and fruit size when 'Hass' trees were going into an 'on' year. In their study, 20–30% of terminal inflorescences were removed between full bloom and about 10% fruit set. They reported that flower pruning in 'on' years increased mean fruit weight and, hence, the value of the crop as a greater percentage of fruit reached export size. However, the treatment had no consistent effect on AB over the 4 years of the study. They concluded that pruning flowering shoots should only be carried out following a distinct 'off' year and when flower intensity was such that a heavy 'on' year could be forecast. They suggested that there was potential to use pruning of flower shoots to correct severe AB patterns, but individual trees should be assessed before treatment.

The different results on AB from selective pruning of flowering shoots of 'Hass' avocado

Table 12.4. Effect of pruning and fruit thinning on alternate bearing and yield of 'Hass' avocado grafted onto 'Topa Topa' rootstocks growing in Spain (from Farré *et al.*, 1987).

Treatments*	I, 1984–85	Mean cumulative yield (kg tree^{-1})
Control	1.19	99.6^{c}
Early light pruning (carried out from 2–8 March)	0.58	129.6^{abc}
Early heavy pruning (carried out from 2–8 March)	0.30	150.0^{a}
Late light pruning (carried out from 18–30 March)	0.96	110.6^{bc}
Late heavy pruning (carried out from 18–30 March)	0.54	145.0^{ab}
Fruit thinned (carried out in late June)	0.86	107.7^{c}

The intensity of yield deviation in successive years (I) is calculated as described by Monselise and Goldschmidt (1982). An I value of 0 = regular bearing and 2 = extreme alternate bearing. Yield data are mean values for 1984 and 1985. Data in columns followed by different letters are significantly different at ($P \leq 0.05$). Yield data are from Farré *et al.* (1987). *Late pruning was timed to coincide with the appearance of the first unopened flowers with early pruning about 3 weeks before that time. Light pruning involved the removal of some of the previous year's summer growth, while heavy pruning included the removal of part of the spring growth.

trees in Spain and South Africa may be due to marked climatic differences between the production regions in each country. However, differences at the time of pruning in relation to tree phenology between the two studies were more likely the greatest contributing factor to the differing results. From the Spanish study (Farré *et al.*, 1987), selective early, heavy pruning about 3 weeks prior to anthesis and before the depletion of reserve carbohydrates (Whiley, 1994) had the greatest effect on reducing AB. Conversely, in the South African study (Roe and Morudu, 1999), pruning occurred later from mid-bloom through to early fruit set after reserve carbohydrates had been committed to flowering (Whiley, 1994). Hence, it is assumed that carbohydrate savings would not be as great, thereby having less of an effect on fruiting in the following year.

In a later study, Roe and Morudu (2000) re-examined pruning of flowering shoots by investigating the time of pruning with respect to flower/fruit set phenology. Pruning was carried out at five different stages from flower bud swell to fruit set by tipping (cutting off 10 cm) all terminals on the tree for 2 consecutive years. None of the treatments significantly increased yield, although there was a trend for the two earliest pruning times (at flower bud swelling and bud break) to marginally increase yield compared to non-pruned trees. The later the pruning was applied, the greater the decline in the cumulative yield over 2 years, which is consistent with a wasted commitment of reserve carbohydrates to flowering and fruiting prior to pruning. Roe and Morudu (2000) also reported that pruning for the most part increased fruit size with a greater percentage falling into the 285–400 g range. The earlier the pruning time, the greater the increase in fruit size. However, pruning of flowering shoots has not been adopted commercially in South Africa. Similar results were reported from New Zealand (P. West, New Zealand, 2023, personal communication), where flower or fruitlet pruning was carried out at three different stages of development (mid-spring, early summer and late summer). There were no statistical differences in the yield among treatments in the first year, but a trend towards lower production with increased fruit size. Similarly, in the second year, there were no statistical differences among treatments, but a trend of increased production, with the earlier removal of reproductive structures being the most productive. Fruit thinning has also been investigated as a means to reduce the extent of AB and to increase fruit size of 'Hass' (Snijder and Stassen, 1997). Preliminary results from South Africa indicate that late spring (November) fruit thinning in combination with regular, light pruning increased yield in one season but had no effect on mean fruit size. Additionally, under a heavy cropping scenario, fruit thinning increased mean fruit weight without affecting yield. However, for trees that carried a light crop, there was no significant effect of fruit thinning on yield or mean fruit size (Snijder and Stassen, 1997). In reality, the large tree size makes hand thinning impractical in production areas with invigorating climates and soils and vigorous rootstock/scion combinations. However, small tree size in experimental high-density plantings (e.g. 3 × 3 m) in Chile, permitting work from ground level in a non-invigorating environment, may allow hand thinning of fruits to reduce or avoid AB problems (Mena *et al.*, 2007), but so far this practice has not been found to be commercially viable.

Time of harvest effects on alternate bearing

Mature avocado fruit do not ripen until harvested and in cool climates may be retained on the tree for up to 6 months or more after attaining acceptable commercial maturity (Schroeder, 1953; see Defilippi, Chapter 15, this volume). This feature has been widely utilized as a 'storage' strategy to take advantage of marketing opportunities. For example, the harvest of mature fruit may be delayed for several weeks or even months if market prices are low, and growers anticipate a significant improvement within the 'tree life' of the crop. However, lipids continue accumulating in fruit well after physiological maturity (Eaks, 1980; Kaiser and Wolstenholme, 1994) and 'tree storage' of fruit will undoubtedly affect the allocation of carbon products within the tree with potential long-term effects on production. In studies with 'Valencia' orange, another fruit crop that may be 'stored' on trees in cool areas for up to 8 months after commercial maturity has been reached, reduced yields of smaller fruit and AB occurred following extended 'on-tree storage' (Hilgeman *et al.*, 1967; Monselise and Goldschmidt, 1982).

Whiley *et al.* (1996a) found that extended 'tree storage' of 'Fuerte' avocado growing in humid subtropical south-east Queensland, Australia induced pronounced AB (in an area with fairly regular cropping) compared to trees where fruit were harvested close to commercial maturity (Fig. 12.13). For example, when fruit were harvested each year after reaching 21% or 24% dry matter (DM), *I* values of 0.06 and 0.08, respectively, were achieved, indicating negligible alternation of cropping. However, when half of the crop was picked at 21% and 30% or at 24% and 30%, the *I* value increased to 0.17 and 0.12, respectively. The most pronounced AB cycle developed when all fruit were left on trees until reaching 30% DM with *I* equal to 0.40 (Fig. 12.13). Additionally, over the 6 years of the study, the cumulative yield from early harvested (21% or 24% DM) trees was 34–37% higher than that from late harvested (30% DM) trees (Whiley *et al.*, 1996a).

In a similar study of shorter duration with 'Hass', Whiley *et al.* (1996b) reported that there were indications of the development of an AB cycle. In that study, fruit were also harvested at different stages of maturity based on fruit DM of 25%, 30% or 35%. In two of the treatments, there was a split harvest with half of the fruit picked at either 25% or 30% DM and the other half at 30% or 35% DM (Fig. 12.14). The *I* values were only calculated for 1993 and 1994 since severe, late season cyclonic winds in 1992 blew a significant quantity of fruit off the trees prior to maturity. Early or split harvest treatments generally showed little indication of AB with *I* values ranging from 0.02 to 0.21. However, when all fruit were allowed to 'hang' on the tree until well after reaching commercial maturity (35% DM), an *I* value of 0.76 was obtained, indicating a pronounced AB cycle had developed (Fig. 12.14). In terms of the effect of harvest maturity on yield, production from trees harvested at 25% DM was approximately 27% higher than that of the other treatments (Whiley *et al.*, 1996b). Kaiser and Wolstenholme (1994) were unable to demonstrate any adverse effect of 'on-tree storage' of 'Hass' in a cool, mesic environment in South Africa on yield or AB. However, the experiment was only conducted for two consecutive seasons and was compromised by theft of fruit from trees.

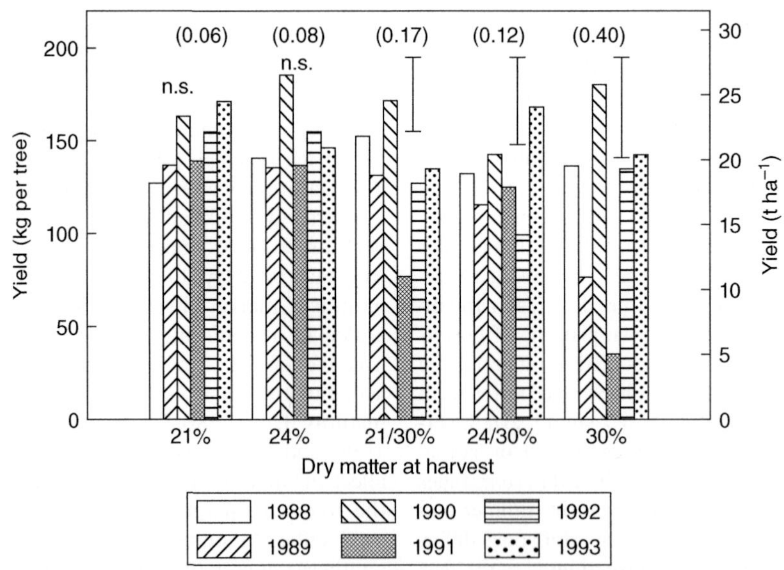

Fig. 12.13. Effect of time of harvest on the sustainability of yield of 'Fuerte' avocado trees growing in south-east Queensland over 6 consecutive years. *I* values (alternate bearing index) for each harvest time based on dry matter values are given in parenthesis. Columns are mean values ($n = 6$) and vertical bars indicate LSDs ($P \leq 0.05$) determined by ANOVA. Reprinted from Whiley *et al.* (1996a), with permission of Elsevier Science (1996).

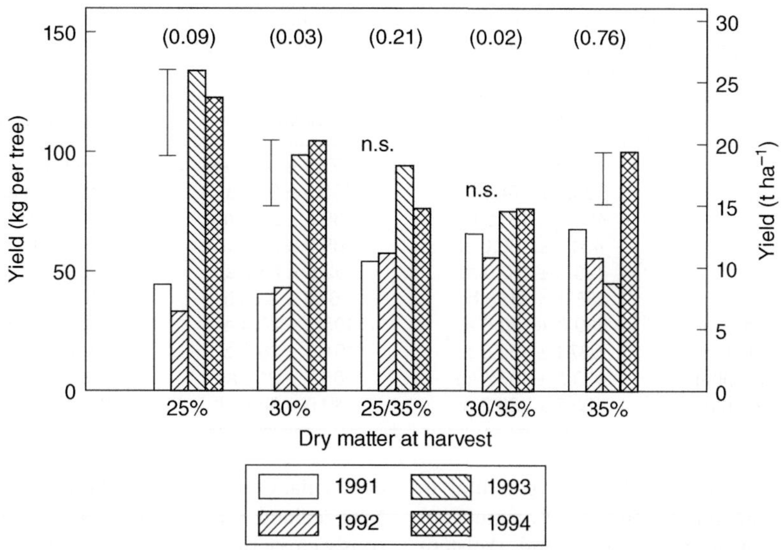

Fig. 12.14. Effect of time of harvest on the sustainability of yield of 'Hass' avocado trees growing in south-east Queensland over 4 consecutive years. *I* values (alternate bearing index) for each harvest time based on dry matter are given in parenthesis. Columns are mean values (n = 6) and vertical bars indicate LSDs ($P \leq 0.05$) determined by ANOVA. Reprinted from Whiley *et al.* (1996b), with permission of Elsevier Science (1996).

'Hass' fruit maturity in the central part of Chile (Panquehue) is usually attained in mid-August, especially when orchards are planted on hillsides. Mena *et al.* (2015b) harvested fruit from trees over 4 years at different times: third week of August, third week of September, third week of October, third week of November, third week of December and third week of January when trees were strip-picked. Concurrently, on a separate group of trees, two harvests were made each year using the following timelines: third week of August and third week of September; third week of August and third week of October; third week of October and third week of November; third week of October and third week of December. For these treatments, the first harvest was size-picked (fruit > 200 g), and the second harvest removed all remaining fruit. Total cumulative yield over 4 years was highest in September (105.5 kg tree^{-1}) for the single harvest and August/September (105.2 kg tree^{-1}) for the split harvest treatments, although only significantly different from the January harvest, which was the lowest yielding treatment (Table 12.5). With respect to fruit size, the January harvest produced the largest fruit, though the least number of fruit per tree, while the earliest harvest (third week of August) resulted in the smallest fruit size. The ABI (Lovatt, 2003) was increased as harvest was delayed, except for the November harvest, which was intermediate in total production.

From the studies of Whiley *et al.* (1996a,b) and Mena *et al.* (2015b), it can be concluded that the management of crop harvest, particularly in relation to the time or maturity at which fruit is removed, can strongly influence both cumulative yield over a number of years and whether or not trees develop strong AB patterns. For example, harvesting fruit as close as possible to minimum commercial maturity standards will be beneficial in maximizing long-term productivity and reducing the severity of AB cycles. However, this strategy needs to be balanced with other criteria such as fruit size and financial returns. Fruit size is an important component of yield with premium prices often paid for larger fruit. In particular, as trees age, there can be a problem with 'Hass' fruit that are at best medium sized, but with a proportion of the crop being too small for profitable marketing (Lahav and Adato, 1990; Köhne, 1992b; Cutting, 1993; Wolstenholme and Whiley, 1995; Cowan,

Table 12.5. Effect of date of harvest on yield (kg), fruit number and mean fruit size per tree over a 4-year period (2011–14) in the central part of Chile. Data in columns followed by different letters are significantly different at ($P \leq 0.05$) (Mena *et al.*, 2015b) used with permission.

Date of harvest	Σ of fruits tree^{-1}	Σ of kg tree^{-1}	Average fruit size (g)
August	449.3 ± 107.2 a	77.3 ± 18.0 ab	176.8 ± 44.0 h
September	585.5 ± 174.6 a	105.5 ± 27.5 a	191.6 ± 54.3 g
October	453.5 ± 131.9 a	89.6 ± 26.8 ab	208.4 ± 56.0 c
November	431.3 ± 160.0 ab	89.3 ± 31.2 ab	214.0 ± 65.4 b
December	429.3 ± 110.0 ab	92.5 ± 23.2 ab	220.7 ± 59.0 b
January	285.1 ± 131.8 b	67.0 ± 28.3 b	243.5 ± 61.1 a
August + September	569.0 ± 128.5 a	105.2 ± 26.4 a	196.3 ± 51.9 f
August + October	537.1 ± 132.0 a	104.3 ± 28.3 a	203.5± 51.2 e
October + November	439.5 ± 107.2 ab	86.7 ± 21.9 ab	211.3 ± 52.5 d
October + December	450.7 ± 138.2 a	91.9 ± 32.6 ab	209.1 ± 51.7 c

1997). Delayed harvest has the potential to bring about small but commercially significant increases in fruit size (Kaiser and Wolstenholme, 1994; Whiley *et al.*, 1996a,b) due to continued cell division in fruit while attached to the tree (Schroeder, 1952; Valmayor, 1967), provided the seed-coat does not senesce prematurely. Thus, while total fruit production may be greater with early harvesting, the volume of premium market-size fruit may be maximized with delayed harvest. Selective size picking, involving only the removal of market-size fruit when commercial maturity is first reached, provides greater opportunity for the remaining fruit to obtain optimum size, while reducing crop load and stress on the tree. In some cases, there may be opportunities to access high-value niche markets by holding fruit late on trees, with yield reduction being more than compensated by the additional returns achieved.

'Hass' has become the most important cultivar grown by countries with mild Mediterranean or subtropical climates or at high elevations in tropical regions such as the Dominican Republic (see Crane *et al.*, Chapter 8, this volume). It is a mid-season maturing cultivar, and fruit can be 'stored' on trees for varying periods after reaching physiological maturity to take advantage of market opportunities, especially in cooler growing areas. Later maturity in cooler areas is also due to later flowering and fruit set. In some instances, delayed harvest of 'Hass', which is normally picked during winter in the warm subtropics, may result in trees simultaneously carrying mature fruit while flowering and setting the next season's crop. For example, where 'Hass' is grown in cooler climates, viz. California, Chile, Israel, southern Australia and New Zealand, it is normal for trees to simultaneously carry two crops for a period of time, mature fruit from the previous season as well as the current season's crop. Adherence to the principles discussed above should help reduce biennial bearing and increase orchard productivity in these regions. However, significant increases in productivity are likely to be found with new varieties that are able to mature fruit in a shorter time than 'Hass', thereby providing a long period for tree recovery before setting the next crop.

Bioregulant modification of alternate bearing

The triazole plant growth inhibitors, paclobutrazol (Cultar®) and uniconazole (Sunny®, Magic®), have been used on avocados for several years, in countries where registration has been obtained (Australia, Chile, Israel, Perú, Colombia, New Zealand and South Africa, but not the USA or Mexico). They block gibberellin biosynthesis (Davis *et al.*, 1988) and are used to increase flower initiation (autumn sprays), increase fruit set (spring sprays), increase fruit size and improve fruit shape (see section on triazoles, this chapter). Uniconazole is less persistent in the plant than paclobutrazol. Both products are being used as an integral part of tree size control in a package of management interventions in ultra-high-density orchards on hillsides in Chile and moderately higher-density orchards in Australia and New Zealand. Tree training, containment pruning, use of bioregulators and in some orchards scoring (mild girdling) have shown promise in mitigating

AB and achieving average yields of over 20 t ha^{-1} in Chile (Gardiazabal *et al.*, 2011a) with similar success in New Zealand (Hermosilla-Madrid and Whiley, 2023). Concepts are continually being modified in light of ongoing experience. New developments can be expected as long as continued bioregulant use is permitted in avocado orchards.

Lopez-Jaminez *et al.* (2011) reported in avocado that the 'on' crop causes correlative inhibition (auxin > cytokinin) of buds that produce the summer and autumn vegetative flushes in California. The result is a reduction of the number of nodes bearing inflorescences next spring and inhibition of spring bud break. Supplying trees with TIBA plus a cytokinin in July (mid-summer) and January (mid-winter) mitigated these adverse effects of the 'on' crop on summer and autumn shoot flushing and spring bud break and enhanced spring flowering.

Prospects for alleviation of alternate bearing

The above-mentioned discussion emphasizes that AB is arguably a major horticultural problem faced by avocado growers. Its extent varies between countries and growing areas, orchards, seasonally, with cultivar and rootstocks. It has been reported as less problematic in some mesic, warm subtropical areas and parts of Mexico. It is clear that certain cultivars and rootstocks are less prone to AB under good orchard management. Thomas (1997), Whiley *et al.* (2013) and Mena *et al.* (2015a) all reported reduction in AB by numbers of 'elite' rootstocks in seedling populations grafted onto 'Hass'. Lovatt (2001) reduced *I* from 0.90 to 0.72 in 20-year-old 'Hass' trees by encouraging the spring growth flush with extra nitrogen in the 'on' crop season. Flower pruning and fruit thinning in an 'on' season are usually helpful but impractical and unpopular with growers in orchards with trees taller than about 4 m. Late harvesting for economic gain will aggravate AB (Whiley *et al.*, 1996a,b; Mena, 2015b), although adverse effects can be reduced by selectively picking half the crop at minimum horticultural fruit maturity.

Currently, in standard (traditional) orchards with tree populations below 300–350 trees ha^{-1}, growers can reduce the amplitude of year-to-year crop variation with a suite of management practices, assuming that the well-known promoters of AB discussed above are countered, as far as practically possible. It is also vital to use best management practices, starting with site selection and preparation and including nutrition, irrigation, pest and disease control, and orchard floor management. The main tools used currently are targeted pruning, use of the bioregulant triazoles where permitted and sometimes girdling. In the humid subtropics, encouragement of spring, and especially summer shoot flushes in an 'on' season by extra N fertilization and pruning, seems logical. Where uniconazole and paclobutrazol are not registered, a range of PBRs are being tested for use in both 'on' and 'off' crop seasons.

For high- and ultra-high-density orchards (greater than 500 and 1000 trees ha^{-1}, respectively) in areas with more stressful climates and soil conditions, success depends on highly intensive orchard management, based on tree training, timely pruning with clear objectives, bioregulant foliar sprays and/or soil applications and sometimes scoring. For ultra-high-density orchards, tree height is ideally kept at about 2–2.5 m, with occasional heavy pruning, which only reduces crop for the following season. Precocity and regularity of bearing help to contain tree size to allotted spaces, aided by root competition, and restricted irrigated soil volume per tree in more arid climates.

Gardiazabal *et al.* (2011a) supplied yield data from a 6-year evaluation of two pioneer 3 × 3 m spaced orchards, in the same general area of Chile. The Llay Llay orchard was planted in August (late winter) 2004, and a 2-m-tall central leader was established in the first season. The first crop was induced the following spring (October) and harvested 24 months after planting. Triazoles were used only in spring to increase fruit set and size, with no special selection of sylleptic lateral shoots or removal of vigorous shoots in the first year. The orchard quickly became overcrowded, necessitating heavy pruning in 2008 and resultant onset of AB. Yields each year were 9.3, 17.7, 4.4, 39.5, 19.1 and 34.0 (estimated) t ha^{-1}.

The Llayquén orchard, planted in January (mid-summer) 2006, was trained to a central leader, and with strict selection of only sylleptic lateral shoots in the first season. Yearly pruning was carried out to maintain initial central leader structure, and plant growth regulators were also used to control vigour, especially after heavy pruning. In this manner, AB was reduced, especially the very high yields in heavy 'on' seasons. Yields were 0.5 t ha^{-1} in 2007 and 18.3, 20.6, 23.9 and 26.0 (estimated) t ha^{-1} in consecutive seasons. The small first crop was due to the later (mid-summer) planting. Triazoles were important in managing AB in these high-density orchards, with summer application leading to increased flowering after an 'on' crop season. High-density orchards required more management of AB than standard spaced orchards (Gardiazabal *et al.*, 2011a).

Long-term solutions to AB will probably result from basic research into avocado genomics and biotechnology, which will facilitate breeding of new cultivars and rootstocks for both the mesic humid subtropical (and tropical) growing regions and the more stressful semi-arid and arid Mediterranean climates. In the shorter term, practically oriented cooperative international breeding research and traditional breeding programmes in various countries, with the release of new crop genetics, may contribute meaningfully to managing AB at the orchard level.

Girdling (Scoring, Cincturing, Ringing)

Girdling is the most commonly used term for an ancient horticultural tree manipulation practice, used mainly to increase flowering and fruiting in excessively vigorous trees. In this sense, it temporarily swings the vegetative: reproductive balance away from vegetative to reproductive growth, depending on the severity of the trunk (or branch or shoot) girdle. Increased fruiting is one of the best ways of preventing excessive tree vigour, which is antagonistic to fruiting. The reverse also applies, so that striking a balance is a key goal of orchard management, complicated by AB. Girdling and cincturing are synonymous in this review and, along with ringing or scoring, are terms that have been widely used in horticulture.

The technique involves the complete severance of phloem on a limb, shoot or trunk of a tree either through the removal of a complete ring of bark (girdling, cincturing or ringing) or by a narrow incision (scoring), without damage to the underlying cambium and xylem tissue (Noel, 1970; Goren *et al.*, 2004) (Fig. 12.15). When successfully carried out, the wound will produce callus tissue bridges and eventually heal, thereby restoring normal physiological functions to the affected part of the tree. The width of the incision varies depending on the time of application and the result being sought.

Girdling has had wide practical application with tree crops but is most commonly used to promote increased flowering and fruiting, although an understanding of likely tree responses is

Fig. 12.15. Scoring is frequently carried out by Californian and Chilean avocado growers to increase yield on treated branches. (a) A handheld pruning saw is used to make a single cut encircling the selected branch. The cut is only bark-deep and does not damage the xylem or cambium tissues (see removed bark patch). Over the course of time, the scoring wound heals and normal physiological activities resume. (b) Multiple scoring events can be made on a branch over a number of years. Photographs supplied by A.W. Whiley, Sunshine Horticultural Services, Queensland, Australia.

necessary, and results are not always predictable. Since the beginning of the modern era of avocado cultivation, the use of girdling for manipulating avocado trees has had a mixed history. For a time, the technique was embraced by some industries, particularly in Israel in the 1960s and 1970s, only to fall from favour within a few years. In recent years, it has found favour in management of some high-density avocado orchards in Chile, where cropping precocity is important to repay high establishment costs and where tree size is severely restricted by various manipulation techniques. Girdling to induce a heavy crop on trees or branches in the season before they are removed is also fairly common. Factors that might cause root stress, such as salinity and drought, have to be taken into account before planning the use of this technique.

Effect of girdling on tree physiology

Goren *et al.* (2004) reviewed the effects of girdling on the physiology of horticultural tree crops. They reported that the main effect of this procedure is the interruption of phloem transport of photoassimilates (carbohydrates) and auxin between the girdled trunk or limb and other parts of the tree, essentially between canopy and roots. An effective girdle usually results in visual changes to the tree, indicating a level of stress. Most commonly reported are yellowing leaves and/or changes to tree phenology such as earlier flowering and premature leaf abscission on treated branches (Swarbrick, 1927; Painter and Brown, 1940; Noel, 1970; Ticho, 1971; Davie *et al.*, 1995; Hackney *et al.*, 1995; Goren *et al.*, 2004). In studies with five avocado cultivars in Israel, Lahav *et al.* (1971a) reported yellowing of leaves on girdled branches that was particularly prominent on 'Hass' trees. However, yellowing was generally not observed on girdled 'Fuerte' trees in humid, subtropical Australia (Trochoulias and O'Neill, 1976). In two Israeli studies with avocados (Lahav *et al.*, 1971a; Tomer, 1977), where girdle treatments were applied to either trunks or single branches, time of flowering was advanced by 2 months or longer compared to untreated trees. The earliness of flowering was greatest following autumn girdling with a much lesser effect from winter treatment (Lahav *et al.*, 1971a). Both studies also reported premature leaf abscission during flowering that was particularly heavy on 'Nabal' (Lahav *et al.*, 1971a; Tomer, 1977). Mature leaf abscission was reduced when new spring shoots were removed as they developed (Tomer, 1977). Most girdling studies with avocado trees increased the number of seedless fruit ('cukes') at harvest (Lahav *et al.*, 1971a; Ticho, 1971; Trochoulias and O'Neill, 1976; Tomer, 1977). It is likely that after this fruit had set, but shortly following ovule fertilization, it was exposed to low minimum temperatures (≤ 10°C; Zamet, 1990) that killed the developing embryo, as found for mango (Sukhvibul *et al.*, 1999). This would be consistent with the earlier flowering (when mean temperatures are cooler) reported for cinctured avocado trees. In other words, girdling increased the chances of retention of stenospermocarpic fruit (with abortive incompletely developed seeds) since flowering is advanced with the likelihood of greater exposure to low temperatures during fruit set.

In correctly girdled trunks or branches, upward transport via the xylem is not affected, and the flow of water, minerals, solutes and other metabolites (but not cytokinins) (Goren *et al.*, 2004) from roots to leaves continues almost as normal (Kurtzman, 1966; Noel, 1970; Goren *et al.*, 2004). However, there have been numerous studies that report the accumulation of carbohydrates and increased auxin concentrations in tissues above the girdle due to temporary phloem dysfunction (Mason and Maskell, 1928a,b; Ticho, 1936; Murneek, 1941; Noel, 1970; Tomer, 1977; Davie *et al.*, 1995). The reduced supply of cytokinins from roots to canopy in girdled trees may be an indirect effect resulting from root starvation (Goren *et al.*, 2004).

In the first reported studies with avocado examining carbohydrate distribution, Ticho (1936) found that the starch concentration in roots of Mexican seedling avocados declined after trees were girdled (down to 8% of that of the control trees), but in branches and leaves distal to the girdle it was higher than in non-treated trees. Later, Tomer (1977), using mature grafted 'Fuerte' and seedling 'Fuerte' trees, found a greater accumulation of carbohydrates (particularly starch) in the leaves, bark and wood of girdled branches compared to control trees. The results were similar for both groups of trees, except differences were greater in seedling trees.

This accumulation of carbohydrates above the girdle can have detrimental effects if excessive, that is, too wide a girdle. Apart from leaf chlorosis associated with an increase in size of starch granules in citrus and apple (Schaffer *et al.*, 1986; Nii, 1992), photosynthesis declines, especially in the absence of fruit sinks. This is usually attributed to sink-feedback inhibition of photosynthesis due to the accumulation of end-product (Goren *et al.*, 2004). The corresponding decrease in carbohydrates below the treated position, especially if prolonged due to a wide trunk girdle, can cause severe root starvation and reduced root growth. These effects can be minimized by narrower girdles (which heal faster) and girdling several branches representing part of the canopy instead of the trunk. The importance of root flushes for normal operation of the avocado growth cycle (Whiley *et al.*, 1988) is self-evident. Girdling of non-vigorous trees, and particularly those infected with Phytophthora root rot, is counterproductive and will have serious detrimental consequences. Cincturing reduces root starch concentration and dry matter accumulation, as first reported for avocado by Ticho (1936).

Lahav *et al.* (1971a) found that leaf chlorosis coincided with reduced nitrogen, calcium, magnesium and manganese concentration in girdled 'Ettinger' trees in Israel for up to 2 years after treatment. They also found that mineral concentration of leaves on girdled branches was depressed in the absence of chlorosis. Tomer (1977) reported that leaf mineral content (nitrogen, phosphorus, potassium and calcium) was depressed in cinctured limbs compared with non-treated branches in the same trees. Leaf yellowing is likely in part due to reduced mineral content of leaves brought about by higher yields and disruption of translocation in girdled branches. Girdled limbs may also be more subject to photoinhibition and photo-oxidation under conditions that severely restrict the utilization of incoming solar radiation, for example, cold, dry winters (Whiley *et al.*, 1999).

Trunk girdling during fruitlet drop increased final fruit set in 'leafy' inflorescences, but not leafless inflorescences, in two citrus cultivars. This was due to a delay in fruitlet abscission and an increase in the quantum efficiency of electron transport through photosystem II (PSII) in girdled 'leafy' flowering shoots, that is, in recently expanded young leaves on the flowering shoot. In citrus, mature leaves and vegetative shoots also support fruit set, but not through improved PSII quantum efficiency. They are therefore not implicated in improving fruit set resulting from trunk girdling (Rivas *et al.*, 2007).

Girdling for increased yield

The first publication on girdling avocado trees is attributed to Coit (1921) in California where increased yield of girdled 'Fuerte' avocado trees was reported. This was followed later by studies that were unable to show beneficial results, which led to the commercial discontinuation of the practice in California during the 1930s.

Following research during the 1960s, girdling of avocados to increase yield found favour with growers in Israel, where results demonstrated consistent benefits following treatment of vigorous low-producing cultivars or individual trees (Lahav *et al.*, 1971b; Ticho, 1971). Lahav *et al.* (1971b) reported that girdling avocado trees for 3 consecutive years gave increased production that ranged from 4.4 to 11.4 t ha^{-1}. The most effective time to girdle was during autumn with a 10–20 mm wide cut being made, the width of the girdle being directly proportional to the diameter of the limb treated. Lahav *et al.* (1971a) found that girdling decreased mean fruit size, which had a marketing advantage for large-fruited cultivars, for example, 'Ettinger' and 'Fuerte'. However, girdling 'Hass' trees increased the number of under-sized fruit, and it was concluded that the treatment was not suitable for this cultivar under the experimental conditions. Girdling also delayed fruit maturity, possibly because of increased fruit set on treated limbs causing greater competition for resources (Lahav *et al.*, 1971a). In a later study in California, Francis (1996) reported that girdling 'Hass' trees significantly improved fruit yield. Girdles were applied in autumn, winter and spring, with all treatments improving yield. However, the winter application produced the greatest yield increase. There is no information from that study on the effect of girdling on mean fruit size.

From the Israeli research, it was recommended that only one-half to two-thirds of limbs on individual trees be treated in any one year as in some instances when trunks were girdled, trees died. Girdling the same limb in consecutive years was not detrimental over the 3 years of the study, but it was thought that the best management strategy was to rotate the limbs with treatment so the same limb would not be girdled for more than 2 consecutive years. Girdling was most beneficial when low-producing, vigorous trees were treated. In most cases, there was no benefit from girdling trees when entering a strong 'on' year of an AB cycle (Lahav *et al.*, 1971b). In theory, however, girdling in the autumn of an 'on' year could help increase flowering the next spring (the 'off' crop year).

Girdling 'Fuerte' trees in humid, subtropical Australia (Trochoulias and O'Neill, 1976) gave similar results to those observed in Israel, with an increase in yield in 3 consecutive years of treatment. However, there was no significant effect on production in the fourth consecutive year of the study. Girdling in autumn also reduced the mean fruit size, which was seen as a marketing advantage for this cultivar (Trochoulias and O'Neill, 1976).

In Israel, Tomer (1977) also reported increased yield in avocado following girdling. The treatment increased the rate of pollen tube growth and the success of ovule penetration, thereby resulting in greater fruit set on girdled branches. It is believed that the elevated carbohydrate content was responsible for the increased rate of pollen tube growth in flowers on girdled branches. Tomer (1977) also reported that girdling inhibited the competitive spring shoot growth from functionally indeterminate flowering shoots, at time when fruit were setting and establishing sink strength that determines their survival (Wolstenholme *et al.*, 1990).

Hackney *et al.* (1995) reported that in an orchard in a cool, humid, subtropical area of South Africa, conventional girdling where a band of trunk bark was removed was too severe, invariably causing leaf chlorosis and a decline in tree health with reduced yield the second year after treatment. They investigated the effect of a single knife cut (scoring) made around the trunks of healthy 2-year-old 'Hass' trees. Scoring was done at 2–4 week intervals throughout the year and the subsequent yield determined. They found that significant increases in yield were obtained from trees scored in autumn (May), winter (June and August) and spring (October and November) with the best result from the spring treatment. Tree health declined when trees were trunk-scored in March and April but there were no visible effects on tree health from the spring treatments.

The potential of trunk or whole-tree girdling to increase yield of avocado has been investigated as a strategy to maximize orchard production in the year prior to thinning trees from crowded orchards. Toerien and Basson (1979) found that autumn girdling of healthy 6-year-old 'Edranol' trees growing in humid, subtropical South Africa increased the subsequent season's yield by 38% over non-girdled trees in the same orchard. Girdling was done low on all main branches of trees with a 9-mm-wide strip of bark removed. The health of girdled trees declined but this was inconsequential since trees were removed following harvest. Köhne (1992a) carried out a similar study on young 'Hass' trees in the year prior to tree thinning. The orchard was planted at 800 trees ha^{-1} at a spacing of 2.5×5 m and when trees were 3 years old, a 5 mm trunk girdle was made on the temporary orchard trees in spring (September). Although the treatment resulted in an increased percentage of small fruit, the 30% increase in yield from girdled trees more than compensated for the number of small fruit, with an overall increase in the volume of exportable fruit (Köhne, 1992a). It was concluded that girdling healthy, young 'Hass' trees in high-density plantings in the year prior to their removal was beneficial for increasing the overall productivity of the orchard. The severity of the girdle in both these studies would normally have led to a detrimental effect on tree health. However, as treated trees were to be removed to prevent orchard crowding, the strategy was useful in maximizing orchard performance.

Girdling to increase fruit size

While autumn girdling has been found to reduce mean fruit size of avocado due to increased numbers of fruit the following season (Lahav *et al.*, 1971a; Trochoulias and O'Neill, 1976), investigations have continued with treatment at different times of the year with the objective of increasing mean fruit size of 'Hass'. This cultivar is notorious for its production of many small fruit with low market value in hot growing areas (Lahav and Adato, 1990; Köhne, 1992b; Köhne and Schutte, 1991; Cutting, 1993; Cowan, 1997).

Following preliminary investigations in South Africa, Davie *et al.* (1995) reported that girdling and scoring resulted in a 35% increase in the mean fruit size of 'Hass'. Branch girdling and scoring were done in early summer (December) when it was established that the trees had set a good crop. However, in a similar experiment the following year with scoring treatments applied earlier in spring (October and November), there was no effect on fruit size, although all scoring treatments significantly increased yield (Davie and Stassen, 1997). The principle behind this technique is to restrict the movement of photoassimilates to other sink regions of the tree and to reduce the amount of new shoot growth, thereby ensuring that fruit on scored limbs experience less competition for photoassimilates. However, this is a complex issue to manage as several factors are involved that can unpredictably affect the outcome. These include tree health (vigour), water status and mineral nutrient supply, feedback inhibition of photosynthesis due to carbohydrate build-up in leaves and leaf/fruit ratios on scored branches at the time of treatment. The latter is perhaps the most significant in terms of extracting the greatest benefit from the treatment but a deficiency in any one of the others may result in no benefit from the procedure. Further research is justified to establish whether repeatable long-term benefits to fruit size are achievable through selective scoring of limbs, on sufficiently vigorous trees in orchards where Phytophthora root rot is not an issue.

Conclusions

In this chapter, some horticultural management procedures that can improve orchard performance have been discussed. Not all strategies will be economically beneficial wherever the crop is grown, and care should be taken when transferring technology from one environment to another. This applies particularly to tree responses in the less invigorating, semi-arid to arid Mediterranean climates as compared with highly invigorating, humid summer rainfall climates with good soils. Understanding the environment and the limitations or excesses it imposes on tree growth is critical for success in determining management strategies from orchard design. In addition to tree responses to various management practices, growers need to consider the cost/benefit relationship, as the cost of management inputs and the returns from fruit can vary significantly within and between avocado-producing countries. An example is the cost of seedling versus clonal rootstocks when tree densities are high. Cloned rootstocks uniformly impart useful characteristics such as Phytophthora root rot resistance but there may be instances, particularly in 'new ground' orchards with negligible disease pressure, where lower cost seedling rootstocks can provide comparable economic performance for the life of the orchard since these rootstocks may be as productive or better than current commercially available cloned rootstocks (Whiley *et al.*, 2013). Additionally, most of the cloned elite rootstocks now available have been selected for salinity tolerance and/or Phytophthora root rot resistance and, in the absence of either or both of these abiotic and biotic stresses, do not always out-yield seedling rootstocks (see Whiley, Chapter 9, this volume).

There are sufficient data available to be confident that avocado trees in most circumstances will respond positively to under-tree mulching and compost application. This can be an expensive orchard operation, and careful consideration is needed in choosing appropriate and readily available materials, and the time of application for local conditions. Plant growth regulators are increasingly playing a role in avocado production. They are commonly used in clonal propagation of rootstocks and to improve fruit set and size through reducing reproductive/vegetative competition (controlling tree vigour) during fruit set and early growth. Soil-applied triazoles are also being effectively used in some countries to manage tree size in high-density planted orchards, resulting in increased production. These products are effective on healthy trees where excess vigour is a problem, but intensive orchard management is required to fully capitalize on the benefits they can potentially deliver. In the shorter term, hormone bioregulator sprays may be commercialized for specific objectives, for example, increased yield and fruit size and partial alleviation of AB.

Canopy management remains one of the most difficult areas in avocado production. Considerable research in Australia, California, Chile, Florida, Israel, Spain and South Africa over the past 30 years has been targeted at developing strategies to contain tree size while sustaining orchard productivity under the prevailing environmental conditions. Although some success has been achieved, there remains no simple solution to this perplexing issue. Furthermore, there are differences in what can be economically done in stressful, compared to mesic, environments. It is likely that successful growers will need to adopt several strategies to contain tree size while maintaining orchard productivity. Since the publication of the first edition of this volume, there has been ongoing, large-scale, innovative development of orchard design in Chile and some other countries, where relatively unique environments have allowed this industry to redefine production systems that introduce new levels of precocity and yield.

Significant AB in avocado remains a production problem, particularly in Mediterranean climates where the annual fruiting cycle is extended, and trees may simultaneously carry two crops for a short period. The severity of alternation can be reduced through harvesting a significant portion of the crop as soon as fruit reach commercial maturity, particularly in 'on' crop years. Rootstocks have also been implicated in alleviating AB (Whiley, Chapter 9, this volume), and ongoing research is required to identify rootstock/scion combinations that will minimize cropping alternation. Bioregulant sprays and soil applications will play an important role in alleviating AB, but in the longer term, improved genetics (both cultivars and rootstocks) are likely to provide the most reliable solution.

Generally, girdling or cincturing of avocados has been particularly successful when used to increase tree yield in the first 1–2 years after starting the treatment, especially where yield is constrained by excessive vigour. However, invariably the results have not been sustainable and in many instances tree health declines as the root systems are weakened through assimilate deprivation. In orchards where Phytophthora root rot, salinity and drought risk are significant problems, girdling is an added threat to long-term tree health and, in reality, should not be practised. Research has clearly shown that girdling reduces the supply of carbohydrates and other phloemtranslocated metabolites, including auxin, to roots and results in less growth (dry matter accumulation). Even partial treatment of trees (selective branch girdling) will decrease carbohydrate supply to roots, thereby reducing their capacity to replace those lost to disease or asphyxiation. Thus, over time, the use of this practice must be carefully assessed against tree vigour (which it reduces) and the disease pressure and effectiveness of Phytophthora root rot control measures. A trunk girdle applied to trees in their final year of production has merit since it has been clearly demonstrated that yield from these trees is increased, and following harvest they will be removed. However, for the success of this strategy, trees should be vigorous and healthy before the girdle is applied. Similarly, where tree size is contained in conventional or high-density orchards through autumn scoring of limbs to enhance subsequent yield prior to their removal, ongoing benefits have been shown to be economically justified. Girdling or scoring has also formed part of the effective intensive management package to constrain tree size and increase precocity and yield with a reduction in the amplitude of AB in some ultra-high-density avocado orchards in Chile. The science of the practice is well researched and stringent criteria are followed when implemented, which preserves tree health.

The need for most of the horticultural management techniques discussed in this chapter reflects the relatively early stage of avocado domestication and the imperatives of modern orcharding. The world's premier avocado cv. 'Hass' is only a 'second generation' chance seedling selection; most rootstocks are still seedlings with a number emanating from a wide range of mostly discarded cultivars. There are nevertheless signs of progress for both scion and rootstock improvement, especially clonal rootstocks for Phytophthora root rot resistance, but also superior 'escape tree' seedling rootstocks able to contribute to higher, more consistent production of orchards. In the future, selection of rootstocks for Phytophthora root rot resistance needs to be accompanied by other features such as reduced tree vigour, precocity and increased productivity for real progress to occur. The impending development of crop genomics and biotechnology should speed up the exploitation of the large avocado gene pool and deliver new cultivars and rootstocks with fewer inherent disadvantages.

References

Adato, I. (1990) Effects of paclobutrazol on avocado (*Persea americana* Mill.) cv. Fuerte. *Scientia Horticulturae* 45, 105–115.

Arpaia, M.L., Tapia, M. and Hofshi, R. (2007) The use of naphthaleneacetic acid (NAA) to control vegetative vigor in avocado trees. *California Avocado Society Yearbook* 90, 131–148.

Bangerth, K.F (2006) Flower induction in perennial fruit trees: still an enigma? *Acta Horticulturae* 727, 177–196.

Bangerth, K.F. (2009) Floral induction in mature, perennial angiosperm fruit trees: similarities and discrepancies with annual/biennial plants and the involvement of plant hormones. *Scientia Horticulturae* 122(2), 153–163.

Barrett, J.E. and Bartuska, C.A. (1982) PP333 effects on stem elongation dependent on site of application. *HortScience* 17, 737–738.

Ben-Ya'acov, A., Michelson, E. and Sela, I. (1995) Development and evaluation of clonal avocado rootstocks in Israel. IV. Experiments with Ettinger cultivar at En-Hachoresh. *Alon Hanotea* 49, 54–60.

Bergh, B.O. (1974) The remarkable avocado flower. *California Avocado Society Yearbook* 57, 40–41.

Bharath, P., Gahir, S. and Raghavendra, S. (2021) Abscisic acid-induced stomatal closure: an important component of defense against abiotic and biotic stress. *Frontiers of Plant Science* 12, 615441. doi:10.3389/fpls.2021.615114.

Biran, D. (1979) Fruitlet abscission and spring growth retardation – their influence on avocado productivity. MSc thesis, The Hebrew University of Jerusalem, Israel. (In Hebrew.)

Blumenfeld, A. and Gazit, S. (1974) Development of seed and seedless avocado fruits. *Journal of the American Society of Horticultural Science* 90(5), 442–448.

Boswell, S.B., Bergh, B.O. and Whitsell, R.H. (1976) Control of sprouts on topworked avocado stumps with NAA [naphthaleneacetic acid] formulations. *HortScience* 11, 113–114.

Broadbent, P. and Baker, K.F. (1974) Behaviour of *Phytophthora cinnamomi* in soils suppressive and conducive to root rot. *Australian Journal of Agriculture Research* 25, 121–137.

Browning, G. (1977) Environmental control of flower-bud development in *Coffee arabica* L. In: Landsberg, J.J. and Cutting, C.V. (eds) *Environmental Effects on Crop Physiology*. Academic Press, London, pp. 321–331.

Cameron, S.H. and Borst, G. (1938) Starch in the avocado tree. *Proceedings of the American Society for Horticultural Science* 36, 255–258.

Cameron, S.H., Mueller, R.T. and Wallace, A. (1952) Nutrient composition and seasonal losses of avocado trees. *California Avocado Society Yearbook* 36, 201–209.

Chandler, W.H. (1958) The avocado. In: *Evergreen Orchards*. Lea and Febiger, Philadelphia, Pennsylvania, pp. 205–228.

Coit, J.E. (1921) The effect of girdling the avocado. *Annual Report of the California Avocado Association* 1920–21, 69–70.

Cowan, A.K. (1997) Why are small Hass fruit small? *South African Avocado Growers' Association Yearbook* 20, 52–54.

Cowan, A.K., Moore-Gordon, C.S., Bertling, I. and Wolstenholme, B.N. (1997) Metabolic control of avocado fruit growth. *Plant Physiology* 114, 511–518.

Cowan, A.K., Taylor, N.J. and van Staden, J. (2005) Hormone homeostasis and induction of the small-fruit phenotype in 'Hass' avocado. *Plant Growth Regulation* 45, 11–19.

Crane, J.H., Schaffer, B., Davenport, T.L. and Balerdi, C. (1992) Rejuvenation of a mature, non-productive 'Lula' and 'Booth 8' avocado grove by topping and tree removal. *Proceedings of the Florida State Horticultural Society* 105, 282–285.

Crane, J.H., Bally, I.S.E., Mosqueda-Vazquez and Tomer, E. (1997) Crop production. In: Litz, R.E. (ed.) *The Mango: Botany, Production and Uses*. CAB International, Wallingford, UK, pp. 203–256.

Crane, J.H., Carrillo, D., Evans, E.A., Gazis, R., Schaffer, B., Ballen, F. and Wasielewski, J. (2020) Recommendations for control and mitigation of laurel wilt and ambrosia beetle vectors in commercial avocado groves in Florida. Horticultural Sciences Department, UF/IFAS Extension. Available at: https://edis.ifas.ufl.edu/publication/HS1360 (accessed 7 July 2023).

Crowley, D. (2008) Managing soils for avocado production and root health. *California Avocado Society Yearbook* 90, 107–130.

Cutting, J.G.M. (1993) The cytokinin complex as related to small fruit in 'Hass' avocado. *South African Avocado Growers' Association Yearbook* 16, 20–21.

Cutting, J.G.M. and Bower, J.P. (1990) Spring vegetative flush removal: the effect on yield, size, fruit mineral composition and quality. *South African Avocado Growers' Association Yearbook* 13, 33–34.

Dahan, Y. Rosenfeld, R., Zadiranov, V. and Irihimovitch, V. (2010) A proposed conserved role for an avocado fw2.2-like gene as a negative regulator of fruit cell division. *Planta* 232, 663–676.

Davenport, T.L. (1982) Avocado growth and development. *Proceedings of the Florida State Horticultural Society* 95, 92–96.

Davie, S.J. and Stassen, P.J.C. (1997) Avocado tree girdling and fruit thinning. *South African Avocado Growers' Association Yearbook* 20, 55–58.

Davie, S.J., Stassen, P.J.C., van der Walt, M. and Snijder, B. (1995) Girdling avocado trees for improved production. *South African Avocado Growers' Association Yearbook* 18, 51–53.

Davis, T.D., Steffens, G.L. and Sankhla, N. (1988) Triazole plant growth regulators. *Horticultural Reviews* 10, 63–105.

Dennis, F.G. and Neilsen, J.C. (1999) Physiological factors affecting biennial bearing in tree fruit: the role of seeds in apple. *HortTechnology* 9, 317–322.

Desta, B. and Amare, G. (2021) Paclobutrazol as a plant growth regulator. *Chemical and Biological Technologies in Agriculture* 8:1. Available at: doi:10.1186/s40538-020-00199-z (accessed 10 June 2023).

Dickson, R.E. (1991) Assimilate distribution and storage. In: Raghavendra, A.S. (ed.) *Physiology of Trees*. Wiley, New York, pp. 51–85.

Downer, J.A., Faber, B. and Menge, J. (2002) Factors affecting root rot control in mulched avocado orchards. *HortTechnology* 12(4), 601–605.

Eaks, I.L. (1980) Respiratory rate, ethylene production, and ripening response of avocado fruit to ethylene or propylene following harvest at different maturities. *Journal of the American Society for Horticultural Science* 105, 744–747.

El-Zeftawi, B.M. and Thornton, L.R. (1975) Effects of rootstocks and fruit stripping on alternate bearing of 'Valencia' orange trees. *Journal of Horticultural Science* 50, 219–226.

Erasmus, H.D. and Brooks, W.H. (1998) Foliar application of uniconazole (Sunny®) to avocado trees to improve fruit size and yield and to change fruit shape. *South African Avocado Growers' Association Yearbook* 21, 52–53.

Ernst, A.A. (1981) Die invloed van mediumdeurlugting, swamdoder en indoolbottersuur op die beworteling van avokadosteggies. *South African Avocado Growers' Association Yearbook* 4, 121–123. (In Afrikaans with English summary.)

Ernst, A.A. (1999) Micro cloning: a multiple cloning technique for avocados using micro containers. *Revista Chapingo Serie Horticultura* 5, 217–220.

Faber, B. (2007) So you want to grow organic avocados? *California Avocado Society Yearbook* 90, 45–46.

Faber, B. (2008) Organic avocado production. *California Avocado Society Yearbook* 91, 35–43.

Faber, B.E., Downer, A.J. and Menge, J.A. (2003) Effects of mulch on avocado and citrus. In: *Proceedings of the V World Avocado Congress*. Vol. II. Granada-Malaga, Spain, pp. 719–724.

Farré, J.M., Hermoso, J.M. and Pliego, F. (1987) Effects of pre-bloom pruning on leaf nutrient status, growth and cropping of the avocado cv Hass. *South African Avocado Growers' Association Yearbook* 10, 71–72.

Fletcher, R.A., Gilley, A., Sankhla, N. and Davis, T.D. (2010) Triazoles as plant growth regulators and stress protectants. *Horticultural Reviews* 24, 55–138.

Francis, H.L. (1996) Girdling trial yield data. *South African Avocado Growers' Association Yearbook* 19, 80.

Gardiazabal, F., Magdahl, C. and Mena, F. (2011a) New avocado planting and management systems in Chile. Internal document, Sociedad Gardiazabal y Mena Ltda. Supplied to B.N. Wolstenholme, Pietermaritzburg, South Africa, p. 10.

Gardiazabal, F., Mena, F., Magdahl, C., Adriazola, C. and Torres, J. (2011b) Neuvos sistemas de poda en arboles adultos de paltos (*Persea americana* Mill.), cv Hass. In: *Proceedings of the VII World Avocado Congress*, Cairns, Australia, p. 8.

Gardiazabal, F., Mena, F., Torres, J. and Pinto, A. (2015a) Efecto de distintas épocas de aplicaciones de Uniconazol al suelo en la producción de Paltos (*Persea americana* Mill.) Cv. Hass. In: *Proceedings of the VIII World Avocado Congress*. Lima, Perú. Available at: https://avocadosource.com/wac8/section_04/gardiazabalfrancisco2015c.pdf (accessed 25 February 2024).

Gardiazabal, F., Mena, F., Torres, J. and Pinto, A. (2015b) Efecto de dosis de Uniconazol y efecto de distintos uniconazoles aplicados al suelo para la producción en paltos (*Persea americana* Mill.) Cv Hass. In: *Proceedings of the VIII World Avocado Congress*. Lima, Peru. Available at: https://avocadosource.com/wac8/section_04/gardiazabalfrancisco2015d.pdf (accessed 25 February 2024).

Gardiazabal, F., Mena, F., Torres, J. and Pinto, A. (2015c) Efecto de la aplicación de Cultar al suelo, sobre la productividad y desarrollo de Paltos (*Persea americana* Mill.) Cv Hass. In: *Proceedings of the VIII World Avocado Congress*. Lima, Perú. Available at: https://avocadosource.com/wac8/section_04/gardiazabalfrancisco2015.pdf (accessed 25 February 2024).

Gardiazabal, F., Mena, F., Torres, J. and Pinto, A. (2015d) Plantaciones de muy alta densidad (2,5 × 1,25m y 1,25 × 1,25m) en paltos (*Persea americana* Mill.) Cv. Hass logran en su primer año de producción más de 40 t/ha. In: *Proceedings of the VIII World Avocado Congress*. Lima, Perú. Available at: https://avocadosource.com/wac8/section_04/gardiazabalfrancisco2015b.pdf (accessed 25 February 2024).

Garner, L.C. and Lovatt, C.J. (2008) The relationship between flower and fruit abscission and alternate bearing in 'Hass' avocado. *Journal of the American Society for Horticultural Science* 133, 3–10.

Gazit, S. and Degani, C. (2002) Reproductive biology. In: Whiley, A.W., Schaffer, B. and Wolstenholme, B.N. (eds) *The Avocado: Botany, Production and Uses*. CAB International, Wallingford, UK, pp. 101–133.

Goren, R., Huberman, M. and Goldschmidt, E.E. (2004) Girdling: physiological and horticultural aspects. *Horticultural Reviews* 30, 1–36.

Gregoriou, C. and Rajkumar, D. (1984) Effect of irrigation and mulching on shoot and root growth of avocado (*Persea americana* Mill.) and mango (*Mangifera indica* L.). *Journal of Horticultural Science* 59, 109–117.

Gur, A., Samish, R.M. and Zamet, D. (1969) A comparison of different planting distances for apple trees. *Israeli Journal of Agricultural Research* 19, 79–86.

Hackney, C.R., Boshoff, M. and Slabbert, M.J. (1995) Increasing yield of young Hass avocado trees using the cincturing technique. *South African Avocado Growers' Association Yearbook* 18, 54–55.

Hadari, M. (2004) A three-dimensional model of the light regime in an avocado orchard. MSc thesis, Israel Institute of Technology, Haifa, Israel, p. 98.

Hardy, J., Partridge, C., Barber, R., Pak, H. and Dixon, J. (2008) Spring: tree management, phenology and flowering. *New Zealand Avocado Growers' Association Guide, plus Avocado Tree Phenology Chart, p.* 36.

Heath, R., Arpaia, M.L. and Mickelbart, M. (2005) Avocado tree physiology: understanding the basis of productivity. In: Lovatt, C.J., Holthe, P.A. and Arpaia, M.L. (eds) *Proceedings of the California Avocado Research Symposium*. UCR, Riverside, California, pp. 87–119.

Hermosilla-Madrid, C. and Whiley, A.W. (2023) High-density (HD) avocado production in New Zealand. In: *Proceedings of the Tenth World Avocado Congress.* Auckland, New Zealand. Available at: https://industry.nzavocado.co.nz/wp-content/uploads/2023/04/1415-Claudia-Hermosilla-.pdf (accessed 29 February 2024).

Hilgeman, R.H., Dunlap, J.A. and Sharples, G.C. (1967) Effect of time of harvest of 'Valencia' oranges on leaf carbohydrate content and subsequent set of fruit. *Proceedings of the American Society for Horticultural Science* 90, 111–116.

Hofshi, R. (1999a) High-density avocado planting – an argument for replanting trees. *Subtropical Fruit News* 7(1), 9–13.

Hofshi, R. (1999b) Should avocado growers consider a 'Nursery Cooperative'. *Subtropical Fruit News* 7(1), 13–15.

Hofshi, R. (1999c) Some economic reasons to consider canopy management. In: Arpaia, M.L. and Hofshi, R. (eds) *Proceedings of Avocado Brainstorming '99*. California Avocado Commission and the University of California, Riverside, California, pp. 45–48.

Ish-Am, G., Cohen, H., Eidelman, E., Regev, I., Winer, L. and Lahav, E. (2007) Achieving larger 'Ettinger' fruits in Western Galilee, Israel. In: *Proceedings VI World Avocado Congress (Actas VI Congreso Mundial del Aguacate)*. Viña Del Mar, Chile. Available at: https://www.avocadosource.com/wac6/en/resumen/3d-117.pdf (accessed 25 February 2024).

Iturrieta, R. (2017) First things first: matching an alternate bearing model to confirmed field phenotypes of avocado (*Persea americana*, Mill.). PhD thesis, University of California, Riverside, California. Available at: https://escholarship.org/uc/item/6fc488ch (accessed 29 February 2024).

Jackson, J.E. (1980) Light interception and utilization by orchard systems. *Horticultural Reviews* 2, 208–267.

Jackson, J.E. (1985) Future fruit orchard design: economic and biology. In: Cannell, M.G.R. and Jackson, J.E. (eds) *Attributes of Trees as Crop Plants*. Institute of Terrestrial Ecology, Abbots Ripton, UK, pp. 441–459.

Janzen, D.H. (1971) Seed predation by animals. *Annual Review of Ecology and Systematics* 2, 465–492.

Kaiser, C. and Wolstenholme, B.N. (1994) Aspects of delayed harvest of 'Hass' avocado (*Persea americana* Mill.) fruit in a cool subtropical climate. II. Fruit size, yield, phenology and whole-tree starch cycling. *Journal of Horticultural Science* 69, 447–457.

Koch, K.E. (1996) Carbohydrate-modulated gene expression in plants. *Annual Review of Plant Physiology and Plant Molecular Biology* 47, 509–540.

Köhne, J.S. (1988) Response of avocado (*Persea americana* Mill.) to the growth regulator paclobutrazol. *Applied Plant Science* 2, 40–43.

Köhne, J.S. (1992a) Increased yield through girdling of young Hass trees prior to thinning. *South African Avocado Growers' Association Yearbook* 15, 68.

Köhne, J.S. (1992b) Increasing 'Hass' fruit size. In: Lovatt, C.J., Holthe, P.A. and Arpaia, M.L. (eds) *Proceedings of the Second World Avocado Congress*. Vol. I. University of California, Riverside, California, p. 242.

Köhne, J.S. and Kremer-Köhne, S. (1987) Vegetative growth and fruit retention in avocado as affected by a new plant growth regulator (paclobutrazol). *South African Avocado Growers' Association Yearbook* 10, 64–66.

Köhne, J.S. and Kremer-Köhne, S. (1989) Comparison of growth regulators paclobutrazol and uniconazole on avocado. *South African Avocado Growers' Association Yearbook* 12, 38–39.

Köhne, J.S. and Kremer-Köhne, S. (1990) Results of a high density avocado planting. *South African Avocado Growers' Association Yearbook* 13, 31–32.

Köhne, J.S. and Kremer-Köhne, S. (1991) Avocado high density planting – a progress report. *South African Avocado Growers' Association Yearbook* 14, 42–43.

Köhne, J.S. and Kremer-Köhne, S. (1992) Yield advantages and control of vegetative growth in a high-density avocado orchard treated with paclobutrazol. In: Lovatt, C.J., Holthe, P.A. and Arpaia, M.L. (eds) *Proceedings of Second World Avocado Congress.* University of California, Riverside, California, pp. 233–235.

Köhne, J.S. and Schutte, J.M. (1991) Increasing 'Hass' fruit size. *South African Avocado Growers' Association Yearbook* 14:38.

Kojima, K., Goto, A. and Nakashima S. (1996) Effects of uniconazole-P on abscission and endogenous ABA, IAA and GA-like substances levels of Satsuma mandarin fruitlet. *Bioscience, Biotechnology and Biochemistry* 60, 901–902.

Kotzé, J.M. (1979) Phases of seasonal growth of the avocado tree. *South African Avocado Growers' Association Research Report* 3, 14–16.

Kurtzman, R.H., Jr (1966) Xylem sap flow as affected by metabolic inhibitors and girdling. *Plant Physiology* 41, 641–646.

Lahav, E. (1999) Issues and strategies for canopy management of avocado orchards in the Jordan valley of Israel. In: Arpaia, M.L. and Hofshi, R. (eds) *Proceedings of Avocado Brainstorming '99*. California Avocado Commission and the University of California, Riverside, California, pp. 49–51.

Lahav, E. and Adato, I. (1990) Increasing the size of 'Hass' avocado fruit. Report of the Institute of Horticultural Science Activities, 1981–1984. Volcani Centre, Bet Dagan, Israel.

Lahav, E., Gefen, B. and Zamet, D. (1971a) The effect of girdling on fruit quality, phenology and mineral analysis of the avocado tree. *California Avocado Society Yearbook* 55, 162–168.

Lahav, E., Gefen, B. and Zamet, D. (1971b) The effect of girdling on productivity of the avocado. *Journal of the American Society for Horticultural Science* 96, 396–398.

Le Lagadec, D. (2010) Five years data collection for clonal and seedling rootstocks with 'Hass' as a scion. *Talking Avocados* 21(4), 36–40.

Leonardi, J. (2001) Progress in canopy management of avocados. In: *Proceedings of the Australian and New Zealand Avocado Growers' Conference 'Vision 2020'*. Conference CD, Australian Avocado Growers' Federation, Brisbane, Australia, p. 11.

Leonardi, J. (2005) New strategies and tools for avocado canopy management. In: *Proceedings of the New Zealand and Australia Avocado Growers' Conference '05*. Tauranga, New Zealand. Available at: https://www.avocado.org.au/wp-content/uploads/2017/02/New-Strategies-Tools-for-Avocado-Canopy-Management-John-Leonardi-AAL.pdf (accessed 22 October 2024).

Leonardi, J. (2008) Canopy management update. *Talking Avocados* 18, 22–28.

Leonardi, J. (2009) The development of canopy management strategies suited to the different growing environments across Australia for increased profitability. Report Av04008. Horticulture Australia Ltd. Available at: https://www.horticulture.com.au/growers/help-your-business-grow/research-reports-publications-fact-sheets-and-more/av04008/ (accessed 22 October 2024).

Lever, B.G. (1986) Cultar – a technical overview. *Acta Horticulturae* 179, 459–466.

Lopez-Jiminez, A., Eyheraguibel, B., Zheng, Y., Khuong, T., Aviles, J. and Lovatt, C.J. (2011) Mitigating alternate bearing of 'Hass' avocado in California. In: *Proceedings of the VII World Avocado Congress*. Cairns, Australia, Program Book, p. 73 (abstract).

Lovatt, C.J. (2001) Properly timed soil-applied nitrogen fertilizer increases yield and fruit size of 'Hass' avocado. *Journal of the American Society for Horticultural Science* 126, 555–559.

Lovatt, C.J. (2003) Eliminating alternate bearing of the 'Hass' avocado: continuing project: year 1 of 4. In: *Proceedings of the California Research Symposium, Ventura 2003*. Available at: https://www.avocadosource.com/ARAC/Symposium_2003/article_11.pdf (accessed 1 March 2024).

Lovatt, C.J. (2005) Plant growth regulators for avocado production. *California Avocado Society Yearbook* 88, 81–91.

Lovatt, C.J. (2010) PGR II – strategies to increase fruit size of 'Hass' avocado. *California Avocado Commission Annual Research Report 2009–2010*. p. 3. Available at: https://www.californiaavocadogrowers.com/sites/default/files/Increase-Fruit-Size-2010-End-of-Year.pdf (accessed 23 October 2024).

Luckwill, L.C. (1980) Hormones and the productivity of fruit trees. *Scientific Horticulture* 31, 60–68.

Mackay, C.K, Hall, J.C., Hofstra, G. and Fletcher R.A. (1990) Uniconazole-induced changes in abscisic acid, total amino acids, and proline in *Phaseolus vulgaris*. *Pesticide Biochemical Physiology* 37, 71–82.

Mason, T.G. and Maskell, E.J. (1928a) Studies on the transport of carbohydrates in the cotton plant. I. A study of diurnal variations in the carbohydrates of leaf, bark and wood and the effects of ringing. *Annals of Botany* 42, 189–253.

Mason, T.G. and Maskell, E.J. (1928b) Studies on the transport of carbohydrates in the cotton plant. II. The factors determining the rate and direction of movements of sugars. *Annals of Botany* 42, 571–636.

Mena, F., Gardiazabal, F., Magdahl, C. and Hofshi, R. (2007) High density orchards: effect of production on tree growth, flowering and crop of Hass avocados (*Persea americana* Mill.) for the year following. In: *Abstracts of the VI World Avocado Congress*. Viña del Mar, Chile. (A PowerPoint presentation in Spanish). Available at: https://www.avocadosource.com/wac6/es/presentacion/3c-113.pdf (accessed 24 October 2024).

Mena, F., Gardiazabal, Torres, J. and Pinto, A. (2015a) Comportamiento poblacional de 200 árboles de Paltos (*Persea americana* Mill.) cv. Hass cultivados en Panquehue, Chile. In: *Proceedings of the VIII World Avocado Congress*. Lima, Perú. Available at: https://avocadosource.com/wac8/section_04/menafrancisco2015.pdf (accessed 6 October 2024).

Mena, F., Gardiazabal, Torres, J. and Pinto, A. (2015b) Efecto de la época de cosecha sobre el rendimiento y calibre de la fruta de paltos (*Persea americana* Mill.) variedad Hass. In: *Proceedings of the VIII World Avocado Congress*. Lima, Perú. Available at: https://avocadosource.com/wac8/section_04/menafrancisco2015b.pdf (accessed 6 October 2024).

Mickelbart, M.V., Bender, G.S., Witney, G.W., Adams, C. and Arpaia, M.L. (2007) Effect of clonal rootstocks on 'Hass' avocado yield components, alternate bearing, and nutrition. *Journal of Horticultural Science and Biotechnology* 82, 460–466.

Mika, A. (1986) Physiological responses of fruit trees to pruning. *Horticultural Reviews* 8, 337–378.

Monselise, S.P. and Goldschmidt, E.E. (1982) Alternate bearing in fruit trees. *Horticultural Reviews* 4, 128–173.

Moore-Gordon, C., Wolstenholme, B.N. and Levin, J. (1996) Effect of composted pinebark mulching on *Persea americana* Mill. cv. Hass fruit growth and yield in a cool subtropical environment. *Journal of the South African Society for Horticultural Science* 6, 23–26.

Moore-Gordon, C., Cowan, A.K. and Wolstenholme, B.N. (1997) Mulching of avocado orchards to increase Hass yield and fruit size and boost financial rewards – a three season summary of research findings. *South African Avocado Growers' Association Yearbook* 20, 46–49.

Moore-Gordon C.S., Cowan A.K., Bertling I., Botha C.E.J. and Cross R.H.M. (1998) Symplastic solute transport and avocado fruit development: a decline in Cytokinin/ABA ratio is related to appearance of the 'Hass' small fruit variant. *Plant Cell Physiology* 39,1027–1038.

Murneek, A.E. (1941) Relative carbohydrates and nitrogen concentration in new tissues produced on ringed branches. *Proceedings of the American Society for Horticultural Science* 38, 133–136.

Newett, S. (1999) Current status of canopy management in Australia. In: Arpaia, M.L. and Hofshi, R. (eds) *Proceedings of Avocado Brainstorming '99*. California Avocado Commission and the University of California, Riverside, California, pp. 56–59.

Newett, S., Whiley, A., Dirou, J., Hofman, P., Ireland, G., *et al.* (2001) Growing the crop. In: Vock, N. (ed.) *Avocado Information Kit*. Queensland Department of Primary Industries, Brisbane, Australia, pp. 3–54.

Nii, N. (1992) Fruiting effects on chloroplast ultrastructure, starch accumulation and localization of chloroplast nucleoids in peach leaves on the girdled shoots. *Scientific Report of the Faculty of Agriculture, Meijo University* 28, 41–48.

Noel, A.R.A. (1970) The girdled tree. *Botanical Review* 36, 162–195.

Noguchi, H. (1987) New plant growth regulators and S–3307D. *Japan Pesticide Information* 51, 15–22.

Opler, P.A., Frankie, G.W. and Baker, H.G. (1976) Rainfall as a factor in the release, timing and synchronisation of anthesis by tropical trees and shrubs. *Journal of Biogeography* 3, 321–326.

Painter, J.H. and Brown, R.T. (1940) Effect of different methods of girdling tung branches. *Proceedings of the American Society for Horticultural Science* 37, 511–514.

Pegg, K.G. and Giblin, F. (2008) Principles of Phytophthora root rot management in established orchards. *Talking Avocados* 19(1), 36–38.

Pegg, K.G. (2010) Pathology challenges in avocado: phytophthora management. In: *Proceedings of the Southern Produce Avocado Conference*. Southern Produce Ltd, Tauranga, New Zealand.

Pegg, K.G. and Whiley, A.W. (1987) Phytophthora control in Australia. *South African Avocado Growers' Association Yearbook* 10, 94–96.

Pegg, K.G., Whiley, A.W., Saranah, J.B. and Glass, R.J. (1985) Control of Phytophthora root rot of avocado with phosphorous acid. *Australasian Plant Pathology* 14(2), 25–29.

Pegg, K.G., Coates, L.M. Korsten, L. and Harding, R.M. (2002) Root, foliar and soilborne diseases. In: Whiley, A.W., Schaffer, B. and Wolstenholme, B.N. (eds) *The Avocado: Botany, Production and Uses*. CAB International, Wallingford, UK, pp. 299–338.

Penter, M.G., Snijder, B., Stassen, P.J.C. and Schäfer, E. (2000) The effect of growth inhibitors on fruit production in Hass avocado trees. *South African Avocado Growers' Association Yearbook* 23, 46–51.

Platt, R.G. (1976) Tree spacing and control of avocado tree growth. In: Sauls, J.W., Phillips, R.L. and Jackson, L.K. (eds) *Proceedings of the First International Tropical Fruit Short Course: The Avocado*. University of Florida, Gainesville, Florida, pp. 54–57.

Platt, R.G., Goodall, G.E., Gustafson, C.D. and Lee, B.W. (undated) Thinning avocado orchards. University of California, Agriculture Extension Leaflet AXT-n3, 8 pp.

Ploetz, R.C., Kendra, P.E., Choudhury, R.A., Rollins, J.A., Campbell, A., *et al.* (2017a) Laurel wilt in natural and agricultural ecosystems: understanding the drivers and scales of complex pathosystems. *Forests* 8:48. Available at: doi:10.3390/f8020048 (accessed 19 September 2023).

Ploetz, R.C., Konkol, J.L., Pérez-Martinez, J.M. and Fernandez, R. (2017b) Management of laurel wilt of avocado, caused by Raffaelea lauricola. *European Journal of Plant Pathology* 149, 133–143. Available at: doi:10.1007/s10658-017-1173-1 (accessed 15 August 2023).

Raviv, M. (2010) Sustainability of organic horticulture. *Horticultural Reviews* 36, 289–333.

Richings E.W., Cripps R.F. and Cowan A.K. (2000) Factors affecting 'Hass' avocado fruit size: carbohydrate, abscisic acid and isoprenoid metabolism in normal and phenotypically small fruit. *Physiologia Plantarum* 109, 81–89.

Rivas, F., Gravina, A. and Agusti, M. (2007) Girdling effects on fruit set and quantum yield efficiency of PSII in two *Citrus* cultivars. *Tree Physiology* 27, 527–535.

Roe, D.J. and Morudu, T.M. (1999) Flower pruning of Hass avocado when going into an 'on' year in an attempt to decrease alternate bearing and increase fruit size. *South African Avocado Growers' Association Yearbook* 22, 84–86.

Roe, D.J. and Morudu, T.M. (2000) Hass avocado yields as affected by dwarfing rootstocks and flower pruning. *South African Avocado Growers' Association Yearbook* 23, 30–32.

Rossouw, T., Robbertse, P.J., Kremer-Köhne, S. and Köhne, J.S. (2000) Effect of gibberellic acid treatments on flowering of avocado. *South African Avocado Growers' Association Yearbook* 23, 43–45.

Salazar-García, S. and Lovatt, C.J. (1997) Use of gibberellic acid to manipulate flowering in the 'Hass' avocado: a preliminary report. In: *Proceedings of the Australian Avocado Growers' Federation and the New Zealand Avocado Growers' Association Conference '97, 'Searching for Quality'*. NZAGA, Rotorua, New Zealand, pp. 106–111.

Salazar-García, S. and Lovatt, C.J. (1998) GA3 application alters flowering phenology of 'Hass' avocado. *Journal of the American Society for Horticultural Science* 123, 791–797.

Salazar-García, S. and Lovatt, C.J. (2000) Use of GA3 to manipulate flowering and yield of 'Hass' avocado. *Journal of the American Society for Horticultural Science* 125, 25–30.

Salazar-García, S., Lord, E.M. and Lovatt, C.J. (1998) Inflorescence and flower development of the 'Hass' avocado *(Persea americana* Mill.) during 'on' and 'off' crop years. *Journal of the American Society for Horticultural Science* 123, 537–544.

Salazar-García, S., Lord, E.M. and Lovatt, C.J. (1999) Inflorescence development of the 'Hass' avocado: commitment to flowering. *Journal of the American Society for Horticultural Science* 124, 478–482.

Salazar-García, W., Cossio-Vargas, L.E., Lovatt, C.J. Gonzales-Duran, I.J.L. and Perez-Barazza, M.H. (2006) Crop load affects vegetative growth flushes and shoot age influences irreversible commitment to flowering of 'Hass' avocado. *HortScience* 41, 1541–1546.

Schaffer, A.A., Liu, K.C., Goldschmidt, E.E., Boyer, C.D. and Goren, R. (1986) Citrus leaf chlorosis induced by sink removal: starch, nitrogen and chloroplast ultrastructure. *Plant Physiology* 124, 111–121.

Schaffer, B. (1995) The environment, the urban jungle and politics versus fruit production in south Florida – with special reference to avocado. In: *Proceedings of the Australian Avocado Growers' Federation Conference*. AAGF, Fremantle, Australia, pp. 127–134.

Schmidt, T., Elfving, D.C., McFerson, J.R. and Whiting, M.D. (2009) Crop load overwhelms effects of gibberellic acid and ethephon on floral initiation in apple. *HortScience* 44, 1900–1906.

Scholefield, P.B., Sedgley, M. and Alexander, D.McE. (1985) Carbohydrate cycling in relation to shoot growth, floral initiation and development and yield in the avocado. *Scientia Horticulturae* 25, 99–110.

Schroeder, C.A. (1952) Floral development, sporogenesis, and embryology in the avocado (*Persea americana*). *Botanical Gazette* 113, 270–278.

Schroeder, C.A. (1953) Growth and development of the Fuerte avocado fruit. *Proceedings of the American Society for Horticultural Science* 61, 103–109.

Schroeder, C.A. (1958) Growth and development of the avocado fruit. *California Avocado Society Yearbook* 42, 114–118.

Shalit, A., Rozman, A., Goldschmidt, A., Alvarrez, J.P., Bowman, J.L., Eshed, Y. and Lifschitz, E. (2009) The flowering hormone florigen functions as a general systemic regulator of growth and termination. *Proceedings of the National Academy of Sciences USA* 106, 8392–8397.

Snijder, B. and Stassen, P.J.C. (1997) Initial results with regard to flower induction and fruit thinning of avocados. *South African Avocado Growers' Association Yearbook* 20, 42–45.

Stassen, P.J.C., Davie, S.J. and Snijder, B. (1995) Principles involved in tree management of higher density avocado orchards. *South African Avocado Growers' Association Yearbook* 18, 47–50.

Stassen, P.J.C., Davie, S.J. and Snijder, B. (1998) Training young 'Hass' avocado trees into a central leader for accommodation in higher density orchards. In: *Proceedings of the World Avocado Congress III*. The Congress, Tel Aviv, Israel, pp. 251–254.

Sukhvibul, N., Whiley, A.W., Vithanage, V., Smith, M.K. and Hetherington, S.E. (1999) Effect of temperature on pollen germination, pollen tube growth and seed development in mango (*Mangifera indica* L.). *Acta Horticulturae* 509, 609–616.

Swarbrick, T. (1927) The healing of wounds in woody stems. II. Contributions to the physiological anatomy of ringed apple shoots. *Journal of Pomology and Horticultural Science* 6, 29–46.

Thomas, G. (1997) Rootstock influence on yield of 'Hass' avocado. In: *Proceedings of the Australian Avocado Growers' Federation and the New Zealand Avocado Growers' Association Conference '97, 'Searching for Quality'*. NZAGA, Rotorua, New Zealand, pp. 138–146.

Thorp, T.G. and Sedgley, M. (1993) Architectural analysis of tree form in a range of avocado cultivars. *Scientia Horticulturae* 53, 85–98.

Ticho, R.J. (1936) The effect of ringing on the carbohydrate and nitrogen content of Mexican avocado seedlings. *California Avocado Society Yearbook* 21, 69–72.

Ticho, R.J. (1971) Girdling, a means to increase avocado fruit production. *California Avocado Society Yearbook* 54, 90–94.

Toerien, J.C. (1999) Integrated management of avocados. In: Arpaia, M.L. and Hofshi, R. (eds) *Proceedings of Avocado Brainstorming '99*. California Avocado Commission and the University of California, Riverside, California, pp. 68–69.

Toerien, J.C. and Basson, A.M. (1979) An investigation into thinning of an avocado orchard. *South African Avocado Growers' Research Report* 3, 59–60.

Tomer, E. (1977) The effect of girdling on flowering, fruit setting and abscission in avocado leaves. PhD thesis, Hebrew University, Rehovot, Israel.

Torres, J. (2017) Efecto de la carga frutal sobre variables fisiológicas en palto (*Persea americana* Mill.) variedad Hass. MSc thesis, Universidad de Chile, Santiago, Chile.

Trochoulias, T. and O'Neill, G.H. (1976) Girdling of 'Fuerte' avocado in subtropical Australia. *Scientia Horticulturae* 5, 239–242.

Turney, J. and Menge, J. (1994) Root health: mulching to control root disease in avocado and citrus. Circular no. 94/2. University of California, Riverside, California.

Valmayor, R.V. (1967) Cellular development of the avocado fruit – blossom to maturity. *Philippine Agriculturist* 50(10), 907–976.

Venning, F.D. and Lincoln, F.B. (1958) Development morphology of the vegetative axis of avocado (*Persea americana* Mill.) and its significance to spacing, pruning practices, and yields of the grove. *Proceedings of the Florida State Horticultural Society* 71, 350–356.

Verreyne, J.S. and Lovatt, C.J. (2009) The effect of crop load on budbreak influences return bloom in alternate bearing 'Pixie' mandarin. *Journal of the American Society for Horticultural Science* 134, 299–307.

Whiley, A.W. (1994) Ecophysiological studies and tree manipulation for maximisation of yield potential in avocado (*Persea americana* Mill.). PhD thesis, University of Natal, Pietermaritzburg, South Africa.

Whiley, A.W. (2001) Adoption of field practices to assist in expanding avocado markets. In: *Proceedings of the Australian and New Zealand Avocado Growers' Conference 'Vision 2020'*. Conference CD, Australian Avocado Growers' Federation, Brisbane, Australia, p. 22.

Whiley, A.W. (2013) Literature review and gap analysis for the development of research plan into irregular bearing. Horticulture Australia Ltd Project No. 12030, p. 79. Available at: https://www.avocado.org.au/wp-content/uploads/2016/12/AV12030-Literature-Review-Gap-Analysis-into-Irregular-Bearing.pdf (accessed 29 February 2024).

Whiley, A.W. and Wolstenholme, B.N. (1990) Carbohydrate management in avocado trees for increased production. *South African Avocado Growers' Association Yearbook* 13, 25–27.

Whiley, A.W., Saranah, J.B., Cull, B.W. and Pegg, K.G. (1988) Manage avocado tree growth cycles for productivity gains. *Queensland Agricultural Journal* 114, 29–36.

Whiley, A.W., Saranah, J.S., Wolstenholme, B.N. and Rasmussen, T.S. (1991) Use of paclobutrazol sprays at mid-anthesis on fruit size and yield of avocado (*Persea americana* Mill. cv Hass). *Journal of Horticultural Science* 66, 593–600.

Whiley, A.W., Hargreaves, P.A., Pegg, K.G., Doogan, V.J., Ruddle, L.J., Saranah, J. and Langdon, P.W. (1995a) Changing sink strengths influence translocation of phosphonate in avocado (*Persea americana* Mill.) trees. *Australian Journal of Agricultural Research* 46, 1079–1090.

Whiley, A.W., Saranah, J.B. and Wolstenholme, B.N. (1995b) Pheno-physiological modelling in avocado – an aid in research planning. In: *Proceedings of the World Avocado Congress III*. The Congress, Tel Aviv, Israel, pp. 71–75.

Whiley, A.W., Rasmussen, T.S., Saranah, J.B. and Wolstenholme, B.N. (1996a) Delayed harvest effects on yield, fruit size and starch cycling in avocado (*Persea americana* Mill.) in two subtropical environments. I. The early-maturing cv. Fuerte. *Scientia Horticulturae* 66, 23–34.

Whiley, A.W., Rasmussen, T.S., Saranah, J.B. and Wolstenholme, B.N. (1996b) Delayed harvest effects on yield, fruit size and starch cycling in avocado (*Persea americana* Mill.) in two subtropical environments. II. The late-maturing cv. Hass. *Scientia Horticulturae* 66, 35–49.

Whiley, A.W., Searle, C., Schaffer, B. and Wolstenholme, B.N. (1999) Cool orchard temperatures or growing trees in containers can inhibit leaf gas exchange of avocado and mango. *Journal of the American Society of Horticultural Science* 124(1), 46–51.

Whiley, A.W., Leonardi, J., Pegg, K.G. and Langdon, P.W. (2001) Use of foliar applications of phosphonate fungicide to control Phytophthora root rot in avocados. In: *Proceedings of the Australian and New Zealand Avocado Growers' Conference 'Vision 2020'*. Conference CD, Australian Avocado Growers' Federation, Brisbane, Australia, p. 13.

Whiley, A.W., Coates, L.M., Hofman, P.J., Dann, E.K., Pegg, K.G. *et al.* (2013) Rootstock improvement for the Australian Avocado Industry – phase 3. Horticulture Australia Ltd Project No. AV08000, p. 154. Available at: https://www.avocado.org.au/wp-content/uploads/2016/12/AV08000-Rootstock-Improvement-for-the-Australian-Avocado-Industry-Phase-3.pdf (accessed 29 February 2024).

Wilkie, J. (2021) Transforming subtropical/tropical tree crop productivity. *Hort Innovation AI13004*, p. 50. Available at: https://www.horticulture.com.au/contentassets/d49b15e3173b4a9184bf7c3f9675babf/ai13004-final-report-complete.pdf (accessed 29 February 2024).

Winkler, A. and Knoche, M. (2021) Xylem, phloem and transpiration flows in developing European plums. *PLOS One* 16(5): e0252085. doi:10.1371/journal.pone.0252085.

Wolstenholme, B.N. (1981) Root, shoot or fruit? *South African Avocado Growers' Association Yearbook* 4, 27–29.

Wolstenholme, B.N. (1986) Energy costs of fruiting as a yield-limiting factor with special reference to avocado. *Acta Horticulturae* 175, 121–126.

Wolstenholme, B.N. (1987) Theoretical and applied aspects of avocado yield as affected by energy budgets and carbon partitioning. *South African Avocado Growers' Association Yearbook* 10, 58–61.

Wolstenholme, B.N. (1991) Making an avocado fruit: energy expensive but mineral cheap. *Avokad* 11, 8–9.

Wolstenholme, B.N. (2009) Alternate bearing in avocado: an overview. In: *Proceedings of the 4th Australia and New Zealand Avocado Growers' Conference*, Cairns, 21–25 July 2009. Conference CD, Australian Avocado Growers' Federation, Brisbane, Australia.

Wolstenholme, B.N. and Sheard, A. (2009) A better understanding of alternate bearing through revised growth model charts. *South African Avocado Growers' Association, Avoinfo* 168, 17–19.

Wolstenholme, B.N. and Whiley, A.W. (1995) Prospects for increasing Hass fruit size – a southern hemisphere perspective. In: *Proceedings of the Australian Avocado Growers' Federation Conference*. AAGF, Fremantle, Australia, pp. 89–102.

Wolstenholme, B.N. and Whiley, A.W. (1998) Strategies for maximizing avocado productivity: an overview. In: *Proceedings of the World Avocado Congress III*. The Congress, Tel Aviv, Israel, pp. 61–70.

Wolstenholme, B.N., Whiley, A.W. and Saranah, J.B. (1990) Manipulating vegetative: reproductive growth in avocado (*Persea americana* Mill.) with paclobutrazol foliar sprays. *Scientia Horticulturae* 41, 315–327.

Wolstenholme, B.N., Moore-Gordon, C. and Ansermino, C.D. (1996) Some pros and cons of mulching avocado orchards. *South African Avocado Growers' Association Yearbook* 19, 87–94.

Wolstenholme, B.N., Moore-Gordon, C.S. and Cowan, A.K. (1998) Mulching of avocado orchards: quo vadis? *South African Avocado Growers' Association Yearbook* 21, 26–28.

Wood, B.W. (2011) Influence of plant bioregulators on pecan flowering and implications for regulation of pistillate flower initiation. *HortScience* 46, 870–877.

Zamet, D.N. (1990) The effect of minimum temperature on avocado yields. *California Avocado Society Yearbook* 74, 247–256.

Zeevaart, J.A. (2008) Leaf-produced floral signals. *Current Opinion in Plant Biology* 11(5), 541–547. doi:10.1016/j.pbi.2008.06.009.

Zheng, Y., Khuong, T. and Lovatt, C.J. (2011) Effect of foliar-applied plant bioregulators on 'Hass' avocado yield. In: *Proceedings of the VII World Avocado Congress*. Avocados Australia Ltd, Cairns, Australia p. 8.

Ziv, D., Zviran, T., Zezak, O., Samach, A. and Irihimovitch, V. (2014) Expression profiling of flowering locus T-like gene in alternate bearing 'Hass' avocado trees suggests a role for PaFT in avocado flower induction. *PLOS One*, 9(10): e110613.

13 Foliar, Fruit and Soilborne Diseases

E.K. Dann[1]* and R. Gazis[2]

[1]*Queensland Alliance for Agriculture and Food Innovation, University of Queensland, Ecosciences Precinct, Dutton Park, Queensland, Australia;*
[2]*Department of Plant Pathology, Tropical Research and Education Center, University of Florida, Homestead, Florida*

Abstract

This chapter provides an in-depth review of diseases affecting avocado production throughout the world, with emphasis on those causing the most severe economic loss. While there are some diseases of global importance, many have more restricted distribution and impact, and several could be considered emerging or intermittent threats. Application of new approaches and technologies for detection, identification and management is presented.

The chapter has been substantially reworked since the second edition, published more than 10 years ago. In that time, some diseases have increased in prevalence and/or economic importance, for example, laurel wilt, black root rot and Fusarium dieback, while others have become less problematic or limited in geographic or financial impact, for example, brown root rot and Duke 6 stem pitting, and have been relegated to a tabular summary. What has not changed since the second edition is the significant constraint to productivity attributable to pathogens, and the various diseases they cause. Phytophthora root rot is still plaguing the avocado industry globally, resulting in diseased trees with reduced yields. Anthracnose and stem-end rot fruit diseases continue to be the predominant cause of rots affecting fruit quality, shelf life and consumer satisfaction, representing increased risk with historically greater volumes of fruit being exported. From time to time, there are localized outbreaks of other pathogens and diseases, and new ones continue to emerge or become recognized as having an impact on productivity. Regardless of the disease, the fundamental principle of integrated pest and disease management, IP(D)M, has been advocated by crop protection specialists for several decades, and remains the preferred strategy for implementation. The concept first arose among entomologists, following problems associated with the widespread use of (and over-reliance on) insecticides in the 1950s and 1960s [for more information, see reviews by Kogan (1998) and Razdan and Gupta (2009)]. The concept and practice were adopted more broadly among plant pathologists and other crop protection specialists following the multidisciplinary Huffaker Project in the 1970s, and are founded on the coordinated use of multiple tactics, including

*Corresponding author: e.dann@uq.edu.au

(eds D. Carrillo, B. Schaffer, A.W. Whiley and B.N. Wolstenholme)
DOI: 10.1079/9781800621824.0013

avoidance, exclusion, eradication, protection, resistance and therapy. Understanding the interaction of plant pathogens with their avocado host is critical to the development of disease management options. Important factors include the presence or absence of a particular pathogen, the susceptibility of the avocado cultivar or rootstock, the optimum environmental conditions for disease development and the level of disease (incidence and severity) that will likely occur in a given situation.

Much has, however, changed in the decade since the second edition of *The Avocado*. Global production of avocado fruit has increased nearly twofold to 8.69 million T across a production area of more than 858,000 Ha, up from 494,000 Ha in 2013 (Food and Agriculture Organization of the United Nations, 2023). Avocados are now growing in extensive monocultures, in a wide range of climates and environments, often far removed from their mesic but diverse centres of origin. Extreme weather events and climate unpredictability in expanded and marginal production regions may exacerbate the incidence and severity of diseases (Cacciola and Gullino, 2019; Cohen and Leach, 2020; Alvarado-Rosales *et al.*, 2022), further challenging management and economic viability. In fact, 2023 was the warmest year on record, with July 2023 being the hottest month in recorded history (World Meteorological Organization, 2024).

Fortunately, at the crop level, the technology and tools available to growers and researchers to predict, detect and manage pests and diseases have improved markedly in the last decade. These include highly sensitive molecular-based identification diagnostics, use of environmental DNA monitoring (e.g. bees and pollen), remote sensing and hyperspectral imaging, real-time monitoring of orchard conditions and pest status, for example, soil moisture and tree sap flow, and storage and transport parameters as fruit track through supply chains.

New approaches and technologies for disease management are being developed and implemented where they are shown to be efficacious and cost-effective, and include disease-resistant (or tolerant) rootstocks, improved fungicides and 'biofungicides', and examples of these are presented throughout this chapter. The digital age has allowed open access to an abundance (perhaps over-abundance) of information shared via different formats, including the latest scientific articles, webinars and electronic newsletters. Some of the older pre-digital literature and reports are also now available online and contain information, which may be relevant and useful as the industry inevitably heads towards more sustainable production and reduced reliance on pesticides.

This chapter has been divided into sections based on which part of the tree is primarily infected: soilborne diseases, root rots; trunk and branch; and fruit and foliage. Minor diseases, including reference to historical records, are summarized under the same categories in Table 13.2 at the end of the chapter. Improvements in the detection and identification via DNA sequencing have prompted extensive changes in the taxonomy and nomenclature of many pathogens causing disease in avocado. The most recent names are used with reference to those previously and still commonly used as synonyms.

Soilborne Diseases, Root Rots

Armillaria root rot

Armillaria, also known as oak root fungus or the honey mushroom, causes trunk and root rot, and has been reported in avocado from several countries. It is mostly a minor problem, except in avocado orchards established on sites where previous susceptible affected hosts were grown, for example, native pine forests in México (Alvarado-Rosales *et al.*, 2022). The fungus has a very wide geographical and host range, including *Pinus* spp., *Eucalyptus* spp. and many fruit tree species (Eskalen and Faber, 2016a; Hartley and van den Berg, 2023).

Symptoms

Armillaria is soilborne, and large roots and the crown of avocado trees may be infected for some time before above ground symptoms become apparent. The decline can be gradual, with reduced vigour, chlorotic leaves and dieback of upper limbs. Alternatively, leaves can wilt and trees die suddenly without defoliating. Young trees often die quickly after infection. White-cream silky mycelial growth occurs along the roots and forms fan-like plaques underneath the bark at the crown. A brown to black decay develops on roots, cambium and wood. Dark structures, pseudosclerotia, can protrude through cracks in the bark, and slender, purplish-brown, shoestring-like rhizomorphs, resembling thin dark roots, often grow along the root surface and into surrounding soil (Fig. 13.1). After rain, clusters of yellow-brown basidiocarps (mushroom-like fruiting bodies) grow under infected trees, often around the base of the trunk (Fig. 13.2). Affected wood becomes soft and brittle. The distribution of dead and diseased trees within orchards is patchy but may exhibit a radial pattern of destruction with successive death of neighbouring trees (Ohr and Zentmyer, 1994a).

Causal agents and epidemiology

Species of *Armillaria* reported from avocado include the *Armillaria mellea* complex, *Armillaria mexicana* and *Armillaria socialis* (Ohr and Zentmyer, 1994a; Alvarado-Rosales *et al.*, 2022). *Desarmillaria tabescens* (Koch *et al.*, 2017; syn. *Armillaria tabescens* and *Clitocybe tabescens*) is listed as causing Clitocybe root rot of avocado (Ohr *et al.*, 2003). More research is warranted to resolve the identity of the species responsible.

Armillaria can survive as pseudosclerotia or mycelial strands on stumps, in root debris or as rhizomorphs buried in soil for many years. It infects healthy avocado trees when roots come into contact with these sources of inoculum. Although spores are produced within the basidiocarps (mushrooms), they are not thought to be important in spreading the disease (Ohr and Zentmyer, 1994a).

Management

Avoid establishing orchards on land where forest or fruit trees were previously affected by *Armillaria*. Remove dead or declining trees and adjacent trees, and thoroughly clean and disinfest all digging tools and equipment used. If replanting into infested sites, remove as much of the woody debris as possible, but it is difficult and probably impractical to completely eliminate rhizomorphs and other residual inoculum (Ohr and Zentmyer, 1994a). *Armillaria* is sensitive to desiccation, so

Fig. 13.1. Yellow-brown basidiocarps (top) of *Desarmillaria tabescens* and thin shoe-string-like rhizomorphs (bottom) recovered from soil.

Fig. 13.2. Honey-coloured fruiting bodies of *Desarmillaria tabescens* at the base of an affected tree.

frequently cultivating replant sites over a period of months to expose infested soil may help with control. Fumigation or other fungicide treatments are unlikely to completely eradicate *Armillaria* from infested soil (Downer and Faber, 2019). Ensure drainage and irrigation are optimal for tree growth. A 6-year study in an infested orchard demonstrated that deep ripping, fresh organic matter (mixed species woodchips) and large planting holes (2 m^3) could improve the survival of grafted peach saplings, but that *Trichoderma* sp. amendment had no effect beyond the postulated suppression from natural populations of *Trichoderma* sp. already present in the woodchip (Downer and Faber, 2019).

Black root rot

Black root rot, previously known as Cylindrocarpon and/or Cylindrocladium root rot, has been reported as a severe disease of avocado in several countries. It is primarily a disease of nursery trees and is often masked by the optimal growing conditions during tree production but becoming a problem in the months and early years after planting in the field. Fungi, now known to be severe pathogens, were first isolated from avocado roots in South Africa in the 1970s (Darvas, 1978, 1979) and Spain in the 1980s (López-Herrera and Melero-Vara, 1991). In Chile, black root rot caused an estimated loss of 22,000 nursery trees in 1994–1995 (Besoain and Piontelli, 1999), and a survey of nurseries in Israel showed a very high incidence, up to 100% of trees, had roots infected with one of the causal organisms (Zilberstein *et al.*, 2007). It has since been reported affecting avocado in Australia (Dann *et al.*, 2012; Parkinson *et al.*, 2017), Italy (Vitale *et al.*, 2012), Colombia (Ramírez Gil and Morales Osorio, 2013) and New Zealand (Herder, 2014; A.W. Whiley, Queensland, Australia, 2024, personal communication).

Symptoms

Severe decline and death of young trees within months of planting may occur, and symptoms are the same as those caused by early and severe infection by *Phytophthora* spp. Leaves wilt, become chlorotic and necrotic, and drop prior to tree death. Trees fail to grow and thrive, and young vegetative shoots may become necrotic and die back (Fig. 13.3). In early stages of symptom development, characteristic discrete lesions up to 5 mm may be observed on feeder roots (Fig. 13.4), which coalesce to rot the entire root. Root systems are considerably reduced and have extensive black necrotic roots with very few white healthy feeder roots (Dann *et al.*, 2012). In nurseries where growing conditions are optimal, above ground symptoms are not usually as severe as those seen in new field plantings; however, root systems may be extensively infected with irreversible necrosis (Fig. 13.5). Necrosis of the crown and lower stem was also reported as being associated with root rot in severe cases (Aiello *et al.*, 2020).

Causal agents and epidemiology

Black root rot is caused by fungi in the Nectriaceae family, which is large and includes pathogens of most agricultural plants. Key causal agents include *Calonectria ilicicola*, *Calonectria pacifica* (previously *Cylindrocladium* sp.), *Dactylonectria macrodidyma*, *Dactylonectria novozelandica*, *Dactylonectria pauciseptata*, *Dactylonectria anthuriicola*

Fig. 13.3. Typical symptoms of black root rot in newly planted orchard.

(previously *Cylindrocarpon* sp.) (Dann *et al.*, 2012; Herder, 2014; Parkinson *et al.*, 2017), *Cylindrocladiella peruviana*, *Pleiocarpon algeriense* associated with crown and stems (Aiello *et al.*, 2020), *Ilyonectria liriodendri*, and many reported under previous names as species of *Cylindrocarpon*, *Neonectria*, *Ilyonectria* and *Cylindrocladium* (Darvas, 1978, 1979; Besoain and Piontelli, 1999; Zilberstein *et al.*, 2007; Vitale *et al.*, 2012; Ramírez Gil and Morales Osorio, 2013). It is likely that many more species may be causal agents under the right conditions.

Nectriaceous fungi are ubiquitous in soil and frequently isolated from necrotic roots, including roots of healthy, mature trees and many of these have been shown to be less pathogenic or non-pathogenic, for example, *Gliocladiopsis* sp., *Fusarium* sp. and *Mariannaea* sp. (López-Herrera and Melero-Vara, 1991; Dann *et al.*, 2012; Parkinson *et al.*, 2017; A. Prabhakaran, Brisbane, Australia, 2019, personal communication). These pathogens spread to orchards through infected nursery material. One study reported the detection of *Dactylonectria* sp. (reported as *Cylindrocarpon* spp.) in potting media and drainage and irrigation water in avocado nurseries (Ramirez-Gil and Osorio, 2021). Their introduction into nurseries and spread is via contaminated soil or potting media or infested plant debris, and water splash from rain or irrigation.

C. ilicicola is most frequently isolated from severely necrotic nursery tree roots, or from declining trees soon after planting, and can be difficult to isolate onto selective media. It produces microsclerotia in root tissue and may survive for many years in soil or root debris. Orange perithecia containing ascospores of *C. ilicicola* are infrequently observed on roots in advanced stages of tree decline (Fig. 13.6).

Fig. 13.4. Characteristic discrete lesions of black root rot may be observed on feeder roots.

Fig. 13.5. Nursery tree showing extensive necrosis of root system caused by *Calonectria ilicicola* with few healthy feeder roots.

Fig. 13.6. Extensive necrosis of root system of a nursery tree which failed to establish. Note small orange-coloured perithecia (arrow) of *Calonectria ilicicola*.

Specific and genus-wide molecular-based detection techniques have assisted diagnosis of black root rot and the causal organism(s) directly from infected plant roots (Parkinson *et al.*, 2019). A draft genome of *Dactylonectria macrodidyma* has been published and may facilitate genome sequencing approaches to detection (Malapi-Wight *et al.*, 2015).

Management

Establishing orchards with nursery trees free of the causal pathogens is critical, and growers should source their trees from reputable and accredited nurseries. Nursery hygiene practices include using steam pasteurized or chemically fumigated potting media (Ramirez-Gil and Osorio, 2021; see Whiley *et al.*, Chapter 9, this volume), treated (disinfested) water for irrigation (Stewart-Wade, 2011), and sanitizing propagation material and tools. Soil solarization utilizing plastic mulches reduced viability of *Calonectria* spp. microsclerotia in soil and propagation media in a nursery situation, even at a depth of 30 cm (Vitale *et al.*, 2013), and this represents a safe and chemical-free option. Keep plants well spaced and do not over-irrigate or use potting media that retains water. Any weak trees and unsold old nursery stock should be discarded. Drenching infected nursery trees with fungicides may suppress black root rot fungi sufficiently to allow the growth of healthy roots and establishment after planting (Prabhakaran and Dann, 2022), but it is not recommended as a regular practice as fungi may develop resistance to specific fungicide active ingredients and treatments will no longer be effective. Orchard practices to minimize loss due to black root rot include choosing sites that are well drained and taking care during the post-planting phase so that trees are not over-irrigated. Use of plastic mulches that are reusable or biodegradable for solarization in the orchard situation may be beneficial and represents an exciting area of continuing research (Vitale *et al.*, 2013). The pathogens causing black root rot in avocado have wide host ranges, including (but not limited to) groundnut, papaya, blueberry, custard apple and grapevine (Parkinson *et al.*, 2017). It is important not to plant avocados in sites where diseased crops were recently grown and removed.

Phytophthora root rot

Phytophthora root rot, caused by *Phytophthora cinnamomi*, is the most serious and important disease of avocado worldwide. It attacks and destroys the fine feeder roots of all ages of trees, including those in nurseries. In 1922, R.D. Rands described *P. cinnamomi* as a pathogen of *Cinnamomum burmanii*, a natural source of the spice cinnamon in the mountains of Western Sumatra. The pathogen was first reported affecting avocado in Puerto Rico in 1929 where it caused severe root rot (Tucker, 1929). *P. cinnamomi* is now known to affect close to 5000 plant species and has severe economic impact across agriculture, horticulture and forestry, as well as devastating environmental effects in natural ecosystems (Jung *et al.*, 2016; Hardham and Blackman, 2018). It has been extensively moved by anthropogenic activities and now affects host plants worldwide.

Symptoms

In the nursery, conditions are usually optimal for plant growth and foliar symptoms of root rot caused by *Phytophthora* (or other soilborne pathogens such as those causing black root rot) can be subtle but may include leaf chlorosis and reduced tree vigour. Examination of roots reveals extensive black necrosis of the feeder roots and saturated potting media due to reduced volume of active roots. If such trees are planted, or if healthy trees are planted into infested soil without prophylactic treatment, they struggle to establish, have weak vegetative growth, drop leaves, shoots die back from the tip, and these are all typical within 12 months from planting. Vegetative suckering from the rootstock may indicate trees attempting to recover, but often the scion variety has died.

In mature trees, infection is mostly limited to the fine white feeder roots, and discolouration and/or dieback of root tips are evident as the initial symptoms of root infection (Fig. 13.7). Necrosis develops, and the pathogen may invade the entire feeder root system, which becomes black and brittle and non-functional (Fig. 13.8). These symptoms may also be seen in roots of healthy-looking trees and demonstrate the ubiquitous distribution of the pathogen and disease. Healthy feeder roots are usually quite

Fig. 13.7. *P. cinnamomi*-infected roots typically die back from the root tips.

Fig. 13.8. Feeder roots affected by Phytophthora root rot become black and brittle.

difficult to find under trees with advanced symptoms of root rot. Beneath such trees, the soil tends to be abnormally wet, as destruction of feeder roots prevents the uptake of water. Larger woody roots are rarely infected and remain functional until the tree dies. The disruption to water and nutrient uptake leads to the typical above ground symptoms. Leaves become chlorotic and wilted, and often have necrotic tips and margins and may abscise prematurely. There is a significant reduction in vegetative growth, resulting in a reduced tree canopy and general decline in tree health and vigour (Fig. 13.9). Entire limbs may die back. If leaves do form, they are small and pale green. Fruit yields decline in affected trees, and sometimes such trees set a large number of fruit that fail to size (Fig. 13.10). Fruit often sunburn and have pepper spot symptoms as a result of the loss of the protective vegetative canopy.

Decline is usually a gradual process if the disease is not adequately managed; however, decline may occur quickly, with symptoms of rapid wilt and death when infected trees are exposed to flooded soils, even for short periods (Ploetz and Schaffer, 1989; Reeksting *et al.*, 2014). These studies also demonstrate the sensitivity of avocado to waterlogging (see Schaffer *et al.*, Chapter 7, this volume).

Causal agent

Phytophthora root rot is primarily caused by the soilborne pathogen, *P. cinnamomi*, which is a member of the phylum Oomycota (oomycetes) and forms fungal-like hyphae, but it is not a true fungus and is more closely related to diatoms and coloured algae. Other *Phytophthora* species such as *P. niederhauserii*, *P. nicotianae* and *P. multivora* as well as the oomycete *Phytopythium vexans* have recently been reported to cause root rot in avocado (Rodríguez-Padrón *et al.*, 2018; Hernández Pérez *et al.*, 2019); however, this section will concentrate on *P. cinnamomi*.

While two distinct mating types (A1 and A2) are required for sexual reproduction, the A2 mating type is the predominant population and most invasive and virulent, causing severe root rot in avocado and other horticultural crops. Recent molecular phylogenetic analyses of A2-type isolates from around the world have demonstrated that there are two dominant clonal lineages with very low genetic diversity, associated with avocado (Shakya *et al.*, 2021; Engelbrecht *et al.*, 2022). The A1 mating type and greater genotypic diversity in *P. cinnamomi* are observed more frequently in isolates recovered from natural vegetation (Engelbrecht *et al.*, 2022) and particularly among isolates from Taiwan and

Fig. 13.9. Spring growth of trees in advanced decline from Phytophthora root rot. Trees are in a poorly drained section of an orchard after a particularly wet summer.

Vietnam, providing strong evidence for a centre of origin in these regions (Shakya *et al.*, 2021).

P. cinnamomi forms three different spore types by asexual reproduction that are involved in disease development or survival. Ovoid to elliptical-shaped sporangia, which produce zoospores with whip-like flagella that facilitates movement in free water, and chlamydospores which are the round survival spores. As *P. cinnamomi* from avocado orchards are predominantly clonal, reproducing asexually, oospores are rare but have been induced in the laboratory in isolates of the A2 mating type by exposure to avocado root extracts or other stimuli (Zentmyer, 1979). The contribution of these selfed oospores to survival of *P. cinnamomi* under field conditions has not been reported for avocado. Interestingly, microscopic examination of *P. cinnamomi*-infected roots and root debris of native Australian forest species, highly susceptible to *Phytophthora*, has confirmed high frequencies of thick-walled, selfed oospores, intercellular encased hyphae and hyphal aggregations. This suggests that these structures may be more important than chlamydospores for long-term survival during the dry summer conditions of south-west Western Australia (Jung *et al.*, 2013).

P. cinnamomi is one of the most easily isolated and identified of all *Phytophthora* spp. It can be readily detected using traditional methods such as direct isolation by plating diseased tissue onto selective media and by baiting with plant tissues from various susceptible species, commonly lupins. The plant tissues are floated or suspended over a slurry of infested soil and root matrix and attract the motile zoospores released from sporangia, colonizing the bait tissue. Different baits and protocols have been developed for various *Phytophthora* species (Martin *et al.*, 2012; Burgess *et al.*, 2021; Sarker *et al.*, 2023). Lesions or symptoms may develop on bait tissues that are examined directly under the microscope and plated onto selective media

Fig. 13.10. Trees with severe Phytophthora root rot frequently flower profusely and set an abundance of small fruit.

for further identification. Morphological features of the sporangia on bait tissues and colony hyphae are characters used to provide initial determination, and micrographs of these may be seen in the review by Martin *et al.* (2012). For example, on V8 juice agar or selective media, *P. cinnamomi* produces clusters of hyphal swellings, giving a distinctive coralloid appearance to the hyphae, and when cultured on potato dextrose agar (PDA), colonies have a distinctive rosette pattern. Chlamydospores form readily in culture on V8 juice agar or broth.

Highly sensitive molecular techniques are now used to accurately identify and even quantify oomycete species from cultures or infected tissues (even if asymptomatic) (Masikane *et al.*, 2019), and a useful online tool and smartphone app have been developed for both morphological and molecular identification of *Phytophthora* species (Abad *et al.*, 2023). Simple field-deployable lateral flow kits, based on immunoassays, are available commercially to detect *Phytophthora* from plant material but have limited sensitivity and specificity. Various PCR protocols are reported; however, they may give unreliable results (false positives or false negatives) and require thorough validation with the target and closely related species (Kunadiya *et al.*, 2017). Metabarcoding of environmental DNA (eDNA) tools are also available, but these only indicate the presence of *Phytophthora* DNA, and do not demonstrate that living, infectious propagules of the pathogen are detected, which is one of the limitations of this technology (Sarker *et al.*, 2023). This highlights that even new technologies are not without their limitations, and due to the ubiquitous nature of *P. cinnamomi*, its presence within an orchard should be presumed and managed accordingly.

Epidemiology

The disease cycle of *P. cinnamomi* with scanning electron micrographs of key spore types and stages and details of interactions at the molecular and cellular level are presented in reviews by Hardham and Blackman (2018) and Pérez-Jiménez (2008), and depicted graphically in Fig. 13.11. Zoospores are the key infectious propagules being produced within and released from sporangia. Sporangia form from mycelium on infected roots or from chlamydospores, when soil pores are full of water. Zoospores are attracted by exudates from roots and swim aided by two flagella towards the root elongation zone, which extends from about 0.4 mm behind the root tip (Aveling and Rijkenberg, 1989). Large numbers of zoospores aggregate and attach to roots and encyst, then germinate typically within 20–30 min to penetrate roots directly via a germ tube, or sometimes forming a swelling like a fungal appressorium (Aveling and Rijkenberg, 1989). *P. cinnamomi* secretes a wide range of cell wall degrading enzymes to digest the main polysaccharides in plant cell walls, enabling penetration and colonization (Hardham and Blackman, 2018). The hyphae grow intra- and intercellularly through the root cortex and into the vascular tissues within 24 h, blocking the xylem thus inhibiting water movement. Along with necrosis of unsuberized feeder roots, this damage causes the typical wilt and decline symptoms associated with Phytophthora root rot. *P. cinnamomi* has been shown to be hemibiotrophic, which means it can initially invade cells and extract nutrients without killing the cells or inducing symptoms (biotrophic phase), then after 12–18 h switches to a necrotrophic phase, killing cells in advance of hyphal colonization along the root (van den Berg *et al.*, 2021). Sporangiophores will emerge from the root surface and another generation of sporangia are produced. The zoospore to zoospore cycle may take

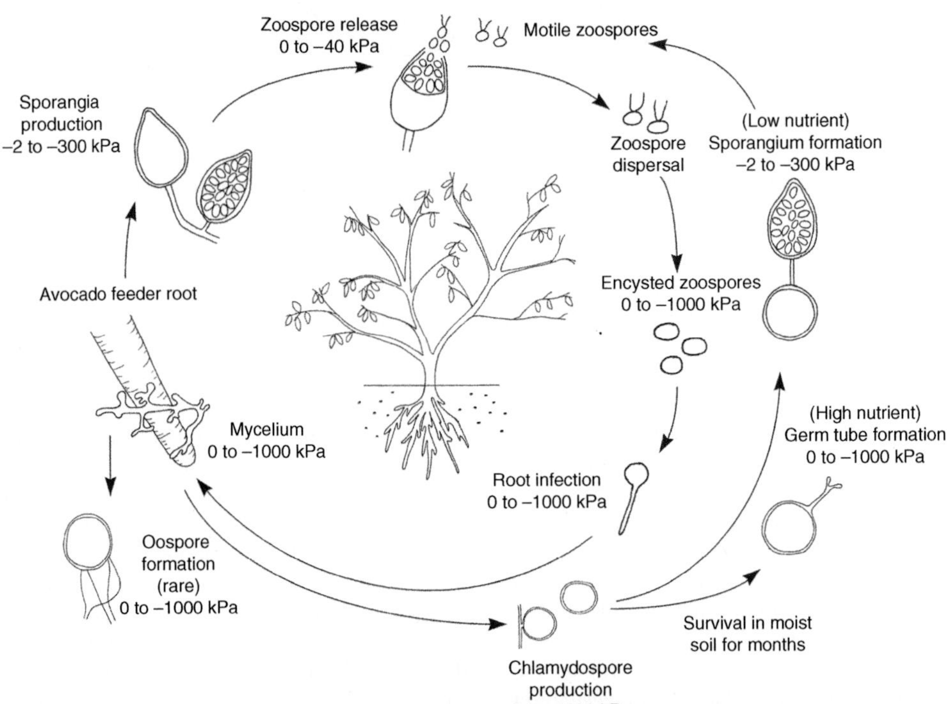

Fig. 13.11. Disease cycle of *Phytophthora cinnamomi* with soil matric potentials for each stage of the pathogen life cycle. The matric potential for zoospore release and dispersal is higher and more narrow than the range over which sporangia are formed. Chlamydospore formation and survival of mycelium occur across a very wide range, including in very dry soil conditions.

as little as 24 h. Chlamydospores and presumably other survival propagules mentioned previously develop in infected decaying roots (Jung *et al.*, 2013).

Soil moisture and temperature are the major environmental factors influencing infection and disease development (Zentmyer, 1981; Menge and Ploetz, 2003; Jolliffe *et al.*, 2024). The influence of soil matric potentials on the life cycle of *P. cinnamomi* is illustrated graphically by Menge and Ploetz (2003), and adapted in Fig. 13.11, and shows that high soil matric potential (−2 to −300 kPa) is required for production of sporangia and zoospore release. However, most other stages, including encystment, infection, mycelial growth and chlamydospore and oospore formation, can occur across a wide range of soil matric potentials, down to −1000 kPa in dry soil. Soil moisture will also mediate the diffusion of root exudates that attract zoospores and stimulate germination of chlamydospores. Zoospore germination, hyphal growth and chlamydospore formation can occur at soil oxygen levels as low as 0.1% and CO_2 levels as high as 20%, conditions that cause anoxia and death of avocado roots (Menge and Ploetz, 2003). Depending on soil type and pore size, well-drained soil may have 15% O_2 and 0.03% CO_2, compared with poorly drained soil, which may have 1% O_2 and 16% CO_2. Relatively short periods of soil saturation (i.e. waterlogging) lead to hypoxic conditions, which in the absence of *P. cinnamomi* will damage or kill avocado roots within days and when *Phytophthora* is present, a more rapid death of roots occurs (see Schaffer *et al.*, Chapter 7, this volume).

Disease development is directly related to soil temperature, given the presence of adequate soil moisture (Zentmyer, 1981). Optimal soil temperature for growth of glasshouse-grown avocado seedlings in non-infested soil was shown to occur at warmer temperatures (21–27°C); however, in the presence of *P. cinnamomi*, temperatures of 9–27°C were conducive to infection and disease development, and plant growth was reduced. Root rot is most severe at lower temperatures (9–21°C) (Menge and Ploetz, 2003) where *P. cinnamomi* grows better than avocado trees but is suppressed at higher temperatures that favour host growth. At a much higher temperature (33°C), plant and root growth were not impaired, but *P. cinnamomi* activity was considerably inhibited. A field study evaluating infection of seedlings planted in infested soil over successive 2-month periods demonstrated minimal infection occurred through cooler winter months (soil temperature at 10 cm ranging from min 4.5°C to max 22°C), with the highest incidence of infected roots during summer and autumn when soil temperatures ranged from a min of 14.5°C to a max of 26°C (Zentmyer, 1981). This was supported by a more recent study quantifying *P. cinnamomi* in roots of mature orchard trees at four seasonal time points. *P. cinnamomi* root quantities determined by quantitative PCR (qPCR) peaked at the late autumn sampling (Jolliffe *et al.*, 2024) and were significantly positively correlated with the number of soil temperature hours at 20–24°C and 20–29°C in the 2 months preceding sampling, corresponding to late summer and autumn (Jolliffe *et al.*, 2024).

Phytophthora root rot develops over a wide range of soil pH (3.5–8.0), but drought and salinity can predispose avocado roots to infection, presumably due to increased secretion of root exudates, resulting from increased stress when growing outside their edaphic range. When roots stressed by high salinity are infected by *P. cinnamomi*, they lose the ability to exclude salt and leaf tips and margins become necrotic, which is a typical symptom of salt and specifically chloride damage (Whiley *et al.*, 1987; Menge and Ploetz, 2003).

Management

The basic principles of integrated disease management are mandatory for successful production when Phytophthora root rot or any other disease is threatening productivity. This is represented simply in a schematic diagram known as the 'Pegg Wheel' and the elements of the integrated management strategy are covered in detail in various reviews (Menge and Ploetz, 2003; Pérez-Jiménez, 2008; Wolstenholme and Sheard, 2010). There is no 'silver bullet' for control of Phytophthora root rot, although new information, approaches and products have become available since publishing the second edition of *The Avocado: Botany, Production and Uses*, which may help to understand and optimize aspects of the fundamental 'Pegg Wheel' components, and a selection of the elements are discussed below (Fig. 13.12).

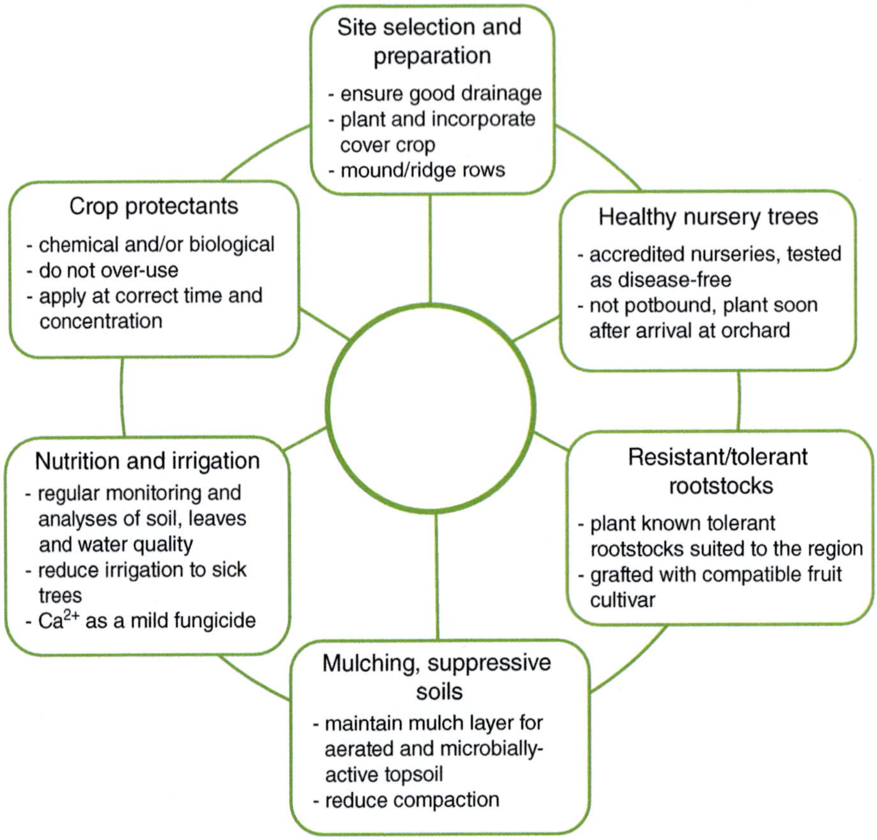

Fig. 13.12. The 'Pegg Wheel' for healthy trees.

Site selection and preparation

It has been known for more than 100 years that avocado is sensitive to waterlogging (Zentmyer, 1955), so irrespective of the potential *P. cinnamomi* threat, sites must be selected and prepared to allow for good internal and surface drainage. Soils with impervious subsoil layers, for example, clay pans or bedrock, even at 80 cm underneath well-drained topsoil, are unlikely to drain sufficiently in those production zones that experience periods of intense rainfall, and these sections of the orchard are typically the first areas to decline from Phytophthora root rot (Broadbent *et al.*, 1989).

A period of fallow or mixed species cover crop (Fish, 2020) prior to planting avocados may be advantageous for establishment, particularly if avocados or other Phytophthora-susceptible hosts were previously grown. A series of rootstock trials revealed higher tree mortality in sites where live (but declining) trees had been removed only 2 months prior to replanting, compared with sites, which had been cleared for a few years (Whiley and Dann, 2021). Subsoil drainage, if required, is best installed during pre-planting earthworks and trees planted on mounds or ridges to ensure roots are growing in well drained, friable, aerated soil. Biofumigation with cruciferous or sorghum species may also reduce *P. cinnamomi* inoculum levels; however, care must be taken to ensure these are thoroughly decomposed prior to planting, as phytotoxicity from freshly incorporated material may occur in young avocado trees. Pelletized formulations of Brassica biofumigant species are available, and a glasshouse study demonstrated the efficacy of this approach when pellets were incorporated into potting mix infested with *P. cinnamomi* a week prior to transplanting oak (*Quercus cerris*) seedlings (Morales-Rodríguez *et al.*, 2016).

Root necrosis was reduced in inoculated plants treated with the pellets, and while *P. cinnamomi* was not completely eradicated from potting mix, infectious propagules were reduced and there was no phytotoxicity during the 3-month duration of the trial (Morales-Rodríguez *et al.*, 2016).

Healthy nursery trees

Nursery trees are a common pathway for the dissemination of *Phytophthora* spp. and have been responsible for the introduction of new species into sensitive ecological sites, causing significant damage (Jung *et al.*, 2016; Rodríguez-Padrón *et al.*, 2019; Ramírez Gil and Morales Osorio, 2021). Sourcing pathogen-free trees from nurseries is the key to the establishment of healthy avocado orchards. Many avocado-producing countries have industry- or government-administered nursery accreditation schemes that provide a level of confidence that nursery trees will be free of soilborne pathogens.

Whether regulated or not, nursery operators must follow strict hygiene procedures, such as disinfestation of irrigation water, pasteurization of potting media, sanitizing wash bays for vehicles entering the nursery complex and footbaths entering each nursery section. Propagation tools and working surfaces should be regularly disinfected. Nursery benches should be totally isolated from soil, either raised high enough off the ground to avoid splash contamination from soil, or by purpose-built systems of coarse gravel or aggregate and sub-surface drainage. Walkways should be kept clean and free of weeds. Budwood should be sourced from healthy orchards and collected at least 1 m from the ground to reduce risk of *P. cinnamomi* contamination by rain splash from the soil. Free draining pots or planter bags along with optimal irrigation are essential. New propagation methods, such as tissue-cultured rootstocks and the use of air pruning pots, should minimize the risk of soilborne pathogens, at least in the early stages (see Whiley *et al.*, Chapter 9, this volume).

The use of fungicides to reduce root damage in nursery trees is not recommended as they mask symptoms and do not eliminate *P. cinnamomi*. Also, their continuous use may select resistant or less sensitive isolates of the pathogen, which become more difficult to control and may be transferred to orchard soils. Weak plants with poor root systems must be destroyed and root systems of neighbouring healthy-looking plants checked. The practice of carrying over nursery stock should be avoided, as this allows pathogens to persist or increase in the nursery environment. Destructive sampling and testing for the presence of *Phytophthora* or other soilborne pathogens should occur frequently across all nursery areas, which will allow quarantine and remedial action to minimize spread. Pre-dispatch inspection that includes root system checks should be undertaken.

Resistant/tolerant varieties

Significant progress has been made since 2013 in evaluating and commercializing rootstocks with superior tolerance or resistance to *P. cinnamomi*. The South African programmes delivered 'Dusa'™ (Merensky 2 cv), 'Latas'™ (Merensky 1 cv) and 'Bounty' about 20 years ago, and 'Dusa'™, in particular, has been very successful in the global market. Two new rootstocks, Leola™ (Merensky 6) and Zerala™ (Merensky 5), were released by the South African industry in 2020, and have superior tolerance to Phytophthora root rot compared to other industry standard rootstocks (Westfalia Fruit, 2021). Zerala™ is reported to have some tolerance to saline conditions. Three rootstock cultivars, 'Zentmyer', 'Uzi' and 'Steddom' with improved tolerance to Phytophthora root rot, have been released from the Californian breeding and selection programme (Menge *et al.*, 2012). Of these, 'Steddom' has comparatively good tolerance to salinity. A highly Phytophthora tolerant rootstock from the Australian selection programme, Procado® (SHSR-04), was commercialized in 2020, and trial data suggest it may be a lower vigour rootstock suited to higher density plantings (Whiley and Dann, 2021). Information from websites of large commercial nurseries report 'Astro' (SR1), recovered from an 'escape' tree in New Zealand, is also being sold as a high yielding rootstock with elevated tolerance to *P. cinnamomi* and ability to thrive in heavier soils (Lynwood Avocado Nursery, 2020). 'Miriam' (VC218) and 'Ben Ya'Acov1' (VC66) from Israel and 'Day' (VC207) from Florida are claimed to have tolerance to both *P. cinnamomi* and salinity (Brokaw Nursery, 2024).

Current understanding of the interaction between avocado roots and *P. cinnamomi*, and defence responses activated in the early stages of infection, has advanced considerably in recent years and has been reviewed by van den Berg *et al.* (2021). The terms 'resistance' and 'tolerance' are frequently used interchangeably; however, tolerance is where a plant is infected by a pathogen, but there is little effect on the plant in terms of symptoms or yields, and resistance is where the pathogen infection and colonization of the plant are reduced or prevented. Quantitative PCR studies show that the amount of *P. cinnamomi* DNA in 'Dusa'™ roots was less than in susceptible R0.12, suggesting that 'partially resistant' is a more accurate descriptor than tolerant (van den Berg *et al.*, 2021). The numbers of zoospores encysting and germinating on roots were previously shown to differ among avocado rootstocks, with more zoospore cysts germinating on susceptible 'Edranol' roots than on tolerant 'Duke 7', 'Martin Grande' and 'G6' (Aveling and Rijkenberg, 1991). This was also observed for susceptible R0.12 compared with resistant R0.06 and was associated with activation of defence enzymes and callose deposits to impede infection and colonization by *P. cinnamomi* in the resistant rootstock. However, other studies implicate lignification in the defence response in 'Dusa'™ (van den Berg *et al.*, 2021). These different observations demonstrate the complexity of the interaction of *P. cinnamomi* with different rootstocks. Several investigations have highlighted differences in multiple defence responses in comparisons between susceptible and resistant rootstocks, although numbers of each type of rootstock in these studies are limited. Nucleotide-binding site-leucine-rich repeat (NLR) proteins of plants mediate recognition of specific pathogens, the first step in the resistance response. These putative resistance genes (PaNLR) were reported for the first time in avocado in response to infection with *P. cinnamomi*, and shown to be strongly upregulated for 24 h in 'Dusa'™ compared with only 6 h in susceptible R0.12 (Fick *et al.*, 2022). Additional information is available in the review of van den Berg *et al.* (2021), which includes a proposed model for the interaction between *P. cinnamomi* and resistant rootstocks in the early stages (up to 24 h) of infection.

There is no doubt that genomic tools, such as single-nucleotide polymorphism (SNP) gene markers (Kuhn *et al.*, 2019), will aid future plant breeding and selection for resistance to *P. cinnamomi*. However, resistance in current (and future) rootstocks provides some protection but will not be the complete answer, and nursery/orchard management still requires an integrated approach as defined by the 'Pegg Wheel'.

Optimal nutrition and irrigation

Nutrition and irrigation must be optimized for healthy tree growth, tailored to seasonal changes in crop load, and ideally be based on leaf analyses and in-orchard water monitoring. A strong summer leaf flush will drive feeder root development during autumn and winter, enabling trees to better cope with stressful periods during flowering and fruit set in spring (see Gil *et al.*, Chapter 11, this volume).

It is difficult to manage irrigation in orchards, where some trees show decline from root rot. These trees with substantially fewer healthy feeder roots for water uptake will be overwatered with further loss of roots if irrigation scheduling is based on water requirements of nearby healthy trees. Remedial action can be taken by replacing sprinkler heads under declining trees with lower output emitters, or in the case of drip irrigation, lines can be pulled back from the irrigated zone. Salinity and chloride toxicity observed as leaf tip burn, which may extend along the leaf margins, is a stress that predisposes avocado to Phytophthora root rot, and may be mitigated by selecting tolerant rootstocks together with careful choice of fertilizers and/or deep leaching irrigation. Additionally, the use of magnetically treated water to drive salts in irrigation water below the root zone may be an option in some situations (see Schaffer *et al.*, Chapter 7, this volume).

Calcium in the form of gypsum, applied to some soils, can reduce Phytophthora root rot and increase yields by improving soil drainage and aeration, creating a more favourable environment for tree growth. It also has mild fungicidal properties. Production of sporangia was reduced in infested field soil amended with gypsum (calcium sulfate, $CaSO_4 \cdot 2H_2O$), and numbers of zoospores produced were reduced in field soil amended with calcium sulfate, calcium nitrate ($Ca(NO_3)_2$) or calcium carbonate ($CaCO_3$) compared with unamended controls (Messenger *et al.*, 2000).

Previous studies had shown that in the absence of *P. cinnamomi*, root and shoot growth was greater in acidic soils with pHs of 5.2 and 5.7, respectively, and decreased as pH increased. However, in pots infested with *P. cinnamomi*, growth was significantly reduced at the lower pHs, and plants had more severe root rot compared with healthier plants and roots at pHs 6.6 and 7.3 (Fernandez Falcon *et al.*, 1984). The high root rot at lower pH was associated with lower leaf and root Ca and higher Mn contents; the authors concluded that favourable soil Ca (and not high pH alone) was likely to be responsible for disease suppression.

Soluble silicon has also been shown to improve health and root growth of trees declining with Phytophthora root rot, but effects were not consistent (Dann and Le, 2017). More recently, in glasshouse trials, a silicate-based amendment reduced root rot and was associated with changes in the bacterial profile of the rhizosphere (Farooq *et al.*, 2022).

Mulching and suppressive soils

The benefits of mulching to suppress Phytophthora root rot have been well documented and reviewed (Wolstenholme *et al.*, 1996; Downer *et al.*, 2002; Crowley, 2007). In its natural environment, the avocado is a tropical/subtropical rainforest tree, where annual replacement of leaves produces a surface mulch of decomposing vegetation. This contributes to high organic content in the top 30 cm of soil, which recycles nutrients in an oxygen-rich environment, thus supporting a vibrant and dense mat of feeder roots. This is also the zone of the greatest microbial diversity and activity, including microorganisms suppressive to *P. cinnamomi*.

Soils collected from orchards have been shown to be suppressive to *P. cinnamomi* (Broadbent and Baker, 1974; Stirling *et al.*, 1992; McDonald *et al.*, 2007), which was attributed to microbial activity, transferable to conducive or non-suppressive soils and which was eliminated by heating or filter sterilizing. Other soils were suppressive in the field, where high populations of *P. cinnamomi* did not adversely affect tree health but were not suppressive to the pathogen in laboratory tests. The authors suggested that suppression could be ephemeral and fluctuate in response to environmental factors, and field effects may be attributed to other abiotic factors (McDonald *et al.*, 2007).

Activity of cellulolytic enzymes, produced by soil microbes to degrade cellulose and related glucans in woody mulches, may be one mechanism for suppression of *P. cinnamomi*, which also contains these structural components in its cell walls (McDonald *et al.*, 2007; Richter *et al.*, 2011). This enzymic suppressive activity is likely to have a greater effect in the upper organic mulch layers with less protective effect in deeper soil layers, although other mechanisms of biological control are clearly still operating.

Maintaining a consistent mulch layer, with careful attention to irrigation so as not to overwater, is important to reduce soil temperature fluctuations, maintain an optimum environment for avocado root growth (Fig. 13.13) and support a diverse and healthy microbial community, including those contributing to suppression

Fig. 13.13. Decomposing avocado leaf litter provides an ideal oxygen-rich environment for feeder root growth.

of *P. cinnamomi*. Genomic techniques such as DNA metabarcoding and microarray chips are tools for characterizing the microbial profile (the 'microbiome') of plant tissues and rhizosphere, and can be useful to investigate effects of treatments on key phyla (Vida *et al.*, 2016; Farooq *et al.*, 2022), although analyses are complex, and results often unique for each soil/tissue sample.

Successful examples of biological control of Phytophthora root rot in avocado using specific antagonists, for example, preparations of *Trichoderma*, *Bacillus* or mycorrhizae under field conditions are rare, with most reports of efficacy from *in vitro* Petri dish assays and greenhouse trials. Simply adding specific microorganisms to disease-conducive soils does not appear as effective as organic matter amendments to manipulate indigenous microorganisms for the suppression of Phytophthora root rot. Perhaps microbial profiling studies will assist with identifying and evaluating a specific set of microorganisms that serve as markers for suppression of *P. cinnamomi* over a range of environments, soil types, management regimes, etc., and which could potentially be introduced for enhanced management.

Crop protectants

Two chemicals have been used successfully for decades to reduce the impact of *P. cinnamomi* in many countries when combined with the more traditional methods of root rot management. The phenylamide pesticide, metalaxyl, and its more active R-enantiomer mefenoxam (or metalaxyl-M) is highly water soluble, moves readily in soils, is absorbed by avocado roots and moves rapidly within the xylem. Metalaxyl has a direct fungistatic activity against oomycetes inhibiting their growth and sporulation (Farih *et al.*, 1981). It will kill some but not all *Phytophthora* inoculum in the soil. However, after multiple applications in some avocado soils, especially those with high organic matter contents, its efficacy is reduced due to rapid biodegradation (Al-Sa'di *et al.*, 2008) and evolution of populations of *P. cinnamomi* that are resistant or less sensitive. It is very effective for short-term use in establishing nursery trees in replant situations.

The phosphonic acid fungicides, for example, fosetyl-Al and potassium phosphonate, are water soluble, readily absorbed and stable within the plant. The metal salts dissociate in plants to the phosphite anion (PO_3-), with fungicidal properties, which moves bidirectionally in xylem and phloem, according to changing physiological sink strengths (Whiley *et al.*, 1995). This facilitates various modes of delivery, the most effective being trunk injections and foliar sprays, but timing of application is critical and must coincide with periods of peak infection and active root growth since reaching and maintaining concentration thresholds in root tissues are critical for protection. Phosphonic acid has multiple modes of action, and these crop protection agents should not be confused with phosphate, PO_4 fertilizers (Dann and McLeod, 2021). Recent field trials have demonstrated that the ammonium phosphonate formulation may be more effective than potassium phosphonate, when applied as a foliar spray, with higher root phosphite concentrations after the same number of foliar sprays (McLeod *et al.*, 2018; McLeod *et al.*, 2021). Loss of sensitivity among isolates of *P. cinnamomi* has been reported; however, phosphonates are still highly effective in avocado when used correctly and as part of integrated management.

There is a paucity of reports on the effects of combined chemical and non-chemical treatments on Phytophthora root rot. However, the benefits of metalaxyl (+ mancozeb) drench and phosphonate injections combined with potassium silicate drench, compost (cow manure and rice hulls with mineral amendment) and leafy organic mulch were demonstrated over a 3-year period, compared with chemicals alone, or organic mulch or compost treatments (Ramírez Gil *et al.*, 2017). Improvement in tree health in the initial stages of the trial was likely due to the fungicide component, which was maintained for the next 6 months, while the organic components became effective. Tree health then gradually declined over the next year or so, as treatments were not reapplied after the initial 3 months and beneficial effects wore off (Ramírez Gil *et al.*, 2017).

Newer anti-oomycete chemistry, including mandipropamid, oxathiapiprolin, fluopicolide and ethaboxam, has been reported to have efficacy in avocado against *P. cinnamomi* (Belisle *et al.*, 2019), and there are likely to be others in the future, but their use must be carefully integrated with an emphasis on the wholistic 'Pegg Wheel' approach.

Verticillium wilt

Verticillium wilt of avocado, previously called asphyxiation or apoplexy, was first described in California (Zentmyer, 1949). The disease, which is usually not serious, is now recognized in several other avocado-producing regions, including Australia, Florida, Israel, México, Chile, Ecuador, South Africa and Spain (Menge and Ploetz, 2003).

Symptoms

Leaves wilt suddenly, turn brown and due to insufficient time to form an abscission layer remain attached to affected branches for several weeks (Figs. 13.14 and 13.15). There is often a characteristic odour associated with newly wilting leaves. In older trees, usually only one or several branches are affected, but sometimes the entire tree will show these symptoms. Young trees may be killed. When bark of affected branches is peeled away, a brown or grey-brown streaking of the vascular tissue is evident (Figs. 13.16a and b). In Mediterranean climates, disease symptoms can occur throughout the year (Dann, 2019); however, in the subtropics, infection occurs during the cooler winter months, but symptoms are not observed until late spring or early summer when environmental stress overcomes dysfunctional xylem tissues, resulting in rapid symptom development.

With warmer temperatures, affected trees usually send out new vigorous shoots below the affected branch within a few months of the initial wilting (Fig. 13.17). Such growth is usually healthy and the trees, if managed correctly, may recover completely. However, in some trees symptoms may reoccur for several years.

Causal agent and epidemiology

Verticillium wilt is caused by the soilborne fungus, *Verticillium dahliae*. This disease often arises in soil where other hosts of *V. dahliae* have grown. They include tomato, aubergine, potato, cotton, grapevine, groundnut, olive and many flower crops, as well as weed species. The fungus survives in the soil for many years as resting

Fig. 13.14. Sudden wilt of leaf flush occurs when the vascular system is blocked by gums and tyloses as the tree responds to the systemic movement of *Verticillium dahliae* conidia.

Fig. 13.15. Desiccated leaves and fruit may remain attached for several weeks in trees affected by Verticillium wilt.

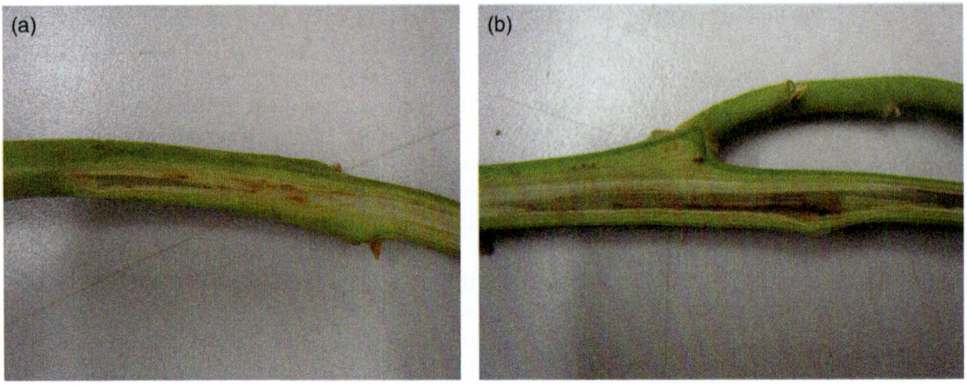

Fig. 13.16. (a) The vascular discolouration caused by Verticillium can be very subtle, (b) or more obvious. The fungus can be isolated from these tissues.

Fig. 13.17. New shoots may form below sections of branches affected by Verticillium wilt.

structures called microsclerotia, which are very small compact masses of thick-walled pigmented cells, budding off from the fungal mycelium. Under the right conditions, the microsclerotia are stimulated to germinate by root exudates, entering the feeder roots directly or through wounds and colonize the water conducting vessels (xylem). The fungus prefers cool (< 25°C), acidic and wet soil conditions, which favour high sap flow and movement of conidia in the tree. Trees respond to infection by producing tyloses and gums that plug vessels and reduce water flow, resulting in external wilt symptoms. Reoccurring symptoms are the result of new root infections.

The fungus is spread by water and wind-blown conidia, in soil particles and decayed plant tissue infested with microsclerotia and resting hyphae, and with budwood and seed from infected trees. In Australia, serious outbreaks of Verticillium wilt have been triggered in spring when major limbs were pruned and/or trees removed in winter due to orchard crowding (Dann *et al.*, 2009). These management practices probably stress or injure feeder roots, thus exacerbating disease. Verticillium wilt is severe in young avocado trees in south-west Western Australia, planted in sites previously growing potatoes or grapes and frequently occurs in summer and early autumn and is not associated with severe pruning or damage to roots (Dann, 2019).

Management

Avocado trees should not be planted on land previously used for susceptible crops or interplanted with such crops. Prune out dead wood once dieback has ceased, and vigorous young shoots have appeared. Pruning will not eliminate the fungus since new infections originate from the roots. Infected wood should not be used as mulch, and prunings should be removed from the orchard. Additionally, do not use organic trash from susceptible crops such as groundnuts as mulch. Tree regrowth following infection should be promoted with careful irrigation and appropriate nutrition. When selecting grafting wood or seed, avoid trees that have displayed symptoms of the disease. Spores can be carried on infected budwood, but such material rarely survives following grafting. In areas where Verticillium wilt is present, canopy management operations and tree thinning should be delayed until warm temperatures occur that do not favour the fungus.

Mexican rootstocks are reported to be more resistant than Guatemalan rootstocks (Zentmyer, 1984). A comprehensive field trial evaluating 'Hass' and 'Ettinger' grafted to 23 rootstocks (mostly from the West Indian race) was undertaken over 5 years in a site in Israel inoculated with potato stems infected with *V. dahliae* (Haberman *et al.*, 2020). Trees grafted with rootstocks VC66, VC26, VC140 and VC162 had the highest proportion with confirmed Verticillium wilt (15–25%), while *V. dahliae* could not be isolated and thus not confirmed from tissues with 'biotic stress' symptoms in several rootstocks, including VC804, 'Dusa', 'Degania' 62, VC96 and VC27 (Haberman *et al.*, 2020).

There are no foliar fungicides that are effective against the disease. Fumigation with chloropicrin or metam sodium is not practical for entire fields but can be used to fumigate sites before replanting.

White root rot (Rosellinia or Dematophora root rot)

White root rot, which is also known as Rosellinia or Dematophora root rot, is a serious disease of avocado in temperate regions such as Spain and Israel. It has a wide host range (and global geographical distribution) and is one of the most destructive diseases of temperate fruit tree crops such as apple, olive and grapevine (Pliego *et al.*, 2012). It was recently reported in Italy (Fiorenza *et al.*, 2021) and in the main growing regions of South Africa (van den Berg *et al.*, 2018; Hartley *et al.*, 2022). It occurs in other countries where it does not cause major losses.

Symptoms

Infected trees do not always show symptoms, and above ground symptoms can develop slowly or rapidly. During the slow expression of symptoms, affected trees stop growing and leaves are wilted and chlorotic, and the canopy is sparse with death of twigs and branches. Symptoms progressively worsen with eventual defoliation and death (Pliego *et al.*, 2012). Alternatively, tree death may be rapid with sudden wilting and desiccation of leaves and shrivelling of fruit, which remain attached and this may occur within a few weeks (Pliego *et al.*, 2012; van den Berg *et al.*, 2018). At the base of the trunk, dark-brown lesions are observed with white mycelial strands in characteristic fan-like formation, which can proliferate between the bark and the wood (Pliego *et al.*, 2012).

Below ground, white cottony strands of mycelium can be seen colonizing and rotting small feeder roots. In later stages, the cambium, larger roots, crown and trunk are affected with the white mycelial fan-like pattern described above, and also into the surrounding soil (Pliego *et al.*, 2012). Later, the fungal tissues melanize and turn green-grey or brown-black and form hard, flattened masses or crusts on roots (sclerotia), brown asexual fruiting bodies (synnemata) bearing conidia and sexual fruiting bodies (perithecia), although the latter are uncommon.

Causal agent and epidemiology

White root rot is caused by the ascomycete fungus *Dematophora necatrix* (syn. *Rosellinia necatrix*). The fungus can survive for long periods in wood, roots and soil as mycelium or sclerotia, and healthy feeder roots from neighbouring trees are infected when they come into contact with these inoculum sources. Hyphae first proliferate on the root surface and penetrate directly, or via natural openings or wounds (Pliego *et al.*, 2012). Hyphae then invade the xylem. The fungus produces secondary metabolites, which are toxic to seedlings, although not necessarily involved with the infection process, but may contribute to disease symptoms as they enter the vascular system (Pliego *et al.*, 2012). Good soil moisture and organic matter levels typically found in well-irrigated and mulched orchards are critical factors for growth of the fungus, and such conditions are conducive to infection (Pliego *et al.*, 2012).

Perithecia (the sexual stage giving rise to ascospores) are uncommon, taking a long time to develop under natural conditions but have been observed on roots of a dead tree in Spain (Pérez-Jiménez *et al.*, 2003). Neither ascospores nor conidia appear to play a role in spreading the disease locally or long distances. Soil contaminated with microsclerotia, infested roots and organic matter are means by which the fungus spreads between orchards. Once present, the fungus spreads rapidly along drip irrigation lines. Detection and identification of *Dematophora necatrix* from roots and soil have been aided in recent years by the development of sensitive nucleic acid-based assays (Arjona-López *et al.*, 2019; Hartley *et al.*, 2022), and non-destructive methods, such as chlorophyll fluorescence, to measure early foliar physiological changes associated with infection (Martínez-Ferri *et al.*, 2015).

Management

Infected trees and those that surround them, even if they are apparently healthy, should be uprooted and destroyed and all stumps and root debris removed from these sites. Affected areas should be isolated from healthy parts of the orchard by trenching or installation of root barriers between trees, to prevent root-to-root spread. Cultivation tools and other equipment should be kept out of infested areas, with irrigation of infected trees discontinued to discourage growth of the pathogen through the soil. Disease-free planting material is essential, and seed, potting media, pots, etc. should be disinfested and tested regularly, for example, via sensitive molecular diagnostic assays mentioned previously, to ensure the absence of the pathogen. Glasshouse and molecular studies have demonstrated that some rootstocks, for example, BG83, may be more tolerant to the disease than others, which may be associated with activation of genes related to salt and stress tolerance (Martínez-Ferri *et al.*, 2015; Zumaquero *et al.*, 2019).

Prior to replanting, expose the soil in these sites to light and air and disconnect irrigation. Soil solarization via covering plots with transparent 75-μm thick polyethylene sheets effectively eliminated viable inoculum in soil. The time taken to achieve this varied from 2 to 6 weeks and was dependent on soil depth and temperature reached in soil under the plastic sheeting (López-Herrera *et al.*, 1999). Chemical fumigation and fungicide drenching may be effective (Pérez-Jiménez, 2006; Arjona-López *et al.*, 2020; Magagula *et al.*, 2021). A large-scale greenhouse trial with containerized trees demonstrated that progression of disease symptoms after *D. necatrix* inoculation was significantly delayed in trees treated with composted organic amendments, including avocado prunings, almond shells or garden waste compared with the unamended control (Bonilla *et al.*, 2015). Addition of the composted avocado prunings and garden waste increased tree growth over the 9-month duration of the trial and was associated with higher soil nutrient content (Bonilla *et al.*, 2015). Several biocontrol agents, including *Trichoderma* sp., *Pseudomonas* sp. and *Bacillus* sp., have shown some efficacy *in vitro* or under greenhouse conditions in controlling *D. necatrix*; however, there are no recent reports of efficacy in orchard situations (Pliego *et al.*, 2012; González-Sánchez *et al.*, 2013; Pliego and Cazorla, 2013; Magagula *et al.*, 2021). Other management approaches, recently reported but requiring practical field assessment, include mild water stress to induce root defence responses (Martínez-Ferri *et al.*, 2019) and deployment of mycoviruses within *D. necatrix* or the related (but non-pathogenic) fungus *Entoleuca* sp. as potential 'virocontrol' agents for white root rot (Velasco *et al.*, 2020; Arjona-López and López-Herrera, 2021; Arjona-López *et al.*, 2021).

Trunk and Branch Diseases

Canker and/or branch dieback

Characteristic 'cankers' are very common on trunks and branches of avocado trees and may be caused by several different oomycete, fungal or bacterial pathogens. They are predominantly caused by entry of these pathogens through wounds created by mechanical damage, animals or wind-driven rain.

Symptoms

The symptoms, associated with canker lesions on trunks or branches induced by several different causal organisms, are very similar. Cankers of varying size typically appear as black, wet, sunken lesions on the lower trunk and branches of all sizes. Sap oozes from the centre of the canker, or from cracks surrounding the canker and dries leaving white to pink-brown powdery C7 sugar residue (Fig. 13.18). On cutting through the bark a distinct red-brown to brown discolouration of the underlying tissue is evident, sometimes confined to the outer bark but often extending into the cambium and sapwood killing phloem and xylem (Figs. 13.19 and 13.20). A sap-filled sac may be obvious when slicing through the canker longitudinally. Symptoms (other than cankers) affecting canopies and branches may differ according to causal organisms and site of infection and are described in more detail in the following sections.

Fig. 13.18. White to pink-brown dried sap residue is frequently observed around trunk or branch cankers.

Causal organisms and epidemiology

PHYTOPHTHORA SPP. *Phytophthora mengei* and *P. cinnamomi* are the most common species causing cankers on major roots, crown, trunk and branches of avocado trees in different parts of the world (Marais, 2004; Eskalen and Faber, 2016c). *P. mengei* was previously regarded as a subgroup of *Phytophthora citricola* (Zentmyer

Fig. 13.19. Trunk canker caused by *Phytophthora mengei* with a brown necrotic lesion in the inner bark. Photograph courtesy of J. A. Menge, University of California, USA.

Fig. 13.20. Cankers can occur on young branches.

et al., 1974) but, following comparative DNA fingerprinting and sequence analyses, was found to constitute a new and separate species named *P. mengei* (Hong *et al.*, 2009). *Phytophthora heveae* has also been reported to cause trunk cankers in avocado (Zentmyer *et al.*, 1976). In California, trunk canker is considered the second most important disease of avocado, after Phytophthora root rot (Eskalen and Faber, 2016c).

Apart from the cankers, trees may show decline and canopy thinning. Unlike root rot caused by *P. cinnamomi*, leaves may retain their normal size as feeder roots remain healthy. Cankers are not always visible above ground but can girdle the trunk just below the soil level, causing wilted and sparse canopies. Trees may appear healthy for several years despite being infected (Marais, 2004). Occasionally trees suddenly die with leaves turning brown but remaining attached. Depending on conditions, cankers may heal and the tree recovers (Eskalen and Faber, 2016c).

Phytophthora spp. causing cankers infect through wounds from orchard equipment, pruning, animals, sucker removal and damage caused by severe storms with strong winds. Disease development is favoured by heavy rainfall and poor drainage, and cankers grow more quickly when sprayed with water from irrigation sprinklers (El-Hamalawi and Menge, 1994). Wet conditions also favour oospore and sporangia/zoospore production of *P. mengei* on wounds, contributing to its survival and dispersal. Studies with greenhouse plants showed that cankers on stems developed more readily after wound inoculation with *P. mengei*, with symptoms increasing in severity with stress imposed by root pruning (El-Hamalawi and Menge, 1994). This was associated with decreased levels of carbohydrate and higher levels of amino acids in bark tissues of stressed (root-pruned) trees.

As with Phytophthora root rot management, trees should be planted on mounds or ridges to facilitate drainage, particularly in high-rainfall areas. Avoid damage to trunk, bark and structural roots, sprinklers should not directly wet trunks (drip irrigation could be considered), and mulches should not be placed up against tree trunks. *Phytophthora* spp. can be easily spread in soil adhering to vehicles and other orchard equipment, pruning tools and foot traffic within nurseries. Hence, sanitation and good nursery and orchard hygiene are important. Pruning wounds from removal of rootstock suckers or low branches should be protected promptly with fungicide. If cankers are not too severe, affected bark tissue can be removed and the wound treated with an effective fungicide. Phosphonate fungicides, applied to trunks either as sprays or paints, are effective for curing as well as protecting avocado trees against canker diseases (Erwin and Ribeiro, 1996). Rootstocks have been reported to vary in their susceptibility to stem cankers caused by *P. mengei* (El-Hamalawi *et al.*, 1994; Eskalen and Faber, 2016c).

FUNGI. Several species within the Botryosphaeriaceae family have been reported to cause branch canker and/or dieback. The disease was formerly known as Dothiorella branch canker/dieback after early reports from California associated *Dothiorella* with branch cankers (Horne, 1934) that was later attributed to *D. gregaria* (Halma and Zentmyer, 1953). Since the early investigations in California, there have been studies from other countries, reporting causal agents within the species of *Neofusicoccum*, *Fusicoccum*, *Dothiorella*, *Lasiodiplodia* and *Diplodia* (Zea-Bonilla *et al.*, 2007; McDonald *et al.*, 2009; McDonald and Eskalen, 2011; Auger *et al.*, 2013; Carrillo *et al.*, 2016; Guarnaccia *et al.*, 2016; Arjona-Girona *et al.*, 2019; Valencia *et al.*, 2019; Liang *et al.*, 2021; Hernández *et al.*, 2023). While members of the Botryosphaeriaceae have been the fungi most frequently reported, other fungal genera have also been isolated from symptomatic tissue. *Diaporthe* spp. (*Phomopsis* spp.) was also isolated from branches with symptomatic cankers (McDonald and Eskalen, 2011; Guarnaccia *et al.*, 2016) and less frequently *Bionectria* sp., *Colletotrichum* spp., *Schizophyllum* sp., *Fusarium* sp., *Eutypella* sp. and *Phaeoacremonium* sp. (McDonald and Eskalen, 2011). *Neopestalotiopsis* spp. and *Neocosmospora perseae* have recently been reported to cause stem lesions and dieback in young avocado trees (Guarnaccia *et al.*, 2018; Fiorenza *et al.*, 2022; Guarnaccia *et al.*, 2022).

Characteristic cankers are observed on branches higher than where Phytophthora cankers are found and bark may be cracked, slightly sunken with a dark discolouration and older necrotic bark tissue may be crumbly. A deep wedge-shaped red-brown discolouration may be visible when the branch is cut transversely through the canker (Fig. 13.21). If a significant amount of vascular tissue is affected, leaves of the branch may wilt suddenly and turn brown but remain attached, causing the branch dieback symptom. Other symptoms include shoot blight and dieback, leaf tip or marginal necrosis, graft dieback of newly planted nursery trees and stem-end rot of ripe fruit (Menge and Ploetz, 2003; Eskalen and Faber, 2016b). The fruiting structures (e.g. pseudothecia or pycnidia) of these fungi may occasionally be observed as small, hard black dots on dead tissue (Menge and Ploetz, 2003; McDonald *et al.*, 2009).

The Botryosphaeriaceae are well known to infect endophytically and exist as latent pathogens of woody hosts without causing symptoms until stress develops (Slippers and Wingfield, 2007). In avocado, branch dieback and canker diseases have often coincided and been linked to abiotic stress conditions, such as severe drought conditions and frost damage, and affect avocado production (Auger *et al.*, 2013; Valencia *et al.*, 2019; Avenot *et al.*, 2021). They are also aggressive wound pathogens in several horticultural woody crops and can infect through damage caused by pruning wounds, split bark from wind, frost or sunburn damage, mechanical injury from orchard equipment and grafting wounds (Fig. 13.22). The incidence and severity of these cankers are greatest in trees subjected to drought stress, nutrient deficiencies, waterlogging and temperature extremes. Spore trapping studies in avocado and macadamia have shown that they are released from fruiting bodies (e.g. pycnidia) on branches and dispersed into tree canopies following rain events during wet winter months in California (Eskalen *et al.*, 2013a) and summer in subtropical Queensland, respectively (Mohankumar *et al.*, 2023). If spore release coincides with pruning, or other wound damage, then infections and cankers may subsequently develop.

Severity of branch canker caused by fungi can be reduced by addressing the following factors: maintain trees under good irrigation regimes to avoid stress; prune dead or infected branches and remove them from the orchard; and conduct canopy pruning operations during dry periods of the year. Additionally, field sprays with fungicides may be beneficial, particularly in high-density orchards that require regular pruning (Twizeyimana *et al.*, 2013b; Eskalen and Faber, 2016b; Avenot *et al.*, 2021). To reduce problems in the nursery or field associated with graft dieback, nurseries should only extract seed from hard green fruit and avoid collecting fruit from the ground underneath trees. Vegetative propagation material (budwood) should not

Fig. 13.21. Wedge-shaped necrosis typical of Dothiorella (Botryosphaeriaceae) branch canker. Photograph courtesy of Akif Eskalen, University of California, USA.

Fig. 13.22. Necrotic graft union caused by infection with Botryosphaeriaceae.

be collected when wet and source blocks should be healthy and well managed. Nurseries should use strict sanitation practices, particularly during grafting. After planting, ensure young trees are not stressed.

BACTERIA. *Pseudomonas* sp. and *Xanthomonas* sp. are reported as causal agents, and sacs of fluid under the canker is characteristic (Korsten and Kotzé, 1987; Cooksey *et al.*, 1993). Causal agents may be epiphytic and enter via wounds or natural openings during conducive conditions and/or when trees are stressed. Severely affected trees may have poor growth, sparse foliage and low yields. Maintain tree health, optimal irrigation and nutrition, particularly boron, since deficiency of this micronutrient can cause bark splitting, which becomes an infection site. If the disease has affected the growth and yield, then the tree should be removed.

UNKNOWN CAUSE 'Black streak' has an unresolved aetiology but possibly linked more recently to Botryosphaeriaceae fungi and/or *Phaeoacremonium* sp. (Horne, 1934; Ohr and Zentmyer, 1994b; Menge and Ploetz, 2003; Carrillo and Eskalen, 2017). Typically cankers of varying size occur on lower trunks and the underside of lower limbs, then later higher in the tree. Foliar symptoms include shortened twig internodes, bunchy terminal leaf growth, small, blotchy, chlorotic leaves, necrotic leaf spots, partial or full defoliation, shoot dieback, early blooms of chlorotic flowers, zinc deficiency, wilted leaves and reduced yield. Affected trees may eventually die. Black streak appears after prolonged periods of environmental or cultural stress, particularly conditions of high salinity and drought. It is managed by maintaining tree health through good fertilizer and irrigation practices (Horne, 1934; Ohr and Zentmyer, 1994b; Menge and Ploetz, 2003; Carrillo and Eskalen, 2017).

Fusarium dieback

Fusarium dieback is a disease complex associated with invasive ambrosia beetles, affecting numerous cultivated and wild woody species across continents. The beetle vectors, members of the *Euwallacea fornicatus* species complex (Coleoptera: Curculionidae: Scolytinae), are commonly called 'invasive shot hole borers' (ISHB), and comprise at least seven genetically and ecologically diverse but morphologically similar lineages of emerging pests [see Smith *et al.* (2019) for the most recent information on taxonomy and nomenclature]. The natural distribution and host range of the *Euwallacea fornicatus* species complex expand across South-east Asia and Oceania (Stouthamer *et al.*, 2017; Smith *et al.*, 2019), and more than 75 plant families (Gomez *et al.*, 2019); however, each beetle species has a preferred host range and geographic distribution (Smith *et al.*, 2019). While species within this complex can infest seemingly healthy trees, they are more often associated with declining trees (Li *et al.*, 2016). In their native region, these beetles are known as pests of several economically important woody crops such as tea (*Camellia sinensis*) and *Acacia mangium* plantings (Hulcr *et al.*, 2017). Outside their native range, this species complex has caused devastation in urban and native forests (Ploetz *et al.*, 2013; Umeda *et al.*, 2016; van Rooyen *et al.*, 2021). Each species carries in their mandibular mycangia (specialized pockets associated with the mouthparts) a similar but unique suite of symbiotic fungi with differential virulence and ecological roles within the beetle life cycle (Kasson *et al.*, 2013; O'Donnell *et al.*, 2015; Freeman *et al.*, 2016; Lynch *et al.*, 2016; Carrillo *et al.*, 2020a). Three species within the *Euwallacea fornicatus* species complex, in conjunction their primary associated *Fusarium* symbionts, are known to infest and reproduce in avocado trees: (i) Tea Shot Hole Borer (TSHB, *Euwallacea perbrevis*); (ii) Kuroshio Shot Hole Borer (KSHB; *Euwallacea kuroshio*) and (iii) Polyphagous Shot Hole Borer (PSHB; *Euwallacea fornicatus*), recent syn. *Euwallacea whitfordiodendrus* (Smith *et al.*, 2019).

Euwallacea is the only genus known to cultivate fungi in the Ambrosia *Fusarium* Clade (AFC), and each of the seven lineages cultivates unique AFC symbionts. The fusaria associated with *Euwallacea* form a monophyletic group within clade 3 of the *Fusarium solani* species complex and contains at least 19 phylogenetically distinct species, not all of which have been formally described (Aoki *et al.* 2021). Those known to be associated with ISHB infesting avocado include *Fusarium euwallaceae*, *Fusarium kuroshium*, *Fusarium obliquiseptatum* and *Fusarium duplospermum* (Aoki *et al.*, 2019, 2021; Freeman *et al.*, 2019). A peculiarity of fusaria in this clade is that, in contrast to the typical banana-shaped macroconidia, most lineages produce unusual apically swollen or clavate (club-shaped) macroconidia that likely represent a potential adaptation for the beetle symbiosis (Kasson *et al.*, 2013; Aoki *et al.*, 2019; Aoki *et al.*, 2021). Due to the promiscuous nature of *Euwallacea* species, and the sympatric distribution of several AFC species and their *Euwallacea* symbionts, there is potential for symbiont switching, which could give rise to more destructive beetle-fungus combinations (Aoki *et al.*, 2021). Although *Graphium* spp. and *Paracremonium pembeum* fungi have also been associated with ISHB (e.g. Freeman *et al.*, 2019), the *Fusarium* are the ones able to colonize the vascular system beyond the beetle galleries and induce symptoms of disease in their hosts (Eskalen *et al.*, 2012; Kasson *et al.*, 2013; Na *et al.*, 2017).

The PSHB and its symbionts have caused the most damage to the avocado industries in California and Israel, although no firm estimate of economic impact is available. The PSHB adult females carry three fungi: *Fusarium euwallaceae*, *Graphium euwallaceae* and *Paracremonium pembeum* (Freeman *et al.*, 2013; Freeman *et al.*, 2016; Lynch *et al.*, 2016). While the first two symbionts have been shown to provide nutrients to different PSHB developmental stages, the role of *Paracremonium pembeum* remains unknown. In Israel, PSHB was first detected in 2009 and by 2016, it had spread to nearly all avocado areas in the country, being present in 60% of the plantings, where a survey of orchards showed 'Hass' being the most susceptible cultivar with the most severe dieback symptom and forcing growers to adopt integrated pest management (IPM) strategies to prevent the proliferation of the beetles within orchards (Mendel *et al.*, 2012, 2017). In California, PSHB was first discovered in 2012 in a backyard avocado tree in the Los Angeles area. By 2015, it was detected in a few avocado orchards in San Diego County (one of the areas with highest avocado production). At the beginning of the PSHB invasion, there was justifiable fear of Fusarium dieback causing extensive mortality to avocado trees, but later observations confirmed that avocado is not a preferred host for PSHB and most infestations occur in smaller tertiary branches, resulting in restricted branch dieback damage (Lynch *et al.*, 2018; Coleman *et al.*, 2019).

Infestations by two other ISHB lineages TSHB and KSHB, and their common *Fusarium* symbionts may similarly cause severe seasonal losses from branch dieback. There are several reports of ISHB and their fungal symbionts associated with avocado in Hawaii and Florida and other countries including South Africa, Australia, Taiwan and México, but these will not be discussed here.

Symptoms and epidemiology

Typical symptoms observed in affected avocado trees include wilting of small diameter branches and discolouration and desiccation of leaves causing a dieback symptom, and often breakage of the branch due to weakness from the beetle galleries. Beetle entrance holes, less than 1 mm in diameter in newly invaded branches, are often covered in white powdery sap exudate or 'sugar volcanos', formed when perseitol sugar in the sap dries out (Eskalen *et al.*, 2013b; Mendel *et al.*, 2017). The outer bark may have a wet, dark and discoloured appearance (Eskalen *et al.*, 2012). Removal of the bark and examination of the wood under the area bored by the beetle reveal a brownish staining of the xylem ('streaking') and necrosis caused by the fungus (Fig. 13.23) (Eskalen *et al.*, 2012; Mendel *et al.*, 2012). Small tubes of compacted sawdust or 'frass tubes' are often observed, resulting from gallery tunnelling. In Israel, trunks and branches of various diameters (from 2 to >30 cm, corresponding to 1- to 30-year-old growth) have been reported to be infested, causing death of branches (Mendel *et al.*, 2012, 2017). However, in California, colonization and reproduction in avocado trees are most often seen in smaller tertiary branches of 4–6 cm diameter, particularly at branch collar or junction regions leading to the death of small diameter branches (Lynch *et al.*, 2018).

Like other Xyleborini ambrosia beetles, species in the ISHB complex have a haplodiploid mating system that allows a single-mated female beetle to start a new infestation. Adult female beetles bore into relatively small diameter branches (less than 5 cm) of avocado trees and build galleries in the sapwood where they deposit

Fig. 13.23. Multiple beetle borer holes in a branch surrounded by red-brown staining. Beetle abdomens can sometimes be seen on close examination. In this case, the beetle is the Tea Shot Hole Borer, *Euwallacea perbrevis*.

the spores of symbiotic fungi (Lynch *et al*., 2016; Mendel *et al*., 2017). The adult beetles and larvae feed on the fungi, which grow within the galleries. Damage to plant hosts is caused by the physical destruction of vascular tissues during gallery formation and by the growth of the fungi. The Fusarium infection causes disruption of water and nutrient movement in the xylem and phloem, which, coupled with extensive beetle boring, can lead to severe tree injury and mortality in some very susceptible hosts such as maple (*Acer* spp.), including boxelder and bigleaf maple (*A. negundo* and *A. macrophyllum*, respectively), plane or sycamore (*Plantanus* spp.), various oaks (*Quercus* spp.), castorbean (*Ricinus communis*), wild tamarind (*Lysiloma latisiliquum*) and *Albizia lebbeck* (Mendel *et al*., 2017; Owens *et al*. 2018; Coleman *et al*., 2019). Contrary to *Harringtonia lauricola* (see laurel wilt section), *Fusarium* species associated with the ISHB complex do not spread very far (<10 cm) from the galleries, although the associated staining of the sapwood may extend further, and the fungus was found to survive in live or dead xylem for more than 2 years (Freeman *et al*., 2019).

Members of the ISHB can be distinguished by molecular methods, using conventional and real-time PCR, and nucleic acid sequencing (Gomez *et al*., 2018; Rugman-Jones *et al*., 2020). The *Fusarium* symbionts can be easily isolated onto semi-selective media (Eskalen *et al*., 2013b; Na *et al*., 2017; Paap *et al*., 2018), and although spores have fairly unique characteristics, species within the ambrosia fusaria group cannot be identified based solely on morphological characteristics (Freeman *et al*., 2013; Kasson *et al*., 2013). PCR and phylogenetic studies utilizing multiple gene regions are required for identification, and good molecular diagnostic methods have been published (Freeman *et al*., 2013; Short *et al*., 2016; Carrillo *et al*., 2020b), including a recent method to detect *Fusarium euwallaceae* directly from woody tissues (de Jager and Roets, 2022).

Management

Management of these ISHB in avocado orchards is restricted to regular surveillance, which may be aided by sticky traps baited with a quercivorol lure (e.g. Dodge *et al*., 2017) and orchard hygiene practices. Regularly check avocado and other susceptible hosts in the orchard vicinity for branch dieback and signs of beetle attack (i.e. pinhole borer holes, dried white sugar residues, sawdust and 'frass' tubes). Sanitation practice, including the removal and destruction of infested branches, represents a cost-effective management approach and is essential to control the population size of the beetle (Lynch *et al*., 2018). Regular pruning to increase light intensity within the orchard canopy is recommended to suppress beetle flight. Depending on the growing region and disease pressure caused by other pathogens, spraying pruning wounds with a fungicide (synthetic or biological) is recommended. Pruned wood should be chipped (<5 cm) and/or solarized prior to removal from the orchard (Lynch *et al*., 2018). There is potential for biocontrol of ISHB by an *Eucosmophorus* sp. parasitoid wasp, shown to have similar patterns of seasonal activity (Liu *et al*., 2022).

Laurel wilt

Laurel wilt is a fungal vascular disease vectored by ambrosia beetles affecting numerous hosts in the Lauraceae family. Even though all lauraceous species tested so far have been documented as susceptible, disease severity varies greatly across genera. Host species that are native to the Americas are particularly susceptible, rapidly developing severe symptoms that result in mortality, while those hosts that originated in tropical Asia can be considered tolerant as only mild and restricted symptoms are induced (Shih *et al.*, 2018). Avocado was first recognized as a host of laurel wilt in 2006 (Mayfield *et al.*, 2008), and its high susceptibility has been demonstrated by artificial inoculation studies and confirmed through field observations. Both the fungus, *Harringtonia lauricola* (formerly *Raffaelea lauricola*), and its primary ambrosia beetle vector *Xyleborus glabratus* were introduced to the USA in 2002 through Port Wentworth (Georgia, USA), but the disease was not reported until 2004 (Fraedrich *et al.*, 2008). The broad geographic range of multiple susceptible hosts, in addition to the abundance and generalist habit of several known beetle vectors, has promoted the unprecedented rapid expansion of laurel wilt across ecosystems. It is present in all 67 counties of Florida (Crane *et al.*, 2020b) and in 11 additional states as far west as eastern Texas (USDA Forest Service, 2023).

Currently, laurel wilt has only been reported in the avocado production areas of Florida and Myanmar (Ploetz *et al.*, 2016). The economic impact of the disease on Florida's avocado industry has been significant with destruction as of 2019 of at least 300,000 fruit-bearing trees and an estimated monetary loss of US$99 million (Evans and Crane, 2009). Economic losses associated with the impact of laurel wilt in Myanmar have not been formally documented but are considered significant (R. Gazis, Homestead, USA, 2023, personal communication). The destructive nature of the disease, in combination with its multiple hosts and potential vectors, makes laurel wilt an imminent threat to avocado production worldwide (Ángel-Restrepo *et al.*, 2022; Kendra *et al.*, 2023). It is threatening major avocado-producing areas in California and México and is likely to expand in Asia as planted area in several countries (i.e. China, Bangladesh and India), where the beetle vector and the pathogenic symbiont are native, is increasing.

Symptoms

Typically, laurel wilt symptoms begin in one section of the tree canopy (i.e. one branch) and progress to other sections over a period of weeks. Branches attacked by the ambrosia beetles are the first to show symptoms, or in the case of root-to-root transmission, branches on the side facing an adjacent infected tree (Fig. 13.24) (Ploetz *et al.*, 2017a; Navia-Urrutia *et al.*, 2022). Conspicuous 'frass' tubes are present at the sites of secondary beetle invasions on branches (Fig. 13.25). Foliar symptoms start with turgor (water pressure) loss followed by foliar wilting ('green wilting'), often on leaves at the top of affected branches, which turns into 'brown wilting' within 2–3 weeks (Fig. 13.26) (Crane *et al.*, 2020b). This arises from the pathogen invading the xylem vessels triggering production of gums and tyloses, which ultimately obstruct sap flow (Inch and Ploetz, 2012). Desiccated leaves often remain attached to branches for several months. Stem and limb dieback follow until eventually the whole tree is compromised (Crane *et al.*, 2020b). Internal symptoms include a blackish-blue-stained streaking of the sapwood (Figs. 13.27 and 13.28), which can also be observed surrounding the entry holes leading to the beetle galleries in redbay trees (Kendra *et al.*, 2013). Discolouration of sapwood can extend to root flares and lateral roots (Ploetz *et al.*, 2012b; Navia-Urrutia *et al.*, 2022). Depending on the initial concentration of pathogen inoculum, tree diameter and canopy structure, most trees succumb to the disease within 2–8 weeks (Cameron *et al.*, 2015; Ploetz *et al.*, 2016; Crane *et al.*, 2019). Although trees may re-sprout from the base, they will very likely eventually die. A few infected but asymptomatic or mildly symptomatic avocado trees, persisting in the landscape for several years ('escape trees'), have been recently identified. However, the mechanisms through which these trees are tolerating the infection and the length of the incubation period remain unknown (R. Gazis, Homestead, USA, 2023 personal communication).

Laurel wilt symptoms may resemble those caused by abiotic factors such as waterlogging, frost damage and fruit load stress (over-cropping), or by other diseases such as Verticillium wilt (Kendra *et al.*, 2013; Ploetz *et al.*, 2017a), as the rapid development of symptoms is similar. While

Fig. 13.24. Laurel wilt-affected trees of the cultivar 'Lula' showing evidence of pathogen transmission through root graft unions between adjacent trees. Tree on the right is showing advanced laurel wilt symptoms, while the one on the left is showing symptoms only in the section facing the adjacent infected tree.

Fig. 13.25. Avocado tree showing signs of ambrosia beetle infestation. Toothpick-like, sawdust protrusions coming from beetle entrance holes.

Fig. 13.26. Avocado tree affected by laurel wilt. Branch showing sections with different disease stages from green wilting to brown wilting and leaf desiccation.

Fig. 13.27. Diagnostic symptom of laurel wilt infection. Pathogen colonization induces the formation of tyloses and gums that result in the occlusion of the vascular system and the discolouration or streaking of the sapwood in diseased branches or main stem observed by removing the bark from a symptomatic limb. Symptom progression from initial green wilting to brown wilting and vascular streaking is typically 2–3 weeks.

Fig. 13.28. Diagnostic symptom of laurel wilt infection. Cross-section of a symptomatic branch showing streaking of the vascular system at different depths. Xylem rings can be completely damaged (appearing as continuous dark-stained rings) or damage can be limited to sections of the xylem (appearing as broken rings).

laurel wilt is relatively easy to identify in Florida where it is widespread, it may be more challenging to identify when it arrives in new production areas as the local biotic and abiotic environmental conditions may affect disease development and symptom expression.

Causal agent

Laurel wilt is caused by the ascomycete *Harringtonia lauricola* (formerly *Raffaelea lauricola*), a fungal nutritional symbiont of the redbay ambrosia beetle, *X. glabratus* (see Fig 14.19) and related species (Fraedrich *et al.*, 2008; Harrington *et al.*, 2008; Carrillo *et al.*, 2014). Female beetles disseminate the pathogen when colonizing seemingly healthy trees, whereas flightless males remain inside the galleries and mate with their sisters (Brar *et al.*, 2013). In suitable hosts, female beetles bore deep into the trunk sapwood, creating a network of interconnected tunnels or galleries. Fungal spores are transported by the beetle in a specialized structure called mycangia and cultivated along the gallery walls serving as the food source for adults and larvae (Kendra *et al.*, 2013). The original invasion into south Florida started with *X. glabratus*, which remains the primary vector of *H. lauricola* in Asia and in natural forests of the south-eastern USA. However, the low recovery of this species in avocado orchards affected by laurel wilt suggests that avocado is not its preferred host (Carrillo *et al.*, 2012; Brar *et al.*, 2013; Ploetz *et al.*, 2017b; Menocal *et al.*, 2018; Rivera *et al.*, 2020). Other ambrosia beetles, native or naturalized, have acquired *H. lauricola* through lateral transfer and now carry the fungus in their mycangia and/or phoretically on the external surface of their exoskeletons (Cruz *et al.*, 2021) and act as secondary vectors of the disease. For more information, see Chapter 14.

H. lauricola (Ophiostomatales: Ascomycota) is a dimorphic fungus, having filamentous (hyphal) and yeast growth. Sexual reproduction within *H. lauricola* has not been reported, and genetic studies suggest that the current population present in the USA is a consequence of a single introduction (Wuest *et al.*, 2017) likely from Taiwan (Dreaden *et al.*, 2019). Efforts to understand the biology of the pathogen, including genomic composition and expression (Ibarra Caballero *et al.*, 2019; Zhang *et al.*, 2020), metabolism and substrate utilization (Joseph *et al.*, 2021), have been reported recently. These studies have found unique traits present in *H. lauricola*, including an expansion of GH18 (chitinases) and of specific secondary metabolism gene clusters, an alternative sulfur assimilation pathway, sensitivity to high pH and cold adaptation, which may explain its superior pathogenicity on lauraceous hosts. Pathogen colonization and movement through the xylem lumen and the vessel elements have been investigated using a GFP-labelled strain, and significant correlations between laurel wilt severity and xylem colonization were demonstrated for avocado and two other Lauraceae hosts (Campbell *et al.*, 2017).

Diagnosis of the disease and causal organism(s) have improved considerably since the second edition of *The Avocado*. Several methods based on the detection of the pathogen have been developed and can be used across lauraceous species (Navia-Urrutia *et al.*, 2022). Due to the plasticity of colony morphology and lack of diagnostic characters that distinguish *H. lauricola* from closely related species associated with avocado, pathogen identification based solely on morphology is not recommended (Parra *et al.*, 2020). Traditional culture-based approaches require the isolation of the pathogen from symptomatic (discoloured) sapwood chips or slivers, using a semi-selective medium followed by DNA extraction and the amplification of two *H. lauricola*-specific microsatellite markers, IFW and CHK (Dreaden *et al.*, 2014). Based on these primers, Parra *et al.* (2020) developed a conventional PCR-based method to detect the pathogen directly from symptomatic sapwood. Other assays include qPCR methods (Dreaden *et al.*, 2014; Jeyaprakash *et al.*, 2014), and loop-mediated isothermal amplification (LAMP) for laboratory (Hamilton *et al.*, 2020) and field use (Hamilton *et al.*, 2021).

Remote sensing approaches for the rapid detection of laurel wilt-affected trees (symptomatic) across large areas have been proposed using aerial imaging (Sankaran *et al.*, 2012; de Castro *et al.*, 2015; Abdulridha *et al.*, 2016; Abdulridha *et al.*, 2018). Although the foliage of laurel wilt-affected trees possesses a unique spectral reflectance signature, discriminatory power is significantly reduced when targeting trees at early stages of disease progress, that is, 'asymptomatic' (<10–14 days after infection). Unfortunately, none of the methods were further improved or developed into a user-friendly interface and consequently these were never implemented by local growers. Finally, laurel wilt sniffing dogs were proposed as an early detection tool. However, besides the lack of practicality, its accuracy and ability to discern false negatives remain doubtful (Mendel *et al.*, 2018). So far, there is not a practical method sensitive enough to detect the pathogen at early stages of infection when the titre (concentration in solution) of the pathogen is low and tissues remain asymptomatic (no vascular damage). Therefore, there is still a critical need to develop field-deployable technologies able to detect molecules secreted by pathogen or host in the early stages of pathogen colonization. Methods to detect the pathogen directly from non-*X. glabratus* ambrosia beetles are critical for surveillance and monitoring programmes. However, this is challenging as the titres of *H. lauricola* in these alternative vectors are very low and may be several-fold lower than in *X. glabratus* (Harrington and Fraedrich, 2010; Cruz *et al.*, 2021).

Epidemiology

Long-distance spread of laurel wilt occurs when ambrosia beetles carrying the pathogen infest and inoculate an avocado tree initiating a new disease outbreak and dispersal by human-mediated transport of infested wood. However, within commercial avocado orchards, it is believed that most tree losses are due to root-to-root transmission of the pathogen, after initial infection of a tree by the ambrosia beetle vector (Ploetz *et al.*, 2017b). Disease spread patterns observed in the field and the consistent isolation of the pathogen from the root system strongly suggest that the pathogen spreads via conidia/spores, yeast and/or hyphal forms carried in xylem fluids through natural root grafts formed between adjacent trees. The pathogen can persist in dead avocado trees or stumps for at least 12 months (Navia-Urrutia *et al.*, 2023). The disease does not spread through fruit as the pathogen does not invade the pulp or seed tissues; however, pedicels, peduncles and hilum tissue are colonized at low frequencies (Ploetz *et al.*, 2012a). Infected scions and contaminated pruning equipment used in commercial avocado production have been discounted as transmission agents (Ploetz *et al.*, 2017a), although greenhouse and field experiments have demonstrated the movement of the pathogen from the inoculated scion into the rootstock (Navia-Urrutia *et al.*, 2023).

Due to the high likelihood of lateral symbiont transfer among ambrosia beetles (discussed above) associated with congeneric symbionts (Skelton *et al.*, 2019), monitoring these beetles for the presence of the pathogen is critical to prevent the establishment of the disease in new areas, as these and other related species are known to occur in major avocado production areas such as Michoacan, México (Ángel-Restrepo *et al.*, 2022). Future epidemics in newly invaded areas will undoubtedly behave differently from what was observed in south Florida. Factors that will influence the likelihood of disease establishment, outbreak expansion and dispersal pace are (i) diversity, density and proximity of susceptible lauraceous species to commercial avocado orchards; (ii) diversity and population density of ambrosia beetles that can adopt *H. lauricola* as a nutritional symbiont; and (iii) orchard spacing and likelihood of root grafting.

Management

Laurel wilt is established so widely in the south-eastern USA (present in 12 states) that eradication of the disease in the USA is not possible. Although eradication of laurel wilt would be difficult in newly invaded natural areas, in agricultural areas where hosts other than avocado are scarce, it may be possible to eradicate new outbreaks if affected trees are destroyed before new generations of the beetle vectors emerge from infected trees and disperse. The rapid and accurate identification of beetles vectoring *H. lauricola* and infected trees and their prompt destruction are crucial. Chipping or burning (when allowed) the infected material should be the preferred disposal option (Spence *et al.*, 2013).

Current strategies to manage laurel wilt disease in avocado orchards are limited and rely on the implementation of cultural practices, including quarantine zones, orchard hygiene (complete removal of infected tree and immediate adjacent trees) and canopy pruning to improve light penetration (Menocal *et al.*, 2022). The systemic nature of the fungal infection represents a critical challenge for management. A recent study demonstrated that once the tree is infected with the pathogen and early symptoms are observed, the tree (main trunk and root system) will remain infected and regrowth of stumped trees will eventually become colonized (Navia-Urrutia *et al.*, 2023). That study supports the recommendation that laurel wilt-infected trees must be completely removed from the planting since the pathogen can not only reinfect the new canopy; infected stumps represent a source of inoculum for adjacent trees due to root-to-root transmission.

Triazole fungicides were found to have the highest efficacy of several synthetic active ingredients tested *in vitro* (Ploetz *et al.*, 2011; Zhou *et al.*, 2018). Formulations containing different concentrations of propiconazole and tebuconazole in addition to other compounds that increase xylem mobility have been tested in potted and field trees (Ploetz *et al.*, 2011; Crane *et al.*, 2015; Ploetz *et al.*, 2017c). However, injections in mature trees with propiconazole do not prevent infection nor symptom development as uptake and movement of the fungicide in mature avocado trees is limited, and it is not uniformly distributed (Ploetz *et al.*, 2017c; Navia-Urrutia *et al.*, 2023). Injections of infected trees with current fungicide formulations are unlikely to succeed as the vascular system is already compromised blocking fungicide translocation, and beetle galleries containing *H. lauricola* fungal gardens extend longitudinally and transversely reaching as far as the heartwood, which potentially results in a constant release of inoculum into the vascular system. Therefore, active ingredients and formulations with lower EC50s, longer half-life and higher xylem mobility need to be explored. Formulations should allow for root uptake and movement within the vascular system of large hardwood trees, as injections and macro-infusion application methods will be cost prohibitive for many production areas. Biologicals have also been tested, but their *in vitro* efficacy has not translated into *in planta* disease control (Pérez-Martínez *et al.*, 2018; Olatinwo and Fraedrich, 2019). Management of *X. glabratus* and other ambrosia beetles vectoring laurel wilt is discussed by Carrillo *et al.*, Chapter 14 of this volume.

Finding genetic resistance to laurel wilt within *Persea americana* is unlikely, since both host and pathogen did not co-evolve. To date, at least 45 avocado cultivars have been reported as susceptible, including the commercially important 'Hass', 'Lamb Hass' and 'Reed', as well as the widely used rootstocks 'Toro Canyon', 'Duke 7' and 'Waldin', and other *Persea* species (i.e. *Persea palustris*, *Persea borbonea*, *Persea humilis* and *Persea indica*). Also other genera within the family (*Lindera*, *Litsea* and *Sassafrass*) succumb to the disease (Navia-Urrutia *et al.*, 2022). Nevertheless, finding a genetic solution that is of commercial value to industry (as scions or rootstocks) should be pursued as the ability to restrict the systemic spread and multiplication of the pathogen could give trees the chance to efficiently compartmentalize the fungus preventing extensive vascular damage. Such material, if discovered, could be incorporated into an IPM programme. Screening work is now being conducted on non-commercial cultivars and hybrid progeny (R. Gazis, Homestead, USA, 2023, personal communication).

Fruit and Foliar Diseases

Anthracnose and other body rots

Anthracnose, caused by fungi in the genus *Colletotrichum*, occurs in all countries where avocados are grown. It is the most serious postharvest disease of avocado fruit and contributes to considerable postharvest losses and consumer dissatisfaction. The key factor with this disease is that fungal infections occur in the field throughout fruit development, yet fruit of the main commercial varieties remain 'clean' and symptomless until ripe, well after fruit have left the orchard. It is essentially the disease that is not seen in orchards yet requires a sound integrated management strategy to reduce losses. The trend towards higher density orchards may exacerbate anthracnose disease, as recently reported (Bustamante *et al.*, 2022).

Symptoms

While symptoms can develop on leaves, flowers, panicles and unripe fruit before harvest under some conditions, the symptoms on fruit developing during ripening are most characteristic. The first symptoms are small, light brown lesions on fruit skin, which enlarge to become dark-brown-black and slightly sunken, circular lesions. These symptoms are obvious on ripe, green-skinned varieties but are very difficult to see in varieties such as 'Hass' which turn dark purple-black when ripe, causing disappointment and frustration in consumers. Pink-to-orange or dark-orange-red coloured spore masses are frequently produced in the centre of the lesions under warm, humid storage conditions (Figs. 13.29 and 13.49). When fruit are cut open or peeled, lesions can be seen penetrating the flesh in a hemispherical shape (Fig. 13.30). Occasionally, spongy but firm tan-coloured lesions occur in the pulp at the base of the ripe fruit (Figs. 13.31a and b). Sometimes infection by Botryosphaeriaceae fungi also causes postharvest rot of fruit (Everett *et al.*, 2007b) and may be seen as a

Fig. 13.29. Advanced anthracnose in a green-skinned variety, with light pink coloured spore mass of *Colletotrichum* spp.

diffuse light-to-dark-grey rot in the flesh of ripe fruit (Fig. 13.32). Disease caused by *Colletotrichum* spp. and Botryosphaeriaceae fungi can also occur at the stem end of fruit, and the symptoms are described in the 'Stem-end rot' section of this chapter.

Dark, spreading circular lesions can also occur on unripe fruit in the field, usually associated with a wound or damage to the peel, for example from insects, which may induce a localized ripening response. Smaller lesions, 1–5 mm, are also observed and are often associated with lenticels (Fitzell, 1987) and fruit affected often abscise prior to harvest. However, small lesions associated with lenticel damage sustained during harvest, packing or cool storage, particularly if fruit are very turgid and/or picked in the rain, are common, and *Colletotrichum* sp. and *Phomopsis* sp. have been isolated from such lesions (Everett *et al.*, 2008). *Colletotrichum* spp. also cause preharvest pepper spot disease of fruit and pedicels and/or peduncles and is observed in orchards primarily associated with stressed trees that have exposed and sunburnt fruit (see 'Pepper spot' section).

Fig. 13.30. Anthracnose lesions caused by *Colletotrichum* spp. infection are typically light brown and penetrate the mesocarp in a hemispherical shape.

Dark lesions can form on the inflorescence, which may cause subsequent death and abortion of fruit (Menge and Ploetz, 2003). More recently, in Queensland, Australia, rapid necrosis and dieback of panicles several weeks after flowering have reduced yields (Fig. 13.33). *Colletotrichum* spp. were consistently isolated from the rachises of thyrses (inflorescences) during early stages of necrosis, and subsequent pathogenicity

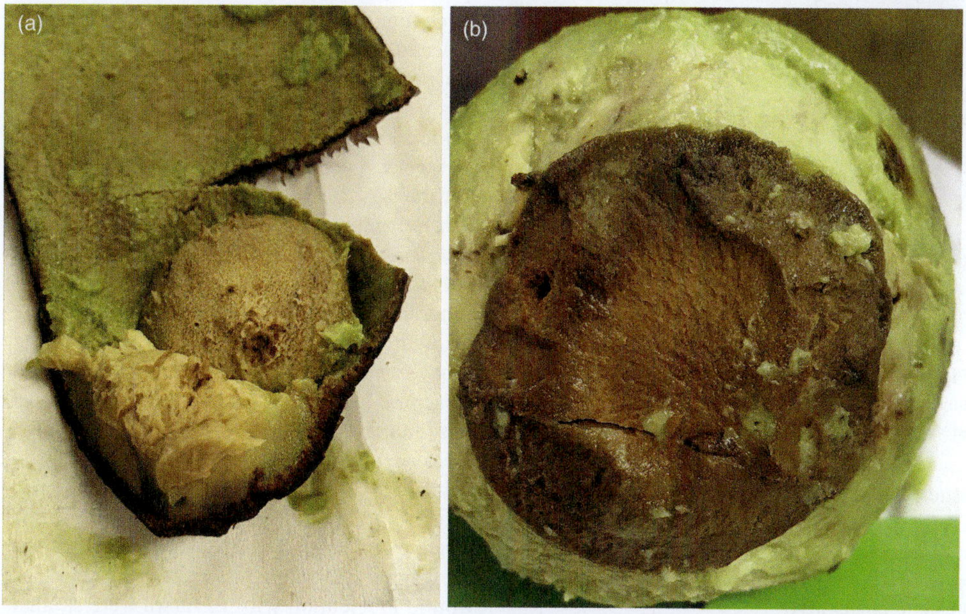

Fig. 13.31a and b. A large, spongy but firm, anthracnose lesion occasionally forms at the base of the fruit.

Fig. 13.32. A dark grey diffuse discolouration of flesh can be caused by infection by fungi within the Botryosphaeriaceae family.

Fig. 13.33. Panicle dieback occurs at flowering or several weeks afterwards, and has been associated with *Colletotrichum* spp.

studies demonstrated its role in this disease (E. Dann, Brisbane, Australia, 2023, personal communication). Other factors such as drought stress and reduced root volume and function are likely to be involved.

Dark-coloured, irregularly shaped lesions can develop on young, soft flush leaves (Fig. 13.34a) or shoots under very wet and humid conditions, and can cause shoot dieback (Pegg and Coates, 1993; Menge and Ploetz, 2003). On older leaves, chlorotic, then brown lesions may be seen (Fig. 13.34b) and *Colletotrichum* spp. can usually be isolated from brown-red advancing margins of lesions associated with tip burn (Menge and Ploetz, 2003), a symptom commonly attributed to salt/chloride or other stress. Small hard black dots, <1 mm are sometimes seen in this desiccated tissue, and these may be perithecia, the sexual fruiting bodies of *Colletotrichum* spp., containing asci and ascospores (Darvas *et al.*, 1987). Other fungi, such as *Diaporthe* sp. and *Pestalotiopsis* sp., also produce similar looking fruiting bodies and microscopy is required to confirm genera.

Causal agents

Anthracnose is caused by species within the *Colletotrichum* species complexes. This genus is responsible for diseases in a wide range of hosts and was voted by plant pathologists among the top-ten fungal plant pathogens globally, in terms of scientific and economic importance (Dean *et al.*, 2012). The genus has undergone extensive taxonomic revisions, aided recently by DNA sequencing techniques, and currently there are three main 'species complexes' associated with avocado. Table 13.1 outlines these complexes and provides examples of the most common species within them, as determined by multi-gene sequence analyses. It is not an exhaustive list of all species reported. It must be noted that while anthracnose has been historically attributed to *Colletotrichum gloeosporioides*, the 'true' *C. gloeosporioides* (classified by molecular taxonomy) is not a common pathogen on tropical fruits (Phoulivong *et al.*, 2010) and is associated more frequently with citrus and other hosts. However, *C. gloeosporioides* was reported as causing anthracnose in Israel, along with seven other species (Sharma *et al.*, 2017). The large numbers of species within the *gloeosporioides* complex highlight the high degree of diversity, and it is not unusual to find more than one species infecting a single piece of fruit. Species may differ in their virulence, for example, lesions on fruit caused by wound inoculation with *C. karstii* were smaller than those caused by *C. aenigma*, *C. alienum* and *C. theobromicola* (Sharma *et al.*, 2017). Limited numbers of isolates from avocado in Sri Lanka

Fig. 13.34. Anthracnose lesions on leaves are not common.

Table 13.1. Examples of common Colletotrichum species within three species complexes causing anthracnose in avocado, as determined by multi-gene sequence analyses.

Complex	Examples of species	References
Colletotrichum gloeosporioides	*fructicola, perseae, siamense, alienum, aenigma, theobromicola, gloeosporioides*	(Weir *et al.*, 2012; Sharma *et al.*, 2017; Giblin *et al.*, 2018; Fuentes-Aragón *et al.*, 2020; Hofer *et al.*, 2021; Soares *et al.*, 2021)
Colletotrichum acutatum	*fioriniae, simmondsii*	(Shivas and Tan, 2009; Fuentes-Aragón *et al.*, 2020; Hofer *et al.*, 2021)
Colletotrichum boninense	*karstii, boninense*	(Silva-Rojas and Ávila-Quezada, 2011; Velázquez-del Valle *et al.*, 2016; Giblin *et al.*, 2018; Hofer *et al.*, 2021; Soares *et al.*, 2021)

and Chile were determined to be *C. gigasporum*, in the *gigasporum* complex (Hunupolagama *et al.*, 2015), and *C. anthrisci*, in the *dematium* complex (Bustamante *et al.*, 2022), respectively.

The geographic distribution of *Colletotrichum* species varies, most notably with *C. fioriniae* and other *acutatum* complex species associated with cooler production regions of New Zealand and south-west Western Australia (Hartill, 1991a; Hartill and Everett, 2002; E. Dann, Brisbane, Australia, 2023, personal communication). Differences in geographical distribution have also been reported for species infecting avocado in Israel (Sharma *et al.*, 2017).

Epidemiology

Infection mainly occurs during extended periods of warm rainy weather, and fruit are susceptible at all stages from fruit set to harvest (Peterson,

1978; Coates *et al.*, 1993a). The disease cycle is depicted in Fig. 13.35. Spore trapping and isolation from different tissues onto selective media showed that conidia (spores) of *Colletotrichum* are produced in large numbers on dead leaves, bark and branches entangled in the tree canopy, on infected mummified fruit still hanging on the tree, and fallen fruit (Darvas *et al.*, 1987; Fitzell, 1987; Sharma *et al.*, 2017) (Figs. 13.35 and 13.36). Perithecia containing asci and ascospores, the sexual stages, are also present on tissues in the orchard (Darvas *et al.*, 1987), observed particularly on leaves where they are more easily seen but their role in causing disease has not been determined. During rainy weather, conidia are washed onto the fruit surface and in the presence of free water, most conidia germinate within 7 h (Parbery, 1981). Each germinated conidium produces a germ tube, then swells at the tip and enlarges, forming an appressorium (Fig. 13.35). The wall of the appressorium thickens and darkens and a central germ pore develops (Parbery, 1981). An infection peg emerges from this germ pore and penetrates the outer wax layer and cuticle of the fruit skin to a depth of less than 1.5 μm (Coates *et al.*, 1993b). Growth of the infection peg is arrested in the cuticular region, where it remains quiescent until fruit ripening (Binyamini and Schiffmann-Nadel, 1972; Prusky *et al.*, 1990; Coates *et al.*, 1993a). The infection at this stage is resistant to extremes of temperature or desiccation, and the fungus remains alive in the peel of the fruit for many months. Flowers may also be infected but remain symptomless (Fischer *et al.*, 2019).

As a fruit ripens, it undergoes physiological and biochemical changes, such as ethylene biosynthesis, antioxidant activity, changes in the cuticle and decline in antifungal secondary metabolites (Prusky *et al.*, 2013). This causes the

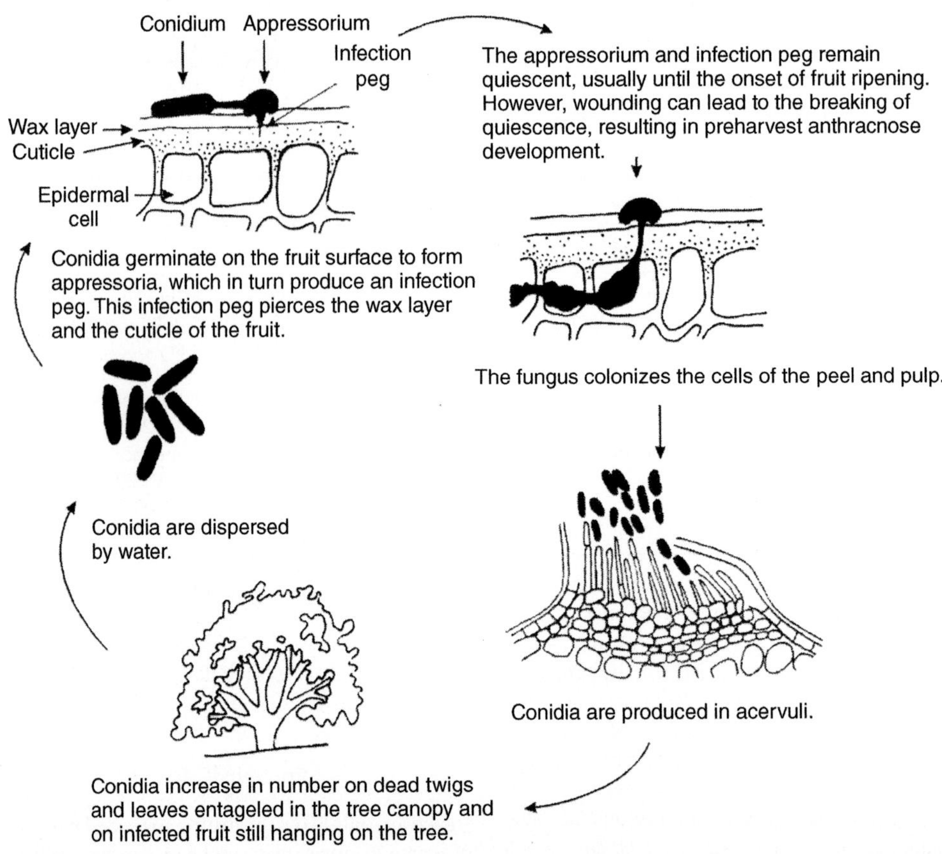

Fig. 13.35. Disease cycle of anthracnose caused by *Colletotrichum* spp.

Fig. 13.36. Mummified fruit still hanging in the canopy covered with conidia of *Colletotrichum* sp.

quiescent phase of the fungus to cease, allowing the resumption of growth. In advanced stages of lesion development, fungal fruiting structures, acervuli, are produced beneath the fruit surface. Eventually, the cuticle and epidermal cell walls are ruptured and conidia are released in a mucilaginous matrix. These are seen as pink-orange-red spore masses on overripe fruit (Figs. 13.29 and 13.49). Conidia are then dispersed by water and the disease cycle continues. After packing, there is no fruit-to-fruit infection or spread.

Management

Successful management of anthracnose and body rots caused by other fungi relies on several pre- and postharvest strategies and is the responsibility of everybody in the supply chain, from growers through to point of sale. Within the orchard, reducing the inoculum (spore) load is achieved by canopy management (pruning) to remove dead twigs and branches and any fruit remaining in the canopy (Hartill, 1991b). Even if fruit is not marketable, it should be picked and dropped to the orchard floor where it will rot more rapidly given the moist conditions under irrigated trees. Such conditions also favour growth of beneficial yeasts and other microorganisms, which help to degrade fruit and may also outcompete *Colletotrichum* sp., as demonstrated in a study with apples (Martin *et al.*, 2021). Similarly, chipping or mulching woody branch material is recommended to encourage rapid decomposition of tissues harbouring spores. Pruning also improves airflow in the canopy, allowing more rapid drying of fruit surfaces after rain events and thus reducing the time for germination of spores and infection of fruit.

Fungicide sprays are a key component to anthracnose management globally, particularly in high-rainfall areas and each country and region will have its own programme of registered products and recommended use patterns. A consideration is the market destination for fruit and their respective maximum residue limits (MRLs) for each pesticide active ingredient. It is critical that spray programmes with protectant fungicides commence early in fruit development and that coverage is maintained by repeated sprays as fruit grow, or the fungicide is washed off. Copper-based formulations have been extensively used around the world for many decades and are still very effective at preventing infection of fruit. There are also several active ingredients, which may have some postinfection activity and these must be used according to label instructions to avoid resistance developing within the pathogen populations, thus ensuring optimum efficacy and prolonging fungicide life. Pressure to reduce the use of traditional pesticides has seen some biological fungicides registered, such as those based on *Bacillus* spp. Evaluations of biological fungicide agents and improved chemical active ingredients or their formulations are continuing. An effective field fungicide programme relies as much on proper use of appropriate spray equipment as on the chemistry and regular checks of calibration, coverage and so on are essential.

Managing fruit nutrient levels can also contribute to reducing anthracnose. Reductions in anthracnose were often correlated with lower nitrogen and higher calcium levels in fruit at harvest (Everett *et al.*, 2007b; Dann *et al.*, 2016), although ratios with other cations, particularly Mg and K, are also likely to influence the relationship with fruit disease (Hofman *et al.*, 2002). Field amendments with soluble silicon formulations have been reported to reduce anthracnose (Dann and Le, 2017). Tree crop load, vigour and rootstock have also been shown to impact fruit anthracnose levels, which may be linked to fruit nutrient status (Hofman *et al.*, 2002; Willingham *et al.*, 2004; Willingham *et al.*, 2006; Dann *et al.*, 2009a; Dann *et al.*, 2016). Other orchard practices include controlling insect damage to fruit to reduce wound-induced anthracnose and rots caused by fungi within the Botryosphaeriaceae family.

Postharvest practices are also important to maintain quality fruit with minimal disease at ripening. Controlled ripening and postharvest temperature management can have a major influence on anthracnose development in harvested fruit, and the protocols for fruit handling may vary depending on cultivar, maturity status (e.g. dry matter content), available infrastructure and market forces. Fungicides with curative or post-infection activity are effective and are applied in packing sheds where available and allowable for destination markets.

Given that *Colletotrichum* spp. infections can be asymptomatic for long periods in fruit (and in other tissues), approaches to detect or estimate the level of infection by *Colletotrichum* spp. in green fruit prior to harvest may be useful as decisions could be made on preharvest sprays, or postharvest storage and market pathways. They could also assist with epidemiological studies to optimize field spray programmes. Treatments to induce premature ripening of fruit or tissue senescence have been effective, with disease symptoms developing in treated tissue (Fischer *et al.*, 2019). A model to predict anthracnose in fruit at harvest, based on quantifying *Colletotrichum* sp. in leaves by the qPCR technique, was developed, and accurately predicted rots for orchards in two out of three regions of New Zealand (Everett *et al.*, 2015).

There have been some exciting new approaches that show promise for management of anthracnose, and postharvest disease caused by other fungi. Most of these are postharvest treatments applied in the packing shed, for example, antifungal essential oils or terpenoid compounds, combined with modified atmosphere packaging or incorporated into biodegradable polymers, edible coatings, natural wax formulations, biocontrol, chitosan and electrolysed oxidizing water. These have been well reviewed by Munhuweyi *et al.* (2020).

Research efforts to deliver efficacious and acceptable management options are ongoing and will utilize increasingly more sophisticated techniques for their discovery and assessment of impact. Studies of the 'microbiome', microbial communities present on or in plant tissues, have recently become popular. Practical applications may include the detection and/or identification of beneficial microbes, leading to effective biocontrol formulations, detection of foodborne (human) pathogens and understanding how environmental or cultural practices in the orchard or along the supply chain could affect the balance of microorganisms on fruit surfaces, contributing to postharvest disease suppression and high-quality fruit (Droby and Wisniewski, 2018).

Avocado sunblotch viroid (ASBVd)

Avocado sunblotch is widely distributed in most avocado-growing countries and causes significant reduction in productivity and losses from unmarketable fruit. It is thought that ASBVd coevolved with avocado and was likely unintentionally introduced to California in germplasm collected from México and Guatemala during the early years of the industry (Saucedo-Carabez *et al.*, 2019). The importance of the disease has diminished somewhat with greater knowledge of its cause and epidemiology. It is considered a minor problem in countries where tree registration programmes exclude the pathogen from propagating material and field tree removal is practiced when the disease is identified.

Symptoms

Symptoms of avocado sunblotch vary considerably, and all cultivars are considered susceptible. Infected trees may be symptomless. When present, the most reliable and characteristic symptom is on fruit, which develop white, yellow or pink sunken streaks originating from the stem end and developing into necrotic black crevices on severely infected fruit (Fig. 13.37). Leaf symptoms in the field are uncommon; however, bleached or chlorotic areas form initially around the leaf midrib and veins and this may progress to complete chlorosis or bleaching with leaves becoming deformed (Fig. 13.38). On green twigs or young stems, narrow longitudinal streaks may occur, first yellow to pink then turning sunken and necrotic. The bark on the trunk or larger branches can have a rectangular cracked appearance, often referred to as 'crocodile skin' or 'alligator bark'. The symptoms of sunblotch may be irregularly distributed throughout an infected tree, and severely affected trees are often stunted, sparsely foliated and develop a sprawling, prostrate architecture.

Fig. 13.37. Avocado sunblotch on 'Hass' fruit.

Fig. 13.38. Leaf chlorosis and deformation caused by avocado sunblotch viroid.

Trees with symptoms can quickly become asymptomatic. Symptomless trees may develop typical symptoms following stress induced by severe pruning as part of canopy management. Yields in symptomatic 'Hass' trees may be reduced by 50–80% (and much of it is unmarketable due to the sunken streaky lesions), while yield reductions of 15–30% may occur in symptomless trees (Saucedo-Carabez *et al.*, 2014).

Causal agent

The disease is caused by Avocado sunblotch viroid (ASBVd), a circular, single-stranded RNA of 246–250 nucleotides and is thus one of the smallest pathogens in the world. More information on recent structure elucidation is provided in the review by Saucedo-Carabez *et al.* (2019). Sequence variants have been categorized according to their association with the different symptom types: those associated with bleached, variegated and asymptomatic tissues have been labelled ASBVd-B, -V, and -SC, respectively (Semancik and Szychowski, 1994). Trees with initially severe leaf bleach symptoms can recover over time and become symptomless carriers but following severe pruning, leaf bleach symptoms may return (Semancik, 2003).

Sunblotch can be detected using molecular methods or by grafting bark patches or razor slash inoculation with sap from infected trees onto indicator hosts such as young Mexican race avocado seedlings (Desjardins *et al.*, 1980). Recent improvements in the methodology for extracting and detecting ASBVd by highly sensitive molecular-based methods, for example, RT-qPCR and high-throughput sequencing, have aided the detection in asymptomatic tissues from trees, honeybees and pollen samples, supporting clean nursery schemes, surveillance and attempts to demonstrate area freedom from the disease (Pretorius *et al.*, 2022; Bonnéry *et al.*, 2023; Roberts *et al.*, 2023).

Epidemiology

ASBVd has a restricted host range and apart from avocado, few species in the Lauraceae family are confirmed as hosts by glasshouse transmission studies, reviewed by Saucedo-Carabez *et al.* (2019). ASBVd is systemic in avocado trees, but its concentration can vary widely between branches. The principal means of transmission are in seed used to propagate rootstocks, in scion material used for grafting, and by natural root grafting. Sunblotch is transmitted, although inefficiently, on sap-contaminated pruning blades, harvesting clippers and injection equipment. While ASBVd has no known insect vector, it is transmitted in pollen (Roberts *et al.*, 2023), but only the developing fruit and seed and not the fruit-bearing tree are infected (Desjardins *et al.*, 1984).

Symptomless carrier trees play an important role in the epidemiology of sunblotch disease. Seed transmission of ASBVd occurs at a much higher frequency (80–100%) in symptomless infected trees than in trees displaying symptoms (up to 5%). Furthermore, plants grown from the seed of a symptomless carrier tree remain symptomless; however, about 5% of seedlings grown from symptom-bearing trees display characteristic symptoms (Wallace and Drake, 1962).

Management

There is no cure for infected trees (whether symptomatic or not), and they should be removed and destroyed, along with all trees within a 15 m radius. The disease can be eliminated from new plantings by nursery certification schemes, which thoroughly pre-test propagation material (seed used for rootstock and budwood) to ensure it is free from the viroid (Saucedo-Carabez *et al.*, 2019). Testing should use the highly sensitive molecular-based indexing methods to ensure the pathogen-free status of propagation material. Orchard sanitation is also important, and pruning tools, harvesting clippers and injection equipment should be thoroughly cleaned and then disinfested when working between trees (Desjardins *et al.*, 1987).

Black spot (Cercospora spot)

Black spot or Cercospora spot of avocado, also known as Pseudocercospora spot or blotch (Pohronezny *et al.*, 1994; Darvas and Kotzé, 1979), is a disease of warm, humid and rainy climates and is found in tropical and subtropical parts of the global production areas. Losses of fruit can be serious, and at one time were reported as high as 69% in unsprayed orchards in the warm subtropics of South Africa (Darvas, 1982). In Colombia, black spot is responsible for downgrading an estimated 48% of fruits to second and third grades (Rodríguez-Polanco *et al.*, 2020). Black spot affects many *Persea* species and all cultivars of *P. americana*.

Symptoms

Symptoms are most conspicuous on leaves and fruit. Characteristic symptoms include lesions that first appear on the lower surface of older leaves as small (1–5 mm) and angular light brown to purple flecks or lesions, usually surrounded by a chlorotic (yellow) halo. Spots eventually become visible on both leaf surfaces (Figs. 13.39a and b). In humid, wet conditions, grey, felty spore-bearing structures of the fungus are clearly visible in the spots on the undersides of leaves (Schoeman and Kallideen, 2018). Individual lesions often coalesce to produce large, brown, dead areas on the leaf. The fungus may survive dormant in leaves and fruit for 3 months [reported in Schoeman and Kallideen (2020)].

On fruit, lesions develop in the field, unlike anthracnose which mostly develops upon fruit ripening. Symptoms first appear on immature fruit as small raised, shiny black flecks or spots, which increase in size as fruit mature, typically up to 6+ mm in diameter and become slightly sunken, brown to black lesions, finally coalescing to irregular black lesions (Fig. 13.40), and often forming white areas of sporulation in the centre (Schoeman and Kallideen, 2018). As lesions dry out cracks may develop, allowing the entry of other pathogens such as *Colletotrichum* spp. (Darvas and Kotzé, 1987). Sometimes infection is temporarily arrested and disease is manifested as minute (<3 mm), raised, black, shiny spots often associated with cracking and corking of lenticels (Darvas, 1982). The disease is usually confined to the peel, but the flesh can be invaded during advanced stages of the disease (Darvas and Kotzé, 1987). Fruit with black spot lesions are downgraded in the packing house and are unacceptable for export (Rodríguez-Polanco *et al.*, 2020). On green twigs and fruit pedicels, dark-brown to black irregular lesions (2–10 mm) can develop causing premature fruit drop.

Fig. 13.39. Black spot (*Cercospora*) symptoms on lower (a) and upper surface (b) of a leaf.

Fig. 13.40. Black spot caused by *Pseudocercospora purpurea* on fruit. The cutaway section shows sub-surface corking and limited penetration into the flesh.

Causal agent and epidemiology

Black spot is caused by the fungus *Pseudocercospora purpurea* (previously known as *Cercospora purpurea*). *P. purpurea* may be extremely difficult to isolate from fruit with advanced sunken lesions as it is slow growing and may be co-isolated and outgrown by *Colletotrichum* spp. (Darvas and Kotzé, 1987). However, once recovered it grows readily on standard media such as PDA. Isolations should be made from fresh young lesions following surface decontamination. It produces a tufted leathery growth, which is initially grey but later becomes brown to black.

Although a teleomorph (sexual stage of the fungus, *Mycosphaerella* sp.) of the pathogen may exist, it is rarely found and considered unimportant in the disease cycle (Menge and Ploetz, 2003). The fungus survives on old lesions present in leaves, and spore trapping studies have shown that multiple releases of spores (conidia) were strongly correlated with rainfall events in spring and summer of wet years, while in dry years the correlation is much weaker. No spores were trapped in dry periods (Schoeman and Kallideen, 2018; Schoeman and Manicom, 2000). Conidia are spread easily by wind, splashing rain, irrigation water, insects and movement of contaminated tools and equipment to initiate infections on new leaves and developing fruit (Schoeman and Kallideen, 2018). Humidity and leaf wetness are likely not significant factors involved with spore release (Schoeman and Kallideen, 2020). Fruit are most susceptible from 2 months after flowering (when greater than about 4 cm in diameter) through to harvest (Rodríguez-Polanco *et al.*, 2020). Penetration can be direct, or through wounds and after penetration, the pathogen remains latent for up to 3 months. Infections early in the season result in the highest disease incidence and severity at harvest.

Management

Timely applications of fungicides are important; the most useful model for commencing sprays includes weather data as well as fruit size and demonstrates that commencing protectant coverage when fruit are about 2.5 cm diameter is critical (Schoeman and Kallideen, 2020). Maintaining protection prior to predicted rainfall events and resulting spore release is necessary. Copper fungicides are very effective (Manicom and Schoeman, 2010), and recent trials with a new product with a lower copper concentration have been encouraging (A. Boyum, World Avocado Congress, Auckland, New Zealand, 2023, personal communication). An aggressive programme with combinations and rotation of protectant and systemic chemistry during the peak infection times reduced black spot incidence in Colombia to less than 10%, translating to a 24% higher economic return compared to growers using conventional programmes (Rodríguez-Polanco *et al.*, 2020). Care must be taken not to overuse these fungicides to avoid development of resistance within *Pseudocercospora* sp. populations, which has been reported in species of this genus affecting other crops, for example, banana (Chong *et al.*, 2021).

Field sprays with *Bacillus subtilis* alone or integrated with copper have been reported to reduce severity of black spot in South Africa (Korsten *et al.*, 1997). Insect control, root rot management and good orchard hygiene, which includes the removal of infected avocado prunings from the orchard, are all beneficial actions that assist with black spot control.

Pepper spot

Pepper spot is a disease of avocado most commonly associated with exposed or sunburnt fruit (Willingham *et al.*, 2000), or following extended warm, wet weather (Schoeman and Manicom, 2001). It is a superficial disease that does not affect the quality of pulp but results in downgrading at the pack shed. Its name comes from the symptoms, which are numerous very small, shiny and raised black spots (0.1–0.5 mm in diameter) on the surface of fruit (Fig. 13.41) and often also seen on pedicels, resembling sprinkling of black pepper. A 'tear stain' pattern of these black lesions on fruit is frequently observed (Fig. 13.42) (Hofer *et al.*, 2021).

Pepper spot is caused by *Colletotrichum* spp., described in this chapter under the 'Anthracnose' section. It is likely that the same species from the *gloeosporioides* complex can cause both anthracnose and pepper spot diseases (Giblin *et al.*, 2010; Giblin *et al.*, 2018; Hofer *et al.*, 2021), but this has not been determined for species within the *acutatum* or *boninense* complexes. The physiological and/or environmental mechanisms underlying infection and expression of the two symptom types, that is, field versus postharvest, have not been elucidated.

Inoculum sources for pepper spot in avocado are most likely the same as those for anthracnose (i.e. conidia produced on dead twigs and leaves entangled in the tree canopy, and washed onto fruit during periods of rain). A combination of temperatures over 18°C and rainfall exceeding a total of 20 mm for a 10- to 14-day period was associated with severe pepper spot development (Schoeman and Manicom, 2001).

Fig. 13.41. Pepper spot frequently occurs on 'Hass' fruit affected by sunburn.

Fig. 13.42. Severe pepper spot can occur in poorly managed orchards, and a 'tear stain' pattern may be observed when multiple infections by *Colletotrichum* sp. occurs from spore-laden drops running down the fruit from the stem end.

Pepper spot can be managed by similar methods as described for anthracnose. Maintaining leaf cover over fruit for sunburn protection is also important and is achieved by keeping trees in good health. Application of sun or stress protectant products to prevent sunburn may also be effective but can be difficult to remove after harvest. Selecting cultivars, known to carry fruit within the canopy, may be an option in regions prone to fruit sunburn.

Scab

Avocado scab causes aesthetic damage to the fruit surface as well as premature abortion of young, infected fruit (Marais, 2004). Several important cultivars such as 'Hass', 'Fuerte', 'Carmen', 'GEM' and 'Lula' are susceptible to the disease (Correa Suárez, 2019; Crane *et al.*, 2020a; Belizaire, 2023). This disease also causes serious postharvest losses because cracks created in the fruit skin provide entry points for other fungal pathogens such as *Colletotrichum* spp. (Pernezny and Marlatt, 2000; Menge and Ploetz, 2003). Significant postharvest losses, up to 50–60%, may occur due to rejection of fruit with scab symptoms in packing houses (Morales-García *et al.*, 2023).

The geographic distribution of *Elsinoe perseae* (the pathogen) and its association with avocado scab symptoms have only been confirmed (i.e. with morphological and genetic identification) in the USA (Florida), Brazil and México (Fan *et al.*, 2017; Correa Suárez, 2019; López Cornejo, 2021; Gañán-Betancur and Gazis, 2023). Caution is suggested when considering the historical distribution records as some of the reports included illustrations depicting symptoms very similar to scab but do not fit the typical symptomatology of the disease. For example, Hartill (1991a) reported for the first time the presence of avocado scab in New Zealand based on fruit symptomatology and fungal isolations mistakenly identified as *Sphaceloma perseae* (asexual state of *E. perseae*). Some years later, this record was sufficient to temporarily suspend exports of fruit to Australia, known to be free of the disease. A significant effort was mounted to accumulate evidence, through molecular assays and testing over 1000 samples with suspect symptoms showing that *E. perseae* was not present in New Zealand and that the 1991 record of avocado scab in the country was incorrect (Everett *et al.*, 2011). That work demonstrated the importance of confirming the causal agent as *E. perseae* to document the presence of avocado scab within a country. Avocado scab is still considered a high priority biosecurity threat for avocado production in Australia and New Zealand and maintaining molecular testing capacity for *E. perseae* is crucial to rapidly screen any suspect samples enabling a rapid response in case of a positive detection (Everett and Siebert, 2018; Parkinson and Geering, 2019).

Symptoms

Symptoms are commonly observed on young host tissue such as new leaves, fruit, young stems including twigs and the fruit peduncle (Pernezny and Marlatt, 2000; Menge and Ploetz, 2003) (Fig. 13.43). On fruit, initial symptoms are scattered across the surface and often observed as raised oval to round lesions, black, brown or purple in colour, giving a sandpaper-like surface (Pohronenzy and Simone,

Fig. 13.43. Developing fruit and pedicel of the cultivar 'Lula' severely affected by scab.

Fig. 13.44. Typical symptoms of scab infection on the avocado cultivar 'Lima Late'.

Fig. 13.45. Avocado fruit of the cultivar 'Lula' severely affected by scab. Lesions have coalesced and generate fissures on the peel. Photo courtesy of L. Gañán-Betancur, Tropical Research and Education Center, University of Florida, Florida, USA.

1994). As lesions enlarge, they can coalesce leading to larger areas of fruit tissue with corky or scab-like appearance (Fig. 13.44). In advanced stages of disease development, several cracks or splits can be observed on the surface of fruit (Fig. 13.45) (Burnett, 1974; Palmateer *et al.*, 2006; López Cornejo, 2021). Early scab symptoms on leaves appear as small circular chlorotic lesions of 1–3 mm in diameter that can be easily observed when turning the leaves against the sunlight and often start along the midrib and veins (Fig. 13.46). As the disease progresses, these lesions turn brown or black and develop as hard corky circular lesions (Palmateer *et al.*, 2006). Infected leaves become distorted as dead tissue is unable to expand (Fig. 13.47). Lesions may coalescence and late symptoms can be similar to shot holes (Pernezny and Marlatt, 2000; Everett *et al.*, 2007a; Correa Suárez, 2019). On avocado fruit peduncles, leaf petioles, twigs and fruit pedicels, scab lesions may also occur appearing as circular black, rough corky tissue.

Given the cork-like appearance of older infected fruit tissue, scab symptoms can be easily confused with similar fruit defects caused by aetiologies other than *E. perseae*. Cork or scab-like lesions on fruit can also be caused by mechanical injuries resulting from fruit rubbing or wind damage. Superficial scraping on the fruit surface after insect feeding, including damage caused by sucking (e.g. thrips) and chewing (e.g. beetles within the Melolonthidae complex) insects, can result in scars that resemble advanced scab symptoms. For example, scab-like lesions observed on 'Hass' fruit in Colombia, resulting after insect feeding, were erroneously attributed to *E. perseae* infection (Lynce-Duque, 2015). This might also be the reason why Ávila-Quezada *et al.* (2003) concluded that there was a direct correlation between scab symptoms and thrips damage in 'Hass' avocado in México. A detailed diagnostic guide has recently been published (Belizaire *et al.*, 2023).

Fig. 13.46. Scab lesions caused by *Elsinoe perseae* on an avocado leaf with concentration of smaller lesions along veins. Photo courtesy of K. Everett, Plant and Food Research, Auckland, New Zealand.

Fig. 13.47. Distorted foliage of the avocado cultivar 'Lula' infected with *Elsinoe perseae*.

Causal agent

Avocado scab is caused by the host-specific ascomycetous fungus, *Elsinoe perseae*, previously known as the basionym (original name) *Sphaceloma perseae*. Pathogen isolation is challenging because its slow growth rate allows fungal contaminants such as saprotrophs and endophytes to overgrow the colonies. Due to the lack of diagnostic characters and their documented high host specificity (Ujat *et al.*, 2023), species of *Elsinoe* are often differentiated according to the host in which the symptoms are observed. However, host identity and symptoms should only be considered a part of the diagnostic process. A second species of *Elsinoe* has recently been isolated from symptomatic avocado fruit in Florida (R. Gazis, Homestead, USA, 2023, personal communication). Therefore, it is important to confirm the identity of the causal agent through morphological and molecular approaches.

There are published molecular techniques for identifying *E. perseae*. A PCR-based assay using the species-specific primers, SpF5 and SpR6, can be used to rapidly identify isolates as *E. perseae* and to detect the pathogen from DNA extracted from symptomatic tissue (Everett *et al.*, 2011). There is currently one reference genome available for *E. perseae* (ASM2944869v1; Gañán-Betancur and Gazis, 2023), which will facilitate the screening of additional detection markers with increased sensitivity.

Epidemiology

The pathogen is spread by the dispersal of spores produced in early stages of infection, which are moved by wind, rain, insects, tools and equipment (Marais, 2004; Palmateer *et al.*, 2006). *Elsinoe perseae* is a pathogen of young tissue. Leaves become resistant 1 month after emergence and fruit become resistant once they are half-grown (Pohronenzy and Simone, 1994). Disease is most severe when susceptible host tissue is exposed to heavy rains or foggy weather with relative humidity above 80% (Menge and Ploetz, 2003). The degree of susceptibility towards scab varies among avocado cultivars (Crane *et al.*, 2020a; Belizaire *et al.*, 2023). Some avocado cultivars can display scab symptoms in leaves, fruit and stems, whereas other cultivars only show disease symptoms on fruit. Some cultivars appear to be resistant to pathogen infection, even when planted near highly susceptible cultivars under high disease pressure. The mechanisms of the resistance to *E. perseae* have not yet been elucidated.

Avocado fruit infected with scab are the main source of inoculum for the following growing season. Therefore, cultural practices, such as elimination of diseased fruit from orchards, are recommended to reduce pathogen inoculum in the field (Palmateer *et al.*, 2006). As the disease is favoured by cool, moist weather and heavy rain and fog, regular pruning to increase aeration within the canopy is recommended to reduce the infection and spread of the pathogen. Currently, chemical management of avocado scab relies largely on sprays of copper-based fungicides (Marais, 2004). Spray applications often start during the first period of flowering, and additional applications are advised during rainy periods when the relative humidity is higher than 70% (Menge and Ploetz, 2003).

Stem-end rot

Stem-end rot is a common postharvest disease of avocado, and although considered less important than anthracnose, can cause serious losses in some situations. It can be caused by a range of fungal pathogens and can be confused with physiological disorders such as vascular browning.

Symptoms

Symptoms usually do not develop until fruit ripens. As the name implies, rotting starts at the stem-end from where it progresses throughout the fruit. External symptoms are obvious in green-skinned varieties as a brown rot with defined margins uniformly progressing from the stem towards the distal end (Fig. 13.48). This may be difficult to see in ripe 'Hass' fruit or other cultivars that turn dark purple/black upon ripening. Mycelial growth is sometimes seen on the surface of skin lesions, particularly during advanced stages of symptom development and under humid storage conditions. Depending on causal agent, pink-orange (*C. gloeosporioides* complex, Fig. 13.48) or dark orange-red (*C. fioriniae*, Fig. 13.49) spore masses or white-grey fluffy mycelium (e.g. Botryosphaeriaceae fungi) may be seen at the stem end, on or around the pedicel abscission scar, although it is not unusual to see both in overripe fruit.

Fig. 13.48. Stem-end rot in a green-skin variety, with pink spore masses on the peel characteristic of infection by *Colletotrichum* sp. within the *C. gloeosporioides* complex.

Internally, symptoms of a brown watery flesh rot progresses rapidly from the stem end and the vascular tissue often becomes discoloured, typically in advance of flesh decay symptoms (Fig. 13.50). A very dark grey or black discolouration is typical of stem-end rot caused by Botryosphaeriaceae fungi (Fig. 13.51).

Causal agents and epidemiology

Fungi in the family Botryosphaeriaceae are commonly responsible for stem-end rot and include those which also cause cankers and branch dieback, reviewed in Hernández *et al.* (2023). Genera within this family have undergone several taxonomic re-classifications and re-naming within the last 10 years. The most recently reported names will be discussed here including species previously known within *Dothiorella*, *Botryosphaeria* and *Fusicoccum* genera. The most common species include *Neofusicoccum luteum*, *Neofusicoccum parvum*, *Neofusicoccum australe*, *Lasiodiplodia theobromae*, *Lasiodiplodia pseudotheobromae* and *Diplodia* spp. (Twizeyimana *et al.*, 2013a; Guarnaccia *et al.*, 2016; Valencia *et al.*, 2019; Hernández *et al.*, 2023). Other genera frequently causing stem-end rot include *Colletotrichum* spp., *Diaporthe* spp. (*Phomopsis* spp.) (Hartill

Fig. 13.49. Dark orange-red spore masses at the stem end of Hass fruit infected by *Colletotrichum fiorinae* (*C. acutatum* complex).

Fig. 13.50. Stem-end rot and vascular discolouration typical of *Colletotrichum* sp. infection.

and Everett, 2002; Twizeyimana *et al.*, 2013a), *Neopestalotiopsis clavispora* (reported as the former name *Pestalotiopsis clavispora*) and other *Pestalotiopsis* spp. (Valencia *et al.*, 2011; Maharachchikumbura *et al.*, 2014). Interestingly, while fungal pathogens of plants rarely cause disease in humans, there is a report of *P. clavispora* infecting the cornea of a gardener's eye, causing keratitis (Monden *et al.*, 2013). The patient had a history of keratitis and eye surgery, and the infection was treated with a medical fungicide.

Fig. 13.51. Dark grey-black symptom of stem-end rot caused by infection by fungi within the Botryosphaeriaceae family.

Many of the stem-end rot pathogens occur as symptomless (endophytic or latent) infections, and they have been isolated from surface-sterilized healthy tissue of avocado terminal shoots and fruit pedicels (Peterson, 1978; Johnson and Kotzé, 1994; Hartill and Everett, 2002). While this colonization may contribute to stem-end rot, the detailed study by Hartill and Everett (2002) suggests that infection of the wound site at harvest by airborne spores or mycelial fragments is more likely. Flower tissues are also abundantly colonized by several fungal genera including known avocado pathogens *Colletotrichum* sp., Botryosphaeriaceae and *Diaporthe* (reported as *Phomopsis* sp.) (Everett *et al.*, 1999). Application of fungicides during flowering reduced stem-end rot severity and also the frequency of isolation of these fungi from diseased fruit. Spores produced by various stem-end rot fungi on dead leaves,

Fig. 13.52. Infection stocking of brown root rot, caused by *Pyrrhoderma noxium*, advancing up the trunk of a tree. The white mycelial margin indicates active fungal growth and is often observed after warm, high-rainfall conditions.

Fig. 13.53. Superficial fungal growth on young branch caused by sooty blotch.

Fig. 13.54. Bacterial soft rot showing complete blackening and breakdown of affected tissues.

twigs and branches are the likely sources of primary infection (Darvas *et al.*, 1987; Hartill and Everett, 2002; Eskalen *et al.*, 2013a). Stem-end rot developing during ripening may be more severe if fruit are harvested during wet conditions, which may be associated with peaks in spore release (Eskalen *et al.*, 2013a) or severe drought, for example, in Chile 2011–2016 (Valencia *et al.*, 2019).

Management

An integrated strategy similar to that described for anthracnose, and other fruit-infecting fungi, will reduce postharvest losses attributed to stem-end rot. Given that many of the stem-end rot fungi also cause branch cankers and dieback, it is important that the latter are well controlled by regular pruning, removing inoculum sources and generally maintaining good tree health, particularly avoiding water stress. Field sprays of fungicides during flowering and throughout fruit development are effective. Harvest fruit, at optimum maturity, avoid harvesting fruit during wet weather and maintain appropriate cool chain management. Postharvest fungicides are also effective and some new treatments applied postharvest such as electrolysed oxidizing water (a sanitizer approved for use in organic agriculture) look promising (Hassan and Dann, 2019).

Table 13.2. Minor diseases of avocado.

Disease by tissue affected	Causal agent/s	Symptoms and management	Reference
Roots/whole tree			
Brown root rot	*Pyrrhoderma noxium* (syn. *Phellinus noxius*) (fungus)	Sudden wilting of leaves, shrivelling of fruit (remain attached), then tree death. Slow progression of consecutive death of trees along row, due to root-to-root spread. A thick brown mycelial crust or 'stocking' usually develops on the outer surface of the roots and may or may not be present advancing up the tree trunk. An actively growing white mycelial margin on the stocking frequently develops during peak fungal growth during wet, warm tropical and subtropical conditions (Fig. 13.52). The wood underneath the stocking is discoloured reddish brown, with white mottled patches. The infection stocking eventually turns black, and in advanced stages of decay the wood becomes light, dry and friable with a honeycombed appearance. As the fungus can survive for many years in root debris in the soil, affected trees and all root material should be removed. Also remove apparently healthy trees on either side of the infected tree, and install root barriers to isolate the affected sites. Replanted trees often die within a couple of years if root debris is not completely removed. Do not mulch with diseased wood. Restricted to avocado in Taiwan and tropical and subtropical regions of Australia, although it has a very wide range of woody hosts so potentially a problem in other countries.	(Ann *et al.*, 2002; Dann *et al.*, 2009b; Garfinkel *et al.*, 2019)
Potato spindle tuber viroid	Potato spindle tuber viroid	Chlorotic and stunted leaves with tip necrosis, defoliation, bunched inflorescence ('witches broom' type symptom), small or seedless fruit, erect vertical branching pattern with weak, slender branches, tree death. Trees testing positive can be symptomless. A likely consequence of intercropping avocados with potatoes. Has a wide host range of mostly vegetables and ornamentals.	(Querci *et al.*, 1995)
Nematodes	*Pratelynchus vulnus* *Radopholus similis* *Meloidogyne hapla*	Stunting, reduction of feeder roots, tree decline.	(Sher, 1959) (Knight, 2001)
Trunk, branches, stems			
Duke 6 stem pitting	Unknown	First observed in South Africa associated with Duke 6 rootstocks but also on 'Hass' and others. Initial symptoms are leaf wilt, wilted leaves then abscise leaving trees severely defoliated and prone to sunburn. Stunting, defoliation, rapid tree decline and collapse of young trees. When bark of the rootstock was peeled away clean indentations or 'pits', approximately 10–15 mm, were observed, with corresponding protuberances (bumps) on the outer bark. Mosaic patterns evident on leaves. Virus or fastidious bacterium suggested. Drought stress triggered the disease and under ideal growing conditions relatively mild symptoms developed. It was graft-transmissible but insect transmission was not demonstrated. The disease has not been detected again and may no longer exist.	(Moll *et al.*, 1987)

Continued

Table 13.2. Continued.

Disease by tissue affected	Causal agent/s	Symptoms and management	Reference
Gall/tumour	*Sphaeropsis tumefaciens* (fungus)	Galls up to 8 cm in diameter of various shapes on branches and stems of old or young trees. Worse in poorly managed orchards.	(Fucikovsky, 1992)
Leaves			
Algal spot or 'red rust disease'	*Cephaleuros virescens* (algae)	Small areas of yellow-green to grey growth on the upper surfaces of leaves, twigs and less frequently fruit. Becomes 'velvety' and turns a rusty-red colour as spore structures form. Favoured by warm temperatures and high rainfall. Can reduce photosynthetic area of leaves, but not usually a problem in well-managed orchards. Canopy management to improve air circulation and copper-based fungicide application assists control.	(Joubert and Rijkenberg, 1971; Vasconcelos *et al*., 2016)
Leaf scorch	*Xylella fastidiosa* (bacterium)	Chlorotic mottling, marginal scorch and deformation of leaves, branch dieback, shortening of internodes, water stress-like symptoms due to xylem blockage. Likely vectored by the glassy-winged sharpshooter *Homalodisca vitripennis,* although other leafhoppers and spittlebugs vector the disease in other crops. Has not become a major problem in avocados but is a listed biosecurity threat for some countries, for example, Australia and New Zealand.	(California Avocado Commission, 2003; Montero-Astúa *et al*., 2008; Geering and Parkinson, 2019)
Tar spot	*Phyllachora gratissima, Ph. perseae* (fungi)	A minor but very common disease in the Americas. Light- to dark-brown lesions on leaves up to 1 cm in diameter that ultimately develop raised black fruiting bodies often surrounded by chlorotic haloes. In severe cases leaves become chlorotic and abscise. Yield not reduced as old leaves are most severely affected. Windy and wet conditions, especially in the spring, favour disease development. When severe it can be managed by removing leaves (inoculum source) from the ground beneath trees and applying fungicide sprays beginning with the appearance of new foliage.	(Hodges, 1969; Menge and Ploetz, 2003)
Vascular streak dieback	*Rhizoctonia theobromae* (syn. *Ceratobasidium theobromae Oncobasidium theobromae*) (fungus)	Observed in seedlings associated with cocoa plantings. Causes vascular necrosis of leaf scars and stems, enlarged lenticels, sickle-shaped leaves with chlorotic flecking and necrosis of leaf margins. White fruiting bodies on leaf scars and stems, white hyphae in xylem vessels.	(Anderson, 1989)
Powdery mildew	*Oidium* sp., *Erysiphe* sp. *Podosphaera perseae-americanae* *Oidium perseae-americanae* (syn. *Pseudoidium perseae-americanae*) (fungi)	Can be severe during dry, cool periods in nurseries and orchards. Dark watery discolouration on the leaf surface, especially along midrib and leaves may be curled or stunted. White, powdery mycelium and spores of the fungus on underside of leaves, which develop into purple-brown blotches. Flush leaves often infected and terminal shoots may be killed. Differential cultivar susceptibility.	(Liberato and Barreto, 2006; Braun and Cook, 2012; Siahaan *et al*., 2016; Suárez *et al*., 2018)

Silver spot	*Mycosphaerella perseae* (fungus)	Mostly confined to older leaves. First appears as small yellow-green spots that later expand into slightly depressed, grey to reddish brown lesions on upper surface, darker brown on lower surface. Lesions can coalesce to cover large areas of leaves. A dark margin may surround lesions and fungal fruiting bodies are seen as small black dots in the old lesions. Symptoms are most severe during wet periods and can cause severe defoliation. Managed by orchard hygiene, canopy pruning for airflow and fungicides.	(Quintero and Sebastian, 1972; Menge and Ploetz, 2003;)
Shoot blight and dieback, stem lesions, graft dieback	*Pestalotiopsis longiseta* *Neopestalotiopsis* spp. (fungi)	Blight on green shoots then necrotic lesions on stems, leaves and dieback of shoots and twigs. Seen on nursery plants and young trees.	(Lin *et al*., 2018; Fiorenza *et al*., 2022)
Fruit			
Sooty blotch	*Stomiopeltis* sp. and *Akaropeltis* sp. (fungi)	Sooty or smoky blotches on the surfaces of branches, stems, leaf veins and fruit, eventually becoming darker and thicker and the affected area turns charcoal black (Fig. 13.53). Infects tissues superficially during warm, wet and humid conditions. If severe, it can cover leaf surfaces and reduce photosynthesis, and result in rejection of fruit.	(Smith *et al*., 1985; Pieterse, 1986)
Sooty mould	*Capnodium* spp. (fungi)	Sooty mould is a general term for the dense, dark mouldy coatings of fungi that live saprophytically on honeydew secretions produced by sap-sucking insects, for example, scales, aphids, planthoppers and flatids. Black, somewhat 'velvety' growth on the surface of fruit, leaves and stems. If severe it can cover leaf surfaces and reduce photosynthesis and downgrade fruit. Managed by controlling insects that produce honeydew and reducing humidity in tree canopies by pruning, etc. Usually washes off fruit in the packing house.	(Du Toit and de Villiers, 1988; Eskalen and Faber, 2016d)
Bacterial soft rot	*Pectobacterium caratovorum* (previously *Erwinia carotovora*) *Pectobacterium aroidearum*	Sunken dark spots or patches on skin, initially with a metallic sheen. Skin develops cracks and breaks apart. Flesh is grey-brown initially then becomes soft and liquified with or without a putrid odour (Fig. 13.54). Occasionally occurs in the field but more common postharvest in fruit harvested in the rain. More prevalent in wet subtropical and tropical production areas, particularly after severe storm and high-wind conditions. Over-mature fruit and green-skin varieties highly susceptible. The bacterium invades through wounds or lesions caused by other diseases. Avoid picking in the rain and prevent skin injury, remove affected fruit from the packing line and use disinfectant wash.	(Volcani, 1959; Allen, 1985; Pathania and Grice, 2021)

References

Abad, Z.G., Burgess, T.I., Redford, A.J., Bienapfl, J.C., Srivastava, S., Mathew, R. and Jennings, K. (2023) IDphy: an international online resource for molecular and morphological identification of *Phytophthora*. *Plant Disease* 107, 987–998.

Abdulridha, J., Ampatzidis, Y., Ehsani, R. and de Castro, A.I. (2018) Evaluating the performance of spectral features and multivariate analysis tools to detect laurel wilt disease and nutritional deficiency in avocado. *Computers and Electronics in Agriculture* 155, 203–211.

Abdulridha, J., Ehsani, R. and de Castro, A. (2016) Detection and differentiation between laurel wilt disease, Phytophthora disease, and salinity damage using a hyperspectral sensing technique. *Agriculture* 6:56.

Aiello, D., Gusella, G., Vitale, A., Guarnaccia, V. and Polizzi, G. (2020) *Cylindrocladiella peruviana* and *Pleiocarpon algeriense* causing stem and crown rot on avocado (*Persea americana*). *European Journal of Plant Pathology* 158, 419–430.

Al-Sa'di, A.M., Drenth, A., Deadman, M.L., Al-Said, F.A., Khan, I. and Aitken, E.A.B. (2008) Association of a second phase of mortality in cucumber seedlings with a rapid rate of metalaxyl biodegradation in greenhouse soils. *Crop Protection* 27, 1110–1117.

Allen, R.N. (1985) *Avocado Diseases. Agfact H6. AB.5*. 2nd edn. Department of Agriculture, New South Wales, Australia, p. 7.

Alvarado-Rosales, D., Elías-Román, R.D., Saavedra-Romero, L.d.L., Michua-Cedillo, J., Ochoa-Ascencio, S. *et al.* (2022) New hosts for *Desarmillaria caespitosa* and *Armillaria mexicana* in Veracruz and Michoacán, México. *Chapingo Magazine Forestry and Environmental Sciences Series* 29, 51–59.

Anderson, R.D. (1989) Avocado, an alternate host for *Oncobasidium theobromae*. *Australasian Plant Pathology* 18, 96–97.

Ángel-Restrepo, M., Parra, P.P., Ochoa-Ascencio, S., Fernández-Pavía, S., Vázquez-Marrufo, G. *et al.* (2022) First look into the ambrosia beetle-fungus symbiosis present in commercial avocado orchards in Michoacán, Mexico. *Environmental Entomology* 51, 385–396.

Ann, P.J., Chang, T.T. and Ko, W.H. (2002) *Phellinus noxius* brown root rot of fruit and ornamental trees in Taiwan. *Plant Disease* 86, 820–826.

Aoki, T., Smith, J.A., Kasson, M.T., Freeman, S., Geiser, D.M., Geering, A.D. and O'Donnell, K. (2019) Three novel Ambrosia Fusarium Clade species producing clavate macroconidia known (*F. floridanum* and *F. obliquiseptatum*) or predicted (*F. tuaranense*) to be farmed by *Euwallacea* spp. (Coleoptera: Scolytinae) on woody hosts. *Mycologia* 111, 919–935.

Aoki, T., Liyanage, P.N., Konkol, J.L., Ploetz, R.C., Smith, J.A. *et al.* (2021) Three novel Ambrosia *Fusarium* Clade species producing multiseptate 'dolphin-shaped' conidia, and an augmented description of *Fusarium kuroshium*. *Mycologia* 113, 1089–1109.

Arjona-Girona, I., Ruano-Rosa, D. and López-Herrera, C. (2019) Identification, pathogenicity and distribution of the causal agents of dieback in avocado orchards in Spain. *Spanish Journal of Agricultural Research* 17:1003.

Arjona-Lopez, J.M., Capote, N. and Lopez-Herrera, C.J. (2019) Improved real-time PCR protocol for the accurate detection and quantification of *Rosellinia necatrix* in avocado orchards. *Plant and Soil* 443, 605–612.

Arjona-López, J.M., Capote, N., Melero-Vara, J.M. and López-Herrera, C.J. (2020) Control of avocado white root rot by chemical treatments with fluazinam in avocado orchards. *Crop Protection* 131, 105100.

Arjona-López, J.M. and López-Herrera, C.J. (2021) *Entoleuca* sp. infected by mycoviruses as potential biocontrol agents of avocado white root rot. *European Journal of Plant Pathology* 159, 409–420.

Arjona-López, J.M., Telengech, P., Suzuki, N. and López-Herrera, C.J. (2021) A moderate level of hypovirulence conferred by a hypovirus in the avocado white root rot fungus, *Rosellinia necatrix*. *Fungal Biology* 125, 69–76.

Auger, J., Palma, F., Pérez, I. and Esterio, M. (2013) First report of *Neofusicoccum australe* (*Botryosphaeria australis*), as a branch dieback pathogen of avocado trees in Chile. *Plant Disease* 97:842.

Aveling, T.A.S. and Rijkenberg, F.H.J. (1989) Behaviour of *Phytophthora cinnamomi* zoospores on roots of four avocado cultivars. *Journal of Phytopathology* 125, 157–164.

Aveling, T.A.S. and Rijkenberg, F.H.J. (1991) A quantitative study of *Phytophthora cinnamomi* zoospore encystment and germination of the roots of four avocado cultivars. *Phytophylactica* 23, 229–231.

Avenot, H.R., Davis, J., Greer, C., Faber, B., Arpaia, M.-L. and Michailides, T.J. (2021) Prevalence, impact and significance of Botryosphaeria fungi in avocado nurseries and new plantings in California. *From the Grove*, Winter 2021, 38–48.

Ávila-Quezada, G., Téliz-Ortiz, D., Mora-Aguilera, G., Vaquera Huerta, H. and Tijerina-Chávez, L. (2003) Spatial and temporal dynamic of scab (*Sphaceloma perseae* Jenk.) on avocado (*Persea americana* Mill.). *Revista Mexicana de Fitopatología* 21, 152–160.

Belisle, R.J., Hao, W., McKee, B., Arpaia, M.L., Manosalva, P. and Adaskaveg, J.E. (2019) New oomycota fungicides with activity against *Phytophthora cinnamomi* and their potential use for managing avocado root rot in California. *Plant Disease* 103, 2024–2032.

Belizaire, C.M. (2023) Characterization of the avocado scab pathogen *Elsinoe perseae* and its sensitivity to synthetic and natural fungicides. M.S., University of Florida. Available at: https://ufl-flvc.primo.exlibrisgroup.com/permalink/01FALSC_UFL/6ad6fc/alma99384164120506597 (accessed 8 October 2024).

Belizaire, C.M., Gañán-Betancur, L. and Gazis, R. (2023) Avocado scab caused by *Elsinoe perseae*: a diagnostic guide. *Plant Health Progress* 25(2), 218-225.

Besoain, X.C. and Piontelli, E.L. (1999) Pudricion negra en raicillas de palto (*Persea americana* Mill) por *Cylindrocarpon destructans:* patogenicidad y aspectos epidemiologicos (Black root rot in avocado plants (*Persea americana* Mill) by *Cylindrocarpon destructans:* pathogenicity and epidemiological aspects). *Boletin Micologico* 14, 41–47.

Binyamini, N. and Schiffmann-Nadel, M. (1972) Latent infection in avocado fruit due to *Colletotrichum gloeosporioides*. *Phytopathology* 62, 592–594.

Bonilla, N., Vida, C., Martínez-Alonso, M., Landa, B.B., Gaju, N., Cazorla, F.M. and De Vicente, A. (2015) Organic amendments to avocado crops induce suppressiveness and influence the composition and activity of soil microbial communities. *Applied and Environmental Microbiology* 81, 3405–3418.

Bonnéry, D.B., Pretorius, L.-S., Jooste, A.E.C., Geering, A.D.W. and Gilligan, C.A. (2023) Rational design of a survey protocol for avocado sunblotch viroid in commercial orchards to demonstrate pest freedom. *PLOS One* 18:e0277725.

Brar, G.S., Capinera, J.L., Kendra, P.E., McLean, S. and Peña, J.E. (2013) Life cycle, development, and culture of *Xyleborus glabratus* (Coleoptera: Curculionidae: Scolytinae). *Florida Entomologist* 96, 1158–1167.

Braun, U. and Cook, R.T.A. (2012) Taxonomic Manual of the Erysiphales (Powdery Mildews). *CBS Biodiversity Series*. CBS-KNAW Fungal Biodiversity Centre, Utrecht, The Netherlands, pp 707.

Broadbent, P. and Baker, K.F. (1974) Behaviour of *Phytophthora cinnamomi* in soils suppressive and conducive to root rot. *Australian Journal of Agricultural Research* 25, 121–137.

Broadbent, P., Trochoulias, T., Baigent, D.R., Abbott, T.S. and Dettmann, E.B. (1989) Effect of soil management on avocados in a krasnozem soil. *Scientia Horticulturae* 38, 87–104.

Brokaw Nursery (2024) *Rootstocks,* online. Available at: https://www.brokawnursery.com/ourproducts/avocado/rootstocks/ (accessed 10 January 2024).

Burgess, T.I., López-Villamor, A., Paap, T., Williams, B., Belhaj, R. *et al.* (2021) Towards a best practice methodology for the detection of *Phytophthora* species in soils. *Plant Pathology* 70, 604–614.

Burnett, H.C. (1974) Avocado scab. Plant Pathology Circular No. 141. Available at: https://ccmedia.fdacs.gov/content/download/11148/file/pp141.pdf (accessed 9 July 2023).

Bustamante, M.I., Osorio-Navarro, C., Fernández, Y., Bourret, T.B., Zamorano, A. and Henríquez-Sáez, J.L. (2022) First record of *Colletotrichum anthrisci* causing anthracnose on avocado fruits in Chile. *Pathogens* 11:1204.

Cacciola, S.O. and Gullino, M.L. (2019) Emerging and re-emerging fungus and oomycete soil-borne plant diseases in Italy. *Phytopathologia Mediterranea* 58, 451–472.

California Avocado Commission, C.M.C.C. (2003) A pest management strategic plan for avocado production in California. Available at: https://ipmdata.ipmcenters.org/documents/pmsps/CAavocado.pdf (accessed 2 July 2023).

Cameron, R.S., Hanula, J., Fraedrich, S. and Bates, C. (2015) Progression and impact of laurel wilt disease within redbay and sassafras populations in southeast Georgia. *Southeastern Naturalist* 14, 650–674.

Campbell, A.S., Ploetz, R.C. and Rollins, J.A. (2017) Comparing avocado, swamp bay and camphortree as hosts of *Raffaelea lauricola* using a green fluorescent protein (GFP)-labeled strain of the pathogen. *Phytopathology* 107, 70–74.

Carrillo, J.D. and Eskalen, A. (2017) Avocado black streak disease revisited: an unsolved mystery. *From the Grove* 7(Fall), 36–38.

Carrillo, D., Duncan, R.E. and Peña, J.E. (2012) Ambrosia beetles (Coleoptera: Curculionidae: Scolytinae) that breed in avocado wood in Florida. *Florida Entomologist* 95, 573–579.

Carrillo, D., Duncan, R.E., Ploetz, J.N., Campbell, A.F., Ploetz, R.C. and Peña, J.E. (2014) Lateral transfer of a phytopathogenic symbiont among native and exotic ambrosia beetles. *Plant Pathology* 63, 54–62.

Carrillo, J.D., Dodge, C., Stouthamer, R. and Eskalen, A. (2020a) Fungal symbionts of the polyphagous and Kuroshio shot hole borers (Coleoptera: Scolytinae, *Euwallacea* spp.) in California can support both ambrosia beetle systems on artificial media. *Symbiosis* 80, 155–168.

Carrillo, J.D., Eskalen, A., Rooney-Latham, S. and Scheck, H.J. (2016) First report of *Neofusicoccum nonquaesitum* causing branch canker and dieback of avocado in California. *Plant Disease* 100:1778.

Carrillo, J.D., Mayorquin, J.S., Stajich, J.E. and Eskalen, A. (2020b) Probe-based multiplex real-time PCR as a diagnostic tool to distinguish distinct fungal symbionts associated with *Euwallacea kuroshio* and *Euwallacea whitfordiodendrus* in California. *Plant Disease* 104, 227–238.

Chong, P., Essoh, J.N., Arango Isaza, R.E., Keizer, P., Stergiopoulos, I. *et al.* (2021) A world-wide analysis of reduced sensitivity to DMI fungicides in the banana pathogen *Pseudocercospora fijiensis*. *Pest Management Science* 77, 3273–3288.

Coates, L.M., Irwin, J.A.G. and Muirhead, I.F. (1993a) The use of a benomyl-resistant mutant to demonstrate latency of *Colletotrichum gloeosporioides* in avocado fruit. *Australian Journal of Agricultural Research* 44, 763–772.

Coates, L.M., Muirhead, I.F., Irwin, J.A.G. and Gowanlock, D.H. (1993b) Initial infection processes by *Colletotrichum gloeosporioides* on avocado fruit. *Mycological Research* 97, 1363–1370.

Cohen, S.P. and Leach, J.E. (2020) High temperature-induced plant disease susceptibility: more than the sum of its parts. *Current Opinion in Plant Biology* 56, 235–241.

Coleman, T.W., Poloni, A.L., Chen, Y., Thu, P.Q., Li, Q. *et al.* (2019) Hardwood injury and mortality associated with two shot hole borers, *Euwallacea* spp., in the invaded region of southern California, USA, and the native region of Southeast Asia. *Annals of Forest Science* 76:61.

Cooksey, D.A., Ohr, H.D., Azad, H.R., Menge, J.A. and Korsten, L. (1993) *Xanthomonas campestris* associated with avocado canker in California. *Plant Disease* 77, 95–99.

Correa Suárez, M. (2019) Identificación molecular y caracterización morfológica de *Sphaceloma perseae* en la franja aguacatera de Michoacán. M.S. in Biological Sciences, Universidad Michoacana de San Nicolas de Hidalgo.

Crane, J.H., Balerdi, C.F. and Maguire, I. (2020a) Avocado growing in the Florida home landscape, online. University of Florida. Available at: https://edis.ifas.ufl.edu/publication/MG213 (accessed 8 October 2024).

Crane, J.H., Carrillo, D., Evans, E.A., Gazis, R., Schaffer, B.A., Ballen, F. and Wasielewski, J. (2020b) *Recommendations for control and mitigation of laurel wilt and ambrosia beetle vectors in commercial avocado groves in Florida: HS1360, 3/2020*, Online. University of Florida. Available at: https://trec.ifas.ufl.edu/media/trecifasufledu/fruitent-daniel-carillo/Crane-et-al-LW-recommendations-EDIS.pdf (accessed 8 October 2024).

Crane, J.H., Ploetz, R.C., White, T., Krogstad, G.C., Prosser, T., Konkal, J. and Wideman, R. (2015) Efficacy of three macroinfused fungicides to control laurel wilt on avocado in Martin and Brevard counties. *Proceedings of the Florida State Horticultural Society* 128, 58–60.

Crane, J.H., Wasielewski, J., Carrillo, D., Evans, E.A., Gazis, R. *et al.* (2019) Effect of the laurel wilt epidemic on Florida's avocado industry. In: *Proceedings of the IX World Avocado Congress*, Medellín, Colombia.

Crowley, D. (2007) Managing soils for avocado production and root health. *California Avocado Society Yearbook* 90, 107–130.

Cruz, L.F., Menocal, O., Kendra, P.E. and Carrillo, D. (2021) Phoretic and internal transport of *Raffaelea lauricola* by different species of ambrosia beetle associated with avocado trees. *Symbiosis* 84, 151–161.

Dann, E. (2019) Verticillium wilt in Western Australia this summer. *Talking Avocados* 30 (Autumn), 52–53.

Dann, E. and McLeod, A. (2021) Phosphonic acid: a long-standing and versatile crop protectant. *Pest Management Science* 77, 2197–2208.

Dann, E., Smith, L. and Pegg, K. (2009) Verticillium more severe in 2009. *Talking Avocados* 20, 32–33.

Dann, E.K., Coates, L.M., Dean, J.R., Smith, L.A., Cooke, A.W. and Pegg, K.G. (2009a) Effect of avocado crop load on postharvest anthracnose and stem end rot, and cations and phenolic acid levels in peel. *The Australasian Plant Pathology Society 2009: Plant Health Management: An Integrated Approach*, APPS, Newcastle, NSW, Australia.

Dann, E.K., Coates, L.M., Pegg, K.G., Dean, J.R., Cooke, A.W. *et al.* (2016) Rootstock selection, nitrogen and calcium influence postharvest disease in avocado. *Acta Horticulturae* 1120, 391–397.

Dann, E.K., Cooke, A.W., Forsberg, L.I., Pegg, K.G., Tan, Y.-P. and Shivas, R.G. (2012) Pathogenicity studies in avocado with three nectriaceous fungi, *Calonectria ilicicola*, *Gliocladiopsis* sp. and *Ilyonectria liriodendri*. *Plant Pathology* 61, 896–902.

Dann, E.K. and Le, D.P. (2017) Effects of silicon amendment on soilborne and fruit diseases of avocado. *Plants* 6:51.

Dann, E.K., Smith, L., Pegg, K., Grose, M. and Pegg, G. (2009b) Report on *Phellinus noxius*, the cause of brown root rot in Australian avocados. *Talking Avocados* 20, 28–34.

Darvas, J.M. (1978) Common root pathogens from avocados. *South African Avocado Growers' Association Research Report* 2, 3–4.

Darvas, J.M. (1979) Ecology of avocado root pathogens. *South African Avocado Growers' Association Research Report* 3, 31–32.

Darvas, J.M. (1982) Etiology and control of some fruit diseases of avocado (*Persea americana* Mill.) at Westfalia Estate. DSc (Agric) thesis, University of Pretoria, South Africa, p. 136.

Darvas, J.M. and Kotze, J.M. (1987) Fungi associated with pre- and postharvest diseases of avocado fruit at Westfalia Estate, South Africa. *Phytophylactica* 19, 83–85.

Darvas, J.M. and Kotzé, J.M. (1979) Cercospora spot of avocado. *South African Avocado Growers' Association Research Report* 3, 38–39.

Darvas, J.M., Kotzé, J.M. and Wehner, F.C. (1987) Field occurrence and control of fungi causing postharvest decay of avocados. *Phytophylactica* 19, 453–455.

De Castro, A.I., Ehsani, R., Ploetz, R.C., Crane, J.H. and Buchanon, S. (2015) Detection of laurel wilt disease in avocado using low altitude aerial imaging. *PlOS One* 10:e0124642.

De Jager, M.M. and Roets, F. (2022) Rapid and cost-effective detection of *Fusarium euwallaceae* from woody tissues. *Plant Pathology* 71, 1712–1720.

Dean, R., Van Kan, J.A.L., Pretorius, Z.A., Hammond-Kosack, K.E., Di Pietro, A. *et al.* (2012) The top 10 fungal pathogens in molecular plant pathology. *Molecular Plant Pathology* 13, 414–430.

Desjardins, P.R., Drake, R.J., Sasaki, P.J., Atkins, E.L. and Bergh, B.O. (1984) Pollen transmission of avocado sunblotch viroid and the fate of the pollen recipient tree. *Phytopathology* 74:845.

Desjardins, P.R., Drake, R.J. and Swiecki, S.A. (1980) Infectivity studies of avocado sunblotch disease causal agent, possibly a viroid rather than a virus. *Plant Disease* 64, 313–315.

Desjardins, P.R., Saski, P.J. and Drake, R.J. (1987) Chemical inactivation of avocado sunblotch viroid on pruning and propagation tools. *California Avocado Society Yearbook* 71, 259–262.

Dodge, C., Coolidge, J., Cooperband, M.F., Cossé, A.A., Carrillo, D. and Stouthamer, R. (2017) Quercivorol as a lure for the polyphagous and Kuroshio shot hole borers, *Euwallacea* spp. nr. *fornicatus* (Coleoptera: Scolytinae), vectors of *Fusarium* dieback. *PeerJ* 5:e3656.

Downer, J. and Faber, B. (2019) Non-chemical control of *Armillaria mellea* infection of *Prunus persica*. *Journal of Plant Science and Phytopathology* 3, 050–055.

Downer, J., Faber, B. and Menge, J. (2002) Factors affecting root rot control in mulched avocado orchards. *HortTechnology* 12, 601–605.

Dreaden, T.J., Davis, J.M., Harmon, C.L., Ploetz, R.C., Palmateer, A.J., Soltis, P.S. and Smith, J.A. (2014) Development of multilocus PCR assays for *Raffaelea lauricola*, causal agent of laurel wilt disease. *Plant Disease* 98, 379–383.

Dreaden, T.J., Hughes, M.A., Ploetz, R.C., Black, A. and Smith, J.A. (2019) Genetic analyses of the laurel wilt pathogen, *Raffaelea lauricola*, in Asia provide clues on the source of the clone that is responsible for the current USA epidemic. *Forests* 10:37.

Droby, S. and Wisniewski, M. (2018) The fruit microbiome: a new frontier for postharvest biocontrol and postharvest biology. *Postharvest Biology and Technology* 140, 107–112.

Du Toit, W.J. and de Villiers, E.A. (1988) Die hartvormige dopluis, Protopulvinaria pyriformis (Cockerell), op avokado's. *South African Avocado Growers' Association Yearbook* 11, 79–80 (In Afrikaans with English summary.)

El-Hamalawi, Z.A. and Menge, J.A. (1994) Avocado trunk canker disease caused by *Phytophthora citricola*: investigation of factors affecting infection and disease development. *Plant Disease* 78, 260–264.

El-Hamalawi, Z.A., Menge, J.A. and Guillemet, F.B. (1994) Comparison of resistance to *Phytophthora citricola* in nineteen avocado rootstocks under greenhouse conditions. *California Avocado Society Yearbook* 78, 121–129.

Engelbrecht, J., Duong, T.A., Paap, T., Hubert, J.M., Hanneman, J.J. and van den Berg, N. (2022) Population genetic analyses of *Phytophthora cinnamomi* reveals three lineages and movement between natural vegetation and avocado orchards in South Africa. *Phytopathology* 112, 1568–1574.

Erwin, D.C. and Ribeiro, O.K. (1996) *Phytophthora Diseases Worldwide*. The American Phytopathological Society, St. Paul, Minnesota.

Eskalen, A., Faber, B. and Bianchi, M. (2013a) Spore trapping and pathogenicity of fungi in the Botryosphaeriaceae and Diaporthaceae associated with avocado branch canker in California. *Plant Disease* 97, 329–332.

Eskalen, A. and Faber, B.A. (2016a) *Armillaria root rot (oak root fungus)*, Online. University of California. Available at: https://ipm.ucanr.edu/agriculture/avocado/armillaria-root-rot-oak-root-fungus/#MANAGEMENT (accessed 22 July 2023).

Eskalen, A. and Faber, B.A. (2016b) *Branch Canker and dieback (formerly Dothiorella canker),* Online. University of California. Available at: https://ipm.ucanr.edu/agriculture/avocado/branch-canker-and-dieback-formerly-dothiorella-canker/ (accessed 22 July 2023).

Eskalen, A. and Faber, B.A. (2016c) *Phytophthora trunk canker and crown rot (formerly Citricola canker),* Online. University of California (accessed 22 July 2023).

Eskalen, A. and Faber, B.A. (2016d) *Sooty mold,* Online. University of California. Available at: https://ipm.ucanr.edu/agriculture/avocado/sooty-mold/#gsc.tab=0 (accessed 22 July 2023).

Eskalen, A., Gonzalez, A., Wang, D.H., Twizeyimana, M., Mayorquin, J.S. and Lynch, S.C. (2012) First report of a *Fusarium* sp. and its vector Tea Shot Hole Borer (*Euwallacea fornicatus*) causing Fusarium dieback on avocado in California. *Plant Disease* 96:1070.

Eskalen, A., Stouthamer, R., Lynch, S.C., Rugman-Jones, P.F., Twizeyimana, M., Gonzalez, A. and Thibault, T. (2013b) Host range of Fusarium dieback and its ambrosia beetle (Coleoptera: Scolytinae) vector in southern California. *Plant Disease* 97, 938–951.

Evans, E.A. and Crane, J.H. (2009) Estimates of the replacement costs of commercial and backyard avocado trees in south Florida: FE825. *EDIS* 2009(10), 3. Available at: https://edis.ifas.ufl.edu/publication/FE825 (accessed 2 July 2023).

Everett, K.R. and Siebert, B. (2018) Exotic plant disease threats to the New Zealand avocado industry and climatic suitability: a review. *New Zealand Plant Protection* 71, 25–38.

Everett, K.R., Fullerton, R.A., Manning, M.A., Rees-George, J. and Hartill, W.F.T. (2007a) Avocado scab is not present in New Zealand. In: *Proceedings of the VI World Avocado Congress*, Viña del Mar, Chile.

Everett, K.R., Boyd, L.M., Pak, H.A. and Cutting, J.G.M. (2007b) Calcium, fungicide sprays and canopy density influence postharvest rots of avocado. *Australasian Plant Pathology* 36, 22–31.

Everett, K.R., Hallett, I.C., Rees-George, J., Chynoweth, R.W. and Pak, H.A. (2008) Avocado lenticel damage: the cause and the effect on fruit quality. *Postharvest Biology and Technology* 48, 383–390.

Everett, K.R., Pushparajah, I.P.S., Rees-George, J., Connolly, P.G. and Vidiella, A. (2015) Towards commercialisation of avocado rot prediction. In: *Proceedings of the VIII World Avocado Congress*, Lima, Peru.

Everett, K.R., Rees-George, J., Pushparajah, I.P.S., Manning, M.A. and Fullerton, R.A. (2011) Molecular identification of *Sphaceloma perseae* (avocado scab) and its absence in New Zealand. *Journal of Phytopathology* 159, 106–113.

Everett, K.R., Stevens, P.S. and Cutting, J.G.M. (1999) Postharvest fruit rots of avocado are reduced by benomyl applications during flowering. *New Zealand Plant Protection* 52, 153–156.

Fan, X.L., Barreto, R.W., Groenewald, J.Z., Bezerra, J.D.P., Pereira, O.L. *et al.* (2017) Phylogeny and taxonomy of the scab and spot anthracnose fungus *Elsinoë* (*Myriangiales, Dothideomycetes*). *Studies in Mycology* 87, 1–41.

Farih, A., Menge, J.A., Tsao, P.H. and Ohr, H.D. (1981) Metalaxyl and efosite aluminium for control of Phytophthora gummosis and root rot on citrus. *Plant Disease* 65, 654–657.

Farooq, Q.U.A., Hardy, G.E.S.J., McComb, J.A., Thomson, P.C. and Burgess, T.I. (2022) Changes to the bacterial microbiome in the rhizosphere and root endosphere of *Persea americana* (avocado) treated with organic mulch and a silicate-based mulch or phosphite, and infested with *Phytophthora cinnamomi*. *Frontiers in Microbiology* 13, 870900.

Fernandez Falcon, M., Fox, R.L. and Trujillo, E.E. (1984) Interactions of soil pH, nutrients and moisture on phytophthora root rot of avocado. *Plant and Soil* 81, 165–176.

Fick, A., Swart, V., Backer, R., Bombarely, A., Engelbrecht, J. and van den Berg, N. (2022) Partially resistant avocado rootstock Dusa® shows prolonged upregulation of nucleotide binding-leucine rich repeat genes in response to *Phytophthora cinnamomi* infection. *Frontiers in Plant Science* 13, 793644.

Fiorenza, A., Aiello, D., Leonardi, G.R., Continella, A. and Polizzi, G. (2021) First report of *Rosellinia necatrix* causing white root rot on avocado in Italy. *Plant Disease* 105:3294.

Fiorenza, A., Gusella, G., Aiello, D., Polizzi, G. and Voglmayr, H. (2022) *Neopestalotiopsis siciliana* sp. nov. and *N. rosae* causing stem lesion and dieback on avocado plants in Italy. *Journal of Fungi* 8:562.

Fischer, I.H., de Moraes, M.F., Firmino, A.C. and Amorim, L. (2019) Detection and epidemiological progress of quiescent avocado diseases. *Ciência Rural* 49: e20180731.

Fish, A. (2020) Avocado cover cropping: what you need to know. *The Avocado Quarterly,* Spring, 24–29.

Fitzell, R.D. (1987) Epidemiology of anthracnose disease of avocados. *South African Avocado Growers' Association Yearbook* 10, 113–116.

Food and Agriculture Organization of the United Nations (2023) Production/Crops and livestock products – metadata. Available at: https://www.fao.org/home/en (accessed 16 October 2023).

Fraedrich, S.W., Harrington, T.C., Rabaglia, R.J., Ulyshen, M.D., Mayfield, A.E. III *et al.* (2008) A fungal symbiont of the redbay ambrosia beetle causes a lethal wilt in redbay and other Lauraceae in the southeastern USA. *Plant Disease* 92, 215–224.

Freeman, S., Miller, G., Protasov, A., Maymon, M., Elazar, M. *et al.* (2019) Aposymbiotic interactions of three ambrosia beetle fungi with avocado trees. *Fungal Ecology* 39, 117–130.

Freeman, S., Sharon, M., Dori-Bachash, M., Maymon, M., Belausov, E. *et al.* (2016) Symbiotic association of three fungal species throughout the life cycle of the ambrosia beetle *Euwallacea* nr. *fornicatus*. *Symbiosis* 68, 115–128.

Freeman, S., Sharon, M., Maymon, M., Mendel, Z., Protasov, A. *et al.* (2013) *Fusarium euwallaceae* sp. nov. – a symbiotic fungus of *Euwallacea* sp., an invasive ambrosia beetle in Israel and California. *Mycologia* 105, 1595–1606.

Fucikovsky, L. (1992) Sphaeropsis tumor of avocado. In: *Proceedings of the Second World Avocado Congress*, pp. 129–132.

Fuentes-Aragón, D., Silva-Rojas, H.V., Guarnaccia, V., Mora-Aguilera, J.A., Aranda-Ocampo, S., Bautista-Martínez, N. and Téliz-Ortíz, D. (2020) *Colletotrichum* species causing anthracnose on avocado fruit in Mexico: current status. *Plant Pathology* 69, 1513–1528.

Gañán-Betancur, L. and Gazis, R. (2023) Genome sequence resource of the avocado scab pathogen *Elsinoe perseae*. *Microbiology Resource Announcements* 12: e00190–23.

Garfinkel, A.R., Cannon, P.G., Klopfenstein, N.B., Stewart, J.E. and Kim, M.-S. (2019) Identification of genetic groups within the invasive brown root rot pathogen, *Pyrrhoderma noxium* (formerly *Phellinus noxius*). In: Reynolds, G.J., Wilhelmi, N.P. and Palacios, P. (eds) *Proceedings of the 66th Western International Forest Disease Work Conference*, Estes Park, Colorado, USA, pp. 141–145.

Geering, A. and Parkinson, L. (2019) *Xylella fastidiosa*, a high priority biosecurity threat for the Australian avocado industry. *Talking Avocados* 30 (Autumn), 49–50.

Giblin, F.R., Coates, L.M. and Irwin, J.A.G. (2010) Pathogenic diversity of avocado and mango isolates of *Colletotrichum gloeosporioides* causing anthracnose and pepper spot in Australia. *Australasian Plant Pathology* 39, 50–62.

Giblin, F.R., Tan, Y.P., Mitchell, R., Coates, L.M., Irwin, J.A.G. and Shivas, R.G. (2018) *Colletotrichum* species associated with pre-and post-harvest diseases of avocado and mango in eastern Australia. *Australasian Plant Pathology* 47, 269–276.

Gomez, D.F., Lin, W., Gao, L. and Li, Y. (2019) New host plant records for the *Euwallacea fornicatus* (Eichhoff) species complex (Coleoptera: Curculionidae: Scolytinae) across its natural and introduced distribution. *Journal of Asia-Pacific Entomology* 22, 338–340.

Gomez, D.F., Skelton, J., Steininger, M.S., Stouthamer, R., Rugman-Jones, P. *et al.* (2018) Species delineation within the *Euwallacea fornicatus* (Coleoptera: Curculionidae) complex revealed by morphometric and phylogenetic analyses. *Insect Systematics and Diversity* 2, 1–11.

Gonzalez-Sanchez, M.A., de Vicente, A., Perez-Garcia, A., Perez-Jimenez, R., Romero, D. and Cazorla, F.M. (2013) Evaluation of the effectiveness of biocontrol bacteria against avocado white root rot occurring under commercial greenhouse plant production conditions. *Biological Control* 67, 94–100.

Guarnaccia, V., Aiello, D., Papadantonakis, N., Polizzi, G. and Gullino, M.L. (2022) First report of branch cankers on avocado *(Persea americana)* caused by *Neocosmospora* (syn. *Fusarium*) *perseae* in Crete (Greece). *Journal of Plant Pathology* 104, 419–420.

Guarnaccia, V., Sandoval-Denis, M., Aiello, D., Polizzi, G. and Crous, P.W. (2018) *Neocosmospora perseae* sp. nov., causing trunk cankers on avocado in Italy. *Fungal Systematics and Evolution* 1, 131–140.

Guarnaccia, V., Vitale, A., Cirvilleri, G., Aiello, D., Susca, A. *et al.* (2016) Characterisation and pathogenicity of fungal species associated with branch cankers and stem-end rot of avocado in Italy. *European Journal of Plant Pathology* 146, 963–976.

Haberman, A., Tsror (Lahkim), L., Lazare, S., Hazanovsky, M., Lebiush, S. *et al.* (2020) Management of Verticillium wilt of avocado using tolerant rootstocks. *Plants* 9:531.

Halma, F.F. and Zentmyer, G.A. (1953) Relative susceptibility of Guatemalan and Mexican avocado varieties to Dothiorella canker. *California Avocado Society Yearbook* 38, 156–158.

Hamilton, J.L., Fraedrich, S.W., Nairn, C.J., Mayfield, A.E. and Villari, C. (2021) A field-portable diagnostic approach confirms laurel wilt disease diagnosis in minutes instead of days. *Arboriculture and Urban Forestry* 47, 98–109.

Hamilton, J.L., Workman, J.N., Nairn, C.J., Fraedrich, S.W. and Villari, C. (2020) Rapid detection of *Raffaelea lauricola* directly from host plant and beetle vector tissues using loop-mediated isothermal amplification. *Plant Disease* 104, 3151–3158.

Hardham, A.R. and Blackman, L.M. (2018) *Phytophthora cinnamomi*. *Molecular Plant Pathology* 19, 260–285.

Harrington, T.C. and Fraedrich, S.W. (2010) Quantification of propagules of the laurel wilt fungus and other mycangial fungi from the redbay ambrosia beetle, *Xyleborus glabratus*. *Phytopathology* 100, 1118–1123.

Harrington, T.C., Fraedrich, S.W. and Aghayeva, D. (2008) *Raffaelea lauricola*, a new ambrosia beetle symbiont and pathogen on the Lauraceae. *Mycotaxon* 104, 399–404.

Hartill, W.F.T. (1991a) Post-harvest diseases of avocado fruits in New Zealand. *New Zealand Journal of Crop and Horticultural Science* 19, 297–304.

Hartill, W.F.T. (1991b) The use of copper sprays and pruning out dead wood to control postharvest rots of avocados. *Scientific Research and Technical Supplement of the New Zealand Avocado Growers Association* 10:5, 7–8.

Hartill, W.F.T. and Everett, K.R. (2002) Inoculum sources and infection pathways of pathogens causing stem-end rots of 'Hass' avocado (*Persea americana*). *New Zealand Journal of Crop and Horticultural Science* 30, 249–260.

Hartley, J., Engelbrecht, J. and van den Berg, N. (2022) Detection and prevalence of *Rosellinia necatrix* in South African avocado orchards. *European Journal of Plant Pathology* 163, 961–978.

Hartley, J. and van den Berg, N. (2023) Armillaria root rot of avocado caused by *Armillaria* species, Online. University of Pretoria, Pretoria, South Africa. Available at: https://www.fabinet.up.ac.za/images/groups/ARP/Armillaria_root_rot_fact_sheet_2023.pdf (accessed 7 July 2023).

Hassan, M.K. and Dann, E. (2019) Effects of treatment with electrolyzed oxidizing water on postharvest diseases of avocado. *Agriculture* 9:241.

Herder, M.C. (2014) Identification of *Ilyonectria* species associated with and determining their role in avocado decline. BAgrSc (Hons), Lincoln University. Available at: https://researcharchive.lincoln.ac.nz/entities/publication/28ef1956-0b6f-4c1a-97c0-21eab203a671 (accessed 17 July 2021).

Hernández, D., García-Pérez, O., Perera, S., González-Carracedo, M.A., Rodríguez-Pérez, A. and Siverio, F. (2023) Fungal pathogens associated with aerial symptoms of avocado (*Persea americana* Mill.) in Tenerife (Canary Islands, Spain) focused on species of the family Botryosphaeriaceae. *Microorganisms* 11:585.

Hernández Pérez, A., Cerna Chávez, E., Delgado Ortiz, J.C., Beltrán Beache, M., Tapia Vargas, L.M. and Ochoa Fuentes, Y.M. (2019) First report of *Phytopythium vexans* causing the 'Avocado sadness' in Michoacan, Mexico. *Phyton-International Journal of Experimental Botany* 88, 11–13.

Hodges, C.S., Jr (1969) A new species of *Phyllachora* on *Persea*. *Mycologia* 61, 838–840.

Hofer, K.M., Braithwaite, M., Braithwaite, L.J., Sorensen, S., Siebert, B. *et al.* (2021) First report of *Colletotrichum fructicola*, *C. perseae*, and *C. siamense* causing anthracnose disease of avocado (*Persea americana*) in New Zealand. *Plant Disease* 105:1564.

Hofman, P.J., Vuthapanich, S., Whiley, A.W., Klieber, A. and Simons, D.H. (2002) Tree yield and fruit minerals concentrations influence 'Hass' avocado fruit quality. *Scientia Horticulturae* 92, 113–123.

Hong, C.X., Gallegly, M.E., Browne, G.T., Bhat, R.G., Richardson, P.A. and Kong, P. (2009) The avocado subgroup of *Phytophthora citricola* constitute a distinct species, *Phytophthora mengei* sp. *Mycologia* 106, 833–840.

Horne, W.T. (1934) Avocado diseases in California. *University of California Agricultural Experiment Station Bulletin* 585:72.

Hulcr, J., Black, A., Prior, K., Chen, C.-Y. and Li, H.-F. (2017) Studies of ambrosia beetles (Coleoptera: Curculionidae) in their native ranges help predict invasion impact. *Florida Entomologist* 100, 257–261.

Hunupolagama, D.M., Wijesundera, R.L.C., Chandrasekharan, N.V., Wijesundera, W.S.S., Kathriarachchi, H.S. and Fernando, T.H.P.S. (2015) Characterization of *Colletotrichum* isolates causing avocado anthracnose and first report of *C. gigasporum* infecting avocado in Sri Lanka. *Plant Pathology and Quarantine* 5, 132–143.

Ibarra Caballero, J.R., Jeon, J., Lee, Y.-H., Fraedrich, S., Klopfenstein, N.B., Kim, M.S. and Stewart, J.E. (2019) Genomic comparisons of the laurel wilt pathogen, *Raffaelea lauricola*, and related tree pathogens highlight an arsenal of pathogenicity related genes. *Fungal Genetics and Biology* 125, 84–92.

Inch, S. and Ploetz, R.C. (2012) Impact of laurel wilt, caused by *Raffaelea lauricola*, on xylem function in avocado, *Persea americana*. *Forest Pathology* 42, 239–245.

Jeyaprakash, A., Davison, D.A. and Schubert, T.S. (2014) Molecular detection of the laurel wilt fungus, *Raffaelea lauricola*. *Plant Disease* 98, 559–564.

Johnson, G.I. and Kotze, J.M. (1994) Avocado stem-end rot. In: Ploetz, R.C., Zentmyer, G.A., Nishijima, W.T., Rohrbach, K.G. and Ohr, H.D. (eds) *Compendium of Tropical Fruit Diseases*. The American Phytopathological Society, St. Paul, Minnesota, pp. 81–83.

Jolliffe, J.B., Dann, E.K., Van Der Rijst, M., Masikane, S.L., Novela, P., Mohale, P. and McLeod, A. (2024) Effect of sampling time, quantification method and tree sample pooling on *Phytophthora cinnamomi* root quantities in South African avocado orchards. *Plant Disease* 108, 725–736.

Joseph, R., Lasa, M., Zhou, Y. and Keyhani, N.O. (2021) Unique attributes of the laurel wilt pathogen, *Raffaelea lauricola*, as revealed by metabolic profiling. *Pathogens* 10:528.

Joubert, J.J. and Rijkenberg, F.H.J. (1971) Parasitic green algae. *Annual Review of Phytopathology* 9, 45–64.

Jung, T., Colquhoun, I.J. and Hardy, G.E.S.J. (2013) New insights into the survival strategy of the invasive soilborne pathogen *Phytophthora cinnamomi* in different natural ecosystems in Western Australia. *Forest Pathology* 43, 266–288.

Jung, T., Orlikowski, L., Henricot, B., Abad-Campos, P., Aday, A.G. *et al.* (2016) Widespread *Phytophthora* infestations in European nurseries put forest, semi-natural and horticultural ecosystems at high risk of Phytophthora diseases. *Forest Pathology* 46, 134–163.

Kasson, M.T., O'Donnell, K., Rooney, A.P., Sink, S., Ploetz, R.C. *et al.* (2013) An inordinate fondness for *Fusarium*: phylogenetic diversity of fusaria cultivated by ambrosia beetles in the genus *Euwallacea* on avocado and other plant hosts. *Fungal Genetics and Biology* 56, 147–157.

Kendra, P.E., Guillén, L., Tabanca, N., Montgomery, W.S., Schnell, E.Q., Deyrup, M.A. and Cloonan, K.R. (2023) Risk assessment of Hass avocado and Mexican Lauraceae for attack by redbay ambrosia beetle (Coleoptera: Curculionidae: Scolytinae). *Agricultural and Forest Entomology* 25, 285–302.

Kendra, P.E., Montgomery, W.S., Niogret, J. and Epsky, N.D. (2013) An uncertain future for American Lauraceae: a lethal threat from redbay ambrosia beetle and laurel wilt disease (a review). *American Journal of Plant Sciences* 4, 727–738.

Knight, K.W.L. (2001) Plant parasitic nematodes associated with six subtropical crops in New Zealand. *New Zealand Journal of Crop and Horticultural Science* 29, 267–275.

Koch, R.A., Wilson, A.W., Séné, O., Henkel, T.W. and Aime, M.C. (2017) Resolved phylogeny and biogeography of the root pathogen *Armillaria* and its gasteroid relative, *Guyanagaster. BMC Evolutionary Biology* 17:33.

Kogan, M. (1998) Integrated pest management: historical perspectives and contemporary developments. *Annual Review of Entomology* 43, 243–270.

Korsten, L., de Villiers, E.E., Wehner, F.C. and Kotze, J.M. (1997) Field sprays of *Bacillus subtilis* and fungicides for control of preharvest fruit disease of avocado in South Africa. *Plant Disease* 81, 455–459.

Korsten, L. and Kotzé, J.M. (1987) Bark canker of avocado, a new disease presumably caused by *Pseudomonas syringae* in South Africa. *Plant Disease* 71:850.

Kuhn, D.N., Livingstone, D.S., Richards, J.H., Manosalva, P., Van den Berg, N. and Chambers, A.H. (2019) Application of genomic tools to avocado (*Persea americana*) breeding: SNP discovery for genotyping and germplasm characterization. *Scientia Horticulturae* 246, 1–11.

Kunadiya, M., White, D., Dunstan, W.A., Hardy, G.E.S.J., Andjic, V. and Burgess, T.I. (2017) Pathways to false-positive diagnoses using molecular genetic detection methods; *Phytophthora cinnamomi* a case study. *FEMS Microbiology Letters* 364, fxn009.

Li, Y., Gu, X., Kasson, M.T., Bateman, C.C., Guo, J. *et al.* (2016) Distribution, host records, and symbiotic fungi of *Euwallacea fornicatus* (Coleoptera: Curculionidae: Scolytinae) in China. *Florida Entomologist* 99, 801–804.

Liang, Y.-P., Wu, C.-J., Tsai, H.-W. and Ni, H.-F. (2021) Avocado branch canker disease caused by *Lasiodiplodia theobromae* and *Lasiodiplodia pseudotheobromae* in Taiwan. *Journal of Taiwan Agricultural Research* 70, 81–97.

Liberato, J.R. and Barreto, R.W. (2006) *Oidium perseae-americanae* sp. nov. on avocado. *Mycotaxon* 98, 189–192.

Lin, C.H., Dong, P.P., Fang, S.Q., Li, M.F., Liu, W.B. and Miao, W.G. (2018) First report of avocado dieback disease caused by *Pestalotiopsis longiseta* in China. *Plant Disease* 102:2660.

Liu, F.L., Rugman-Jones, P., Liao, Y.C., Husein, D., Liang, H.H., Tuan, S.J. and Stouthamer, R. (2022) Seasonal dynamics of flight phenology of the *Euwallacea fornicatus* species complex and an associated parasitoid wasp in avocado groves in Taiwan. *Journal of Economic Entomology* 115, 1901–1910.

López Cornejo, C.I. (2021) Aislamiento e identificación morfológica y molecular del agente causal de la roña del aguacate (*Persea americana* Mill.), en Michoacán y Jalisco. M.S. in Biological Sciences, Universidad Michoacana de San Nicolas de Hidalgo.

Lopez-Herrera, C.J. and Melero-Vara, J.M. (1991) Diseases of avocado caused by soil fungi in the southern Mediterranean coast of Spain. In: *Proceedings of the Second World Avocado Congress,* California, USA.

Lopez-Herrera, C.J., Perez-Jimenez, R.M., Basallote-Ureba, M.J., Zea-Bonilla, T. and Melero-Vara, J.M. (1999) Loss of viability of *Dematophora necatrix* in solarized soils. *European Journal of Plant Pathology* 105, 571–576.

Lynce-Duque, D. (2015) Mirids produce pre-harvest spotting on avocado fruits, affecting sales. In: *Proceedings of the VIII World Avocado Congress*, Lima, Peru.

Lynch, S.C., Carrillo, J.D., Stouthamer, R. and Eskalen, A. (2018) Severity of Fusarium dieback–shot hole borers analyzed. *From the Grove* 8, 46–50.

Lynch, S.C., Twizeyimana, M., Mayorquin, J.S., Wang, D.H., Na, F. *et al*. (2016) Identification, pathogenicity and abundance of *Paracremonium pembeum* sp. nov. and *Graphium euwallaceae* sp. nov. – two newly discovered mycangial associates of the polyphagous shot hole borer (*Euwallacea* sp.) in California. *Mycologia* 108, 313–329.

Lynwood Avocado Nursery (2020) *Rootstock summary,* Online. Available at: https://www.lynwoodavocado.co.nz/root-stock-summary (accessed 10 January 2024).

Magagula, P., Taylor, N., Swart, V. and van den Berg, N. (2021) Efficacy of potential control agents against *Rosellinia necatrix* and their physiological impact on avocado. *Plant Disease* 105, 3385–3396.

Maharachchikumbura, S.S.N., Hyde, K.D., Groenewald, J.Z., Xu, J. and Crous, P.W. (2014) *Pestalotiopsis* revisited. *Studies in Mycology* 79, 121–186.

Malapi-Wight, M., Salago-Salazar, C., Demers, J., Veltri, D. and Crouch, J.A. (2015) Draft genome sequence of *Dactylonectria macrodidyma*, a plant pathogenic fungus in the Nectriaceae. *Genome Announcements* 3:e00278–15.

Manicom, B.Q. and Schoeman, M.H. (2010) Evaluation of fungicides for *Pseudocercospora* control – final report. *South African Avocado Growers' Association Yearbook* 22, 70–72.

Marais, L.J. (2004) Avocado diseases of major importance worldwide and their management. In: Naqvi, S.A.M.H. (ed.) *Diseases of Fruits and Vegetables: Volume II: Diagnosis and Management*. Kluwer Academic Publishers, The Netherlands, 1–37.

Martin, F.N., Abad, Z.G., Balci, Y. and Ivors, K. (2012) Identification and detection of *Phytophthora*: reviewing our progress, identifying our needs. *Plant Disease* 96, 1080–1103.

Martin, P.L., King, W.L., Bell, T.H. and Peter, K.A. (2021) The decay and fungal succession of apples with bitter rot across a vegetation diversity gradient. *Phytobiomes Journal* 6, 26–34.

Martínez-Ferri, E., Moreno-Ortega, G., van den Berg, N. and Pliego, C. (2019) Mild water stress-induced priming enhance tolerance to *Rosellinia necatrix* in susceptible avocado rootstocks. *BMC Plant Biology* 19:458.

Martínez-Ferri, E., Zumaquero, A., Ariza, M.T., Barceló, A. and Pliego, C. (2015) Nondestructive detection of white root rot disease in avocado rootstocks by leaf chlorophyll fluorescence. *Plant Disease* 100, 49–58.

Masikane, S., Jolliffe, J., Swart, L. and McLeod, A. (2019) Novel approaches and methods for quantifying *Phytophthora cinnamomi* in avocado tree roots. *FEMS Microbiology Letters* 366, fnz199.

Mayfield, A.E., III, Smith, J.A., Hughes, M. and Dreaden, T.J. (2008) First report of laurel wilt disease caused by a *Raffaelea* sp. on avocado in Florida. *Plant Disease* 92:976.

McDonald, V. and Eskalen, A. (2011) Botryosphaeriaceae species associated with avocado branch cankers in California. *Plant Disease* 95, 1465–1473.

McDonald, V., Lynch, S. and Eskalen, A. (2009) First report of *Neofusicoccum australe*, *N. luteum*, and *N. parvum* associated with avocado branch canker in California. *Plant Disease* 93:967.

McDonald, V., Pond, E., Crowley, D., McKee, B. and Menge, J. (2007) Selection for and evaluation of an avocado orchard soil microbially suppressive to *Phytophthora cinnamomi*. *Plant and Soil* 299, 17–28.

McLeod, A., Masikane, S.L., Novela, P., Ma, J., Mohale, P. *et al*. (2018) Quantification of root phosphite concentrations for evaluating the potential of foliar phosphonate sprays for the management of avocado root rot. *Crop Protection* 103, 87–97.

McLeod, A., Moyo, L., van der Westhuyzen, J.G., Novela, P., Modisane, J.G. and Mohale, P. (2021) Evaluating ammonium- and potassium phosphonate foliar sprays over two seasons based on root phosphite concentrations for the preventative management of Phytophthora root rot. *South African Avocado Growers' Association Yearbook* 44, 72–79.

Mendel, J., Burns, C., Kallifatidis, B., Evans, E., Crane, J., Furton, K.G. and Mills, D. (2018) Agri-dogs: using canines for earlier detection of laurel wilt disease affecting avocado trees in South Florida. *HortTechnology* 28, 109–116.

Mendel, Z., Protasov, A., Maoz, Y., Maymon, M., Miller, G., Elazar, M. and Freeman, S. (2017) The role of *Euwallacea* nr. *fornicatus* (Coleoptera: Scolytinae) in the wilt syndrome of avocado trees in Israel. *Phytoparasitica* 45, 341–359.

Mendel, Z., Protasov, A., Sharon, M., Zveibil, A., Ben-Yehuda, S. *et al.* (2012) An Asian ambrosia beetle *Euwallacea fornicatus* and its novel symbiotic fungus *Fusarium* sp. pose a serious threat to the Israeli avocado industry. *Phytoparasitica* 40, 235–238.

Menge, J.A., Douhan, G.W., McKee, B., Pond, E., Bender, G.S. and Faber, B. (2012) Three new avocado rootstock cultivars tolerant to Phytophthora root rot: 'Zentmyer', 'Uzi', and 'Steddom'. *HortScience* 47, 1191–1194.

Menge, J.A. and Ploetz, R.C. (2003) Diseases of avocado. In: Ploetz, R.C. (ed.) *Diseases of Tropical Fruit Crops*. CABI Publishing, Wallingford, UK, 35–71.

Menocal, O., Kendra, P.E., Montgomery, W.S., Crane, J.H. and Carrillo, D. (2018) Vertical distribution and daily flight periodicity of ambrosia beetles (Coleoptera: Curculionidae) in Florida avocado orchards affected by laurel wilt. *Journal of Economic Entomology* 111, 1190–1196.

Menocal, O., Kendra, P.E., Padilla, A., Chagas, P.C., Chagas, E.A., Crane, J.H. and Carrillo, D. (2022) Influence of canopy cover and meteorological factors on the abundance of bark and ambrosia beetles (Coleoptera: Curculionidae) in avocado orchards affected by laurel wilt. *Agronomy* 12, 547. doi:10.3390/agronomy12030547

Messenger, B.J., Menge, J.A. and Pond, E. (2000) Effects of gypsum on zoospores and sporangia of *Phytophthora cinnamomi* in field soil. *Plant Disease* 84, 617–621.

Mohankumar, V., Dann, E.K. and Akinsanmi, O.A. (2023) Seasonal dynamics of inoculum of Botryosphaeriaceae in macadamia orchards in Australia. *Plant Pathology* 72, 1160–1170.

Moll, J.N., Grech, N.M. and van Vuuren, S.P. (1987) A lethal, transmissible stem-pitting of avocados associated with Duke 6 rootstocks. *South African Avocado Growers' Association Yearbook* 10, 122–123.

Monden, Y., Yamamoto, S., Yamakawa, R., Sunada, A., Asari, S., Makimura, K. and Inoue, Y. (2013) First case of fungal keratitis caused by *Pestalotiopsis clavispora*. *Clinical Ophthalmology* 7, 2261–2264.

Montero-Astúa, M., Saborio-R, G., Chacon-Diaz, C., Garita, L., Villalobos, W. *et al.* (2008) First report of *Xylella fastidiosa* in avocado in Costa Rica. *Plant Disease* 92:175.

Morales-García, J.L., López-Cornejo, C.I., Pedraza-Santos, M.E., Chávez-Bárcenas, A.T., Esquivel-Miguel, E., García-Morales, S. and Pineda-Guillermo, S. (2023) Morpho-molecular identification of the causal agent of avocado scab in Michoacán. *Mexican Journal of Phytopathology* 41, 182–202.

Morales-Rodríguez, C., Vettraino, A.M. and Vannini, A. (2016) Efficacy of biofumigation with *Brassica carinata* commercial pellets (BioFence) to control vegetative and reproductive structures of *Phytophthora cinnamomi*. *Plant Disease* 100, 324–330.

Munhuweyi, K., Mpai, S. and Sivakumar, D. (2020) Extension of avocado fruit postharvest quality using non-chemical treatments. *Agronomy* 10:212.

Na, F., Carrillo, J.D., Mayorquin, J.S., Ndinga-Muniania, C., Stajich, J.E. *et al.* (2017) Two novel fungal symbionts *Fusarium kuroshium* sp. nov. and *Graphium kuroshium* sp. nov. of Kuroshio shot hole borer (*Euwallacea* sp. nr. *fornicatus*) cause Fusarium dieback on woody host species in California. *Plant Disease* 102, 1154–1164.

Navia-Urrutia, M., Sánchez-Pinzón, L., Parra, P.P. and Gazis, R. (2022) A diagnostic guide for laurel wilt disease in avocado. *Plant Health Progress* 23, 345–354.

Navia-Urrutia, M., Sendoya-Corrales, C.A., Crane, J.H. and Gazis, R. (2023) Reevaluation of the application method and efficacy of propiconazole in controlling laurel wilt in commercial avocado orchards in south Florida. *HortTechnology* 33, 425–436.

O'Donnell, K., Sink, S., Libeskind-Hadas, R., Hulcr, J., Kasson, M.T. *et al.* (2015) Discordant phylogenies suggest repeated host shifts in the Fusarium-Euwallacea ambrosia beetle mutualism. *Fungal Genetics and Biology* 82, 277–290.

Ohr, H.D., Coffey, M.D. and McMillan Jr., R.T. (2003) *Diseases of avocado (Persea americana* Miller*)*, Online. The American Phytopathological Society. Available at: https://www.apsnet.org/edcenter/resources/commonnames/Pages/Avocado.aspx (accessed 22 July 2023).

Ohr, H.D. and Zentmyer, G.A. (1994a) Armillaria root rot. In: Ploetz, R.C., Zentmyer, G.A., Nishijima, W.T., Rohrbach, K.G. and Ohr, H.D. (eds) *Compendium of Tropical Fruit Diseases*. APS Press, St. Paul, Minnesota.

Ohr, H.D. and Zentmyer, G.A. (1994b) Avocado black streak. In: Ploetz, R.C., Zentmyer, G.A., Nishijima, W.T., Rohrbach, K.G. and Ohr, H.D. (eds) *Compendium of Tropical Fruit Diseases*. APS Press, St. Paul, Minnesota.

Olatinwo, R. and Fraedrich, S. (2019) An *Acaromyces* species associated with bark beetles from southern pine has inhibitory properties against *Raffaelea lauricola*, the causal pathogen of Laurel wilt disease of Redbay. *Plant Health Progress* 20, 220–228.

Owens, D., Cruz, L.F., Montgomery, W.S., Narvaez, T.I., Schnell, E.Q. *et al.* (2018) Host range expansion and increasing damage potential of *Euwallacea* nr. *fornicatus* (Coleoptera: Curculionidae) in Florida. *Florida Entomologist* 101(2), 229–236.

Paap, T., de Beer, Z.W., Migliorini, D., Nel, W.J. and Wingfield, M.J. (2018) The polyphagous shot hole borer (PSHB) and its fungal symbiont *Fusarium euwallaceae*: a new invasion in South Africa. *Australasian Plant Pathology* 47, 231–237.

Palmateer, A.J., Ploetz, R.C. and Harmon, P.F. (2006) 2006 Florida Plant Disease Management Guide: Avocado (Persea americana), Online. University of Florida.. Available at: https://doi.org/10.32473/edis-pg134-2006 (accessed 6 July 2023).

Parbery, D.G. (1981) Biology of anthracnose on leaf surfaces. In: Blakeman, J.P. (ed.) *Microbial Ecology of the Phylloplane*. Academic Press, London, pp. 135–154.

Parkinson, L. and Geering, A. (2019) Biosecurity capacity building for the Australian avocado industry. *Talking Avocados* 30, 54–55.

Parkinson, L.E., Le, D.P. and Dann, E.K. (2019) Development of three loop-mediated isothermal amplification (LAMP) assays for the rapid detection of *Calonectria ilicicola*, *Dactylonectria macrodidyma*, and the *Dactylonectria* genus in avocado roots. *Plant Disease* 103, 1865–1875.

Parkinson, L.E., Shivas, R.G. and Dann, E.K. (2017) Pathogenicity of Nectriaceous fungi on avocado in Australia. *Phytopathology* 107, 1479–1485.

Parra, P.P., Dantes, W., Sandford, A., de la Torre, C., Pérez, J. *et al.* (2020) Rapid detection of the laurel wilt pathogen in sapwood of Lauraceae hosts. *Plant Health Progress* 21, 356–364.

Pathania, N. and Grice, K. (2021) Investigating the cause of soft rot in avocado. In: *Proceedings of the 2021 Australasian Plant Pathology Society Biennial Conference*, online.

Pegg, K. and Coates, L.M. (1993) Avocado. In: Persley, D. (ed.) *Diseases of Fruit Crops*. Department of Primary Industries, Brisbane, Australia, pp. 20–24.

Pérez-Jiménez, R. (2008) Significant avocado diseases caused by fungi and oomycetes. *The European Journal of Plant Science and Biotechnology* 2, 1–24.

Pérez-Jiménez, R.M. (2006) A review of the biology and pathogenicity of *Rosellinia necatrix*– the cause of white root rot disease of fruit trees and other plants. *Journal of Phytopathology* 154, 257–266.

Pérez-Jiménez, R.M., Zea-Bonilla, T. and López-Herrera, C.J. (2003) Studies of *Rosellinia necatrix* perithecia found in nature on avocado roots. *Journal of Phytopathology* 151, 660–664.

Pérez-Martínez, J., Ploetz, R.C. and Konkol, J.L. (2018) Significant *in vitro* antagonism of the laurel wilt pathogen by endophytic fungi from the xylem of avocado does not predict their ability to control the disease. *Plant Pathology* 67, 1768–1776.

Pernezny, K. and Marlatt, R. (2000) *Diseases of avocado in Florida,* online. University of Florida. Available at: https://plantpath.ifas.ufl.edu/misc/media/factsheets/pp0021.pdf (accessed 2 July 2023).

Peterson, R.A. (1978) Susceptibility of Fuerte avocado fruit at various stages of growth to infection by anthracnose and stem-end rot fungi. *Australian Journal of Experimental Agriculture and Animal Husbandry* 18, 158–160.

Phoulivong, S., Cai, L., Chen, H., McKenzie, E.H.C., Abd-Elsalam, K., Chukeatirote, E. and Hyde, K.D. (2010) *Colletotrichum gloeosporioides* is not a common pathogen on tropical fruits. *Fungal Diversity* 44, 33–43.

Pieterse, C.L. (1986) Afkeuringsfaktore by uitvoeravokado's. *South African Avocado Growers' Association Yearbook* 9:14. (In Afrikaans.)

Pliego, C. and Cazorla, F.M. (2013) Biocontrol of tree root diseases. In: de Bruijn, F.J. (ed.) *Molecular Microbial Ecology of the Rhizosphere*. John Wiley & Sons, Inc.

Pliego, C., López-Herrera, C., Ramos, C. and Cazorla, F.M. (2012) Developing tools to unravel the biological secrets of *Rosellinia necatrix*, an emergent threat to woody crops. *Molecular Plant Pathology* 13, 226–239.

Ploetz, R.C., Hughes, M.A., Kendra, P.E., Fraedrich, S.W., Carrillo, D. *et al.* (2017a) Recovery plan for laurel wilt of avocado, caused by *Raffaelea lauricola*. *Plant Health Progress* 18, 51–77.

Ploetz, R.C., Hulcr, J., Wingfield, M.J. and de Beer, Z.W. (2013) Destructive tree diseases associated with ambrosia and bark beetles: black swan events in tree pathology? *Plant Disease* 97, 856–872.

Ploetz, R.C., Inch, S.A., Perez Martinez, J.M. and White, T.L. Jr. (2012a) Systemic infection of avocado, *Persea americana*, by *Raffaelea lauricola*, does not progress into fruit pulp or seed. *Journal of Phytopathology* 160, 491–495.

Ploetz, R.C., Kendra, P.E., Choudhury, R.A., Rollins, J.A., Campbell, A. *et al.* (2017b) Laurel wilt in natural and agricultural ecosystems: understanding the drivers and scales of complex pathosystems. *Forests* 8:48.

Ploetz, R.C., Konkol, J.L., Pérez-Martínez, J.M. and Fernandez, R. (2017c) Management of laurel wilt of avocado, caused by *Raffaelea lauricola*. *European Journal of Plant Pathology* 149, 133–143.

Ploetz, R.C., Perez-Martinez, J.M., Evans, E.A. and Inch, A.J. (2011) Toward fungicidal management of laurel wilt of avocado. *Plant Disease* 95, 977–982.

Ploetz, R.C., Pérez-Martínez, J.M., Smith, J.A., Hughes, M., Dreaden, T.J., Inch, S.A. and Fu, Y. (2012b) Responses of avocado to laurel wilt, caused by *Raffaelea lauricola*. *Plant Pathology* 61, 801–808.

Ploetz, R.C. and Schaffer, B. (1989) Effects of flooding and Phytophthora root rot on net gas exchange and growth of avocado. *Phytopathology* 79, 204–208.

Ploetz, R.C., Thant, Y.Y., Hughes, M.A., Dreaden, T.J., Konkol, J.L. *et al.* (2016) Laurel wilt, caused by *Raffaelea lauricola*, is detected for the first time outside the southeastern United States. *Plant Disease* 100:2166.

Pohronenzy, K.L. and Simone, G.W. (1994) Scab. In: Ploetz, R.C., Zentmyer, G.A., Nishijima, W.T., Rohrbach, K.G. and Ohr, H.D. (eds) *Compendium of Tropical Fruit Diseases*. APS Press, St. Paul, Minnesota, p. 81.

Pohronezny, K.L., Simone, G.W. and Kotzé, J.M. (1994) Pseudocercospora spot (blotch). In: Ploetz, R.C., Zentmyer, G. A., Nishijima, W.T., Rohrbach, K.G. and Ohr, H.D. (eds) *Compendium of Tropical Fruit Diseases*. APS Press, St. Paul, Minnesota, pp. 79–80.

Prabhakaran, A.D. and Dann, E.K. (2022) Evaluation of fungicide soil drench treatments to manage black root rot disease of avocado. *Plant Disease* 106, 2026–2030.

Pretorius, L.-S., Chandra, K.A., Jooste, A.E.C., Motaung, L.C., Parkinson, L.E. and Geering, A.D.W. (2022) Adaptation of a filter paper method for RNA template preparation for the detection of avocado sunblotch viroid by reverse transcription qPCR. *Journal of Virological Methods* 114455.

Prusky, D., Alkan, N., Mengiste, T. and Fluhr, R. (2013) Quiescent and necrotrophic lifestyle choice during postharvest disease development. *Annual Review of Phytopathology* 51, 155–176.

Prusky, D., Plumbley, R.A. and Kobiler, I. (1990) Germinated appressoria: the latent stage of *Colletotrichum gloeosporioides* in unripe avocado fruits. *Phytoparasitica* 18:74.

Querci, M., Owens, R.A., Vargas, C. and Salazar, L.F. (1995) Detection of potato spindle tuber viroid in avocado growing in Peru. *Plant Disease* 79, 196–202.

Quintero, E.J. and Sebastian, R.C. (1972) A Mycosphaerella leaf spot disease on avocado and its control. *California Avocado Society Yearbook* 55, 148–151.

Ramírez-Gil, J.G., Castañeda-Sánchez, D.A. and Morales-Osorio, J.G. (2017) Production of avocado trees infected with *Phytophthora cinnamomi* under different management regimes. *Plant Pathology* 66, 623–632.

Ramírez Gil, G. and Morales Osorio, J.G. (2013) Primer informe de *Cylindrocarpon destructans* (Zinss) Scholten afectando plántulas de aguacate (*Persea americana* Mill.) en Colombia (First report of *Cylindrocarpon destructans* (Zinss) Scholten affecting avocado (*Persea americana* Mill) seedling in Colombia). *Revista de Protección Vegetal* 28, 27–35.

Ramírez Gil, G. and Morales Osorio, J.G. (2021) Source of inoculum of pathogens, the origin of disorders and diseases management in avocado nurseries. *Australasian Plant Pathology* 50, 457–468.

Razdan, V.K. and Gupta, S. (2009) Integrated disease management: concepts and practices. In: Peshin, R. and Dhawan, A.K. (eds) *Integrated Pest Management: Innovation-Development Process*. Springer, Dordrecht, The Netherlands, pp. 369–389.

Reeksting, B.J., Taylor, N.J. and van den Berg, N. (2014) Flooding and *Phytophthora cinnamomi*: effects on photosynthesis and chlorophyll fluorescence in shoots of non-grafted *Persea americana* (Mill.) rootstocks differing in tolerance to Phytophthora root rot. *South African Journal of Botany* 95, 40–53.

Richter, B.S., Ivors, K., Shi, W. and Benson, D.M. (2011) Cellulase activity as a mechanism for suppression of Phytophthora root rot in mulches. *Phytopathology* 101, 223–230.

Rivera, M.J., Martini, X., Conover, D., Mafra-Neto, A., Carrillo, D. and Stelinski, L.L. (2020) Evaluation of semiochemical based push-pull strategy for population suppression of ambrosia beetle vectors of laurel wilt disease in avocado. *Scientific Reports* 10:2670.

Roberts, J.M.K., Jooste, A.E.C., Pretorius, L.-S. and Geering, A.D.W. (2023) Surveillance for avocado sunblotch viroid utilising the European honey bee (*Apis mellifera*). *Phytopathology* 113, 559–566.

Rodríguez-Padrón, C., Rodríguez, A. and Siverio, F. (2019) Survey in nurseries and irrigation water reservoirs as sources of oomycetes found in avocado orchards in the Canary Islands. *Plant Disease* 103, 1264–1274.

Rodríguez-Padrón, C., Siverio, F. and Pérez-Sierra, A. and Rodriguez, A. (2018) Isolation and pathogenicity of *Phytophthora* species and *Phytopythium vexans* recovered from avocado orchards in the Canary Islands, including *Phytophthora niederhauserii* as a new pathogen of avocado. *Phytopathologia Mediterranea* 57, 89–106.

Rodríguez-Polanco, E., Reina-Noreña, J.A., Tamayo-Molano, P.J., Rodríguez-Polanco, L.A. and Varón-Devia, E.H. (2020) Validation of black spot [*Pseudocercospora purpurea* (Cooke) Deighton] management strategies in avocado crops in northern Tolima (Colombia). *Revista Colombiana de Ciencias Hortícolas* 14, 178–191.

Rugman-Jones, P.F., Au, M., Ebrahimi, V., Eskalen, A., Gillett, C.P.D.T. *et al.* (2020) One becomes two: second species of the *Euwallacea fornicatus* (Coleoptera: Curculionidae: Scolytinae) species complex is established on two Hawaiian Islands. *PeerJ* 8:e9987.

Sankaran, S., Ehsani, R., Inch, S.A. and Ploetz, R.C. (2012) Evaluation of visible-near infrared reflectance spectra of avocado leaves as a non-destructive sensing tool for detection of laurel wilt. *Plant Disease* 96, 1683–1689.

Sarker, S.R., McComb, J., Hardy, G.E.S.J. and Burgess, T.I. (2023) Sample volume affects the number of *Phytophthora* and *Phytopythium* species detected by soil baiting. *European Journal of Plant Pathology* 166, 303–313.

Saucedo-Carabez, J.R., Téliz-Ortiz, D., Ochoa-Ascencio, S., Ochoa-Martínez, D., Vallejo-Pérez, M.R. and Beltrán-Peña, H. (2014) Effect of avocado sunblotch viroid (ASBVd) on avocado yield in Michoacan, Mexico. *European Journal of Plant Pathology* 138, 799–805.

Saucedo Carabez, J.R., Téliz Ortiz, D., Vallejo Pérez, M.R. and Beltrán Peña, H. (2019) The avocado sunblotch viroid: an invisible foe of avocado. *Viruses* 11:491.

Schoeman, M.H. and Kallideen, R. (2018) Cercospora spot on avocado - a preliminary report on the relook at the epidemiology of the pathogen. *South African Avocado Growers' Association Research Report* 41, 86–91.

Schoeman, M.H. and Kallideen, R. (2020) Cercospora spot of avocado - validation of spray timing: final results. *South African Avocado Growers' Association Yearbook* 43, 64–69.

Schoeman, M.H. and Manicom, B.Q. (2000) Report on the epidemiology of pepper spot on 'Hass' avocado. *South African Avocado Growers' Association Yearbook* 23, 95–97.

Schoeman, M.H. and Manicom, B.Q. (2001) Further work on the epidemiology of pepper spot. *South African Avocado Growers' Association Yearbook* 24, 29, 31–32.

Semancik, J.S. (2003) Avocado sunblotch viroid. In: Hadidi, A., Flores, R., Randles, J. and Semancik, J. (eds) *Viroids: Properties, Detection, Diseases and their Control*. CSIRO Publishing, Clayton, Australia, pp. 171–177. Available at: https://ebooks.publish.csiro.au/content/viroids#tab-info.

Semancik, J.S. and Szychowski, J.A. (1994) Avocado sunblotch disease- a persistent viroid infection in which variants are associated with differential symptoms. *Journal of General Virology* 75, 1543–1549.

Shakya, S.K., Grünwald, N.J., Fieland, V.J., Knaus, B.J., Weiland, J.E. *et al.* (2021) Phylogeography of the wide-host range panglobal plant pathogen *Phytophthora cinnamomi*. *Molecular Ecology* 30, 5164–5178.

Sharma, G., Maymon, M. and Freeman, S. (2017) Epidemiology, pathology and identification of *Colletotrichum* including a novel species associated with avocado (*Persea americana*) anthracnose in Israel. *Scientific Reports* 7:15839.

Sher, S.A. (1959) Nematodes in avocados. *California Avocado Society Yearbook* 43, 91–93.

Shih, H.-H., Wuest, C.E., Fraedrich, S.W., Harrington, T.C. and Chen, C.-Y. (2018) Assessing the susceptibility of Asian species of Lauraceae to the laurel wilt pathogen *Raffaelea lauricola*. *Taiwan Journal of Forest Science* 33, 173–184.

Shivas, R.G. and Tan, Y.P. (2009) A taxonomic re-assessment of *Colletotrichum acutatum*, introducing *C. fioriniae* comb. et stat. nov and *C. simmondsii* sp. nov. *Fungal Diversity* 39, 111–122.

Short, D.P.G., O'Donnell, K., Stajich, J.E., Hulcr, J., Kijimoto, T. *et al*. (2016) PCR multiplexes discriminate *Fusarium* symbionts of invasive *Euwallacea* ambrosia beetles that inflict damage on numerous tree species throughout the United States. *Plant Disease* 101, 233–240.

Siahaan, S.A.S., Hidayat, I., Kramadibrata, K., Meeboon, J. and Takamatsu, S. (2016) *Podosphaera perseae-americanae*, a new powdery mildew species on *Persea americana* (avocado) from Indonesia. *Mycoscience* 57, 417–421.

Silva-Rojas, H. and Ávila-Quezada, G. (2011) Phylogenetic and morphological identification of *Colletotrichum boninense*: a novel causal agent of anthracnose in avocado. *Plant Pathology* 60, 899–908.

Skelton, J., Johnson, A.J., Jusino, M.A., Bateman, C.C., Li, Y. and Hulcr, J. (2019) A selective fungal transport organ (mycangium) maintains coarse phylogenetic congruence between fungus-farming ambrosia beetles and their symbionts. *Proceedings of the Royal Society B* 286, 20182127.

Slippers, B. and Wingfield, M.J. (2007) Botryosphaeriaceae as endophytes and latent pathogens of woody plants: diversity, ecology and impact. *Fungal Biology Reviews* 21, 90–106.

Smith, E.M., Kotzé, J.M. and Wehner, F.C. (1985) Sooty blotch of avocado caused by *Akaropeltopsis* spp. *Phytophylactica* 17, 101–102.

Smith, S.M., Gomez, D.F., Beaver, R.A., Hulcr, J. and Cognato, A.I. (2019) Reassessment of the species in the *Euwallacea fornicatus* (Coleoptera: Curculionidae: Scolytinae) complex after the rediscovery of the 'lost' type specimen. *Insects* 10(9), 261.

Soares, M.G.O., Alves, E., Silveira, A.L., Pereira, F.D. and Guimarães, S.S.C. (2021) *Colletotrichum siamense* is the main aetiological agent of anthracnose of avocado in south-eastern Brazil. *Plant Pathology* 70, 154–166.

Spence, D.J., Smith, J.A., Ploetz, R., Hulcr, J. and Stelinski, L.L. (2013) Effect of chipping on emergence of the redbay ambrosia beetle (Coleoptera: Curculionidae: Scolytinae) and recovery of the laurel wilt pathogen from infested wood chips. *Journal of Economic Entomology* 106, 2093–2100.

Stewart-Wade, S.M. (2011) Plant pathogens in recycled irrigation water in commercial plant nurseries and greenhouses: their detection and management. *Irrigation Science* 29, 267–297.

Stirling, A.M., Hayward, A.C. and Pegg, K.G. (1992) Evaluation of the biological control potential of bacteria isolated from a soil suppressive to *Phytophthora cinnamomi*. *Australasian Plant Pathology* 21, 133–142.

Stouthamer, R., Rugman-Jones, P., Thu, P.Q., Eskalen, A., Thibault, T. *et al*. (2017) Tracing the origin of a cryptic invader: phylogeography of the *Euwallacea fornicatus* (Coleoptera: Curculionidae: Scolytinae) species complex. *Agricultural and Forest Entomology* 19, 366–375.

Suárez, S.N., Saucedo, J.R., Sanahuja, G. and López, P. (2018) First report of a putative new *Erysiphe* sp. causing powdery mildew on avocado in Florida. *Plant Disease* 102:448.

Tucker, C.M. (1929) Report of the plant pathologist. *Report of the Puerto Rico Agricultural Experiment Station* 1928, 29–35.

Twizeyimana, M., Forster, H., McDonald, V., Wang, D.H., Adaskaveg, J.E. and Eskalen, A. (2013a) Identification of pathogenicity of fungal pathogens associated with stem-end rot of avocado in California. *Plant Disease* 97, 1580–1584.

Twizeyimana, M., McDonald, V., Mayorquin, J.S., Wang, D.H., Na, F., Akgül, D.S. and Eskalen, A. (2013b) Effect of fungicide application on the management of avocado branch canker (formerly Dothiorella canker) in California. *Plant Disease* 97, 897–902.

Ujat, A.H., Ono, T., Hattori, Y. and Nakashima, C. (2023) Re-examination of several *Elsinoë* species reported from Japan. *Mycobiology* 51, 122–138.

Umeda, C., Eskalen, A. and Paine, T.D. (2016) Polyphagous shot hole borer and Fusarium Dieback in California. In: Paine, T.D. and Lieutier, F. (eds) *Insects and Diseases of Mediterranean Forest Systems*. Springer International Publishing, Cham, Switzerland, pp. 757–768.

USDA Forest Service. (2023) Laurel Wilt Public Dashboard, online. Available at: https://www.arcgis.com/apps/dashboards/d43391c8fdb741b597e6ccf1236d2a02 (accessed 2 July 2023).

Valencia, A.L., Gil, P.M., Latorre, B.A. and Rosales, I.M. (2019) Characterization and pathogenicity of Botryosphaeriaceae species obtained from avocado trees with branch canker and dieback and from avocado fruit with stem end rot in Chile. *Plant Disease* 103, 996–1005.

Valencia, A.L., Torres, R. and Latorre, B.A. (2011) First report of *Pestalotiopsis clavispora* and *Pestalotiopsis* spp. causing postharvest stem end rot of avocado in Chile. *Plant Disease* 95:492.

van den Berg, N., Hartley, J., Engelbrecht, J., Mufamadi, Z., van Rooyen, Z. and Mavuso, Z. (2018) First report of white root rot caused by *Rosellinia necatrix* on *Persea americana* in South Africa. *Plant Disease* 102:1850.

van den Berg, N., Swart, V., Backer, R., Fick, A., Wienk, R., Engelbrecht, J. and Prabhu, S.A. (2021) Advances in understanding defense mechanisms in *Persea americana* against *Phytophthora cinnamomi*. *Frontiers in Plant Science* 12, 636339.

van Rooyen, E., Paap, T., de Beer, W., Townsend, G., Fell, S. *et al.* (2021) The polyphagous shot hole borer beetle: current status of a perfect invader in South Africa. *South African Journal of Science* 117, 1–10.

Vasconcelos, C.V., Pereira, F.T., dos Santos Galvão, C. and Costa Carvalho, D.D. (2016) Occurrence of algal leaf spot (*Cephaleuros virescens* Kunze) on avocado in Goiás State, Brazil. *Summa Phytopathologica* 42:108.

Velasco, L., López-Herrera, C. and Cretazzo, E. (2020) Two novel partitiviruses that accumulate differentially in *Rosellinia necatrix* and *Entoleuca* sp. infecting avocado. *Virus Research* 285, 198020.

Velázquez-del Valle, M.G., Campos-Martínez, A., Flores-Moctezuma, H.E., Suárez-Rodríguez, R., Ramírez-Trujillo, J.A. and Hernández-Lauzardo, A.N. (2016) First report of avocado anthracnose caused by *Colletotrichum karstii* in México. *Plant Disease* 100:534.

Vida, C., Bonilla, N., de Vicente, A. and Cazorla, F.M. (2016) Microbial profiling of a suppressiveness-induced agricultural soil amended with composted almond shells. *Frontiers in Microbiology* 7:4.

Vitale, A., Aiello, D., Guarnaccia, V., Perrone, G., Stea, G. and Polizzi, G. (2012) First report of root rot caused by *Ilyonectria* (= *Neonectria*) *macrodidyma* on avocado (*Persea americana*) in Italy. *Journal of Phytopathology* 160, 156–159.

Vitale, A., Castello, I., D'Emilio, A., Mazzarella, R., Perrone, G., Epifani, F. and Polizzi, G. (2013) Short-term effects of soil solarization in suppressing *Calonectria* microsclerotia. *Plant and Soil* 368, 603–617.

Volcani, Z. (1959) Bacterial soft rot of avocado fruit. *Bulletin of the Research Council of Israel* 7D, 39–42.

Wallace, J.M. and Drake, R.J. (1962) A high rate of seed transmission of avocado sunblotch virus from symptomless trees and the origin of such trees. *Phytopathology* 52, 237–241.

Weir, B.S., Johnston, P.R. and Damm, U. (2012) The *Colletotrichum gloeosporioides* species complex. *Studies in Mycology* 73, 115–80.

Westfalia Fruit. (2021) Our Rootstocks, online. Available at: https://www.westfaliafruit.com/for-growers/our-rootstocks/ (accessed 10 January 2024).

Whiley, A.W., Hargreaves, P.A., Pegg, K.G., Doogan, V.J., Ruddle, L.J., Saranah, J.B. and Langdon, P.W. (1995) Changing sink strengths influence translocation of phosphonate in avocado (*Persea americana* Mill.) trees. *Australian Journal of Agricultural Research* 46, 1079–1090.

Whiley, A.W., Pegg, K.G., Saranah, J.B. and Langdon, P.W. (1987) Influence of Phytophthora root rot on mineral nutrient concentrations in avocado leaves. *Australian Journal of Experimental Agriculture* 27, 123–127.

Whiley, T. and Dann, L. (2021) Procado®: a new Australian avocado rootstock. *Talking Avocados* 32, 57–64.

Willingham, S.L., Coates, L.M., Cooke, A.W. and Dean, J.R. (2004) Tree vigour influences disease susceptibility of 'Hass' avocado fruits. *Australasian Plant Pathology* 33, 17–21.

Willingham, S.L., Cooke, A.W., Coates, L.M. and Pegg, K.G. (2000) Pepper spot: a new preharvest Colletotrichum disease of avocado cv. Hass. *Australasian Plant Pathology* 29, 151–151.

Willingham, S.L., Pegg, K.G., Anderson, J.M., Cooke, A.W., Dean, J.R., Giblin, F.R. and Coates, L.M. (2006) Effects of rootstock and nitrogen fertiliser on postharvest anthracnose development in Hass avocado. *Australasian Plant Pathology* 35, 619–629.

Wolstenholme, B.N., Moore-Gordon, C. and Ansermino, S.D. (1996) Some pros and cons of mulching avocado orchards. *South African Avocado Growers' Association Yearbook* 19, 87–91.

Wolstenholme, B.N. and Sheard, A. (2010) Integrated management of Phytophthora root rot the 'Pegg wheel' updated. *South African Avocado Growers' Association Avoinfo Newsletter* 175, 11–15.

World Meteorological Organization (2024) *State of the Global Climate 2023*. World Meteorological Organization, Geneva, Switzerland, p. 47. Available at: https://wmo.int/sites/default/files/2023-11/WMO%20Provisional%20State%20of%20the%20Global%20Climate%202023.pdf (accessed 21 March 2024).

Wuest, C.E., Harrington, T.C., Fraedrich, S.W., Yun, H.Y. and Lu, S.S. (2017) Genetic variation in native populations of the laurel wilt pathogen, *Raffaelea lauricola*, in Taiwan and Japan and the introduced population in the United States. *Plant Disease* 101, 619–628.

Zea-Bonilla, T., González-Sánchez, M.A., Martín-Sánchez, P.M. and Pérez-Jiménez, R.M. (2007) Avocado dieback caused by *Neofusicoccum parvum* in the Andalucia region, Spain. *Plant Disease* 91:1052.

Zentmyer, G.A. (1949) Verticillium wilt of avocado. *Phytopathology* 39, 677–682.

Zentmyer, G.A. (1955) Diseases of the avocado. *California Avocado Society Yearbook* 39, 44–58.

Zentmyer, G.A. (1979) Stimulation of sexual reproduction in the A2 mating type of *Phytophthora cinnamomi* by a substance in avocado roots. *Phytopathology* 69, 1129–1131.

Zentmyer, G.A. (1981) The effect of temperature on growth and pathogenesis of *Phytophthora cinnamomi* and on growth of its avocado host. *Phytopathology* 71, 925–928.

Zentmyer, G.A. (1984) Avocado diseases. *Tropical Pest Management* 30, 388–400.

Zentmyer, G.A., Jefferson, L., Huckman, C.J. and Yung, C.H. (1974) Studies of *Phytophthora citricola* isolated from *Persea americana*. *Mycologia* 66, 830–845.

Zentmyer, G.A., Klure, L.J. and O'Neal, E. (1976) Trunk canker of avocado caused by *Phytophthora heveae*. *California Avocado Society Yearbook* 60, 169–170.

Zhang, Y., Zhang, J., Vanderpool, D., Smith, J.A. and Rollins, J.A. (2020) Genomic and transcriptomic insights into *Raffaelea lauricola* pathogenesis. *BMC Genomics* 21:570.

Zhou, Y., Avery, P.B., Carrillo, D., Duncan, R.H., Lukowsky, A., Cave, R.D. and Keyhani, N.O. (2018) Identification of the Achilles heels of the laurel wilt pathogen and its beetle vector. *Applied Microbiology and Biotechnology* 102, 5673–5684.

Zilberstein, M., Noy, M., Levy, E., Elkind, G., Zeidan, M., Teverovski, E. and Ben Ze'ev, I. (2007) Wilting disease of young avocado trees caused by *Neonectria radicicola* in Israel. In: *Proceedings of the 6th World Avocado Congress*, Viña del Mar, Chile.

Zumaquero, A., Martínez-Ferri, E., Matas Arroyo, A., Reeksting, B., Olivier, N.A. *et al*. (2019) *Rosellinia necatrix* infection induces differential gene expression between tolerant and susceptible avocado rootstocks. *PLOS One* 14:e0212359.

14 Insect and Mite Pests

D. Carrillo[1,*], M.S. Hoddle[2], M. Aluja[3] and J.E. Peña[1]
[1]*Tropical Research and Education Center, University of Florida, Homestead, USA;* [2]*Department of Entomology, University of California, Riverside, USA and Center for Invasive Species Research, University of California, Riverside, USA;* [3]*Instituto de Ecologia - INECOL, Xalapa, Mexico*

Abstract
This chapter provides a comprehensive review of avocado arthropod pests, including mites and insects from the native range of avocado and areas where it has been introduced. A special emphasis is placed on quarantinable and key economic pests that limit avocado production. As avocado production expands to new geographic areas, new pests emerge threatening the sustainability of avocado production and causing economic damage.

Multiple arthropod species attack and cause economic damage to avocados. The majority are arthropods indigenous to the neotropics that specialize in feeding on avocado and other Lauraceous plants. Many others are polyphagous (feeding on many different plant species) arthropods that feed on avocados in growing areas around the world where the plant has been introduced. In their native range, avocados are affected by a complex of fruit and stem borers, and external feeders that damage leaves, fruit and stems. These arthropods are often considered quarantine pests subject to specific mitigation tactics to prevent their introduction into other avocado-producing areas. In Asia, Africa, Oceania and Europe, where avocados have been introduced, the main pests are primarily native polyphagous insects that feed on avocados. A few arthropod species of American origin, primarily mites, have been introduced in these areas.

In Mexico, part of the native habitat of avocado, Hernandez *et al.* (2000), Salgado (1993) and SENASICA (2018) provide a good compilation of major pest complexes. The main regulated pests are the big avocado seed weevil, *Heilipus lauri* Boheman; the small avocado seed weevils *Conotrachelus aguacatae* Barber and *C. perseae* Barber; the branch borer weevils, *Copturus aguacatae* Kissinger (= *Macrocopturus aguacatae*) (all Coleoptera: Curculionidae) and the avocado seed moth, *Stenoma catenifer* Walsingham (Lepidoptera: Depressariidae). Other economically important pests include the persea mite (*Oligonychus persea* Tuttle, Baker and Abbatiello) and avocado thrips (*Scirtothrips perseae* Nakahara).

In the past decade, avocado production in Colombia evolved from primarily small farming units to a large industry focused on international exports. The fruit borers, *S. catenifer, H. lauri* and *Heilipus trifasciatus* Fabricius; and the stemborer, *Copturomimus hustachei* Kissinger (=*C. perseae*), are regulated quarantine pests. Other major pests of economic importance in Colombia include

*Correspondence: dancar@ufl.edu

(eds D. Carrillo, B. Schaffer, A.W. Whiley and B.N. Wolstenholme)
DOI: 10.1079/9781800621824.0014

Astaena pygidialis Kirsch (Coleoptera: Scarabaeidae), *Frankliniella gardeniae* Moulton (Thysanoptera: Thripidae), *Monalonion velezangeli* Carvalho & Costa (Hemiptera: Miridae) and *Bruggmanniella persea* Gagne (Diptera: Cecidomyiidae) (Carabalí Muñoz *et al.*, 2021).

In Brazil, *S. catenifer* and the stem borer *Heilipus catagraphus* Germar (Coleoptera: Curculionidae) are quarantine pests and significant constraints to avocado production (Hohmann *et al.*, 2000, 2003; Nava *et al.*, 2005a, 2006, 2007; Furtado-Moreira, 2022). *Heilus freyreissi* Boheman (Coleoptera: Curculionidae) has also been reported to feed on tender lateral branches, the central leaf vein, inflorescences, peduncles and young fruit (Reis *et al.*, 2018). Chile is free of quarantine pests that are present in other avocado-producing countries but has several economically important species, including thrips (*Heliothrips haemorrhoidalis* Bouché), mealybugs (*Pseudococcus calceolariae* Maskell and *Pseudococcus longispinus* Targioni-Tozzetti), scales (*Hemiberlesia lataniae* Signoret, *Aspidiotus nerii* Bouché, *Hemiberlesia rapax* Comstock and *Saissetia oleae* Olivier) and mites (*Oligonychus yothersi* McGregor) (Luppichini *et al.*, 2013).

In Peru, *S. catenifer* is the main quarantine pest. Other economic important pests include bagworms (*Oiketicus kirbyi* Guilding), caterpillars (*Oxydia vesulia* Cramer), stem borers (*Cratosomus* sp.) thrips (*H. haemorrhoidalis*), mites (*Oligonychus punicae* Hirst, *Oligonychus beeri* Estebanes & Baker and *O. yothersi*), whiteflies (*Aleurodicus coccois* Curtis, *A. coccolobae*, *Palareyrodes sp.* and *Aleurotrachellus* sp. *Aleurodicus juleikae* Bondar), mealybugs (*Planococcus citri* and *P. longispinus*) and scales (*Ceroplastes* sp., *S. oleae* Oliver, *Protopulvinaria pyriformis* Cockerell, *Fiorinia fioriniae* Targioni-Tozzetti, *Coccus hesperidum* L., *Hemiberlesia cianophylli* and *H. lataniae*) (Nuñez, 2008; Marin and Mocarro, 2019; personal communication Ms Guillermo Sánchez V.).

In Ecuador, *S. catenifer* is also the main quarantine pest. Economically important pests include mites (*O. yothersi*), bagworms (*O. kirbyi*) and leaf miners (*Gracilaria persea* Busck) (Agrocalidad, 2024).

In California, USA, all serious pests are exotic leaf feeders native to Mexico. These pests include the red-banded whitefly (*Tetraleurodes perseae* Nakahara), persea mite (*O. perseae*), avocado thrips (*S. perseae*), avocado lace bug (*Pseudacysta persea*) and *Neohydatothrips burungae* Hood. Of these five foliar pests, *T. perseae*, *O. perseae* and *S. perseae* were species new to science at the time of first detection in the USA (Hoddle, 2004). Other pests of California avocados are of minor significance because they are biologically controlled by predators, parasites and parasitoids (Dreistadt, 2008).

In Florida, USA, ambrosia beetles that vector fungal pathogens are primary pests of avocado. Of particular importance are the exotic redbay ambrosia beetle (*Xyleborus glabratus* Eichhoff) and native ambrosia beetles (*Xyleborus bispinatus* and *Xyleborus volvulus*) because of their role as vectors of the fungus (*Harringtonia lauricola*) that causes a lethal disease called laurel wilt (Carrillo *et al.*, 2014; Cruz *et al.*, 2018, 2019, 2021). Other pests of importance include avocado lace bug *P. persea* Heidemann, mites (*O. yothersi*), mirids (*Daghbertus* spp.) and red-banded thrips *Selenothrips rubrocinctus* Giard.

In Hawaii, USA, *P. persea* was found in 2019 on the island of O'ahu and later spread and became the main pest of avocado trees on multiple islands (Matsunaga and Silva, 2020). Few insects have been documented on Texas avocados, and none has been severe enough to warrant control measures (Nesbitt *et al.*, 2013).

In the Dominican Republic, thrips (*H. haemorrhoidalis*), gall midges (*B. persea*) and a large complex of Hemipterans, including avocado lace bug (*P. persea*), various mealybugs, whiteflies and scales, are primary pests of avocado (Valdez *et al.*, 2016). Of particular concern in the Caribbean is the recent detection of avocado trees infested by the mango stem borer, *Batocera rufomaculata* De Geer, an insect native to Eurasia, which is a destructive pest of mango and several other fruit trees.

The pest complex in the Old World includes a few New World species that were probably accidentally introduced with avocado plants and many endemic arthropods that feed on them. In Europe, where Spain is the leading avocado producer, *O. perseae*, *O. punicae*, *P. pyriformis*, *Nipaecoccus nipae* Maskell, *H. haemorrhoidalis* and ant species *Linepithema humile* Mayr, *Lasius grandis* Forel and *Pheidole pallidura* Nylander are the main avocado pests (Gil and Aranda, 2021).

Scant information is available from Asian countries where avocado cultivation is relatively recent. In Indonesia oriental fruit fly (*Bactrocera dorsalis* Hendel), red mite (*Tetranychus cinnabarinus* Boisd), Cricula silkmoth (*Cricula trifenestrata* Helfer) and twig borers (*Xyleborus coffeae* Wurth) are reported as major avocado pests. In Israel, some local pests have moved on to avocado (e.g. *S. rubrocinctus*), while others are introduced (e.g. *O. perseae*) (Wysoki *et al.*, 2002; EFSA, 2021). Lists of avocado pests in Israel were prepared by Avidov and Harpaz (1969), Wysoki and Izhar (1978), and Swirski *et al.* (1986, 1991b). The number of known avocado pests in Israel increased from 15 to more than 96 species due to the expansion of avocado cultivation within the country (Bragard *et al.*, 2021).

The most important pests of avocado in sub-Saharan Africa are tephritid fruit flies (*B. dorsalis* and *Ceratitis* spp.) and the false codling moth, *Thaumatotibia leucotreta* Meyrick (Lepidoptera: Tortricidae) (Toukem *et al.*, 2020). In South Africa, sucking insects are regarded as the most important pests. These include mirid bugs (e.g. *Taylorilygus* spp.), various stink bug species (Hemiptera: Pentatomidae) and coreids, including the coconut bug, *Pseudotheraptus wayi* Brown. Other pests include palm fiorinia scale, *F. fioriniae*, which infests avocado leaves and fruit, affecting marketability (De Villiers and Van den Berg, 1987). The long-tailed mealybug *P. longispunus* is also present in avocado orchards in South Africa. Thrips (*S. rubrocinctus* and *H. haemorrhoidalis*), the false codling moth, *T. leucotreta* and several fruit fly species (*Ceratitis* spp.) are considered important pests of avocado in South Africa (Grové, 2022). In Tanzania, thrips have been recorded feeding on cultivated avocados that are exported to Europe (Hoddle and Mound, 2020).

Stevens (1999) considered that the geographical isolation of New Zealand has helped exclude many exotic pests of avocados that are present in other countries. The main pests of New Zealand avocados are several species of native leafrollers, the invasive greenhouse thrips (*H. haemorroidalis*), latania scale (*H. lataniae* Signoret) and six spotted mites (*Eotetranychus sexmaculatus* Riley). In Australia, mirids (*Helopeltis* sp.), Queensland fruit fly *Bactrocera tryoni* Froggatt, *S. rubrocinctus*, red-shouldered leaf beetle (*Monolepta australis* Jacoby), swarming leaf beetles (*Rhyparida* spp.) and leaf rollers (e.g. *Homona spargotis* Meyrick) are important pests.

This chapter describes the most common pests of avocado in different growing areas of the world and summarizes information on management tactics.

Mites

Tetranychidae

Persea mite

Oligonychus perseae (persea mite, previously misidentified as *Oligonychus peruvianus* McGregor) is native to Mexico (Lara *et al.*, 2017) and invaded Costa Rica (1974), California (1990), Israel (2001), Hawaii (2002), Spain (in 2004), Portugal (2005), Florida (2007), Italy (2015), Morocco (2018) and France (2019). Avocado is considered its main host, but it can feed on around 20 different hosts in 17 botanical families (EPPO, 2024).

Persea mites (Fig. 14.1) are found in avocado orchards year-round, and avocado cultivars vary significantly in their susceptibility to infestation (Kerguelen and Hoddle, 2000). Mites prefer mature leaves. In California, populations begin to build in April–June, peak in July–August and decline in September–October (Aponte and McMurtry, 1997b; Kerguelen and Hoddle 1999a,b; Hoddle *et al.*, 2000). An additional peak of infestation is observed in Israel in October–November (Maoz *et al.*, 2011). Persea mite can be present year-round in the Canary Islands (Spain) with one or two peaks from April to July (Torres *et al.*, 2022). Temperature is an important driver of *O. perseae* population dynamics (Salinas, 1992; Ramirez *et al.*, 1993). *O. perseae* proliferates in areas with moderate temperatures and is sensitive to high temperatures above 35°C and low humidity. Abrupt population crashes have been observed under these conditions over several consecutive days.

Mites live, feed and breed on the undersides of avocado leaves in small colonies within the protection of circular silken nests (Aponte and McMurtry, 1997a). Characteristic circular necrotic spots result from feeding (Fig. 14.1). Once necrotic feeding damage on the undersides of leaves approaches ~10% (or approximately 100 mites per leaf), trees begin to drop leaves (Kerguelen and Hoddle, 1999a; Hoddle *et al.*, 2000).

Fig. 14.1. Persea mite, *Oligonychus perseae* Tuttle. A. Adult females, B. Damage, C. Persea mite nests along leaf veins, D. *Neoseiulus californicus* predators inside Persea mite nest. Photo credits: Marta Montserrat Larrosa, Applied Ecology Lab, IHSM 'La Mayora' – UMA – CSIC (with permission).

Maoz *et al.* (2011) recommended that professional pest control experts adopt an action threshold of 50–100 mites per leaf in Israel but cautioned that more research is needed to refine the action threshold. Lara and Hoddle (2015) developed a fixed-size binomial sampling model for California avocados with guidelines on sample tree and leaf selection within blocks of trees. This model requires sampling of ~75 leaves per hectare (30 per acre) and has a ~125-mite action threshold (50-mites per acre). Hernandez *et al.* (2000) suggested the removal of fallen foliage infested with mites from the ground below infested trees as a cultural control method. However, this tactic is not efficacious (Takano-Lee and Hoddle, 2002).

According to Hernandez *et al.* (2000), no known prey-specific predators of *O. perseae* have been reported in Mexico. Consequently, management of persea mite has focused on three major avenues: (i) biocontrol with commercially available predatory mites, (ii) enhancement of naturally occurring predator populations via resource subsidies and (iii) miticides.

Screening of six species of predatory mites in the field (Hoddle *et al.*, 1999) indicated that *Neoseiulus californicus* McGregor (Acari: Phytoseiidae) could provide control equivalent to pesticide applications (Hoddle *et al.*, 1999; Kerguelen and Hoddle, 1999b). However, this was 13–14 times more expensive than pesticides (Hoddle *et al.*, 2000). A modified mist blower was very effective at delivering predators to avocado trees (Takano-Lee and Hoddle, 2001), but this application technology was not sufficiently efficient to reduce costs associated with predator inoculations of mite-infested trees. Generalist predator populations build in response to increasing *O. perseae* populations, but control is often inadequate (Yee *et al.*, 2001a).

In Spain, an omnivorous phytoseiid mite, *Eusieus stipulatus* Athias-Henriot, can be enhanced in avocado orchards via the artificial provision of pollen from deliberately planted maize plants. Increased populations of *E. stipulatus* were correlated with lower *O. perseae* populations (González-Fernández *et al.*, 2009). Additionally, airborne pollen from surrounding olive orchards was correlated with increased populations of *E. stipulatus* in avocado orchards. Subsequent laboratory experiments indicated that olive pollen was suitable for *E. stipulatus* development (González-Fernández *et al.*, 2009). However, Montserrat *et al.* (2013a) reported that high temperatures decreased *E. stipulatus* abundance despite pollen supplementation. Moreover, Montserrat *et al.* (2013b) concluded that *O. perseae* has a much wider range of temperature tolerances with positive per capita growth rates than the predatory mites, and therefore, predator mites cannot respond to *O. perseae* population growth when environmental conditions are unfavourable for predators but still suitable for *O. perseae*. In Israel, *Euseius scutalis* Athias-Henriot, previously described as *Euseius rubini* Swirski and Amitai, was effective in reducing persea mite populations. *E. scutalis* responds positively to resource subsidization via windblown pollen from patches of Rhodes grass or artificial application of pollen to trees with an electrostatic sprayer (Maoz *et al.*, 2009). Field surveys concluded that pollen subsidization generally promoted higher predator populations which were correlated with lower pest densities, but this was not significantly different from control treatments that did not receive pollen (Maoz *et al.*, 2009). Significant practical and economic challenges confront the use of pollen for enhancing predator mite populations in avocado orchards.

In California, the standard industry control practice for persea mite is the application of abamectin combined with horticultural oil. Abamectin and horticultural oil are also used in combination with control populations of avocado thrips (*S. perseae*). Consequently, pest populations in avocado orchards can be exposed to two abamectin applications per year. Abamectin application has a very long period (~10–12 weeks) of activity against targeted avocado pests because, in combination with highly refined horticultural oil, it provides translaminar activity for abamectin, thereby protecting this product from UV degradation (Morse *et al.*, 2000), but resistance development is a concern (Humeres and Morse, 2005). To combat this potential problem, in California, insecticide rotation programmes to manage resistance development are encouraged. In Mexico, the use of hot-pepper extracts failed to provide significant mite control (Reyes *et al.*, 1995). The best chemical control was obtained with organophosphates and the poorest control with sulfur applications (Andrade, 1988). Martinez (1989) suggested that *O. perseae* has developed resistance to organophosphates in the states of Morelos and Michoacán in Mexico because of the overuse of broad-spectrum contact insecticides.

Persea mite is not considered a pest in Florida, where it has been detected periodically but populations do not establish or spread, suggesting that persea mite requires a Mediterranean climate to build damaging populations (D. Carrillo, personal observation).

Avocado red mite

In Florida, Colombia, Ecuador, Brazil and Chile, avocado red mite (*Oligonychus yothersi*) is a common pest of avocados (Vargas and Rodriguez, 2008) (Fig 14.2). Feeding is first confined to the upper surface of the leaves along the midrib, then along secondary leaf veins. The areas along the veins become reddish-brown. During heavy infestations, leaves can be covered with cast exoskeletons shed by feeding mites. Leaf damage is regularly observed from October through February, causing up to a 30% reduction of photosynthetic activity (Schaffer *et al.*, 1986). Leaves affected by red mite feeding regularly drop earlier (45–60 days after infestation) than non-infested leaves. According to Rioja *et al.* (2016, 2021), avocado shoots infested by *O. yothersi* increase the attraction of micro-coleopterans, including natural enemies of this mite such as *Parastethorus histrio* (Chazeau) and *Oligota pygmaea* (Solier). Red mite is a sporadic pest, occasionally observed in some orchards but is seldom problematic in others. Periodic inspections for red mite are recommended during December, January and February (in the northern hemisphere). Control measures may begin when populations reach six or more mites per leaf (Peña and Johnson, 1999). Duration of life stages varies from 14 to 15 days. Females are

Fig. 14.2. Avocado red mite, *Oligonychus yothersi* McGregor. A. Avocado red mite colony, B. Adult female mites.

capable of laying 40–50 eggs during their lifespan (Peña and Johnson, 1999). In Chile, the lifespan is completed in 27 days, with highest density peaks observed between January and May and with faster development on mature rather than young leaves (Vargas and Rodriguez, 2008; Rioja *et al.*, 2019).

In Florida, few miticides are registered for use on avocados when fruit is present. Sulfur or oil emulsion sprays are recommended (Peña and Johnson, 1999). In Chile, mineral oil is considered effective against eggs and motile stages (Vargas and Rodriguez, 2008). In Peru, entomopathogenic bacteria and fungi suppressed laboratory and field populations of this mite (Parco-Quinchori *et al.*, 2023). Natural enemies of *O. yothersi* in Chile include *Stethorus histrio* Chazeau, the staphylinid beetle *(Oligota pygmea* Solier) and the phytoseiid (*Cydnodromus picanus* Ragusa) (Vargas and Rodriguez, 2008). Severe outbreaks of red mite have been observed after the use of neonicotinoid insecticides when applied as foliar sprays or delivered through irrigation systems.

Avocado brown mite

Avocado brown mite (*Oligonychus punicae*) is present in the Americas, Asia and Africa but has only been reported as an avocado pest mite in Mexico, California, Peru and Chile. Hernandez *et al.* (2000) suggested that *O. punicae* originated in the south of Mexico and northern Guatemala. *O. punicae* are found on the upper leaf surface, causing leaf bronzing and reduction of photosynthetic activity. It is considered a minor avocado pest in southern California that flares up occasionally in dry periods when road dust is heavy on trees, or excessive ash from wildfires is deposited on leaves (McMurtry, 1985a; Aponte and McMurtry, 1997a,b; Dreistadt, 2008).

In Mexico, different avocado cultivars show varying susceptibilities to *O. punicae* infestations. Reyes and Salgado (1994) demonstrated that the avocado cultivars '175 PLS', '54PLS', '131PLS', '120PLS', '18PLS', '137PLS' and '30PLS' show some tolerance against *O. punicae*, while Colin and Rubi (1992) indicated that 'Colinmex' and '148PLS' were moderately susceptible. The cultivars 'Rincon' and 'Fuerte' are considered tolerant to brown mite Gallegos (1983). One brown mite generation can be completed in 15.4 days under laboratory conditions at 22°C (Hernandez *et al.*, 2000). Avocado brown mite densities are reduced during periods of low ambient temperature and high humidity. Coria-Avalos (1993) recommended sulfur and propargite applications 3–4 times during the dry season between December and May in Mexico for control of brown mite. The phytoseiid mites, *Amblyseius limonicus* Garman and McGregor, *Typhlodromus floridanus* Muma and the coccinellid *Stethorus picipes* Casey, exhibit good predatory potential against *O. punicae* (McMurtry and Johnson, 1966; Jeppson *et al.*, 1975; McMurtry, 1985a,b; Dreistadt, 2008).

Tea red mite

Tea red mite (*Oligonychus coffeae* Nietner) is widely distributed in the Americas, Asia and Africa but has only been reported infesting avocado in South Africa, Australia and India (Waite and Pinese, 1991; Srikumar *et al.*, 2015). It is known to attack avocados, cashews, citrus, coffee, guava, mango, tea, cotton, grape, mulberry, protea, rubber, strawberry and about 30 other plant species (Meyer, 1987). *O. coffeae* prefers the adaxial surface of mature leaves but will be found on both leaf surfaces during severe infestations (Meyer, 1987). Injury is first seen as yellowish spotting along the midrib and veins and occasionally on the petioles. As the mites continue to feed, these patches darken, until the entire leaf becomes deeply bronzed, necrotic and abscissed (Meyer, 1987; De Villiers, 1998).

In Australia, tea red mite is a common pest of avocados in eastern states, particularly when knockdown insecticides are used. Being surface dwellers, heavy summer rains may reduce populations by washing mites from leaves.

Das (1955) reported that the life cycle of *O. coffeae* on tea in India takes 9–12 days in summer and about 30 days in winter. Under these conditions, 22 generations can occur annually. Tea red mites are spread by wind, animals, farm implements and labourers (Meyer, 1987).

Many natural enemies, mostly predacious mites, are associated with *O. coffeae*. In Australia, the main predators in coastal areas are coccinellids, in particular *Stethorus* spp., while in drier inland areas, phytoseiid mites are important (Waite and Pinese, 1991). Coccinellids, especially *Stethorus madecassus* Chazeau in the Malagasy Republic, were found feeding on eggs and other stages of *O. coffeae*. The larvae of *Chrysopa* spp. also attack the active stages of this mite (Blommers and Guttierrez, 1975).

In South Africa, chemical control of the tea red mite is usually not necessary since natural enemies keep population densities low. In Australia, approximately 20% leaf bronzing can be tolerated before control action needs to be taken (Waite and Pinese, 1991).

Sixspotted mite

Sixspotted mite (*Eotetranychus sexmaculatus*) was introduced into Western Australia, Australia in 1986, apparently on budwood imported illegally from California, where it is a minor pest (Dreistadt, 2008). *E. sexmaculatus* has been present in New Zealand since the 1950s, but it only became a serious pest in the 1990s, causing leaf drop and reducing productivity and yield. *Stethorus* sp. was the most common predator observed actively feeding on *E. sexmaculatus* early in the season. Phytoseiids were more abundant later in the season but there was a lack of correlation between these predators and pest populations. The acaricides abamectin, milbemectin and fenpyroximate effectively reduce mite numbers (Steven, 2004).

Jamieson and Stevens (2007) determined *E. sexmaculatus* development, longevity and fecundity at constant temperatures. Total development from egg to adult ranged from 29.6 days at 18°C to 11 days at 30°C. Mean adult longevity ranged from 18.8 days at 30°C to 41.4 days at 18°C. Jamieson and Stevens (2007) estimated that 202.8 degree-days above a threshold temperature of 12.2°C were required to complete development from egg to adult.

Leaf drop associated with high populations of *E. sexmaculatus* represents an extra stress factor just before and during flowering. Accordingly, monitoring and control measures are recommended before flowering. Mite populations often decline naturally in midsummer (Avogreen® Manual, 2010).

Eriophyidae

Avocado bud mite

In Brazil, avocado bud mite, *Tegolophus perseaflorae* (Keifer), injures flowers and contributes to flower drop. Similarly, in Florida, *T. perseaflorae* is observed in developing flower and leaf buds. Peña and Denmark (1996) related the presence of this mite to excessive flower drop and fruit deformation. Avocado bud mites cause necrotic spots and subcircular and irregular openings on apical leaves. Feeding by this mite may cause fruit deformation and discolouration. The adult avocado bud mite has a yellowish appearance. In Florida, avocado bud mite populations increase from March to May (Peña and Johnson, 1999), but its life cycle has not been determined.

Other mite species

In Peru, *Paratetranychus* sp. and *Panonychus citri* McGregor are problematic when avocados are grown together with citrus trees (Wolfe *et al.*, 1969).

In Israel, *Tetranychus* species are rarely found in avocado orchards. However, in the early 1980s, the carmine mite, *Tetranychus cinnabarinus* Boisduval and two-spotted mite (*Tetranychus urticae* Koch) were found causing leaf damage in avocado orchards adjacent to cotton fields, which were treated aerially with synthetic pyrethroids. Occasionally, the oriental spider mite (*Eutetranychus orientalis* Klein) causes considerable damage to the leaves of 'Pinkerton', especially in arid areas (Swirski *et al.*, 1988).

The broad mite, *Polyphagotarsonemus latus* (Banks) (Tarsonemidae), sometimes attacks avocado seedlings in glasshouses, causing rolling and browning of young leaves, and inflicts severe harm to apical buds which reduces development of shoots by lateral buds, and results in dwarfing of the seedlings (Waite and Pinese, 1991; Swirski *et al.*, 1998). There is also some indication that *Tuckerella pavoniformis* (Acari: Tuckerellidae) feeding can damage avocado fruit (Escobar-Garcia *et al.*, 2022).

The eriophyid mite, *Calepitrimerus muesbecki* Keifer, feeds inside vegetative buds causing bronzing and virus-like chlorosis damage to the foliage of avocado in Guatemala (Velasquez and Santizo, 1992).

Insects

Termites

Termites are occasional pests of avocado in different production areas. Africa has a large number of termite species, with many of these present in areas where avocados are grown. Termites live in nests and feed on cellulose in wood, grass, humus and other plant matter. Termites mainly destroy roots and lower stems of young avocado trees but may also girdle and kill large trees (Fig. 14.3). Damage is mostly to plants with low vigour or new transplants. Termites are more destructive during the first year after planting. Trees become less susceptible once they are well established (Webb, 1974). The taxonomy, biology, distribution and ecology of termites in Africa were studied by Coaton and Sheasby (e.g. Coaton, 1971, 1974; Coaton and Sheasby, 1972, 1978) and Skaife (1956). Ruelle (1985) provided a synopsis of the termites of Southern Africa. In Chile, termites are observed mostly on trees over 20 years old (Luppichini and Ripa, 2008; Ripa and Luppichini, 2008b). Three groups of termites are economically important in avocado orchards, all in the family Termitidae.

Fig. 14.3. Termite damage to avocado.

1. Large fungus growers in the genera *Macrotermes* and *Odontotermes*.

2. Small fungus growers in the genera *Microtermes*, *Allodontotermes* and *Ancistrotermes*.

3. Carton nest builders of the genus *Microcerotermes*. Members of this group are of less importance than those of the other two groups.

Natural enemies of termites include ants, especially in the genera *Dorylus* and *Anoplolepis*. Fungi and parasitic nematodes play an important role in causing mortality of termite nests. Anteaters dig open and can destroy entire nests, while various bird species scratch open the protective clay passageways to expose termites which they then eat. Various animals, including insects, birds and mammals, feed on winged termites that emerge after spring rains. This natural mortality plays a large role in limiting the establishment of new nests (Gouse *et al.*, 1998). According to Gouse *et al.* (1998), it is often difficult to find nests of *Odontotermes badius* Havil, especially after the soil has been prepared. Termite nests can occur at a depth of 300 mm to more than 3 m below the soil surface (Hartwig, 1966). Termite mounds are usually a slightly raised area, 100–300 mm high and approximately 1.5 m in diameter. The nest of *O. badius* can be detected from discarded soil particles cemented together to form small compact heaps close to the nest. During humid weather, numerous small white mushrooms can be seen on these sites for a few days.

Some nests of *Odontotermes* spp. are characterized by having air pipes or 'chimneys' of clay. *Macrotermes* spp. construct above ground heaps with a thick wall. *Ancistrotermes* and *Microtermes* make small underground nests spread over a wide area, with no above ground signs to indicate nest presence. The nests of *Microcerotermes* spp. are completely underground (Gouse *et al.*, 1998).

Winged adults, called alates, fly out of the nests during late afternoon after the spring rains, usually coming out of passages that open directly into the central nest cavity. These passages are most suitable for fumigation (Gouse *et al.*, 1998).

Hemipterans

Whiteflies and blackflies

In Chile, *Trialeurodes vaporariorum* (Westwood) affects avocado seedlings and grafting material growing in greenhouses (Ripa *et al.*, 2008a). In Mexico, *Tetraleurodes* spp. nymphs and adults are found on the underside of leaves and are more abundant during June–November. During heavy attacks, leaves from the lower tree canopy deteriorate, and the injury may cause defoliation. *Paraleyrodes perseae* (Quaintance) is found close to the midvein of leaves. Its feeding causes chlorotic circular leaf spots. Heavy damage may be associated with low flower production and defoliation. Population peaks are observed from June–November (Coria-Avalos, 1993). In Cuba, *Aleurodicus cordini* Back and *Aleurovitreus insignis* (Bondar) are avocado pests (Pérez-Artiles *et al.*, 2020). Biocontrol agents attacking these whitefly species include *Baccha clavata* F., *Baccha parvicornis* Loew (Syrphidae), *Chrysopa* sp. (Chrysopidae), *Isodromus iceryae* How. and *Carthasis distinctus* Harris (Nabidae). In Florida and Hawaii, *Aleurotrachelus anonae* Corbett was recently identified infesting avocados (D. Carrillo, personal observation).

In Israel, the Japanese bayberry whitefly (*Parabemisia myricae* Kuwana) was discovered in 1978, causing heavy damage to avocado and citrus trees (Sternlicht, 1979). Local predacious mites in the family Phytoseiidae, lacewings (Neuroptera), ladybeetles (Coccinellidae), predacious bugs (Anthocoridae) and parasitic wasps did not effectively control the pest. Thus, the parasitoid *Eretmocerus debachi* Rose and Rosen (Aphelinidae) was imported from California, successfully established, and eventually controlled *P. myricae* (Rose and DeBach, 1992). Other exotic natural enemies, *Eretmocerus* sp., *Encarsia* sp. (Aphelinidae) (from Japan), ladybird beetles (*Nephaspis amnicola* Wingo) and *Delphastus pusillus* Le Conte (from Hawaii), the beetle *Cybocephalus binotatus* Grouvelle (Cybocephalidae), and the fungus *Aschersonia aleyrodis* Webber were released in numerous avocado and citrus orchards but probably did not become established (Swirski *et al.*, 1987). Outbreaks *P. myricae* were recorded in 1992 in some avocado orchards in Western Galilee, Israel. These outbreaks were probably caused by drift of baits containing insecticide (malathion), which were applied aerially against the Mediterranean fruit fly in adjacent citrus orchards and adversely affected *E. debachi*. This parasitoid was reestablished, and *P. myricae* was brought under control again (Swirski *et al.*, 1998).

The red-banded whitefly (*Trialeurodes perseae*), native to Mexico, infests avocados in California, where it was discovered in 1982 (Hoddle, 2006). *T. perseae* is probably univoltine (has one brood of offspring per year) with adult densities showing distinct peaks during August. Partial life tables determined by Hoddle (2006) indicated that survivorship from settled first and second instars to emerged adult whiteflies ranged from 34% to 37%. At 25°C, complete development is achieved in 43–46 days. In California, the aphelinid *Cales noacki* Howard causes 30–100% parasitism (Hoddle, 2006). In Mexico, the parasitoid *Eretmocerus perseae* Rose and Zolnerowich was reared from *T. perseae* (Rose and Zolnerowich, 2004).

The citrus blackfly (*Aleurocanthus woglumi* Ashby) is a minor pest of avocado. According to Bedford (1998b), the native range of citrus blackfly is India, Ceylon and the Philippines. It also occurs in the West Indies, Costa Rica, Panama, Mexico and Kenya. Citrus blackfly breeds on 75 different plant species, which include avocado, banana, citrus, coffee, grapes, mango and persimmon (Bosman, 1959; see Clausen and Berry, 1932; Ebeling, 1959; Bedford and Thomas, 1965; and Bedford, 1998b for information on natural enemies).

Aphids, mealybugs and scale insects

Cotton aphid

The cotton aphid (*Aphis gossypii* Glover) has a worldwide distribution (Stoetzel and Davidson, 1974; Schwarz, 1998). In Israel, it is found on young foliage and occasionally on inflorescences of two avocado cultivars, 'Nabal' and 'Horshim' (Swirski *et al.*, 1991a). According to Avidov and Harpaz (1969), the damage caused by the cotton aphid is characterized by foliar and floral discolouration and deformities caused by the aphid's toxic saliva, resulting in stunted development of infested plants, growth of extensive layers of sooty mould on the excreted honeydew and transmission of viral diseases.

The nymphs are greenish to brownish with a distinctive appearance, spotted with an exudation of powdery wax (Van den Berg and De Villiers, 1987). Winged and apterous females are found, but males are absent. Depending on their host plant, they vary from yellow-green to green to almost black. The body length is 1.2–2.0 mm (Avidov and Harpaz, 1969).

Both winged and apterous females produce live young aphids. The females may live for 2–3 weeks and produce two or more offspring daily (Hill, 1975). Ebeling (1959) refers to a nymphal period of 3–20 days.

The aphelinid, *Aphelinus* sp., attacks *A. gossypii* in South Africa (Prinsloo and Neser, 1994).

Green peach aphid

The green peach aphid (*Myzus persicae* Sulzer) is a cosmopolitan species and was recorded infesting avocado in South Africa (Millar, 1994). *M. persicae* feeds on the flowers and young leaves of avocado. The damage is caused by sucking sap, producing honeydew, inducing black sooty mould and fouling the plants with cast skins (Daiber, 1992).

Aphelinus abdominalis Dalman and *Aphelinus asychis* Walker have been reported parasitizing *M. persicae* (Prinsloo and Neser, 1994). According to Daiber (1992), the green peach aphid is attacked by ladybird beetles and fungal diseases. Control of the green peach aphid on avocado is seldom necessary.

Brown citrus aphid

The brown citrus aphid (*Toxoptera aurantii* Boyer de Fonscolombe) was first recorded in Malawi in 1910 (Theobald, 1914). This species has been recorded in other African countries, Asia, Australasia, Central and South America, southern USA and southern Europe. In Africa, it has been reported from most countries (Millar, 1994).

Hill (1975) reported distortion of young leaves with black sooty mould on the upper surface of leaves and stems. Nymphs of *T. aurantii* are brown. Alate and apterous viviparous females are brown to black, and the body length varies from 1.4 mm to 2.0 mm. The median veins of the forewings of the alate females found in Africa have only one branch (Avidov and Harpaz, 1969), while those of *Toxoptera citricidus* Kirkaldy have two branches.

Cottony cushion scale

Icerya purchasi Maskell (Monophlebidae), also known as the Australian bug, the cottony cushion scale or fluted scale, is known to be present in South Africa since 1873 (Trimen, in Lounsbury, 1897). Australia is the native home of this insect, which now has a worldwide distribution, attacking a wide range of plants (Lounsbury, 1897; Munro and Fouché, 1936).

Heavy infestations of the Australian bug can cause defoliation, fruit drop and a decrease in the vitality of the host plant (Cilliers, 1978). Large quantities of honeydew are produced, which attracts ants and can disrupt the natural or biological control of other pests. Because of the classical biological control that occurred (*vide* Lounsbury, 1940), *I. purchasi* is seldom found on avocado in South Africa.

The introduction of the vedalia ladybird, *Novius cardinalis* Mulsant (= Rodolia cardinalis, Coleoptera: Coccinellidae), was the first case of classical biological control of an insect pest in South Africa (Lounsbury, 1897, 1940). *Cryptochetum iceryae* Williston (Diptera: Cryptochetidae) was also introduced and released in South Africa in an attempt to control *I. purchasi* (Hattingh and Cilliers, 1998). *C. iceryae* and *Rodolia iceryae* Janson were introduced to Israel for the control of *I. purchasi* (Blumberg *et al.*, 1989; Blumberg and Mendel, 1991; Mendel and Blumberg, 1991; Blumberg *et al.*, 1992).

Natural enemies indigenous to Africa include the coccinellids, *Rodolia ferruginea* Weise and *R. iceryae*. Another coccinellid, the Australian *Cryptolaemus montrouzieri* Mulsant, introduced to South Africa from California for the control of *P. citri* Maskell (Cilliers, 1978), is one of the predators that also feed on *I. purchasi* in Australia (Smith *et al.*, 1997). Important natural enemies of *I. purchasi* in Australia include the hymenopterous parasitoids, *C. iceryae* and *Ophilosia* spp., and predators such as lacewings and a coccinellid (Smith *et al.*, 1997).

Long-tailed mealybug

Long-tailed mealybug (*Pseudococcus longispinus*) has a cosmopolitan distribution (Ebeling, 1959; Hattingh *et al.*, 1998; Ripa *et al.*, 2008d) and is a potentially serious pest in all areas where avocados are cultivated. The life cycle of *P. longispinus* takes about 4 weeks in summer and several months in winter. Three overlapping generations occur per year (De Villiers *et al.*, 1987a). Both sexes of the mealybug pass through three nymphal instars. The adult male long-tailed mealybug has one pair of wings and two long waxy anal filaments (Ebeling, 1959). The long-tailed mealybug is viviparous.

Mealybugs are usually kept at low numbers by natural enemies. Most outbreaks are associated with chemical disruption of their natural enemies. Furthermore, ants are attracted to the honeydew that mealybugs excrete. By protecting them against natural enemies, ants can also help to cause a mealybug outbreak. In Israel during 1960s and 1970s, avocado orchards adjacent to cotton fields were heavily damaged by *P. longispinus*. Drift of broad-spectrum insecticides from aerially sprayed fields killed natural enemies and resulted in severe mealybug outbreaks. Additionally, females of the honeydew moth (*Cryptoblabes gnidiella* Milliere) were attracted to the honeydew of the mealybug and their caterpillars nibbled at the fruit. The problem was solved by avoiding broad-spectrum insecticide applications within 200 m of avocado orchards and by releasing the parasitic wasps *Arhopoideus peregrinus* Compere (= *Hungariella peregrina* Compere), introduced to Israel in 1954, and *Anagyrus fusciventris* Girault, introduced from Australia in 1971 (Swirski *et al.*, 1980). In Chile, the parasitoids *Coccophagus gurneyi* Compere, *Pseudaphycis* spp., *nr angelicus* Howard and *Tetracnemoidea brevicornis* Girault were observed in the region of Valparaiso, with *C. gurneyi* considered one of the most successful parasitoids (Ripa *et al.*, 2008d).

Nipaecoccus nipae

In Cuba and Florida, the coconut mealybug *N. nipae* Maskell =*Pseudococcus nipae* feeds on young avocado leaves and flower buds (Fig. 14.4). Several biological control agents control this mealybug species, including *Scymnus bahamicus* Csy (Coccinellidae), *Lobodiplosis pseudococci* Felt, *Pseudoaphycus utilis* Timberlake, *Lecanicillium lecanii* Zimmerman, *Empusa fresenii* Nowak and *Botrytis rileyi* Farlow (Cañizares, 1975).

Planococcus lilacinti

Planococcus lilacinti Cockerell infests young shoots of avocado and fruit pedicels in the Philippines. Heavy infestations may cause fruit to drop but *P. lilacinti* is generally of minor concern (Cendana *et al.*, 1984).

Fig. 14.4. The coconut mealybug, *Niapecocus nipae*, feeding on avocado leaves and leaf buds.

White wax scale

White wax scale (*Ceroplastes destructor* Newstead) has been collected from avocado in South Africa (Munro and Fouché, 1936; Cilliers, 1998) but is not listed as a pest in De Villiers (2001a). In Australia, white wax scale (*Gascardia destructor* Newstead), Indian white wax scale (*Ceroplastes ceriferus* Fabricius) and pink wax scale (*Ceroplastes rubens* Maskell) are minor pests and rarely warrant control actions (Waite and Pinese, 1991) (Fig. 14.5).

Soft brown scale

Soft brown scale (*Coccus hesperidum* L.) is a cosmopolitan pest of many tropical and subtropical plants (Annecke and Georgala, 1978), including avocado (Ebeling, 1959) and granadilla (De Villiers and Van den Berg, 1987). Direct loss of avocado fruit due to soft brown scale appears to be small. However, fruit may be stained by sooty mould deposits, causing downgrading. Management of this insect has been studied in detail by Bodenheimer (1951), Annecke (1959, 1964, 1965, 1966), Searle (1964), Avidov and Harpaz (1969), Annecke and Georgala (1978), Krause *et al.* (1996), Bedford (1998a) and De Villiers (2001a).

Mango shield scale

Mango shield scale (*Milviscutulus mangiferae* Green) feeds on several host plants, including avocado, citrus, guava, mango and persimmon. It has been reported in many countries, including Egypt, Kenya, Madagascar, Mauritius, Seychelles, South Africa, Sri Lanka and Zanzibar (De Lotto, 1957; Ben-Dov, 1971; Ben-Dov *et al.*, 1976; Kamburov, 1987). Mango shield scale is found on leaves and occasionally on avocado fruit. Damage is mainly caused by large amounts of honeydew excreted by the insects on the leaves and fruit, which causes thick layers of sooty mould. Natural enemies found in South Africa are the parasitoids *Coccophagus pulvinariae* Compere, *Scutellista cyanea* Motschulsky, *Aprostocetus* sp., *Tetrastychus* sp. and an unidentified predatory beetle (Kamburov, 1987; Daneel *et al.*, 1998). According to Merrill (1953), mango shield scale is controlled in Florida during rainy periods by the fungus *Cephalosporium lecanii* (Zimmerman) Viegas. Chemical control does not seem necessary, since its natural enemies keep mango shield scale in check.

Fig. 14.5. The pink wax scale *Ceroplastes rubens* Maskell.

Heart-shaped scale or pyriform scale

In South Africa, heart-shaped scale (*Protopulvinaria pyriformis* Cockerell) was first recorded on avocado in 1916 at Pietermaritzburg, KwaZulu-Natal, South Africa (Brain, 1920) (Fig. 14.6). It also occurs in Israel (Wysoki, 1987), Asia, the Mediterranean, and South and North America (Du Toit and De Villiers, 1988). *P. pyriformis* feeds on avocado (mainly on 'Hass'), guava, the ornamental plant *Aphanamixis polystachya* (= *A. moora rohituka*) (De Villiers, 1981) and citrus (Bedford, 1978a). A preference for certain avocado cultivars has been observed (De Villiers and Van den Berg, 1987; Wysoki, 1987). According to De Villiers (1989), a steady increase in the occurrence of heart-shaped scale was reported from 1986 to 1989 in parts of South Africa. In Israel, *P. pyriformis* was first recorded in 1980 (Ben-Dov and Amitai, 1980) and is now widespread. In Chile, *P. pyriformis* is considered a minor pest, sometimes needing insecticide applications to reduce honeydew produced by the scale that induces sooty mould on the fruit (Ripa *et al.*, 2008c). Heart-shaped scale is one of the most damaging insects affecting avocado in Peru (Wolfe *et al.*, 1969) and sometimes causes serious damage to avocado in Florida (Ebeling, 1959). 'Nabal' is the most susceptible cultivar, followed by 'Ein Vered', 'Reed', 'Hass' and 'Fuerte'; whereas 'Ettinger' trees are sometimes attacked when located close to

Fig. 14.6. Heart-shaped scale, *Protopulvinaria pyriformis* Cockerell. A. Heart-shaped scales along veins in the underside of the leaf, B. Sooty mould accumulation on leaves resulting from heart-shaped scale infestation.

infested trees of more susceptible cultivars. In laboratory trials, *P. pyriformis* completed development and reproduced less on 'Nabal' and 'Ein Vered' seedlings than on 'Hass'. Survival of *P. pyriformis* was rather poor on 'Fuerte', 'Horshim' and 'Wurz' and higher on 'Ettinger', 'Pinkerton' and 'Reed' (De Meijer *et al.*, 1989). Two generations of *P. pyriformis* are observed annually in Israel, a winter generation in November–December and a summer generation in June–July (Hadar, 1993).

Heart-shaped scale secretes excessive quantities of honeydew on which sooty mould grows and stains the fruit and leaves (Du Toit and De Villiers, 1990; Steyn *et al.*, 1994; De Villiers, 2001b). Leaf drop can occur with heavy infestations (De Villiers, 2001b). The life cycle of heart-shaped scale on avocado seedlings was determined in the laboratory at 27°C and 70% RH (Du Toit *et al.*, 1994). Mean duration of the first instar (which includes the crawler stage) was 10.5 days. The second instar lasted 17.5 days, the third instar 24.5 days and the adult female stage (including the ring stage) 45.5 days. Two generations of heart-shaped scale on avocado occur per year in Nelspruit, South Africa (De Villiers, 1989, 2001b). Peak numbers of first instar larvae occurred in November and March and adults peaked in October and March. The winter generation lasted about 7 months from egg to adult and the summer generation lasted 5 months. *P. pyriformis* feeds on the ventral side of the leaves. Fruit is normally not attacked. In descending order, avocado cultivars preferred by heart-shaped scale in South Africa are 'Hass', 'Collinson', 'Ryan', 'Fuerte' and 'Edranol' (De Villiers and Robertson, 1988).

The following natural enemies of heart-shaped scale have been identified in South Africa (Robertson and De Villiers, 1986): *Metaphycus galbus* Annecke, *Metaphycus helvolus* Compere, *Metaphycus stanleyi* Compere (Encyrtidae), *Coccophagus basalis* Compere, *C. pulvinariae* Compere (Aphelididae) and *Tetrastichus* sp. (Eulophidae). Two hyperparasitoids, *Cheiloneurus cyononotus* Waterston (Encyrtidae) and *Marietta javensis* Howard (Aphelinidae), were also found. Predators include *Chilocorus angolensis* Crotch and *Hyperaspis senegalensis hottentota* Mulsant (De Villiers, 2001b).

Du Toit *et al.* (1991) reported that 70% parasitism of heart-shaped scale occurred in the adult female stage, 4% in the second instar and 14% in the third instar. *Metaphycus* spp. were dominant parasitoids of *P. pyriformis* during late summer and autumn but were replaced in winter by *Coccophagus* spp. and *Tetrastichus* sp. According to Du Toit *et al.* (1991), *Marietta javensis* Howard had a negative effect on the parasitoid complex and attacks the two *Metaphycus* species, *M. stanleyi* and *Metaphycus swirskii* Annecke & Mynhardt, in Israel. Since local and introduced parasitoids do not sufficiently control heart-shaped scale populations, oils are used targeting nymphal stages but not the ovipositing females (Swirski *et al.*, 1988).

In Israel, parasitic wasps, *Microterys flavus* Howard, *Metaphycus flavus* Howard (Encyrtidae) and *Coccophagus lycimnia* Walker (Aphelinidae); predators, *Chilocorus bipustulatus* Linnaeus, *Oenopia (Synharmonia) conglobata* Linnaeus, *Scymnus flavicollis* Redtenbacher (ladybeetles, Coccinellidae), *Anisochrysa carnea* Stephens (green lacewings, Chrysopidae), spiders; and the fungus, *Verticillium lecanii* Zimmerman (Hadar, 1993), were unable to keep populations of *P. pyriformis* below the economic threshold. Thus, efforts were made to import various natural enemies. *M. swirskii* (imported from Kenya), initially the most abundant parasite, was soon replaced by *M. stanleyi* (imported from the USA, South Africa and Spain), which is today the dominant natural enemy of *P. pyriformis* in Israel. *M. helvolus* (imported from the USA) is rare in avocado orchards, while *M. galbus* (from South Africa) probably failed to become established. Two ladybeetle species were imported from Spain: *C. montrouzieri* is sporadically found in avocado orchards, whereas *Nephus peyerimhoffi* Sicard is rare. Two secondary parasites, *M. javensis* and *Pachyneuron concolour* Förster, reduce the efficiency of *Metaphycus* spp. (Blumberg and Blumberg, 1991; Hadar, 1993). In Peru, biological control can be achieved by releasing the wasp, *Mi. flavius*. The oviposition, host marking and host stage preference of various species of *Metaphycus* were studied by Hadar (1993).

Mineral oil is recommended to control three nymphal stages of *P. pyriformis*. Oil sprays are likely to produce satisfactory results if applied (northern hemisphere) in December–January (or January–February) and July–August, when *P. pyriformis* populations consist mainly of young nymphal stages (Hadar, 1993). Only heavily infested trees are treated with broad-spectrum insecticides (Wolfe *et al.*, 1969). In Colombia, *Bombacoccus aguacatae* Kondo has been observed, infesting avocado branches and twigs (Kondo, 2010).

Saissetia sp.

A single outbreak of *Saissetia* sp. was reported in 1997 from a South African avocado orchard (Grové, 2001). This coccid feeds on avocado leaves and twigs and produces vast amounts of honeydew, which attracts ants and serves as a substrate for sooty mould fungi. In Chile, *Saissetia oleae* is considered a pest of citrus more than of avocado, although heavy infestations have been observed in some avocado orchards (Ripa *et al.*, 2008b). Natural enemies include parasitoids (*Encarsia* sp. (Aphelidinidae), *Aprostocetus ceroplastae* Girault (Eulophidae), *Metaphycus* sp. (Encyrtidae)) and the predator *C. angolensis* Crotch (Coccinellidae).

Avocado scale

Avocado scale (*Fiorinia fioriniae*) has been found in Florida, Hawaii and the West Indies attacking leaves and fruit (Ebeling, 1959). It is of little economic importance in South Africa where it is regarded as a minor pest (Erichsen and Schoeman, 1992; Claassens, 1994). Small brown spots develop where nymphs feed, which on 'Hass' fruit fade as fruit colour develops (Claassens, 1994). Female scales are oval and shield shaped, with a thin, translucent covering. Avocado scale is brownish-yellow to orange-brown and 1.0–1.3 mm long (De Villiers and Van den Berg, 1987). A morphological description and illustrations of the first instar of *F. fioriniae* were provided by Howell and Tippens (1977).

Latania scale

Latania scale or palm scale (*H. lataniae* Signoret) was recorded on avocado in Australia, Israel, South Africa, Chile and the USA (California and Florida) (Quayle, 1938; Ebeling, 1959; Avidov and Harpaz, 1969; Howard and Oliver, 1985; Swaine *et al.*, 1985; De Villiers and Van den Berg, 1987; Vargas and Rodriguez, 2008). According to Dekle (1976), it has more than 370 host plants in Florida. Davidson and Miller (1990) stated that *H. latania* feeds on ornamental plants in 78 families and 278 genera. It feeds on both tropical and subtropical plants (Quayle, 1938; Ebeling, 1959) (Fig 14.7).

According to Ebeling (1959), *H. lataniae* causes damage to 'Fuerte' avocado fruit and possibly to thin-skinned avocado cultivars. Damage is characterized by nodules adhering to the inside of the peel with corresponding depressions in the flesh of ripe fruit. Latania scale attacks branches, twigs, leaves and fruit. Infested fruit is degraded or culled. In South Africa, heavy infestations were observed on 'Hass'. A heavier latania scale infestation on 'Hass' fruit hanging near the ground was observed (Steyn, 1954).

Natural enemies in South Africa include aphelinid parasitoids, *Aphytis lingnanensis* Compere and *Aphytis africanus* Quednau; and encyrtids, *Habrolepis obscura* Compere and Annecke, and *Comperiella lemniscata* Compere and Annecke. Predators are coccinellids, *Rhyzobius lophanthae* Blaisdell, and *Chilocorus* sp., as well as a cybocephalid, *Cybocephalus* sp. (De Villiers and Van den Berg, 1987; Daneel, 1998). Natural enemies in Chile include *Aphytis* sp. and *R. lophanthae* (Vargas and Rodriguez, 2008). The predatory mites, *Cheletomimus berlesei* (Cheyletidae) and

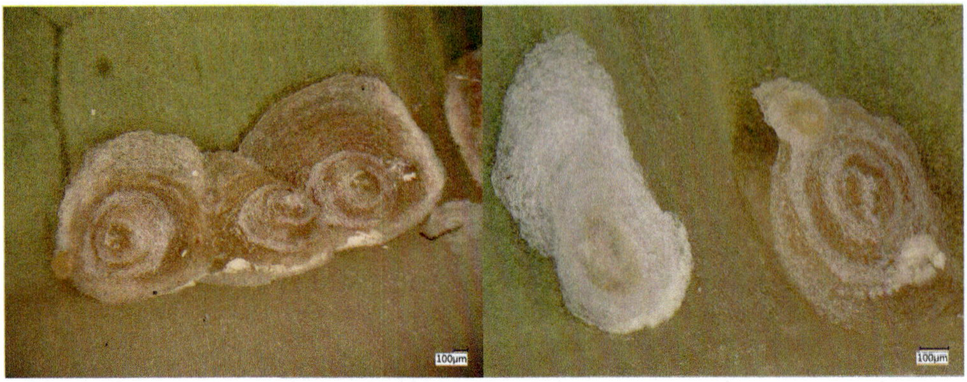

Fig. 14.7. Latania scale, *Hemiberlesia lataniae* Signoret.

Hemisarcoptes spp. (Hemisarcoptidae), feed on *H. lataniae* crawlers (Gerson and Zor, 1973). *Dentifibula obtusilobae* Felt is a cecidomyid predator of latania scale. Evans and Prior (1990) stated that armoured scale insects, particularly in the humid tropics, are subject to periodic and often devastating attacks by highly adapted fungal pathogens. The fungal pathogen *Nectria flammea* (Tulasne and C. Tulasne) Rossman and Samuels was isolated from *H. lataniae*.

Spanish red scale

Spanish red scale (*Chrysomphalus dictyospermi* Morgan) feeds on avocado twigs, branches, leaves and fruit (Van den Berg *et al.*, 1999). Avocado twigs and branches severely infested with *C. dictyospermi* become roughened and cracked. Leaf drop may also occur. Spanish red scale is occasionally found in California, but in Florida, it is a pest of avocado (Quayle, 1938; Ebeling, 1959).

Spanish red scale females are circular, slightly convex, 1.5–2.0 mm in diameter, light brown or red-brown and the covering may be readily separated from the body, even during moulting. This distinguishes it from the red scale (*Aonidiella aurantii* Maskell) (Quayle, 1938). Spanish red scale reproduces parthenogenetically or sexually (Ebeling, 1959). According to Quayle (1938), there can be three to four generations per year in southern California and five to six in southern Florida.

Parasitoids recorded from *C. dictyospermi* in the USA include *Aphytis chrysomphali* Mercet, *Prospaltella aurantii* Howard, *Thysanus merceti* Malenotti and *Thysanus flavopalliata* Ashmead. The coccinellids, *Rhizobius debilis* Blackburn and *Lindorus lophanthae* Blaisdell, seemed more important than the parasitoids (Ebeling, 1959).

In South Africa, Spanish red scale is not considered a major pest on avocado, and several parasitic wasps attack this insect. Chemical control is not necessary (De Villiers, 2001a). Natural enemies of Spanish red scale in South Africa include the parasitic wasps, *Aphytis* sp. *proclia* Walker group, *Signiphora perpauca* Girault, *Encarsia citrina* Craw, *Signiphora flavella* Girault; the green lacewing (*Chrysopa oblatis* Banks), and the coccinellid, *Rhizobius satellus* Blackburn.

Other scale insects

In Peru, *Selenaspidus articulatus* Morgan and *Unaspis citri* Comstock regularly attack avocado when orchards are located close to citrus orchards. In Mexico, 14 species of diaspidid scales: *Agrallaspis howardi* Cockerell, *Acutaspis albopicta* Cockerell, *Aspidiotus spinosus* Comstock, *Chrysomphalum aonidum* L., *Chrysomphalum dyctiospermi* Morgan, *Diaspis cacoccois* Lichtenstein, *Hemiberlesia diffinis* Newstead, *H. lataniae* Signoret, *H. rapax* Comstock, *Melanaspis aliena* Newstead, *Myxetaspis personata* Comstock, *Pinnaspis strachani* Cooley, *Quadraspidiotus perniciosus* Comstock and *Velataspis dentate* Hoke; three Coccidae: *S. oleae* Oliver, *C. hesperidum* L. and *Pulvinaria simulans* Cockerell; and the mealybugs: *P. citri* Risso, *Dysmicoccus brevipes* Cockerell, *Ferrisia virgata* Cockerell and *N. nipae* are considered the most frequent Coccoidea attacking avocado. In Chile, *Aspidiotus nerii* is considered an important pest of avocado that affects fruit quality (Vargas *et al.*, 2008). In California, imported 'Hass' avocado fruit from Mexico was found infested with *Abgrallaspis aguacatae* (Evans *et al.*, 2009; Morse *et al.*, 2009; Rugman-Jones *et al.*, 2010) and methods to detect six diaspidid species, including *A. aguacatae*, were developed (Rugman-Jones *et al.*, 2009). In Mexico, 16 species of armoured scales were identified, 12 from the subfamily Aspidiotinae and four from Diaspidinae associated with avocado in central Mexico (Lazaro-Castellanos *et al.*, 2023). Forty species of associated natural enemies were also recorded.

Avocado planthopper

Avocado planthopper (*Decipha viridis* Synave) is a minor pest of both avocado and macadamia (M.A. Van den Berg, Agricultural Research Council, South Africa, 2001, unpublished data). Nymphs and adults suck sap from the leaves and apparently also from the fruit stalks. In South Africa, large amounts of honeydew are produced, which support growth of sooty mould. *D. viridis* represented 18.0% and 2.2% of Hemiptera found on avocado and macadamia trees, respectively (M.A. Van den Berg, Agricultural Research Council, South Africa, 2001, unpublished data). The adult *D. viridis* is light green with a blackish spot on the distal edge. Eggs are laid in oval-shaped masses of about 5 mm in diameter. Nymphs are light green with white and yellow stripes on their backs. Long white waxy filaments protrude from the tips of their abdomens. The life cycle of avocado planthopper takes about 6 weeks in summer. There may be up to six generations per year (M.A. Van den Berg, Agricultural Research Council, South Africa, 2001, unpublished data).

Parathiscia sp. nr. truncata

In South Africa, both adults and nymphs of *Parathiscia* sp. nr. *truncata* feed on avocado and macadamia. Adults are mustard green and are up to 8 mm in length. The wings are folded over the abdomen when at rest, giving them a flattened appearance. The nymphs are green. In surveys in South Africa, *P.* sp. nr. *truncata* represented 2.7% and 2.9% of the total number of Hemiptera collected from avocado and macadamia, respectively (M.A. Van den Berg, Agricultural Research Council, South Africa, 2001, unpublished data).

Mottled avocado bug

Both adults and nymphs of mottled avocado bug (*Parapioxys jucundus* Distant) feed on avocado and macadamia. This species is probably a phloem feeder. It was more abundant in avocado (3.1%) than in macadamia (0.5%) (M.A. Van den Berg, Agricultural Research Council, South Africa, 2001, unpublished data). The adult is 15 mm long and 7 mm wide, with a mottled appearance varying from orange-green through green to a brownish colour with a bright red abdomen. The nymphs are dark brown with long anal filaments.

Leafhoppers

In Mexico, *Idona minuenda* Ball is a common pest of avocados grown in more temperate areas with frequent rainfall. In addition to *I. minuenda*, Pinedo-Escatel and Dmitriev (2019) listed eight other leafhopper species associated with avocado in Mexico. *I. minuenda* population peaks are observed in Mexico during July–October. Nymphs feed on leaves causing a grey colour injury on the upperside. When young leaves are attacked, they turn yellow and remain underdeveloped. This species is suspected of transmitting a virus. *Anagrus raygilli* Triapitsyn was collected parasitizing *I. minuenda* in Mexico (Triapitsyn, 2000).

In South Africa, citrus leafhopper (*Penthimiola bella* Stål) is a pest of citrus. This species is probably one of the causative agents of protrusions on 'Hass' avocado fruit from mature trees (Du Toit *et al.*, 1993; Bruwer, 1996). Feeding on avocado has not been proven. Pretorius (1971) reported life history traits of *P. bella* on citrus: eggs are laid singly through a minute slit and placed in a superficial envelope formed by the tissue of either leaves or fruit. There are five nymphal instars that last approximately 25–63 days at 20–27°C. Adult *P. bella* are brownish, mottled and measure 3.0–3.9 mm in length. Eggs of *P. bella* are attacked by seven Chalcidoidea species (Annecke, 1965), which may parasitize more than 50% of the eggs (Annecke, 1964).

Avocado psyllids

Trioza anceps

Trioza anceps Tuthill is 2–5 mm long and is a stenophagous (having few hosts) species that causes galls on avocado leaves and is widely distributed in Mexico and Central America, where it attacks regional non-commercial avocado cultivars (Hollis and Martin, 1997). However, *T. anceps* is seldom observed on commercial cultivars such as 'Hass' or 'Fuerte'. Heavy infestations characterized by the presence of brown galls may cause defoliation (Hernandez *et al.*, 2000).

Trioza perseae

Trioza perseae is reported from Peru (Chavez *et al.*, 1985) and Venezuela. It is approximately 2.5 mm long, yellowish, with dark bands on the abdomen and thoracic regions; the wings are almost transparent, with reduced venation (Chavez *et al.*, 1985). Females deposit eggs on the tender flush, preferring the leaf underside, close to the midvein. Eggs have a pedicel and whitish colouration and are difficult to see with the naked eye. The emerging nymph is yellowish with inconspicuous body segmentation. Most feeding occurs on the leaf underside, but some nymphs and adults could be found on the leaf upperside. *T. perseae* galls are formed on the leaf underside, while the adaxial surface shows depressions surrounded by a yellowish circle. This psyllid is regularly found in localities more than 900 m above sea level. Chavez *et al.* (1985) reported several unidentified Eulophidae and Pteromalidae species as possible parasitoids of *T. perseae*. Hollis and Martin (1997) provided an identification key to different psyllids associated with avocado. Other psyllid species that infest avocado are *Trioza aguacatae* Hollis and Martin, deforming shoots and leaves, and *Trioza godoyae* Hollis and Martin, causing leaf rolling (Hollis and Martin, 1997; Hernandez *et al.*, 2000). Around 58 species of psyllids are associated with host plants in the Lauraceae family (Hollis and Martin, 1997).

Froghoppers

Broom (1970) described the life history of froghoppers (*Metcalfiella monogramma* Germar, Hemiptera: Cercopidae), which is a common pest in Mexico. Injury is caused by the deposition of eggs into young branches, causing death of tissue. The most important injury is caused by the feeding of nymphs and adults.

Tip wilters

Many coreids that damage tropical and subtropical crops are known as tip wilters because their feeding often causes wilting of new shoots. Tip wilters are mostly stout, dull brown to black, and robust medium to large insects. As a defence mechanism, some species eject a foul-smelling secretion (Van den Berg, 1998). Adults and nymphs suck the sap of succulent new growth. The saliva injected by tip wilters into plant tissue can cause wilting and dieback of the new shoots beyond the point of feeding (Brain, 1929). As a result of this dieback, secondary axillary buds start to grow, which may cause dense sprouting (Smit 1964a,b). Their attacks are most serious during shoot growth flushing, especially on young trees (Hartwig and De Lange 1978). In South Africa, the most important species attacking tropical and subtropical crops, including avocado, are *Anoplocnemis curvipes* F., *Leptoglossus australis* F. and *Pseudotheraptus wayi* Brown (Vanderplank, 1958; Van den Berg, 1998).

Large black tip wilter

Anoplocnemis curvipes adults are dull grey-brown and about 25 mm in length. The femora of the males are thickened and noticeably bent, with a thorn-like spine. The eggs are grey-brown and barrel shaped and are deposited in single rows of 6–12 eggs on branches and leaves of their host plants. After the egg is opened, the nymph rests halfway out and moults into the second instar, leaving its first cast skin with the egg burster in the eggshell (Smit, 1964a,b). The second to final (fifth) instar nymphs look like the adult, except for the absence of wings.

There are 3–6 overlapping generations of *A. curvipes* annually in the Nelspruit area of South Africa (Van den Berg, 1998). According to Hartwig and De Lange (1978), females exhibit maternal care of the eggs and young. Females are the overwintering stage.

The egg parasitoid, *Gryon* sp. (Hymenoptera: Scelionidae), has been reported from Nelspruit causing less than 15% parasitism of *A. curvipes* (Van den Berg, 1998). *Hermya* sp., a tachinid fly, parasitized about 2% of *A. curvipes* adults in the Nelspruit area (Van den Berg, 1998). Hartwig and De Lange (1978) found that *Pheidole megacephala* F. preys on up to 96% of young *A. curvipes* nymphs in Pretoria North, South Africa.

Leaf-footed plant bug

Leaf-footed plant bug (*L. australis*, previously known as *Leptoglossus membranaceus* F.) is dark brown to black with orange spots on the ventral parts of the abdomen and thorax. An orange spot is also present on each of the hind flattened tibiae. *L. australis* is not as robust as most of other tip wilters. *L. australis* occurs in South Africa (Bedford, 1978b), Zimbabwe (Hall and Ford, 1933), Kenya (Ondieki, 1975), Australia (Murray, 1976), India (Jadhav *et al.*, 1980) and Japan (Yasuda and Kinjou, 1983). *L. australis* has been recorded feeding on avocado and mango shoots (Van den Berg, Agricultural Research Council, South Africa, 2001, unpublished data). Jadhav *et al.* (1980) reared *L. australis* on pomegranate and reported on its biology. Eggs are laid in strips along the twigs, individual females lay up to 60 eggs in their lifetime. The egg stage lasts 5–9 days and five nymphal stages 27–31 days. The life cycle takes 63–74 days. Adults live up to 41 days.

Coconut bug

The coconut bug (*P. wayi*) is endemic to Africa and can cause extensive damage to avocados and various other subtropical crops (Viljoen, 1986; Bruwer, 1987; De Villiers and Van den Berg, 1987; De Villiers, 1990a, 1992; Dupont, 1993; Joubert and Neethling, 1994). According to De Villiers (1990a), nymphs and adults of *P. wayi* suck sap from the avocado fruit. About 2 days after feeding on a mature avocado fruit, the lesion can be recognized as a patch slightly darker than the rest of the skin and resembles a bruise. As the lesion ages, it enlarges to approximately 8 mm in diameter, becomes indented and turns brown to black. Lesions may sometimes become tumescent and wart-like. Mature fruit can become malformed and asymmetrical if *P. wayi* feeds on a young fruitlet. Internally, the lesions display a brown stain that can penetrate to a depth of 10 mm. A typical hard clot forms that is easily removed together with the skin. Coconut bugs do not cause rotting of the fruit flesh (De Villiers, 1990b). A packing house survey in the Nelspruit/Hazyview area of South Africa by Dennill and Erasmus (1991) indicated that damage by *P. wayi* during the 1990 season amounted to 4.7% of all avocado fruit. A similar study the following year indicated a figure of 2.8% (Erichsen and Schoeman, 1992).

Eggs are laid singly, scattered over the fruit, small twigs and blossom stems. They are creamy white to light brown, oval and approximately 1.5 mm long. Just before hatching, the colour changes to a darker red-brown. The egg stage lasts approximately 7 days (Bruwer, 1992; De Villiers, 1992). Newly hatched nymphs are 5 mm in length, 1.5 mm in width and light brown. Five instars occur during which the nymphs are morphologically very similar with their characteristically long antennae. Adult coconut bugs are strong fliers with well-developed wings. The dorsal side of *P. wayi* is red-brown, while the ventral side is lighter in colour. Adults are approximately 15 mm in length and 4 mm in width, with males slightly smaller than females (De Villiers, 1992). The female lays on average 80 eggs; the life cycle from egg to adult lasting 31–48 days, depending on temperature (Bruwer, 1992). The adult lifespan of *P. wayi* ranges from 73 to 84 days (Lever, 1969). According to Way (1953), nine or more generations per year can occur on coconuts in East Africa and Zanzibar.

In East Africa and its coastal islands, where *P. wayi* is a serious pest of coconuts, the bug is controlled by a tree-nesting ant (*Oecophylla longinoda* Latreille) (Way, 1951). In South Africa, two egg parasitoids of *P. wayi*, *Anastatus* sp. (Eupelmidae) and *Trissolcus* sp. (Scelionide), cause 58% and 26% of parasitism, respectively (Bruwer, 1992).

Schoeman *et al.* (2012) point out that monitoring coconut bug in avocado orchards is difficult due to the reticent nature and high mobility of these insects. This pest tends to move from macadamia to avocado orchards in late summer, preferring more mature avocado fruit.

Green coconut bug

The green coconut bug (*Homoeocerus* sp. nr. *elongatus* Dallas) is up to 15 mm long and 5 mm wide. It has almost the same size and general form as the coconut bug (*P. wayi*) but is pale green in colour. It is indigenous to South Africa and probably causes the same damage to avocado as *P. wayi*. It has been found in small numbers in avocado (Van den Berg *et al.*, 1999). In Taiwan, the coreid fruitspotting bug (*Paradasynus spinosus* Hsiao), which breeds on *Magnolia* spp., damages avocado fruit similarly to Australian and South African coreids (Hung and Jong, 1997). Bagging fruit to prevent bug access is recommended.

Fruit spotting bugs

Fruit spotting bugs, *Amblypelta nitida* Stal and *A. lutescens lutescens* Distant (Coreidae), are major and often devastating pests of avocados in Queensland and northern New South Wales, Australia. They attack fruit from the time it sets in September–October until the end of April. Bugs of all stages can cause damage, but avocado is not a suitable host for *A. l. lutescens* immature development. Adults migrating into orchards from adjacent breeding areas cause the most damage (Lindsay *et al.*, 2019). Severe damage usually occurs in orchards situated close to rainforests.

Egg parasitoids, *Gryon* sp. (Scelionidae), *Ooencyrtus caurus* Huang and Noyes (Encyrtidae), and *Anastatus* sp. (Eupelmidae), may parasitize up to 90% of eggs but the immigrant adults, which result from unparasitized eggs, cause severe damage. Because immigration is constant through the season, individual crops may require protection every 2 weeks. Orchards more distant from rainforests may require only occasional controls. Thin-skinned cultivars, especially 'Fuerte', appear to suffer the most, but the thick-skinned 'Hass' can be just as severely affected without showing the severe lesions and cracks observed on 'Fuerte' (Fay and Huwer, 1993; Waite and Huwer, 1998).

Stinkbugs

Stinkbugs (Pentatomidae) that attack avocado are primarily indigenous to Southern Africa, except for *Nezara viridula* L., which is a cosmopolitan species. In Israel, the variegated caper bug (*Stenozygum colouratum* Klug) generally develops on wild caper bush (*Capparis spinosa*), sometimes migrating to avocado orchards (mainly 'Hass' but also 'Fuerte' and 'Ettinger'). Damage is accompanied by heavy secretion of perseitol from the fruit and the appearance of black spots (Izhar *et al.*, 1990b). Feeding by stinkbugs on small developing avocado fruit causes many to drop, while larger fruit are often blemished. Growth points on which feeding takes place may wither and die. The biology, life history and control of stinkbugs discussed in this chapter are similar for the different species. Natural enemies consist of egg (Hymenoptera) and adult parasitoids (Diptera: Tachinidae), and predators such as ants. Tachinids may occasionally parasitize nymphs but complete their development in the adult stink bug.

Powdery stinkbug

Powdery or woolly stinkbug (*Atelocera raptoria* Germar) is indigenous to Southern Africa. It is brown, sprinkled with red-brown and with a whitish wax. The ventral sides of the thorax and abdomen are mostly covered with a whitish waxy substance (Van den Berg *et al.*, 2000)

Grey-brown stinkbug

In South Africa, the dorsal side of adult grey-brown stinkbug (*Coenomorpha nervosa* Dallas) is brown with grey-black dots forming patterns over the head, thorax, abdomen and wings. *C. nervosa* adults are up to 14 mm long, and the ventral side of the abdomen and thorax is off-white to a whitish-brown. One of the characteristics of *C. nervosa* is the width of the abdomen, which is broader than the thorax, characterized by grey-black lines that accentuate each segment (Van den Berg *et al.*, 2000). Fairly large numbers of *C. nervosa* were recorded on avocado, feeding on fruit or on the bark (Joubert and Claassens, 1994; Van den Berg *et al.*, 2000).

Spotted stinkbug

Spotted stinkbug (*Farnya* sp.) feeds on avocado and macadamia in South Africa (Van den Berg *et al.*, 2000). *Farnya* sp. are green-brown to greyish-brown with yellow-brown spots and up to 14 mm long. At a constant temperature of 25°C and 75% relative humidity, *Farnya* sp. eggs last on average 6.8 days and the immature development (egg to adult) takes 48 days (Bruwer, 1992). Egg parasitism by two *Trissolcus* spp. (Scelionidae) and a *Pachyneuron* sp. (Pteromalidae) at

Nelspruit, South Africa accounted for 72.6% and 4.5% parasitism, respectively (Bruwer, 1992).

Yellow-edged stinkbug

Small numbers of the yellow-edged stinkbug (*Nezara pallidoconspersa* Stal) were collected on avocado and macadamia (De Villiers, 1986). Eggs of *N. pallidoconspersa* are yellowish-white and laid in clusters on the leaves averaging about 70 eggs. Adult females of *N. pallidoconspersa* are up to 18 mm long, and males are 11–14 mm long. Adult *N. pallidoconspersa* are green, with the edge of the abdomen yellowish with brown lines that accentuate each segment. *N. pallidoconspersa* also have yellowish hindwings, which are striking when they fly. In hotter climates, such as in Nelspruit, South Africa, *N. pallidoconspersa* breed throughout the year on wild castor-oil plants (*Ricinus communis*), which should be removed from areas where avocados are grown.

Natural enemies of *N. pallidoconspersa* include the egg parasitoid, *Trissolcus* sp. and two unidentified tachinid parasitoids that attack adult stinkbugs (De Villiers, 1986).

Small green stinkbug

Adults of small green stinkbug (*Nezara prunasis* Dallas) are often found in relatively large numbers in avocado (Joubert and Claassens, 1994). However, no nymphs were recorded suggesting that breeding takes place on other plants. Adults of *N. prunasis* are green, from 9 mm to 11 mm long. At the beginning of winter, they congregate in sheltered areas such as barns or even on the roofs of houses.

Green vegetable bug

Nezara viridula, the green vegetable bug, is a cosmopolitan species that attacks over 200 cultivated plants including avocado. Dennill and Dupont (1992) reported that feeding by *N. viridula* causes protrusions on avocado fruit. Joubert and Claassens (1994) reported that *N. viridula* was the most abundant stinkbug on avocado in South Africa.

Monalonion velezangeli

M. velezangeli is a polyphagous pest of fruit trees including avocado, guava and coffee. It attacks vegetative buds, inflorescences, flowers and fruit and is considered one of the most harmful pests of avocado cv. Hass in Colombia. It can cause yield losses between 50% and 100% (Londoño Zuluaga, 2014). Adults are red with black wings, and females are generally larger than males (Fig. 14.8). Eggs are inserted into the plant tissue, preferably on young stems and are laid singly or in groups of two or three. The incubation period is approximately 23 days. *M. velezangeli* has five nymphal stages that are orange in colour with red and black marks on the head, abdomen, legs and antenna (Carvalho and Costa, 1988). The immature development from egg to adult is approximately 27 days, and the adults live for 10–12 days (Londoño Zuluaga, 2014).

Fig. 14.8. A. *Monalonion velezangeli* Carvalho & Costa from Colombia; Damage to B. stems, C. branches and D. fruit by *Monalonion* sp. In Brazil. Photo credits: A = Instituto ENTOMA-Naturavision, B–D = Dr Graceille Furtado (Brazil) (with permission).

Damage on vegetative shoots is characterized by punctures inducing plant exudates that first appear as red dots and later turn into white dust-like structures. Feeding damage on fruit first appears as brown circular spots with an oily appearance, which later turns black with a white exudate over the wound (Fig. 14.8). Monitoring for damage on stems, branches or fruit is recommended, especially during flowering. The presence of *M. velezangeli* and its damage are more abundant on the flowering and fruiting branches than on the vegetative ones, as well as in the medium and upper strata of the trees (Torres Jaimes *et al.*, 2012). *M. velezangeli* prefers dense and shaded canopies. Therefore, appropriate tree spacing and pruning to allow light into the canopy are important cultural practices against this pest. The entomopathogenic fungi *Beauveria bassiana* and several conventional insecticides have been used successfully to manage *M. velezangeli* (Londoño Zuluaga, 2014).

Mirids

Avocado bug

In South Africa, the avocado bug (*Taylorilygus* sp.) is brown to black and up to 3 mm long. Two dirty white spots are present on the base of the wings. The first two nymphal instars are green. Adults and nymphs feed on avocado flowers, young avocado fruit and presumably also on young leaves. Avocado flowers seem to be preferred to fruit. Because of the large numbers of flowers produced, feeding on them is of minor importance. Damage to avocado fruit is caused within the first few weeks after fruit set, leading to protrusions on larger fruit, which are only visible about a month after feeding (Bruwer, 1998). Lesions on avocado fruit caused by *Taylorilygus* sp. are pimply elevations on the fruit surface (Du Toit *et al.*, 1993) and are colloquially known as 'Vosknoppe'. Damage can be severe, especially on 'Hass' fruit in old orchards. A 30% cull of 'Hass' fruit on a farm in the Kiepersol area of South Africa and a 50% cull on a farm at Wartburg, KwaZulu-Natal, South Africa were reported. Brown corky scar tissue can be found in the centre of the lesions. Surveys shortly after fruit set should be conducted, and control actions are granted if large *Taylorilygus* sp. populations are observed (Van den Berg *et al.*, 1999).

Other mirids

In Florida, a number of mirid species (*Daghbertus fasciatus* Reuter, *Daghbertus olivaceous* Reuter, *Rhinacloa* sp.) feed and insert their eggs on opening buds, leaves, flowers and small fruit (Fig. 14.9). Attacks affect flowers and recently set fruit, causing them to drop. Wounds on fruit may serve as a point of entry for decay organisms. Mirid species are green-brown and comparatively small (1 cm long). Mirids usually appear during bloom and the early fruit-setting stage. It is suggested that weeds and grass in and around the orchard be mowed as closely as practical to reduce refuges for mirids. Mirid populations are most common from January through April (northern hemisphere) when avocado flowers are fully open (Peña *et al.*, 2003). *Leiophron* nr. *fumipennis* Loan has been found parasitizing mirid species in Florida. Spray applications targeting mirids during flowering should be made in the late afternoon to reduce losses of honeybees (Peña and Johnson, 1999).

In the Philippines, the mirid bugs, *Helopeltis bakeri* Pop. and *Helopeltis collaris* Donovan, attack avocado shoots and fruit, causing significant damage (Cendana *et al.*, 1984).

Avocado lace bug

Pseudacysta perseae (Fig. 14.10) was described in 1908 as *Acysta perseae* from Florida specimens and was considered a minor pest of avocado for several years. However, persistent population outbreaks of *P. perseae* observed since the mid-1990s in Florida and

Fig. 14.9. *Daghbertus fasciatus* Reuter.

Fig. 14.10. Avocado lace bug colony, *Pseudacysta perseae* Heidemann.

the Caribbean, and in the last decade in Hawaii and California, have made this insect one of the most important pests of avocado (Abud-Antun, 1991; Medina-Gaud *et al.*, 1991). *P. perseae* is found in Florida, Hawaii, Texas, California and Georgia in the USA, in the Caribbean (e.g. Barbados, Bermuda, Dominican Republic, Guadeloupe, Jamaica, Puerto Rico, Martinique, US Virgin Islands), Central America (Guatemala and Panama), northern South America (Venezuela and French Guyana) and in Mexico (Mead and Peña, 1991; Rugman-Jones *et al.*, 2012). Other hosts for this pest are red bay (*Persea borbonia* L.) and camphortree (*Cinnamomum camphora* L). The life cycle of *P. perseae* was reported by Abud-Antum (1991) to require 22 days from the egg stage to adult. The most complete description of adults and late instar nymphs was elaborated by Heidemann (1908).

P. perseae infests the lower surface of the foliage, causing chlorosis, necrosis and severe defoliation of avocado, probably reducing yields. Avocado lace bug usually lives in colonies, depositing eggs upright in irregular rows in clusters on the lower leaf surface. This opens an avenue of penetration for the anthracnose fungus, *Colletotrichum gloeosporioides* (Mead and Peña, 1991). Since the avocado lace bug was previously not considered an important pest, it is suggested that, in recent times, suitable natural enemies were decimated by applications of pesticides or by some other type of environmental disequilibrium.

In Florida, avocado lace bug population densities increase during the dry season (November–February) and decline during spring and summer (Peña *et al.*, 1998). The cultivars 'Waldin', 'Booth 8' and 'Loretta' had the highest natural infestation levels. The most susceptible cultivar appears to be 'Booth 8' with 20–28% of the leaf area damaged. Photosynthesis was reduced by 50% when the leaves sustained 40% damage (Peña *et al.*, 1998). Trees of 'Simmonds', with 100% of their leaves infested, exhibited an early leaf drop and no fruit set. In contrast, a West Indian × Guatemalan hybrid was scarcely affected by *P. perseae*.

Major biological control agents of *P. perseae* in Florida are two egg parasitoids, an *Oligosita* sp. (Trichogrammatidae) and the mymarid, *Eyrthmelus klopomor* Triapitsyn (Peña *et al.*, 2009). Predators include *Stethoconus prafectus* Distant (Heteroptera: Miridae) (Peña *et al.*, 2007; Henry *et al.*, 2009; Holguin *et al.*, 2009) and *Tingidoletes praelonga* Gagné (Diptera: Cecidomyiidae) (Gagné *et al.*, 2008). Peña *et al.* (2012) reported that natural enemies in Florida caused 16–90% mortality of *P. perseae*. Effective use of entomopathogenic fungi against *P. persea* has also been reported in Florida and Cuba (Peña *et al.*, 1996; Romero and Ravelo, 2020).

Ants

Ants are avocado pests primarily because they disrupt biological control of other pests. Ants protect phloem-sucking insects from natural enemies, especially in young avocado trees where mealybugs and other honeydew producers are occasional pests. However, ants sometimes chew crop twigs and tender bark, damage irrigation tubing or annoy workers. In Peru, *Atta sexdens* L. causes severe damage by cutting new avocado leaf flushes and consuming leaves. In Mexico, *Atta mexicana* Smith cuts the leaves into irregular pieces, the damage frequently starting at the leaf apex. The central vein of the leaf remains intact. The severity of the damage depends on the time of the year when the infestation occurs. *A. mexicana* is a variable-size ant, 3–14 mm long, with polymorphic markings. It is reddish to dark brown in colour. The workers are polymorphic, with spines on the pronotum, mesonotum and propodeum.

In South Africa, the driver or red ant (*Dorylus helvolus L.*) causes damage to avocado fruit hanging or lying on the ground under the trees. Parts of the fruit peel are damaged, and deep holes may be eaten in the fruit pulp (Erichsen and Schoeman, 1992).

In Spain, *L. humile* Mayr, *L. grandis* Forel and *P. pallidura* Nylander are considered avocado pests (Gil and Aranda, 2021). In addition, red imported fire ants (*Solenopsis Invicta* Buren) were found to be girdling and killing young avocado trees in California.

Thrips

Avocado Thrips

Avocado thrips, *S. perseae* (Thysanoptera: Thripidae), was found damaging avocado fruit and foliage in California in 1996. This pest has a native range that extends from Mexico to Guatemala (Hoddle *et al.*, 2002a). The population that invaded California originated from Coatepec-Harinas in Mexico (Rugman-Jones *et al.*, 2008). Economic damage is caused by *S. perseae* larvae and adults feeding on the skin of immature fruit (Fig. 14.11), and losses (i.e. fruit damage and increased management costs) from this pest can amount to several million dollars ($US) per year (Hoddle *et al.*, 2003a). Scarring can be severe enough to render the entire fruit surface brown with a characteristic 'alligator skin' appearance, which is downgraded in packing houses. Fruit, 25–44 mm long, are highly preferred for oviposition by *S. perseae*, but females will also oviposit

Fig. 14.11. The avocado thrips, *Scirtothrips perseae* and the damage it inflicts on flowers, fruitlets, leaves and mature fruit. Photo credits: Ing. Hector Guillen (with permission).

into fruit pedicels and the top and the undersides of immature leaves (Hoddle, 2002a,b,c). Leaves heavily damaged by adult and larval feeding become cup-shaped and bronzed, and in rare instances *S. perseae*-damaged leaves may abscise prematurely (Nakahara, 1997). Avocado thrips larvae and adults can build to high densities from spring through early summer when temperatures are moderate, and abundant flush growth is available for feeding and oviposition (Hoddle, 2002a).

Female *S. perseae* lay eggs singly into soft plant tissue. Eggs are kidney-shaped and whitish-yellow. Upon hatching, developing thrips pass through two active immature stages. Larvae pupate in leaf litter under trees, in cracks and crevices in the bark or within persea mite nests on leaves. Studies on the pupation biology of *S. perseae* on potted avocado trees in the laboratory indicated that 77% of second instar larvae fall from plants to pupate in the soil, while the remaining 23% pupated in cracks on tree branches (M. Hoddle, University of California, Riverside, 2012, unpublished data). Field surveys of leaf litter indicate that 94% of pupating thrips are found in the upper 5.1 cm (Yee *et al.*, 2001a).

The developmental and reproductive biology of *S. perseae* has been studied in the laboratory at five constant temperatures (Hoddle, 2002b). At 20°C, *S. perseae* exhibited the greatest larval to adult survivorship rates, and mated females produced a greater proportion of female offspring than those at higher or lower temperatures (Hoddle, 2002b). These data suggest that *S. perseae* likely evolved with avocados at high elevations with cool to mild year-round temperatures, which is supported by collection records in the home range of this pest (Hoddle *et al.*, 2002a). California surveys indicate that *S. perseae* damage significantly correlates with orchard distance from the coast, with hotter inland orchards suffering less thrips damage when compared to orchards that are situated in cooler coastal areas (Hoddle, 2002b).

S. perseae population increases positively correlate with new flush growth on trees. Peaks of 3–5 larvae per leaf during fruit set are correlated with 6–15% scarring on fruit. When 0.5–1.5 larvae are found on small fruit (< 1.5 cm in length), 22–51% economic scarring of the fruit surface is likely to develop. Fruit is no longer vulnerable to scarring when it is 3 cm in length. Consequently, the recommended action threshold for managing *S. perseae* is an average of 3–5 larvae per leaf when fruit sizes vulnerable to feeding are present (Yee *et al.*, 2001b).

Cultural control efforts at managing *S. perseae* in avocado orchards have focused primarily on controlling late second instars, propupae and pupae in the soil. One strategy for increasing thrips pupae mortality is to use composted organic yard waste. Composted mulch applied to a depth of approximately 30 cm and spread to the edge of the canopy can reduce *S. perseae* emergence rates by ~50% compared with non-treated control plots. The exact mechanism for suppression is unknown but could be due to an abundance of generalist arthropod predators that colonize the mulch and opportunistically exploit *S. perseae*. The release of secondary plant compounds from decaying mulch could also be antagonistic to *S. perseae*, along with entomopathogenic nematodes (e.g. *Steinernema* spp.) or fungi (e.g. *B. bassiana*) that have been isolated from mulch. In addition to *S. perseae* control, mulching assists with controlling Phytophthora root rot, improving soil quality and water retention, and weed suppression (Hoddle *et al.*, 2002b).

Several insecticides have been rigorously evaluated under field conditions for efficacy against *S. perseae* and to determine application rates and frequencies. Aerial applications by helicopter of sabadilla (a botanical pesticide made from the ground seeds of *Schoenocaulon officinale*) and abamectin provided effective control of *S. perseae* but different rates and spray volumes are required for small and large avocado trees (Yee *et al.*, 2001d). Abamectin suppressed *S. perseae* densities for 1–3 months after application compared with control treatments. Reductions in populations were sometimes delayed until 20–23 days after treatment even when coverage was high (Yee *et al.*, 2001d). Abamectin appears to paralyse *S. perseae* and intoxicated thrips enter a persistent quasi-moribund state before dying. The residual activity of abamectin increases when combined with horticultural oil or a surfactant that assists with translaminar movement that protects the active ingredient (AI) from photodegradation on the leaf surface. Similar results are obtained with spinosad when applied with horticultural oil or surfactants to avocado trees (~4.5 m tall trees) for *S. perseae* control (Tollerup and Morse, 2006). Resistance development by *S. perseae* to different insecticides is anticipated if they are overused or not rotated with products that have different modes of action. For example, *S. perseae* exhibited increased tolerance to sabadilla in orchards treated repeatedly with this product (Humeres and Morse, 2006).

Systemic neonicotinoids applied to potted nursery avocado trees showed varying efficacy against *S. perseae*, with imidacloprid being the most effective insecticide, providing at least 70% mortality of thrips for 14 weeks (Byrne *et al.*, 2007). However, chemigation of large trees (20–25 years old) with imidacloprid applied at a high rate (560 g AI l^{-1}) and low rate (280 g AI l^{-1}) was not effective in protecting young leaves from *S. perseae*. It appears that imidacloprid uptake is not sufficiently fast to keep pace with developing young leaves that need protection against *S. perseae* (Byrne *et al.*, 2010).

Classical biological control of *S. perseae* has not been attempted in California because foreign exploration efforts failed to find specialist host-specific natural enemies in the home range of this pest (Hoddle *et al.*, 2002a). Resident natural enemies can respond to increasing *S. perseae* populations in unsprayed orchards, with the predatory thrips *Aeolothrips kuwanaii* Moulton, *Ae. fasciatus* Linnaeus, *Leptothrips mali* Fitch and *Franklinothrips orizabensis* Johansen being the most important (Oevering *et al.*, 2002). However, augmentative releases of *F. orizabensis* and green lacewing larvae (*Chrysoperla carnea* (Stephens) Neuroptera: Chrysopidae) failed to suppress *S. perseae* populations compared with control treatments (Hoddle, 2003a,b; Hoddle and Robinson, 2004; Hoddle *et al.*, 2000, 2001, 2004).

S. perseae and its natural enemies, particularly the predator thrips *F. orizabensis*, can be monitored in avocado orchards using coloured sticky cards. Yellow sticky cards (peak spectral reflectance at 570 nm) are most attractive to *S. perseae*, and white cards (peak reflectance at 430 nm) capture more *F. orizabensis* (Hoddle *et al.*, 2002c).

Citrus thrips

African citrus thrips (*Scirtothrips aurantii* Faurè) was reported from avocado, citrus, guava and lychee in South Africa during the early 1920s (Faurè, 1929). Its presence has been confirmed outside the African continent in Cape Verde, Yemen, Mauritius, Reunion (Gilbert and Bedford, 1998) and Australia (Morris and Mound, 2004). Both the adults and larvae of *S. aurantii* cause lesions when feeding on avocado, citrus, guava, macadamia and mango fruit. It is a very important pest of citrus (Bedford, 1943) and mango, but it seldom does damage of economic importance to avocado. *S. aurantii* has been recorded on over 60 plant species (Faurè, 1929; Bedford, 1943), including tropical and subtropical crops and indigenous Acacia species. The male is smaller than the female. The presence of a comb of dark hairs on the ventral side of the hind femur and dark-hooked drepana (Faurè, 1929) on the end of the abdomen is diagnostic for *S. aurantii* males. The life cycle is completed in 18 days in summer and 44 days during winter (Bedford, 1943). Diapause does not take place. The number of generations per year was estimated to be around nine on citrus in Rustenburg, South Africa (Bedford, 1943). Schwartz (1983), Grout and Richards (1992), and Grout (1994) stated that predacious mites play an important role in regulating *S. aurantii* populations. A eulophid parasitoid, *Goetheana incerta* Annecke, parasitizes *S. aurantii* larvae (Grout and Stephen, 1995).

Red-banded thrips

Red-banded thrips (*S. rubrocinctus* Giard) has a pantropical distribution (Hill, 1975) and is present in many avocado-producing areas worldwide (Fig 14.12). Red-banded thrips feed on the ventral leaf surfaces. In Florida and Australia, infested leaves are spotted on the upper surface and usually have dark reddish-brown faecal pellets. Damaged areas turn a 'rusty' colour with numerous small, shiny black spots of excreta. The edges of infested leaves curl up. Heavy feeding on fruit causes a russetted appearance, cracking and decay. In greenhouses, red-banded thrips damage fruit and tend to feed on the larger and more mature fruit. They are found most frequently where two fruit are in contact or where a leaf contacts the fruit.

Eggs are inserted into the leaf tissue. They are white, kidney-shaped and about 0.25 mm long (Hill, 1975). The first and second nymphal stages are yellow, with a bright red band around

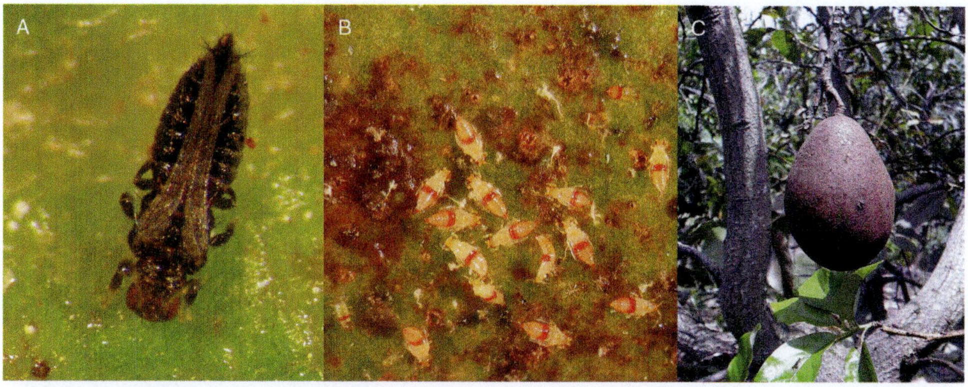

Fig. 14.12. Red-banded thrips, *Selenothrips rubrocinctus* Giard.

the base of the abdomen. When fully grown, the second instar is about 1 mm long. The tip of the nymph's abdomen is turned up and carries a drop of excreta on anal setae (Hill, 1975). The propupa is yellowish with red eyes and a red band across the first three abdominal segments (Hill, 1975; Steyn, 2001b). The adult female is dark brown and just over 1 mm long. Males are smaller and rare (Hill, 1975). Reproduction by the red-banded thrips is parthenogenetic (Avidov and Harpaz, 1969). Females live about 7 weeks and lay an average of 25 eggs, which hatch in 12–18 days (Hill, 1975). The nymphal stage lasts 6–10 days, and the propupa and pupal stage lasts 3–6 days.

Prior to implementing control measures, observations should be made on leaves and fruit for red-banded thrips during summer and fall for any developing infestation. Frequent fruit inspections are recommended, beginning in early August and continuing until harvest. Peña and Johnson (1999) and Waite and Pinese (1991) evaluated several insecticides for control of *S. rubrocinctus* in Florida and Australia, respectively. The predator *Franklinothrips vespiformis* D.L. Crawford is considered a predator of *S. rubrocinctus* Giard, as is *Leptothrips macroocellatus* Watson (Johansen and Mojica-Guzman, 1996).

Frankliniella spp.

In Mexico, *Frankliniella chamulae* Johansen and *Frankliniella bruneri* Watson are found on avocado flowers in Uruapan. There are unconfirmed reports that they cause injury to the fruit. Coria-Avalos (1993) reported that *Frankliniella* spp. and *Liothrips perseae* Watson were becoming the most important avocado pests in the Michoacán state of Mexico. However, the author did not specify which species injure fruit or leaves.

Western flower thrips (*Frankliniella occidentalis* Pergande), native to California, has been recorded infesting avocado inflorescences near infested mango orchards in the south of Israel. Many different thrips species were also found on avocado inflorescences, including *Thrips tabaci* Lindeman and *Thrips major* Uzel (M. Wysoki, R. zur Strassen and W. Kuzlicky, Agricultural Research Organization, the Volcani Center, Israel, 2011, unpublished data). In Chile, *F. occidentalis* has been observed in some cases in very high densities in the flowers of avocado, with no adverse effects (Ripa and Larral, 2008). A similar observation has been made in California (Hoddle pers. obs.)

Black vine thrips

In Israel, the black vine thrips (*Retithrips syriacus* Mayet) was reported causing damage to avocado fruit and leaves. The most heavily infested cultivars are 'Horshim', 'TX-531', '4102', '4203' and 'T-142'. The thrips can be readily controlled with sabadilla (Izhar *et al.*, 1992).

Greenhouse thrips

Greenhouse thrips (*H. haemorrhoidalis* Bouché) is native to Brazil and Peru and reproduces sexually (Mound and Monteiro, 1997). Invasive populations are parthenogenetic. This thrips is an economically important pest of avocados in

Israel, South Africa, Spain, Dominican Republic, Peru and Chile (Bodenheimer, 1951; Ebeling, 1959; De Villiers and Van den Berg, 1987; De Villiers, 1990b; McMurtry *et al.*, 1991; Larral *et al.*, 2008). The importance of this pest in California avocado orchards has diminished significantly, possibly because of natural enemies and improved IPM practices that reduced use of broad-spectrum insecticides (McMurty *et al.*, 1991; McMurtry, 1992). Effective cultural control of this pest has been achieved in California and simply requires thinning of fruit bunches by manipulating harvest dates to reduce contact points between fruit where greenhouse thrips larvae and adults reside in these well-protected areas (Phillips *et al.*, 1992).

In Israel, large populations of *H. haemorrhoidalis* usually start on trees of Mexican cultivars or 'Benik' and 'Hass'. From these infestation foci, thrips disperse to 'Fuerte', 'Nabal' and 'Ettinger' varieties. Heavy damage to 'Ardit' was recorded, and 'Stuart', 'Nahlat' and 'Horshim' are also susceptible cultivars (Izhar *et al.*, 1990a). Mortality rates of this thrips are high at relative humidities below 60% when temperatures are above 27°C (Rivnay, 1935). Thus, it seems that outbreaks of greenhouse thrips at certain sites and years are influenced by climatic factors. Greenhouse thrips is found on avocado trees across all seasons, with peak infestations occurring in autumn. During especially warm winters, populations may reach considerable numbers.

The body of adult greenhouse thrips is black, except for the legs, wings and antennae, which are white (De Villiers, 1990b). They are about 1.25 mm long and most of them are females that can reproduce parthenogenetically. The female lives for about 7 weeks and lays an average of 25 eggs (Avidov and Harpaz, 1969). The eggs are white, kidney-shaped and about 0.3 mm long. They are inserted singly into the leaf tissue just beneath the epidermis. They increase in size and become considerably swollen shortly before hatching. The optimum temperature range for oviposition is between 20 and 28°C (Ebeling, 1959). Most countries have four generations per year, but as many as 6–7 generations per year have been recorded in California. The life cycle, from egg to adult, lasts about 8 weeks (Hill, 1975).

Various parasitoids and predators have been recorded attacking the greenhouse thrips. A small pirate bug (*Orius tripoborus* Hesse) feeds on nymphs of *H. haemorrhoidalis* on avocado fruit in South Africa. This anthocorid may be partly responsible for the rarity of thrips outbreaks (Dennill and Erasmus, 1992a,b). A tiny parasitoid, *Thripobius semiluteus* Boucek, was found attacking nymphs of the greenhouse thrips feeding on *Croton* spp. in South Africa (Steyn, 1996). This parasitoid was introduced into California in 1986 and into Israel in 1991. The female endoparasitoid oviposits into first- or second-stage nymphs of *H. haemorrhoidalis*. In the laboratory, the development of the parasite is completed at 23°C within 22–25 days (McMurtry *et al.*, 1991). This parasitoid became established in some avocado orchards in California where up to 60% parasitism occurred within 2 years after their release (McMurty *et al.*, 1991). In California and Chile, the egg parasitoid, *Megaphragma mymaripenne* Timberlake, and the imported larval parasitoid, *Thripobius semiluteus* Boucek are important natural enemies of greenhouse thrips (Larral *et al.*, 2008). They successfully control the thrips population in Israel and Chile (Wysoki *et al.*, 1997). Natural enemies of greenhouse thrips recorded in Israel include the predacious mites, *Typhlodromus athiasae* Porath and Swirski, *Amblyseius swirskii* Athias-Henriot (Phytoseiidae); the thrips *Franklinothrips megalops* Trybom; and spiders and predacious bugs of the family Anthocoridae. However, natural enemies residing in orchards may be inefficient in curbing the pest populations (Wysoki *et al.*, 1997).

Greenhouse thrips mostly occur sporadically and usually only in small numbers; chemical control measures are seldom necessary. Chemical control should only be applied when absolutely necessary to maximize the benefit of natural enemies in avocado orchards (Steyn, 2001a,b). The predator *Franklinothrips vespiformis* Crawford is considered a predator of greenhouse thrips as is *F. orizabensis* (Hoddle, 2003a).

Other thrips species

Scirtothrips hansoni Mound & Hoddle, a new species described from Costa Rica, Ecuador and Colombia from the young leaves of avocado, is closely related to *S. perseae* and also to *S. astrictus* from Costa Rica (Mound and Hoddle, 2016). *Frankliniella gardenia* Moulton was also reported from Colombia and Mexico as a significant pest of avocado (Carabalí Muñoz *et al.*, 2021).

In Mexico, larvae and adults of *Scirtothrips aguacatae* Johansen and Mojica and *Scirothrips kupae* Johansen and Mojica (the identity of these thrips has been questioned and *S. aguacatae* has been synonymized with *S. perseae* (Hoddle *et al.*, 2008)) are found on immature leaves and rarely on flowers; they cause damage to the epidermis of young fruit.

Lepidoptera

Avocado seed moth

The avocado seed moth (*Stenoma catenifer*) is a specialist herbivore of plants in the Lauraceae family (Cervantes Peredo *et al.*, 1999). This moth is native to neotropical areas and is considered one of the most important pests of avocados in Mexico and Central and South America (Wysoki *et al.*, 2002) (Fig 14.13). It has been recorded attacking avocados in Mexico (Muñiz Velez, 1958; Mendez-Villa, 1961; Wolfenbarger and Colburn, 1966; Arellano, 1975), El Salvador (Wolfenbarger and Colburn, 1966), Guatemala, Honduras (Sasscer, 1921), Venezuela (Boscán de Martínez and Godoy, 1984), Guyana (Cervantes Peredo *et al.*, 1999), Peru (Wysoki *et al.*, 2002), Colombia (Carabalí Muñoz *et al.*, 2021), Brazil (Hohmann *et al.*, 2003) and Argentina (Wysoki *et al.*, 2002).

The primary economic loss is caused by direct damage by larvae feeding on fruit which may induce premature fruit drop. Damaged fruit are characterized by white exudates of the C_7 sugar alcohol, perseitol, which oozes from the entrances

Fig. 14.13. The avocado seed moth (*Stenoma catenifer* Walsingham). A. Adult female moth, B. Moth captured on monitoring pheromone trap, C. Larvae feeding on avocado seed, D. Damage on avocado fruit. Photo credits A = Dr Sangmi Lee, Insect Collections of ASU Natural History Collections, Arizona State University (ASU), B–D = Dr Gracielle Furtado (Brazil) (with permission).

of feeding tunnels and accumulations of frass at tunnel entrances on the fruit surface. On average, about 53% of tunnel entrances open in the bottom third of fruit, 40% are in the central region of fruit and only about 7% of holes are found in the top third of the fruit near the pedicel (Hoddle and Hoddle, 2008a). Within the avocado canopy, the level of fruit by *S. catenifer* infestation was significantly higher at the top than in the middle and bottom of the trees. Fruit infestation rates tend to be higher on the side of avocado trees facing the east (sunrise) (Vacari *et al.*, 2021). In addition to avocados and other plants in the genus *Persea*, *S. catenifer* has been reared from additional hosts in the Lauraceae family, including fruit of *Nectandra megapotamica* in Brazil (Link and Link, 2008) and *Chlorocardium rodiei* in Guyana (Cervantes Peredo *et al.*, 1999).

Crop losses of 60% and 80% have been recorded in Brazil and Venezuela, respectively, despite intensive broad-spectrum pesticide use (Boscán de Martínez and Godoy, 1982; Nava *et al.*, 2005a). Similarly, in Mexico, up to 95% of fruit can be attacked and monthly insecticide treatments are recommended for control (Wysoki *et al.*, 2002). In Guatemala, some commercial 'Hass' orchards can suffer up to 45% fruit infestation levels, despite rotating monthly applications of malathion and endosulfan (Hoddle and Hoddle, 2008a). In addition to fruit, *S. catenifer* larvae can attack and kill young twigs and stems (Wolfenbarger and Colburn, 1979), and heavy infestations of stem-mining larvae can kill young avocado trees (Cervantes Peredo *et al.*, 1999).

The adult is a ~1.25 cm long, light-brown-coloured moth with a distinctive 'C' pattern and black spots at the distal margin of the forewings (Fig. 14.13). Moths mate readily after emergence and can live up to 10–15 days. Females do not need to feed on carbohydrates to sustain their reproductive output (Milano *et al.*, 2010). The life cycle of this insect has been well studied (Boscán de Martinez and Godoy, 1984; Cervantes Peredo *et al.*, 1999) and degree-day estimates for developmental times of different life stages are available (Nava *et al.*, 2005b).

The egg is 0.6 mm long, oval and whitish-green, and ecloses in ~5–6 days; larvae penetrate fruit and consume pulp and the seed (Fig. 14.13). Females lay eggs on avocado branches, fruit, between the fruit and the fruit pedicel and on the pedicel (Hoddle and Hoddle, 2008a). There are five larval instars and one pupal stage. Early instars (first and second) are whitish, while late instars are reddish-brown (Fig. 14.13). The ventral side of mature fifth instar larvae is a distinctive light blue colour. Larval stages last ~16–21 days. Most (> 90%) of fifth instar larvae will exit the seed or fruit and walk for ~24 h searching for pupation sites. Larvae then enter a quiescent phase within a lightly spun silk bivouac, and about 24 h later they enter the pupal stage. Larvae that exit seeds and fall to the ground pupate in the top 0.5–2.0 cm of the soil (Boscán de Martínez and Godoy, 1984). The pupal stage lasts ~11–19 days; young pupae (< 8 h old) are turquoise in colour; older pupae are dark brown. The entire life cycle lasts ~44–49 days, and the sex ratio is 1:1 female:male. All developmental times presented here are estimates and vary with temperature. Adult moths are nocturnal and hide among weeds and in the soil on the orchard floor during the day (Hoddle and Hoddle, 2008c).

Trichogrammatids can parasitize up to 60% of *S. catenifer* eggs in Brazil, which is insufficient to prevent substantial economic losses (Hohmann *et al.*, 2003), but augmentative field releases of *Trichogramma pretiosum* may be feasible (Nava *et al.*, 2007). A variety of braconid and ichneumonid parasitoid species have been recorded attacking *S. catenifer* larvae (Cervantes Peredo *et al.*, 1999; Hohmann *et al.*, 2003; Nava and Parra, 2005), with *Apanteles* sp. (Braconidae) inflicting ~30–40% mortality in Venezuela and Brazil (Boscán de Martínez and Godoy, 1982; Nava and Parra, 2005). In Guatemala, a gregarious *Apanteles* sp. parasitized ~22–53% of larvae (Hoddle and Hoddle, 2008a,b,c). Gregarious *Apanteles* sp. in Guatemala produce, on average, 8–9 offspring per *S. catenifer* larva and the sex ratio is 47% female (Hoddle and Hoddle, 2008a). A sympatric solitary *Macrocentrus* sp. (Braconidae) can parasitize 0.4–15% of larvae in Guatemala (Hoddle and Hoddle, 2008a,b,c). *Hogna* sp. spiders (Lycosidae) are likely important predators of *S. catenifer* life stages in the soil. Laboratory studies indicate that *Hogna* sp. will readily attack and consume mature fifth instar larvae that have exited seeds and are searching for pupation sites, pupae and adult moths (Hoddle and Hoddle, 2008c).

The sex pheromone of *S. catenifer* is a highly unsaturated aldehyde, (9Z)–9,13-tetradecadien-

11-ynal, and at the time of identification, a new class of natural product (Millar *et al.*, 2008). Synthesis instructions for the pheromone are available (Hoddle *et al.*, 2009a; Zou and Millar, 2010). The best dispensers for releasing the pheromone in wing traps are 11 × 5 mm grey rubber septa and 11 × 5 mm white rubber septa (Hoddle *et al.*, 2011). The pheromone has a longevity of at least 4 weeks (Hoddle *et al.*, 2009a). Traps placed at ~0.15 m, ~1.75 m, and ~4.0 m above the ground catch statistically equal numbers of male *S. catenifer*, suggesting that trap height is not important, and traps should be placed at heights that are convenient for inspection (Hoddle *et al.*, 2011). However, Cruz-Lopez *et al.* (2020) reported that traps placed 4 and 6 m above the ground captured significantly more *S. catenifer* moths than traps that were 2 m high.

In Guatemala, peak captures of male moths in pheromone traps occurred between 2:30 am and 4:30 am and ceased abruptly after 5:30 am (Hoddle *et al.*, 2009a). Mark-recapture studies of male *S. catenifer* in pheromone traps indicated that, on average, males flew 67 m in one night. However, this is likely an underestimate and these studies need to be repeated (Hoddle *et al.*, 2011).

In Mexico, Velázquez-Martínez *et al.* (2022) used pheromone-baited traps to study the population dynamics of *S. catenifer* in 'Hass' avocado orchards and determined two population peaks. The first was in June–August, during the fruiting stage, and the second was in December–January, which coincided with the end of the fruiting and flowering stages.

The *S. catenifer* sex pheromone can be used to detect and monitor this pest in countries with endemic populations that are exporting fresh avocados (e.g. Mexico, Colombia, Ecuador, Chile and Peru) and for quarantine detection and incursion monitoring in countries receiving avocado imports from high-risk areas that have commercial avocado industries (e.g. the USA, Israel and Spain).

The importance of proactive monitoring for *S. catenifer* cannot be overstated. This pest has demonstrated an ability to invade and establish in new areas. The invasion threat from importation of fresh avocados was realized in 2000 when *S. catenifer* established in the Galápagos Islands after being accidentally imported inside fruit that originated in mainland Ecuador (Landry and Roque-Albelo, 2003). It has been estimated that larvae damage ~90% of the biannual crop, which has effectively eliminated a locally produced food on at least two islands, Santa Cruz and San Cristóbal, in the Galápagos. Additionally, larval to adult survivorship rates are high (62%) and parasitism is very low (< 5%) in the Galápagos. Of the *S. catenifer* larvae that were parasitized, parasitoids failed to emerge from cocoons, suggesting opportunistic exploitation of an unsuitable host (Hoddle, 2009).

In all countries with resident *S. catenifer* populations, chemical control is predominantly achieved through the heavy and regular use of broad-spectrum insecticides to minimize fruit infestation. In areas of Peru with endemic *S. catenifer* populations, avocados that are not treated with insecticides can reach 100% infestation levels. Even when natural enemy activity is high, as observed in Guatemala, larval parasitoids do not emerge from *S. catenifer* until the pest has reached the fifth and final instar and larvae have exited fruit. By this time, fruit have significant damage and are unmarketable, and the subsequent death of larvae through parasitoid activity occurs too late to reduce economic losses.

Bark borers

In South Africa, the bark borers, *Salagena obsolescens* Hampson and *Salagena transversa* Walker, feed on avocado, guava, lychee, macadamia, pecan, wild fig (*Ficus* sp.), bushwillows (*Combretum zeyheri* and *C. collinum*), jacket plum (*Pappea capensis*) and water berry (*Syzygium cordatum*) (De Villiers, 1973). *Eugenia* sp. is also a host of *S. obsolescens* (Pinhey, 1975). According to Pinhey (1975), *S. obsolescens* has been recorded from East Africa, Mozambique, South Africa, West Africa and Zimbabwe.

Salagena spp. eggs are cream coloured and laid in clusters on the bark of branches. The young larva feeds on the bark and later tunnels into the wood for shelter. The larva is light brown to dark brown and almost without hair. The final stage can be up to 30 mm in length. The pupa is dark brown and develops in the tunnel, its head facing the opening. The adult of *S. obsolescens* has a variegated pale and dark-brown thorax (Pinhey, 1975). The abdomen is grey with prominent black or brown dorsal tufts near the base, followed by reddish-brown ones. The forewing is cream colour, with rows of black rings encircling reddish-brown spots. The forewing is 15–23 mm long.

Moths of *Salagena* spp. are active at night, and a female can lay about 270 eggs (De Villiers, 1973). The larva feeds on the bark of trees under a frass-covered silken web. This cover is enlarged when the larva grows larger and may encircle a branch. The larva may then girdle the bark and kill a branch. The older larva tunnels into the wood, usually where branches fork. The tunnel serves as a shelter for the larva and is up to 70 mm long and about 5 mm in diameter when the larva is fully grown (De Villiers, 1973).

Two parasitic wasps of these bark borers have been recorded. One of these has been identified as *Genaemirum* sp. (Ichneumonidae).

Avocado leaf roller and ivy leaf roller

The major caterpillar pests in Australia are the avocado leaf roller (*Homona spargotis* Meyrick) and the ivy leaf roller (*Cryptoptila immersana* Walker). These pests web leaves together and feed on them, but the more serious damage is caused when they chew into fruit or web leaves to fruit (Waite and Pinese, 1991).

Natural enemies recorded for *C. immersana* include the predatory larva of the syrphid fly (*Melanostoma agrolas* Walker) as well as the parasitoids *Goniozus* sp. (Bethylidae), *Sympiesis* sp. (Eulophidae) and *Phytodietus celsissimae* Turner (Ichneumonidae). A trichogrammatid also parasitizes the eggs (Waite and Pinese, 1991).

Cryptaspasma sp. nr. lugubris

Cryptaspasma lugubris Felder and Rogenhofer was recorded from 'Hass' avocados in Michoacán, Mexico, where larvae were initially confused with *S. catenifer*; it is also known from cultivars other than 'Hass' in Puerto Rico (Brown and Brown, 2004). *Cryptaspasma perseana* Gilligan and Brown was reported as a potential pest of avocado fruit in Mexico and Guatemala (Gilligan *et al.*, 2011). Other species of *Cryptaspasma* have been recorded from fruit of *Quercus* sp., *Macadamia integrifolia* Maiden and Betche, *Beilschmidia pendula* (Sw.) Hemsl., *P. borbonia* L. and *P. palustris* (Raf.) Sarg. (the latter three host species are in the Lauraceae family) (Brown and Brown, 2004). It was hypothesized that *Cryptaspasma* infest hard seeds that are exposed in fallen fruit (Brown and Brown, 2004). Consequently, *Cryptaspasma* would not be classified as pests of on-tree avocado fruit. Laboratory studies indicated that *C.* sp. nr. *lugubris* would oviposit on fruit and seeds, but any avocado substrate preference was overwhelmed by this moth's predilection for ovipositing on plastic surfaces in rearing cages where choice studies were conducted.

Rearing studies of field-collected fruit infested with *C.* sp. nr. *lugubris* indicated that ~30% of larvae are parasitized by *Pseudophanerotoma* sp. (Hymenoptera: Braconidae), and this is the first rearing record for any member of this parasitoid genus (Hoddle and Hoddle, 2008d). In the laboratory, egg masses of *C.* sp. nr. *lugubris* average ~21 eggs and require ~12 days to hatch at ~22°C. Larval to adult survivorship in the laboratory is 72%, and the sex ratio was 54% female (Hoddle and Hoddle, 2008d). Coloured photographs of all life stages of *C.* sp. nr. *lugubris* and its natural enemies are available (Hoddle, 2010b).

Avocado fruit destroyer

The avocado fruit destroyer (*Histura perseavora* Brown) is an avocado-feeding tortricid, first collected and subsequently described from material collected in Guatemala (Brown and Hoddle, 2010). This is a poorly known genus with just eight described species from South America and one from Panama. Species of *Histura* are extraordinarily rare, and the life history and host plants for members of this genus are poorly known. The record presented here is the first time a species of *Histura* from Guatemala has been documented. This is the second host plant record for a member of this genus, the other species being *H. costaricana* Razowski also reared from avocados in Costa Rica.

Adult moths are small, ~5.5 mm in length, and are grey-brown in colour. Larvae are ~6–7 mm long and creamy-pink with conspicuous dark spots and can be easily confused with *S. catenifer* (*H. perseavora* larvae are significantly smaller than *S. catenifer* larvae). Pupae are ~5–6 mm in length, light brown and typically tortricoid in shape with prominent dorsal spines. A larval parasitoid, *Chelonus* sp. (Hymenoptera: Braconidae), was reared from *H. perseavora* larvae (Brown and Hoddle, 2010).

In Guatemala, *H. perseavora* was only associated with avocados grown at elevations higher than 1450 m, and adult moths were reared from larvae infesting fruit of 'Hass' and other cultivars. This moth is sympatric with *S. catenifer* at elevations from ~1450 to 1600 m above sea level. Above 1800 m, *H. perseavora* is more commonly encountered in fruit than *S. catenifer* (Hoddle and Hoddle, 2008d; Brown and Hoddle, 2010).

The major damage to avocado fruit is caused by larvae boring into fruit pedicels and the uppermost parts of the fruit close to the pedicel. Larvae were also reared from young green twigs. This preferred feeding area in the fruit, which is very close to the fruit pedicel, is in direct contrast to *S. catenifer*, which concentrates its feeding efforts in the middle to bottom third of fruit (i.e. distal to the pedicel) (Hoddle and Hoddle, 2008a). *H. perseavora* and *S. catenifer* may reduce competition by feeding preferentially in different parts of the fruit and by having evolved different climatic preferences that become evident with changes in altitude where one species or the other is more commonly encountered.

Damage is characterized by white exudates of perseitol oozing from the feeding tunnel exit holes that often have accumulations of fine granulated frass (for colour photos of all life stages and feeding damage, see Brown and Hoddle (2010)). Work is needed to determine the severity of this pest in orchards, to better delineate its range in Central America and possibly Mexico and to identify its sex pheromone.

Avocado bark borer

The avocado bark borer (*Aegeria* sp.) feeds beneath the bark of avocado trees in the Philippines. Sap may ooze from the wound and branches may be weakened so that they snap in strong winds. The pest is of minor importance (Cendana *et al.*, 1984). Quimio and Cayetano (1985) report controlling this insect with insecticidess applied at the onset of the infestations.

False codling moth

False codling moth (*Cryptophlebia leucotreta* Meyrick) was reported as a minor pest of avocado in South Africa (De Villiers *et al.*, 1987b; Joubert and Du Toit, 1993). When *C. leucotreta* infests avocado, a tiny hole is visible where the larva penetrated the skin of the fruit. This is usually surrounded by a white powdery substance presumed to be perseitol. A small feeding tunnel is made just beneath the skin, but large larvae are seldom found. Kok (in Milne, 1973) mentioned that only once has an adult false codling moth been reared, in this case from an avocado seed. However, 'Pinkerton' avocados interplanted with maize were extensively damaged by false codling moth after the maize was sprayed with cypermethrin (Joubert and Du Toit, 1993). Up to 50 false codling moth eggs were counted on a single fruit and an estimated 80% of the avocado fruit were damaged.

The developmental stages have been described in detail by several authors, the most comprehensive by Gunn (1921), Daiber (1979) and Schwartz (1981). Eggs are pearl white and translucent, turning slightly reddish with a black spot shortly before the larva emerges. The egg is oval, flattened and with a reticulate sculpture, measuring 0.77 × 0.60 mm. Eggs are laid singly on fruit, nuts and, according to Gunn (1921), also on guava leaves. The first instar larva is cream-white with a dark-brown head and ~1.5 mm long. The full-grown larva is pinkish-red with a brown head and is 12–15 mm in length. The pupa, which is dark brown, is formed in a silken cocoon in the soil. The adult is inconspicuous with mottled dark grey coloration. Males are smaller than females and have a greyish anal tuft of scales, densely packed grey-white hairs on the hind tibia and a scent gland near the anal angle of each hindwing.

According to Newton (1998), the sex ratio of field populations is close to unity. Females mate shortly after emergence and pre-oviposition lasts 5–6 days in the field and 1–2 days in laboratory cultures. Multiple mating takes place in both sexes (Schwartz, 1981). Daiber (1980) found an average fecundity of 456 eggs at a constant 25°C and 87 eggs per female at 15°C. Five days after moth emergence, a peak of 29 eggs per female was observed at 20°C. Peak egg laying occurs during the first night of the oviposition period which is normally 2 days after emergence. Five generations may develop per year in the Pretoria area (Daiber, 1980). Adults live 2–3 weeks in the field and are not attracted to light traps (Gunn, 1921; Ripley *et al.*, 1939; Catling and Aschenborn, 1978). A female pheromone was isolated by Read *et al.* (1968, 1974) but it was later concluded (Newton and Mastro, 1989; Newton *et al.*, 1993) that sex pheromones comprise several structural and optical isomers.

The broad range of host plants, together with mild tropical and subtropical winters, ensure that the pest is an all-year-round threat to crop hosts in most areas of its distribution (Newton, 1998). Ullyett and Bishop (1938) obtained strong evidence that, in some areas, moths originating in indigenous hosts give rise to infestations in avocado orchards at the beginning of the season (Newton, 1988).

Catling and Aschenborn (1974) reported egg parasitism by *Trichogrammatoidea cryptophlebiae* Nagaraja. The egg-larval parasitoid, *Chelonus curvimaculatus* Cameron (Braconidae) was reported by Searle (1964) and Broodryk (1969). A number of larval parasitoids were recorded in Southern Africa, namely the ichneumonid *Apophua leucotretae* Wilkinson, the braconids *Agathis bishopi* Nixon, *Agathis leucotreta* Nixon, *Bassus* sp. and *Phanerotoma curvicarinata* Cameron, the chalcid *Oxycoryphe edax* Waterston, and an unidentified tachinid (Ford, 1934; Ullyett, 1939; Thompson, 1946; Begemann and Schoeman, 1999). *Rhynochorus albopunctatus* Stål and *Orius* sp. and *P. megacephala* F. have been recorded as predators of the false codling moth (Omer-Cooper, 1939; Steyn, 1954; Newton, 1998).

Controlling false codling moth is extremely difficult because eggs are laid continually when fruit are on the trees. Furthermore, the hatching larvae bore into the fruit or husk of the nuts within a few hours (Newton, 1998). Orchard sanitation, recommended by Fuller (1901) and Gunn (1921), is still one of the most important methods of reducing damage by the false codling moth. In South Africa, the granulovirus product Cryptogran®, the active ingredient being *Cryptophlebia leucotreta*, was registered for use on avocado. Access to the US market for South African 'Hass' will require mitigation treatments (Louw *et al.*, 2012). Cold disinfestation of 22 days would be required to destroy all potential 'Hass' fruit infestations of false codling moth.

Western avocado leaf roller

The western avocado leaf roller (*Amorbia cuneana* Walsingham) is a sporadic pest of avocados in California and Mexico. *A. cuneana* larva rolls and feeds on the folded edge of the leaves; the injury can be extended to superficial feeding on fruit, frequently when two fruit are close together, causing economic losses to growers (Bailey and Olsen, 1990; Coria-Avalos, 1993).

Adult female moths lay light green, oval-shaped, overlapping flat masses of 5–100 eggs. Eggs are generally laid on upper leaf surfaces close to the midrib. Eggs hatch within 13–15 days. Newly hatched larvae are yellowish green and gradually turn darker green as they reach maturity. Larvae roll and tie leaves together with a silken web. This forms a shelter or nest in which the larva feeds on leaves and fruit and is protected from pesticide applications. Larvae go through five instars and pupate in leaf rolls. The pupal stage lasts about 17 days. Adult moths are bell shaped, with a wingspan of about 2.5 cm. Adults are active at night. Female moths deposit 400–500 eggs during 2–3 weeks (Faber and Phillips, 2001).

Young *A. cuneana* larvae typically feed only on the surface of avocado leaves, leaving a thin brown membrane or skeleton of leaf veins. Mature larvae consume the whole leaf. Mature avocado trees can tolerate considerable leaf damage by *A. cuneana* larvae without severely affecting tree growth or fruit yield. Fruit damage may occur where larvae web leaves against fruit or where webs are made between touching fruit. Larvae may feed on fruit skin and cause scarring, which causes severe downgrading or culling (Faber and Phillips, 2001).

In California, Hoffman *et al.* (1983) tested two components of the sex pheromone of *A. cuneana* (E,Z)-10,12 and (E,E)-10,12 tetradecadien-1-ol acetates (McDonough *et al.*, 1982) and determined that optimum trap catches were obtained with an isomer content of 29–82% EZ (as a percentage of EE + EZ) and dosages of 0.06–1.7 mg per rubber septum. Later, Bailey *et al.* (1988) tested a combination of two components (E,Z)-10,12 and (E,E)-10,12 tetradecadien-1-ol acetates in a 1:1 ratio and in a 9:1 ratio to trap populations of *A. cuneana*. The difference in moth response led them to believe that populations represented different species or different races of *A. cuneana*. Faber and Phillips (2001) suggested monitoring larvae in late spring by looking for leaf rolls in young foliage and feeding damage in mature fruit. There are no established thresholds for chemical control.

The egg, larval and pupal stages of *A. cuneana* are parasitized by various beneficial insects. The most important egg parasitoid is *Trichogramma platneri* Nagarkatti. Results of a preliminary study by Oatman and Platner (1985) showed at least 87% parasitization of *A. cuneana* eggs. Fleschner *et al.* (1957) reported *Elachertus proteoteraris* Howard and an ichneumonid parasitizing the larvae of *A. cuneana*, and the tachinid, *Phorocera erecta* Coquillett, parasitizing the pupa.

Apple leaf roller

The apple leaf roller (*Tortrix capensana* Walker) is an indigenous species of the Ethiopian region (Pinhey, 1975), and it occurs throughout Africa (Begemann *et al.*, 1998). Feeding on avocado fruit is mostly done on the side of the fruit. In most cases only the peel of the fruit is fed on, which results in it being downgraded.

Eggs are laid in oval masses on the dorsal leaf surface. The individual eggs overlap and are attached with a clear secretion (Begemann *et al.*, 1998). Freshly laid eggs are a dull yellow-cream colour, which changes to a yellow-orange before hatching. The newly hatched larva is pale yellow with a black head. As the larva ages, the body darkens to a yellow-green. Five instars occur in males and six in females which reach 20–25 mm in length. Pupation takes place in the tree in leaves webbed together. The newly formed pupa is light green and becomes brown. Female pupae are, on average, 11 mm long and males are 9 mm long. Both male and female adults are brown, with varying patterns of a darker or lighter colour on the forewing (Begemann *et al.*, 1998).

Development takes, on average, 10 days in summer and 21 days in winter. At 26°C, male larvae complete their development in 18–29 days, female larvae in 18–29 days and the pupal stage 8–12 days. The duration from egg to egg is 41–43 days. Six generations of *T. capensana* occur annually in citrus orchards in South Africa (Begemann *et al.*, 1998).

Several natural enemies, mostly larval parasitoids, keep apple leaf roller in check in avocado orchards in South Africa (De Villiers, 2001a). Barrow (1977) reared the following parasitoids from *T. capensana* larvae: *Apanteles* sp. (Braconidae), *Goniozus* sp. (Bethylidae) and *Pediobius amaurocoelus* (Westwood) (Eulophidae). Pupal parasitoids are *Nemorilla afra* Curran (Tachinidae), *Theronia* sp. (Ichneumonidae) and *Brachymeria boranensis* Masi (Chalcididae) (Barrow, 1977). The bacterium, *Bacillus thuringiensis* var. *kurstaki*, is registered in many countries for control of leaf rollers.

Citrus leaf roller

The citrus leaf roller (*Archips occidentalis* Walsingham) larva spins leaves or fruit together to form a protective niche where it feeds. Damage to the fruit is characterized by large epidermal lesions, which lead to the degrading of the fruit (Van den Berg *et al.*, 1999). A survey in the Nelspruit-Hazyview region of South Africa indicated that citrus and apple leaf rollers damaged 0.34% of avocado fruit, which represented a calculated loss of ~US $14,000 for the season (Erichsen and Schoeman, 1992).

The eggs of the citrus leaf roller are laid on the leaves in compact, oval-shaped masses of between 20 and 176 eggs (Newton and Catling, 1998). A protective white membrane covers the egg masses, while individual eggs are glued together in an overlapping pattern. The eggs are dull, yellow-grey, changing to orange before hatching. The remains of the egg masses stay on the leaves long after they have hatched. When they hatch, the larvae are pale orange with a darkly pigmented brown head capsule (Newton and Catling, 1998). They later turn pale green. On hatching, the larvae crawl directly to young leaves and fruit or disperse by spinning thin threads. Webs are spun over the immediate feeding area (Newton and Catling, 1998). Fruit may be spun together, while young leaves are rolled and webbed together, each with a larva hidden inside. Larvae reach a length of 20–25 mm (Van den Berg, *et al.*, 1999). According to Newton and Catling (1998), the insect usually pupates in the rolled leaves; the pupa is dark brown. The adult's anterior wings are brown with two oblique bands and a darker area at the apex (Newton and Catling, 1998). Avocado fruit is damaged in South Africa mainly from January to April (Newton and Anderson, 1985; Van den Berg *et al.*, 1999). The larva is sensitive to direct light and only feeds under protected conditions. Young leaves are rolled up by spinning threads. Only portions of young leaves are eaten, rarely the entire leaf (Newton and Catling, 1998). On avocado, the larvae usually damage the lateral portions of fruit where two fruit hang next to each other or a leaf is webbed to a fruit. When disturbed, the larvae quickly wriggle backwards or drop to the ground on a silken thread.

Several parasitoids keep the citrus leaf roller in check on avocado (Van den Berg *et al.*, 1999). Parasitoids reared from citrus (Newton and Catling, 1998) include *Brachymeria microlinea* Walker, *Brachymeria boranensis* Masi (Chalcididae), *Apanteles* spp. (Braconidae) and *Pristomerus* sp. (Ichneumonidae) (Prinsloo, 1984; Newton and Catling, 1998).

As the citrus leaf roller parasitoids keep the pest in check, chemical intervention is usually unnecessary. However, if necessary, the bacterial insecticide *Bacillus thuringiensis* var. *kurstaki* WP can be effective when the larvae are still in their early instars.

Other leafrollers

The brown-headed leafrollers (*Ctenopseustis obliquana* Walker and *C. herana* Felder and Rogenhofer) are the major pests of avocados in New Zealand. These moths are endemic to New Zealand, where a complex of six leaf roller species that feed on both foliage and fruit can cause up to 30% rejection of fruit for export from unsprayed orchards. On average, seven insecticide applications per year are made to most commercial orchards to combat these pests (Stevens *et al.*, 1998). Frequent applications of *Bacillus thuringiensis* have been shown to provide acceptable control. Despite the increased cost associated with the additional sprays, the advantages gained through being able to pick fruit immediately after spraying outweigh this cost (Stevens, 1997).

Caloptilia perseae Busck was first described on avocado in Florida in 1920 (Busck, 1920). In Mexico, it is considered a pest in the states of Oaxaca and Veracruz (Gallegos, 1983). The adult is 2–3 mm in length with whitish-cream-coloured wings; the larvae mine the tender leaves, causing deformation; however, they very seldom cause defoliation. In general, this is considered an unimportant pest. There is no knowledge of its life cycle or natural enemies (Gallegos, 1983). This species was recently reported from the Canary Islands, Spain (Gaston *et al.*, 2018). In addition, *Apanteles* n.sp. (Braconidae) and *Sympiesis dolichogaster* Ashmead (Eulophidae) have been reported parasitizing *C. persea* in Cuba.

In Peru, *Phyllocnistis* spp. (Lepidoptera: Gracillariidae) can reduce tree vigour by mining in the leaves, while in Florida, *Phyllocnistis* spp. is considered a minor pest problem.

Arctopoda maculosa

Arctopoda maculosa Butler is a native species of Chile, affecting primarily 'Chileno', 'Mexicola' and 'Champion' avocados. It feeds on the leaves only and its densities are considered low. Adults are 13 mm in length, and the larva is 2 cm long (Ripa, 2008b).

Citrus looper

The citrus looper (*Ascotis selenaria recipocaria* Walker) was recorded first on avocado in South Africa (De Villiers, 2001c). This looper was also recorded in equatorial Africa, Cameroon, Malawi, Uganda, South Africa, Zaire, Zambia and Zimbabwe. The citrus looper has a broad range of host plants, including citrus, avocado, coffee (robusta and arabica) and macadamia. Other hosts include boxwood, kei apple, pepper tree (*Schinus molle*), *Acacia* sp., *Pappea* sp., *Rhus* sp., wattle, willow and *Ziziphus* sp. (Pinhey, 1975). It is considered to be of minor importance on avocado in certain parts of Africa.

Schoeman (1998) describes the stages as follows: the eggs are blue-green and oval, 0.79 mm in length and 0.48 mm in width. They are laid in groups of 10–80, but sometimes up to 260. The eggs are deposited beneath loose bark or in cracks of dead twigs. The larvae are slender and twig-like, measuring 25–50 mm long. Mature larvae drop from trees and enter the soil to a depth of about 20–30 mm, where they pupate. Prepupae are olive-green and grey mottled and 16–22 mm in length. Pupae are 16–21 mm long, greenish at first, changing to mahogany-brown. Adult moths are white and brown with a 45–48 mm wingspan. Moths are active at night but rest on tree trunks and branches during the day. The larvae feed on the skin of young and mature avocado fruit. Young larvae gnaw the fruit's skin superficially, while full-grown larvae cause deeper lesions, slightly penetrating the fruit flesh (De Villiers and Van den Berg, 1987).

In South Africa, Erichsen and Schoeman (1994) found the greatest damage by the citrus looper on 'Edranol' followed by 'Pinkerton', 'Hass', 'Fuerte' and 'Ryan'.

Natural enemies include braconid larval parasitoids, including two *Rogas* spp., *Cardiochiles* sp., as well as the ichneumonid *Charops* sp. According to Schoeman (1998), more than 90% parasitism occurred due to these two species at Zebediela Citrus Estate in South Africa. Dipterous parasitoids include *Peribaea cervina* Mesnil, *Exorista sorbillans* Wiedemann, *Sturmia imberbis* Fallén Wiedemann, *Pales coerulea* Jaennike and *Muscina stabulans* Fallén. An entomopathogenic fungus (*Beauveria bassiana*) also attacks the larvae (Schoeman, 1998).

Avocado loopers

Several loopers, *Epimeces detexta* Walker (Fig. 14.14), *E. matronaria* Guenée, *Anacamptodes defectaria* Guenée and *Oxydia vesulia transponens* Walker, feed on avocado leaves in Florida. The most common of these is *Epimeces detexta*. This looper is the larva of a medium-sized grey or greyish-white moth. Young larvae are 0.6 cm or less in size and usually grey or greyish black and grow rapidly to 3–4 cm in length. Older larvae are generally tan or greenish-yellow in colour. Larvae feed on flower panicles, even fruit, but prefer the tender growth in the upper part of the tree. Looper infestations appear seasonal and are more severe in spring and summer, generally becoming less of a problem in fall and winter (Peña *et al.*, 1996). The adult moth is short lived and mates and oviposits soon after emergence from the pupa. Eggs are laid in narrow, elongated masses, preferentially on needles of Australian pine (*Casuarina* sp.) and hatch in about 5 days. The larvae grow rapidly and pupate 17–22 days after eggs hatch. The pupal stage can last 10 days. Thus, a full generation is expected to last 34–37 days. Pupae drop to the ground and the adult emerges in 12 days to restart the cycle. Some avocados are culled because of damage from feeding on the fruit by two or three small caterpillars.

Bacillus thuringiensis is sold under several trade names and has an effect on certain types of caterpillars. It is cleared for use on avocados in many producing countries. Native natural enemies of *E. detexta* include the predators *Calleida decora* Fabricius and *Podisus maculiventris* Say. *Alcaerrhynchus grandis* Dallas, the parasitoid *Parapanteles* sp. and *Trichospilus diatreae* Cherian are reported as parasitoids of *E. matronaria*, *A. defectaria* and *O.v. transponens*, respectively (Peña *et al.*, 1996). In Florida, several attempts to introduce exotic biological control agents, including *Telenomus* sp. and *Trichogramma platneri* Nagarkatti, have failed.

Other loopers

Ectropis sabulosa Warren (Geometridae) can be a serious pest in north Queensland (Australia), where it can completely defoliate trees, exposing fruit, resulting in sunburn. It can also cause severe damage to fruit when it feeds on the skin. Infestations are more severe near windbreaks. Control can be obtained by spraying a few rows near the windbreak, thus sparing the parasitoid

Fig. 14.14. The avocado looper, *Epimeces detexta* Walker. A. Adult moth, B. Looper. Photo Credits A = Dr Sangmi Lee, Insect Collections of ASU Natural History Collections, Arizona State University (ASU), B = Dr Jorge Pena (USA, Florida) (with permission).

Apanteles sp. nr. *Vitripennis*, which normally suppresses *E. sabulosa* and prevents outbreaks (Waite and Pinese, 1991).

In Australia, the grey looper (*Cleora inflexaria* Snellen), the brown looper (*Lophodes sinistraria* Guenée), the bizarre looper (*Eucyclodes pieroides* Walker (Geometridae)) and the orange fruit borer (*Isotenes miserana* Walker (Tortricidae)) are all minor pests which occasionally attack the fruit, causing holes and excavations. In Peru, *Sabulodes caberata* Guenée is common in orchards near the jungle. Although it may cause some defoliation, it does not attack frequently enough to justify pesticide applications.

Giant looper

In Israel, the giant looper (*Boarmia selenaria* Denis and Schiffermüller (Geometridae)) is the most important avocado pest in the regions where cotton is grown. It produces five generations a year, of which the most destructive are the first (spring) and second (early summer) (Wysoki and lzhar, 1986). Its reproductive behaviour was studied by Hadar (1983).

A long list of the natural enemies of the giant looper in Israel was compiled in earlier publications (Wysoki and lzhar, 1980). Spiders are very common in avocado orchards and contribute to some extent to the biocontrol of the pest (Mansour *et al.*, 1985). The wasp, *Apanteles cerialis* Nixon (Braconidae), oviposits in young caterpillars of the giant looper. Observations in avocado orchards revealed that parasitism rates were highest in October and November, reaching 70% (Wysoki and lzhar, 1981). Females of tachinid flies (Tachinidae), such as *Compsilura concinnata* Meigen and *Exorista nr. sorbillans* Wiedemann, attack advanced stages of the giant looper caterpillars; the parasitism rate peaks in late summer or autumn. Since local egg parasites of the giant looper have not been recorded in Israel, two exotic wasps, *Ooencyrtus ennomophagus* Yoshimoto (Encyrtidae) and *Telenomus alsophilae* Viereck (Scelionidae), were introduced but failed to attack eggs of the giant looper (Wysoki and lzhar, 1980).

Infestations are controlled, when necessary, by preparations containing *Bacillus thuringiensis (B.t.)* var. *kurstaki* Berliner, but these preparations are toxic only to young caterpillars. Timing of control measures is based on traps baited with virgin females (attracting males) and scouting the orchards for young caterpillars. Since mass production of giant looper virgin females for monitoring purposes is laborious and expensive, efforts were made to replace the virgin females with synthetic pheromone. The chemical compounds (Z,Z)-6,9-*cis*-3S,4R epoxynonadecadiene and (Z,Z,Z)-3,6,9-nonadecatriene were identified as sex pheromone components (Becker *et al.*, 1990). Bioassays performed by electroantennography (EAG) in a wind tunnel gave positive results, but in field tests, males were not sufficiently attracted to these two compounds (Becker *et al.*, 1990).

Other lepidopterous defoliators

In Mexico, *Copaxa multifenestrata* Heinrich-Shaffer (Saturnidae) is considered a specific pest of avocado. It feeds on the older leaves and is not economically important. By contrast, *Papilio garanas garanas* Hübner (Papilionidae) feeds on tender avocado leaves and may cause some economic damage (Hernandez *et al.*, 2000).

The confetti worm (*Pyrrhopyge chalybea* Scudder (Hesperiidae)) is a neotropical species that creates holes in leaves and eventually causes defoliation. The adult female deposits eggs on the foliage, eclosing within 20 days; the neonate larva chews a circular hole in the leaf, folding and hiding in the damaged area; the larva is active at night. Two generations are observed yearly; the first occurs in March, and the second in August. In Mexico, Gallegos (1983) reports that *Trichogramma fasciatus* spp. have been collected parasitizing eggs of *Pyrrohopyge chalybea* and Eupelmidae *Anastatus* spp. parasitizing its larvae.

The pine emperor moth (*Imbrasia cytherea* (F.) (Saturnidae)) is an important defoliator of *Pinus* spp. It also attacks a large number of indigenous plants (PPRI, 1970). The subspecies, *Imbrasia cytherea clarki* Geertsema, feeds on avocado leaves in Africa (Van den Berg, 1973). The most important natural enemies of *I. cytherea clarki* include the eupelmid egg parasitoid *Mesocomys pulchriceps* Cameron, three other egg parasitoids and larval and larval-pupal parasitoids. The most effective predator of *I. cytherea clarki* was found to be the Cape Chacma baboon (*Papio ursinus* Kerr). Other important predators include the vervet monkey (*Cercopithecus aethiops* L.) and birds such as the Cape raven (*Corvultur albicollis* Latham). Furthermore, viruses and three species of fungi were found to attack the larvae and pupae (Van den Berg, 1974).

In Brazil, *Papilio scamer scamer* Boiusduval and *Saurita cassra* L. can cause serious injury to the leaves. Both are brown-coloured moths, but *Papilio scamer scamer* is approximately 80 mm long, while *Saurita cassra* is more than half that size (30 mm long). *Papilio scamer scamer* caterpillars are whitish-brown and change to a green colour during their later instars. Larval instars of *S. cassra* are predominantly dark in colour (Medina *et al.*, 1978). Fornazier *et al.* (1997) suggest that the avocado cultivars 'Ouro Verde', 'Primavera' and 'Briosqui' are not as prone to attack by an unidentified geometrid as other cultivars. *I. cytherea clarki* has been found from Zimbabwe to KwaZulu-Natal in South Africa. The other subspecies, *I. cytherea cytherea* F., occurs further south in the Eastern and Western Cape Provinces (Geertsema, 1971). *I. cytherea clarki* larvae are present from March to October with damage occurring during the year's driest and coldest months. The egg stage lasts from 26 days at 20°C up to 84 days at 14°C. In the field, at ~15°C, the five larval instars are completed in about 100 days. Pupation takes place in the soil beneath the trees. The pupal stage lasts about 8 months (Van den Berg, 1975).

Bagworms (Psychidae)

Oiketicus kirbyi (Guildeman) is a common bagworm in South American avocado orchards and a key pest in Peru (Rhainds *et al.*, 2010). Eggs are deposited within the cases, eclosing in 30 days. Female larvae are 60–70 mm long, and male larvae are 45–50 mm in length. The larva remains within the case and moves around, selecting new feeding places. The pupa develops within the case, and when the adults emerge, the apterous female remains within the case where the male fertilizes it. Alate males are attracted by female sexual pheromones. Later, the female deposits the eggs within the same case where she spent her life and dies after the eggs are deposited (Sanchez, 1983).

Larval feeding on leaves causes almost uniform, 1-cm-diameter holes. Likewise, Saunders' case moth (*Oiketicus elongatus* Saunders (Psychidae)) occasionally defoliates portions of individual trees. *Chirotica* spp. (Ichneumonidae) is reported parasitizing *O. elongates*.

Diptera

Gall midges

The gall midge (*Asphondylia websteri* Felt) is a polyphagous fly that attacks immature fruiting bodies of native and exotic crop plants in the south-western USA (Barnes, 1946; Gagné, 1994; Gagné and Woods, 1988). There are over 250 described species of *Asphondylia* and many known undescribed species. *A. websteri* is cosmopolitan in distribution and the majority of species appear to be mono or stenophagous (Gagné, 2004). The native range of *A. websteri* is uncertain, but south-western USA and northern Mexico are thought to be areas constituting the natural range in North America (Gagné and Woods, 1988).

This galling fly was first found attacking small (< 2.0 cm in length) avocado fruit in commercial 'Hass' orchards in Guatemala (Hoddle, 2008). Immature *A. websteri* feed in developing fruit. Female flies lay one egg per fruit. Larvae complete development within the fruit and create an escape tunnel in the side of the fruit from which the adult fly can emerge. The emergence tunnel is created by a 'spatula', a structure on the head region of third instar larvae that is used to scrape away plant material until a thin exocarp wall is formed at the distal end of the tunnel. Pupation occurs within the central cavity of the deformed fruit. When the fully developed adult midge is within the pupal case, the pupa advances to the end of the escape tunnel using abdominal spicules to gain traction within the tunnel. The pupa opens the necrotic entrance in the exocarp wall to the outside by applying its pupal horns to the closed end of the tunnel and rotating its body about the longitudinal axis, thereby cutting open the exit hole (Gagné *et al.*, 2004). Once the circular tunnel entrance is open, the pupa advances partway out of the tunnel exit, and the adult then emerges from the protruding pupal case. Typically, the deformed fruit dehisces shortly after the emergence of the adult fly and drops from the tree. Subsequently, abscissed fruit with protruding midge pupal cases can be collected from the ground.

As larvae develop and feed internally, infested young 'Hass' avocado fruit become misshapen, acquiring a thin and elongated appearance. This shape is extremely characteristic of young avocado fruit infested by *A. websteri*. Internally, the young fruit is hollowed out by larval feeding. Gagné *et al.* (2004) also reported *Bruggmanniella perseae* Gagne, attacking avocado fruit in Colombia and Costa Rica (Fig. 14.15). This species was recently reported from the Dominican Republic. *A. websteri* appears to be a pest mainly of small avocado fruit that are set during the off-bloom period (August–September) in Guatemala. Depending on the site and the year, off-bloom fruit damaged by *A. websteri* can range from 21% to 85%, and < 5% of fruit from the principal bloom (December–February) appear to be attacked (Hoddle, 2008; Hoddle *et al.*, 2009b).

Larvae of *A. websteri* were found to be heavily parasitized by eight species of hymenopterous parasitoids: (1) *Torymus solidaginis* Huber, (2) *Torymus bedeguaris* Linnaeus (group) (both Hymenoptera: Torymidae), (3) *Rileya* sp. (Hymenoptera: Eurytomidae), (4) *Paragaleopsomyia* sp. (Hymenoptera: Eulophidae), (5–7) three species of *Galeopsomyia* (Hymenoptera: Eulophidae) and (8) *Lyrcus* sp. (Hymenoptera: Pteromelidae). The parasitoid fauna associated with *A. websteri* in Guatemala and its impact can vary greatly between years (Hoddle *et al.*, 2009b). In addition to parasitoids, larvae of *A. websteri* were attacked by *Epipona guerini* Saussure (Hymenoptera: Vespidae: Polistinae). Adult wasps were observed foraging in study orchards, cutting open infested fruit with their mandibles, and extracting and consuming *A. websteri* larvae. The sides of fruit from which *A. websteri* were removed by *E. guerini* were characterized by large circular holes (Hoddle, 2008).

A. websteri is the second galling midge now known to attack young avocado fruit. *B.perseae* Gagné (Diptera: Cecidomyiidae) has been recorded

Fig. 14.15. A. The gall midge, *Bruggmanniella perseae* Gagne in Colombia, B. *Bruggmanniella* sp. in Brazil. Photo credits: A = Instituto ENTOMA-Naturavision, B = Dr Grazielle Furtado (Brazil) (with permission).

causing very similar damage in Costa Rica and Colombia, where it is considered a severe pest (Gagné *et al.*, 2004) (Fig. 14.15). Part of the native range of *A. websteri* encompasses southern California (Gagné and Woods, 1988), which puts populations of this midge in sympatry with the largest 'Hass'-avocado-growing areas in the USA (~26,500 ha). Curiously, this midge has not been recorded attacking young 'Hass' fruit or other avocado cultivars in southern California. There are five possible reasons for this. First, *A. websteri* does attack developing fruit over the single fruit set period April–June in southern California. Damaged fruit could have gone unnoticed because infestations are inconsequential, possibly due to natural enemy activity, and attacks do not adversely affect expected crop yields. Second, significant damage does occur but has simply gone unrecognized because deformed fruit are aborted, and this has been considered a normal developmental occurrence for 'Hass' fruit during fruit set. Third, the phenology of early 'Hass' fruit development does not coincide with reproductively active populations of *A. websteri*, so this insect has never emerged as a serious avocado pest in southern California. Fourth, avocados are not attacked in southern California because they are not a preferred host when more highly favoured native plants are present, when avocados are setting fruit that would be vulnerable to attack. Fifth, A. websteri is a cryptic species complex, and *A. websteri* attacking avocado fruit in Guatemala is a species that is morphologically indistinguishable from those that are native to California. This cryptic species hypothesis should be investigated through the use of DNA analyses. The possible association of *A. websteri* with avocados, in particular the commercially important 'Hass' cultivar in southern California warrants future research, and molecular analyses of *A. websteri* from California and Guatemala should be conducted to confirm species identities (Hoddle, 2010c).

Fruit flies

Healthy, hard-skinned avocados (independent of cultivar or origin) are resistant to the attack of fruit flies under field conditions. Consequently, this group of insects cannot be strictly considered pests of avocado, although small lesions may occur in certain cultivars from abortive stings. Nevertheless, there are quarantine concerns among importing countries, which should be addressed on a case-by-case basis following the procedures described by Aluja and Mangan (2008). The mere presence of regulated species of fruit flies in avocado-growing regions often triggers quarantine restrictions severely limiting the export of avocado fruit. Due to the unsightly star-shaped lesions resulting from some fruit fly stings in some avocado cultivars, routine control measures are undertaken in certain countries, and there are restrictions on exports from host countries.

Literature on the status of avocado as a host of fruit flies is anecdotal, marred with methodological flaws or inaccurate. The problem is compounded by the fact that authors rarely specify the avocado cultivar they worked with or the fruit fly taxonomist who performed the identification of the fruit fly species involved (Aluja *et al.*, 2004; Aluja and Mangan, 2008). Nevertheless, economically important fruit fly species have been wrongly but repeatedly included in listings of insects known to infest avocado (e.g. Wasbauer, 1972; White and Elson-Harris, 1992).

In general, field infestations in any *Persea* species are rare and invariably restricted to soft-skinned cultivars or damaged fruit (e.g. Willard *et al.*, 1929, Bush, 1957). Cultivars with hard-skinned fruit (e.g. 'Hass', 'Fuerte' and 'Sharwil') are resistant to fruit fly attack if the fruit is not damaged prior to attack or if trees are not stressed (Armstrong *et al.*, 1983; Liquido *et al.*, 1995; Aluja *et al.*, 2004; De Graaf, 2009). In the case of artificial, forced infestations in Hass avocados by *Anastrepha ludens* (Loew), shortly after eggs are inserted, commercially ripe fruit respond by generating defensive fruit tissues, causing neoplasia and hindering egg eclosion (Aluja *et al.*, 2004). In a recent study aimed at identifying the molecular mechanisms behind the neoplasia observed in Hass avocados, Aluja and collaborators (Aluja *et al.*, 2023) found that 'Oviposition (and possibly bacteria on the eggs' surface) induces a plant hypersensitive response, triggering a chitin receptor, producing an oxidative burst and synthesizing phytoalexins. In addition, cell wall modification and polyphenols biosynthesis occur, leading to polymerization in the neoplastic tissue surrounding the eggs'. Eggs are enclosed by a hardened corky tissue called 'calluses' that fully develops weeks after egg insertion. These calluses kill or asphyxiate the eggs (Aluja *et al.*, 2004) and are also formed when other insects sting the fruit (Horticulture Innovation Australia, 2018). A similar phenomenon has been documented in 'Sharwil' avocados (Armstrong *et al.*, 1983; Liquido *et al.*, 1995). When eggs are laid at the junction of the pedicel and the fruit, larvae are then able to develop and pupate, yielding viable adults. Nevertheless, such sporadic events do not signify that fruit flies can be considered pests of 'Sharwil'' avocados.

Ceratitis spp.

The only reliable references about *Ceratitis* spp. on avocado are an old paper by Willard *et al.* (1929) and one by De Graaf (2009). Willard *et al.* (1929) reported that only some soft-skinned cultivars of the Guatemalan race (e.g. 'Itzamna', 'Chabil' and 'Kashlan') were infested at very low densities under field conditions (e.g. two out of 180 'Itzamna' fruit harvested). All other cultivars studied were not infested under natural field conditions, although some could be infested under artificial laboratory conditions (fruit exposed to large numbers of gravid females in glass jars). Willard *et al.* (1929) concluded that avocado cultivars with soft-skinned fruit showed a significant degree of resistance to attack by *Ceratitis capitata* (Mediterranean fruit fly). De Graaf (2009) experimentally documented that 'Hass' fruit are not hosts of *C. capitata*. However, laboratory artificial infestations with the fruit fly *Ceratitis rosa* (Natal fruit fly) and *cosyra* (marula fruit fly) were successful (Aluja and Mangan, 2008). The fact that fruit fly larvae do not develop in fruit of commercial avocado cultivars under normal orchard practices has also been reported by De Villiers and Van den Berg (1987) and Brink *et al.* (1997). In Hawaii, Armstrong *et al.* (1983) showed that 'Sharwil' avocado fruit could only be infested with *C. capitata* if the flies were confined to artificially damaged fruit in cages on the tree. These authors concluded that avocados harvested with the stem attached and protected from postharvest fruit fly infestations are not at risk of fruit flies being exported. However, avocado fruit can be damaged by the fruit fly's ovipositor during egg laying. These holes develop into cracks or star-shaped lesions (Du Toit *et al.*, 1993), which require control measures to reduce the potential export of fruit flies.

Queensland fruit fly

In Australia, the Queensland fruit fly (*B. tryoni* Froggatt) occasionally lays eggs in 'Fuerte' avocado fruit. However, eggs are laid too superficially and do not hatch into viable larvae, only causing small cosmetic damage consisting of star-shaped cracks in the skin (Smith, 1973). Although this fly will sting hard-skinned cultivars including 'Hass', the damage is usually not visible and is of no consequence. However, in 'Fuerte', 'Wurtz' and 'Rincon', the scars are quite noticeable and chemical sprays or baiting may be required to prevent excessive damage. Protein bait sprays are effective and the preferred option over the use of conventional insecticides (Waite and Pinese, 1991).

Oriental fruit fly

Fruit flies within the *Bactrocera dorsalis* Hendel complex have been questionably regarded as avocado pests mostly in Africa, Asia and Oceania. For instance, various cultivars with thin-skinned fruit are grown throughout the Philippine islands and can apparently be infested with fruit fly larvae (G.K. Waite, Nambour, Australia, 1999, personal communication). Reports by Armstrong *et al.* (1983) indicated that commercially mature 'Sharwil' avocados grown in Hawaii were resistant to infestations by the Mediterranean fruit fly (*Ceratitis capitata* Wiedemann), the melon fly (*Bactrocera cucurbitae* Coquillet) and the oriental fruit fly (*B. dorsalis*). Contradicting these findings, plant inspectors found 'Sharwil' avocados infested with larvae of *B. dorsalis* (APHIS, 1992). Shortly thereafter, Liquido *et al.* (1995) confirmed that 'commercially mature' 'Sharwil' avocados can serve as hosts for *B. dorsalis* and that females were capable of ovipositing on lesions surrounding the pedicel of stressed fruit. This lesion, called 'ringneck', appears under severe drought conditions and is caused by an incomplete abscission and callus formation at the pedicel (see Schaffer *et al.*, Chapter 7 of this volume for a discussion of ringneck). Under non-drought conditions, ripe avocado fruit detaches at the pedicel and fall to the ground. However, fruit with 'ringneck' remain attached to the tree and are thus susceptible to oriental fruit fly attack. The infestability of 'Sharwil' avocado fruit by *B. dorsalis* appears to depend on the hardness of the fruit skin. However, this cultivar was considered a poor host by Liquido *et al.* (1995) and Follett (2009). Similarly, 'Malama' Avocados, also grown in Hawaii, are considered poor hosts of *B. dorsalis* (Follet *et al.*, 2022). Hamacek *et al.* (2005) also suggested that avocados (Hass, Lamb Hass, Shepard and Wurtz cultivars) are poor hosts for fruit flies in Australia and control treatments are seldom required if fruit are harvested while still hard. *B. dorsalis* (=*B. invadens*) is believed to have invaded Africa from the Indian subcontinent and was reported from Kenya in 2003. It has spread rapidly to many African countries, with mango, loquat, banana and guava the preferred hosts, and papaya and questionably avocado less preferred. It is highly mobile and dominates indigenous fruit flies. It is considered a great threat to South Africa (Grové *et al.*, 2009). Jessup (1994) found no third instar *B. tryoni* larvae infesting 'Sharwil' fruit after they were dipped in 500 ppm benomyl at 46°C. Jang (1996) estimated that heat treating 'Sharwil' fruit at 40°C for 24 h reduced the oriental fruit fly population by 99.5–100%. A cold mitigating treatment to provide phytosanitary security to importing countries (not less than 18 days at 2.5°C or lower) was proposed for 'Hass' avocados, should this become necessary (Ware *et al.*, 2012).

Anastrepha spp.

The old literature (e.g. Rust, 1918) on the host status of avocado to flies within *Anastrepha* is inaccurate and unreliable (Aluja *et al.*, 2004; Aluja and Mangan, 2008). Aluja *et al.* (2004) showed that 'Hass' avocados are not natural hosts of four species of *Anastrepha* (*A. ludens*, *Anastrepha obliqua* (Macquart), *Anastrepha serpentina* (Wiedemann) and *Anastrepha striata* (Schiner)). Of 5200 fruits forcibly exposed to the oviposition activity of 26,000 gravid, laboratory-reared and wild *A. ludens*, *A. obliqua*, *A. serpentina* and *A. striata* females with a high oviposition drive, only four were infested by *A. ludens* (laboratory and wild origin), and of the 22 larvae and 19 pupae recovered (three larvae were found dead), no adult ever emerged. *A. ludens* pupae from 'Hass' avocados were almost half the size of pupae from the natural host (grapefruit), and the few fruit that were forcibly infested were from diseased trees that were also water stressed (infestations occurred in the middle of the dry season). Hennessey *et al.* (1995) reported that, of 25 *P. americana* cultivars tested, only five (20%) were infested in caged conditions by *A. suspensa* (Loew) (Caribbean fruit fly) and few adults emerged under highly artificial laboratory conditions (only portions of fruit were used to test egg hatch and larval development). These authors attributed this phenomenon to the sugar, estragole, which may be lethal to eggs and developing larvae (Hennessey and Schnell, 2001). Even though *A. ludens* females can probably overcome a hostile microenvironment for larval development by laying more eggs per clutch (Díaz-Fleischer and Aluja, 2003), 'Hass' avocados clearly represent an inadequate rearing medium for flies in the genus *Anastrepha* (Aluja *et al.*, 2004; Aluja and Mangan, 2008).

Aluja *et al.* (2010) found that keeping artificially infested 'Hass' fruit at 5°C prevented larvae from developing into viable pupae or adults. By contrast, Carrillo *et al.* (2017) found that green-skinned avocados do not tolerate methyl bromide fumigation against fruit flies.

Coleoptera

Copturus aguacatae

The avocado stem weevil, *Coptorus aguacatae* Kissinger, causes economic losses in multiple states in Mexico (Engstrand *et al.*, 2010). Fruit losses range from 20% to 80%, depending on the control tactics used. Canseco (1971) noted some confusion with the distribution of the species *C. persea* and *C. aguacatae*. The first species is found in Guatemala, and the latter in Mexico. *C. aguacatae* can also be confused with *Copturus sapotae*. However, the major difference is the shape of the aedeagus. Canseco (1971) provides a detailed description of the different stages of *C. aguacatae*.

C. aguacatae is a stout reddish-brown univoltine weevil, 4 mm in length with diurnal activity. The female deposits eggs on younger branches but has been observed damaging older shoots. The egg stage lasts 10–12 days, while larvae live up to 117 days. Pupae develop in approximately 17–19 days, and the adults live for approximately 34–44 days. The generation time is 169–192 days (Hernandez *et al.*, 2000). Extensive taxonomical descriptions of different stages of the weevil are provided by Kissinger (1957) and Muñiz (1960). Major activity is observed in June–July. However, Garcia *et al.* (1986) demonstrated that adults are observed from May to November in the warm regions of Yautepec, Mexico. Eggs can be collected between July and January and pupae between June and February, while in the more temperate area of Tetela del Volcan, Mexico, adults are found between June and February, eggs are observed in September, larvae are found throughout the year and pupae from May to December. Garcia *et al.* (1986) suspect some of the weevils could be diapausing as larvae from January to April. Canseco (1971) reported that *C. aguacatae* spends the winter as an adult, hiding in the fallen leaves. During the spring, they move to the tree foliage and start feeding on leaves. Later, the copula starts, and as soon as the fruit are formed, they start feeding on them. The emergence of the adults coincides with the first rains. Eggs are inserted into the fruit, and after each egg deposition, the female seals the site with a gelatinous substance. Each female will deposit about 70 eggs. The eggs eclose within 7–10 days, and the larva bores through the pulp and seed. When the larva reaches the seed, the galleries within the seed are sinuous. Close to the end of the larval stage, they bore through the pulp again and fall to the soil surface. They pupate in the soil at 2.5–5 cm from the soil surface. The pupal stage can last 15–30 days. The number of overlapping generations is considered to be two (Canseco, 1971).

Gallegos (1983) reported that most of the infestations are found in small orchards where regional cultivars are grown without any type of pest management, while Coria-Avalos (1993) indicated that this pest can be found in orchards located in warm climates. Damage is concentrated in the upper half of the tree canopy (Salazar-García and Bolio-García, 1992). Meléndez-Gonzales and Espinosa-Garcia (2018) compared GC-MS metabolic profiles of cambial tissue from avocado cv. Hass trees infested and non-infested by *C. aguacatae*. They found a set of 16 highly correlated terpenoids, showing increasing proportions in borer trees as response to *C. aguacatae* larvae. In addition, low proportions of isoobtusilactone A and a-humulene were found in non-bored trees.

To control this weevil, Hernandez *et al.* (2000) recommend the removal of infested branches, followed by applications of contact insecticides. Insecticides can also be applied to the soil to control the larvae when they fall to the ground or to control the adults after they emerge (Canseco, 1971).

Some chalcidoid wasps have been observed as active biocontrol agents of these weevils. However, the species have not been identified until now. Huerta *et al.* (1990) collected *Camponotus cerberulus* Emery within the galleries made by *C. aguacatae* and speculated that *C. cerberulus* could be preying on *C. aguacatae*. The same authors indicated that the species *Oncophanes* (Braconidae), *Euderus* (Eulophidae), *Erythmelus* (Mymaridae) and *Eurydinoteloides* (Pteromalidae) could be associated with *C. aguacatae*. A dose of 1500 infective juveniles per ml of *Steinernema bibiones* caused 10% mortality, while *Heterorhabditis heliothidis* caused 45–75% larval mortality under laboratory conditions (Huerta *et al.*, 1990). More recently, Dedios-Avila *et al.* (2021) reported 11 species of parasitic Hymenoptera found associated with avocado stem weevil in Nayarit, Mexico. The most abundant was *Heterospilus bruchi* (Viereck) (Braconidae). In addition, three species of entomopathogenic fungi (*Metarhizium pingshaense, Metarhizium anisopliae* and *B. bassiana*) were identified. Also in Nayarit, Dedios-Avila *et al.* (2016) reported *Neocatolaccus tylodermae* (Ashmead) (Pteromalidae) as a parasitoid of the avocado stem weevil.

Heilipus spp.

The genus *Heilipus* is represented by 92 species in the Americas with 39 recorded from North and Central America and 53 registered for South America. Currently eight species of *Heilipus* attack avocados. In Florida, *Heilipus squamosus* LeConte (= *H. apiatus*) was reported in the mid-20th century boring into avocado stems and killing young trees (Anzaldo and Diaz-Grizales, 2022). In Mexico, the large avocado borers, *Heilipus albopictus* Champion and *H. lauri*

Bohemann, cause damage to stems and fruit, respectively (Castañeda-Vildózola *et al.*, 2017) (Fig 14.16). In Central America, *H. lauri* is reported from Costa Rica, Guatemala, Nicaragua; the stem borer, *Heilipus elegans* Guerin, is known from Costa Rica, while the fruit borers, *H. trifasciatus* Fabricius and *Heilipus pittieri* Barber, have been collected in Nicaragua, Costa Rica and Panama. *H. lauri*, *H.* nr. *pittieri*, *H. trifasciatus* and *H. elegans* are reported from Colombia (Rubio *et al.*, 2009; Arévalo Peñaranda *et al.*, 2012). *Heilipus rufipes* Perty and *H. catagraphus* Germar were reported infesting avocados Brazil (Castañeda-Vildozola *et al.*, 2010; Vanin and Bená, 2015).

Heilipus lauri

The large borer weevil, *H. lauri*, damages up to 80% of avocado fruit, causing extensive fruit drop. Its distribution reportedly includes Mexico, Costa Rica, Guatemala, Nicaragua and Colombia. *H. lauri* is a pest of significant quarantine concern. In Mexico, the elevational range of *H. lauri* is 570–2100 m above sea level (Ortega-Licona *et al.*, 2019). The dark-brown (with two incomplete yellow bands or spots on the elytra) adults are 14–17 mm long and emerge from the fallen fruit (Ebeling, 1959; Salgado and Bautista, 1993) (Fig. 14.16 a & b). There is an evident sexual dimorphism; females are larger than males and have a longer, thinner more curved rostrum (Ortega-Licona *et al.*, 2019). Hoddle *et al.* (2022) conducted flight mill studies to measure the dispersal capacity of *H. lauri* with respect to gender, mating status and size on the single and repeat flight capabilities of weevils. Gender, mating status and size did not significantly affect measured flight parameters. The total distances flown, flight velocity and mean maximum flight bout distances and durations significantly declined as weevil age increased and when weevils engaged in repeated sequential flights. Survivorship rates were significantly reduced as the number of successive flights undertaken increased (Hoddle *et al.*, 2022).

Fig. 14.16. A. *Heilipus lauri* Bohemann, B. *Heilipus albopictus* Champion, C. *Heilipus catagraphus*, D. Damage by *H. catagraphus* Germar Photo credits: A–B = Instituto ENTOMA-Naturavision, C–D = Dr Gracielle Furtado (Brazil) (with permission).

Female weevils oviposit through the epidermis into the pulp. To accomplish this, females chew an oviposition hole into which the ovipositor is inserted. The small oval eggs are 1 mm long and change from pale green to cream colour. The female deposits 1–2 eggs per fruit, and a total of 36 eggs per month can be laid. Twelve to 15 days after oviposition, the legless larva bores through the pulp into the seed, where it feeds and spends its larval and pupal stages (Ebeling, 1959). The larva has four instars that last approximately 54–63 days (Anonymous, 1984; Castañeda-Vildozola *et al.*, 2016). The pupal stage lasts 15 days. Emerging adults feed on foliage and live up to 4 months. In the case of fallen fruit, the larvae sometimes leave the fruit and enter the soil to pupate, but the majority of weevil larvae pupate within the seed. Ortega-Licona *et al.* (2019) recorded between one and four oviposition perforations per fruit, but all seeds had only one larva inside. The highest percentages of damage occurred on the east-facing quadrant of the tree. Larval infestations may cause premature fruit drop. The adult feeds on leaves, buds, sprouts and fruit (Ebeling, 1959). Further information on development is described by Salgado and Bautista (1993). Romero-Frías *et al.* (2019) recorded four male-specific semiochemicals (grandisol, papayanol, papayanal and grandisal in a ratio of 94.8:3.0:1.8:0.4) that may have potential for monitoring and controlling populations of this pest. Repeated collection and destruction of infested fruit is critical for the management of *H. lauiri*. Clavijo and Holguin (2020) evaluated the pathogenicity of commercial entomopathogenic fungal strains on *H. lauri* in Colombia and determined that *B. bassiana strains* have the greatest control potential over this pest. However, this control option is probably not feasible given the difficulty of delivering fungal spores to highly protected areas where eggs are laid and larvae feed.

Heilipus catagraphus

In Brazil, *H. catagraphus* Germar attacks avocado branches and is considered the second most important pest of the crop after *S. catenifer*. Adults are dark in colour with characteristic lateral beige markings (Fig. 14.16 c). According to Lourencao *et al.* (1984), adult *H. catagraphus* can damage the fruit skin and pulp, but larvae have not been reported to cause seed damage, suggesting that this species is restricted to reproducing on woody plant parts. This type of fruit damage is generally seen at the beginning of the rainy season (October) (Furtado-Moreira, 2022) (Fig. 14.16 d). Adults bore into the trunk to oviposit, causing large, conspicuous wounds. Larvae develop in the trunk and oozing tree secretions indicate damaged areas. Large infestations can kill young avocado trees (Furtado-Moreira, 2022). Management of *H. catagraphus* requires cultural practices such as pruning and destroying infested branches.

Conotrachelus perseae, Conotrachelus aguacatae and Conotrachelus serpentines

Conotrachelus is a New-World weevil genus with approximately 1200 described species (O'brien and Wibmer, 1982). Three *Conotrachelus* species (Fig. 14.17) that feed on avocado are known from Mexico and Central America: *C. perseae* Barber, *C. aguacatae* Barber and *C. serpentines* Klug (Chamorro and Barclay, 2018). The first two are of economic importance because they

Fig. 14.17. *Conotrachelus* sp. A. *Conotrachelus perseae* Barber (Frontal view), B. *Conotrachelus perseae* (lateral view), C. *Conotrachelus serpentines* Klug. Photo credits: Dr M. Lourdes Chamorro, USDA, ARS, SEL – Specimens from Department of Entomology, Smithsonian National Museum of Natural History (USNM).

are direct pests of fruit and are considered quarantinable pests by most avocado-importing countries. Despite their economic importance, these species are poorly studied. Whitehead (1979) provided taxonomic keys and diagnoses, expanding the information by Barber (1919). Muñiz and Barrera (1958) provide a taxonomic key that includes species of *Conotrachelus* collected from tropical and subtropical America.

Conotrachelus aguacatae (seed weevil)

The small, shiny, almost black weevil (*C. aguacatae*) can cause up to 85% fruit loss. It is found in the areas of Queretaro, Michoacán, Jalisco, Puebla, Morelos and Guanajuato, Mexico. Adult *C. aguacatae* are 7 mm long and copulate immediately after emergence. Adults are nocturnal and remain hidden on folded leaves or in any crevices in the trunk. The adults can attack newly formed fruit, and the females usually deposit 1–4 eggs per fruit, but sometimes as many as 70. Eggs are whitish and eclose in 7–10 days. Larvae develop in seeds for 20–35 days. The number of larvae per fruit can be as high as three or four. The larvae leave the fruit to pupate in the soil at about a 5 cm depth. The life cycle from egg to adult is 42–75 days. The first generation starts in January and February and lasts approximately 10 weeks. The second generation starts in April and ends in June or July. The female prefers to oviposit in the stylar area of the fruit. Higher levels of fruit damage occur in the lower tree canopy than in the upper canopy (Hernandez *et al.*, 2000).

To control *C. aguacatae*, Hernandez *et al.* (2000) recommended the collection of fallen fruit, weed removal and tilling the soil to expose pupae. The only known biological control agent is *B. bassiana*, an entomopathogenic fungus, infecting larvae and pupae.

Copturomimus sp.

In Venezuela and Colombia, *Copturomimus hustachei* Kissinger, 1957 (= *Copturomimus perseae*), a small greyish weevil, 3.7–5 mm long, with a black transverse band on the elytra causes damage to branches and stems of avocado (Moreno-Gaviria *et al.*, 2023). Both larvae and adults bore into branches and stems. The total time from egg to adult was 86.78 ± 25.41 days; the egg stage was 9.29 ± 1.22 days, the larval stage lasted 66.18 ± 10.86 days and the pupal stage 13.08 ± 3.99 days. Adult longevity ranged between 15 and 49 days (Moreno-Gaviria *et al.*, 2023). The legless, whitish curculionid larva is 10–12 mm long. The damage is concentrated on the branches; infested branches show white frass exudates from the galleries made by this insect; affected parts show a slow dieback, and the tree eventually dies (Saldarriaga, 1977; Avilan *et al.*, 1997). Collecting and burning infested branches is the control method suggested by Avilan *et al.* (1997).

Fruit tree weevil

The fruit tree weevil (*Naupactus xanthographus* Germar) is reported from Chile, Brazil, Argentina, Paraguay and Uruguay. Adults consume avocado leaves, and larvae feed on avocado roots, weakening the trees. Temperatures above 13.5°C promote weevil emergence, and females oviposit on leaves, in cracks in broken branches and under any material used to support grafts (Ripa and Luppichini, 2008a). Natural enemies include the egg parasitoids, *Fidiobia asina* Loiacano, the adult parasitoid *Centistes* spp. and the egg predator, *Megatoma* spp. (Dermaptera) (Ripa, 2008a).

Monolepta apicalis

Since early 1990, the avocado beetle (*Monolepta apicalis* Sahlberg) (Chrysomelidae) has been a pest of avocado orchards, mostly at higher altitudes in the Kiepersol region of South Africa. The infestations were more severe in orchards near eucalyptus plantations and natural vegetation. *M. apicalis* was also reported in two other areas in the Mpumalanga Province of South Africa (Erichsen and Schoeman, 1992, 1993; Claassens, 2001).

M. apicalis is a sporadic pest in South Africa. The intensity of damage differed considerably according to the location of the orchards. Some of the young orchards in Kiepersol, situated in the relatively high mountainous areas, had a total crop loss due to feeding by the beetles. *M. apicalis* feeds on avocado twigs, leaves, leaf stalks and fruit. In certain instances, the epidermis of the fruit was completely removed, causing a brown blemish covering the entire fruit. Lighter damage appeared as darkish holes of ~2 mm in diameter. Very young leaves are often eaten completely, while older leaves are mainly eaten from the ventral sides, resulting in a skeletonized appearance (Claassens, 2001). *M. apicalis* was also recorded from Namibia and Central Africa (G.L. Prinsloo, Plant Protection Research Institute, Pretoria, South Africa, 1993, personal communication). *Bridelia macrantha*, lychee and macadamia trees, *Hibiscus trionum*, *Triumfetta rhomboidae*, *Sida rhomboidea* and *Protea* spp. were also reported as *M. apicali* hosts (Erichsen and Schoeman, 1992; Claassens, 2001).

The adult *M. apicalis* is bright red with shiny black elytra. The insects vary from 3 mm to 5 mm in length. The life history of *M. apicalis* has not been studied in South Africa. In Australia, the closely related *M. australis* (Jacoby) lays eggs in the soil. The larvae feed on grass roots and are ~5 mm when fully grown. They pupate in the soil. Their life cycle takes about 2 months. Three to four generations occur annually. Adults emerge after rain (Fay and De Faveri, 1990).

M. apicalis is active in avocado orchards in the early mornings. During the warmer parts of the day, the insect seeks shade underneath leaves and on shady stems. When disturbed, *M. apicalis* will hide behind leaves or drop and fly away.

In Kiepersol, South Africa, *M. apicalis* was noticed in a heavily infested orchard, where 50–100 insects per young tree were counted. During 1998, the insect was still present but in very small numbers. Crop loss due to *M. apicalis* varied on individual trees from about 5%, up to 100% (Erichsen *et al.*, 1993). *Monolepta* spp. are especially a threat because they produce an aggregating pheromone and form swarms that can invade an orchard and cause extensive damage in a short period (Fay and De Faveri, 1990).

Scouting for beetles should commence in November. This should occur in the early morning or late afternoon, during cool conditions when the beetles are more likely to be found on rather than under the leaves. Observation should be made of damage to leaves and fruit and the presence of beetles on the tree (Erichsen, 1993). The beetles are attracted to eucalyptus windbreaks, especially *Eucalyptus torreliana* in Australia, which act as sentinel trees so that growers can control swarms on these trees before they invade the orchard. Fay and De Faveri (1990) found that numerous individual beetles present at flowering (i.e. not in a swarming phase) were actually beneficial for pollination. When swarms invade, there is no alternative but to spray, and action must be taken quickly to prevent damage.

The red-shouldered leaf beetle (*M. australis* Jacoby (Chrysomelidae)) is a serious pest in Australia and can defoliate trees when swarms descend on an orchard. Individual beetles are no problem; however, when they swarm and get into a feeding frenzy, trees can be stripped in a matter of hours. If fruit are present, their skin will be removed. In Australia, the black swarming leaf beetle and the brown swarming leaf beetle (*Rhyparida* spp.) cause damage similar to that of *Monolepta* (Waite and Pinese, 1991). *Monolepta bifasciata* Hornstedt (Coleoptera, Chrysomelidae) occasionally damages foliage in the Philippines (Cendana *et al.*, 1984).

Apate monachus F.

Apate monachus F. (Bostrichidae) attacks avocados in different South American and Caribbean countries. The female of *A. monachus* oviposits on dead branches and trunks. Eggs are laid randomly on the trees and eclose in 7–8 days. The number of eggs per female is 64–106. Larval development lasts 67 days, the pupal stage lasts 8–9 days and adults live for up to 40 days. They are active during the night, and the two sexes are found copulating within the galleries during the day. The complete cycle lasts 87–190 days. In Brazil, *Apate terebrans* Pallas and *Acanthoderes jaspidea* Germar affect branches by ovipositing or sometimes by larval boring and adult feeding (Medina *et al.*, 1978). Descriptions of adults are provided by Medina *et al.* (1978). The adults attack healthy plants and use their mandibles to construct galleries within branches and the trunk. Injured branches are recognized because of the yellowing of the leaves; these branches regularly break easily.

Bethylidae (Hymenoptera) and *Clyptodoryctes* (Braconidae) are registered as natural enemies of *A. monachus*.

Ambrosia beetles

Ambrosia beetles (Coleoptera: Curculionidae: Scolytinae) have recently emerged as important pests of avocado. A few species in the genera *Xyleborus* and *Euwallacea* have been found to vector plant pathogens to avocado (Carrillo *et al.*, 2012,

2014, 2016). Other genera reported from avocado include *Xylosandrus* sp. *Corthylus*, *Aegeria*, *Hypothenemus*, *Monarthrum* and *Polycaon* (Ebeling, 1959; Ngoan *et al.*, 1976; Schonherr and Pedrosa-Macedo, 1981; Daehler and Dudley, 2002; Peña, 2003; Perez-De la Cruz *et al.*, 2009; Mendel *et al.*, 2010; Carrillo *et al.*, 2012; Mendel *et al.*, 2023). In general, only adult female ambrosia beetles initiate infestation by boring tunnels in the trunks or branches of the host plant, and they use the host plant to cultivate symbiotic fungi. Males are flightless, remain in brood galleries and arise from parthenogenesis (Ngoan *et al.*, 1976).

Euwallacea shot hole borers

The tea shot hole borer, *Euwallacea perbrevis* (Schedl), was introduced into Florida in 2002, and since 2010, increasing damage has been recorded in avocado plantations and native species (Carrillo *et al.*, 2012, 2016; Owens *et al.*, 2018) (Fig. 14.18). In 2012, beetles matching the same morphology were found in Israel attacking avocados but were later found to be genetically different and named the polyphagous shot hole borer *Euwallacea fornicates* Eichhoff (PSHB) (Mendel *et al.*, 2012; Stouthamer *et al.*, 2017). The PSHB was recorded in 2003 in California causing damage to avocado and several other hosts (Eskalen *et al.*, 2012). In 2019, PSHB was also reported infesting a backyard tree in South Africa (van den Berg *et al.*, 2019). Another species commonly known as the Kuroshio shot hole borer, *Euwallacea Kuroshio* Gomez and Hulcr, was also introduced into California and Mexico, causing severe damage to various native willows in riparian habitats (Boland, 2016; García-Avila *et al.*, 2016).

Shot hole borers are vectors of several fungi species in the genus *Fusarium* that colonize the xylem cells. *Euwallacea* beetles bore and tunnel into stems and branches of healthy trees and cause damage through inoculation of *Fusarium* spp. carried by the beetles (O'Donnell *et al.*, 2016). The typical injury to avocado is wilting branches due to a vascular infection by the fungus introduced by the beetle. Beetle entry holes in branches or stems are characterized by copious perseitol exudates called 'sugar volcanos'. Carrillo *et al.* (2015) reported that *Euwallacea* beetles are attracted to lures containing quercivorol. Several management tactics have been developed for these beetles. However, effective management of shot hole borers has been achieved in affected areas primarily through sanitation (removal and destruction of infested branches). Mitigating shot hole borers in the landscape and natural areas remains challenging.

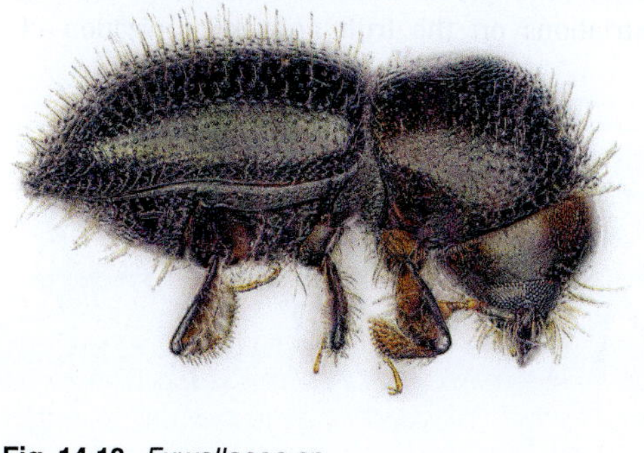

Fig. 14.18. *Euwallacea* sp.

Xylosandrus

The black twig borer, *Xylosandrus compactus* Einchhoff is another ambrosia beetle that attacks healthy avocados. *X. compactus* is reported from tropical Africa, Europe, South-east Asia, New Zealand, tropical Pacific islands (including Micronesia), the Caribbean (Dominican Republic and Puerto Rico), South America (including Brazil, Guyana and Peru), the USA (Hawaii and Florida) and Israel (Wood, 2007; Mendel *et al.*, 2023).

In Florida, lateral tree branches are regularly attacked, but the main trunk is seldom attacked, and it is considered a minor pest. However, Dixon *et al.* (2003) reported that avocado sustained considerable losses from *X. compactus* according to observations in the 1950s and 1960s.

The granulate ambrosia beetle, *Xylosandrus crassiusculus* Motschulsky, also infests weakened avocado trees in Florida (Carrillo *et al.*, 2012). This beetle is considered a secondary pest; however, controlled studies determined that this beetle can transmit the fungus *H. lauricola*, the causal agent of laurel wilt, to avocado trees (Carrillo *et al.*, 2014) (see Dann and Gazis, Chapter 13 of this volume for a discussion of *H. lauricola*).

Xyleborus spp.

The redbay ambrosia beetle (*X. glabratus* Eichhoff) is an invasive wood-boring insect in the USA and the primary vector of laurel wilt (LW), a new disease of trees in the Lauraceae family.

The causal agent, *H. lauricola*, is an exotic fungal pathogen brought into the USA by the X. *glabratus*. This beetle is rare in avocado orchards, but the pathogen is now vectored by various species of native and exotic ambrosia beetles that have acquired *H. lauricola* via horizontal transfer (Carrillo *et al.*, 2014) (Fig 14.19). The redbay ambrosia beetle is native to Asia (i.e. Taiwan, Japan and India). In the USA, it was first discovered at Port Wentworth, Georgia in 2002 (Rabaglia *et al.*, 2006). After killing millions of native Lauraceous trees in South Carolina, Georgia and Florida, that is, *P. borbonia* (L.) Spreng, laurel wilt was detected in commercial avocado orchards in south Florida in 2012. Initial efforts focused on characterizing the ambrosia beetle communities in avocado orchards and native forests affected by the disease. Carrillo *et al.* (2012) discovered that *X. glabratus* does not breed naturally in avocado trees in Florida, but native ambrosia beetles have acquired the laurel wilt pathogen and play a key role as secondary vectors (Carrillo *et al.*, 2014). This finding has been fundamental to understanding the epidemiology of laurel wilt in avocado, and a key to designing a strategy to control a complex of ambrosia beetle species (*X. bispinatus, X. volvulus, X. affinis, Xyleborinus saxesenii* and *Xylosandrus crassiusculus*) that act as secondary vectors of the disease.

In response to laurel wilt, avocado growers increased the number of applications of chemical insecticides from approximately two to six per year. However, field trials made it apparent that broadcast applications with insecticides do not reduce ambrosia beetle populations because breeding sites inside the xylem of host trees, where most of the population develops, are not controlled (Peña *et al.*, 2011; Carrillo *et al.*, 2013). Therefore, initial efforts are focused on the early detection of disease symptoms followed by prompt removal and destruction of infected trees, and conventional insecticides are only recommended to eliminate beetle colonies exposed after rouging and shredding infected trees.

Subsequent efforts are focused on identifying and implementing effective biological and cultural control practices. Surveys for natural enemies have not detected parasitoids or predators of ambrosia beetles associated with laurel wilt (Peña *et al.*, 2015). However, using entomopathogenic fungi has provided an effective and environmentally friendly biological alternative to chemical insecticides (Carrillo *et al.*, 2015). Entomopathogenic fungi are applied using seasonality data to target periods of a high prevalence of ambrosia beetles in flight. More recent recommendations for regular orchard pruning stem from a better understanding of the flight biology of ambrosia beetles and the effect of pruning on the orchard light environment on suppressing ambrosia beetle movement (Menocal *et al.*, 2022).

As of 2023, approximately 300,00 mature tropical avocado trees, worth an estimated US$99 million, have been lost to laurel wilt. Approximately 70% of the lost avocado trees are being replanted, indicating that avocado growers are now more confident in managing this disease. However, despite significant progress in the biological understanding of the laurel wilt pathogen and ambrosia beetle vectors, additional practical control eludes growers and the scientific community.

Other beetles

Scarab beetles

Astaena pygidialis Kirsch, the main chafer beetle, causes damage to avocado fruit and young leaves and is considered a key pest in Colombia (Holguin and Mira, 2021) (Fig. 14.20). Adults chew the green skin of small fruit causing striations on the fruit and skeletonization of

Fig. 14.19. A. *Xyleborus glabratus* Eichhoff, B. *Xyleborus bispinatus*, C. *Xyleborus volvulus*.

young leaves by chewing the leaf tissue between veins. Interestingly, the authors demonstrated that *A. pygidialis* and not *Phyllophaga obsoleta* Blanchard is the species damaging avocados in some Colombian regions. Adults of *A. pygidialis* emerge from the soil during the rainy season, usually in March and April (Londoño Zuluaga, 2020). Other species reported by Holguin and Mira (2021) in Colombia include *Anomala cincta* Say (Coleoptera: Rutelinae), *Charioderma xylina* Blanchard and *Strigoderma* sp. (all Coleoptera: Rutelinae) associated with flowers, *Isonychus* sp. (Coleoptera: Melolonthinae) with flower buds, *Cyclocephala fulgurata* Burmeister (Coleoptera: Dynastinae) with tender buds, stems and flowers, and *Astaena valida* Burmeister and *Plectris pavida* (Burmeister) (both Coleoptera: Melolonthinae) with mature leaves.

In Peru, *Oncideres poecila* Bates (Coleoptera: Scarabaeidae) causes girdling of branches. *O. poecila* deposits its eggs on fallen branches; collection of fallen branches is recommended to reduce the pest problem. *Macrodactylus mexicanus* Burmeister, a univoltine, highly polyphagous insect, is associated with avocado and vegetation found in the hot and temperate areas of central and western Mexico (Ramirez-Ponce *et al.*, 2009).

Long-horned beetles

Mallodon dasystomus Say (Coleoptera: Cerambycidae: Prioninae) is a pest of hardwood trees in the southern USA that recently began attacking trees in avocado orchards in Miami-Dade County, Florida (Fig. 14.21). Females oviposit near the base of stressed host trees or near

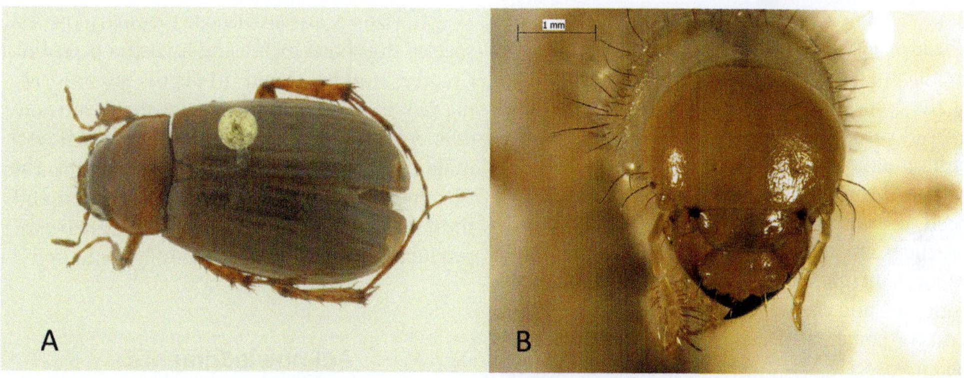

Fig. 14.20. The main chafer beetle *Astaena pygidialis* Kirsch. A. Adult beetle, B. Larvae. Photo credits: Dr Jhon César Neita (with permission).

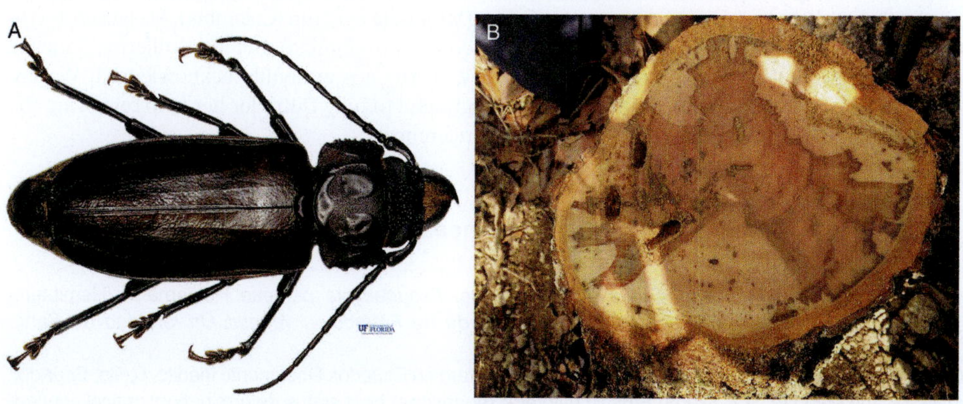

Fig. 14.21. The hardwood stump borer *Mallodon dasystomus* Say. A. Adult beetle, B. Excavated galleries.

wounds where wood is exposed. The larvae feed briefly under the bark and then tunnel through the sapwood and heartwood, damaging the integrity of the tree structure. The infestation begins in lower cut branches or the main trunk, especially in areas with sunscald injury. Larvae bore into the sapwood and heartwood making large tunnels, moving downwards to the base of the tree (Mankin *et al.*, 2018). Of particular concern is the recent detection of avocado trees in the Caribbean (US Virgin Islands and Puerto Rico) infested by the mango stem borer, *Batocera rufomaculata* De Geer. This is a highly destructive pest of mango and several other fruit trees in Asia and Africa. Preliminary observations suggest that avocados are severely affected.

Conclusions

The avocado pest complex is dynamic due to the significant expansion of the geographical range of the crop. The majority of the known pests of avocado are native to the New World. Still, new pests emerge frequently due to new associations with arthropods in areas where avocados have been introduced. Avocado pest management practices are affected not only by domestic and export fruit markets but also by consumer health concerns and cosmetic appearance of the fruit. Costs of pest control, current lack of information about insect biology and control, and lack of registration for the new generation of pesticides complicate the continuation or development of pest management programmes for avocado. From the information compiled in this chapter, it can be deduced that major producer countries such as Mexico, Colombia, Peru, Chile, Dominican Republic, South Africa, Israel, Australia and the USA maintain and develop avocado pest management programmes. Other countries, such as Indonesia and Myanmar, are either beginning to work towards this goal or accumulating basic information on the biology and life histories of the pests and descriptions of damage inflicted to the crop, with some emphasis on pest management, that is, economic injury levels, monitoring, sampling and biological and autocidal control.

The large differences in the scope and breadth of information provided by various avocado-growing countries may be based on economics. Countries where avocado is commercially grown on a large scale place more emphasis on gathering information on the basic biology of specific pests, and responses to grower needs are observed. In countries where avocados are mainly grown in small farms and backyards primarily for domestic consumption, very little research has been aimed at rational pest management.

Horizon scans are needed for emerging pest species that have increasing invasion potential. Greater numbers and in-depth surveys are needed in countries where avocados are endemic, especially throughout Mexico and Guatemala, to better detail and understand the extremely rich biodiversity and potential threats these unknown 'wild card' species pose to avocado-producing areas.

Acknowledgments

Information and images provided by Dr Gracielle Furtado-Moreira (Brazil), Engineer Hector Guillen (Mexico), Dr Marta M. Monserrat (Spain), Dr Claudia Holguin (Colombia), Ms Edison Torrado-Leon (Colombia) and Ms Guillermo Sánchez V. (Peru) are gratefully acknowledged. We are grateful to Jully Dutra for her help reviewing the references.

References

Abud-Antum, A. (1991) Presence of avocado lacebug, *Pseudacysta perseae* Heidemann (Hemiptera: Tingidae) in Dominican Republic. *Primera Jornada de Proteccion Vegetal, Universidad de Santo Domingo, Dominican Republic* p. 4 (abstract).

Agrocalidad. (2024) Listado oficial de plagas del aguacatero en Ecuador. Documento inédito, Quito, Ecuador.

Aluja, M. and Mangan, R.L. (2008) Fruit fly (Diptera: Tephritidae) host status determination: critical conceptual, methodological, and regulatory considerations. *Annual Review of Entomology* 53, 473–502.

Aluja, M., Diaz-Fleuscher, F. and Arredondo, J. (2004) Nonhost status of commercial *Persea americana* 'Hass' to *Anastrepha ludens, Anastrepha obliqua, Anastrepha serpentina* and *Anastrepha striata* (Diptera: Tephritidae) in Mexico. *Journal of Economic Entomology* 97, 293–309.

Aluja, M., Díaz-Fleischer, F., Arredondo, J., Valle-Mora, J. and Rull, J. (2010) Effect of cold storage on larval and adult *Anastrepha ludens* (Diptera: Tephritidae) viability in commercially ripe, artificially infested *Persea americana* 'Hass'. *Journal of Economic Entomology* 103, 2000–2008.

Aluja, M., Vázquez-Rosas-Landa, M., Cerqueda-García, D., Monribot-Villanueva, J.L., Altúzar-Molina, A., *et al.* (2023) Assessment of the molecular responses of an ancient angiosperm against atypical insect oviposition: the case of Hass avocados and the tephritid fly *Anastrepha ludens*. *International Journal of Molecular Sciences* 24(3), 2060.

Andrade, G.S. (1988) Control Quimico de la Arana Cristalina *Oligonychus Perseae*, Tutle, Baker and Abatiello en el Cultivo del Aguacatero. Thesis, Universidad Michoacána de San Nicolas Hidalgo, Uruapan, Michoacán, Mexico.

Annecke, D.P. (1959) The effect of parathion and ants on *Coccus hesperidum* L. (Coccidae: Hemiptera) and its natural enemies. *Journal of the Entomological Society of Southern Africa* 22, 245–275.

Annecke, D.P. (1964) The encyrtid and aphelinid parasites (Hymenoptera: Chalcidoidea) of soft brown scale, *Coccus hesperidum* Linnaeus (Hemiptera: Coccidae) in South Africa. *Entomology Memoir, Department of Agricultural Technical Services, Republic of South Africa* 7, 1–73.

Annecke, D.P. (1965) A new species of *Centroda* Foerster (Hymenoptera: Aphelinidae) parasitic in the eggs of a cicadelid injurious to citrus in South Africa. *South African Journal of Agricultural Science* 8, 1133–1138.

Annecke, D.P. (1966) Biological studies on the immature stages of soft brown scale, *Coccus hesperidum* Linnaeus (Homoptera: Coccidae). *South African Journal of Agricultural Science* 9, 205–228.

Annecke, D.P. and Georgala, M.B. (1978) Soft brown scale *Coccus hesperidum* L. In: Bedford, E.C.G. (ed.) *Citrus Pests in the Republic of South Africa.* Science Bulletin of the Department of Agricultural Technical Services, Republic of South Africa No. 391, p. 253.

Anonymous (1984) *Manual Fitosanitario del Aguacate*. Bayer de Mexico, p. 38.

Anzaldo, S.S. and Díaz-Grisales, V. (2022) *Heilipus squamosus* (LeConte, 1824): clarification of the correct name for the 'avocado tree girdler' with updates on its biology and distribution (Coleoptera, Curculionidae, Molytinae, Hylobiini). *Biodiversity Data Journal* 10: e85499.

APHIS (1992) 7 CFR Part 318 (Docket No. 92-081-1). 'Sharwil' avocados from Hawaii. *Federal Register* 57, 31306–31308.

Aponte, O. and McMurtry, J. (1997a) Damage on 'Hass' avocado leaves, webbing, and nesting behavior of *Oligonychus perseae* (Acari: Tetranychidae). *Experimental and Applied Acarology* 21, 265–272.

Aponte, O. and McMurtry, J. (1997b) Biology, life table and mating behavior of *Oligonychus perseae* (Acari: Tetranychidae). *International Journal of Acarology* 23, 199–207.

Arellano, P.G. (1975) Key for the identification of larvae of avocado stone borers in Mexico. *Folia Entomologica Mexicana* 31–32, 127–131.

Arévalo Peñaranda, E., Caicedo, A.M., González, M. and Jaramillo D. (2012) Resultados piloto de vigilancia fitosanitaria escama verde (*Coccus viridis* (Green) (Hemiptera: Coccidae)), pasador del fruto (*Stenoma catenifer* Walsingham (Lepidoptera: Elaschistidae)) y barrenadores (*Heilipus lauri* Boheman, *Heilipus* sp. *cerca pittieri* Barber, *Heilipus trifasciatus* (Fabricius), *Heilipus elegans* Guérin-Méneville (Coleoptera: Curculionidae)) ENANTIOQUIA. Available at: chrome-extension://efaidnbmnnnibpcajpcglclefindmkaj/https://www.ica.gov.co/areas/agricola/servicios/epidemiologia-agricola/boletines/departamentales/2012/boletin_aguacate_antioquia.aspx (accessed 12 February 2024).

Armstrong, W.J., Mitchell, W.C. and Parais, G.J. (1983) Resistance of 'Sharwil' avocados at harvest maturity to infestation by three fruit fly species (Diptera: Tephritidae) in Hawaii. *Journal of Economic Entomology* 76, 119–121.

Avidov, Z. and Harpaz, I. (1969) *Plant Pests in Israel*. Israel University Press, Jerusalem, p. 549.

Avilan, L., Leal, F. and Bautista, D. (1997) *El Aguacatero*. Espasando, Venezuela, p. 63.

Avogreen™ Manual (2010) Version 1.0 May 2010. Available at: http://www.nzavocado.co.nz/Avo%20Green/WEB%20AG%20Manual%2010May.pdf (accessed 16 March 2011).

Bailey, J.B. and Olsen, K.N. (1990) Chemical control of *Amorbia*, an insect pest of avocado and citrus. *California Agriculture* 44, 10–19.

Bailey, J.B., Hoffman, M.P. and Olsen, K.N. (1988) Blacklight monitoring on two avocado insect pests. *California Agriculture* 42, 26–27.

Barber, H.S. (1919) Avocado seed weevils. *Proceedings of the Entomological Society of Washington* 21, 53–60.

Barnes, H.F. (1946) *Gall Midges of Economic Importance. Vol.* II. *Gall Midges of Fodder Crops*. Crosby Lockwood and Son Ltd, London, p 852.

Barrow, M.R. (1977) The biology, economic significance and control of *Tortrix capensana* Wlk. (Lepidoptera: Tortricidae) on citrus at Zebediela Estate. MSc thesis, University of Natal, South Africa.

Becker, D., Cyjon, R., Cosse, A., Moore, I., Kimmel, T. and Wysoki, M. (1990) Identification and enantioselective synthesis of (Z,Z)-6,9-cis-3S,4R-epoxynonadecadiene, the major sex pheromone component of *Boarmia selenaria. Tetrahedron Letters* 31, 4923–4926.

Bedford, E.C.G. (1943) The biology and economic importance of the South African citrus thrips, *Scirtothrips aurantii* Faurè. *Publications of the University of Pretoria, South Africa, Series* 2, *Natural Science* 7, 1–68.

Bedford, E.C.G. (1978a) Other scale insects of minor importance. In: Bedford, E.C.G. (ed.) *Citrus Pests in the Republic of South Africa*. Science Bulletin, Department of Agricultural Technical Services, Republic of South Africa No. 391, p. 130.

Bedford, E.C.G. (1978b) Tip wilters and shield bugs on citrus. In: Bedford, E.C.G. (ed.) *Citrus Pests in the Republic of South Africa*. Science Bulletin, Republic of South Africa, Department of Agricultural Technical Services No. 391, pp. 135–136.

Bedford, E.C.G. (1998a) Methods of controlling citrus pests. In: Bedford, E.C.G., Van den Berg, M.A. and De Villiers, E.A. (eds) *Citrus Pests in the Republic of South Africa*. ARC Institute for Tropical and Subtropical Crops, Nelspruit, South Africa, pp. 8–14.

Bedford, E.C.G. (1998b) Citrus blackfly *Aleurocanthus woglumi* Ashby. In: Bedford, E.C.G., Van den Berg, M.A. and De Villiers, E.A. (eds) *Citrus Pests in the Republic of South Africa*. ARC Institute for Tropical and Subtropical Crops, Nelspruit, South Africa, pp. 102–104.

Bedford, E.C.G. and Thomas, E.D. (1965) Biological control of the citrus blackfly, *Aleurocanthus woglumi* (Ashby) (Homoptera: Aleyrodidae) in South Africa. *Journal of the Entomological Society of Southern Africa* 28, 117–132.

Begemann, G. and Schoeman, A. (1999) The phenology of *Helicoverpa armigera* (Hübner) (Lepidoptera: Noctuidae), *Tortrix capsensana* (Walker) and *Cryptophlebia leucotreta* (Meyrick) (Lepidoptera: Tortricidae) on citrus at Zebediela, South Africa. *African Entomology* 7, 131–148.

Begemann, G.J., Barrow, M.R and Bedford, E. (1998) Apple leaf roller *Tortrix capensana* (Walker). In: Bedford, E.C.G., Van den Berg, M.A. and De Villiers, E. (eds) *Citrus Pests in the Republic of South Africa*. ARC Institute for Tropical and Subtropical Crops, Nelspruit, South Africa, p. 288

Ben-Dov, Y. (1971) An annotated list of the soft scale insects (Homoptera: Coccidae) of Israel. *Israel Journal of Entomology* 6, 23–34.

Ben-Dov, Y. and Amitai, S. (1980) The pyriform scale, *Protopulvinaria pyriformis* Cockerell, in Israel. *Alon Hanotea* 34, 797–798. (In Hebrew.)

Ben-Dov, Y., Williams, M.L. and Ray, Jr, C.H. (1976) Taxonomy of the mango shield scale *Protopulvinaria mangiferae* Green (Homoptera: Coccidae). *Israel Journal of Entomology* 10, 1–17.

Blommers, L. and Gutierrez, J. (1975) Les tétranyques vivant sur agrumes et avocatiers dans la région de Tamatave (Madagascar – est) et quelques-uns de leurs prédateurs. *Fruits* 30, 191–200.

Blumberg, D. and Blumberg, O. (1991) The pyriform scale, *Protopulvinaria pyriformis*, and its common parasitoid, *Metaphycus stanleyi*, on avocado and *Hedera helix. Alon Hanotea* 45, 265–269. (In Hebrew with English summary.)

Blumberg, D. and Mendel, Z. (1991) Improvement of the biological control of the cottony cushion scale *Icerya purchasi* by the parasitic fly *Cryptochaetum iceryae*. *Hassadeh* 71, 1107–1112. (In Hebrew with English summary.)

Blumberg, D., Mendel, Z., Assael, F. and Rasis, A. (1989) Outbreaks of the cottony cushion scale *Icerya purchasi* and acclimatization of the parasitic fly *Cryptochaetum iceryae* in Israel. *Alon Hanotea* 43, 395–400. (In Hebrew with English summary.)

Blumberg, D., Mendel, Z. and Assael, F. (1992) Improvement of the biological control of the cottony cushion scale *Icerya purchasi*, by the parasitic fly *Cryptochaetum iceryae. Phytoparasitica* 20, 72–73.

Bodenheimer, F.S. (1951) *Citrus Entomology in the Middle East*. Dr. W. Junk, The Hague, The Netherlands.

Boland, J.M. (2016) The devastating impact of an invasive ambrosia beetle on the riparian habitats of the Tijuana River Valley, California (2167–9843). PeerJ. Available at: doi: 10.7717/peerj.2141. PMID: 27366644; PMCID: PMC4924130.

Boscán de Martínez, N. and Godoy, F.J. (1982) *Apanteles* sp. (Hymenoptera: Braconidae) parasito del taladrador del agaucate *Stenoma catenifer* Walsingham (Lepidoptera: Stenomatidae) en Venezuela. *Agronomia Tropical* 32, 205–208.

Boscán de Martinez, N. and Godoy, F.J (1984) Observaciones preliminaries sobre la biologia de *Stenoma catenifer* Walsingham (Lepidoptera: Stenomiidae) taladrador del Aguacate (*Persea americana* Mill.). *Agronomia Tropical* 34, 205–208.

Bosman, T. (1959) Blackfly can destroy entire crop. *Farming in South Africa* 35, 6–27.

Bragard, C., Dehnen-Schmutz, K., Di Serio, F., Gonthier, P., Jacques, M.A., *et al.* (2021) Commodity risk assessment of *Persea americana* from Israel. *EFSA Journal* 19(2): e06354. doi:10.2903/j.efsa.2021.6354.

Brain, C.K. (1920) The Coccidae of South Africa. *Bulletin of Entomological Research* 11, 1–42.

Brain, C.K. (1929) *Insect Pests and their Control in South Africa.* Nasionale Pers Beperk, Cape Town, South Africa, p. 468.

Brink, T., Steyn, W.P. and De Beer, M. (1997) Artificial exposure of different avocado cultivars to fruit flies. *South African Avocado Growers' Association Yearbook* 20, 78–79.

Broodryk, S.W. (1969) The biology of *Chelonus (Microchelonus) curvimaculatus* Cameron (Hymenoptera: Braconidae). *Journal of the Entomological Society of Southern Africa* 32, 169–189.

Broom, E. (1970) *El Aguacate*. Comision Nacional de Fruticultura, S.A.G. Mexico, pp. 172–264.

Brown, J.W. and Brown, R.L. (2004) A new species of *Cryptaspasma* Walsingham (Lepidoptera: Tortricidae: Olethreutinae) from Central America, the Caribbean, and southwestern United States, with a catalogue of the world fauna of Microcorsini. *Proceedings of the Entomological Society of Washington* 106, 288–297.

Brown, J.W. and Hoddle, M.S. (2010) A new species of *Histura* Razowski (Lepidoptera: Tortricidae: Polyorthini) from Guatemala attacking avocados (*Persea americana*) (Lauraceae). *Proceedings of the Entomological Society of Washington* 112, 10–21.

Bruwer, I.J. (1987) Stinkbugs in macadamia orchards. *Institute for Tropical and Subtropical Crops, Information Bulletin* 178, 2–4.

Bruwer, I.J. (1992) The influence of various hemipteran species on macadamia and some factors which can limit nut damage. PhD thesis, University of Stellenbosch, Stellenbosch, South Africa, p. 184.

Bruwer, I.J. (1996) Die Hemiptera-plaagkompleks op avokado's in die Kiepersol-gebied. *South African Avocado Growers' Association Yearbook* 19, 33–35. (In Afrikaans.)

Bruwer, I.J. (1998) Bio-ekologiese inligting van suigbesies op avokado. *South African Avocado Growers' Association Yearbook* 20, 73–77. (In Afrikaans.)

Busck, A. (1920) A new Gracillaria injurious to avocado (Lepidoptera). *Canadian Entomologist* 52:239.

Bush, G. (1957) Some notes on the susceptibility of avocados in Mexico to attack by the Mexican fruit fly. *Journal of the Rio Grande Valley Horticultural Society* 6, 75–95.

Byrne, F.J., Toscano, N.C., Urena, A.A. and Morse, J.G. (2007) Toxicity of systemic neonicotinoid insecticides to avocado thrips in nursery avocado trees. *Pest Management Science* 63, 860–866.

Byrne, F.J., Humeres, E.C., Urena, A.A., Hoddle, M.S. and Morse, J.G. (2010) Field evaluation of systemic imidacloprid for the management of avocado thrips and avocado lace bug in California avocado groves. *Pest Management Science* 66, 1129–1136. doi:10.1002/ps.1991.

Cañizares, J. (1975) *Los Aguacateros*. Edicion Revolucionaria, Instituto Cubano del Libro, La Habana, Cuba, pp. 239–258.

Canseco, F.D. (1971) Biologia y control del barrenador pequeno del hueso del aguacate (*Conotrachelus aguacatae*) Barber en la region de Colima, Queretaro. *Secretaria de Agricultura y Ganaderia, Direccion General de Sanidad Vegetal* 24, 37–48.

Carabalí Muñoz, A., Caicedo Vallejo, A.M. and Holguín, C.M. (2021) *Guía para el reconocimiento y manejo de las principales plagas de aguacate cv. Hass en Colombia.* Corporación Colombiana de Investigación Agropecuaria - AGROSAVIA; ICA, Mosquera, Colombia. Available at: https://doi.org/10.21930/agrosavia.nbook.7404913 (accessed 25 November 2024).

Carrillo, D., Duncan, R.E. and Peña, J.E. (2012) Ambrosia beetles (Coleoptera: Curculionidae: Scolytinae) that breed in avocado wood in Florida. *Florida Entomologist* 95(3), 573–579.

Carrillo, D., Crane, J.H. and Peña, J.E. (2013) Potential of contact insecticides to control Xyleborus glabratus (Coleoptera: Curculionidae: Scolytinae), a vector of laurel wilt in avocados. *Journal of Economic Entomology* 106: 2286-2295.

Carrillo, D., Duncan, R.E., Ploetz, J., Campbell, A., Ploetz, R. and Peña, J.E. (2014) Lateral transfer of a phytopathogenic symbiont among native and exotic ambrosia beetles. *Plant Pathology* 63(1), 54–62.

Carrillo, D., Cruz, L., Kendra, P., Narváez, T., Montgomery, W., Monterroso, A. and Cooperband, M. (2016) Distribution, pest status and fungal associates of Euwallacea nr. Fornicatus in Florida avocado groves. *Insects* 7(4), 55.

Carrillo, D., Narvaez, T., Cossé, A.A., Stouthamer, R. and Cooperband, M. (2015) Attraction of *Euwallacea nr. Fornicatus* (Coleoptera: Curculionidae: Scolytinae) to lures containing quercivorol. *Florida Entomologist* 98(2), 780–782.

Carrillo, D., Roda, A., Sarmiento, C., Monterroso, A., Wei, X., *et al.* (2017) Impact of oriental fruit fly postharvest treatments on Avocado. *American Journal of Plant Sciences* 8(3), 549.

Carvalho, J.M. and Costa, L.A. (1988) Mirídeos neotropicais. CCXCVII: Duas novas espécies do gênero Monalonion Herrich-Schaffer (Hemiptera). *Revista Brasileira de Biologia* 48(4), 893–896.

Castañeda-Vildozola, A., Peña, J.E. and Equihua-Martinez, A. (2010) Avocado weevils of the genus *Heilipus*. In: *Proceedings of the Potential Invasive Pests Workshop 10–14 October 2010, Miami,* p. 21 (Abstract).

Castañeda-Vildózola, Á., Palacios-Torres, R.E., Sánchez-Pale, J.R., Franco-Mora, O., Valdez-Carrasco, J. and Equihua-Martínez, A. (2017) Sympatry of two species of Heilipus Germar, 1824 (Coleoptera: Curculionidae) infesting avocado (*Persea americana* Mill.) in central Mexico. *The Coleopterists Bulletin* 71(2), 361–363.

Castañeda-Vildózola, Á., González-Hernández, H., Equihua-Martínez, A., Valdez-Carrasco, J., Peña, J.E., Cazado, L.E. and Franco-Mora, O. (2016) Head capsule width is useful for determining larval instar in *Heilipus lauri* (Coleoptera: Curculionidae). *Florida Entomologist* 99(4), 822–825.

Catling, H.D. and Aschenborn, H. (1974) Population studies of the false codling moth, *Cryptophlebia leucotreta* Meyrick, on citrus in the Transvaal. *Phytophylactica* 6, 31–38.

Catling, H.D. and Aschenborn, H. (1978) False codling moth *Cryptophlebia leucotreta* (Meyrick). In: Bedford, E.C.G. (ed.) *Citrus Pests in the Republic of South Africa*. Science Bulletin, *Republic of South Africa*, Department of Agricultural Technical Services, Pretoria, South Africa, pp. 165–170.

Cendana, S.M., Gabriel, B.P. and Magallona, E.D. (1984) *Insect Pests of Fruit Crops in the Philippines*. Department of Entomology, University of the Philippines, Los Banos.

Cervantes Peredo, L., Lyal, C.H.C. and Brown, V.K. (1999) The stenomatine moth, *Stenoma catenifer* Walsingham: a pre-dispersal seed predator of Greenheart (*Chlorocardium rodiei* [Schomb.] Rohwer, Richter and van de Werff) in Guyana. *Journal of Natural History* 33, 531–542.

Chamorro, M. and Barclay, M. (2018) On the identity of a U.S. intercepted *Conotrachelus* Dejean (Coleoptera: Curculionidae) with avocado (*Persea americana*). *Biodiversity Data Journal* 6: e26362. doi:10.3897/BDJ.6.e26362.

Chavez, H., Pereira, C. and Gonzalez, R. (1985) Chicharrita del aguacate *Trioza perseae* Tuthill (Homoptera: Psyllidae) plaga potencial del aguacate en zones altas del estado de Lara. *Boletin de Entomologia Venezolana* 4, 17–24.

Cilliers, C.J. (1978) Australian bug *Icerya purchasi* Maskell. In: Bedford, E.C.G. (ed.) *Citrus Pests in the Republic of South Africa*. Science Bulletin, Department of Agricultural Technical Services, Republic of South Africa No. 391, pp. 85–87.

Cilliers, C.J. (1998) Wax scales on citrus. In: Bedford, E.C.G., Van den Berg, M.A. and De Villiers, E.A. (eds) *Citrus Pests in the Republic of South Africa*. ARC Institute for Tropical and Subtropical Crops, Nelspruit, South Africa, pp. 129–131.

Claassens, V.E. (1994) *Fiorinia*-dopluis op avokadovrugte. *Information Bulletin, Institute for Tropical and Subtropical Crops, Nelspruit* 260:12. (In Afrikaans.)

Claassens, V.E. (2001) Avocado beetle *Monolepta apicalis*. In: Van den Berg, M.A., De Villiers, E.A. and Joubert, P.H. (eds) *Pests and Beneficials of Tropical and Non-Citrus Subtropical Crops in South Africa*. ARC Institute for Tropical and Subtropical Crops, Nelspruit, South Africa, pp. 271–274.

Clausen, C.P. and Berry, P.C. (1932) The citrus blackfly in Asia and the importation of its natural enemies into tropical America. *USDA Technical Bulletin* 320:59.

Clavijo, A.P. and Holguin, C.M. (2020) Pathogenicity of commercial entomopathogenic fungal strains on the avocado seed borer (ASB), *Heilipus lauri* (Coleoptera: Curculionidae) under laboratory conditions. *International Journal of Tropical Insect Science* 40(4), 1059–1067.

Coaton, W.G.H. (1971) Five new termite genera from South West Africa (Isoptera: Termitidae). *Cimbebasia Series* A 2(1), 1–34.

Coaton, W.G.H. (1974) Order Isoptera. In: Coaton, W.G.H. (ed.) *Status of the Taxonomy of the Hexapoda of Southern Africa*. Department of Agricultural Technical Services, South Africa, *Entomological Memoir* 38, pp. 22–29, 124.

Coaton, W.G.H. and Sheasby, J.L. (1972) Preliminary report on a survey of the termites (Isoptera) of South West Africa. *Cimbebasia Memoir* 2, 1–129.

Coaton, W.G.H. and Sheasby, J.L. (1978) National survey of the Isoptera of Southern Africa. 14. The genus *Stolotermes* Hagen (Termopsidae: Stolotermitinae). *Cimbebasia* A 3, 208–213.

Colin, J.G. and Rubi, M. (1992) Determinacion de la susceptibilidad de ocho selecciones de aguacate al ataque de araña roja y trips. *Centro de Investigaciones Cientificas y Tecnologicas del Aguacate en el Estado de Mexico, Memoria,* Coatepec Harinas, Mexico, pp. 149–155.

Coria-Avalos, V. (1993) Principales plagas del aguacate en Michoacán. Secretaria de Agricultura, Instituto Nacional de Investigaciones Forestales y Agropecuarias. *Centro de Investigaciones del Pacifico Centro, Michoacán, Mexico*, p. 19, 20.

Cruz-López, L., Vázquez, M. and Castillo, A. (2020) Effectiveness of the sex pheromone for monitoring Stenoma catenifer (Lepidoptera: Elachistidae) in avocado orchards. *Neotropical Entomology* 49, 332–336.

Cruz, L.F., Rocio, S., Duran, L.G., Menocal, O.G., Garcia-Avila, D.J. and Carrillo, D. (2018) Developmental biology of *Xyleborus bispinatus* (Coleoptera: Curculionidae) reared on an artificial medium and fungal cultivation of symbiotic fungi in the beetle's galleries. *Fungal Ecology* 35, 116–126.

Cruz, L.F., Menocal O.G., Mantilla J., Ibarra L. and Carrillo D. (2019) Biology and fungal associates of *Xyleborus volvulus* (Coleoptera: Curculionidae). *Applied and Environmental Microbiology* 85(19): e01190–19, 11.

Cruz, L.F., Menocal, O., Kendra, P.E. and Carrillo, D. (2021) Phoretic and internal transport of *Raffaelea lauricola* by different species of ambrosia beetle associated with avocado trees. *Symbiosis* 84(560), 151–161.

Daehler, C. and Dudley, N. (2002) Impact of the black twig borer, an introduced insect pest, on *Acacia koa* in the Hawaiian Islands. *Micronesica Supplement* 6, 35–53.

Daiber, C.C. (1979) A study of the biology of the false codling moth [*Cryptophlebia leucotreta* (Meyr.)]: the larva. *Phytophylactica* 11, 141–144.

Daiber, C.C. (1980) A study of the biology of the false codling moth *Cryptophlebia leucotreta* (Meyr.): the adult and generations throughout the year. *Phytophylactica* 12, 187–193.

Daiber, K.C. (1992) Insects and nematodes that attack cole crops in Southern Africa – a research review. *Zeitschrift für Pflanzenkrankheiten und Pflanzenschutz* 99, 430–440.

Daneel, M.S. (1998) Palm scale. In: De Villiers, E.A. (ed.) *The Cultivation of Mangoes*. ARC Institute for Tropical and Subtropical Crops, Nelspruit, South Africa, p. 216.

Das, S.C. (1955) Bionomics of the tea red spider, *Oligonychus coffeae* Nietner. *Bulletin of Entomological Research* 50, 265–274.

Davidson, J.A. and Miller, D.R. (1990) Ornamental plants. In: Rosen, D. (ed.) *Armoured Scale Insects, their Biology, Natural Enemies and Control*, Vol. 4B. Cornell University Press, Ithaca, New York, pp. 603–632.

De Graaf, J. (2009) Host status of avocado ('Hass') to *Ceratitis capitata*, *Ceratitis rosa*, and *Ceratitis cosyra* (Diptera: Tephritidae) in South Africa. *Journal of Economic Entomology* 102, 1448–1459.

Dekle, G.W. (1976) *Florida Armored Scale Insects*. Florida Department of Agriculture and Consumer Services, Gainesville, Florida, p. 345.

De Lotto, G. (1957) On some Ethiopian species of the genus *Coccus* (Homoptera: Coccoidea: Coccidea). *Journal of the Entomological Society of Southern Africa* 22, 150–173.

De Meijer, A.H., Wysoki, M., Swirski, E., Blumberg, D. and Izhar, Y. (1989) Susceptibility of avocado cultivars to the pyriform *scale, Protopulvinaria pyriformis* Cockerell (Homoptera: Coccidae). *Agriculture, Ecosystems and Environment* 25, 75–82.

Dennill, G.B. and Dupont, F.M.A. (1992) An ecological study of the damage done to avocado fruit by Hemiptera in the Hazyview region, South Africa. *South African Avocado Growers' Association Yearbook* 15, 57–61.

Dennill, G.B. and Erasmus, M.J. (1991) A packhouse survey of insect damage to avocados in the Nelspruit/Hazyview area during 1990. *South African Avocado Growers' Association Yearbook* 14, 79–82.

Dennill, G.B. and Erasmus, M.J. (1992a) The insect pests of avocado fruits – increasing pest complex and changing pest status. *Journal of the Entomological Society of Southern Africa* 55, 51–57.

Dennill, G.B. and Erasmus, M.J. (1992b) Basis for a practical technique for monitoring thrips in avocado orchards. *Crop Protection* 11, 89–91.

De Villiers, E.A. (1973) Die voorkoms van basboorders (*Salagena* sp.) op lietsjie en ander vrugtebome. *The Citrus and Sub-Tropical Fruit Journal* 477, 17–19. (In Afrikaans.)

De Villiers, E.A. (1981) Heart-shaped scale in avocados. *Farming in South Africa*, Leaflet H.4. Department of Agricultural Technical Services, Pretoria, South Africa, p. 2.

De Villiers, E.A (1986) Bugs on macadamia nuts. *Farming in South Africa*. *Macadamia* H.1. Department of Agricultural Technical Services, Pretoria, South Africa, p. 3.

De Villiers, E.A. (1989) Seisoensvoorkoms van verskillende stadia van die hartvorminge dopluis. *South African Avocado Growers' Association Yearbook* 12, 58–59. (In Afrikaans.)

De Villiers, E.A. (1990a) Coconut bugs on avocados. *Farming in South Africa, Avocado Series* H.5.Department of Agricultural Technical Services, Pretoria, South Africa, p. 2.

De Villiers, E.A. (1990b) Blaaspootjies by avokados. *Farming in South Africa. Avocado Series* H.3. Department of Agricultural Technical Services, Pretoria, South Africa, p.1. (In Afrikaans.)

De Villiers, E.A. (1992) Coconut bug on guavas. *Farming in South Africa, Guavas Series* H.3. Department of Agricultural Technical Services, Pretoria, South Africa, p. 2.

De Villiers, E.A. (1998) Rooiteemyt slaan toe op avokado's. *Information Bulletin, Institute for Tropical and Subtropical Crops* 299:38. (In Afrikaans.)

De Villiers, E.A. (ed.) (2001a) *The Cultivation of Avocado*. ARC Institute for Tropical and Subtropical Crops, Nelspruit, South Africa, p. 370.

De Villiers, E.A. (2001b) Heart-shaped scale, *Protopulvinaria pyriformis* (Cockerell). In: Van den Berg, M.A., De Villiers, E.A. and Joubert, P.H. (eds) *Pests and Beneficial Arhropods of Tropical and Non-Citrus Subtropical Crops in South Africa*. ARC Institute for Tropical and Subtropical Crops, Nelspruit, South Africa, pp. 170–174.

De Villiers, E.A. (2001c) Citrus looper *Asscotis reciprocaria reciprocaria* (Walker). In: Van den Berg, M.A., De Villiers, E.A. and Joubert, P.H. (eds) *Pests and Beneficial Arthropods of Tropical and Non-citrus Subtropical Crop in South Africa*. ARC Institute for Tropical and Subtropical Crops, Nelspruit, p. 525.

De Villiers, E.A. and Robertson, C.M. (1988) Chemiese beheer van hartvormige dopluis op avokado's. *Subtropica* 9, 9–12. (In Afrikaans.)

De Villiers, E.A. and Van den Berg, M.A. (1987) Avocado insects of South Africa. *South African Avocado Growers' Association Yearbook* 10, 75–79.

De Villiers, E.A., Bedford, E.C.G. and Petty, G.T. (1987a) Mealybugs and Australian bug. In: Myburgh, A.C. (ed.) *Crop Pests in Southern Africa, Volume 2: Citrus and Other Subtropicals*. Plant Protection Research Institute, Department of Agriculture and Water Supply, Bulletin, Pretoria, South Africa 411, pp. 38–42.

De Villiers, E.A., Newton, P.J., Petty, G.T. and Myburgh, A.C. (1987b) Fruit and leaf-eating caterpillars. In: Myburgh, A.C. (ed.) *Crop Pests in Southern Africa, Vol. 2, Citrus and Other Subtropicals*. Plant Protection Research Institute Bulletin, Pretoria, South Africa 411, pp. 82–90.

Dedios-Avila, N., Cambero-Campos, J., Coronado-Blanco, J., Carvajal-Cazola, C., Rios-Velasco, C. and Pena-Sandoval, G. (2016) First register of *Neocatolaccus tylodermae* Ashmead in México as parasitoid of *Copturus aguacatae* Kissinger. *Southwestern Entomologist* 41(4), 1203–1206.

Dedios-Avila, N., Estrada-Virgen, O., Coronado-Blanco, J., Rios-Velasco, C., Bustillos-Rodríguez, C. and Cambero-Campos, J. (2021) Parasitoids and entomopathogens associated with the avocado stem weevil, *Copturus aguacatae* (Coleoptera: Curculionidae) in Nayarit, Mexico. *Entomological News* 130(1), 61–69.

Díaz-Fleischer, F. and Aluja, M. (2003) Clutch size in frugivorous insects as a function of host firmness: the case of the tephritid fly *Anastrepha ludens. Ecological Entomology* 28, 268–277.

Dixon, W., Woodruff, R. and Foltz, J. (2003) Black twig borer, *Xylosandrus compactus* (Eichoff) (Insecta: Coleoptera: Curculionidae: Scolytinae). *University of Florida, Entomology and Nematology Department, EENY* 5 2011. Available at: http://edis.ifas.ufl.edu/in577 (accessed 10 February 2011).

Dreistadt, S.H. (2008) *Integrated Pest Management for Avocados*. Statewide Integrated Pest Management Program, University of California Agriculture and Natural Resources Publication, Mountain View, California 3503, p. 222.

Dupont, F.M. (1993) An ecological study of and development of a monitoring technique for the damage done to avocado fruits by Hemiptera in the Hazyview region of South Africa. *South African Avocado Growers' Association Yearbook* 16, 107–112.

Du Toit, W.J. and De Villiers, E.A. (1988) Die hartvormige dopluis, *Protopulvinaria pyriformis* (Cockerell), op avokado's. *South African Avocado Growers' Association Yearbook* 11, 79–80. (In Afrikaans.)

Du Toit, W.J. and De Villiers, E.A. (1990) Effects of insect growth regulators on the development of heart-shaped scale (Hemiptera: Coccidae) on avocados in South Africa. *Tests of Agrochemicals and Cultivars. No. 11, Annals of Applied Biology* 116(Suppl.), 4–5.

Du Toit, W.J., Schutte, M.S. and Steyn, W.P. (1991) Parasitoids of the heart-shaped scale, *Protopulvinaria pyriformis* (Cockerell) (Hemiptera: Coccidae) on avocados in South Africa. *South African Avocado Growers' Association Yearbook* 14, 74–77.

Du Toit, W.J., Steyn, W.P. and de Beer, M.S. (1993) Occurrence of protrusions on avocado fruit and the causative agent. *South African Avocado Growers' Association Yearbook* 16, 100–102.

Du Toit, W.J., Steyn, W.P. and de Beer, M.S. (1994) Integrated control of avocado pests with special reference to the heart-shaped scale. *Final Report 1993/94, Project No. TS 5510/03, ITSC, Nelspruit*.

Ebeling, E. (1959) *Subtropical Fruit Pests*. University of California Press, Riverside, p. 436.

EFSA Panel on Plant Health (PLH), Bragard, C., Dehnen-Schmutz K., Di Serio F., Gonthier P., *et al.* (2021) Commodity risk assessment of Persea americana from Israel. *EFSA Journal* 19(2): e06354. doi:10.2903/j.efsa.2021.6354. PMID: 33552299.

Engstrand, R.C., Cibrián-Tovar, J., Cibrián-Jaramillo, A. and Kolokotronis, S.O. (2010) Genetic variation in avocado stem weevils *Copturus aguacatae* (Coleoptera: Curculionidae) in Mexico. *Mitochondrial DNA* 1, 38–43. doi:10.3109/19401736.2010.536226.

EPPO (2024) *Oligonychus perseae* (OLIGPA). Available at: http://www.gd.eppo.int/taxon/OLIGPA/hosts (accessed 12 February 2024).

Erichsen, C. (1993) Comparison of and scouting for beetle and looper damage to avocado leaves and fruit. *South African Avocado Growers' Association Yearbook* 16, 125–127.

Erichsen, C. and Schoeman, A. (1992) Economic losses due to insect pests on avocado fruit in the Nelspruit/Hazyview region of South Africa during 1991. *South African Avocado Growers' Association Yearbook* 15, 49–54.

Erichsen, C. and Schoeman, A.S. (1993) *Monolepta apicalis* Sahlberg (Coleoptera: Chrysomelidae): a new addition to the avocado pest-complex. *South African Avocado Growers' Association Yearbook* 16, 122–124.

Erichsen, C. and Schoeman, A.S. (1994) Identification and potential pest-status of looper on avocado. *South African Avocado Growers' Association Yearbook* 17, 113–116.

Erichsen, C., McGeoch, M.A. and Schoeman, A.S. (1993) Invasion of orchards by the avocado beetle *Monolepta apicalis* Sahlberg (Coleoptera: Chrysomelidae): assessment of damage to leaves and fruit. *South African Avocado Growers' Association Yearbook* 16, 118–122.

Escobar-Garcia, H.A., Beard, J.J. and Ochoa, R. (2022) Report of *Tuckerella pavoniformis* (Acari: Tuckerellidae) on Mamey, *Mammea americana* (Calophyllaceae), in Northwestern Peru. *Insects* 13:473.

Eskalen, A., Gonzalez, A., Wang, D., Twizeyimana, M., Mayorquin, J. and Lynch, S. (2012) First report of a *Fusarium* sp. and its vector tea shot hole borer (*Euwallacea fornicatus*) causing Fusarium dieback on avocado in California. *Plant Disease* 96(7), 1070–1070.

Evans, H.C. and Prior, C. (1990) Entomopathogenic fungi. In: Rosen, D. (ed.). *Armoured Scale Insects, their Biology, Natural Enemies and Control, Vol.* 4B. Elsevier, New York, pp. 3–17.

Evans, G., Watson, G. and Miller, D. (2009) A new armoured scale (Hemiptera: Coccoidea: Diaspididae) found on avocado fruit from Mexico and a key to the species of armored scales found on avocado world-wide. *Zootaxa*, 1991, 57–68.

Faber, B.A. and Philips, P.A. (2001) *Pests of Avocado. UC Pest Management Guidelines*. University of California, State Wide Integrated Pest Management Project. Available at: http://www.imp.ucdavis.edu/PMG/selectnewpest.avo.html (accessed 10 February 2011).

Faurè, J.C. (1929) The South African citrus thrips and five other new species of *Scirtothrips* Shull. *Bulletin of the Faculty Agriculture, Transvaal University College* 18, 1–18.

Fay, H.A.C. and De Faveri, S.G. (1990) Invasion of flowering avocado trees by *Monolepta australis* Jacoby (Coleoptera: Chrysomelidae) and its impact on fruit-set in north Queensland. *Crop Protection* 9, 392–394.

Fay, H.A.C. and Huwer, R.K. (1993) Egg parasitoids collected from *Amblypelta lutescens lutescens* (Distant) (Hemiptera: Coreidae) in north Queensland. *Journal of the Australian Entomological Society* 32, 365–367.

Fleschner, C., Ricker, D. and Johnson, H. (1957) Parasites of amorbioa and the omnivorous loopers in avocado orchards. *California Avocado Yearbook* 41, 107–117.

Follett, P.A. (2009) Puncture resistance in 'Sharwil' avocado to oriental fruit fly and Mediterranean fruit fly (Diptera: Tephritidae) oviposition. *Journal of Economic Entomology* 102, 921–926.

Follett, P.A., Asmus, G., Hamilton, L. and Stockton, D. (2022) Susceptibility of 'Malama' Avocados to Bactrocera dorsalis (Diptera: Tephritidae) in Hawai'i and a proposed systems approach for market access. *Journal of Economic Entomology* 115(6), 1894–1900.

Ford, W.K. (1934) Some observations on the bionomics of the false codling moth – *Argyroploce leucotreta*, Meyr. (Family Eucosmidae) – in Southern Rhodesia. *Publications of the British South Africa Company* 3, 9–34.

Fornazier, M., Pratissoli, D. and Teixeira, C. (1997) Susceptibilidade de cultivares de abacateiro a um geometrideo desfolhador. *Resumos do 16 Congresso Brasileiro de Entomologia, Salvador, Bahia*, Brazil, p. 370 (Abstract).

Fuller, C. (1901) The Natal codling moth, *Carpocapsa* sp. *First report of Government Entomologist 1899–1900*. Department of Agriculture, Natal, South Africa, pp. 48–51.

Furtado-Moreira, G. (2022) Manejo Integrado de Pragas. In: Costa-Sampaio, A. and Whately, M.C. (eds) *Abacaticultura sustentável*. Atena Editora. Available at: https://doi.org/10.22533/at.ed.6432227048 (accessed 25 November 2024).

Gagné, R.J. (1994) *The Gall Midges of the Neotropical Region*. Cornell University Press, Ithaca, New York.

Gagné, R.J. (2004) A catalogue of the *Cecidomyiidae* (Diptera) of the World. *Memoires of the Entomological Society of Washington*. No. 23.

Gagné, R.J. and Woods, W.M. (1988) Native American plant hosts of *Asphondylia websteri* (Diptera: Cecidomyiidae). *Annals of the Entomological Society of America* 81, 447–448.

Gagné, R.J., Posadam, F. and Gil, Z.N. (2004) A new species of *Bruggmanniella* (Diptera: Cecidomyiidae) aborting young fruit of avocado *Persea americana* in Colombia and Costa Rica. *Proceedings of the Entomological Society of Washington* 106, 547–553.

Gagné, R.J., Peña, J.E. and Acevedo, F.E. (2008) A new lestodiplosine (Diptera: Cecidomyiidae) preying on the avocado lace bug, *Pseudacysta perseae* (Heteroptera: Tingidae) in Southern Florida. *Florida Entomologist* 91, 43–48.

Gallegos, R. (1983) *Algunos Aspectos del Aguacate y su Produccion en Michoacán*. Universidad Autonoma de Chapingo, Grupo Editorial Gaceta, Mexico City, Mexico. p. 311.

García-Avila, C.D., Trujillo-Arriaga, F.J., López-Buenfil, J.A., González-Gómez, R., Carrillo, D., *et al.* (2016) First report of *Euwallacea nr. fornicatus* (Coleoptera: Curculionidae) in Mexico. *Florida Entomologist* 99(3), 555–556.

Garcia, J.C., Mujica, E., Vazquez, L. and Aranda, E. (1986) Fenologia de Copturus aguacatae Kissinger (Coleoptera: Curculionidae) en tres regions del estado de Morelos. *Resumenes, XXI Congreso Nacional de Entomologia, 16–19 Marzo 1986, Monterrey, Nuevo Leon, Mexico*, pp. 61–62.

Gaston, J., Revilla, T. and Vives Moreno, A. (2018) Contribution to the knowledge of Microlepidoptera, with new mentions for Spain and other contributions of interest (Insecta: Lepidoptera). *Shilap-Revista De Lepidopterologia* 46(182), 221–233.

Geertsema, H. (1971) Description of the stages of two subspecies of the pine emperor moth, *Nudaurelia cytherea* (Fabr.) (Lepidoptera: Saturniidae). *Annals of the University of Stellenbosch*, Stellenbosch, South Africa, p. 37.

Gerson, U. and Zor, Y. (1973) The armoured scale insects of avocado in Israel. *Journal of Natural History* 7, 513–533.

Gil, A.M. and Aranda, G. (2021) Aguacate, Guia de gestion intagrada de plagas. Ministerio de Agricultura, Pesca y Alimentacion, Secretaria General Tecnica, Centro de Publicaciones, Madrid Espana.

Gilbert, M.J. and Bedford, E.C.G. (1998) Citrus thrips *Scirtothrips aurantii* Faurè. In: Bedford, E.C.G., Van den Berg, M.A. and De Villiers, E.A. (eds) *Citrus Pests in the Republic of South Africa*. ARC Institute for Tropical and Subtropical Crops, Nelspruit, South Africa, pp. 164–183.

Gilligan, T.M., Brown, J.W. and Hoddle, M.S. (2011) A new avocado pest in Central America (Lepidoptera: Tortricidae) with a key to the Lepidoptera larvae threatening avocados in California. *Zootaxa* 3137, 31–45.

González-Fernández, J.J., De la Peña, F., Hormaza, J.I., Boyero, J.R., Vela, J.M., *et al.* (2009) Alternative food improves the combined effect of an omnivore and a predator on biological pest control. A case study in avocado orchards. *Bulletin of Entomological Research* 99, 433–444.

Gouse, H.H., Du Toit, D.M. and Vorster, A.M. (1998) Termites as pests in citrus orchards. In: Bedford, E.C.G., Van den Berg, M.A. and De Villiers, E.A. (eds) *Citrus Pests in the Republic of South Africa*. ARC Institute for Tropical and Subtropical Crops, Nelspruit, South Africa, pp. 85–86.

Grout, T.G. (1994) The distribution of phytoseiid mites (Acari: Phytoseiidae) on citrus in southern Africa and their possible value as predators of citrus thrips (Thysanoptera: Thripidae). *Experimental and Applied Acarology* 18, 61–71.

Grout, T.G. and Richards, G.I. (1992) *Euseius addoenensis addoensis*, an effective predator of citrus thrips, *Scirtothrips aurantii*, in the Eastern Cape Province of South Africa. *Experimental and Applied Acarology* 15, 1–13.

Grout, T.G. and Stephen, P.R. (1995) *Goetheana incerta* parasitizing citrus thrips in southern Africa. *Citrus Journal* 5, 30–32.

Grové, T. (2001) A soft scale *Saissetia* sp. In: Van den Berg, M.A., De Villiers, E.A. and Joubert, P.H. (eds) *Pests and Beneficial Arthropods of Tropical and Non-Citrus Subtropical Crops in South Africa*. ARC Institute of Tropical and Subtropical Crops, Nelspruit, South Africa, pp. 220–221.

Grové T. (2022) Progress towards an eco-friendly insect pest management approach in subtropical agro-ecosystems (South Africa). *CABI Agriculture and Bioscience* 3:44.

Grové, T., Joubert, P.H. and De Beer, M.S. (2009) Invasive fruit fly species in Africa and the possible threat to the subtropical fruit industry in South Africa. *Avoinfo, South Africa Avocado Growers' Association* 167, 9–12.

Gunn, D. (1921) The false codling moth (*Argyroploce leucotreta* Meyr.). *Science Bulletin, Department of Agriculture and Forestry, Union of South Africa* 21, 1–28.

Hadar, D. (1983) Reproductive behavior and control of the giant looper, *Boarmia (Ascotis) selenaria* Schiff. (Lepidoptera, Geometridae), a pest of avocado in Israel. MSc thesis, The Hebrew University of Jerusalem, Israel, p. 75. (In Hebrew with English summary.)

Hadar, D. (1993) Population dynamics of the pyriform scale, *Protopulvinaria pyriformis* Cockerell, and its natural enemies in avocado groves. PhD thesis, The Hebrew University of Jerusalem, Israel, p. 14. (In Hebrew with English summary.)

Hall, W.J. and Ford, W.K. (1933) Notes on some citrus insects in southern Rhodesia. *Mazoe Citrus Experimental Station Publication* 2a, 1–51.

Hamacek, E., Hall, E., Leach, P., Cavallaro, J., Wyatt, P., *et al.* (2005) Development of a non-chemical treatment system for avocado against Queensland fruit fly for interstate trade. In: *Proceedings of the New Zealand and Australia Avocado Growers' Conference 05*, 20–22 September 2005, Tauranga, New Zealand, p. 12.

Hartwig, E.K. (1966) The nest and control of *Odontotermes lactericius* (Haviland) (Termitidae: Isoptera). *South African Journal of Agricultural Science* 9, 407–418.

Hartwig, E.K. and de Lange, H.C. (1978) Observation on three coreids, *Carlisis wahlbergi* Stål, *Anoplocnemis curvipes* Fabr. and *Elasmopoda valga* L. *Technical Communication, Department of Agricultural Technical Services, Republic of South Africa* 141:12.

Hattingh, V. and Cilliers, C.J. (1998) Australian bug *Icerya purchasi* Maskell. In: Bedford, E.C.G., Van den Berg, M.A. and De Villiers, E.A. (eds) *Citrus Pests in the Republic of South Africa*. ARC–Institute for Tropical and Subtropical Crops, Nelspruit, South Africa, pp. 108–112.

Hattingh, V., Cilliers, C.J. and Bedford, E.C.G. (1998) Citrus mealybugs. In: Bedford, E.C.G., Van den Berg, M.A. and De Villiers, E.A. (eds) *Citrus Pests in the Republic of South Africa*. ARC Institute for Tropical and Subtropical Crops, Nelspruit, South Africa, pp. 112–120.

Heidemann, O. (1908) Two new species of North American Tingidae. *Proceedings of the Entomological Society of Washington* 10, 103–108.

Hennessey, M.K. and Schnell, R.J. (2001) Resistance of immature mango fruits to Caribbean fruit fly (Diptera: Tephritidae). *Florida Entomologist* 84, 318–319.

Hennessey, M.K, Knight Jr, R.J. and Schnell, R.J. (1995) Antibiosis to Caribbean fruit fly in avocado germplasm. *HortScience* 30, 1061–1062.

Henry, T.J., Peña, J.E., Long, D. and Acevedo, F. (2009) *Stethoconus praefectus* (Hemiptera: Miridae): first North American records of an old world predatory plant bug preying on avocado lace bug, *Pseudacysta perseae* (Hemiptera: Tingidae) in Florida. *Proceedings of the Entomological Society of Washington* 111, 98–105.

Hernandez, H.G., Johansen, R., Gazca, L., Equihua, A., Salinas, A., *et al.* (2000) Plagas del aguacate. In: Teliz, D. (ed.) *El Aguacate y su Manejo Integrado*. Ediciones Mundi-Prensa, Mexico City, Mexico, pp. 117–136.

Hill, D. (1975) *Agricultural Insect Pests of the Tropics and their Control*. Cambridge University Press, London, p. 516.

Hoddle, M.S. (2002a) Oviposition preferences of *Scirtothrips perseae* Nakahara (Thysanoptera: Thripidae) in Southern California avocado orchards. *Pan Pacific Entomologist* 78, 177–183.

Hoddle, M.S. (2002b) Developmental and reproductive biology of *Scirtothrips perseae* (Thysanoptera: Thripidae): a new avocado pest in California. *Bulletin of Entomological Research* 92, 279–285.

Hoddle, M.S. (2002c) Phenology of *Scirtothrips perseae* Nakahara (Thysanoptera: Thripidae) and associated natural enemies in Southern California avocado orchards. *California Avocado Society Yearbook* 86, 127–145.

Hoddle, M.S. (2003a) The effect of prey species and environmental complexity on the functional response of *Franklinothrips orizabensis*: a test of the fractal foraging model. *Ecological Entomology* 28, 309–318.

Hoddle, M.S. (2003b) Predation behaviors of *Franklinothrips orizabensis* (Thysanoptera: Aeolothripidae) towards *Scirtothrips perseae* and *Heliothrips haemorrhoidalis* (Thysanoptera: Thripidae). *Biological Control* 27, 323–328.

Hoddle, M.S. (2004) Invasions of leaf feeding arthropods: why are so many new pests attacking California-grown avocados? *California Avocado Society Yearbook* 87, 65–81.

Hoddle, M.S. (2006) Phenology, life tables, and reproductive biology of *Tetraleurodes perseae* (Hemiptera: Aleyrodidae) on California avocados. *Annals of the Entomological Society of America* 99, 553–559.
Hoddle, M.S. (2008) First record of *Asphondylia websteri* (Diptera: Cecidomyiidae) infesting Hass avocados. *Florida Entomologist* 91, 501–503.
Hoddle, M.S. (2009) The avocado seed moth, *Stenoma catenifer* Walsingham (Lepidoptera: Elachistidae). Available at: http://www.biocontrol.ucr.edu/Stenoma/Stenoma.html (accessed 6 August 2010).
Hoddle, M.S. (2010a) The biology and management of persea mite, *Oligonychus perseae* Tuttle, Baker, and Abatiello (Acari: Tetranychidae). Available at: http://www.biocontrol.ucr.edu/mite1.html (accessed 5 August 2010).
Hoddle, M.S. (2010b) *Cryptaspasma* sp. (Lepidoptera: Tortricidae: Olethreutinae). Available at: http://www.biocontrol.ucr.edu/Cryptaspasma/Cryptaspasma.html (accessed 6 August 2010).
Hoddle, M.S. (2010c) *Asphondylia websteri* (Diptera: Cecidomyiidae) attacking 'Hass' avocados. Available at: http://www.biocontrol.ucr.edu/Asphondylia/Asphondylia.html (accessed 6 August 2010).
Hoddle, M.S. and Hoddle, C.D. (2008a) Bioecology of *Stenoma catenifer* (Lepidoptera: Elachistidae) and associated larval parasitoids reared from Hass avocados in Guatemala. *Journal of Economic Entomology* 101, 692–698.
Hoddle, M.S. and Hoddle, C.D. (2008b) Survey results for Lepidoptera and associated parasitoids attacking Hass and non-Hass avocados in Guatemala. *Journal of Economic Entomology* 101, 692–698.
Hoddle, M.S. and Hoddle, C.D. (2008c) Aspects of the field ecology of *Stenoma catenifer* (Lepidoptera: Elachistidae) infesting Hass avocados in Guatemala. *Florida Entomologist* 91, 693–694.
Hoddle, M. and Hoddle, C.D. (2008d) Lepidoptera and associated parasitoids attacking Hass and non-Hass avocados in Guatemala. *Journal of Economic Entomology* 101, 1310–1316.
Hoddle, M.S. and Robinson, L. (2004) Evaluation of factors influencing augmentative releases of *Chrysoperla carnea*, for control of *Scirtothrips perseae* in California avoacado orchards. *Biological Control* 31, 268–275.
Hoddle, M.S., Aponte, O., Kerguelen, V. and Heraty, J. (1999) Biological control of *Oligonychus perseae* (Acari: Tetranychidae) on avocado: I. Evaluating release timings, recovery, and efficacy of six commercially available phytoseiids. *International Journal of Acarology* 25, 211–219.
Hoddle, M.S., Robinson, L. and Virzi, J. (2000) Biological control of *Oligonychus perseae* on avocado: III. Evaluating the efficacy of varying release rates and release frequency of *Neoseiulus californicus* (Acari: Phytoseiidae). *International Journal of Acarology* 26, 203–214.
Hoddle, M.S., Jones, J., Oishi, K., Morgan, D. and Robinson, L. (2001) Evaluation of diets for the development and reproduction of *Franklinothrips orizabensis* (Thysanoptera: Aeolothripidae). *Bulletin of Entomological Research* 91, 273–280.
Hoddle, M.S., Nakahara, S. and Phillips, P.A. (2002a) Foreign exploration for *Scirtothrips perseae* Nakahara (Thysanoptera: Thripidae) and associated natural enemies on avocado (*Persea americana* Miller). *Biological Control* 24, 251–265.
Hoddle, M.S., Morse, J.G., Phillips, P.A., Faber, B.A. and Jetter, K.M. (2002b) Avocado thrips: new challenge for growers. *California Agriculture* 56, 103–107.
Hoddle, M.S., Jetter, K.M. and Morse, J.G. (2003) The economic impact of *Scirtothrips perseae* Nakahara (Thysanoptera: Thripidae) on California avocado production. *Crop Protection* 22, 485–493.
Hoddle, M.S., Oevering, P., Phillips, P.A. and Faber, B.A. (2004) Evaluation of augmentative releases of *Franklinothrips orizabensis* for control of *Scirtothrips perseae* in California avocado orchards. *Biological Control* 30, 456–465.
Hoddle, M.S., Mound, L.A., Rugman-Jones, P.F. and Stouthamer, R. (2008) Synonomy of five *Scirtothrips* species (Thysanoptera : Thripidae) described from avocados (*Persea americana*) in Mexico. *Florida Entomologist* 91, 16–21.
Hoddle, M.S. and Mound, L.A. (2020) Surveys for thrips (Thysanoptera) associated with avocados (*Persea americana*) in Mbeya, Tanzania. *African Entomology* 28(1), 195–197.
Hoddle, M.S., Millar, J.G., Hoddle, C.D., Zou, Y. and McElfresh, J.S. (2009a) Synthesis and field evaluation of the sex pheromone of *Stenoma catenifer* (Lepidoptera: Elachistidae). *Journal of Economic Entomology* 102, 1460–1467.
Hoddle, M.S., Mottern, J.L. and Hoddle. C.D. (2009b) Observations on *Asphondylia websteri* (Diptera: Cecidomyiidae) infesting Hass avocados in Guatemala. *Florida Entomologist* 92, 673–674.
Hoddle, M.S., Millar, J.G., Hoddle, C.D., Zou, Y., McElfresh, J.S. and Lesch, S.M. (2011) Field optimization of the sex pheromone of *Stenoma catenifer* (Lepidoptera: Elachistidae): evaluation of lure types, trap height, male flight distances, and number of traps needed per avocado orchard for detection. *Bulletin of Entomological Research* 101, 145–152.

Hoddle, M.S., Hoddle, C.D., Milosavljević, I., Estrada-Venegas, E.G. and Equihua-Martínez, A. (2022) Assessment of age, gender, mating status, and size on single and repeat flight capabilities of *Heilipus lauri* Boheman (Coleoptera: Curculionidae). *Journal of Insect Science* 22(2), 3.

Hoffman, M.P., McDonough, L.M. and Bailey, J.B. (1983) Field test of the sex pheromone of *Amorbia cuneana* (Walsingham) (Lepidoptera: Tortricidae). *Environmental Entomology* 12, 1387–1390.

Hohmann, C.L., Dos Santos, W.J. and Menegium A.M. (2000) Avaliação de técnicas de manejo para o contrôle da broca-do-abacate, *Stenoma catenifer* Wals. (Lepidoptera: Oecophoridae). *Revista Brasileira Fruticultura Jaboticabal* 22, 359–363.

Hohmann, C.L., Meneguim, A.M., Andrade, E.A., Garcia de Novaes, T. and Zandona, C. (2003) The avocado fruit borer, *Stenoma catenifer* Wals. (Lepidoptera: Elachistidae): egg and damage distribution and parasitism. *Revista Brasileira Fruticultura Jaboticabal* 25, 432–435.

Holguin, C.M., Peña, J.E., Henry, T.J. and Acevedo, F. (2009) Biology of *Stethoconus praefectus* (Distant) (Heteroptera: Miridae), a newly established predator of the avocado lace bug, *Pseudacysta perseae* (Heteroptera: Tingidae), in Florida. *Florida Entomologist* 92, 54–57.

Holguin, C.M. and Mira, R.H. (2021) Report of *Astaena pygidialis* Kirsch (Coleoptera: Scarabaeidae), the main chafer beetle causing damage to avocado fruit and young leaves in Antioquia Department, Colombia. *Florida Entomologist* 104(1), 36–41.

Hollis, D. and Martin, J. (1997) Jumping plantlice (Hemiptera: Psylloidea) attacking avocado pear trees, *Persea americana*, in the New World, with a review of Lauraceae-feeding among psylloids. *Bulletin of Entomological Research* 87, 471–480.

Horticulture Innovation Australia (2018) *Avocado Fruit Quality Problem Solver, 1st edn.* Applied Horticultural Research, Eveleigh, Australia, p. 29. Available online: https://avocado.org.au/wp-content/uploads/2018/05/Avocado-Fruit-Quality-Problem-Solver.pdf (accessed on 15 April 2024).

Howard, F.W. and Oliver, A.D. (1985) Armored scale insects (Homoptera: Diaspididae) of Louisiana. *Bulletin, Louisiana Agricultural Experiment Station* 762. p. 125.

Howell, J.O. and Tippens, H.H. (1977) Descriptions of first instars of nominal type-species of eight disputed tribes. *Annals of the Entomological Society of America* 70, 119–135.

Huerta, A., Trujillo, J., Equihua, A. and Carrillo, J. (1990) Enemigos naturales y evaluacion Preliminar de dos nematodos para biocontrol de *Copturus aguacatae* (Coleoptera:Curculionidae) en Atlixco, Puebla, Mexico. *Agrociencia, Proteccion Vegetal* 1, 47–57.

Humeres, E.C. and Morse, J.G. (2005) Baseline susceptibility of persea mite (Acari: Tetranychidae) to abamectin and milbemectin in avocado groves in Southern California. *Experimental and Applied Acarology* 36, 51–59.

Humeres, E.C. and Morse, J.G. (2006) Resistance of avocado thrips (Thysanoptera: Thripidae) to sabadilla, a botanically derived bait. *Pest Management Science* 62, 886–889.

Hung, S.C. and Jong, T.M. (1997) Control of the fruitspotting bug on avocados. *Agriculture World* 163, 54–56. (In Chinese.)

Izhar, Y., Swirski, E., Ben-Yehuda, S., Wysoki, M. and Hadar, D. (1990a) Susceptibility of the avocado cultivar 'Ardit' to the greenhouse thrips, *Heliothrips haemorrhoidalis* (Bouché) (Thysanoptera: Thripidae) in Israel. *Alon Hanotea* 44, 1009–1014. (In Hebrew with English summary.)

Izhar, Y., Wysoki, M., Swirski, E. and Amitai, S. (1990b) The variegated caper bug, *Stenozygum colouratura* (Klug) (Rhynchota: Pentatomidae), and its damage to avocado and persimmon. *Hassadeh* 70, 1244–1245. (In Hebrew with English summary.)

Izhar, Y., Ben-Yehuda, S., Swirski, E., Wysoki, M. and Dagan, M. (1992) Occurrence of the black vine thrips, *Retithrips syriacus* Mayet, on avocado in Israel and trials in controlling the pest by insecticides of plant origin. *Alon Hanotea* 46, 523–529. (In Hebrew with English summary.)

Jadhav, L.D., Kadam, M.V., Ajri, D.S. and Pokharkar, R.N. (1980) Observations on the biology of leaf-footed plant bug *Leptoglossus membranaceus* Fabricius, a pest of pomegranate. *Bulletin of Entomology* 21, 79–82.

Jamieson, L.E. and Stevens, P.S. (2007) Development rates, longevity and fecundity of six-spotted mite (*Eotetranychus sexmaculatus*) at constant temperatures. *New Zealand Plant Protection* 60, 72–77.

Jang, E.B. (1996) Systems approach to quarantine security: postharvest application of sequential mortality in the Hawaiian grown 'Sharwil' avocado system. *Journal of Economic Entomology* 89, 950–956.

Jeppson, L.R., Keifer, H.H. and Baker, E.F. (1975) *Mites Injurious to Economic Plants*. University of California Press, Berkeley, California, p. 614.

Jessup, A. (1994) Quarantine disinfestaion of 'Hass' avocados against *Bactrocera tryoni* (Diptera: Tephritidae) with a hot fungicide dip followed by cold storage. *Journal of Economic Entomology* 87, 127–130.

Johansen, R. and Mojica-Guzman, A. (1996) Reconsideracion del concepto de depredador y parasitoide en Tisanopteros Mexicanos (Insecta) de interes en control biologico natural. *Folia Entomologica Mexicana* 97, 21–38.

Joubert, P.H. and Claassens, V.E. (1994) Stink bug damage on avocado fruit in the Nelspruit region. *South African Avocado Growers' Association Yearbook* 17, 106–108.

Joubert, P.H. and Du Toit, W.J. (1993) False codling moth damage to avocados. *Institute for Tropical and Subtropical Crops Information Bulletin* 252:7.

Joubert, P.H. and Neethling, C. (1994) Kernel damage to pecan nuts in the Nelspruit area of South Africa. *Journal of the Southern African Society for Horticultural Sciences* 4, 42–44.

Kamburov, S. (1987) The mango shield scale *Protopulvinaria mangiferae* Green (Homoptera: Coccidae), a new pest on mango (*Mangifera indica)*. *Citrus Journal* 635, 10–11.

Kerguelen, V. and Hoddle, M.S. (1999a) Measuring mite feeding damage on avocado leaves with automated image analysis software. *Florida Entomologist* 82, 119–122.

Kerguelen V. and Hoddle, M.S. (1999b) Biological control of *Oligonychus perseae* (Acari: Tetranychidae) on avocado: II. Evaluating the efficacy of *Galendromus helveolus* and *Neoseiulus californicus* (Acari: Phytoseiidae). *International Journal of Acarology* 25, 221–229.

Kerguelen, V. and Hoddle, M.S. (2000) Comparison of the susceptibility of several cultivars of avocado to the persea mite, *Oligonychus perseae* (Acari: Tetranychidae). *Scientia Horticulturae* 84, 101–114.

Kissinger, D.G. (1957) Description of a new *Copturus* pest of avocado from Mexico (Coleoptera: Curculionidae: Zygopinae). *Acta Zoologica Mexicana* 2, 1–8.

Kondo, T. (2010) Description of a new coccid (Hemiptera: Coccidae) on avocado (*Persea americana* Mill.) from Colombia, South America. *ZooKeys* 42, 37–45.

Krause, M., Nel, A. and Van Zyl, K. (1996) *A Guide to the Use of Pesticides and Fungicides in the Republic of South Africa*. Government Printer, National Department of Agriculture, Pretoria, p. 320.

Landry, B. and Roque-Albelo, L. (2003) Presence of *Stenoma catenifer* Walsingham (Lepidoptera, Elachistidae, Stenomatinae), the avocado seed moth in the Galápagos. *Noticias de Galápagos* 62, 15–16.

Lara, J.R., Rugman-Jones, P.F., Stouthamer, R. and Hoddle, M.S. (2017) Population genetics of *Oligonychus Perseae* (Acari: Tetranychidae) collected from avocados in Mexico and California. *Florida Entomologist* 100(3), 616–626. doi:10.1653/024.100.0320.

Lara, J.R. and Hoddle, M.S. (2015) Comparison and field validation of binomial sampling plans for *Oligonychus Perseae* (Acari: Tetranychidae) on Hass avocado in Southern California. *Journal of Economic Entomology* 108(4), 2074–2089.

Larral, P., Ripa, R., Montenegro, J. and Veliz, P. (2008) Greenhouse thrips, *Heliothrips haemorrhoidalis* (Bouché) In: Ripa, R. and Larral, P. (eds) *Manejo de plagas en paltos y cítricos.* Instituto de Investigaciones Agropecuarias, Ministerio de Agricultura, La Cruz, Chile, Colleccion de Libros INIA 23, pp. 207–219.

Lazaro-Castellanos, C., González-Hernández, H., Romero-Nápoles, J., Ortega-Arenas, L., Equihua-Martínez, A., and Ochoa-Ascencio, S. (2023) Armored scale insects (Hemiptera: Diaspididae) associated with avocados (*Persea americana*) in México's main production areas, with keys for identification and natural enemies. *Zootaxa*, 5357(2), 241–269. doi:10.11646/zootaxa.5357.2.4.

Lever, R.G.A.W. (1969) Pests of the coconut palm (*Pseudothe.raptus wayi*). *Food and Agricultural Organization* 77, 38–42.

Lindsay, K.R., Zalucki, M.P., Newton, I.R. and Furlong, M.J. (2019) Spatial and temporal dynamics of the native banana-spotting bug, *Amblypelta lutescens lutescens* (Hemiptera: Coreidae), in avocado crops in north Queensland, Australia. *Journal of Economic Entomology* 112(4), 1812–1820.

Link, D. and Link, F.M. (2008) Identificação de plantas hospedeiras da broca do abacate, *Stenoma catenifer* Walsingham (Lepidoptera: Elachistidae) no Rio Grande do Sul. *Neotropical Entomology* 37, 342–344.

Liquido, N., Chan Jr., H.T. and McQuate, G.T (1995) Hawaiian Tephritid fruit flies (Diptera); integrity of the infestation-free quarantine procedure for 'Sharwil' avocado. *Journal of Economic Entomology* 88, 85–96.

Londoño Zuluaga, M.E. (2014) Manejo integrado de *Monalonion velezangeli* en aguacate. Corp. Colombiana de Investigación Agrop. Available at: https://repository.agrosavia.co/bitstream/handle/20.500.12324/1931/Ver_Documento_1931.pdf?sequence=1&isAllowed=y (accessed 16 October 2024).

Londoño Zuluaga, M.E. (2020) *Insectos: Cucarrones marceños. En Actualizac tecnológica y buenas prácticas agrícolas cultivo de aguacate (pp. 228–284)*. Corp Colombiana de Investigación Agrop. Available at: https://repository.agrosavia.co/bitstream/handle/20.500.12324/33242/50483_1.pdf?sequence=1&i (accessed 16 October 2024).

Lounsbury, C.P. (1897) *Report of the Government Entomologist for the year 1896*. Cape of Good Hope, South Africa, p. 153.

Lounsbury, C.P. (1940) The pioneer period of economic entomology in South Africa. *Journal of the Entomological Society of Southern Africa* 3, 9–29.

Lourencao, A., Rossetto, C. and Soares, N. (1984) Ocorrencia de adultos de *Heilipus catagraphus* Germar 1824 (Coleoptera: Curculionidae) danificando frutos de abacateiro. *Bragrantia Campinas* 43, 245–253.

Louw, E., Shandukanu, E. and Mzimba, O. (2012) Determining the host status of 'Hass' avocado fruit to false codling moth. In: *Proceedings of the SAAGA Research Symposium Abstracts,* 16 February, Tzaneen, South Africa, p. 86.

Luppichini, P. and Ripa, R. (2008) Chilean termite, drywood termite. In: Ripa, R. and Larral, P. (eds) *Manejo de Plagas en Paltos y Citricos*. Instituto de Investigaciones Agropecuarias, Ministerio de Agricultura, La Cruz, Chile, Collection de Libros INIA 23, pp. 223–226.

Luppichini, P., Olivares, N. and Montenegro, J. (2013) Guía de Campo: Plagas del Palto y sus enemigos naturales. 118 p. *Boletín INIA* 239. ISSN 0717-4829. Available at: https://biblioteca.inia.cl/handle/20.500.14001/7578?-show=full (accessed 16 October 2024).

Mankin, R.W., Burman, H., Menocal, O. and Carrillo, D. (2018) Acoustic detection of *Mallodon dasystomus* (Coleoptera: Cerambycidae) in *Persea americana* (Laurales: Lauraceae) branch stumps. *Florida Entomologist* 101(2), 321–323,

Mansour, F., Wysoki, M. and Whitcomb, W.U. (1985) Spiders inhabiting avocado orchards and their role as natural enemies of *Boarmia selenaria* Schiff. (Lepidoptera:Geometridae) larvae in Israel. *Acta Oecologica* 6, 315–321.

Maoz, Y., Gal, S., Abrahams J., Gan-Mor, S., Coll, M. and Palevsky, E. (2009) Pollen provisioning enhances *Eusieus scutalis* (Phytoseiidae) populations and improves control of *Oligonychus perseae* (Tetranychidae). In: Mason, P.G., Gillespie, D.R. and Vincent, C. (eds) *Proceedings of the 3rd International Symposium on the Biological Control of Arthropods*, Christchurch, New Zealand, pp. 339–346. USDA Forest Service Publication FHTET-2008-06.

Maoz, Y., Gal, S., Zilberstein, M., Izhar, Y., Alchnatis, V., Coll, M. and Palevsky, E. (2011) Determining an economic injury level for persea mite *Oligonychus perseae*, a new pest of avocado in Israel. *Entomologia Experimentalis et Applicata* 138, 110–116.

Marin, J.A. and Mocarro, G.E. (2019) Fluctuacion poblacional de los principales insectos fitofagos en el cultivo del palto (*Persea americana* Mill) Var. 'Hass' en la etapa de fructificacion, Distrito de Olmos (Lambayeque). Tesis de Grado Universidad Nacional Pedro Ruiz Gallo, Olmos, Peru.

Martinez, B.R. (1989) El cultivo del aguacate en Mexico, sus problemas fitosanitarios, soluciones y perspectivas. *Fruticultura de Michoacán* 2, 7–13.

Matsuna, J.N. and Silva, J. (2020) Avocado lace bug *Pseudacysta perseae* (Heidemann) (Hemiptera: Tingidae). Hawaii Department of Agriculture New Pest Advisory No. 20-01. https://hdoa.hawaii.gov/ pi/ files/2020/04/ Avocado-lace-bug-NPA-3-31-20-FINAL.pdf (accessed 25 November 2024).

McDonough, L.M., Hoffman, M., Bierl-Leonhardt, B., Smithisler, C., Bailey, J. and Davis, H. (1982) Sex pheromone of the avocado pest, *Amorbia cuneana* (Walsingham) (Lepidoptera: Tortricidae): structure and synthesis. *Journal of Chemical Ecology* 8, 255–265.

McMurtry, J.A. (1985a) Avocado. In: Helle, W. and Sebellis, M.W. (eds) *Spider Mites, Their Biology, Natural Enemies and Control. World Crop Pests, Volume* IB. Elsevier, Amsterdam, The Netherlands, pp. 327–332.

McMurtry, J.A. (1985b) Studies on a *Euseius* species complex in avocado in Mexico, Central America, with a description of new species (Acari: Phytoseiidae). *Acarologia* 26, 107–106.

McMurtry, J.A. and Johnson, H. (1966) An ecological study of the spider mite *Oligonychus punicae* Hirst and its natural enemies. *Hilgardia* 37, 363–402.

McMurtry, J.A., Johnson, H.G. and Newberger, S.J. (1991) Imported parasite of greenhouse thrips established on California avocado. *California Agriculture* 45, 31–32.

McMurtry, J.A. (1992) The role of exotic natural enemies in the biological control of insect and mite pests of avocado in California. In: *Proceedings of the Second World Avocado Congress, 21–26 April 1991, Orange, California, USA*, pp. 247–252.

Mead, F. and Peña, J.E. (1991) Avocado lace bug, *Pseudacysta perseae* (Hemiptera: Tingidae). *Florida Department Agriculture and Consumer Services, Division of Plant Industry Entomology Circular* 346, p. 4.

Medina, J.C., Bleinroth, E.W., Sojiro, J. and Leite, W. (1978) *Frutas Tropicais 1. Abacate.* Istituto de Tecnologia de Alimentos, Campinas, Brazil, pp. 63–68.

Medina-Gaud, S., Segarra, A. and Franqui, R. (1991) The avocado lacewing bug *Pseudacysta perseae* (Heidemann)(Hemiptera: Tingidae). *Journal of Agriculture, University of Puerto Rico* 75, 185–188.

Meléndez-González, C. and Espinosa-García, F.J. (2018) Metabolic profiling of *Persea americana* cv. Hass branch volatiles reveals seasonal chemical changes associated to the avocado branch borer, *Copturus aguacatae. Scientia Horticulturae* 240, 116–124.

Mendel, Z., Protasov, A., Sharon, M., Zveibil, A., Yehuda, S.B., *et al.* (2012) An Asian ambrosia beetle *Euwallacea fornicatus* and its novel symbiotic fungus *Fusarium* sp. pose a serious threat to the Israeli avocado industry. *Phytoparasitica* 40, 235–238.

Mendel, Z., Freeman, S., Hulcr, J., Recht, E., Zagron, G., *et al.* (2023) Black twig borer and other ambrosia and bark beetles in Israel. *Phytoparasitica* 51, 1–9.

Mendel, Z. and Blumberg, D. (1991) Colonization trials with *Cryptochetum iceryae* and *Rodolia iceryae* for improvement of biological control of *Icerya purchasi* in Israel. *Biological Control* 1, 68–74.

Mendel, Z., Protasov, A., Freeman, S., Ben-Yehuda, S., Noy, M. and Hamburger, M. (2010) Ambrosia beetle *Euwallacea fornicatus* a new pest in avocado orchards. *Pamphlet, Ministry of Agriculture and Rural Development*, Bet Dagan, Israel, p. 2.

Mendez-Villa, M. (1961) Medidas practicas para combater los barrenadores del huesco del aguacate. *Fitofilo* 14, 4–7.

Menocal, O.G., Kendra, P.E., Padilla, A., Cardoso Chagas, P., Alves Chagas, E., Crane, J.H and Carrillo, D. (2022) Influence of canopy cover and meteorological factors on the abundance of bark and ambrosia beetles in avocado orchards affected by laurel wilt. *Agronomy* 12(3), 547. doi:10.3390/agronomy12030547.

Merrill, G.B. (1953) A revision of the scale-insects of Florida. *Bulletin of Standard Plant Breeding, Florida* 1:143.

Meyer, M.K.P. (1987) African Tetranychidae (Acari: Prostigmata), with reference to the world genera. *South African Department of Agriculture and Water Supply, Entomology Memoir* 69:175.

Milano, P., Filho, E.B., Parra, J.R.P., Oda, M.L. and Cônsoli, F.L. (2010) Efieot de alimentação de fase adulta na reprodução e longevidade de species de Noctuidae, Crambidae, Tortricidae e Elachistidae. *Neotropical Entomology* 39, 172–180.

Millar, I.M. (1994) A Catalogue of the Aphids (Homoptera: Aphididae) of Sub-Saharan Africa. *Plant Protection Handbook* 4:130.

Millar, J.G., Hoddle, M.S., McElfresh, J.S., Zou, Y. and Hoddle. C.D. (2008) (9Z)-9, 13-tetradecadiene-11-ynal, the sex pheromone of the avocado seed moth, *Stenoma catenifer. Tetrahedron Letters* 49, 4820–4823.

Milne, D.L. (1973) Bagrada bug, *Bagrada hilaris*. In: Milne, D.L. (ed.) Insect pests of subtropical fruit. *Republic of South Africa, Department of Agricultural Technical Services, Entomological Memoir* 33, p. 42.

Montserrat, M., Guzmán, C., Sahún, R.M., Belda, J.E. and Hormaza, J.I. (2013a) Pollen supply promotes, but high temperatures demote, predatory mite abundance in avocado orchards. *Agriculture, Ecosystems & Environment* 164, 155–161.

Montserrat, M., Sahún, R.M. and Guzmán, C. (2013b) Can climate change jeopardize predator control of invasive herbivore species? A case study in avocado agro-ecosystems in Spain. *Experimental and Applied Acarology* 59(1–2), 27–42.

Moreno-Gaviria, J., Díaz-Grisales, V. and Carabalí-Muñoz, A. (2023) Biology of *Copturomimus hustachei* Kissinger, 1957 (Coleoptera: Curculionidae), a stem and branch borer of avocado (*Persea americana*) in Colombia. *Revista Colombiana de Entomología*, 49(1), e12215. Available at: https://doi.org/10.25100/socolen.v49i1.12215 (accessed 25 November 2024).

Morris, D.C. and Mound L.A. (2004) Molecular relationships between populations of South African citrus thrips (*Scirtothrips aurantii* Faure) in South Africa and Queensland, Australia. *Australian Journal of Entomology* 43, 353–358.

Morse, J.G., Hoddle, M.S. and Urena, A.A. (2000) Persea mite pesticide efficacy trial. *California Avocado Society Yearbook* 84, 127–137.

Morse, J., Rugman-Jones, P., Watson, G., Robinson, L., Bi, T. and Stouthamer, R. (2009) High levels of exotic armored scales on imported avocados raise concerns regarding USDA-APHIS' Phytosanitary Risk Assessment. *Journal of Economic Entomology* 102, 855–867.

Mound, L.A. and Monteiro, R.C. (1997) A review of the genus Heliothrips (Thysanoptera; Thripidae), with a new sister-species of the Greenhouse Thrips from South Eastern Brazil. *Journal of the New York Entomological Society* 105(3/4), 154–160.

Mound, L. and Hoddle, M.S. (2016) The *Scirtothrips perseae* species-group (Thysanoptera), with one new species from avocado, *Persea americana. Zootaxa* 4079(3), 388–392.

Muñiz Velez, R. (1958) La oruga barrenadora del hueso del aguacate, *Stenoma catenifer* Walsingham (Lepidoptera: Stenomatidae). *Memoria del Primer Congreso Nacional de Entomologia y Fitopatologia*, pp. 170–174.

Muñiz, R. (1960) *Copturus aguacatae* Kissinger: plaga del agucatero (*Persea gratissima* Gaertn.) en Mexico. *Fitofilo* 12, 42–48.

Muñiz, R. and Barrera, A. (1958) Clave para los barrenadores de las ramas del aguacatero (*Persea gratissima* Gaertn.) en la America Tropical y Subtropical (Coleoptera: Curculionidae: Zygopinae). *Resumenes, XXI Congreso Nacional de Entomologia, 16–19 Marzo, 1986, Monterrey, Nuevo Leon, Mexico*, pp. 61–62.

Munro, H.K. and Fouché, F.A. (1936) A list of the scale insects and mealybugs (Coccidae) and their host plants in South Africa. *Department of Agriculture and Forestry, Union of South Africa, Bulletin* 158, 1–104.

Murray, D.A.H. (1976) Insect pests of passion fruit. *Queensland Agricultural Journal* 102, 145–151.

Nakahara, S. (1997) *Scirtothrips perseae* (Thysanoptera:Thripidae), a new species infesting avocado in southern California. *Insecta Mundi* 11, 189–192.

Nava, D.E. and Parra, J.R.P. (2005) Biologia de *Stenoma catenifer* Walsingham (Lepidoptera: Elachistidae) em dieta natural e artificial e estabelecimento de um sistema de criação. *Neotropical Entomology* 34, 751–759.

Nava, D.E., Parra, J.R.P., Costa, V.A., Guerra, T.M. and Cônsoli, F.L. (2005a) Population dynamics of *Stenoma catenifer* (Lepidoptera: Elachistidae) and related larval parasitoids in Minas Gerais, Brazil. *Florida Entomologist* 88, 441–446.

Nava, D.E., de Lara Haddad, M. and Parra, J.R.P. (2005b) Exigências térmicas, estimative de número de geroções de *Stenoma catenifer* e comprovação do modelo em campo. *Pesquisa Agropecuaria Brasileira* 40, 961–967.

Nava, D.E., Parra, J.R.P., Bento, J.M.A., Diez-Rodriguez, G.I. and Haddad, M.L. (2006) Distribuição vertical, danos e controle cultural de *Stenoma catenifer* Walsingham (Lepidoptera: Elachistidae) em pomar de abacate. *Neotropical Entomology* 35, 516–522.

Nava, D.E., Takahashi, K.M. and Para, J.R.P. (2007) Linhagens de *Trichogramma* e *Trichogrammatoidea* para controle de *Stenoma catenifer*. *Pesquisa Agropecuaria Brasileira Brasília* 42, 9–16.

Nesbitt, M,. Stein, L.A. and Kamas, J. (2013) *Texas Fruit and Nut Production: Avocados.* AgriLife Extension, Texas A&M System, College Station, TX. Available at: https://agrilifelearn.tamu.edu/s/product/texas-fruit-and-nut-production-avocados/01t4x000004OfhxAAC.

Newton, P.J. (1988) Inversely density-dependent egg parasitism in patchy distributions of the citrus pest *Cryptophlebia leucotreta* (Lepidoptera: Tortricidae) and its agricultural efficiency. *Journal of Applied Ecology* 25, 145–162.

Newton, P.J. (1998) False codling moth, *Cryptophlebia leucotreta* Meyrick. In: Bedford, E.C.G., Van den Berg, M.A. and De Villiers, E.A. (eds) *Citrus Pests in the Republic of South Africa*. Department of Agricultural Technical Services Science Bulletin No. 391, pp. 192–200.

Newton, P.J. and Anderson, T. (1985) Crop loss due to false codling moth, *Cryptophlebia leucotreta* Meyr. (Lepidoptera: Tortricidae), and other factors, on different navel orange selections at Citrusdal. *Citrus and Subtropical Fruit Research Institute, Information Bulletin* 159, 22–24.

Newton, P.J. and Catling, H.D. (1998) Citrus leaf roller *Archips* (= *Cacoecia*) *occidentalis* Walshingham.In: Bedford, E.C.G., Van den Berg, M.A. and De Villiers, E.A. (eds) *Citrus Pests in the Republic of South Africa*. ARC Institute for Tropical and Subtropical Crops, Nelspruit, South Africa, pp. 207–210.

Newton, P.J. and Mastro, V.C. (1989) Field evaluations of commercially available traps and synthetic sex pheromone lures of the false codling moth, *Cryptophlebia leucotreta* Meyr. (Lepidoptera: Tortricidae). *Tropical Pest Management* 35, 100–104.

Newton, P.J., Thomas, C.D., Mastro, V.C. and Schwalbe, C.P. (1993) Improved two-component blend of the synthetic female sex pheromone of *Cryptophlebia leucotreta*, and identification of an attractant for *C. peltastica*. *Entomologia Experimentalis et Applicata* 66, 75–82.

Ngoan, N., Wilkinson, R., Short, D., Moses, C. and Mangold, J. (1976) Biology of an introduced ambrosia beetle *Xylosandrus compactus*, in Florida. *Annals of the Entomological Society of America* 69, 872–876.

Nuñez, S.E. (2008) *Plagas de paltos y cítricos en Perú* [en línea]. Quillota: Version-Producciones Graficas Ltda. Colección Libros INIA - Instituto de Investigaciones Agropecuarias. no. 23. Disponible en: https://hdl.handle.net/20.500.14001/3525 (Consultado: 12 febrero 2024).

Oatman, E.R. and Platner, G. (1985) Biological control of two avocado pests. *California Agriculture* 39, 21–23.

O'Brien, C. and Wibmer, G.J. (1982) Annotated checklist of weevils (Curculionidae sensu lato) of North America and the West Indies (Coleoptera: Curculionidae). *Memoirs of the American Entomological Institute* 34, 1–382.

O'Donnell, K., Libeskind-Hadas, R., Hulcr, J., Bateman, C., Kasson, M.T., *et al.* (2016) Invasive Asian Fusarium–Euwallacea ambrosia beetle mutualists pose a serious threat to forests, urban landscapes and the avocado industry. *Phytoparasitica* 44, 435–442.

Oevering, P., Phillips, P.A. and Faber, B.A. (2002) Natural enemies associated with avocado thrips in Ventura County avocado groves: results of a pilot study and year one of a three year study. *California Avocado Society Yearbook* 86, 105–126.

Omer-Cooper, J. (1939) *Remarks on False Codling Moth.* Rhodes University, Grahamstown. Mimeograph pp. 1–17.

Ondieki, J.J. (1975) Diseases and pests of passion fruit in Kenya. *Acta Horticulturae* 49, 291–293.

Ortega-Licona, A., Equihua-Martínez, A., Castañeda-Vildózola, Á., Estrada-Venegas, E.G. and Sánchez-Escudero, J. (2019) New contributions to the knowledge of Heilipus lauri Barber (Coleoptera: Curculionidae) in Mexico. *Proceedings of the Entomological Society of Washington* 121(3), 484–490.

Owens, D., Cruz, L.F., Montgomery, W.S., Narvaez, T.I., Schnell, E.Q., *et al.* (2018) Host range expansion and increasing damage potential of *Euwallacea* nr. *fornicatus* (Coleoptera: Curculionidae) in Florida. *Florida Entomologist* 101(2), 229–236.

Parco-Quinchori, J., Valverde-Rodriguez, A., Cornejo, A., Briceño, H., Barrionuevo, L. and Romero, J. (2023). Efficiency of entomopathogenic bacteria and fungi on *Oligonychus yothersi* in vitro and on *Persea americana* Mill. plants. Revista de la Facultad de Agronomía, 40(4), e234033. doi:10.47280/RevFacAgron(LUZ).v40.n4.02

Peña, J.E. (2003) Pests of avocado in Florida. In: *Proceedings of the V Congreso Mundial del Aguacate, 19–24 October 2003, Granada-Malaga, Spain, Acta*, 3, pp. 487–494.

Peña, J.E. and Denmark, H. (1996) An eriophyid *Tegolophus perseaflorae* (Acari: Eriophyidae) new to Florida and the USA. *Florida Entomologist* 79, 74–76.

Peña, J.E. and Johnson, F. (1999) Insect management of avocados. Insect Management Guide, Department of Entomology and Nematology, Florida Cooperative Extension Service, University of Florida. Available at: http://edis.ifas.ufl.edu/Body_IG068 (accessed 5 March 2012).

Peña, J.E., Baranowski, R. and Nadel, H. (1996) Pest management of tropical fruit trees in Florida. In: Rosen, D., Bennett, F. and Capinera, J. (eds) *Pest Management in the Subtropics-Integrated Pest Management-A Florida Perspective.* Intercept, Andover, UK, pp. 349–370.

Peña, J.E., Sunhari, S., Hunsberger, A., Duncan, R. and Schaffer, B. (1998) Monitoring, damage, natural enemies and control of avocado lacebug, *Pseudacysta perseae* (Hemiptera: Tingidae). *Proceedings of the Florida State Horticultural Society* 111, 330–334.

Peña, J.E., Glenn, H. and Baranowski, R.M. (2003) Dynamics and sampling of mirids (Hemiptera: Miridae) in avocado in Florida. In: Proceedings of the V Congreso Mundial del Aguacate, 19–24 October 2003, Granada, Malaga, Spain, Acta, 3, pp. 495–501.

Peña, J.E., Duncan, R., Roltsch, W., Gagné, R. and Agudelo, F. (2007) Natural enemies of the avocado lace bug, *Pseudacysta perseae* (Heteroptera: Tingidae) in Florida, USA. In: *Proceeding of the VI World Avocado Congress*, Viña del Mar, Chile. Available at: http://www.avocadosource.com/wac6/en/extenso/2a–48.pdf (accessed 9 March 2012).

Peña, J.E., Triapitsyn, S.V., Long, D., Evans, G. and Roltsch, W. (2009) First record of *Erythmelus klopomor* (Hymenoptera: Mymaridae) as a parasitoid of the avocado Lace bug, *Pseudacysta perseae* (Heteroptera:Tingidae). *Florida Entomologist* 92, 394–395.

Peña, J.E., Duncan, R., Crane, J., Capinera, J., Mclean, S., *et al.* (2011) Chemical control of the redbay ambrosia beetle, *Xyleborus glabratus. Florida Entomologist* 94, 882–896.

Peña, J.E., Duncan, R., Roltsch, W. and Carrillo, D. (2012) Mortality factors of the avocado laceubug, *Pseudacysta perseae* (Heteroptera:Tingidae) in Florida. *Florida Entomologist* 95, 179–182.

Peña, J.E., Weihman, S.W., McLean, S., Cave, R.D., Carrillo, D., *et al.* (2015) Predators and parasitoids associated with Scolytinae in Persea species (Laurales: Lauraceae) and other Lauraceae in Florida and Taiwan. *Florida Entomologist* 98(3), 903–911.

Pérez-Artiles, L., Hernández-Espinosa, D., Rodríguez-Tapia, J.L., Borges-Soto, M. and Fernández-Argudín, M. (2020) First record of the *Aleurovitreus insignis* Bondar (Hemiptera: Aleyrodidae) on avocado (Persea americana Mill.) in Cuba. *Ctro. Agr.* 47(2): 89–94.

Perez-de la Cruz, M., Equihua, A., Romero-Napoles, J., Sanchez, S., Garcia-Lopez, E. and Bravo-Mojica, H. (2009) Escolitidos (Coleoptera: Scolytidae) asociados al agroecosystema cacao en Tabasco, Mexico. *Neotropical Entomology* 38, 602–609.

Phillips, P.A., Bekey, R.S. and Blair Bailey J. (1992) An effective strategy for the management of Greenhouse Thrips in 'Hass' avocado orchards. *California Avocado Society Yearbook* 76, 99–107.

Pinedo-Escatel, J.A. and Dmitriev, D. (2019) Redescription of the Dikraneurini leafhopper *Dikrella mella* Ruppel & DeLong, 1952 (Hemiptera, Cicadellidae) with a synoptic checklist of leafhoppers on avocado trees in Mexico. *ZooKeys* 857, 17–27.doi:10.3897/zookeys.857.33910.

Pinhey, E.C.G. (1975) *Moths of Southern Africa*. Tafelberg Publishers Ltd, Cape Town, South Africa, p. 273.

PPRI (Plant Protection Research Institute) (1970) Check list of insects on forest trees and shrubs in South Africa. *Department of Agricultural Technical Services, South Africa, Entomological Memoir* 21:81.

Pretorius, N.H.S. (1971) Die biologie en morfologie van die sitrusbladspringer *Penthimia bella* Stål (Hemiptera). Unpublished M.Sc. thesis, University of Pretoria, South Africa, p. 118. (In Afrikaans.)

Prinsloo, G.L. (1984) An illustrated guide to the parasitic wasps associated with citrus pests in the Republic of South Africa. *Republic of South Africa, Department of Agriculture Science Bulletin* No. 402, 1–119.

Prinsloo, G.L. and Neser, O.C. (1994) The southern African species of *Aphelinus* Dalman (Hymenoptera: Aphelinidae), parasitoids of aphids (Homoptera: Aphidoidea). *Journal of African Zoology* 108, 143–162.

Quayle, H.J. (1938) *Insects of Citrus and Other Subtropical Fruits*. Comstock Publishing Company, Inc., Ithaca, New York, p. 583.

Quimio, A.J. and Cayetano, C. (1985) Control of the avocado bark borer (*Aegeria* sp.) with Chlorpyrifos and methomyl (Philippines). In: *Proceedings of the Annual Convention of the Pest Control Council of the Philippines*, La Trinidad, Benguet, Philippines, p. 1.

Rabaglia, R.J., Dole, S.A. and Cognato, A.I. (2006) Review of American *Xyleborina* (Coleoptera: Curculionidae: Scolytinae) occurring north of Mexico, with an illustrated key. *Annals of the Entomological Society of America* 99, 1034–1056.

Ramirez, M., Otero-Colina, G., Lopez-Collado, J. and Resendiz, B. (1993) Constantes termicas y tablas de vida de *Oligonychus perseae* (Trombidiformes: Tetranychidae). *Agrociencia Serie Proteccion Vegetal* 4, 67–80.

Ramirez-Ponce, A., Allende, J. and Moron, M. (2009) Fauna de coleopteros lamelicomos de Santiago Xiacui, Sierra Norte, Oaxaca. *Acta Zoologica Mexicana* 25, 323–343.

Read, J.S., Warren, F.L. and Hewitt, P.H. (1968) Identification of the sex pheromone of the false codling moth (*Argyroploce leucotreta*). *Chemical Communications,* 792–793.

Read, J.S., Hewitt, P.H., Warren, F.L. and Myburgh, A.C. (1974) Isolation of the sex pheromone of the moth *Argyroploce leucotreta*. *Journal of Insect Physiology* 20, 441–450.

Reis, D.S., Machado, L.G., Silva, E.M. and Fernandes, F.L. (2018) First report of *Heilus freyreissi* (Coleoptera: Curculionidae) attacking avocado and associated with *Colletotricum* sp. in Brazil. *Florida Entomologist* 101(1), 317–320.

Reyes, J.C. and Salgado, M. (1994) Tolerancia de algunas selecciones de agucate al Attaque de araña roja (*O. punicae)* y thrips (*Scirtothrips* sp). *Memorias, Centro de Investigaciones Cientificas y Tecnologicas del Aguacate en el Estado de Mexico, Fundacion Salvador Sanchez Colin*, Coatepec Harinas, Mexico, pp. 175–178.

Reyes, J.C., Rubi, M. and Aguilar, J. (1995) Manejo organico en el cultivo del aguacate. *Memorias Fundacion Salvador Sanchez Colin*, Coatepec Harinas, Mexico, pp. 29–37.

Rhainds, M. and Cabrera–La Rosa, J.C. (2010) Oiketicus kirbyi (Lepidoptera, Psychidae), a key pest in Peruvian orchards of avocado. *International Journal of Pest Management* 56(2), 103–107.

Rioja, T., Tello, V., Zarzar, M., Cardemil, A. and Ceballos, R. (2019) Avocado 'Hass' leaf age affects life table parameters of *Oligonychus yothersi* (McGregor)(Acari: Tetranychidae) under laboratory conditions. *Chilean Journal of Agricultural Research* 79(4), 557–564.

Rioja, T., Ceballos, R., Holuigue, L. and Vargas, R. (2016) Different population densities and continuous feeding by Oligonychus Yothersi (McGregor)(Acari: Tetranychidae) affect the emissions of herbivore-induced plant volatiles on avocado (*Persea Americana* Mill. Cv. Hass) shoots under semi-field conditions. *International Journal of Acarology* 42(6), 310–318.

Rioja, T., Ceballos, R., Fernandez, N. and Curkovic, T. (2021) Autumnal recruitment of mite's predators by herbivore-induced plant volatiles in avocado (*Persea americana* Mill. Cv. Hass) infested with *Oligonychus yothersi* (Acari: Tetranychidae). *International Journal of Acarology* 47(4), 352–360. doi:10.1080/01647954.2021.1913220.

Ripa, R. (2008a) Fruit tree weevil *Naupactus xanthographus* (Germar) In: Ripa, R. and Larral, P. (eds) *Manejo de Plagas en Paltos y Citricos*. Instituto de Investigaciones Agropecuarias, Ministerio de Agricultura, La Cruz, Chile, Collection de Libros INIA 23, pp. 155–158.

Ripa, R. (2008b) Mariposa del cachito, *Arctopoda maculosa* Lepidoptera: Oecophoridae. In: Ripa, R. and Larral, P. (eds) *Manejo de Plagas en Paltos y Citricos*. Instituto de Investigaciones Agropecuarias, Ministerio de Agricultura, La Cruz, Chile, Colleccion de Libros INIA 23, pp. 235–236.

Ripa, R. and Larral, P. (eds) (2008) *Manejo de Plagas en Paltos y Citricos*. Colección Libros, INIA, 23, Instituto de Investigaciones Agropecuarias, Ministerio de Agricultura, Chile, p. 399.

Ripa, R. and Luppichini, P. (2008a) Fuller rose weevil, *Pantomorus cervinus* (Boheman) In: Ripa, R. and Larral, P. (eds) *Manejo de Plagas en Paltos y Citricos*. Instituto de Investigaciones Agropecuarias, Ministerio de Agricultura, La Cruz, Chile, Collecion de Libros INIA 23, pp. 158–162.

Ripa, R. and Luppichini, P. (2008b) Subterranean termite *Reticulitermes flavipes* Kollar In: Ripa, R. and Larral, P. (eds) *Manejo de Plagas en Paltos y Cítricos*. Instituto de Investigaciones Agropecuarias, Ministerio de Agricultura, La Cruz, Chile, Collection de Libros INIA 23, pp. 224–226.

Ripa, R., Larral, P. and Luppichini, P. (2008a) Greenhouse whitefly, *Trialeurodes vaporariorum* (Westwood) In: Ripa, R. and Larral, P. (eds) *Manejo de Plagas en Paltos y Cítricos*. Instituto de Investigaciones Agropecuarias, Ministerio de Agricultura, La Cruz, Chile, Colleccion de Libros INIA 23, pp. 120–121.

Ripa, R., Larral, P., Rojas, S. and Rodrigues, F. (2008b) Black scale, olive black scale, *Saissetia oleae* (Olivier) In: Ripa, R. and Larral, P. (eds) *Manejo de Plagas en Paltos y Cítricos*. Instituto de Investigaciones Agropecuarias, Ministerio de Agricultura, La Cruz, Chile, Colleccion de Libros INIA 23, pp. 135–143.

Ripa, R., Larral, P. and Rojas, S. (2008c) Pyriform scale, *Protopulvinaria pyriformis* (Cockerell) In: Ripa, R. and Larral, P. (eds) *Manejo de Plagas en Paltos y Cítricos*. Instituto de Investigaciones Agropecuarias, Ministerio de Agricultura, La Cruz, Chile, Colleccion de Libros INIA 23, pp. 135–143.

Ripa, R., Larral, P., Luppichini, P., Guajardo, V. and Rojas, S. (2008d) Longtailed mealybug *Pseudococcus longispinus* (Targioni and Tozzetti) In: Ripa, R. and Larral, P. (eds) *Manejo de Plagas en Paltos y Cítricos*. Instituto de Investigaciones Agropecuarias, Ministerio de Agricultura, La Cruz, Chile, Colleccion de Libros INIA 23, pp. 180–191.

Ripley, L.B., Hepburn, G.A. and Dick, J. (1939) Mass breeding of the false codling moth, *Argyroploce leucotreta* Meyrick, in artificial media. *Plant Industry Series No. 53. Entomology. Department of Agriculture and Forestry, Union of South Africa, Science Bulletin* 207, 1–18.

Rivnay, E. (1935) Ecological studies of the greenhouse thrips *Heliothrips haemorrhoidalis* in Palestine. *Bulletin of Entomological Research* 25, 267–278.

Robertson, C.M. and De Villiers, E.A. (1986) Parasites of avocado pest bite the dust. *Citrus and Subtropical Fruit Research Institute, Information Bulletin* 168, 9.

Romero-Frías, A.A., Sinuco, D.C. and Bento, J.M.S. (2019) Male-specific volatiles released by the big avocado seed weevil *Heilipus lauri* Boheman (Coleoptera: Curculionidae). *Journal of the Brazilian Chemical Society* 30, 158–163.

Romero, L.M.M. and Ravelo, V.H.G. (2020) The avocado lace bug: *Pseudacysta perseae* (Heid.)(Hemiptera: Tingidae). Bioecology and biological control in Cuba. *Centro Agricola* 47(Suppl.1), 59–62.

Rose, M. and DeBach, P. (1992) Biological control of *Parabemisia myricae* (Kuwana) (Homoptera: Aleyrodidae) in California. *Israel Journal of Entomology* 25–26, 73–95.

Rose, M. and Zolnerowich, G. (2004) *Eretmocerus persea* n.sp., (Hymenoptera: Chalcidoidea:Aphelinidae) reared from *Tetraleourodes perseae* Nakahara (Homoptera: Aleyrodidae:Aleyrodinae) living on avocado in Mexico. *Vedalia* 11, 45–52.

Rubio, J.D., Posada, F., Osorio, O.I., Vallejo, L. and Lopez, J.C. (2009) First record of *Heilipus elegans* Guerin-Meneville (Coleoptera: Curculionidae) attacking the avocado tree stem in Colombia. *Revista Universidad Ciencias Aplicadas y Ambientales Actualidad y Divulgacion Cientifica* 12, 59–68.

Ruelle, J.E. (1985) Order Isoptera (Termites). In: Scholtz, C.H. and Holm, E. (eds) *Insects of Southern Africa*. Butterworths, Durban, South Africa, pp. 53–61.

Rugman-Jones, P.F., Hoddle, M.S. and Stouthamer, R. (2008) Population genetics of *Scirtothrips perseae*: tracing the origin of a recently introduced exotic pest of Californian avocado orchards, using mitochondrial and microsatellite DNA markers. *Entomologia Experimentalis et Applicata* 124, 101–115.

Rugman-Jones, P.F., Morse, J. and Stouthamer, R. (2009) Rapid molecular identification of armored scale insects (Hemiptera: Diaspididae) on Mexican 'Hass'. *Journal of Economic Entomology* 102, 1948–1953.

Rugman-Jones, P.F., Andersen, J., Morse, J., Normark, B. and Stouthamer, R. (2010) Molecular phylogenetic placement of the recently described armored scale insect *Abgrallaspis aguacatae* and several congeners (Hemiptera: Diaspididae). *Annals of the Entomological Society of America* 103, 30–38.

Rugman-Jones, P.F., Hoddle, M.S., Phillips, P.A., Jeong, G. and Stouthamer, R. (2012) Strong genetic structure among populations of the invasive avocado pest *Pseudacysta perseae* (Heidemann)(Hemiptera: Tingidae) reveals the source of introduced populations. *Biological Invasions* 14, 1079–1100.

Rust, E.W. (1918) *Anastrepha fraterculus* Wied. (Trypetidae) – a severe menace to the southern United States. *Journal of Economic Entomology* 11, 457–467.

Salazar-Garcia, S. and Bolio-Garcia, J.M. (1992) Damage caused by avocado branch weevil (*Copturus aguacatae* Kiss.) in Fuerte avocado trees in Atlixco, Puebla, Mexico. In: *Proceedings of the Second World Avocado Congress,* Riverside, California, p. 269 (abstract).

Saldarriaga, A. (1977) Las plagas del aguacate y su control. *Instituto Colombiano Agropecuario, Separata Curso Frutales, Compendio* 20:70.

Salgado, M. (1993) Problemas fitosanitarios del aguacate en Coatepec Harinas. *Fundacion Salvador Sanchez Colin, Memorias*, Coatepec Harina, Mexico, pp. 191–212.

Salgado, M. and Bautista, N. (1993) El barrenador grande del hueso del aguacate (*Heilipus lauri* Bohemann) en Ixtapan de la Sal, Mexico. *Memorias, Fundacion Salvador Sanchez Colin, Memorias*, Coatepec Harina, Mexico. pp. 225–231.

Salinas, A.P. (1992) Ciclo biologico de la araña cristalina del aguacatero *Oligonychus perseae* (Tuttle, Baker and Abatiello) (Prostigmata:Tetranychidae). Professional thesis, Universidad Autonoma de Chapingo, Mexico.

Sanchez, H.M. (1983) *Plagas y Enfermedades de los Frutales*. Ed. Pueblo y Educacion, Havana, Cuba, p. 195.

Sasscer, E.R. (1921) Important insects collected on imported nursery stock in 1920. *Journal of Economic Entomology* 14, 353–355.

Schaffer, B., Peña, J., Lara, S.P. and Buisson, D. (1986) Net photosynthesis, transpiration and stomatal conductance of avocado leaves infested by avocado red mite. *Proceedings of the Interamerican Society of Tropical Agriculture* 30, 73–82.

Schoeman, O.P. (1998) Citrus looper. In: Bedford, E.C.G., Van den Berg, M.A. and De Villiers, E.A. (eds) *Citrus Pests in the Republic of South Africa*. ARC Institute for Tropical and Subtropical Crops, Nelspruit, South Africa, pp. 212–216.

Schoeman, P.S., Grové, T. and Mohlala, R. (2012) New guidelines for the management of the coconut bug (*Pseudotheraptus wayi*) in avocados. In: *Proceedings of the SAAGA Research Symposium Abstracts*, Tzaneen, South Africa, p. 11.

Schonherr, Von J. and Pedrosa-Macedo, J.H. (1981) Scolytoidea in den Aufforstungen Brasiliens Ein Beitrag zur Kenntnis der Borkendkafer Sudamerikas. *Zeitschfrit fur Angewandte Entomologie* 92, 48–61.

Schwartz, A. (1981) 'n Bydrae tot die biologie en beheer van die valskodlingmot *Cryptophlebia leucotreta* Meyr. (Lepidoptera: Eucosmidae) op nawels. PhD thesis, University of Stellenbosch, Stellenbosch, South Africa. (In Afrikaans.)

Schwartz, A. (1983) Sitrusblaaspootjie: Wat van biologiese beheer? *Subtropica* 4, 14. (In Afrikaans.)

Schwarz, R.E. (1998) Citrus aphids. In: Bedford, E.C.G., Van den Berg, M.A. and De Villiers, E.A. (eds) *Citrus Pests in the Republic of South Africa*. ARC Institute for Tropical and Subtropical Crops, Nelspruit, South Africa, p. 288.

Searle, C.M. and St, L. (1964) A list of the insect enemies of citrus pests in Southern Africa. *Technical Communication Department of Agricultural Technical Services, South Africa* 30, 1–18.

SENASICA (2018) Manual de Identificación de las Principales Plagas del Aguacate en México. *Dirección General de Sanidad Vegetal – Centro Nacional de Referencia Fitoranitaria, Tecámac, México,* 33 p.

Skaife, S.H. (1956) *Dwellers in Darkness: An Introduction to the Study of Termites*. Longmans Green, London, p. 134.

Smit, B. (1964a) A further chapter in the history of entomology in South Africa. *Journal of the Entomological Society of Southern Africa* 23, 3–15.

Smit, B. (1964b) *Insects of Southern Africa. How to Control Them.* Oxford University Press, Johannesburg, p. 399.

Smith, D. (1973) Insect pests of avocados. *Queensland Agricultural Journal* 99, 645–653.

Smith, D., Beattie, G.A.C. and Broadley, R. (eds) (1997) *Citrus Pests and Their Natural Enemies: Integrated Pest Management in Australia*. Queensland Department of Primary Industries Information Series Q 197030, Brisbane, Australia, p. 272.

Srikumar, K.K., Kumar, B.S. and Radhakrishnan, B. (2015) Record of new host plant of red spider mite, Oligonychus Coffeae (Nietner). *Bulletin of Entomological Research* 50, 265–274.

Sternlicht, M. (1979) *Parabemisia myricae* (Homoptera: Aleyrodidae) – a new pest in Israel. *Hassadeh* 59, 1830–1831. (In Hebrew.)

Steven, D. (2004) Control of six spotted mite *Eotetranychus sexmaculatus*. *New Zealand Avocado Growers' Association Annual Research Report* 4, 13–24.

Stevens, P. (1999) The impact of new pests on New Zealand avocado production. In: Arpaia, M. and Hofshi, R. (eds) *Proceedings of Avocado Brainstorming '99*, California Avocado Commission, University of California, Riverside, p. 119.

Stevens, P.S. (1997) Avo-Green – an option for reducing pesticide use. In: *Proceedings of the Australian, New Zealand Avocado Conference,* Rotorua, New Zealand, pp. 112–118.

Stevens, P.S., McKenna, C.E. and Steven, D. (1998) Pest management for avocados in New Zealand. In: *Proceedings of the World Avocado Congress III 22–28 October 1995, Tel Aviv Israel*, pp. 429–432.

Steyn, J.J. (1954) The pugnacious ant (*Anoplolepis custodiens* Smith) and its relation to the control of citrus scales at Letaba. *Memoirs of the Entomological Society of Southern Africa* 3, 1–96.

Steyn, W.P. (1996) First record of *Thripobius semiluteus* Boucek, (Hymenoptera: Eulophidae), as a parasitoid of pine-tree thrips nymphs in South Africa. *Subtropica* 17(9), 14–16.

Steyn, W.P. (2001a) Greenhouse thrips *Heliothrips haemorrhoidalis* (Bouché). In: Van den Berg, M.A., De Villiers, E.A. and Joubert, P.H. (2001) *Pests and Beneficial Arthropods of Tropical and Non-Citrus Subtropical Crops in South Africa*. ARC Institute of Tropical and Subtropical Crops, Nelspruit, pp. 228–231.

Steyn, W.P. (2001b) Red-banded thrips *Selenothrips rubrocinctus* (Giard). In: Van den Berg, M.A., De Villiers, E.A. and Joubert, P.H. (2001) *Pests and Beneficial Arthropods of Tropical and non-Citrus Subtropical Crops in South Africa*. ARC Institute of Tropical and Subtropical Crops, Nelspruit, pp. 239–240.

Steyn, W.P., Du Toit, W.J. and De Villiers, E.A. (1994) Effect of insect growth regulator, CGA 211446, on the third instar of the heart-shaped scale on avocados. *Subtropica* 15(4), 15–18.

Stoetzel, M.B. and Davidson, J.A. (1974) Biology, morphology and taxonomy of immature stages of 9 species in the Aspidioti. *Annals of the Entomological Society of America* 67, 475–509.

Stouthamer, R., Rugman-Jones, P.F., Thu, P.Q., Eskalen, A., Thibault, T., *et al.* (2017) Tracing the origin of a cryptic invader: phylogeography of the *Euwallacea fornicatus* (Coleoptera: Curculionidae: Scolytinae) species complex. *Agricultural and Forest Entomology* 19(2017), 366–375.

Swaine, G., Ironside, A. and Yarrow, W.H.T. (1985) *Insect Pests of Fruit and Vegetables*. Queensland Department of Primary Industries, Information Series Q 183021, Brisbane, p. 176.

Swirski, E., Izhar, Y., Wysoki, M. Gurevitz, E. and Greenberg, S. (1980) Integrated control of the long-tailed mealybug *Pseudococcus longispinus* [Hom: Pseudosoccidae], in avocado plantations in Israel. *Entomophaga* 15, 415–426.

Swirski, E., Wysoki, M. and Izhar, Y. (1986) Biological and integrated control of subtropical fruit pests in Israel. *Alon Hanotea* 40, 664–674. (In Hebrew with English summary.)

Swirski, E., Blumberg, D., Wysoki, M. and Izhar, Y. (1987) Biological control of the Japanese bayberry whitefly, *Parabemisia myricae* (Kuwana) (Homoptera Aleyrodidae), in Israel. *Israel Journal of Entomology* 21, 11–18.

Swirski, E., Wysoki, M. and Izhar, Y. (1988) Integrated pest management in the avocado orchards of Israel. *Applied Agricultural Research* 3, 1–7.

Swirski, E., Izhar, Y. and Wysoki, M. (1991a) Appearance of *Aphis gossypii Glover* and *Aphis spiraecola* Patch (Rhynchota: Aphidoidea) on avocado, persimmon and macadamia. *Alon Hanotea* 45, 413–416. (In Hebrew with English summary.)

Swirski., E., Wysoki, M. and Izhar, Y. (1991b) Twelve years survey of avocado pests and their natural enemies (1978–1990). *Alon Hanotea* 46, 65–103. (In Hebrew.)

Swirski, E., Wysoki, M. and Izhar, Y. (1998) Avocado pests in Israel. In: *Proceedings of the World Avocado Congress III, 22–27 October 1995, Tel-Aviv, Israel*, pp. 419–428.

Takano-Lee, M. and Hoddle, M.S. (2001) Biological control of *Oligonychus perseae* (Acari: Tetranychidae) on avocado: IV. Evaluating the efficacy of a modified mistblower to mechanically dispense *Neoseiulus californicus* (Acari: Phytoseiidae). *International Journal of Acarology* 27, 157–169.

Takano-Lee, M. and Hoddle, M.S. (2002) *Oligonychus perseae* (Acari: Tetranychidae) population responses to cultural control attempts in an avocado orchard. *Florida Entomologist* 85, 216–226.

Theobald, F.V. (1914) African Aphididae. *Bulletin of Entomological Research* 4, 313–337.

Thompson, W.R. (1946) *A Catalogue of the Parasites and Predators of Insect Pests. Sec. 1. Parasite host catalogue. Part 8. Parasites of the Lepidoptera (N-P)*. Imperial Agricultural Bureaux, Belleville, Canada, pp. 386–523.

Tollerup, K.E. and Morse, J.G. (2006) The effect of horticultural spray oil and surfactants on the residual efficacy of spinosad against avocado thrips, *Scirtothrips perseae* (Thysanoptera: Thripidae). *Journal of Agriculture and Urban Entomology* 22, 127–131.

Torres, E., Hernández-Suárez, E., Alvarez-Acosta, C. and Ferragut, F. (2022) *Oligonychus Perseae* Tuttle, Baker & Abbatiello (Acari: Tetranychidae) population dynamics and associated Phytoseiid Mites (Acari: Phytoseiidae) in avocado orchards in the Canary Islands (Spain). *International Journal of Acarology* 48(7), 551–563.

Torres Jaimes, L.F., Correa Londoño, G.A., Cartagena Valenzuela, J.R., Monsalve García, D.A. and Londoño Zuluaga, M.E. (2012) Relationship of *Monalonion velezangeli* Carvalho & Costa (Hemiptera: Miridae) with the phenology of avocado (*Persea americana* Mill., cv. Hass). *Revista Facultad Nacional de Agronomía Medellín* 65(2), 6659–6665.

Toukem, N.K., Yusuf, A.A., Dubois, T., Abdel-Rahman, E.M., Adan, M.S. and Mohamed, S.A. (2020) Landscape vegetation productivity influences population dynamics of key pests in small avocado farms in Kenya. *Insects* 11:424. doi:10.3390/insects11070424.

Triapitsyn, S. (2000) A new *Anagrus* (Hymenoptera: Mymaridae), egg parasitoid of *Idona minuenda* (Homoptera: Cicadellidae), a pest of avocado in Mexico. *Folia Entomologica Mexicana* 110, 89–94.

Ullyett, G.C. (1939) Parasites of the false codling moth (*Argyroploce leucotreta* Meyr.) in South Africa. In: *Proceedings of the Entomological Conference*, Union Buildings, Pretoria, Mimeograph, pp. 54–55.

Ullyett, G.C. and Bishop, H.J. (1938) Report on the false codling moth experiment carried out at Martindale, C.P. 1938. *Report of the Parasite Laboratory, Division of Entomology, Pretoria*, p. 23.

Vacari, A.M., Damato, F., Dami, B.G., de Lima, M.L.F., Lima, L.S.U., *et al.* (2021) Within-canopy distribution of *Stenoma catenifer* (Lepidoptera: Elachistidae) infestation in avocado orchards. *Journal of Insect Science*, 21(5), 5.

Valdez, F.J., Matos, L.A. and Alvarez, P.A. (2016) *Indice de plagas y enfermedades de importancia economica en la Republica Dominicana.* Ministerio de Agricultura, Oficina de Ejecución de Proyectos (OEP), República Dominicana.

Van den Berg, M.A. (1973) Host plants of three saturniids and the degree of defoliation they cause to *Pinus patula* Schlechtd. & Cham. *Phytophylactica* 5, 65–70.

Van den Berg, M.A. (1974) Natural enemies and diseases of the forset insect *Nudaurelia cytherea clarki* Geertsema (Lepidoptera: Saturniidae). *Phytophylactica* 6, 39–44.

Van den Berg, M.A. (1975) Bio-ecological studies on forest pests 3. *Nudaurelia cytherea clarki* Geertsema (Lepidoptera: Saturniidae). *Forestry in South Africa* 17, 1–16.

Van den Berg, M.A. (1998) Tip wilters and stink bugs on citrus. In: Bedford, E.C.G., Van den Berg, M.A. and De Villiers, E.A. (eds) *Citrus Pests in the Republic of South Africa*. ARC Institute for Tropical and Subtropical Crops, Nelspruit, South Africa, pp. 161–163.

Van den Berg, M.A., Newton, P.J., Deacon, V.E. and Crause, C. (1987) Dispersal of *Trichogrammatoidea cryptophlebiae* (Hymenoptera: Trichogrammatidae), an egg parasitoid of the false codling moth, *Cryptophlebia leucotreta* (Lepidoptera: Tortricidae), in an empty habitat. *Phytophylactica* 19, 515–516.

Van den Berg, M.A., De Villiers, E.A. and Joubert, P.H. (1999) *Identification Manual for Avocado Pests*. ARC Institute for Tropical and Subtropical Crops, Nelspruit, p. 53.

Van den Berg, M.A., Steyn, W.P. and Greenland, J. (2000) Hemiptera occurring on avocado trees in the Mpumalanga Loweveld of South Africa. *Africant Plant Protection* 6, 29–33.

Van Den Berg, N., Du Toit, M., Morgan, S.W., Fourie, G. and De Beer, Z.W. (2019) First report of *Fusarium euwallaceae* causing necrotic lesions on *Persea americana* in South Africa. *Plant Disease* 103, 1774. doi:10.1094/PDIS-10-18-1818-PD.

Vanderplank, F.L. (1958) Studies on the coconut pest, *Pseudotheraptus wayi* Brown (Coreidae), in Zanzibar. 1. A method of assessing the damage caused by the insect. *Bulletin of Entomological Research* 49, 559–584.

Vanin, S.A. and Bená, D. de C. (2015) A new species of Heilipus Germar (Coleoptera: Curculionidae: Molytinae) associated with commercial species of Annonaceae in Brazil, and comments on other species of the genus causing damage to avocado trees in Brazil. *Zootaxa* 3905(4), 541–556.

Vargas, R. and Rodriguez, S. (2008) Avocado red mite, *Oligonychus yothersi* (McGregor). In: Ripa, R. and Larral, P. (eds) *Manejo de Plagas en Paltos y Cítricos*. Instituto de Investigaciones Agropecuarias, Ministerio de Agricultura, La Cruz, Chile, Colleccion de Libros INIA 23, pp. 239–246.

Vargas, R., Ripa, R. and Rodriguez, S. (2008) Oleander scale, *Aspidiotus neri* (Bouché). In: Ripa, R. and Larral, P. (eds) *Manejo de Plagas en Paltos y Cítricos*. Instituto de Investigaciones Agropecuarias, Ministerio de Agricultura, La Cruz, Chile, Colleccion de Libros INIA 23, pp. 171–177.

Velasquez, M. and Santizo, L. (1992) El acaro *Calepitrimerus muesebecki*, plaga del follaje del aguacate. Nota Tecnico Cientifica 14. Instituto de ciencia and tecnologia agricola (Guatemala), Disciplina de proteccion vegetal, 2 pp.

Velázquez-Martínez, G.C., González-Hernández, H., Equihua-Martínez, A., Lomeli-Flores, J.R., Rojas, J.C. and Lopez-Collado, J. (2022) Population dynamics of *Stenoma catenifer* Walsingham (Lepidoptera: Depressariidae) on Hass avocado orchards in México. *Journal of Asia-Pacific Entomology* 25(1), 101866.

Viljoen, H.M. (1986) Kokosneutstinkbesie – 'n potensiële plaag op avokado's. *South African Avocado Growers' Association Yearbook* 9, 72–74. (In Afrikaans.)

Waite, G.K. and Huwer, R.K. (1998) Host plants and their role in the ecology of the fruitspotting bugs *Amblypelta nitida* Stal and *Amblypelta lutescens lutescens* (Distant) (Hemiptera: Coreidae). *Journal of the Australian Entomological Society* 37, 340–349.

Waite, G.K. and Pinese, B. (1991) Pests. In: Broadley, R.H. (ed.) *Avocado Pests and Disorders*. Queenstand Department of Primary Industries Information Series, Brisbane, Australia, Q190013, p. 74.

Ware, A.B., du Toit, C.L.N., Ekasi, S., Mohamed, S.A. and Nderitu, P.W. (2012) Cold tolerance and disinfestations of *Bactrocera invadens* (Diptera: Tephritidae) in 'Hass' avocado. In: *Proceedings of the SAAGA Research Symposium Abstracts*, Tzaneen, South Africa, p. 10.

Wasbauer, M.S. (1972) An annotated host catalog of the fruit flies of America north of Mexico (Diptera: Tephritidae). *Occasional Paper, California Department of Agriculture Bureau of Entomology* 19, 1–172.

Way, M.J. (1951) An insect pest of coconuts and its relationship to certain ant species. *Nature* 168:302.

Way, M.J. (1953) Studies on *Theraptus* sp. (Coreidae); the cause of the gumming diseases of coconuts in East Africa. *Bulletin of Entomological Research* 44, 657–667.

Webb, D.V.V. (1974) Forest and timber entomology in the Republic of South Africa. *Department of Agriculture Technical Services, Republic of South Africa Entomology Memoirs* 34:21.

White, I.M. and Elson-Harris, M.M. (1992) *Fruit Flies of Economic Significance: Their Identification and Bionomics*. CAB International, Wallingford, UK, p. 601.

Whitehead, D. (1979) Recognition characters and distribution records for species of *Conotrachelus* (Coleoptera: Curculionidae) that damage avocado fruits in Mexico and Central America. *Proceedings of the Entomological Society of Washington* 8, 105–107.

Willard, H.F., Mason, A.C. and Fullaway, D.T. (1929) Susceptibility of avocados of the Guatemala race to attack by the Mediterranean fruit fly in Hawaii. *The Hawaiian Forester and Agriculturist* 26, 171–176.

Wolfe, H., Franciosi, R., Van Oordt, E. and Figueroa, R. (1969) El culivo del palto en el Perú. *Ministerio de Agricultura y Pesqueria, Estacion Experimental Agricola de La Molina, Lima, Perú. Boletin Tecnico* 73:40.

Wolfenbarger, D.O. and Colburn, B. (1966) Recent observations on some avocado pests in Mexico and El Salvador. *Proceedings of the Florida State Horticultural Society* 79, 335–337.

Wolfenbarger, D.O. and Colburn, B. (1979) The *Stenoma catenifer*, a serious avocado pest. *Proceedings of the Florida State Horticultural Society* 92, 275.

Wood, S.L. (2007) Bark and ambrosia beetles of south America (Coleoptera, Scolytidae). Books by Faculty of the Monte L. Bean Life Science Museum, Brigham Young University, Provo, USA, p. 900. ISBN 0-8425-2635-8.

Wysoki, M. (1987) A bibliography of the pyriform scale, *Protopulvinaria pyriformis* (Cockerell), 1894 (Homoptera: Coccidae), up to 1986. *Phytoparasitica* 15(1), 73–77.

Wysoki, M. and Izhar, Y. (1978) A list of arthropod pests of avocado and pecan trees in Israel. *Phytoparasitica* 6, 89–93.

Wysoki, M. and Izhar, Y. (1980) The natural enemies of *Boarmia (Ascotis) selenaria* Schiff. (Lepidoptera: Geometridae) in Israel. *Acta Oecologica, Oecologia Applicata* 1, 283–290.

Wysoki, M. and Izhar, Y. (1981) Biological data on *Apanteles cerialis Nixon* (Hymenoptera: Braconidae), a parasite of *Boarmia (Ascotis) selenaria Schiff.* (Lepidoptera: Geometridae). *Phytoparasitica* 9, 19–25.

Wysoki, M. and Izhar, Y. (1986) Fluctuation of the male population of *Boarmia (Ascotis) selenaria* (Lepidoptera: Geometridae) estimated by virgin female baited traps. *Acta Oecologia Applicatta* 7, 251–259.

Wysoki, M., Kuzlitzky, W., Izhar, Y., Swirski, E., Ben-Yehuda, S., Hadar, D. and Rene, S. (1997) Successful acclimatization of *Thripobius semiluteus*, a parasitoid of *Heliothrips haemorrhoidalis* (Bouché) in Israel. *Phytoparasitica* 25:155.

Wysoki, M., Van den Berg, M.A., Ish Am, G., Gazit, S., Peña, J.E. and Waite, G.K. (2002) Pests and pollinators. In: Peña, J.E., Sharp, J.L. and Wysoki, M. (eds) *Tropical Fruit Pests and Pollinators*. CAB International, Wallingford, UK, pp. 223–293.

Yasuda, K. and Kinjou, T. (1983) Seasonal prevalence of the leaf-footed bug, *Leptoglossus australis* Fabricius (Heteroptera: Coreidae) on fruit of the wild host, *Diplocyclos palmatus* C. Jeffrey, and development of immature stages at various constant temperatures. *Proceedings of the Association for Plant Protection of Kyushu* 29, 89–91.

Yee, W., Phillips, P.A., Rogers, J.L. and Faber, B.A. (2001a) Phenology of arthropod pests and associated natural predators on avocado leaves, fruit, and in leaf litter in southern California. *Environmental Entomology* 30, 892–898.

Yee, W., Phillips, P.A., Faber, B.A. and Rogers, J.L. (2001b) Relationships between *Scirtothrips perseae* (Thysanoptera: Thripidae) populations on avocado leaves, fruit, and scarring damage on fruit. *Environmental Entomology* 30, 932–938.

Yee, W., Phillips, P.A. and Faber, B.A. (2001c) Effects of aerial spray volume, coverage, and abamectin on *Scirtothrips perseae* (Thysanoptera: Thripidae). *Journal of Economic Entomology* 94, 1090–1096.

Zou, Y. and Millar, J.G. (2010) Improved synthesis of (9Z)-9,13-tetradecadien-11-ynal, the sex pheromone of the avocado seed moth, *Stenoma catenifer*. *Tetrahedron Letters* 51, 1336–1337.

15 Harvesting, Packing, Postharvest Technology, Transport and Processing

P.J. Hofman[1], J. Bower[2], A. Woolf[3], B.G. Defilippi[4], D. Olivares[4] and M. García-Rojas[4]

[1]*Maroochy Research Station, Department of Agriculture, Fisheries and Forestry, Queensland, Australia;* [2]*British Columbia, Canada;* [3]*The New Zealand Institute for Plant and Food Research Limited, New Zealand;* [4]*Instituto de Investigaciones Agropecuarias (INIA La Platina), Chile*

Abstract

This chapter reviews postharvest technology and practices (including processing) for avocado fruit. The effect of production practices on postharvest performance is also discussed.

The harvest and postharvest phases account for up to 60% of all costs in horticultural production and marketing systems, and up to 50% of all fruit is not consumed (Kader, 2005; Elik *et al.*, 2019). Postharvest losses of fruits and vegetables are estimated to be 5–20% in developed countries and 20–50% in developing countries (Bantayehu *et al.*, 2017). These postharvest losses are observed at harvesting, during packing, during transportation, in wholesale and retail markets, and because of delays at different stages of handling (Kitinoja *et al.*, 2010). In general, the risk of quality loss is higher when the time from harvest to consumption is longer. Thus, the trend towards greater distances between production sites and the consumer and storage of fruit to overcome localized market oversupply has increased the emphasis on postharvest technology and practices. Additionally, the increasing demand for quality and reduced chemical use (particularly postharvest), increasing interest in convenience foods, and concerns about carbon footprints are placing additional demands on postharvest technologies in the horticultural sector. There is also a growing awareness of the interactions among all parts of the production and postharvest systems in relation to fruit quality. In general, maximum product quality is determined preharvest, and postharvest systems are designed to minimize quality loss during handling and distribution. Greater knowledge and understanding of these interactions lead to new developments in supply chain systems to address the complex issues involved in horticultural production and marketing.

This chapter reviews postharvest technology and practices (including processing) for avocado (*Persea americana* Mill.) fruit. The effect of production practices on postharvest performance is also discussed, recognizing that this section intersects with other chapters in this book. Postharvest diseases are discussed by Dann and

*Correspondence: bdefilip@inia.cl

 The Avocado: Botany, Production and Uses (eds D. Carrillo, B. Schaffer, A.W. Whiley and B.N. Wolstenholme)
DOI: 10.1079/9781800621824.0015

Gazis (Chapter 13, this volume), and in-field control of quarantine insect pests is discussed by Carrillo *et al.* (Chapter 14, this volume). Earlier reviews (Lewis, 1978; Lee, 1981; Bower and Cutting, 1988; Hofman and Smith, 1994; Milne, 1994; Hofman *et al.*, 2002a; Yahia and Woolf, 2011) have also discussed postharvest aspects of avocado production. This chapter provides updated information on quality parameters, sensory perception and maturity, production and postharvest factors affecting quality defects, disinfestation and storage (including preconditioning), prediction of outturn quality and processing.

Quality

Quality and loss assessment

Quality can be defined as 'fitness for purpose; a product state that meets the expectations of the customer/consumer' (Hewett, 2006). Acceptable quality may be considered as the state the purchaser considers 'value for money'. It reflects the individual's needs and perceptions, which may vary between grower, agent, retailer and consumer. In most cases, quality is influenced by external and internal appearance, flavour, texture and value (price). The important quality characteristics of avocado are described in Table 15.1.

Many production and postharvest factors can influence avocado quality. Maximum potential quality is determined at harvest, whereas postharvest technologies aim to maintain quality but cannot improve it (Hewett, 2006). Therefore, it is important for members of the production-to-consumer continuum to be aware of each other's role in producing and maintaining quality. Regularly assessing quality at various stages during production helps ensure that customer/consumer preferences are met, identifies practices causing unacceptable quality loss and helps set research and development priorities and encourage practice improvements (Ledger *et al.*, 2006).

Field defects detracting from external appearance can typically be removed during harvesting and sorting in the packhouse, but defects not obvious at packing are more difficult to manage. Mechanical damage during harvesting and packing may not be obvious until several days later. Flesh defects are often not visible externally, particularly for black-skin cultivars such as 'Hass' but have a significant impact on value and consumer satisfaction. Up to 40% of 'Hass' fruit sampled from Australian retail chain store shelves can have at least 10% of the flesh affected by defects, with rot and flesh bruising being the most common (Hofman and Ledger, 1999, 2001a; Anonymous, 2009). Providing 'ripe for tonight' fruit for the consumer increases the risk of flesh defects caused by bruising and overripeness but also significantly increases sales.

Outturn quality monitoring of South African avocados in European markets (Bezuidenhout, 1992) and New Zealand fruit in the USA (Dixon *et al.*, 2003b) led to the conclusion that fruit age and transit temperature were important in problems caused by fruit rot. In Australian retail surveys, grower/packer practices and the number of days and holding conditions from packing to arrival at the retail store were related to the incidence of rot (Hofman and Ledger, 2001b). Little flesh bruising appears to occur during picking and packing (Hofman, 2002), but bruising is likely to increase once the fruit starts to ripen.

Table 15.1. Typical quality parameters valued by different members in the avocado supply chain

Grower	Packer	Agent	Retailer	Consumer
Size	Size	Size	Size	Size
Shape	Gradable defects (% pack out)	Consistent/predictable level of defects	Shape	Shape
Skin defects	Freedom from pests	Uniform fruit size within the entire pallet	Skin colour	Skin markings
Maturity		Predictable ripening	Absence of skin defects	Ripeness stage
		Maturity	Absence of disease	Flesh defects
		Absence of disease	Ripeness stage	Flavour
		Freedom from pests	Maturity	Texture

Quality parameters

Clear, unambiguous descriptions of quality defects for growers, packers and marketers are required to avoid confusion and misdiagnosis of causes and solutions. Swarts (1985), Fitzell *et al.* (1995) and Anonymous (2004) present illustrations of common disorders, but the nomenclature is not consistent (Table 15.2). White *et al.* (2009) developed an avocado quality assessment manual to help standardize nomenclature and rating systems. In most cases, disorder names are based on the physical appearance of the defect rather than potential causes. Various organizations have suggested quality descriptors and standards. The Organisation for Economic Co-operation and Development (OECD) has established standards for imported avocados (Anonymous, 2004). Country industry organizations often produce generic grade standards, which are sometimes refined for specific chains.

The important quality parameters are discussed in the following section. Fruit size, shape and field defects such as sunburn, wind rub, insect damage and nutrient deficiencies are not discussed. The causes of the defects are summarized in the sections on 'Effects of Production Practices on Postharvest Quality' and 'Effects of Postharvest Practices on Quality'.

Skin colour and brightness

Ripe fruit skin colour is generally not a problem with green-skinned avocado cultivars such as 'Edranol' and 'Fuerte', apart from field blemishes such as sunburn and damage from impacts and diseases. These cultivars often change from bright green at harvest to a slightly darker green when ripe. In these avocado cultivars, differences in colour were not observed during cold storage and shelf life (Olivares *et al.*, 2020).

Black-skinned or dark-skinned cultivars such as 'Hass' and 'Gem' change from green to purplish-black during ripening (Cho *et al.*, 2021) mainly due to a slight loss of chlorophyll in the first few days of ripening followed by an increase in the concentration of only one anthocyanin, cynanidin-3-O-glucoside (Cox *et al.*, 2004; Ashton *et al.*, 2006). However, the change from purple to black is not completely explained by the increase in anthocyanin concentration, and other changes later in softening may be involved. The fact that the skin turns purple/black as these avocado varieties ripen postharvest has made skin colour change an indicator of ripeness and has been used by handlers and consumers as a guide (Donetti and Terry, 2011; Rivera *et al.*, 2017; Cho *et al.*, 2021). 'Hass' fruit isconsidered ripe by Australian consumers when fully purple/black; however, thereis great variability in fruit skin color during the ripening and ready-to-eat(RTE) stage (Fig. 15.1). The delay in skin coloration with ripening is observed more in fruit ripened at lower temperatures (e.g. 15°C; Hopkirk *et al.*, 1994; Cox *et al.*, 2004) or fruit treated with 1-methylcyclopropene (Olivares *et al.*, 2022). These fruit are often not consumed until fully purple/black, by which stage they can be overripe and have more diseases and disorders (Ledger and Barker, 1995) because these increase as the fruit softens (Hopkirk *et al.*, 1994).

Options for improving the skin colour of ripe 'Hass' fruit are discussed in the sections 'Effects of Production Practices on Postharvest Quality' and 'Ripening Temperature'.

Black areas on the skin

These defects are typically caused by one or more of the following:

- Mechanical damage during harvest and packing results in skin spotting, which is shown as black areas of less than 1 mm in diameter with discrete, well-defined margins (White *et al.*, 2009) (Fig 15.1). This defect is also called nodule damage or lenticel damage and typically becomes obvious only 1–2 days after harvesting/packing. In rough-skinned 'Hass', skin spotting is typically confined to the nodules because these are more likely to be damaged during picking and packing. Skin spotting is also associated with damage to the lenticels (pores that allow gas exchange between the environment and internal tissue) that become non-functional as the fruit expands. This occurs more easily if the cells within the lenticel cavity are more turgid because of, for example, harvesting during or after rain (Everett *et al.*, 2008) (Fig 15.2). Presumably, skin spotting on smooth-skinned cultivars is closely associated with lenticel damage.

Table 15.2. Common names for avocado postharvest defects

Defect name[a]	Alternate names	Description
Skin spotting	Lenticel damage, nodule damage	Brown or black areas on the skin, less than 3 mm diameter. Mostly associated with damage to the nodules on 'Hass', and possibly to the lenticels in smooth-skinned green cultivars.
Discrete patches		Brown or black areas of the skin with very discreet margins, greater than 3 mm diameter, and generally sunken.
	Dusky cold	Brown or black areas of the skin but with diffuse margins. Often occurring over larger areas of the skin, and with less intense colour compared with discrete patches. Not sunken.
External rot	Fuzzy patches or 'measles' when lesions are very small	Brown or black areas on the skin surface with a poorly defined margin.
Flesh bruising		Discoloured flesh with relatively distinct margins. Typically associated with the skin.
Diffuse flesh discoloration	Diffuse discoloration, mesocarp discoloration, flesh greying	Grey or grey-brown areas of discoloured flesh with a diffuse margin. Typically starts near the base of the seed.
Vascular browning	Vascular streaking	Browning of the vascular bundles. Often associated with stem-end rot.
Stem-end rot		Light discoloured areas of flesh starting from the stem end, typically softer than the surrounding flesh.
Body rot		Discrete, often firm symmetrical areas of discoloured flesh on the body of the fruit.
Seed cavity browning		Browning of the flesh directly adjacent to the seed.
Tissue breakdown		A slimy breakdown of the bright green flesh directly under the skin. The affected tissue looks dark green and shiny.
Stringy vascular		Thickening of the vascular bundles that run longitudinally through the fruit.
Vascular leaching		Browning of the flesh around the vascular bundles, generally a diffuse appearance.
Stones in the flesh	Fruit spotting bug damage	Hard lumps or 'stones' in the flesh under the skin.
Pulp spot		Grey spots along the margins of the vascular bundles. Most often associated with 'Fuerte'.

[a]Based on White *et al.* (2009), with permission.

Fig. 15.1. Black areas on the skin: skin spotting on 'Hass' generally caused by mechanical damage during harvest and packing, and more common when fruit are harvested during wet weather. Photograph from P. Hofman, with permission from the Department of Employment, Economic Development and Innovation, Queensland.

- Low storage temperatures (generally below approximately 3–4°C for longer than approximately 10 days) cause chilling damage, resulting in areas of black skin greater than 1 mm in diameter with a well-defined, discrete margin (Fig 15.3). Excessive heat treatment (HT) can also cause discrete patches, but the affected areas typically have more regular margins than those caused by chilling injury (CI)-inducing temperatures. Dusky cold is likely a less severe form of skin CI that occurs at marginally cold temperatures and shorter durations.
- Disease causing black areas with less defined margins. These are less obvious in black-skinned cultivars because disease severity increases as the fruit ripens, and the ripe fruit skin colour masks the discoloured

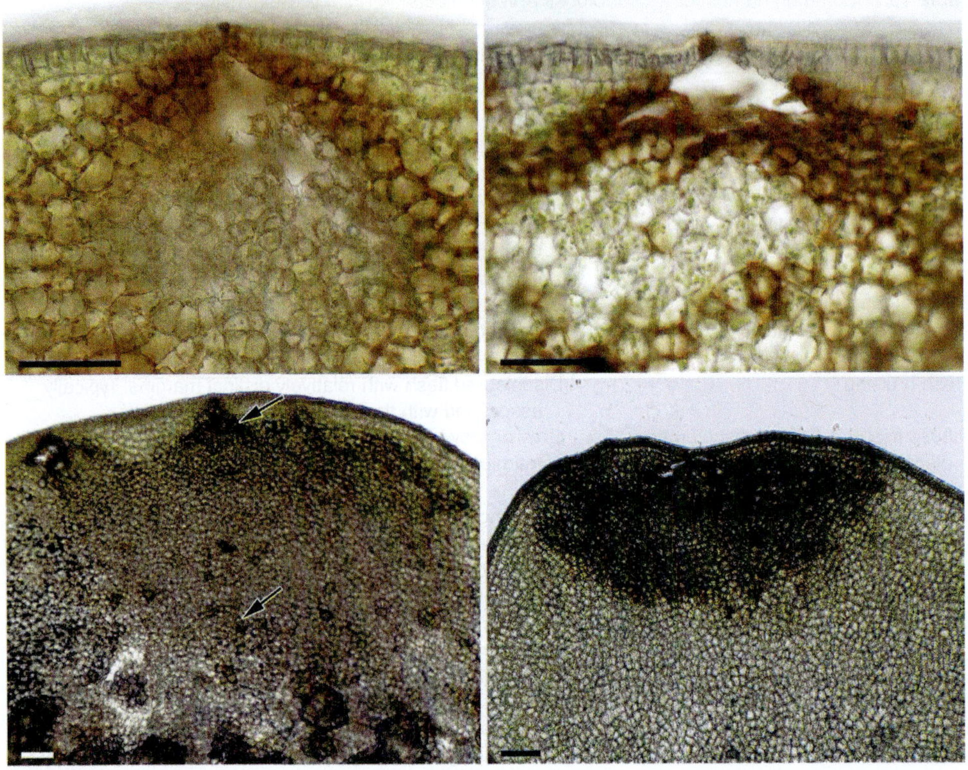

Fig. 15.2. Top: lenticel on the nodule of 'Hass' skin 2 h and 8 h after simulated damage caused by harvesting and packing, showing brown discoloration. Bottom: larger view of a nodule showing discoloration 30 min and 24 h after treatment. Bars = 0.1 mm. Photograph from Everett *et al.* (2008), with permission.

skin. However, disease skin lesions are sometimes obvious in unripe fruit when disease pressure is high, and storage times are longer. These defects are sometimes called fuzzy patches or measles (Everett *et al.*, 2008).

- In recent years, a new defect known as a black spot has been observed in the fruit epicarp. Black spot is characterized by brown to black spots 2–3 cm in diameter on the exocarp but not on the mesocarp tissue. The symptoms are visible only after storage and are caused by physical, physiological or phytopathological damage (Uarrota *et al.*, 2020). This defect is typically confused with mechanical damage and anthracnose, a disease caused by a fungus that produces dark, irregular spots on the epicarp but also compromises the mesocarp of the fruit. Moreover, other disorders, such as chilling injury during storage and lenticel breakdown, mimic black spot symptoms. Recent research distinguishes black spots from anthracnose and indicates that the possible factors associated with the incidence of black spots on the epicarp of avocados are produced by cold storage for more than 10 days combined with a drastic decrease in the enzymatic defence system of the epicarp (Uarrota *et al.*, 2020; Fuentealba *et al.*, 2022). The causal factor of black spot is still unclear and is currently being investigated.

Discoloured flesh

BRUISING. Fruit tissue damage results from external forces that cause physical changes in texture and/or chemical changes in colour, taste and odour. This damage is known as bruising and typically manifests as a dark-grey flesh and light-coloured discoloration, often associated with hairline cracking of the flesh and a well-defined area close to the impact site (Mazhar *et al.*, 2015, Wedding *et al.*, 2018). In softer fruit, the injury starts near the skin and progresses towards the seed, whereas in firm fruit, the bruise is typically halfway between the skin and the seed (Wedding *et al.*, 2018). Mechanical impacts on firmer fruit can cause cracking of the flesh and slight browning of the affected areas (Fig 15.4). These symptoms are relatively infrequent, with one survey detecting bruising of more than 10% of the flesh in only approximately 1% of fruit sampled at the end of the packing (Hofman, 2002). Flesh bruising occurs more easily as fruit soften, requiring lower drop heights to achieve similar bruising severity (Arpaia *et al.*, 1987; Ledger, 1995), particularly when fruit are handled in tray batches or individually (Hofman and Sandoval, 2002). The relationships between bruising, energy absorbed by the fruit and the maturity of 'Hass' avocado fruit have been confirmed by Arpaia *et al.* (2006). Identifying practices contributing to flesh bruising is a prerequisite to reducing the incidence and severity of flesh bruising in fruit on the retail shelf.

Fig. 15.3. Black areas on the skin: discrete patches on 'Hass', usually caused by storage below 3–4°C for longer than about 10 days. Photograph from P. Hofman, with permission from the Department of Employment, Economic Development and Innovation, Queensland, Australia.

Fig. 15.4. Flesh bruising on softening fruit (left) and on firmer fruit (right) resulting in cracking of the flesh. Photograph from P. Hofman, with permission from the Department of Employment, Economic Development and Innovation, Queensland, Australia.

Bruising in these softer fruits results in discrete areas of brown discoloration and is often associated with cavities in the flesh. The risk of bruising will likely increase as the fruit softens and when trays are removed from the pallet and fruit are removed from the trays.

Physiological disorders

There are several physiological flesh disorders, most of which are associated with cold storage (after 2–4 weeks) or excessively cold temperatures before harvest. Most of these disorders involve flesh browning, which is at least partly catalysed by polyphenol oxidase (PPO) following some form of cell dysfunction (Bower and Cutting, 1988). Disorder susceptibility differs according to cultivar (Plumbley *et al.*, 1989; Milne, 1994), maturity (Plumbley *et al.*, 1989), stage of ripeness (Kosiyachinda and Young, 1976; Hopkirk *et al.*, 1994), storage temperature (van Rooyen and Bower, 2006), humidity and storage duration (Bower and Cutting, 1987), ripening temperatures (Hopkirk *et al.*, 1994) and production factors (see 'Effects of Production Practices on Postharvest Quality' in this chapter).

Diffuse discoloration is also called mesocarp discoloration or grey pulp. It appears as a discoloration with ill-defined margins (Fig 15.5) and is typically associated with extended cold storage, particularly at temperatures above 4–6°C (van Rooyen and Bower, 2006). This disorder is characterized by a grey-brown discoloration of the pulp, which intensifies upon cutting the fruit and exposing the cut surfaces to the air. Initial studies associated this disorder with chilling injury (Chaplin *et al.*, 1982); however, discoloration was

Fig. 15.5. Physiological disorder: diffuse discoloration. Photograph from P. Hofman, with permission from the Department of Employment, Economic Development and Innovation, Queensland, Australia.

Fig. 15.6. Physiological disorder: seed cavity browning. Photograph from P. Hofman, with permission from the Department of Employment, Economic Development and Innovation, Queensland, Australia.

also found in unstored fruit maintained at non-chilling temperatures (van Rooyen and Bower, 2005). Mesocarp discoloration after cold storage in avocado correlates with increases in electrical conductivity (EC) and PPO activity (Hershkovitz *et al.*, 2005).

Seed cavity browning is a browning of the seed cavity only, with little penetration into the flesh (Fig 15.6). It can be caused by mild impact damage or from physiological causes.

Vascular browning has been described as mesocarp discoloration with a hardening of the vascular strands and off-flavour development (Bill *et al.*, 2014) (Fig 15.7) and can be associated with long cold storage times (>2 weeks) and/or incorrect storage temperatures (<5°C). It is also often associated with stem-end rot, except that the origin of vascular browning in stem-end rot is from the stem-end of the fruit, whereas physiological vascular browning (due to CI) starts at the distal end of the fruit. Uneven ripening, resulting in hard sections in the flesh, is occasionally a problem, particularly in less mature fruit in certain cultivars. This may be caused by earlier ripening of the proximal end compared with the distal end or by slower ripening of other portions of the flesh (often close to the seed at the proximal end). This disorder has

Fig. 15.7. Physiological disorder: vascular browning. Photograph from P. Hofman, with permission from the Department of Employment, Economic Development and Innovation, Queensland, Australia.

been associated with lower concentrations of 1-aminocyclopropane-1-carboxylic acid (ACC) synthase in the less ripe areas (Offer, 1986), high ripening temperatures (Hopkirk *et al.*, 1994) and waxing of the fruit. It may also reflect calcium (Ca) gradients in the flesh.

The presence of 'stony' layers in avocado flesh has been described by Schroeder (1981), which he called sclerocarpelosis. These layers can form throughout the entire flesh due to the hardening of the stone cells (sclereids) in the flesh. The affected flesh is a tan-to-grey colour and has a gritty texture. It has been observed only in 'Hass', and there appeared to be no correlation with fruit size or position on the tree. The disorder was more common in fruit from trees severely affected by frosts.

Pulp spot, also known as low-temperature disorder, is another significant physiological disorder of avocados that can develop during storage (Bill *et al.*, 2014, Olivares *et al.*, 2020). High incidences of pulp spot have been commonly observed in the cultivar 'Fuerte'. This disorder is characterized by black or grey spots along the margins of the vascular bundles and distributed randomly throughout the fruit. The incidence and severity of this disorder are strongly associated with cultivation practices; mainly linked to calcium, potassium and magnesium ratios; and highly variable across seasons. The symptoms develop after storage and increase with exposure to air (Olivares *et al.*, 2020).

DISEASES. The postharvest life of avocados is affected by fungal pathogens; therefore, postharvest diseases become a major constraint for successful storage and shipment. Anthracnose and stem-end rot are two major postharvest diseases that cause serious losses during export (Bill *et al.*, 2014). Anthracnose fruit rot is caused by the fungal pathogens *Colletotrichum gloeosporioides* and, to a lesser extent, *Colletotrichum acutatum*. Stem-end rot is caused by a range of fungal pathogens, primarily anamorphs of *Botryosphaeria* spp. (Obianom and Sivakumar, 2018). For further details on these two diseases, see Dann and Gazis, Chapter 13 in this volume.

Eating quality

Avocado eating quality is characterized not only by its unique texture, exquisite taste and aroma, and nutritional profile but also by the numerous health benefits that it provides (Hurtado-Fernandez *et al.*, 2018; Fuentealba and Hernández, Chapter 16, this volume). Texture is a significant indicator of avocado quality and of concern to the consumer (Maftoonazad and Ramaswamy, 2008; Toivonen and Brummell, 2008; Landahl *et al.*, 2009). Avocados undergo drastic changes in texture (Landahl *et al.*, 2009; Li *et al.*, 2010). Chen *et al.* (2009) stated that the oil content is a key component in the texture of avocados, which Hofman *et al.* (2002a) identified as contributing to 'smoothness'. Despite the relation between texture and oil content, Chen *et al.* (2009) discovered that an increase in the oil content over the harvest period did not manifest as a change in the texture. However, storage temperature, oxygen and carbon dioxide concentrations and wounding directly affect texture (Maftoonazad and Ramaswamy, 2008). The relationship between texture and firmness can be extended to the 'strength' of avocados, that is, the ability of the fruit to withstand loading during storage. Firmness can be described as the force necessary to attain a previously defined deformation during textural evaluation (Landahl *et al.*, 2009). As avocados ripen, their texture, firmness and strength are reduced (Kassim *et al.*, 2013). Texture is mainly influenced by the stage of ripeness, which also affects firmness and oil content (with a higher content reflecting increased maturity), which increases flesh creaminess and smoothness.

The existing research indicates that the content and composition of fatty acids and other nutrient components and the physical and chemical properties of avocado undergo relevant changes during ripening after storage (Wang *et al.*, 2018). Simultaneously, changes in the nutrient composition of the flesh inevitably lead to changes in its textural characteristics (Ma *et al.*, 2020).

Texture can also be affected by uneven ripening (firm lumps in the flesh), stones in the flesh (caused by insect damage) and stringy vascular bundles (caused by thickening of the vascular bundles or fibres in the flesh).

Flavour is influenced by cultivar and maturity. Changes in fruit composition associated with fruit development and ripening in avocado include changes in concentrations of oil, organic acids, sugars and the accumulation of volatile compounds (Obenland *et al.*, 2012).

These changes contribute to the overall quality perceived by the consumer. Although avocado flavour is highly associated with fatty acid content and texture, flavour composition is a complex attribute in which the combination of sugars, acids and volatiles plays a primary role (Lewinsohn *et al.*, 2001). With regard to oil content, lipid composition is the most important component of avocado, comprising 60–80% of the dry matter of the mesocarp, and is responsible for the taste that consumers demand (Marangoni *et al.*, 1996; Requejo-Tapia *et al.*, 1999). Oil content is also involved in defining the texture and nutritional properties of the fruit (Yousef and Hassaneine, 2010). Additionally, due to its major role in overall flavour, oil content was used to define a maturity index (harvest index) almost a century ago. The most important fatty acids present in the avocado mesocarp are long-chain fatty acids (C16 or higher), including oleic (75%), palmitic (8.6%), linoleic (8.7%) and palmitoleic acid (content lower than 1.4%) (Schwartz *et al.*, 2007). The content of each fatty acid varies as the fruit matures and is cultivar dependent. For example, the percentage of oleic acid increased during growth of the Fuerte cultivar in Japan, and the percentage of linolenic acid decreased (Inoue and Tateishi, 1995). Because fatty acids are synthesized or accumulated during fruit growth and not after harvest, their final concentration at harvest is modified by preharvest factors that affect tree physiology. Sugars are primary metabolites involved in several cell functions, including signalling, and as energy sources (e.g. respiration) and cell wall components. However, they are also involved in overall flavour as the sugar's sucrose, glucose and fructose are responsible for sweetness in fruit, with some contribution from sorbitol in apples (Knee, 1993; Baldwin, 2002). Avocado has been found to contain the C_7 sugar mannoheptulose and its reduced form perseitol, which are present in higher concentrations at earlier stages of fruit development and exhibit a steady reduction until maturity. Similarly, during postharvest, a continuous reduction of these sugars in the flesh has been described. Therefore, their role in avocado has been mainly related to storage reserves and respiratory metabolism (Liu *et al.*, 1999b, 2002). Interestingly, studies suggest that D-mannoheptulose benefits human health due to its inhibitory action on insulin secretion, impact on obesity and anticancer activity. These functional attributes are presently highly valued by consumers. Malic and citric acids are the most abundant organic acids in avocado, followed by ascorbic acid (Cresswell, 1970). Because organic acids are respiration substrates, their levels typically decrease during the climacteric stage of ripening. Despite the importance of organic acids in the overall flavour of fruit, their role in avocado would be minor; however, little information is available in the literature. The aroma properties of fruit depend on the combination of volatiles produced and the concentrations of individual volatile compounds. Aroma volatiles play an important role in determining the organoleptic properties of a fruit, primarily because aroma is one of the most important qualities demanded by consumers, and it also interacts with and affects other flavour attributes mentioned previously such as sweetness and acidity (Baldwin, 2002). There are few studies on aroma in avocados. Terpenes comprise more than 80% of the volatile profile, followed by heptanal and hexanal aldehydes (Pino *et al.*, 2004; Sinyinda and Gramshaw, 2004). However, as shown in other fruit species, the final profile will be affected by the genotype, growing conditions and maturity stage (fruit age), among other factors. The most recent report available in the literature by Obenland and collaborators (2012) showed that avocado has aroma-volatile compounds that play a role in determining the characteristic flavour of 'Hass' avocado, as represented by a creamy, smooth and buttery texture with a minimum richness and grassy notes. Thus, aroma compounds would be a complementary attribute to oil content, contributing to the overall quality of avocado fruit.

Maturity

Avocado is an unusual fruit in that it can be held on the tree for very long periods after reaching physiological maturity (defined as the ability to ripen after harvest). In some environments, fruit can be held on the tree for one year or more after physiological maturity. It is commercially important to identify the minimum maturity (i.e. the maturity standard) that ensures an acceptable eating experience and minimal fruit disorders while allowing early harvest to take advantage of higher early-season prices. Harvesting fruit at very early maturity can lead to poor fruit quality due to shrivelling; a rubbery texture; ripe, stringy vascular tissue; insipid flavour and increased rot (Lee *et al.*, 1983; Pak *et al.*, 2003a).

Maturity and sensory perception

The most frequently referenced early avocado maturity evaluations used basic taste tests and descriptions (Lewis, 1978; Lee, 1981). More formal consumer evaluations using a smaller number of panellists, a hedonic scale and 'Hass' fruit sampled over the season indicated that 23% dry matter (DM) was required to achieve 'like slightly' responses (Arpaia *et al.*, 2003a). Studies with more than 100 consumers and fruit harvested from several regions to provide a 20–40% DM range indicated similar results, and the 'liking' score and intent to purchase increased with increasing % DM (Gamble *et al.*, 2010). A DM of greater than 26% was needed to achieve a 70–89% chance of purchase. Chen *et al.* (2009) found no effect of harvest date on flavour score for 'Sharwil', possibly because the lowest % DM of the tested fruit was 21.9% (11.9% oil content), which was probably adequate to provide good flavour.

Full sensory descriptions of avocado flavour were performed in New Zealand (Yahia and Woolf, 2011) and California (Obenland *et al.*, 2012). New Zealand 'Hass' were described by attributes of 'hay/dried grass', 'green' and 'nutty' odours; 'woody/pine', 'savoury' and 'canned pea' flavours; 'bitter' taste; and the texture terms 'oil release', 'water release' and 'fibrousness'. Californian descriptors (carried out for 'Hass' from California, Peru, Mexico and Chile) used terms such as 'creaminess', 'richness', 'grassiness', 'blandness', 'nuttiness' and 'butteriness'. Changes in volatiles were also examined using gas chromatography–mass spectrometry (GC-MS), some of which increased and others decreased with harvest date. These tools should allow better definition of optimum harvest maturity for 'Hass' and other cultivars.

Ethylene inhibitors affect volatile compounds differently in many fruits. In avocado, Pereira *et al.* (2013) demonstrated that the ethylene inhibitor 1-MCP delayed ripening and mainly affected the flavour of the fruit in the middle stage of ripening. Studies in climacteric fruits have shown that the production of aroma volatiles is mainly derived from fatty acids modulated by ethylene during ripening (Yahia, 1994; Defilippi *et al.*, 2009). These free fatty acids, or those liberated by lipase activity and further metabolized by β-oxidative enzymes and/or lipoxygenase (LOX), are generally regarded as being the main precursors of esters, alcohols and aldehydes (Fellman *et al.*, 2000). These compounds change as the fruit develop; for example, short-chain aldehydes are predominant in immature avocados, particularly hexanal and E-2-hexenal compounds; their concentrations then decrease during maturation, contributing to the decline of the herbaceous aroma (El-Mageed, 2007; Obenland *et al.*, 2012). García-Rojas *et al.* (2016) demonstrated that changes in the linolenic acid concentration, a primary substrate for volatile aroma biosynthesis, and the relative expression of *PamLOX*, the key enzyme in the lipoxygenase pathway, responded to ethylene modulation during storage in 'Hass' avocado, which modified the final production of hexanal as ripening progressed to the ready-to-eat stage.

Oil content and dry matter

The oil concentration of avocado fruit increases during fruit development and is a significant determinant of the eating and nutritional quality of the fruit (Hofman *et al.*, 2002). It is one of the early maturity indices (Lee *et al.*, 1983) and is still the most reliable indicator of maturity (Kaiser, 1994). However, the cost and difficulty of measuring oil concentrations have resulted in a search for alternatives. Per cent dry matter (% DM) is strongly related to oil content and quality (Lee *et al.*, 1983; Brown, 1984; Ranney, 1991). The oil content and % DM of 'Hass' from a wide range of growing areas in New Zealand, Australia, California and Hawaii showed a remarkably strong correlation ($r^2 = 0.96$) (e.g. Fig. 15.2), particularly given the wide range of rainfall, soil types and climates (Hofman *et al.*, 2008a; Chen *et al.*, 2009; Woolf *et al.*, 2009). Generally, the oil content is approximately 11 units less than % DM (e.g. 20% total oil content equates to approximately 31% DM). Because of this strong relationship, % DM is used routinely as the maturity index in Australia, Israel, New Zealand, Chile and the USA, whereas South Africa uses the reciprocal, maximum flesh moisture content, pioneered by Swarts (1978). Legal % DM maturity indices in several countries were summarized by Hofman *et al.* (2002a). Recent changes include increasing the Australian 'Hass' minimum standard from 21% to 23% ('Shepard' remains the same at 21%).

The prediction of the earliest harvest date based on changes in % DM over time has been investigated (Ranney, 1991) but was found to be unsatisfactory because of the large influence of production location (even over small distances) on the time to attain minimum % DM (Coggins, 1984). Oil accumulates until harvest; thus, within a location, late-season fruit have much higher concentrations of oil than early-season fruit, ranging between 9% and 14% (Hernández *et al.*, 2016). Structural lipids are part of the cell membrane (phospholipids and glycolipids), and storage lipids (triglycerides) are found in the idioblasts (Requejo-Tapia, 1999). The fatty acid profile results from adaptation to the environment (Blakey, 2011), and fatty acids have recently been proposed as potential biomarkers to distinguish avocado fruit-growing areas (Donetti and Terry, 2014). Avocados grown in cooler climates present a higher proportion of monounsaturated to saturated fatty acids (Requejo-Tapia, 1999). Minimal changes in the fatty acid composition have been reported postharvest (Luza *et al.*, 1989). However, an increase in oil content was observed during storage and ripening. This increase is related to postharvest dehydration and lipid recovery from partial cell wall breakdown (Pedreschi *et al.*, 2016, 2019).

Dry matter content (DMC), the gold standard avocado harvest index, which is highly correlated with oil content, is extensively used to harvest 'Hass' avocados to comply with the minimum standards to guarantee good eating quality. Non-destructive tools such as near-infrared spectroscopy (NIRS) are being implemented online in packing houses in an attempt to better segregate 'Hass' avocados at the gate and reduce the observed postharvest ripening heterogeneity in long-distance markets (Pedreschi *et al.*, 2014).

Maturity is commonly indexed by the related measures of oil, DMC or moisture content (Parodi *et al.*, 2007). Oil content and the related DMC increase with maturation time and are related to flesh creaminess and smoothness and thus the eating quality of ripened fruit (Hofman *et al.*, 2013). Arpaia *et al.* (2003a) and Gamble *et al.* (2010) established avocado fruit DMC specifications by cultivar in the context of consumer acceptance (i.e. eating quality). A (cultivar-specific) minimum DMC specification has been set in most significant production regions to ensure a minimum maturity before harvest and acceptable eating quality of ripened fruit (Magwaza and Tesfay, 2015). In some regions, a maximum DMC specification has also been established to reduce physiological disorders in stored fruit (Subedi *et al.*, 2020).

DMC has been accepted worldwide as an indicator of harvest maturity, and a minimum percentage of DM, standardized for each avocado cultivar, will ensure the continuation of the ripening process after harvest (OECD, 2004). The % DM is highly and proportionally correlated with the percentage of oil in 'Hass' avocado. Moreover, the DM content has been linked with the potential marketability and intention to buy avocado fruit (Rivera *et al.*, 2017).

Other indices

Attempts have been made to develop simpler and more reliable maturity indices. Plumbley *et al.* (1989) investigated the change in the fruit growth rate to predict the earliest harvest date based on the assumption that the growth rate would be minimal at maturity. However, this method was unreliable because of variability among cultivars and production locations. Additionally, smaller fruit may have higher % DM than larger fruit (Sippel *et al.*, 1992; Hofman and Jobin-Décor, 1999), probably due to earlier horticultural maturity caused by premature seedcoat senescence.

The activity of the softening enzymes cellulase and pectin methylesterase (PME) has been investigated as a maturity index based on their involvement in the softening of avocado fruit. The PME activity on the day of harvest decreased with increasing fruit maturity. Additionally, PME activity decreased more rapidly after harvest in less mature fruit than in more mature fruit (Zauberman and Schiffmann-Nadel, 1972). On this basis, it was suggested that PME activity might be a maturity indicator. However,

Kaiser *et al.* (1995) found that PME activity was too variable between and within replications in two seasons. In addition, cellulase activity increased from approximately 40 units g^{-1} at harvest to 210 units g^{-1} when ripe, but the increase in activity was inconsistent between harvests in two seasons and was also considered unsuitable as a maturity index (Kaiser *et al.*, 1995).

Fruit firmness has been used for many years as a measure of avocado ripeness. Considerable efforts have been made to develop firmness assessment units that can be used in packing lines (Peleg *et al.*, 1990). These include the Sinclair IQ™ Firmness Tester and Aweta Acoustic Firmness Sensor (Arpaia *et al.*, 2006). These are based on concepts such as acoustic responses to a mechanical impulse and ultrasonic wave transmission over the fruit surface, respectively (Self *et al.*, 1994; Mizrach and Flitsanov, 1999). However, these devices were not developed to determine horticultural maturity but only to assess the stage of ripeness of harvested fruit, as indicated by fruit softening. They have limited use as maturity indicators.

The skin colour of 'Hass' and 'Gem' avocados changes from green to purple/black during ripening (Guerrero and Benavides, 2014; Magwaza and Tesfay, 2015). The fact that the skin turns purple/black as these avocado varieties ripen postharvest has made skin colour change an indicator of ripeness (Cho *et al.*, 2020, 2021). Arzate-Vázquez *et al.* (2011) applied image processing to classify the ripeness of 'Hass' avocado during the ripening process and reported that the colour and texture features evaluated from the images showed an adequate correlation with avocado firmness. Karimi *et al.* (2012) applied hyperspectral imaging to classify the colour of 'Hass' avocado subjected to different treatments and reported that it could classify ripeness with more than 92% accuracy. Machine learning techniques can be applied to colour change in agro-food sectors and can find patterns in a large data set (Cho *et al.*, 2020). Cho *et al.* (2021) evaluated the accuracy of machine learning and deep learning models developed by combining colour features (L*a*b* and YUV) extracted from RGB images and firmness information of 'Hass' avocados. They suggest that an artificial neural network (ANN) based on colour components provides a useful model for estimating firmness in 'Hass' avocado during storage. The correlation coefficients between the colour features (L*a*b* and YUV) and firmness of 'Hass' avocados ranged from −0.72 to −0.83. The L*a*b* and YUV values of 'Hass' avocado skin colour successfully estimated the firmness of 'Hass' avocados ($R^2 = 0.94$, RMSE = 0.38 and RPD = 4.03). However, more experiments will be needed under different storage conditions, such as temperature and humidity ranges, to achieve more comprehensive estimations in the future (Cho *et al.*, 2021).

Sugar content is the most important quality parameter used to indicate the sweetness of many horticultural products by research laboratories and industry to determine marketing standards (Magwaza and Opara, 2015). Non-structural sugars (sucrose, glucose and fructose) and sugar alcohols (e.g. sorbitol and mannitol) constitute the majority (approximately 85%) of total soluble solids in many fruits (Magwaza and Opara, 2015). In avocado, the disaccharide sucrose and its component hexoses, fructose, glucose and seven carbon (C_7) sugars (D-mannoheptulose, its reduced form polyol and perseitol) are dominant sugars constituting at least 98% of the fruit total soluble sugars (Liu *et al.*, 1999a,b; Cowan, 2004). C_7 sugars have been found to comprise more than half of avocado fruit total soluble sugars; the balance is accounted for by the more common hexose sugars, glucose and fructose (Liu *et al.*, 1999b). As a result, avocado fruit maturity has recently been related to the concentration of mesocarp C_7 sugars. Liu *et al.* (1999b) showed that as avocado fruit grew in size, the flesh accumulated proportionally higher levels of total soluble solids, coinciding with the increase in dry weight. The authors also showed that the C_7 sugars, D-mannoheptulose and perseitol, were the major non-structural carbohydrates present in the fruit tissues of 'Hass' avocado and as such appeared to play an essential role in fruit growth. Total soluble sugars were a major biomass component for the young fruit during the early rapid development stage and stopped accumulating in fruit tissues when the fruit reached the minimum maturity stage (20.8% DM for 'Hass' avocado). At this point, the flesh begins to accumulate oil, which may explain the cessation of sugar accumulation. These results and those reported by Tesfay *et al.* (2012a,b) suggest the potential for using C_7 sugars and total soluble sugars to quantify avocado fruit maturity. However, given that sugar accumulation stops after reaching minimum maturity, the concentration of this chemical parameter cannot be used commercially to quantify maturity beyond the minimum maturity stage of 20.8% DM. In addition, analysis of individual sugar concentrations requires time-consuming sample preparation and analysis using high-performance liquid chromatography or GC-MS and hence is not practical in commercial setups (Magwaza and Tesfay, 2015).

Factors affecting maturity indices

In some countries, one standard is used for several cultivars and production locations. However, cultivar, cultivation practices, season and micro- and macro-environments can influence the relationship between fruit oil content, % DM and fruit quality (Lahav and Kalmar, 1977; Lee *et al.*, 1983; Coggins, 1984; Kruger *et al.*, 1995). For example, Stahl (1933) noted large differences in % DM between two locations in Florida and attributed the differences to high rainfall in one of the districts, causing a decrease in % DM. Thus, Coggins (1984) suggested that production location influences % DM to such an extent that it is an unreliable predictor of the earliest acceptable harvest date. However, Ranney (1991) concluded that season and locality effects on % DM were not significant in California. When oil content is used as a maturity index, the influence of soil moisture or rain can be reduced by determination on a dry rather than a wet basis (Kruger *et al.*, 1995). To overcome some of these influences, more specific maturity standards can be applied. In South Africa, export regulations require the % DM of all cultivars to be > 20% to prevent uneven ripening and shrivelling, except for 'Hass' and 'Ryan' (>23%) and 'Edranol' (25%). Fruit from the cooler areas of KwaZulu-Natal reach higher oil concentrations before other fruit maturity parameters are satisfied. Therefore, it is recommended that the fruit have minimum DMs of 25% ('Fuerte'), 28% ('Edranol') and 30% ('Hass' and 'Ryan') for export (Kaiser and Wolstenholme, 1994b).

Fruit sampling procedures can also affect the accuracy of a maturity test. The % DM can be affected by position in the canopy and flowering date (Plumbley *et al.*, 1989; Vuthapanich, 2001) and by storage conditions between harvest and % DM determination (Hofman and Jobin-Décor, 1999). Sampling fruit of different sizes, damaged fruit (Hofman and Jobin-Décor, 1999) and the portion of the flesh used (Schroeder, 1986) may also affect the % DM result. As flowering can last up to 6–8 weeks, fruit age at harvest will vary correspondingly. In instances where only one maturity index is used, the previously mentioned factors require that the maturity standard be sufficiently conservative to prevent marketing fruit that attain the standard but do not ripen to acceptable quality. However, this will invariably result in some fruit not being marketed as soon as commercial maturity is reached. Thus, greater consideration should be given to adopting different maturity standards for each cultivar and for production districts where the growing conditions are very distinct (Kaiser, 1994). The simultaneous use of two indices (e.g. 21% DM and fruit capable of ripening without shrivelling) will further reduce the risk of marketing immature fruit. This is pertinent in view of the multiple maturity indices used for other fruits, for example, citrus and apple. For example, a proportion of the observed variability in fruit quality and ripening is due to the broad range of conditions in which the trees are grown, particularly with respect to environmental factors such as the temperature during the growing season (Arpaia, 1994; Sams, 1999; Woolf and Ferguson, 2000) and level of sun exposure (Woolf *et al.*, 1999, 2000).

Fruit-ripening behaviour and postharvest quality can be associated with different cultivation practices, including irrigation management (Bower, 1988), and with the composition of macronutrients and micronutrients such as calcium (Witney *et al.*, 1990; Thorp *et al.*, 1997), nitrogen (Arpaia *et al.*, 1996) and zinc (Vorster and Bezuidenhout, 1988) within the fruit. Based on our findings, we can postulate that predicting the postharvest behaviour of 'Hass' avocado using a single preharvest variable such as DM or calcium content could be a misleading simplification of reality because several factors, including climate/environment, horticultural management and physiological variables directly (e.g. via fruit mineral content) or indirectly (e.g. via metabolites from photosynthesis), influence the ripening behaviour of 'Hass' avocado fruit. Our study showed that the cumulative degree days, DMC, fruit calcium content, fruit firmness at harvest, irrigation management at bloom, mean minimum temperature and leaf area index are important variables significantly influencing the overall ripening behaviour of 'Hass' avocado fruit. Therefore, the next step will be to identify quantitative thresholds for each preharvest variable affecting the attributes determining quality, which would be a major contribution to growers and exporters for guaranteeing an optimum and consistent avocado at the consumer level (Rivera *et al.*, 2017).

Maximum maturity standards

Late-maturing cultivars are often 'stored' on the tree, particularly in cool climates, to supply high-priced markets when fruit availability is low. These fruit are often more prone to disease (Vuthapanich *et al.*, 1995b; Hofman *et al.*, 2000), rancidity (Erickson *et al.*, 1970) and faster poststorage ripening. A maximum maturity standard may help reduce losses when fruit are harvested when they are too mature. However, the presence of off-flavours is not always related to oil content and appears to vary with season and district (Erickson *et al.*, 1970). Additionally, the % DM does not always increase with later harvests (Hofman *et al.*, 2000, 2008b) and varies with production location and season (Vuthapanich, 2001), suggesting that this is not a good late-maturity standard. Fuchs and Zauberman (1987) found that cellulase and PME were higher and lower, respectively, with late harvest and may be used as indicators of the impact of 'late hanging' on the risk of ripening during storage. The authors did not consider the effect of late harvest on flavour in their study.

The seven carbon (C_7) sugars, perseitol and mannoheptulose, are the main sugar forms in avocado fruit, and higher concentrations during ripening may reduce physiological flesh disorders (Meyer and Terry, 2010). Sugar concentrations in the flesh vary with maturity and decreases in sugar concentrations may reduce energy reserves and the quality of late-harvested fruit (Bertling and Bower, 2005). However, perseitol concentrations varied inconsistently with harvest date over two seasons, whereas mannoheptulose concentrations declined in a consistent manner (Lallu *et al.*, 2005; Burdon *et al.*, 2007). This suggests that mannoheptulose may reflect a decline in fruit quality that is common in late-harvested fruit, but this needs to be confirmed. Soluble solids, total sugars, sucrose, fructose and glucose do not vary consistently with harvest date (Lallu *et al.*, 2005). The water content of the seed, as estimated by nuclear magnetic resonance (NMR), and the degree of germination of the seed are also unlikely to provide an estimate of the quality of late-harvested fruit (Lallu *et al.*, 2005).

Non-destructive assessment

Due to considerable variation among fruit in an orchard or even within the same tree, destructive sampling using % DM, moisture content or even oil content can be considered to be no more than a general guide. Sample size and procedures become critical and may vary between operators. The net effect is that at the transition from immature to mature, failure to accurately predict the effects of maturity, such as the potential to ripen normally, will result in fruit of variable quality in the market. This is particularly important to prepackers who require consistent ripening across and between cartons to simplify their logistics. The disadvantage of these methods, apart from their labour-intensive and destructive nature, is that growers face the risk of delivering poor-quality fruit to the market due to fruit quality variation within a batch (Olarewaje *et al.*, 2016). Poor-quality fruit resulting from immature fruit in a batch may lead to uneven ripening and consumer dissatisfaction, affecting the future purchase of the fruit (Kruger and Lemmer, 2007; Kruger and Magwaza, 2012). Intravarietal differences in fruit composition, as well as difficulties in controlling and forecasting avocado fruit-ripening patterns, have been recently reported as major causes of concern among consumers and importers (Donetti and Terry, 2014). Non-destructive techniques for maturity testing (which relate to ripening) would be highly advantageous, particularly if installed on the packing line, allowing for sorting of fruit into more homogeneous classes than presently possible. Non-destructive techniques detect fruit quality parameters due to differences in optical characteristics, sound and density. Several non-destructive methods for maturity evaluation and quality parameters of avocado fruit have been developed (Olarewaje *et al.*, 2016). These methods include NMR (Chen *et al.*, 1993), vibrational energy (Peleg *et al.*, 1990), acoustic sensors (Galili *et al.*, 1998), Vis/NIRS (Schmilovitch *et al.*, 2001; Clark *et al.*, 2003) and ultrasonic excitation. For example, Chen *et al.* (1993) and Kim *et al.* (1999) showed that nuclear magnetic resonance (NMR) has the potential to estimate the mesocarp oil content of avocado fruit. Developments in NMR equipment design indicate the potential for in-line maturity testing in packhouses (Kim *et al.*, 1999). This technology may also have potential for non-destructive detection of disorders (Bergh *et al.*, 1989; Sanches *et al.*, 2003; Mazhar *et al.*, 2011). Mizrach and Flitsanov (1995a,b) showed that DM and oil content were correlated with ultrasonic parameters of avocado fruit, indicating the potential of the ultrasonic system for non-destructive prediction of avocado maturity and ripening patterns. Other non-destructive techniques for quantifying avocado fruit maturity parameters include ultrasound (Flitsanov *et al.*, 2000; Mizrach, 2008) and visible to near-infrared (Vis/NIR) spectroscopy (Vis/NIRS, Clark *et al.*, 2003; Wedding *et al.*, 2013).

NIRS is used commercially in Japan for in-line estimation of sweetness in apples, peaches and pears (Kawano, 1994) and as a commercial test for oil content in oilseeds (Williams and Sobering, 1993). Near-infrared measurement is rapid, and equipment costs are lower than for NMR, making NIRS a possible alternative in packing lines as a non-destructive indicator of maturity. However, the use of the technology does not appear as advanced with avocados compared with other fruit industries. This may be partially due to a lack of clarity over what to test. Sugars are often the target where NIRS is successfully used for other fruit crops. In other commodities, oils have been used, and proteins have been estimated in animal feeds. Clark *et al.* (2003) estimated the % DM in 'Hass' avocado fruit with reasonable accuracy. Blakey *et al.* (2009) conducted similar work but attempted to go further by determining water content and relating this to ripening. Based on this and additional work by Blakey (unpublished results), it appears that NIRS could separate fruit into broad groups with respect to % DM, water content and ripening potential under commercial conditions. The difficulty in utilizing the technology relates to the speed of scanning for use on a packing line due to the multiple wavelengths needed for accurate results as well as optimal penetration depth into the fruit. It is necessary to measure compounds in the fruit flesh without interference from the rind. The roughness of some avocado exocarps (such as 'Hass') may cause wavelength scatter and reflection before adequate penetration, which could vary from fruit to fruit. Regardless, validation using large sample numbers has shown the technique to have promise (Blakey, 2011; van Rooyen, 2011; Wedding *et al.*, 2011).

The potential for other non-destructive maturity techniques appears limited. A number of authors cite work relating to fruit skin colour changes. For example, a patented apparatus for non-destructive maturity determination of peaches was developed based on spectral changes due to chlorophyll loss in the fruit, which was linked to internal quality related to maturity (Ziosi *et al.*, 2008). However, the best results were obtained for fruit that had a visible colour change related to maturity. There is no evidence that this can be used for avocado.

In conclusion, % DM has served relatively well as a maturity standard for avocado fruit, but problems with the sometimes-poor relationship between % DM and ripe fruit quality justify continued work on maturity standards. Improved technologies such as more efficient per cent oil analysis techniques may allow more cost-effective use of this parameter. NIRS may also allow in-line assessment of maturity, with the potential to predict the time for fruit to ripen (Clark *et al.*, 2003; Blakey *et al.*, 2009; van Rooyen, 2011). Techniques that rely on fruit softness do not have application for maturity determination but are useful to predict the number of days for fruit to fully ripen once softening has commenced. Compositional changes in late-harvested fruit are less consistent than those during early harvest, making it more difficult to develop a reliable late-maturity standard. The latest harvest date based on district and season, which considers the risk of increased rancidity and disorders (Erickson *et al.*, 1970; Hofman *et al.*, 2000), may be the best option.

Effects of Production Practices on Postharvest Quality

There is a growing awareness of the significance of preharvest/postharvest interactions on horticultural products, as highlighted by several reviews (Monselise and Goren, 1987; Beverly *et al.*, 1993; Arpaia, 1994; Hofman and Smith, 1994; Hewett, 2006). Hofman *et al.* (2002a) described these interactions in some detail.

Fruit-ripening behaviour and postharvest quality can be associated with different environmental factors such as temperature, light availability, the level of sun exposure and location (Arpaia, 1994; Sams, 1999; Woolf *et al.*, 1999; Hernández *et al.*, 2016), which are associated with different cultivation practices such as irrigation management and in turn affect shoot vigour and nutritional status of the plant (Bower, 1988; Thorp *et al.*, 1997; Hernández *et al.*, 2016). Rivera *et al.* (2017) postulated that predicting the postharvest behaviour of 'Hass' avocado using a single preharvest variable, such as DM or calcium content, could be a misleading simplification of reality because several factors, including climate/environment, horticultural management and physiological variables directly (e.g. via fruit mineral content) or indirectly (e.g. via metabolites from photosynthesis), influence the ripening behaviour of 'Hass' avocado fruit.

Skin colour

Large variation in skin colour occurs in 'Hass' fruit ripened under the same conditions, but there is no understanding of the causes. The skin colour of ripe 'Hass' fruit can vary from tree to tree (Hofman and Jobin-Décor, 1999; Marques *et al.*, 2006) and be affected by production location (Vuthapanich, 2001), GA and girdling (Espindola-Barquera *et al.*, 2006) and possibly by sun exposure (Woolf *et al.*, 1999a; Cox *et al.*, 2004). Harvesting during hot conditions and delaying precooling for 24 h can result in darker fruit (Considine *et al.*, 2005). Additionally, very mature 'Hass' fruit can develop a darker colour on the tree (Hofman and Jobin-Décor, 1999; Vuthapanich, 2001), which is not associated with ripening (Cox *et al.*, 2004). These later harvested fruit can also have darker skin colour (lower chroma and hue angle) at eating soft (Hofman and Jobin-Décor, 1999).

Skin colour in many fruits is influenced by light and plant nutrition, particularly nitrogen (N) (Monselise and Goren, 1987; Hofman and Smith, 1994), but there is little understanding of these effects on avocado (apart from sunburn on exposed fruit). One of the common effects of high N on ripe fruit skin colour is either higher fruit skin chlorophyll concentrations at harvest or reduced chlorophyll loss during ripening. The dominant role of anthocyanin production in the development of black skin colour in 'Hass' (Cox *et al.*, 2004) suggests a relatively minor role of chlorophyll and N nutrition on the skin colour of ripe 'Hass' fruit.

Eating quality

Avocado eating quality is largely determined by oil content (Lee, 1981), which can be influenced by harvest date (see the 'Maturity' section of this chapter). However, cultivars with higher oil content do not necessarily have better eating quality, indicating that other fruit components are involved. Subtropical cultivars with a nutty flavour distinctly differ from the less oily and slightly sweeter tropical cultivars.

There is scant reference in the literature to the importance of texture in eating quality, except for the role of oil in the 'smoothness' of the flesh. However, hard lumps and firm, 'rubbery-textured' flesh around the seed can affect eating quality. Prolonged periods of inadequate rain during early fruit growth and/or high rainfall (> 500 mm in 10 days) just prior to maturity have been associated with this disorder in 'Fuerte' (Sanewski, 1984), and insect attack from spotting bugs can result in 'stones' in the flesh just under the seed (Waite and Martinez Barrera, 2002). The vascular strands in the flesh occasionally become stringy, particularly following long storage.

Ripening

There can be considerable intrinsic variation in fruit shelf life (days from harvest to eating ripe at ambient temperature), particularly if fruits are not stored before ripening. For example, Smith *et al.* (1992) noted that 'Hass' fruit harvested at the same time from one orchard block can take 4–26 days to ripen, and Hopkirk *et al.* (1994) reported a range of 5–12 days after cool storage at 6°C for 7 days. Maturity at harvest appears to be a factor in this variability. More mature fruit (as determined by % DM and harvest date) typically ripen more quickly, but this is not universally observed (Vuthapanich, 2001). The fact that larger fruit often ripen more quickly may result from their greater maturity (Eaks, 1980; Continella *et al.*, 1992). However, this may not be the case for cultivars that have a long flowering period such as 'Pinkerton' (Sippel *et al.*, 1992), and in some seasons, 'Hass' (Hofman and Jobin-Décor, 1999). In these cases, fruit size is a poor indicator of % DM. Fruit from higher-yielding trees can ripen more slowly (Hofman *et al.*, 2002c) but not in all cases (Cutting and Vorster, 1991).

Fruit mineral concentrations also influence ripening. Calcium infiltration delays softening in avocado (Tingwa and Young, 1974; Eaks, 1985), and fruit with higher flesh Ca concentrations at harvest often ripen more slowly (Witney *et al.*, 1990; Cutting *et al.*, 1992; Hofman *et al.*, 2002c). These effects occur partly through a delay in respiration and ethylene production (the climacteric state) (Eaks, 1985) and a general slowing of senescence. Higher concentrations of other divalent cations (e.g. manganese (Mn), cobalt (Co) and magnesium (Mg) in tomato) can have beneficial effects, but these are generally less than those of Ca (Wills and Tirmazi, 1979). Lower N (Abou-Aziz *et al.*, 1975; Arpaia *et al.*, 1996) and higher boron (B) (Smith *et al.*, 1997) have also been associated with longer shelf life.

Plant growth regulators (PGRs) other than ethylene can influence ripening. Cultar® (paclobutrazol) application at flowering can delay ripening (Vuthapanich *et al.*, 1995a), but this may occur through delayed maturity or higher Ca concentrations in the fruit (Vuthapanich, 2001).

Factors influencing the ripening rate can affect ripe fruit quality. For example, when harvested from a single orchard and stored and ripened under the same conditions, fruit that ripened more quickly had less rotting than those taking longer to ripen (Hopkirk *et al.*, 1994). However, ripe fruit with higher flesh Ca concentrations at harvest can ripen more slowly and have less disease (Hofman *et al.*, 2002c). This apparent contradiction may be related to the high Ca concentration reducing disease growth (Penter and Stassen, 2000; Manganaris *et al.*, 2005) sufficiently to prevent the longer ripening times from increasing disease severity in the ripe fruit.

Fruit diseases, physiological disorders and mechanical damage

Storage conditions are critical to the onset of postharvest diseases and disorders. They include the time associated with postharvest handling, temperature and gas composition, as well as the degree of fruit maturity at harvest time (Burdon *et al.*, 2013; Hernández *et al.*, 2016; Pedreschi *et al.*, 2016). The development of fruit diseases identified at postharvest stages typically begins as preharvest quiescent processes. This condition poses a challenge for their management because disease presence is identified when fruit begin to soften or rot; these symptoms generally occur at the market stage when it is not possible to control the problem (Hartill and Everett, 2002; Sharma *et al.*, 2017).

Susceptibility to fruit diseases and disorders can increase with increased maturity (Cutting and Wolstenholme, 1992; Kruger *et al.*, 2004; van Rooyen and Bower, 2005), warmer orchard temperatures, higher rainfall and older trees (Kruger *et al.*, 2004). There is a general relationship between higher crop load (Hofman *et al.*, 2002c, 2005; Kruger *et al.*, 2004) (Fig. 15.3), more vigorous trees (Vorster *et al.*, 1989) and smaller fruit (Hofman *et al.*, 2002c) having fewer diseases and disorders. Freeze injury susceptibility can be more severe in larger fruit (Fuchs *et al.*, 1975). The time of fruit set can also affect disorder development. After 4 weeks of cold storage, early set 'Pinkerton' fruit had more postharvest physiological disorders such as lenticel damage, skin discoloration and grey pulp than late-set fruit (Sippel *et al.*, 1995). Additionally, discrete patches on the skin from CI can decrease with maturity (Vorster *et al.*, 1987). However, the effect of maturity on these disorders can vary with production location (Vuthapanich, 2001).

Calcium is the mineral most strongly implicated in physiological fruit disorders, probably because of its role in the cell wall, membrane structure and function and as a catalyst and cofactor for a number of enzymatic reactions (Poovaiah *et al.*, 1988). The relationship between high fruit Ca concentrations and reduced flesh disorders has been demonstrated in many fruits (Shear, 1975), including avocado (Thorp *et al.*, 1997; Hofman *et al.*, 2002a,c; Willingham *et al.*, 2004). However, in some cases, fruit Ca concentration does not appear to be strongly related to flesh disorders (van Rooyen and Bower, 2005), possibly because of small differences in fruit Ca concentration and/or other factors having a more dominant effect.

Calcium availability during the first 12 weeks of fruit growth is important to ensure adequate Ca for good cell wall and cell membrane formation during the early stages of cell division and cell expansion. However, it can be difficult to increase fruit Ca concentrations through soil or foliar applications because of the behaviour of Ca in the soil, the competition between vegetative growth and fruit for Ca uptake, and the poor movement of Ca across the fruit skin. Soils with low cation exchange capacity (common in tropical/subtropical avocado-growing areas) have a reduced capacity to retain Ca in the upper root zone where maximum root nutrient uptake occurs. In these cases, less soluble forms of Ca (e.g. granular gypsum) or smaller, more frequent Ca applications of more soluble forms (e.g. microfine gypsum) are preferred (Hofman *et al.*, 2007). Maintaining adequate vegetative/reproductive balance can increase the proportion of Ca translocated to the fruit in the xylem stream (van Rooyen and Bower, 2005).

Potassium (K) and Mg have also been implicated in disorders of various fruits (Ferguson, 1980), including avocado, often through ratios such as (Mg+K)/Ca or (Ca+Mg)/K (Witney *et al.*, 1990; Vuthapanich, 2001; Willingham *et al.*, 2006; Hofman *et al.*, 2007). These relationships are not always observed. Koen *et al.* (1990) identified a positive relationship between proximal fruit K concentrations and grey pulp but a strong negative correlation between leaf and soil (Mg+Ca)/K and grey pulp. These ratios also did not appear to be significant in internal disorders of 'Pinkerton' fruit (van Rooyen and Bower, 2005).

High fruit N is often associated with increased flesh disorders and rotting; 'Pinkerton' fruit in South Africa obtained from trees grown in mesic areas on fairly fertile soils with high organic matter content often had more flesh disorders after storage than those from trees grown on more sandy soil in a slightly cooler climate (van Rooyen and Bower, 2005). Fruit from trees on richer soils also had higher flesh N concentrations, and there were strong correlations between flesh N concentrations and disorders. Similar relations between higher fruit N concentrations and increased occurrences of rots and disorders have been observed in 'Hass' (Marques *et al.*, 2003; Willingham *et al.*, 2006). Soil N application trials have had mixed effects on fruit disorders, sometimes increasing ripe fruit disorders (Arpaia *et al.*, 1996) and, in other cases, having little effect (Willingham *et al.*, 2006). This may reflect the existing soil and tree N status at the start of the trials. Higher N could affect fruit quality by increasing vegetative growth and competition with the fruit, thereby reducing fruit Ca concentrations (Shear and Faust, 1975) or carbohydrate reserves, reducing cell wall thickness and increasing fruit softening during storage (van Rooyen and Bower, 2005). Regardless, the 'Pinkerton' fruit quality problem, which threatens exports from South Africa, has been brought under control by reducing N fertilization on soils rich in organic matter (high N mineralization) and by imposing maximum and minimum fruit pulp N limits (Snijder *et al.*, 2002; Kruger and Magwaza, 2010). Fruit Ca concentration is more easily controlled indirectly by controlling shoot vigour through N fertilization.

With regard to other nutrients, low distal fruit phosphorus (P) has been associated with high avocado grey pulp incidence after cold storage (Koen *et al.*, 1990), and high fruit and leaf zinc (Zn) concentrations have been associated with lower pulp spot incidence in 'Fuerte' (Vorster and Benzuidenhout, 1988; Bezuidenhout and Vorster, 1991). High Fe and Zn and low Mn and copper (Cu) concentrations have also been associated with increased pulp disorders (van Rooyen and Bower, 2005).

Fruit mineral concentrations may sometimes be more easily manipulated by factors other than soil or foliar application. For example, differences in fruit disorders from adjacent 'Hass' trees on seedling rootstock (Hofman *et al.*, 2002c) and frequent effects of rootstock on scion fruit internal disorders and diseases have been identified, and these are often associated with rootstock effects on fruit flesh and skin mineral concentrations (Marques *et al.*, 2003;

Willingham *et al.*, 2006). The influence of high N is likely manifested through tree vigour. The effects of crop load and tree vigour may also manifest through fruit size and fruit mineral concentrations (Hofman *et al.*, 2002c). Factors that increase tree vigour apart from N, such as reduced *Phytophthora* root rot pressure, can also increase flesh diseases and disorders (Willingham *et al.*, 2004). Bower and van Lelyveld (1985) found that different irrigation schedules altered PPO activity in harvested 'Hass' fruit flesh. However, Vuthapanich (2001) noted that irrigation regimes had little influence on 'Hass' fruit quality.

Field temperatures close to harvest may also influence disorders. For example, sun exposure of avocado fruit on the tree resulted in high skin and flesh temperatures and a reduced incidence of postharvest heat injury and CI (Woolf *et al.*, 2000). This may explain the variation between fruit in skin CI (discrete patches), which is often observed during storage below approximately 4°C. Field temperatures above ~30°C have been associated with reduced rot and flesh discoloration in ripe 'Hass' fruit (Considine *et al.*, 2005). Flesh discoloration and rot can also increase if fruit are held above 30°C between harvest and precooling (Arpaia *et al.*, 1992), and this is more likely to occur at higher ambient temperatures during harvest.

Other fruit constituents during fruit growth may also influence postharvest performance. Van Rooyen and Bower (2006) found that mesocarp flesh discoloration in 'Pinkerton' was strongly related to the orchard of origin. This could be partially explained by the concentration of critical minerals in the fruit, but PPO activity was higher in fruit from 'poor fruit' orchards. The C_7 sugars in the mesocarp may also be important. The avocado mesocarp contains high concentrations of the C_7 sugars mannoheptulose and perseitol, whereas the C_6 sugars fructose, glucose and sucrose are present in very low concentrations (Liu *et al.*, 2002; Tesfay *et al.*, 2010). The C_7 sugar concentrations decrease significantly during ripening, possibly indicating a role in the ripening process (Liu *et al.*, 2002). They may act as an energy source (Bertling *et al.*, 2008; Meyer and Terry, 2010), and the double bond in mannoheptulose provides powerful water-soluble antioxidant properties, which may be a first line of defence against the negative effects of reactive oxygen species (ROS) (Tesfay *et al.*, 2010). The C_7 sugar concentrations in the flesh were somewhat lower in immature fruit and again in very mature fruit, which may explain the changes in the incidence of internal disorders in fruit harvested at different stages of maturity. Furthermore, it is possible that greater vegetative vigour reduces C_7 sugar allocation to fruit, thus reducing antioxidant capacity and increasing susceptibility to flesh disorders. Hence, production conditions that optimize sugar reserves in fruit at harvest may reduce flesh quality loss during storage and ripening. However, attempts to increase fruit sugar reserves through B and Zn applications and girdling were unsuccessful (Bertling *et al.*, 2008).

Effects of Postharvest Practices on Quality

The impact of practices from harvest onwards on postharvest quality is discussed below and summarized in Table 15.3.

Harvesting and transport to the packhouse

When to harvest

Because of the effect of maturity on disorder susceptibility, the date of harvest is influenced by the minimum accepted maturity, market prices and the required storage or transport time (see the 'Effects of Production Practices on Postharvest Quality' and 'Effects of Postharvest Practices on Quality' sections in this chapter). The timing of the latest harvest is also influenced by the increased risks of pests and diseases, weather damage and off-flavours (although little work has been done in this latter area). In some growing environments, trees may also be forced into alternate cropping patterns if fruit remain on the tree in the next cropping cycle (Whiley, 2002). Other factors include the risk of low field temperatures that cause chilling and freezing injury, increasing fruit size with maturity and labour/machinery availability.

The time of day for picking should also be considered. Temperatures above 30°C between

Table 15.3. Postharvest factors that can reduce ripe fruit quality

Condition/causes	Prevention	References
Fewer days to ripen (no storage or after storage)		
Higher temperatures		Kok *et al.*, 2010; Lütge *et al.*, 2010
Exposure to ethylene, possibly during storage	Avoid ethylene exposure at all times (less than 0.05 $\mu l\ l^{-1}$)	Hofman *et al.*, 1995
More mature fruit		Burdon *et al.*, 2005
More moisture loss, particularly during early ripening	Higher humidity systems. 1-MCP, as moisture loss likely acts through ethylene	Bower, 2005; Burdon *et al.*, 2005; Blakey *et al.*, 2009
Skin colour ('Hass')		
Early-season fruit develop less black at ripeness	Delay harvest	Cox *et al.*, 2004
Lower ripening temperatures, particularly with low maturity fruit	Ripen at approximately 20°C	Hopkirk *et al.*, 1994; Cox *et al.*, 2004
Late-season fruit already coloured (black/purple) but not ripe		Cox *et al.*, 2004
Skin spotting (lenticel damage)		
Harvesting and packing practices	Reduce mechanical damage from impact, abrasion and compression	Hofman, 2005
Harvesting wet fruit (rain, dew)	Delay harvest for 48 h after rain	Pak *et al.*, 2003b
	Delay packing of wet fruit	Everett *et al.*, 2008
Less mature fruit are more susceptible	More care to reduce mechanical damage during harvest and packing of early-season fruit	Arpaia *et al.*, 1987; van Rooyen and Bezuidenhout, 2010
Later season fruit have more skin spotting after storage		Mandemaker, 2004
Discrete patches		
Storing 'Hass' at 0–3°C for longer than approximately 10 days. There will be time x temperature effects, and even short times (a few days) can result in discrete patches	Store above approximately 4°C. Alternatively, use low-temperature conditioning treatments prior to storing at low temperatures (0–3°C). Decrease water loss during cooling	Hofman *et al.*, 2003; Bower and Magwaza, 2004; Dixon *et al.*, 2004b; Bower, 2005; Dixon *et al.*, 2008; Hofman *et al.*, 2010; van Rooyen and Bezuidenhout, 2010
Excessive skin spotting can increase discrete patches development after cold storage	Reduce skin spotting by reducing mechanical damage	Hofman, 2005
Production location	Unknown causes	Dixon *et al.*, 2008
Excessive moisture loss. More than 2–5% moisture loss at 2°C or 6–8% at 5.5°C	Reduce moisture loss by limiting forced-air cooling, using high humidity cold storage or MAP	Donkin and Cutting, 1994; Bower and Magwaza, 2004; Bower, 2005; Dixon *et al.*, 2008
Inappropriate maturity	Risk of discrete patches higher in early and late maturity	Bower and Palpi, 2006
Warmer temperatures at harvest. New Zealand fruit harvested in summer had more discrete patches than those harvested in autumn, and South African fruit harvested in winter had less than those harvested earlier or later	Winter harvest temperatures may provide some low-temperature conditioning before harvest	Toerien, 1986; Bower and Palpi, 2006; Dixon *et al.*, 2008

Continued

Table 15.3. Continued.

Condition/causes	Prevention	References
Shaded fruit	Sun-exposed side of fruit will have significantly fewer discrete patches, although overall fruit quality not as significantly affected. May be less effective in winter	Woolf *et al.*, 1999a, 2000
Flesh diseases (excluding use of fungicides)		
Excessive skin spotting	Do not allow fruit to drop more than 30 cm on to hard surfaces or other fruit during picking and packing. Minimize rubbing of fruit against hard surfaces or against other fruit	Mandemaker *et al.*, 2006: Not always the case; Everett *et al.*, 2008
Discrete patches. Flesh diseases and discrete patches increase with lower (less than approximately 3°C) storage temperatures. However, eliminating discrete patches using cold temperature conditioning reduces disease	Prevent discrete patches by using high storage temperatures or low-temperature conditioning	Hofman *et al.*, 2003; Dixon *et al.*, 2008
Increased ripening temperatures. Flesh diseases are higher when ripened at 20°C compared with 15–16°C	Ripen at 15–16°C. May reduce purple skin colour in early maturity fruit	Hopkirk *et al.*, 1994
Longer days to ripen. Not using ethylene. Using 1-MCP	Use ethylene to increase ripening rate. Reduce 1-MCP concentration or do not use 1-MCP	Dixon *et al.*, 2003a; Adkins *et al.*, 2005
Using ethylene after cold storage	Do not use ethylene if fruit are stored longer than 2–3 weeks	Arpaia *et al.*, 2006 Ethylene provides no additional benefit: Dixon *et al.*, 2003a; Hofman *et al.*, 2003
Delays between picking and cold storage	Place fruit in cold storage within 24 h of harvest	Dixon *et al.*, 2005; Kok *et al.*, 2010; Lütge *et al.*, 2010
Excessively long cold storage period	Match cold storage duration to fruit inoculum load/vitality. Strong increase in rot incidence with storage, particularly around the 4-week point. Step-down temperatures during transport	Vorster *et al.*, 1987; Dixon *et al.*, 2003b; Mandemaker, 2004
Short-term cold storage (4–16 days at 5°C) can reduce rots, primarily by reducing time to ripening and synchronizing ripening within a line		Hopkirk *et al.*, 1994; Hofman *et al.*, 2003; Arpaia *et al.*, 2006
Early and late maturity	Use mid-maturity fruit for more challenging scenarios such as long-distance transport	Dixon *et al.*, 2003b

Continued

Table 15.3. Continued.

Condition/causes	Prevention	References
Fruit overripe	Consume slightly firmer as rot severity increases significantly from just ripe to overripe	Everett, 2001; Elmsly and Dixon, 2008
CO_2 concentration during CA storage	Lower CO_2 concentrations reduce diseases in New Zealand but increase diseases in South Africa. May depend on fungal species	Lemmer *et al*., 2007; Burdon *et al*., 2008
High humidity during storage and ripening		Dixon *et al*., 2004a
Diffuse discoloration (mesocarp discoloration)		
Late-maturity fruit	Do not store late-maturity fruit for too long. This fruit may start ripening at the low storage temperature, which is a trigger to diffuse discoloration	van Rooyen and Bower, 2005; Woolf *et al*., 2005
Storage above 5°C	Store < 5°C to prevent the start of ripening. This needs balancing with susceptibility of fruit to discrete patches (i.e. the balance of internal and external chilling disorders). Use step-down temperatures during shipping	Bezuidenhout *et al*., 1992; van Rooyen and Bower, 2005
Exposure to ethylene before or during storage	Prevent exposure to ethylene	Pesis *et al*., 2002
Excessive delays between harvest and the start of storage	Reduce fruit temperature to the storage temperature within 24 h of harvest	Kok *et al*., 2010; Lütge *et al*., 2010
Lack of use of 1-MCP	Use 1-MCP to minimize the effects of ethylene and prevent the start of ripening during storage	Woolf *et al*., 2005; Kruger and Lemmer, 2007
Absence of controlled atmospheres	Used controlled atmospheres to prevent the start of ripening	Burdon *et al*., 2008
Excessive moisture loss before or during storage	Minimize delay between harvest and storage. Maintain greater than 90% RH during storage. Use fruit coatings or modified atmospheres	Bower, 2005; Kok *et al*., 2010; Lütge *et al*., 2010

harvest and the start of cooling can increase rot and flesh disorders (Arpaia *et al.*, 1992), and fruit can ripen more quickly and have less disease when ripe if harvested at field temperatures above 30°C (Considine *et al.*, 2005). However, in New Zealand's relatively cool spring and summer environment, mid-afternoon harvest did not consistently reduce flesh disease, possibly because the high soil moisture content did not result in sufficiently different fruit moisture contents between early morning and mid-afternoon (Elmsly *et al.*, 2007). The slightly higher disease in fruit harvested in the early morning during one year was suggested to result from higher fruit turgidity and increased mechanical damage during picking and packing (Elmsly *et al.*, 2007).

Rain just before harvest can increase fruit turgidity and susceptibility to mechanical damage.

For example, picking wet 'Fuerte' fruit can increase skin CI and pulp spot, particularly early in the season (Duvenhage, 1993). Skin spotting in 'Hass' can also increase if fruit is picked wet because of increased cell turgidity (Pak *et al.*, 2003b; Everett *et al.*, 2008), and increased skin spotting can increase skin CI if storage temperatures are too low (Hofman *et al.*, 2003; Hofman, 2005). Additionally, vascular browning in late-maturing fruit increases if harvested wet (Duvenhage, 1993). Fruit diseases can increase with as little as 5 mm of rain 24 h before harvest and rain is significantly correlated with increased skin spotting (Pak *et al.*, 2003b). Rain for more than 48 h or longer before harvest still increased skin spotting but reduced disease, presumably because the disease inoculum was washed away from the fruit. This may explain the variation in the impact of harvesting during rain on fruit rot observed by other researchers (Darvas, 1982; Duvenhage, 1993). Delaying harvest for 24–72 h after rain reduced skin spotting to prerain levels.

How to harvest

Fruit can be harvested either by manual clipping or 'plucking' (pulling the fruit from the tree, which results in a stem scar on the fruit where the stem is removed; sometimes called 'snap' picking). In the past, international markets preferred clipped fruit with the stem attached but are increasingly accepting plucked fruit.

Fruit can be clipped using secateurs (also called pruning shears or clippers) or picking poles, with or without standard or self-propelled mechanical ladders ('cherrypickers' or hydroladders). However, clipping is slower than plucking (Hofshi, 2002), and the sanitation of clippers may be important to reduce the spread of tree diseases such as sunblotch. The remaining stem can also cause mechanical damage to adjacent fruit during picking, transport and packing. Plucking may not be suitable with some cultivars (e.g. 'Shepard') because the skin can be torn at the stem insertion point, resulting in exposed flesh.

Plucking does not significantly affect weight loss or ripening in a number of cultivars (Eaks, 1973; Köhne and Kremer-Kohne, 1995). Research in New Zealand and California suggests that large differences can occur between the quality of plucked and clipped fruit and that plucked fruit from humid or high rainfall areas or seasons are likely to have more stem rot than clipped fruit (Woolf *et al.*, 1999b). Therefore, plucked fruit from these areas should be ethylene treated and preferably ripened at 15°C to reduce the risk of rot in ripe fruit. Clipping is advised for extended storage for long-distance transport. In some cases in South Africa, an anti-fungal compound is applied to the cut pedicel surface to reduce the risk of stem-end rot.

Skin spotting increases significantly and gradually as fruit progresses through harvest to packing (Hofman, 2005) (Table 15.4), most likely due to mechanical damage. For example, dropping fruit from as little as 0.1 m during harvest or on the pack line can cause skin spotting (Mandemaker *et al.*, 2006). However, rubbing against other fruit (Everett *et al.*, 2008) and solid surfaces and compression forces in bins during transport to the packhouse may also contribute to spotting. Practices reducing these forces will reduce skin spotting in the packed product and likely reduce discrete patches because of the relationship between skin spotting and discrete patches (Hofman, 2005). Picking and packing procedures do not normally cause significant flesh bruising (Hofman, 2002) but can increase flesh diseases, particularly with the dropping fruit from heights greater than 0.3 m (Mandemaker *et al.*, 2006).

From harvest to packhouse

Fruit are generally placed in 400 kg (Australia) or 360–380 kg (Israel) field bins or 18.5 kg (South Africa and South America) crates for transport to the packhouse. Bin size is a compromise between handling efficiency and machinery requirements for larger containers. Container depth is a compromise between excessive compression damage from the fruit above and having a larger percentage of the fruit on the top layer in shallow containers, which are more prone to bouncing during transport. Zauberman *et al.* (1969) noted that the main cause of fruit injury during transport was contact with the sides or the bottom of wooden bins. Therefore, the percentage of injured fruit in bulk bins (13.6%) was less than that in field boxes (33.4%), as relatively fewer fruit contacted the bins. Injury incidence was reduced by 50% when bins were lined with canvas. These effects will

Table 15.4. The effect of picking and packing on skin spotting severity (% of the skin affected) of 'Hass' avocado fruit. Samples were taken either directly from the tree (off-tree), from bins in the field or in the shed, after brushing, or at the end of the packing line. Assessments were performed either several days after harvest (for no storage) or at removal from cold storage. From Hofman *et al.* (2010), with permission.

Where sampled	No storage	5°C, 20 days	6°C, 3 days and 1°C, 17 days
Off-tree	1.8f[a]	2.5f	3.3f
Bin field	8.1de	4.7ef	12.0cd
Bin shed	13.1c	5.0ef	11.3cd
After brush	14.8bc	3.4f	17.8b
End packline	23.3a	5.6ef	26.8a

[a]Numbers with different letters indicate that the treatments produced significantly different responses ($P \leq 0.05$).

vary with the box material and design (e.g. wooden slat containers versus plastic, solid-walled containers).

Fruit should be placed in the shade immediately after picking as excess heating will cause dehydration, sunburn and reduced quality. Covering the field bins with leaves was found to be more beneficial in reducing flesh discoloration and disease after storage than covering with a space blanket or paper or leaving the bins uncovered (Arpaia *et al.*, 1992). Fruit should be brought into the packhouse within 2 h of picking (Hardy *et al.*, 1995). Bin trailers or trucks transporting the fruit should have good suspension (and preferably airbags) and travel at a speed that prevents the top fruit from bouncing.

Receiving, sorting, packing and precooling

Traceability and quality assurance

The OECD has produced international guidelines for quality standards for avocados (Anonymous, 2004). This guide includes varietal characteristics, quality definitions and tolerance levels for mechanical injuries, sunburn, diseases and rots, insect damage, frost damage in orchards and CI. Fruit size and packaging requirements are also given. Many avocado-producing countries produce generic standards and tolerance levels for several grades of fruit.

Quality and contaminant management systems may require a record system to track from which block or trees each bin or tray of fruit was harvested (traceability). The system should include regular assessment of fruit on arrival at the packhouse (receival assessment) and of packed trays to ensure adherence to the required quality standards. Additionally, the assessment of second-grade and rejected fruit provides valuable indications of causes for downgrading, which allows remedial practices to be implemented to improve the percentage of premium-grade fruit. These systems, incorporated with block/tree-to-tray traceability, allow problems (such as excessive or unapproved pesticide detections) to be traced back to the site of the problem and relevant action to be taken. It also allows records of tree or block yield and fruit quality, which is important for the ongoing improvement of tree and fruit performance. These systems can also be used to motivate producers or picking teams (through financial incentives) to deliver high-quality produce.

Many importers in important world markets require producers to be accredited with quality management and traceability systems. For example, most importers now require GlobalGap accreditation, which forces producers to not only have traceability of all orchard history and production practices but also produce in accordance with internationally accepted norms of good agricultural practice. A number of other systems to enhance and ensure quality are also used, some of which are specific to retail outlets (Vorster, 2005). At arrival in the packhouse, fruit samples should be evaluated immediately for maturity, blemishes and disease and pest incidence.

Removing field heat before packing

Harvested fruit should be packed and cooled as soon as possible. Arpaia *et al.* (1992) recommended that fruit should be cooled within 6 h of harvest and that fruit temperature should be reduced to below 30°C if the delay is greater than 6 h. Where long storage life is needed, there are advantages to precooling field bins to below 16–20°C (Hardy *et al.*, 1995) before sorting and packing. However, the holding temperature should be above the dew point to prevent condensation on the fruit, which would result in wet fruit being packed. In New Zealand, fruit should be cooled within 24 h of harvest (Dixon *et al.*, 2005), reflecting the lower temperatures during harvest.

Cleaning, sorting, grading and packing

Fruit damage caused during picking and packing can result in damage that may not be obvious for several days (e.g. skin spotting). This is particularly important for export fruit, as transport and marketing costs will be incurred before defects are obvious. High fruit turgidity and rough handling increase the potential for damage to the nodules, causing skin spotting. This can be reduced by delaying packing to reduce fruit turgidity (Everett *et al.*, 2008). It is important to minimize all potential points of fruit impact and rubbing.

Field residues may be removed using water, with or without brushing. Brush type, number and speed can be particularly important with regard to skin spotting in 'Hass' if fruits rub against each other. High-pressure water blasting, developed to remove insect pests (Jamieson *et al.*, 2000), can be used for more difficult deposits such as pollen (Mandemaker, 2003) and may reduce or remove the need for brushes. Fungicide treatment (if tolerated in the target market) is applied after fruit cleaning. Insecticides can be applied against fruit flies, but these are becoming less acceptable. Fruit is also waxed in South Africa, typically before sorting for defects.

Sorting to remove blemished fruit is generally done by visual inspection; however, optical sorting systems are becoming more cost-effective, particularly in countries with high labour costs. It is also possible that NIRS units could be employed to grade maturity, shelf life and/or internal defects (Blakey *et al.*, 2009).

Weight-grading systems are now standard in all packing lines. A 10% average deviation by number or weight is permissible, provided that no fruit is more than one code size larger or smaller than the code size group indicated. The carton size varies with customer requirements in various markets, but the common sizes are 6 kg in Australia, 4 kg in Israel and South Africa and 11.3 kg in California. Fruit is generally subject to different packing patterns dependent upon size. In some countries, packing sheets (carton inserts or liners) are used to position and hold fruit. Lower grade and small fruit are often packed in bulk cartons (loose fill), typically of 10–20 kg. The importance of net weight varies with market, depending on whether fruit is normally sold by weight or piece. To ensure correct net weight, a common practice is to slightly overpack to account for water loss during shipping.

Cartons should be adequately ventilated to allow efficient and rapid cooling while maintaining the mechanical strength required for transport and handling. Approximately, 9% of the free-flow area (area of holes relative to the total carton area) is sufficient to achieve satisfactory cooling, and there is no advantage in increasing this percentage (Haas and Felsenstein, 1985). Fruit should not be packed wet because of the increased risk of skin spotting and disease.

Pallet and stacking design are a compromise between transport requirements, stability of the pallet, and the need for even and efficient airflow over all fruit in the pallet. Pallet size should be such that shipping container space is fully utilized. The carton stacking/alignment configuration should optimize airflow, considering the direction of airflow (horizontal and/or vertical). For example, significant improvements in fruit cooling are possible by ensuring that the holes in the cartons align to maximize airflow (Boelema, 1987) and that practices to improve pallet stability (pallet wraps, etc.) do not impede air movement through the pallet. The so-called 'ideal' pallet for shipping container transport uses techniques such as strapping and cover sheets or link sheets to stabilize the pallets and prevent toppling (Hardy *et al.*, 1995).

Precooling

Cooling as soon as possible after harvest has the most significant effects on long-term storage potential (Yearsley *et al.*, 2002; Kok *et al.*, 2010; Lütge *et al.*, 2010). Hence, fruit temperatures should be reduced to near the final storage temperature as soon as possible after harvest, particularly if long-term storage is needed.

Traditional room cooling involves placing pallets in the cold room with no additional manipulation of cold airflow. This results in air flowing mostly past the outside of the package walls, resulting in slow and uneven cooling in palletized loads. Maintaining at least 5 cm gaps between pallets can provide more even cooling, and air-stacking the cartons (gaps between cartons on the pallet) can increase air movement further but will require restacking before pallet dispatch.

Rapid and uniform cooling can be achieved with forced-air or pressure cooling. The ideal forced-air cooling design is tunnel cooling, which consists of two rows of pallets with a tunnel (or plenum) between them. The top and end of the tunnel are covered with a tarpaulin to provide a good air seal between the two pallet rows, and the pallets within each row are stacked tightly together to prevent air from leaking past the pallets (Watkins and Ledger, 1990). Other sources of air leakage, such as through the pallet slats, are also minimized. All pallets within the row should be the same height, and all cartons in both rows should have similar free-flow areas. In addition, dunnage bags can be placed in gaps to force air movement in the desired direction. A fan creates negative pressure inside the tunnel, thus sucking cold air from the outside horizontally through the cartons and across the fruit into the tunnel. Airflow across the fruit in the front pallets will be greater than the fruit in the back pallets if the tunnel width and the spacing between pallets of adjacent rows are too small. The design should cool the pallets as efficiently as possible, as evenly as possible in all cartons, and with minimal water loss from the fruit.

Avocados can be forced-air cooled to storage or transport temperature in approximately 8 h with airflows of 1.0 and 0.5 $l\ s^{-1}\ kg^{-1}$, respectively (Watkins and Ledger, 1990). Faster cooling times may not be practical, as the pressure drop across the pallet required for higher airflows is very high. The airflow rate and distribution are important in achieving even fruit temperatures, but an excessive airflow rate can cause increased weight loss and skin CI from wind chill.

Excessive water loss after harvest can increase the risk of discrete patches on the skin and increase the ripening rate (Bower and Magwaza, 2004) and ripe fruit rot (Lallu *et al.*, 2004; Dixon *et al.*, 2005). Most fruit water loss occurs during the initial cooling phase, and the amount can be estimated using psychometric tables. Hence, cooling units should be large enough to minimize the temperature differential between the air and the cooling coil (which reduces moisture removed from the air by condensation). Well-designed systems should maintain a relative humidity above ~85%. Additional moisture can be added to the air by spraying a water mist into the airstream on the downstream side of the cooling coil. Hydrocooling for the removal of field heat and initial precooling have been found to be useful by some packers but are not widely used. This may require additional packhouse equipment such as an air knife to dry the fruit. Waxing of fruit in the packhouse also reduces water loss (Lütge *et al.*, 2010) but has cost and logistical implications (see the section 'Controlled and modified atmospheres' in this chapter).

Ripening

General

Bower and Cutting (1988) provided a comprehensive review of avocado fruit-ripening physiology. In broad terms, ripening of a fruit is a process whereby physiological and physical changes take place that make it attractive and palatable for consumption (Rhodes, 1980). These changes normally only occur once a fruit has reached maturity during the growth and development phase of the tree. Because the final stages of fruit life involve senescence, considerable anabolic and catabolic processes take place during ripening. This is an ordered process requiring large amounts of energy and the maintenance of membrane integrity, failure of which may cause various ripening disorders (Bower and Cutting, 1988; Wills *et al.*, 2007). The ripening process in avocado fruit is unusual in that softening and most other ripening characteristics do not normally occur in fruit firmly attached to the tree but are initiated only after harvest. The most obvious changes during ripening are softening, and in some cultivars, exocarp (skin, rind) colour change (although these colour changes and ripening are not directly linked), but many physiological changes accompany the ripening process. On a macro level, the two major phenomena are the respiratory and ethylene climacterics (Biale and Young, 1971). These are major, easily measurable changes. However, many structural and physiological changes, both catabolic and anabolic (Seymour and Tucker, 1993), also occur, which results in final softening and organoleptic changes. These processes involve enzymatic and structural changes as well as the production of a range of volatiles (Obenland *et al.*, 2012).

Onset of ripening

A number of factors can be measured during the early stages of ripening. Dallman *et al.* (1988) reported that Golgi bodies and plasma membranes increased in buoyant density during ripening of avocado, whereas thylakoids and mitochondrial membranes showed no change. Ultrastructural changes occurring during ripening have been observed using transmission electron microscopy (Platt-Aloia and Thompson, 1981), and some changes in EC (Montoya *et al.*, 1994) have also been reported. However, most of the literature on ripening processes in avocado has dealt with physiological and biochemical changes (Biale and Young, 1971) at the enzymatic (Pesis *et al.*, 1978) and molecular levels (Christoffersen *et al.*, 1984). More than 30 different enzyme activities have been quantified in avocado fruit, and of these, cellulase, polygalacturonase (PG), PPO and cytochrome P450 have been the most studied. Lipid peroxidation may be regarded as one of the earliest detectable processes in avocado fruit ripening (Meir *et al.*, 1991). Cytochrome P450 has been suggested as a ripening-related enzyme in avocado (Bozak *et al.*, 1990). The gene encoding this enzyme is translated in the early preclimacteric events leading to the onset of the climacteric (Starrett and Laties, 1993), and it is induced by both ethylene pulsing and wounding. In contrast, Bozak *et al.* (1990) reported that cellulase, PG and the ethylene-forming enzyme ACC oxidase were not involved in initiating the climacteric in avocado.

Although these are some of the earliest detectable changes in ripening, other factors initiate the process. A certain level of fruit maturity at harvest is essential for ripening to proceed. Although immature fruit will show a number of the changes associated with the initiation of ripening (Adato and Gazit, 1974), fruit do not continue to true ripeness, showing only some softness due to dehydration.

It is suggested that the initiation of the ripening process is related to a changing balance between PGRs and fruit water loss. Abscisic acid has been shown to stimulate ethylene synthesis (Vendrell and Palomer, 1998). Bower and Cutting (1988) suggested that water stress, abscisic acid and ethylene are important in the early stages of ripening. Blakey *et al.* (2009) found that the infusion of abscisic acid or water could modify the initiation and rate of ripening. Once fruit are harvested, a net loss of water occurs and can be associated with enhanced abscisic acid concentrations (Bower, 1985) due to water stress. This change in abscisic acid may stimulate the ethylene pathway, and other ripening processes may then be activated. A certain stress threshold may be necessary to achieve activation. This may help explain why fruit from water-stressed trees appear to ripen faster (Bower and Cutting, 1988), as does fruit considered more mature as measured by a greater % DM (lower water content) (Adato and Gazit, 1974). Adato and Gazit (1974) also noted that fruit subject to greater postharvest water loss ripens more rapidly. However, the process is likely to be more complex. The role of other PGRs is uncertain, but they may contribute as promoters or inhibitors (Bower and Cutting, 1988). Furthermore, C_7 sugars, in particular mannoheptulose and perseitol, have been noted as possible inhibitors of ripening (Liu *et al.*, 2002). Liu *et al.* (2002) also found that these sugars declined with ripening and that an apparent threshold level of approximately 20 mg g^{-1} DM was required before ripening was initiated. Ripening also occurred in fruit left on the tree if the fruit pedicel was girdled. Again, this process resulted in a decline in the C_7 sugars to approximately the same level as that found to be associated with fruit ripening in harvested fruit.

Softening

Softening of avocado fruit has been extensively studied. Softening is commonly associated with cell wall disassembly in climacteric fruits as a consequence of modifications in the polysaccharides of the primary cell wall and their interactions (Brummell, 2006; Defilippi *et al.*, 2018). The plant cell wall is composed of cellulose and hemicelluloses embedded in a matrix of pectin. Pectins are the most abundant polysaccharides in the cell wall matrix and middle lamella, and Ca^{2+}–pectin crosslinks are a major factor in determining the physical and structural properties of fruit (Billy *et al.*, 2008; Michailidis *et al.*, 2017; Defilippi *et al.*, 2018, Olmedo *et al.*, 2021). The preharvest history of the fruit with respect to Ca deposition, particularly during the early stages of cell division, may also impact fruit softening (Bower, 1978). In avocado, low levels of Ca^{2+} have been associated with rapid softening after harvest (Tingwa and Young, 1974; Wills and Tirmazi, 1982). Witney *et al.* (1990) showed a significant proportional relationship between fruit Ca concentration and days of ripening. Moreover, postharvest treatments, including calcium chloride infiltration in avocado fruits, have been shown to delay the ripening process compared with untreated fruit; however, this effect is variety dependent (Wills and Tirmazi, 1982). The process of degrading the cell walls requires a number of enzymes operating in a prescribed order and should therefore be seen as a complex and integrated process (Ali *et al.*, 2004). Changes in pectins such as solubilization and depolymerization have been proposed as the primary causes for fruit softening during ripening as a result of cell wall loosening and disintegration (O'Donoghue and Huber, 1992; Jeong *et al.*, 2002; Brummell, 2006).

Each of the cell wall components has its own enzyme system (due to the different substrate molecules), but overall, the most important enzymes are cellulase, PG and PME. A correlation between cellulase activity and softening in avocado was shown in the late 1970s (Pesis *et al.*, 1978) and appears to be critical because cellulase activity is required for the functioning of a number of other enzymes (Hatfield and Nevins, 1986). Since then, there has been extensive work to study the induction (Starrett and Laties, 1993), activity (Tucker *et al.*, 1985; Kanellis *et al.*, 1990; O'donoghue *et al.*, 1993) and gene expression, isolation and characterization (Christoffersen *et al.*, 1984; Dallman *et al.*, 1989; Cass *et al.*, 1990; Dopico *et al.*, 1993) of cellulase in relation to the softening process. When ethylene binding is blocked by 1-methylcyclopropene (1-MCP), cellulase activity is inhibited (Jeong and Huber, 2004). The removal of ethylene from the cold storage atmosphere delayed the appearance of mRNAs linked to cellulase and PG (Buse and Laties, 1993).

In firm, commercially mature fruit, cellulase activity increases from very low levels at harvest. In more mature fruit, higher cellulase activity can be detected at harvest when the fruit is still firm (Fuchs and Zauberman, 1987). Low storage temperatures delay ripening and retard the appearance of mRNAs for cellulase, PG and EFE (Dopico *et al.*, 1993). In a storage atmosphere of 2.5% O_2, softening and the rise in cellulase and PG activities were prevented (Kanellis *et al.*, 1989), and even the addition of 100 $\mu l\ l^{-1}$ ethylene did not alter the suppressive effects of low O_2. Thus, there is strong evidence for a relationship between cellulase and avocado softening, but there are also other hydrolytic enzymes and unknown factors involved.

Several other hydrolytic enzymes implicated in cell wall structure changes such as PG (Zauberman and Schiffmann-Nadel, 1972; Kanellis *et al.*, 1991), PME (Zauberman and Schiffmann-Nadel, 1972) and xylanase (Ronen *et al.*, 1991) have also been studied. It has been suggested that the solubilization and depolymerization of cell wall polysaccharides during avocado fruit ripening involves the concerted action of PG and PME (Wakabayashi *et al.*, 2000). The purpose of PME is to prepare the pectin substrate for solubilization by PG. The activity of PME appears to vary during avocado fruit ripening (Awad and Young, 1979). PG activity increases during ripening but follows that of cellulase (Awad and Young, 1979), and high levels of PG activity observed in avocado have been related to the extensive solubilization and depolymerization of pectins that take place during avocado ripening (Awad and Young, 1979; O'Donoghue and Huber, 1992; Jeong *et al.*, 2002). The evidence shows that PG and PME are not exclusively responsible for pectin degradation (Fischer and Bennett, 1991). β-Galactosidase activity, for example, has been associated with a marked loss of galactose units, facilitating pectin solubilization during avocado ripening (Tateishi *et al.*, 2007). Previous studies have identified three β-galactosidase isoforms in avocado cultivars such as 'Fuerte' AV-GAL I, II and III. Among them, AV-GAL III produced a substantial release of galactose units from avocado cell wall polysaccharides, and its enzyme activity was found to be positively associated with fruit softening (Tateishi *et al.*, 2007). Defilippi *et al.* (2018) indicated that PG activity was correlated with a decrease in galacturonic acid in avocado cell wall material.

Galactose, a main component of rhamnogalacturonan type-I side chains (Buffetto *et al.*, 2015), presented major changes during ripening. It was possible to correlate the increased β-galactosidase with a reduction in galactose content in the cell wall (Defilippi *et al.*, 2018). In particular, the reduction in galactose could be linked to Na_2CO_3-soluble pectins, which showed the largest changes in galactose content.

It is presumed that the overall process during ripening is the initiation of cellulase activity, which partially modifies the cell wall structure, followed by PME action, and finally, PG activity responsible for softening (Jeong *et al.*, 2002). There is evidence that PG is also linked to ethylene, as it appears to be inhibited by the application of 1-MCP.

Respiration

As previously indicated, avocado fruit show strong climacteric activity during ripening, including changes in respiration. The high respiration rate and rapid ripening are challenges in the extended storage required for long-distance shipping from many exporting countries to their major markets. The many anabolic changes that take place during the initiation of the ripening process imply a high energy demand and may explain the rapid increase in respiratory activity. Although avocado fruit has a high lipid content, there is evidence that this is not the primary substrate for respiration. Blanke (1991) stated that the respiratory quotient is approximately 1, indicating that the substrate is carbohydrate. The C_7 sugars mannoheptulose and perseitol are the main sugars in avocado fruit, and their rapid decline during ripening suggests that they may be the primary energy source (Tesfay, 2009; Meyer and Terry, 2010). As the total respiratory carbohydrates are finite at the time of harvest, both pre- and postharvest conditions may play a role in the shelf life of the fruit. This also provides a challenge for long-distance shippers.

Ripening temperature

Hopkirk *et al.* (1994) found that stem-end rot, body rot, uneven ripening and vascular browning in 'Hass' fruit increased as the ripening temperature increased from 20°C to 30°C, and the severity was generally the least at 15°C. Similar results were obtained by Arpaia *et al.* (2003b) when comparing ripening temperatures of 20°C and 25°C, and only 24 h at 25°C was sufficient to increase disease incidence. However, the average skin colour is less purple–black when ripened at 15°C compared with 20–25°C [colour also shows a slightly higher hue angle and higher L (lightness) value] as a result of lower total anthocyanin and cyanidin-3-glucoside concentrations (Cox *et al.*, 2004). Consumers often assume that 'Hass' fruit with green skin colour are not ripe, so the fruit is left to ripen further, thus increasing the risk of disease or other disorders in the flesh. This risk is increased further in less mature fruit because these fruit often do not attain a purple/black colour when eating soft. Thus, the benefits of lower ripening temperatures reducing diseases can be lost on these less mature fruit unless consumers understand that fruit softness is the major determinant of 'Hass' ripeness rather than skin colour. The use of lower ripening temperatures to reduce diseases and intermittent higher temperatures to stimulate darker skin colour in ripe 'Hass' needs further investigation. Additionally, there is little understanding of the effects of ripening temperatures on diseases in other avocado cultivars.

Ethylene

Ethylene is the hormone that modulates a wide range of responses during fruit ripening and senescence in climacteric fruits such as avocado (Khan, 2006). In avocado, the 'peak' in the production of ethylene during ripening has been specifically associated with an increase in the respiratory rate, softening of the flesh, colour change and development of aroma compounds (Zamorano *et al.*, 1994; Kassim *et al.*, 2013).

Ethylene physiology in avocado has been reviewed by Hofman *et al.* (2002a), and practical guidelines for ethylene ripening are provided by Ledger *et al.* (2011). Regarding commercial application, accidental exposure of avocados to ethylene and its analogues (e.g. from adjacent ripening rooms of avocados or other climacteric fruits, particularly bananas and passion fruit; exhaust fumes from internal compression engines or wound ethylene produced from damaged/ripening fruit) can cause premature ripening, and very low doses just before/during cold storage contribute to physiological flesh disorders (see above).

Ethylene can be removed by oxidizing mechanisms such as the use of potassium permanganate (either in sachets or in ethylene scrubbing units in cool stores or containers) catalytic oxidizers or by ozone-based systems (Reid, 2002). A palladium-based catalyst has been developed as part of an alternative ethylene scrubbing system (Terry *et al.*, 2007). However, 1-MCP (see section '1-MCP' in this chapter) is arguably more effective at reducing the effects of unintentional exposure to ethylene or its analogues.

Induction of ripening using ethylene provides considerable commercial advantages by reducing the time to ripen, reducing variability in ripening among fruit, and reducing disease severity in eating soft fruit (Arpaia *et al.*, 2004; Adkins *et al.*, 2005). Ethylene ripening has been widely used in Australia for more than 20 years to provide partly ripened fruit for the retail shelf and has recently become more common in the USA and EU. Ethylene can be applied from gas bottles as either a trickle or shot system (Table 15.5). A number of automatic ethylene gas control systems are now available (CRS, 2011) to maintain ethylene concentrations within required limits.

Ethylene gas can also be generated by catalytic conversion of ethanol in ethylene generators (e.g. CG, 2011). In most cases, these mimic the shot system, but automated measurement and control systems can also be applied to these units . Ethylene is explosive at concentrations of 3–30% in air (Reid, 2002). Explosion risk can be reduced by using a bottled mixture of ethylene and an inert gas (CO_2 or nitrogen; 'Ripegas') or by ensuring that the volume of ethanol in the ethylene generator is less than that required to reach explosive concentrations in the ripening room. In addition, automated measuring and dosing systems should activate alarms and automated shut-off valves well before concentrations approach explosive levels.

The high respiration rate of ripening avocados can increase CO_2 concentrations in the ripening room (up to 5.3%; Ledger, 2007), particularly if the room is full and has poor ventilation. It should also be noted that health and safety recommendations suggest a limit of 3% under 10 min for CO_2 exposure and that concentrations do not exceed 4% (Glatte *et al.*, 1967). Ethylene treatment of 10 μl l^{-1} for 48 h reduced the average ripening time and variability in ripening (Arpaia *et al.*, 2004). Lower concentrations had less effect, and there was little difference at 100 μl l^{-1}. High CO_2 delayed ripening, particularly with 10 μl l^{-1} ethylene, because of a competitive effect of CO_2 on ethylene action. Increasing ethylene concentrations reduced the high CO_2 effects. Carbon dioxide concentrations should be kept below 1–2% when using 10 μl l^{-1} ethylene and below 2.5% when using 100 μl l^{-1} ethylene (Arpaia *et al.*, 2004). Opening the

Table 15.5. Systems for applying ethylene to fruit

Process	Comments	Advantages/disadvantages
Trickle	Constantly trickling of ethylene from a gas bottle into the ripening room to maintain approximately 10 μl l^{-1} (10 ppm) ethylene (Arpaia *et al.*, 2004).	Good control, more convenient. Can be more expensive to set up. Requires room ventilation to prevent CO_2 accumulation.
Shot	Injecting sufficient ethylene gas every 6–8 h at a concentration of approximately 100 ppm. Leakage from the cold room reduces concentrations over time. The room door is left open for approximately 15 min every 8 h before ethylene is reinjected.	Cheaper to set up. Less control. Risk of CO_2 build-up because of inadequate ventilation.
Ethylene generators	Generates ethylene by catalytic breakdown of ethanol. Dosing can be automated.	Removes reliance on bottled ethylene.
Dipping fruit in Ethephon®[a]	Applied as a solution before packing. Ethephon is converted to ethylene inside the fruit.	Cheaper for small batches but not practical for larger batches. May not be registered in some countries for avocados.

[a]2-Chloroethylphosphonic acid degrades to ethylene within the higher pH of the tissue.

ripening room doors for approximately 15 min every 6–8 h or using extractor fans that can be activated via a CO_2 monitor (CRS, 2011) will maintain CO_2 concentrations below the recommended levels.

There is little benefit of ethylene gassing when fruit are stored for longer than 14 days (Dixon *et al.*, 2003a; Arpaia *et al.*, 2006), except when ripening at 15°C (Hofman *et al.*, 2003).

1-Methylcyclopropene (1-MCP)

The gas 1-MCP is a potent inhibitor of ethylene-related senescence, presumably via inhibition of ethylene action (Feng *et al.*, 2000b; Hofman *et al.*, 2001). The binding is irreversible, but inhibition can be lost due to the synthesis of new receptors (Sisler *et al.*, 1996).

1-MCP is a powerful inhibitor of ripening in avocado, with concentrations as low as 30–70 nl l^{-1} delaying ripening for as many as 10 days (Feng *et al.*, 2000a). 1-MCP can delay ripening of a range of cultivars (e.g. 'Ettinger', 'Hass', 'Reed' and 'Fuerte') as well as reduce cell wall softening enzyme activity (PG and cellulase), climacteric ethylene production and the respiration rate (Feng *et al.*, 2000a; Jeong *et al.*, 2003). A combination of waxing and 1-MCP treatment can delay ripening even longer in 'Tower II' avocados (Jeong *et al.*, 2003).

Although 1-MCP binding to receptors is mostly irreversible, recovery during ripening is generally observed in most fruit, although less so in avocado, even after long periods in storage (Woolf *et al.*, 2005). This difficulty in 'restarting' ripening may be due to 1-MCP being absorbed into the oil of the fruit, which acts as a 1-MCP reservoir allowing slow release and binding to new receptors (Dauny *et al.*, 2003). Using controlled atmosphere (CA) treatments of 2% O_2 and 20% CO_2 following storage can restart and synchronize ripening following 1-MCP treatment and storage (Roets *et al.*, 2009).

One possible application of 1-MCP is to arrest avocado ripening at a mid-ripe stage (Adkins *et al.*, 2005). The concept is that fruit could be ripened to a near-ripe stage and then distributed with less susceptibility to bruising. Although 1-MCP was very effective in delaying ripening of 'Hass' when applied prior to the start of ripening or ethylene treatment, application after fruit started to soften was not effective and showed little potential to control ripening when used after ethylene treatment. However, Zhang *et al.* (2011) showed that there was a reduction in fruit sensitivity to 1-MCP at the mid-climacteric stage, and application of low O_2 (2% for 12 h prior to 1-MCP exposure) partially restored 1-MCP sensitivity.

The main commercial purpose of 1-MCP is to delay ripening/softening during cold storage, which has the ancillary benefit of reducing flesh CI (e.g. diffuse flesh discoloration, the key physiological problem for long-stored avocados) (Pesis *et al.*, 2002; Woolf *et al.*, 2005). Pesis *et al.* (2002) first showed the effect of 1-MCP on flesh CI in 'Ettinger' avocados. Fruit treated at 5°C with 100–300 nl l^{-1} 1-MCP for 24–48 h and then stored for 4 weeks had less diffuse flesh discoloration and lower PPO activity than non-1-MCP-treated fruit. The response of 'Hass' fruit to 1-MCP treatment at 6°C compared with 15°C, concentrations of 50–1000 nl l^{-1} and durations of 6, 12 or 24 h before storage for 4 and 7 weeks showed that 1-MCP treatment at typical industry storage temperatures of approximately 6°C is as effective as 15°C, and an overnight treatment of 18–24 h was a suitable and practical duration (Woolf *et al.*, 2005). Olivares *et al.* (2022) demonstrated that the delayed application of 1–MCP during storage is effective if applied before 21 days of cold storage; after this period, the fruit does not respond to the treatment. The application of 1–MCP during cold storage causes a delay in the ripening process of avocado due to an inhibition of ethylene metabolism, mainly by decreasing the expression levels of genes related to critical enzymes involved in ethylene synthesis, *PamACS1*, *PamACS2* and *PamACO*, and of the receptors *PamETR1* and *PamCTR1*. However, that study showed that this delay, at least at the transcriptional level, lasted close to 21 d after harvest, and at that stage, transcripts involved in ethylene biosynthesis/perception increased their expression concomitant with a decrease in maturity and physiological attributes.

Green-skinned cultivars (e.g. 'Ettinger' and 'Pinkerton') exhibit 1-MCP treatment responses similar to the dark-skinned cultivars with reduced flesh CI and reduced PPO, peroxidase and chlorophyllase activities (Hershkovitz *et al.*, 2005). Olivares *et al.* (2020) studied 'Edranol' and 'Fuerte' avocados, and the application of 1-MCP was equally effective in both cultivars with regard to delayed softening, delayed ethylene and CO_2 production during cold storage and downregulated expression of genes involved in ethylene biosynthesis.

Thus, 1-MCP is a powerful tool for retarding ripening (softening and skin coloration) and reducing flesh CI disorders following storage. However, the increased development of rots (Adkins *et al.*, 2005) and the slower and more inconsistent ripening (Kruger and Lemmer, 2007) need to be considered, particularly in countries where rot pressure is higher due to wet growing environments.

Other chemical treatments

Other chemical treatments are being investigated to manipulate ripening and fruit quality. Pesis *et al.* (1998) showed that fumigation with acetaldehyde delayed ripening of 'Fuerte' fruit by retarding softening and reducing ethylene production. Holding avocado fruit in ethanol-saturated air can also delay ripening but may result in internal disorders (Ritenour *et al.*, 1997). The application of methyl jasmonate to 'Fuerte' and 'Hass' avocados reduced CI severity and incidence during storage at 2°C for 4–10 weeks (Meir *et al.*, 1996). These treatments have not received wide commercial application.

Disinfestation

Fumigation of avocado with methyl bromide (MeBr) is effective against a wide range of insects, including fruit flies. However, external and sometimes internal fruit quality can decline, and shelf life is reduced by 2–4 days. Thus, MeBr is not used as a routine disinfestation treatment.

Irradiation is a well-accepted treatment for fruit fly disinfestation, and the development of X-ray and electron beam-type irradiation systems (which eliminate the need for transport and disposal of radioactive materials) was hoped to provide effective disinfestation for avocados. However, Akamine and Goo (1971) showed that avocados suffer surface and internal damage at 50 Gy, and 150 Gy X-ray treatment of non-ripe 'Hass' followed by 3 weeks of storage at 5°C resulted in significant fruit quality loss, including delayed ripening, increased disease incidence and off-odours (Arpaia *et al.*, 2003b). These results, and the pressure to move to non-chemical disinfestation treatments, have led to the development of successful physical treatments.

Avocados are relatively intolerant of postharvest HTs, with a limit of approximately 1 h at 42°C, 6 h at 40°C and 12 h at 38°C for 'Hass' (Woolf *et al.*, 1995). Damage can be induced in hot water by as little as 3 min at 50°C (Woolf and Lay-Yee, 1997), and 40°C water treatments for 8 h result in severe rind blackening in 'Fuerte' (Donkin and Wolstenholme, 1995). These limits are well below the level of effective treatment for heat disinfestation of fruit flies (Sinclair and Lindgren, 1955), and most external pests (such as thrips) require approximately 7–10 min in 50°C water. Even the application of thermotolerance-inducing pretreatments was not effective at preventing the loss of ripe fruit quality (Woolf, unpublished data), although external heat damage can be eliminated (Woolf and Lay-Yee, 1997).

Low-temperature disinfestation against fruit flies is a widely accepted treatment that is relatively simple to apply (storage of fruit at less than 1°C for 12–16 days, depending on fruit fly species) but causes external CI on 'Hass' (mainly discrete skin patches) under most growing conditions. Heat or low-temperature conditioning (LTC) treatments (described in more detail in the 'Preconditioning' section of this chapter) show potential to induce chilling tolerance.

Hot-air pretreatments of 'Sharwil' fruit at 37–38°C for 17 to 18 h reduced external CI after storage at 1.1°C for 14 days for insect disinfestation (Sanxter *et al.*, 1994). However, the average external appearance of the heat-treated fruit was only rated as 'marginally acceptable' and below a marketable rating. Later work showed that treatment at 38°C for 8–12 h more effectively reduced 'Sharwil' CI during disinfestation at pulp temperatures of less than 2.2°C for 16 days (Nishijima *et al.*, 1995). Hofman *et al.* (2002b) applied a range of hot water treatments (HWTs) to 'Hass' fruit prior to a Queensland fruit fly disinfesting cold treatment (1°C for 16 days). The optimum HWTs were 40–42°C for 30 min. However, heat pretreatment/low-temperature disinfestation has not been commercialized to the best of our knowledge.

LTC (application of temperatures close to chilling-inducing temperatures) is very effective in reducing external chilling injury in 'Hass' fruit. The optimum treatment combination (approximately 6°C for 3–5 days) for Australian fruit achieved effective disinfestation of 30,000 fruit fly larvae after 16 days at 1°C with negligible skin CI (Hofman *et al.*, 2003). This is used commercially for exports to New Zealand and is effective on South African (van Rooyen and Bezuidenhout, 2010) and Peruvian (L. Gonzales, Peru, 2009, personal communication) fruit.

High-pressure water washing (or water blasting) was originally developed for insect removal from citrus fruit in South Africa in the 1970s (Honiball *et al.*, 1979). This system used rotating brushes and relied on random movement of fruit with treatment times of generally 10–15 seconds. An alternative system developed in New Zealand orients the fruit, applies jets impinging from three directions, and treats for a much shorter (generally 0.5–1 s) exposure (Jamieson *et al.*, 2000). This technology was modified to target the removal of leaf roller egg rafts but was found to remove all crawling insects (leafrollers, thrips and mites) and other contaminants such as pine (*Pinus radiata*) pollen, bird droppings and spray residues.

Storage

Temperature and humidity

Temperature is the single most important factor affecting fruit storage because it affects the reaction rates of biological processes. The challenge is to slow natural fruit deterioration down by lowering temperatures without causing CI. Avocado, because of its subtropical and tropical origins, is typically stored for extended periods at 4–7°C. Typical avocado fruit CI symptoms include internal flesh discoloration (diffuse discoloration, vascular browning and pulp spot, particularly in 'Fuerte'), discrete patches on the skin (external CI), chlorotic appearance of the skin ('Ettinger' in Israel), failure to ripen, increased susceptibility to microorganism attack, and in very long storage periods, outer flesh blackening, stringy vascular bundles and vascular leaching (Woolf *et al.*, 2004). In general, flesh discoloration is more common at relatively higher storage temperatures of over 5°C (Bower *et al.*, 2000; Vuthapanich, 2001; Dixon *et al.*, 2008; van Rooyen and Bezuidenhout, 2010), and skin blackening and pitting (discrete patches) are more common at 0–2°C (Vuthapanich, 2001; Dixon *et al.*, 2008). However, on rare occasions, external CI as discrete patches can be observed in 'Hass' after long storage periods at 5–6°C (A. Woolf and A. White, unpublished data; P.J. Hofman, unpublished data). Hence, selecting an optimum storage temperature is a balance between minimizing external (typically occurring at lower temperatures) and internal (typically occurring at higher temperatures) CI symptoms.

Maintenance of the cold chain once storage has begun is very important. There is good evidence that breaks in the cold chain enhance premature softening and increase fruit disorders (Kok *et al.*, 2010; Lütge *et al.*, 2010). The effects of breaks in the cold chain include using more refrigeration energy to recool fruit; condensation on the fruit, which can increase microorganism growth; increased CI; weakening of cartons; and reduced storage life. Eksteen and Bester (1987) provide some practical suggestions in relation to the cold chain, including harvesting, precooling, air circulation, carton ventilation, container loading and storage temperature recommendations.

Research and commercial experience in South Africa have shown that fruit are more sensitive to CI during the first week in storage (Toerien, 1986). Hence, a regime that gradually reduces the temperature to 7.5°C for the first week of storage, then to 5.5°C for 2 weeks and 3.5°C for the last week, or similar stepped-down temperature regimes, have been used commercially for sea freight to European markets (Eksteen and Bester, 1987; Vorster *et al.*, 1987, 1990). Later work indicated that 'Fuerte' from the colder areas in South Africa does not benefit from this regime of reducing temperature (Donkin *et al.*, 1995). The South African industry also recommends that fruit harvested from colder areas and later in the season from warmer areas be stored at lower temperatures because of increased tolerance to CI (Vorster *et al.*, 1990). The fact that cultivar and production conditions can influence CI susceptibility (Zauberman *et al.*, 1973, 1977) suggests that improved storage performance can be obtained by determining specific cold storage conditions for cultivar and production location.

Past storage efforts have typically concentrated on alleviating internal disorders during storage at 5–7°C and preventing lower temperatures because of the risk of skin discoloration. However, recent results indicate that skin damage during low-temperature storage can be avoided. Zauberman and Jobin-Décor (1995) noted better outturn quality of 'Hass' after storage at 2°C for 4 weeks than at 5 or 8°C, and Jessup (1991) successfully stored 'Hass' at 1°C for 30 days following dipping in 0.5% benomyl at 50°C for 3 min. Cold preconditioning (see the 'Preconditioning' section of this chapter) allows cold disinfestation at 1°C for 16 days without skin damage and results in fewer flesh defects than no storage (Hofman *et al.*, 2003). Semicommercial trials using sea freight containers modified to improve temperature uniformity in conjunction with a conditioning treatment resulted in acceptable quality following 22 days with a delivery air set temperature of 1°C (van Rooyen and Bezuidenhout, 2010). At least 40 days of cold storage may be possible using conditioning and 2°C storage (Hofman *et al.*, 2010). Further reductions in the risk of skin damage during low-temperature storage could be achieved by minimizing skin spotting caused during harvesting and packing, as increased skin spotting results in more discrete patches following cold storage (Hofman, 2005).

Both pre- and postharvest water stress can reduce fruit quality and ripening (Bower, 1985; Wills *et al.*, 2007). While some water loss after harvest may be needed to stimulate ripening (see the section on 'Ripening' in this chapter), excessive loss can be deleterious to long-term storage by increasing skin CI and internal disorders (Bower and Bertling, 2008). Skin CI starts to appear above approximately 2% water loss (Dixon *et al.*, 2008) or approximately 5% water loss during storage (Bower, 2005). These results were obtained using modified humidity bags with sufficient permeability to minimize CO_2 or O_2 changes. However, they can also increase rot in ripe fruit, partly by increasing ripening times (Dixon *et al.*, 2004a). None the less, reducing water loss in combination with minimizing skin spotting from harvest to packing and combined with cold conditioning treatments (Hofman *et al.*, 2003; Hofman, 2005) could eliminate the risk of skin damage during storage at 1–2°C.

Relatively little work has been done on the storage potential of ripening or ripe fruit. Early research has suggested that 'Hass' fruit in the climacteric rise or at the climacteric peak are most sensitive to CI, but fully ripe California fruit have been stored for 6–7 weeks at 2°C without significant quality loss (Kosiyachinda and Young, 1976). Additionally, fruit treated with ethylene immediately after harvest have been stored for 53 days at 2°C and have acceptable eating quality (Young and Kosiyachinda, 1976). These results are unexpected given the role of ethylene in flesh disorders during cold storage. Additionally, 'Hass' fruit ripened to a firmness of 2 (0 = hard, 5 = eating soft) developed significant diffuse discoloration after 7 days at 2°C, and the severity increased when riper fruit was stored (Campbell *et al.*, 2011).

Controlled and modified atmospheres

Decreasing the O_2 and/or increasing the CO_2 concentration around the fruit reduces the respiration rate and ethylene production and action, resulting in retarded ripening and reduced CI (Yahia, 2006; Thompson, 2010). However, if the O_2 concentration is too low (dependent on cultivar, storage temperature, fruit maturity and ripeness stage), anaerobic respiration will result in the production of ethanol and acetaldehydes, off-flavours and physiological disorders (Bender *et al.*, 2000). Atmosphere modification can be active or passive or a combination of the two. Surface coatings also provide a modified atmosphere (MA) inside the fruit.

CA storage controls O_2 and CO_2 concentrations in the container by actively monitoring these gases and injecting N_2 or bleeding air into the container as needed. It can extend storage life and reduce CI in avocado (Hatton and Reeder, 1972; Truter and Eksteen, 1987; Faubion *et al.*, 1992; Meir *et al.*, 1995). Storage times for 'Hass' of at least 10 weeks at 7°C have been obtained under experimental conditions using 2% O_2 and at least 4% CO_2 (Smith *et al.*, 1992; Jordan and Smith, 1993), and more than 9 weeks of storage have been obtained with 2% O_2 and 5% CO_2 at 5.0–5.5°C (Faubion *et al.*, 1992). Carbon dioxide concentrations of 4–10% had relatively little impact on storage life or outturn quality (Smith *et al.*, 1992), although Faubion *et al.* (1992) reported that concentrations above 5% increased flesh discoloration, particularly in early-season fruit. With Israeli 'Hass', 8% CO_2 resulted in less flesh chilling damage compared with 3% CO_2 and with either 3% or 21% O_2 (Meir *et al.*, 1995). There was no mention of treatment effects on disease. CA conditions often change the relative rate of 'Hass' fruit softening and skin darkening after removal, but skin colour differences at eating soft may be minimal (Meir *et al.*, 1995; Burdon *et al.*, 2008). Similar effects have been observed with mango (Johnson and Hofman, 2009) and are largely attributed to the high CO_2 rather than the low O_2 concentrations.

CA treatment is mainly used on 'Hass' avocado, but there are a significant number of green skin varieties such as 'Edranol' and 'Fuerte' that have important roles in several countries. Olivares *et al.* (2020) evaluated the effects of CA during long-term storage of avocados harvested at two maturity stages. They showed that ripening was delayed, and quality was maintained compared with regulated air storage. Also, CA was more effective than 1-MCP in reducing most of the external and internal damages evaluated, but the effectiveness within each disorder was influenced by the maturity stage at harvest and evaluation time.

Commercial use of CA for long-distance transport is now common, although competition with 1-MCP at similar cost may reduce its use. CA shipping containers control temperature, O_2 and CO_2 with ethylene scrubbing as a further option. Standard CA storage uses constant CO_2 and O_2 concentrations. Dynamic CA (DCA) is a relatively new development that aims to reduce O_2 concentrations as low as possible without causing fruit damage from anaerobiosis. By monitoring fruit stress, a lower O_2 concentration can be maintained, which results in reductions in softening and diseases in the ripe fruit (Burdon and Lallu, 2008). The most practical means to date for monitoring fruit stress has been chlorophyll fluorescence. This was first commercialized as the HarvestWatch™ sensor and is now available through Isolcell (Isolcell, 2011) as the fluorescence interactive response monitor (FIRM). Achieving an O_2 concentration of approximately 2% and repeatedly stressing the fruit for very short durations during storage leads to improved fruit quality without the need for high CO_2 concentrations (Burdon *et al.*, 2008), which can lead to delayed ripening and increased incidences of fruit diseases (Yearsley and Lallu, 2001). In contrast, other treatments that improve storage life (such as 1-MCP) can increase shelf life (time to ripen after storage) and fruit diseases.

In more passive systems, such as the MaXtend® system (Maxtend, 2011), fruit respiration reduces O_2 concentrations and the concentration is monitored and manipulated by venting the container as needed. Options in addition to O_2 control include flushing with N_2 at the start of storage to rapidly establish the desired atmosphere, monitoring CO_2 and venting with air, passive CO_2 (controlled rate) absorption and ethylene scrubbing. These systems can be attached to standard, good condition reefer containers, eliminating the need for dedicated CA containers. Avocados have been sea exported from Australia, South Africa, Chile and Mexico using MaXtend®.

Optimum CA conditions may be influenced by cultivar and disease profile. In New Zealand, 5% CO_2 with DCA delayed ripening and increased body and stem-end rots in 'Hass' compared with DCA only (Burdon *et al.*, 2008). However, in South Africa, increasing the CO_2 concentration from 6% (with 4% O_2) to 10% (with 3% O_2) reduced both stem-end rot and anthracnose incidence in the ripe fruit, particularly when combined with SmartFresh℠ (Lemmer *et al.*, 2007).

The efficiency of CA is influenced by fruit characteristics, agronomic management, temperature and shipping times, among other factors. An important factor is the number of days before the fruit enters the CA from harvest. Burdon *et al.* (2008) showed the effect of delays of 1, 5, 10 or 15 d after harvest on the establishment of CA on the quality of 'Hass' avocados. The results showed that delaying the establishment of CA did not affect the development of skin colour during storage, ripening time after storage or incidence of rot in ripe fruit. Therefore, delaying CA does not always result in the loss of fruit quality.

Modern CA containers can be electronically programmed to create active/intelligent atmospheres that can monitor and appropriately modify the storage atmosphere and temperature en route. Furthermore, the CA chambers can be manipulated to change from storage conditions into ripening rooms by altering the atmospheric temperature and gas composition (Burdon *et al.*, 2017).

MA storage relies on fruit respiration and the gas permeability of the barrier (e.g. a plastic bag) to develop an MA around the fruit. The potential for an MA to extend the storage life of avocado has been demonstrated under laboratory conditions (Aharoni *et al.*, 1968; Naiman *et al.*, 1995; Meir *et al.*, 1997), but commercial application is currently limited. Careful maintenance of the cold chain is essential with current wraps because higher temperatures will increase respiration and result in detrimental gas concentrations in the bag. Additionally, high-quality fruit should be used to prevent ethylene production associated with premature ripening of some of the fruit and from fruit wounds and disease.

Surface coatings are a variant of MA in which spraying or dipping the fruit in suitable solutions and emulsions provides a coating that alters the gas permeability of the fruit (Amarante and Banks, 2001). Waxing of fruit is used commercially in South Africa and Israel to supply several European markets, primarily because of consumer preference for glossy fruit and to decrease water loss as well as prevent premature softening during shipping (Kok *et al.*, 2010; Lütge *et al.*, 2010). However, ripening disorders may develop because of altered gas exchange (Durand *et al.*, 1984). Application rates are critical to success. If the fruit coating is too thick, gas permeability will be too restricted, resulting in disorder development, whereas the treatment will be ineffective if too little is applied (Bower *et al.*, 2003). The benefits of surface coatings are their ease of application, glossy appearance and lack of disposal problems, which occur with plastic wraps. However, the same challenges apply as in MA in that the coating should have the permeability characteristics to create the desired O_2 and CO_2 atmospheres in the fruit (and hopefully retard water loss) but be sufficiently robust to accommodate fruit-to-fruit variability in respiration and fluctuations in storage temperatures. Further development of surface coatings for avocado is warranted (Johnston and Banks, 1998) as formulations vary considerably in effectiveness (Bower *et al.*, 2003). Furthermore, restrictions by some countries on using certain components have created additional challenges.

Preconditioning

Preconditioning refers to exposing plants to mild stress conditions that confer tolerance to more stressful subsequent conditions and is a common phenomenon in plants (Kushad and Yelenosky, 1987; Palonen *et al.*, 2000). This response can be used to manipulate postharvest quality and/or potential disinfestation scenarios. HT and LTC have been examined extensively, and LTC is used commercially. Both treatments increase the tolerance of avocados to low storage temperatures and thus have applications for extending storage life and for cold disinfestation, at least for fruit flies.

Heat

Avocado fruit are relatively intolerant of heat. However, short-duration HTs result in the induction of tolerance to both low (Sanxter *et al.*, 1994; Woolf *et al.*, 1995; Florissen *et al.*, 1996; Roman-Mares and Yahia, 2000) and high (Woolf and Lay-Yee, 1997) temperatures (Fig. 15.4). Interestingly, fruit exposure to high temperatures in the field (e.g. due to sun exposure) does not result in the same damage as that induced by long-duration HTs. For example, sun exposure can result in fruit temperatures up to 50°C (Woolf and Ferguson, 2000) but does not cause external heat damage. Indeed, these occasional high temperatures in the field are accompanied by the induction of specific proteins (called heat shock proteins; HSPs) in diurnal cycles (Woolf *et al.*, 1999a, 2000), which might be responsible for the induced thermotolerance to postharvest treatments.

In general, 38°C hot-air treatment for approximately 8–12 h is optimal for the induction of tolerance to subsequent low-temperature storage (Nishijima *et al.*, 1995; Woolf *et al.*, 1995). Hot water pretreatment (38°C) of 'Hass' for 60 min was the most effective treatment for reducing CI when stored at 0.5°C for 4 weeks (Woolf, 1997). Additionally, Hofman *et al.* (2002b) showed that external CI in 'Hass' after disinfestation for 16 days at 1°C can be minimized by hot water conditioning at 40–42°C for 30 min. HTs did not eliminate chilling injury after more than 4 weeks at 0–1°C (A. Woolf, Auckland, New Zealand, 2011, unpublished data).

A hot water treatment (38°C) used in conjunction with a CA of 4 kPa O_2 and 6 kPa CO_2 prior to storage for 30 days at 5°C induced the development of soluble sugars (sucrose and galactinol) and stress-related enzymes (Escribano and Mitcham, 2014). The integration of metabolomics and proteomics data revealed a positive correlation between reduced glycolytic throughput and induced protein degradation with heat shock treatment.

HTs induce the formation of HSPs in many plants, which may have a protective effect on other proteins and enzyme systems during subsequent HT or cold storage. In avocado fruit, Florissen *et al.* (1996) noted that HSP concentrations started to increase with treatment times longer than 4 h at 38°C, which coincided with the minimum HT conditions required to reduce CI in storage. Messenger RNAs encoding HSPs reached a maximum at 38°C for 6–10 h, with little formation below 38°C or at 42°C for 6 h (Woolf and Lay-Yee, 1997). The reduction in CI was inversely related to the increased levels of a specific HSP RNA (Woolf *et al.*, 1998).

Subsequent work explored the effect of time delays between HWT and hot or cold stress treatment (Woolf *et al.*, 2004). A 3–24 h delay before a 50°C HWT resulted in reduced heat damage for 5- and 20-min pretreatments, whereas a loss of heat tolerance was observed with delays up to 5 days. Delays of 1–5 days resulted in reduced chilling tolerance, but the effect of less than a 24 h delay was less clear for the 5- and 20-min treatments. Levels of HSP17 and 70 homologous RNAs were induced by a 38°C pretreatment and delayed increased RNA levels with a maximum induction at 6 h. These patterns were correlated with the induced tolerance (i.e. increased HSP levels in the first 24 h) and then a decline (between 1 and 5 days). However, this pattern was less closely associated with a loss in tolerance. These patterns of induction and loss of thermotolerance could be exploited, for example, by delaying the time between pretreatment and disinfestation for a period when the insect loses tolerance faster than the fruit.

The correlations between HSPs and cold tolerance suggest that cultivars with a strong HSP response during HT may be more tolerant to CI after conditioning. If so, HSPs may represent a rapid screening tool to identify cultivars with better storage potential. They may also help predict outturn quality after conditioning and cold fruit storage from different production locations and seasons. However, large-scale commercial adoption of HTs has not yet occurred, most likely because of the high capital and energy costs.

Although HTs may reduce CI and control decay, they can also seriously damage fruit sensory quality, causing deterioration of flavour and aroma, resulting in consumer rejection (Escribano and Mitcham, 2014). Therefore, the integration of postharvest HTs into commercial postharvest chains should be applied with caution.

Low-temperature conditioning

LTC is the induction of tolerance to low temperatures by applying temperatures close to chilling temperatures. This technique is used successfully to allow cold disinfestation of 'Hass' (see 'Disinfestation'). The step-down temperature regimes developed in South Africa for sea freight (e.g. 7.5°C for the first week, 5.5°C for the second and third week and 3.5°C for the last week) (Eksteen and Bester, 1987; Vorster *et al.*, 1990) are arguably another example of a conditioning treatment. Combining the LTC treatment with storage temperatures of 1–2°C can significantly reduce the risk of internal flesh disorders with little risk of skin CI, thus potentially reducing the need for CA and 1-MCP for long-term cold storage and sea freight transport (Hofman *et al.*, 2010; van Rooyen and Bezuidenhout, 2010).

Atmospheric gases

Pretreatment with short exposure to high CO_2 or high N_2 (low O_2) before storage (called pulse or shock treatment) has been investigated in a number of fruits as a low-cost alternative to CA storage. Concentrations of 20–25% CO_2 for 3 days for 'Fuerte' (Truter and Eksteen, 1987), 5–25% for 2 days for 'Fuerte' (Allwood and Cutting, 1994) and 20% for 7 days for 'Hass' (Saucedo Veloz *et al.*, 1991) have generally provided better cold storage outturns than cold storage without pretreatment, but the quality is not as good as that with CA storage. Fruit ripening can be delayed, and internal disorders and diseases can be reduced by pulsing fruit with CO_2 during storage. However, Allwood and Cutting (1994) found little significant effect of pulsing 'Fuerte' fruit with 10% or 20% CO_2 for 2 days before storage at 5.5°C, and observed a delay in ripening after removal from storage at 3.5°C. Allwood and Wolstenholme (1995) also reported little benefit of CO_2 pretreatment for 'Fuerte' fruit.

Pulsing fruit with N_2 (effectively reducing O_2 concentrations) may have commercial potential (Pesis *et al.*, 1994). Oxygen concentrations from 0.2% to 4.2% for 2 days were more effective in delaying ripening after storage at 3.5°C or

5.5°C than pulsing with CO_2 (Allwood and Cutting, 1994). Treatment of 'Fuerte' with 3% O_2 for 1 day at 17°C significantly reduced CI after 3 weeks at 2°C (Pesis *et al.*, 1994).

Various techniques have been used for the application of pulsing treatments. Truter and Eksteen (1987) flushed fruit containers with 25% CO_2 and then sealed them. After 3 days at 5.5°C, the CO_2 and O_2 concentrations were 35% and 1%, respectively. Allwood and Cutting (1994) flushed sealed containers of fruit with premixed gas at the desired concentration daily for a total treatment time of 2 days. However, some of these reports did not indicate the temperature at which pulsing was conducted. Pulsing treatments are not used commercially to any significant extent.

Combination treatments

Storing fruit can help improve domestic marketing opportunities by influencing the supply–demand balance and can reduce reliance on imports. However, storage costs and loss of quality should not undermine the increased returns. Trials combining 1-MCP (applied as SmartFresh^{SM}) and CA with monthly reapplication of fungicide to control disease resulted in up to 3-month storage of late-season fruit (Lemmer *et al.*, 2007). It was possible to store 'Hass' fruit for up to 56 days when 1-MCP was used in conjunction with 1°C storage (Kok *et al.*, 2011). However, similar treatment of early-season fruit should be avoided to reduce the risk of CI. Further trials with conditioning, lower temperature storage, CA and 1-MCP may provide additional storage duration improvements.

Predicting outturn quality

Library trays

Library trays are used by many fruit industries worldwide but are more common for avocados, even though there is little published information on their use (e.g. Pak *et al.*, 2005a). Typically, trays of fruit are taken during the packing process and held for a specified time (generally 4 weeks) at standard storage temperatures, and fruit are removed for ripening at a constant temperature. More emphasis is placed on fruit quality assessment at the ripe stage, but external fruit quality may be assessed at removal from storage (Dixon and Pak, 2002; Pak *et al.*, 2005b). Given the difficulty of standardizing storage and ripening conditions, assessing fruit at a standard stage of ripeness and the assessment standards/systems, it is best that an industry uses a single (or minimal number) of laboratories to carry out storage and fruit quality assessments. For example, the New Zealand avocado industry has developed an integrated library system, which is outlined in the 'Avocado Industry Council Best Practice Guides' (Anonymous, 2007). This system aims to provide feedback to growers on fruit quality, pinpoint problems/issues, provide an industry-wide snapshot during the season and provide research direction.

Some key considerations in establishing a library tray system are the aim(s) of the system and who will be responsible for responding to the results and/or recommendations. For example, what is done with information that shows that one grower has significant fruit quality problems, and what if this occurs on one occasion or is consistent over many harvests/seasons? Is this information used as a strong inducement, or can it be used to determine what factors are causing problems and thus lead to improvement of preharvest and/or harvest/packing systems? Clearly, linking library tray results with other information leads to greater usefulness of the data (as long as they are analysed). This could/should include orchard information, spray diaries, harvesting conditions (e.g. rainfall) and if possible, outturn/in-market fruit quality assessments or feedback.

Fruit composition

Due to the apparently strong correlations of various mineral elements with internal quality, Snijder *et al.* (2002) identified critical fruit Ca and N concentrations several months prior to harvest that were related to storage potential. Exporters in South Africa have used the recommendations from Snijder *et al.* (2002) to set a 'Hass' fruit Ca concentration of more than 1000 mg kg^{-1} in November (approximately the end of the main cell division period) and N of less than 1% in January (showing a declining trend for N) as an export decision making tool (Kruger and Magwaza, 2010). Fruit N is also a primary driver in internal disorders of 'Pinkerton' (van Rooyen and Bower, 2005).

Non-destructive assessment

Online testing of quality parameters during packing is the ultimate goal of non-destructive quality assessment techniques. Good progress has been made in estimating % DM or water content using NIRS (Clark *et al.*, 2003; Blakey *et al.*, 2009; Wedding *et al.*, 2011), which in turn may be used to predict the ripening rate and some quality parameters. However, to be even more commercially significant (and thus more likely to be implemented given the costs), non-destructive assessments will need to predict the incidence of various disorders. The difficulty is to determine what parameters to measure, as most avocado disorders (except some mechanical damage) will not be evident at the time of harvest. Therefore, biochemical parameters will need to be determined. The concentration of sugars, particularly C_7 sugars, may be useful indicators (Bertling and Bower, 2005; Bertling and Tesfay, 2011), which again makes NIRS a potential choice for assessment. Magnetic resonance imaging (MRI) can accurately estimate oil content (Clark *et al.*, 1997) and potentially assess internal disorders such as flesh bruising (Sanches *et al.*, 2003; Mazhar *et al.*, 2011), but both cost and suitable equipment for use on packing lines, together with the decision as to what needs to be measured, preclude widespread use in the near future. Considerable use is made of various softness testing equipment using acoustic or pressure sensors, but these are only useful for predicting the time of ripeness once the fruit has started softening and are not useful at harvest.

Retail

Surveys of 'Hass' fruit from the retail shelf regularly find that 20–40% of the ripe fruit have at least 10% of the flesh showing defects (Hofman and Ledger, 2001a,b). Rot and bruises account for most defects, with lesser contributions from diffuse discoloration and vascular browning. A good understanding is required of the typical ripeness stage when consumers cut the fruit (e.g. Gamble *et al.*, 2010), as flesh diseases (Everett, 2001), flesh bruises (Mazhar *et al.*, 2011) and possibly other flesh defects increase as fruit soften and ripen. Consumer surveys suggest that less than one in five fruit should have no more than 10% of the flesh with defects; otherwise, sales will be impacted at typical retail prices.

Consumers increasingly prefer to purchase ripe fruit rather than unripe fruit. Hard, green 'Hass' fruit presented at the wholesale market or on the retail shelf generally do not sell as quickly as ripening fruit (Arpaia *et al.*, 2005). Selling 'ripe tonight' or ready-to-eat fruit increased sales by 200–300%, and the increased preference for ripe fruit was supported by observations that ripe fruit on the retail shelf sell more quickly than less ripe fruit. In Australia, approximately 60% of avocados sold by wholesalers and 90% sold by retailers are at or past the 'sprung' stage (the neck of the fruit yields to firm pressure), and ripening fruit are increasingly being sold in the USA and Europe.

The increasing demand for near-ripe fruit may increase the risk of flesh damage on the retail shelf. Ripening fruit is often stored by the wholesaler or retailer to balance supply and demand. Fruit have been considered more susceptible to CI when cool-stored in the early climacteric stages (Kosiyachinda and Young, 1976). Campbell *et al.* (2011) found that diffuse discoloration was significantly higher when fruit at the sprung to softening stage were held at 2°C for 7 days compared with when they were held at the rubbery to sprung stage; hence, susceptibility to diffuse discoloration was lower in the less ripe fruit. Additionally, from the sprung to softening stage onwards, diffuse discoloration was greater when stored below 5°C compared with 5–8°C. Hence, storage should focus on hard, green fruit, as this also reduces disease in ripe fruit compared with storing ripening fruit (Arpaia *et al.*, 2006). For example, hard fruit stored for as little as 4 days at 5°C had less rotting when ripe compared with storing softening or firm ripe fruit for 4 days before full ripening.

Unripe fruits are relatively resistant to bruising but become very susceptible as they soften (Arpaia *et al.*, 1987; Ledger and Barker, 1995). Little flesh bruising was observed in Australian 'Hass' fruit during harvesting and packing (Hofman, 2002) and during handling at the ripening and distribution centre, but bruising increased significantly on arrival at the retail store and particularly on the retail shelf (Mazhar *et al.*, 2011). This coincides with greater handling of individual trays and fruit compared with whole pallets as well as of riper fruit.

Systems need to be developed to minimize flesh bruising when near-ripe fruit are presented to the consumer. Preliminary results suggest that tray liners that hold fruit in individual compartments significantly reduce flesh bruising compared with no liners or volume filling (Arpaia *et al.*, 2006), but these liners result in less fruit per tray. Accurate prediction of days to ripen and placing fruit in protective packaging, such as transparent 'clamshells', can increase consumer confidence in the product and reduce the impulse to squeeze fruit to determine ripeness. However, disadvantages may include resistance to perceived excessive use of packaging (even if recyclable), reduction in fruit per pallet, and cost.

The greater consumer demand for fruit that will be ripe in 1–3 days after purchase has increased the use of postharvest ethylene treatments to induce more predictable ripening and more uniform ripening within a batch (Arpaia *et al.*, 2004). More consistent ripeness in prepacks can be achieved by grading for firmness during ripening. Estimating firmness by very gentle whole-hand squeezing can accurately rate firmness with little fruit damage, but this is time-consuming. In-line non-destructive systems such as the Sinclair IQ™ Firmness Tester and the Aweta Acoustic Firmness Sensor (Arpaia *et al.*, 2006) can estimate fruit firmness, but they are less accurate on firm fruit, and some manual resorting may be needed. These systems could be used to estimate the days to eating soft, but factors such as cultivar, maturity and storage conditions affect ripening behaviour; hence, predicting the time to eating soft based on firmness alone, particularly in the earlier stages of ripening, can be inaccurate.

An alternative system for determining ripeness has been developed by ripeSense® (www.ripesense.com). This system uses a clamshell with an 'intelligent' sticker or a direct-to-fruit sticker, which detects specific volatiles released from the ripening fruit. The colour of the sticker changes from red to yellow as the fruit ripens, and the consumer can choose to eat fruit at a less or more ripe stage.

Processing

Processing is an increasingly important component of the avocado industry worldwide because of the growing demand for convenience, the popularity of avocado, the increasing popularity of Mexican cuisine, new processing technologies, and knowledge of the health benefits of avocado products. High-pressure processing (HPP) is a relatively new technology that has resulted in highly successful commercialization of processed avocado products with long shelf lives. The use of HPP has grown significantly over the last 5 years, and guacamole and avocado pastes are flagship products for the HPP industry.

Approximately, 90% of processed product is sold to hotels and restaurants, mainly for use as guacamole (Kurlaender, 1996). There is currently a small market for minimally processed (MP) avocado halves and slices, but this market is expected to expand with the increasing demand for convenience foods. Avocado oil is used in the cosmetics industry and increasingly as a high-quality food and cooking oil (for more information about the uses of avocado fruit, see Fuentealba and Hernández, Chapter 16, this volume).

The major factors affecting the storage of MP products are senescence of the flesh, microbial spoilage and browning. For purees, microbial spoilage, browning and rancidity are the major problems, whereas for oils, rancidity is the major problem. Typical control measures are storage at low temperatures, inert atmospheres and using antioxidants and acids. Suitable cultivar selection can also be important.

For high-quality processed products, a minimum maturity of 25% DM is recommended. This ensures a more 'nutty' flavour, a greener colour and a more viscous puree (Kurlaender, 1996). For oil production, later maturity increases yield. Cooler growing conditions will increase the concentration of desirable monounsaturated oils (Kaiser and Wolstenholme, 1994a). Fruit must be ripe for purees and MP products, and uniform ripening of batches of fruit increases production efficiency. In California, this is achieved by cooling the fruit to 5°C for 24 h, then warming at 20–22°C, ethylene-treating and ripening them (Kurlaender, 1996).

Minimal processing

Minimal processing requires avocados to be ripened, the skin and seed to be removed, and storage at reduced temperature with some form of wrapping or packaging. However, deterioration of the product due to microbial growth (Ohlsson, 1994) and flesh discoloration is a significant problem (Gerdes and Parrino-Lowe, 1995). Browning can be delayed using SO_2, but this chemical is becoming unacceptable to the consumer. Alternative antibrowning agents and mixtures, such as 1% tetrasodium pyrophosphate and 0.2% cysteine (at a water activity of 0.80 and pH 5.5), may have some potential to reduce browning (Dorantesalvarez *et al.*, 1998). Dipping fruit halves or chunks in solutions of citric or ascorbic acid can also be used. For example, Nikapitiya and Lileperuma (2003) successfully used 1000 mg l^{-1} citric acid and 200 mg l^{-1} ascorbic acid to inhibit fruit discoloration. A patented process using a combination of ascorbate and calcium ions is also claimed to be useful for reducing flesh discoloration. However, such additives can alter fruit flavour. Although heating of the fruit produces a bitter taste (Ben-Et *et al.*, 1973), Bower and Dennison (2005) demonstrated that adverse effects did not occur provided that fruit temperature can be maintained below 80°C in the water dip. Furthermore, phenolic compounds were leached from the cut surfaces and probably contributed to a decreased potential for browning. MA packaging of ripe 'Fuerte' halves in either air, vacuum or a mixture of 5% CO_2, 15% O_2 or 80% CO_2 did not reduce browning sufficiently (Gerdes and Parrino-Lowe, 1995), although this may have been due to anaerobic conditions, which can also cause browning (Spalding and Marousky, 1981). Undurraga *et al.* (2007) found the same MA conditions used by Gerdes and Parrino-Lowe (1995) to be successful, although Bower and Dennison (2003, 2005) found no advantage with MA. Some processors use vacuum packaging, but this is not favoured by others due to the possibility of anaerobic bacterial growth and food safety concerns. Storage temperature is normally used to decrease the rate of product deterioration. Lower temperatures can retard softening, senescence and microbial spoilage. Acidification of the fruit to ~4.5 pH with ascorbic acid and citric acid dips can further reduce microbial spoilage (Pao and Petracek, 1997). Regardless, the chilled product has a relatively short shelf life. Undurraga *et al.* (2007) were only able to maintain the product at an acceptable quality for 11 days when stored at 3°°C.

Freezing of ripe halves has been attempted as an alternative to fresh fruit, but the thawed product was inferior to the fresh fruit because the texture changed from 'creamy' to 'watery' (Scudamore-Smith, 1984). Despite this, a number of commercial frozen products are available. Although the texture is not as good as the fresh product, an acceptable product can be produced, providing freezing is done optimally (Bower and Dennison, 2005). Although such products may not be considered suitable for replacing fresh avocado eaten alone, they are suitable for the catering trade when mixed in salads or as a spread for fast foods such as burgers or sandwiches.

Work is continuing on improving the quality of MP fruit, but the advent of HPP may replace MP as the preferred technology as it solves most of the problems encountered with MP, although cost is a factor to be considered.

High-pressure processing

A new technology developed and commercialized over the last 10 years is HPP, also known as ultrahigh pressure processing (UHP) or high hydrostatic pressure (HHP). The production of long-storing guacamole is one of the recent commercial success stories (Torres and Velazquez, 2005) for HPP, and the increase in sales and number of processing plants is a testament to its commercial efficacy. HPP involves the use of extreme pressures (200–600 MPa or up to 900 MPa for 1–10 min). The treatment is applied on a batch basis, typically in the water phase in 300–500 l tubes. For food products (such as guacamole), the product is prepackaged (e.g. in a plastic or foil pouch). Capital costs are high at approximately US$1 million per machine, and fruit processing, pulp blending and bagging equipment are also needed. Additionally, the system must operate in a food-safe environment, thus generally necessitating a multi-million-dollar investment.

As with nearly all food products, maintenance of the cool chain of HPP-treated avocado is critical to maintain quality. Temperatures of 5°C or lower will result in significantly longer shelf life than higher temperatures (20–25°C; López-Malo *et al.*, 1998), and even temperatures of 8°C will result in reduced quality compared to 4°C (Woolf *et al.*, Auckland, New Zealand, 2012, unpublished data). Clearly, the selection of the packaging/film type will also significantly influence product quality as the atmosphere generated will influence degradation (Soliva-Fortuny *et al.*, 2002).

It is generally believed that the main benefits of HPP are inactivation of microorganisms and enzymes in foods with minimal effects on colour, flavour and nutritional content (Hjelmqwist, 2002). This is achieved by denaturation of the protein structure, particularly by enzymes, which can retard browning reactions (Weemaes *et al.*, 1999).

A range of commercial products have been developed using HPP technology, including guacamole and avocado pulp (with no or minimal additives). The extension of shelf life for these products is remarkable, up to or exceeding 50 days at ~ 4°C. The high flavour quality (very close to fresh) and shelf-life durations are well beyond what can be achieved using other processing technologies such as freezing. López-Malo *et al.* (1998) demonstrated that in HPP-treated purees, the presence of yeasts and moulds was quite low (< 10 cfu g^{-1}) even after 100 days at 5°C. Slight colour changes occurred during storage but were generally not evident to the eye (López-Malo *et al.*, 1998).

In avocado pulp, PPO activity decreased with increasing pressure and decreasing pH (López-Malo *et al.*, 1998). Pressures of 345, 517 and 689 MPa resulted in residual enzyme activities of ~75%, 50% and 25%, respectively. Jacobo-Velázquez and Hernández-Brenes (2010) also reported a 50% reduction in PPO activity immediately after treatment at 600 MPa; enzyme activity recovered after 10–15 days and then again decreased to ~ 40% by day 45. The same authors found a similar reduction in LOX activity after an HPP treatment of 600 MPa (Jacobo-Velázquez and Hernández-Brenes, 2010). Palou *et al.* (2000) observed similar results when guacamole was treated for 5 min at 689 MPa. They observed that longer treatment times (e.g. 15–20 min) resulted in a greater reduction in PPO activity (~25–20%) and complete inactivation of LOX and that multiple cycles of HPP (two and three cycles for 5 min at 689 MPa) further reduced PPO activity to as low as 15%. However, the commercial viability of such long-term treatments or multiple cycles is doubtful.

Although it is generally believed that the efficacy of HPP is due to reduced enzyme activity, it should be noted that HPP can induce stabilization and/or activation of some enzymes. The nature of the product (e.g. pulp or whole fruit, pH and temperature) may also influence enzyme activities (Eisenmengera and Reyes-De-Corcuera, 2009). There has not been an extensive examination of other enzymes in avocado products (most emphasis has been placed on PPO and LOX), and these other enzymes may be important in product quality.

Oil

Avocado oil is generally considered a high-health oil, in part because it contains large quantities of the monounsaturated oil oleic acid (Kurlaender, 1996; see; Fuentealba and Hernández, Chapter 16, this volume). In addition, a number of 'healthful' phytochemicals are fat-soluble, including pigments such as chlorophylls and carotenoids and other compounds such as tocopherols (principally α-tocopherol) and sterols (principally β-sitosterol) (Woolf *et al.*, 2009). Avocado oil is favoured for use in the pharmaceutical and cosmetic industries but is also very stable and suitable for cooking and frying because of its fatty acid profile and very high smoke point.

Chemical or high-temperature extraction

Extraction of oil from avocado is significantly more difficult than extraction from olives, and thus, a range of 'harsh' techniques have been employed in the past to maximize yield. Before extraction, the seed (and sometimes the skin) is generally removed. Solvent extraction of the fresh pulp with ethanol:hexane (3:2) provides high yields and better-quality oil (Jacobsberg, 1988). Drying the flesh before extraction increases yield; temperatures of 80–90°C or even 120–130°C can be used. However, drying results in significantly lower-quality oil due to oxidation, concomitant off-flavours and destruction of a range of beneficial phytochemicals. Temperature may be reduced towards the end of drying to reduce the risk of these effects. Sun drying has also been investigated as a low-cost option (Southwell *et al.*, 1990). The oil is removed from the dried product by solvent extraction with yields of up to 90% of the total oil content (Jacobsberg, 1988). Centrifugation can also be used, and to maximize yield, avocado pulp is first heated by the addition of hot water (e.g. to approximately 70°C; Werman and Neeman,

1987), but this also results in lower oil quality, and the oil requires refining (see below).

Low-cost extraction techniques have been described for use in developing countries (Bizimana *et al.*, 1993). At the other end of the technology spectrum is supercritical extraction. This uses highly pressurized CO_2 (35 MPa), but the oil recovered is yellow, suggesting damage during extraction (Botha, 2004).

Refining

Because of the relatively harsh nature of the above-mentioned extraction techniques, the oil must be 'cleaned up', typically using three steps depending on the extraction process, desired oil quality and target oil product. Refining, bleaching and deodorizing (RBD) involve deacidification to remove free fatty acids, removal of pigments such as chlorophyll with Fuller's earth and charcoal, and deodorization by steam and vacuum treatment (Jacobsberg, 1988), respectively. This results in a stable product (oxidized fatty acids removed), but it is yellow (due to removal of the chlorophyll) with little flavour (the deodorizing step). These steps result in significant losses of up to 30% by volume during refining (C. Nathan, New Zealand, 2009, personal communication).

Cold-pressed avocado oil

Although cold-pressed avocado oil was examined many decades ago (Werman and Neeman, 1987), it has only become a significant commercial product since approximately 2000 (Wong *et al.*, 2008). The main use for this oil is culinary, with applications similar to those of extra virgin olive oil (drizzling on salads, dipping and high-temperature frying). Extraction is carried out using processes similar to olive oil, that is, tissue grinding using a hammer mill (or similar technique), malaxing (mixing) for approximately 1 h, then separation of solid, oil and water phases using a horizontal centrifuge (decanter). The best practice requires subsequent centrifugation of the oil to minimize the water content (Wong *et al.*, 2008). The seed should be removed before processing. The amount of skin in the malaxer can affect oil quality (Wong *et al.*, 2011). To achieve high-quality oil, it is important to use high-quality fruit with minimal to no rot/decay and free from internal physiological disorders such as diffuse flesh discoloration. Achieving commercially acceptable yields throughout the harvest season is a challenge, particularly early in the season, due to the lower % DM/oil content. Although Ca salts and exogenous cell wall enzymes can assist in oil release (Dominguez *et al.*, 1994), their commercial application has not been successful.

The high quality of cold-pressed avocado oil allows use in some high-end cosmetic applications, which may be refined only with the Fuller's earth and charcoal refining step to remove pigments. Reviews of avocado oil (with the greatest emphasis on cold-pressed avocado oil) have been recently published (Woolf *et al.*, 2007, 2009) and include summaries on commercial practicability.

Conclusions

Considerable progress has been made over the last 10 years in our understanding of avocado fruit responses after harvest, leading to improvements in commercial practices. Detailed sensory analysis has provided clear descriptors of the sensory characteristics, and more detailed studies have confirmed maturity standards and identified consumer tolerances to defects and ripeness. The increased commercial significance of avocado oil has resulted in a better understanding of its characteristics and processing requirements, although considerable information about oil processing techniques is kept proprietary by the industry. There is also a better understanding of the factors contributing to postharvest defects, including skin damage and flesh disorders. This has led to improved harvesting and storage practices. The greater availability of CA containers, the potential for DCA and a better understanding of 1-MCP application and its limitations have improved the level and consistency of quality in export markets. Demonstration of the potential for preconditioning to minimize CI has paved the way for cold disinfestation, potentially lower storage temperatures and less reliance on CA and 1-MCP to minimize flesh disorders during storage. Appropriate treatment combinations may further increase storage life, but detailed comparisons of these combinations relative to single treatments are required to verify efficacy.

An improved understanding of the effect of production factors (particularly fruit mineral concentrations) on fruit quality and shelf life has allowed better targeting of fruit batches to markets based on storage potential. Additional factors, such as C_7 sugar reserves, may provide additional indicators of fruit quality, but further research is needed. Non-destructive assessment of internal quality is gaining momentum. NIRS can predict % DM and potentially days to ripen. Further research and technological developments, including technologies such as MRI, may lead to accurate estimation of outturn quality using in-line packhouse systems, thus providing improved consistency for the consumer.

Providing close-to-ripe or ripe fruit to the consumer ('ripe tonight') is becoming increasingly common and typically increases consumer sales, although it adds complexity because of reduced shelf life and increased risk of flesh bruising. Improved systems to estimate days to eating soft and maximum shelf life and ways of minimizing bruising during handling will help increase consistent consumer satisfaction.

Finally, improved integration and information are required to minimize the time from harvest or packing to retail sale due to disruptions in product flow. Duration of 20 days for normal domestic marketing can be reduced by improving the integration of harvest and sale, thus reducing the risk of short-term oversupply. Improved information technology and the desire of all people in the supply chain to continually improve the quality of fruit offered to consumers are required to optimize quality and storage. In the longer term, new cultivars and rootstocks are needed with selection for improved fruit quality as an important criterion.

Note

The mesocarp and endocarp together constitute the whole fruit flesh (pulp). The term 'mesocarp discoloration' is well known commercially. Wherever the word mesocarp is used, it is intended to indicate the combined botanically correct mesocarp plus endocarp.

References

Abou-Aziz, A.B., Desauki, I. and El-Tanahy, M.M. (1975) Effect of nitrogen nutrition and storage temperature on avocado fruit behaviour during storage. *Gartenbauwissenschaft* 40, 223–227.

Adato, I. and Gazit, S. (1974) Postharvest response of avocado fruits of different maturity to delayed ethylene treatments. *Plant Physiology* 53, 899–902.

Adkins, M.F., Hofman, P.J., Stubbings, B.A. and Macnish, A.J. (2005) Manipulating avocado fruit ripening with 1-methylcyclopropene. *Postharvest Biology and Technology* 35, 33–42.

Aharoni, Y., Schiffmann-Nadel, M. and Zauberman, G. (1968) Effect of gradually decreasing temperatures and polyethylene wraps on the ripening and respiration of avocado fruit. *Israel Journal of Agricultural Research* 18, 77–82.

Akamine, E. and Goo, T. (1971) Respiration restoration of gamma-irradiated fresh fruits. *Journal of Food Science* 36, 1074–1077.

Ali, Z.M., Chin, L.H. and Lazan, H. (2004) A comparative study on wall degrading enzymes, pectin modifications and softening during ripening of selected tropical fruits. *Plant Science* 167, 317–327.

Allwood, M.E. and Cutting, J.G.M. (1994) Progress report: gas treatment of 'Fuerte' avocados to reduce cold damage and increase storage life. *South African Avocado Growers' Association Yearbook* 17, 22–26.

Allwood, M.E. and Wolstenholme, B.N. (1995) Modified atmosphere shock treatment and an orchard mulching trial for improving Fuerte fruit quality. *South African Avocado Growers' Association Yearbook* 18, 85–88.

Amarante, C. and Banks, N.H. (2001) Postharvest physiology and quality of coated fruits and vegetables. *Horticultural Reviews* 26, 161–238.

Anonymous (2004) *International Standardisation of Fruit and Vegetables: Avocados*. Organisation for Economic Co-Operation and Development, Paris, France.

Anonymous (2007) *Fruit Quality Best Practice Manual*. Avocado Industry Council Ltd, Tauranga, New Zealand. Available at: http://www.nzavocado.co.nz/Best%20practice/best%20practice%202007.pdf (accessed 5 March 2012).

Anonymous (2009) Overview of the avocado supply chain improvement program: project AV06006. *Talking Avocados* 20(3), 25–33.

Arpaia, M.L. (1994) Preharvest factors influencing postharvest quality of tropical and subtropical fruit. *HortScience* 29, 982–985.

Arpaia, M.L., Mitchell, F.G., Katz, P.M. and Mayer, G. (1987) Susceptibility of avocado fruit to mechanical damage as influenced by variety, maturity and stage of ripeness. *South African Avocado Growers' Association Yearbook* 10, 149–151.

Arpaia, M.L., Meyer, J.L., Witney, G.W., Bender, G.S., Stottlemyer, D.S. and Robinson, P.R. (1996) The Cashin Creek nitrogen fertilizer trial – what did we learn? *California Avocado Society Yearbook* 80, 85–98.

Arpaia, M.L., Ontai, S.L. and Reints, J.S. (1992) Protecting the postharvest quality of avocado fruit. *California Avocado Society Yearbook* 76, 93–97.

Arpaia, M.L., Boreham, D., Collin, S., Fjeld, K. and Sievert, J. (2003a) Hass maturity project. In: Witney, G. (ed.) *Proceedings of the California Avocado Research Symposium*. California Avocado Commission, Irvine, CA, pp. 120–124.

Arpaia, M.L., Sievert, J., Collin, S., Fjeld, K., Coetzer, M. *et al.* (2003b) Avocado postharvest quality. In: Witney, G. (ed.) *Proceedings of the California Avocado Research Symposium*. California Avocado Commission, Irvine, CA, pp. 125–139.

Arpaia, M.L., Sievert, J., Collin, S., Fjeld, K., Stottlemyer, D. *et al.* (2004) Avocado postharvest quality. In: Witney, G. (ed.) *Proceedings of the California Avocado Research Symposium*. California Avocado Commission, Irvine, CA, pp. 112–125.

Arpaia, M.L., Requejo, C., Woolf, A., White, A., Thompson, J.F. *et al.* (2005) Avocado postharvest quality. In: *Proceedings of the California Avocado Research Symposium*. California Avocado Commission, Irvine, CA, pp. 120–133.

Arpaia, M.L., Requejo-Jackman, C., Woolf, A., White, A., Thompson, J.F. *et al.* (2006) Avocado postharvest quality. In: Witney, G. (ed.) *Proceedings of the California Avocado Research Symposium*. California Avocado Commission, Irvine, California, pp. 143–155.

Arzate-Vázquez, I., Chanona-Pérez, J.J., Perea-Flores, M.D.J., Calderón-Domínguez, G., Moreno-Armendáriz, M.A., Calvo, H. and Gutiérrez-López, G. (2011) Image processing applied to classification of avocado variety Hass (*Persea americana Mill.*) during the ripening process. *Food and Bioprocess Technology* 4, 1307–1313.

Ashton, O.B.O., Wong, M., McGhie, T.K., Vather, R., Wang, Y. *et al.* (2006) Pigments in avocado tissue and oil. *Journal of Agricultural and Food Chemistry* 54, 10151–10158.

Awad, M. and Young, R.E. (1979) Postharvest variation in cellulase, polygalacturonase and pectinmethylesterase in avocado (*Persea americana* Mill. cv. Fuerte) fruits in relation to respiration and ethylene production. *Plant Physiology* 4, 306–308.

Baldwin, E.A. (2002) Fruit flavor: volatile metabolism and consumer perceptions. In: Knee, M. (ed.) *Fruit Quality and its Biological Basis*. Sheffield Academic Press, Sheffield, UK, pp. 89–106.

Bantayehu, M., Alemayehu, M., Abera, M. and Solomon, B. (2017) Postharvest losses assessment of tropical fruits in the market chain of North-Western Ethiopia. *Food Science and Quality Management* 66.

Bender, R.J., Brecht, J.K., Sargent, S.A. and Huber, D.J. (2000) Mango tolerance to reduced oxygen levels in controlled atmosphere storage. *Journal of the American Society for Horticultural Science* 125, 707–713.

Ben-Et, G., Dolev, A. and Tatarsky, D. (1973) Compounds contributing to heat-induced bitter off-flavour in avocado. *Journal of Food Science* 38, 546–547.

Bergh, B., Kumamoto, J. and Chen, P. (1989) Determining maturity in whole avocados. *California Avocado Society Yearbook* 73, 173–176.

Bertling, I. and Bower, J.P. (2005) Sugars as energy sources – is there a link to avocado fruit quality? *South African Avocado Growers' Association Yearbook* 28, 24–27.

Bertling, I. and Tesfay, S.Z. (2011) Avocado sugars – key to postharvest shelf life? In: *Proceedings of the VII World Avocado Congress*. Avocados Australia Ltd, Brisbane, Australia. Available at: http://worldavocadocongress2011.com/userfiles/file/Isa%20Bertling%201420-1440.pdf (accessed 23 January 2012).

Bertling, I., Tesfay, S.Z. and Bower, J.P. (2008) The search for means to improve the C7 sugar pool in 'Hass' avocado. *South African Avocado Growers' Association Yearbook* 31, 12–15.

Beverly, R.B., Latimer, J.G. and Smittle, D.A. (1993) Preharvest physiological and cultural effects on postharvest quality. In: Shewfelt, R. and Prussia, S. (eds) *Postharvest Handling: A Systems Approach*. Academic Press, San Diego, USA, pp. 74–93.

Bezuidenhout, J., Vorster, L.L. and Toerien, J.C. (1992) Temperature management – the basis for successful export of South African 'Fuerte' avocados. In: *Proceedings of the II World Avocado Congress*, Orange, California, pp. 427–433.

Bezuidenhout, J.J. (1992) Analysis of transit temperature and fruit condition of South African export avocados. *South African Avocado Growers' Association Yearbook* 15, 39–40.

Bezuidenhout, J.J. and Vorster, L.L. (1991) The relation between fruit quality and climatic and soil factors. *South African Growers' Association Yearbook* 14, 40–41.

Biale, J.B. and Young, R.E. (1971) The avocado pear. In: Hulme, A.C. (ed.) *The Biochemistry of Fruits and their Products, Vol.* 2. Academic Press, London, pp. 1–63.

Bill, M., Sivakumar, D., Thompson, A.K. and Korsten, L. (2014) Avocado fruit quality management during the postharvest supply chain. *Food Reviews International* 30, 169–202.

Billy, L., Mehinagic, E, Royer, G., Renard, C., Arvisenet, G., Prost, C. and Jourjon, F. (2008) Relationship between texture and pectin composition of two apple cultivars during storage. *Postharvest Biology and Technology* 47, 315–324.

Bizimana, V., Breene, W.M. and Csallany, A.S. (1993) Avocado oil extraction with appropriate technology for developing countries. *Journal of the American Oil Chemists' Society* 70, 821–822.

Blakey, R.J. (2011) Management of avocado postharvest physiology. PhD thesis, University of KwaZulu-Natal, South Africa.

Blakey, R.J., Bower, J.P. and Bertling, I. (2009) Influence of water and ABA supply on the ripening pattern of avocado (*Persea americana* Mill.) fruit and the prediction of water content using near infrared spectroscopy. *Postharvest Biology and Technology* 53, 72–76.

Blanke, M. (1991) Respiration of apple and avocado fruits. *Postharvest News and Information* 2, 429–436.

Boelema, T. (1987) Long-distance transport of avocados. *South African Avocado Growers' Association Yearbook* 10, 153–156.

Botha, B.M. (2004) Supercritical fluid extraction of avocado oil. *South African Avocado Growers' Association Yearbook* 27, 24–27.

Bower, J.P. (1978) Ecophysiological studies of three cultivars of *Persea americana* (Mill.) emphasising photosynthesis and internal water relations. MSc thesis, University of Natal, South Africa.

Bower, J.P. (1985) Some aspects of water relations on avocado (*Persea americana* Mill.) tree and fruit physiology. PhD thesis, University of Natal, South Africa.

Bower, J.P. (1988) Pre and postharvest measures for long-term storage of avocados. *South African Avocado Growers' Association Yearbook* 11, 68–72.

Bower, J.P. (2005) The effect of coatings and packaging on fruit quality in avocado cv. Hass stored at low temperature for phytosanitary purposes. *South African Avocado Growers' Association Yearbook* 28, 29–31.

Bower, J.P. and van Lelyveld, L.J. (1985) The effect of stress history and container ventilation on avocado fruit polyphenol oxidase activity. *Journal of Horticultural Science* 60, 545–547.

Bower, J.P. and Cutting, J.G.M. (1987) Some factors affecting post-harvest quality in avocado fruit. *South African Avocado Growers' Association Yearbook* 10, 143–146.

Bower, J.P. and Cutting, J.G. (1988) Avocado fruit development and ripening physiology. *Horticultural Reviews* 10, 229–271.

Bower, J.P. and Dennison, M.T. (2003) Progress in the development of avocado products. *South African Avocado Growers' Association Yearbook* 26, 35–39.

Bower, J.P. and Magwaza, L.S. (2004) Effect of coatings and packaging on external and internal quality with emphasis on 'cold injury'. *South African Avocado Growers' Association Yearbook* 27, 35–39.

Bower, J.P. and Dennison, M.T. (2005) A process to prevent browning of frozen avocado halves and chunks. *South African Avocado Growers' Association Yearbook* 28, 40–41.

Bower, J.P. and Palpi, G. (2006) Effect of fruit coatings and packaging on chilling injury of 'Hass' avocados. *South African Avocado Growers' Association Yearbook* 29, 69–72.

Bower, J.P. and Bertling, I. (2008) Shelf life extension and reduction of chilling injury of 'Hass' avocados stored at ultra-low temperatures. *Acta Horticulturae* 768, 189–194.

Bower, J.P., van Rooyen, Z. and Allwood, G.A. (2000) Progress in identifying the primary causes of mesocarp discoloration in *cv.* Pinkerton. *South African Avocado Growers' Association Yearbook* 23, 15–20.

Bower, J.P., Dennison, M.T. and Fowler, K. (2003) Avocado and mango cold storage damage as related to water loss control. *Acta Horticulturae* 628, 401–406.

Bozak, K.R., Yu, H., Sirevag, R. and Christoffersen, R.E. (1990) Sequence analysis of ripening-related cytochrome-p-450 cDNA from avocado. *Proceedings of the National Academy of Sciences (USA)* 87, 3904–3908.

Brown, B.I. (1984) Market maturity indices and sensory properties of avocados grown in Queensland. *Food Technology in Australia* 37, 474–476.

Brummell, D.A. (2006) Cell wall disassembly in ripening fruit. *Functional Plant Biology* 33, 103–119.

Buffetto, F., Cornuault, V., Rydahl, M.G., Ropartz, D., Alvarado, C. *et al.* (2015) The deconstruction of pectic rhamnogalacturonan I unmasks the occurrence of a novel arabinogalactan oligosaccharide epitope. *Plant and Cell Physiology* 56, 2181–2196. doi:10.1093/pcp/pcv128.

Burdon, J. and Lallu, N. (2008) Dynamic controlled atmosphere storage of avocados: technology for managing exports? *New Zealand Avocado Growers' Association Annual Research Report* 8, 123–126.

Burdon, J., Lallu, N., Haynes, G., Francis, K., Pidakala, P. and Billing, D. (2005) Mode of action of water loss on fruit quality of 'Hass' avocados. In: *Proceedings of the New Zealand and Australia Avocado Growers Conference '05*. Available at: http://www.avocadosource.com/journals/ausnz/ausnz_2005/burdonj2005.pdf (accessed 21 October 2011).

Burdon, J., Lallu, N., Haynes, G., Pidakala, P., Wilcocks, P. *et al.* (2007) Carbohydrate of late season 'Hass' avocado fruit. *New Zealand Avocado Growers' Association Annual Research Report* 7, 97–102.

Burdon, J., Lallu, N., Haynes, G., McDermott, K. and Billing, D. (2008) The effect of delays in establishment of a static or dynamic controlled atmosphere on the quality of 'Hass' avocado fruit. *Postharvest Biology and Technology* 49, 61–68.

Burdon, J., Billing, D. and Pidakala, P. (2017) Avoiding chilling damage in 'Hass' avocado fruit by controlled atmosphere storage at higher temperature. *HortScience* 52, 1107–1110.

Buse, E.L. and Laties, G.G. (1993) Ethylene-mediated posttranscriptional regulation in ripening avocado (*Persea americana*) mesocarp discs. *Plant Physiology* 102, 417–423.

Campbell, T., Marques, J. and Hofman, P. (2011) Cold storage of partially ripened 'Hass'. *Talking Avocados* 22(3), 24–25.

Cass, L.G., Kirven, K.A. and Christoffersen, R.E. (1990) Isolation and characterization of a cellulase gene family member expressed. *Molecular and General Genetics* 223, 76–86.

CG (2011) Catalytic generators: the ripening advantage. Catalytic Generators, LLC. Available at: http://www.catalyticgenerators.com (accessed 29 May 2011).

Chaplin, G.R., Wills, R.B.H. and Graham, D. (1982) Objective measurement of chilling injury in the mesocarp of stored avocado. *HortScience* 17, 238–239.

Chen, N.J., Wall, M.M., Paull, R.E. and Follett, P.A. (2009) Variation in 'Sharwil' avocado maturity during the harvest season and resistance to fruit fly infestation. *HortScience* 44, 1655–1661.

Chen, P., McCarthy, M.J., Kauten, R., Sarig, Y. and Han, S. (1993) Maturity evaluation of avocados by NMR methods. *Journal of Agricultural Engineering Research* 55, 177–187.

Cho, B.H., Koyama, K., Olivares-Díaz, E. and Koseki, S. (2020) Determination of 'Hass' avocado ripeness during storage based on smartphone image and machine learning model. *Food and Bioprocess Technology* 13, 1579–1587.

Cho, B.H., Koyama, K. and Koseki, S. (2021) Determination of 'Hass' avocado ripeness during storage by a smartphone camera using artificial neural network and support vector regression. *Journal of Food Measurement and Characterization* 15, 2021–2030.

Christoffersen, R.E., Tucker, M.L. and Laties, G.G. (1984) Cellulase gene expression in ripening avocado fruit: the accumulation of cellulase mRNA and protein as demonstrated by cDNA hybridisation and immunodetection. *Plant Molecular Biology* 3, 385–391.

Clark, C.J., Hockings, P.D., Joyce, D.C. and Mazucco, R.A. (1997) Application of magnetic resonance imaging to pre- and post-harvest studies of fruits and vegetables. *Postharvest Biology and Technology* 11, 1–21.

Clark, C.J., McGlone, V.A., Requejo, C., White, A. and Woolf, A.B. (2003) Dry matter determination in 'Hass' avocado by NIR spectroscopy. *Postharvest Biology and Technology* 29, 301–308.

Coggins, Jr., C.W. (1984) Feasibility of using fruit size and percentage dry weight to predict maturity. *California Avocado Society Yearbook* 68, 145–160.

Considine, M., McCarthy, A., Speijers, J. and Tan, S.C. (2005) Separating high harvest temperature effects on postharvest avocado quality. *Acta Horticulturae* 687, 167–173.

Continella, G., La Rosa, G. and Larosa, G. (1992) Observations on the maturity process of six avocado cultivars in Sicily. *Acta Horticulturae* 296, 69–73.

Cowan, A.K. (2004) Metabolic control of avocado fruit growth: 3-hydroxy-3-methylglutaryl coenzyme a reductase, active oxygen species and the role of C7 sugars. *South African Journal of Botany* 70, 75–82.

Cox, K.A., McGhie, T.K., White, A. and Woolf, A.B. (2004) Skin colour and pigment changes during ripening of 'Hass' avocado fruit. *Postharvest Biology and Technology* 31, 287–294.

Cresswell, A. (1970) *The Biochemistry of Fruits and Their Products, 1st edn.* University of Michigan, Ann Arbor, MI, pp. 172–175.

CRS (2011) Ethylene systems. Controlled ripening systems Australia. Available at: http://www.ethylenesystems.com.au (accessed 29 May 2011).

Cutting, J.G.M. and Vorster, L.L. (1991) Difference in fruit ripening and quality from trees with different yield – a progress report. *South African Avocado Growers' Association Yearbook* 14, 56.

Cutting, J.G.M. and Wolstenholme, B.N. (1992) Maturity and water loss effects on avocado (*Persea americana* Mill.) postharvest physiology in cool environments. *Journal of Horticultural Science* 67, 560–575.

Cutting, J.G.M., Wolstenholme, B.N. and Hardy, J. (1992) Increasing relative maturity alters the base mineral composition and phenolic concentration in avocado (*Persea americana* Mill.) fruit. *South African Avocado Growers' Association Yearbook* 15, 64–67.

Dallman, T.F., Eaks, I.L., Thomson, W.W. and Nothnagel, E.A. (1988) Postharvest density changes in membranes from ripening avocado fruits. *Journal of the American Society for Horticultural Science* 113, 729–733.

Dallman, T.F., Thomson, W.W., Eaks, I.L. and Nothnagel, E.A. (1989) Expression and transport of cellulase in avocado mesocarp during ripening. *Protoplasma* 151, 33–46.

Darvas, J.M. (1982) Etiology and control of some fruit diseases of avocado (*Persea americana*). DSc thesis, University of Pretoria, South Africa.

Dauny, P.T., Joyce, D.C. and Gamby, C. (2003) 1-Methylcyclopropene influx and efflux in 'Cox' apple and 'Hass' avocado fruit. *Postharvest Biology and Technology* 29, 101–105.

Defilippi, B.G., Manríquez, D., Luengwilai, K. and González-Agüero, M. (2009) Aroma volatiles: biosynthesis and mechanisms of modulation during fruit ripening. *Advances in Botanical Research* 50, 1–37.

Defilippi, B.G., Ejsmentewicz, T., Covarrubias, M.P., Gudenschwager, O. and Campos-Vargas, R. (2018) Changes in cell wall pectins and their relation to postharvest mesocarp softening of 'Hass' avocados (*Persea americana* Mill.). *Plant Physiology and Biochemistry* 128, 142–151.

Dixon, J. and Pak, H.A. (2002) Analysis of packhouse library tray data from 2001/2002 season. *New Zealand Avocado Growers' Association Annual Research Report* 2. Available at: http://www.avocadosource.com/journals/nzaga/nzaga_2002/nzaga_2002_10.pdf (accessed 16 October 2024).

Dixon, J., Elmsley, T.A., Smith, D.B. and Pak, H.A. (2005) Increasing pick to pack times increases ripe rots in 'Hass' avocados. *New Zealand Avocado Growers' Association Annual Research Report* 5, 43–50.

Dixon, J., Elmsly, T.A. and Cutting, J.G.M. (2003a) The effect of ethylene treatment on fruit quality and ripening for late season 'Hass' avocados stored up to six weeks. *New Zealand Avocado Growers' Association Annual Research Report* 3, 80–85.

Dixon, J., Pak, H.A., Mandemaker, A.J., Smith, D.B., Elmsly, T.A. and Cutting, J.G.M. (2003b) Fruit age management: the key to successful long distance export of New Zealand avocados. *New Zealand Avocado Growers' Association Annual Research Report* 3, 60–65.

Dixon, J., Smith, D. and Elmsly, T.A. (2004a) Quality of Hass avocado (*Persea americana* Mill.) fruit after high humidity storage in polyethylene bags. *New Zealand Avocado Growers' Association Annual Research Report* 4, 54–60.

Dixon, J., Smith, D. and Elmsly, T.A. (2004b) Fruit age, storage temperature and maturity effects on Hass avocado fruit quality and ripening. *New Zealand Avocado Growers' Association Annual Research Report* 4, 47–53.

Dixon, J., Elmsley, T.A., Smith, D.B., Mandemaker, A.J. and Pak, H.A. (2008) Low temperature limits of New Zealand Hass avocado fruit. *New Zealand Avocado Growers' Association Annual Research Report* 8, 111–122.

Dominguez, H., Nunez, M.J. and Lema, J.M. (1994) Enzymatic pretreatment to enhance oil extraction from fruits and oilseeds: a review. *Food Chemistry* 49, 271–286.

Donetti, M. and Terry, L. A. (2011) Investigation of skin colour changes as non-destructive parameter of fruit ripeness of imported "Hass" avocado fruit.. In: *Proceedings of the IV International Conference Postharvest Unlimited* 2011 945, pp. 189–196.

Donetti, M. and Terry, L. A. (2014) Biochemical markers defining growing area and ripening stage of imported avocado fruit cv. Hass. *Journal of Food Composition and Analysis* 34(1), 90–98.

Donkin, D.J. and Cutting, J.G.M. (1994) Rind moisture loss – the cause of cold damage in harvested avocado fruit? *South African Avocado Growers' Association Yearbook* 17, 28–30.

Donkin, D.J. and Wolstenholme, B.N. (1995) Postharvest heat treatments with a view to reducing chilling injury in Fuerte avocado fruit. *South African Avocado Growers' Association Yearbook* 18, 80–84.

Donkin, D.J., Mans, C.C., Slabbert, M.J., Levin, J. and Wolstenholme, B.N. (1995) 'Stepped down' temperature regimes for Fuerte fruit grown in the Kwazulu/Natal Midlands: do they reduce the incidence of physiological disorders? *South African Avocado Growers' Association Yearbook* 18, 89–93.

Dopico, B., Lowe, A.L., Wilson, I.D., Merodico, C. and Grierson, D. (1993) Cloning and characterization of avocado fruit mRNA and their expression. *Plant Molecular Biology* 21, 437–449.

Dorantesalvarez, L., Paradadorantes, L., Ortizmoreno, A., Santiagopineda, T., Chiraltboix, A. and Barbosa-canovas, G. (1998) Effect of anti-browning compounds on the quality of minimally processed avocados. *Food Science and Technology International* 4, 107–113.

Durand, B.J., Orcan, L., Yanko, U., Zauberman, G. and Fuchs, Y. (1984) Effect of waxing on moisture loss and ripening of 'Fuerte' avocado fruit. *HortScience* 19, 421–422.

Duvenhage, J.A. (1993) The influence of wet picking on postharvest diseases and disorders of avocado fruit. *South African Avocado Growers' Association Yearbook* 16, 77–79.

Eaks, I.L. (1973) Effect of clip vs. snap harvest of avocados on ripening and weight loss. *Journal of the American Society for Horticultural Science* 98, 106–108.

Eaks, I.L. (1980) Respiratory rate, ethylene production, and ripening response of avocado fruit to ethylene or propylene following harvest at different maturities. *Journal of the American Society for Horticultural Science* 105, 744–747.

Eaks, I.L. (1985) Effect of calcium on ripening, respiratory rate, ethylene production, and quality of avocado fruit. *Journal of the American Society for Horticultural Science* 110, 145–148.

Eisenmengera, M.J. and Reyes-De-Corcuera, J.I. (2009) High pressure enhancement of enzymes – a review. *Enzyme and Microbial Technology* 45, 331–347.

Eksteen, G.J. and Bester, J.M. (1987) Storage and transport of avocados – practical considerations for the South African export situation. *South African Avocado Growers' Association Yearbook* 10, 157–159.

Elik, A., Yanik, D.K., Istanbullu, Y., Guzelsoy, N.A., Yavuz, A. and Gogus, F. (2019) Strategies to reduce post-harvest losses for fruits and vegetables. *Strategies* 5(3), 29–39.

El-Mageed, M.A.A. (2007) Development of volatile compounds of avocado and casimiro a during fruit maturation. *Arab Universities Journal of Agricultural Sciences* 15, 89–100.

Elmsly, T.A. and Dixon, J. (2008) Growth rates of ripe rot fungi at different temperatures. *New Zealand Avocado Growers' Association Annual Research Report* 8, 77–84.

Elmsly, T.A., Mandemaker, A.J., Dixon, E.M., Smith, D.B. and Dixon, J. (2007) Influence of time of harvest on 'Hass' avocado ripe fruit quality. *New Zealand Avocado Growers' Association Annual Research Report* 7, 91–96.

Erickson, L.C., Eaks, I.L. and Porter, G.G. (1970) Over-maturity in 'Hass' avocados. *California Avocado Society Yearbook* 54, 62–65.

Escribano, S. and Mitcham, E.J. (2014) Progress in heat treatments. *Stewart Postharvest Review* 3:1.

Espindola-Barquera, M.C., Cano-Medrano, R. and Rodriguez-Alcazar, J. (2006) Fruit set responses to gibberellic acid, nitrogen and girdling on 'Hass' avocado. *Acta Horticulturae* 727, 409–413.

Everett, K.R. (2001) Stem-end rots: infection of ripening fruit. *New Zealand Avocado Growers' Association Annual Research Report* 1, 50–59.

Everett, K.R., Hallett, I.C., Rees-George, J., Chynoweth, R.W. and Pak, H.A. (2008) Avocado lenticel damage: the cause and the effect on fruit quality. *Postharvest Biology and Technology* 48, 383–390.

Faubion, D.F., Mitchell, F.G., Mayer, G. and Arpaia, M.L. (1992) Response of 'Hass' avocado to postharvest storage in controlled atmosphere conditions. In: *Proceedings of the II World Avocado Congress*, Orange, California, pp. 467–472.

Fellman, J.K., Miller, T.W., Mattinson, D.S. and Mattheis, J.P. (2000) Factors that influence biosynthesis of volatile flavour compound in apple fruits. *HortScience* 35, 1026–1033.

Feng, X.Q., Apelbaum, A., Sisler, E.C. and Goren, R. (2000a) Control of ethylene responses in avocado fruit with 1-methylcyclopropene. *Postharvest Biology and Technology* 20, 143–150.

Feng, X.Q., Apelbaum, A., Sisler, E.C. and Goren, R. (2000b) Control of ethylene responses in avocado fruit with 1-methylcyclopropene. *Postharvest Biology and Technology* 20, 143–150.

Ferguson, I.B. (1980) Mineral composition of fruit and vegetables in relation to storage life. *CSIRO Food Research Quarterly* 40, 94–100.

Fischer, R. and Bennett, A. (1991) Role of cell wall hydrolases in fruit ripening. *Annual Review of Plant Physiology and Plant Molecular Biology* 42, 675–703. doi:10.1146/annurev.pp. 42.060191.003331.

Fitzell, R.D., Coates, L.M., Ledger, S.N., Yuen, C.M.C., Tan, S.C. and Joyce, D.C. (1995) Avocados. In: Coates, L.M., Cooke, A.W., Persley, D.M., Beattie, B. and Wade, N. (eds) *Postharvest Diseases of Horticultural Produce, Vol. 2. Tropical Fruit*. Queensland Department of Primary Industries, Brisbane, Australia, pp. 11–17.

Flitsanov, U., Mizrach, A., Liberzon, A., Akerman, M. and Zauberman, G. (2000) Measurement of avocado softening at various temperatures using ultrasound. *Postharvest Biology and Technology* 20(3), 279–286.

Florissen, P., Ekman, J.S., Blumenthal, C., McGlasson, W.B., Conroy, J. and Holford, P. (1996) The effects of short heat-treatments on the induction of chilling injury in avocado fruit (*Persea americana* Mill). *Postharvest Biology and Technology* 8, 129–141.

Fuchs, Y. and Zauberman, G. (1987) Physiological aspects of delayed marketing of avocado fruit. *South African Avocado Growers' Association Yearbook* 10, 132–135.

Fuchs, Y., Zauberman, G. and Yanko, U. (1975) Freeze injuries in avocado fruit. *HortScience* 10, 64–65.

Fuentealba, C., Vidal, J., Zulueta, C., Ponce, E., Uarrota, V., Defilippi, B.G. and Pedreschi, R. (2022) Controlled atmosphere storage alleviates Hass avocado black spot disorder. *Horticulturae* 8(5), 369.

Galili, N., Shmulevich, I. and Benichou, N. (1998) Acoustic testing of avocado for fruit ripeness evaluation. *Transactions of the ASAE* 41(2), 399.

Gamble, J., Harker, F.R., Jaeger, S.R., White, A., Bava, C. *et al.* (2010) The impact of dry matter, ripeness and internal defects on consumer perceptions of avocado quality and intentions to purchase. *Postharvest Biology and Technology* 57, 35–43.

García-Rojas, M., Morgan, A., Gudenschwager, O., Zamudio, S., Campos-Vargas, R., González-Agüero, M. and Defilippi, B.G. (2016) Biosynthesis of fatty acids-derived volatiles in 'Hass' avocado are modulated by ethylene and storage conditions during ripening. *Scientia Horticulturae* 202, 91–98.

Gerdes, D.L. and Parrino-Lowe, V. (1995) Modified atmosphere packaging (MAP) of Fuerte avocado halves. *Lebensmittel-Wissenschaft und Technologie* 28, 12–16.

Glatte, J.H.A., Motsay, G.J. and Welsh, B.E. (1967) *Carbon Dioxide Tolerance Studies*. School of Aerospace Medicine, Brooks Air Force Base, San Antonio, Texas. Available at: http://archive.rubicon-foundation.org/6045 (accessed 2 June 2011).

Guerrero, E.R. and Benavides, G.M. (2014) Automated system for classifying Hass avocados based on image processing techniques. In: *Proceedings of the I.E. Colombian Conference on Communications and Computing (COLCOM),* 4–6, Bogota, pp.1–6.

Haas, E. and Felsenstein, G. (1985) Factors affecting the cooling rate of avocados packed in corrugated cartons. In: *Proceedings of Meetings of Commissions D1, D2, D3 of the International Institute of Refrigeration*, pp. 291–299.

Hardy, J., Finnemore, H.J., Partridge, C. and Joubert, J. (1995) *SAAGA Avocado Exporter's Packing Guide*. Tzaneen, South Africa, p. 23.

Hartill, W.F.T. and Everett, K.R. (2002) Inoculum sources and infection pathways of pathogens causing stem-end rots of 'Hass' avocado (*Persea Americana*). *New Zealand Journal of Crop and Horticultural Science* 30(4), 249–260.

Hatfield, R. and Nevins, D.J. (1986) Characterization of the hydrolytic activity of avocado cellulase. *Plant and Cell Physiology* 27, 541–552.

Hatton, T.T. and Reeder, W.F. (1972) Relationship of bloom date to the size and oil content of Booth avocados. *The Citrus Industry* 53, 20–21.

Hernández, I., Fuentealba, C., Olaeta, J.A., Lurie, S., Defilippi, B.G, Campos-Vargas, R. and Pedreschi, R. (2016) Factors associated with postharvest ripening heterogeneity of 'Hass' avocados (*Persea americana* Mill). *Fruits* 71, 259–268.

Hershkovitz, V., Saguy, S.I. and Pesis, E. (2005) Postharvest application of 1-MCP to improve the quality of various avocado cultivars. *Postharvest Biology and Technology* 37, 252–264.

Hewett, E. (2006) An overview of preharvest factors influencing postharvest quality of horticultural products. *International Journal of Postharvest Technology and Innovation* 1, 4–15.

Hjelmqwist, J. (2002) New technology for maintaining freshness of foods. *Alimentacion, Equipos y Tecnologia* 21, 45–49.

Hofman, P.J. (2002) Bruising of 'Hass' avocado from harvest to the packhouse. In: *Final Report for Project AV02015*. Horticulture Australia Ltd, Sydney, Australia, pp. 1–10.

Hofman, P.J. (2005) Skin damage to Hass avocados can cause quality problems down the track. *Talking Avocados* 16(2), 24–26.

Hofman, P.J. and Smith, L.G. (1994) Preharvest effects on postharvest quality of subtropical and tropical fruit. In: Champ, B.R., Highley, E. and Johnson, G.I. (eds) *Postharvest Handling of Tropical Fruits. ACIAR Proceedings No. 50.* ACIAR, Canberra, Australia, pp. 261–268.

Hofman, P.J. and Jobin-Decor, M. (1999) Effect of fruit sampling and handling procedures on the percentage dry matter, fruit mass, ripening and skin colour of 'Hass' avocado. *Journal of Horticultural Science and Biotechnology* 74, 277–282.

Hofman, P.J. and Ledger, S.N. (1999) Retail surveys show little quality improvement. *Talking Avocados* 10(4), 22–23.

Hofman, P.J. and Ledger, S.N. (2001a) Rots and bruising main quality problems. *Talking Avocados* 12(1), 20–22.

Hofman, P.J. and Ledger, S.N. (2001b) Reducing avocado defects at retail level. In: *Final Report for Project AV99007*. Horticulture Australia Ltd, Sydney, Australia, pp 1–32.

Hofman, P.J., McLauchlan, R.L. and Smith, L.G. (1995) Sensitivity of avocado fruit to ethylene. In: *Proceedings of the III World Avocado Congress*, Tel Aviv, Israel, pp. 335–339.

Hofman, P.J., Jobin-Decor, M. and Giles, J. (2000) Percentage of dry matter and oil content are not reliable indicators of fruit maturity or quality in late-harvested 'Hass' avocado. *HortScience* 35, 694–695.

Hofman, P.J., Jobin-Décor, M., Meiburg, G.F., Macnish, A.J. and Joyce, D.C. (2001) Ripening and quality responses of avocado, custard apple, mango and papaya fruit to 1-methylcyclopropene. *Australian Journal of Experimental Agriculture* 41, 567–572.

Hofman, P.J., Fuchs, Y. and Milne, D.L. (2002a) Harvesting, packing, postharvest technology, transport and processing. In: Whiley, A.W., Schaffer, B. and Wolstenholme, B.N. (eds) *The Avocado: Botany, Production and Uses*. CAB International, Wallingford, UK, pp. 363–402.

Hofman, P.J., Stubbings, B.A., Adkins, M.F., Meiburg, G.F. and Woolf, A.B. (2002b) Hot water treatments improve 'Hass' avocado fruit quality after cold disinfestation. *Postharvest Biology and Technology* 24, 183–192.

Hofman, P.J., Vuthapanich, S., Whiley, A.W., Klieber, A. and Simons, D. (2002c) Tree yield and fruit minerals concentrations influence 'Hass' avocado fruit quality. *Scientia Horticulturae* 92, 113–123.

Hofman, P.J., Stubbings, B.A., Adkins, M.F., Corcoran, R.J., White, A. and Woolf, A.B. (2003) Low temperature conditioning before cold disinfestation improves 'Hass' avocado fruit quality. *Postharvest Biology and Technology* 28, 123–133.

Hofman, P.J., Marques, J.R., Searle, C., Stubbings, B. and Moody, P. (2005) Improving avocado fruit quality through tree nutrition – present knowledge. In: *Proceedings of the New Zealand and Australia Avocado Growers' Conference Avocado '05*. Tauranga, New Zealand. pp, 1–15. Available at: http://www.avocadosource.com/Journals/AUSNZ/AUSNZ_2005/HofmanPeter2005.pdf (accessed 17 October 2024).

Hofman, P.J., Marques, J.R., Stubbings, B. and Searle, C. (2007) Optimising the postharvest qualities of Hass avocado through improved calcium nutrition. In: *Final Report for Project AV02009*. Horticulture Australia Ltd, Sydney, Australia, pp. 1–66.

Hofman, P.J., Requejo-Jackman, C., Olsson, S., Stubbings, B. and Woolf, A. (2008a) The relation between percent dry matter and oil content, and the challenges of a commercial oil industry in Australia. *Talking Avocados* 19(3), 13–17.

Hofman, P.J., Requejo-Jackman, C., Olsson, S., Stubbings, B. and Woolf, A. (2008b) Changes in dry matter and oil content of avocados from several production regions in Queensland. *Talking Avocados* 19(1), 28–32.

Hofman, P.J., Marques, J.R., Stubbings, B., Dann, E. and Coates, L. (2010) Using lower storage temperatures to improve Hass avocado fruit quality after storage. *Talking Avocados* 21(3), 26–29.

Hofman, P.J., Bower, J. and Woolf, A. (2013) Harvesting, packing, postharvest technology, transport and processing. In: Schaffer, B., Wolstenholme, B.N. and Whiley, A.W. (eds) *The Avocado: Botany, Production and Uses*. CAB International, Wallingford, UK, pp. 489–540.

Hofshi, R. (2002) Should the California avocado industry consider 'snap' harvesting? *AvoResearch* 2, 1–4.

Honiball, F., Giliomee, J.H. and Randall, J.H. (1979) Mechanical control of red scale *Aonidiella auranti* (Mask) on harvested oranges. *Citrus and Subtropical Fruit Journal* 549, 17–18.

Hopkirk, G., White, A., Beever, D.J. and Forbes, S.K. (1994) Influence of postharvest temperatures and the rate of fruit ripening on internal postharvest rots and disorders of New Zealand 'Hass' avocado fruit. *New Zealand Journal of Crop and Horticultural Science* 22, 305–311.

Hurtado-Fernández, E., Fernández-Gutiérrez, A. and Carrasco-Pancorbo, A. (2018) Avocado fruit – *Persea americana*. In: *Exotic Fruits*. Academic Press, Cambridge, MA, pp. 37–48.

Inoue, H. and Tateishi, A. (1995) Ripening and fatty acid composition of avocado in Japan. In: *Proceedings of the World Avocado Congress III*, Tel Aviv, Israel, pp. 366–369.

Isolcell. (2011) DCA: Dynamic controlled atmosphere. Available at: http://www.isolcell.it/en/Isolcell-nitrogen-generators/dca.html (accessed 8 December 2011).

Jacobo-Velázquez, D.A. and Hernández-Brenes, C. (2010) Biochemical changes during the storage of high hydrostatic pressure processed avocado paste. *Journal of Food Science* 75, 264–270.

Jacobsberg, B. (1988) Avocado oil: a literature survey. *Belgian Journal of Food Chemistry and Biotechnology* 43, 115–124.

Jamieson, L.E., Whiting, D.C., Woolf, A.B., White, A. and McDonald, R.M. (2000) Water-blasting avocados to remove leafroller eggs. In: *Proceedings of the 53rd New Zealand Plant Protection Conference*, pp. 371–374.

Jeong, J. and Huber, D.J. (2004) Suppression of avocado (*Persea americana* Mill.) fruit softening and changes in cell wall matrix polysaccharides and enzyme activities: differential responses to 1- MCP and delayed ethylene application. *Journal of the American Society for Horticultural Science* 129, 752–759.

Jeong, J., Huber, D.J. and Sargent, S.A. (2002) Influence of 1-methylcyclopropene (1-MCP) on ripening and cell-wall matrix polysaccharides of avocado (*Persea americana*) fruit. *Postharvest Biology and Technology* 25, 241–256.

Jeong, J., Huber, D.J. and Sargent, S.A. (2003) Delay of avocado (*Persea americana*) fruit ripening by 1-methylcyclopropene and wax treatments. *Postharvest Biology and Technology* 28, 247–257.

Jessup, A.J. (1991) High-temperature dip and low temperatures for storage and disinfestation of avocados. *HortScience* 26:1420.

Johnson, G.I. and Hofman, P.J. (2009) Postharvest technology and quarantine treatments. In: Litz, R.E. (ed.) *The Mango: Botany, Production and Uses*. CAB International, Wallingford, UK, pp. 529–605.

Johnston, J.W. and Banks, N.H. (1998) Selection of a surface coating and optimisation of its concentration for use on Hass avocado (*Persea americana* Mill.) fruit. *New Zealand Journal of Crop and Horticultural Science* 26, 143–151.

Jordan, R.A. and Smith, L.G. (1993) The responses of avocado and mango to storage atmosphere composition. In: *CA '93, Proceedings of the Sixth International Controlled Atmosphere Research Conference*. Northeast Regional Agricultural Engineering Service, Cooperative Extension, USA, pp. 629–638.

Kader, A.A. (2005) Increasing food availability by reducing postharvest losses of fresh produce. *Acta Horticulturae* 682, 2169–2176.

Kaiser, C. (1994) Evaluation of maturity standards in avocado fruit. *Subtropica* 15, 18–20.

Kaiser, C. and Wolstenholme, B.N. (1994a) Aspects of delayed harvest of 'Hass' avocado (*Persea americana* Mill.) fruit in a cool subtropical climate. I. Fruit lipid and fatty acid accumulation. *Journal of Horticultural Science* 69, 437–445.

Kaiser, C. and Wolstenholme, B.N. (1994b) Aspects of delayed harvest of 'Hass' avocado (*Persea americana* Mill.) fruit in a cool subtropical climate. II. Fruit size, yield, phenology and whole-tree starch cycling. *Journal of Horticultural Science* 69, 447–457.

Kaiser, C., Wolstenholme, B.N. and Levin, J. (1995) Towards improving maturity standards for 'Fuerte' avocado (*Persea americana* Mill.) fruit in a cool subtropical climate. In: *Proceedings of the III World Avocado Congress*, Tel Aviv, Israel, pp. 277–284.

Kanellis, A.K., Fuchs, Y., Zauberman, G. and Roubelakis-Angelakis, A.K. (1990) Differential expression of avocado cx and pg isoenzymes during ripening. In: *Proceedings of the XXIIIrd International Horticultural Congress,* Firenze, Italy, p. 690.

Kanellis, A.K., Solomos, T. and Mattoo, A.K. (1989) Hydrolytic enzyme activities and protein pattern of avocado fruit ripened. *Plant Physiology* 90, 259–266.

Kanellis, A.K., Solomos, T. and Roubelakis-Angelakis, A.K. (1991) Suppression of cellulase and polygalacturonase and induction of alcohol. *Plant Physiology* 96, 269–274.

Karimi, Y., Maftoonazad, N., Ramaswamy, H.S., Prasher, S.O. and Marcotte, M. (2012) Application of hyperspectral technique for color classification avocados subjected to different treatments. *Food and Bioprocess Technology* 5, 252–264.

Kassim, A., Workneh, T.S. and Bezuidenhout, C.N. (2013) A review on postharvest handling of avocado fruit. *African Journal of Agricultural Research* 8(21), 2385–2402.

Kawano, S. (1994) Present condition of nondestructive quality evaluation of fruits and vegetables in Japan. *JARQ, Japan Agricultural Research Quarterly* 28, 212–216.

Kim, S.M., Chen, P., McCarthy, M.J. and Zion, B. (1999) Fruit internal quality evaluation using on-line nuclear magnetic resonance sensors. *Journal of Agricultural Engineering Research* 74, 293–301.

Kitinoja, L., Saran, S., Roy, S.K. and Kader, A.A. (2011) Postharvest technology for developing countries: challenges and opportunities in research, outreach and advocacy. *Journal of the Science of Food and Agriculture* 91, 597–603.

Knee, M. (1993) Apple. In: Seymour, G.B., Taylor, J.E. and Tucker, G.A. (ed.) *Biochemistry of Fruit Ripening*. Chapman and Hall, NY, pp. 325–346.

Koen, T.J., du Plessis, S.F. and Terblanche, J.H. (1990) Nutritional factors involved in physiological post-harvest fruit disorders of avocados (cv. Fuerte). *Acta Horticulturae* 275, 543–550.
Köhne, J.S. and Kremer-Kohne, S. (1995) Picking Hass avocados without pedicels. *South African Avocado Growers' Association Yearbook* 18:66.
Kok, R.D., Bower, J.P. and Bertling, I. (2010) Low temperature shipping and cold chain management of 'Hass' avocados: an opportunity to reduce shipping costs. *South African Avocado Growers' Association Yearbook* 33, 33–37.
Kok, R.D., Bower, J.P. and Bertling, I. (2011) Ultra-low temperature shipping and cold chain management of 'Hass' avocados: investigation into reducing shipping costs. *South African Avocado Growers' Association Yearbook* 34, 13–19.
Kosiyachinda, S. and Young, R.E. (1976) Chilling sensitivity of avocado fruit at different stages of the respiratory climacteric. *Journal of the American Society for Horticultural Science* 101, 665–667.
Kruger, F.J. and Magwaza, L. (2010) Avocado fruit mineral content norms made easy. *Avoinfo* 170:18.
Kruger, F.J. and Lemmer, D. (2007) Commercial ripening trials with South African avocados in the United Kingdom. *South African Avocado Growers' Association Yearbook* 30, 12–15.
Kruger, F.J. and Magwaza, L.S. (2012) Does orchard soil moisture content at the time of harvest influence the post-storage ripening pattern of 'Hass' avocado fruit. *South African Avocado Growers Association* 35, 47–53.
Kruger, F.J., Stassen, P.J.C. and Snijders, B. (1995) A preliminary study on variation in the maturity parameters of avocados from the Kiepersol/Hazyview area. *South African Avocado Growers' Association Yearbook* 18, 67–73.
Kruger, F.J., Snijder, B., Mathumbu, J.M., Lemmer, D. and Malumane, R. (2004) Establishing appropriate maturity and fruit mineral content norms for the main avocado export cultivars. *South African Avocado Growers' Association Yearbook* 27, 5–10.
Kurlaender, A. (1996) Avocados. In: Somogyi, L.P., Barrett, D.M. and Hui, Y.H. (eds) *Processing Fruits: Science and Technology, Vol. 2, Major Processed Products*. Technomic Publishing Co., Lancaster, UK, pp. 445–457.
Kushad, M.M. and Yelenosky, G. (1987) Evaluation of polyamine and proline levels during low temperature acclimation of citrus. *Plant Physiology* 84, 692–695.
Lahav, E. and Kalmar, D. (1977) Water requirements of avocado in Israel. II. Influence on yield, fruit growth and oil content. *Australian Journal of Agricultural Research* 28, 869–877.
Lallu, N., Burdon, J., Haynes, G., Francis, K., Boldingh, H. *et al.* (2005) Is the poor quality of late season New Zealand 'Hass' avocado fruit related to fruit characteristics at harvest? *New Zealand Avocado Growers' Association Annual Research Report* 7, 57–66.
Lallu, N., Punter, M., Haynes, G., Pidakala, P. and Burdon, J. (2004) Role of water loss in ripening of 'Hass' avocados. *New Zealand Avocado Growers' Association Annual Research Report* 4, 70–79.
Landahl, S., Meyer, M.D. and Terry, L.A. (2009) Spatial and temporal analysis of textural and biochemical changes of imported avocado cv. Hass during fruit ripening. *Journal of Agricultural and Food Chemistry* 57(15), 7039–7047.
Ledger, S. (1995) *Hass Bruising*. Horticulture Postharvest Group, Queensland Department of Primary Industries, Brisbane, Australia.
Ledger, S.N. (2007) Managing mango ripening and storage rooms. In: *Proceedings of The Amistar 6th Australian Mango Conference*. Australian Mango Industry Association, Brisbane, Australia, pp. 26–29. Available at: http://www.mangoes.net.au/downloads/File/PDF/industry_conference/mango_conference_07_proceedings.pdf (accessed 20 January 2010).
Ledger, S., Barker, L., Campbell, T., Hofman, P. and Marques, J. (2011) *Avocado Ripening Manual*. Department of Employment, Economic Development and Innovation, Brisbane, Australia.
Ledger, S.N. and Barker, L.R. (1995) Black avocados – the inside story. In: *Proceedings of the Australian Avocado Growers' Federation Conference 'The Way Ahead'*, pp. 71–77.
Ledger, S.N., Campbell, T.P., Holmes, R.J., Campbell, J.A. and Barker, L.R. (2006) Improving knowledge and practices in fresh produce chains. *Acta Horticulturae* 712, 105–112.
Lee, S.K. (1981) A review and background of the avocado maturity standard. *California Avocado Society Yearbook* 65, 101–109.
Lee, S.K., Young, R.E., Schiffman, P.M. and Coggins, C.W. (1983) Maturity studies of avocado fruit based on picking dates and dry weight. *Journal of the American Society for Horticultural Science* 108, 390–394.

Lemmer, D., Malumane, T.R., Ntandane, J. and Kruger, F.J. (2007) Extending the storage life of late season 'Hass' avocados to increase availability of pre-packed ready-to-eat fruit during the off-season. *South African Avocado Growers' Association Yearbook* 30, 7–11.

Lewinsohn, E., Schalechet, F., Wilkinson, J., Matsui, K., Tadmor, Y. *et al.* (2001) Enhanced levels of the aroma and flavor compound S-Linalool by metabolic engineering of the terpenoid pathway in tomato fruits. *Plant Physiology* 127, 1256–1265.

Lewis, C.E. (1978) The maturity of avocados – a general review. *Journal of the Science of Food and Agriculture* 29, 857–866.

Li, X., Xu, C., Korban, S.S. and Chen, K. (2010) Regulatory mechanisms of textural changes in ripening fruits. *Critical Reviews in Plant Sciences* 29(4), 222–243.

Liu, X., Robinson, P.W., Madore, M.A., Witney, G.W. and Arpaia, M.L. (1999a) 'Hass' avocado carbohydrate fluctuations. I. Growth and phenology. *Journal of the American Society for Horticultural Science* 124(6), 671–675.

Liu, X., Robinson, P.W., Madore, M.A., Witney, G.W. and Arpaia, M.L. (1999b) 'Hass' avocado carbohydrate fluctuations. II. Fruit growth and ripening. *Journal of the American Society for Horticultural Sciences* 124, 676–681.

Liu, X., Sievert, J., Arpaia, M.L. and Madore, M.A. (2002) Postulated physiological roles of the seven-carbon sugars, mannoheptulose, and perseitol in avocado. *Journal of the American Society for Horticultural Science* 127, 108–114.

López-Malo, A., Palou, E., Barbosa-Canovas, G.V., Welti-Chanes, J. and Swanson, B.G. (1998) Polyphenoloxidase activity and color changes during storage of high hydrostatic pressure treated avocado puree. *Food Research International* 31, 549–556.

Lütge, A., Bower, J.P. and Bertling, I. (2010) An investigation into ultra-low temperature shipping and cold chain management of 'Fuerte' avocados-an opportunity to reduce shipping costs. *South African Avocado Grower' Association Yearbook* 33, 39–43.

Luza, J.G., Lizana, L.A. and Masson, L. (1989 Comparative lipids evolution during cold storage of three avocado cultivars. In: *Proceedings of the Symposium on Tropical Fruit in International Trade* 269, 153–160.

Ma, H., Liu, Y., Tu, X., Deng, D., Ye, J., Yang, C. and Li, L. (2020) Yun optimization of test conditions for TPA texture properties of avocado flesh. In: *IOP Conference Series: Earth and Environmental Science* (526(1), 012058). IOP Publishing.

Maftoonazad, N. and Ramaswamy, H.S. (2008) Effect of pectin-based coating on the kinetics of quality change associated with stored avocados. *Journal of Food Processing and Preservation* 32(4), 621–643.

Magwaza, L.S. and Opara, U.L. (2015) Analytical methods for determination of sugars and sweetness of horticultural products – a review. *Scientia Horticulturae* 184, 179–192.

Magwaza, L.S. and Tesfay, S.Z. (2015) A review of destructive and non-destructive methods for determining avocado fruit maturity. *Food and Bioprocess Technology* 8(10), 1995–2011.

Mandemaker, A.J. (2003) Effect of postharvest water blasting on fruit pesticide residues. *New Zealand Avocado Growers' Association Annual Research Report* 3, 66–70.

Mandemaker, A.J. (2004) The quality of New Zealand avocados in the Australian market. *New Zealand Avocado Growers' Association Annual Research Report* 4, 61–69.

Mandemaker, A.J., Elmsley, T.A. and Smith, D.B. (2006) Effects on drop heights and the fruit harvesting methods on the quality of 'Hass' avocados. *New Zealand Avocado Growers' Association Annual Research Report* 6, 97–104.

Manganaris, G.A., Vasilakakis, M., Mignani, I., Diamantidis, G. and Tzavella-Klonari, K. (2005) The effect of preharvest calcium sprays on quality attributes, physicochemical aspects of cell wall components and susceptibility to brown rot of peach fruits (*Prunus persica* L. cv. Andross). *Scientia Horticulturae* 107, 43–50.

Marangoni, A.G., Palma, T. and Stanley, D.W. (1996) Membrane effects in postharvest physiology. *Postharvest Biology and Technology* 7, 193–217.

Marques, J.R., Hofman, P.J. and Wearing, A.H. (2003) Rootstocks influence 'Hass' avocado fruit quality and fruit minerals. *Journal of Horticultural Science and Biotechnology* 78, 673–679.

Marques, J.R., Hofman, P.J. and Wearing, A.H. (2006) Between-tree variation in fruit quality and fruit mineral concentrations of Hass avocados. *Australian Journal of Experimental Agriculture* 46, 1195–1201.

Maxtend. (2011) Atmosphere control systems. Available at: http://www.maxtend.com.au (accessed 31 May 2011).

Mazhar, M., Joyce, D., Hofman, P., Collins, R., Brereton, G., Cowin, G. and Gupta, M. (2011) Bruising in the Queensland supply chains of Hass avocado fruit. In: *Proceedings of the VII World Avocado Congress*, Cairns, Australia. Available at: http://www.avocadosource.com/wac7/section_11/mazharmuhammad2011.pdf (accessed 14 December 2011).

Mazhar, M., Joyce, D., Cowin, G., Brereton, I., Hofman, P., Collins, R. and Gupta, M. (2015) Non-destructive 1H-MRI assessment of flesh bruising in avocado (*Persea americana M.*) cv. Hass. *Postharvest Biology and Technology* 100, 33–40.

Meir, S., Akerman, M., Fuchs, Y. and Zauberman, G. (1995) Further studies on the controlled atmosphere storage of avocados. *Postharvest Biology and Technology* 5, 323–330.

Meir, S., Philosoph-Hadas, S., Zauberman, G., Fuchs, Y., Akerman, M. and Aharoni, N. (1991) Increased formation of fluorescent lipid-peroxidation products in peels precedes other signs of ripening. *Journal of the American Society for Horticultural Science* 116, 823–826.

Meir, S., Philosoph-Hadas, S., Lurie, S., Droby, S., Akerman, M. *et al.* (1996) Reduction of chilling injury in stored avocado, grapefruit, and bell pepper by methyl jasmonate. *Canadian Journal of Botany* 74, 870–874.

Meir, S., Naiman, D., Akerman, M., Hyman, J.Y., Zauberman, G. and Fuchs, Y. (1997) Prolonged storage of 'Hass' avocado fruit using modified atmosphere packaging. *Postharvest Biology and Technology* 12, 51–60.

Meyer, M.D. and Terry, L.A. (2010) Fatty acid and sugar composition of avocado, cv. Hass, in response to treatment with an ethylene scavenger or 1-methylcyclopropene to extend storage life. *Food Chemistry* 121, 1203–1210.

Michailidis, M., Karagiannis, E., Tanou, G., Karamanoli, K., Lazaridou, A., Matsi, T. and Molassiotis, A. (2017) Metabolomic and physico-chemical approach unravel dynamic regulation of calcium in sweet cherry fruit physiology. *Plant Physiology and Biochemistry* 116, 68–79.

Milne, D.L. (1994) Postharvest handling of avocado, mango and lychee for export from South Africa. In: Champ, B.R., Highley, E. and Johnson, G.I. (eds) *Postharvest Handling of Tropical Fruits. ACIAR Proceedings No. 50*. ACIAR, Canberra, Australia, pp. 73–89.

Mizrach, A. (2008) Ultrasonic technology for quality evaluation of fresh fruit and vegetables in pre-and postharvest processes. *Postharvest Biology and Technology* 48(3), 315–330.

Mizrach, A. and Flitsanov, U. (1999) Nondestructive ultrasonic determination of avocado softening process. *Journal of Food Engineering* 40, 139–144.

Mizrach, A. and Flitsanov, U. (1995a) Predicting avocado shelf life by an ultrasonic non-destructive method. *Acta Horticulturae* 421, 111–119.

Mizrach, A. and Flitsanov, U. (1995b) Ultrasonic device for avocado shelf-life predicting and maturity detection. In: *Proceedings of the World Avocado Congress* 3, pp. 300–306.

Monselise, S.P. and Goren, R. (1987) Preharvest growing conditions and postharvest behavior of subtropical and temperate zone fruits. *HortScience* 22, 1185–1189.

Montoya, M.M., De La Plaza, J.L. and Lopex-Rodriguez, V. (1994) Electrical conductivity of avocado fruit during cold storage and ripening. *Food Science and Technology, Lebensmittel-Wissenschaft, Technologie* 27, 34–38.

Naiman, D., Akerman, M., Hyman, J., Zauberman, G., Fuchs, Y. and Meir, S. (1995) Prolonged modified atmosphere storage of Hass avocados. *Alon Hanotea* 49, 462–470.

Nikapitiya, C. and Lileperuma, C.K. (2003) Quality of avocado slices as affected by the stage of ripeness, anti-browning agents and packaging. *Tropical Agricultural Research* 15, 156–176.

Nishijima, K.A., Chan, Jr., H.T., Sanxter, S.S. and Linse, E.S. (1995) Reduced heat shock period of 'Sharwil' avocado for cold tolerance in quarantine cold treatment. *HortScience* 30, 1052–1053.

O´Donoghue, E.M. and Huber, D.J. (1992) Modification of matrix polysaccharides during avocado (*Persea americana*) fruit ripening: an assessment of the role of Cx-cellulase. *Physiologia Plantarum* 86(1), 33–42.

O'Donoghue, E.M., Huber, D.J. and Timpa, J.D. (1993) Action of cx-cellulase on avocado cell wall polymers. *Acta Horticulturae* 343, 212–214.

Obenland, D.M., Collin, S., Sievert, J., Negm, F. and Arpaia, M.L. (2012) Influence of maturity and ripening on aroma volatiles and flavour in 'Hass' avocado. *Postharvest Biology and Technology* 71, 41–50.

Obianom, C. and Sivakumar, D. (2018) Differential response to combined prochloraz and thyme oil drench treatment in avocados against the control of anthracnose and stem-end rot. *Phytoparasitica* 46, 273–281.

Offer, R.A. (1986) Maturation and ripening of avocado fruit. MSc thesis, the Hebrew University of Jerusalem, Israel.

Ohlsson, T. (1994) Minimal processing – preservation method of the future: an overview. *Trends in Food Science and Technology* 51, 341–345.

Olarewaju, O.O., Bertling, I. and Magwaza, L.S. (2016) Non-destructive evaluation of avocado fruit maturity using near infrared spectroscopy and PLS regression models. *Scientia Horticulturae* 199, 229–236.

Olivares, D., Alvarez, E., Véliz, D., García-Rojas, M., Díaz, C. and Defilippi, B.G. (2020) Effects of 1-Methylcyclopropene and controlled atmosphere on ethylene synthesis and quality attributes of avocado cvs. Edranol and Fuerte. *Journal of Food Quality* 2020, 1–14.

Olivares, D., García-Rojas, M., Ulloa, P.A., Riveros, A., Pedreschi, R., Campos-Vargas, R. and Defilippi, B.G. (2022) Response mechanisms of 'Hass' avocado to sequential 1–methylcyclopropene applications at different maturity stages during cold storage. *Plants* 11(13), 1781.

Olmedo, P., Zepeda, B., Rojas, B., Silva-Sanzana, C., Delgado-Rioseco, J. *et al.* (2021) Cell wall calcium and hemicellulose have a role in the fruit firmness during storage of blueberry (*Vaccinium* spp.). *Plants* 10:553. doi:10.3390/plants10030553.

Pak, H.A., Dixon, J. and Cutting, J.G.M. (2003a) Influence of early season maturity on fruit quality in New Zealand 'Hass' avocados. *New Zealand Avocado Growers' Association Annual Research Report* 3, 54–65.

Pak, H.A., Dixon, J., Smith, D.B., Elmsly, T.A. and Cutting, J.G.M. (2003b) Impact of rainfall prior to harvest on ripe fruit quality of 'Hass' avocados in New Zealand. *New Zealand Avocado Growers' Association Annual Research Report* 3, 22–31.

Pak, H., Dixon, J. and Cutting, J.G. (2005a) Library trays – a valuable tool for managing fruit quality. In: *Proceedings of the New Zealand and Australia Avocado Growers' Conference '05*. Available at: http://www.avocadosource.com/Journals/AUSNZ/AUSNZ-2005/PakHenry2005.pdf (accessed 7 April 2010).

Pak, H.A., Dixon, J. and Cutting, J.G.M. (2005b) Postharvest quality outturn. In: *Proceedings of the New Zealand and Australia Avocado Grower's Conference '05*. Available at: http://www.avocadosource.com/Journals/AUSNZ/AUSNZ_2005/PakHenry2005.pdf (accessed 21 July 2011).

Palonen, P., Buszard, D. and Donnelly, D. (2000) Changes in carbohydrates and freezing tolerance during cold acclimation of red raspberry cultivars grown *in vitro* and *in vivo*. *Physiologia Plantarum* 110, 393–401.

Palou, E., Hernandez-Salgado, C., Lopez-Malo, A., Barbosa-Canovas, G.V., Swanson, B.G. and Welti-Chanes, J. (2000) High pressure processed guacamole. *Innovative Food Science and Emerging Technologies* 1, 69–75.

Pao, S. and Petracek, P.D. (1997) Shelf life extension of peeled oranges by citric acid treatment. *Food Microbiology* 14, 458–497.

Parodi, G., Sanchez, M. and Daga, W. (2007) Correlation of oil content, dry matter and pulp moisture as harvest indicators in Hass avocado fruit (*Persea americana Mill*) grown under two conditions of orchards in Chincha-Peru. In: *Proceedings of the VI World Avocado Congress*, Chile, pp. 12–16.

Pedreschi, R., Uarrota, V., Fuentealba, C., Alvaro, J.E., Olmedo, P. *et al.* (2019) Primary metabolism in avocado fruit. *Frontiers in Plant Science* 10:795.

Pedreschi, R., Muñoz, P., Robledo, P., Becerra, C., Defilippi, B.G. *et al.* (2014) Metabolomics analysis of postharvest ripening heterogeneity of Hass avocadoes. *Postharvest Biology and Technology* 92, 172–179.

Pedreschi, R., Hollak, S., Harkema, H., Otma, E., Robledo, P. *et al.* (2016) Impact of postharvest ripening strategies on Hass avocado fatty acid profiles. *South African Journal of Botany* 103, 32–35.

Peleg, K., Ben-Hanan, U. and Hinga, S. (1990) Classification of avocado by firmness and maturity. *Journal of Texture Studies* 21, 123–139.

Penter, M.G. and Stassen, P.J.C. (2000) The effect of pre- and postharvest calcium applications on the postharvest quality of Pinkerton avocados. *South African Avocado Growers' Association Yearbook* 23, 1–7.

Pereira, M.E., Tieman, D.M., Sargent, S.A. and Klee, H.J. (2013) Volatile profiles of ripening West Indian and Guatemalan-West Indian avocado cultivars as affected by aqueous 1-methylcyclopropene. *Postharvest Biology and Technology* 80, 37–46.

Pesis, E., Fuchs, Y. and Zauberman, G. (1978) Cellulase activity and fruit softening in avocado. *Plant Physiology* 61, 416–419.

Pesis, E., Marinansky, R., Zauberman, G. and Fuchs, Y. (1994) Prestorage low-oxygen atmosphere treatment reduces chilling injury symptoms in 'Fuerte' avocado fruit. *HortScience* 29, 1042–1046.

Pesis, E., Faiman, D. and Dori, S. (1998) Postharvest effects of acetaldehyde vapour on ripening-related enzyme activity in avocado fruit. *Postharvest Biology and Technology* 13, 245–253.

Pesis, E., Ackerman, M., Ben-Arie, R., Feygenberg, O., Feng, X. *et al.* (2002) Ethylene involvement in chilling injury symptoms of avocado during cold storage. *Postharvest Biology and Technology* 24, 171–181.

Pino, J.A., Marbot, R., Rosado, A. and Fuentes, V. (2004) Volatile components of avocado cv. Moro grown in Cuba. *Journal of Essentials Oil Research* 16, 130–140.

Platt-Aloia, K.A. and Thompson, W.W. (1981) Ultrastructure of the mesocarp of mature avocado fruit and changes. *Annals of Botany* 48, 451–465.

Plumbley, R.A., Jeger, M.J. and Carbon, W. (1989) Optimal picking date of avocado in Dominica, West Indies, in relation to flowering, fruit growth and chilling. *Tropical Agriculture (Trinidad)* 66, 129–134.

Poovaiah, B.W., Glenn, G.M. and Reddy, A.S.N. (1988) Calcium and fruit softening: physiology and biochemistry. *Horticultural Reviews* 10, 107–152.

Ranney, C. (1991) Relationship between physiological maturity and percent dry matter of avocados. *California Avocado Society Yearbook* 75, 71–85.

Reid, M.S. (2002) Ethylene in postharvest technology. In: Kader, A.A. (ed.) *Postharvest Technology of Horticultural Crops, 3rd edn*. University of California, Oakland, CA, pp. 149–162.

Requejo-Tapia, L.C., Woolf, A.B., Roughan, G., Schroeder, R., Young, H. and White, A. (1999) Seasonal changes in lipid content and fatty acid composition of 'Hass' avocados. *Avocado Postharvest Research* (1998/99), pp. 1–29.

Rhodes, N.J.C. (1980) Respiration and senescence of plant organs. In: Davies, D.D. (ed.) *The Biochemistry of Plants – A Comprehensive Treatise, Vol.* 2 – *Metabolism and Respiration*. Academic Press, New York, pp. 419–462.

Ritenour, M.A., Mangrich, M.E., Beaulieu, J.C., Rab, A. and Saltveit, M.E. (1997) Ethanol effects on ripening of climacteric fruit. *Postharvest Biology and Technology* 12, 35–42.

Rivera, S.A., Ferreyra, R., Robledo, P., Selles, G., Arpaia, M.L., Saavedra, J. and Defilippi, B.G. (2017) 'Identification of pre-harvest factors determining postharvest ripening behaviors in ´Hass´ avocado under long term storage'. *Scientia Horticulturae* 216, 29–37.

Roets, N.R.J., Lemmer, D. and Kruger, F.J. (2009) Evaluation of different atmospheric conditioning treatments to synchronize the ripening of SmartFresh[SM] 'Hass' avocado fruit. *South African Avocado Growers' Association Yearbook* 32, 42–45.

Roman-Mares, A. and Yahia, E.M. (2000) Hot air treatment (38°C for 360 min and 40°C for 120 min) effect on the postharvest quality, chilling injury and lipid content in 'Hass' avocado fruit. In: Artes, F., Gil, M.I. and Conesa, M.A. (eds) *Improving Postharvest Technology of Fruits, Vegetables and Ornamentals*. International Institute of Refrigeration, Spain, pp. 491–497.

Ronen, R., Zauberman, G., Akerma, M., Weksler, A., Rot, I. and Fuchs, Y. (1991) Xylanase and xylosidase activities in avocado fruit. *Plant Physiology* 95, 961–964.

Sams, C.E. (1999) Preharvest factors affecting postharvest texture. *Postharvest Biology and Technology* 15, 249–254.

Sanches, J., Biscegli, C.I., Durigan, J.F., Simoes, M.L. and da Silva, W.T. (2003) Diagnosis of mechanical injuries in avocados by magnetic resonance imaging. In: *Proceedings of the V World Avocado Congress*, Uruapan, Mexico, pp. 695–700.

Sanewski, G.M. (1984) Firm flesh – severe avocado disorder. *Queensland Fruit and Vegetable News* 55:114.

Sanxter, S.S., Nishijima, K.A. and Chan, Jr., H.T. (1994) Heat-treating 'Sharwil' avocado for cold tolerance in quarantine cold treatments. *HortScience* 29, 1166–1168.

Saucedo Veloz, C., Aceves Vega, E. and Mena Nevarez, G. (1991) Prolonging the duration of cold storage and marketing of Hass avocado fruits by treatment with high concentrations of CO_2. *Proceedings of the Interamerican Society for Tropical Horticulture* 35, 297–303.

Schmilovitch, Z.E., Mizrach, A., Hoffman, A., Egozi, H. and Fuchs, Y. (2000) Determination of mango physiological indices by near-infrared spectrometry. *Postharvest Biology and Technology* 19(3), 245–252.

Schroeder, C.A. (1981) Sclerocarpelosis in avocado fruit. *California Avocado Society Yearbook* 65, 125–132.

Schroeder, C.A. (1986) Physiological gradient in avocado Fruit. *California Avocado Society Yearbook* 69, 137–143.

Schwartz, M., Olaeta, J.A., Undurraga, P. and Costa, V. (2007) Mejoramiento del rendimiento de extracción de aceite de Palta (Aguacate). In: *Proceedings of the VI World Avocado Congress (Actas VI Congreso Mundial del Aguacate)*.

Scudamore-Smith, P.D. (1984) The utilization of avocado as frozen savoury spread. *Food Technology in Australia* 36, 375–378.

Self, G.K., Ordozgoiti, E., Povey, M.J.W. and Wainwright, H. (1994) Ultrasonic evaluation of ripening avocado flesh. *Postharvest Biology and Technology* 4, 111–116.

Seymour, G.B. and Tucker, G.A. (1993) Avocado. In: Seymour, G.B., Taylor, J.E. and Tucker, G.A. (eds) *Biochemistry of Fruit Ripening*. Chapman and Hall, London, pp. 53–81.

Sharma, G., Maymon, M. and Freeman, S. (2017) Epidemiology, pathology and identification of *Colletotrichum* including a novel species associated with avocado (*Persea americana*) anthracnose in Israel. *Scientific Reports* 7(1), 15839.

Shear, C.B. (1975) Calcium-related disorders of fruits and vegetables. *HortScience* 10, 361–365.

Shear, C.B. and Faust, M. (1975) Preharvest nutrition and postharvest physiology of apples. In: Haard, N.K. and Salunkhe, D.K. (eds) *Postharvest Biology and Handling of Fruit and Vegetables*. Avi Publishing, Westport, Connecticut, pp. 35–42.

Sinclair, W. and Lindgren, D. (1955) Vapor heat sterilization of California citrus and avocado fruits against fruit-fly insects. *Journal of Economic Entomology* 48, 133–138.

Sippel, A.D., Conradie, W. and Claassens, N.J.F. (1992) Growth rates of Pinkerton avocado fruit. *South African Avocado Growers' Association Yearbook* 15, 72–74.

Sippel, A.D., Holmes, M.A. and Claassens, N.J.F. (1995) Manipulation of fruit-set period of the Pinkerton avocado and its effect on fruit and tree characteristics. *South African Avocado Growers' Association Yearbook* 18, 77–79.

Sisler, E.C., Dupille, E. and Serek, M. (1996) Effects of 1-methylcyclopropene and methylcyclopropene on ethylene binding and ethylene action on cut carnation. *Journal of Plant Growth Regulation* 18, 79–86.

Smith, L.G., Jordan, R.A., Hofman, P.J., Jobin-Décor, M., Zauberman, G. and Ledger, S.N. (1992) Long term storage of avocados. In: *Proceedings of the Conference 92 'Management into the Future', Australian Avocado Growers' Federation Inc.*, Queensland Department of Primary Industries, Brisbane, Australia.

Smith, T.E., Hofman, P.J., Stephenson, R.A., Asher, C.J. and Hetherington, S.E. (1997) Improving boron nutrition improves 'Hass' avocado fruit size and quality. In: Cutting, J. (ed.) *Proceedings of the Conference '97; Searching for Quality. Joint Conference of the Australian Avocado Growers' Federation and the New Zealand Avocado Growers' Association*, Tauranga, New Zealand, pp. 131–137.

Snijder, B., Penter, M.G., Mathumbu, J.M. and Kruger, F.J. (2002) Further refinement of Pinkerton export parameters. *South African Avocado Growers' Association Yearbook* 25, 51–55.

Soliva-Fortuny, R.C., Elez-Martinez, P., Sebastian-Caldero, M. and Martin-Belloso, O. (2002) Kinetics of polyphenol oxidase activity inhibition and browning of avocado puree preserved by combined methods. *Journal of Food Engineering* 55, 131–137.

Southwell, K.H., Harris, R.V. and Swetman, A.A. (1990) Extraction and refining of oil obtained from dried avocado fruit using a small expeller. *Tropical Science* 30, 121–131.

Spalding, D.H. and Marousky, F.J. (1981) Injury to avocados by insufficient oxygen and excessive carbon dioxide. *Proceedings of the Florida State Horticultural Society* 94, 299–301.

Stahl, A. (1933) Changes in composition of Florida avocados in relation to maturity. *University of Florida, Agricultural Experiment Station Bulletin* 259, 1–61.

Starrett, D.A. and Laties, G.G. (1993) Ethylene and wound-induced gene expression in the preclimacteric phase of ripening avocado fruit and mesocarp discs. *Plant Physiology* 103, 227–234.

Swarts, D.H. (1978) The no-nonsense determination of oil content for avocados. *Citrus and Subtropical Fruit Research Institute Information Bulletin* 42:4.

Swarts, D.H. (1985) Postharvest problems in avocados. *Farming in South Africa* 13/1985, 1–4.

Tateishi, A., Shiba, H., Ogihara, J., Isobe, K., Nomura, K., Watanabe, K. and Inoue, H. (2007) Differential expression and ethylene regulation of β-galactosidase genes and isozymes isolated from avocado (*Persea americana* Mill.) fruit. *Postharvest Biology and Technology* 45, 56–65. doi:10.1016/j.postharvbio.2007.01.009.

Terry, L.A., Ilkenhans, T., Poulston, S., Rowsell, L. and Smith, A.W.J. (2007) Development of new palladium-promoted ethylene scavenger. *Postharvest Biology and Technology* 45, 214–220.

Tesfay, S.Z., Bertling, I. and Bower, J.P. (2012a) D-mannoheptulose and perseitol in 'Hass' avocado: metabolism in seed and mesocarp tissue. *South African Journal of Botany* 79, 159–165.

Tesfay, S.Z., Bertling, I., Bower, J.P. and Lovatt, C. (2012b) The quest for the function of 'Hass' avocado carbohydrates: clues from fruit and seed development as well as seed germination. *Australian Journal of Botany* 60(1), 79–86.

Tesfay, S.Z. (2009) Special carbohydrates of avocado- their functions as sources of energy and anti-oxidants. PhD thesis, University of KwaZulu-Natal, South Africa.

Tesfay, S.Z., Bertling, I. and Bower, J.P. (2010) Anti-oxidant levels in various tissues during the maturation of 'Hass' avocado (*Persea americana* Mill.). *Journal of Horticultural Science and Biotechnology* 85, 106–112.

Thompson, A.K. (2010) *Controlled Atmosphere Storage of Fruit and Vegetables,* 2nd edn. CAB International, Wallingford, UK.

Thorp, T.G., Hutching, D., Lowe, T. and Marsh, K.B. (1997) Survey of fruit mineral concentrations and post-harvest quality of New Zealand-grown 'Hass' avocado (*Persea americana* Mill.). *New Zealand Journal of Crop and Horticultural Science* 25, 251–260.

Tingwa, P.O. and Young, R.E. (1974) The effect of calcium on the ripening of avocado (*Persea americana* Mill.) fruits. *Journal of the American Society for Horticultural Science* 99, 540–542.

Toerien, J.C. (1986) Temperature control of avocados for sea export. *South African Avocado Growers' Association Yearbook* 9, 31–32.

Toivonen, P.M. and Brummell, D.A. (2008) Biochemical bases of appearance and texture changes in fresh-cut fruit and vegetables. *Postharvest Biology and Technology* 48(1), 1–14.

Torres, J.A. and Velazquez, G. (2005) Commercial opportunities and research challenges in the high pressure processing of foods. *Journal of Food Engineering* 67, 95–112.

Truter, A.B. and Eksteen, G.J. (1987) Controlled and modified atmospheres to extend storage life of avocados. *South African Avocado Growers' Association Yearbook* 10, 151–153.

Tucker, M.L., Christoffersen, R.E., Woll, L. and Laties, G.G. (1985) Induction of cellulase by ethylene in avocado fruit. In: Roberts, J.A. and Tucker, G.A. (eds) *Ethylene and Plant Development*. Butterworth, London, pp. 93–100.

Uarrota, V.G., Hernández, I., Ponce Guequen, E., Vidal Cruz, J., Fuentealba, C., Defilippi, B.G. and Pedreschi, R. (2020) Unravelling factors associated with 'blackspot' disorder in stored Hass avocado (*Persea americana* Mill.) fruit. *The Journal of Horticultural Science and Biotechnology* 95(6), 804–815.

Undurraga, P., Olaeta, J.A. and Olivares, C. (2007) Evaluation of three kinds of packing material in Edranol avocados (*Persea americana* Mill.) as flesh-cut products. In: *Proceedings of the VI World Avocado Congress*, Viña del Mar, Chile. Available at: http://www.avocadosource.com/wac6/en/resumen/4b-200.pdf (accessed 20 January 2012).

Van Rooyen, Z. (2011) New developments in horticultural research at Westfalia, South Africa. In: *Proceedings of the VII World Avocado Congress*, Cairns, Australia. Available at: http://www.avocadosource.com/wac7/section_09/vanrooyenz2011.pdf (accessed 6 February 2012).

Van Rooyen, Z. and Bower, J.P. (2005) The role of fruit mineral composition on fruit softness and mesocarp discolouration in 'Pinkerton' avocado (*Persea americana* Mill.). *Journal of Horticultural Science and Biotechnology* 80, 793–799.

Van Rooyen, Z. and Bower, J.P. (2006) Effects of storage temperature, harvest date and fruit origin on post-harvest physiology and the severity of mesocarp discolouration in 'Pinkerton' avocado (*Persea americana* Mill.). *Journal of Horticultural Science and Biotechnology* 31, 89–98.

Van Rooyen, Z. and Bezuidenhout, J. (2010) Semi-commercial trials to determine the risk of shipping South African 'Hass' at 1°C. *South African Avocado Growers' Association Yearbook* 33, 27–31.

Vendrell, M. and Palomer, X. (1998) Hormonal control of fruit ripening in climateric fruits. *Acta Horticulturae* 463, 325–334.

Vorster, L.L. (2005) South African avocado industry update. *South African Avocado Growers' Association Yearbook* 88, 55–60.

Vorster, L.L. and Benzuidenhout, J.J. (1988) Does zinc play a role in reducing pulp spot? *South African Avocado Growers' Association Yearbook* 11:60.

Vorster, L.L., Toerien, J.C. and Bezuidenhout, J.J. (1987) A storage temperature regime for South African export avocados. *South African Avocado Growers' Association Yearbook* 10, 146–149.

Vorster, L.L., Toerien, J.C. and Bezuidenhout, J.J. (1989) Factors involved in fruit quality. *Yearbook – South African Avocado Growers' Association Yearbook* 12, 76–78.

Vorster, L.L., Toerien, J.C. and Bezuidenhout, J.J. (1990) Temperature management of avocados – an integrated approach. *South African Avocado Growers' Association Yearbook* 13, 43–46.

Vuthapanich, S. (2001) Preharvest practices affecting postharvest quality and minerals composition of 'Hass' avocado fruit. PhD thesis, University of Queensland, Australia.

Vuthapanich, S., Hofman, P.J., Whiley, A.W., Klieber, A. and Simons, D.H. (1995a) Effects of irrigation and foliar Cultar® on fruit yield and quality of 'Hass' avocado fruit. In: *Proceedings of the III World Avocado Congress*, Tel Aviv, Israel, pp. 311–315.

Vuthapanich, S., Klieber, A., Hofman, P.J., Whiley, A.W. and Simons, D.H. (1995b) Effects of production location in south-east Queensland on postharvest quality of 'Hass' avocado fruit. In: *Proceedings of the III World Avocado Congress*, Tel Aviv, pp. 316–320.

Waite, G.K. and Martinez Barrera, R. (2002) Insect and mite pests. In: Whiley, A.W., Schaffer, B. and Wolstenholme, B.N. (eds) *The Avocado: Botany, Production and Uses*. CAB International, Wallingford, UK, pp. 231–258.

Wakabayashi, K., Chun, J., Huber, D.J. and Chun, J.P. (2000) Extensive solubilization and depolymerization of cell wall polysaccharides during avocado (*Persea americana*) ripening involves concerted action of polygalacturonase and pectinmethylesterase. *Physiologia Plantarum* 108, 345–352.

Wang, J.S., Wang, A.B., Zang, X.P., Tan, L., Ge, Y. *et al.* (2018) Physical and oxidative stability of functional avocado oil high internal phase emulsions collaborative formulated using citrus nanofibers and tannic acid. *Food Hydrocolloids* 82, 248–257.

Watkins, J.B. and Ledger, S.N. (1990) *Forced-Air Cooling, 2nd edn*. Queensland Department of Primary Industries, Brisbane, Australia.

Wedding, B.B., Wright, C., Grauf, S., Gadek, P. and White, R.D. (2019) The application of FT-NIRS for the detection of bruises and the prediction of rot susceptibility of 'Hass' avocado fruit. *Journal of the Science of Food and Agriculture* 99(4), 1880–1887.

Wedding, B.B., Wright, C., Grauf, S., White, R.D., Tilse, B. and Gadek, P. (2013) Effects of seasonal variability on FT-NIR prediction of dry matter content for whole Hass avocado fruit. *Postharvest Biology and Technology* 75, 9–16.

Wedding, B.B., White, R.D., Grauf, S., Wright, C., Tilse, B., Hofman, P. and Gadek, P.A. (2011) Non-destructive prediction of 'Hass' avocado dry matter via FT-NIR spectroscopy. *Journal of the Science of Food and Agriculture* 90, 6–10.

Weemaes, C., Ludikhuyze, L., van den Broeck, I. and Hendrickx, M.E. (1999) Kinetic study of antibrowning agents and pressure inactivation of avocado polyphenoloxidase. *Journal of Food Science* 64, 823–827.

Werman, M.J. and Neeman, I. (1987) Avocado oil production and chemical characteristics. *Journal of the American Oil Chemists' Society* 64, 229–232.

Whiley, A.W. (2002) Crop management. In: Whiley, A.W., Schaffer, B. and Wolstenholme, B.N. (eds) *The Avocado: Botany, Production and Uses*. CAB International, Wallingford, UK, pp. 231–258.

White, A., Woolf, A.B., Hofman, P.J. and Arpaia, M.L. (2009) *The International Avocado Quality Manual*. Plant and Food Research, Auckland, New Zealand.

Williams, P.C. and Sobering, D.C. (1993) Comparison of commercial near infrared transmittance and reflectance instruments for analysis of whole grains and seeds. *Journal of Near Infrared Spectroscopy* 1, 25–32.

Willingham, S.L., Coates, L.M., Cooke, A.W. and Dean, J.R. (2004) Tree vigour influences disease susceptibility of 'Hass' avocado fruits. *Australasian Plant Pathology* 33, 17–21.

Willingham, S.L., Pegg, K., Anderson, J.M., Cooke, A.W., Dean, J., Giblin, F.R. and Coates, L. (2006) Effects of rootstock and nitrogen fertiliser on postharvest anthracnose development in Hass avocado. *Australasian Plant Pathology* 35, 619–629.

Wills, R.B.H. and Tirmazi, S.I.H. (1979) Effect of calcium and other minerals on ripening of tomatoes. *Australian Journal of Plant Physiology* 6, 221–227.

Wills, R.B.H. and Tirmazi, S.I.H. (1982) Inhibition of ripening of avocados with calcium. *Scientia Horticulturae* 16, 323–330.

Wills, R.B.H., McGlasson, W.B., Graham, D. and Joyce, D. (2007) *Postharvest – An Introduction to the Physiology and Handling of Fruit, Vegetables and Ornamentals*. University of New South Wales Press, Sydney, Australia.

Witney, G.W., Hofman, P.J. and Wolstenholme, B.N. (1990) Effect of cultivar, tree vigour and fruit position on calcium accumulation in avocado fruits. *Scientia Horticulturae* 44, 269–278.

Wong, M., Ashton, O.B.O., McGhie, T., Requejo-Jackman, C., Wang, Y. and Woolf, A.B. (2011) Influence of proportion of skin present during malaxing on pigment composition of cold pressed avocado oil. *Journal of the American Oil Chemists Society* 88, 1373–1378.

Wong, M., Ashton, O.B.O., Requejo-Jackman, C., McGhie, T.K., White, A. *et al.* (2008) Avocado oil – the colour of quality. In: Culver, C.A. and Wrolstad, R.E. (eds) *Color Quality of Fresh and Processed Foods*. American Chemical Society, Washington, DC, pp. 328–349.

Woolf, A.B. (1997) Reduction of chilling injury in stored Hass avocado fruit by 38°C water treatments. *HortScience* 32, 1247–1251.

Woolf, A.B. and Lay-Yee, M. (1997) Pretreatments at 38°C of 'Hass' avocado confer thermotolerance to 50°C hot water treatments. *HortScience* 32, 705–708.

Woolf, A.B. and Ferguson, I.B. (2000) Postharvest responses to high fruit temperatures in the field. *Postharvest Biology and Technology* 21, 7–20.

Woolf, A.B., Watkins, C.B., Bowen, J.G., Maindonald, J.H., Lay-Yee, M. and Ferguson, I.B. (1995) Reducing external chilling injury in stored 'Hass' avocados with dry heat treatments. *Journal of the American Society for Horticultural Science* 120, 1050–1056.

Woolf, A.B., Ball, S., White, A., Spooner, K.J., Bowen, J.H., Lay-Yee, M. and Ferguson, I.B. (1998) Hot air and hot water treatments reduce chilling injury of avocados during storage. In: *Proceedings of the III World Avocado Congress*, Tel Aviv, Israel, pp. 348–353.

Woolf, A.B., Bowen, J.H. and Ferguson, I.B. (1999a) Preharvest exposure to the sun influences postharvest responses of 'Hass' avocado fruit. *Postharvest Biology and Technology* 15, 143–153.

Woolf, A.B., Fergunson, I.B., Requejo-Tapia, L., Boyd, L., Laing, W.A. and White, A. (1999b) Impact of sun exposure on harvest quality of Hass avocado fruit. *Revista Chapingo Serie Horticultura* 5, 353–358.

Woolf, A.B., White, A., Sievert, J.R. and Arpaia, M.L. (1999c) Summary of New Zealand and Californian experience with 'snap' picking. In: Arpaia, M.L. and Hofshi, R. (eds) *Proceedings of the Avocado Brainstorming '99*. University of California, Riverside, California, pp. 161–162.

Woolf, A.B., Wexler, A., Prusky, D., Kobiler, E. and Lurie, S. (2000) Direct sunlight influences postharvest temperature responses and ripening of five avocado cultivars. *Journal of the American Society for Horticultural Science* 125, 370–376.

Woolf, A.B., Bowen, J.H., Ball, S., Durand, S., Laidlaw, W.G. and Ferguson, I.B. (2004) A delay between a 38°C pretreatment and damaging high and low temperature treatments influences pretreatment efficacy in 'Hass' avocados. *Postharvest Biology and Technology* 34, 143–153.

Woolf, A.B., Requejo-Tapia, C., Cox, K.A., Jackman, R.C., Gunson, A., Arpaia, M.L. and White, A. (2005) 1-MCP reduces physiological storage disorders of 'Hass' avocados. *Postharvest Biology and Technology* 35, 43–60.

Woolf, A., Requejo-Jackman, C., Lund, C., McGhie, T., Olsson, S. *et al.* (2007) Avocado oils and other niche culinary oils in New Zealand. In: O'Connor, C.J. (ed.) *Handbook of Australasian Edible Oils*. American Oil Chemists Society, Auckland, pp. 12–37.

Woolf, A., Wong, M., Eyres, L., McGhie, T., Lund, C. *et al.* (2009) Avocado oil. In: Moreau, R.A. and Kamal-Eldin, A. (eds) *Gourmet and Health-Promoting Speciality Oils*. AOCS Press, Urbana, IL, pp. 73–125.

Yahia, E.M. (1994) Apple flavor. *Horticultural Reviews* 16, 197–234.

Yahia, E.M. (2006) Modified and controlled atmospheres for tropical fruits. *Stewart Postharvest Review* 5(6), 1–10.

Yahia, E.M. and Woolf, A. (2011) Avocado (*Persea americana* Mill.). In: Yahia, E.M. (ed.) *Postharvest Biology and Technology of Tropical and Sub-tropical Fruits*. Woodhead Publishing, Urbana, Illinois, pp. 73–126.

Yearsley, C. and Lallu, N. (2001) Symptoms of controlled atmosphere damage in avocados. *New Zealand Avocado Growers' Association Annual Research Report* 1, 22–25.

Yearsley, C., Lallu, N., Burdon, J., Billing, D., Punter, M. and Osman, S. (2002) Effects of prepacking holding temperatures on shelf life quality of Hass avocados. *New Zealand Avocado Growers' Association Annual Research Report* 2, 1–12.

Young, R.E. and Kosiyachinda, S. (1976) Low temperature storage of ripe avocado fruit. *California Avocado Society Yearbook* 59, 73–76.

Yousef, A.R.M. and Hassaneine, M.M.M. (2010) Influence of different harvest dates and ripening periods on fruit quality and oil characteristics of 'Fuerte' avocados. *Agriculture and Biology Journal of North America* 1, 1223–1230.

Zamorano, J., Dopico, B., Lowe, A.L., Wilson, I.D., Grierson, D. and Merodio, C. (1994) Effect of low temperature storage and ethylene removal on ripening and gene expression changes in avocado fruit. *Postharvest Biology and Technology* 4(4), 331–342.

Zauberman, G. and Jobin-Décor, M. (1995) Avocado (*Persea americana* Mill.) quality changes in response to low-temperature storage. *Postharvest Biology and Technology* 5, 235–243.

Zauberman, G. and Schiffmann-Nadel, M. (1972) Pectinmethylesterase and polygalacturonase in avocado fruit at various stages of development. *Plant Physiology* 49, 864–865.

Zauberman, G., Schiffmann-Nadel, M. and Yanko, U. (1973) Susceptibility to chilling injury of three avocado cultivars at various stages of ripening. *HortScience* 8, 511–513.

Zauberman, G., Schiffmann-Nadel, M. and Yanko, U. (1977) The response of avocado fruits to different storage temperatures. *HortScience* 12, 353–354.

Zauberman, G., Schiffmann-Nadel, M., Yanko, U., Sarig, I. and Alper, I. (1969) Factors causing injury during transportation of avocado pears to packing. *The National and University Institute of Agriculture, The Volcani Institute, Israel*, 656, 1–11.

Zhang, Z.K., Huber, D.J. and Rao, J.P. (2011) Ripening delay of mid-climacteric avocado fruit in response to elevated doses of 1-methylcyclopropene and high hypoxia-mediated reduction in internal ethylene concentration. *Postharvest Biology and Technology* 60, 83–91.

Ziosi, V., Noferini, M., Fiori, G., Tadiello, A., Trainotti, L., Casadoro, G. and Costa, G. (2008) A new index based on Vis spectroscopy to characterize the progression of ripening in peach fruit. *Postharvest Biology and Technology* 49, 319–329.

16 Uses and Human Nutrition

C. Fuentealba[1,*] and I. Hernández[1]
[1]*School of Agronomy, Faculty of Agronomical and Food Sciences, Pontifical Catholic University of Valparaiso, Chile*

Abstract

This chapter comprehensively reviews avocado's rich history and consumption patterns, dating back to pre-Columbian times, highlighting its role as a staple food and ingredient in various dishes (see Galinda-Tovar *et al.*, Chapter 2, this volume). In addition to its traditional uses, avocado is increasingly processed into products such as dehydrated pulp, frozen pulp and guacamole, driven by the need to utilize surplus and damaged fruit effectively. Furthermore, this chapter highlights avocado processing, focusing on the development of new healthy products and innovative applications, such as the increasing popularity of avocado oil extracted using advanced technologies for its health benefits and cosmetic applications. The fruit composition, high in unsaturated fatty acids, fibre and antioxidants, offers numerous health advantages, including improved cardiovascular health. Additionally, avocado by-products, including the seed and peel, are increasingly studied for their potential in pharmaceutical, cosmetic and industrial applications, contributing to a more sustainable use of this fruit.

History of Avocado Consumption

The earliest use of avocado was assumed to be as an energy-rich food eaten fresh. Early Americans would have been attracted to *Persea* since it contains far more oil than any other New World fruit, as well as two to three times more protein (Coe, 1994). Anderson (2003) wrote, '(Avocado) must have been one of the major staples and nutrition sources in pre-Columbian times'. The earliest archaeological evidence finds it among other early crops in the process of domestication. Among the fruit and vegetable crops consumed by the early inhabitants of Tehuacan, avocado was found along with pepper (*Capsicum annuum*), pumpkin (*Cucurbita mixta*), amaranth (*Amaranthus* spp.), bean (*Phaseolus* spp.), maize (*Zea mays*), black sapote (*Diospyros digyna*), white sapote (*Casimiroa* sp.), canistel (*Pouteria campechiana*) and so on. (MacNeish, 1964). In more tropical areas (e.g. San Isidro, Colombia), evidence of the West Indian race avocado was found along with an edible palm species (*Acrocomia aculeate*) and species with edible roots (Piperno and Holst, 1998), while in the Mayan zone avocado, evidence was found along with *Z. mays*, *C. annuum*, *Cuburbita* sp., chicozapote (*Manilkara zapota*) and cacao (*Theobroma cacao*) (Turner and Miksicek, 1984). It is impossible to know at what point avocados went from being a fruit eaten fresh out of hand to an ingredient in salsas, such as the well-known

*Correspondence: claudia.fuentealba@pucv.cl

 The Avocado: Botany, Production and Uses (eds D. Carrillo, B. Schaffer, A.W. Whiley and B.N. Wolstenholme)
DOI: 10.1079/9781800621824.0016

Table 16.1. Composition of the edible portion of 100 g avocados (USDA, 2023).

Nutritional composition	Value per 100 g
Water (g)	73.2
Protein (g)	2.0
Lipids (g)	14.7
SFA (g)	2.1
MUFA (g)	9.8
PUFA (g)	1.8
Total carbohydrate (g)	8.53
Fibre (g)	6.7
Sugars (g)	0.7
Minerals	
Calcium (mg)	12
Magnesium (mg)	29
Phosphorus (mg)	52
Potassium (mg)	485
Sodium (mg)	7
Vitamins	
Vit A (IU)	146
Vit C (mg)	10
Vit E (mg)	2.1
Vit K (μg)	21
Folate (μg)	81

guacamole (which is derived from the word 'ahuaca-mulli' in Nahuatl) and a spread on maize tortillas. Both uses are thought to be pre-Columbian in origin (Coe, 1994). Early Spanish accounts include references to their enjoyment of the new-found fruit, sweetened with sugar as a sweetmeat and savoury with salt and/or pepper (Coe, 1994).

We can catch a glimpse of pre-Columbian uses of avocado and the hybridization of pre-Columbian cuisines with Old World cuisines, in the current cuisines of the Americas and the Caribbean Basin, throughout which avocado is still an important food. Latin America has the world's highest per capita consumption of *Persea americana* fruit (FAO, 2020). In its indigenous home, Mexico, avocado is found in many dishes other than guacamole. It is used to make salsa verde (green sauce) along with cilantro, onions and tomatillos and is even used as part of a cold soup with chicken broth. It also frequently tops black beans and enchiladas and is found in and accompanying the many variations of taco-like delicacies found throughout Mexico, such as empanadas, flautas, panuchos and salbutes. Indeed, avocado is ubiquitous on Mexican plates as a condiment and rich, cooling element to contrast with the abundant spicy chilies. In addition to the flesh of the fruit, Mexicans use the anise-flavoured leaves of the Mexican race as a spice for barbacoa (grilled and/or stewed meat) and for beans. In eastern South America, the large 'palta' halves are frequently stuffed with ceviche, picadillo (a mixture of ground meat and peppers) or chicken and are also included in salads.

Throughout the Caribbean, avocados are eaten in various forms, sliced as a condiment with soups, in fruit salads and chilled 'pear' (as the fruit is frequently called in Caribbean English) soup (Harris, 1991). In Brazil, and throughout most of Asia, avocados are consumed sweet, in cubes mixed with sugar and condensed milk, milkshakes, ice cream and mousses (Morton, 1987; Smith *et al.*, 1992; Nguyen, 2007). In Java, pureed avocado pulp is mixed with sweetened coffee, while in Indonesia avocado-chocolate shakes are reported (Morton, 1987). One of the most unique and curious uses of avocado was as an ingredient in the thick liqueur known as advocaat, made in the past by Dutch populations in Suriname and Recife, Brazil. The drink continues to be made in The Netherlands with the same name but uses egg yolks instead of avocados.

Processed Avocado Products

Whereas most, if not all, traditional culinary uses of avocado are restricted to the fruit flesh, industrial food applications have been sought to put surplus and damaged fruit to use and create less perishable, more easily transportable food products. Processed avocado products include dehydrated, frozen and vacuum-packed pulp, as well as refrigerated ready-to-eat guacamole. The market for these products appears to be growing (FAO, 2018; FAO, 2020).

Edible oil extracted from avocado pulp may be an important part of the future of this fruit. Although the oil has been produced for many years, extraction techniques were based on harsh technologies (e.g. organic solvents and/or high temperatures). New extraction technologies, such as the deodorization process (>200°C), can significantly decrease heat-sensitive bioactive substances in vegetable oils such as phytosterols, phenols and carotenoids (Tan, 2019).

Avocado oil with this type of extraction is commonly used in the manufacture of cosmetic products due to its high skin penetration properties (Eyres *et al.*, 2001).

Consumers are now more aware of the differences between refined and unrefined edible oil from the perspective of quality and health benefits. They prefer oils from green and sustainable extraction, along with minimal oil processing (Tan, 2019). According to Wong *et al.* (2010), virgin avocado oil can be extracted from the mesocarp of healthy fruit or fruit with some degree of physiological damage and rots, using mechanical or natural techniques at low temperatures. Not many years ago, New Zealand started commercial production of avocado oil extracted by cold pressing (Wong *et al.*, 2008). But cold pressing is not the only method used for the production of virgin avocado oil; methods such as expeller-pressed extraction, subcritical CO_2 extraction (SCO_2), aqueous enzymatic extraction and ultrasound-assisted aqueous extraction (UAAE) have also been studied (see Hofman *et al.*, Chapter 15, this volume, for a discussion of oil extraction techniques). Nevertheless, the efficiency and reproducibility of these methods need to be further evaluated (Tan, 2019).

Other processed avocado products have become relevant in the market, such as dehydrated pulp, frozen pulp, vacuum-packed pulp and ready-to-eat refrigerated guacamole, among others. The marketing of processed avocado products is a great challenge in terms of quality preservation due to the high susceptibility of these products to browning by the polyphenoloxidase enzyme and degradation reactions by peroxidase (Martin, 1991). Various physical methods such as freezing, high hydrostatic pressures, low-oxygen atmospheres, freeze drying and heat treatments have been used for many years to control enzymatic browning in avocado processed products, such as pulps or guacamole, with heat treatment being the most widely used (Toledo and Aguirre, 2017). According to Ben-Et *et al.* (1973), heat treatment is effective in inactivating the polyphenoloxidase enzyme, but it can change the texture and develop a bitter taste in guacamole, which is attributed to the high lipid content of avocado. In addition to physical treatments, different chemical methods, such as acidifying agents, chelating agents and antioxidant compounds, have been used to inhibit enzymatic browning in different processed avocado products (Toledo and Aguirre, 2017). Over the years, different additives and additive mixtures have successfully reduced browning but have also shown possible negative effects on human health, such as sulfites, for people potentially exacerbating respiratory problems such as asthma (Vally and Misso, 2012). Daiuto *et al.* (2011) tested the addition of ascorbic acid for the preservation of guacamole. They observed that ascorbic acid efficiently preserved guacamole stored in slow or fast freezing, but after the thawing process, the product lost important biochemical and sensory parameters. Subsequently, López-Ramírez and Duarte-Sierra (2020) developed a hydrocolloid of non-animal origin that, together with ascorbic acid and heat treatments, protects avocado pulp from enzymatic browning and the fruit is not affected in texture and roughness.

Different processed products, including pulps in different formats, avocado honey, avocado jelly and avocado puree as a fat substitute in oatmeal cookies, have been developed to preserve avocado's nutritional properties (Wekwete and Navder, 2005; Rodríguez *et al.*, 2019; López-Ramírez and Duarte-Sierra, 2020). The dairy industry has embraced the trend of incorporating dairy products into fruit beverages, giving rise to the concept of 'juiceceuticals'. These products combine dairies with fruit, providing consumers with a distinctive blend of taste and health benefits (Khurana and Kanawjia, 2007). Because instant milk powder and dry ice cream mixes do not contain phytonutrients, a highly nutritious avocado powder smoothie was developed by Pandey *et al.* (2020), which provided higher content of protein, fat, ash, crude fibre, phenols, flavonoids and β-carotene, compared to a milk-only smoothie.

Composition and Human Nutrition

Avocado has been recognized for its health benefits for quite a long time. The edible part of avocado is the pulp (mesocarp); its composition depends on the variety, grade of ripening, cultivation conditions and climate (Bhuyan *et al.*, 2019). The nutritional composition of avocado, according to the United States Department of Agriculture, is shown in Table 16.1 (USDA, 2023). The main component is water, followed by lipids and carbohydrates. However, the most remarkable characteristic of avocado is its high content of unsaturated fatty acids, comparable with olive oil. Moreover, vitamins (A, C, E) and phenolic compounds present in avocados function as antioxidants that reduce the amounts of free radicals and reactive oxygen species (Stephen and Radhakrishnan, 2022).

The lipid content in avocados is higher than in other fruit, which contributes to most of the calories (Bhuyan *et al.*, 2019). Avocado oils are high in monounsaturated fatty acids (MUFA), where oleic acid is the main fatty acid (45% of total fatty acids), followed by saturated fatty acids (SFA) and polyunsaturated fatty acids (PUFA), with palmitic and linoleic acid the most abundant SFA and PUFA, respectively. West Indian race cultivars typically have a lower oil content in their pulp compared to Mexican and Guatemalan races, as well as their hybrids, rendering them less oily and potentially healthier in terms of fat intake. Nonetheless, it is important to note that the lipid content in avocado, regardless of the variety, consists mainly of healthy fats like monounsaturated fatty acids, which are considered beneficial for cardiovascular health when consumed in moderation. The lipid fraction of avocado fruit also contains polyhydroxylated fatty alcohol derivates, such as acetogenins, which are more polar than fatty acids due to their hydroxyl groups (Salazar-López *et al.*, 2020). Acetogenin-rich extract from avocado pulp can be used as an alternative treatment for cancer and rheumatoid arthritis due to its antioxidant and anti-inflammatory activities (Hassan *et al.*, 2022).

The most abundant carbohydrate in avocado fruit is insoluble fibre (approximately 70% of total fibre). A high-fibre diet is often linked with a healthy digestive system by improving bowel movement (Bhuyan *et al.*, 2019). Moreover, avocado intake has been associated with decreased risk of cardiovascular and coronary heart diseases (CVD) (Pacheco *et al.*, 2022) due to its high-unsaturated fatty acid (MUFA and PUFA) contents, low SFA and phytosterols, and dietary fibre that may play a secondary role in lowering cholesterol (Dreher and Davenport, 2013). Likewise, substituting other food carbohydrate sources with avocados may enhance glucose control and decrease biomarkers of cardiometabolic risk in overweight or obese adults with insulin resistance (Zhang *et al.*, 2022). In addition, consuming fresh avocado could stimulate improved colonic microbiota health in overweight or obese adults by promoting healthier microflora and faecal metabolites (Dreher *et al.*, 2021). However, the meta-analysis of Conceição *et al.* (2022) showed that the consumption of avocado pulp did not change the body weight of people with excess weight.

The main mineral present in avocado pulp is potassium. A low-sodium and high-potassium diet has been shown to protect against CVD by regulating muscle activity and promoting blood pressure control in adults (Dreher and Davenport, 2013; Bhuyan *et al.*, 2019). Avocados also contain other minerals such as phosphorus, magnesium, calcium, sodium, iron and zinc, and the pulp is very rich in antioxidant vitamins. The intake of vitamin C and E may contribute to vascular health, arterial plaque stabilization and slowing atherosclerotic progression. These effects may be greater in at-risk populations such as smokers and people with hypercholesterolemia or hypertension, are overweight or have type 2 diabetes (Dreher and Davenport, 2013). Avocado pulp also contains folate (vitamin B) and phylloquinone (vitamin K1). Folate has antioxidant activity, which may play a role in preventing stomach, pancreatic, colon and cervical cancers (Stephen and Radhakrishnan, 2022). Phylloquinone is a cofactor for enzymes that facilitate coagulation activity and enhance calcium absorption (Dreher and Davenport, 2013).

Avocado pulp has a significant concentration of carotenoids, which are potent free radical scavengers. Carotenes and xanthophylls are subclasses of carotenoids. The main carotenes are α- and β-carotene, which have provitamin A activity. However, the most abundant carotenoid in avocados is the xanthophyll lutein (approximately 70% of total carotenoids). In addition to the colour contribution of lutein (and zeaxanthin), the xanthophylls have been shown to suppress blood vessel damage by decreasing the amount of oxidized low-density lipoproteins (LDL) and slowing the progression of cataracts (Bhuyan *et al.*, 2019).

Phytosterols and phenolic compounds are secondary metabolites present in avocados. Both compounds have extensive health benefits such as antifungal, anticancer, antioxidant, antidiabetic and antimicrobial properties, improving cardiovascular health and weight management (Stephen and Radhakrishnan, 2022; Bhuyan *et al.*, 2019). Phytosterols are compounds very similar to cholesterol, which inhibit the intestinal absorption of cholesterol (Duarte *et al.*, 2016). The most abundant phytosterol in avocado is β-sitosterol, followed by campesterol and stigmasterol. In addition to its cholesterol-lowering activity, β-sitosterol suppresses carcinogenesis and strengthens the immune system by enhancing lymphocytes proliferation and natural killer cell activity (Duarte *et al.*, 2016; Bhuyan *et al.*, 2019). The most relevant phenolic compounds in avocados are the phenolic acids, flavonoids and tannins; however, the phenolic content is known to be influenced by the degree of avocado ripeness (López-Cobo *et al.*, 2016). Coumaric acid is the main hydroxycinnamic acid, followed by ferulic acid, caffeic acid, sinapic acid and *trans*-cinnamic acid, which are mainly conjugated with sugars and quinic acid to form conjugated hydroxycinnamic acids such as 5-feruloylquinic acid, *p*-coumaric acid glucoside, ferulic acid glucoside and *p*-coumaric acid rutinoside among other hydroxycinnamic acids (Salazar-López *et al.*, 2020). The flavonoids reported in avocado fruit include flavan-3-ols (catechin, epicatechin and epicatechin gallate), flavonols (rutin, isorhamnetin, narirutin and quercetin) and flavanonols (taxifolin). Both phenolic acids and flavonoids are potent free radical stabilizers, and their high content in avocados make it the fruit with the highest lipophilic antioxidant capacity, which could help to reduce serum lipid peroxidation and promote vascular health (Dreher and Davenport, 2013; Salazar-López *et al.*, 2020). Elsadek and Alquraishi (2023) recently reported an anxiolytic effect of avocado powder and juice in male albino rats, which is related to its bioactive flavonoids and polyphenolic compounds that could positively affect mental hormones. Likewise, Cheng *et al.* (2021) reported that older adults who consume avocado fruit have better cognitive function than non-consumers.

Novel Uses from Non-edible Parts

Avocado industrialization has increased significantly in recent years driven by the creation of new and functional products from avocado pulp. While this new trend allows the consumer to obtain the diverse organoleptic as well as functional properties of this fruit in different less-perishable formats, it also unfortunately leads to a significant amount of avocado by-products, such as the seed and peel, which could end up as waste. The avocado market data for 2022 revealed that 8.4 million tonnes of fresh avocados were produced worldwide, meaning that approximately 1.3 million tonnes of avocado seeds were discarded as waste. Several investigations have characterized avocado seed wastes and explored their utilization. The most recent review by Setyawan *et al.* (2021) focused on extraction methods and analysis of avocado seed phytochemicals, including their chemical and functional properties. Other research ventured into using the avocado seed in various applications, such as in health (Dabas *et al.*, 2013; Jimenez *et al.*, 2020; Ochoa-Zarzosa *et al.*, 2021), in the production of biofuels, to adsorb pollutants (Colombo and Papetti, 2019), as an antioxidant and antimicrobial (Yasir *et al.*, 2010; Bahru *et al.*, 2019), in the food and cosmetic industry and to produce bioplastics (Charles *et al.*, 2022).

There are efforts to use avocado fruit that does not meet quality standards for fresh consumption. Several components of the fruit, such as the peel, seed and defatted pulp, are commonly discarded, contributing to environmental contamination. However, these components are rich in carbohydrates (seed 42–81%, shell 43–81%), lipids (seed 3–15 %; shell 2–9 %), proteins (seed 0.14–9%; shell 0.17–8%), fibres (seed 2–4.2%; shell 1.3–55%), minerals (seed 1.3–4.3%; shell 1.5–6.0%) and several bioactive compounds such as hydroxycinnamic acids, hydroxybenzoic acids, organic acids, phenolic alcohols derivatives, flavonoids, proanthocyanidins, terpenoids, alkaloids, saponins, acetogenins, phytosterols, and other polar and non-polar compounds (Salazar-López *et al.* 2020). Therefore, the avocado fruit that is nowadays discarded is rich in compounds with very important nutritional, pharmaceutical and industrial applications.

Both avocado peel and seed have high contents of phenolic compounds, with phenolic acids, flavonoids and tannins being most relevant (Tremocoldi *et al.*, 2018). Phenolic acids and flavonoids in avocado fruit are potent free radical stabilizers; epicatechin and catechin aglycones have been reported to be the main flavonoids, but in lower content in the pulp (0.0004–0.05 mg/g FW epicatechin) and higher content in the seeds (10.2–11.06 mg/g DW epicatechin and 3.6–8.1 mg/g DW catechin) and peel (30.4–40.2 mg/g DW epicatechin and 0.15 mg/g DW catechin) (Salazar-López *et al.*, 2020). Epicatechin and catechin can stabilize peroxyl radicals, superoxide radicals and hypochlorous reactive species. However, catechin is 1.3, 2.5 and 1.6 times more effective than epicatechin at stabilizing peroxyl, superoxide and hypochlorous radicals, respectively (Tremocoldi *et al.*, 2018). These compounds present in avocado by-products have a high antioxidant capacity. Furthermore, avocado by-products contain other compounds such as triterpenoids, phytosterols, avocatins, polyhydroxylated fatty alcohols, fatty acids and persenins, each possessing various bioactive properties that are beneficial to health (Salazar-López *et al.*, 2020). Rodríguez-Carpena *et al.* (2011) demonstrated that avocado seed and peel extracts were able to inhibit oxidative reactions and colour deterioration of raw pork during storage (at 4°C for 15 days) by reducing thiobarbituric acid reactive substances (TBARS) and the formation of protein carbonyls. Another study evaluated the effect of a pig diet supplemented with 30% avocado paste on meat quality. Meat from pigs fed with the supplemented diet had lower lipid concentration and more unsaturated fats, less malonaldehyde (MDA) during refrigerated storage (4°C, 12 days), less discoloration and fewer protein carbonyls (Hernández-López *et al.*, 2016). Although studies on the safety of using these avocado by-products are lacking, the available evidence suggests that diets supplemented with avocado or avocado peel and seed extracts improve meat quality and increase shelf life.

The avocado seed is rich in starch (Salazar-López *et al.*, 2020). In the food industry, starch is widely used as a thickening agent because, with increased temperature, it absorbs water and swells, thickening the liquid (Mason, 2009). Cornelia and Christianti (2018) observed that starch from avocado seeds loses viscosity during storage, but they managed to reverse this by crosslinking the starch molecules using sodium tripolyphosphate. Native and chemically modified avocado seed starch presents antioxidant and antimicrobial properties, which could be also considered a sustainable substitute for plastic packaging in the food industry (Hassan *et al.*, 2023). Moreover, Wang *et al.* (2023) reported that avocado seed starch exhibits higher amylose content and larger granule size compared with maize starch, which renders it a promising source of resistant starch. Resistant starch refers to starch that resists digestion in the small intestine, offering potential health benefits such as improved gut health and blood glucose control.

Avocado seed and peel meals have also been investigated. In addition to their high phytochemical content, avocado shells contain 12 different fatty acids: three monounsaturated (MUFA; oleic, palmitoleic and heptadecenoic acids), two polyunsaturated (PUFA; linolenic and linoleic acids) and seven saturated (SFA; myristic, palmitic, margaric, stearic, capric, lauric and pentadecanoic acids). MUFA (571.32–4814.77 mg/kg FW) were the most abundant, followed by SFA (242.73–1741.22 mg/kg FW) and PUFA (212.88–1683.70 mg/kg FW) (Morais *et al.*, 2017). Avocado peels could be used to produce baking flour with low fatty acid content, which increases the palatability of bakery products and provides important health-beneficial phytochemicals, potentially substituting part of wheat flour in various bakery products (Ramos-Aguilar *et al.*, 2021). Avocado kernel-based flour has desirable quantities of protein (8%), moisture (14%) and carbohydrates (75%) required of a baking flour (Mahawan *et al.*, 2015). However, flours contain a quality attribute called falling number, which affects the texture of the baked product. In avocado kernel flour, the falling number is far from the desirable range (Mahawan *et al.*, 2015).

For all avocado by-products that require a drying process, such as flour or dehydrated avocado, the drying conditions are optimized to retain the bioactive characteristics of avocado by-products. In a study conducted by Saavedra *et al.* (2017), the drying time was longer for avocado seeds compared to the peel, due to the higher moisture content of the seed. Through the statistical technique, 'response surface methodology' used to optimize the drying time and temperature, 62.82% and 54.81% of the initial phenolic content was retained in dehydrated peels and seeds, respectively, showing very high concentrations compared with other dehydrated by-products.

The cosmetic industry seeks to use natural and minimally processed products. In 1989, two patents related to the cosmetic uses of avocado seed were submitted in the USA. One consisted of a scalp treatment to prevent hair loss, using macerated avocado seeds and other ingredients in mineral oil. The other patent consisted of a silica gel activated with avocado seed extract to treat skin stretch marks and reduce skin lesions (Ruiseco, 1989; Moy, 1999). These properties were corroborated by Laksmiani *et al.* (2020), who found that flavonoids in avocado seed extracts have the potential to lighten the skin, due to the inhibition of tyrosinase. This enzyme oxidizes tyrosine to produce melanin and other pigments. In addition to lightening the skin, avocado seed extracts (Ettinger cultivar) have demonstrated healing qualities for skin repeatedly exposed to UVB radiation, because the polyhydroxylated fatty acids extracted from the seed do not absorb UVB light, so they are believed to act differently than sunscreens (Rosenblat *et al.*, 2011). In addition to the potential use of avocado by-products to improve health and well-being, other studies concluded that fruit components, mainly the seed, could have a positive impact on different industries and the environment through the generation of biomass that can be used to produce energy (Colombo and Papetti, 2019).

The fuel and food industries have explored using avocado seed as biomass to generate energy. Today, 80% of the world's energy use comes from coal, natural gas and oil. There are substantial efforts to find new renewable resources that deliver a more sustainable solution to energy demand (Perea-Moreno *et al.*, 2016). Biomass is one of the most studied renewable energy sources, due to its wide extension worldwide and its availability from agricultural or industrial waste products that can be burned in waste conversion plants to produce electricity without chemical processing (Manzano-Agugliaro, 2007). One of the main controversies for using combustion is the generation of carbon dioxide (CO_2). However, the production of energy from biomass does not contribute to the increase of greenhouse gases because the balance of CO_2 emissions to the atmosphere is neutral. In fact, any CO_2 generated during biomass combustion is fixed by plants through photosynthesis; consequently, the amount of CO_2 present in the atmosphere does not increase (Perea-Moreno *et al.*, 2016). It is due to the increasing cost of wood biomass that other biomass sources such as nut shells (González *et al.*, 2005; Haykiri-Acma *et al.*, 2012), maize husks (Xiong *et al.*, 2008) or olive pits (Garcia Sanz-Calcedo *et al.*, 2011) have been considered. Avocado waste could be used as biomass. According to Perea-Moreno *et al.* (2016), avocado seed has higher heating values. However, factors such as moisture, chlorine and ash content must be considered when using biofuel in an energy conversion process and in the associated steps of this industry's process line, especially in combustion.

The plastic industry has also explored the use of the avocado seed. Large-scale plastic production between the 1950s and 2010s introduced about 6.3 billion tonnes of plastic into the world, 80% of which ended up in landfills or the environment (Rhodes, 2018). The use of avocado seeds to produce bioplastic has great potential due to its high starch content. In a study by Ginting *et al.* (2015), avocado seed starch was used as raw material for bioplastic production, along with chitosan as filler and glycerol (in water) as plasticizer. They observed that increasing the amount of filler and the gelatinization temperature resulted in a stronger bioplastic. However, Lubis *et al.* (2018) and Sartika *et al.* (2018) used microcrystalline cellulose as filler and glycerol as plasticizer and observed that lower amounts of plasticizer and higher amounts of filler increased the strength of the avocado seed starch.

Many studies indicate that processing avocado waste can be valuable in elaborating new avocado-based by-products, contributing directly to the emerging bioeconomy markets based on food, cosmetic and industrial applications.

Conclusions

The avocado, with its high nutrient density and oil content and high levels of lipids (mostly monounsaturated), proteins, vitamins and beneficial phytochemicals, may be part of the solutions to a number of health problems that are now almost an epidemic in the developed world. The industry of processed avocado products goes beyond developing a product with different formats but rather seeks to offer products that contain the multiple properties of this fruit satisfying the diverse needs of today's consumers. In addition, there is a growing niche for products developed from avocado waste, contributing to the emerging bioeconomy while improving the sustainability of the avocado market.

References

Anderson, E.N. (2003) *Those Who Bring the Flowers: Maya Ethnobotany in Quintana Roo, Mexico*. Ecosur, San Cristobal, Mexico.

Bahru, T.B., Tadele, Z.H. and Ajebe, E.G. (2019) A review on avocado seed: functionality, composition, antioxidant and antimicrobial properties. *Chemical Science International Journal* 27(2), 1–10.

Ben-Et, G., Dolev, A. and Tatarsky, D. (1973) Compounds contributing to heat-induced bitter off-flavor in avocado. *Journal of Food Science* 38, 546–547.

Bhuyan, D.J., Alsherbiny, M.A., Perera, S., Low, M., Basu, A. *et al.* (2019) The odyssey of bioactive compounds in avocado (*Persea americana*) and their health benefits. *Antioxidants* 8:426.

Charles, A.C., Dadmohammadi, Y. and Abbaspourrad, A. (2022) Food and cosmetic applications of the avocado seed: a review. *Food & Function* 13:6894.

Cheng, F.W., Ford, N.A. and Taylor, M.K. (2021) US older adults that consume avocado or guacamole have better cognition than non-consumers: national health and nutrition examination survey 2011–2014. *Frontiers in Nutrition* 8, 746453.

Coe, S.D. (1994) *America's First Cuisines*. University of Texas Press, Austin, Texas.

Colombo, R. and Papetti, A. (2019) Avocado (*Persea Americana* Mill.) by-products and their impact: from bioactive compounds to biomass energy and sorbent material for removing contaminants. A review. *International Journal of Food Science & Technology* 54(4), 943–951.

Conceição, A.R., Fraiz, G.M., Rocha, D.M.U.P. and Bressan, J. (2022) Can avocado intake improve weight loss in adults with excess weight? A systematic review and meta-analysis of randomized controlled trials. *Nutrition Research* 102, 45–58.

Cornelia, M. and Christianti, A. (2018) Utilization of modified starch from avocado (*Persea americana* Mill.) seed in cream soup production. *IOP Conference Series: Earth and Environmental Science* 102, 012074.

Dabas, D., Shegog R.M., Ziegler, G.R. and Lambert, J.D. (2013) Avocado (*Persea americana*) seed as a source of bioactive phytochemicals. *Current Pharmaceutical Design* 19(34), 6133–6140.

Daiuto, E.R., Vieites, R.L., Simon, J.W., Carvalho, L.R. and Pegoretti, C. (2011) Avaliações sensoriais, bioquímicas e microbiológicas do guacamole, um produto à base de abacate, sob armazenamento a frio e comadição de ácido ascórbico. *Semina-ciencias Agrarias. Londrina: Universidade Estadual de Londrina (UEL)* 32(2), 599–611.

Dreher, M.L. and Davenport, A.J. (2013) Hass avocado composition and potential health effects. *Critical Reviews in Food Science and Nutrition* 53, 738–750.

Dreher, M.L., Cheng, F.W. and Ford, N.A. (2021) A comprehensive review of Hass avocado clinical trials, observational studies, and biological mechanisms. *Nutrients* 13:4376.

Duarte, P.F., Chaves, M.A., Borges, C.D. and Mendonça, C.R.B. (2016) Avocado: characteristics, health benefits and uses. *Ciência Rural* 46(4), 747–754.

Elsadek, M.F. and Alquraishi, M.I. (2023) Investigating the avocado (*Persea americana*) fruit's anti-anxiety potentials in rat models. *Journal of King Saud University – Science* 35, 102504.

Eyres, L., Sherpa, N. and Hendricks, G. (2001) Avocado oil: a new edible oil from Australasia. *Lipid Technology*, 13, 84–88.

FAO (2018) *Food Outlook: Biannual Report on Global Food Markets*. FAO, Québec City, Canada.

FAO (2020) *Major Tropical Fruits: Market Review 2020*. Rome. Available at: https://www.fao.org/economic/est/est-commodities/tropical-fruits/en/ (accessed 28 September 2022).

Garcia Sanz-Calcedo, J., Cuadros-Blazquez, F., Ruiz-Celma, A. and Lopez-Rodriguez, F. (2011) Efficiency of a biomass boiler in a hospital center. Use of olive crushing stones to produce heat. *DYNA* 86(3), 343–349.

Ginting, M.H.S., Tarigan, M.F.R. and Singgih, A.M. (2015) Effect of gelatinization temperature and chitosan on mechanical properties of bioplastics from avocado seed starch (*Persea americana* Mill.) with plasticizer glycerol. *The International Journal of Engineering and Science* 4, 36–43.

González, J.F., González-García, C.M., Ramiro, A., Gañán, J., González, J. and Sabio, E. (2005) Use of almond residues for domestic heating. Study of the combustion parameters in a mural boiler. *Fuel Processing Technology* 86(12), 1351–1368.

Harris, J.B. (1991) *Sky Juice and Flying Fish: Traditional Caribbean Cooking*. Simon and Schuster, New York.

Hassan, M., Moneim, S.A., Mahmoud, E.A. and Mohamed, D.A. (2022) Antioxidant, anti-cancer and anti-arthritic activities of acetogenin-rich extract of avocado pulp. *Egyptian Journal of Chemistry* 65(2), 539–550.

Hassan, S., Kabir, M.H., Mazhar, D., Imran, Z., Rehman, A., *et al.* (2023) Chemical processing and various applications of starch extracted from avocado seeds: a step towards sustainable food waste management. *Starch* 76, 2300114.

Haykiri-Acma, H., Yaman, S. and Kucukbayrak. S. (2012) Combustion characteristics of sodium-free pyrolytic char from hazelnut shell. *Fuel Processing Technology* 96, 169–174.

Hernández-López, S.H., Rodríguez-Carpena, J.G., Lemus-Flores, C., Grageola-Nuñez, F. and Estévez, M. (2016) Avocado waste for finishing pigs: impact on muscle composition and oxidative stability during chilled storage. *Meat Science* 116, 186–192.

Jimenez, P., Garcia, P., Quitral, V., Vasquez, K., Parra-Ruiz, C. *et al.* (2020) Pulp, leaf, peel and seed of avocado fruit: a review of bioactive compounds and healthy benefits. *Food Reviews International* 37(6), 619–655.

Khurana, H.K. and Kanawjia, S.K. (2007) Recent trends in development of fermented milks. *Current Nutrition & Food Science* 3, 91–108.

Laksmiani, N., Sanjaya, I. and Leliqia, N. (2020) The activity of avocado (*Persea americana* Mill.) seed extract containing catechin as a skin lightening agent. *Journal of Pharmacy & Pharmacognosy Research* 8, 449–456.

López-Cobo, A., Gómez-Caravaca, A.M., Pasini, F., Caboni, M.F., Segura-Carretero, A. and Fernández-Gutiérrez, A. (2016) HPLC-DAD-ESI-QTOF-MS and HPLC-FLD-MS as valuable tools for the determination of phenolic and other polar compounds in the edible part and by-products of avocado. *LWT - Food Science and Technology* 73, 505–513.

López-Ramírez, A.M. and Duarte-Sierra, A. (2020) Avocado jelly: formulation and optimization of an avocado gel using hydrocolloids. *International Journal of Gastronomy and Food Science* 21, 100234.

Lubis, M., Harahap, M.B., Ginting, M.H.S., Sartika, M. and Azmi, H. (2018) Production of bioplastic from avocado seed starch reinforced with microcrystalline cellulose from sugar palm fibers. *Journal of Engineering Science and Technology* 13(2), 381–393.

MacNeish, R.S. (1964) Ancient Mesoamerican civilization. *Science* 143, 531–537.

Mahawan, M.A., Tenorio, M.F.N., Gomez, J. and Bronce, R.A. (2015) Characterization of flour from avocado seed kernel. *Asia Pacific Journal of Multidisciplinary Research* 3, 34–40.

Manzano-Agugliaro, F. (2007) Gasification of greenhouse residues for obtaining electrical energy in the south of Spain: localization by GIS. *Interciencia* 32(2), 131–136.

Martin, Z.J. (1991) *Processamento: produtos, características e utilização*. In: Teixeira, C.G., Bleinroth, E.W., Castro, J.V., Martin, Z.J., Tango, J.S., *et al.* (eds) *Abacate: cultura, matéria prima, processamento e aspectos econômicos.* 2nd edn. ITAL, Campinas, pp. 148–155.

Mason, W.R. (2009) *Starch Use in Foods*. In: BeMiller, J.N. (ed.) *Starch: Chemistry and Technology*, 3rd edn. Elsevier Inc., San Diego, pp. 745–795.

Morais, D.R., Rotta, E.M., Sargi, S.C., Bonafe, E.G., Suzuki, R.M., Souza, N.E. and Visentainer, J.V. (2017) Proximate composition, mineral contents and fatty acid composition of the different parts and dried peels of tropical fruits cultivated in Brazil. *Journal of the Brazilian Chemical Society* 28, 308–318.

Morton, J.F. (1987) *Fruits of Warm Climates*. Wipf and Stock Publishers, Eugene, Oregon, pp. 91–102.

Moy, L.S. (1999) Cosmetic formulation and method for amelioration of skin keratoses and striae distensae. US 5928659. U.S. Patent and Trademark Office. https://patentimages.storage.googleapis.com/0f/8b/74/747150de91c023/US5928659.pdf (accessed on October 10, 2022).

Nguyen, A. (2007) Like butter. *Savuer* 104, 76–87.

Ochoa-Zarzosa, A., Báez-Magaña, M., Guzmán-Rodríguez, J.J., Flores-Alvarez, L.J., Lara-Márquez, M. *et al.* (2021) Bioactive molecules from native Mexican avocado fruit (*Persea americana* var. drymifolia): a review. *Plant Foods for Human Nutrition* 76, 133–142.

Pacheco, L.S., Li, Y., Rimm, E.B., Manson, J.E., Sun, Q. *et al.* (2022) Avocado consumption and risk of cardiovascular disease in us adults. *Journal of the American Heart Association* 11: e024014.

Pandey, S.A., Shiby Varghese, K., Kumar Chauhan, A. and Singh, M. (2020) Development of phytonutrient enriched avocado milkshake powder and its quality evaluation. *Indian Journal of Dairy Science* 73(6), 556–565.

Perea-Moreno, A.J., Aguilera-Ureña, M.J. and Manzano-Agugliaro, F. (2016) Fuel properties of avocado stone. *Fuel* 186, 358–364.

Piperno, D.R. and Holst, I. (1998) The presence of starch grains on prehistoric tools from the humid tropics: indications of early tuber use and agriculture in Panama. *Journal of Archaeological Science* 25, 765–776.

Ramos-Aguilar, A.L., Ornelas-Paz, J., Tapia-Vargas, L.M., Gardea-Béjar, A.A., Yahia, E.M. *et al.* (2021) Effect of cultivar on the content of selected phytochemicals in avocado peels. *Food Research International* 140, 110024.

Rhodes, C.J. (2018) Plastic pollution and potential solutions. *Science Progress* 101(3), 207–260.

Rodríguez, I., Cámara-Martos, F., Flores, J.M. and Serrano, S. (2019) Spanish avocado (*Persea americana* Mill.) honey: authentication based on its composition criteria, mineral content and sensory attributes. *LWT - Food Science and Technology* 111, 561–572.

Rodriguez-Carpena, J.G., Morcuende, D., Andrade, M.J., Kylli, P. and Estevez, M. (2011) Avocado (*Persea americana* Mill.) phenolics, in vitro antioxidant and antimicrobial activities, and inhibition of lipid and protein oxidation in porcine patties. *Journal of Agricultural and Food Chemistry* 59(10), 5625–5635.

Rosenblat, G., Meretski, S., Segal, J., Tarshis, M., Schroeder, A., Gilead, A.Z. and Hochberg, M. (2011) Polyhydroxylated fatty alcohols derived from avocado suppress inflammatory response and provide non-sunscreen protection against UV induced damage in skin cells. *Archives of Dermatological Research* 303, 239–246.

Ruiseco, M.G. (1989) Oil based scalp treatment composition. US4849214A. U.S. Patent and Trademark Office. https://patentimages.storage.googleapis.com/18/aa/8a/af0c047c073f1a/US4849214.pdf (accessed on October 10, 2022).

Saavedra, J., Córdova, A., Navarro, R., Díaz-Calderón, P., Fuentealba, C. *et al.* (2017) Industrial avocado waste: functional compounds preservation by convective drying process. *Journal of Food Engineering* 198, 81–90.

Salazar-López, N.J., Domínguez-Avila, J.A., Yahia, E.M., Belmonte-Herrera, B.H., Wall-Medrano, A., Montalvo-González, E. and González-Aguilar, G. (2020) Avocado fruit and by-products as potential sources of bioactive compounds. *Food Research International* 138, 109774.

Sartika, M., Lubis, M., Harahap, M.B., Afrida, E. and Ginting, M.H.S. (2018) Production of bioplastic from avocado seed starch as matrix and microcrystalline cellulose from sugar palm fibers with Schweizer's reagent as solvent. *Asian Journal of Chemistry* 30(5), 1051–1056.

Setyawan, H.Y., Sukardi, S. and Puriwangi, C.A. (2021) Phytochemicals properties of avocado seed: a review. *IOP Conference Series: Earth and Environmental Science* 733, 012090.

Smith, N.J.H., Williams, J.T., Plucknett, D.L. and Talbot, J.P. (1992) *Tropical Forests and Their Crops*. Cornell University Press, Ithaca, New York.

Stephen, J. and Radhakrishnan, M. (2022) Avocado (*Persea americana* Mill.) fruit: nutritional value, handling and processing techniques, and health benefits. *Journal of Food Processing and Preservation* 46: e17207.

Tan, C.X. (2019) Virgin avocado oil: an emerging source of functional fruit oil. *Journal of Functional Foods* 54, 381–392.

Toledo, L. and Aguirre, C. (2017) Enzymatic browning in avocado (*Persea americana*) revisited: history, advances, and future perspectives. *Critical Reviews in Food Science and Nutrition* 57, 3860–3872.

Tremocoldi, M.A., Rosalen, P.L., Franchin, M., Massarioli, P., Denny, C. *et al.* (2018) Exploration of avocado by-products as natural sources of bioactive compounds. *PLOS One* 13: e0192577.

Turner, B.L. and Miksicek, C.H. (1984) Economic plant species associated with prehistoric agriculture in the Maya lowlands. *Economic Botany* 38, 179–193.

USDA (U. S. Department of Agriculture), Agricultural Research Service. (2023) Avocados, raw, all commercial varieties. FoodData Central. Available at: https://fdc.nal.usda.gov/fdc-app.html#/food-details/171705/nutrients (accessed 18 January 2023).

Vally, H. and Misso, N.L.A. (2012) Adverse reactions to the sulphite additives. *Gastroenterology and Hepatology from Bed to Bench* 5, 16–23.

Wang, J., Li, Y., Jin, Z. and Cheng, Y. (2023) Physicochemical, morphological, and functional properties of starches isolated from avocado seeds, a potential source for resistant starch. *Biomolecules* 12:1121.

Wekwete, B. and Navder, K.P. (2005) Effect of avocado puree as a fat replacer on the physical, textural and sensory properties of oatmeal cookies. *Journal of the Academy of Nutrition and Dietetics* 105:47.
Wong, M., Ashton, O., Requejo-Jackman, C., McGhie, T., White, A., Eyres, L. and Woolf, A. (2008) Avocado oil: the color of quality. *ACS Symposium Series*, Vol. 983, pp. 328–349.
Wong, M., Requejo-Jackman, C. and Woolf, A. (2010) What is unrefined, extra virgin coldpressed avocado oil? Available at: https://www.aocs.org/stay-informed/readinform/featured-articles/what-is-unrefined-extra-virgin-cold-pressed-avocado-oilapril-2010 (accessed 28 September 2022).
Xiong, S., Burvall, J., Örberg, H., Kalen, G., Thyrel, M. and Öhman, M. (2008) Slagging characteristics during combustion of corn stovers with and without kaolin and calcite. *Energy Fuels* 22, 3465–3470.
Yasir, M., Das, S. and Kharya, M. (2010) The phytochemical and pharmacological profile of *Persea americana* Mill. *Pharmacognosy Reviews* 4(7), 77–84.
Zhang, X., Xiao, D., Guzman, G., Edirisinghe, I. and Burton-Freeman, B. (2022) Avocado consumption for 12 weeks and cardiometabolic risk factors: a randomized controlled trial in adults with overweight or obesity and insulin resistance. *The Journal of Nutrition* 152, 1851–1861.

Index

H

Printed and bound by CPI Group (UK) Ltd, Croydon, CR0 4YY

13/09/2026

14961111-0007